CRATERS, COSMOS, AND CHRONICLES
A New Theory of Earth

CRATERS, COSMOS, AND CHRONICLES

A New Theory of Earth

Herbert R. Shaw

Stanford University Press
Stanford, California
1994

Stanford University Press
Stanford, California
© 1994 by the Board of Trustees
of the Leland Stanford Junior University
Printed in the United States of America

CIP data appear at the end of the book

Stanford University Press publications are
distributed exclusively by Stanford University
Press within the United States, Canada, and
Mexico; they are distributed exclusively by
Cambridge University Press throughout the
rest of the world.

To Andrea, who shares my adventure,

To Carol, who made it possible,

To Lorraine and Deborah, who co-opted my surprise,
   and

To all of the "children of chaos," who inspire me,

In memory of Máire — artist and mother — who made of chaos its fruition, a truly transcendent gift.

Foreword

William Glen

Meteorites from the shooting gallery of space, scaled from the daily patter of pebbles and dust to the crushing impacts of mountain-size rocks every thirty million years or so, have pelted the Earth and shaped its history from the moment of its violent birth. Shaw deciphers this chronicle of impacts from the cratering record of the planets and satellites, and astonishingly ties that chronicle to the dynamics of the Earth's interior — and thus to its tectonic movements and volcanism, its magnetic-field behavior and fossil record, and a host of other phenomena, all linked to the resonances of the Cosmos.

Shaw's pioneering application of the nonlinear dynamics of chaos science to a range of rapidly moving fields has birthed a major new theory that explains patterns of cosmic imprinting in the solid Earth body. Arresting in breadth and predictive promise, the theory calls for immediate tests of established data, and for further tests as new data are spawned by the host of research programs spanned by its vast conceptual reach.

Formulating this new overarching scheme of earth dynamics and history required that Shaw leap beyond the broad range of disciplines within his own fields of geology and geophysics to master both chaos theory and celestial mechanics. And having taken that leap, he has faced what other authors of such multidisciplinary syntheses — melds of the endemic and the exotic — encounter in their own professional communities: preferred seating on the hot griddle of tradition.

In 1986, while serving as an editor of *Eos, Transactions of the American Geophysical Union*, I was faced with the seemingly impossible task of finding referees for Shaw's manuscript on "The Periodic Structure of the Natural Record and Nonlinear Dynamics" — which was in some ways a precursor to this book. The task was a small nightmare, first because just getting the gist of the manuscript

Foreword

was a struggle for me, and second because we found ourselves almost in a referee vacuum: the first scientist reviewer I selected became lost in the chaos theory of the article, and the chaos theorist was daunted by the science. At that time a bridge of communication between chaos science and conventional mathematics had not yet been built, let alone one to the earth sciences. A minuscule number of scientists had become familiar with chaos theory, and fewer still were applying it.

In 1987, after a year-long search for referees, the greatly delayed article finally appeared; glowingly reviewed, it was widely hailed as an "object of study around which graduate seminars should be organized nationally." The *Eos* editorial staff felt vindicated: the paper had exceeded the journal's length limit by more than anyone could recall seeing before, and the staff members, in spite of very positive referee reports, had suffered deep qualms about publishing a paper that was beyond their ken. Nonlinear dynamics seemed an ordinance from an alien realm.

But things have changed since 1986. Recent years have seen chaos theory applied to science at a rate seemingly unprecedented in the history of ideas. Symposia, workshops, and study sessions now abound, and a burgeoning literature pervades fields from astronomy to cardiology to meteorology and beyond — even into the social sciences. Old data are being focused through the lens of "chaos" to reveal new patterns of meaning everywhere.

Chaos science has taken form over the last three decades through a confluence of ideas whose histories have been rent with uncertainties about priority and precise channels of intellectual development. The cardinal virtue of the methods that have evolved under the rubric "chaos theory" or "chaos science" is their capacity to facilitate the cognition of patterns of order in the fluctuations of complex dynamical systems — for example, in the movement of fluids or the behavior of planetary rings. Storehouses of data describing such fluctuations had long been analyzed by conventional methods and had come to be regarded confidently as either rigorously periodic or hopelessly random. Randomness had virtually been defined by the jumbled, seemingly patternless cascades and torrents of turbulent systems! Thus did the pioneers of chaos theory, who claimed order where all before had deemed disorder, meet a hostile reception. Even common sense seemed to speak against their claim. And why would anyone name a new "order-seeking" science "chaos"?

A most difficult charge of science has always been to find order in dynamical systems. Indeed, the very process of scientific analysis, by whatever means, is concerned with the identification and isolation of the pieces or parts of complex systems, such that one might examine the individual pieces in the hope of discovering their properties, geneses, and combinatorial propensities, and thereby predict the behavior of the whole. But the work of science since Newton's day has been done under the assumption that a flight of sparrows in Barcelona could not figure in the formation of clouds over Bombay. Long experience had taught that vanishingly small variables could, and even must, be overlooked if one expected to make any progress in deciphering the complexities of the real world.

Foreword

Chaos science, like other systems of analysis, also seeks order in the apparent disorder of dynamical systems, but it undertakes the search in new ways. "Chaos" has grown from, and has in turn engendered, a repertoire of initially arcane ideas and vocabulary that daily move more into the mainstream. The "nonlinear dynamics" of chaos science treats changes in complex systems: it describes how a small change in a single variable, such as a fractional rise in temperature in a very localized weather system, could produce a global weather response out of all proportion to its size. Within weather systems, as in all nonlinear-dynamical systems, are loci that have come to be called "strange attractors." Around such strange attractors the system orients itself over time as it attempts to move into a set of balanced states. But note that the system moves toward the entire set of balanced states, not to a single state, rendering our characterization of its behavior all the more elusive. (There are exceptions, typically in very simple cases: as an example, the swinging pendulum that settles steadily toward a single position as its energy is dissipated.)

Mathematically, the strange attractor appears as a computer-generated "portrait of order" within the chaos of the system as a whole. Herbert Shaw, and others, have proposed that these artifactual strange attractors of computer experimentation simulate the organic complexity of the natural world. This constitutes an order of transformation that even some chaos theorists of mathematical bent would resist, much as many traditional mathematicians resist chaos theory itself. New geometric measures of the strange-attractor-centered portraits of order, called "fractals," inevitably appear in such systems; fractals show similar patterns of complexity at smaller and smaller scales, like the dendritic branching pattern of a river system that is exhibited as one moves from the river to its streams to its brooks. Ranges of objects with such fractal geometry, which appear similar at different scales, are said to possess "self-similarity." Another example of self-similarity is the pattern by which the capillaries of the vascular system join small veins, and the small veins larger veins, and so forth. These ideas and others, comprising a new set of cognitive modes, now allow chaos theorists to view dynamical systems differently, and to demonstrate how the sparrows of Barcelona do indeed affect the clouds over Bombay. It was to this new perspective on nature — reshaped through a novel language — that Shaw was "attracted" in 1981. By that time he had felt frustration for fifteen years, occasioned by the complacency and traditionalism that had muffled his cries about the nonlinear world he had found in his own studies. A brief look at his career and publications — up to that time — topically diverse and profoundly theoretical — is revealing of how acutely attuned he has been to a broad array of fundamental problems in science. What we also discover is that in his own work, as early as the 1960's, he had already foreseen, without benefit of the language or the computer expositions that would emerge later, some of the fundaments of chaos theory.

In Shaw's college years there was little of the maverick that would emerge later. Straight from a rural high school of only 100 students, he was awestruck by the worlds opening before him at the University of California, Berkeley. Looking

Foreword

singlemindedly ahead, he played the grade game to the hilt and was named to Phi Beta Kappa in geology in his junior year. Shaw stayed on at Berkeley to complete the Ph.D. — having posted only "A" grades across his entire university record! During his long career at the U.S. Geological Survey in Menlo Park, he served three times as Visiting Professor at Berkeley, won the 3rd Ernst Cloos Award of Johns Hopkins University, and became a Fellow of the American Geophysical Union and the Geological Society of America.

In the earliest years of Shaw's career he devised a method of measuring the ease of flow in rock melts (magmas) in which there were known concentrations of water. That study, completed in 1963, garnered an Award for Outstanding Achievement. He then measured and calculated the rates at which water molecules could diffuse through glasses at high temperatures. In each instance, his pioneering measurements of dynamical systems initiated what became an established program of research in rock-melt processes. During this same period (1961–64), by breaking a barrier of experimental limitation, he also devised a method for the osmotic control and measurement of the chemical activity of hydrogen in experiments at high temperatures and pressures. Hans Eugster, the revered experimentalist, came to refer to "Shaw's offbeat originality." Even today, Shaw's demonstration of the mixing properties of hydrogen and water gases may be historically unique.

On a divergent foray during the hydrogen experiments, involving statistical mechanics and his own solution theory for the behavior of water in rock melts, Shaw wrote a paper that attempted to predict how water would interact in such conditions. The noted isotope chemist G. J. Wasserburg of Cal Tech, in reviewing that paper, was surprised by Shaw's mastery of a theory in which he had never had a formal course.

In 1965 Shaw published on many dynamic features of rock melts, work that was eventually to lead to a full awakening of his fascination with nonlinear dynamics. And field work during the 1960's fueled his interests in the predictive powers of the new discoveries of magnetic striping on the sea floor and, later, plate tectonics — interests that he saw were intimately related to the origin and upward migration of melts (magma transport) in the Earth's mantle and crust. In the early 1970's, with Robert L. Smith of the U.S.G.S., Shaw formulated a quantitative scheme by which they predicted the total heat stored in the rock melts of the continental crust of the conterminous U.S. That contribution became "the exploration bible" in the geothermal-energy industry during the late 1970's.

With scientists of the Hawaiian Volcano Observatory (HVO) in 1965, Shaw performed flow experiments with molten basalt. These experiments refocused his attention from assumptions about the simple, proportional properties of materials to a fascination with their nonlinear flow characteristics (rheologies). The complex aspects of flow were powerfully displayed to Shaw that year during an eruption that partially filled Makaopuhi pit crater on Kilauea's East Rift Zone, creating a 100-meter-deep lava lake. The inadequacy of the conventional linear suppositions, which he believes have historically hindered rheological flow analysis,

Foreword

became abundantly clear. In those phenomena — as he remarked two decades later — he had seen a microcosm of 4.5 billion years of nonlinear rheological evolution in the Earth and planets!

Shaw and his HVO team "ice-fished" by drilling through the solidified crust of the lava lake and, after repeated failures, immersed a rotator in the molten lava to measure resistance to flow. Their rotator was a scaled-up version of the device that James Joule had used more than a hundred years earlier in England to demonstrate the mechanical equivalent of heat. But where Joule's device was put to work in a small container of water in the laboratory, Shaw's was in a great cauldron of incandescent lava — apparently the only time in history that magma has been directly measured rheologically.

By 1968, Shaw and the rheologist Irving Gruntfest had developed a program to investigate the Earth's tide by means of rheological experiments, computer simulations of tidal deformation, and field measurements of flexing across tectonic-plate margins by the use of lasers — but the proposal for that study went unsupported. During the period 1966–68, Shaw also made — and published in 1969 — the first full-fledged, experimental attempt to evaluate thermomechanical deformations and instabilities of rock-forming materials; from that base he made predictions for earthquakes, mantle flow, and dissipative melting in the Earth's mantle.

After another decade of what Shaw described as "banging his head against the walls of geophysical conservatism and complacency," he was more than ready to seize upon the concept of "deterministic chaos" or "chaos science" that had evolved in several different domains by 1981. He had discovered important parts of it himself in his experimental studies, much as Edward Lorenz had discovered the butterfly (or Barcelona sparrows) effect in meteorological computer experiments — both in the 1960's!

I first chanced on Shaw in 1984 in the wee hours, at a water fountain in Building 2 of the Survey in Menlo Park, California — both of us seeking refuge from sunlit workaday distractions. Full of the fervor of grant-proposal writing I told him of my new historical study of the unfolding debates over mass extinctions, which were moving at a frenetic pace from the impact hypothesis itself to the alternative volcanist theory to periodicity of extinctions to astronomical hypotheses and on to a host of geologic effects implied by the new ideas. I had found what I had sought for three years: an important, current, theoretical debate to study in vivo, ethnographically. It was an opportunity to do a longitudinal, internal, scientific history, documenting the behaviors of the scientists and their products as the debates unfolded. This, as became clearer to me in my dialogue with Herb, was my own hope of going beyond the all too frequent, linearly convergent descriptions of history.

I wondered at the time if Shaw was fixed by my enthusiasm or by the captivating character of the debated ideas — perhaps both. But that was the "turning point" — as he has remarked — of that decade in his career. The rest is the history that led directly to this book. So easily set aflame, he burned like phos-

Foreword

phorus, upping the ante on his normal ten-hour workday to twelve, and within a year he had doubled the piles of literature in his impossibly cluttered office. To peruse those piles is to subtend an astonishingly broad disciplinary arc.

Shaw went on to master the methods of chaos science, and through them turned the sleeve of his favorite conceptual garment — volcanism — inside out, discovering that the sleeve's drab exterior hid a wondrous polychromatic lining. During the next half-decade he raced about, turning out the sleeves of mountain building, earthquakes, extinctions of life, the flux of Earth-crossing space objects, the distribution of impact craters, and other phenomena. He shows here how this diversity of dynamical garments, all so different in origin and appearance, share a host of subtle, but fundamental, chaotic properties — a common sleeve lining — that is the key to their deepest nature. It is a book which — more than one reviewer has noted, but Archie Roy, the celebrated celestial dynamicist of Glasgow said best — "will arouse strong feelings, either enthusiastically for or scathingly against, but that is the fate of all new work that threatens traditional views."

Has Shaw shown a way to answer Edna St. Vincent Millay:

> Upon this gifted age rains from the
> sky a meteoric shower of facts...
> they lie unquestioned, uncombined.
> Wisdom enough to leech us of our ills
> is daily spun, but there exists
> no loom to weave it into fabric.

Contents

Foreword by William Glen — vii

Figures, Appendixes, Chapter Notes — xvi

Prologue: Reverie, Obsession, and Algorithmic Natural Selection in Stalking the Nonlinear Paradigm in Science — xxi

Acknowledgments — xxxix

A Note on Terms, Conventions, and Cross-References — xli

Postscript Added in Proof — xliii

Introduction — 1

The Legend of the Beginning Written in the Earth 6 / Back to the Future: Some Lessons from the Satellites 8 / Some Terms and Concepts, and a "Travel Advisory" 12 / The Random State: Central Dogma of Science 16 / The First Transformation of Thought 18 / The Second Transformation of Thought 26 / Impact Cratering: The Coordinating Database of Earth History 30 / The Celestial System of Geographic Coordinates 31 / Cratering Patterns and Geodesy: Distributions of the Mass and Shape of the Geoid 33 / Cratering Patterns and Paleo-magnetism 37 / The Celestial Reference Frame and Geocentric Axial Dipole Hypotheses: A Paradox of Poles 42 / Cratering Patterns of Late Cretaceous and Younger Ages 48 / Recapitulation of the Celestial Reference Frame Hypothesis 49 / The Celestial Reference Frame Hypothesis and the Evolution of the Biosphere 53 / General Plan of the Book 56

Contents

	Table of Equations	**60**
	Figures: The Basic Illustrations	**61**
1.	**Geographic Patterns of Impacts on Earth**	**118**
2.	**Temporal Patterns of Impacts on Earth**	**126**
3.	**Geometric Complexity and Nonlinear Processes in the Solar System**	**132**

The Solar System as a Fractal Object 135 / Shower Cycles of Impactor-Impactee Bombardments in the Solar System 139

4.	**Chaotic Orbits and Nonlinear Resonances**	**142**

The Asteroid Belt 144 / Analogies Between the Asteroid Belt, the Kuiper Belt, and the Oort Cloud 150

5.	**Large- and Small-Scale Chaotic Crises, Intermittency, Universality, and the Big Picture**	**156**
6.	**Spin-Orbit Resonances and Proximal Flight Control of Objects in the Celestial Reference Frame**	**165**

Spin-Orbit Resonances with Earth 169 / Ordering of the Planets by Resonance with the "Spin" of the Solar System 178 / Satellitic Orbits and the Celestial Reference Frame 194

7.	**Correlations with and Consequences of the Celestial Reference Frame: In Situ Astrogeology**	**203**

Cratering Patterns, Iridium Anomalies (and Shocked Quartz), Flood-Basalt Provinces, and "Plant Pathology" 204 / Kinematic Correlations with Impact Origins of Flood-Basalt Provinces 211 / The Geomagnetic Signature of a Celestially Coupled Earth 217 / Geomagnetic Reversals and Coupling Between Interior Crises of the CRF and the Core Dynamo 227 / Interior Crises in Numerical Studies of the Rikitake Two-Disk Dynamo 232

8.	**Critical Self-Organization, Universality, Chaotic Crises, and the Celestial Reference Frame**	**235**

Crisis-Network Resonances of External and Internal "Satellitic" Planetary Spaces 238 / Energy Budgets and Straw Models: Self-Similar Cycles

Coupled Between the Sun and Earth 245 / The Celestial Reference Frame and Paleogeography 259 / Schematic Paleogeographic History of the Earth Following Formation of the Moon by a Collisional Mechanism 276 / Nonlinear Biostratigraphic Codification and the Celestial Reference Frame 283 / Intermittent Biological Crises, and Background Extinction vis-à-vis Mass Extinction 292

9. **Summary of Conclusions** — 307

   Epilogue: Copernican Craters and the Phases of the Moon — 319

   Appendixes: The Supplementary Illustrations — 333

   Faraday, on the Fate of Hypotheses — 381

   Chapter Notes — 385

   References Cited — 567

   Index — 645

Figures, Appendixes, Chapter Notes

Figures: The Basic Illustrations

1. Precambrian craters: Atlantic view	62
2. North polar view of composite cratering by reported age	63
3. North polar view of crater ages 50–100 Ma	64
4. North polar view of impact loci and continents	66
5. North polar projection, transformed axes	68
6. Phanerozoic craters by age and minimum diameter	69
7. Mesozoic and Cenozoic craters by age and minimum diameter	70
8. Cratering frequencies, geomagnetic reversals, and magma production	71
9. Distributions of cumulative planetary mass and fictive density in the Solar System	72
10. Orbital and rotational energy distribution of Solar System	76
11. Multifractal domains of the Solar System	78
12. Nonlinear orbital computations for asteroids	80
13. Frequency distributions and orbital structure of Asteroid Belt	81
14. Numerical meaning of Devil's Staircase	85
15. Orbital characteristics of cometary reservoirs	86
16. Monte Carlo delivery of comets to inner Solar System	87
17. Oscillations of a forced magnetoelastic ribbon	88
18. Ferromagnetic resonance and spin waves in YIG spheres	90
19. Approach scenarios for near-Earth objects	91
20. Spin-orbit resonances for Earth	94
21. Generalized "spin-orbit" resonances of the Solar System	95
22. 3-D perturbations of asteroids	97

Figures, Appendixes, Chapter Notes

23.	North polar view of nodal great circles of craters	98
24.	Distribution of K/T iridium anomalies	99
25.	Flood basalt provinces viewed from North Pole	100
26.	Mercator projection of impact cratering nodes and swaths, and flood basalt provinces	101
27.	North polar projection of cratering nodes and magnetic flux into the core	102
28.	Geomagnetic patterns and global cratering nodes and swaths	103
29.	Chaotic motion in a thermosyphon loop	104
30.	Mass-impact and geomagnetic-reversal frequencies	104
31.	Mean times between impacts and between geomagnetic reversals	106
32.	Rikitake two-disk dynamo and chaotic crises	108
33.	Impacts of large objects with the Earth following origin of the Moon	110
34.	Paleogeography and cratering pattern	112
35.	Frequencies of occurrence of geologic periods	113
36.	Biological diversity during Phanerozoic	114
37.	Copernican cratering of the Moon and Earth's Phanerozoic cratering record	116

Appendixes: The Supplementary Illustrations

A1.	North polar equal-area projection of new craters	334
A2.	Cratering pattern for ages less than 50 Ma	336
A3.	Phanerozoic to modern impact trajectories	336
A4.	Northern Hemisphere cratering pattern	340
A5.	Time distribution of cratering and geomagnetic, magmatic, and plate-tectonic processes	341
A6.	Nonlinear resonance grid corresponding to a generalized Titius-Bode relationship for the planets	342
A7.	Density of fractal planetary nebula relative to assumed nebular mass and assumed number of planets	346
A8.	Fractal size distributions of planets, satellites, and asteroids	348
A9.	Chaotic-resonance relations for asteroids relative to Kirkwood Gaps and Devil's Staircase	349
A10.	Pattern of major Mercurian impact basins	350
A11.	Generalized resonance diagram for the Solar System	351
A12.	Index maps of global "hot spots," cratering nodes, swaths, and nodal great circles	352
A13.	North polar projection of cratering nodes, nodal great circles, "hot spots," and "hot spot" great circles	353
A14.	Great Circles through global "hot spots" and their antipodes	355
A15.	"Hot spot" great circles in equatorial equal-area projection viewed from Americas	356

Figures, Appendixes, Chapter Notes

A16. North polar equal-area projections of "hot-spots" and their antipodes — 357
A17. Orbital/cratering great circles compared with "hot spot" antipodal great circles — 361
A18. Great circles and small circles through antipodes of global "hot spots," compared with Phanerozoic crater pattern — 362
A19. Apparent polar-wander paths during the Phanerozoic — 364
A20. North polar equal-area projections of African polar-wander paths — 366
A21. North American and Eurasian apparent polar-wander paths — 368
A22. Africa, India, North America, and Eurasia apparent polar-wander paths compared with cratering and impact patterns — 369
A23. Crater patterns 150 Ma compared with North American apparent polar-wander path — 372
A24. Volcanic "hot spots" on Io — 373
A25. Comparison of cratering and/or "hot spot" patterns for Earth, Moon, Mercury, and Io — 375
A26. Schematic pre-Phanerozoic continent distribution — 376
A27. Global crater distribution compared with dominantly Precambrian cratonic terranes — 378

Topics of Chapter Notes

P.1: Mathematical characterization of feedback systems — 385
P.2: Henri Poincaré and a brief history of nonlinear dynamics — 387
P.3: Attractors — 388
P.4: Universality parameters — 396
P.5: Strain-energy feedback — 401
P.6: Myth of the Laplacian ideal — 404
P.7: Michael Faraday, the candle, and experimental dynamics — 410
I.1: Rawal, self-similar "Oort Clouds," and cosmochemistry — 414
I.2: Algorithmic complexity and computational effort — 415
I.3: Pulsars, quasars, blazars, and spatiotemporal scaling in systems of coupled oscillators — 417
I.4: Literature sources of terminology and classification in nonlinear dynamics — 420
I.5: Rheology, feedback, and deformation fields — 421
I.6: Phase space — 422
I.7: Virtual geomagnetic pole (VGP): positions, paths, and patterns — 422
I.8: Uniformitarianism of James Hutton — 423
1.1: K/T craters and the age of the Popigai structure, Siberia — 426
1.2: Low-angle trajectories, grazing encounters, bolides, skipping impacts, and aligned sets of multiple craters — 427
2.1: Orbital focusing and delivery rates of meteoroids to Earth — 429
3.1: Dynamical resonance and concepts of synchronization — 429

Figures, Appendixes, Chapter Notes

3.2: Binary stars, planetary nebulas, and the Solar System	433
3.3: Orbital structure of the Solar System	436
4.1: Commensurability and nonlinear resonance	438
4.2: Discovery of Ceres, and the evolution of cosmological concepts	440
4.3: Comet sources, solar-galactic dynamics, and cosmochemistry	454
5.1: State space and the numerical scaling structure of the Cosmos	456
5.2: Self-organized tuning of feedback systems in Nature	458
5.3: Dynamical dimensions of phase-space data and attractor states	459
6.1: "Fuzzy" routes for capture of satellites by primary bodies	464
6.2: Synchronous lunar spin-orbit resonance condition	476
6.3: Extra-Solar "Oort Clouds," the Galaxy, and cosmochemistry	477
6.4: Center of mass (barycenter) and planetary dynamics	479
6.5: Ambiguity in celestial dynamics, and programs of search for new objects in the Solar System	482
6.6: n-body interactions in the Solar System	486
6.7: Asteroid taxonomy and orbital-dynamic uncertainty	487
6.8: Planetary resonances of Titus-Bode type and planet stabilities	488
6.9: New horizons in terrestrial-cosmological research	492
6.10: Astronomical discovery and the status of orbital theory	494
6.11: Geocentric system of spin-orbit resonances	497
7.1: Geographic invariances in geophysical patterns	498
7.2: Geophysical fluid dynamics, viscosity, and dissipative structures	501
7.3: Dynamo theory, "fast" dynamos, and chaos	503
7.4: Self-organized criticality, coupled oscillators, and chaotic crises	503
7.5: Gaian-symbiogenetic concepts of terrestrial-organic evolution	504
8.1: Terrestrial recurrence patterns, and paleomagnetic-paleontologic time scales	505
8.2: Impact dynamics and the core dynamo	514
8.3: Magmas and plasmas	516
8.4: Solar activity	517
8.5: Coupling between solar activity, mass impacts, and Earth processes	518
8.6: Direct and antipodal patterns of "hot spots" and mass impacts	519
8.7: Protoplanet collisions with proto-Earth	520
8.8: Crustal foundering and the cooling of magma oceans	520
8.9: Collisional origin of the Moon, and Earth's protomantle	525
8.10: Supercontinent cycles	526
8.11: Ambiguities in the record of glaciation, tillites vs. impactites	527
8.12: Ager's "catastrophes," and the language of biostratigraphy	529
8.13: Geometric progressions, power-law behavior, and logarithmic scaling in Nature	533
8.14: Clams vs. brachiopods, natural selection, competition for survival, Darwin's "wedge," and nonlinear dynamics	537

xix

Figures, Appendixes, Chapter Notes

8.15: "Uncertainty-of-precedence principle" and the origin of species, vortex trees, "mutational events," and critical self-organization in Nature (Cosmos, galaxies, planetary nebulas, dynamos, stream networks, and living organisms)	539
9.1: Recent enhancements in discoveries of near-Earth objects (NEO)s	547
9.2: T-Tauri, FU-Orionis types of stellar evolution, binary stars, and the early Solar System	548
E.1: Story of the Beginning, and cosmic world order	551
A.1: North American Great Fireball Procession (GFBP) of 9 February, 1913	557
A.2: Nonlinear clockworks, and the periodic structure of planetary systems	559
A.3: Geologic boundary events and abrupt changes in the trajectories of polar wander paths (geodynamic crises)	561
A.4: Jules Verne, chaotic dynamics, and mass impacts as excitations of sudden changes in Earth's parameters of rotation and nutation	561

PROLOGUE

Reverie, Obsession, and Algorithmic Natural Selection in Stalking the Nonlinear Paradigm in Science

> *Life is — when all is said and done — essentially a matter of research.*
> — A shavian saying

Among the most notable recent events in the history of science, two stand out: (1) the revolution in biology associated with the breaking of the genetic code and an enunciation of the general principles of biological organization (Judson, 1979); and (2) the revolution in geology associated with the breaking of the geomagnetic code (or time scale) and the enunciation of the general principles of plate-tectonic organization (Glen, 1982). Although in both cases the breakthroughs that opened the floodgates were particular events, leaping forth at particular times, the new disciplinary fields that the two provoked had each had their precursors, and both new fields experienced an interval of greatly accelerated activity in the 1960's. This was a period notable not only for these major paradigmatic shifts in biology and geology but also for the inception of a third, involving both physics and mathematics. Though this one — which has come to be associated with names such as nonlinear dynamics and deterministic chaos — would be operative by the 1970's, it would not be generally recognized for another two decades, and even now it is not fully accepted in the strongholds of disciplinary conservatism. Significantly, the two principal chronicles cited above (the works by Judson and Glen, respectively; see their prefatory statements) were both initiated roughly a decade after the first peak in the acceleration of results produced by the new paradigms — and the rates of publication within the new fields, in contrast with their derivatives, the accelerations, have in both instances undoubtedly continued to climb (though I have not compiled their statistics).

The acceleration of general activity in the Earth sciences, and in a number of other fields, actually followed that in the biological sciences by several years, thus manifesting a form of time-delayed resonance — a phenomenon that is important in the discussion of coupled nonlinear-dynamical effects between mass impacts and the internally resonant behaviors of geological processes, an essential theme

Prologue

of this book. [Curiously — but perhaps not surprisingly, in view of the synergistic (nonlinear) context of multidisciplinary resonances noted both here and subsequently in the present volume — paradigm shifts, whether permanent or transitory, appeared to have got under way in this same timeframe within many other disciplinary fields of study, examples being astronomy (discoveries of many new phenomena; cf. Notes I.3, I.5, 4.1, 4.2), unified field theories (the inception of ideas found in the most advanced present-day theories; see Note 4.2), linguistics ("transformational grammar"), anthropology ("structuralism"), and the motivational areas of humanistic psychology and cultural change (for instance the aftermaths we are still experiencing from the Vietnam debacle and the "lost" generation of the 1960's, as in the transformational — "consciousness-raising" — movements of the 1970's, etc.).]

Parallels between sociological-psychological phenomena and physical phenomena are conspicuous to anyone who, as I have done, has studied the behaviors of feedback systems in diverse contexts **(Note P.1)**, including: feedback in the rheology of materials (e.g., Gruntfest, 1963; Gruntfest et al., 1964; Shaw, 1969; Gruntfest and Shaw, 1974); "industrial dynamics" and feedback in supranational social dynamics and other global contexts (as developed by Forrester, 1961, 1973; cf. Meadows and Meadows, 1973; Cole et al., 1973; Shaw, 1982), and particularly feedback theory as applied to questions of the stability of radioactive-waste isolation strategies (Shaw, 1978, 1981; Shaw et al., 1980b), the transport of magma from the mantle to the root systems of volcanic eruptions (see Notes P.1, P.5, I.5, 8.3; and Shaw, 1969, 1973, 1980, 1987a, 1988b; Shaw and Jackson, 1973; Shaw et al., 1980a; Shaw and Chouet, 1991), and the evolution of the Earth in the Cosmos (Shaw, 1970, 1983a, b, 1988a; Shaw et al., 1971; Gruntfest and Shaw, 1974). The practices of science represent the genetic crossover, as well as the coupling mechanisms, by which the behavioral aspects of these phenomenologies are, in some instances, perpetuated and, in other instances, understood. This places the activities of science and the history of science in the unique position of representing not only our knowledge of the patterns of behavior in our physical environment, but also the patterns of organization in human behavior and our knowledge of *those* patterns.

Far from being detached and objective, scientists are caught up in *all* of these worlds. Research in science is a bellwether, a system of self-organized activity (scientific programs) within a larger system of self-organized activity (social programs), the whole connected, by means of a system of more specific oscillating actions (the "bells and whistles" of our scientific research tools), with yet another system of self-organized activity (the physical and biological environment). Those who do scientific research in such a context — the young idealists of favored specialties, and the established experts of those specialties — deserve our sympathy, for the crossover nexus among these systems is, figuratively speaking, subject to the greatest stress, while simultaneously offering the greatest potential for benchmark change (a role metaphorically analogous to the self-organizational processes and events that accompany "crossing over" during the meiotic stage of biological reproduction). In the context of the present work, we see scientists acting as agents in the passing on of genetic information for social and cultural change, in somewhat the same way that mass impacts — as I interpret them — act as agents in the passing on of genetic information, from the Solar System and the universe at large, for terrestrial physical and biological change, which is the subject of this book.

Peak events and periodicities in such feedback systems are notoriously punctuated and complex, because the outcomes rarely have obvious proportional (*linear*) relationships to any particular mechanisms, subsystems, or arithmetic combinations of mechanisms and subsystems. And any one or more of the "parts" of a feedback system, as well as any of its output functions, can be either continuous or discontinuous (*singular*) in space and/or in time. In short, multiple-feedback systems often display both unintuited and unpredictable behaviors. Equations or sets of equations that might be written to describe feedback systems generally cannot be solved analytically (i.e., by closed-form mathematical manipulations that give unique numerical answers). In this computer age, we might rephrase this to say that any digital and/or analog data streams that are produced by implementing computer-algorithmic descriptions of feedback-system responses — including their naturally occurring analogues (e.g., the propagation of a solitary wave in the ocean, the orbital motion of an asteroid in the Solar System) — to a given set of input conditions are generally nonunique. This is not to say that specific solutions (stable solutions) never exist in *any* type of feedback system, but rather that, in general, we cannot *count on* finding solutions prior to studying the possible numerical behaviors of the system — and some types of feedback systems may never yield stable solutions. (For some examples, see Note P.1.)

Although it is possible to construct artificial examples of directly proportional feedback (as in feedback control of a simple laboratory device, or of an especially simple industrial process), most feedback systems, and especially natural ones, involve one or more mechanisms and/or processes in which the output is not proportional to the input. Usually there is some sort of amplification and/or suppression of one or more particular motions or signals. These are typically called *nonlinear* effects, and systems of coupled nonlinear effects often are called, generically, *nonlinear feedback systems,* or simply *nonlinear systems* (cf. Notes P.2 and P.3, referred to in the following discussion). In seismic phenomena, for example, the response function of the global tectonic engine consists, in part, of sets of discontinuous events (earthquakes) and peak events (great earthquakes) that would appear to be more discontinuous (singular) than the background of the smallest events. Yet, large earthquakes can sometimes occur as paired events — or even as multiple events of similar or increasing magnitudes (cf. Shaw, 1987c). Analogously, the ongoing revolution in the Earth sciences is showing evidence of a following "seismic" wave of discovery of even greater import than plate tectonics (cf. Glen 1975, 1982), as represented by the current studies of relationships between terrestrial and extraterrestrial phenomena, initiated by the sudden onset (actually the rediscovery) in 1979–80 of the so-called *Impact-Extinction Debates* (see Note 8.1; and Glen, 1990, 1992, 1994).

In parallel with the recent histories of biology and geology, space exploration and planetological studies also have burgeoned since the 1960's, and astronomers have made new discovery after new discovery, in breathtaking profusion. Astronomical *masers*, an intragalactic phenomenon of coherent microwave radiation, were discovered in 1965 (Maran, 1992, pp. 414ff; Cohen, 1992; Elitzur, 1992; cf. Note I.5); *pulsars*, a stellar phenomenon closely related to the oscillatory phenomena discussed in the present work, were discovered in 1967 (see Pasachoff, 1991, p. 474; and Note I.3); and quasi-stellar objects and related radio sources, *quasars*, became established as an amazing new type of astronomical phenomenon during the 1960's (e.g., Pasachoff, 1991, Chap. 32; Kaufmann, 1991, Chap. 27; cf. Notes I.3, I.5, 4.1, 4.2). The data and concepts of stellar evolution — including the concept of *gravitational lenses* and theories of *dark matter* that are intrinsically related to the nature of mass distributions in the universe [see the readable reviews by Kristian (1992) and Mateo (1992)] — have advanced almost too rapidly for us to follow them (e.g., Pasachoff, 1991; Kaufmann, 1991; Maran, 1992). And in the realm of unifying theories of the universe, the so-called *superstring theory* in physics (see Note 4.2), presently in its heyday, really had its inception in 1968 (see Kaku and Trainer, 1987, pp. 87ff). Eclipsed by all of these revolutionary discoveries, and central to the main theme of this book, were two breakthroughs in nonlinear dynamics, also made during the 1960's, neither of which initially received much attention from the scientific community, even in the areas of study where they were applied. One was represented by the enunciation of the physical principle called *sensitive dependence on initial conditions* (Lorenz, 1963, 1964), a principle that can be characterized, simplistically, by a scenario in which the flapping, for example, of the wings of a Malaysian butterfly leads, by stages, to a tornado in Alabama **(Note P.2)**. The other was an awakening to principles of thermomechanical feedback in rheology, as first advanced by I. J. Gruntfest in 1963 and subsequently explored by me during that decade, both experimentally and theoretically, relative to rheological behaviors in the Earth at elevated temperatures and pressures — principles that included within their many implications variants of the Lorenz effect, cast in the analogous form of *exponential divergence from initial conditions* (Gruntfest 1963; Gruntfest et al., 1964; Gruntfest and Shaw, 1974; Shaw, 1969, 1991). (It is perhaps worth repeating that within the most general context of nonlinear dynamics, which would include cultural, social, and psychological phenomena, the unusual activity and fecundity of the 1960's, on so many fronts, would seem to have represented a generalized chaotic crisis, expressed by scintillating chaotic bursts in many different realms of human endeavor; cf. Notes 5.1–5.3.)

According to the interdisciplinary principles of nonlinear dynamics advanced herein, none of these actual or latent events of revolutionary change in scientific research happened independently, and none was without precedent (cf. Note 8.15).

Poincaré, for example, in his studies of mathematical methods of prediction as applied to celestial mechanics, had already enunciated the central notion of sensitive dependence on initial conditions (Note P.2) more than 60 years before the same idea was independently discovered in computer experiments by Edward Lorenz, a meteorologist studying methods of global weather forecasting. In contrast with the period during which Poincaré did his research, however, recent developments have occurred within a particularly confused and confusing, but not random, global context of interacting physical-biological-psychological-sociological-cultural-political change, a condition perhaps best described by the phrase *chaotic ferment,* a ferment sometimes carried to the point of *foment* (including personal and political dissent, the inciting of riots, rebellion, etc.), which finds its nonlinear-dynamical parallels in the phenomena of *chaotic crises* and *chaotic bursts* (cf. Chap. 5).

All of these developments illustrate the operation of a new and subsuming mathematical paradigm — a paradigm that in principle, and in combination with Poincaré's latent revision of mathematical physics, created the seeds of destruction of the authoritarian grip that traditional mathematics has for so long held on science. And in the process, ironically, the new paradigm opened mathematics to previously unimagined vistas of relevance to relationships among all scientific (and social-scientific) disciplines. This shift in mathematical paradigm often is attributed to a theorem published by Kurt Gödel in 1931, called Gödel's Theorem, or the Incompleteness Theorem, or *the principle of mathematical incompleteness.* Underpinning that work, however, is the encompassing idea called *the principle of self-reference* (see Gödel, 1931; Nagel and Newman, 1958; Spencer-Brown, 1972; Hofstadter, 1979; Chaitin, 1982). Gödel's work, of course — as is the case in any recursive endeavor like scientific "advancement" (cf. Shaw, 1994) — was undertaken in response to the work of others, in this case the idealistic pursuit of mathematical completeness advanced in the *Principia Mathematica* by Bertrand Russell (1872–1970) and Alfred North Whitehead (1861–1947). [I would add that no scientist has ever lacked for indirect precedents (at least not so indirect as most of us like to pretend), or has ever worked productively in the proverbial vacuum or ivory tower, images that some reflection on the dynamics of research would indicate are pure fictions. The romantic image of Alfred Russel Wallace (co-founder with Charles Darwin of the concept of natural selection; cf. Chap. 8 and Notes 8.14 and 8.15) — one that would have him discovering full-blown the principles of organic evolution by virtue of his isolated and heroic research in the head-hunting wilds of Borneo and New Guinea, as amazing and intrepid as that was — is clearly far from the greater truth of historical and contextual precedent (cf. Lull, 1949, Chap. 1; Raup and Stanley, 1978, pp. 432ff; Mayr, 1982, Chap. 8, 1991, Chaps. 1–3). The dynamical crossover with, and implication of, such an *uncertainty-of-precedence principle* (see Note 8.15) is explored further in Shaw (1994).]

The idealistic concept of a "complete mathematics," in the wake of the mathematical foment stirred up by the reputed allegations of the *Principia Mathematica,* was first, and finally, posed as a formal challenge to mathematicians by

Prologue

David Hilbert (1862–1943) in his *twenty-three problems* (Hilbert's Program), the solutions to which, he proposed, would fulfill a system of self-consistent proofs (the Truth Machine) for all of mathematics, including arithmetic (see Kramer, 1970, p. 686; Hofstadter, 1979, pp. 23f; Hodges, 1983, pp. 90ff, 107, 111ff; Casti, 1990, pp. 329ff, 367ff). Gödel's Theorem demolished that vaunted ideal of mathematical authority (e.g., Kramer, 1970, p. 686) — thankfully, we might say, because the parallel between Hilbert's ideal and the specter of political autocracy conjures haunting overtones, particularly when arrayed before the backdrop of the peer-review system of governance in science (cf. Glen, 1989; and see the case in point concerning Kronecker's treatment of Cantor, cited in Note P.7).

Advances in applied physics during the same period enhance this historical theme. During the past three decades, lasers, masers, solitons (e.g., the so-called *stadium wave*, "the wave" performed by onlookers at a stadium event), and so on became familiar nonlinear-dynamical phenomena of technological importance. I mention such things because they demonstrate ways in which coherently focused phenomena and events can be generated from feedbacks among multimode phenomena and events. My own computer experiments in nonlinear dynamics have demonstrated analogous numerical effects in which many-dimensional interactions can generate patterns with few-dimensional and even fixed-point geometries (cf. Shaw and Doherty, 1983; Gu et al., 1984; Kaneko, 1986, 1990; Crutchfield and Kaneko, 1987; Shaw, 1988a, b; Willeboordse, 1992). It is difficult to avoid the conclusion that just as the subjects and objects of scientific research follow these principles of nonlinear interactions, so, too, do the phenomenologies of scientific research.

Scientific research organizes itself according to nonlinear-dynamical principles — the principles of self-organized chaos — that are largely unrecognized by the individual researchers and would seem to be recognized even less by those administrators and/or scientist-administrators who plan programs of research dedicated to strictly limited goals (cf. Ruthen, 1993). Thankfully, a knowledge of this phenomenology need not change anyone's individual research bent, but in some circumstances — perhaps in the context of large-scale programs of government-sponsored research (thus with imposed goals) carried out through the auspices of, say, the U.S.G.S. or N.I.H. — a knowledge of the principles and behavioral patterns demonstrated by chaotic-systems research might offer a more effective basis for the self-organization of compatible research strategies, and probably a morale boost as well (e.g., Peters, 1987).

This book exemplifies the study of effectual relationships within a large contextual system (here, the Cosmos) pursued from the overt standpoint of nonlinear systems theories and principles of nonlinear feedback phenomena. Such research follows its own lead, so to speak, in a manner analogous to the way in which the recursion algorithms of nonlinear dynamics themselves operate. An apparent result from one facet of the study is reentered into the study in the recognition that its essential status may be expressed, and/or may be revealed, by the possible effects it has on other facets or related inquiries (as is the case in the

Prologue

neurophysiological model of phasic reentrant signaling and neuronal group selection advanced by Edelman, 1978, 1985, 1987, 1989; cf. Skarda and Freeman, 1987; Freeman, 1991, 1992; Rose, 1992). Such a strategy can be pursued, however, only if no arbitrary bounds have been placed on the range or validity of subjects included in the study.

I began the present study with the spontaneous notion that meteoroid impacts on Earth should be patterned in time and space, a thought preconditioned by a decade of studying nonlinear numerical algorithms. From this vague beginning, I came quickly to the implications that impact phenomena not only on the Earth but also on other planets and satellites of the Solar System should be: (1) related in a demonstrable way to the orbital and spin-orbital characteristics of the planets and satellites, as well as to the behavior of the vast population of smaller objects (asteroids, comets, etc.) that have been preconditioned in their trajectories of approach, and thus their impacts on the larger objects in the Solar System, by the same nonlinear orbital principles; (2) related to the *consequences* of such impacts on Solar System objects (especially for those planets and/or satellites analogous to Earth in tectonic fecundity) via their coupled nonlinear effects on the mass distributions and geodetic features of the impacted objects through the agencies of tectonic, magmatic, geomagnetic, meteorological-oceanographic, and (on Earth) biological interactions; and (3) related by the orbital and geodynamical categories of all such actions — as mediated by (1) and (2) — to, and within, an encompassing self-organized context that is characteristic of the Solar System as a whole, its interactions with the Milky Way Galaxy, and the interactions of the Galaxy with the Cosmos as a whole. In such a self-referential loop of feedback processes, *no one process is independent of the others*; failing to appreciate that fact is the first and foremost mental block that must be overcome in seeking an understanding of the role of mass impacts in Earth history. Meteoroid impacts are not simply cosmic events independently imposed *on* Earth. They are events that have evolved *with* the Earth, and with all other influences affecting (collectively) their own pre-impact orbital histories in the solar-galactic-cosmological context of evolution.

My original title for the book was *A Celestial Reference Frame for Terrestrial Processes*. This working title reflected my central hypothesis that a *celestial reference frame* (CRF) can be identified by which this outline of nonlinear effects can be examined and, at least in part, demonstrated. The CRF confirms the initial assumption that a pattern should exist in the record of impacts on Earth, a pattern that would link Earth's geographic system of coordinates to a celestial system of geometric coordinates (the *celestial sphere* of astronomy and celestial dynamics). Hence, for the first time in the history of geological and astronomical research, a direct kinematic-dynamic linkage between the internal dynamical evolution of a planet (Earth) and its astronomical setting is postulated, and is explored relative to the observed phenomenologies of both the Earth and the Solar System. Nonlinear dynamics represents both the perceptive device (the *lens*) through which the problem is revealed and explored and the cohesiveness (the *glue*) that holds the diverse phenomenologies together in common patterns of behavior and dimen-

Prologue

sional effects (*universality principles*; cf. Notes P.4 and I.8). What follows recounts how I came to hold this viewpoint.

About twelve years ago a part-time word-processing assistant named Raynelle Feinstein was working on a manuscript of mine that concerned geologic feedback systems. One day, she casually mentioned to me that it reminded her of an article by Douglas Hofstadter she had just seen in *Scientific American*. This turned out to be his regular column, Metamagical Themas, for November 1981 (Hofstadter, 1981). That particular column, a now-famous article on the "new" theories of nonlinear dynamics, was titled "Strange Attractors: Mathematical Models Delicately Poised Between Order and Chaos" (**Note P.3**). Hofstadter was in a unique position to evaluate the impact of these discoveries because he had just two years previously published his monumental work *Gödel, Escher, Bach: An Eternal Golden Braid*, a book that builds its case from the central notion of *Strange Loops*. For Hofstadter, the Strange Loop is a global or orchestral form of generalized feedback-systems effect that contains within its own complex hierarchies the essence of its own creation.

I had already studied Hofstadter's treatment of the Strange Loop because I had come to believe that dissipative feedback operates in the universe at large, and the Strange Loop seemed to me to be describing how I believed natural feedback systems work. The prototype for his definition was the "Canon per Tonos" in the "Musical Offering" of J. S. Bach to King Frederick of Prussia, which Bach had based on his visit to Frederick's Court in 1747. The same, seemingly paradoxical, type of effect occurs visually in works by M. C. Escher, and mathematically in Gödel's Theorem. Visual constructs of this character are often called optical illusions, and given this common usage for seemingly anomalous visual effects, it makes equally good sense to call the analogous mathematical and auditory effects, respectively, *mathematical illusions* and *auditory illusions* (cf. Shepard, 1964). According to Deutsch (1991, 1992a, b), the phenomenon of auditory illusions is related to characteristic regimes of mental processing that simultaneously engage the nature of language perception—implying that the way one perceives auditory illusions (e.g., in musical forms) is partly dependent on one's native language, dialect, and personal idiosyncrasies of language recognition.

The implication of language in the perception of illusory patterns of sound, I suggest, may have an even greater contextual meaning—particularly when it is applied to the notions of visual and mathematical illusions. To me, these are fascinating questions that probably have great import for sensory and cognitive perception in general. Unfortunately, they cannot be dealt with here. It would appear, however, that one's aptitude for (general perception of) mathematical languages, as well as one's perception of geometric (i.e., mathematical) constructions, is grounded in the relationship between linguistics (in the broadest sense) and neurophysiological information processing (cf. Edelman, 1978, 1985, 1987, 1989; Shafto, 1985a, b; Shaw, 1987c, p. 5, 1994; Levine, 1988; Rose, 1992).

In microcosm, each of these types of description—mathematical, visual, and auditory—represents a system in which self-reference is essential. In a more global context, all three would appear to form sets of interacting perceptions wherein the mental machinery of linguistic information processing plays a central role. The same types of nonlinear relationships apply to the neurophysiological system in general — as documented, for example, by Skarda and Freeman (1987) and Freeman (1991, 1992) — providing a richness and subtlety of communicative complexity that both includes and goes beyond David Bohm's concept of "the implicate order" (in the archaic, but expressive, sense of being folded together; cf. Bohm and Peat, 1987, Chap. 4) by invoking sensitive dependence on initial conditions, the essential principle of chaotic dynamics [i.e., paradoxically, the quantum-dynamical precedent of Bohm's theory — itself often seemingly paradoxical — is fundamentally periodic in nature, hence realms of behavior must exist between those of quantum dynamics in the strict sense and the behaviors of "classical chaos" in the modern nonlinear-dynamical context (cf. Berry, 1987; Ford et al., 1990)]. In mathematical set theory, self-reference is represented by those categories of sets wherein each set contains itself as a member of its set. Such sets are therefore never-ending, "eternal," or eternally cyclical. This self-referential nature is intrinsic to mathematical description, for we can never construct a complete mathematics. Nor can we construct one that ends in an all-encompassing theorem that describes any and all conceivable sets and hence stands uniquely "above" and separate from membership in any one of them and has no peers or superiors.

I can conclude only that the universe itself is of this nature, because no matter how it is defined — whether in Newtonian, Einsteinian, quantum-dynamical, or chaotic models — there is no way to encapsulate it as a unique existence that we could somehow stand outside of and describe as an entity, or as anything complete (cf. de Vaucouleurs, 1970). There is always some way to expand it, split it, cluster it, cycle it, or otherwise see in it aspects of incompleteness, whether I imagine myself to be in space, in time, or in spacetime. If this were not so, then it would be possible to say that there is such a thing as an all-powerful mathematics capable of circumscribing the universe, hence circumscribing mathematics itself.

It is not surprising that at this point we come to the limit of words, or of any form of communicating by parts. If "holism" characterizes a system of interacting wholes, then we are left with nothing but the "ism," as Hofstadter (1979, p. 254) has put it. In Zen, this "ism" only implicates and, lacking any existential pointer, points beyond form to (indicates) that which cannot be given any definition. Even the phrase "a whole that cannot be broken into parts at all" seems ruled out as the ismic meaning because it implies that there is something called "all of it"; hence it hints at completeness (the word "void" is often used for the indefinable whole). A discussion of parallel nature by Spencer-Brown (1972, p. 105) suggests a simpler way to put it, one that is consistent with my discussion of self-referential phenomena. The following is what I take to be Spencer-Brown's meaning, in terms that are concordant with my own views: *The universe is that which can observe only aspects of itself, and which is always and eternally self-involved in the process of self-referential observation.* This evasion, if it is one, contains notions

Prologue

of recurring cyclical processing, self-similarity, self-organization, and, within all that, the implication of universality among the agencies and modalities of observation (cf. Notes P.1–P.4). In short, such a viewpoint is consistent with nonlinear dynamics and with the attempts made in this book to see our cosmological environment and its history in such a light.

In haiku, or Zen, or the Tao of Pooh, the *ineffable transfinite* (*IT*) is simply accepted and experienced (see Note E.1) — meaning *mindfully digested*, as the mental analogue of metabolism, a *metabolism of impressions* in the sense of Ouspensky (1949, pp. 181ff), rather than being *thought about*, in the sense of being endlessly chewed upon "like a dog with a bone" — and thereby allowed to *be* (cf. Spencer-Brown, 1972; Capra, 1977; Hofstadter, 1979; Hoff, 1982; Dauben, 1983; Shaw, 1988c, 1994).

Thus the Strange Loop, as developed by Hofstadter, represented for me an excellent symbol for the concept of cyclical feedback, one that I had come to view as the essential ingredient of evolving processes. It also formed a bridge that permitted me to formulate the necessary modulations, as in canons and fugues, by which to arrive at a reconciliation with the nature of recursive effects that I later found to be beautifully displayed by the never-ending varieties of algorithms of nonlinear dynamics, especially by the eternally cyclical circle maps (i.e., phase-space plots of clock-like character, such as the sine-circle map of Table 1, Sec. D; see Note P.3) of coupled-oscillator systems that figure prominently in the present work. Together with the concept of *chaotic crises*, as I apply it to terrestrial-cosmological phenomena herein, and the relationship between that concept and *self-organized criticality*, the recursive nature of nonlinear dynamics has come to symbolize for me Nature's "musical offering" to the Cosmos. Hofstadter's 1981 description of the emergent, holistic field of nonlinear dynamics was filled, apparently for him as well as for me, with both excitement and insight. It is well worth reading, and rereading, especially in the context of his book, even though, today, the relevant literature has grown in little more than a decade from a few articles on particular aspects of nonlinear dynamics to many thousands of articles, hundreds of books, and several new journals dedicated specifically to its furtherance.

This incident, my reading of Hofstadter's description of *strange attractors*, reignited what has been a central theme, if not an obsession, of my research life — both officially and unofficially. How I had missed Robert May's article in *Nature* five years earlier is a mystery — though the reason I had not heard about the much earlier work of Edward Lorenz in computational meteorology is less obscure (Lorenz, 1963, 1964; May, 1976). For those days saw the beginning of specialization run amok in science, and though I had tried to follow developments in several aspects of the natural sciences, my efforts apparently overlooked meteorology. The lessons in these incidents, and their ramifications for the course of my research, are too many to describe. During that same period, Earth Science was being taught a lesson of inestimable value concerning the interdisciplinary outlook, but one that apparently was not taken to heart. The success of "The New Global Tectonics" had vindicated essential aspects of Alfred Wegener's much earlier notions of Continental Drift. And Wegener, along with some others working in structural geology — Hans Cloos and Ernst Cloos immediately come to mind — had shown that interdisciplinary analogies, such as those between meteorology and geology, can be powerful indeed. I had been preoccupied with forms of nonlinear systems analysis that had proved productive in materials science and rheology (the latter being, literally, the study of flow, particularly in relation to the material makeup of a system). If I had also been aware of what Lorenz was doing in the 1960's I would the sooner have found a more general forum for descriptions of the nonlinearities of magmatic and tectonic interactions that had come to fascinate me in my first forays into studies of viscous flow and diffusive phenomena, which also began in the early 1960's (Shaw, 1963, 1965, 1972, 1974). This work, which had first concerned the behavior of highly viscous silicic melt systems, formed the basis for contributions to the origin and evolution of the granitic rocks and of large-scale continental rhyolitic volcanism (the evolution of caldera-forming and ash-flow-forming crustal magma chambers) that many of my colleagues were working on at that time — and that many of them are working on even now because of the richness, complexity, and importance of the continental magmatic history.

Extending the scope of my studies to include the rheology of basaltic magma, beginning in 1965 with a program of field measurements of ponded basaltic lava in Hawaii (Shaw et al., 1968), lent great impetus to my fascination with nonlinearity, for compared with the ranges of behaviors I had seen in other magmatic silicate systems, the rich rheologies of basaltic lavas typical of Hawaii and other oceanic systems of the world, as well as in continental basaltic systems, "compresses" them within a relatively small range of temperatures between the solidus and the liquidus (cf. Notes P.5, I.5, and 8.3). These studies, and analog experiments conducted in parallel with them, became the testing ground for an evolving viewpoint in which thermomechanical feedback was the central theme. The equations involved were highly nonlinear and unstable, and, in my impressionable mind, dissipation soon was seen as the fundamental organizing — as well as disorganizing — principle of Earth processes (Shaw, 1969). This viewpoint was augmented by an affiliation with Irving Gruntfest, then at the General Electric Research Laboratory near Valley Forge, Pennsylvania. His influence on my thinking was profound, and perhaps even too great, in the sense that the focus of my attention on the materials and rheological literature delayed my discovery of equally tillable fields being exploited in other areas of applied mathematics — such as the aforementioned work by Edward Lorenz in meteorology (Lorenz, 1963, 1964), that by Robert May in the dynamics of biological populations (May, 1976), and the work done by several persons at Los Alamos that led eventually to the pivotal studies of *mathematical universality* carried out by Mitchell Feigenbaum (Feigenbaum, 1979a, b, 1980). Universality is the principle of common effect that runs as a unifying theme through the application of nonlinear dynamics to every kind of phenomenology, regardless of material makeup or scales of time or size **(Note P.4)**.

During this period I was posted in Washington, D.C., but that great national center turned out to be a conceptual backwater relative to what was happening elsewhere in science. When I returned to the West, in 1974, as I later found out, I was only a stone's throw from institutions where revolutionary work on two seemingly different fronts was poised to be carried out by the end of the decade. The new paradigm that Hofstadter described in 1981 was, and would continue to be, greatly aided and abetted by a cadre of brash young graduate students at the University of California at Santa Cruz (see Gleick, 1987, on "The Dynamical Systems Collective," pp. 241ff). And some astounding proposals of great import to the future of geological studies — and to the writing of this book — were about to be put on the table at the University of California at Berkeley by Luis Alvarez and others at the Lawrence Berkeley Laboratory.

Robert Shaw, working at UC Santa Cruz, was about to begin the research that eventually led to the publication in 1981 of a remarkable paper titled "Strange Attractors, Chaotic Behavior, and Information Flow," one of the first, if not the first, detailed discussions of the relationship between the world of nonlinear recursion experiments and Claude Shannon's well-known numerical concepts of information flow. [In the process, he had abandoned a nearly completed Ph.D. thesis on superconductivity in favor of the fascinations he had discovered in the exploration of attractors (see Note P.3), using an abandoned analog computer he had found in the basement of the Physics Department.] Shaw later constructed a model of the dripping faucet and studied its behavior as a paradigm for a great variety of problems in nonlinear intermittency (R. Shaw, 1984). Many of these applications are relevant to geophysics: one example is the intermittency of volcanic eruptions, and another is the intermittent flow of the solar wind leading to the formation of "plasmoids" in Earth's magnetotail (see my review: Shaw, 1991).

Rob Shaw was joined in 1977 and 1978 by other, now famous, members of that "Collective" (or "Chaos Cabal," as some UC Santa Cruz associates called it). Additional members of the core group, initially impelled by Shaw's courageous and now almost legendary experimental adventures, included Doyne Farmer, Norman Packard, and James Crutchfield, each of whom has gone on to produce research in nonlinear dynamics of fundamental importance. Farmer had become fascinated by the challenge of beating the gambler's odds in roulette, a game based on a mechanical system that he clearly saw was subject to the rules of dynamics, but subject in an extremely complex and nonlinear way that, like the playing of dice, was generally assumed to be governed by the "random rules of chance." The story behind that ambition, as well as some of the behind-the-scenes goings-on of the Chaos Cabal, has been told in the sometimes hilarious book called *The Eudaemonic Pie*, by Thomas Bass (1985). Farmer's interest in predicting and/or forecasting the outcomes of nonlinear systems behavior, especially in forms based on nonlinear time-series analysis, was influential at the time, and has continued to this day in increasingly sophisticated formats, a methodology to which both Packard and Crutchfield have contributed significantly (see Farmer et al., 1980; Packard et al., 1980; Crutchfield et al., 1986; Crutchfield and Kaneko, 1987; Farmer and Sidorowich, 1988).

All of this was poised to happen, coincidentally, about the time I accepted an assignment as Visiting Professor in Geology at Berkeley for the academic year 1974–75. Though I was now less than a hundred miles from the Santa Cruz campus, I was still light-years away insofar as my awareness of the work there was concerned. My affiliation with UC Berkeley, where much earlier I had done my graduate studies, continued from 1976 through 1980 in the form of intermittent graduate seminars in volcanology. It was during this period that Luis Alvarez and his colleagues at the Lawrence Berkeley Laboratory, of that same campus, were about to converge on the idea that came to motivate a major portion of the present work, an idea that is herein referred to as the Impact-Extinction Hypothesis. The present status of that hypothesis owes much to the enthusiastic work of Luis' son Walter Alvarez, and to Walter's subsequent alignments with a far-sighted and influential group of paleontologists led by David Raup and J. J. Sepkoski, Jr. (see L. Alvarez et al., 1979, 1980; W. Alvarez et al., 1982, 1989, 1991; Sepkoski, 1982, 1989, 1990; Raup and Sepkoski, 1984, 1986; W. Alvarez, 1986, 1990; Raup, 1986a, b, 1988a, b, 1990, 1991; Sepkoski and Raup, 1986; L. Alvarez, 1987; W. Alvarez and Asaro, 1990; and the concise summaries of this history in Glen, 1990 and 1994).

Ironically, I had heard no more than distant hints of any of these goings on until Raynelle Feinstein opened my eyes to what I had been missing in the field of nonlinear dynamics. I have never had an opportunity to thank her, because she left soon after that and I lost track of her whereabouts. If by good fortune this book comes to be read by someone who knows her, perhaps that person will be kind enough to point out this Prologue as a note of my appreciation to someone who, with neither the "credentials" nor the relevance of a formal research role, taught me a great lesson and gave me a great gift, for such is what I perceive to be the impact, the power, and the importance of nonlinear dynamics for science — and far beyond that, for humankind.

By way of the students at Berkeley, in 1980, I had heard about what Luis Alvarez was doing, but frankly I had dismissed it as being remote from my main interests in geodynamics — another major irony, in view of the central thrust of the present work. In 1984, however, I met William Glen, and my research world was dramatically expanded. He brought to life for me the importance and excitement of the Impact-Extinction Hypothesis, as well as alternative theories of mass extinctions, with all of their implications for paleontology and Earth history. I was ready to listen, because by then I had already pursued three years of intensive exploration of nonlinear-dynamical algorithms and how they might impact my concepts of feedback systems and terrestrial evolution. An earlier sojourn into studies of extraterrestrial influences on terrestrial phenomena had contributed to my readiness.

Prologue

During the years 1968 through 1974 I had been caught up in an investigation of the dissipative character of the Earth's tides and their possible influences on geodynamics (Shaw, 1970; Shaw et al., 1971; Gruntfest and Shaw, 1974; Wones and Shaw, 1975; cf. Greeley and Schneid, 1991, p. 998). This I owed in large part to Irv Gruntfest, and to his influence on my realization that small dissipative effects can be amplified by feedback processes into exponentially great departures from thermomechanical equilibrium, a realization that later found its counterpart in the phrase "sensitive dependence on initial conditions," which Lorenz had dramatically demonstrated in his attempts to compute models of the weather. This phenomenon has come to be known popularly as "the butterfly effect," a phrase also attributed to Lorenz (see Gleick, 1987, on "The Butterfly Effect," pp. 9ff). Although the hypothesis of tidal dissipation as a major influence on magmatic phenomena in the Earth has never gained great credence in geology or geophysics, it awakened in my mind a graphic realization that Earth processes do not take place in a vacuum, as it were, as if the Earth moved through space in a bubble or isolation tank, shielded from everything but the Sun's radiation. Such an expanded viewpoint was later vindicated indirectly through the medium of the planetological discoveries made during the international program of space exploration (cf. Rothery, 1992).

In early 1979, active volcanic eruptions were discovered occurring on Io, initially identified in a photograph taken by the U.S. Voyager missions to Jupiter (Morrison, 1982a, b) that happened to catch an eruption plume in the act (see Strom and Schneider, 1982; Rothery, 1992, Plates 2–5). On Io, a satellite of Jupiter about the size of our Moon, near-surface sulfur-rich compounds enhance volcanic-plume activity (Kieffer, 1982). That same year, Stanton Peale and coworkers proposed that volcanism on Io probably was caused by the melting of silicate materials in its interior due to the dissipation of mechanical energy related to tidal interaction with Jupiter (Peale et al., 1979). The molten silicate material (magma), in rising toward the surface, much as it does in the Earth (cf. Shaw, 1980), melts and volatilizes sulfur and other low-melting-temperature compounds that had previously been deposited at and near Io's surface. The role of tidal deformations in the thermal histories of several of Jupiter's satellites was soon widely accepted by many planetologists (Cassen et al., 1982; cf. Pearl and Sinton, 1982; Greeley and Schneid, 1991, p. 998; Rothery, 1992, pp. 23f).

The tidal effect on Io, though maintained by a complicated system of resonances with Europa, resonances between Europa and Ganymede, and, ironically, the tide raised by Io on Jupiter itself, is easily accepted, because of the relatively large and permanent tidal bulge raised by the gravitational influence of nearby Jupiter on the synchronously rotating satellite. (Io's period of rotation is, on average, the same as its orbital period, so that Io maintains a synchronous rotational orientation relative to Jupiter, like Earth's Moon relative to Earth; see Spin-Orbit Resonances with Earth, Chap. 6; cf. Notes P.5, 6.2, 6.4.) On a homogeneous Io, this tidal bulge would have a constant amplitude of about 8 km (a huge amplitude compared with the tides raised on the Earth by the Moon, or on the

Moon by the Earth; cf. Kaula, 1968, pp. 198ff) relative to the equilibrium figure of tidally unperturbed rotation. But because of the effects of these interactions between Io, Europa, Ganymede, and Jupiter, Io's orbit is much more eccentric than would be so in the unperturbed case. As a result, the tidal bulge oscillates in amplitude during each spin-orbit cycle, and the resulting cycle of strain pumps thermal energy into Io (dissipation of tidal-strain energy) in proportion to the imperfection of its elastic properties, as measured by its average *quality factor*, Q. (An analogous effect is produced in the Moon by the Earth, and although dissipation is now minuscule by comparison with that in Io, it was larger during the Moon's early history, and even now it is sufficient to influence low-amplitude seismicity in the Moon; cf. Wones and Shaw, 1975; Wilhelms, 1987, p. 269.) Furthermore, Io contains easily volatilized substances, as witness the sulfur compounds that so colorfully adorn its surface (Kieffer, 1982; Strom and Schneider, 1982; Schaber, 1982), and the dissipated thermal energy is thus more than enough to account for significant volcanic activity; theoretically, it is enough for wholesale melting of virtually the entire satellite (Cassen et al., 1982, p. 99; cf. Pearl and Sinton, 1982).

The analogous arguments that Gruntfest and I had advanced for the Earth, initially in 1968 but not published until 1974, and that David Wones and I had advanced in 1975 for the Moon's early history, had been unconvincing to geophysicists, on the grounds that the energy dissipated in the "solid" Earth by the lunar tide, and vice versa, was considered to be negligible. Few appreciated the fact that, like the inhomogeneous "deposition" of tidal energy in Io, the dissipations of tidal energy in both the Earth and the Moon are extremely localized and focused by *strain-energy feedback* **(Note P.5)**. Now, with the notion of *sensitive dependence on initial conditions* — one of the very effects that we had attempted to apply to the tidal process — dynamically realized, the door is once again open to a contemplation of significant coupling between dissipative processes in the Moon and dissipative processes deep in Earth's interior. [Slowing of the Earth's rotation by the lunar tide has been attributed by geophysicists almost entirely to the existence of the Earth's shallow seas, following the lead of Harold Jeffreys (1920, 1970; cf. Munk and MacDonald, 1960, p. 200), who, in turn, had extrapolated the remarkable study by G. I. Taylor (1919) of tidal friction in the Irish Sea to a highly uncertain estimate of global tidal dissipation — an effect clearly inapplicable to the tidal deceleration of the Moon's rotation, or to those rotational decelerations known to have occurred in other bodies of the Solar System lacking oceans or dense atmospheres.]

The study of volcanism on Io in relation to its orbital setting and tidal dissipation is an example of a conceptual resonance with the seemingly never-ending implications of the hypothesis advanced in the present book. Reexamination of other aspects of the tidal phenomenon was stimulated by a recent paper by Yamaji (1991) that interprets the distribution of "hot spots" on Io in terms of possible fracture patterns induced by its tidal history. Kinematically, Yamaji's hypothesis is a variant of the idea advanced by Gruntfest and Shaw (1974) for the

Earth and by Wones and Shaw (1975) for the Moon (cf. Klein, 1976; Dzurisin, 1980; Weems and Perry, 1989; Greeley and Schneid, 1991, p. 998), in the sense that the consequence of protracted tidal deformation, in addition to contributing directly to the melting of the interior, is to induce flexural instabilities that afford molten material access to the surface, by mechanisms of extensional fracture. In Appendixes 24 and 25 I investigate the parallelism between this process for the Earth and that for Io, and find a remarkable similarity in the distributions of so-called hot spots on the two bodies. [The formal term "hot spot," as it is used in concepts of plate kinematics on Earth, denotes a site of persistently stationary or propagating volcanic venting that delivers magma to the surface from a melting anomaly presumed to exist at some uncertain depth in the mantle, a depth that may range from as little as 60 km beneath a site such as Hawaii to as great — in some hypotheses — as the depth of the core-mantle boundary (cf. Morgan, 1971, 1972; Shaw and Jackson, 1973).]

In view of the main thesis of this book, which concerns the coupling between impact dynamics and geodynamics, the parallelism between tidally induced melting in the Earth and that in Io would further suggest that there should be coupling between impact dynamics and tidal dynamics, as well, in both objects. Twenty years ago this notion might have been received as an outrageous, if not "irresponsible," conjecture. Now, however, it would appear to be but another of the many "coincidences" that seem to occur unsought in the course of examining the phenomenology of the terrestrial-cosmological "connection machine" through the lens of nonlinear dynamics. *It hardly matters if this correlation proves to be correct in detail*, because it is inevitable that many heretofore unknown and unsought correlations having even greater implications will be uncovered, simply as a byproduct of our recognizing Strange Loops (Hofstadter, 1979) involving the Earth and its cosmological environment, a recognition that can occur only within the recursive universe that nonlinear dynamics will have brought to our attention.

Some of my remarks in this Prologue (and this volume) have raised issues of seemingly great and, for some persons, even esoteric or "arcane" difficulty. But this has been the fate of the natural sciences at least since these issues of self-referential nature have been raised by mathematicians like Cantor, Gödel, Spencer-Brown, Mandelbrot, Hofstadter, and others, and in fact since Poincaré pointed to the impossibility of preserving the Laplacian ideal of perfect determinism, in which the trajectory of every particle of the universe could, in principle, be calculated once and for all, given sufficiently precise descriptive data and sufficiently powerful computing machines (**Note P.6**).

Still, the ideas in this book are not as difficult as a "schooled" attempt at understanding them would suggest. My hope is that interested readers will bear with me, and will allow their innate ability for understanding, which is intrinsic to the cognitive mechanisms of mind, to come to the fore as they proceed (see Skarda and Freeman, 1987; Freeman 1991, 1992). A form of contextual integration then emerges even in those instances where analytical and/or numerical integrations are impossible — and thus just as nonrealizable by the expert mathematical practitioner as they are by the mathematically untutored scientist (**Note P.7**). It may help to remember that the history of studies in mechanics and dynamics has shown that when reality is limited only to that which can be subjected to closed-form solutions, or to computer-aided proofs of such logical closures, mathematics becomes a handicap. Mathematical proof as a criterion for the revelation of truth, in the physically existential sense (not to be equated with "existentialism"), becomes a block rather than a mechanism by which we can better experience the world (Spencer-Brown, 1972, p. 101). This does not mean that quantitative measurements and mathematical methods of analysis must be discarded. The differential and integral calculus are retained as "tools of the trade" along with arithmetic and ordinary algebraic manipulations, just as the recognition of and uses of *fractal geometry* have not obviated the practical value of Euclidean geometry. Classical methods in dynamics are simply augmented by other criteria, among which are the notions, partially described herein, of such things as *mathematical universality*, *universality parameters*, *fractal self-similarity*, and related properties of *self-organized systems* and *chaotic crises*, which underscore the importance of *power-law statistics* and combinatorial forms of geometric progressions. Aside from the possible intimidation of an unfamiliar terminology, many of these concepts, in practice, are very simple indeed.

What I am attempting to convey is a different way of looking at the phenomena of nature, but it is a way that is not new either to the evolution of my thoughts about the Earth in the Cosmos or to the thoughts of many who have preceded me (e.g., Note I.8). Nonetheless, many geologists, astronomers, and cosmologists see the language of nonlinear dynamics and chaotic-systems theories as something new to, and even alien to, the more traditional developments in these fields. The other day a colleague opined that I probably didn't "get into the field" to look at rocks and do "normal" geological studies anymore, now that my thoughts were "off in space." I suggested that he might take note that all of us are de facto in outer space, and that, whether we study rocks in the field or a computer screen in the laboratory, we are inevitably studying matters of a cosmological nature. He had to concede that this is true. Nonetheless he walked away looking a bit nonplussed. What we see around us manifests the organizational principles that are intrinsic to the makeup of the Cosmos, for how could it be otherwise unless we claim to be phenomenologically separable and separated from that which we observe?

This book is thus about an involved reality wherein we are part of a universe caught in the act of observing itself. And such self-observations are but a form of "seeing" that acknowledges the reality of processes that heretofore some of us would have called exotic, catastrophic, or arcanely irrelevant to the human condition. But we are caught up in, and are the product of, all this excitement; the recursive nature of our own evolution is part and parcel with that which produced and eventually "killed" the dinosaurs. Our seeing of the evolutionary fates that have transpired on Earth in its responses to the cosmic environment may be no guarantee that we can avoid them, but just as the opening reference to Bach and his

Prologue

"endlessly rising canon" (Hofstadter, 1981, p. 10) reveals but an example, individually masterful though it may be, of our common ability to mimic something even more common in the structure of the universe, so too may our ability to see the eternal in the fabric of our past be just the offering we need if we are to perceive the nature of our choices among all possible futures. When in time we come to see all this drama as just another aspect of ourselves, perhaps we will cease to view the universe as a threatening place that snuffs out life like candles before the winds of "cosmic catastrophes." For it is absurd to think that a mechanism that has learned of itself to endlessly recreate itself could, by some quirk of final analytical capability, learn to implement, or even to forecast, its own demise. Such an outcome, wrought within a self-referential universe by a form of being that sees but aspects of itself, is not foreseeable. And this book, for me, chronicles only some few fragments taken from the fieldnotes of that seeing.

Acknowledgments

Although the content of this work may appear to address subjects remote from the mission of the U.S. Geological Survey, as some may think that mission is or should be delimited, I am fortunately heir to a great precedent set by Eugene M. Shoemaker, who, as time and circumstances have shown, has seen clearly and correctly the practical vestments and romantic vistas of planetary "geology." I am, along with the rest of the world, in his debt. But I owe too much to far too many others to mention them all individually. I can only hope that they will recognize, in those instances where their influence surfaces in the present work, my abiding gratitude. My debt to Irving J. Gruntfest will be evident in his conspicuous contributions to the science of nonlinear dynamics (e.g., Note P.5). I can never adequately thank Robert L. Smith for his thirty years of instruction in science, behaviorism, and natural philosophy, and for being an unwavering model of personal and scientific integrity. Whatever knowledge I have gained in my career concerning the phenomena of volcanic evolution and igneous processes in the Earth I owe largely to him, to my graduate supervisor at UC Berkeley and lifelong friend Charles Meyer (1915–1987), and to my mentor and colleague in the study of oceanic volcanism E. Dale Jackson (1925–1978).

My debt to Bill Glen has been mentioned and will be evident throughout — beyond measure. Both he and Bill Carver, as representatives of Stanford University Press, have been indefatigable in their editorial assistance as well as in their constant encouragement, stimulating discussions, and moral support. Reviews of a first draft of the manuscript by Digby McLaren (Royal Society of Canada, Ottawa), Ralph Abraham (Department of Mathematics, University of California, Santa Cruz), Archie Roy (Department of Physics and Astronomy, University of Glasgow), Victor Clube (Department of Physics, University of Oxford), and Bernard Chouet (U.S.G.S., Menlo Park) were invaluable in giving me the courage

Acknowledgments

and perspective to continue in this adventure. In my discussions of paleomagnetism, I have benefited immeasurably from "tutorials" by Jonathan Glen (University of California, Santa Cruz) concerning recent developments in that field. My interest in possible antipodal effects of mass impacts was stimulated initially by discussions with Jonathan Hagstrum (U.S.G.S., Menlo Park) during my review of his manuscripts on that subject. I am also grateful to Arthur Grantz (U.S.G.S., Menlo Park) for sharing with me his ideas on the circumstances of origin of the Avak structure, Alaska, the relic of an impact crater of probable Cretaceous-Tertiary age.

The moral support of my daughter Andrea Maxa, who shares in my dedication of this book, has been inspirational by virtue of — among many empathies — her own adventures in writing, and the "struggle to get published" that goes with the territory. And I am grateful to my nieces Deborah Shaw and Lorraine Dale for sharing with me their mutual excitements concerning humanistic applications of chaos theory — and Debbie provided strategic assistance during proofreading of the galleys. Without the support of Nancy Blair, Ellen White, Anna Tellez and all the staff of the U.S.G.S. Library in Menlo Park, this book could never have been written — and I owe them even more for their morale-boosting good humor than for their expertise and promptness in carrying out every literature search and obtaining every obscure interlibrary loan that I requested of them. This holds also for the help of my project associate Anne E. Gartner and the support of many other colleagues in the U.S.G.S. Aida Larsen provided invaluable assistance in the checking of page proofs and in preparation of the Index, an excruciating task that the serendipity of her expert knowledge greatly lightened. The enthusiasm that she — with the additional, and spontaneous, support of her husband, Lee — brought to this project provided the critical-state tuning and impetus needed at that stage to carry this work to its completion. I am especially grateful to Bernard Chouet, not only for the many years of our collaborations on the application of nonlinear dynamics to seismic phenomena, but also for taking the time from a period of intensive seismic monitoring of the active volcanoes of the world to read this book and to keep in touch with frequent words of encouragement, and to Paula Chouet for the spiritual strength that extends to include me, and everyone else who knows her, in the security of her gracious presence. Finally, I am eternally grateful to Carol-J. K., who was my phantom muse throughout this longest of vigils — the seemingly endless abdication of all human contact while in the grip of that demonic force which assumes all control in those of us who would aspire to be authors.

H.R.S.
Menlo Park
June 1993

A Note on Terms, Conventions, and Cross-References

In the Prologue I have italicized words, terms, and passages that I judge to be either new or unfamiliar to a largely geological/astrophysical readership. I have tried to sustain this practice throughout the text, either where such words are newly introduced, or where there has been such a long break between one appearance and another that a reminder seemed appropriate. Much of the material in the Chapter Notes (pp. 385–564) elucidates various of these terms and phrases. Otherwise I have used italics occasionally to emphasize words or passages, and for a few emphatic quotations; such usages should be obvious in context.

I use *Cosmos* as the form of the word when there is an implication of special structural order, whether linear or nonlinear, and *cosmos* as the generic form when no particular structural implication is intended for the arrangements of astronomical objects in space and time (cf. Notes 3.3, 4.2, and 7.5). Similarly, *Galaxy* and *Solar System* refer explicitly or implicitly, and respectively, to the Milky Way Galaxy and the Sun-centered stellar-planetary system. I use *galaxy* and *solar system* if the context is generic with respect to cosmological processes resembling or analogous to the Galaxy and Solar System.

The Chapter Notes following the text and Appendixes are intended to offer perspectives, from more than one disciplinary viewpoint, on the concepts discussed in the text; they are cited in text in boldface type where definitions and/or elaborations of a current topic first seemed to be indicated (e.g., **Note 7.3**). They are also cited throughout the text in plain type where cross-referencing may be helpful, or where ancillary background material may simply enhance the multidisciplinary nature of the theme.

Every work cited in the text is given in full in the References Cited (pp. 567–641). Where text citations indicate page numbers as well, in the works cited, I have employed "f" (following a page number) to indicate that the citation carries over

Note on Terms, Conventions, and Cross-References

to the following page, and "ff" to indicate that it carries over to the following few pages (which will be evident in the context of the cited reference). On those few occasions when the citation refers to a substantial number of pages, but not to an entire section or chapter of a book, I have given the inclusive pages (e.g., pp. 1103–17).

I have placed the principal text illustrations (cited, for example, as Fig. 32) together as a unit immediately following the Introduction and Table 1, partly to indicate something of the flavor of the book's thesis, but chiefly because most of the figures are referred to on many occasions in the Introduction, as well as elsewhere in the text, in a variety of contexts, and would be difficult to find repeatedly and intermittently if scattered in the usual fashion. The Appendixes (pp. 333–79) present a number of supplementary illustrations, cited in the text as, for example, "Appendix 3" (boldface type is used only where a particular Appendix, or group of Appendixes, is first indicated in support of a given topic in the text, e.g., **Appendix 3**, or **Appendixes 12–15**).

Postscript Added in Proof

Though this work is far from being comprehensive, it would be even less so if I failed to mention three happenings that occurred while it was being typeset and proofed:

(1) Derek Ager — mentor to us all in biostratigraphical insight — died in Swansea, Wales, in 1992 during the final stages of production of his last work, *The New Catastrophism: The Importance of the Rare Event in Geological History* (Cambridge University Press, 231 pp., 1993), which he dedicated to the Department of Geology, University College of Swansea (opened in 1920 and, sadly, closed in 1990), and "with gratitude to all the medical staff who kept me alive long enough to finish writing it." My admiration for Derek Ager, so evident in the present volume, would have been even more apparent had Ager (1993) been available to me during its preparation.

(2) On March 25, 1993, the comet-searching astronomer team of Carolyn Shoemaker, her husband Eugene Shoemaker, and David Levy — working at the Palomar Mountain Observatory in California, with confirmatory observations by James Scotti at Kitt Peak Observatory in Arizona — discovered what looked like a "squashed comet," which turned out to be a string of twenty or so separate cometary nuclei, comae, and tails arranged much like a "string of pearls" moving in a skewed and eccentric jovian orbit spiraling toward a "splashdown" on the far side of Jupiter during July 1994. That object, collectively, has been named comet Shoemaker-Levy 9, a source of great excitement in the astronomical community because it will offer the first "live" witnessing of an impact of a comet with a planet (see "Jupiter Watch: The Celestial Necklace Breaks" by Charlene Anderson in *The Planetary Report, v. 14,* pp. 8 and 9, January/February 1994). The individual post-fragmentation comet nuclei — separated by irregular and evolving distances that average roughly 20,000 km (see H. A. Weaver et al., *Science, v. 263,*

Postscript

p. 787–791, 11 February 1994) — have been estimated to be only about a kilometer or so in diameter, but the potential effect of the multiple impact event on Jupiter's atmosphere has galvanized many astrophysical modeling teams. In the present context, two aspects of this event are noteworthy. First, the approach scenario for this array of impactors resembles two types of multiple impact events on Earth described in the present work: (a) the Great Fireball Procession — with an average length exceeding, at any one time, $\sim$ 1000 km — that was tracked across eastern North America and the Atlantic Ocean for more than 4000 km at an estimated altitude of one hundred kilometers or so during the late evening (about 9 PM EST) of 9 February 1913, an event that was essentially forgotten in the post-Sputnik (post-1957) literature and is resurrected here (see the Introduction, Chaps. 6 and 9, Appendix 3, and Notes 1.2 and A.1), and (b) a hypothetical train of Moon-orbiting natural satellites (see the Epilogue and Fig. 37; cf. Chap. 1 and Note 1.1) as a source of impacts at the Cretaceous/Tertiary boundary — i.e., my *K/T swath*. Second, the effects of a spatially distributed and protracted multiple collision with Jupiter's density-stratified gaseous atmosphere, at a theoretical escape speed (entry speed; see Note 6.1) of nearly 60 km/sec (without accounting for frictional braking in Jupiter's outermost tenuous atmosphere), may reveal novel aspects of strain-rate feedback instabilities discussed here in Note P.5. Because of the inverse temperature-dependence of the "viscosity" of gases (cf. Notes P.5 and 7.2), thermal feedback associated with intense shearing of a gas phase may cause it to display transient (irreversible-thermodynamic) liquid-like or solid-like properties, suggesting that such a feedback mechanism might — accompanying and following compressional shock-wave effects — nucleate a vortical, "black-hole-like," cascade of chemical reactions that could have far-reaching consequences [e.g., spin-up of the cometary "string of pearls" to a braided "twister" of graphite, diamond, and hydrocarbon compounds (?); cf. Notes I.1 and 1.2] — a scenario that is eerily reminiscent of scenes in the novel *2010* by Arthur C. Clarke (1982): cf. (i) my interpretation of the Tunguska air-burst event in Siberia on June 30, 1908 (see Note 1.2), (ii) the "butterfly effect" (see Note P.2), and (iii) my discussion of a "ghost-binary" star, and the remote possibility that Jupiter could, even now, conceivably become a low-mass white dwarf star (see Chap. 3, Notes 3.2 and 9.2, and Appendix 7).

(3) W. J. Broad, writing in the *New York Times*, Tuesday, 25 January 1994 (p. B5), and J. K. Beatty, writing in the February 1994 issue of *Sky and Telescope* (v. 87, pp. 26 and 27), describe information long held secret by the U.S. Department of Defense concerning meteoroid impacts with Earth's atmosphere ("air-burst events"). Now declassified, these data (cf. my proposal "Project SPACE-STRAP" in Notes 6.5 and 6.10) will soon be documented in a report by Edward Tagliaferri and others, which will appear as a chapter in T. Gehrels, ed., *Hazards Due to Comets and Asteroids* (University of Arizona Press, 1994). That report allegedly documents the annual incidence of tens to hundreds of natural air-burst events (rather than the single event usually cited; cf. Chapman and Morrison, 1989, p. 278) with explosive yields often exceeding that of the Hiroshima atomic

Postscript

bomb (cf. Chap. 9 here) — phenomena that the present volume reveals to be "normal" aspects of the near-Earth objects, including natural satellites, that have intermittently impacted Earth over geologic time, sometimes in far greater numbers and/or with orders of magnitude greater explosive powers (cf. item 2 above).

H.R.S.
Menlo Park
March 1994

CRATERS, COSMOS, AND CHRONICLES
A New Theory of Earth

> *Pythagoras, if he could but be with us, would (I hope) smile indulgently upon our endeavors. But I think that he would be inclined to say that he knew that the Universe would turn out to be harmonious, for harmony was for him an axiom, a definition of the way in which he chose to organize his experience of the world.*
>
> —M. W. Ovenden, 1975
>   *Bode's Law — Truth or Consequences?*

Introduction

What is it that I intend to argue in this volume? Fundamentally, that the records of mass impacts on Earth, and on the Moon and the planets in general, describe persistent patterns, and that these patterns are generated by a figurative, nonlinear hourglass of actual and potential impactors flying about the Solar System in highly organized (critically self-organized) regimes, in diverse groupings of heliocentric (Sun-centered), geocentric (Earth-centered), and other, similarly localized, orbiting-object reservoirs. In short, I argue that these patterns and orbits do not conform—as has been supposed—to the notion of "random models." A random model is one that is *maximally complex*: the entity modeled is totally lacking in any simplifying hierarchies, codes, or structures relative to the total number of bits of information that an algorithm intended to describe the entity, as a whole and in all its detail, would require. The random model presents the entity either as observed, thus wholly unanalyzed, or as the observer chooses to observe it (a sometimes subtle distinction), in either case often implying preconceptions about the nature of organization and/or blind faith in standard statistical methodologies.

The principles of *algorithmic complexity* (Chaitin, 1979, 1982, 1987a, b, 1988) tell us that we often cannot know in a given case—i.e., for a given expenditure of effort (cf. Nicolis and Prigogine, 1989, Secs. 1.5 and 4.9; Ruelle, 1991, Chap. 22)—whether the complexity of some aspect of nature really represents a class of entirely uncomputable structures (equivalent to undecidable propositions in mathematics; cf. Post, 1965; Ruelle, 1991), or whether we have just been unwilling or unable to search for an algorithm that is capable of reducing the maximally complex description to a simpler, more redundant, hence in some measure periodic, description. According to the *incompressibility rule* of algorithmic-complexity theory, a given set of data remains random *simply by default*—as a matter of conditional classification — in an open-ended, often fallow, quest for a simplifying algorithm. This is far different from stating that a given type of data is random *in principle*, thus that there is no point in our looking for a simplifying algorithm because, in principle, it is impossible to find one.

Two of this century's landmark discoveries — of the genetic code in biology and the geomagnetic time scale in geology — were examples of unexpected and dramatic reductions in the algorithmic complexities of natural phenomena. In these two instances, suddenly far less information was required to describe a given biological structure or a given geological history, and some structures and histories that had never been described at all could now be given quantitative expression. An analogous shift occurred in astrodynamics during the period 1990–92 (see Belbruno, 1992), when an entirely new type of algorithm was put to the test for the computation of fuel-efficient orbital-transfer routes for travel from the Earth to the Moon. Almost everyone had assumed that the type of transfer orbit sought by Belbruno did not exist, because "if it did, someone would have found it by now" during the several decades of computations of space-vehicle transfer orbits by the many specialists in the Soviet and American space programs (see Note 6.1). I argue that analogous transfer orbits, similarly unsought until now, must exist for natural objects.

The so-called *"fuzzy route"* discovered by Belbruno would appear to exploit *fractal* properties of finite-width ("fuzzy") boundaries between the spheres of attraction of celestial objects. In other words, the gravitational-potential wells of objects like the Earth, Moon, planets, and asteroids are analogous to, or equivalent to, fractal *basins of attraction* in nonlinear dynamics (cf. Notes P.3 and 6.1). Accepting the existence of fractal potential wells totally transforms the problem of calculating a probability distribution for the orbital-transfer problem — hence for the capture of natural satellites — just as it is likely to explain the orbital complexities of (1) the systems of satellites of the planets (e.g., Cruikshank et al., 1982; Jewitt, 1982; Thomas and Veverka, 1982; Greenberg and Brahic, 1984; Esposito and Colwell, 1989; Kolvoord et al., 1990; Esposito, 1991, 1992; Horn and Russell, 1991; Porco, 1991; Colwell and Esposito, 1993), (2) the reservoirs of asteroids, hybrid asteroid-comet mixtures, and comets (e.g., Wisdom, 1987a, b, c; Weissman et al., 1989; Torbett and Smoluchowski, 1990), (3) the planetary system itself (e.g., Laskar, 1989; Mallove, 1989; Wisdom, 1990; Kerr, 1992c; Milani and Nobili, 1992; Peterson, 1992; Sussman and Wisdom, 1992), and (4) the Solar System in the Galaxy (e.g., Stern, 1987, 1991).

The hourglass I invoke (e.g., Fig. 11) symbolizes an open-system nonlinear process called *self-organized criticality* (see Chap. 5 and Notes P.2, I.2, 5.1–5.3; cf. Shaw, 1987b; Bak et al., 1988; Shaw and Chouet, 1988, 1989, 1991; Chen and Bak, 1989; Kadanoff et al., 1989; Bak and Chen, 1989, 1991; Bak and Tang, 1989), that extends, in principle, beyond the Solar System into the interstellar reaches of the Cosmos, implicating, along the way, interstellar processes, galactic and intergalactic processes, processes of supergalactic clustering, and so on. Within the inner Solar System, critical-state processes sustain a sporadically continuous

Introduction

source of showers of asteroids, comets, and diverse near-Earth objects (NEOs; cf. Steel, 1991; Matthews, 1992; Kerr, 1992g). A fractionally small, but poorly known, population of these objects intermittently comes to Earth according to both ballistic and resonant geocentric trajectories, emerging terminally from systems of natural Earth satellites — i.e., systems comprising the time-dependent reservoir of surviving prior captures in the Earth-Moon vicinity — that fluctuate in numbers and sizes and interact resonantly with longitudinal mass anomalies in the Earth (see Chap. 6; cf. Allan, 1967a, b, 1971).

Although these objects typically are smaller than tens to hundreds of meters in diameter — the measurements that would be recorded if the geocentric reservoir were sampled now (i.e., in situ, as a population of natural Earth satellites) or at any arbitrary time in the past (e.g., as a population of the smallest impact craters discernible on the Earth's surface during a geologically short interval) — they nonetheless pose potentially catastrophic implications for human culture (compare the discussions of Notes 1.2 and 9.1 with the global patterns illustrated in Appendixes 1–4). And geologically speaking — i.e., as a result of the naturally selective influence of the geocentric reservoir on the spatial organization of the far less frequent to rare single or multiple "large-body" impactors with maximum individual diameters as great as tens of kilometers — the same suites of phenomena have posed intermittent catastrophic implications for the extinctions and originations of Earth's biological populations in general. The large-body impacts are the "rare" events that correlate with massive extinctions in the fossil record, which in turn — by both observation and hypothesis — mark the most dramatic biostratigraphic discontinuities in the geologic record, such as the Cretaceous/Tertiary and Permian/Triassic boundaries at, respectively, 65 million and 245 million years before the present (Ma), according to Harland et al. (1990).

The past and present threats to human culture, as well as to the biosphere in general, have been, and are, heightened by the relatively high incidence of grazing and/or clustered impacts characteristic of objects in geocentric or near-geocentric orbits (see Appendix 3; cf. Notes 9.1 and A.1). This general process has written, on the surfaces of all planetary and satellitic objects in the Solar System, spatiotemporal patterns of impact craters and other impact-related scars that record the nonlinear thermomechanical histories of these thus smitten objects (see Note P.5). As a result, a kind of impulsive "ratcheting" of surface and interior motions has been effected by impact-tectonism and rheological feedback with the crustal veneer, the mantle, and the core, representing a form of orbital-ballistic telegraphy that has punctuated planetary surfaces and interiors much in the manner of a nonlinear laser printer. In the Earth, these printed patterns can then be transcribed through the media of paleontology, isotope geology, and geochronology into the languages of biostratigraphy, magnetostratigraphy, and magmatic (volcanic and plutonic) stratigraphy, becoming what we call the *Geologic Column* and the *Geologic Time Scale* (e.g., Harland et al., 1990; cf. Harland et al., 1967, 1982; Shoemaker, 1961; Shoemaker and Hackman, 1962; Baldwin, 1971, 1985; Brush, 1986; Wilhelms, 1987; Stothers, 1992). This record might in fact be better described as a multidimensional calibration chart, or code, that memorizes the spatiotemporal patterning of nonlinear geological-cosmological processes.

These ideas of course have their precursors. In the realm of celestial mechanics and the orbital evolution of the Solar System, more is owed to the vision of Michael W. Ovenden (1926–1987) than has been acknowledged in the general literature of planetology and astronomy. He foresaw, I believe, the ineffable forms toward which the trajectories of *celestial strange attractors* (see Notes P.3 and 6.1) may flow—though he did not call them that. He attempted to elucidate a theory of orbital interaction that can now be compared with the more advanced outcomes of specialties that derive, at least in part, from concepts of nonlinear dynamics and chaos science—including implications for the geometric fabric of orbital evolution that now would be called *fractal* in nature (see Chaps. 3 and 6). In *Bode's Law—Truth or Consequences* (1975), Ovenden revealed a number of insights that are still incompletely appreciated [some are recapitulated in Roy (1988a, Chaps. 1, 5, and 8)]. In that work (loc. cit., p. 485) he called one of his evocative insights *The Principle of Planetary Claustrophobia*, "namely that, in any system, the planets will spend most of their time as far away from each other as possible." This concept is quite similar to the *principle of frustration*, a term that has been invoked by the practitioners of a new interdisciplinary field of studies—a field closely related to nonlinear dynamics and *deterministic chaos* (cf. Note P.3)—called *spin-glass theory,* a figurative name for a theory of complexity that is being applied to studies as seemingly different as materials science (studies of the structural states of actual glasses, electromagnetic-spin domain structures, etc.) and the fabrics of complex behaviors in the social sciences, wherein the anthropomorphic term "frustration" becomes literal (cf. Palmer, 1986; Carroll et al., 1987; Dewdney, 1987; Kinzel, 1987; Stein, 1989; Suarez et al., 1990; Gilbert, 1991; Mosekilde et al., 1991).

In its more general ramifications, the notion of frustration is consistent with my conclusion that the Solar System, and cosmological structures in general, are *attractor systems*: they are regions of cosmological space within which special trajectories of material motion have converged to a locus, such that there is a complex distribution of resonances, quasiperiodic behaviors, stable attractors, and other forms of ordered and/or clustered states that in ensemble occupy a spatial volume that neither shrinks to a fixed number of points—in reality or in phase space—nor diverges to infinity. Such a complex state is analogous to a *fractal gravitational-potential well* in my general notion of *celestial attractors* (see Notes 6.1–6.9). Because of the principle of *metric universality*, which operates at all scales of motion (see Note P.4), attractor systems are recognizably similar throughout the Cosmos. [Within the spatial volume of any cosmological attractor, however, including the Solar System, nearby trajectories that are not caught in special resonances or fixed-point behavior generally exhibit *exponential divergence*, as is the case in the nonlinear-dynamical structure, introduced above, called a strange attractor (see Notes P.3 and I.4). Some authors refer to such forms of generally chaotic but almost periodic phenomenology as *"stable chaos"* (e.g.,

Milani and Nobili, 1992; Murray, 1992; cf. Laskar, 1989; Sussman and Wisdom, 1992).]

Ovenden (1975, p. 495) also concluded, as do I, that Newton's Laws of Motion are, in effect, tautologies that have defined how we have chosen to organize and represent our experience of the world — rather than discoveries of an immutable construct of nature (cf. Milgrom, 1983; Shaw, 1983b, c; Schneider and Terzian, 1984; Lindley, 1992a; Milgrom and Sanders, 1993). Their celebrated capability for predicting trajectories of orbital motions is really but a form of superbly designed mimicry, for Newton's laws, by the happy coincidence of an opportune merger, simply wed Galileo's experiments on falling objects to the systematics of geometric and kinematic measurements that Kepler had demonstrated to be characteristic of planetary motions (see Gamow, 1962; cf. Shaw, 1994). Paraphrased in the language of nonlinear dynamics, "we mimic the characteristics of the strange attractor within which we live, for that is our experience, and some have made this experience numerically quantitative the better to find our way around within the material parameters of this attractor space." The so-called "inertial frame" in Newtonian mechanics, according to this view, is an artifact of an overarching simplification that reconciles the kinematics with a self-consistent pattern of "forces" (cf. Shaw, 1994). Such a view, therefore, is consistent with the next level of sophistication in our mimicry of the world, a geometric description embodied in the general theory of relativity (cf. Sexl and Sexl, 1979, Fig. 1.2), but it is a view that overlooks the essential nature of *nonlinear-dissipative-feedback processes* — viz. *nonintegrability* and *sensitive dependence on initial conditions* (see Notes P.1 and P.2). A resolution of the relationship between nonlinear dynamics and theories of relativity — rather, the inability to solve numerically the evolving examples of such a relationship — should likewise resolve the conceptual disparities between relativity, as a paradox of an essentially Newtonian continuum, and the particle-wave duality (the ideological bifurcation, or schizophrenia, of *dynamical frustration*) in quantum dynamics. Such a resolution might be called a theory of the relativity of relativity, or a *Theory of Nothing* (see Notes 4.2 and E.1; cf. Rotman, 1993a, b).

But let us move on now, to an examination of the evidence underlying the patterns of impacts on Earth, beginning with some thoughts on how it all might have begun. It is clear from what fragmentary data we have on the earliest periods of Earth history, from the surface history of the Moon and current theories of its origin, and from common-sense deductions and intuitions that the principal features of the Earth, as we know it, were products of cataclysms unimaginably greater in magnitude than, but not different in kind from (cf. Note I.8), anything the Earth has witnessed since. This conclusion seems virtually self-evident, despite the fact that during even my own relatively brief career I have seen the dominating theories of the Earth change in emphasis and scope several times — sometimes coming full circle to an earlier theory (the early history of the Earth and origin of the Moon being a case in point, which, I think, will be seen as the present theme unfolds). And yet I have noticed one constant during this period — and that is the general validity of a concept that, to me, represents the spirit of what James Hutton (1726–1797) intended by the principle of Uniformitarianism (see Note I.8). As we will learn from the experiments and principles of nonlinear dynamics, the geologic past can never be unraveled with anything like the rigor the "exact sciences" would demand, but if there is anything about that geologic past that we can state with confidence, it is that our sense of how the world must have worked *then*, at least insofar as we can express it in a way that rings of truth and universality, must be based on our *experience* of the way the world works *now*. In this context, "experience" embraces more *potential* information than it does presently established "knowledge," because to date we have not done a very comprehensive or very faithful job of documenting the actuality of our experience (cf. Notes P.4, P.6, and I.2). And that, I testify, from my own sampling of our experiences in field observations, in theory, and in laboratory, field, and computer experimentation, is what the (my) excitement over nonlinear dynamics and chaos theory is really all about.

Thus advised—and assured that we will be returning to the contemplation of these perspectives from time to time—let us continue with the exploration of my theme.

The Legend of the Beginning Written in the Earth

In the beginning there was Proto-Earth, born in the death throes of its short-lived and then-nameless parent-star Genesis, binary companion to a Protostar (the one now called Sun) in the pre-main-sequence pair of stars that became the Solar System. Not long after its birth, as these things go, Proto-Earth narrowly escaped death by disintegration during the violent birth of the Moon, Earth's now healed, and healing, offspring. As a Phoenix rising from that unimaginable cataclysm, Earth evolved from Proto-Earth, and survived a traumatic early life of mammoth but lesser cataclysms. And each trauma left a scar upon Earth's face—and on the mirror of the Moon. And each one testified to the character of Earth's resiliency, which rendered it immune to smaller catastrophes during its greater lifetime. And each new catastrophe gave Earth a greater and more tenacious vibrancy—until the fecundity founded in it multiplied and grew, eventually to paint the great tapestry of color and integral form that now obscures the brutal scars of youth. And in time that youthful violence was all but forgotten, in the more subtle stirrings of microbes and molds and macroorganisms, whose evolution was punctuated intermittently, now and now again, recurrently and persistently, by the impacts of meteoroids (meteorites, bolides, and comets)—and is assailed and threatened today even by its human progeny. Having discovered the secret of the power of the universe, the children of Earth have tried to usurp even the power of the Sun, and—in the foreboding of their nightmares—perhaps ultimately to terminate even their own brief terrestrial story in the burst of a supernova, the death cycle of stars that is also rebirth. Thus might darkened Genesis, the binary companion to our Protostar—the ultimate source of all the chaos so long ago

Introduction

unleashed onto and within Earth and the other planets — also be called Siva.

And also thus, from the cyclical gallery of space, has Earth been visited with a chaotic pattern of meteoroids, pelting its sky and its face with frequencies and dimensions that have ranged from a continual mist and daily patter of dust-size and pebble-size objects to intermittent colossal blasts far more terrible than the most fearsome earthquakes or volcanic eruptions imaginable. And these great blasts were dealt Earth by the impacts of exotic blocks of Gibraltar-size to Malta-size rock, which traversed space at speeds of tens of kilometers per second and arrived on Earth hundreds of thousands to tens of millions of years apart, accentuating an incessant, syncopated percussion that has struck this small cosmic outcrop in a persistently ringing nonlinear cacophony. Like an ensemble of massively muted bells, this meteoritic shower activity has pummeled and shaped the fundaments of Earth's history for all of its four and a half billion years. Such is the rhythmic intricacy and dynamical involvement of impacts from space. Nor are these impacts but random events that do local if massive damage in space and time but are then quickly lost to history, as if they had been no more than the bad dreams of Jupiter and Pluto, gods of our outer and inner weather. They are integral elements in a larger process, one that has imprinted trajectories of the limitless extraterrestrial influence on Earth's face and has wound that influence into a patterned cocoon of dynamical involvements that envelop and penetrate Earth and all of its companions. Some of these trajectories vary in weave at finer and finer scales, while others cross again and again at the same places, driving spikes of involvement deep into the heart of the globe, with recurring and spatially invariant emphases. And thus has been made, from an intermittency of traumas, and Earth's manifold responses to trauma, a bestiary of forms that has evolved to that time when this story was first told, when the storyteller had finally stopped to look back, to recapitulate the way come — and to contemplate the future course.

This story is told in a manner that simulates the spoken legend, both to suggest the panoply of topics to be taken up and to set the stage for the descriptions that follow — descriptions written more in the style of twentieth-century science. I tell the story, as well, to honor the great traditions of the more indigenous Californians, who learned about living in peace with nonlinear Nature long before, and for far longer than, the scientific-technologic mind of European culture has existed — i.e., long before Europe sent to American shores and beyond a chaotic burst of what that culture calls civilization, to forever change the face of Asian-American evolution. (For what, otherwise, might the Maya and the Inca and the others have come to be, and to contemplate?) I tell it to honor even older traditions, those that represent our resonances with a Cosmos that twentieth-century science may yet come to recognize, and perhaps even to acknowledge (cf. Capra, 1977; Gould, 1984a; Stewart, 1989; Davies, 1992). As interim epitome, I defer to the story of Umai told by Theodora Kroeber in *The Inland Whale* (Kroeber, 1971, pp. 91ff; originally published in 1959), one of the few attempts to honor California's past and to learn from it. [Kroeber reprised that attempt in her biography of Ishi

(Kroeber, 1961), which she dedicated to her husband Alfred Louis Kroeber, Professor of Anthropology at the University of California, who had been both mentor and student to Ishi.] The legends of the Inland Whale reveal an awareness of, and a unity with, a cosmology that has become the more familiar to me as I have learned something of the nature of nonlinear dynamics and deterministic chaos.

Back to the Future: Some Lessons from the Satellites

There is a general penchant in geology and planetology — and now in paleontology — to see the cratering patterns on the planets, satellites, and larger asteroids of the Solar System as the results of a random process (cf. Notes 6.1 and 8.15). A remarkably comprehensive study of the cratering histories of three satellites of Saturn by Lissauer et al. (1988) illustrates several of the difficulties that have led to this belief. In the absence of a chronological framework for the cratering history of any of the three objects, or of any other satellites or planets (as yet, we have no "moon rocks" from such places!), it is difficult to discern evidence of spatially or temporally ordered sequences of impacts across their landscapes. The study of Rhea, the second largest of Saturn's satellites (diameter ~1528 km, orbital semimajor axis ~527,000 km), is a case in point. Rhea is only slightly larger than Iapetus, and both are a little less than one-third the diameter of Titan, Saturn's largest satellite, which, at a diameter of 5150 km, is almost equal in size to Ganymede — the largest satellite of Jupiter, and the largest satellite in the Solar System — one and one-half times the diameter of Earth's Moon, and more than double the diameter of Pluto, *a planet* (for data on the planets and the satellites of the planets, see Considine, 1983; Davies et al., 1989; Anon. 1991; Pasachoff, 1991, Appendix 4; Maran, 1992, by planet and/or by satellite name; and, for an up-to-date review of the less well known Pluto-Charon system, see Stern, 1992). Rhea's orbital distance from Saturn is roughly four-tenths that of Titan (Titan's semimajor axis is ~1,221,900 km), and Iapetus's distance is roughly three times that of Titan and nearly seven times that of Rhea. Mimas, the third satellite studied by Lissauer et al. (1988; cf. Dermott, 1971), is much smaller (diameter ~390 km) and is only ~185,500 km from Saturn (roughly one-third the distance to Rhea, equivalent to halfway between the Earth and the Moon). Lissauer et al. concluded that the cratering pattern of Rhea is somewhat more uniform than a random distribution would have it, which they interpreted to mean that random clusters of craters were preferentially modified or obliterated by later cratering events, but not enough to destroy the underlying random character of the pattern. An interesting assessment.

The relative sizes and distances of these satellites of Saturn are mentioned to emphasize a point raised in the Prologue concerning the complexity of resonance phenomena of planetary satellites, as demonstrated there in terms of the tidal behavior of Jupiter's satellite Io. Present knowledge of the richness of the resonance phenomena displayed by the satellites of the planets illustrates two points of

Introduction

crucial importance to the theme of this book: (1) the orbital histories of the satellites are very complex, but they are systematically ordered according to resonant hierarchies of mutual interactions with both their primary planets and their satellite "families"; (2) by the same token, systems of smaller satellitic objects (e.g., satellites of satellites, particulate rings, dust bands, and so on) are resonantly related to the systems of the more primary satellitic objects, and the interactions and/or collisions of these smaller objects with the more primary satellites will have been modulated by complex systems of resonance overlaps and other nonlinear phenomena (see the discussion of "ring arcs" and related interactions of the satellitic bodies of Neptune by Porco, 1991; cf. B. A. Smith et al., 1989; Ross and Schubert, 1990; Esposito, 1991, 1992; Colwell and Esposito, 1993).

In other words, the distribution of objects in the Solar System is *self-similar*, but this self-similarity is no longer — and really never was — statistically either random *or* uniform, because the chaotic evolution of, first, the solar-planetary nebula, then the system of protoplanets and planetesimals, and finally the "modern," terrestrially coeval, system of planetary, satellitic, asteroidal, relict planetesimal, and cometary structures constitutes the progressive development of increasing self-organization that has manifested itself in several different hierarchical regimes of chaotic order. The Solar System became progressively clotted (or "curdled," in the terminology of Mandelbrot, 1983, p. 89), and its present complement of objects, including the Sun itself, can be grouped into several different object-size classes, depending on one's purpose, each of which presents a roughly consistent fractal dimension or subset of dimensions (e.g., Appendix 8; cf. Rawal, 1991, 1992). Examples of such classes might be: (1) the Sun and the major planets, from Mercury to Pluto, plus the "minor planet" Ceres (the largest asteroid), and a subset of the largest satellites of the planets; (2) the intermediate-size satellites of the planets, perhaps including the largest asteroids; (3) the small but individually recognizable satellites of the planets; (4) the asteroids as a separate class by itself (see Appendix 8, Inset); (5) the comets as a separate class by itself; and (6) all objects below the limits of optical and/or other contemporary methods of resolution of individual objects. Class (6) is simply a catchall for a host of extremely complex and incompletely recognized structures, such as the zodiacal dust, other meteoritic dusts, meteoroid and asteroid "streams" (cf. Alfvén, 1971a, b; Baxter and Thompson, 1971; Danielsson, 1971; Lindblad and Southworth, 1971; Drummond, 1981, 1991; Olsson-Steel, 1989; McIntosh, 1991), satellite rings and ring arcs (cf. Jewitt, 1982; Greenberg and Brahic, 1984; Esposito and Colwell, 1989; Kolvoord et al., 1990; Esposito, 1991, 1992; Horn and Russell, 1991; Porco, 1991; Colwell and Esposito, 1993), etc., that are limited in minimum object sizes only by the particle phenomena of the solar wind, cosmic plasmas, and subatomic particle fluxes of various kinds (cf. Alfvén and Arrhenius, 1975, 1976; Alfvén, 1980, 1981, 1984). Taken as a whole, it can be seen that we are dealing with a *spectrum* of object classes, probably none of which is really separable, in the dynamical sense, from all the others [and this includes the "giant" comets of Clube and

Napier (1990, pp. 136–46) and/or the relict planetesimals of Stern (1991)]. It may be noticed that this sort of descriptive classification resembles the qualitative definitions of *chaotic attractors* and *strange attractors* outlined in the Prologue (cf. Note P.3).

Hierarchical fractal structures of the kind just described typically are the nonlinear-dynamical resultants of compound resonances, some of which are evolving in both periodic and chaotic modes in a manner analogous to that of a "structured gas" [e.g., see the description of a multifractal gas, a form of strange attractor analogous to those displayed by cigarette smoke and Jupiter's atmosphere — or, for that matter, by the acoustic, magma-transport-related multifractal attractors of seismic tremor (Shaw and Chouet, 1989, 1991; Chouet and Shaw, 1991) — as described by Hoover and Moran (1992; cf. Ruelle, 1980)]. Barring the stabilizing effect of one or a few overriding resonances (as is the case in the *celestial reference frame hypothesis* of the present volume for Earth's history of extraterrestrial impact phenomena), cratering records in the Solar System will be so dimensionally complex — and so heavily overwritten — that the palimpsest of their long-term histories will be indistinguishable from those of overlapping "random walks," at least at the level of discernment provided by standard statistical tests (cf. Sugihara and May, 1990; Tsonis and Elsner, 1992), *even though they are printed by a meticulously ordered system of orbital resonances*. By standard statistical test I mean any test that is based on the statistics of *independent* (chance) events, such as the statistics of numerical sequences generated from a series of coin tosses, or from a random-number generator.

Patterns generated by processes governed by the *idealized* statistics of chance events differ from patterns generated by *actual* processes of self-organized chaos. The difference, in the latter instance, stems from the biasing or "weighting" of the statistics by chaotically self-selected number sequences that are ordered in precise, if highly mixed and complicated, ways by nonlinear interactions (see May, 1976; Shaw, 1987b, 1988a, 1991; Kaplan, 1987; Sugihara and May, 1990). A major premise of the present work is that *all* natural systems involve such weightings in their numerical structures — the reason being simply that (1) processes based on the probabilities of chance events are processes consisting of some combination of *independent* mechanisms each of which is ordered by the probabilities of chance events, but (2) no dissipative system in nature can exist independently of nonlinear interactions with other systems, an *axiomatic principle that simply testifies to the ubiquity of nonlinear interactions and dissipative systems in the universe*, hence (3) no dissipative system can be constituted of linear superpositions or combinations of independent mechanisms, and therefore (4) because any actual, even if arbitrary, system of interactions in the universe is dissipative in some sense, *there is no such thing in nature as an independent system* (cf. Shaw, 1987b).

Even without such reasoning, one need but examine the exquisite details of the orbital computations and photographic documentation that go into the making of new discoveries of objects in the Solar System, not to mention the long, cold nights of watching for them and the days, or years, of waiting for that perfect photographic plate, to see how great is the role of meticulously executed order in the delivery of, and discoveries of, Earth-crossing objects. Nonlinear dynamics does nothing more than extend such organizational principles to include sensitive interactions as ancillary mechanisms that influence the trajectories and rates of delivery of impactors to a planet such as the Earth (cf. Everhart, 1979; Wisdom, 1985; Greenberg and Nolan, 1989; see Notes 6.5–6.8 for commentaries on the effects of chaotic orbits on the discovery circumstances of Solar System objects, and their role in complex resonance phenomena; for more on the optical searches for asteroids and comets, see particularly Shoemaker et al., 1989, 1990, writing in the great tradition of the search for a trans-Neptunian planet and the discovery of Pluto by Clyde W. Tombaugh on 18 February, 1930, documented in a series of photographic plates beginning on 21 January, 1930; see Tombaugh and Moore, 1980, pp. 125ff).

The complexities of natural interactions are so great — the fabric of complexity is so interwoven among all of the phenomenological fields of the terrestrial sciences, hence with celestial dynamics as well — that statistical techniques of analysis for distinguishing the degrees of interactions of diversely related systems have been slow to evolve. Such statistical techniques now exist in rudimentary forms and are being energetically pursued in several fields of the natural sciences, beginning with the diverse forms of studies that convolved into what has come to be called *fractal geometry* (see the diversity of fundamental studies cited in Mandelbrot, 1977, 1983). Significant progress is being made in the statistical study of chaotic systems, notwithstanding the fact that applications of fractal methods of statistical analysis are (1) difficult, (2) sometimes seemingly contradictory or ambiguous relative to the results of standard statistical techniques (which are incapable of revealing hidden numerical structures), and (3) sometimes — particularly at this stage in their development — subject to great controversy (cf. Shaw, 1987b; Shaw and Chouet, 1989; Chouet and Shaw, 1991; Lorenz, 1991; Tsonis and Elsner, 1992).

All of this points to the fundamental importance of my conveying my own ideas concerning the characteristics of *natural* processes, as outlined above, vis-à-vis those of *unnatural* (independent) processes. This is necessary both to set the stage and to demonstrate the rationale by which I began to formulate the evidence upon which I have based my conclusion not only that the record of impact cratering on Earth — not to mention similar evidence subsequently presented for the Moon, Mercury, and Io (see the Prologue and Epilogue; cf. Appendix 25; and, respectively, Wilhelms, 1987; Vilas et al., 1988; Morrison, 1982a) — is ordered, but that the ordering process has been so selective within particular spatial and temporal windows that the geographical precision in some subsets of cratering sequences demands the direct intervention of geocentrically coordinated nonlinear-resonance phenomena (the spin-orbit resonances and related nonlinear orbital phenomena introduced in Chap. 6, Figs. 19 and 20). This means in turn that the Earth, at least since the time of the putative impact that attended the formation of the Moon (and the proto-Earth/Earth transition), has possessed from time to time — and, I would argue, possesses even now as we speak — a subsystem of natural satellites, singly and collectively quite small compared to the Moon, that has fluctuated fractally both in numbers and in time-dependent size distributions but has nonetheless been a persistent feature of terrestrial dynamics for more than four billion years (**Note I.1**).

Some Terms and Concepts, and a "Travel Advisory"

The word *nonlinear* applied to results of general feedback-systems research, or as a qualifier to the class of problems in physics and physics-related fields of research called dynamics — which categorically includes all physical processes in the natural sciences — has, in my experience, been a block to communication concerning dynamical phenomena in general (cf. Notes P.1 and P.6). *Nonlinear dynamics* and its related phenomenologies have been "mystified" in the literature by an unfamiliarity of most scientists, including physicists, with the pattern-generating complexity of feedback-systems effects. One source of confusion is that most of us have been trained to think that mechanical problems can be dealt with effectively in a linear or approximately linear fashion (see comments by May, 1976; cf. Shaw, 1987b, 1988c, 1991). In attempting to break with that tradition, however, it is difficult to present the mathematical ideas of nonlinear dynamics in both a reasonably accurate and understandable way — i.e., one is criticized either for being too vague and imprecise or for being too rigorous and incomprehensible [cf. reviews of chaos theory in the popular media (e.g., Gleick, 1987; Pool, 1989; Peterson, 1992; Peterson and Ezzell, 1992) vis-à-vis textbooks that emphasize mathematical rigor (e.g., Guckenheimer and Holmes, 1983; Devaney, 1986; cf. Guckenheimer, 1991)]. Accordingly, my comments here are very general ones. (The ironical aspect is developed further in the Prologue, along lines that stem from the concept of *mathematical incompleteness*.)

It is also my experience that most people really do understand the difference between linear and nonlinear phenomena — *except* when attempts are made to give them rigorous definitions. Broadly speaking, a linear system is one in which the whole is the sum of the parts. There is little or no *synergy* in linear systems (cf. Haken, 1978). Nonlinear systems are everything else. Synergy thrives in nonlinear systems — and vice versa — wherein the whole is greater than the sum of the parts.

The words "linear" and "nonlinear," especially when used as qualifiers of physical processes involving feedback control, can be ambiguous, therefore misleading. In order to be clear about whether a system is really linear or nonlinear, relative to the context of nonlinear processes in general, as ironic or as circular as this reasoning may seem, it is necessary to reduce the interacting parts of a system to some set of fundamental and irreducible mechanisms (cf. Hofstadter, 1979; Richardson, 1991, pp. 77ff; Morrison, 1991, Chap. 18). But this is impossible in an absolute sense, because science has never found — and is incapable of finding, on grounds analogous to the principle of mathematical incompleteness (see discussion in the Prologue) — a finite set of absolutely fundamental matter entities (e.g., subatomic particles) and/or a finite set of absolutely fundamental mechanisms of interactions among such entities. Science can always find ways either to split and/or to lump the properties and/or dynamical states of matter. Thus, when we describe some macroscopic system as epitomizing the case of linear interactions, what we are really describing is a system in which the innate feedbacks among multifaceted phenomena that may be disguised within the system have resulted in a quasisteady and proportional balance of inputs to the system and outputs by the system. An example is the notion of *local equilibrium* in irreversible thermodynamics, characterized by conditions of *microscopic reversibility* and the so-called *Onsager reciprocity relations* (cf. Castellan, 1971, p. 777; Prigogine et al., 1972; Prigogine and Stengers, 1984, pp. 137ff). Prigogine and Stengers (loc. cit.) classify such systems under the heading of *linear thermodynamics*.

This ambiguity of the linear-nonlinear duality is perhaps best exemplified by the simple mechanical clock, which Morrison (1991, pp. 92ff) has already described. The clock mechanism puts out what appears to be a perfectly linear string of ticks and tocks, the spacings between which are proportional, on average, to inputs of mechanical energy to the clock mechanism (from springs, pendulums, batteries, solar cells, etc.). The system is "linear" in a sense analogous to the Prigogine-Stengers definition, but, as Morrison points out, the mechanical clock — when inspected closely — is really a horse of another color. The outputs of cyclic loops of local energy exchanges within the system, each with its attendant local dissipation of mechanical energy, actually are not proportional to their inputs. By both criteria (nonproportional input-output relations, and dissipation) the innocuous-seeming clock — the standard reference for notions of regularity and order in the world — *is a nonlinear feedback system!* In the terminology used elsewhere in the present volume (see Notes I.5 and 6.1), the mechanical clock is really a *pumped system operating far from equilibrium* (though not nearly as far from equilibrium as other pumped systems, such as the laser; see Note I.5). The escapement mechanism of the clock (see Morrison, 1991, Fig. 8.1) represents a form of cyclical *ramp function*, or *integrate-and-fire mechanism*, which, in other contexts, is a form of nonlinear coupled-oscillator system analogous to the sine-circle function of the present volume [e.g., compare the patterns of phase-locked resonances illustrated by Allen (1983) with those illustrated in Fig. 14 and Appendix 9 ("Devil's Staircases" for asteroids) of the present volume, as well as in Shaw (1987b, Figs. 1 and 2; 1991, Figs. 8.32–8.36)]. Such considerations, combined with numerical-dynamic experimentation based on the sine-circle function (**Table 1, Sec. D**) form the rationale for the central assertion of my present thesis that, on sufficiently close examination, there are no linear or nearly linear (truly integrable) dynamical systems in nature (including celestial mechanics, which — with some significant exceptions — was founded on linear, closed-system conceptions of nature; cf. Prigogine et al., 1991).

In 1972, *Physics Today* published a very short and innocuous-looking review paper (in two parts) — written by Ilya Prigogine, Gregoire Nicolis, and Agnes Babloyantz, titled "Thermodynamics of Evolution" — which contained ideas of cosmological scope (cf. Krinsky, 1984; Prigogine and Stengers, 1984; Nicolis and Prigogine, 1989). That paper was presented as an elucidation of the principles of self-organization as applied to biological processes, but given suitable substitutions of celestial and geological mechanisms for biological mechanisms — and neglecting for the moment the loaded issue, or mystique, of perfect self-replication — their description of dissipative processes is directly applicable to the concepts of self-organization advanced in the present volume for natural processes in general. It holds, in particular, for the concepts I have advocated for the role of thermal-feedback processes, and of localized dissipation, in the volcano-tectonic structural organization of the Earth. It also holds, with suitable shifts in scaling, for the structural organization of the Solar System, the Galaxy, and universal systems of galaxies (cf. Seiden and Gerola, 1982; Shaw, 1983a, b, c; Rubin, 1983, 1984; Schneider and Terzian, 1984; Hunter and Gallagher, 1989; Barnes and Hernquist, 1992; Flam, 1992a, b). The following statement from their paper (Prigogine et al., 1972, p. 24, Inset) paraphrases my own conclusions, as described previously and in the present work, concerning the celestial-geological-biological *connection machine*, a term borrowed from the computer realm of parallel processing systems (e.g., Coffey et al., 1990, 1991; Kerr, 1993c; but notice the conceptual distinctions between the systems described in these references and those referred to in the following quotation from Prigogine et al., loc. cit.): *"One of our main points here shall be that an increase in dissipation is possible for nonlinear systems driven far from equilibrium. Such systems may be subject to a succession of unstable transitions that lead to spatial order and to increasing entropy production."* Variations on this theme appear throughout the present work (e.g., Notes P.1–P.7, I.2, I.4, 4.1, 4.2, 5.1, 6.1).

In principle, we can understand everything there is to know about linear systems to arbitrary precision; such systems are truly conservative in lacking any form of energy dissipation and are subject to ordinary operations of differential and integral calculus, and/or to equivalent operations of difference equations if the data are discontinuous (i.e., linearly discontinuous data can be converted to continuous ordinary differential and integral equations by some form of averaging). By contrast, nonlinear systems generally do not have proportional responses or closed-form analytical solutions. And even when they *can* be expressed as a system of equations, those equations typically have multiple solutions. A simple example is a quadratic equation (e.g., an equation graphically represented by a parabola) that gives two possible solutions for a single value of the function. As a result, the response function is nonunique and generally is not invertible (the results of forward calculations differ from those of backward calculations). [Kaplan (1987), in my experience — and especially in conjunction with the seminal articles by May (1976), Feigenbaum (1980), and Hofstadter (1981) — gives the clearest elucidation of the methodology, and the numerical consequences, of the recursive mappings of quadratic functions (and "one-humped" functions in general) onto themselves.] Even in such a simple prototype of nonlinear feedback systems, there are virtually unlimited sets of bifurcations, as well as more complex behavior — and, within such a bifurcational maze, there is a multiplicity of routes by which to traverse the system, routes that in the general case become so mixed up by periodic and aperiodic intervals that they are not recoverable once they are traversed. *Chaos* is the general term that has been adopted in nonlinear dynamics for complex periodic-aperiodic mixtures of states — a complexity that rivals that of the random state but clearly is not random (cf. Note I.2). Morrison (1991) gives valuable insights into such quadratic models as the exemplar of nonlinear-feedback-systems behavior (cf. Shaw, 1987a, 1988b).

In order to be rigorous about what nonlinear dynamics is all about, then, one would evidently have to start by describing what it is not — or what linear dynamics *is* all about. That in itself is a tall order (cf. Note P.6), one that is rarely mentioned in the literature of nonlinear dynamics (but see Morrison, 1991, p. 197). In a qualitative sense, a fundamental distinction between linear and nonlinear systems is that a linear system is, in principle, entirely knowable (algorithmically computable), while nonlinear systems are in some respects unknowable (uncomputable) — thus cannot be solved completely and once and for all for at least some, if not most, parameter values. Nonlinear systems are, in this respect, elusive in nature and can only be understood *experientially* by exploring their properties for selected ranges of parameter values. This exploration, fortunately, is greatly aided by certain numerical invariances and patterns that are encountered over and over in different systems of equations. Searches for such patterns and their *universality parameters* represent one of the major goals of nonlinear dynamics (see Note P.4). Ultimately, linear systems are those systems for which a solution holds no surprise. Nonlinear systems are everything else, and because the linear system is an idealization that does not actually exist, virtually everything is a nonlinear system.

A corollary to these remarks is that absolutely definitive statements about nonlinear dynamics can rarely be made. This may seem an evasion, but in context it is not. *Surprise* is an essential ingredient of nonlinear dynamics, because surprise is the hallmark of new information (cf. R. Shaw, 1981). If nonlinear dynamics could be exhaustively defined, it would hold no more surprises — and we would stand to learn nothing more from it. So all this boils down to living with uncertainty and paradox, to living with enigmas wrapped in conundrums, because if I could tell you exactly what nonlinear dynamics is — vis-à-vis all expectable states of a system — it would not be nonlinear dynamics. And yet I assert that, despite such imprecision of meaning, the patterns of nonlinear-dynamical phenomena are increasingly familiar the more they are explored, and their familiarity has to do with our own makeup as nonlinear scanning devices. But this is no fluke or accident of nature, because the neurophysiology of our perceptive and cognitive apparatuses operates by the very principles that those apparatuses would attempt to define as nonlinear behavior (e.g., Skarda and Freeman, 1987; Freeman, 1991, 1992; Andreyev et al., 1992). For example, as much as some scientists would try to deny it, everyone recognizes the difference between "linear thinking" and "nonlinear thinking." Nonlinear dynamics is as simple — and as mysterious — as whatever that distinction means. This enigmatic character is both the puzzle and the power of the *nonlinear-dynamical lens* by which the world can be scanned for recognizability.

I will have more to say about all this from time to time. But for now, perhaps, a "travel advisory," concerning routes and rates of progress through the book, may be appropriate. I would ask of the reader as much forbearance as can be brought to the contemplation of my proposals, for it took me, too, a long time to see what now seems obvious — even though I had conditioned myself to look for it — and I cannot fairly ask for a more immediate acceptance from my readers. Overstreet (1949) phrased best what I would ask, that, to the greatest extent possible, one should exercise one's "ability to hold a suspended judgment," should stand at some remove from the subject matter, and hold in abeyance the way one sees it, in relation to the constructs of nonlinear dynamics. Such distancing, I have found, is necessary that debate not hide the truth among the facts by the practice of logic, for "logic" is often but a disguise for contempt before the fact (cf. Spencer-Brown, 1972). Paradox, or the illusion of paradox, is often the real ruler of the scientific mind. I suggest therefore that the reader look for and emphasize the simplicities wherever possible — even when the going seems tough — for there is no adequately common language by which to convey what I wish to say. It is the capacity to see the glimmer of meaning that "simple" would convey here that distinguishes the concept of "chaos," as understood by the modern-day student of nonlinear dynamics, from the "chaos" of the Bible, which would threaten anarchy and reinforce our holds on clocks and/or "perfect chance" as hedges against the forebodings of strange undertones and overtones of apocalyptic ills of a future either sought, by some, or given unasked, by the practices of our cherished motto, "the scientific method."

The Random State: Central Dogma of Science

Until recently, nearly everyone working in the earth and planetary sciences would have ignored or dismissed my fanciful picture of the patterning of impacts by meteoritic objects, whatever the source of these objects or the character of their impingement on Earth's surface. It has been the common belief of science that crater distributions on the planets and satellites are "random." Randomness has been the accepted standard of the normal or null model in the natural sciences, as well, and it continues to rule as the central dogma in neo-Darwinian models of molecular genetics (e.g., Burnet, 1976; Gould, 1989c; cf. Shaw, 1987b; Gillespie, 1991; and Notes 8.14 and 8.15 in the present volume). If one intended to show that there is a pattern in some precinct or phenomenon of nature, then one had to "prove" that there is some exceptional reason for the pattern. An underlying theme of the present work is that this dictum has it backwards, that, on the contrary, multidimensional and nonlinear Nature — left to her own devices — *can do nothing else but* generate organized complexity by perfectly ordinary mechanisms of self-organization. The random state is *not* the natural state of matter, nor is it the natural state of waves, ethers, or "empty" space. And yet this central dogma of randomness has been, and to some extent is likely to persist, and to remain a paradox to the human mind (we often find, in spite of ourselves — I speak here from my own habits of mind — that we tacitly accept it, even when we know better, as evidenced by such phrases as "the random sample," "the random choice," "the random state," etc.), and the root of the paradox is that one truth is unavoidable — *all of our percepts, precepts, and concepts are based on self-reference*. Historically, we have based our concepts of spatial measure, for example, on units proportional to ourselves — relative to an individual, a kingdom, or the "world" — and we have based our concepts of time, whether Newtonian or Einsteinian, on units relative to the narrow confines of our own spatial history. Our calibration of time as a system of measurement is expressed in units of our movements relative to the circumferences of our horizons — man scattered about the Earth as center, Earth moving about the Sun as center.

Our beliefs about length and time are deeply embedded in the cognitive mechanisms by which we have come to define order, or its absence, and few have noticed that our self-referential mechanisms of measurement and analysis deny us "free choice" in defining, in any absolute sense — and despite our efforts to impose absolute, hence arbitrary, standards — measures of length independently from measures of time, and vice versa, in our descriptions of nature. [In other words, length is invoked or implied in our measures of time, and time is invoked or implied in our measures of length — thereby establishing an effective length-time feedback loop (or, in special relativity, even a *folded-feedback loop*; cf. Shaw, 1987a, Fig. 51.14) of self-referential observations.] Our measures of distances, areas, and volumes in space that neglect the time-dependency err, with respect to more realistic measures in the domains of spacetime, in a manner that is analogous to the way planar measures of distances and areas on a flat Earth would err in the context of modern *geodesy*, the science that studies "the figure of the Earth," meaning its geometric-gravitational configuration. As a field of study, geodesy was revolutionized by the application of artificial-satellite gravity and topographic "surveying" methodologies (e.g., King-Hele, 1976, 1992; cf. Anderson and Casenave, 1986; Carter and Robertson, 1986; J. Davis et al., 1989; D. E. Smith et al., 1990; Dixon, 1991; Hughes, 1993; and Notes 6.5 and 6.10 of the present volume). Observations and measurements in spacetime, then, represent aspects of the universal self-organization, hence these acts are not separable from the "things" being observed and measured. For if the random as null model were truly to describe the fundamental nature of the universe, and if our cells and the evolution of our sensory apparatus — by which the observations and measurements of our worlds are made — have sprung from that universe, as presumably they have, then our concepts of space and time, too, would have to be artifacts of that randomness. In that case, order — if it is to be viewed as something separate and distinct from randomness — would be the province of a different universe, not the random one but one of special creation, a universe made by man or by god, or by man as god. But just as our concept of space cannot really be separated from our concept of time, neither can our concept of randomness be separated from our concept of order. What is more, we cannot separate our concept of space and time from our concept of randomness and order. The concept of an "existence" for each of the measures *length, time, randomness,* and *order* — i.e., a distinctness that is separable from their *relationships* — has no real meaning, earnestly though we have tried to impose real meaning onto each of them in isolation from the others.

Our views of order and disorder, determinism and chance, are caught in a "Strange Loop" of consistently illogical logic (Hofstadter, 1979). Paradox and contradiction seem to hound our reasoning at every turn. Relatively few theorists have come to terms with self-contradiction, yet self-contradiction is inevitable in every system of self-reference. We are confronted with and affronted by the liar's dilemma (Epimenides' Paradox, or liar's paradox) even as we revere the vast might of the mathematical principles we have brought to bear on natural phenomena. Yet the evolution of mathematics has only confirmed the central dilemma and the paradoxical foundation on which mathematics and logic are based. If few have noticed the contradictions in world models based either on randomness or on order, fewer still have come to realize that no law of nature requires that this paradox be solved. It is only a belief in the need "to prove limiting or ultimate truths" that sustains the grip of any paradox.

Those who have offered resolutions — rather than solutions — to the dilemma have done so by reference to expressions of a continuum of relativities spanning the limiting idealizations. Something shown to be disorder in one context may be shown to be order in another, just as "perfectly random chance" in one context, as in the throwing of dice, can be shown to be consistent with "perfect determinism" in another (see Ford, 1983; cf. Gould, 1989c). Resolutions that express such a continuum of relativities are transformational in character, and at least two transformations, or major reconfigurations, have occurred in the symphonic constructs of human thought over the last few decades. The effects of these reconfigurations began to take hold very quietly but then built rapidly to the crescendo that we are beginning to perceive throughout all scientific disciplines and science-related technologies and practices — including principles of codification and communication, principles of optical transmission and measurement, techniques of characterization of subatomic motions, principles and practices of molecular engineering and medicine, and principles and practices of celestial dynamics and cosmology.

The First Transformation of Thought

The random state is that which can neither be encompassed nor structurally described. It is simply the ineffable. It is the "maximally complex," in the language favored by some mathematicians (see Chaitin, 1987a, b, 1988; Nicolis and Prigogine, 1989; Ruelle, 1991). And this means simply that it can be described only in terms of itself. If anything else, meaning something more complex, can be shown to encompass a given set of numerical data (i.e., leaving that set intact and identifiable as such relative to the expanded context), then that sequence is said to have been "compressed" and is no longer a model of randomness. For example, a macromolecule may be so complex that in isolation there is no obvious way to describe it in terms of fewer entities than the quantity of individual atomic elements (atoms, ions, etc.) it contains. In that isolated context such a molecule is an *algorithmically random structure*, an "aperiodic crystal" as Schrödinger called it [see Schrödinger (1992, pp. 60ff); the work cited is a republication of selected writings by Erwin Schrödinger (1887–1961), including his little book *What Is Life?* first published in 1944, which represented some largely untutored thoughts of a mathematical physicist about biology, thoughts which proved to be catalytically influential in the development of the genetic code; cf. Judson (1979)]. But in a larger context, such as in a mixture of macromolecules, it may be found that certain sequences of atoms, and/or geometric configurations of atoms, may be recognized as recurring entities (code-like or word-like structures), and in that context, not only is the description of the mixture itself much simpler than its totality of atoms suggests, but the macromolecule previously defined as a random or aperiodic structure (by default, for want of any way to reduce its description) may be found to contain subentities that recur in other macromolecules. The macromolecule is thereby *symbolically compressed*, and fewer bits of information are required to describe its role in the mixture than are represented by the number of different atoms it contains. It has become something codified, hence ordered. In other words, something that is symbolically incompressible (*random by default*) in one context may become compressible (*ordered*) in another. Randomness is a contextually relative concept, and it has no useful meaning as a general (algorithmically unqualified) description of spatial and/or temporal structures in nature. Were a given mechanism or process, such as the coin toss, shown to be thus codified, and even though it had been believed to be the very "standard" of a random process, then it will have relinquished its random character, for it would now have a describable substructure **(Note I.2)**.

An idealist would, presumably, rationalize such a demonstration by arguing that even though there is no such thing, physically, as a truly "fair" coin toss, the abstract notion of an idealized random-number generator—numbers chosen purely by "chance"—can still be applied to the statistical analysis of physical processes. This approach is as dynamically misleading as the notion that the universe contains somewhere within it "a perfect clock"—or one that is perfect but is somehow "external" to the universe—representing an absolute and invariant measure of time. Once, and not that long ago, the motions of the planets were seen to represent such perfection. We put our faith in a "perfect orrery" (PO), a mechanical clockwork of orbital motion so precisely fitted that it could be counted on as the absolute measure of time. With increased precision of measurements in astronomy and in particle physics, it was eventually realized that the "atomic clock" was a more convenient and more precise measure of time. Suddenly (but subtly, for most persons still think of time as something meted out by a mechanical clock, or by the solar day), we had shifted our standard of perfection from the ideal deterministic clockwork to the "ideal" game of chance. The perfection now became a statistical one based essentially on the notion of a perfect random-number generator, the "perfect coin-flipping machine" (PCFM) of atomic decay [see Dalrymple (1991, pp. 80ff); but compare the discussion of improvements in "accuracy" by Thompson (1993) with discussions of "quantum chaology" (Berry, 1987), quantum chaos (e.g., Ford et al., 1990; Jensen, 1992), etc., in Notes P.4 and 4.2; cf. Bohm and Peat (1987) and the discussion of Bohm's "implicate order," quantum dynamics, and "classical chaos" in the Prologue]. In fact, the PO and the PCFM are only two among an unlimited variety of cosmological mechanisms by which we might "tell time" (actually, by which we can empirically define time, because there is no universal law or principle that tells us what "time" actually is in any absolute sense; it is really only a derivative notion from our observations and measurements of objects in cyclical motion). If the circumstances of technology and position in space were somewhat different, it is not hard to imagine a civilization that would use the signal from a pulsar **(Note I.3)** as its "absolute" measure of time.

From the nonlinear-dynamical viewpoint I advocate here, both the absolute mechanical clockwork (the PO) and the absolute probabilistic "clockwork" (the PCFM) amount to looking at nature as though "the tail is wagging the dog." Only the behavior of the natural system itself (its properties of nonlinear recursion under specified conditions of observation) identifies the contextual meaning of time. Even though, in practice, time is defined as if such things as POs and PCFMs exist, our calibrations of time are subject to measurements made on systems that are now known to operate in nonlinear and even chaotic patterns (e.g., the chaotic motions of the planets; cf. Laskar, 1989; Wisdom, 1990; Milani and Nobili, 1992; Sussman and Wisdom, 1992). The distinction between the conventional and the actual meanings of time is of no great moment in the immediate activities of social and cultural life, but it does have significance relative to our views of the world and our beliefs and strategies concerning how the world works; hence it does influence our beliefs about how the organizational behaviors of societies and cultures are manifested — e.g., in their political beliefs, the waging of war, and so on (cf. Note 8.15).

We have, in fact, attempted to force our perceptions of the world into a linear conception of time. And we have succeeded in this, up to a point, but we have incurred a debt that is now being collected at the cost of major reorientations in our methods of doing science and in our perceptions of our ecological role in the planetary environment. Our idealizations have attempted to impose beliefs about the dynamical meaning of time, beliefs that impose on our operational definition of time something it is not, an absolute concept. The fact that there are conflicting viewpoints concerning the nature of the perfectly ordered state and the perfectly random state brings us to the nexus of the dynamical-mathematical reconciliation that will be addressed in this book. If any sort of nonlinear, nonuniform interaction with the environment can be shown to exist, then a clock cannot be perfect, and a coin toss cannot be fair; hence neither the notion of a perfectly independent periodic process nor the notion of a perfectly independent random process is valid [nor is there such a thing — once we remove the artificial stipulation of container walls (cf. Nicolis and Prigogine, 1989, Fig. 1) — as a perfectly random gas in Nature, a gas that corresponds perfectly to the kinetic theory of gases; see Hoover and Moran, 1992]. Once interaction is allowed — and there is no way to exclude it from a nondelimited world — there is an ever-increasing certainty that some form of characteristic structure will appear in the outcomes, and the longer the sequences we can construct and recognize, the "sooner" (cyclically speaking) a structure will become demonstrable. And because such a world has no beginning and no end, all natural sequences — meaning those that are not bounded, either by artificial container walls or by theoretical constraints — are of indeterminate length, and within them can be found, given sufficient effort, compressible (codifiable) nonlinear structures (cf. Note I.2).

This view of nature differs from one based on the "laws of probability," because probability — as in the forms of statistical averaging that have made the laws of thermodynamics appear to be standards of certainty in a random world, and that have made the coin toss appear to be the standard of reference for "exact" measurements of time — is based on the idea that an immutable mechanism, such as the fair coin toss, can exist independently of, and in isolation from, interaction with a dynamically heterogeneous world. (By the same token, the idealized clockwork would imply the existence of a mechanism that operates in an invariant manner *no matter what else is happening in the system*; both ideas are equivalent to suggesting that some mechanism by which processes *within* the universe can be measured can exist "outside" the universe, an idea obviated by the *principles of incompleteness and self-reference* that were discussed in the Prologue relative to our conception of the universe.)

Except for the relative numbers of "bits" of information that exist within a substructure, the coin toss becomes conceptually no different from the orbital processes by which we measure "years" that can be compressed into other units called "days," "hours," and so on. It is only in the dimensional scaling of a process that we can see the process in a different light. We can measure time with a pendulum, with an hourglass, or with an "atomic clock," and departures from uniformity in each of these mechanisms depend on how the scale of the dynamics in each case can be perturbed by interactions with the rest of the world relative to the number of iterations of the mechanism that we translate into a measure of time. The dripping of water through a standardized orifice from an effectively infinite reservoir (relative to the duration in time that we judge to be significant) would represent an analogous clock capable of arbitrarily great precision. Yet, when the distributions of drops from the orifice are examined in detail, we find that they describe recursive patterns that have all of the structural complexities and simplicities that are found in the study of nonlinear dynamics (e.g., R. Shaw, 1984; Wu and Schelly, 1989; Cahalan et al., 1990; Shaw, 1991). Certain phenomena to be discussed later, called *self-organized criticality* and *chaotic crises*, are directly related to the dynamics of the dripping faucet and to the motion of sand grains in an hourglass. All of these mechanisms are of mixed characteristics, just as detailed observations of the coin toss, or of the workings of an "atomic clock," would be found to be under appropriate shifts in relative scaling—in the latter case involving nonlinear quantum-chaotic effects [Thompson's (1993) discussion of an "unexpected hurdle" discovered in recent attempts to improve the "accuracy" of atomic clocks raises nonlinear-dynamical considerations of the sorts discussed, for example, by Pietronero (1986), Berry (1987), Richards (1988), Ford et al. (1990), Jensen (1992), and Parmenter and Yu (1992)].

To a small but intelligent creature with a lifespan of the order of ten Earth-person seconds and a size far smaller than a grain of sand, measurements of the same standards of length and time by which we have "compressed" the world—by the very act of establishing standards of measurement—even if scaled downward in linear proportion to size and lifespan (our units divided, say, by a factor of ten to one hundred million), would look as if they were random-number strings. The inverse also would be true, at least until we developed a cryptography to decipher the historical record of that creature's culture (but look at the effort we humans have had to put into deciphering the ancient language codes of our own species, codes in which the objects and methods used as standards of measurement are fundamentally like our own in nature, requiring no significant dimensional transformation). Even when decoded, the creature's system of measurement would appear to be nonlinear, noninteger, and greatly distorted relative to ours, and its self-referential perceptions of the celestial sphere would also differ greatly from ours. Even if the creature were to have recognized that its measure of length was cyclical in some nonlinear sense, lengths based on the (fractal) perimeter of its world would be much greater than ours, relative to a common standard of length, and the comparison would not be reproducible, because from place to place the lengths measured likely would differ from day to day relative to ours, perhaps even from minute to minute, or from second to second (one of our days, rescaled in terms of creature-units of age and individual longevity, would be equivalent to nearly ten thousand Earth-person years, while one of our minutes would correspond to about six Earth-person generations, and one of our seconds to about one Earth-person "decade").

Averaging undoubtedly would figure prominently in this creature's concepts of space and time, given an adequate history to define a chronology—for statistical averaging is important in our own systems of measurements, where standard "meter-sticks" and "days" are readily identified and easily measured, especially when we contemplate phenomena that are relatively very large or very small. The relationships between measurements made, on the one hand, in the small-scale creature-culture just described and, on the other, in human-culture would involve what we now call "fractal" or "multifractal" dimensional scaling—scaling that involves a multitude of, perhaps even a continuum of, noninteger exponents as proportionality factors in transformations between arithmetic and geometric series [I refer here to the difference between (1) counting by integers in *arithmetic series* involving multiples of the units of a standard of measurement and (2) "counting" by means of *geometric series* involving noninteger powers of the units of a standard of measurement, representing so-called *fractal statistics*, or *"power-law" statistics*].

The world of large organic molecules, proteins, viruses, and bacteria, viewed at high magnification, is metaphorically analogous to the "creature-culture" just described. The metaphor applies even with regard to "intelligence." Certainly, viruses cannot be described as lacking intelligence, in view of their structural organization and ability to insinuate themselves into the life cycle of the biological cell (electron micrographs show them to have "space-vehicle-like" morphologies, down to the intricacy of prong-like "landing pods"; see Jacob and Wollman, 1974, p. 39; Sampson, 1984, pp. 97ff; Levine, 1992, Chap. 2). Not that long ago (prior to 1953), before the crystallographic structure of DNA was first fully described by Watson and Crick (see *The Eighth Day of Creation* by Horace Judson, 1979), organic molecules of the living cell were seen as vague, "infinitely long" structures that lacked characteristic repeat units. Proteins and "genes" — even chromosomes — were suggested to be of this "aperiodic" character (Schrödinger, 1992, pp. 60–61; cf. Judson, 1979, pp. 244f). They were called "aperiodic crystals," the structures of which reasonably could be called random — at least according to the mathematical criterion of compressibility described here — until their substructures were defined and codified. But once the substructures called nucleotides and amino acids were identified, the periodic structure of living materials became "compressed" to a much lower state of complexity. These revelations constituted a veritable crescendo of collapsing uncertainty, as dramatic in its way, and in terms of the number of symbolic entities needed to describe a system, as the collapse of a portion of a galaxy into a "black hole." Both types of phenomenology became "ordered," not by the discovery of any new and immutable mathematical principles, analogous to the axiomatic foundation of probability theory, but simply by the exercise of existing criteria leading to the formulation and documentation of more perceptive *mechanisms of recognition*.

The scientific views, respectively, of biological, geological, and astrophysical uncertainty have undergone major advances in little more than three decades — a time span virtually coincident with that of my own scientific career. But when I think about the distinctions between these views, I wonder if we are not perhaps too quick to congratulate ourselves on the sophistication of our discoveries in these fields. The language of molecular biology, for example, is now intricate in the extreme — unintelligibly so to anyone but the trained expert in its semantics (see any current issue of *Science*) — but what primitive notions might not still be masked by such sophistications? Consider the terminology of molecular biology in the light of the hypothetical "creature-culture" described above. Let us say that such a creature is analogous to a virus. In the terminology of viral studies, the word-root "phage" figures prominently in technical nomenclatures. Yet the history of that usage is rather less sophisticated than the semantics might imply, for "phage" — both in the dictionary and in its technical connotation — means simply "one that eats," a primitive notion indeed. In the word "bacterio*phage*," we see that the primitive criterion of eating has been joined to the name of a class of organisms to produce a rather specialized, even if efficiently compact, technical term (cf. Lin et al., 1984; Sampson, 1984, p. 98).

The way this term was established also involves rather primitive processes and observations. Many viruses are so small that they would pass through all of the filtering devices available in the earlier days of bacterial studies. Thus, one of the classes of viruses came to be known as the "filterable viruses," a term seen frequently in technical articles until a few years ago. Since the filterable viruses could not be isolated and characterized using the optical microscope (before the electron microscope was perfected), one way to determine their presence and to characterize their behavior was to ascertain whether the filtrate would "eat" a given bacterial culture; hence the class of viruses called bacteriophages. Now, of course, the mechanisms of viral "eating" have been studied extensively (e.g., Sampson, 1984, Fig. 6.2), and they have been related to the molecular mechanisms by which the genetic material of the viral cell enters into and interacts with the host cell (a process so visually horrific that I sometimes think it must have inspired the imagery of the motion picture "Alien"; ironically, the mechanisms of submicroscopic viral interventions seem far more terrifying than the film, and more profoundly disturbing than they would otherwise be, if they are viewed as a form of attack by a totally alien culture).

In a sense, it might be said that our difficulties and uncertainties in understanding the virus-induced diseases of the human species relate to the difficulties of dimensional transformations between the two "cultures." Part of the problem, curiously enough, is that it may represent in biology the same bias that has dominated the outlook of most researchers in geology and geophysics. Viruses are treated as "alien" entities, by both media characterization and medical characterization, just as the impacts of meteoritic objects and the effects of other extraterrestrial objects on Earth have been treated as "alien" phenomena. A new paradigm — in which the "external cause" is seen as but another aspect, or "effect," of an extended feedback system of "internal mechanisms" in an appropriately expanded universe — places both types of problem in the same light, one in which techniques of dynamical scaling are applied in a search for common effects that reflect mechanisms operating independently of scale (*dynamical universalities*). This is the strategy explored in the present work for the dynamical history of the Earth.

If the same strategy were applied to the molecular biology of immunological research, the analogous viewpoint would see bacterial and viral "infections" simply as forms of a larger context, one in which the immune system is not treated uniquely in terms of mechanisms of defense against an attacking alien force (a strategic war uniquely waged by antibodies, antigens, and so on). I, for one, would be inclined to explore the larger context within which both the immune system of a host and the "viral field" — the latter has heretofore been characterized almost exclusively as an alien "military" force deployed, initially at least, from a phase space external to the host — would be examined for those aspects by which the host systems demonstrate characteristic dynamical patterns of involvement with their global genetic environment. These patterns are likely to be analogous to, or identical to, strange attractors, attractors that may be manifested in subtle and fractal forms never before sought out, or even considered, in the ecosphere and global biosphere. Such phenomenologies would include nonlinear resonances and chaotic mechanisms of contextual self-organization that encompass both the viral world and the host world.

Such a view would therefore see the human immune system as extending beyond the individual to the global biologic community, just as the analysis of geodynamics that I offer here sees the manifestations of earthquakes, volcanic eruptions, weather, and climate as extending beyond Earth to include, at least, the Solar System and the Milky Way Galaxy. The universal context does not render us, either as individuals or as species, "immune" from the actions and interactions of "local" systems (e.g., involving the "untimely" deaths, extinctions, and other deleterious phenomena that affect individual organisms and species on Earth; and involving analogous phenomena that affect the lifetimes and ultimate fates, as individuals or "species," of asteroids, satellites, comets, and planets — and the Sun itself — in the Solar System). Yet the implicit medical goal of making us immune to all "disease" is tantamount to a goal in which Earth is made "immune" to all meteoritic impacts, plasma storms, and other phenomena imposed from "outside," as evidenced by an internal "cure" represented by the symptomatic absences of major earthquakes, volcanic eruptions, tornadoes, hurricanes, and floods. The absurdity of such an aspiration is self-evident in the geological context, and this book seeks to make that absurdity self-evident in the planetary and cosmological contexts, as well. (Compare the above remarks with Raup, 1991.)

In the end, the relativities, uncertainties, and codifications that have been applied to the numerical aspects of Earth processes differ little from those applied to the characterization of the universe at large. The dynamical information contained in the cosmos is compressed into units made up of stars and planetary nebulas, which can in turn be compressed into units called planetesimals, which can be compressed into major and minor planets, satellites, comets, and shooting stars, which can be compressed into the units of length and time that provide the measures by which we have defined the nature of complexity and compression. *That which has constituted our concept of the random is subject to relativity, cyclicity, and compressibility within a universal and self-referential framework that operates fundamentally in the same way at all levels of observation and description* (cf. Note I.8). The "truth" of this assertion is circular, in the sense that it is based on the nature of self-reference that pervades all known concepts in science, including mathematics, nonlinear dynamics, biological genetics, and cosmology, thereby subsuming whatever other scientific viewpoints we might conceivably bring to bear without revoking every known principle of science and mathematics. By its very nature, this conclusion must apply to any principle capable of being enunciated by the human cognitive mechanism. Furthermore, the self-referential context would deny the possibility advanced by some theorists (see Ruelle, 1991, p. 3) that an era might be coming in which computers "will take the initiative" and surpass the human cognitive potential in the mathematical context — suggesting, perhaps, that a totally new and different paradigm of mathematical organization could be applied to nature. The viewpoint expressed in the present work would imply that the evolution of a paradigm that can see the universe in a totally different light is incompatible with the most basic conclusions of mathematics, because computers are self-referential extensions of the minds of their designers, hence are incapable of drawing a boundary between themselves and their designers on the basis of their own logic *and then jumping across it* (jumping out of themselves, as it were — on the basis of a system of rules developed within themselves — into a different "skin" that represents a construction made from a new and independent set of rules; cf. Spencer-Brown, 1972).

We cannot know, either a priori or once and for all, what is incompressible, hence random. Our concept of the "fair coin toss" is based on limited data and idealized assumptions. The number sequences of a "random number table" are "random" only within sets that are small compared to the repeat period of the recursion algorithm that generated it. These sets are but "molecules" of a more complex, more protean, number set. To me, it seems more reasonable to think of what we have called the random state as a *Gödelian state*, one that is but an admission of natural incompleteness and our own fallibility (cf. Chaitin, 1982). To believe that something, by its very nature, is immutably random is to say, "I do not know what it is, and there is no way that I can ever know what it is, but it is nonetheless the foundation of all my principles." It is to say that the universe is a logical duality: (a) it is forever mysterious; and (b) it is axiomatically defined. It is to believe in an absolute god of reality that is forever defined by an axiom I have made. And this is a form of statement that only a god could make (cf. Rotman, 1993b).

Neither the phenomena of geology nor those of cosmology are random. They are Gödelian. This conviction is based not simply on many years spent searching for ways of compressing the complex. Rather, it springs from the observation that the universe of experience is both nonlinear and dissipative, meaning that every describable object, and every mechanism of interaction between objects, involves some form of conversion of the energy of position and motion into heat energy, and that the dissipation of energy is accompanied by the contraction of volume of phase space, where the "volume of phase space" is that which encompasses those states of motion of the objects that the objects might be imagined to have "visited" in the absence of dissipation.

The Second Transformation of Thought

Beginning with Poincaré at the turn of the century, and picking up the pace in the 1960's when computer experiments by Edward Lorenz demonstrated what is now called *sensitive dependence on initial conditions*, and/or *exponential diver-*

gence from initial conditions, scientists have begun to look at the subject matter of their disciplines within the nonlinear-dynamical context. An effect directly analogous to sensitive dependence on initial conditions surfaced in my own experimental studies of dissipative (viscous) flows, also during the 1960's, confirming theoretical conclusions derived by Gruntfest in 1963, the same year that Lorenz first published on what is now called "deterministic chaos." My own experimental confirmation of thermal feedback was first published in 1969 (cf. Nicolis and Prigogine, 1989, Sec. 2.2; Shaw, 1991, Figs. 8.1–8.9). Dissipative feedback, called over the years by such other names as "thermal feedback," "thermomechanical feedback," "thermal runaway," and so on, demonstrates, directly and explicitly, the principle of exponential divergence from initial conditions (cf. Notes P.1 and P.5; and Gruntfest, 1963; Gruntfest et al., 1964; Shaw, 1969; Gruntfest and Shaw, 1974; Anderson and Perkins, 1975; Wones and Shaw, 1975; Nelson, 1981; Shaw, 1991).

A contextually more meaningful name for dissipative feedback might be "entropic feedback," because the nature of the effects produced — expressed in terms of material states, mechanisms, and processes — is sensitively dependent on diffusive transport and chemical reactions (transfers and transformations of mechanical work, heat, and mass), which in turn are sensitively dependent on gradients of velocity, temperature, and chemical potentials relative to neighboring material states, mechanisms, and processes (cf. Prigogine et al., 1972, 1991; Prigogine, 1978; Krinsky, 1984; Prigogine and Stengers, 1984; Nicolis and Prigogine, 1989). In such a context it is easier to see that the nature of entropic transport is relative, both spatially and temporally. Some mechanisms and processes move toward more and more aperiodic and codifiably complex states (states of high entropy, or uncertainty), others move toward more and more periodic and codifiably ordered states ("negentropic" states of relatively increased redundancy), and still others oscillate relative to the two tendencies, the distinctions between directions and oscillations being scale-dependent, hence relative at all scales of time and size.

The universe of experience is ruled by processes that rarely, and in detail never, boil down to direct proportional parts — parts that can simply be summed to describe the net effects of those processes. This is equally true for those processes that involve "fundamental structural units" — such as specific atoms, molecules, or groups of atoms and/or molecules described by the formalism of the crystallographic "unit cell" — and/or "fundamental particles," such as electrons. In the presence of interactions there are always forms of dissipation that modify the structures, material makeup, and energies of the interacting parts. And interactions are ruled by mechanisms and processes that cannot sustain themselves in perpetual motion in the absence of interactions with other mechanisms and processes, regardless of the scales of length and time by which the mechanisms and processes are described. This is more than just a statement of the "Second Law of Thermodynamics." The universe is entropic, but it does not "run down." If it *could* run down, we would be left with the same contradiction that the random model presents: a state that is believed to be defined absolutely (as in "absolute zero"), but one that the same belief can state only in the form of an unrealizable axiom.

The following, then, are the central ideas of this book: (1) geological history and cosmological history involve the same principles of organization, (2) terrestrial mechanisms and processes are coupled dynamically (entropically) with cosmological mechanisms and processes, (3) all these mechanisms and processes are invoked by and invoke nonrandom patterns in nature, and (4) nonlinear interactive feedback involving sensitive dependence on prior conditions is the governing principle of evolution in the Cosmos, just as it is in the Earth (cf. Note I.8). These principles, in turn, are composed of those principles of nonlinear dynamics currently subsumed within the phenomena of "deterministc chaos" **(Note I.4)**.

Nonlinear dynamics, in its more complex manifestations, concerns the details of the synthetic and analytic characters of ordinary phenomena that have evolved and/or recur in such a way as to be recognizable, hence quantifiable, in both the descriptive and taxonomic senses. Put in these terms, it is evident that nonlinear dynamics deals with the common issue of what is or is not recognizable, hence familiar, as a categorically similar form or configuration within an array of complex configurations that, although different, are also recognizable in the same sense. It is not surprising that these are the same criteria by which living and fossil organic forms — flowers, birds, trilobites, people, whatever — are recognized and classified as the same and/or different, because organic forms originate within nonlinear dynamical systems made up contextually of the same types of dynamical mechanisms and processes mentioned above (nonlinear periodic-aperiodic arrays of trajectories and cells, attractors, strange attractors, critical-state processes, chaotic crises, and the like). This metaphorical comparison also underscores the taxonomic problem associated with the concept of the random state. An unrecognizable fossil form represents a geometric form, or array of geometric forms, that can be placed in only one taxonomic category: the category of unrecognizable fossil forms. This category has no meaningful taxonomic status — other than the criterion of unrecognizability — because its membership can be recognized only in terms of its enigmatic status. However, the ability to conditionally group such fossils (e.g., *"Enigmata jardineii"*) according to even vague resemblances to known fossil assemblages "compresses" the complexity of the global system of taxonomic classification — and if the *Enigmata* subsequently become recognized as fragments of a particular body part of an already classified fossil group, then of course the algorithmic complexity of the taxonomic classification of enigmatic fossils automatically collapses to something far simpler (cases in point concern controversies over the complexity of the Precambrian and early Cambrian biota vis-à-vis more modern biota; cf. Note 8.14; and Knoll, 1991; Foote and Gould, 1992; Knoll and Walter, 1992; Levinton, 1992; Morris, 1993). I would also note that simply the recognition of complicated-looking objects in nature as fossil forms in and of itself places them within the self-referential framework of *Gödelian arrays,* thereby accomplishing in one fell swoop a huge compression of complexity relative to the world at large.

My thesis is explored in several different ways in the chapters of this book. It proposes that the historic array of Earth impactors, deciphered from the cratering record of this and other planets and the Moon, has recognizable attributes and substructures that depart significantly from a maximally complex or random configuration. Geometric patterns in this historic array are shown to exist, and they are shown to be compressed in terms of both spatial and temporal recurrences, thus forming recognizable subsets of the full record, hence are spatiotemporally ordered in the sense described above. This geometric "compression" is then described in relation to familiar patterns and concepts of orbital dynamics. It is also described in terms of recurrent dynamical relationships between the spatiotemporal character of the cratering record, mantle dynamics, volcanism and/or patterns of "hot spots," paleogeographic features of the Earth's crust, patterns of geomagnetic reversals and clustering of virtual geomagnetic poles, the fossil and biostratigraphic records, and a host of other earthly processes and their products. The recurrent (generally cyclical) character of these processes testifies, even if taken alone and without regard to relationships, to my opening declaration that none of these terrestrial phenomena conforms — even in its most general spatiotemporal characteristics — to the notion of randomness, in the sense either of endogenous effects or of exogenous coupling with the cratering record. The cosmic imprint in many quarters of the solid Earth body, however, can often be read only through the lens of nonlinear-dynamical concepts and analyses that allow the demonstration of effects produced by systems of terrestrial and extraterrestrial coupled oscillations, thus giving systematic meaning to irregular and "catastrophic" events of intermittent character. In this light the sudden and/or episodic events of, for instance, volcanism and geomagnetic reversals are seen to be the "normal" behaviors of chaotic systems rather than the chance happenings of random and independent magmatic and geomagnetic source phenomena.

The views expressed in these introductory remarks are largely my own. Additional insights into the concepts discussed can be found in a number of the references cited in this volume, such as Prigogine et al. (1972, 1991), Spencer-Brown (1972), Ovenden (1975), Berry (1978, 1987), Ford (1978, 1983, 1987), Prigogine (1978), Hofstadter (1979), Chaitin (1979, 1982, 1987a, b, 1988), Schlegel (1980), R. Shaw (1981), Krinsky (1984), Prigogine and Stengers (1984), Crutchfield et al. (1986), Shaw (1986), Lima-de-Faria (1988), Gould (1989a, b, c), Nicolis and Prigogine (1989), Ford et al. (1990), Goldberger et al. (1990), Graham and Spencer (1990), Wisdom (1990), Casti (1990, 1991), Abraham et al. (1991a, b), Raup (1991), Ruelle (1991), Eigen (1992), Schrödinger (1992), and Shaw (1994). Excellent tutorials on nonlinear dynamics and chaos theory are given in, for example, Abraham and Shaw (1982–88, 1987), Baker and Gollub (1990), Middleton (1991), and Tufillaro et al. (1992), including computer software that enables one to obtain firsthand experience concerning the behavior of nonlinear-recursion algorithms, such as the quadratic function, the sine-circle function (see Table I, Sec. D), the Lorenz equations, and other algorithms. Software is offered in both IBM-compatible and Macintosh-compatible formats by Baker and Gollub (by prepaid order), and is included with the printed text by Middleton in IBM-compatible format, and by Tufillaro et al. in Macintosh-compatible format.

The rest of the Introduction previews some of the observations and interpretations that are documented more fully in the chapters that follow.

Impact Cratering: The Coordinating Database of Earth History

Pursuant to these investigations, I have studied the geographical and temporal distribution of the known impact craters on Earth, finding — contrary to common belief but in accord with nonlinear-dynamical prediction — that it is not random. My reexamination of the cratering record reveals the persistence over time of at least three spatially invariant clusters of craters. Some of the characteristics of this pattern apply to very small young craters (ages < 10 Ma; diameters $\ll 1$ km) as well as to older large and small craters (ages from ~ 10 Ma to ~ 2000 Ma; diameters from ~ 1 to > 100 km). Hence the invariance is not simply an artifact of the long-term survival of large craters in relatively undisturbed cratonic terranes. On the contrary, the stability of certain special terranes since early in Earth's history appears to be a consequence of the long-term spatial invariance of the cratering record! I conclude that this relationship is a manifestation of Earth's history of coupled geodynamic and orbital-dynamic resonances, one of several categorical types of resonance relationships that form an important part of the central theme of this book. A catalog of 121 terrestrial impact craters was the principal data base used in Figures 1 through 3 and Figure 6 (Grieve, 1982, 1987; Grieve and Robertson, 1987; cf. Grieve and Dence, 1979; Grieve and Robertson, 1979; Grieve et al., 1988). Both the figures and the text discussion are based on this data set. Grieve (1991) and Grieve and Pesonen (1992) have amended this list to include several additional craters, and have given new or revised ages for several others. Their amendments, if anything, bolster the points I have made. A number of other sites have been proposed as candidates for impact craters, and some of these are in the process of being documented (cf. Fig. 3: the global distribution of supplementary craters and craters with revised ages is illustrated in Appendix 1).

Most of the craters thus documented are of Phanerozoic age (≤ 570 Ma; Harland et al., 1990); roughly 10% of the craters in the catalogs of Grieve and coworkers are undated, and another 5% or so are either of Precambrian age (> 570 Ma) or have age uncertainties that span the Cambrian-Precambrian transition but place them close to that transition. The global invariances in the cratering history are shown by mutual overlaps of cratering patterns among four different age groups of craters encompassing the Phanerozoic Eon (Fig. 2). Patterns within each of the age groups differ in their local distributions and regional trends, but all of the progressively younger patterns intersect or overlap older ones at characteristic geographic locations. Although the regions of overlap may span more than 20° of arc for any two age groups, the locus of common overlap among the patterns of all four Phanerozoic age groups within distinct clusters is restricted to roughly 5° in latitude and longitude (i.e., less than the expanse of the large asterisks that mark these positions in Fig. 2). The common center of each of the overlapping sets is called a *cratering node*. The three cratering nodes, which have remained invariant in relative geographic position throughout the Phanerozoic Eon, are called *Phanerozoic cratering nodes*.

The Celestial System of Geographic Coordinates

I emphasize in this book that the planetary system of geographic coordinates reflects positions and orientations relative to the fixed stars (e.g., Polaris, currently the "North Star"). Although this reference frame is expressed in the present-day system of geographic coordinates, it looks outward from the Earth toward the fixed stars as markers of an external coordinate system on which the motion of Earth's coordinate system can be traced. This convention differs from the traditional inward-looking perspective wherein coordinate frames for relative motions of markers on the Earth's surface depend on detailed knowledge of differential motions between the continents, oceans, underlying mantle, core, and Earth's rotational and geomagnetic axes (the spin-axis and the mean dipolar axis of Earth's magnetic field, where compass "north" — corresponding at the present time, according to a confusing convention, to the south *magnetic pole* (see Gubbins and Coe, 1993, p. 51; but see Butler, 1992, Fig. 1.5) — is the direction of the dipolar field axis most nearly aligned with the orientation of the North Geographic Pole of the spin axis). At the present *epoch* (representing the instant in time that is chosen as the *astronomical date* of reference, which is not necessarily the present time), celestial north is that direction relative to which the Earth rotates on its axis and revolves about the Sun in the clockwise direction, as seen from directly south of Earth's South Pole. This is the *direct* or *prograde* sense of motion. The reverse sense of rotation or revolution is called *retrograde* motion. These are relative directions; for example, the Sun only appears to revolve around the Earth in the counterclockwise direction, whereas relative to the center of mass (CM) of the Solar System it too revolves in the direct or clockwise sense (the Sun's position relative to the CM, however, is complex, and its angular momentum relative to the CM has both positive and negative terms; see Jose, 1965).

At present, the northward projection of Earth's spin axis points in the approximate direction of the North Star, or "Pole Star," Polaris. This approximation to the true direction is accurate to only about 1° of arc, a deviation that varies with time. The precession of Earth's equatorial plane relative to the plane of its orbit causes the Earth's spin axis to trace out a circle of ~23.5° radius relative to the fixed stars. The central point of this circle falls on the projection of a normal to the ecliptic plane (the plane defined by the sequence of astronomical eclipses during the year, hence by definition the plane of the Earth's orbit). About 4000 years ago (Roy and Clarke, 1988, Sect. 10.9) the northward projection of the spin axis pointed to the vicinity of the star γ Draconis, and roughly 12,000 years hence it will project to within a few degrees of α Lyrae (Vega), returning to the vicinity of Polaris in about 26,000 years (the approximate sidereal period of precession which, relative to orbital relationships involving solar-terrestrial cycles, consists of several different "harmonic" components; cf. Berger et al., 1992).

The background of stars upon which the precessing path of the Earth's spin axis is traced is called the *celestial sphere*. The celestial sphere is a coordinate system defined by a fictitious sphere of infinite radius on the inside of which are projected the positions of the fixed stars and the geocentric coordinate system defined at a given epoch, such as the present-day geographic poles and Equator (Bate et al., 1971, Chap. 2; Taff, 1985, Chap. 3; Roy, 1988a, Chap. 2; Roy and Clarke, 1988, Chap. 5). Thus, the celestial north and south poles and the celestial equator are the projections of Earth's North and South poles and Equator on the celestial sphere. Accordingly, the invariance of the three Phanerozoic cratering nodes (PCNs) inferred from the Earth's cratering history is equivalent to the existence of three different virtual poles mapped on the celestial sphere, poles that have maintained a constant angular orientation relative to each other and to Earth's spin axis. This relationship is inferred from the constancy of the cratering pattern throughout the Phanerozoic Eon (the nodal invariance already described) and from relationships between the orbital configurations of impacting objects (Chap. 6) and the mass distribution in the Earth that governs both the orientation of its spin axis and the conditions of spin-orbit resonances that have governed the patterns of impacts (Chap. 8). These additional celestial poles, and the system of spin-orbit resonances that governed the dominant pattern of impacts (representing many, but not all, impactors that have struck the Earth), thus have been locked — by this interpretation — to the position and motion of the Earth's rotational axis for the duration of this invariance. My interpretation of the cratering pattern suggests that this duration has been at least as long as the Phanerozoic Eon (570 Ma) — and according to the hypothesis I advance later concerning deep mass anomalies in the Earth (see discussion of Fig. 33), the same invariance has persisted, in a qualitative sense, probably since the time of the formation of the Moon ($\sim$4 Ga).

According to this hypothesis, cratering invariance was qualitatively established in its present form a few hundred million years after the formation of the Moon (i.e., after about 4.4 Ga and earlier than 4 Ga; see Dalrymple, 1991, p. 401). The set of geographic coordinates defined by the three PCNs, together with Earth's polar-equatorial coordinates and other invariances of cratering patterns to be described, is referred to as the Celestial Reference Frame (CRF) for geodynamic processes. Therefore, this reference frame provides positional information in both latitude and longitude for any geodynamic processes that can be shown to have a known kinematic history relative to the vicinities of the PCNs and other geographic elements of the CRF (cf. Courtillot and Besse, 1987, p. 1146, Note 18). I emphasize that the meaning of "celestial reference frame," as applied in this

Introduction

context, *does not imply that a coordinate system based on the distribution of impact craters is fixed relative to the fixed stars,* any more than that the Earth's spin axis is fixed relative to the fixed stars. It implies only that the CRF (by hypothesis) is fixed relative to the principal axial moments of inertia that are determined by Earth's distribution of mass — the distribution that determines the existence of, and the nonlinear dynamics of, the system of spin-orbit resonances of satellitic objects (natural and artificial) — hence that the CRF is fixed relative to the same effects that determine the orientation and stability of the spin axis (i.e., satellitic orbits are subject to precession, and to any other phenomena that affect the Earth's rotational stability). The word "celestial" distinguishes this type of coordinate system — one mapped on the celestial sphere in the same sense that the celestial equator and the precession of the equinoxes are mapped on the celestial sphere — from those pinned to internal properties of the Earth, properties that depend on relative nonlinear-dynamical relationships, the kinematics of which are not linearly synchronized and, in some instances, may involve exponential divergences in rates and/or directions of relative motions.

Cratering Patterns and Geodesy: Distributions of the Mass and Shape of the Geoid

I have used the following arbitrary groupings of crater ages to establish the age-dependence of spatial patterns: (1) < 50 Ma, (2) 50–100 Ma, (3) > 100–250 Ma, (4) > 250–600 Ma, and (5) > 600 Ma. These age ranges were chosen to emphasize, respectively, (1) the later post-Cretaceous/Tertiary (K/T) cratering history; (2) the history immediately preceding and following the K/T "boundary"; (3) the mid-Cretaceous to late-Permian history [including the possible Triassic/Jurassic (208 Ma) impact event of Bice et al. (1992; cf. Sepkoski, 1990; Hodych and Dunning, 1992; Trieloff and Jessberger, 1992; Benton, 1993; Rogers et al., 1993) and the Permian/Triassic (245 Ma) extinction event; cf. Notes 1.1 and 8.1]; (4) the several Paleozoic extinction events (cf. Fig. 6); and (5) the paucity of information for Precambrian cratering events. (Curiously — with some conspicuous exceptions constituted by the few very large Precambrian cratering events on Earth — the nature and timing of the transition between the Precambrian and Phanerozoic histories is not entirely unlike a tectonic transition that has recently been discovered on Venus; see Note 8.8; cf. Note I.5.) If warranted by future discoveries of new craters — a process that is likely to continue for many years to come — the above classification can be refined, especially for the Paleozoic Era, where there are at present insufficient numbers of documented craters at the proper biostratigraphic intervals for a useful examination of the cratering record vis-à-vis, for example, the late Devonian (Frasnian-Famennian, 367 Ma) extinction event of McLaren (1970, 1982, 1983, 1986); but see Appendix 3a.

The three Phanerozoic cratering nodes (PCNs) are based on age groups (1) through (4). Together, they crudely delineate a distorted small circle on a Lambert Azimuthal Equal-Area Projection with a diameter of about 130° of arc and a pole

33

at about 45°N, 160°E on the celestial sphere relative to the Greenwich Meridian and Earth's Equator (= celestial equator; see Fig. 5). The approximate geographical centers of the three principal PCNs (Figs. 1–5) are located on the celestial sphere at about 47°N, 92°W (North America), 58°N, 36°E (Eurasia), and 21°S, 128°E (Australia). The small circle connecting the three PCNs is called the *Phanerozoic cratering swath*, where "swath" refers to a band of roughly ± 25° on either side of the circle. At its greatest limit the Phanerozoic cratering swath approximates a great circle (Fig. 5) that divides the Earth into a North Pacific–Arctic hemisphere and a South Atlantic–Indian Ocean–Antarctic hemisphere.

The invariance of the three PCNs, including the Phanerozoic cratering swaths that define them (plus another small-circle locus described below), establishes the celestial reference frame (CRF) as the putatively fundamental geodynamic coordinate system. That is, the CRF comprises a set of markers that move in constant angular relationships with the Earth's Equator and rotational axis. But, because there is no absolute reference frame for motions in the universe, not even the coordinate system based on the "fixed stars," all apparent invariances are relative in some way. The concept of nonlinear spin-orbit resonances, upon which I base the theoretical explanation of the PCNs and CRF, states that there is a system of both stable and unstable orbits in the Earth and/or Earth-Moon vicinity that is generally coupled in the nonlinear periodic and/or chaotic sense with a system of mass anomalies within the Earth. Things are not all that simple, however, because the external variations of Earth's gravitational field owing to the Moon, the Sun, and the other planets are also active participants in these resonances, and if the Earth were a homogeneous spheroid of rotation, they would constitute the dominant perturbing functions. But in fact the Earth departs significantly from this symmetrical figure, and there are several types of well-known gravity anomalies — the ones existing now and/or those that have existed in similar forms at different times in the past — that influence the Earth-Moon orbital interaction and the orbital perturbations of both artificial satellites and such natural satellites, in addition to the Moon, as may exist (see Figs. 20 and 21; and Appendix 11).

In addition to the dominating effect of the Earth's equatorial bulge, artificial-satellite geodesy has demonstrated that satellite orbital stabilities are significantly influenced by higher-order mass anomalies (e.g., Kaula, 1968; Stacey, 1969; King-Hele, 1976, 1992; Hager et al., 1985; Anderson and Casenave, 1986; Bowin, 1992). I argue, partly following Allan (1965, 1967a, b, 1971), that the longitudinal distribution of such anomalies is crucially important to the existence and configurations of spin-orbit resonances (see Figs. 20 and 21; also Coffey et al., 1986; Friesen et al., 1992). In the present work I describe a significant body of circumstantial evidence (see Figs. 19, 20, and 37; cf. Appendixes 1–4) that argues strongly for the intermittent existence of small (diameters generally much less than 10 km) *natural Earth satellites* in ordered sets of preferred orbital configurations aligned, on average, with the PCN-CRF system of invariances. A number of descriptions of prehistoric and historic fireball trajectories are consistent with, and

Introduction

sometimes identical with, the same orbital parameters (see Appendix 3a and Note A.1; and Chant, 1913a, b; Mebane, 1955, 1956; LaPaz, 1956; Baker, 1958; Cassidy et al., 1965; Bagby, 1969; Schultz and Lianza, 1992).

If I accept the existence of spin-orbit resonances and preferred-approach scenarios for impactors, which I do, it becomes evident that the spatial invariance of such patterns over a given time frame must be a function of the evolution of the mass distribution in the Earth over a longer time frame. In detail, the patterns of orbital resonances at different distances from the Earth are likely to reflect different aspects of the mass anomalies. It may be, for example, that satellites inside the geosynchronous distance (see below) evolve differently than do satellites between that distance and the lunar distance. The more distant resonances are influenced by differing aspects of the gravitational field, especially as the relative effects of the terrestrial anomalies approach parity with various combinations of the lunar-solar influences. Such locations in the vicinity of the Earth-Moon system are complex or "fuzzy" and resemble the properties of the *fractal-attractor basin boundaries* discussed in Chapter 5 (cf. Shaw, 1991; Belbruno, 1992). The *geosynchronous distance* is that orbital distance where the mean orbital period of a satellite is the same as Earth's period of rotation. This occurs at about 6.61 Earth radii ($\sim$42,000 km from the center of the Earth, or $\sim$36,000 km altitude; see Fig. 20 and Allan, 1967a; Friesen et al., 1992). The lunar orbit is at a geocentric distance of about $\sim$384,400 km (semimajor axis of the Moon's orbit; see Anon., 1991, p. F2; Pasachoff, 1991, Appendix 4), a distance approximately equal to 60.3 Earth radii, or roughly nine times the geosynchronous altitude.

In order to simplify and coordinate the orbital-geodynamic discussion in the text, I follow a model (developed in Chap. 8) according to which the currently dominant mass distribution of the Earth was acquired shortly after the formation of the Moon (between roughly 4.4 and 4.2 Ga; cf. Dalrymple, 1991) and has since been modified only in minor respects by plate-tectonic "circulations," because tectonic feedback has been guided by the existence and magnitudes of the original anomalies. In the main, the effects of plate subduction have been distributed in a broad belt or girdle that more or less coincides with the meridional character of the early-formed deep mass anomaly (see Fig. 33 and Appendix 26; cf. the "lithospheric graveyards" of Engebretson et al., 1992, Fig. 1; Richards and Engebretson, 1992). According to this model (Fig. 33), the more robust features of PCN-CRF invariance (their mean orbital-resonance states) have persisted throughout most of Earth's history, whereas certain features related to *interior orbital resonances* (those orbits inside the geosynchronous distance) might have been more sensitive to effects related to near-surface plate motions and/or continental drift and may have differed to some degree from the mean resonance states. However, the central thesis of my work holds that a system of geodynamic feedback, between the cumulative energy of impacts and the net motions of plate tectonics and continental drift, has persisted throughout geologic history.

According to my general model, the long-term invariance of spin-orbit resonances rests within the mean kinematic effects of the distributions of all cratering patterns on Earth's tectonic history. This distribution over time is governed by two main effects: (1) the nonlinear spin-orbit effects, already outlined, as guiding factors in the positioning and coordination of cratering patterns; and (2) the nonlinear-orbital resonance phenomena (periodic and/or chaotic) between the sources of potential impacting objects and the rest of the Solar System, particularly Jupiter, the inner planets, and the Sun, as guiding factors in governing the approach scenarios of impactors and the conditions of transfer into spin-orbit resonances with Earth and/or the Earth-Moon system (see Chaps. 5–7). The second effect is of major importance to the distribution on the celestial sphere of approach scenarios, hence it is important to the "selection effects" by which sets of spin-orbit resonances are most likely to be populated relative to the long-term system of mass anomalies that is maintained by the tectonic feedback mentioned above (see Fig. 19). Alternatively, if the CRF hypothesis were to be disproved in this general form, a global pattern of mass anomalies that has changed systematically with time would be needed to explain invariances by mechanisms of spin-orbit resonance. In that event, the PCNs would represent a moving reference frame that has tracked the centroids of the evolving mass anomalies. For example, the PCNs might have varied in position relative to each other and to the Earth's spin axis as the continents changed their relative positions. In this simplistic case, the PCNs would constitute resonances, not with the figure of the Earth as a whole, but with a special distribution of near-surface mass anomalies correlated almost exclusively with the motions of only three unique cratonic terranes: North America (Laurentia?), Eurasia (Baltica?), and northwestern Australia.

I view the generalized case as far the more likely one from the standpoint of its simplicity and consistency with a model of integrated nonlinear self-organization of the global system. In that context, all forms of resonances — which include (1) the global system of spin-orbit resonances governed by a mass distribution that retains nearly the same symmetries over time, (2) the pattern of coordinated impacts that is globally locked to the same system and that modulates the magmatic-tectonic engine, (3) the global system of lithospheric circulations ("tectonic resonances") that, in a quasi-steady sense, maintain or enhance the crude meridional symmetry of the global mass distribution that, in turn, maintains (4) the relationship between the spin-orbit resonance conditions and the pattern of impacts — automatically satisfy the requirements for a common system of coupled feedback mechanisms which, according to the nonlinear-dynamical phenomenology of coupled-oscillator sytems, is precisely the type of system most likely to produce the forms of mixed periodic-chaotic resonances just described (see Chaps. 7 and 8). This is a cyclic system, hence it is consistent with the nonlinear intermittency (see Chap. 5, and Note 5.1) of magmatic-tectonic "events" in Earth's history and with the episodic character of plate tectonics (cf. Glen, 1975, 1982; Seyfert and Sirkin, 1979; Cox and Hart, 1986; Anderson et al., 1992), continental drift, "suspect terranes" (e.g., Howell, 1985, 1989; Moores, 1991; Elliott and Gray, 1992), "accretionary continents" (e.g., P. F. Hoffman, 1988, 1991; Zonenshain et al., 1991; de Wit et al., 1992; cf. Ernst, 1988), and "supercontinent cycles" (e.g., Nance et al., 1986, 1988; Hartnady, 1991; Dalziel, 1991, 1992a, b; Murphy and Nance, 1991, 1992).

Cratering Patterns and Paleomagnetism

These alternative bases for *celestial reference frame (CRF) invariances* constrain existing concepts of paleogeographic reconstructions, motions of geologic terranes ("suspect terranes," "exotic terranes," etc.; see Howell, 1989), and the evolution of *paleomagnetic pole positions*, depending on whether one or another form of geodynamic invariance has persisted over time. The standard assumption, or central dogma, of paleomagnetic studies is that, statistically speaking, a paleomagnetic pole position can be assumed to be identical with the Earth's spin axis. This is called the *geocentric axial dipole* hypothesis (Butler, 1992, Chaps. 1 and 10). To the extent that paleomagnetic poles have moved relative to the Phanerozoic cratering nodes (PCNs), the geocentric axial dipole hypothesis would appear to be contradicted by PCN-CRF invariance, according to the generalized long-term model. The specialized, or moving-reference-frame, explanation for PCN-CRF invariance would imply that the PCNs have moved with a unique subset of the continents, hence they would have traced the same relative motions as those of continental drift — at least for that special subset of three principal cratonic terranes. Because the specialized model therefore offers nothing new to test relative to traditional views of polar-wander paths, the description below follows the subsequent text in postulating the generalized PCN-CRF model of long-term invariance of PCNs relative to Earth's spin axis.

In order to be explicit about the distinction between this model and the more traditional views of polar wander (cf. Goldreich and Toomre, 1969; Irving, 1982; Irving and Irving, 1982; Gordon et al., 1984; Andrews, 1985; Gordon and Jurdy, 1986; Gordon, 1987; Piper, 1987; Sager and Pringle, 1988), I shall designate the position of the true North Pole of planetary rotation, relative to the mean distribution of mass in the Earth, as the *mean North Pole of whole-Earth rotation* — the reason being that a more explicit phrase is needed to take the place of "spin axis," because most workers automatically equate "spin axis" and "magnetic dipole axis" over times longer than about 10^5 years, the period of time most workers assume is sufficient to average out the "secular" variations in the geomagnetic field (cf. Butler, 1992, Figs. 1.7–1.10 and pp. 161ff; but see Courtillot and Le Mouël, 1984, 1985, 1988; Mankinen et al., 1985; Prévot et al., 1985; Champion et al., 1988; Coe and Prévot, 1989; Jin and Jin, 1989; Courtillot et al., 1992; Thouveny and Creer, 1992). This "whole-Earth" rotational pole is a meaningful reference only for those structural features of the Earth that either have been locked rigidly to the spin axis throughout its history or have been locked in some form of 1:1 toroidal convective resonance with the spin axis. Whichever may be the case, the PCN-CRF invariance represents such a structure to the extent that it is in resonance with a generally invariant pattern of mean anomalous mass distributions throughout the Earth (i.e., the anomalous mass could be either largely rigid or largely fluid and nonetheless satisfy this condition if the internal motions reflect the feedback resonance conditions outlined above).

In the usual approach taken for paleomagnetic interpretations, the paleomagnetic pole is assumed to represent the position of the mean North Pole of whole-Earth rotation at a given time ("true north") — despite the fact that during historic time, relative to a given location on the Earth, the north magnetic pole has deviated to a significant extent, and in a variable and sometimes systematic manner from the whole-Earth rotational pole (cf. Jin and Jin, 1989). [In England, for example, the historic record extends back to ca. 1600, and during this time the range of vertical deviation (inclination) of a compass needle spanned a difference of $\sim 10°$, and the horizontal deviation (declination) spanned a difference as large as $\sim 35°$ (e.g., Butler, 1992, Fig. 1.7). The usual assumption that such variations have averaged to zero over time spans of the order of $\sim 10^5$ years (see above) is suspect in the light of theories of nonlinear periodicities (see Notes P.3 and 3.1; cf. Shaw, 1987b, Fig. 1; 1991, Fig. 8.32).] According to this common practice, and despite an inadequate understanding of nonlinear effects over time spans from $10^2–10^5$ years — the durations of the "historic" and "secular" variations — the mappable traces of the paleomagnetic pole positions across different continents are called *apparent polar wander paths* (APWPs), where "polar" is assumed to refer, in common, to the mean dipole axis of the geomagnetic field and to the mean spin axis.

The justification for this seemingly ambiguous state of affairs is that over periods of up to tens of thousands of years — the so-called "secular variation" of the geomagnetic field (e.g., based on radiocarbon dating of archaeological sites and suitable lake deposits; cf. Butler, 1992, Chap. 1) — variations such as those of the historic record "average out" approximately to a geomagnetic dipole axis that is "indistinguishable" from the whole-Earth rotational axis. To me, this conclusion would appear to be based on a rather questionable combination of other assumptions, such as: (1) an assumption that a small number of cycles of deviation from "true north" in the most recent and best studied geomagnetic records is a representative sample of the geomagnetic record in general (e.g., Butler, 1992, Fig. 1.9), and (2) an assumption that the general motion of the geomagnetic dipole axis over longer times ($>$ 100,000 years) defines the relative motion of the spin axis because the variations are *assumed* to represent a "random walk" about a mean geocentric-axial-dipole position (e.g., Butler, 1992, p. 10; but see the discussion of paleomagnetic "subchrons" with durations $\leq$ 100,000 years in Champion et al., 1988; cf. Mankinen et al., 1985; Prévot et al., 1985; Courtillot and Le Mouël, 1988; Coe and Prévot, 1989; Courtillot et al., 1992; Thouveny and Creer, 1992). Thus the concept of a "random null model" — which implicitly ignores the possibility of systematic effects such as those revealed by the work of Champion et al. and Coe and Prévot — is once again seen to form the standard-of-reference for the physical interpretation of a terrestrial phenomenon, as discussed earlier in other contexts.

Introduction

Although the above convention follows "good scientific practice" in the statistics of data management, my own bias—based on principles of nonlinear dynamics—would be to treat such variations as systematic and ordered in some cyclic way (assuming that measurement "noise" can be accounted for by independent testing and/or by the methods of statistical nonlinear-dynamical analysis themselves), whether or not standard statistical tests are capable of distinguishing that order or not. [See Sugihara and May (1990) for a discussion of contrasts between samples of chaotic data and samples of white noise, where both are "uncorrelated" (indistinguishable) on the basis of statistical tests normally used to evaluate the existence of order and/or "determinism" in scientific data; these authors rarely used the word "random."] In point of fact, the present-day geomagnetic dipole axis is an *inclined axial dipole*, or more generally a *precessing inclined axial dipole*, that may or may not, over a sufficiently long time, average to a *geocentric axial dipole*. Even if such an average state existed, the dynamical behavior represented by the motion of an inclined dipolar axis could be of considerable dynamical interest (e.g., compare this point of view with the inclined axis of the electromagnetic source of radio waves in the *lighthouse model* of the stellar object called a *pulsar*, as described in Note I.3).

On the evidence that continents have moved relative to each other over time, in consequence of the combined effects of tectonic plate motions and continental drift (the two can differ, depending on interpretations of plate boundaries, exotic terranes, and so on), APWPs have traditionally been used to test paleogeographic reconstructions according to the net combinations of rotations required to define an APWP common to all of the continents (cf. Appendixes 20 and 21). The standard method of such reconstructions is based on the definition of appropriate *Euler poles* of rotation, where the Euler pole at a given position on the Earth's surface is defined by a line, or virtual axis of rotation, drawn through the center of the Earth and about which given continents or other selected spherical areas of the Earth's surface are rotated along small-circle trajectories normal to that pole (see Cox and Hart, 1986; Butler, 1992). Hence these reconstructions are referred to as *paleomagnetic Euler pole* models. Although the method is a general one in spherical geometry, those who apply such data to paleogeographic reconstructions sometimes do so without pointing out ambiguities and alternative interpretations. However, the near-identity of reconstructions of intricate APWPs from two or more different continents—where such paths involve "cusps" and "hairpins" with very large angular changes of direction (but see Hagstrum, 1993)—is sometimes convincing evidence that, at least with respect to those surficial markers represented by geomagnetic pole positions (even if not relative to the whole-Earth axis of rotation), paleomagnetic criteria of coordinated continental motion are difficult to refute (cf. Appendixes 19–21; and Butler, 1992, Chap. 10).

Acknowledging the fact that there have been major relative motions of continents and tectonic plates throughout Earth's history, I emphasize in the present work that it may be valid to assume that the outer shells of the Earth (specifically the lithospheric and crustal layers) have experienced net displacements relative to at least three internal coordinate frames, one locked to a set of relatively invariant markers in the sublithospheric mantle, another locked to the spin axis, and a third locked to the mean motion of an inclined axial dipole — a coordinate frame that is not locked either to the spin axis or to the sublithospheric mantle. The term "locked" is used loosely to indicate that most of the relative motion of large areas of the Earth's surface can be systematically coordinated and conserved simply by rotational translations relative to these coordinate frames, assuming that appropriate allowances can be made for the production of new lithosphere and crust along the global system of oceanic ridges and for the destruction of older lithosphere and crust along the global system of subduction zones (e.g., Morgan, 1981, 1983; Engebretson et al., 1992). Three-dimensional (radial and azimuthal) displacements relative to the first two coordinate frames must occur within dominantly solid-state petrologic regimes between the order of 100 km depth and the core-mantle boundary (CMB) at ~2900 km depth (cf. Jeanloz, 1990, Fig. 1). If it were possible to define a net coordinate frame (see Goldreich and Toomre, 1969) — for example, by suitable rotations of geomagnetic pole positions about a set of Euler poles combined with a suitable rotation relative to a deep-mantle reference frame — then the result could provide a suitable test of the net relative motion with respect to the whole-Earth pole of rotation. But such a test is only possible if there is some way to identify and describe the actual orientation of the whole-Earth pole of rotation over the same spans of geologic time, relative to either the geomagnetic reference frame or the deep-mantle reference frame, or relative to yet another, more general, reference frame that has known relationships to all other systems of coordinated motions. Such a test would constitute a measure of so-called *true polar wander* (TPW) — meaning the net rotation of the system of angular motions of the respective paleomagnetic and deep-mantle coordinated frames relative to the Earth's axis of rotation, hence relative to the Earth's principal axial moments of inertia. True polar wander is sometimes called the "roll" of the mantle relative to the spin axis (Hargraves and Duncan, 1973; cf. Gold, 1955; Andrews, 1985), a phenomenon conceptually similar to the one that I later invoke on a much grander scale for the interpretation of the origin of mass anomalies early in Earth's history (see Chap. 8, Fig. 33; cf. Bostrom, 1971, 1992; Andrews, 1985; Schultz, 1985; Runcorn, 1987; Sager and Bleil, 1987; Besse and Courtillot, 1991; Duncan and Richards, 1991; Cadek and Ricard, 1992; Ricard et al., 1991, 1992; Spada et al., 1992).

True polar wander paths (TPWPs) would be established on the basis of the paleomagnetic record (assuming that a suitable deep-mantle reference frame can be established; see the following sections of this chapter) if the geomagnetic axial dipole were, in fact, aligned with the spin axis. This is the so-called *geocentric axial dipole hypothesis* — called the *GAD hypothesis* by Butler (1992, Chap. 10). The GAD hypothesis automatically satisfies the conditions described above for the delineation of TPWPs. Because many researchers currently working in the field of paleomagnetism make that assumption, there are now a number of determinations of so-called TPWPs (see Appendixes 5 and 20; and Andrews,

1985; Gordon, 1987; Courtillot and Besse, 1987; Besse and Courtillot, 1991). The validity of the GAD hypothesis rests largely on the verifiability of the position of the mean axial geomagnetic dipole, relative to the whole-Earth pole of rotation, for durations of geologic time far longer than that of the "secular variation" discussed above (i.e., $\gg$ 100,000 years). The validity of the TPWPs rests partly on such a verification, partly on verification of the existence and coherence over geologic time of a deep-mantle reference frame, and partly on verification of the statistically derived differences between net paleomagnetic "wander" over geologic time (reconstructed APWPs) and the net "wander" of the "markers" that define the deep-mantle reference frame. In detail, these are difficult and statistically demanding problems in spherical geometry and kinematics (see the discussion of analogous problems by Pulliam and Stark, 1993), and the examples worked out in the references just cited remain hypothetical in nature.

In the light of this brief summary of difficulties in the measurement of "true" polar wander, it should now be evident that the existence of a cratering reference frame — the PCNs of the *celestial reference frame hypothesis* (CRFH) — adds yet another set of criteria for the establishment of TPWPs. For, if future work substantiates the relationships described in the present study, the PCN-CRF coordinate system would establish at least three "geographic poles" (the PCNs) as reference markers that are locked to the long-term motion of the mean axis of whole-Earth rotation — whereas until now there has been no direct way to establish that history, except by the implication of the putative null model of statistically random motions in the core that would maintain an alignment between the mean axis of the dipole field and that of the Earth's net rotation. [Nonlinear dynamics would suggest, on the other hand, that *inclined resonant axial dipoles* and/or *inclined chaotic axial dipoles* are the more general cases in a heterogeneous rotating body, in a manner analogous to rotationally induced fluid structures in the Sun and the large outer planets, in the lighthouse model of pulsars (Note I.3), and in the systems of external orbital resonances described in the present study; see Chaps. 8 and 9.]

To the extent that APWPs require significant relative motions between the three cratonic blocks containing the PCNs, the geocentric axial dipole hypothesis and the present hypothesis (CRFH) would appear to be in conflict, for the mean PCN-CRF is, by definition, locked to the spin axis, and it and the PCNs move together. The two might be reconciled if the reconstructed APWPs from different continents can be accommodated by rotations that do not greatly change the relative positions of the PCNs, positions that at present are located within rather large angular tolerances (see Chap. 1; and Appendixes 20–23). Nonetheless, the extensive body of data on paleomagnetic pole positions and their reconstructions as APWPs and putative TPWPs has demonstrated a remarkable fact concerning terrestrial evolution as codified by the "Geologic Column" (e.g., by the geological time scale of Harland et al., 1990). That is, polar wander paths are characterized by dramatic changes in direction (major azimuth swings) that often fall within the time intervals of major geologic-biostratigraphic boundary events (see Appendixes 19–21; cf. Butler, 1992, Figs. 10.6 and 10.9; Hagstrum, 1993). In the geomagnetic literature, these major azimuth swings are typically called *cusps* (e.g., Gordon et al., 1984; Butler, 1992, p. 251) or *hairpins* (e.g., Irving and Irving, 1982; Courtillot and Besse, 1987; Besse and Courtillot, 1991). A general correlation between APWP cusps and major divisions of the Geological Column was first pointed out to me by Jonathan Glen (spoken comm., 7 February, 1992). This observation does not resolve the APWP-TPWP relativity, but it is consistent with the implications of the CRFH with regard to a time-delay model of impact-dynamic/geodynamic coupling (see Chap. 7, Figs. 30 and 31). It is therefore qualitatively consistent with major perturbations of the geodynamo at such times (times of geomagnetic cusps, broadly delineated by the geologic boundaries), suggesting the possibility — one already implied on other grounds (see Chap. 7) — that orientations of the dynamo field correlate with impact-mediated geodynamics of the "solid" Earth.

The Celestial Reference Frame and Geocentric Axial Dipole Hypotheses: A Paradox of Poles

In principle, the motions implied by putative "polar wander paths" (APWPs and/or TPWPs) represent rotational instabilities in the Earth of types that are analogous to those I appeal to in the celestial reference frame hypothesis (CRFH) for the original organization of mass distribution in the Earth associated with the origin of the Moon and the formation of the core early in Earth's history — an organization that is required, at least in some form, to satisfy the Phanerozoic-cratering-node/celestial-reference-frame (PCN-CRF) invariances (see Chap. 8, Fig. 33). The essential difference between the CRFH and *geocentric axial dipole* (GAD) hypotheses is that the CRFH refers to a systematically coordinated ("self-organized") system of coupled processes — each of which is "oscillatory" in the general sense — whereas GAD hypotheses derive from the assumption that sets of spatially related but largely independent geodynamic systems "wander" about in more or less "random walks" relative to a statistically averaged alignment between a magnetic dipole axis and the Earth's spin axis.

The CRFH postulates systems of global geodynamic resonances that express feedback coupling between impact dynamics and the geodynamics of the deep mantle, lithosphere, and crustal layers of the Earth — including resonances with the core dynamo. These resonances are self-tuned, as it were, by mechanisms of critical self-organization such that the constant angular relationships between the spin axis and PCN-CRF invariances are preserved — and also in a manner such that the observed sets of relationships that describe the systematic migrations of the paleomagnetic field relative to plate tectonics, continental drift, and convective motions in the mantle (including the magmatic source mechanisms that give rise to the global fabric of volcanism) are satisfied. In short, the global fabric of all these mechanisms and processes metaphorically represents a sort of self-fulfilling (cybernetic) "prophecy." Polar "wander," instead of being a sort of random walk,

Introduction

then, represents a resonantly coordinated sequence of geographic trajectories of geochronologically "stable" paleomagnetic field orientations that expresses the net involvement of the CRF system, the generalized fabric of tectonic-magmatic phenomenologies, and the time-averaged fabric of geomagnetic polarity oscillations and net polarity migrations that are coordinated with the resonantly coupled system as a whole. Expressed in the form of geographic trajectories of a magnetic dipole axis, the polarity migration is generally inclined to the spin axis, and this effective, or apparent, inclination varies systematically over several different time scales — e.g., those that are characteristic of secular variations, geomagnetic excursions, patterns of virtual geomagnetic pole (VGP) paths (see Note I.7), and patterns of quasistable polarity reversals — in response to the global feedback of the resonantly coupled CRF-geodynamic system (in such a system no single subsystem is causative, because the behavior exists only by virtue of the nonlinear synchronization of the whole; cf. Freeman, 1992).

In other words, the CRFH would appeal to only one general form of nonlinear-dynamical process, one consisting of a *system of coupled oscillators* that is subject to the types of phenomena that are described under the rubric of *deterministic chaos*, to explain both the originating mechanisms and the relative kinematics of all terrestrial reference frames — i.e., the magnetic, volcanic ("hotspot"), plate-tectonic, deep-mantle, and celestial reference frames. "True polar wander" (TPW), on the other hand — at least in the form in which it is currently promulgated in the geophysical literature — would appeal to several different types of essentially independent, or stochastically coupled, mechanisms for its existence. Three of these mechanisms are given further attention here so that the paleomagnetic theme — which, in my view, constitutes one of the two major categories of generalized tests of the CRF hypothesis, the other being the theory and observations of near-Earth objects and natural Earth satellites (these two categories, respectively, reflecting the "orbital" mechanisms of Earth's internal and external, periodic and chaotic, nonlinear-resonance phenomena) — may be kept in mind while the evidence pertaining to Earth's cratering history is described in subsequent chapters.

Mechanism (1) is identified with a statistically averaged axis of geomagnetic "spin" (the GAD hypothesis) that wanders "randomly" within tens of degrees of deviation from the whole-Earth spin axis over time scales ranging from hundreds of years to tens of thousands of years, but, beyond that time frame, also represents an axis that is generally locked to Earth's geographic pole of rotation. [If this were, in fact, a demonstrable invariance, it would be seen in the context of the CRFH as a special case of a nonlinear-dynamical chaotic resonance that expresses the general stability of a dominant strange attractor, i.e., a single chaotic mode of a family of possible attractor and/or strange-attractor states of the geodynamo — a state of the geodynamo that would be somewhat analogous to the Moon's present state of spin-orbit resonance among many possible spin-orbit resonances relative to Earth (cf. Alfvén and Arrhenius, 1969).] Mechanism (2) is identified with the existence in the mantle of a global process that leaves a set of traces on the Earth's surface (e.g., volcanic "hot spots") that migrate relative to that surface but, at any given time, maintain invariant angular relationships relative to each other (in the same sense that the PCNs maintain a constant angular separation over time) and that is not necessarily causally connected with fundamental mechanisms of the core dynamo — a global process that is largely independent of the paleomagnetic process (some workers argue that there is a thermal connection between the two systems; cf. Chap. 8). Mechanism (3) is the global geodynamic process of whole-Earth rotation that permits large and abrupt (geologically "instantaneous") changes in the position of the spin axis relative to the continents, lithospheric plates, and deep mantle — the latter motions being the consequences of plate tectonics and mantle convection — while permitting instantaneous reversals of the geomagnetic field (representing the net polarity of magnetohydrodynamic core motions) without affecting the stability (meaning the geologically instantaneous identity of the magnetic axis and the net-Earth spin axis) of the geocentric axial dipole (GAD), thereby constituting a geologically frozen synchronous resonance of the fluid core with the relatively nonfluid portions of the Earth's geodetic system. Despite the many, de facto, hints — or outright assumptions — of resonant behavior of the core dynamo, few researchers have viewed the phenomenological behavior of the whole, "solid" Earth (including the inner core) plus geodynamo as the manifestation of a globally interactive (self-organized) nonlinear-dynamical system.

Mechanism (3) above would appear to originate from an idea put forward by Gold (1955), which, in order to be generally operative in the Earth as it is presently constituted, would have two implications. First, Earth's radial and longitudinal mass distribution is so nearly uniform, and/or is so nearly symmetrically balanced (as in the distribution of weights required to balance the wheels on a car), that an almost imperceptible shift in mass, or addition of mass at some point on its surface (Gold's "beetle"; cf. Goldreich and Toomre, 1969; Andrews, 1985, Fig. 7), would be sufficient — assuming that there is negligible resistance to the compensating deformation implied by the redistribution of mass relative to the spin axis — to reorient Earth's principal axial moments of inertia, thereby reorienting the spin axis relative to any preexisting system of relatively invariant radial and longitudinal distributions of markers. Second, the rate of yielding of Earth materials to small stress differences is sufficiently rapid, and/or the time-dependent thresholds of yielding to small stress differences are so low (cf. Shaw, 1969, Fig. 10), that any displacements of mass will be compensated in a steady way by balancing shifts in the relative position of the spin axis. For example, if a "beetle" were placed on the Equator and it walked very slowly northward along a meridian, it would always remain on the Equator because of the compensating shift of the axis of rotation.

The difficulty with this notion, according to the dynamical principles emphasized in the present volume, and the nonlinear rheological principles of my earlier studies (e.g., Shaw, 1969, 1991), is that the Earth is so heterogeneous — e.g., with regard to the long-term bias in the deep-mantle distribution of mass, the biases in the crust and shallow-mantle distributions of mass, and the long-term time constants of rheological responses — that any mechanisms that could cause significant

Introduction

changes in the magnitudes of inertial effects related to plate tectonics would be highly nonuniform as well as highly nonlinear; see, for example, discussions of the rheology of mantle instabilities in Notes P.5, I.5, 8.3, and 8.8.

The present model (Chap. 8) proposes that the Earth has inherited a large-scale mass heterogeneity from its early history, and that large shifts in its axis of rotation relative to its deep-mass configuration have been minimized since early in the Archean Eon. Furthermore, this keel-like stabilization has acted to guide the recirculation of mass owing to plate motions within a wide and crudely delimited but relatively invariant spherical segment through the Earth, one that is more or less bounded by the meridional band of dominant present-day geophysical anomalies (see Figs. 28 and 33; cf. Engebretson et al., 1992, Fig. 1). Although the rates and directional changes of putative TPWPs are not inconsistent with rates of relative plate motions (i.e., the velocity vector of the effective "center of mass" of the net transport conceivably could change instantaneously in direction, and the range of speeds of polar wander would then be comparable to speeds of plate motion; see Besse and Courtillot, 1991; and Appendix 5 herein), the amplitudes of likely excursions of the Earth's spin axis from a relatively invariant position — according to the PCN-CRF system — would be constrained, perhaps in an oscillatory manner, within a range of roughly $\pm 20°$ from its present orientation (matching the positional uncertainty of PCN locations; see Fig. 2). Gold's (1955) conjecture would hold only if the convective processes in the Earth were spherically symmetric and if circulation times (and stress relaxation times) as a function of radial position were short enough to effectively homogenize large chemical heterogeneities that originated early in Earth's history (relative to an equilibrium gradient of radial stratification). By contrast, the present model envisions nonlinear cycles of dissipatively focused deformation and flow, such that the system as a whole operates in a quasisteady manner far from equilibrium **(Note I.5)**.

In most discussions of TPWPs based on the GAD hypothesis, the deep-mantle reference frame is identified with the invariance of a global pattern of volcanic systems called "hot spots." This form of mantle reference frame is therefore usually referred to as the *"hotspot reference frame"* (HSRF). The existence of "hot spots" is now habitually "explained" (meaning by rote, or by mandate) as the manifestation of a system of advective mantle plumes ("thermal plumes") rising from the vicinity of the core-mantle boundary. Such "thermal plumes" are viewed by many as the principal mechanism for focused mantle melting at all "hot spot" locations (meaning wherever it is kinematically expedient to place them, even at locations along the systems of more "dendritically" localized mid-ocean ridges). A number of problems associated with this model are addressed in Chapter 8. For example, recent mapping of shear-wave attenuation in the upper 400 km of the Earth (Anderson, 1991a, b; Anderson et al., 1992) has shown that "hot spots" are not really spot-like from a thermal point of view (cf. Shaw and Jackson, 1973; D. L. Anderson, 1990; Fukao, 1992; Shearer and Masters, 1992), even in the context of broad "plume heads" invoked by many authors to explain widely distributed volcanic phenomena.

Conceivably, an invariant HSRF could be interpreted as a system of advective mantle plumes that represent a form of convective resonance relative to core-mantle coupling of global heat flow, where the net motion — kinematically, dynamically, and thermodynamically — would be correlated with the analogous properties of the core dynamo. By contrast, the CRFH correlates all of these motions with orbital-dynamic action and cumulative impact energy as the modulator of the coupled geodynamic effects, including the analogous effects of core-mantle coupling. The difference is that in the CRFH both the GAD and the HSRF are permissible constructs, *but neither one is essential to the model*. Other forms of resonances in the geodynamo and in mantle-magmatic transport are implied by the CRFH, but they are likely to be of greater complexity than the simplistic GAD model with respect to possible periodic and/or chaotic attractor structures in *phase space* (**Note I.6**). CRFH mechanisms are intermittent in detail, but they can be expected to produce long-term invariances (attractors with fixed-point or limit-cycle behaviors, or localized systems of more complex but invariant loci, such as strange attractors of specific forms) of global geomagnetic and magmatic signatures (cf. variability of Hawaiian magmatic propagation episodes within the spot-like invariances assumed to exist in conventional "plume" models, as discussed in Shaw, 1973, 1988a, 1991; Shaw and Jackson, 1973; Jackson et al., 1975; Jackson, 1976; Shaw et al., 1980a). The most obvious difference, of course, between the generalized CRFH-HSRF model and a generalized GAD-HSRF model (even one that invokes the existence of nonlinear resonances) is that the former involves mechanisms that are coupled with and modulated by phenomena external to the Earth (mass impacts and their associated mechanisms of origin and integral involvement with the Earth as well as with the Solar System), whereas the latter appeals to the more restrictive properties, mechanisms, and energies originating largely within the dynamical limitations of the core and core-mantle interactions.

Rather than measuring trajectories of motion of tectonic plates relative to the assumptions (a) that the orientation of the paleomagnetic pole is always identical to that of the spin axis (GAD) and (b) that there is a long-term and invariant geodynamic reference frame fixed in the deep mantle (HSRF), the CRF implies that there is an alternative interpretation in which the pole of the spin axis and the poles of the CRF maintain constant relative angular orientations in time. In this context, the mean effective paleomagnetic dipole and the HSRF both move in coupled and compensatory ways relative to the PCN-CRF system as well as relative to each other. According to the CRFH, if the HSRF were truly fixed relative to the deep mantle, then it also would probably be fixed relative to the ellipsoid of the three principal rotational moments of inertia. If so, the HSRF and PCN-CRF systems would be crudely locked together in a form of relative angular invariance. Accordingly, the spin axis, CRF, and HSRF could only move together about a pole common to all three of them, and that would be the pole that represents the normal to the plane of Earth's greatest moment of inertia (i.e., such a pole would coincide with the normal to the plane of the equatorial bulge, which constrains the mean orientation of the spin axis). Paleomagnetic pole positions in models based on the CRFH would be constrained mainly by the relationship between impact dynamics and the dynamics of the core, as coupled through the intermediaries of magmatism and mantle convection (cf. Chap. 8, Figs. 27, 28, 30, and 31). The possibilities for nonlinear fluid resonances of core motions produced by such a coupling mechanism are more varied, both temporally and spatially, than they are in conventional models, but this does not preclude the existence of either stationary or moving inclined axial dipole fields characterized by long-term stability (e.g., analogous to the stability of electromagnetic sources in pulsars, or to stable vortical invariances in the Sun, or to the stability of Jupiter's Great Red Spot; cf. Note I.3; Chap. 8).

The alternatives provided by CRFH models would appear to conform better to the meaning of *true polar wander* than does the widely accepted current interpretation in which "true" polar wander is calculated as the difference between globally averaged apparent polar wander (APW) and coordinates determined from the HSRF (see Andrews, 1985; Courtillot and Besse, 1987; Besse and Courtillot, 1991). A new class of motion would be required to describe the paleomagnetically averaged paths of inclined axial dipole fields, one that would represent the resolved dipolar components of long-term *virtual geomagnetic pole* (VGP) paths (**Note I.7**). Accordingly, patterns shown by clustering of VGPs and trajectories of VGP paths should provide sensitive tests of dynamical relationships between the history of mass impacts and their effects on the core dynamo, recording both transient disturbances (such as *chaotic crises,* as discussed in Chaps. 6 and 8) and the evolution of intermittent clusters of quasisteady apparent dipole field orientations designated here by the phrase *characteristic paleomagnetic poles of epoch* (CPPEs), using "epoch" in the same sense as it is used in astronomy, for the time datum of an event. These CPPEs, in turn, might correspond to any of the following types of phenomena: short-term field excursions, nondipole effects, or geologically stable (i.e., stratigraphically demonstrable) paleomagnetic field orientations. In the last of the three cases, CPPEs and stratigraphically established *paleomagnetic pole positions* (PPPs) become identical. VGPs, CPPEs, and PPPs differ primarily with regard to characteristic, though relative and variable, time frames of measurements of geomagnetic field orientations, in the general order of relative transience: VGP = "instantaneous" (usually unstable, but possibly metastable), CPPE = quasisteady (metastable), and PPP = steady (conditionally stable; but see the distinctions between the time scales of paleomagnetic "excursions" and "subchrons" discussed in Champion et al., 1988; cf. Courtillot and Le Mouël, 1988; Coe and Prévot, 1989; Thouveny and Creer, 1992).

Paleomagnetic interpretations based on the celestial reference frame hypothesis (CRFH) resurrect questions concerning deviations of the orientation of the averaged dipole field of the geodynamo relative to the Earth's spin axis—and possibly relative to the orbital parameters of Earth's motion in the Solar System (cf. Champion et al., 1988; Berger et al., 1992). Since the advent of notions of fixed mantle reference frames for kinematic reconstructions of plate-tectonic motions (e.g., Morgan, 1971, 1972, 1981, 1983), the corollary assumption of invariance between the Earth's spin axis and the averaged axial dipole has been widely adopted in geomagnetic studies (e.g., Andrews, 1985; Courtillot and Besse, 1987; Gordon, 1987; Besse and Courtillot, 1991; Duncan, 1991; Duncan and Richards, 1991). The CRF may make it possible to test this assumption in new ways, some of which have been mentioned, and to evaluate questions of variability in the fixity of the hot spot reference frame (HSRF) or other mantle reference frames (see Molnar and Atwater, 1973; Molnar and Francheteau, 1975; Molnar et al., 1975, 1988; Molnar and Stock, 1987; DeMets et al., 1990; Cande and Kent, 1992; DeMets, 1993; and cf. Shaw and Jackson, 1973; Jackson and Shaw, 1975; Jackson et al., 1975; Jackson, 1976; Shaw et al., 1980a). This is one example among many in both geomagnetism and plate tectonics where the CRFH and concepts of CRF-geodynamic resonances may shed a new and revisionary light on relationships between the internal dynamics of the Earth and the dynamics of the Earth in the Solar System.

Cratering Patterns of Late Cretaceous and Younger Ages

Craters with ages from 50 to 100 Ma (Group 2, Fig. 2) form a geographical subset delineated by ~180° of arc of a second circular swath that is characteristic of at least the mid-Cretaceous to early Tertiary cratering history and is most clearly demonstrated at the Cretaceous-Tertiary (K/T) boundary (Appendix 2 compares this pattern with a similar one for craters in Group 1, Fig. 2, i.e., those with ages less than about 50 Ma). This pattern is referred to as the *K/T cratering swath*. This trend is represented by a distorted small circle located predominantly in the Northern Hemisphere of north-polar equal-area projections, with its plane almost perpendicular to the plane of the Phanerozoic cratering swath and its pole at roughly 40° N, 40° W (Fig. 3). The two principal clusters of craters in this group are nearly coincident with the Phanerozoic cratering nodes (PCNs) of the Northern Hemisphere. Southward extrapolations of the limbs of this circular arc are commensurate with the pattern of K/T iridium anomalies (Alvarez et al., 1982) plotted on the same projection (Fig. 24). The net pattern of the K/T cratering swath resembles a canted crown or auroral halo circling the Northern Hemisphere in such a way that its northern limit intersects the Siberian Arctic region near the Laptev Sea and its southern limit crosses the South Atlantic Ocean near the Tropic of Capricorn (i.e., near 23.5° S) about halfway between Africa and South America. Together, the K/T and Phanerozoic cratering swaths form an approximately orthogonal double crown "draped" almost symmetrically over the geographic North Pole (cf. Figs. 1 and 5).

If the K/T cratering swath had been produced by a single event, then the "shot pattern" of K/T impactors and their putative signatures would have represented either an almost simultaneous global "cloudburst" of impacts or a closely timed sequence of geographically coordinated events (for more on iridium anomalies, shocked quartz, tektites, diamonds, etc., see Alvarez et al., 1982; Hsü et al., 1982; Smit, 1982; Tschudy et al., 1984; Tschudy and Tschudy, 1986; Lerbekmo et al.,

Introduction

1987; Xu et al., 1989, Chap. 4; Bohor, 1990; Doehne and Margolis, 1990; Huffman et al., 1990; Orth et al., 1990; Rocchia et al., 1990; Sharpton and Grieve, 1990; Sharpton et al., 1990, 1991, 1992; Sigurdsson, 1990; Sweet et al., 1990; Thomas, 1990; Ward, 1990a, b; Wolbach et al., 1990; Wolfe, 1990, 1991; Izett et al., 1991a, b; Robin et al., 1991; Roddy et al., 1991; Sigurdsson et al., 1991a, b; Wang et al., 1991; Zhou et al., 1991; Carlisle, 1992; Gilmour et al., 1992; Shoemaker and Izett, 1992; Swisher et al., 1992; Wang, 1992). The existing pattern suggests that the cratering record of K/T impacts is still incompletely documented, even though the arcuate pattern of known large craters in the Northern Hemisphere is itself highly redundant relative to concepts of single-impact dynamics (e.g., Hut et al., 1987, 1991; Shoemaker and Izett, 1992; cf. Bohor, 1992). Starting in the Americas, the northernmost arc of K/T craters can be traced from the Chicxulub structure in Yucatan across the Gulf of Mexico and across areas of central and western U.S. (e.g., Manson, Iowa) until it crosses the Arctic Ocean near Point Barrow, Alaska. The arc then passes within about 20° of the North Pole, crosses the Laptev Sea, and connects east of Popigai Crater, Russia (north-central Siberia), with another cluster of craters that crosses Eurasia into the Mediterranean region. The northern arc is informally referred to as the "Glen Line," acknowledging the prior recognition of the trans-American-Eurasian connection by William Glen (spoken communication, 6 June, 1991; the trans-Arctic connection itself had been pointed out earlier by Sharpton and Burke, 1987). Other impact locations that might exist along a completed K/T circle would be found along an arc that crosses northwestern Africa, the Gulf of Guinea, the South Atlantic (roughly between Ascension and St. Helena Islands), central South America, and portions of the southern Caribbean, meeting the arc of Central and North American craters again in Yucatan.

Recapitulation of the Celestial Reference Frame Hypothesis

My present treatment of the impact-cratering history of the Earth departs from prior descriptions primarily in its concept of spin-orbit resonances and in positing the persistent though intermittent existence of systems of natural Earth satellites. I have concluded that to the degree the cratering patterns have remained invariant, the guidance mechanisms that have controlled the final approach of objects impacting on Earth are necessarily constrained to fit orbital trajectories that intersected the Earth at highly selective values of latitude and longitude. This precision on a rotating Earth over times of the orders of 10^8 to 10^9 years virtually demands a geocentric system of satellitic impactors. It may be, however, that such a geocentric proximal-reservoir of impactors is filled only during transient intervals prior to multiple impact events (the dynamical process is intermittent on a spectrum of time and size scales given by the absolute and relative chronological records of impact craters on the Earth and Moon).

In general terms, the CRFH states that the cratering pattern on Earth constitutes the imprint of impactors that during some interval of time prior to impact moved in geocentric orbits consistent with that pattern and impacted the Earth at subsequent intervals of time with a precision analogous to that of a nonlinear laser printer. The hypothesis explains the CRF as a self-organized nonlinear system of extraterrestrial objects ("meteoroids" of mixed pedigrees) that have been captured in the vicinity of the Earth-Moon system (i.e., at a distance of ~ 1 Astronomical Unit from the Sun). The apparent small-circle distributions of cratering on Earth may be the products of three or more great-circle intersections of nonlinearly precessing satellitic orbits, or more direct "shot patterns" from multiple objects orbiting the Moon in similar resonances (cf. Figs. 19, 23, and 37). The CRF therefore documents the integral history of a long-lived but intermittently filled and emptied proximal reservoir of objects derived from diverse sources in the Solar System and beyond (cf. Clube, 1985, 1992; Clube and Napier, 1986, 1990). Individual histories of objects and the resulting mineralogical and geochemical signatures of their meteoritic fragments represent storage and mixing over different durations for different objects in the reservoir (cf. Safronov and Zvjagina, 1969; Bandermann, 1971; Dohnanyi, 1971; Gehrels, 1971a, 1979; Kiang, 1971; Lindblad and Southworth, 1971; Roosen, 1971; Bowell et al., 1979; Kresák, 1979; Carusi and Valsecchi, 1979, 1985; Shoemaker et al., 1979, 1989; Van Flandern et al., 1979; Wasson and Wetherill, 1979; Brown et al., 1983; Donnison and Sugden, 1984; Chapman, 1986, 1990; Turcotte, 1986; Taylor et al., 1987; Kerridge and Matthews, 1988; Stöffler et al., 1988; Bell, 1989; Bell et al., 1989; Binzel, 1989; Binzel et al., 1989b, 1991, 1992; Chapman et al., 1989; D. Davis et al., 1989; French et al., 1989; Fujiwara et al., 1989; Gaffey et al., 1989; Gradie et al., 1989; Hoffmann, 1989a, b; Ip, 1989; Lipschutz et al., 1989; McKay et al., 1989; Nobili, 1989; Ruzmaikina et al., 1989; Scott et al., 1989; Sykes et al., 1989; Tedesco et al., 1989; Tholen, 1989; Valsecchi et al., 1989; Weidenschilling et al., 1989; Weissman et al., 1989; Wetherill, 1989b; Safronov, 1990; Dones, 1991; Drummond, 1991; Kresák, 1991; Swindle et al., 1991; Yoshikawa, 1991; Ipatov, 1992; Melosh et al., 1992; Binzel and Xu, 1993; Hoffmann et al., 1993). This proposition gives rise to a heterogeneous system with size-dependent and property-dependent residence times. The complex and independently documented record of small meteoritic objects — most of which are not considered in this work, which deals mainly with the larger, individually documented meteor craters (but see Chap. 7; and Note 6.8) — therefore represents a mixed population of objects, some of which passed through the CRF staging process while others came directly from the orbital vicinities of their parent bodies (asteroids, comets, intragalactic material) and/or from meteoroid streams that resulted from the fragmentation, within the Asteroid Belt and/or by diverse mechanisms in the vicinity of the Sun, of the larger objects from those sources (cf. Alfvén, 1971a, b; Baxter and Thompson, 1971; Danielsson, 1971; Lindblad and Southworth, 1971; Drummond, 1981, 1991; Fujiwara et al., 1989; Olsson-Steel, 1989; McIntosh, 1991).

I have concluded that the CRF reservoir has evolved as a product of: (1) nonlinear-dynamical interactions within and between other object reservoirs in the

Introduction

Solar System (Asteroid Belt, Kuiper Belt, and Oort Clouds ± comet-like objects from "Oort Clouds" produced in other star-forming regions of the Milky Way Galaxy), (2) nonlinear-dynamical periodic and chaotic resonances of these reservoirs with a system of planetary resonances in the Solar System at large, and (3) the nonlinear-dynamical evolution of a system of spin-orbit resonances in the Earth-Moon vicinity, with or without an intervening stage of mean-motion orbital resonances with the planets and Sun, producing a geocentric system of recurring orbital trajectories mapped on the celestial sphere. The CRFH also implies that latitudes and longitudes identified in this reference frame are fixed relative to each other and to the rotational figure of the Earth that defines the celestial sphere. The cratering patterns therefore map the immediate heavens as seen from any position on the Earth's surface, depending only on the orientation and motion of the Earth's equatorial great circle and its poles, hence on the dynamical figure of the Earth and on the Precession of the Equinoxes. The figure of the Earth in turn is a function of deviations of moments of inertia (mass distributions) from perfect spheroidal symmetry (implicitly including secondary motions of lithospheric plates and continental fragments as well as the gravitational effects of the Moon and Sun) that represent the net longitudinal mass anomalies of a system of spin-orbit resonances ordered according to nonlinear periodic-chaotic dynamical mechanisms (cf. Figs. 19–21 and 23).

Conceivable variations on the theme of the CRFH would allow the CRF to be oriented according to four possible types of orbital systems, the first of which, for want of information to constrain other possibilities, is used to illustrate the CRF process throughout most of the book. The four orbital-system types are each identified by the nature of the central mass: (1) the Earth is the primary of a single geocentric spin-orbit reservoir of small satellitic objects, (2) the Earth and Moon together form the central mass of a modified "geocentric" system in which the dual or coordinated binary action leads to the formation of a single satellitic system coordinated with both, (3) the process in type 2 evolves in such a way that the Earth and Moon each act as primaries of separate spin-orbit systems, where the paired system interacts as a complex coupled nonlinear oscillator composed of two subsystems, each of which is also a coupled nonlinear oscillator (Earth + satellites; Moon + satellites), implying complex mutual influences on the timing of impacts on either or both the Earth and the Moon from either or both subsystems, and (4) an ersatz "geocentric" system formed by specialized long-term mean motion resonances of diverse small objects within 2 Astronomical Units of the Sun that have interacted with the Sun ± Jupiter, with Jupiter ± other outer planets ± Earth and/or Earth ± Moon ± other inner planets, in such a manner as to produce invariant sets of great-circle impact trajectories on both the Earth and Moon over at least the latest ~1 Ga of their histories.

The action of type 4, though conceivable, would have to allow for effects of the 10^4-yr to 10^5-yr precessions of all the orbital and rotational motions involved, as well as the $\sim 10^9$-yr orbital evolution of the Earth-Moon system itself. The complexity of such a process stands in marked contrast with the more focused actions of types 1, 2, and 3, which are intrinsic to the evolution of the Earth-Moon system as a whole and therefore would generate CRFs in a mutually coordinated manner. The longevity of the CRF depends on the extent to which it has remained in balance with the system of torques associated with changing distributions of mass in the Earth and/or Moon. In this respect, the CRF's longevity is analogous to that of the Moon itself, particularly with regard to the system of torques, for example, whereby the lunar spin-orbit resonance has remained in a state such that the same opposite hemispheres always face, respectively, toward and away from the Earth — giving rise to a nonvarying view of the nearside of the Moon as seen from Earth (i.e., the Moon rotates on its axis just fast enough that it performs one full rotation during one full revolution in its orbit around the Earth). This is a form of spin-orbit resonance — or perhaps I should say "orbit-spin resonance," because I am referring to the conditions that attended the evolution of the synchronous rotation of the Moon with respect to its own orbital period about the Earth, rather than to the conditions attending the orbital period of a satellite relative to Earth's period of rotation (where the latter is the usual meaning of spin-orbit resonance in the present work; see Fig. 20; cf. Alfvén and Arrhenius, 1969).

Analogously, the triaxial figure of the rotating Earth (see Fig. 33, Inset) conforms to a long-lived system of both near-surface and deep anomalies of mass distributions that stemmed originally from the dynamical effects of early impacting objects, and that therefore has controlled the orientation and dynamical properties of the CRF reservoir that subsequently evolved with it (this is the type 1 system, which for the most part I emphasize in this work). The spatiotemporal patterns of impacts from the CRF reservoir have in turn mediated the internal dynamics of the Earth by their influences on the inception and subsequent timing of magmatic processes that in turn have mediated the inceptions and timing of continental drift, plate tectonics, core motions, geodynamo polarities, and their feedbacks with mantle convection. This cyclic-feedback process has thus controlled the subsequent evolution of the mass distributions, moments of inertia, and torques responsible for the continued orientation of the PCN-CRF system. CRF invariance consequently testifies to a quasisteady invariance of Earth's mass anomalies, and to a quasisteady balance of geophysical processes and patterns related to mass transport.

The principal mass anomalies of both the Earth and Moon, as well as their present-day orientations, appear to have been established during the early epochs of activity ($\geq 4.2 \pm 0.2$ Ga) following formation of the Moon and the Earth's core (Fig. 33). These intervals of Earth and lunar history were accompanied by inertial reorientations of their dynamical figures whereby the equatorial great circles and pole positions were inverted during two or more stages of global tectonic revolutions at scales that dwarf all subsequent activity on both the Earth and Moon (see Fig. 33). The earliest of these tectonic revolutions were associated with the formation of the Procellarum Basin on the Moon and an analogous basin on Earth, which appears to have been the ancestral Pacific Basin prior to condensation of the earliest gaseous atmosphere, which would have followed the megaevent of lunar

Introduction

formation (recent hypotheses — as well as some older, forgotten, hypotheses — favor a collisional origin involving a nearly Mars-sized protoplanet with the proto-Earth; cf. Daly, 1946; Hartmann and Davis, 1975; Wetherill, 1985, 1990; Brush, 1986; Stevenson, 1987; Chapman and Morrison, 1989, p. 165; Benz and Cameron, 1990; Cameron and Benz, 1991; Kaufmann, 1991, Fig. 9–22; Baldwin and Wilhelms, 1992; Spera and Stark, 1993).

The Celestial Reference Frame Hypothesis and the Evolution of the Biosphere

Having compared the terrestrial cratering record with that of the Moon, as documented by Wilhelms (1987), I have concluded that the global feedback loop mediated by the CRF system, and possibly involving lunar orbits as well as Earth orbits, was established sometime between the end of the Moon's Pre-Nectarian Period (≥ 3.9 Ga) and the beginning of its Copernican Period (~ 1.1 Ga; Wilhelms, 1987, Fig. 14.3). The result was a universal spacetime frame for the development of the CRF system and for the nonlinear periodicities that evolved from its interactions with Earth's tectonosphere (including the dynamics of the core), "magmasphere" (cf. Note 8.3), glacier-ocean-atmosphere-magnetosphere, and biosphere. The record of this nonlinear network of globally coupled oscillations is written in the chronological history of the igneous rocks, in the paleogeography of continental and plate motions, in the surface traces of dynamo action in the core relative to the CRF (virtual geomagnetic pole paths, flux lobes at the surface of the core, spatiotemporal frequencies of polarity excursions and reversals, apparent polar wander paths, etc.; see Figs. 27–32 and Appendixes 19–22), and in the self-similar scaling of the small to large catastrophes that represent the hierarchical spectrum of discontinuous "happenings" recorded by the biostratigraphic column in both space and time (cf. Ager, 1984; Raup, 1991).

The scaling of catastrophes in the biostratigraphic column constitutes a nonlinear code within which the histories of geodynamic processes are recorded, and the cracking of this *biostratigraphic code* therefore represents the cracking of both the *satellite resonance code* and the *macropaleontologic code* of biological evolution (cf. McLaren, 1970, 1982, 1983, 1986; Shaw, 1970, 1987b, 1988a; Shaw et al., 1971; Pannella, 1972; Ager, 1980, 1984; Williams, 1981, 1989a, b; House, 1985; Raup, 1990; Berger et al., 1992; Kerr, 1992f; Winograd et al., 1992; D. G. Smith, 1994). These code crackings are possible because the hierarchical ordering, in both space and time, of the event horizons that give structure to these codes correlates directly with taxonomic originations and extinctions in the fossil record, which themselves correlate directly with the CRF system (cf. Fig. 36). The CRF, in turn, codifies the impact horizons by which the biostratigraphic column is directly and indirectly calibrated, both by the chronology of impacts and by the chronologies of coupled magmatic, tectonic, geomagnetic, and sedimentological effects. Thus the theme of the present work holds that the CRF system both symbolizes and acts to implement the organizational structure of a generalized terrestrial code within which other codes have evolved to display synergistically compatible forms (cf. Note 8.1). In the language of nonlinear dynamics (Note I.4), the Geologic Column comprises the multivalent inscription of the multidimensional phase-space data of terrestrial evolution, the universal phase portrait of the Earth (Note 5.1). The numerical framework of this generalized code is expressed in the present work in terms of categorical structures (systems of attractors) classified as to their representational regimes of *nonlinear periodicity* and *chaos*— including quasiperiodic, nonlinear-periodic (mode-locked or phase-locked), and chaotic *resonances*, *frustrations* (Notes 3.1 and 5.3), and *crises* (Notes P.3 and 5.1)—and their *universality parameters* (Note P.4). Some of these structures are illustrated and discussed in Chapters 3–8 and illustrated in Figures 11–18, 20–22, 29–32, 35, and 36, and Appendixes 6–11. They imply that geologic time series of diverse physical, chemical, and biological processes will reflect similar degrees of simplification, or *algorithmic compression*, in the sense discussed earlier, relative to their maximum *algorithmic complexity* (Note I.2).

In this general context, I have previously explored the relative complexities of different natural time series—including geological, biological, impact-cratering, and Galactic phenomena—(1) by constructing *phase portraits* (Note P.3) from the temporal data (Shaw, 1985b), and (2) by simulating the time series, to arbitrary precision, by means of power-law sequences of events derived from combinations of simple geometric progressions (Shaw, 1988a, Table 1 and Fig. 25). The resulting patterns were of similar simplicity in that they all involved about the same degree of algorithmic compression—qualitatively as well as in terms of Shannon's criterion of redundancy (see Shannon and Weaver, 1949; Pierce, 1980; Shaw, 1994)—an outcome that is consistent with the hierarchical simplicity of the Geologic Column, as expressed in terms of the numbers of subdivisions that are used to describe it. The distribution of the numbers of chronostratic subdivisions of the Phanerozoic Eon, in turn, resembles the distribution of the numbers of large, intermediate, and small Phanerozoic impact craters, as discussed in Chapters 2 and 8 and in the Epilogue (cf. Figs. 6–8, and Appendixes 1–5, 16–19, and 22–25).

According to the "GTS 89 definitive time scale" of Harland et al. (1990, Fig. 1.7; pp. 19ff in this reference define the chronostratic names and their hierarchical classification), the Phanerozoic Eon is subdivided into $\sim$3 eras (giving an average of $\sim$190 m.y. between their delimiting boundaries; e.g., corresponding, nominally, to the few largest Phanerozoic impact events), $\sim$11 periods, counting the Tertiary as a period rather than a sub-era (giving $\sim$52 m.y. between boundaries; e.g., corresponding, nominally, to a like proportion of large impact events), $\sim$43 epochs (giving $\sim$13 m.y. between boundaries, corresponding to a like proportion of intermediate-size impact events), and $\sim$130 stages ($\sim$4.4 m.y. between boundaries, corresponding to the preponderance of small impact events). [It is worth noting that these "average periods" are all reasonably close to rational-number multiples of the 26-m.y. extinction period of Raup and Sepkoski (1984)—i.e., $\sim$22, $\sim$7, 2, 1/2, and $\sim$1/6, respectively, as the ratios for eons, eras, periods, epochs, and stages—as is consistent with the notion that they all reflect a common system of nonlinear resonances. This would suggest that the 26-m.y. extinction period is analogous to a chronostratic sub-period.] The range of qualitative impact magnitudes would correspond to craters ranging between the order of 200 km in diameter (e.g., Chicxulub, Yucatan; cf. Notes 1.1, 1.2, and 2.1) and 10 km in diameter. Grieve (1991, Table 1, cf. Fig. 4) lists 132 documented craters, excluding Chicxulub, Yucatan, and a few other craters included in the present work (see Appendix 1). [Relationships between the numbers, diameters, and ages of craters listed by Grieve (1982, Table 2) are illustrated in Shaw (1988a, Figs. 11–13). Figure 13B in Shaw (ibid.) shows that the numbers of craters are almost directly (though inversely) proportional to their diameters for diameters greater than 10 km. If Grieve's 1982 catalog is between 10% and 20% complete (see the discussions of completeness in Chap. 2 and the Epilogue), the numbers of craters of sizes between 10 km and 200 km closely approximate the chronostratic proportions given above.]

The average times between boundary events in the Geologic Column, of course, do not reflect the highly nonuniform proportions at different levels in the hierarchy, a feature that the geologic record shares in common with the mean periodicities (*winding numbers*; see Notes P.4, 3.1, and 4.1) of phase-locked nonlinear-dynamical recursion phenomena (*nonlinear clock mechanisms*) relative to the irregular distributions of states in phase space (cf. Note 8.1; and Shaw, 1987b, Fig. 1; 1991, Fig. 8.32). It is worth noting also that both types of record are based on imperfect data that have been improved — each in terms of the relative numerical "completeness" of the respective records — at about the same rates during the last few decades (compare the discussions of discovery rates of asteroids and comets vis-à-vis the rates of astronomical discoveries in Chaps. 4 and 6; cf. Notes 4.2 and 9.1). Harland et al. (1967) displayed 72 biostratigraphic stages (cf. Raup et al., 1973; Gould and Calloway, 1980) in the Phanerozoic Eon, which is just over half the number displayed by Harland et al. (1990) some twenty years later [i.e., over a time span that roughly approximates the doubling time of asteroid discoveries (about 16 years; see Bowell et al., 1989) and the doubling time of astronomical discoveries in general (see de Vaucouleurs, 1970)]. Although the total *effort* (see Note I.2) spent in the documentation of Earth's cratering record is probably much less than that devoted to improvements in the precision of the biostratigraphic record (cf. McLaren, 1970, 1982, 1983, 1986; Ager, 1980, 1984; McLaren and Goodfellow, 1990), the *relative* growth in the number of known craters between about 1970 and 1990 has crudely paralleled the rate of refinement of the Geologic Column — as measured by the ratio of the data categories, (1) crater size vs. number of craters, and (2) time duration between bracketing events at a given biostragraphic level vs. the number of subdivisions of the Geologic Column at that level [e.g., the world maps of crater distributions shown by Grieve (1991, Fig. 1; cf. Appendix 27 of the present volume) that compare the status of knowledge in 1972 with that in 1991 indicate that crater discoveries have approximately doubled in the last twenty years].

The celestial reference frame hypothesis (CRFH) suggests that the geological column constitutes a nonlinear time series of geodynamic happenings representing a hierarchical system of local and global "catastrophes" that mark both the chaotic intermittencies of terrestrial processes and their resonances with the record of impact-cratering processes, hence with the nonlinear dynamics of orbital evolution in the Solar System. This is, generically speaking, a fractal or multifractal time series, and the paleomagnetic record is a key intermediary, in several respects, to the documentation of this chaotic geologic time series of self-similar catastrophes (see Chaps. 7 and 8). A feature not previously emphasized in work by others is the fact that sharp changes in the azimuths of APWPs and TPWPs (cusps, hairpins, etc.) correlate at several different (self-similar?) scales with major geologic time intervals (cf. Appendix 19). Such correlations offer the possibility of establishing a more complete *geomagnetic code* by which to establish a generalized translation among the other codes identifying chaotic processes of terrestrial evolution (e.g., a *volcano-tectonic code*, a *seismotectonic code,* and so on; cf. Jackson and Shaw, 1975; Jackson et al., 1975; Smith and Shaw, 1975; R. L. Smith, 1979; Shaw, 1980; Shaw et al., 1980a; Smith and Luedke, 1984; Shaw, 1987a, b, c, 1988a, b; Shaw and Chouet, 1988, 1989, 1991).

The principles of nonlinear dynamics applied within the context of the CRFH are not alien to traditional geological and paleontological concepts. On the contrary, they represent a nonlinear revision of Hutton's principle of uniformitarianism **(Note I.8)** and Darwin's principle of organic evolution (cf. Notes I.8 and 8.12–8.15). In major respects, the role of CRF impacts within the revised context of global dynamics can be viewed as a restatement of Darwin's metaphor for the action of "10,000 wedges," a restatement symbolizing the coordination of the biological circumstances and environmental mechanisms involved in the origin and survival (or extinction) of species by natural selection, rephrased in the language of nonlinear dynamics.

General Plan of the Book

The topics I have outlined here are considered in the light of advances in nonlinear dynamics and deterministic chaos, as applied to developments in the documentation of Earth history furnished by researches supporting or opposing the Impact-Extinction Hypothesis (IEH) identified with the landmark papers of Alvarez et al. (1980) and Raup and Sepkoski (1984, 1986). An earlier proposal by McLaren (1970) for the idea of an impact-related extinction late in the Devonian Period (the Frasnian-Famennian boundary event at ~367 Ma, according to Harland et al., 1990), which emphasized the criteria needed for biostratigraphic codification, is particularly relevant from the perspective of nonlinear dynamics advanced here and in Shaw (1987b, 1988a; cf. Kaneko, 1990). McLaren (1983) presents a carefully reasoned overview of biologic-extinction phenomena in Earth history and their possible causes that is broadly consistent with viewpoints expressed in the present work. Contrasts between the IEH and opposing arguments that would relate extinctions to large-scale volcanic eruptions are debated, respec-

Introduction

tively, by Alvarez and Asaro (1990) and Courtillot (1990). The case for endogenous causations of extinctions favored by most paleontologists and geologists in the past—as discussed cogently by, among many others, McLean (1978, 1988), Russell (1982), Schopf (1982), Stanley (1984), Officer and Drake (1985), Hallam (1987, 1988, 1989), Rampino and Volk (1988), Rampino et al. (1988), Vogt (1989), etc.—was given impetus recently by Officer et al. (1987, 1992), Campbell et al. (1992), and Czamanske et al. (1992); cf. Baksi and Farrar (1991), Schmitt et al. (1991), Erwin and Vogel (1992). Concise reviews of the history of the IEH and related viewpoints pro and con are given by Glen (1990, 1994) and by McLaren and Goodfellow (1990). The classical issues of cause and effect, however—which, to the proponents of this or that "cause," are drawn sharply in black and white—are not specifically resolved in coupled nonlinear systems with delayed feedback (e.g., because the distinction between causal mechanisms and effectual mechanisms may be indeterminate in such systems, where sharp demarcations become fuzzy, and a black-and-white pattern turns to gray in the phase space of fractal probability distributions; cf. Notes P.3, 5.1, and 6.1). It is this aspect of terrestrial-extraterrestrial phenomenologies that is emphasized throughout the present work (see especially Chaps. 7 and 8, and Notes 8.1 and 8.11).

In the chapters immediately following I examine the record of impact craters on Earth for evidence of patterns that may represent vestiges or "fossil imprints" of spatiotemporal processes in the Solar System that have mediated the distal and proximal stages of orbital histories of the impactors in a manner consistent with predicted effects of nonlinear dynamics. I find a simple spatial pattern in the cratering history of the Earth, a pattern consisting of three spatial nodes that have persisted since the late Precambrian. The nodes mark three widely separated points where impact loci have been coincident at all ages along a crude and swath-like small circle in equal-area projections. With age there are departures of the spatial distribution from these nodes, but a single almost circular swath is sufficient to fit the positions of the nodes during the Phanerozoic.

During the late Cretaceous and early Tertiary, wherein the data are better focused, owing to the great quantity of studies of the K/T transition interval at ~65 Ma, another approximately small-circle swath, deployed at a high angle in the opposite hemisphere, intersects the Phanerozoic cratering swath near two of the Phanerozoic cratering nodes (PCNs). Viewed from the North Pole, the Phanerozoic swath passes through North America, Eurasia, and Australia, while the Cretaceous/Tertiary (K/T) swath passes through North America, Eurasia, Africa, and South America. I initially thought that one more node should have existed at earlier times, giving two paired nodes along two great circles, but only the three PCNs (North American, Eurasian, and Australian) are clearly demonstrable for all age groups, perhaps because the "fourth" PCN was obliterated by the effects of plate tectonics, seafloor spreading and subduction, and continental breakup and/or continental-shelf submergence during post-Paleozoic time [see, for example, the proposed craters on the Falkland Plateau (Rampino, 1992) discussed in Note 8.11].

If the idea of additional, symmetrically paired, PCNs were to be pursued in the light of a much-expanded future catalog of impact craters, it would appear to be conceivable — taken at face value — that three additional PCNs might be found. The places to look for them would be in the vicinity of southern South America (paired with the North American PCN), in a vicinity southeast of South Africa (paired with the Eurasian PCN), and in the vicinity of eastern Siberia and/or northern China (paired with the Australian PCN); these relationships can be seen more easily in the Mercator projections of Figures 26 and 28. Later discussion of geomagnetic patterns (Chap. 7, Fig. 28), however, would suggest that there might be only one "missing" PCN, such that, at least until the later Mesozoic, there would have been two pairs of PCNs meridionally aligned, respectively, with the Americas and with Australia-Eurasia. The present location of the "fourth" PCN, on this assumption, would lie somewhere in the vicinity of southern South America; hence the evidence for it would be complicated by numerous effects attending continental breakup and the opening of the South Atlantic (cf. Note 8.11).

The K/T swath is dominated by impact localities in the Northern Hemisphere, and to the degree that this pattern is correlated with iridium anomalies, shocked-quartz localities, biostratigraphic breaks, and other sharply defined events of K/T age, the K/T event appears to have been largely a Northern Hemisphere event. The results of Huffman et al. (1990) for DSDP (Deep-Sea Drilling Project) Site 527 in the South Atlantic, and of Zhou et al. (1991) for DSDP Site 596 in the South Pacific, may qualify this conclusion, depending on an evaluation of mechanisms of global geochemical and mineralogical dispersal. Certainly, searches for cratering events in the Southern Hemisphere, made difficult by the vast spread of its oceans, are indicated by questions of nodal symmetries.

Using as evidence the descriptions by McLaren (1970, 1982) and McLaren and Goodfellow (1990) of the localities of a major biostratigraphic break during the late Devonian (the Frasnian-Famennian extinction at 367 Ma on the time scale of Harland et al., 1990), we may surmise that the mean position of the Phanerozoic swath would have been rotated 20 degrees or so farther east (in North Polar view) to include localities in China during that time interval. Such a rotation can be accommodated without significantly affecting an intersection with the three nodal points. Similar variations, or oscillations, about the mean position of the Phanerozoic swath are expectable as functions of age, but they do not greatly affect the positions of the PCNs. To the precision that paleogeographic correlations presently permit (roughly 10 to 20 degrees in latitude and longitude), the PCNs have remained fixed relative to each other, thereby constraining relative surface motions associated with continental drift and plate tectonics at least during the Phanerozoic, and possibly during the latest 1 to 2 Ga of Earth's history. In this sense, the PCNs represent reference points ("benchmarks"), for both latitude and longitude, that are analogous to those of celestial navigation — in the same sense that the orientation of the geographic North Pole is fixed relative to the North Star, Polaris. The secular pattern of cratering as a whole is referred to here as a celestial reference frame (CRF) for the surface history of the Earth, against which relative reference frames of continental reconstructions, polar wander, seafloor spreading, tracks of volcanic "propagation," patterns of virtual geomagnetic pole locations, and core flux patterns can be compared and/or correlated. Some of these implications are pointed out as they are encountered in the discussion.

I begin by showing a set of Lambert Azimuthal Equal Area Projections that illustrate my introductory remarks. From there, I consider those aspects of other observations, interpreted in the light of nonlinear dynamics, that lend credence to the possibly outrageous idea of a CRF for geologic processes. The phenomena considered are highly interdisciplinary, and they necessitate consideration of feedback coupling in both the near and far fields of space and time. It is therefore difficult to present the progression of ideas in a linear way. I try nonetheless to adhere to the following general sequence of topics: (1) the paleogeographic patterns of impacts on Earth; (2) the corresponding history of temporal frequencies of impacts; (3) the general dynamical basis for Solar System structure and source regimes of potential impactors, including the contrasting implications of chaotic dynamics vis-à-vis classical interpretations of orbital histories of different classes of objects in the Solar System; (4) concepts of orbit-orbit and/or spin-orbit resonances that would be required if the CRF is to be correlated with secular invariances of impacts with the putative spatiotemporal precision; (5) the dynamics of conspicuous geophysical processes, such as plate tectonics, mantle convection, magmatism, geomagnetism, and the behavior of the core dynamo, that are consistent with the notion of systematic and intermittently continuous impacting of the Earth's surface at both linearly distributed and invariant loci; (6) an introductory discussion of the paleogeographic revisions that might be required if the cratering patterns are to be reconciled with existing models of continental drift and plate tectonics (if the CRF is invariant, then paleogeographic reconstructions must be consistent with a new longitudinal constraint as well as with those in latitude, and physical implications apply especially to the inception and directions of seafloor spreading and to "hot spot reference frames," both of which are coupled in current models with global patterns of magmatism and geomagnetic core-mantle interactions); and (7) a brief introduction to the implications for biostratigraphic codification that is simultaneously consistent both with the intermittency of impacting and with nonlinear dynamics (feedback coupling between nonlinear rates of growth and decay or, generally speaking, between birthrates and deathrates, is implicated in these concepts, just as it must be in the evolution of impacting objects in the Solar System).

Table I. Examples of Equations with Regimes of Nonlinear Resonances and Chaos

A. Accelerations Due to Nonspherical Mass Distributions In Primary; cf. Bate et al. (1971):

$$\mathbf{a} = \nabla\phi = \frac{\partial\phi}{\partial x}\mathbf{I} + \frac{\partial\phi}{\partial y}\mathbf{J} + \frac{\partial\phi}{\partial z}\mathbf{K} \qquad (1)$$

$$\phi = \frac{\mu}{r}\left[1 - \sum_{n=2}^{\infty} J_n \left(\frac{r_e}{r}\right)^n P_n \sin L\right], \qquad (2)$$

where:
a = acceleration
ϕ = potential function
∇ = vector differential operator ($\mathbf{I}$, $\mathbf{J}$, $\mathbf{K}$, are the unit vectors for orthogonal coordinates x, y, z).
μ = gravitational parameter = $G(M+m)$
G = gravitational constant
M = mass of primary
m = mass of satellite
r_e = radius of primary
r = distance to satellite
J_n = harmonic coefficients (geoid)
P = Legendre polynomials (higher than 1st)
L = geocentric latitude
$\sin L = \dfrac{z}{r}$

B. The Damped Driven Pendulum; cf. Baker and Gollub (1990):

$$d^2\theta/dt^2 + (1/\lambda)d\theta/dt + \sin\theta = k\cos(\omega_d t), \qquad (3)$$

where:
θ = angle of swing
t = time
λ = damping (quality) coefficient
ω_d = angular velocity of forcing, adjustable relative to natural frequency (tuning)
k = forcing amplitude

C. Dissipative Standard Map (Kicked Rotor); cf. Jensen (1987):

$$X_{n+1} = X_n + Y_{n+1} \qquad (4a)$$
$$Y_{n+1} = \lambda Y_n + k\sin X_n, \qquad (4b)$$

where X and Y, respectively, are the angle of the rotor and angular velocity (modulo 2π) at iteration steps n and n+1, and λ and k are the same as for the pendulum.

D. Sine-Circle Map; cf. Jensen et al. (1984) and Fein et al. (1985):

$$X_{n+1} = X_n + \Omega - (K/2\pi)\sin(2\pi X_n), \qquad (5)$$

where X is the iterated angle on the circle (modulo 1), Ω is a constant phase difference (bias), and K is the amplitude of forcing. The average number of cycles per iteration is the relative frequency, W, called the Winding Number, given by: W = Limit (large N) of (Total Rotation From Initial State)/N. At K = 1, all possible rational frequencies are present (complete Devil's Staircase) and is limit of the phase-locked nonlinear resonances + quasiperiodic cycles (irrational frequencies). Phase-locking (mode-locking) is directly analogous to resonant commensurabilities of orbital mechanics. Above the value K = 1, trajectories are noninvertible and iterated maps become chaotic with two or more values of W possible. In two dimensions, the sine-circle map is closely related to the dissipative standard map.

Figures: The Basic Illustrations

The Figures illustrate the principal ideas in the text, as summarized in the Introduction, and are arranged by chapter in the following sequences:

Chapter	Figures
1	1–5
2	6–8
3	9–11
4	12–16
5	17–18
6	19–23
7	24–32
8	33–36
Epilogue	37

The primary citation of each Figure in the text is shown in boldface type, for instance **(Figure 5)**.

Additional illustrations are given in a series of twenty-seven numbered Appendixes (see Contents), each number corresponding to an illustration that is cited in the text by Appendix number, to avoid confusion with the basic illustrations and to emphasize the ancillary nature of the appended information. Relevant chapter numbers for each of the appendixes are indicated in the individual Appendix headings, and primary citations of the Appendixes in the text also are shown in boldface type, for instance **(Appendix 3)**.

Figure 1

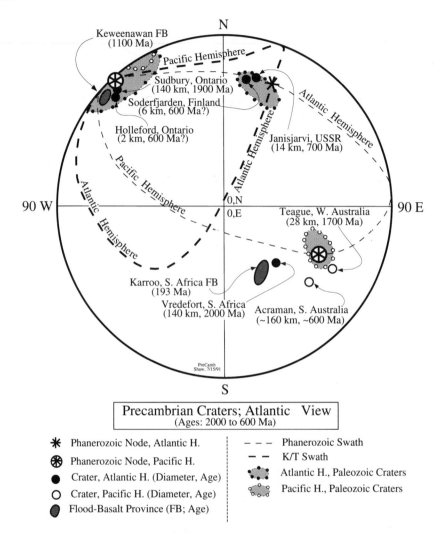

Figure 1 (Chapter 1). Lambert Azimuthal Equal-Area Projection of Earth viewed from above the Atlantic (0°N, 0°W), showing known impact craters with ages ≥ 600 Ma (solid symbols are craters on front hemisphere; open symbols, on back hemisphere). Also shown are Paleozoic impact loci, Phanerozoic nodes, and trends of global swaths (see text and Figs. 2–4). Crater locations and ages are catalogued by Grieve (1982) and Grieve and Robertson (1987), except for the Acraman Crater, South Australia, which is described by Gostin et al. (1986, 1989) and Williams (1986). Two flood-basalt provinces are shown for reference; locations, areas, and age spans are discussed in Rampino and Stothers (1988, Table 1), White and McKenzie (1989, Table 2), and Hutchinson et al. (1990, Table 3).

Figure 2

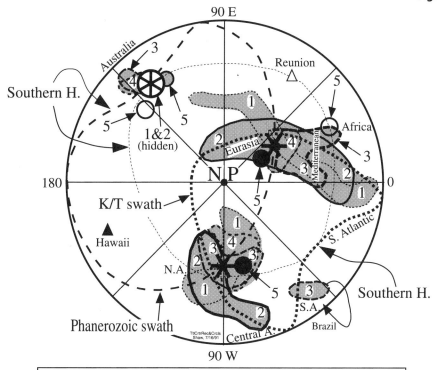

Composite Cratering by Reported Age, North Polar View
(Intervals were chosen to place K/T and P/T well within a group;
Era, Period, & Stage names from Harland et al., 1990, Fig. 1.7)

1. < 50 Ma (Cenozoic younger than Ypresian).
2. 50 to 100 Ma (Ceno. Ypresian to Cret. late Albian).
3. > 100 to 250 Ma (Cret. late Albian to Perm. Capitanian).
4. > 250 to 600 Ma (Perm. Capitanian to Vendian Smalfjord).
5. > 600 to 2000 Ma (Vendian Smalfjord to early Animikean).

● -- Precambrian, two loc. each): ● = N. Hemis.; ○ = S. Hemis.
✶ ⊕ -- Nodes of mutual overlap of all age groups (PCNs).

Figure 2 (Chapter 1). Lambert Azimuthal Equal-Area Projection of Earth viewed from North Pole, showing approximate loci of impact cratering of all age groups (shaded areas, numbered in legend, and enclosed by dashed and solid closed curves). Solid curves enclose age group 2, which includes the K/T subgroup. Open and filled circles are isolated craters of age group 5 (filled symbols, upper hemisphere; open symbols, lower hemisphere). The positions of Hawaii (Kilauea) and Reunion are shown for reference (filled and open triangles, respectively). Heavy short dashes show the trend of the K/T swath; heavy long dashes, the trend of the Phanerozoic swath (see text and Figs. 3 and 4). The light dashed line passing through Reunion and the Australian cratering node represents the low-latitude limit of a small-circle swath that could exist in the Southern Hemisphere but is poorly constrained by the data (there is no information to confirm or deny this or other crater patterns at high southern latitudes).

Figure 3

Figure 3 (Chapter 1). North polar projection as in Figure 2, showing the distribution of craters of age group 2 and a subgroup of craters with ages close to or at the K/T boundary (~65 Ma). Crater data are mainly from Grieve (1982) and Grieve and Robertson (1987); the latter is essentially the G-R Catalog described in the text. Additions and revisions to the G-R Catalog in age group 2 are from the following sources:

The location of Chicxulub, Yucatan, its discovery circumstances, and its inferred K/T age are discussed in Hildebrand and Penfield (1990), Hildebrand and Boynton (1990a, b), and Hildebrand et al. (1991).

The K/T age of Popigai is from Deino et al. (1991); that of Kara from Nazarov et al. (1991a, b) and Badjukov et al. (1991); cf. Note 1.1.

Montagnais Crater (age ~50 Ma; diameter $\geq$ 45 km), described in Jansa et al. (1990), is located on the continental shelf near its edge, ~200 km south of Halifax, Nova Scotia. Also see Poag et al. (1992), who describe evidence from DSDP Site 612 of a late Eocene "bolide event" near the east coast of the United States.

Avak structure, located at Point Barrow, Alaska, was discovered in the course of a geophysical survey; it has been described by A. Grantz (written comm., 26 June, 1991; cf. Grantz and Mullen, 1991; Kirschner et al., 1992).

The queried impact site in the Columbia Basin, Caribbean, was proposed by Hildebrand and Boynton (1991).

The "Unidentified Arcuate Scarps" NW of Hawaii, and the vicinity of the Hess Rise, are proposed here to be potential targets in the ongoing search for deep-ocean impact sites, none of which so far proposed have been documented; the scarps were discovered during a U.S. Exclusive Economic Zones bathymetric survey (R. I. Tilling, written comm., 30 June, 1991; Holcomb et al., 1991).

Figure 3

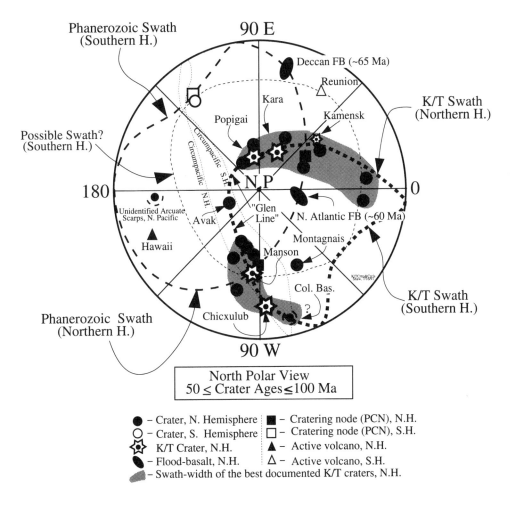

Figure 4

Figure 4 (Chapter 1). North polar projection as in Figs. 2 and 3, showing cratering nodes and swaths against outlines of the continents and the locations of eight flood-basalt provinces, numbered below from youngest to oldest (the numbers here correspond to the numerals in the illustration):

(1) Columbia River (CRB) Province, northwestern U.S. (Shaw and Swanson, 1970), with peak volume emplacement at ~17 to 15 Ma (Baksi, 1988, 1990a, b).

(2) Ethiopian Province, with age range from ~35 to 25 Ma (Hutchinson et al., 1990, Table 3; Rampino and Stothers, 1988, Table 1).

(3) British-Arctic Tertiary Province, with age range from ~62 to 56 Ma (Rampino and Stothers, 1988, Table 1; White and McKenzie, 1989, Table 2; Hutchinson et al., 1990, Table 3).

(4) Deccan Traps, India, with age range from ~67 to 60 Ma (citations in 3, above, and Baksi, 1990a, b; Baksi and Farrar, 1991).

(5) Serra Geral (Paraná) Province, South America, age range ~135–120 Ma (citations in 4; and Harry and Sawyer, 1992; Rennc ct al., 1992).

(6) Karroo Province, South Africa, with age range from ~200 to 175 Ma (citations in 4).

(7) Siberian Traps, Russia, with age range from ~250 to 240 Ma (Rampino and Stothers, 1988, Table 1; Hutchinson et al., 1990, Table 3; Baksi, 1990a, b; Baksi and Farrar, 1991; I. H. Campbell et al., 1992; Czamanske et al., 1992).

(8) Keweenawan Volcanic Province, central and northeastern U.S. and Canada, age ~1.1 Ga (Hutchinson et al., 1990, Table 3).

The McLaren Line is based on outcrop localities for the late Devonian (Frasnian-Famennian) extinction events described in McLaren (1970, 1982, 1983) and McLaren and Goodfellow (1990); see also Appendix 3a. The assigned age for this boundary in Harland et al. (1990) is 367 Ma. The outcrop pattern has no necessary relation to impact craters, but there are six craters in the numbered list of Grieve and Robertson (1987) within reasonable proximity to the McLaren Line and/or the Phanerozoic swath (swath-width ± 15° or so). The six craters, identified by [#, name, Lat.-Long., diameter (km), reported age (Ma)], are the following (cf. Appendix 3a): [**#16**, Charlevoix, Quebec, 47°32′N–70°18′W, 46, 360±25]; [**#27**, Flynn Creek, Tennessee, 36°17′N–85°40′W, 3.8, 360±20]; [**#42**, Kaluga, Russia, 54°30′N–36°15′E, 15, 380±10]; [**#63**, Mishina Gora, Russia, 58°40′N–28°00′E, 2.5, <360]; [**#73**, Piccaninny, Western Australia, 17°32′S–128°25′E, 7, <360]; [**#90**, Siljan, Sweden, 61°02′N–14°52′E, 52, 368±1].

Figure 4

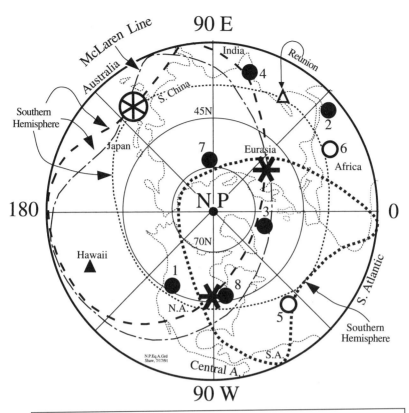

NORTH POLAR VIEW OF IMPACT LOCI AND CONTINENTS

▪▪▪▪▪ K/T Swath.

— — Phanerozoic Swath.

——- McLaren Line [based on sites of Devonian (F-F) extinctions in Canada-US, Central Europe, Spain, Morocco, and China].

▪▪▪▪▪▪▪▪ Southern Hemisphere Locus [based on sites passing through Australia Node, S. Africa, & southern S. America; sites nearer the S. Polar vicinity are not known but possible].

● Flood-Basalt Province, Northern Hemisphere

○ Flood-Basalt Province, Southern Hemisphere

⟩ (Numbers identified in caption)

Figure 5

Figure 5 (Chapter 1). Transformation of axes of North Polar Equal-Area Projection from present-day Greenwich Standard **(a)** to a pole, NNP', located in (a) at ~45°N–160°E and replotted as the north pole of the transformed projection **(b)**. The dash-dot line and three cratering nodes in (a) are represented by the heavy circle and asterisks in the transformed projection (b), where they all now plot within the new northern hemisphere. The shaded great-circle section in (a) is the equatorial section in (b). Because the shaded great-circle section in (a) is inclined at 45° to the Standard Equator, its perimeter [the equatorial section of the transformed projection in (b)] passes through the projected positions of the new poles, NNP' and NSP', plotted in (a) as the filled and open circles, respectively. In the equatorial section of the new projection [shaded area in (a)], the point NNP' (filled circle) lies above the position of the point 180° E'/W' plotted on the dashed line in the lower hemisphere of (a), and the point NSP' (open circle) lies below the point 0° E'/W' plotted on the solid line in (a). The unshaded area in (b) shows the extent of departure from a great circle of the small circle (heavy line) passing through the transformed positions of the three Phanerozoic cratering nodes (asterisks), a departure less readily seen in the distorted approximation represented by the difference between the dash-dot line and shaded area in (a). The heavy circle in (b) is a geographic parallel in the transformed coordinate system at a latitude of about 25° N'.

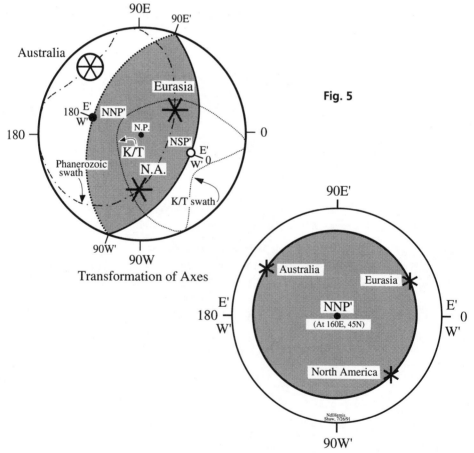

68

Distributions of craters within the individual age groups of Figure 2 locally can extend as much as ± 45° from the average trend of the Phanerozoic nodal swath passing through the asterisks in either (a) or (b). Evidently, at any given time (see patterns in Fig. 2), craters may cluster along either small or great-circle swaths departing significantly from the average Phanerozoic trend. Thus, spatial patterns more localized in time can trend at angles ranging between nearly normal to and subparallel to the mean nodal circle in (b) and can fall along arcs of either small circles or great circles (see discussions of Fig. 23, Chap. 7; Fig. 37, Epilogue). The K/T swath of Figure 3 [fine dotted line in (a)] is an example of a small circle that is roughly normal to the plane of the small circle passing through all three cratering nodes [dash-dot line in (a); heavy circle in (b)]. All projections suggest that the patterns of cratering on Earth are biased toward the Northern Hemisphere of the present-day Standard Coordinate System. Whether this global bias is an artifact of the present predominance of continental area in the Northern Hemisphere or is a property of the mechanisms of spin-orbit resonances discussed in Chap. 7 will depend in part on eventual documentation of cratering densities in Antarctica and in the southern oceans.

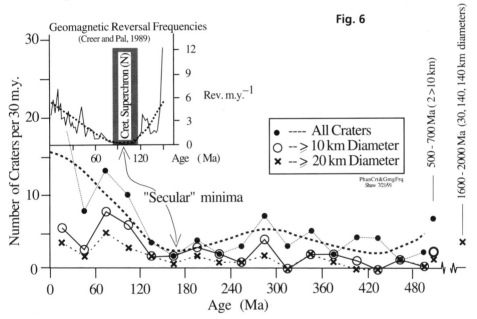

Figure 6 (Chapter 2). Histogram of Phanerozoic craters plotted against age and minimum diameters (points are plotted at centers of 30-m.y. bins and connected by straight-line segments to aid visualization). Craters of Cambrian and older ages are represented by counts of the indicated diameters and age ranges at the right of the diagram.

Inset: General relationship between crater frequencies and geomagnetic reversal frequencies over the latest 150 Ma (discussed in Chap. 7). The sources of crater ages and diameters are given in the captions of Figs. 1 and 3; geomagnetic data are from Creer and Pal (1989, Fig. 2); cf. Merrill and McFadden (1990, Fig. 1). Periodicities in the temporal data of reversal frequencies are discussed by Creer and Pal (1989), Negi and Tiwari (1983), Stothers (1986), and Mazaud and Laj (1991). Reversal mechanisms are reviewed by Merrill and McFadden (1988, 1990); see Jacobs (1992) for a state-of-knowledge summary. Concepts of nonlinear coupling between cratering frequencies and geomagnetic reversal frequencies are discussed in Chap. 7 (see Figs. 30 and 31).

Figure 7

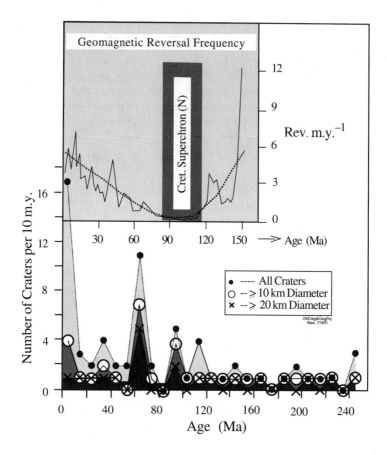

Figure 7 (Chapter 2). Histogram of Mesozoic and Cenozoic craters plotted against age and minimum diameters (points are plotted at centers of 10-m.y. bins and connected by straight-line segments to aid visualization).

Inset: Comparison with analogous history of geomagnetic reversal frequencies (see caption of Fig. 6; and Appendix 5).

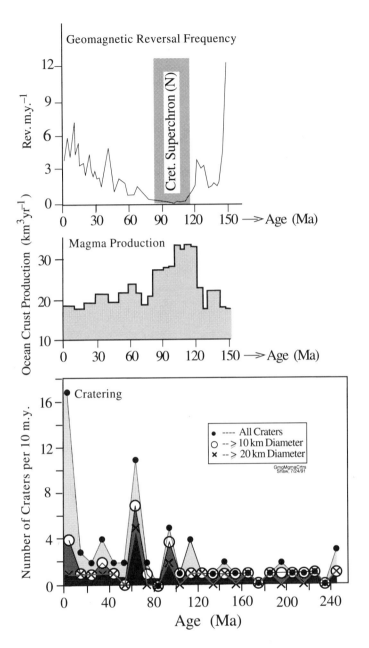

Figure 8 (Chapter 2). Histograms of cratering frequencies and geomagnetic-reversal frequencies from Figure 7 compared with generalized histogram of global magma production during the latest 150 Ma (data from Larson, 1991a, Fig. 1; cf. Larson, 1991b). Nonlinear coupling between impact frequencies, geomagnetic-reversal frequencies, and magma-production rates are discussed in Chaps. 7 and 8 (see Figs. 30 and 31; and Appendix 5).

Figure 9

Figure 9 (Chapter 3). Two-part illustration showing the distributions of cumulative planetary mass (a) and *fictive density* (b) in the Solar System (see subheadings a and b below), where fictive density refers to cumulative mass, Σm_E (starting with Mercury, Me), normalized by Earth mass (5.9742×10^{27} gm) and by the cube of the radius (R) equal to a given planetary distance from the Sun in Astronomical Units (AU). [1 AU is the mean orbital distance of Earth from the Sun ($\sim 149{,}600{,}000$ km).] The value R^3 is proportional to the volume of a sphere with radius equal to planetary distance (i.e., $R^3 = 3V_R/4\pi$, where V_R is spherical volume out to radius R from the Sun). The sphere is used as standard of reference because the topology of the protoplanetary nebula is not known. The ratio $(\Sigma m_E)/(R^3)$ as a measure of fictive density is proportional to nebular density only if the original mass-distance relation resembled the present one (arguments in the text suggest that a catastrophic redistribution of mass-energy occurred immediately prior to or accompanying planetary accretion; see Notes 3.2 and 9.2; see also Appendixes 6–8).

(a) Logarithm of cumulative mass vs. logarithm of mean orbital distances for the planets, Asteroid Belt, Kuiper Belt, and Oort Comet Clouds, where the Kuiper Belt is represented by distances from the vicinity of Neptune (~ 30.06 AU) to the order of 10^3 AU, and the Inner Oort Cloud is represented by distances from $\sim 10^3$ AU to $\sim 2 \times 10^4$ AU, the beginning of the Outer (or classical) Oort Cloud (both the Kuiper and Oort vicinities are generally considered to represent cometary reservoirs). The masses of objects beyond Neptune are relatively uncertain and/or conjectural. Prior to the discovery of Pluto's large "satellite" Charon in 1978 (Kaufmann, 1991, p. 320), the mass of Pluto was given as ~ 0.1 Earth mass (m_E) or greater but has subsequently been revised downward to a value as small as $\sim 0.002\, m_E$ (Kaufmann, 1991, p. 130; Pasachoff, 1991, Appendix 3); *The Astronomical Almanac for the Year 1992* (Anon., 1991, p. E88) gives a value of 0.015×10^{24} kg $= 0.00251 m_E$. The original mass of the Asteroid Belt is unknown; it has been estimated at values either similar to that of the present estimated mass or far larger, depending on models of planetary evolution. Bertotti and Farinella (1990, p. 339) give a value of $\sim 5 \times 10^{-4}\, m_E$.

The fine dotted lines with positive slopes represent various power-law extrapolations

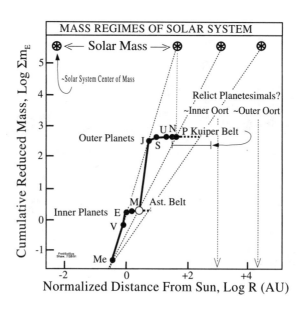

Figure 9

for original mass distributions. The steepest slope is one in which the cumulative mass values at the Earth and Jupiter positions are about the same as at present—i.e., this line is approximately equivalent to the "constant-density trend" shown in the Inset of (b)—and intersect a value corresponding to the present mass of the Sun at roughly the Pluto distance (if one or more other massive objects intervened between the Earth and Jupiter, then the original mass of Jupiter would have been smaller, unless it formed at a greater distance than its present one of ~5.2 AU; cf. Appendix 7). The dotted lines with smaller slopes represent hypothetical possibilities for primordial distributions in which most of the mass is at distances approaching those of the Inner and Outer Oort Clouds, respectively. In any distribution, steps would correspond to fractal aggregation over different distance intervals (e.g., near the outer limit of the Kuiper interval for the intermediate slope). Estimates of possible masses in the outer regions are discussed by Weissman (1990b). Stern (1991) later predicted that there should be a large number of 10^3-km size objects at distances beyond Pluto (the "plutons" of Weidenschilling, 1991)—though their existence has not been established (but see Note 6.9)—and the implications for the total mass and fractal distributions of mass for the planetary nebula are not known (an original distribution averaging to somewhere near the "constant-density trend" of the Inset in (b) might be implied, unless a hierarchical distribution of mass had already developed as a property of the pre-planetary nebula).

(b) Logarithm of normalized fictive density vs. logarithm of orbital distances. This graph gives a crude evaluation of the fractal dimensions of the present-day planetary distribution, following the demonstration of power-law density distributions in the cosmos by de Vaucouleurs (1970), as elaborated by Mandelbrot (1983) in terms of fractal dimen-
(*cont.*)

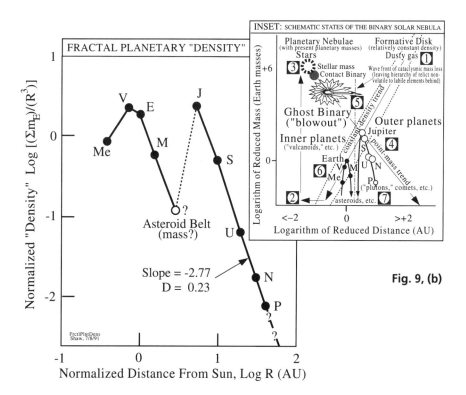

Fig. 9, (b)

Figure 9

(*Fig. 9, cont.*)
sions. The same approach has been shown to apply to the fractal dimensions of magma distributions in the Earth by Shaw and Chouet (1991). Analogous constructions are discussed in Alfvén and Arrhenius (1975, p. 15ff; 1976, p. 29ff) and in Bertotti and Farinella (1990, p. 338ff). From this plot we can estimate the mass of the Asteroid Belt by extrapolating the line through Earth and Mars to an assumed orbital distance for the center of mass of the asteroids. Taking the position of Ceres as a first approximation gives a mass of the order of 0.1 m_E for the asteroids (cf. Chapman and Davis, 1975), which is 200 times larger than the value (0.0005 m_E) cited in (a); the values for the masses of the three largest asteroids, Ceres, Pallas, and Vesta, given in *The Astronomical Almanac for the Year 1992* (Anon., 1991, p. K7), sum to 0.0003 m_E.

Reversing the above procedure and calculating a value of orbital distance for the Asteroid Belt from a value of 0.0005 m_E gives R = 2.71 AU, compared with 2.767 for Ceres (see Fig. 13a). Thus, a small change in the orbital distance of the assumed center of mass for the asteroids produces a large change in estimates of mass, as compared with the present-day planetary distribution. Trial and error calculations of this sort demonstrate the great sensitivity of calculated mass to fractal density-radius relationships of the above type (see the discussion in Appendix 7), and, conversely, the relative insensitivity of orbital distances relative to an assumed total mass of the asteroids. Some discussions in the literature based on models of orbital evolution have proposed the past existence of much larger planetary objects at the position of the Asteroid Belt (see Ovenden, 1972, 1973, 1976; Ovenden and Byl, 1978; Van Flandern, 1978). Many astrophysicists appear to favor a small mass at this location as an intrinsic outcome of planetesimal accretion (cf. Wetherill, 1989b, c, 1990, 1991b). It is evident from the figure (and Inset), however, that the discontinuity between the bipartite fractal or multifractal trends constitutes a transition of major significance in the evolution of the Solar System (cf. Fig. 10, Appendix 7, Notes 3.2 and 9.2, and discussions in the text of Chap. 3).

Inset: Schematic diagram (not to scale) illustrating conjectural stages in the evolution of a binary primordial solar system that approximate the model discussed in Chapter 3 and Notes 3.2 and 9.2. The sequence of highlighted numbers illustrates stages of this process. The evolution begins (**Stage 1**) after a disk-like region of star-forming gases and particles has coalesced within the Galaxy to form a disk-like nebula (**Stage 2**) that has a relatively uniform density, compared to later stages, just prior to gravitational collapse into a bifurcating instability that resembles the biological process of *gastrulation and fission* (see Note 8.15) thereby giving birth to **Stage 3**, a pre-main-sequence contact-binary star surrounded by "leftover" centers of planetary nucleation (see Note 9.2). During binary-star formation and approach to the main-sequence stage of stellar evolution (cf. Notes 3.2 and 9.2), the planetary nebula that nucleated in the outer parts of the stellar disk evolves in essentially the standard fashion (cf. Wetherill, 1989c, 1990, 1991b). But this process — beginning during Stage 3 — necessarily involves additional instabilities in the mass-distance relations, as expressed by the changing fractal geometry of the mass distribution. These post-star-formation instabilities represent completion of the transition between the volume-filling mode (where D = 3) of Stages 1 and 2, corresponding to the state of constant density vs. radial distance (the region of positive slope labeled "constant-density trend") and the point-mass-like mode (where D = 0) of **Stage 4**, corresponding to the condensed-state idealization of a central star plus a, by comparison, low-mass stable planetary system (the line of negative slope labeled "point-mass trend"). The implication of this complex transition for mass transport is illustrated schematically by the "radial mass distribution function" of Appendix 7b (cf. de Vaucouleurs, 1970; Shaw and Chouet, 1991). At approximately this juncture — some time after the initial accretionary coalescence of the planetary centers of

Figure 9

mass—the locus of the contact-binary companion of the central star experiences yet another event in the general sequence of vortical instabilities related to the Roche condition (cf. Appendix 6, Inset C) in which jets of high-velocity plasma and gas stream out through the partially formed or forming planetary system (shown by the swirling arrows labeled "Ghost Binary," as in Fig. 10), constituting the transition between Stages 3 and 4. **Stage 5** represents the motion of a shock wave (vertical line of long and short dashes near the position of the present-day Asteroid Belt) passing through the system of planetary nuclei which strips, or blasts, much of the lighter mass (the volatile elements and some portion of the labile elements; cf. Lipschutz and Woolum, 1988) from the region that later becomes the locus of the present-day terrestrial planets. **Stage 6** and **Stage 7** represent the complex of phenomena that include (a) the collisional regimes of the growing plantesimals (e.g., Wetherill, 1985, 1988, 1989a, b, c), (b) the collisional formation of Mercury's mass distribution and core (e.g., Cameron et al., 1988), (c) the collisional formation of the Moon (e.g., Cameron and Benz, 1991) and asteroids (cf. Note 6.8), (d) the formation of outer-solar-system objects, the "plutons" and comets (cf. Duncan et al., 1987, 1988; Stern, 1991), and (e) the chaotic "mixups" that effected the present-day orbital properties of the inner planets, the Pluto-Charon pair, and the Uranus-Neptune pair (cf. Stern, 1992, 1993), as suggested by the "strange-attractor-like" distribution (cf. Notes P.3 and 6.1) of orbital and rotational resonance states illustrated in Figure 10 and Appendix 6 (Inset C).

Figure 10

Figure 10 (Chapter 3). Two-part illustration showing the logarithms of orbital and rotational kinetic energies plotted against (a) the logarithms of object masses, and (b) the logarithms of object masses normalized relative to mean radii and distances from the center of mass of the Solar System (excluding the satellites of the planets, the Asteroid Belt, and the various cometary reservoirs). Data sources are given in Figure 9a (see also Fish, 1967; Kaula, 1968). [Note: These diagrams were originally drawn before the mass of Pluto was revised downward, as cited in Figure 9a (see *The Astronomical Almanac for the Year 1992*, Anon., 1991, p. E88). The new value would shift the plotted points for Pluto offscale to the left. The original points were retained to show where the more massive state would appear in the diagrams (a value of ~0.1 m_E was used; cf. Considine, 1983, p. 2249). In any case, Pluto (and the Pluto-Charon binary) are even more anomalous than are the other nonlinearities displayed.]

(a) Logarithm of the orbital and rotational kinetic energies (K.E.) of the Solar System objects vs. the logarithm of mass. The value of K.E. for the Sun is a crude estimate based on its motion about the center of mass of the Solar System (Jose, 1965; Landscheidt, 1983, 1988). The discontinuity seen in Figure 9b is also evident in the trends of orbital K.E. (solid circles). Subjective trends through the two sets of data (inner planets vis-à-vis the outer planets, except Pluto) intersect at a mass roughly an order of magnitude smaller than that of the Sun, labeled "Ghost" Binary (mass of Sun = 1.9891×10^{33} gm; Pasachoff, 1991, Appendix 2). This label represents an inference that the Solar System retains a dynamical memory of a primordial stage when there existed either a closely bound binary companion of the Sun or a massive primordial planetary nebula that bifurcated and lost mass from the inner regions (cf. Notes 3.2 and 9.2; and Appendix 7). This inference implies a dynamical history analogous to T Tauri, FU Orionis-like objects (see Weissman and Wasson, 1990; Bertotti and Farinella, 1990, p. 340). The trend of rotational K.E. (open circles) converges with orbital K.E. at a vicinity several orders of magnitude higher in K.E. and at least twice the mass of the Sun. This is the inferred protostellar system that evolved through T Tauri stages to become the planetary nebula. Recent reviews of observations on stellar evolution and astrophysical hypotheses appear to support this interpretation (see Kaufmann, 1991; Pasachoff, 1991; Maran, 1992).

(b) Logarithm of orbital and rotational kinetic energies (K.E.) of Solar System objects vs. the logarithm of mass-length ratios (m/R, m/r), where R is the approximate orbital distance of the objects from the Solar System center of mass, and r is the radius of the object. The upper sets of points (solid circles and circles with crosses) represent orbital K.E. normalized by R (solid circles) and by r (circles with crosses); the lower sets of points (open circles and crosses) represent rotational K.E. normalized by R (open circles) and by r (crosses). These conventions expand the kinetic-energy spectrum, reflecting a diversity of effects in planetary evolution (the divergences and clustering effects are analogous to *fractal singularity spectra* (see Note P.4; and Shaw and Chouet, 1989). Sets and subsets of trends in this plot converge to the locus of the protostellar system in (a). The precise alignment of the uppermost trend is an artifact of the Newtonian definition of orbital K.E.

Figure 10

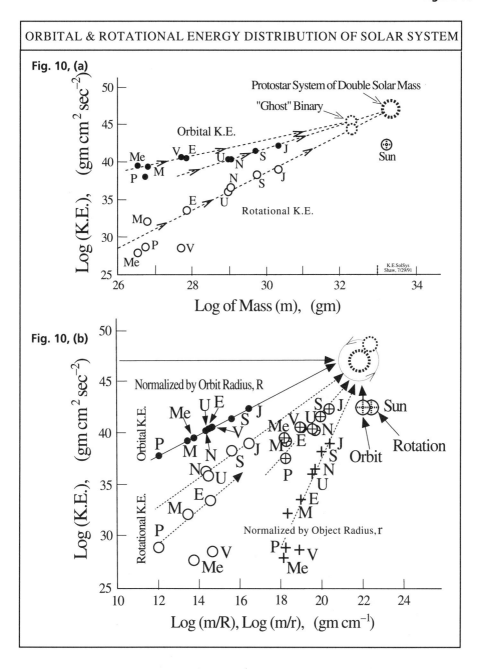

Figure 11

Figure 11 (Chapter 3). Schematic perspective of the Solar System, showing positions of potential impactor reservoirs relative to planets and a fictive density, M_*/R^3, based on a reference mass, M_*, equal to the mass of the Sun (1.9891×10^{33} gm). This "density" convention differs from that in Figure 9b because of an assumed invariance of the mass of the system; i.e., it is equivalent to the idealization of a point mass, hence the fractal dimension is zero (corresponding to the heavy solid line with a slope of -3 extending from the position marked "Sun" through the planetary sequence shown by solid circles; the open circle is the position of the asteroid Ceres). The "Sun" position is plotted at a finite distance on the abscissa, because the actual body of the Sun revolves around the Solar System Center of Mass (CM) at distances, R, that vary from near zero to $2.19\,R_* \cong 1.524 \times 10^6$ km, where $R_* \cong 0.696 \times 10^6$ km is the Sun's radius (cf. Landscheidt, 1983). Thus, the position of the Sun oscillates relative to the CM within $\sim 10^{-2}$ AU (one Astronomical Unit $\cong 1.496 \times 10^8$ km). The heavy dashed line with a slope of -3 shows the effect of doubling the constant mass (distances $10R_*$ and $100R_*$ are shown for reference). The reentrant dashed and solid line with solid circles and one open circle at the bottom of the shaded region (forming an "arrow" pointing toward the Sun) is the fractal distribution of Figure 9b.

The hourglass symbols and direction lines with arrows at the right indicate general possibilities for either inward or outward cascades ("sand-pile slumps") of objects between different vicinities and/or reservoirs, including extrasolar space (representing ejection of objects from the Solar System and capture of interstellar objects by the Solar System; see Clube, 1985, 1989; Clube and Napier, 1984a, b; Napier, 1985, 1989; Duncan et al., 1987; Stern, 1987, 1988). The asteroid categories Aten, Apollo, and Amor represent various near-Earth and/or Earth-crossing values of semimajor axes, a, and perihelia, q: Atens are asteroids that plot generally within 1 AU from the Sun; Apollos have values of a generally greater than 1 AU but values of q generally less than 1 AU (Earth-crossers); and Amors have values of q within a narrow range slightly greater than 1 AU (see Binzel et al., 1989). More detailed views of asteroid distributions are shown in Figure 13. Any of these categories could be a source of near-Earth asteroids (NEAs) or similarly evolved objects or fragments of objects; the term "near-Earth object" (NEO) acknowledges the existence of hybrid sources (e.g., asteroidal and cometary). There appears to be no rigorous dynamical distinction between these and other loosely labeled object reservoirs, hence any of them might contribute to low relative-velocity collisions and/or to convergent scenarios as the proximal source of intermittently captured natural Earth satellites (NESs); see Chapter 6 and Figure 19.

Nonconventional capture scenarios for NESs are envisioned to occur, owing to a diversity of chaotic and mode-locked nonlinear resonances involving groups of objects approaching the orbital vicinity of the inner planets (NEAs, NEOs, and FMOs, the last referring to "fast-moving objects" that exist in great numbers but usually are of undocumented origins and orbital parameters; see the discussion in Chap. 5, Fig. 12; cf. Wisdom, 1985, 1987a, b, c; Tatum, 1988, 1991; Binzel, 1991; Binzel et al., 1991; Pike, 1991; Ipatov, 1992). All of these objects are subject to perturbations by longitudinal gravitational coupling (oscillating mass anomalies) involving the Sun ± inner planets, or by the inner planets ± Sun ± outer planets (see Figs. 12 through 14 and 19 through 21, and discussions in the text). The nonlinear oscillations of the Sun relative to the CM represent both a large gravitational anomaly and a major source of torques acting on solar-system objects, especially the inner planets, asteroids, and new comets. Forces related to Sun-CM-Planet orbital oscillations are balanced by major dissipative motions within the Sun, which, in turn, are related to the electromagnetic-convective properties of the solar dynamo (see Chap. 8; and Wood, 1972; Landscheidt, 1983, 1988; Maran, 1992, p. 848).

Figure 11

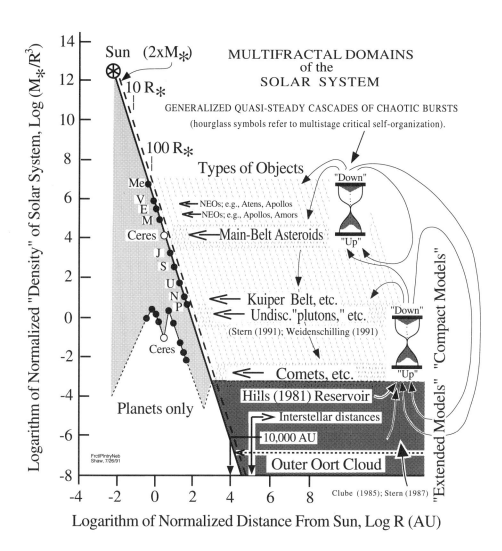

Figure 12

Figure 12 (Chapter 4). Examples of nonlinear orbital computations for the asteroids, redrawn and modified after Wisdom (1987a, b, c):

(a) Oscillatory evolution of eccentricities as a function of time, with occasional spikes or "bursts" that represent potential Mars- or Earth-crossing trajectories. Note the bistable character of the time series, which is typical of diverse forms of nonlinearly coupled oscillators subject to chaotic crises (cf. Shaw, 1991).

(b) Poincaré section of regular and chaotic regions of a conservative Hamiltonian system showing simple (concentric circular structures at left) and complex (intricate shaded cauliform structure on right) loci of recurrences surrounded by more regular parameters of motion (outer dashed lines). Points in the diagram represent instantaneous values of eccentricity, e, given by the distance from the origin (0,0) to a given point. In such a *phase portrait* the area of phase space is conserved, and the ordering of the orbital functions, X and Y, are described in terms of the regions of phase space "visited" by a given temporal sequence of computational results [compare the temporal sequences in (a), where the ranges of amplitudes oscillate within a generally unvarying range (on average) and neither expand globally nor decay to limit cycles or fixed points of vanishing phase-space areas]. By contrast, in a dissipative system, trajectories of motion evolve, and the two vicinities would be analogous to what are called loci of attraction or *attractor basins*. In that case arbitrary trajectories are attracted to ("gravitate" toward) or are repelled away from the vicinities of attractors. In the vicinities of chaotic attractors, trajectories of motion can be caught up in *chaotic crises* (see Chap. 5) in which they experience abrupt and amplified changes of motion or are ejected from one attractor basin to the vicinity of some other attractor basin as system parameters are varied; see the discussion of Figures 17 and 18 in Chapter 5 (cf. Shaw, 1991, Figs. 8.13–8.20). In either case, eccentricities jump about with time in the irregular but essentially bistable manner shown in (a) [in nonlinear dynamics, *bistability* refers to motions of trajectories on a two-well potential surface, or *manifold* (cf. Abraham and Shaw, 1982–88)].

(c) Example of computations in which asteroids in the vicinity of a particular nonlinear-resonance condition (representing the 3:1 Kirkwood Gap in the Asteroid Belt) are cleared out (ejected) from that vicinity by chaotic behavior, phenomenologically fitting the notion of chaotic crises in dissipative systems that repel trajectories from the vicinity (unshaded) of an unstable periodic attractor in a chaotic region (cf. Chap. 5; and Wisdom, 1987a, p. 121 and Fig. 9). If the periodic attractor were stable it would represent a condition of nonlinear resonance, which would be selectively attracting to asteroids.

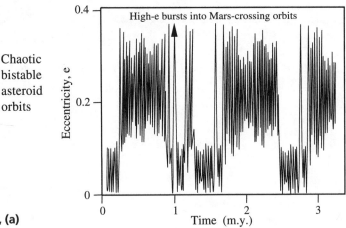

Chaotic bistable asteroid orbits

Fig. 12, (a)

Figure 13

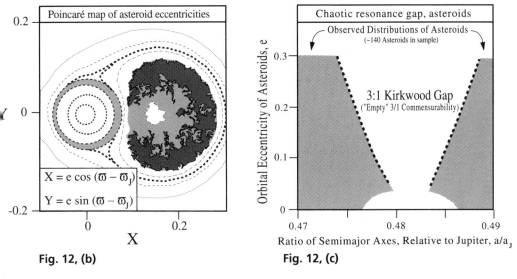

Fig. 12, (b) Fig. 12, (c)

Figure 13 (Chapter 4). Generalized diagrams illustrating the frequency distributions and orbital structure of the Asteroid Belt based on ~4000 named asteroids (data from Binzel, 1989, Fig. 2; Binzel et al., 1989a):

(a) Photopositive ("night-view") histogram of numbers of asteroids as a function of approximate distance from the Sun (populated regions are shown in white as though viewed against the night sky; the name asteroid derives from the Greek for "star-like," referring to the starry asterism relative to the planets). The taxonomic labels are commonly used names for different parts of the Asteroid Belt, but they do not necessarily indicate common genetic origin in terms of orbital evolution, petrology, or chemistry. "Empty" resonances, first established by Daniel Kirkwood in 1857 as the locations of numerous gaps, are now called *Kirkwood Gaps* in his honor (see Kirkwood, 1876, p. 365; cf. Kirkwood, 1891). The existence of the gaps has been given plausible explanations in terms of nonlinear dynamics applied to orbital computations (see Fig. 12, and Wisdom, 1987a, b, c; Froeschlé and Greenberg, 1989; Nobili, 1989; Scholl et al., 1989; Yoshikawa, 1991; Ipatov, 1992). The conventional astronomical symbols ("canonical symbols"; see Bate et al., 1971, p. 41) placed above the distance scale identify the approximate locations of Earth (1 AU), Mars (1.524 AU), and Jupiter (5.203 AU) relative to the asteroid distribution. Notice the spread of values for the near-Earth Aten-Apollo-Amor subset, as if this region represents a transient spillover from the peak distribution, as is shown in (b) from the orbital perspective (cf. Inset 1). See Figure 11 and the discussions in Chapters 4–6 for nonlinear "delivery" of asteroids to Earth and ideas concerning self-organized "sand-pile slumps" of objects, fragments, and groups of fragments derived from the "central heap" (e.g., "rubble-pile" asteroids from the MBA and/or from outlying resonances; cf. Greenberg and Nolan, 1989; Chapman, 1990; Binzel et al., 1991) resulting from collisional processes and stream-like focusing of these objects — together with objects from cometary sources (see the Insets) — in the region of the inner planets (cf. Alfvén, 1971a, b; Bandermann, 1971; Baxter and Thompson, 1971; Danielsson, 1971; Dohnanyi, 1971; Lindblad and Southworth, 1971; Trulsen, 1971; Alfvén and Arrhenius, 1975, 1976; Brown et al., 1983; Taylor et al., 1987; Olsson-Steel, 1989; Drummond, 1991; McIntosh, 1991; Steel, 1991; Shoemaker and Izett, 1992). Figure 14 and Appendix 9 illustrate the nonlinear-dynamical resonances of a two-parameter one-

(*cont.*)

Figure 13

(*Fig. 13, cont.*)
dimensional algorithm (the sine-circle map of Table 1, Sec. D) that comes remarkably close to simulating numerical resonances in the vicinity of the asteroids (cf. Fig. 21 and Appendix 11). By that analogy, near-Earth resonances (whether partially filled or "empty") comprising locations correlated with the Aten-Apollo-Amor taxonomic groups would correspond to small-number fractions (high Kirkwood ratios) near and beyond 1/8 (8:1 Kirkwood), as indicated by the symbolic Kirkwood ratio labeled "1:0" (see the "Devil's Staircases" of Fig. 14 and Appendix 9; see also Notes 6.5, 6.7, and 6.9).

(b) Schematic snapshot of asteroid positions at a specific time, as in a wide-angle freeze-frame photograph viewed from a position looking down on the ecliptic plane (schematically modified and redrawn from the illustration on the back cover of Binzel et al., 1989). The date and orientation relative to the direction of the Vernal Equinox are shown at the bottom; the positions of the planets at that instant are indicated by tick marks on the orbits. The density of clustering in the MBA is exaggerated for emphasis. On the other hand, the number of impactors within the orbits of Mars and Earth that pose significant present and future "cosmic hazards" for mankind is far greater than would be implied by the number of objects shown (see Notes 1.2, 6.9, 9.1, A.1, and Appendix 3; cf. Tatum, 1988, 1991; Binzel, 1991; Pike, 1991; Steel, 1991; Yeomans, 1991b). The scale in (b) is reduced from (a) by the ratio 1/3 [12 AU across in (b), compared with 4 AU for the length along the scale bar between Earth and Jupiter in (a)]. The Insets show three schematic diagrams that illustrate the scaling of asteroidal-to-cometary orbits in the Solar System and why there are ambiguities concerning the observational evidence for the origin and evolution of the asteroids vs. the comets, thereby relating the "local" scale of Figure 13 to the more encompassing scales of Figure 15.

Inset 1: Schematic diagram of the extended domain of the "asteroids." The shaded region corresponds to the conventional Asteroid Belt [its less well known peripheral regions, inside the orbit of the Earth and outside the orbit of Jupiter, are shown by lighter shading (see Weissman et al., 1989, Fig. 1)]. The nearly circular orbit (short dashes) labeled Ce, for the largest asteroid Ceres, is shown for cross-reference with the MBA of Figure 13a. The elliptical trajectories (heavy dashes) were drawn to resemble the orbits of four objects that are neither "asteroids" nor "comets" — in the sense that their known physical and orbital properties do not correspond to classical definitions nor do they show definitive visual aspects of either asteroidal or cometary character — as follows [the numbers preceding the names indicate chronological sequence in the catalog of objects formally called "asteroids" (see Chap. 4; and Bowell et al., 1989)]: (a) 2201 Oljato, with a semimajor axis a of ~ 2.18 AU and eccentricity e of ~ 0.71 (both evolving chaotically; cf. Weissman et al., 1989, Fig. 5), has its present perihelion inside the orbit of Venus, hence it crosses the orbits of all of the terrestrial planets except Mercury, and has physical properties recent observers have called "enigmatic" relative to either asteroids or comets (McFadden et al., 1993); (b) 944 Hidalgo, with $a \cong 5.76$, $e \cong 0.66$, and perihelion beyond the orbit of Mars, is one of the objects traditionally called "an asteroid with cometary orbit," which also includes at least some of the Earth-crossing objects in the Apollo group shown in Figure 13a (cf. Kaula, 1968, Fig. 5.9; Greenberg and Nolan, 1989); (c) 5145 Pholus, discovered in 1992, and resembling 2060 Chiron somewhat in its orbital parameters, rather large size relative to most of the asteroids and the usual expectations for Solar System comets (both Chiron and Pholus apparently have diameters between 100 and 200 km; cf. Stern, 1991; Cowen, 1992; Kerr, 1992e; and Note 6.9 of the present volume), and mystique (cf. French et al., 1989; Hoffmann et al., 1993), has an orbital eccentricity similar to those of Oljato and Hidalgo but with a perihelion beyond Jupiter and an aphelion beyond Neptune, as well as physical properties that, according to Hoffmann et al., "do not provide a clue for either a cometary or an asteroidal origin"; (d) an analogously ambiguous object with an orbit that is hypothetically analogous to that of comet Halley, the

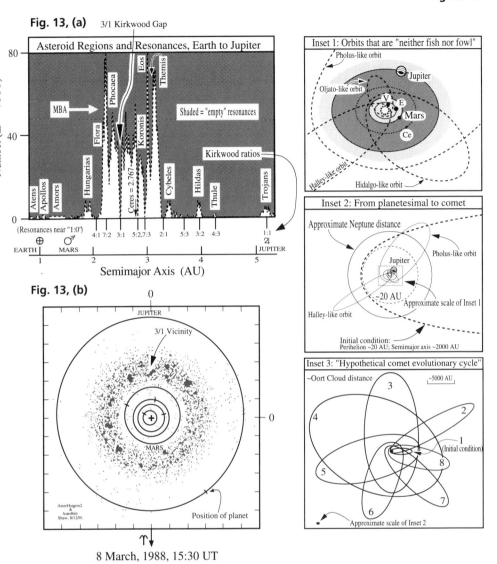

Figure 13

best-known "short-period comet" (see Yeomans, 1991, Table 10.1), hence with a perihelion $q \cong 0.6$ AU (i.e., inside the orbits of both the Earth and Venus, and similar to that of Oljato), a Pluto-crossing aphelion (i.e., in the outermost part of the planetary region, hence in the vicinity of the Kuiper Belt; cf. Figs. 9a and 11), and an eccentricity $e > 0.95$, where $e = 1$ is the parabolic limit of an "infinite" semimajor axis at the threshold of the hyperbolic regime of ejected objects or objects originating from extrasolar sources. [Pasachoff (1991, pp. 32f) illustrates the relationships between elliptical shapes, eccentricity e, and conical sections.]

Inset 2: Planetary-system-scale perspective (i.e., reduced by a factor of about one-seventh from Inset 1) of the Oljato-, Hildago-, Pholus-, and Halley-like orbits (solid ellipses) relative to the assumed initial conditions (the partial trajectory shown by the heavy dashes) of a planetesimal that was originally formed in the outer planetary region (i.e., the example chosen was assumed to have originally formed in a nearly circular orbit near the

Figure 13

ecliptic plane in the Uranus-Neptune vicinity within what is sometimes called the "Kuiper Disk") and that subsequently evolved to become a typical Hills-Cloud or Oort-Cloud comet owing to the perturbing orbital influences of the outer planets, passing stars, and the Galactic tide [see Fig. 15b herein, as modified from Fig. 3 of Duncan et al. (1987); cf. their Fig. 4, and Duncan et al., 1988]. The high ellipticity of the "initial condition" shown was a "trick" used by Duncan et al. (1987) to shorten the computer time of Monte Carlo simulations, because exploratory results had shown that most objects evolved in approximately the same way from the Uranus-Neptune vicinity (i.e., at low orbital inclination) until the semimajor axis a reached a value of at least 2000 AU at the original perihelion of $q \cong 20$ AU (cf. Fig. 15, and the discussion of chaotic orbital evolution in Chap. 4).

Inset 3: Caricature of the orbital evolution of a cometary object according to the Monte Carlo simulations of Duncan et al. (1987). The sequence of ellipses corresponds more or less to the following relationships documented in Figure 15b: (a) beginning with the initial state illustrated in Inset 2 (ellipse #1 in this Inset), the object initially expands in semimajor axis a at constant perihelion q to the orbit shown by ellipse #2 (i.e., the closest proximity to the Sun per orbital cycle remains almost constant until the semimajor axis grows to the order of $\sim 10^4$ AU); (b) the perihelion value q then begins to grow at approximately constant semimajor axis a (ellipses #3 and #4) until q approaches the same order of magnitude as a, hence with decreasing ellipticity [i.e., if, or when, q approaches parity with a, then the object has attained a nearly circular orbit at the Oort Cloud distance — but computations suggest that the timescale to attain this condition is of the order of the lifetime of the Solar System, therefore this Monte Carlo simulation cannot differentiate between objects evolved to the Oort Cloud state and objects formed there and/or captured from the Galaxy or from the "Oort Clouds" of other stars (cf. Shoemaker and Wolfe, 1984; Clube, 1985; Stern, 1987)]; (c) when q has grown to large values, but before it has reached $q = a$, the orbital states begin to show indications of recursive behavior — meaning that some states are returned to the vicinities of earlier states (e.g., ellipses #5–#8), as indicated by Figure 15b — and the system as a whole could conceivably evolve to an orbital state resembling the initial one (see the discussion of Fig. 15 in Chaps. 4 and 5 vis-à-vis *chaotic crises*). In other words, the cometary evolution suggested by this Inset resembles that of an expanding and contracting elliptical spring coupled to another oscillating system (the stellar-galactic oscillator) so that the cometary orbit responds chaotically (including opportunities for nonlinear resonances and/or chaotic crises) — i.e., somewhat like a rubber band excited to vibrate chaotically between isotropic and anisotropic stretched and relaxed modes — as it precesses slowly about the planetary locus. Because it is known that at least some objects in the Asteroid Belt proper have *de*volved from objects of cometary character (e.g., Weissman et al., 1989; Wetherill, 1991a; cf. Note 6.9 of the present volume), according to the visual criteria by which comets derived their name (see Clube and Napier, 1982b; Yeomans, 1991), this model suggests that it is *impossible* to define a unique cometary source vs. a unique asteroidal source in the planetary nebula (and in the present-day asteroids and comets) if there was any overlap at all in their "original" states (see, for example, the binary-star evolutionary model of the present volume, as discussed in Chap. 3 and Notes 3.2 and 9.2). Consider also the humbling spatial perspective shown by the relative sizes of Inset 2 (lower right corner) and Inset 1 (the small dot at the center of this Inset, roughly one-thousandth the scale of Inset 1) relative to the Solar System *sphere of influence* (compare the astronomical concept of the "sphere of influence" with the nonlinear-dynamical concept of the "basin of attraction," as discussed in Note 6.1) for the orbital evolution of planetesimals that have become what we now call asteroids ("minor planets"), comets, and other Solar System objects [such as the satellites of the planets, the "vulcanoids" of Leake et al. (1987), and the "plutons" of Stern (1991); see Weidenschilling (1991) and Notes 6.5 and 6.9 of the present volume].

Figure 14

Figure 14 (Chapter 4). Diagram illustrating the numerical meaning of the *Devil's Staircase* (see Jensen et al., 1984; Bak, 1986), based on implementation of the sine-circle algorithm of Table 1, Eq. 5 (see Shaw, 1987b, 1991, for other types of sine-circle plots and applications to natural periodicities). The abscissa is the *bias* term, Ω, an independently assigned variable in Eq. 5, and the ordinate is the derived resonance, or nonlinear mode-locked periodicity, called the *winding number*, W, produced at a given value of Ω for an assigned value of nonlinear coupling strength, K (in the present example, $K = 1$). In the literature, Ω is sometimes called the "bare winding number" and W the "dressed winding number." The value $K = 1$ is the critical value of the coupling parameter at the transition to chaos ("edge of chaos"), where all possible values of rational fractions are produced when the recursion is performed (hypothetically to unlimited precision).

Inset: Magnification of the small rectangle in the lower diagram, showing the self-similarity and fractal nature of the distribution of rational fractions (see Jensen et al., 1984, for derivation of the fractal dimension of the completed set).

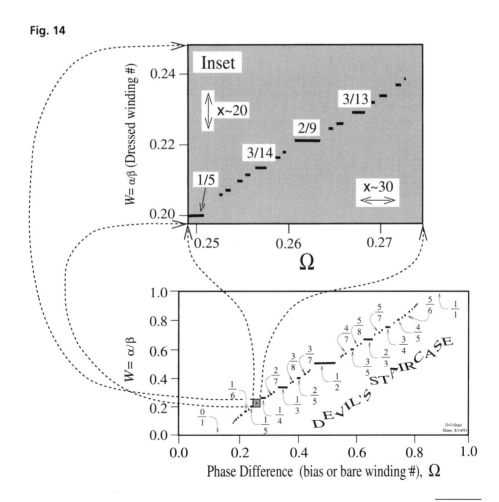

Figure 15

Figure 15 (Chapter 4). Illustrations of the nonlinear evolution of orbital characteristics of cometary reservoirs:

(a) Example of chaotic motions in the Kuiper Disk, modified and redrawn from computer simulations by Torbett and Smoluchowski (1990, Fig. 3); effects of 10-m.y. integrations of orbital perturbations of objects with initial orbital parameters as shown.

(b) Results of a hybrid-integration Monte Carlo simulation of the evolution of a "typical" comet assumed to have formed originally in the outer planetary region and to have evolved to a present state through planetary perturbations, stellar encounters, and the Galactic tide (modified and redrawn from Duncan et al., 1987, Fig. 3). Their simulation was started at a value of the semimajor axis, $a = 2000$ AU (open square), because repeated experiments showed that most objects beginning with an initial perihelion of $q = 20$ AU [i.e., in a region just outside the present orbit of Uranus, near the conventional inner limit of the Kuiper Disk; cf. Figs. 9 and 11, and Bailey (1990)] passed through this state during a simulated lifetime of 4.5 b.y. In this time frame there was one transient return to the vicinity $a = 3000$ AU, $q = 20$ AU (open circle) during chaotic oscillations within the shaded regions at values of semimajor axes of the order of $a \sim 10^4$ AU and perihelion distances in the range $20 \leq q \leq 10^4$ AU. Thus there were pendulum-like chaotic oscillations of both a and q between the limits 20 and 10^4 AU with one or more possible flips between the dominating ranges (potential wells) at either approximately constant perihelion q or semimajor axis a. [Fig. 13 (Inset 3) illustrates approximately one global cycle of the nonlinear pendulum-like orbital oscillations of the evolving a vs. q oscillations — as distinct from the orbital-precessional cycles of specific objects. That diagram should be compared with the nonlinear-periodic and chaotic phase-space patterns of the driven and damped nonlinear pendulum, as studied, for example, by Miles (1962, 1984), Gwinn and Westervelt (1986), and Tritton (1986, 1988, 1989). See Kaula (1968, pp. 207f and 226) and Wisdom (1987a, p. 111) for formal analogies between the equations of motion of the pendulum and of orbital mechanics (cf. Table 1, Secs. A and B, of the present volume; and Baker and Gollub, 1990). See also the comparative discussion of "spheres of influence," "fuzzy orbits," and "basins of attraction" in Note 6.1 (cf. Notes 3.1, 4.1, 5.1, 5.2).]

Fig. 15, (a)

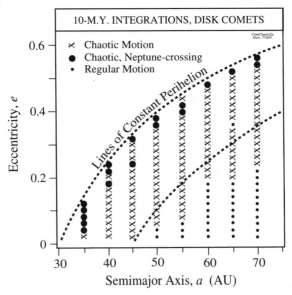

Figure 16

Fig. 15, (b)

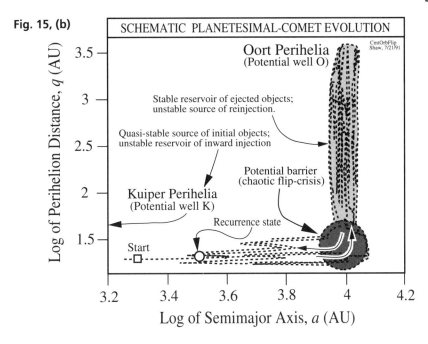

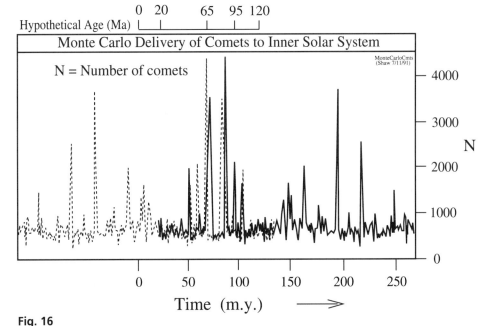

Fig. 16

Figure 16 (Chapter 4). An example of the results of Monte Carlo simulation of the flux of new comets from the Oort Cloud between 10,000 and 40,000 AU into the inner Solar System with perihelia $q < 10$ AU, including perturbations by both the Galactic tidal field and by passing stars (modified and redrawn from Heisler, 1990, Fig. 1). The ordinate is the

Figure 17

(Fig. 16, cont.)
number of new comets per m.y. entering the inner Solar System at values of $q < 10$ AU from a total population of $\sim 10^7$ comets at 10^4 AU. Normalized to an Oort Cloud containing a total of $\sim 10^{11}$ comets, the number of new comets per m.y. is nearly 10^4 times the ordinate, or up to nearly 40 *per year*! The abscissa represents the time of evolution of intermittent pulses of new comets shown by the time series. The dashed curve shows the time reversal of this series, as though plotted against age from an arbitrary zero age. Zero age was chosen so that there is an approximate match of the intermittency peaks, suggesting a very crude or noisy invertibility of the time series.

The numbers used for the populations of comets at the Oort Cloud distances are based on genetic hypotheses (see Hills, 1981; Weissman, 1990b). In principle, such a nonlinear evolution is neither integrable nor invertible. I have noted, however, simply by inspection, that time series resulting from nonlinear processes are sometimes crudely invertible in chaotic domains well beyond regimes of nonlinear resonances (cf. Shaw, 1991, Figs. 8.2, 8.16, 8.18, 8.38, 8.43).

Figure 17 (Chapter 5). Return-map and time-series data showing recurrence patterns of nonlinear oscillations of a forced magnetoelastic ribbon, schematically redrawn from Ditto et al. (1989, Figs. 2 and 3). Values of X in (a) were derived from readings of voltage in (b):

(a) Poincaré section based on values of voltage readings of a Fotonic Sensor (Ditto et al., 1989, Fig. 1). Data-state vectors, X, are related to voltage readings, V, of the time series at five different values of delay times relative to a progression of times measured by multiples of a fixed sampling time interval; the first and fourth values in each set of five values are plotted, representing the projection of the data to the plane X_1–X_4 in a five-dimensional phase space. The ribbon was clamped to favor asymmetric buckling relative to the vertical; the vertical was at ~ 4.1 volt. The open squares represent an unstable period-9 attractor within the phase space of the chaotic attractor (dots). Thus, in this experiment, oscillations occurred mainly within one potential well, with occasional transients (*chaotic bursts*) to the second well on the opposite side of the vertical position ($V > 4.1$ volt), depending on amplitude and critical frequency, f_c, of the time-varying magnetic field (the critical value f_c was inferred to be in the range $0.834 < f < 0.835$ Hz). The time between bursts, τ, increased as a power-law function of proximity to the critical frequency $|f_c - f|$. Similar transient bursting phenomena should exist in the vicinities of other unstable periodic attractor states, presumably even within one potential well (cf. Rollins and Hunt, 1984; Carroll et al., 1987).

(b) Voltage output of the Fotonic Sensor vs. time at two values of frequency, one above (top) and one below (bottom) the critical frequency. The shaded area represents the approximate attractor basin of the dominant potential well. Note that there is an ambiguity between a time series that is so near f_c that the time between bursts exceeds the duration of the experiment, and another ambiguity so far from f_c that the bursting time is too short to be distinguishable from steady chaotic oscillations. In order to establish τ vs. proximity to a given unstable periodic attractor, bursting times should be measured over a continuous range above and below the critical value (cf. Carroll et al., 1987).

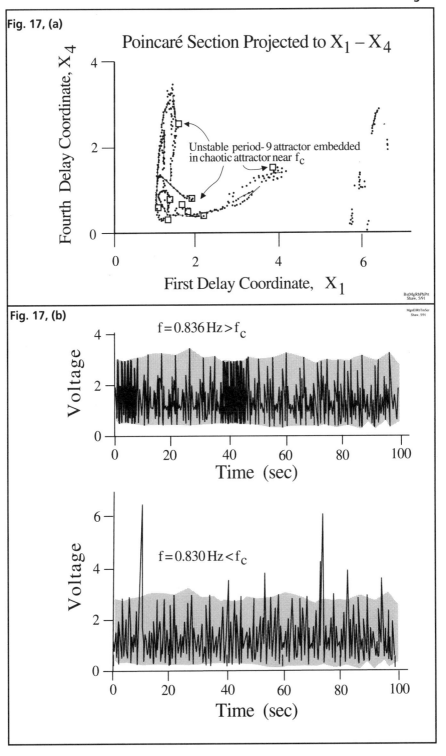

Figure 17

Figure 18

Figure 18 (Chapter 5). Phase-space plot (a) and time series (b) of radio-frequency (rf) magnetic excitation of ferromagnetic resonance and spin waves in yttrium iron garnet (YIG) spheres held in a DC field perpendicular to the excitation source (see also individual descriptions below; diagrams are schematically redrawn from the experimental results of Carroll et al., 1987, Figs. 1 and 3). The system was tuned to the vicinity of transition from a ferromagnetic resonance line, where the orientation of spins precess uniformly about the DC field, to a regime of magnetic spin-wave modes (first-order "Suhl instability"), where energy is transferred from the uniform mode into bistable oscillatory modes of alternating polarities; two or more magnetc spin waves interact as coupled oscillators, giving a "difference frequency" that modulates the spin-wave signal. These "auto-oscillations," with frequencies in the range 5 to 400 kHz, were recorded and plotted in a manner similar to that in Figure 17. Rather than varying the driving frequency, however, the microwave power was turned on at $t = 0$, and the durations of the resulting chaotic transients were plotted as a function of microwave power above the Suhl instability (Carroll et al., 1987, Fig. 2; the results indicate that the spectrum of durations of chaotic bursting reflects the existence of two periodic attractors in this experiment).

(a) Phase-space plot of trajectories of chaotic bursting relative to the vicinity of a quasiperiodic attractor for one set of conditions (shaded region at center); in this case a three-dimensional phase space was sufficient to reveal the general character of the chaotic vis-à-vis quasiperiodic trajectories (compare the discussions of fractal dimensions in Carroll et al., 1987, and Ditto et al., 1989; see also the text and Fig. 17a).

(b) Time series showing alternating output voltage vs. duration in the transient regime (unshaded region) and the decay to the steady-state quasiperiodic attractor (shaded region).

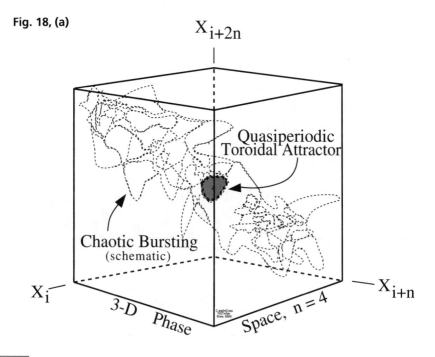

Fig. 18, (a)

Fig. 18, (b)

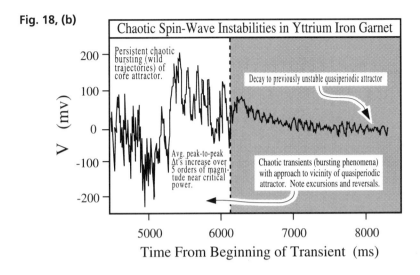

Figure 19 (Chapter 6). Schematic diagram illustrating the diversity of possible approach scenarios and dynamical effects (bottom and left center) influencing conditions for collision, capture, or flyby of a near-Earth object (NEO), according to the standard orbital mechanics of artificial satellites and space probes (see Bate et al., 1971, pp. 374ff; Taff, 1985; Roy, 1988; Roy and Clarke, 1988). Configurations were chosen arbitrarily to highlight possible high-inclination approach scenarios. Such trajectories represent transfer orbits that might culminate in an event resembling the Great Fireball Procession (GFBP) of February 9, 1913 (Chant 1913a, b; Mebane, 1955, 1956; LaPaz, 1956). The GFBP was visually tracked by many observers in Canada, in the eastern U.S., and in Bermuda along an approximate great-circle arc with coordinates given by Chant (1913a, Table I, Figs. 1 and 2). This trace is compared in Appendix 3a with the Phanerozoic and K/T cratering swaths and great-circle trajectories of Figure 23 (Chap. 6). The diagram on the right (above the Inset) illustrates capture from high-inclination heliocentric orbits (the shaded area labeled "Cone of High Inclination Approaches") into diverse forms of geocentric spin-orbit resonances of natural Earth satellites (NESs), according to the theory of Allan (1967a, b; 1971). These approach scenarios are analogous to the proposal by Alfvén and Arrhenius (1969, 1975, 1976) for resonance-level states progressively occupied during the evolution of the lunar orbit by retrograde noncataclysmic capture. [Notice that the convention for the geocentric orbital inclination is relative to the orbital trajectory of interest, and that the phase of the Earth's orientation is relative to the 26,000-year precession of the equinoxes. The convention for the "approach cone," as shown, was drawn to be consistent with the discussion of Alfvén and Arrhenius (1969), whereas that in the Inset (approximately its inverse) is the usual convention for artificial satellites (see Bate et al., 1971, pp. 141ff). Over geologic time, of course, if the "approach cone" is approximately fixed relative to the ecliptic plane of the Solar System, it precesses about the 90° position (which corresponds, in the present discussion, to the Earth's geographic North Pole).] Their proposal offered a rheologically

(*cont.*)

Figure 19

(*Fig. 19, cont.*)

plausible alternative to orbital-history calculations in which disruption of the Moon by a close approach inside the Roche limit was implied (cf. G. H. Darwin, 1907, 1962; Gerstenkorn, 1955, 1967, 1969; MacDonald, 1964; Williams, 1972, 1989; Brosche and Sündermann, 1978, 1982; Lambeck, 1980; Walker and Zahnle, 1986; Sridhar and Tremaine, 1992). The Alfvén-Arrhenius model of lunar orbital history is relevant either to a capture or to a collison model for the origin of the Moon (see Chap. 9). Collision models are currently the subject of active debate among workers in cosmology, planetology, geophysics, and geochemistry (cf. Wetherill, 1985, 1990; Boss, 1986; Boss and Peale, 1986; Benz et al., 1986, 1987, 1989; Ringwood, 1986; Stevenson, 1987; Cameron et al., 1988; Benz and Cameron, 1990; Cameron and Benz, 1991). Rheological effects described by Gruntfest (1963), Shaw et al. (1968), Shaw (1969), and Gruntfest and Shaw (1974), and intermittent convection-dissipation-melting mechanisms in the Earth and Moon (Shaw, 1970, 1973, 1980, 1985a, 1987a), Shaw et al. (1971), and Wones and Shaw (1975) favor nonlinear-resonance events during the evolution of the lunar orbit because of the complex nonlinear time-dependences of the tidal phase lags; the paleontological evidence of Pannella (1972) also suggests an episodic history for the lunar orbit. Analogous orbital-rotational effects are known to exist for some of the satellites of Jupiter (see the Prologue; and Cassen et al., 1982; Rothery, 1992).

Inset: Histogram showing the distribution of orbital inclinations in the geocentric reference frame of the many thousands of artificial-satellitic objects (artificial satellites and their debris) that were older than ten years as of 1990 (modified from Coffey et al., 1991, Fig. 1; see also Taff, 1985, Chap. 5, for the amount of "space garbage" that existed in low Earth orbit in the mid-1980's). The distribution reveals several attributes of Earth-orbiting satellites, artificial *or* natural: (1) nonlinear-resonance states ("frozen orbits") are relatively stable, hence useful for technological purposes *and* for purposes of identifying search strategies for natural objects [to my knowledge — other than the implications of the pioneering studies by the English space scientist R. R. Allan (1965, 1967a, b, 1969, 1971) — this property of frozen orbits has not previously been exploited in astronomical searches for natural Earth satellites (NESs); see also Note 6.5], (2) artificial objects "gravitate" to resonance states even if they were not initially placed precisely into a "frozen orbit," (3) there is a virtual infinity of potential nonlinear resonances (see the discussion of Fig. 20 in Chap. 6; see also Figs. 11 and 21, and Appendix 11), and (4) the range of observed inclinations documented in this histogram is nearly identical to the range of equivalent inclinations shown in the diagram immediately above it, *an illustration that I had drawn about a year before I had begun to inquire about artificial-satellite monitoring activities and had encountered the work by Shannon Coffey and his coworkers from which this histogram was taken* (cf. Coffey et al., 1986, 1990, 1991). [I contacted Coffey — whose name had been given to me by John B. Clark, Chief, Applications Division, Headquarters Air Force Space Command, Peterson Air Force Base, Colorado — by telephone at the Naval Research Laboratory, Washington, DC, early in 1992 with the idea that we might collaborate on applying nonlinear-dynamical resonance techniques to a new monitoring scheme for NESs, employing the existing data of artificial-satellite orbits as a "filter," so to speak, in a manner analogous to that used in another pioneering study, in this instance by the American scientist John Bagby (1969). Coffey confirmed his interest in such a collaboration in a letter dated 25 March, 1992, a letter in which he described the research cited above on "frozen-orbit" techniques and some ideas on applying it to problems of mutual interest. Completion of the present volume has postponed implementation of that mutual project, which I had outlined in a Gilbert Fellowship Program Proposal to the U.S. Geological Survey, submitted on 30 March 1992 and granted in August 1992.]

Figure 19

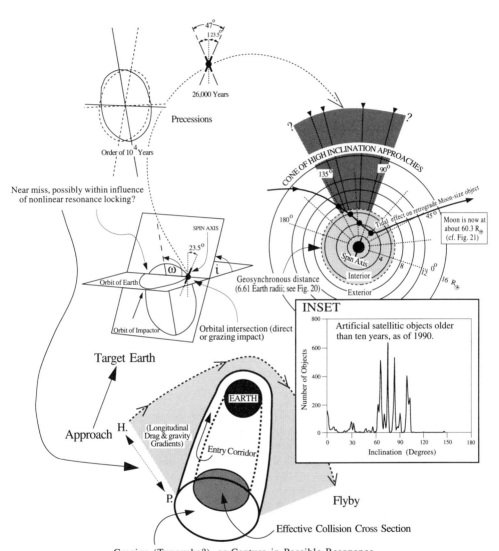

93

Figure 20

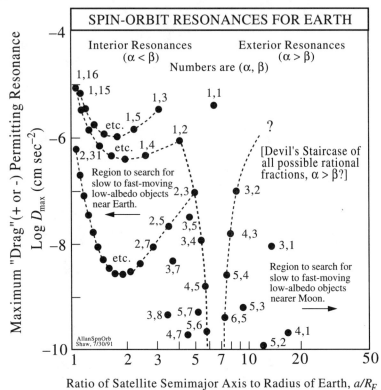

Figure 20 (Chapter 6). Illustration of resonance effects due to the longitude-dependence of the Earth's gravitational field (*longitude-dependent spin-orbit resonance*), modified and redrawn from R. R. Allan, 1967a, Table 1 and Fig. 3; see also Allan, 1967b). The ordinate gives the estimated maximum acceleration, positive or negative, that would drive the satellite out of its resonance potential well. Note the high strength of close-in interior resonances, such as $\alpha/\beta = 1/16$, where *interior* refers to resonances in which the orbital frequency, β, of the satellite exceeds the surface-rotation frequency, α, of the primary; analogously, *exterior* resonances are those for which $\alpha > \beta$. The condition $\alpha = \beta$, $\alpha/\beta = 1$ occurs at the synchronous distance equivalent to $a/R_E \cong 6.6108$, or $a \cong 6.6108\,R_E \cong 42{,}164$ km, giving an altitude of ~(42,164–6378) = 35,786 km, where R_E is Earth's mean radius, and a is the mean semimajor axis of an essentially circular satellite orbit (i.e., the synchronous distance is ~6.61 Earth radii from the center of rotation and revolution and 5.61 Earth radii altitude "above" the mean distance of the Earth's surface from that center). In this reference frame, the Earth's surface is at $a/R_E = 1$, but the calculations obviously do not apply near this limit because of perturbations of the other five orbital elements in addition to a (see Allan, 1967b). The inclination of the satellite orbit in the same frame for a given resonance is a function of α and β; see the magnitude and sign of the "inclination function," F_{Imp} (*I*) of Allan (1967a, Figs. 1 and 2). Allan (1967a, Table I) lists values of I_{opt} that represent values of inclination at the maximum of F_{Imp}; these inclinations are in the range 70° to 90° relative to the equatorial plane of rotation and revolution for the greatest disparities between α and β shown in the figure (both interior and exterior resonances).

Figure 21

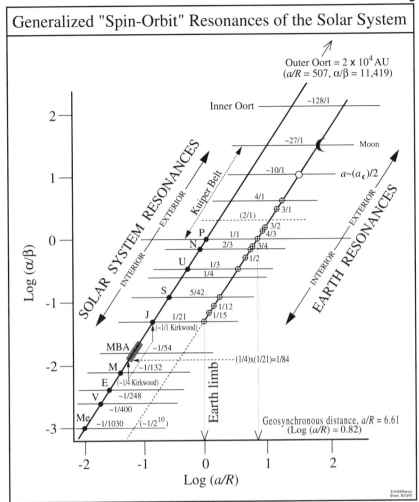

Fig. 21

Figure 21 (Chapter 6). Schematic illustration of the idea of longitude-dependent spin-orbit resonances applied to orbital resonances in the Solar System (SS). The solid line with slope of 3/2 on the right represents the relationships of Figure 20 extrapolated to the position of the Moon. Resonances α/β are shown by the light horizontal lines, where the integers correspond to α and β relative to a given definition of the "primary" radius (i.e., the Earth radius for the heavy reference line on the right; the radial distance from the center of the SS to a chosen "primary" for the heavy reference line on the left). The datum "primary" for the lefthand resonance line relative to the center of the SS is arbitrary; it is chosen to be the Pluto distance in the case shown (see the text for transformation to the Jupiter distance). The planetary resonances are based on Keplerian orbital distances, therefore $a = R$ at the synchronous resonance $\alpha = \beta$ for any planetary orbit chosen as the reference datum; this is not true if the SS datum is referred to the rotation of the Sun, or to the revolution of the Sun around the SS Center of Mass (CM); see Appendix 7a (Inset). This distinction occurs

(*cont.*)

Figure 21

(Fig. 21, cont.)

because the presumption of a Keplerian relation between semimajor axes and orbital periods (the 3/2 slope of the heavy reference lines in the figure) does not hold for condensed-body rotation, nor does it hold for the orbit of the Sun about the CM (cf. Jose, 1965; Landscheidt, 1983, 1988).

The latter relations in particular reveal the essentially dissipative character of the organization of orbital motions in the SS (and in planetary nebulae in general) when there is strong coupling between "internal" and "external" trajectories of the motion (represented by the internal and external motions of the Sun in this case; see the Inset, Appendix 7a). Since this coupling is necessarily reflected in the orbital evolution of the seemingly "decoupled" bodies (i.e., the planets relative to the CM), the SS as a whole can be formally viewed as a dissipative system, and the evolution of the orbits of the planets in that context must represent attractors of the system, whether the basins of attraction are periodic or chaotic (cf. Figs. 12–14 and Appendix 9). Accepting the evidence of Laskar (1988, 1989, 1990), Laskar et al. (1992) and Sussman and Wisdom (1992) that the SS is formally in the chaotic regime, it would follow that planetary orbits are subject to multiple possibilities of unstable periodic orbits relative to given attractor states, hence are subject to diverse forms of chaotic crises (see Figs. 12, 17, 18, and 31, and discussions in the text).

The same reasoning follows for the asteroids, the satellites of the planets, and the cometary reservoirs (see the text for discussion of correspondences between asteroid resonances and planetary resonances; e.g., as is shown relative to the position of the main belt of asteroids in the figure by the dashed horizontal line for resonance $\alpha/\beta = 1/84$). The theoretical reasoning for this figure is based directly on the implications of nonlinear dynamics for Allan-type spin-orbit resonances (Fig. 20); it would also appear to be a logical extension, taking dissipation into account, of the work of Roy and Ovenden (1954, 1955) on orbital resonance states (*The Mirror Theorem*) and the work of Ovenden (1972, 1975, 1976) and Ovenden et al. (1974) on *The Principle of Least Interaction Action*. If these relationships can be modified to take account of the orbital evolution of the planets relative to the evolution of dissipative coupling in the Sun-CM-planet system, they may provide a basis for development of fully coupled models of dissipative self-organization in celestial mechanics (cf. Roy, 1988a).

Figure 22

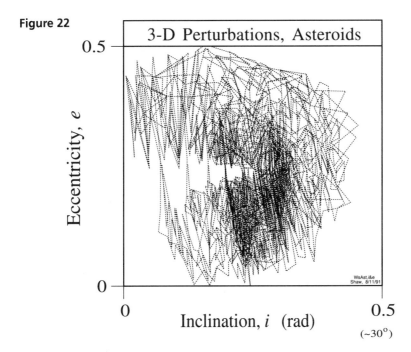

Figure 22 (Chapter 6). Schematic illustration of eccentricity-inclination oscillations (e vs. i), modified and redrawn from Wisdom (1987a). The dashed trajectories represent chaotic variations in e at constant i, in i at constant e, or both simultaneously. If Δe is large enough, the asteroid "kicks out" of its resonance potential well and may become Earth- and/or Mars-crossing. This could happen within a wide range of small to large inclinations.

Figure 23

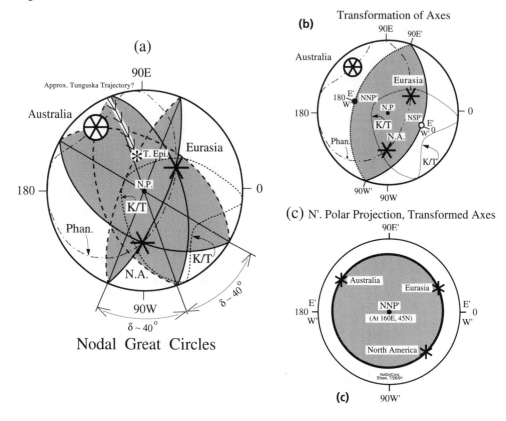

Figure 23 (Chapter 6). (a) North polar view of three great-circle sections, each section passing through one of the three pairs of cratering nodes; the Phanerozoic and K/T cratering swaths are shown, respectively, as the curves with light dashes and dots and light fine dashes. The range of azimuths of the intersection with the equatorial plane has a range of nearly 90°, or roughly twice the range of the hypothetical "capture cone" of incoming trajectories in Figure 19 (righthand diagram). This is the approximately correct relationship for the variation of the capture cone in the ecliptic reference frame relative to the precession of the equinoxes (i.e., the average occupancy of potential spin-orbit resonances repeatedly sweeps through a geocentric range of inclinations roughly twice the initial ecliptic inclinations of the capture cone during the ~26,000-year precessional cycle of the equatorial plane relative to the ecliptic plane).

The transformed projection (b) and small circle containing the three nodes (c) are from Figure 5, shown here for reference. The light path with arrows in (a), a crude inference for the orientation of the trajectory of the Tunguska object of 30 June, 1908, is based on narratives in Krinov (1963b) and in Clube and Napier (1982b, p. 140). The latter description states that the object "entered the atmosphere over western China." This constrains the path approximately as shown in (a) for the documented location of the epicenter, indicated by T. Epi. (e.g., Krinov, 1963b; Rocchia et al., 1990). Krinov's (1963b, Figs. 1 and 2) maps of the proximal trajectory before final disintegration appear to indicate a more easterly direction, but the basis for that trajectory is not explicitly stated. If both apparent trajectories were accurate, the orbit would have had to rotate significantly during the object's passage through the atmosphere; this can occur during re-entry of artificial satellites (e.g., Arho, 1971; see also the discussion in Rocchia et al., 1990), but another explanation seems more likely for the Tunguska object. [After I had completed the illustrations for the present volume, Chyba et al. (1993) published an analysis of the proximal stages of the Tunguska "explosion," interpreting the angle of descent as much steeper than previously supposed (cf. Melosh, 1993a). This posed a paradox relative to the above description of the Tunguska approach trajectory, one that I have attempted to resolve by slightly modifying the scenario postulated by Chyba et al. (1993), as discussed in Chapter Note 1.2.]

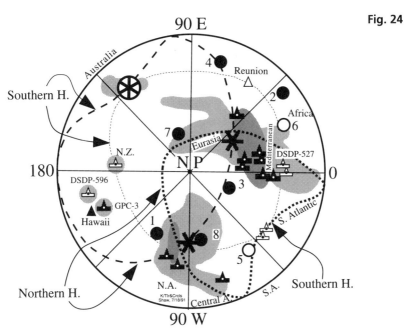

Fig. 24

Distribution of K/T Iridium Anomalies

⛰ - Iridium Anomaly with Associated Shocked Quartz in Vicinity, Northern Hemisphere.

△ - Iridium Anomaly with Associated Shocked Quartz in Vicinity, Southern Hemisphere.

● - Locations of Flood-Basalt Provinces, N. Hemisphere (identified by number, as in text).

○ - Locations of Flood-Basalt Provinces, S. Hemisphere (identified by number, as in text).

[Tektite strewnfields of K/T age and younger correlate with CRF nodes and/or circles; see Glass (1982), Smit et al. (1991), Izett et al. (1991)].

Figure 24 (Chapter 7). North polar projection, showing the distributions of known K/T iridium anomalies against a backdrop of the Phanerozoic cratered regions of Figure 2 (light shading). Localities of iridium anomalies are based mainly on the data in Alvarez et al. (1982, Fig. 1 and Table 1). Data for DSDP Sites 527 and 596, respectively, are from Huffman et al. (1990) and Zhou et al. (1991). Numbered solid and open circles are flood-basalt provinces as identified in the caption of Figure 4. Regions in which shocked quartz of K/T age is generally prevalent are indicated by darker shading. In general, closed symbols refer to the Northern Hemisphere and open symbols to the Southern Hemisphere.

99

Figure 25

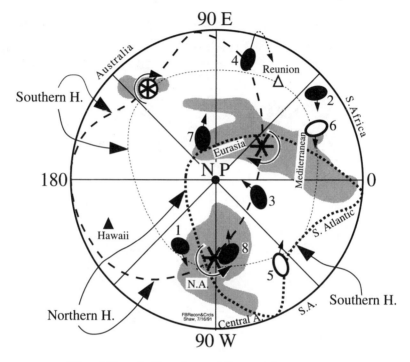

Flood-Basalt Provinces Viewed from North Pole

(Arrows indicate rotation sense for backtracking relative to the nodal points as invariant; at large distances from nodes, motion depends on local rotations of substructures of continental blocks.)

Figure 25 (Chapter 7). North polar projection of Figure 24, in which backtracking directions (not vectors) are indicated relative to their present-day positions. In general, these directions would bring the original geographic locations of these volcanic provinces into better alignment with the patterns of impact craters shown by the shaded regions, the Phanerozoic swath (long heavy dashes) and the K/T swath (short heavy dashes). For the identification of numbered basalt provinces, see the discussion in the text and the caption of Figure 4.

Figure 26

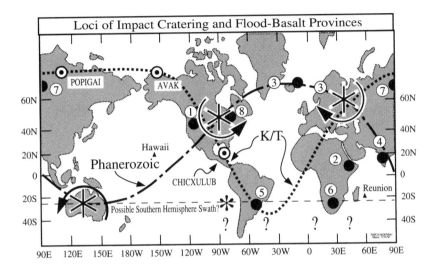

Figure 26 (Chapter 7). Mercator projection showing the transformed patterns of cratering nodes and swaths and locations of flood-basalt provinces (filled circles numbered according to Fig. 4). Cratering nodes shown by the large heavy asterisks are surrounded by a segment of a circle with an arrow showing the backtracking sense of rotation implied by standard paleogeographic reconstructions and assumed global geographic invariance of the cratering nodes (this applies only to the vicinity of each node and to those portions of the adjacent continental structures that have maintained rigid-body rotational relationships with them; i.e., the sense of rotation does not apply to exotic blocks or to accretionary terranes in general). The small asterisk just west of central South America identifies the location of a cluster of virtual geomagnetic poles (VGPs) as determined by J. Glen at the University of California, Santa Cruz (comm. of unpubl. data, 26 December, 1991); compare this location with Figure 28 and the discussion in text. The heavy dotted horizontal line near 25° S identifies the lower limit of high southern latitudes within which undiscovered cratering patterns could exist.

Figure 27

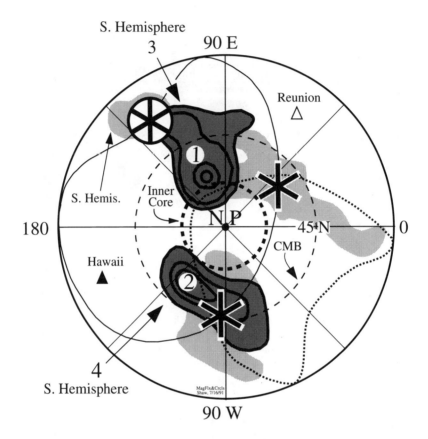

Figure 27 (Chapter 7). North polar projection comparing (1) the global pattern of cratering nodes (heavy asterisks), (2) the dominant regions of cratering (light shading), and (3) the Phanerozoic and K/T cratering swaths — shown respectively by the light solid curve and medium-weight short dashes — with the patterns of the radial component of geomagnetic flux into the core, as mapped on the surface of the core by Gubbins and Bloxham (1987, Fig. 2); the positions of flux lobes 1 through 4 on their map are shown schematically by the dark shading, heavy contours, circled numbers, and numbers with arrows (see the Mercator projection in Fig. 28 for the locations of flux lobes 3 and 4 in the Southern Hemisphere). Note the general coincidence of, on the one hand, the distribution of flux lobes and cratering patterns, and, on the other hand, the position of the cratering nodes in the Northern Hemisphere within the interval between the projected inner-core/outer-core boundary (the heavy dashed circle centered on the North Pole) and the projected core-mantle boundary (the lighter-weight dashed circle labeled CMB that crudely circumscribes the flux lobes and cratering nodes in the Northern Hemisphere).

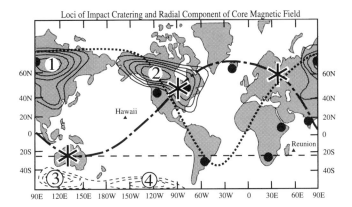
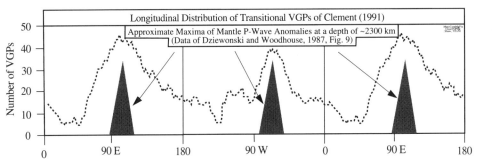

Figure 28 (Chapter 7). Comparison of geomagnetic patterns (numbers and solid and dashed contours at top; histogram at bottom) and mantle P-wave anomalies (large filled triangles at bottom) with the Mercator projection of global cratering nodes and swaths of Figure 26 (top). The geomagnetic data (top) are those of Figure 27 (data from Gubbins and Bloxham, 1987); the histogram of virtual geomagnetic poles (VGPs) (bottom) is based on the data of Clement (1991). The large filled triangles (bottom) show the crude meridional configuration of deep-mantle P-wave anomalies (data of Dziewonski and Woodhouse, 1987) that align with the VGP maxima and are suggested here to represent the present-day signature of early-formed mass anomalies in the Earth (see Fig. 33, as discussed in Chap. 8; cf. Laj et al., 1991, 1992).

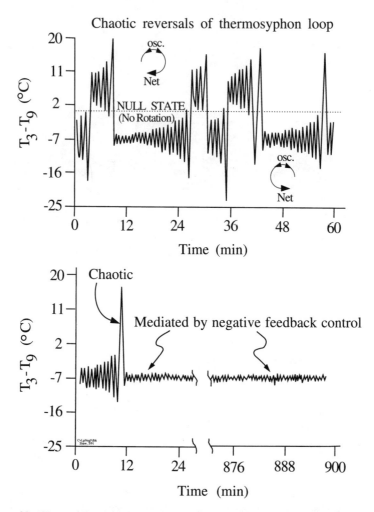

Figure 29 (Chapter 7). Illustration of chaotic oscillations and reversals of flow in a vertically oriented convecting thermosyphon loop (top) heated at the bottom and cooled at the top; from Shaw (1991, Fig. 8.23) as modified from Singer et al. (1991). The bottom diagram illustrates the suppression of chaotic motion by negative feedback. The ordinate represents the difference between temperatures at the 3 o'clock and 9 o'clock positions in the vertically oriented cycle (12 o'clock at top); changes of sign indicate reversals of net flow directions. It should be noted that a suppressed signal of this type could also be a form of so-called stable chaos (cf. Milani and Nobili, 1992; Murray, 1992) or represent chaotic intermittency with long intermissions between chaotic bursts (see the discussion in Chap. 7; cf. Robbins, 1977; Gorman et al., 1984; Widmann et al., 1989; Shaw, 1991).

Figure 30 (Chapter 7). Relationship between mass-impact frequencies and geomagnetic-reversal frequencies (data sources identified in figure):
 (a) Ratios of geomagnetic-reversal frequencies to cratering frequencies (per m.y.). Note the approximate 30-m.y. peak-to-peak "period" (relative winding number), which is characteristic of the coupled system (see the discussion in Chap. 7).

Figure 30

Fig. 30, (a)

Fig. 30, (b)

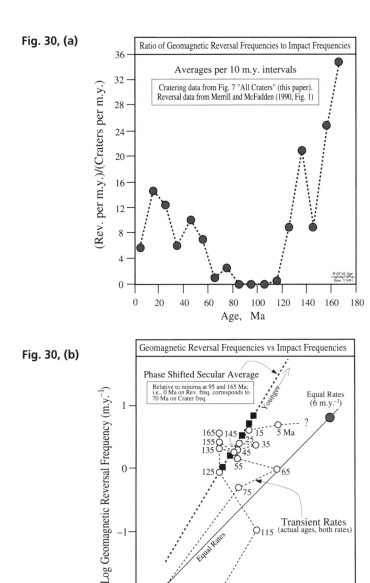

(b) Intricacies of the individual frequency variations with age (numbered circles). The fine dotted line tracks transient variations in the respective frequencies; the solid line at 45° is the trend of equal rates, and the large filled circle on that line gives a value of 6 per m.y. as a possible extrapolation since 5 Ma. See Chapter 7 for a discussion of the difference between transient and steady equal-rate curves and the "phase-shifted" curve (heavy dashed line). Note that the irregularities in (b) are ordered in (a) by "self-normalization," or rhythmic mutual compensation (delayed feedback coupling; cf. Freeman, 1992, pp. 467ff).

Figure 31

Figure 31 (Chapter 7). Logarithm of mean times (τ) between impacts and between geomagnetic reversals vs. logarithm of proximity to a characteristic frequency, where $1/\tau \cong 0$. This proximity is assumed to be given by the absolute value of a frequency difference that can be approximated from the age difference between a point on the curve of mean impact and reversal frequencies of Figure 6 and the respective "Secular Minima" of that figure, as given by $|f_{Crit.} - f| \cong |(t_{Crit.} - t)/(t \times t_{Crit.})|$. The age values for the minima are 95 Ma and 165 Ma for geomagnetic reversals and impacts, respectively. Values of τ represent the reciprocals of events per m.y. read from the curves of mean values, where the approximate ranges of values are shown by vertical bars through the symbols (open circles for geomagnetic reversals; filled circles for impacts). Subjective power-law trends through the data are indicated by solid lines and the indicated slopes (power-law exponents). The dotted line represents the results of Ditto et al. (1989) for times between transient phase-space trajectories describing large-amplitude spikes ("bursts") in the vibrational motion of a magnetically driven magnetoelastic ribbon (see Chap. 5, Fig. 17). In that case, a burst represents a transient escape from one to another potential well of the bistable system, where values of τ are measured in seconds and vibration-frequency differences in Hz. The critical frequency is one where the vibration may be stable within one well for some set of chaotic conditions that is near the phase-space points of a periodic attractor that is unstable under the conditions of those chaotic motions. The physical interpretation for geomagnetic phenomena is that a polarity reversal of either sign represents a chaotic flip from one generalized bipolar potential well to the other (or between higher-order potential wells of nondipolar conditions). A literal analogy with the magnetoelastic ribbon would represent a 180° excursion that returned to the first well during the burst. However, it is evident that bursts can occur relative to more than one kind of unstable periodic attractor state in the same physical system (cf. Fig. 18, and Carroll et al., 1987), and in that case the notion of a simple return flip (reversal) is not mandatory.

The analogy for impact events is less direct but is interpreted, for example, as applying to either of two possible dynamical regimes: (1) transient ejection from an asteroid potential well (cf. Fig. 12, and Wisdom, 1987a, b, c) with enhanced probability of capture and/or collision with Earth, and (2) ejection from one of the spin-orbit resonances of Figure 20 in such a way that an impact occurs directly from it or from an "avalanche" of perturbations that activates an impact event from any of the other resonance states.

Inset: The same type of calculation and reasoning is applied to the two-disk dynamo data of Ito (1980) and Hoshi and Kono (1988), as shown in Figure 32.

Figure 31

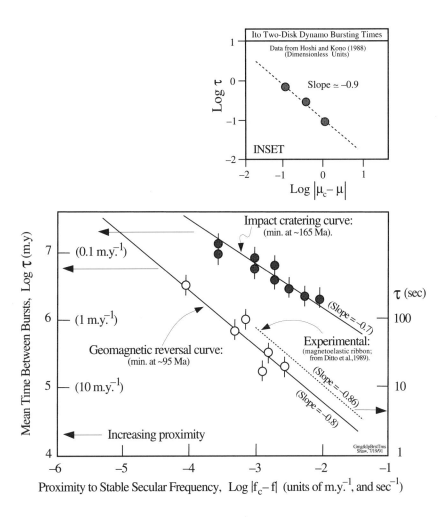

Figure 32

Figure 32 (Chapter 7). Results of an experimental study of the Rikitake two-disk dynamo by Hoshi and Kono (1988), interpreted according to the concept of chaotic crises (illustrations are redrawn and modified from data in Hoshi and Kono, 1988):

(a) Phase-space diagram showing the nonlinear periodic and chaotic regimes, and the *Minimum Entropy Regime* (MER) of Ito (1980). The MER is interpreted as the vicinity of one or more unstable periodic attractors within the chaotic regime. Distance along the heavy dashed line at a constant value of coupling parameter k is used as the measure of proximity to a critical frequency, as plotted in the Inset of Figure 31 (see discussion in text).

(b) Diagram showing how the reversal frequency approaches zero at critical values of the ratio, μ, between the mechanical and electromagnetic time scales that represent the analogues of the critical frequencies for chaotic crises illustrated in Figures 17 and 18.

(c) Phase portrait in the vicinity of the MER of Ito (1980), showing how chaotic orbits circulate around the vicinities of two unstable fixed points, N (normal) and R (reversed). The parameters X_1 and X_2 are the electrical currents of the two disks, which pass through various (+) and (−) states rather than simply oscillating between the (+,+) and (−,−) fixed points (modified schematically from Hoshi and Kono, 1988, Fig. 13). The family of phase-space trajectories constitutes a fractal set called a *strange attractor*, which resembles the chaotic regime of the *Lorenz attractor* (Lorenz, 1963, 1964, 1979; cf. Shaw, 1991, pp. 125ff). Comparison with Figures 17 and 18 indicates how a chaotic trajectory approaching the vicinity of N or R (or MER in general) can be kicked out to large excursions representing a chaotic burst, with or without eventual decay to an actual stable fixed point (see the discussion of Fig. 18 in Chap. 5).

(d) Samples of typical reversal histories (modified from Hoshi and Kono, 1988, Fig. 12); the large spike on the right is a typical transient burst, but the system does not necessarily return to the previous polarity. The study of convection in a thermosyphon loop, as shown in Figure 29, resembles this behavior and is even more instructive with regard to possible reversal sequences related to cyclic vortical and/or helical flows in the core dynamo. Thus it is important to note that in actual fluid flows, chaotic behavior and crises are not restricted to the turbulent regime as determined by the usual stability criteria. Robbins (1977) makes this point in reference to the geodynamo, and Widmann et al. (1989, Figs. 3–7) demonstrate examples of chaotic histories in both the laminar and turbulent regimes (see also Chillingworth and Holmes, 1980). And such observations, among others, imply that chaos and turbulence — despite discussions in the literature of nonlinear dynamics that speak loosely of "routes to chaos" and "routes to turbulence" as if they were identical phenomena — are not necessarily synonymous (e.g., Shaw, 1991, Figs. 8.1–8.9; cf. Gollub and Swinney, 1975; Gollub and Benson, 1980; Moon and Li, 1985a, b; Ditto et al., 1990a; Moon, 1992).

Figure 32

Fig. 32, (a)

Fig. 32, (b)

Fig. 32, (c)

Fig. 32, (d)

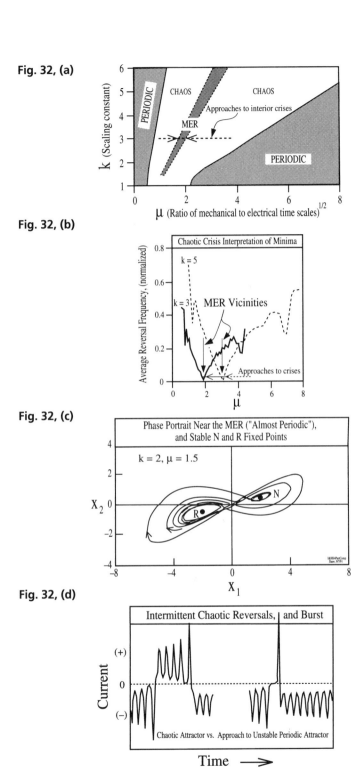

Figure 33

Figure 33 (Chapter 8). Schematic evolution of mass anomalies and major geodetic "scars" in and on the Earth due to the primordial history of impacts of 100-km-scale and larger objects with the Earth following the origin of the Moon. The sequence, simplified to two principal stages, is analogous to and follows the arguments advanced by Schultz (1985) and Runcorn (1987) for rotations of the figure of a body (Mars and Moon, respectively) relative to the spin axis due to large changes in moments of inertia induced by major mass displacements caused by cratering in equatorial localities. [Note that, like a gyroscope, the spin axis of the Earth remains approximately invariant relative to the inertial frame of the Earth-Moon-planet system while the internal mass distribution undergoes a reconfiguration relative to it (e.g., material originally located within the equatorial bulge migrates toward one of the poles, depending on the distribution of the mass deficiency that initiates an "overturn," while other material that resided at higher latitudes migrates toward the Equator to compensate the revised configuration of the bulge).] Mass is removed from equatorial regions and deposited at higher latitudes in two stages. Most of this ejecta material subsequently became caught up in plate-tectonic mixing effects. [Later bombardment episodes of smaller scales in the early Precambrian — but still immense compared to the Phanerozoic cratering record — should have created massive Precambrian ejecta deposits, deposits that have never been definitely identified in the geologic record, and the same observation applies to the ejecta deposits of the largest Phanerozoic craters. Recently, some workers have proposed that some portion of those terrestrial deposits that were formerly mapped as glacial tillites, diamictites, etc., may, in fact, represent the "missing" impact-ejecta deposits in the geologic record; cf. Oberbeck et al. (1992), Marshall and Oberbeck (1992), Aggarwal and Oberbeck (1992), Rampino (1992).]

After each stage of mass displacement, some radial interval of the Earth, presumed to extend from the surface to great depth in what is now the mantle, is envisioned to have reoriented itself so that the axis of the greatest moment of inertia becomes realigned with the spin axis. In the schematic idealization, the equatorial plane that had previously contained the principal axis of smallest moment of inertia suddenly became the plane that contained the axis of greatest moment of inertia, owing to the mass removed from the equatorial vicinity by impact excavation. Since the greatest effect is associated with the most intense cratering, some form of longitudinal mass anomaly also is associated with the poleward reorientation.

Stage One of this process is envisioned to have begun during the Pre-Nectarian time of lunar chronology (prior to ~4 Ga), when Procellarum and other immense lunar basins were formed (see Wilhelms, 1987, Fig. 14.3) — i.e., if the Moon was the product of the collision of a Mars-sized planetesimal with the proto-Earth, then both the newly formed Moon and the Earth body that survived from the mangled state of the proto-Earth must have been subjected to massive infall of the residual fractions of orbiting debris during more or less the same period of time (e.g., Boss, 1986, Fig. 4; Kerr, 1989; Kaufmann, 1991, Fig. 9.22; cf. Stewart, 1988). The analogously immense "basin" on the early Earth is suggested to represent the earliest expression of the present Pacific Basin at a time when core formation had already been largely completed (but see the discussion of chronological uncertainties in Chap. 8; cf. Ringwood, 1986, 1990; Ringwood et al., 1990). The repetition of this general process in Stage Two involved shallower depths, hence the previous stratification due to igneous layering processes favored a radial stratification of mass, resulting in a large mass anomaly in the deep mantle [i.e., the less-dense upper part of the Stage-One igneous fractionation process in the "magma ocean" that filled the Pacific Basin (Oceanus Pacificus) was rotationally displaced during Stage Two so that the pre-Stage-One mantle was superimposed on the dense ultramafic residuum of Stage-One igneous fractionation, result-

Figure 33

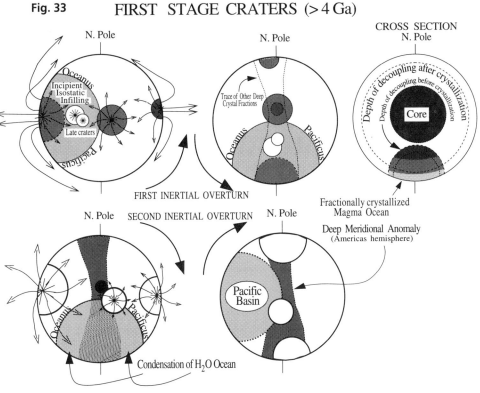

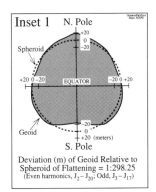

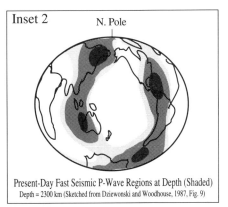

ing in a radial net mass excess in that region, but this was later modified by the modern type of plate-tectonic processes (see the discussion in Chap. 8)].

Because of the equatorial biases of both stages, the reoriented Earth retained a meridional mass anomaly that I suggest is correlated with present-day fast P-wave anomalies in the deep mantle (e.g., Dziewonski and Woodhouse, 1987) and with the familiar

(*cont.*)

Figure 34

(*Fig. 33, cont.*)

"pear-shaped" figure of the Earth derived from satellite geodesy (King-Hele, 1976), as shown in **Inset 1**, and with the present-day fast P-wave anomalies in the deep mantle (e.g., Dziewonski and Woodhouse, 1987; Laj et al., 1991), as shown in **Inset 2**. The resulting meridional mass anomaly, as subsequently modified at shallower depths by processes of plate tectonics, is envisioned to be the source of the longitudinal gravity anomalies responsible for spin-orbit resonances of present-day satellites, natural and artificial. Therefore, this process has mediated the eventual correlations of impact-cratering patterns, patterns associated with core motions (e.g., Figs. 27 and 28), and patterns of plate motions and continental drift relative to both the cratering patterns and patterns of core processes. In essence, the cratering nodes shown in several different types of map projections throughout the present work (the CRF nodes) are envisioned to correlate with a similarly invariant reference frame for the evolution of mass distributions in the Earth. This is necessary because the mechanism of spin-orbit resonances that represents the essential control on the disposition of the resulting natural Earth satellites (the CRF NES mechanism represented by Figs. 19 and 20), that in turn is proposed as the control of the cratering patterns, is coupled with anomalies of the Earth's gravity field, hence with the evolution of mass distributions schematically described in this set of diagrams. The generalized centering "anomaly" for both the satellitic and internal geodynamic processes is referred to in the text as the *meridional geodetic keel* (MGK) of the Earth.

Figure 34 (Chapter 8). Comparison of the paleogeographic reconstruction by Dietz and Holden (1970) with the nodal invariance suggested by the cratering pattern (cf. Note 8.10; and Norton and Sclater, 1979; Bambach et al., 1980; Ziegler et al., 1982; Duncan and Hargraves, 1984; Royer and Chang, 1991; Smith and Livermore, 1991). The senses of rotation shown relative to each node refer specifically only to those immediate vicinities and to those portions of the Earth's surface rocks that are rotationally coordinated with the nodes (see text of Chap. 8 for discussion; cf. Appendix 26).

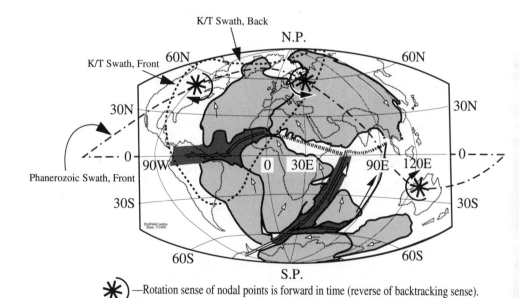

—Rotation sense of nodal points is forward in time (reverse of backtracking sense). Direction arrows for translation are according to model of Dietz and Holden (1970).

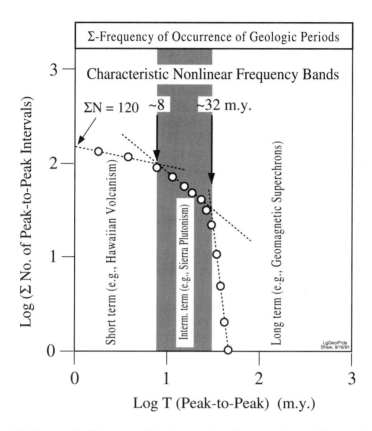

Figure 35 (Chapter 8). Histogram of the frequencies of geologic "periods" determined by counting peak-to-peak intervals in diverse records (geophysical, geochemical, and paleontological) and comparing them with synthetic histograms made up of known geometric progressions, as derived in Shaw (1988a, Table 1 and Fig. 25; see the discussion in Chap. 8).

Figure 36

Figure 36 (Chapter 8). Evolution of numerical biologic diversity during portions of the Phanerozoic Eon, as demonstrated by genera of clams and brachiopods, and by bird families, the three groups chosen to test the power-law character of cumulative trends and the correlation of inflections and discontinuities with geologic boundary events.

(a) Clam and brachiopod diversity data from Gould and Calloway (1980). The "fictive" diversity is simply the sum of genera by geologic Stage; hence like a maximum thermometer it can show only increases. The purpose of this device is to test long-term trends not revealed by the incremental standing diversity that existed within each Stage (i.e., the actual processes of origination and extinction affecting the diversity are not revealed by either the incremental or the cumulative numbers; see McLaren, 1986). The cumulative curves, however, indicate "sudden" events by departures from straight lines of steady-state power-law growth (light dashed lines) compared with the data for each lineage (heavy solid line, brachiopods; heavy dotted line, clams; open circles, their sums). Arrows show that the inflections and breaks in these curves correlate closely with geologic boundaries known from other criteria to represent important extinction events (see the discussion in Chap. 8).

(b) Standing diversity of avian families based on data from Fisher (1967). Although the fossil record of birds is poor, the data reveal characteristic power-law trends that have two grossly different rate regimes, occurring prior to and after the K/T boundary, as shown by the abrupt change in slope for all families (the large open circle marking the discontinuity in the trend of filled circles) and in the abrupt onset and rapid rise of the perching birds (open triangles) and the relative decline in growth rates of the pelicans, cormorants, etc. (open squares). The ages in parentheses correspond to the time scale of Harland et al. (1990).

Figure 36

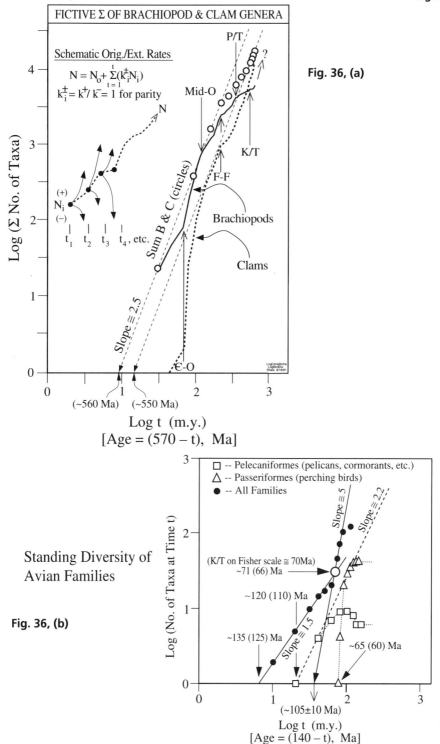

Fig. 36, (a)

Standing Diversity of Avian Families

Fig. 36, (b)

Figure 37

Figure 37 (Epilogue). The cratering history of the Moon during the Copernican Period (Wilhelms, 1987, Fig. 14.3; age ≤ ~1100 Ma) compared with the cratering nodes, nodal swaths, and nodal great circles of Earth's Phanerozoic cratering record.

The upper lefthand diagram is an equal-area projection of the Moon viewed from the side facing the Earth (see Notes 6.2 and 6.4; cf. Kaula, 1968, pp. 203ff, for an analysis of the spin-orbit coupling required to keep one face of the Moon always pointing toward Earth). Only the craters with diameters ≥ ~10 km (Wilhelms, 1987, Plates 11A, B and Table 13.1) are shown (filled circles, nearside; open circles, farside); the numbers by the circles are the diameters (in km) given in Wilhelms (1987, Table 13.1). The locations of Copernicus and Tycho (the two largest lunar craters of this Period) are labeled for reference; Hayn and Bel'kovich K also are indicated because they are on opposite sides and fall near the arc of a great circle passing from the nearside to the farside; four craters, including Hayn, are aligned along a great circle on the nearside (heavy solid lines) and at least two others, including Bel'kovich K, are aligned with the same great circle on the farside (dotted lines). Arcs of great circles are shown wherever three or more craters were found to align conspicuously on the stereonet (the arcs shown are examples and do not represent a complete search for all sets of aligned craters). A nearside-farside bias is shown by the numbers and distributions of craters (39 nearside craters, 19 farside craters); the nearside craters tend to be clustered within an area representing about 90° of latitude and longitude, skewed somewhat to the Northern and Western hemispheres.

Below the plot of lunar craters is the north polar projection of Earth from Figure 23a, showing the three cratering nodes (large asterisks), the Nodal Great Circles, and the Phanerozoic and K/T nodal swaths. The approximate trajectory of Tunguska is shown for orientation, and for comparison with its position in the equatorial projection of the **Inset** (upper right). The Inset shows the Earth as viewed from the Atlantic side, centered on the Greenwich Meridian; the Phanerozoic swath is tracked by circle symbols, and the K/T swath is tracked by square symbols (filled symbols indicate the Atlantic side; open symbols, the Pacific side). The projections of the Nodal Great Circles in the Inset are shown by heavy meridional lines (solid on Atlantic side; dashed on Pacific side); note that in this projection, two of the cratering nodes are in the far hemisphere, as shown by the asterisks inscribed within open circles. The shaded areas between two of the Nodal Great Circles roughly coincide with the 90°W–90°E meridional anomaly of Figures 28 and 33.

The orientation of Earth's spin axis in the Inset is approximately as it would look relative to the Moon at a compatible epoch indicated by the angular relationships shown at right center. The tilt of Earth's spin axis relative to the lunar orbit varies between roughly 18.3° and 28.7°, depending on the rotation of the Moon's orbital plane relative to the ecliptic ("rotation of the line of nodes"), with a period of about 18.6 years (Bate et al., 1971, p. 326); the Moon's orbit around the Earth is inclined about 5.2° to the ecliptic, compared to 23.5° for Earth's Equator relative to the ecliptic (see the Epilogue for discussion of relative cratering frequencies and what might happen if objects in lunar orbit were released toward Earth).

Figure 37

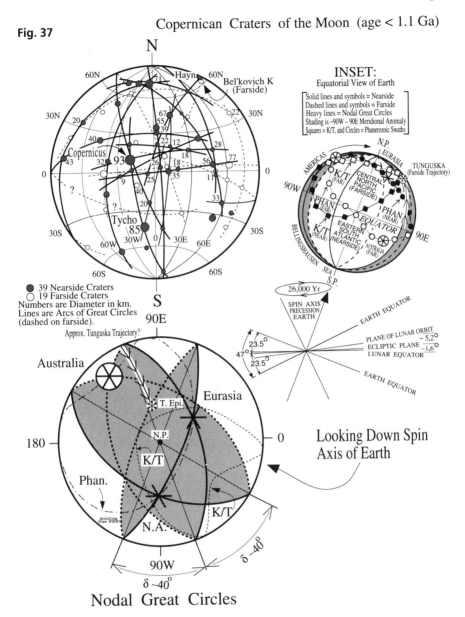

CHAPTER 1

Geographic Patterns of Impacts on Earth

The patterns of impacts on Earth outlined in the Introduction are shown in the equal-area map projections of **Figures 1 through 4**, which are largely self-explanatory. The catalog of terrestrial impact craters used previously in Shaw (1988a) was that of Grieve (1982). It was revised and updated by R. A. F. Grieve and P. B. Robertson, and I refer to the revised list as the G-R Catalog (written comm. from Grieve, 1986; Grieve and Robertson, 1987). The catalog that appears in Grieve (1987) is identical with the G-R Catalog, except for a discrepancy in the age of Sääksjärvi Crater, Finland (see below). A few additional craters, shown in updates of the G-R Catalog (Grieve, 1991; Grieve and Pesonen, 1992), were not received in time to be included in Figures 1–4, but the new and revised data only substantiate the general patterns shown in these figures (see **Appendix 1** for craters supplementary to those in Figs. 1–4).

Grieve (1982) gave the age of Sääksjärvi Crater as 490 Ma without reported uncertainty. The 1986 G-R Catalog, and Grieve (1991), give the age as 514 ± 12 Ma, but Grieve (1987) reported Sääksjärvi to have a maximum age of 330 Ma. Müller et al. (1990) report ^{40}Ar–^{39}Ar ages for Dellen (Sweden), Sääksjärvi (Finland), and Jänisjärvi (Russia) — all on the Baltic shield (cf. Appendix 27 herein) — that are identical (including estimated errors) to those given in the Grieve catalogs (1982–91), except for Sääksjärvi. It is difficult to tell from the report by Müller et al. (loc. cit.) what age they considered to be most representative for the Sääksjärvi Crater, because (1) in their Abstract they give it an "isochron" age of 560 ± 12 Ma, (2) on p. 1 they quote Bottomley et al. (1977) as concluding that the maximum crater age of Sääksjärvi is ~330 Ma (which may explain the value listed in Grieve, 1987), (3) in Table 1 they report the age as 514 Ma without indicated uncertainty, and (4) on p. 10 they conclude that 560 ± 16 Ma (presumably an average from release spectra such as their Fig. 9) is the "best value" for the impact event.

Geographic Patterns of Impacts

Apparently all we can conclude is that the Sääksjärvi impact event "probably" occurred between 500 and 600 Ma, but may have been much younger. Unfortunately, this is not an unusual example of confusion in the reporting of impact ages (see **Note 1.1**), and it would seem that a review and recompilation of all age determinations, including descriptions of the sampling histories, sample states, and analytical laboratories is needed (Grieve's tabulations, for example, do not indicate literature sources for the reported ages).

Other discoveries, not included in the G-R Catalog, and especially those related to investigations of possible K/T impact sites, are reported at appropriate places in the text and/or in the figure captions. Only six craters with ages of about 600 Ma or greater are listed in the G-R Catalog; these are shown in **Figure 1**, to which I have added the ~600-Ma, ~160-km-diameter Lake Acraman impact site in South Australia described by Williams (1986) and by Gostin et al. (1986, 1989), and the ~65-Ma, ~200-km-diameter Chicxulub Crater, Yucatan (see Note 1.1; and Hildebrand and Penfield, 1990; Hildebrand and Boynton, 1990a, b, 1991; Hildebrand et al., 1991; Penfield and Camargo, 1991; Pope et al., 1991a, b; Kring et al., 1991; Sharpton et al., 1992; Swisher et al., 1992). A bibliography of original literature sources for many of the terrestrial impact structures in the G-R Catalog is given by Grolier (1985); cf. Grieve and Robertson (1979), Grieve and Dence (1979), and Grieve et al. (1988).

For perspective, Figure 1 indicates the areal loci of Paleozoic craters and the locations of two contiguous flood-basalt fields. Although the Karroo field in South Africa is of only Mesozoic age, it is shown because its location anticipates later illustrations of a general association of flood-basalt provinces with impact loci (cf. Stothers and Rampino, 1990; Rampino, 1992). This association poses one of many correlations, pointed out throughout the text, that may seem paradoxical but are consistent with the major theme that patterns in a number of different magmatic, tectonic, and geomagnetic phenomena are remarkably similar to the putative self-organized invariances of the cratering pattern. In this instance, how can an association between the flood-basalt province and the cratering pattern, which has persisted in a relatively invariant way for the order of 10^9 years, be reconciled with paleogeographic reconstructions? A partial resolution, one that does not do major violence to existing concepts of continental drift, will be discussed later (cf. Fig. 34 and Appendix 26).

With respect to the documented impact loci, the Atlantic view of Figure 1 shows a general asymmetry between the Northern and Southern hemispheres, and the asymmetry is more conspicuous in the Phanerozoic record illustrated in Figures 2 and 3. Though the Southern Hemisphere is largely ocean, the K/T impact locus does cross South America (Fig. 3), yet the G-R Catalog shows only three craters there with diameters greater than 1 km, one undated and two possibly of Mesozoic age [Chicxulub, at the tip of the Yucatan Peninsula—which, with the northern tip of Cuba and the southern tip of Florida, roughly marks the transition between the Gulf of Mexico and the Caribbean Sea (see the paleogeographic discussion in Chap. 8)—is the only documented crater in Mexico, occurring where the northward trace of the K/T swath first makes tangential contact with North America (e.g., Figs. 4 and 26)]. South Africa is of interest for the reasons mentioned above, and the good exposures and mineral-exploration activities there would seem to favor the discovery of impact craters (cf. Appendix 27). Still, no Phanerozoic craters are documented. In Central Africa, too, only one is known (Bosumtwi, Ghana; ~6°N, ~10 km diameter, ~1 Ma), but several are known in North Africa near Mediterranean latitudes. The Central and North African localities are consistent with both the general Phanerozoic loci and the K/T loci. The paucity of impact localities in South America may be related to difficulties of identification. If so, the position of the K/T locus, which has persisted in much the same form during at least the latest 100 m.y. (see Fig. 3 and Appendix 2), suggests where to take a closer look.

A generalized composite projection for the patterns of Phanerozoic craters, as viewed from the North Pole, is shown in **Figure 2**. Because of the northern bias in the patterns, North Polar equal-area projections are used for most of the subsequent map illustrations. Positions of the Phanerozoic and K/T swaths on the more familiar Mercator projections are shown in later illustrations (e.g., Figs. 26 and 28). The numbered shaded areas outlined by solid and dashed lines in Figure 2 show the distributions of impact craters of different ages, as summarized in the Introduction and discussed below (cf. Appendix 27). The numbered loci were chosen subjectively to indicate the similarities in cratering patterns over four broad intervals of geologic time — viz. the Paleozoic (4), the lower three-fifths of the Mesozoic (3), the middle Cretaceous to lower Eocene (2), and the upper five-sixths of the Cenozoic (1). The radiometrically calibrated age spans of these four categories of cratering patterns — e.g., with reference to the chronometric time scale of Harland et al. (1990) — are, from youngest to oldest: (1) < 50 Ma, (2) 50–100 Ma, (3) > 100–250 Ma, (4) > 250–600 Ma. Another category, (5), consisting of Precambrian craters with ages greater than 600 Ma, is represented by only four craters in Figure 2; two of them (large solid circles) are in the Northern Hemisphere near the large asterisks that mark the Phanerozoic cratering nodes (PCNs), and the other two (heavy open circles) are in, respectively, Australia and South Africa (cf. Appendix 27). It is worth noting that all but one of these four Precambrian craters is in close proximity to a PCN, and the fourth one is in a vicinity — speaking very roughly — that I mentioned in the Introduction as a possible candidate for a pre-rifting PCN in the present-day South-Atlantic-Indian-Ocean region of the globe (see the discussions of Figs. 26–28 in Chap. 7). The light dashed line in Figure 2 is indicated as a possible cratering swath that would be consistent with the existing very sparse distribution of Southern Hemisphere craters (cf. Appendix 18).

The distribution of craters with ages between 50 and 100 Ma is shown in more detail in **Figure 3** (the analogous distribution for younger craters is shown in **Appendix 2**). Those craters that have published ages at or close to 65 Ma are identified by a star-like symbol. The Montagnais Crater of Jansa et al. (1990) on the continental shelf of eastern North America widens the western arc of the 50–

100 Ma swath if it is included with this age group, but the age they report, 50 ± 0.8 Ma, falls right on my somewhat arbitrary division between Age Groups 1 and 2 of Figure 2 (cf. Appendix 2). Also shown in Figure 3 are the Deccan flood-basalt province in India (~65 Ma) and the North Atlantic flood-basalt province (~60 Ma). The North American to Eurasian portion of the heavy dashed swath in Figure 3 is labeled the "Glen Line." This name acknowledges an event that stimulated the present thesis that the terrestrial pattern of craters is spatially ordered in a manner that resembles the precision, geologically speaking, of a laser printer. The idea of projecting a trend between the Chicxulub (Yucatan) and Manson (Iowa) craters across the Arctic Ocean to Eurasia occurred to William Glen (spoken comm., 11 June, 1991) and me mutually while we were discussing new information presented at the 22nd Lunar and Planetary Science Conference (18–22 March, 1991, Houston, Texas). Several important papers dealing with new interpretations of K/T boundary events were given there. An especially important one for my own conception of a possible global cratering pattern was the report by Deino et al. (1991) revising the age of the ~100-km-diameter Popigai Crater in north-central Siberia. In the G-R Catalog, the reported age of Popigai is 39 ± 9 Ma [changed to 35 ± 5 Ma in Grieve (1991) and Grieve and Pesonen (1992)]. However, Deino et al. (1991), using laser-fusion $^{40}Ar/^{39}Ar$ techniques on selected fragments of glass from a "melt-breccia" (suevite), reported revised ages of 65.4 ± 0.3 Ma ("plateau age") and 66.3 ± 0.4 Ma ("total fusion age"). The earlier ~40-Ma age for Popigai, combined with its size, which — prior to the discovery of Chicxulub, Yucatan — shared with Manicouagan, Quebec (~100 km, 212 ± 2 Ma according to Grieve, 1991) the record for the largest Phanerozoic crater in the G-R Catalog, had long puzzled me, because it seemed anomalous in the time-size distribution of cratering I had previously examined (Shaw, 1988a). The possibility that Popigai was actually of K/T age placed an entirely new light on my perception of the K/T cratering pattern (see Note 1.1). [Manicouagan was redated at 212 Ma (without reported uncertainty) by Trieloff and Jessberger (1992), and 214 ± 1 Ma by Hodych and Dunning (1992).]

The Kara structure, not far from Popigai, was listed in the G-R Catalog as two craters, Kara (60 km in diameter, 57 ± 9 Ma) and Ust-Kara (25 km in diameter, 57 ± 9 Ma), but Nazarov et al. (1991a) have reinterpreted the structure as one crater with a rim-to-rim diameter of about 120 km and an age in the range 65–70 Ma [Nazarov et al. (1991b) give the age as 65.7 Ma without reported uncertainty, but Trieloff and Jessberger (1992) report new age determinations of 69.3–71.7 Ma for four "impact melts" from "Kara Crater"]. Despite these ambiguities in the interpretation of the Kara/Ust-Kara complex (cf. Koeberl et al., 1990; Badjukov et al., 1991), there is some indication of more than one type of impact product, yet only one coherent crater morphology. This is not the only indication of nearly isochronous double or multiple impacts (see Notes 1.1 and 1.2), both at the same locality and at several widely separated localities, in the chronological vicinity of the K/T transition interval (cf. Hut et al., 1987, 1991; Shaw, 1988a, Figs. 14–16; Kerr, 1993b). The 60-km Kamensk structure and the 3-km Gusev structure in southern Russia share the same coordinates (48° 20' N, 40° 15' E, not far from the Eurasian nodal point of Fig. 2) and the same age (65 Ma, without reported uncertainty), as reported in the G-R Catalog (cf. Appendix 27, Inset).

The Avak structure, shown in Figure 3 between the Eurasian and North American clusters, was brought to my attention by Arthur Grantz of the U.S.G.S. (spoken and written communications, 26 June, 1991; cf. Grantz and Mullen, 1991). Grantz interprets the structure as the lower distal portion of a larger, now eroded impact crater having a projected original diameter greater than 12 km by some unknown factor. He believes the age to be upper Cretaceous to lower Tertiary, but the structural and stratigraphic evidence (Kirschner et al., 1992) does not allow the age estimate to be narrowed conclusively. The Avak structure, located at Point Barrow, Alaska, at the continental margin with the Arctic Ocean, would reasonably bridge the arcuate K/T swaths of North America and Eurasia.

The 35-km-diameter Manson Crater, in Iowa (Kunk et al., 1989; Hartung et al., 1990; Hartung and Anderson, 1991), and the very large Chicxulub Crater, in Yucatan (Hildebrand and Boynton, 1990a, b, 1991; Hildebrand and Penfield, 1990; Hildebrand et al., 1991; Penfield and Camargo, 1991), both of K/T age (see Sharpton et al., 1992; Swisher et al., 1992; Kerr, 1993b), are the principal North American K/T localities of Figure 3. The diameter of Chicxulub has been somewhat controversial, but most workers give estimates of ~200 km (cf. Penfield et al., 1991; Pope et al., 1991a, b). Sharpton et al. (1991) originally favored a size more like that of the Manicouagan Crater, Quebec (~100 km), but Sharpton et al. (1992) apparently concur with the larger estimate (for additional discussion of the age and diameter of Chicxulub, and the age of Popigai, see Note 1.1). The queried symbol in Figure 3 for the Colombia Basin is based on a speculation by Hildebrand and Boynton (1991), and the same symbol northwest of Hawaii is based on an observation of high-relief arcuate scarps on the ocean floor identified during a side-scan sonar *GLORIA* (*G*eologic *LO*ng *R*ange *I*nclined *A*sdic) survey in April–May, 1991 (location pointed out by R. I. Tilling, U.S.G.S., 30 June, 1991). If the Colombia Basin were to be verified as an impact site, it might be found to be even larger than Chicxulub. This would make the Gulf of Mexico–Caribbean region, with its threefold (or multifold?) suspiciously basin-like bathymetric subdivisions — obviously modified by complex plate-tectonic effects — the general site (protoscar) of impacts larger than any of the other known Phanerozoic events, conceivably representing impact basins marginal to the Procellarum-like Pacific basining events postulated here (Chap. 8) to have occurred shortly after the formation of the Moon, very early in Earth's history. The puzzling feature northwest of Hawaii is near the southeast end of the Hess Rise in a region of complicated topography on the Cretaceous ocean crust (remnants of latest Cretaceous or younger multiple-impact scars on the ocean floor?), located about 12° east of the Hawaiian-Emperor Bend (the H-E Bend is at ~33° N, ~172° E).

The K/T swath of Figure 3 therefore would appear to consist of at least three ~100-km or larger craters (Popigai, Kara/Ust-Kara, and Chicxulub; but see Note 1.1), at least two additional sites with crater diameters exceeding 25 km

(Kamensk/Gusev and Manson), and several others of uncertain size and K/T age. Notably, at least two of these sites are composite in character (Kara/Ust-Kara and Kamensk/Gusev). The Avak structure, at Point Barrow, Alaska (Kirschner et al., 1992) — which bridges the K/T swath across the Arctic Circle between North America and Siberia (Fig. 3) — shows evidence of a low angle of incidence ($< 5°$ from horizontal) on its final approach to impact (A. Grantz, written comm., 3 December, 1991; Grantz and Mullen, 1991). Low to grazing angles of incidence are evident in a number of planetological studies of crater morphologies (see Schultz and Gault, 1990a, b), and, though they have many possible explanations, they are especially conspicuous in the approach scenarios emphasized in the present work (see Figs. 19 and 20; cf. **Appendix 3** and Note A.1). Schultz and Gault (1990b) demonstrate that at impact angles of less than five degrees, an impactor may "ricochet" in such a way that much of it may remain more or less intact (e.g., closely clustered) — the fragments retaining velocities only slightly reduced from the initial impact velocity (cf. Baldwin and Sheaffer, 1971; Melosh et al., 1992; Schultz and Gault, 1992). At impact angles between five and ten degrees from horizontal a large impactor is likely to break up into several large chunks. The same authors state that a 10-km-diameter impactor arriving at 20 km/sec could ricochet with attendant production of numerous 0.1 to 1-km-diameter fragments traveling at some reduced fraction of the original speed (**Note 1.2**). This mechanism, tantamount to frictional braking of the population of fragments, is a conspicuous mechanism for the capture of some portion of the fragments into geocentric orbits (natural Earth satellites), as discussed in Chapter 6 (see Fig. 19; cf. Gault and Schultz, 1990).

During the time span of the 50–100 Ma group of Figure 3, i.e., occurring between the middle Cretaceous and early Tertiary Periods, there were 20 or more large impact events. The number of original impactors (approach events from space) related to these impact events, however, could have been substantially fewer than 20, according to the above discussion of low-angle events. (It is even conceivable that a single previous event of geocentric capture — or combined geocentric-selenocentric capture — of one or a few relatively large satellitic objects could have set the stage for a closely timed burst of multiple events on Earth that culminated in the system of terrestrial processes now identified, in net consequences, as the K/T boundary event; e.g., see the hypothetical scenario described in the Epilogue.) This 50–100-m.y. interval represents more than 20% of the craters of all ages and diameters > 1 km in the G-R Catalog, a fraction of the total number of impacts that is more than twice the corresponding fraction of geologic time ($< 10\%$) represented by the impact record as a whole. Moreover, this group is clustered within an $\sim$180-degree horseshoe-shaped swath in the Northern Hemisphere and has mixed oceanic and continental source materials in its collective ejecta. Because of the large geographic extent, as well as the great violence, of the dominantly Northern-Hemisphere impacts spanning the K/T boundary, the finding of a global distribution of the resulting geochemical and fine-grained mineralogical anomalies would occasion little surprise, depending on the interpretation of ocean-atmosphere dynamics during that interval of time. Whereas only a few years ago everyone was looking for *the* elusive K/T impact site (the so-called "smoking gun" of the K/T extinction), there now seems to be a plethora of such impact sites (the metaphor "smoking Gatling gun" perhaps is more apt in view of the spin-orbital mechanisms proposed in this volume for the coordination of impacts; see Chaps. 6–9; cf. the Epilogue, Fig. 37; and **Appendix 4**). Time will tell what similar concentrations of efforts may do for the definitions of the other impact loci outlined here. Notably, however, the general patterns of Figures 2 and 3 would remain about the same whether they were based on Grieve's 1982 listing of impact craters, on the 1987 G-R Catalog (as supplemented by me from the geological literature), or on Grieve's 1991 revision of crater data (cf. Appendixes 1–4).

The literature offers a number of discussions of "multiple impact events" of different characteristics and their possible relationships to "comet showers" (e.g., Hut et al., 1987, 1991; Shaw, 1988a; Weissman, 1990a, b). But the pattern of Figure 3, by contrast, would be a "cloudburst" relative to the secular variations of impact frequencies. Subsequently, I will suggest that this general characterization is more than a metaphor. Crisis-like "bursting" phenomena are typical in the types of chaotic intermittency (see Chap. 5, and Note 5.1) that appear to fit recent descriptions of the dynamics of source reservoirs of impactors in the extended Solar System (including the Oort Cloud). This mechanism, which also applies to orbital processes that may be coupled with the timing and distribution of impact loci on Earth, is universal. Nonlinear crises in both terrestrial and extraterrestrial systems, as well as in laboratory experiments, are illustrated in the following chapters.

Figure 4 summarizes the impact loci discussed above and also shows the distribution of major flood-basalt provinces against a sketch outline of the Northern Hemisphere continents (cf. the Mercator projection in Appendix 12 showing the global distribution of flood basalts and volcanic "hot spots"). The *McLaren Line* in Figure 4 (light short and long dashes) shows the result of rotating the composite Phanerozoic swath (medium-heavy long dashes) to accommodate the late Devonian (Frasnian-Fammenian) extinction localities of McLaren (1970) and McLaren and Goodfellow (1990), as mentioned in the Introduction (a few craters that correlate spatially with the McLaren Line and are of similar ages are shown in Appendix 3a). A transformation of axes is shown in **Figure 5**, where an idealized small circle (Fig. 5b) is drawn through the three nodes about a pole of rotation at 45° N, 160° E in North Polar projection (Fig. 5a). The equatorial great circle of the transformed projection and its north (filled circle) and south (open circle) poles are shown in Figure 5a as the poles labeled NNP' and NSP', respectively. Because of the 45° section, NNP' and NSP' coincide with the due east (E') and west (W') positions on the great circle. If the orientation of the great circle and the positions of the NNP' and NSP' poles in Figure 5a were reversed and rotated about 20° W, they would correspond roughly to the transformed coordinates of a somewhat smaller circle drawn through the K/T swath. Together, these two small circles, intersecting a bit east of the North Pole, form the Northern Hemisphere "pitched roof," "crown," or pair of "canted halos" of the Phanerozoic cratering pattern (cf. Figure 37, Inset; and Appendix 15). Later, I return to the possible implications of these orientations and to those of three great circles (the nodal great circles, NGCs, used in later discussions; cf. Appendixes 1–4) that are drawn through pairs of the Phanerozoic cratering nodes (PCNs) of the celestial reference frame (CRF) — where the NGCs (see Fig. 23) are extensions of the stereographic relationships outlined in Figure 5a and are later used in discussions of the orbital and/or spin-orbital mechanisms that may describe the proximal trajectories of impactors that have been engaged by the Earth-Moon system in one way or another (cf. Figs. 19, 23, and 37; and Appendixes 1–4, 12–18, and 22).

CHAPTER 2

Temporal Patterns of Impacts on Earth

The time distribution of craters in the G-R Catalog, supplemented by the newly discovered craters and revised ages plotted in Figures 1 through 3, is shown in Figures 6 through 8 (cf. **Appendix 5**). **Figure 6** shows running counts of craters in 30-m.y. bins starting at zero age (0–30 Ma, 31–60 Ma, etc.). **Figure 7** shows the latest 240 Ma of the same record, plotted in 10-m.y. bins, compared with the record of paleomagnetic reversal frequencies of Creer and Pal (1989, Fig. 2), and **Figure 8** compares the two records in Figure 7 with the histogram of ocean-crust production of Larson (1991a). The ages listed in the G-R Catalog (except for those that have been specifically revised) are taken at face value. Where the age is listed as less than a given value, that value is used, in the absence of other information. (Craters lacking age data, of course, are not included.)

The effects of these uncertainties in age distributions cannot be explicitly evaluated, but in order to partially offset this deficiency, three sets of counts are given: (1) all craters ≥ 1 km in diameter; (2) craters ≥ 10 km in diameter; and (3) craters ≥ 20 km in diameter. Age uncertainties should affect the three categories somewhat differently, yet the three time series are reasonably similar in form. The secular average trend (heavy dashed line) was guided generally by the total count, yet it follows the longer time trends (> 60 m.y.) for the larger craters; at the youngest ages there may be a real shift toward smaller objects (cf. Shoemaker, 1983; Shaw, 1988a; Barlow, 1990; Shoemaker et al., 1990; Weissman, 1990a). The inset in Figure 6 compares the impact record with that of the geomagnetic-reversal frequencies (Creer and Pal, 1989) that will be used later in discussing the geophysical implications of cratering loci and frequencies (cf. Figs. 30 and 31; and Appendix 5).

Secular averages of impact frequencies based on the curves in Figure 6 appear to represent on the order of 10% of the potential number of craters of 10-km or greater diameter that could have punctuated the continents. This estimate, based partly on discussions in Shoemaker (1983) and Shaw (1988a), is roughly consistent, as a gross average, with the results of Shoemaker et al. (1990) for astronomical estimates of all potential impactors that may exist at the present time (constructing that 1990 estimate, largely from the results of their own astronomical observations, was an undertaking of truly heroic proportions). When comparisons between the geological record of impact craters and the astronomical estimates of uniformly distributed impact rates (meaning rates normalized by the Earth's total area or by land area) are examined in relation to sizes of craters and impactors, however, it is found that the ratio of potential impacts to documented impacts exceeds the above estimate of ten to one for craters smaller than about 10 km, but the ratio varies erratically with the size and physical properties of the impactors, hence with the diameters of documented craters on Earth. According to the following discussion, the "completeness" of the cratering catalog for the latest 600 m.y., as judged only by comparison with the astronomical evaluation of the apparent availability of impactors by Shoemaker et al. (1990), would appear to increase from a low value of far less than 10% for the smallest craters to a value that is ostensibly 100% complete for the largest craters. Factors are then pointed out which would suggest that, on analogous grounds of incompleteness of the "sampling" of the heavens by the standard practices of astronomical observation, the astronomical estimates of numbers of potential Earth impactors — asteroids, comets, and objects without established relationships to specific source reservoirs (cf. Fig. 13) — are minimum values that probably are exceeded on average, and that may be greatly exceeded during maxima of intermittent delivery of Earth-crossing objects owing to nonlinear resonances related to the chaotic dynamics of orbital motions in the Solar System.

The Abstract of Shoemaker et al. (1990) states that the estimated collision rate of impactors with the Earth is consistent with rates deduced from the geologic record for craters > 20 km formed during the latest 120 Ma. However, this span of time is not representative of the geological record of impacts because it contains the exceptional "burst" of events at the K/T boundary and, more generally, all of the many events between 50 and 100 Ma, as well as those at the youngest ages, when the rate of small cratering events may have increased relative to the Phanerozoic average for all events (cf. Fig. 3). Assuming that their estimate of ~ 0.7 m.y.$^{-1}$ for asteroidal impactors ≥ 2 km in diameter equates approximately with craters ≥ 20 km in diameter (see Shoemaker, 1983, for a discussion of scaling laws), the integrated number of craters over 120 m.y. would be 84. If it is assumed that only one-third of these could produce geologically detected craters on land (because of the great preponderance of sea surface), then the number is reduced to 28 (**Note 2.1**). The uniformly normalized number of impacts on land (28) is about twice the number shown in Figure 6 for the latest 120 Ma, and is essentially the same as the count (29) for craters ≥ 20 km of all ages in the G-R Catalog. It would appear, therefore, that for intermediate-size craters, the astronomical estimate by Shoemaker et al. (1990) is roughly double the highest cratering rate in the augmented record of the latest 100 m.y., and it is many times the average geologically documented Phanerozoic cratering rate.

Discounting Precambrian ages and taking the total count of craters ≥ 20 km to represent the latest 600 Ma gives an average rate of 29 ÷ 600 ≅ 0.05 m.y.$^{-1}$ (~7% of the astronomically inferred rate, or ~21% of that rate normalized to land area). Alternatively, if the astronomically estimated rate were assumed to be constant during the Phanerozoic, it would predict ~420 craters ≥ 20 km during that time, or about 140 on land (more than five times the observed number). The discrepancy increases, as would be expected, for smaller impactors and smaller crater diameters. When we compare the astronomically inferred rate of 4.3 ± 2.6 m.y.$^{-1}$, representing objects ≥ 1 km in diameter producing craters ≥ ~10 km in diameter, with the data of Figure 6, the astronomical rate predicts a total of ~2580 ± 1020 craters ≥ ~10 km during the Phanerozoic, or ~860 ± 340 normalized to the Earth's land area, whereas the total count of craters ≥ 10 km in diameter in the G-R Catalog is 46 (~5 ± 2% of the number calculated from the astronomical rate).

Shoemaker et al. (1990) concluded that comets (but see Fig. 13, Insets) probably are the most important contributors to the formation of craters > 50 km in diameter, and they gave a rate of the order of 0.01 m.y.$^{-1}$ for cometary nuclei ≥ 10 km in diameter (those producing the largest terrestrial craters). This estimate would give ~6 craters ≥ 100 km during the Phanerozoic, in apparent agreement with the augmented record discussed here. As expected, the agreement between the rates derived on the basis of astronomical observations and inferences, on the one hand, and the rates derived from the geological documentation of craters, on the other, appears to greatly improve as crater diameter increases. But to some unknown extent, however, this agreement may be fortuitous. The literature on discoveries of large craters on Earth would suggest that, even in that case, the record may be far from complete. For example, three large craters [~600 Ma, ~160 km, at Acraman, South Australia (Gostin et al., 1986; Williams, 1986); ~365 Ma, > 70 km, at Taihu, China (Wang, 1992; cf. Wang et al., 1991); and ~65 Ma, ~200 km, at Chicxulub, Yucatan, Mexico (Hildebrand and Boynton, 1990a, b; Hildebrand and Penfield, 1990; Swisher et al., 1992)] have been discovered during just the last few years, two of them approximately equal in size to the largest documented Precambrian craters (~1,850 Ma, ~200 km, at Sudbury, Canada (Grieve, 1991, Table 1); ~2,000 Ma, ~140 km, at Vredefort, South Africa (Grieve, ibid.)]. Within a period of about five years, the catalog of Earth's largest impact craters thus almost doubled — and it may double again within the next few years (cf. Oberbeck et al., 1992; Rampino, 1992) — suggesting that the astronomical estimate by Shoemaker et al. (1990) of collision rates for the largest impactors probably is too low relative to the geological record of impacts.

Weissman's (1990a) review of cometary sources is more or less consistent with Shoemaker et al. (1990) concerning the completeness of the terrestrial cratering record. However, because Weissman emphasizes cometary objects as the principal sources of impacts, the discrepancies between the astronomical and

Temporal Patterns of Impacts

geological estimates of cratering rates probably are much greater than would be indicated from the general statements made by these workers, especially at intermediate and small crater diameters. These discrepancies relate in part to the fact that, since there are so few craters of large size, any changes in estimates of the numbers and ratios of asteroidal and cometary sources of the larger impactors may greatly modify the estimated ratio of potential impactors to geologically observed craters. Furthermore, astronomical measurements of the frequencies of comets crossing Earth's orbit are not well established compared to our knowledge of the orbital distributions of asteroids (cf. Shoemaker et al., 1979, 1982, 1989, 1990; Wetherill and Shoemaker, 1982; Shoemaker and Wolfe, 1982, 1986; Weissman, 1982, 1990a, b; Shoemaker, 1983, 1984; Wetherill and Chapman, 1988; Weissman et al., 1989; Wetherill, 1989a, b, 1991a; Safronov, 1990).

This situation is understandable, particularly in the context of the present volume, for three reasons: (1) the distinction between asteroids and comets is nonunique (see Fig. 13), because some asteroids are thought to represent the nuclei of "dead" comets, meaning comets that have experienced orbital decay and capture within the Asteroid Belt (cf. Wasson and Wetherill, 1979; Weissman et al., 1989; Lipschutz et al., 1989; Wetherill, 1991a); (2) the inner regions of the Solar System (i.e., near and inside Earth's orbital distance from the Sun) are apparently capable of "storing" large amounts of cometary material of virtually unknown orbital parameters and size distributions, much of which may be in high-inclination orbits such as the approach scenarios of Figure 19 (cf. Clube, 1982, 1989b, 1992; Duncan et al., 1987, 1988; Stern, 1987; Olsson-Steel, 1989; Steel, 1991); and (3) the amount, size distribution, and temporal intermittencies (irregularities of timing relative to long-term — meaning astronomically "secular" — averages) of cometary material entering the inner Solar System during comet "showers" are poorly known, in the sense that they are based largely on theoretical considerations. That is, uncertainties may be related to (a) assumptions concerning intermittent disturbances of the inner Oort Cloud (sometimes called the Hills Cloud) by various kinds of galactic perturbations (e.g., by the frequency of passing stars, or by galactic tides; cf. Weissman, 1982, 1986, 1990a, b; Duncan et al., 1987, 1988; Heisler, 1990) or (b) assumptions concerning predicted but so far undocumented contributions of comet-like materials from outside the Solar-System comet reservoirs (cf. Clube, 1967, 1978, 1982, 1985, 1989b, 1992; Clube and Napier, 1982a, b, 1984a, b, 1986, 1990; Napier, 1985, 1989; Stern, 1986, 1987, 1988, 1991; Stern and Shull, 1988; Bailey et al., 1990).

Our knowledge of cometary sources and the frequencies of "new" comets, meaning those that enter the inner Solar System for the first time, is discussed more fully in Chapters 4–7. Weissman (1990a) concludes, in agreement with my own evaluation of the cratering history — e.g., in Shaw (1988a) — that we are not presently at a time of peak impact rates by large objects (at least compared to the K/T rate). The interpretation of the cratering record in the present volume is consistent with that conclusion. However, others would qualify this assessment in the following way. The so-called *"giant comet hypothesis"* would suggest that cometary events related to galactic phenomena have occurred in the recent past, and that they have profoundly affected the inner Solar System, hence have also affected the historical record of comet observations. Clube and Napier (1990, pp. 136–46), for example, suggest that comets as large as 50 km in diameter enter the inner Solar System at intervals averaging about 100,000 years apart "during a high-risk period" when the Solar System is crossing one of the spiral arms of the Milky Way Galaxy (cf. Schmidt-Kaler, 1975; Rubin, 1983, 1984; Clube, 1989b; Olsson-Steel, 1989). The Solar System is just emerging from the vicinity of the Orion arm of the Galaxy (Schmidt-Kaler, 1975; Spitzer, 1982; Shaw, 1988a, Fig. 21b), which implies—according to the Clube-Napier model—that a peak of giant comet activity may have occurred as little as three to five million years ago, and that we are still in a period of enhanced risk. The 100,000-year type of intermittency, and related periodicities, are thought by Clube and Napier to be at least partially responsible for the existence of, as well as the periodicities of, the Pleistocene glaciation cycles (notably, an approximate 10^5-year cycle is similar to that of Milankovich-like oscillations; cf. Milankovich, 1969; Sergin, 1980; Benzi et al., 1982; Held, 1982; Ghil, 1985; Peltier, 1987; Berger et al., 1992; Kerr, 1992f; Winograd et al., 1992).

Although the giant comet hypothesis is not widely accepted among students of cometary phenomena, neither it nor the other concepts of galactic sources of material and/or physical influences on the comet flux, cited above, can be ruled out of consideration—if for no other reason than that our state of knowledge of cometary dynamics is heavily influenced by theoretical bias. And even though the coupling of cometary phenomena with the dynamics of terrestrial glaciation might appear to conflict with accepted correlations between glacial cycles and Earth's orbital parameters (see citations above), arguments such as those mentioned in the Prologue concerning coupling between impact dynamics and tidal dynamics would suggest that—from the standpoint of the nonlinear dynamics of coupled oscillations—coupling between the periodicities of cometary, and other, impact phenomena and the astronomical periodicities of Earth's paleoclimatic phenomena would be expectable consequences of the viewpoint advocated in this volume (cf. discussion in Shaw, 1988a, concerning relationships between galactic processes and periodicities relative to Earth processes and periodicities).

Shoemaker (1983) and Shaw (1988a) discuss geological factors of crater preservation (cf. Appendix 27; and Blatt and Jones, 1975; Derry, 1980; Kieffer and Simonds, 1980) and problems of comparing estimates of present-day rates with those of the geological record. Shaw (1988a) concluded, in agreement with the preceding discussion, that the cratering rate is probably episodic, and that the present-day rate for small objects probably is higher than the Phanerozoic average. Stothers (1992, Fig. 2) illustrates this effect in terms of the cratering history of the Moon, suggesting that there is a correlation between the episodicity of lunar cratering and the episodicity of Earth's tectonic evolution—both observations being consistent with the hypothesis of the present volume (see the Epilogue, and Fig. 37). The message of the studies by Shoemaker et al. (1990)—combined with the implication that there are sources of impactors, as well as orbital trajectories of approach of impactors to the vicinity of Earth, that have been neither identified nor acknowledged by the astronomical community in general — leads me to conclude that there probably have been far more potential impactors in Earth-crossing orbits over geologic time than have been accounted for even by the evidence for bursts of clustered events in the geologic record. This conclusion will be reinforced by observations (given in later chapters) concerning the remarkable growth histories of discoveries of both asteroids and comets since the inception of astronomical documentations. It seems clear to me, therefore, that we stand to discover many more additional craters of all sizes, throughout the geologic column.

Figures 7 and 8 illustrate crater counts at intervals of 10 Ma over the latest 250 Ma. Both figures illustrate the relationship between crater counts and the record of geomagnetic reversals during the latest 150 Ma, and Figure 8 compares global magma production, as represented by the net production of ocean crust (Larson, 1991a, b; Fuller and Weeks, 1992), with both the geomagnetic signatures and the impact signatures. Time delays between fluctuations of impact frequencies, geo-magnetic-reversal frequencies, and rates of magma generation and transport are discussed in Chapter 8 (cf. Appendix 5), where relationships to rates and periodicities revealed by the history of polar-wander paths are also illustrated. In the later discussions, I evaluate all these relationships in terms of chaotic bursting phenomena in the dynamics of magmatic and paleomagnetic processes coupled with the orbital-terrestrial dynamics that govern incidences of bursts of impact events in the geologic column.

In the following two chapters I will examine the background structure of the Solar System and the source reservoirs of impactors, as interpreted in the light of nonlinear dynamics, since this information is essential to an understanding of the context of the *celestial reference frame hypothesis* (CRFH) by which the history of impacting on Earth is traced.

But before that is done, perhaps it is time once again both to warn and to reassure the reader (and indeed myself) that the seeming abyss of difficulty represented by the unfamiliar material that bedecks the next few chapters will disgorge, as well, by bits, pieces, and glimmers, the reader's own innate understanding — much as it may seem that none of that has occurred. Whatever can be gathered together through this stretch will inform a reading of the later chapters. I urge, then, the patience and indulgence that one might bring to the assembly of an intricate jigsaw puzzle, or to the cultivation of a garden. All is not lost, even when the prospect that a picture will ever emerge, or that transplanted azaleas will ever bloom, seems hopeless. Just when the patterns seem most irretrievably confused, the right piece *finds itself*, and the picture laid before us emerges with a suddenness and a satisfaction as surprising as any season of azalea blossoms.

CHAPTER 3

Geometric Complexity and Nonlinear Processes in the Solar System

Several research groups have by now established large-scale computational programs to study the nonlinear dynamics of orbital processes in particular parts of the Solar System, while other groups are studying in analogous ways the nonlinear patterns demonstrated by systems of artificial satellites orbiting the Earth, and space vehicles probing the interplanetary environment (e.g., Coffey et al., 1986, 1990, 1991; Friesen et al., 1992; Belbruno, 1992). Wisdom (1987a, b, c) has greatly elucidated the dynamics of the Asteroid Belt in the general context of chaotic behavior in the Solar System (cf. Berry, 1978; Wisdom, 1990; and the demonstrations of Solar System chaos by Sussman and Wisdom, 1992). Prior to the study by Sussman and Wisdom (1992), a series of orbital computations of chaotic instabilities of the planets themselves were carried out by a number of workers, including Wisdom (1987a, b, c, 1990), Laskar (1988, 1989, 1990; cf. Berger et al., 1992), and Laskar et al. (1992). Milani (1989), Milani et al. (1989), and Milani and Nobili (1992) have carried out complementary studies. Other aspects of the nonlinear dynamics of the Solar System have been considered by Tittemore (1990) and by Stern (1991). There are many studies of cometary reservoirs, but those of Duncan et al. (1987) and Torbett and Smoluchowski (1990) illustrate particularly important nonlinear effects. Altogether, these studies bring new light to the structures of storage domains of small to large objects in the Solar System (from < 1-km meteoroids to > 1000-km relict planetesimals; cf. Ip, 1989; Bailey, 1990; Weissman, 1990b; Weidenschilling, 1991; Stern, 1991).

The Solar System extends to more than 10^4 AU from the Sun. [The standard unit of length in planetary astronomy is the Astronomical Unit (AU), where 1 AU is the gravitationally normalized mean distance between the Earth and Sun = $1.49597870 \times 10^{11}$ meters, or roughly 150 million kilometers (see the Astronomical Almanac for the Year 1992, Anon., 1991, p. K6).] Indeed, it can be demonstrated that the outer fringes of the cometary cloud (the Oort Cloud) that surrounds the Solar System extend to a distance of $\sim 10^5$ AU, a distance where stellar interactions become more conspicuous (Stern and Shull, 1988). Impactors have been thought to originate (1) in the Asteroid Belt, which extends overall from inside Earth's orbit (< 1 AU) to beyond Jupiter (> 5.2 AU), and/or (2) in several different cometary reservoirs of much greater dimensions. The latter include the Kuiper Belt, or Disk, at the order of 30 to 500 AU, and the inner and outer Oort Clouds, ranging from the order of 10^2 AU to $\sim 10^5$ AU, with the inner limit of the outer Oort Cloud conventionally placed at about 2×10^4 AU (see Bailey, 1990; Weissman, 1990b). All of these cometary reservoirs, though subdivided by convention, together represent a dynamically continuous, if distributionally complex, region of the Solar System. Stern (1991) has added a new wrinkle to this complex realm. He argues that resonances **(Note 3.1)** between planets and their satellites near and beyond Uranus (~ 20–50 AU) imply that the outer part of the Solar System should include a large population of 10^3-km-size objects surviving from early stages of planet formation [this is the approximate size of Ceres, in the Asteroid Belt (cf. Bowell et al., 1979; Tedesco, 1989), and of a number of satellites of the outer planets (Pasachoff, 1991, Appendix 4) — and is roughly comparable to the diameters of Pluto ($\sim 2,300$ km) and Pluto's "satellite" Charon ($\sim 1,200$ km); see Stern (1992, Table 1)]. There is no observational evidence as yet for such "independent" objects, but the issue remains open, especially at distances within the Kuiper Belt and beyond [cf. Fig. 13 (Insets), and Notes 6.5 and 6.9].

There are in fact abundant indications of chaotic orbits and resonances at virtually all spatiotemporal scales in the Solar System. The calculations by Laskar (1988, 1989, 1990) have demonstrated that even Earth's orbit is unstable on the remarkably short time scale of 5 m.y., as determined by integrations over 200 m.y., and therefore is chaotic. Such calculations concern the divergence time scale for departures from stable orbital parameters, and do not mean that Earth's mean orbital diameter, eccentricity, or inclination will change significantly over 5 m.y. (cf. Laskar, 1989; Milani, 1989; Wisdom, 1990; Milani and Nobili, 1992; Berger et al., 1992; Laskar et al., 1992; Sussman and Wisdom, 1992). The simple classical picture of a stable progression of planetary orbits is now seen to be *characteristically* in a state of flux. In light of these discoveries and the patterns identified in the previous chapter, I suggest further that impacts on Earth often have occurred as a consequence of nonlinear resonances with objects in the Earth's vicinity, rather than only by chance collisions with asteroids or comets coming directly from their source reservoirs.

But deducing the dynamical complexity of the Solar System is neither a new enterprise nor one demanding large-scale computations. Kaula (1968) provided numerous hints pointing toward these later results, and Alfvén and Arrhenius (1975, 1976) examined many dynamical problems implying resonant interactions at all scales in the Solar System. Nonlinear resonances and invariances of many kinds have been known for a long time, beginning with the famous but previously unexplained Titius-Bode relation for the spacing of the planets (e.g., Roth, 1962,

p. 20; Kaula, 1968, p. 222; Nieto, 1972; Tombaugh and Moore, 1980, p. 46). This relationship, previously viewed as one of the great mysteries of cosmogony (e.g., Struve et al., 1962; Berlage, 1968; Sexl and Sexl, 1979; Henbest, 1981), now looks as if it represents numerical invariances characteristic of the universality parameters of nonlinear dynamics (cf. Figs. 9 and 10; Notes 4.2, 6.8; and **Appendix 6**). The fractal complexity of the Solar System relative to the origin and evolution of solar-planetary dynamics is elucidated in the next two sections of this chapter.

I have relied on background material from many different sources for the viewpoints expressed in the following discussions. Some of the textual sources I have used in mechanics and nonlinear dynamics are listed below. Numerous works by Ralph H. Abraham and coworkers have supplied me with a perspective on the mathematical foundations of nonlinear dynamics that emphasizes its historical roots and spans developments from classical mechanics to recent theories of chaotic systems and manifolds (configurations of complex surfaces in phase space; see Note P.3), with applications. Principal among these works are the well-known textbooks of Abraham and Marsden (1987) and Abraham et al. (1988), and the impeccably illustrated set of books by Abraham and C. D. Shaw (1982–88) that trace the evolution of developments in bifurcation theory and chaotic dynamics in an exciting sequence of graphics that emphasizes the visualization of trajectories of motion in phase space in a way that stimulates application to virtually any field of research, including neurophysiology, psychology, and sociology (cf. Dewdney, 1987; Skarda and Freeman, 1987; Zeeman, 1987; Rapp et al., 1987, 1988; Glass and Mackey, 1988; Holden, 1988; Alper, 1989; Pool, 1989; F. D. Abraham, 1990; Goldberger et al., 1990; Pickover, 1990; Suarez et al., 1990; Abraham et al., 1991b; Lewis and Glass, 1991; Freeman, 1991, 1992; Andreyev et al., 1992; Garfinkel et al., 1992; Larnder et al., 1992; Peterson and Ezzell, 1992).

A sampling of other useful texts and symposium volumes in nonlinear dynamics is as follows (and I highly recommend the technique of "comparison shopping" in perusing such a rapidly growing literature): Jorna (1978), Campbell and Rose (1983), Horton et al. (1983), Guckenheimer and Holmes (1983), Lichtenberg and Lieberman (1983), Hao Bai-Lin (1984, 1987, 1990), Bergé et al. (1984), Cvitanovic (1984), Devaney (1986, 1989, 1991), Holden (1986), Thompson and Stewart (1986), Berry et al. (1987), Moon (1987), Flaschka and Chirikov, 1988; Schuster (1988), Jackson (1989–90), Baker and Gollub (1990), Campbell (1990), Middleton (1991), and Tulfillaro et al. (1992). Excellent semipopular reviews are given by Crutchfield et al. (1986), Gleick (1987), Jensen (1987), Stewart (1989), and Schroeder (1991). An entrée to historically fundamental studies in Russia, as well as in other countries, can be found in the compilation of articles derived from the Soviet-American Conference on Chaos (Campbell, 1990), which includes articles by Aranson and coworkers, Berman, Chernikov, Khanin, Neishtadt, Sinai, and Zaslavsky. Historically, studies of coupled-oscillator systems, both theoretically and experimentally, central to the present theme, owing much to the work of A. N. Kolmogorov and V. I. Arnold (e.g., the so-called *KAM theorem*, upon which the more realistic applications of Hamiltonian dynamics rest, grew first from the work of Kolmogorov, was added to by Arnold, and was developed in its present form by J. Moser; see Moser, 1978; Lichtenberg and Lieberman, 1983; cf. Note P.3 of the present volume; and Meiss, 1987; Schmidt, 1987; Smith and Spiegel, 1987). A. M. Liapunov, B. V. Chirikov, and other Russian scientists contributed what are now central ideas in studies of nonlinear-dynamical behavior and criteria of divergence rates in studies of sensitive dependence on initial conditions (see Liapunov, 1947; Chirikov, 1979; cf. Feit, 1978; Wolf, 1986; Nese, 1989).

The Solar System as a Fractal Object

Fractal geometric complexity and nonlinear-dynamical processes go hand-in-hand. Chaos implies orbital complexity; it does not imply the loss of rigorous order parameters, but rather means that a new kind of classification of geometric regimes in the Solar System is possible (see Note I.2; and Shaw, 1987b). The background structure of the Solar System as seen in this light is summarized in **Figures 9 and 10**, and a schematic overview of source locations and interactions between reservoirs of impacting objects is shown in Figure 11.

In **Figure 9**, the cumulative mass of planets inside a radius R from the Sun (Fig. 9a) is divided by the effective volume of a sphere of that radius (excluding the mass of the Sun), giving the fictive density of planetary mass within that volume (Fig. 9b). The logarithmic plot of these data (Fig. 9b) shows an offset in mean densities between the inner and outer planets. Because the lines in Figure 9b have noninteger limiting slopes, the mass distributions are of *fractal* character. This relationship was discovered long ago for stars and galaxies by de Vaucouleurs (1970); the slope in Figure 9b is similar to his value for the distribution of stars. Mandelbrot (1977, p. 111; 1983, p. 85) discusses the fractal implications of de Vaucouleurs' observations, as well as those of other cosmological phenomena (cf. Seiden and Gerola, 1982; Rubin, 1983, 1984; Shaw, 1983a, b, c, 1988a, 1991, Fig. 8.43; Schneider and Terzian, 1984; Coniglio, 1986; Lucchin, 1986; Pietronero and Kupers, 1986; Hunter and Gallagher, 1989; Shaw and Chouet, 1991; Barnes and Hernquist, 1992; Luo and Schramm, 1992; Flam, 1992a, b, 1993).

The fractal dimension of the density distribution is a power-law exponent or proportionality constant analogous to the Titius-Bode relation, but it includes the mass of the bodies as well as their distances. These relationships fail beyond limits that may represent different resonances among the planets — and breaks in, or transitional states of, these resonances appear to be associated with the general positions of the Asteroid Belt, the Kuiper Belt, and the Oort Cloud. There is no reason a priori why there should not be another offset beyond Pluto in Figure 9, and the objects predicted by Stern (1991) that span the Kuiper Belt and Oort Cloud distances might represent such a resonance step (cf. Stern, 1992, 1993). Weissman (1990b), in reviewing estimates of the possible masses in the outer belts of objects, finds that one Earth mass of objects could exist in the Kuiper Belt, and that the mass of the Oort Cloud conceivably could exceed that of the planetary system as a whole. These estimates preceded the results by Stern (1991), which demonstrated how remarkably incomplete is our knowledge of the evolution, and present distribution, of mass in the Solar System (cf. Fig. 9, Inset; and **Appendix 7**). Weidenschilling (1991) has coined the term "plutons" for the 10^3-km objects (small planets) predicted by Stern (1991), a term that presumably derives from their putative positions at distances near and beyond Pluto (the netherworld connotation is appropriate only if we think of the outer reaches of the Solar System as being more remote or "deeper in space," a spatial inversion that reconciles this usage with that in geology, where the term "pluton" refers literally to an igneous body that originated at depth in the Earth and "never saw the light of day" while in the molten or partially molten state). All of these models, and the present discussion, should be compared with the "standard model" of nebular accretion elucidated in a recent review by Wetherill (1990; cf. the scenarios illustrated in the Inset of Fig. 9b and discussed in Notes 3.2 and 9.2), as well as with the varieties of open-system models that invoke inputs of mass and/or energy from surrounding regions of the Galaxy (cf. Clube, 1967, 1978, 1982, 1985, 1989a, b, 1992; Clube and Napier, 1982a, b, 1984a, b, 1986, 1990; Weissman, 1982, 1986, 1990a, b; Napier, 1985, 1989; Stern, 1986, 1987, 1988, 1991; Duncan et al., 1987, 1988; Stern and Shull, 1988; Bailey et al., 1990; Heisler, 1990).

Figure 10 shows the distributions of orbital and rotational kinetic energies (K.E.) in the Solar System. Some of the misalignments are explained by known effects, such as gravitational coupling and tidal dissipation (e.g., the Earth-Moon system would be shifted to somewhat higher values of K.E.), but the data points are generally "scrambled" in some complex manner (cf. Fig. 9b, Inset; and Appendix 7). The dense terrestrial planets (+ Pluto) tend to fit a distribution different from that of the large, less-dense, outer planets (in some cases there are overlaps of data points between Earth, Uranus, and Neptune). The precise fit of the uppermost curve in Figure 10b is tautological, in the sense that it defines K.E. according to Newton's laws of motion. The other curves are all fractal deviations from this reference state.

In the data sets of Figure 10, one regularity persists: the distance-mass-energy progressions for the outer planets, which are monotonic for the sequence (from outside in): Pluto, (Uranus + Neptune), Saturn, Jupiter. The distances decrease with increasing mass and K.E., except for the case of Neptune and Uranus, which are reversed in distance but have similar diameters and mass. Combined, Neptune and Uranus represent about 30% of Saturn's mass and 10% of Jupiter's. The Neptune/Uranus similarity suggests that the two have been sensitive to nonlinear effects, as is the case with smaller binary objects such as the paired objects Pluto/Charon. The suggestion of anomalous dynamical states represented by the Neptune/Uranus pair is also supported by the fact that both of these large planets, each more than ten times Earth's mass, have large axial tilts (see Stern, 1991, 1992). [Charon was discovered in 1978 by J. W. Christy of the U.S. Naval Observatory (see Pasachoff, 1991, pp. 272ff). Although Charon is listed as a satellite of Pluto in catalogs of the satellites of the planets, its diameter is roughly

Geometric Complexity and Nonlinear Processes

half that of Pluto—a very large size ratio compared with the size ratios of other satellites relative to their primaries (the sizes and masses of Pluto and Charon are not known with great precision, and their chemistries and internal structures are still largely speculative; see the review by Stern, 1992)—and the Pluto/Charon pair might qualify more appropriately as a "binary planet" by analogy with binary stars (see Stern, 1992, 1993; cf. Strahler, 1971, pp. 84f; Kaufmann, 1991, Sec. 19-6 and pp. 448f; Pasachoff, 1991, Chap. 22; Maran, 1992).]

In contrast with the outer planets, still referring to Figure 10, the terrestrial planets (i.e., those planets resembling Earth in composition, including Mercury, Venus, Earth, Mars, and whatever fraction of the Asteroid Belt has inner-Solar-System affinities rather than cometary affinities; cf. Fig. 13, and Notes I.1, 4.3, 6.9), vary irregularly in nearly all respects. Except for Mars and the Asteroid Belt, mass and energy decrease inward, the reverse of the pattern of the outer planets. Earth, the most massive, roughly bisects the distance between the Sun and the inner edge of the Main-Belt Asteroids (MBAs). Such combinations of regular and irregular motions are characteristic of a system operating near a transition between chaos and regimes of both quasiperiodicity and phase-locked resonance (cf. Appendix 6).

Bistability or multistability, suggested by the fractal plots of Figure 9, also is reflected in the bipartite fractal (or multifractal) trends of orbital K.E. between the terrestrial planets (+ Pluto) and the outer planets in Figure 10a (solid circles). These trends intersect at a mass roughly one-tenth that of the Sun, a virtual state that I refer to as the "Ghost Binary" in allusion to a hypothetical prior stage of the solar nebula when a second star, smaller than the Sun, occupied that position in mass-energy space **(Note 3.2)**. This state, of course, did not survive, but, rather, left a relict signature in the properties of the terrestrial planets, the Asteroid Belt, and Pluto (as well as the "plutons" predicted by Stern, 1991). Accordingly, the protoplanets could have represented the retrograde products of the disruption of this second star in the original binary pair; alternatively, the mass-energy state of the putative second star could have represented a massive nebular disk rather than a condensed star (cf. Notes 3.2 and 9.2). In either scenario, the system essentially turned itself inside-out, leaving behind the lithophilic terrestrial planets as relicts of the less volatile states of the stellar disk materials (cf. Fig. 9, Inset; Appendixes 6 and 7; and Petersons, 1984; Sagan and Druyan, 1989; Ringwood, 1986, 1989, 1990; Ringwood et al., 1990; Drake, 1990). Pluto and the "plutons" of Stern's (1991) model would mark the inner edge of the "loss sphere" (outer edge of survival) of the binary star's condensible components (following the catastrophic disruption event), and accordingly all of the small-object belts would have acquired a "mixed pedigree," right from the start, relative to the Solar System as it has been sampled geologically by meteoritic materials (cf. Note 3.2). It is impossible to ascertain at this stage—because of the uncertainties in our knowledge of the dynamics of binary-star systems (but see Note 9.2)—what relative contributions were made by each of the original stellar objects (and/or what prior history those materials had experienced in the Galaxy). Complex mixing must have taken place over a significant fraction of the uncertain binary lifetime — a lifetime that was probably short relative to the total age of the Solar System as a stellar object, but one that had a profound effect on the chemical and mass-energy structure of the planetary system in the forms we see it today — first by turbulent vorticity and later by chaotic orbital instabilities of types analogous to those discussed by Wisdom (1987a, b, c), Duncan et al. (1987), and Torbett and Smoluchowski (1990).

According to the hypothesis that the Solar System originated as a binary-star system, its present structure is chaotically, and fractally, heterogeneous in chemical mass-energy, and in the orbital/rotational parameters of the early planetesimals and those of the evolved states of planets, satellites, and relict planetesimals. This does not mean that the Solar System was "randomized," however, because ordered domains exist — as the product of self-organized processes — and are predicted to be of mixed nonlinear periodic-chaotic signatures (*pattern periodicities* in the sense of Shaw, 1987a, b; 1988a, b). For early times, this model implies *collisional exchanges* between planet-size objects, which is consistent with the eventual hypothesized collisional origins of the Moon and Mercury, and the collisional demise of other infant planets involved in such collisions now favored by many workers (e.g., Wetherill, 1985, 1990; Ringwood, 1986; Stevenson, 1987; Stewart, 1988; Melosh, 1989; cf. Daly, 1946; Hartmann and Davis, 1975; Baldwin and Wilhelms, 1992; Spera and Stark, 1993). My model of an early-binary stage of the Solar System (Fig. 9b, Inset) differs from the so-called *Nemesis hypothesis* in which a distant companion star to the Sun has been — and putatively continues to be — responsible for periodic perturbations of the Oort Cloud of comets and for comet showers on Earth (cf. Whitmire and Jackson, 1984; Hut, 1986; Hills, 1981, 1984a, b, 1985, 1986; Muller, 1986, 1988; Raup, 1986b; Tremaine, 1986). Perlmutter et al. (1990) review the history and prospects of the search for "Nemesis."

The conference Protostars and Planets III (Tucson, Arizona, 5–9 March, 1990) revealed that certain stages of stellar evolution broadly fit the properties of Figures 9 and 10 (see Weissman and Wasson, 1990, for a summary of that conference). Pre-main-sequence stars called T-Tauri stars exhibit extended protoplanetary disks and high mass-loss rates, and the Sun was of this type (cf. Notes 3.2 and 9.2). Some of the T-Tauri stars, those called FU-Orionis type stars, can undergo great bursts of both mass loss and enhanced luminosity, both processes probably influenced by catastrophic instabilities and/or infall of disk material (cf. Hartmann and Kenyon, 1985; Sargent, 1989).

Chaotic oscillations of a double-star system (or massive-disk star) would be particularly susceptible to interior chaotic crises of the type later identified in this volume for particular values of critical driving frequencies (see Chap. 5; and Figs. 17 and 18). As the driving frequency approaches a characteristic natural frequency of the system, phase-space trajectories (and actual ones) experience chaotic bursts (see Note P.3). In this form of chaotic intermittency (see Gwinn and Westervelt, 1986; cf. Chap. 5, and Note 5.1), the bursts of unstable activity are more widely spaced in time and more violent the closer the system approaches a critical frequency (there can be more than one; cf. Carroll et al., 1987). This phenomenon has been experimentally verified by Ditto et al. (1989, 1990a), Savage et al. (1990), and Sommerer et al. (1991) for the buckling of a magnetoelastic ribbon with strain-dependent Young's modulus (representing a class of sensitive coupled oscillators within the general category of multi-well potential problems), following a concept of phase-space crises posited by Grebogi et al. (1982, 1983, 1986, 1987a, b, c; cf. Gwinn and Westervelt, 1985, 1986; Sommerer et al., 1991; Sommerer and Grebogi, 1992). This concept is universal in nature and applies at all scales of orbital evolution, including collisional-impact phenomena, and in other terrestrial processes, such as the instabilities of the core dynamo mentioned in the Introduction and discussed in Chapter 7. (I will return in Chapter 5 to the provocative implications of the experimental demonstrations of chaotic crises.)

Shower Cycles of Impactor-Impactee Bombardments in the Solar System

A schematic diagram of source regions of potential impactors is shown in **Figure 11** as a form of "orbital circulatory system." Objects within any particular orbital reservoir are "lost" by collisions (plus other accelerations) among themselves and/or with other objects, and are "gained" by transfers involving the same and/or analogous processes from other reservoirs; the latter reservoirs may be either inside or outside the Solar System. Early large-object collisions and/or close encounters of the protoplanetary nebula with other stellar systems would have influenced the initial states of all reservoirs. The coordinates of Figure 11 are scaled up from Figure 9, and object classes are labeled according to the studies cited above. The important Aten, Apollo, and Amor types of objects (conspicuous as potential sources of near-Earth meteoroids; see Fig. 13) are discussed more fully in Gehrels (1971a), Gehrels (1979), Binzel et al. (1989a), and especially Shoemaker et al. (1989, 1990).

Because the types of transfers indicated in Figure 11 are many, and ambiguous, it is difficult to be specific about terms such as "injection," "ejection," "capture," "orbital transfer," "aggregation," etc. The term injection, as employed here, means simply transfer from one orbital state to another (as is done with artificial satellites). Thus in Figure 11, "injection" refers to a transfer from outer-orbital reservoirs to inner ones, or vice versa, and "reinjection" refers to the subsequent inverse *return* to outer reservoirs from inner ones, or vice versa. "Ejection" usually refers to removal from the Solar System altogether, but it might refer to a removal from one planetary influence and a "capture" by some other. Ambiguities of terminology are largely unavoidable in a general context, such as that of Figure 11, in view of the kinematic reversibility shown there—and this is particularly true of the outer reservoirs of cometary objects, which, because they involve sensitive interactions with the Galaxy, are especially ambiguous in dynamical behavior (cf. Oort, 1950; Oort and Schmidt, 1951; Clube, 1967, 1978, 1982, 1985, 1989a, b, 1992; Hills, 1975, 1981, 1984a, b, 1985, 1986; Clube and

Napier, 1982a, b, 1984a, b, 1986, 1990; Napier, 1985, 1989; Stern, 1986, 1987, 1988, 1991; Tremaine, 1986; Weissman, 1986, 1990a, b, 1991b; Bailey, 1987, 1990; Duncan et al., 1987, 1988; Stern and Shull, 1988; Bailey et al., 1990; Heisler, 1990).

The underlying theme of Figure 11 is *self-organized criticality* (see Note P.4), implying that there are quasisteady cascades of chaotically intermittent "shower activity" (cf. Fig. 13). Particular domains are chaotic, but the system as a whole oscillates near the borders of chaos, giving power-law distributions of frequencies that, by and large, are self-similar at all scales (Shaw, 1987b, 1991; Bak et al., 1988; Shaw and Chouet, 1988, 1989, 1991; Bak and Tang, 1989; Chen and Bak, 1989; Kadanoff et al., 1989). Frequencies of collisions would scale inversely with object sizes and orbit sizes (relative to Earth's orbit) to be generally consistent with the energy relations in Figure 10. The individual exceptions that occur depend on Poissonian tails of power-law averages, as demonstrated in the phenomenology of chaotic crises (e.g., Ditto et al., 1989, Figs. 4 and 5; see the later chapters in this volume). In the general scheme of Figure 11, no absolute meaning can be accorded an object class such as the Asteroid Belt or Oort Cloud, because such a class is defined by its relative frequencies of exchanges with other object classes (e.g., Fig. 13, Insets; and Fig. 15). Relativity in these definitions makes the classification of cratering events on Earth difficult, because there is no unambiguous way to discriminate between a "comet" and an "asteroid," especially in cases where nonlinear mechanisms are important to the proximal states of Earth-crossing impactors. At the present state of knowledge, we can say that some undetermined number of asteroids were of cometary origin (Weissman et al., 1989), and I would surmise that some objects now in the cometary reservoirs could have originated in the Asteroid Belt (cf. Ovenden and Byl, 1978; Van Flandern, 1978; Duncan et al., 1987), while some undetermined number of objects in both of these reservoirs may have come from sources external to the Solar System (e.g., Stern, 1987; Papaelias, 1991; cf. Chap. 2). Unfortunately, the literature of planetology sometimes conveys a level of certainty concerning the sources of Earth-crossing objects that is not supported by closer examination of alternative scenarios.

The symbolic hourglass in Figure 11 is the prototype for self-similar chaotic showers. Particles in the hourglasses (simulating objects in the Solar System) impact "surfaces" (the general "planetary surfaces" at more interior positions in the Solar System) below the necks or throttle points that sustain the flow at a critical state, such that there are quasisteady "failure-cascades" that simulate comet showers, asteroid swarms, multiple terrestrial impacts, pulsatory tectonic motions, and, on Earth, biological extinctions. Each scale of the hourglass process represents some different orbital "level" of the object reservoirs, each of which has a characteristic power-law distribution of self-similar sizes and frequencies such that the average rate of delivery of impacting objects at a particular position (e.g., Earth) is scaled hierarchically in proportion to the total mass flux. Those who are familiar with nonlinear dynamics may note that the flow of sand grains within

Geometric Complexity and Nonlinear Processes

an hourglass represents both fluid-like and solid-like intermittency (cf. Chap. 5, and Note 5.1), and that the "dripping" of grains resembles the dynamics of the dripping faucet (R. Shaw, 1984; Wu and Schelly, 1989; Cahalan et al., 1990; Shaw, 1991), whereas slumps of the accumulating sandpile that the grains impact symbolize the dynamics of intermittent failure of a solid substrate (Bak et al., 1988; Bak and Tang, 1989; Kadanoff et al., 1989; Held et al., 1990). Each scale of motion in Figure 11 could be viewed as a microcosm, in the sense that each impacted object has its own "local" reservoir of impacting objects. But, like the hourglass, this local reservoir has a "global" aspect by which it must be continuously or periodically reset or refilled. The resetting (refilling) process is the function of the total circulatory balance of Figure 11, and it depends on whether the system operates at a steady state of replenishment from outside the Solar System or is a system that has decayed gradually with time from an original state that contained such large numbers of objects that the decay-rate has been small over time spans of hundreds of millions of years. At the present state of knowledge, there is no general agreement concerning these alternatives **(Note 3.3)**.

CHAPTER 4

Chaotic Orbits and Nonlinear Resonances

It is well known that Johannes Kepler (1571–1630) was fascinated by the question of numerical regularities in the universe (e.g., Haase, 1975; Barut, 1986), and that notion had been a commonly held worldview at least since the time of Pythagoras (see the tribute to Michael W. Ovenden early in, and immediately preceding, the Introduction). This implicit sense of order is reflected in the roots of the words "cosmos" and "cosmology," which invoked a harmonious image of the universe, but "harmonious" was meant in the linear sense of harmonic functions — the "pleasing" music of the spheres — rather than in the nonlinear sense emphasized in the present volume. In making such an anthropomorphic judgment concerning cosmic structure, it was not recognized — or was recognized and went unappreciated — that all but a minuscule fraction of the numerically ordered states available to the universe were excluded by fiat (see the discussion of the wheel of fortune in Note 4.2). Everything nonharmonic was referred to in terms, or epithets, employing words such as "dissonance," "discord," "unnatural," "pathological," and so on, because most persons did not recognize — by the same criterion, the criterion of harmony read by the machinery of sensory perception — that the most awe-inspiring musical sounds, spectral colors, and architectural forms, of either human or nonhuman construction (both being natural, a fact that often goes either unrecognized or unacknowledged) are derivatives of the irrational and transcendental numbers, namely the relationships between the numbers *pi* and the *golden mean* (see Universal Order, Geometry, and Number, in Note P.4).

The nonlinearly resonant and chaotic order that I emphasize in the present volume does not really change the fundamental implication that the word Cosmos stands for a numerically ordered universe. It just expands on that theme to include the implications of the systematic nonlinear processes that characterize a physically *realistic* universe. During the "more modern" era of nineteenth-century cosmological studies, the astronomer Daniel Kirkwood stands out as a scientist who, like Kepler, devoted much of his career to the search for numerical regularities in the universe. Although he is mainly remembered for his studies of the asteroids, and for first identifying the relatively depopulated orbital intervals of the Asteroid Belt now called the Kirkwood Gaps, he also wrote many papers on other forms of commensurabilities, ranging from relationships between the orbital and rotational periods of the planets to the structure of the Solar System and the incidences of meteors and comets (e.g., Kirkwood, 1849, 1860, 1867, 1871, 1872, 1876, 1888, and 1891). A number of his ideas and observations anticipated relationships discussed in the present volume — as did those of Michael Ovenden (see Ovenden, 1972, 1973, 1975, 1976; Ovenden and Byl, 1978; Ovenden et al., 1974; and Roy and Ovenden, 1954, 1955).

Kaula (1968, pp. 222ff) summarizes several types of commensurabilities and resonances long known to exist in the Solar System (**Note 4.1**). His Table 5.6 (after Roy and Ovenden, 1954, 1955) shows that the number of observed commensurabilities within a given bandwidth of numerical uncertainty, relative to an exact rational fraction, is at least double the number that could be attributed to chance (see Roy, 1988a, Chaps. 1, 5, and 8, for a recapitulation and a more extensive discussion of commensurabilities in celestial dynamics). What is more, the ratio of the number of commensurabilities observed to the number attributable to chance increases with decreasing bandwidth! This increase is important in view of the stepwidths of rational fractions in nonlinear phase-locking shown by laboratory and computer experiments. Such a progression is called a Devil's Staircase (e.g., Bak, 1986; Shaw, 1987b). The width of a phase-locked step in a completed staircase of resonances of the sine-circle map (see Table 1, Sec. D; and Fig. 14), counted relative to integer numbers of 2π cycles, for example, is greatest at the ratios of relative phase *0/1* or *1/1* (depending on one's viewpoint concerning whether the motion is normalized relative to the numerator or the denominator), is next largest at *1/2* (or *2/1*), next largest at *1/3* and *2/3* (or *3/1* and *3/2*), etc. — where the latter pairs of ratios are symmetrically distributed about the central ratio *1/2* (or *2/1*) — and it decreases with decreasing separation between adjacent fractions, where the integers stand for the number of 2π cycles of one oscillator vs. the other.

No matter how subdivided a rational sequence becomes, however, the chance of hitting any rational fraction with vanishing uncertainty, or with vanishing resonance, approaches zero, because there are uncountable numbers of irrational numbers between any two rational numbers (see Dauben, 1983; Shaw, 1987b). *This is the opposite of what is observed.* The demonstration that the relative numbers of commensurabilities increase with increasing resolution testifies to increasingly complex but rigorous nonlinear order (cf. Ovenden, 1975). In the limit, where there is an approach to noninvertibility and nonintegrability, such a system operates near the transition to chaotic dynamics. Alternatively, an irrational frequency ratio (a quasiperiodic orbit) testifies directly to nonlinear resonance, because it can occur by chance only if there is zero uncertainty in the orbital parameters (Shaw, 1987b). This relationship is one form of "uncertainty principle" of nonlinear dynamics. An analogous form of nonlinear uncertainty concerns the "widths" or "fuzzinesses" of fractal attractor-basin boundaries (cf. McDonald et al., 1985; Gwinn and Westervelt, 1986; Shaw, 1991, p. 121; Belbruno, 1992).

The Asteroid Belt

Studies of the Asteroid Belt by Wisdom (1982, 1983, 1987a, b, c; 1990) demonstrate that the Kirkwood Gaps in the distributions of asteroids, first established by Kirkwood (1876, 1888), can be explained by nonlinear effects in which particular commensurabilities between asteroid orbital periods and the period of Jupiter ($\sim$11.86 years) are unstable relative to neighboring chaotic orbits ($\pm$ quasiperiodic and/or higher-order resonances). Some results of the computations by Wisdom (1987a) are summarized in **Figure 12**. Taking Jupiter's orbital period as $T_J = 1$ (and $f_J = 1/T_J = 1$), because Jupiter's orbital period is chosen as the common divisor by which the commensurability is to be tested, we can define commensurabilities in the Kirkwood convention by $T_J/T_A = 1/T_A = f_A/f_J = f_A$, where T_A and f_A are the reduced orbital period and frequency of an asteroid. Major gaps within the main-belt asteroids (MBA) — as shown by a histogram of about 4000 asteroids given by Binzel (1989, Fig. 2) — are at the commensurabilty ratios of 3/1, 5/2, and 7/3, where the MBA itself is delineated by larger gaps at the 4/1 and 2/1 commensurabilities (see Fig. 13; cf. Yoshikawa, 1991; Ipatov, 1992). The reciprocals of these ratios vary in crude, and nonlinear, proportion to the semimajor axes of the orbits (where the semimajor axis is half the major axis of an elliptical orbit and corresponds roughly to the orbital distance from the Sun for low-eccentricity — meaning nearly circular — orbits).

An asteroid in a synchronous orbit with Jupiter, with a commensurability ratio of *1/1*, has a semimajor axis approximately equal to Jupiter's orbital distance of $\sim$5.2 AU from the Sun. Asteroids that have semimajor axes approximately equal to the orbital distances of Mars (at $\sim$1.52 AU from the Sun) and Earth (at 1 AU from the Sun) have relatively large Kirkwood ratios ($\gg$ 4/1). As the Kirkwood ratios increase, so does the probability that an asteroid from the MBA may become a "Mars-crosser" or an "Earth-crosser" sometime during its chaotic evolution (if its orbit crosses Earth's orbit during this evolution, of course, it has also crossed that of Mars). The asteroid groups called Atens and Amors (Figs. 11 and 13) include those asteroids — by taxonomic classification, whatever their orbital history — that now have orbits, respectively, that lie entirely inside and outside of Earth's orbit (see McFadden et al., 1989, Table 1; cf. Gradie et al., 1989). The orbits of the Apollo asteroids, on the other hand, cross Earth's orbit, and many also cross the orbit of Mars — and conceivably even the orbits of all the other asteroids (i.e., the Apollo group is dynamically heterogeneous, and the orbits of some of its membership strongly resemble the orbits of short-period comets, perhaps representing the late stages of orbital decay of objects from the cometary reservoirs; see Figs. 11, 13, and 15, and Note 6.7; cf. Kaula, 1968, Fig. 5.9;

Chaotic Orbits and Nonlinear Resonances

Kresák, 1979, 1991; Drummond, 1981, 1991; Weissman et al., 1989, Fig. 1 and Table III; McIntosh, 1991; Meech, 1991; Wetherill, 1991a).

With the recognition that asteroids could have been "kicked out" of the Kirkwood Gaps — reflecting the notion that they can be ejected from the vicinity of any resonance state by *chaotic bursts* (see Chap. 5; and Notes 5.1 and 6.1) — it is conceivable that the asteroids in the vicinity of Mars and Earth could have come from anywhere within the Asteroid Belt (cf. Wisdom, 1982, 1983, 1985, 1987a, b, c), or beyond (see Fig. 13, Insets; and Note 6.7). The "asteroids" 2060 Chiron (not to be confused with Pluto's satellite Charon) and 2201 Oljato (Weissman et al., 1989, pp. 908f) are cases in point — where the two-part classification by number and name indicates that these are among the class of formally accepted "asteroids" with well-determined orbital elements (see Bowell et al., 1979, 1989; Pilcher, 1979, 1989; Tedesco, 1989). 2201 Oljato is an $\sim$2-km-diameter near-Earth object that is "one of the strongest candidates for an extinct comet among the near-Earth asteroids" (Weissman, loc. cit., p. 909). 2060 Chiron is an "outer-Solar-System asteroid" (see Fig. 13, Insets; and Weissman et al., loc. cit., Fig. 1 and Table III) which, like many Apollo asteroids, has a comet-like or degenerate comet-like orbit (cf. Kaula, 1968, Fig. 5.9; McFadden et al., 1989, Table 1), but it is not classed with the Apollo-type objects because it lacks the Earth-crossing criterion — indicating how arbitrary the taxonomic classifications can be relative to orbital evolution (cf. Fig. 13, Insets; and Note 6.7). Chiron's orbital semimajor axis is $\sim$13.7 AU, and its distance from the Sun at perihelion passage is $\sim$8 AU, between Jupiter and Saturn (its next perihelion passage will be in 1996, when we will be hearing a lot more about this unusual "asteroid"). Chiron apparently is quite large (perhaps $\sim$200 km in diameter), and it may in fact still be a comet, at least according to the criteria of comet-like outbursts and a visible coma (French et al., 1989, pp. 479ff. and p. 486; Weissman, 1991a; Hoffmann et al., 1993).

The chaotic regimes in Wisdom's computations strongly resemble those of the damped driven pendulum and other coupled nonlinear oscillators (see Table 1; and studies by Miles, 1962, 1984; Beckert et al., 1985; Gwinn and Westervelt, 1985, 1986; Tritton, 1986, 1988, 1989; Henderson et al., 1991; Matsuzaki and Furuta, 1991). Figure 12a, redrawn from Wisdom (1987a, Fig. 5), is a simulation of the orbital eccentricity of an asteroid near the 3/1 gap over an interval of a few million years. This oscillatory time series, which alternates intermittently between oscillatory regimes with relatively low orbital eccentricities and oscillatory regimes with intermediate to high eccentricities, constitutes a form of bistable intermittency that is common to many types of coupled nonlinear oscillators that display phenomena of chaotic bursts (cf. Chap. 5 and Note 5.1; and Shaw, 1991). Wisdom (1987a, p. 118) observes: "There are bursts of irregular high-eccentricity behavior interspersed with intervals of irregular low-eccentricity behavior, with an occasional eccentricity spike." The "spikes" represent what I refer to as chaotic bursts. The irregularities at lower amplitudes might be viewed as high-frequency bursts or as chaotic excursions that intermittently escape from one potential well to enter the other in a quasisteady manner. The smaller bursts imply possible

Mars-crossing excursions ($e > 0.3$), while the more widely spaced, and larger, spikes may lead to Earth-crossing orbits ($e > 0.6$). I shall demonstrate later (Chap. 5; cf. Notes 5.1 and 7.4) that the timing of spike-like bursts in coupled oscillators are power-law functions of the difference between chaotic orbital frequencies and the critical frequency for unstable periodic orbits dynamically embedded within the chaotic orbital motions. In that case, the critical frequency could be a function of the orbital parameters of Jupiter, of Jupiter + Jovian satellites, and/or of other orbital parameters of the Solar System that affect the motions of the asteroids. This interpretation assumes that the dynamics are dissipative, and that the orbits move on chaotic attractors (cf. Notes P.3 and I.4).

A section of a phase portrait for the case of Figure 12a is shown in Figure 12b. The distance from the origin (0,0) to any point is a measure of the asteroid's orbital eccentricity, and the coordinates describe the asteroid's motion as a function of the mean longitudinal differences between the asteroid and Jupiter. The circles near the origin correspond to regular motion, and the irregularly shaded regions correspond to chaotic motions. The cortex-like involutions of the greatest density of points (darkest shading), which I have crudely outlined, correspond to the irregularities in the upper part of Figure 12a. This structure is a connected fractal version of the densities of points shown in the computer plot by Wisdom (1987a, Fig.7). The density of points in Wisdom's diagram is a measure of recurrence frequencies, because each point in his diagram represents one cycle of the mean longitudes (the Poincaré section is stroboscopic; cf. Note P.3). Such involuted-convoluted structures, resembling a section of cerebral cortex, are characteristic of the informational content of chaotic regimes. The "fractal coastline length" tends to fill that area of the plot, suggesting a fractal dimension near 2 for points around the periphery of the atoll-shaped section, or 3 (like the "surface area" of the brain) in a three-dimensional phase space. These results — assuming that the distribution of points in Wisdom's diagram are not artifacts produced by the drafting of the figure — are reminiscent of organized complexity in adjoining regimes of dissipative attractors analogous to those produced by the sine-circle map (Table 1, Sec. D; cf. Wisdom, 1990).

Figure 12c illustrates the vacant 3/1 commensurability in the Asteroid Belt. This relationship fits well with dynamical phenomena that are characteristic of an interior crisis. The crisis and subsequent bursting effects occur in a vicinity where a chaotic attractor, called the "core attractor," enters the "sphere of influence" of an unstable periodic attractor interior to it in phase space (see the discussion of later demonstrations, as portrayed in Figs. 17 and 18). The periodic attractor would represent the filled state of an ordered 3/1 resonance condition, which fits with the fact that some of the commensurability positions in the Asteroid Belt as a whole, illustrated in **Figure 13**, are peaks in the distribution of asteroids rather than gaps — shown schematically in Figure 13a on the basis of a population of about 4000 asteroids from Binzel (1989, Fig. 2). The spatial distribution of these 4000 asteroids is sketched pictorially in Figure 13b.

It is instructive here to examine the growth rate of asteroid discoveries, which has been increasing exponentially since the discovery of Ceres — first, and largest, in the list of the numbered asteroids (cf. Bowell et al., 1979; Tedesco, 1989) — in 1801, with a doubling time of about 16 years over 12 generations (Bowell et al., 1989, Table 1), and there is no reason to expect it suddenly to "saturate" at a fixed number **(Note 4.2)**. [This doubling time of 16 years apparently corresponds to the time constant of astronomical discovery rates in general (see de Vaucouleurs, 1970), as discussed in Chapter 8 (cf. Note 8.15).] It is also quite clear from the tabulations of asteroid optical magnitudes and diameters (Bowell et al., 1979; Tedesco, 1989) that the record of named asteroids is far from complete at diameters less than about ten to twenty kilometers, and there is no minimum limit on these sizes, apparently down to dust-size particles (cf. Sykes et al., 1989). (An overview of the fractal size distributions of various classes of Solar System objects, including the asteroids, is given in **Appendix 8**.) If the growth rate for asteroid discoveries is a function of continually improving techniques of documentation, then the number-size relation probably is a power-law or fractal function. At the present rate the count would hypothetically exceed *65,000 documented* (numbered and named) asteroids in about four more generations (year ~2055). The fractal dimension of 2.37 estimated from the data in Appendix 8 (Inset) suggests that this number would consist of objects with minimum diameters of 5 to 10 km.

The observational growth factor of $\sim 2\times$ every 16 years (mean value) derived by Bowell et al. (1989, Table 1) emphasizes the discussion in Chapter 2 concerning discrepancies between the low documented crater counts and the astronomically inferred potential-impact rates. In that same analysis (Bowell et al., 1989, Table 1), the projected number of "discovered," but not fully documented and named, asteroids has already reached about 65,000 objects. There are also many tens of thousands of incompletely documented "asteroids" representing one-time sightings of probable asteroids or miscellaneous unidentified objects, such as "fast-moving objects" with unknown orbital characteristics (comet fragments, etc.), within the records of individual observatories (e.g., Tatum, 1988). Binzel et al. (1992, p. 779) have estimated that the number of near-Earth asteroids (NEAs), consisting of the Aten, Apollo, and Amor asteroids (plus any asteroids with orbital properties of hybrid characteristics; cf. Fig. 13), is roughly 10,000 objects with diameters larger than 0.5 km (equivalent to potential craters with diameters greater than several kilometers), of which only 200 have been discovered. The number 10,000 for NEAs can be compared with the total known population of asteroids by examination of the relative peak heights in the histogram of Figure 13a, which would suggest that the above estimates of very large numbers of undiscovered asteroids are, if anything, conservative. Binzel et al. (loc. cit.) also point out that the equivalent sampling of main-belt asteroids (MBAs) has a size cutoff at least four times larger than that of the NEAs (i.e., corresponding to diameters of main-belt objects greater than 2 km), and imply that even the dedicated studies of MBAs with diameters as large as 10 km or more may be incomplete. This statement, too, would appear to support my remarks concerning the incompleteness of asteroid discoveries (see Appendix 8).

These observations are of major importance to the types of discoveries that could be consistent with the *celestial reference frame hypothesis* (CRFH) of the present volume, especially because they would represent primarily the populations of objects between 1 and 10 km in diameter, rather than those of the larger diameters, and these are the diameters of impactors that are generally correlated with terrestrial craters between about 10 and 100 km in diameter. It is therefore likely that estimates of potential terrestrial impacts based on mechanisms of the Wisdom type are likely to increase during the next few decades (cf. Chap. 2). The history of discoveries of the named asteroids is documented with regard to physical properties by Bowell et al. (1979) and Tedesco (1989; asteroid numbers 1–3318, inclusive), and with regard to the circumstances of discovery by Pilcher (1979, 1989; asteroids 1–4044, inclusive).

The "photo-negative" histogram in Figure 13a emphasizes the telescopically "empty" states in the Asteroid Belt (empty is dark, filled is light; the reverse of Fig.12c). Kirkwood ratios are shown at the bottom, together with the orbital distances from the Sun and canonical symbols for Earth, Mars, and Jupiter. The taxonomic names shown by the labels identify characteristic groups of asteroids associated with the illustrated Kirkwood Gaps, but the taxonomy here does not necessarily correspond to identical genetic histories that might be implied, for example, by the context of Figure 11 (cf. Binzel et al., 1989a), a problem sometimes encountered in paleontological vis-à-vis biological taxonomies as well — meaning simply that taxonomic hierarchies and genetic hierarchies in the natural sciences often do not match up. Figure 13b shows a "snapshot" of the Asteroid Belt at the date and time of the Asteroids II conference (Tucson, Arizona, 15:30 UT, 8 March, 1988), drawn schematically from the back cover of Binzel et al. (1989a). The Kirkwood Gaps are blurred by orbital eccentricity (if the orbits were concentric circles, the Gaps would be conspicuous). The ordered clustering and filamentary structures seen in Figure 13b reflect instantaneous densities of points, but the "continuous" patches are misleading because the "holes" in them are not portrayed. Notice the scatter of points within the foci of the inner planets and the clusters or "swarms" of points crossing Jupiter's orbit (planetary positions are shown by tick marks). This illustration emphasizes the nonlinear organization discussed relative to Figure 12b and lends some realism to the symbolic hourglass of Figure 11. The "granular" appearance resembles the hourglass in the sense that objects scatter both outward and inward (i.e., depending on which way the hourglass is "turned" by the effects of self-organized orbital forcing phenomena), in steady-state balance with the MBA (the "central heap"), and in conjunction with long-term net replacements of objects from elsewhere (e.g., from cometary reservoirs and/or from outside the Solar System; cf. Chap. 2, and Fig. 13).

The Devil's Staircase of **Figure 14** presents a comparison between the resonances of Figures 12 and 13 and the phase-locked resonances of dissipative coupled oscillators, such as the damped driven pendulum and the sine-circle map (Table 1, Sec. D; cf. Note 4.1). The Staircase corresponds to a condition near the edge of chaos, a condition where there is complete "filling" of all possible rational

Chaotic Orbits and Nonlinear Resonances

frequencies (cf. Bak, 1986; Shaw, 1987b, Figs. 1 and 2). The sine-circle function can be called "the simplest coupled oscillator" because the $2p$, π, cycles of the circle represent a natural frequency, and the nonlinear recursion steps represent the relative frequency of another oscillator coupled with it according to assigned values of W and K (Table 1, Eq. 5). Figures 13 and 14 are analogous if Jupiter is taken as representing the circle (fundamental oscillator) and the positions of the asteroids are taken as representing the recursive iterations of an oscillator coupled with it (see **Appendix 9**). The ratios in Figure 13a are inverted in Figure 14, as previously discussed (i.e., the 3/1 Kirkwood Gap corresponds to the 1/3 step in Fig. 14; cf. Note 4.1). Resonance states in Figure 14 are labeled α/β to fit the convention of Allan (1967a) used later to describe spin-orbit resonances (see Chap. 6, Fig. 20; and Note 6.1).

Now we can compare the asteroid resonance conditions with the stepwidths in Figure 14, as illustrated in Appendix 9 (see Note A.2). The "empty" resonances of the Asteroid Belt (Kirkwood Gaps) are represented by the (inverted) circle-map resonances 1/4, 1/3, 2/5, 3/7, 1/2, and possibly 3/5 (the first and last of these resonance conditions occur in sparsely populated regions of the Asteroid Belt). Partly filled resonances, in this inverted convention, occur near 2/3, 3/4, and 1/1. The peaks labeled Hungarias and Cybeles in Figure 13a, which occur at $< 1/4$ and $> 1/2$, are ambiguous, because the neighboring ratios have small stepwidths (see Inset in Fig. 14; and Appendix 9). This comparison suggests that the Asteroid Belt might be modeled in a manner analogous to an hourglass in which the particles ("asteroids") are "binned" preferentially as some function of the nonlinear-resonance conditions of a circle map (e.g., Appendix 9). In such a model, the bins are being simultaneously filled and emptied by the net responses of the nonlinear-resonant and chaotic dynamics of the Asteroid Belt, as well as by the distribution of bursting frequencies in other object-belts of the Solar System that can contribute to any of the orbital "bins" in the vicinity of the Asteroid Belt. The result is that instead of one symmetrical pile of objects, as in the usual hourglass, the accumulation of objects would resemble a hopper-like system in which a series of bins are filled and emptied at different rates, but where the system of "baffles" between the bins is leaky, permitting some exchange of material between them (to be more realistic, these imaginary "bins" would have to be filled with bursting popcorn, so that the analogous effect would also permit objects to hop across one or more intervening bins). The population of objects in a given "bin" would depend on the attracting vs. repelling forces acting in that vicinity (self-induced effects) as well as on the net balance of forces acting on the system as a whole.

The "bin model" resembles the motions of a "tectonic landscape" that is being both uplifted and eroded by the net effects of many dynamical phenomena simultaneously acting on it and within it. In the absence of any standard of comparison for the steady-state profile of the "asteroid valley and ridge province" of the Solar System, however, we cannot say whether the system is (1) generally filling, (2) generally emptying, or (3) intermittently filling and emptying at time-dependent/sign-reversing rates characteristic of each of the different orbital resonance states. The resonance states, in turn, are functions of the relative interactions between (a) pairs of individual asteroids, (b) clusters of asteroids, (c) orbital bands of clustered asteroids, (d) clustered bands of clustered asteroids, and (e) the crudely self-similar structure of the Asteroid Belt as a whole relative to general chaotic motions of the planets at orbital distances between the Sun and Jupiter (as modulated, damped, and/or enhanced by the motions of the other planets and object belts beyond Jupiter; see Fig. 13 and the discussion of Fig. 21; cf. Appendix 11). Such a description is reminiscent of the self-similar fractal structure of a Devil's Staircase and/or a *strange attractor* (cf. Appendix 9 and Note P.3) — and it is also reminiscent of the filling and emptying of electron shells among the various valence levels of a multivalent chemically reactive atom (cf. Richards, 1988). By the landscape analogy, relatively small populations in the Hildas, Thule, and Trojan vicinities (Fig. 13a) might be interpreted as residual "hillocks" of a denuded landscape, because there are no apparent occupancies at 3/4, 4/5, etc. I am more inclined, however, to view these locations — as I do the possible existence of resonances in the vicinity of Earth (e.g., Atens, Apollos, and Amors) — as being analogous to the detrital products of the watershed of an active mountain range, the latter represented by the MBA of Figure 13a, where the geomorphic profile across the range and pediment surfaces locally waxes and wanes according to the varying rates of tectonic and hydrologic processes (cf. Shoemaker et al., 1989, and other reviews in Binzel et al., 1989).

The peak "topologic survivorship" (the MBA) represents orbital frequencies much higher than that of Jupiter and much lower than those of Mars and Earth, suggesting resonance conditions that may be consistent with systems of complex multiple oscillators (cf. Carroll, 1987; Arecchi, 1991; Brown and Chua, 1991; Frouzakis et al., 1991; Murali and Lakshmanan, 1991, 1992). For example, Ceres is at ~2.8 AU, which is near the bisector (2.85 AU) of the difference between the Earth-Jupiter bisector (3.1 AU) and the Sun-Jupiter bisector (2.6 AU). This orbital position reflects a weighted net distribution of resonances involving Jupiter, the other planets, and the Sun (see Fig. 21). Resonances that are synchronous with Jupiter ($\alpha \cong \beta$), such as those of the Trojan asteroids, could represent strong orbital stabilities or incomplete erosion or both vs. resupply (cf. clusters in Fig. 13b); detailed descriptions of objects in the Trojan vicinity are given by Shoemaker et al. (1989). According to the model in Figure 11, and the above discussion, such balances are general possibilities for all sites of the Asteroid Belt — and beyond (see Fig. 13, Insets) — meaning that Figure 13 is a portrayal of a kinetic landscape.

Analogies Between the Asteroid Belt, the Kuiper Belt, and the Oort Cloud

The properties of the Asteroid Belt are richly documented (Binzel et al., 1989). By contrast, the documentation of the Kuiper Belt, Oort Cloud, and other poorly defined regions of the outer Solar System is based mainly on inferences from (1) orbital distinctions between "new" and "old" comets and short-period and long-period comets, etc. (e.g., Oort, 1950; Kaula, 1968, pp. 231ff; Bailey et al., 1990), (2) theories of planetary accretion and relics of preexisting or existing stellar structures around the Sun (e.g., Kuiper, 1951; Bailey, 1987; Stern, 1987, 1988; Stern and Shull, 1988; Aumann and Good, 1990; Rawal, 1991, 1992), or (3) models of interactions between the planets and the distal regions of the system between about 30 AU and $\sim 10^4$ AU, relative to the Galactic disk and to perturbing stars (e.g., Hills, 1981; Weissman, 1986, 1990b, 1991b; Duncan et al., 1987, 1988; Heisler, 1990; Bailey, 1990; Bailey et al., 1990). The distal reaches of the Solar System, though little known, show indications of resonant and chaotic motions (Duncan et al., 1987; Torbett and Smoluchowski, 1990; Weissman, 1990a, b; Stern, 1991; Weidenschilling, 1991). This is the realm of comets, but "comets" are proving as difficult to define precisely as is the unqualified term "asteroid" (e.g., Fig. 13, Insets; cf. Bailey, 1990). The work by Stern (1991) suggests that large planetary objects ($\sim 10^3$ km in diameter) should exist beyond 30 AU (these objects differ from the "giant comets" thought to originate outside the Solar System; see Clube and Napier, 1986, 1990; Clube, 1989b, 1992). Stern's model, combined with that suggested by Figures 9 and 10, would imply a spectrum of relict objects some of which would be of terrestrial affinities, hence perhaps without comas and typical comet trails. Presumably, such objects would be indistinguishable in their eventual orbits from Apollo-family asteroids (asteroids with cometary orbits; see Weissman et al., 1989). Similarly, the Kuiper Belt, the inner Oort Cloud (Hills Cloud, or "inner cometary reservoir"), and the outer or "classical" Oort Cloud (which is the region originally predicted by Oort, 1950, hence is properly identified with his name) are ill-defined except by orbital criteria and knowledge of perturbing forces between 10^3 and 10^4 AU (cf. Duncan et al., 1987; Bailey, 1990; Weissman, 1990b).

Judging from the complexity, the contiguous and/or overlapping orbital distances, and the confusion of theories and opinions about all these cometary sources, I would opine that, in general, "the cometary source" within the Solar System can be more simply viewed as a self-organized system of attractors that interacts with the Solar System as a whole in essentially the same way as does the Asteroid Belt (cf. Fig. 21), where the two are "diffusively" intermingled across a fuzzy attractor-basin boundary in the vicinity of the outer planets and the Kuiper Belt (see Fig. 13, Insets; cf. Weissman et al., 1989, Fig. 1; Bailey, 1990). But it is clear that many workers would also appeal to sources beyond the sphere of influence of the Solar System for at least some part of the comet flux. This latter viewpoint has been espoused for many years by S. V. M. Clube (see Clube, 1967, 1978, 1982, 1985, 1989b, 1992; Clube and Napier, 1982a, b, 1984a, b, 1986, 1990; Olsson-Steel, 1989; Bailey et al., 1990), and it is also given support, on the basis of different arguments, by Stern (1987, 1988). Within the nonlinear-dynamical context of the present volume, I would therefore suggest—as is typical in complex systems of coupled oscillators with numerous attractor basins—that the system of cometary attractors within the Solar System also is diffusively intermingled with the more general basins of attraction of cometary objects within the Galaxy, whether that entails direct production within molecular cloud complexes (the "comet factories" of Clube, 1985) or capture from the "Oort Clouds" of other stars (Stern, 1987) during interactions with the Solar System Oort Cloud **(Note 4.3)**.

Duncan et al. (1987), Torbett and Smoluchowski (1990), Heisler (1990), and Weissman (1990a, b) discuss some of the processes that can influence cometary reservoirs. Chaotic behavior of cometary orbits in these reservoirs is illustrated in **Figure 15**, which is based on the work of Torbett and Smoluchowski (1990) and Duncan et al. (1987). Figure 15a, redrawn from Torbett and Smoluchowski (loc. cit., Fig. 3), shows chaotic effects in the relationship between semimajor axes and eccentricities for a sample of comets in the inner part of the Kuiper Belt. Although Figure 15a is plotted differently from Figure 12, the results it presents for comets are similar to chaotic effects in the Asteroid Belt. Chaotic motions (crosses and filled circles) tend to "open up" the phase space, permitting some objects to be ejected from one position and reinjected into another position at smaller Neptune-crossing distances (filled circles). This ejection-reinjection process is suggestive of the phenomena of interior crises already discussed relative to the Asteroid Belt (Fig. 12). Another portion of the Kuiper Belt in Figure 15a (small dots) retains the properties of regularity that were originally assigned as starting conditions by Torbett and Smoluchowski. Such a result, combining regular and chaotic motions, thus fits a general criterion for a system poised near the "edge" of chaos (cf. Note 4.2; and Ruthen, 1993). Transition tendencies for only one set of initial a and e values are shown, but other calculations by the same authors show large increases in e from low-e initial states during a simulated evolution time of 10 m.y., thus yielding patterns that resemble the phenomena of Figure 12.

Figure 15b shows results from Duncan et al. (1987, Fig. 3) based on the Monte Carlo evolution of a typical comet in a generalized comet cloud over the lifetime of the Solar System. The initially assigned perihelion, q, shown by the square is near the Uranus distance (as is the case for Fig. 15a), but has a semimajor axis, a, of about 2000 AU. In other simulations, objects not originally accreted to or bound to the protoplanets inside about 100 AU maintained constant perihelia for initial values of $a < 2000$ AU; thus the longer computation was shortened by starting with that value of a. The semimajor axis, a, in Figure 15b increases in an oscillatory fashion to about 10^4 AU at nearly constant perihelion, and at about 10^4 AU there is a dynamical transition that increases the comet's perihelion in an oscillatory manner until it is within a factor of three or so of the semimajor axis (meaning that the extremely elongate elliptical orbits that existed prior to this transition opened out to somewhat more circular orbits). In the case shown, the comet returns to the previous type of motion at least once (as is indicated by the label "Recurrence state" in Fig. 15b), owing to perturbations by the Galactic tide. In caricature, the evolution would look something like the motion of a rubber band that gyrates around the planetary system, alternately in an elongate mode of relatively close approach and in a more open mode remote from the planets (e.g., Fig. 13, Inset 3).

The long-term temporal evolution of perihelion, q, and semimajor axis, a, of the orbital motion of Figure 15b qualitatively resembles that of the bistable oscillation shown in Figure 12a, in that the values of both of these parameters oscillate in time between characteristic limits. The nonlinear driven and damped spherical pendulum shows analogous motions, the orbital properties of which, in phase space, have been worked out by Miles (1984); see also Tritton (1986, 1988, Fig. 24.4, 1989). Such oscillators, in the chaotic regime, experience unpredictable changes in orbital trajectories as well as demonstrating the existence of interior crises that show even more erratic behavior — consisting of high-amplitude bursts or spikes, as in Figure 12a. Each of these chaotic systems — asteroids, spherical pendulums, and comets — illustrates the same types of behaviors, despite the fact that the forcing and damping functions, and the scales of motion, are vastly different. In the Asteroid Belt, chaotic motion is driven by the perturbing influence of Jupiter's gravitational field (and/or that of the planetary motions in general), while the spherical pendulum in the laboratory is perturbed by periodic displacements of the point of suspension, and the cometary motions of Figure 15b are perturbed by the periodic influence of the Galactic tide as the main forcing term (Duncan et al., 1987). Stated differently, the Galactic tide, by analogy with Jupiter in the asteroid case, is (like the sine-circle map) the forcing frequency or controlling natural frequency, and the orbital frequencies of the comet (like the motions of an asteroid, or of a pendulum bob) represent the iterated function. In each instance, other effects of forcing and damping contribute to the orbital complexity. For example, comets are affected by complicated positive and negative accelerations related to the dynamics of the central region of the Solar System, as well as by nearby stars (see Table 1, Eq. 1). I return to the discussion of coupled oscillators in the next chapter, where experimental examples of chaotic crises are illustrated.

The calculations of Duncan et al. (1987) also illustrate the evolution of orbital inclinations. In the calculated cometary evolution, the ecliptic of the Solar System is inclined about 60° to the Galactic plane. An evolution dominated by the Galactic tide therefore produces large inclinations (relative to the ecliptic plane) in the outer regions, which subsequently evolve to small values at distances less than 1000 AU (Duncan et al., 1987, Fig. 5). According to the model, then, a large range of inclinations should exist for objects coming from the Oort Cloud, including individual long-period comets and showers of shorter-period comets from beyond 10^3 AU. Such correlations, however, are uncertain, and Duncan et al. (1987) did not consider the secular evolution of the angular relationship between the ecliptic plane of the Solar System and the Galactic plane (an evolution that is also nonlinear) over the lifetime of the Solar System. Wisdom (1987a, b, c), exploring the e-i relationship for the Asteroid Belt, found mutually dependent chaotic variations that enhance the instabilities already discussed (cf. Fig. 22). To judge from Wisdom's results, the e-i relationship — hence also the e-i-q-a relationships of the cometary evolution in Figure 15b (and Fig. 13, Insets) — may be an important discriminant between chaotic, quasiperiodic, and phase-locked regimes.

Perturbations due to passing stars were found by Duncan et al. (1987) to be less important for the control of the secular behavior than is the Galactic tide. Stars did, however, control the activation of comet showers from an evolved inner reservoir by disrupting the region beyond about 2×10^4 AU (the classic Oort reservoir), giving rise to cascades analogous to those of Figure 11. In the eventual state of the model system, about 80% of the "surviving" objects were in the inner cloud, about 20% in the outer cloud. The survivors represent about 40% of the initial objects; of the others, about 5% entered the inner Solar System, and the rest (roughly 55% of the initial objects) were either ejected by planetary and/or stellar effects or evolved to distances $\gg 10^5$ AU. But because the actual initial population in such a model (as well as in all other models of cometary reservoirs) is unknown, the average incidence of "new" comets is highly uncertain.

A time series based on the results of Monte Carlo simulations of the background comet flux and comet showers by Heisler (1990, Fig. 1), with modifications, is shown in **Figure 16**. The numbers reflect the assumed population of initial objects, where N is the number per m.y. calculated on the basis of a reservoir containing about 10^7 objects with $a \cong 10^4$ AU. Estimates four orders of magnitude larger were used for the *total* number of objects at all values of a (Heisler, 1990, Fig. 1b), giving an uncertain average of the order of ten comets per year and peak rates (showers) of the order of 100 comets per year — in each case referring to comets with perihelia < 10 AU (cf. Hills, 1981; Weissman, 1990a, b). Inside 10 AU, complications of the sorts previously discussed in relation to the basins of attraction of the Asteroid and Kuiper Belts introduce additional ambiguities concerning the fates of individual comets. Possible effects in this sort of calculation of the comet flux by the objects predicted by Stern (1991) have not, so far as I know, been evaluated.

The dashed curve in Figure 16 illustrates a phenomenon I have noticed in chaotic time series that retain some semblance of invertibility. This phenomenon is illustrated by reversing the solid curve, but shifted in such a way that there is a crude match between the curve forms of forward and reverse progressions. This manipulation is directly analogous to the so-called *Mirror Theorem* of Roy and Ovenden (1955; cf. Roy, 1988a, pp. 119f, 141ff). The Mirror Theorem identifies two configurations of many-body systems in which the future states of the system are reflected across the position of generalized "conjunctions" where all of the bodies are either (1) arranged in a plane, with velocity vectors perpendicular to that plane, or (2) arranged along an axis, with velocity vectors perpendicular to that axis. To the degree that a system approximates such configurations, it is characterized by a sort of degenerate periodicity $2\Delta t$, where Δt represents the one-way recurrence time of the particular configuration. That is, the system "winds up" to the mirror state in one orbital direction during Δt (in a manner that is literally analogous to the motions that define the *winding numbers*, W, in phase plots such as Fig. 14; cf. Note 3.1) and "unwinds" identically backward from that state in the reverse direction (across the "mirror"), so that the whole sequence takes $2\Delta t$. Because the configurations represented by Figure 16 were originally "randomized," the result is a spectrum of "pseudoperiods" reflected across the "origin," or inversion point, at roughly $t \cong 80$ m.y. (and age $\cong 80$ Ma), giving pseudoperiods, $2\Delta t$, ranging from ~ 20 m.y. or so to ~ 240 m.y. or more, the latter resembling various rotation periods of the Galaxy (cf. Shaw, 1988a). By coincidence, the maximum shower activity would be at an "age" near 65 Ma (upper scale) if the pattern were crudely symmetrical about the maximum amplitudes, and if it were plotted backward from an absolute starting time at an age of 120 Ma or so (e.g., between the minima shown in Figure 7 for the paleomagnetic and cratering time series, respectively). In general, of course, there is no reason to expect that repeated runs of a Monte Carlo simulation, or even repeated runs of an integration of a deterministic nonlinear function with precisely known starting conditions, would reproduce the same history in any but the crudest of forms, unless nonlinear mode-locked resonances and/or *stochastic resonances* occur (cf. Benzi et al., 1982; Meiss, 1987; McNamara and Wiesenfeld, 1989; Schleich and McClintock, 1989; Irwin et al., 1990; Kloeden et al., 1991).

CHAPTER 5

Large- and Small-Scale Chaotic Crises, Intermittency, Universality, and the Big Picture

The "big picture" I invoke refers not to size but rather to the generality of the *phase space* (see Note P.3) within which dynamical systems can be mapped (**Note 5.1**). I have already used this idea in order to metaphorically illustrate the relationships between simple nonlinear-dynamical algorithms and conceptions of cosmological structure (e.g., as in the brief overview of metric universality, universal order, and number theory in Note P.4). We will need that sort of generality in this chapter in order to see the connections between the simplest ideas of chaotic and periodic attractors and the potentially very complex phenomena of nonlinear-dynamical "crises" that can arise from their topological interactions in phase space.

The phenomena of *chaotic intermittency* and *chaotic crises* (see the discussion in Note 5.1) in the Solar System, at both large and small scales, is focused by my conjectural interpretation of the cometary evolution shown in Figure 15b (cf. Fig. 13, Inset 3). However, the nature of chaotic crises is first illustrated in this chapter by describing two analogous laboratory experiments, ordered at different scales, that demonstrate the scale invariance of critical phenomena, a concept that is equally relevant to problems of orbital resonances, spin-orbit resonances, and geophysical phenomena, as well as their correlations, because there is no fundamental distinction in nonlinear-dynamical phenomenology between the scales of the laboratory and those of the Earth, the Solar System, the Galaxy, or the Cosmos. An analogous observation concerning the universe as a system of critically self-organized *cellular automata* is made by Chen and Bak (1989, p. 302; cf. Notes I.4, 7.3, and 7.4 of the present volume).

The reason that such nonlinear-dynamical invariances operate in the universe is simple (cf. Note P.4): *nonlinear dynamics deals with numerical and geometric relationships that are independent of mass, length, and time*. Trajectories in phase space have in common certain structural properties, properties that depend only on resemblances among nonlinear (feedback-implicit) functions, rather than on material states and/or spatiotemporal scaling. Because this premise says that the properties of attractors are "universal," the nature of chaotic crises must also be universal, the one caveat being that I am assuming the existence of attractors, hence a dissipative phase space rather than a conservative one. For other discussions of dissipative vis-à-vis conservative phase-space phenomena and crises, see, for example, Miles (1962, 1984), Helleman (1978, 1980, 1983), Grebogi et al. (1982, 1983, 1987a, b, c), Shaw and Doherty (1983), Bergé et al. (1984), Gu et al. (1984), Abraham and Stewart (1986), Gwinn and Westervelt (1986), Moon (1987), Hao Bai-Lin (1990), Campbell (1990), Shaw (1987b, 1991), and Sommerer and Grebogi (1992).

Experimental data from Ditto et al. (1989), shown in **Figure 17**, describe the oscillatory behavior of a magnetoelastic ribbon clamped at its base and driven to and fro in nearly vertical orientation by a variable magnetic field. With vertical symmetry, this situation would correspond to a typical two-well potential problem, but in this case, because of a slightly canted orientation, the oscillations are biased to one side. That is, for conditions of forcing whereby a symmetrical system would oscillate alternately between the two potential wells with equal — but nonlinearly intermittent, hence pointwise distinct — cumulative durations in each well (in a manner analogous to the time-series record of Figure 12a), the asymmetric magnetoelastic ribbon oscillates chaotically only within the lower of the two potential wells. But above a critical forcing condition (meaning, in this case, below a critical frequency of the imposed magnetic field), the flexing of the ribbon eventually crosses the vertical position and oscillates for a time in the less stable potential well before returning to chaotic oscillations within the more stable well.

The magnetoelastic material used by Ditto et al. (1989, Fig. 1) has stress-dependent elastic compliances that introduce nonlinear feedback between the stress and strain fields and amplify the motion so that the horizontal component at the position of a motion sensor can be tuned to very large displacements (Savage et al., 1990). The ribbon flips about in complex patterns somewhat like an inverted pendulum with flexible suspension (analogous to the behavior of the damped-driven spherical pendulum described by Miles, 1962, 1984; cf. Beckert et al., 1985; Gwinn and Westervelt, 1985, 1986; Tritton, 1986, 1988, 1989; Henderson et al., 1991; Matsuzaki and Furuta, 1991). This behavior is shown by the V-shaped or nearly L-shaped Poincaré sections of Figure 17a plotted on the X_1–X_4 plane of a five-dimensional phase space chosen to remove self-intersections of trajectories (three dimensions were sufficient to describe the dynamics, which had a fractal correlation dimension of 2.3 ± 0.1). Horizontal displacements (X-readings) were measured from the voltage output of a Fotonic sensor. The vertical position corresponds to the reading $X = 4.1$ V (volts). Smaller values of X (those < 4.1 V) correspond to flexing in the biased direction (giving points in phase space within the more stable potential well); positions within the other potential well are shown by much larger values of X. Intermediate points between the two potential wells are sparse, because the trajectories of motion that cross the transition interval between the two wells are rarely "caught in the act," so to speak, by the method of stroboscopic sampling typically used in vibration problems (see Note P.3). Transient oscillations within the less stable well are shown by the clusters or "bursts" of points near $X_1 = 6$ (the vertical axis of the phase plot, the X_4-axis, apparently is skewed to smaller values by the timing of the readings relative to the asymmetry and nonlinearity).

Time series of voltage readings just above and just below the critical frequency are shown in Figure 17b. The upper curve, $(f > f_c)$, which is typical of chaotic intermittency within the more stable potential well, represents irregular bucklings on the biased side of vertical. This pattern of intermittency, representing the oscillatory behavior within the region of phase space that corresponds to the dominant potential well and basin of attraction, represents the "core attractor" of the chaotic motion. In a more symmetrically designed experiment, the chaotic core attractor might involve both potential wells, but by intentionally introducing the asymmetry shown in Figure 17, the conditions under which a chaotic crisis is manifested by a burst of motion "uphill" across the potential divide are dramatized. The lower curve in Figure 17b, $(f < f_c)$, shows that as the forcing frequency was lowered, thus establishing a new chaotic pattern and time scale of intermittency, the ribbon would suddenly cross the vertical, stay there briefly, and then return to the side of easier initial buckling. According to the predictions of Grebogi et al. (1982, 1983, 1986, 1987a, b, c), an unstable *burst* of this kind occurs when the forcing term (*tuning parameter*) causes the chaotic trajectories to enter a region of phase space near the location of an *unstable periodic attractor*. This periodic attractor is located interior to the basin of attraction of the initial chaotic attractor (the *core attractor*) in the embedding space, but it does not lie on the core attractor, which, in this case, represents a more excited state of chaotic motion. Because of these relationships, crises that conform to the topological relationships just described are called *interior crises* (cf. Note 5.1). [In the general case of multiple chaotic/periodic attractor states — such as the bands in bifurcation sequences, as illustrated in May (1976), Crutchfield et al. (1986), Jensen (1987), or Shaw (1987b) — there can be unlimited numbers of core attractors, and periodic attractors that are unstable relative to them, where the relevant pairs of states capable of producing chaotic crises also form hierarchies of relatively increasing activation, or *tuning* **(Note 5.2)**; cf. Grebogi et al. (1982, 1983, 1986, 1987a, b, c), Gwinn and Westervelt (1986), Carroll et al. (1987).]

Coincident with the approach of a chaotic trajectory to the unstable manifold of the periodic attractor (its "sphere of influence"), there is a figurative collision between the respective basins of attraction (core attractor vs. unstable periodic attractor). Attractor basins, excluding those that are degenerate, often have fractal boundaries. If the periodic structure is robust, the chaotic core attractor can shift to values beyond "tangential" contact, thereby increasing the probability of bursts by increasing the fractal interactions and lowering the mean time between bursts, which in the case of Figure 17 occurs as f decreases farther and farther below f_c (for $f > f_c$, the trajectories are still on the core attractor and the system has not yet "gone critical"). The bursts are intermittent chaotic transients, but they can recur persistently near the critical condition. Well below f_c — but not yet within the realm of influence of some other attractor basin — the system eventually locks onto the new stability condition (i.e., the previously unstable periodic attractor has become stable). By contrast with the above description of the interior crisis, a *boundary crisis* refers to a situation in which a chaotic attractor is abruptly terminated at a very sharp transition to a periodic attractor (all attractors have transients, but in the boundary crisis the previously stable chaotic attractor is quickly "destroyed").

Chaotic crises are somewhat analogous to transient overstabilities (overshoot and undershoot) of mechanical or thermodynamic equilibrium points, but they are amplified by the chaotic motion and fractal indeterminism of interior basins of attraction ("roughness," "interlamination," etc.). A qualitative explanation for the suddenness of the bursts can be envisioned from the locations of the unstable periodic points in Figure 17a, which are shown by the open squares. If the squares are imagined to be analogous to the pins of a pinball machine and the core attractor is imagined to represent chaotic motion of balls unconstrained by these pins but in the same general space (e.g., magnetically suspended in a plane above the tips of the pins, both planes being imperfect or fractal), then as the two imperfect surfaces come together, initially there will be only a few hits of the most protuberant pins, and a few of the balls will experience sudden "wild" trajectories. As the two surfaces "overlap," the frequencies will increase, and the net amplitudes of the "wildness" will decrease, until there is eventually an average nonlinear periodic variability of hits on all pins, in a manner analogous to an evaluation of the average winding number (W) of iterations in a sine-circle map (Table 1, Eq. 5; and Fig. 14). This imaginary system offers a more appropriate analogy for three-dimensional "celestial pinball" than does the sine-circle map, because the spheres of influence of the chaotically moving balls (celestial objects) vary fractally in size relative to the fractal variations of the pin heights, where the latter represent the unstable periodic orbits with which the chaotic objects interact (the pins are analogous to a distribution of gravitational potential wells, collision cross sections, and so on, of the periodic objects; cf. Isomäki et al., 1985; Hénon, 1988; Bleher et al., 1989; Whelan et al., 1990; Rega et al., 1991; Hoover and Moran, 1992).

Ditto et al. (1989) demonstrated experimentally that the average time between bursts is a power-law function of the proximity of the tuning frequency, f, to a critical frequency, f_c, where the critical frequency of the coupled oscillator is that frequency at which the core attractor and unstable periodic attractor become tangent to each other (their "collision" point). This experimental result — together with additional work by Ditto et al. (1990a, b), Savage et al. (1990), and Sommerer et al. (1991b) — confirmed the theoretical power-law dependence between bursting times and critical-point proximity (tuning) derived by Grebogi and coworkers for chaotic crises in general, which they based on interpretation of phase-space relationships (e.g., Grebogi et al., 1986, 1987c; Sommerer et al., 1991a; Sommerer and Grebogi, 1992). The frequency of bursts, $1/\tau$ (where τ is the average time intermission between bursts), logarithmically approaches zero as the driving frequency logarithmically approaches the critical frequency. The slope defines a critical exponent, γ, that measures the fractal density of recurrences in phase space (e.g., Ditto et al., 1989, Figs. 4–6). Geophysical examples are given in Figure 31 (described in Chap. 7), where data from terrestrial cratering are compared with data from geomagnetic-reversal frequencies, both treated as chaotic crises of a type that is approximately self-similar with the experimental system of Ditto et al. (1989) — meaning that the complexity of the natural systems apparently is reduced by self-organization to a simplicity consistent with the few degrees of freedom of laboratory coupled oscillators ($\leq$ three degrees of freedom in the case of the magnetoelastic ribbon of Ditto et al., 1989).

Figure 18 is a sketch of the analogous chaotic bursting found to operate at crystallographic scales of time and size in ferromagnetic resonance and spin-wave instabilities of yttrium iron garnet (YIG), as modified from Carroll et al. (1987). Here, X and V are the voltage responses of a YIG sphere suspended in a direct-current field and excited by a microwave power source perpendicular to the field. Nonlinear-periodic resonances exist at some values of the driving power, while chaotic behaviors ("spin-wave instabilities") exist at others. The tuning of this system to a chaotic interval in the vicinity of quasiperiodic resonances produced chaotic bursts relative to a critical driving power, just as the magnetoelastic experiment did relative to a critical frequency. The initial chaotic response experienced a sudden jump to higher intermittent output voltages that persisted for transient, but systematic, intervals of time. Examples of this type of transient response are shown by the dashed trajectories of Figure 18a and by the left side of Figure 18b, which represent culminating episodes of the chaotic crisis prior to the eventual decay to a smaller, previously unstable, toroidal quasiperiodic attractor (shaded areas in both diagrams) interior to the chaotic core attractor (trajectories of the quasistable core attractor are not shown; they would resemble the bursting trajectories, but with smaller amplitudes and different frequency structure). It is notable that the chaotic behavior illustrated by Figure 18, like that of Earth's magnetic field, involves voltage oscillations of several types, including: (1) chaotic oscillations of the same sign ("field excursions"), (2) transient excursions from one sign to the other followed by return to the original polarity ("polarity excursions"), and (3) major bands of polarity reversals of varying styles and durations (intervals of rapid polarity reversals, protracted intervals of the same polarity but with small to large excursions, and so on).

I have described these experiments in order to illustrate that: (1) the crisis mechanism is universal over almost unlimited changes of scale in length and time (cf. Figs. 17, 18, and 31); (2) bursting phenomena can occur at many different values of the tuning parameters of systems with multiple bands of chaotic/periodic states, and the different unstable periodic attractors show self-similar shifts in the power-law distributions of bursting times characteristic of each state (see Fig. 2 in Carroll et al., 1987); (3) chaotic bursting phenomena of the same categorical type (interior crises) occur in systems of diverse properties, including solid-state oscillators (Fig. 18), fluid-convection systems (Fig. 29), geomagnetic-reversal phenomena (Figs. 31 and 32), and astronomical phenomena (Figs. 12 and 31); and (4) the universal relationships are established not only by theory (Grebogi et al., 1982, 1983, 1986, 1987c; Gwinn and Westervelt, 1986; Sommerer et al., 1991a, b; Sommerer and Grebogi, 1992), but also by diverse types of mechanical, fluid-dynamical, magnetoelastic, and electromagnetic experiments (e.g., Gorman et al., 1984; Rollins and Hunt, 1984; Gwinn and Westervelt, 1985, 1986; Carroll et al., 1987; Ditto et al., 1989, 1990a, b; Widmann et al., 1989; Savage et al., 1990; Hunt, 1991; Shaw, 1991; Singer et al., 1991; Sommerer et al., 1991b; Garfinkel et al., 1992) and by observation of natural systems (cf. Fig. 31; and Shaw, 1991, Figs. 8.37–8.43). Furthermore, the power-law scaling of bursts in chaotic crises phenomenologically parallels that found in experimental and theoretical studies of self-organized criticality in virtually *all* types of systems (e.g., Bak et al., 1988; Kadanoff et al., 1989; Bak and Tang, 1989; Chen and Bak, 1989; Held et al., 1990; Babcock and Westervelt, 1990). Chaotic crises and self-organized criticality are but two different ways of describing dynamical phenomena that occur in the neighborhood of the transition between nonlinear phase-locked periodicities ($\pm$ quasiperiodicities) and chaos (see the "strange nonchaotic attractor" of Ditto et al., 1990a; cf. Shaw, 1987b, 1991; Shaw and Chouet, 1988, 1989, 1991; Olinger and Sreenivasan, 1988).

The parallelisms between chaotic crises, self-organized criticality, and critical golden-mean transitional quasiperiodic/chaotic states — e.g., the *critical golden-mean nonlinearity* (CGMN) condition discussed by Shaw (1987b) and Shaw and Chouet (1988, 1989) — clarify one of the seeming dilemmas in nonlinear-dynamical discussions of fractal dimensions and "degrees of freedom" in natural systems (cf. Lorenz, 1984, 1991; Lauterborn and Holzfuss, 1986, 1991; Tsonis and Elsner, 1990; Chouet and Shaw, 1991; Pierrehumbert, 1991). A self-organized critical cascade in a natural system effectively has an infinite number of degrees of freedom (e.g., an "infinite" hourglass, sandpile, or magnetic-domain structure in which there are events of all possible sizes), yet the sizes of events and their durations and intermissions are neither random nor Poissonian, but, rather, accord with power-law scaling just like that of chaotic crises as functions of proximity to a critical condition in finite-dimensional attractor systems **(Note 5.3)**.

Furthermore, the singularity spectrum of the one-dimensional sine-circle function — when it is set at $K = 1$, and with W, the coupled winding number (see Table 1, Sec. D, Eq. 5), tuned, either by arbitrary tuning in the laboratory or by *cybernetic self-tuning* in Nature (a form of "autopoiesis"; see Jantsch, 1980, p. 7; Shaw and Chouet, 1989, p. 197), to the quasiperiodic (irrational) value equal to the *inverse of the golden mean* (see Note P.4) — is directly analogous to such a self-organized critical cascade (cf. Ostlund et al., 1983; Bohr et al., 1984; Jensen et al. 1984, 1985; Fein et al., 1985; Halsey et al., 1986; Sreenivasan and Meneveau, 1986; Gwinn and Westervelt, 1987; Meneveau and Sreenivasan, 1987; Shaw,

1987b, 1991; Newell et al., 1988; Olinger and Sreenivasan, 1988; Shaw and Chouet, 1988, 1989, 1991). The spectrum is one-dimensional in the sense that it is restricted to motion on the circle, but for the above values of K and W the system is *frustrated* (cf. Notes 3.1 and 5.3) by attempting to generate simultaneously all of the possible phase-locked frequencies (a few of which are shown in Fig. 14) while also trying to focus on an exact quasifrequency equal to the irrational number given by the inverse of the golden mean (Note P.4; Shaw, 1987b), all the while balancing itself exactly at the onset of noninvertibility and chaos (thus representing a condition *poised at the "edge" of chaos*; cf. Note 4.2; and Shaw and Chouet, 1988, 1989, 1991; Ruthen, 1993).

Since no system can do all these things at once, the result is analogous to a sampling of the characteristics of all three properties. The computer generates logarithmically scaled (power-law) bundles of all possible frequencies normalized by the golden mean as the fundamental universality parameter (see Fein et al., 1985), thus approximating the value of the irrational number with a system of rational numbers, and thereby satisfying the arrays of phase-locked nonlinear resonances, as well. In doing so, the power-law scaling of frequencies also simulates the behavior of all possible chaotic interior crises in the vicinities of the infinite number of unstable phase-locked periods (recall that all possible rational fractions are possible at $K = 1$). These phase-locked resonances are unstable because the stable frequency is theoretically quasiperiodic (irrational). In effect, the computer attempts to satisfy all of these frustrating conditions by sampling them in turn over hundreds of thousands of iterations. The computation proceeds as if one visits the phased-locked regime for a while, then the chaotic regime, then another part of the phase-locked regime, etc., thus as if the tuning parameters were cycling back and forth across the periodic-chaotic transition. This is what self-organized criticality does "automatically" in nature, by virtue of the unlimited cyclicity over all possible scales of time and size (i.e., the perpetual cyclic avalanching process operates like a cellular automaton; cf. Bak et al., 1988; Kadanoff et al., 1989). In effect, natural self-organized systems operate in just the same way, as I imagine it, that an entirely self-designed and self-tuned computer algorithm would operate — assuming that a "special-purpose computer" to carry out the computation could be catalyzed ("mutated") in some way as the offspring of a general-purpose computer (cf. Note 8.15).

Natural spectra of self-similar periodicities are therefore functions of tuning to a (multilevel) critical vicinity, rather than being functions of the fractal "dimensionality," or "degrees of freedom," of any specific deterministic dynamical system of the type used for purposes of illustrating the nature of—and the universal structural regimes of—nonlinear phenomena. Evidently, this is why a two-parameter cyclic function (e.g., the sine-circle equation of Table 1) can be used to simulate the multifractal characteristics of coupled oscillating flows (e.g., Jensen et al., 1984, 1985; Bohr et al., 1984; Halsey et al., 1986; Olinger and Sreenivasan, 1988), infinite-dimensional electronic oscillators (e.g., Gwinn and Westervelt, 1987), and seismic tremor in volcanic systems, or in the Sun, etc. (e.g.,

Chaotic Crises, Intermittency, and Universality

Kurths and Herzel, 1987; Shaw and Chouet, 1989; Shaw, 1991; Chouet and Shaw, 1991; Shaw and Chouet, 1991). We cannot define a unique core attractor or an unstable periodic manifold in complex natural systems (e.g., in the systems of Figs. 11 and 21; cf. controversies in the literature concerning the existence of attractors in the Earth's weather-climate system, as discussed by Nicolis and Nicolis, 1984; Grassberger, 1986; Vallis, 1986; Essex et al., 1987; Shaw and Moore, 1988; Tsonis and Elsner, 1990; Lorenz, 1991; Pierrehumbert, 1991), because there are so many attractor basins that we cannot even enumerate them (i.e., there are *uncountable infinities* of attractor states in the universe; cf. "A Linguistic Digression" in Chap. 8; and the *wheel of fortune* in Shaw, 1987b), much less identify specific attractor states, and they are continually interacting and changing at many different scales of time and size. But we *can* say that the aggregate effect is as if we were describing an array of attractors each of which is operating in the vicinity of a critical frequency, therefore with power-law distributions of bursting times of all magnitudes. [A caveat here: natural and experimental critical-state processes of a given type—such as sandpile cascades, acoustic-cavitation cascades, seismic-tremor cascades, comet-shower cascades, etc.—each have truncation effects (rolloffs) at both the small and large ends of their spectra, and therefore they each depart from the universal scaling relationships near those characteristic cutoff dimensions.]

Knowledge of the universal properties of nonlinear-dynamical systems is useful because, if there is evidence that an interactive system is taking part in a general critical-state process of large or effectively unlimited scale [e.g., systems such as the Earth's core, the "magmasphere" (cf. Note 8.3), the glacier-ocean-atmosphere-magnetosphere, or the biosphere—ultimately including the "heliosphere," which embraces all processes figuratively and/or dimensionally implicated within the critically self-organized Solar System illustrated diagrammatically in Figs. 11 and 21], then it may be possible to identify particular frequency bands or bundles with particular processes. [Some evidence for such *universal scaling* of frequency bands or bundles is given by Shaw and Chouet (1991, p. 10,205) in discussing the phenomenology of magma-related seismicity and magma transport in Hawaii, where the self-similarity appears to extend over a range of 15 decades in frequency (roughly 1 Hz to $< 10^{-15}$ Hz, or periods from about 1 sec to $\sim$30 million years!); an analogous form of universality recently was identified in studies of nucleotide frequencies in the molecular makeup of the biological cell (see Amato, 1992), where DNA sequences hundreds of thousands of nucleotides in length may be relevant to the information content of just one particular gene (cf. Note P.4).]

Returning to the more specific notion of chaotic crises and bursting spectra, we can infer that the role of self-organization in infinite-dimensional systems effectively ensures that the system runs the gamut of all possible critical states; hence over a sufficiently long time, and comparable scales of observation, the system will sample all possible interior crises and bursting times. An experimental alternative to the somewhat abstract numerical properties of the sine-circle function could be designed, as either an analog or digital attractor system, by arranging a feedback loop of self-tuning that cyclically visits the positions of all possible periodic attractors within the field of chaotic states (this sort of experiment would be the inverse of so-called "controlled chaos" wherein a negative feedback circuit is strategically placed in a system in such a way that it opposes any tendencies for the development of chaotic transients; cf. Singer et al., 1991; and Fig. 29).

An example of how these relationships can be applied to a portion of the generalized orbital flux of objects in the interplanetary "circulatory system" of Figure 11 is given by the interpretation of data from Duncan et al. (1987) shown in Figure 15b (cf. Fig. 13, Inset 3). The geometric similarity between the chaotic instabilities of Figure 17a and the orbital parameters of Figure 15b is evident. In mirror image, X_1 in Figure 17a would correspond figuratively (relative to a three-dimensional potential well scaled to Fig. 15b) to variations of the semimajor axis a at constant perihelion q, and X_4 would correspond to variations of q at constant a. Although the general implications are similar, the bursting modes indicate ways that trajectories can be influenced by deterministic chaos with or without random effects. In this sense the orbital instabilities would resemble those of a nonlinear spherical pendulum, where the motion is unstable and erratic when a forcing frequency is near to the natural (critical) frequency (e.g., Miles, 1984). The timing of bursts would occur as a power-law function of proximity to a critical Galactic-tidal frequency and/or to a critical amplitude and frequency of stellar perturbations in the neighborhoods of possible cometary resonances. The bursting mechanism predicts self-similar distributions of cometary events in orbital parameters and/or abruptness as functions of proximities to critical frequencies. The same mechanism applies equally to new comets, comet showers, and old comets, as well as to transition frequencies relevant to distinctions between comet and asteroid domains and their orbital evolution (cf. Note 6.7).

CHAPTER 6

Spin-Orbit Resonances and Proximal Flight Control of CRF Objects

Nonlinear phenomena underscore the complexities that confront the geologist, planetologist, or astrodynamicist seeking to discover the nature of orbital histories related to the impact patterns on a planet (their *celestial reference frames, CRFs*). What are the scaling properties and characteristic frequencies of objects emanating from a reservoir sufficiently local to satisfy the orbital precision of an ordered CRF when a "local" reservoir may in fact, according to the model of Figure 11, correspond to several different scales of orbital phenomena (e.g., the Asteroid Belt as a whole, or some dynamical subset of it; comet reservoirs in general, or a subclass of them such as the Kuiper Belt) and may be integrally dependent on relationships between object reservoirs extending throughout the Solar System? To satisfy the properties of Earth's CRF—as I have described it in the Introduction to the present volume, and as it is illustrated in Chapters 1 and 2 (see Figs. 1–8; and Appendixes 1–5)—there must be a proximal stage that acts, at least transiently, as a holding reservoir that is analogous to a flight pattern in the vicinity of an airport. Its orbital properties must be such that the timing of impacts is resonantly coupled with ("synchronized with," in the general sense of Blekhman, 1988) the planet's orbital and/or rotational periods. Mean-motion resonances (e.g., the "orbit-orbit" resonances of Froeschlé and Greenberg, 1989; cf. Note 4.1) between the orbits of potential impactors and the Earth, which would satisfy these requirements in the seasonal sense, lead to a zonal distribution of impacts for low-inclination orbits, and to some sort of sectorially banded distribution of impacts for high-inclination orbits. This form of nonlinear resonance might provide significant controls in the latitudes and longitudes of impactors if the precision of orbital synchronization is high (as is the case with periodic meteor showers that recur in the same locality on the same day of the year; see Kippenhahn, 1990; Yeomans, 1991a).

A "random" or variable set of trajectories would yield scattered shot patterns smeared out by the planet's spin, hence forming small-circle distributions of craters — an alignment I have seen only rarely in my examinations of cratering patterns on the Earth, Moon, and Mercury (the lunar cratering pattern is shown in Fig. 37, as discussed in the Epilogue; major Mercurian impact basins are shown in **Appendix 10**; cf. Appendix 25). Hypothetically, coincidence at an exact position and time on the Earth's surface could occur if the average period of a nonlinear orbit (its winding number, W, as described in Notes 3.1 and 4.1) were (1) exactly synchronous with a specific place and time of day (like a synchronous communications satellite), (2) an exact-integer multiple of the Earth's rotation period (an integer multiple of the yearly period) over times short compared with the Earth's precessional period, (3) an exact-integer multiple of the Earth's precessional period ($\sim$26,000 years), or (4) an integer multiple of any orbital period of the Earth's motion that places it in exactly the same orientation relative to the orbital period of the motion of an intermittent source of impactors after some different but integer number of cycles of the object source (cf. Notes 3.1 and 4.1). Such guidance mechanisms relative to the Earth would imply some combination of (a) mean-motion resonances of potential impactors with, at least, the inner planets and the Sun, (b) capture of impactors in spin-orbit resonances with the Earth-Moon system, and/or (c) capture of impactors in spin-orbit resonances with the Earth itself. The relevance of these types of resonance conditions for Earth impactors is indicated by several of the types of nonlinear phenomena previously documented — e.g., by Wisdom (1982, 1983, 1985, 1987a, b, c) for the Asteroid Belt (Fig. 12) and by the diverse nonlinear phenomena in cometary reservoirs (see Fig. 15; and Duncan et al., 1987, 1988; Torbett and Smoluchowski, 1990) and in comet-asteroid dynamics [cf. Fig. 13 (Insets) and Note 6.9; and Weissman et al., 1989; Froeschlé, 1990] — and by the concepts of spin-orbit resonances developed systematically by Allan (1967a, b).

Some hypothetical approach scenarios are sketched in **Figure 19**; factors affecting collision or capture are indicated schematically in the lower portion of the diagram. Artificial-satellite guidance illustrates the intricacy of hitting the right transfer orbits or reentry conditions even under optimal conditions of adjustable parameters and instantaneous thrust control. The conditions necessary for collision with a "random" orbit from the Asteroid Belt are given perspective by Figure 13b, where the diameter of the Sun is $\sim 10^{-2}$ AU and the diameter of Earth is $\sim 10^{-4}$ AU (much smaller than any of the dots). If the orbital scale of Figure 13b were expanded to accommodate possible sources of new comets, the size of the region now shown in the diagram — including the Sun, Jupiter, the Asteroid Belt, and the inner planets — would be reduced to an invisible speck, and the modified scale would be $\sim 2 \times 10^4$ AU instead of about 10 AU (more than a thousandfold reduction of the size of the inner Solar System, relative to the expanded diagram; cf. Fig. 13, Insets). Theoretical treatments of the orbital convergences of new comets within the inner Solar System, for perihelia inside 2 AU, typically invoke probabilistic methods, such as those of Shoemaker and Wolfe (1984) and Duncan et al. (1987), combined with statistical estimates of the comet flux based on: (1) actual observations (e.g., Shoemaker et al., 1989, 1990) or (2) Monte Carlo simulations of background and shower frequencies (e.g., Heisler, 1990; Weissman, 1990a, b), as illustrated by Figures 15b and 16.

An alternative to reliance on statistical estimates based on astronomical observations or on Monte Carlo simulations for the present population of comets inside 2 AU, as was done earlier for asteroids, begins with a consideration of the growth rate of comet discoveries. Between A.D. 1680 and 1980, discoveries of comets increased at a crudely exponential rate, a rate much like that of asteroid discoveries, as cited earlier, but slower, corresponding to roughly twice the doubling time of asteroid discoveries. If we ignore ancient records and begin our count with a "first" discovery in 1680, we will have 1024 comets after ten doubling generations of 30 years each. This is roughly the number in the discovery source I draw upon (Bailey et al., 1990, Fig. 7.1). Yeomans (1991a) gives an annotated list of some hundreds of "naked eye" observations prior to 1700 A.D., but that list presumably includes many short-period comets later rediscovered by telescope. In view of the interest since 1980 (the beginning of the debates on the Impact-Extinction Hypothesis), this rate probably has significantly increased. A current population of $\sim 10^3$ comets estimated in this fashion can be compared with the Monte Carlo prediction of Figure 16, which gives an average just under 1000 m.y.$^{-1}$ for a source of 10^7 comets near 10^4 AU and perihelia < 10 AU (Heisler, 1990, Fig. 1). This comparison would imply that the mean residence time for an instantaneous population of 1000 comets inside 10 AU is $\sim 10^6$ years.

Astrophysical estimates of the total numbers of comets in source reservoirs for all distances $> \sim 10^3$ AU — i.e., within both the Hills and the classical Oort reservoirs — is more like 10^{11} to 10^{12} (cf. Weissman, 1990a, b). Normalized to these higher numbers, the Monte Carlo results would predict between 10 and 100 new comets *per year* and showers of up to several hundred per year over intervals longer than 10^6 years (cf. Heisler, 1990, Fig. 1)! At these rates of influx, residence times as small as 10^3 years would give populations within 10 AU at any one time of 10^4 to 10^5 comets and peak populations (times of "showers") approaching 10^6 comets. The lower of these numbers (10^4) would be reached in about three more generations (~ 100 years) of astronomical searches at the past rate of increase in discovery rate, and in a much shorter time if the rate picks up to match that of asteroid discoveries (cf. Note 4.2). If these different extrapolations are mutually compatible, the steady-state-average number of comets ($10^4 < N < 10^6$) would theoretically exist now, waiting to be discovered. In view of orbital focusing, even 10^4 comets cycling in and out of the interval within 2 AU from the Sun (the discovery interval) probably would imply great displays of debris trails, a consequence that would appear to be consistent with interpretations of meteor showers and their radiants (see below) as being caused by orbital intersections of the Earth with such trails (e.g., Kippenhahn, 1990; Yeomans, 1991a). If additional material has emerged from the breakup of "giant comets" $\sim 10^2$ km in diameter (cf. Clube, 1989b, 1992; Olsson-Steel, 1989; Clube and Napier, 1990), the potential numbers and sizes of near-Earth objects of cometary origin would be increased by a large but unknown factor (cf. Steel, 1991).

An impressive variety of "guidance effects" are attributed to astronomical phenomena observed within 2 AU of the Sun (e.g., Binzel et al., 1990; Clube, 1989b, 1992; Bailey et al., 1990). Meteor streams and showers, as mentioned above, represent highly focused phenomena—witness the remarkably periodic recurrences of meteor *radiants* (points in the sky from which meteor showers appear to emanate; see Kippenhahn, 1990, pp. 167ff; Yeomans, 1991a, pp. 191ff). The residual velocities of such debris would imply divergence from special orbits unless they *either* experience zero relative acceleration during breakup *or* are confined to the vicinities of short-period comet and/or asteroid orbits by gravitational focusing. To judge from the discussion of the Asteroid Belt and the likelihood of near-resonant and/or chaotic conditions in the Solar System, it would seem likely that most *repetitively observable* objects either are near a resonance state or are approaching or leaving one. This places an organizational emphasis on the interpretations of possible trajectories in Figure 19.

The cone of high-inclination approaches shown at the upper right of Figure 19 represents possible orbital resonances with Earth's spin (capture scenarios). The inclinations in the figure are based partly on the positions of the cratering nodes of Figures 1–5 and partly on inclinations characteristic of exterior resonances in the theory of spin-orbit resonances by Allan (1967a, b), outlined below (Fig. 20). Capture scenarios are discounted by most geophysicists and astrophysicists because they have not allowed for the possibility of organized resonance conditions during orbital approach, and because they assume that the *entry corridor* (the quasiparabolic "funnel" around the Earth shown at the bottom of Figure 19) is too narrow to be "hit" by "random" trajectories of approach, as compared either with the collision cross section of the Earth or with hyperbolic orbits that correspond to *flybys* (the collision cross section and the region of hyperbolic orbits corresponding to flybys are represented by dark and light shaded areas, respectively, at the bottom left of Fig. 19; these relationships are discussed in Bate et al., 1971).

This seeming conflict of interpretations can be resolved, at least to some extent, by realizing that a number of different factors exist by which incoming objects may be orbitally perturbed and eventually captured in an orbit about the Earth and/or Moon **(Note 6.1)**, such as: (1) the Earth's gravitational *sphere of influence*, which is—according to my interpretation of the universe as a dissipative system—analogous to its orbital *basin of attraction* (see Note P.3), is of the order of 10^6 km or more (larger for a *fractal basin of attraction*; see Note 6.1), or roughly 1% or so of the Earth-Sun distance (e.g., Bate et al., 1971, p. 40; Duncan et al., 1987, p. 1335), (2) the perimeter of the sphere of influence is dynamically and topologically complex (a fractal boundary), being affected by resonances between the Earth, the other planets, and the Sun (see Notes 6.4 and 8.4; and Jose, 1965; Wood, 1972; Landscheidt, 1983, 1988; Laskar et al., 1992), as well as representing the neighborhood of the so-called Earth-Sun *fuzzy boundary* representing a complex region of uncertain balance between the competing gravita-

Spin-Orbit Resonances and Flight Control of Objects

tional attractions of the Earth and the Sun, in concert with the Moon and other planets, on a "third body" (Belbruno, 1992, Fig. 2), (3) the net effect of solar-plasma drag, cosmic-dust drag, and atmospheric drag (cf. Notes 1.2, 3.3, 6.1; and Allan, 1967a, b, 1969, 1971; Greenberg, 1978, 1982; Pollack et al., 1979; Pollack and Fanale, 1982; Greenberg and Brahic, 1984; Atreya et al., 1989; Cox et al., 1991; Toon et al., 1992; Zahnle et al., 1992) can be an important influence on the trajectories and residence times spent by an object inside of the *"uncertainty width"* of the fractal ("fuzzy') resonance boundaries of a planet's sphere of influence (attractor basin), hence on the probability of the object's capture within the gravitational potential well of that planet (i.e., I am referring here to the *fractal basin boundaries of planetary attractor states*, as in the finite-width basin boundaries of fractal attractors in general; e.g., McDonald et al., 1985, Fig. 5; Moon and Li, 1985a, b; Shaw, 1991, Sec. 8.3, Fig. 8.20; and Notes 5.1 and 6.1 in the present volume), and (4) low-angle and/or grazing encounters with the surface of the Earth and/or the Moon (cf. Gault and Schultz, 1990) can act as effective braking mechanisms for objects in hyperbolic, flyby orbits (bottom of Fig. 19), such that the fragments—several of which may have diameters that are significant fractions of the diameter of the original object—may be captured into a family of geocentric orbits (cf. Schultz and Gault, 1990a, b, 1992; Schultz and Lianza, 1992; and Appendix 3). [If the integrated mass of the matter penetrated, or otherwise affecting drag coefficients, during one orbital passage of an object intruding into the sphere of influence of a celestial body, is approximately equal to the mass of that object, direct capture may occur (cf. Pollock et al., 1979; Pollack and Fanale, 1982; Stern, 1991, p. 275; Toon et al., 1992). Many new comets orbiting the Sun over geologic time were likely, by such a mechanism alone, to have dissipated precisely enough energy to be brought into a form of mean-motion resonance permitting eventual capture (in their entirety or, as is more likely, in the form of fragments and residual debris caused by the Sun's action during flyby; cf. Hut et al., 1991) by the Earth-Moon system, a likelihood enhanced by the hierarchy of possible spin-orbit resonances with Earth that are affected by atmospheric drag (see Fig. 20).]

Spin-Orbit Resonances with Earth

Nonlinear dynamics is a two-edged sword. The same reasoning that argues for the prevalence of chaotic or resonant orbital parameters makes it impossible to calculate an exact approach trajectory by integrating the equations of motion. At the edge of dissipative chaos, integrability breaks down in the absence of information that permits orbital corrections (artificial satellites and space probes, for example, usually have capability for in-flight correction). However, the circumstances that would accommodate the existence of small satellitic objects in spin-orbit resonances with Earth can be outlined in principle (see Note 6.1). Their theoretical presence poses two questions. First, if objects 0.1–10 km in diameter exist in high-inclination orbits at distances intermediate between those of the

Earth and Moon (exterior resonances) or at less than about 30,000 km from the Earth (interior resonances), what is the likelihood that they would be documented by the strict observational criteria applied to discoveries of new asteroids (see Bowell et al., 1989; Shoemaker et al., 1990)? Second, if we became convinced by observational evidence that they do not exist at the present time, more difficult questions are, why not — and how do we evaluate the possibility that they existed in the past, for example, according to the same principles of chaotic intermittency explored in Chapters 4 and 5 concerning the conditions that have affected the occupancies of asteroids in the Kirkwood Gaps of the Asteroid Belt (see Notes 3.1, 4.1, 5.1, 6.7)?

These questions are accentuated by the diversity of objects within 2 AU of the Sun (zodiacal dust, meteor streams and radiants, etc.; cf. Alfvén, 1971a, b; Baxter and Thompson, 1971; Danielsson, 1971; Lindblad and Southworth, 1971; Drummond, 1981, 1991; Olsson-Steel, 1989; McIntosh, 1991) and by the fact that most planets *do* have satellitic objects (cf. Kaula, 1968; Alfvén and Arrhenius, 1975, 1976). A case in point is the recent documentation by Porco (1991) of six new satellites and their resonances, accompanied by a complex system of particle rings and ring arcs around Neptune (cf. B. A. Smith et al., 1989). Some asteroids seem to have satellites (e.g., Van Flandern et al., 1979; Weidenschilling et al., 1989) and/or to be binary (cf. Anon., 1993), or to consist of aggregates of more than two objects — and these possibilities have even been raised for comets (e.g., Van Flandern, 1981). Binzel et al. (1991, p. 94) show a remarkable illustration of a self-gravitating or self-adhering "contact binary" asteroid, recently catalogued as the near-Earth asteroid 4769 Castalia (note, incidentally, that this number indicates that nearly 800 new asteroids have been added to the catalog since the compilation two years previously by Binzel, 1989; see Fig. 13 here). Other composite asteroids mentioned by these authors are described as "rubble piles" of fragmented pieces (cf. Fujiwara et al., 1989; Chapman, 1990). Binary and/or aggregated asteroids (and/or comets) are obvious candidates for multiple capture and/or multiple impact events (see Note 6.1). [An interesting mechanism of *ballistic agglomeration* — a mechanism that predicts power-law rates of growth, as would be implied by the *universal model of critical self-organization* discussed in Chapter 5 (cf. Notes 7.3 and 7.4) — that may be relevant in forming aggregated asteroids in the inner Solar System and cometary aggregates in the Oort Cloud, or in the Galaxy (cf. Clube, 1985), is described by Carnevale et al. (1990; cf. Maddox, 1990). Alfvén (1971a, b) had previously suggested an analogous mechanism for particle aggregation in space (cf. Lindblad and Southworth, 1971; Hoffmann, 1989b).]

Despite these interesting observations, however, the astronomical aspect of the *celestial reference frame hypothesis* (CRFH) is faced with an observational problem almost as severe as that needed to confirm Stern's (1991) prediction of small "planets" beyond the outer planets (see Note 6.5). This is so because the CRFH predicts objects so small and so close to the Earth that they may be almost as difficult to locate (cf. Bagby, 1956; 1966, 1967, 1969; Baker, 1958; Tombaugh et al., 1959; Tombaugh 1961a, b; Kordylewski, 1961; Frank et al., 1990; Binzel,

1991; Yeomans, 1991b; Steel, 1991). The Earth-Moon system is minuscule by comparison with the Solar System, even by comparison with that part of the Solar System inside of Jupiter (e.g., the ratio of celestial volume inside the lunar distance from Earth, relative to the celestial volume inside Jupiter's orbit, is about 10^{-10}), but it covers the entire celestial sphere with respect to the field of view from Earth, and this view confronts a backdrop containing myriad small and/or faint objects that confuse the issue. It is difficult, and often impossible, to differentiate between transient geocentric and heliocentric orbits in one-time sightings that produce instantaneous traces (small arcs) on the celestial sphere (e.g., calculated values of apparent perihelion can be similar to possible values of perigee and/or apogee, especially for eccentric high-inclination geocentric orbits; cf. Aksnes, 1971). An example might be a high-inclination trajectory of a one-time tracking of a meteor that fits an asteroidal orbit but might be a geocentric orbit with a similar trace on the celestial sphere. If, then, a calculated reappearance based on assumed asteroidal parameters does not occur (as it would not if the object is in fact in a geocentric orbit), the sighting might be classed with meteors, be unreported, be classed with unknown objects (objects with unrecognized orbital characteristic and/or objects that move so fast relative to the field of view that nothing can be said about their trajectories, where the latter are sometimes recorded within a generic group called "fast-moving objects"; cf. Tatum, 1988), or it may be classed with unrecovered possible asteroids. The validity of new discoveries usually is based on repeated appearances and orbital parameters that are carefully worked out by more than one observer. On that basis, unless all possible geocentric orbits are specifically included in a computer search routine, small geocentric satellitic objects would not be documented as such.

Bowell et al. (1989) discuss the criteria that have been established, by consensus, for the discovery of asteroids, and give statistical data for the total number of such "discoveries." In this context, formal discovery is constituted by some number of reported observations, by different observers, that all appear to refer to the same object, indicated by the agreement of orbital parameters computed by different workers. This is a very stringent requirement, one that is not likely to be met by many of the sightings of real, but nonlinearly perturbed, asteroids and other near-Earth objects (see Note 6.5). Among roughly 4800 well-documented numbered asteroids (including the 4000 cited by Binzel, 1989, Fig. 2, and another 800 indicated by the discussion of Binzel et al., 1991), there have been, in fact, about four "discoveries" per object. If the same ratio applies to other objects (i.e., if only one-fourth of the objects tracked by fewer than four "discoverers" prove in time to constitute acceptable discoveries), then the number of incompletely documented (but perhaps real) objects can be obtained from the statistical data of Bowell et al. (1989, p. 25). At the time of writing by these authors, this number would be given as follows: (total "discoveries" = 49,466) − (rediscoveries of numbered asteroids = 14,970) = 34,496 ÷ 4 = 8624 objects, or roughly two to three times the number of fully documented and officially numbered asteroids documented at that time (i.e., depending on whether reference is made to the 4096 numbered asteroids of Bowell et al., 1989, Table 1, or the 3318 named asteroids in the tabulation of Tedesco, 1989). Because the ratio of discoveries per object, now put at 4:1, may be smaller if "less conventional" orbits are "rediscovered" less frequently, this estimate must be considered minimal.

On the basis of the preceding argument, we can be sure that there are not just the 4800 confirmed asteroids but at least 10^4 asteroids (cf. extrapolations in Table 1 of Bowell et al., 1989). This number of asteroids represents the sum of those that are fully documented plus those that are *partially* documented (in current practice, *partially documented asteroids* are recorded by the year plus two added letters, such as 1991 BA; cf. Binzel 1991; Binzel et al., 1991, p. 91) — and this means that there has to be some incredible number of *un*discovered asteroids out there, plus an incredible number of *un*identified near-Earth objects, too (cf. Tatum, 1988; Steel, 1991; Scotti ct al., 1991). The following statement, from Steel (1991), gives dramatic emphasis to this discussion: *"The 'annual event' for the Earth is apparently a 20-kiloton airburst (Chapman and Morrison, 1989). Although the chance of an individual object like 1991 BA hitting the Earth is only 10^{-10}–10^{-7} a year (depending on its orbital parameters), there are believed to be 10^9 such NEOs larger than 10 metres across. This implies that,* **each day, at least one passes by the Earth closer than did 1991 BA, but escapes detection.** *Asteroid 1991 BA was discovered only because the sole instrument capable of detecting and tracking it was pointed in the right direction at the right time, and because the operators were skilled and experienced enough to recognize and follow it."* [Boldface type has been added to emphasize the increasing estimates of the frequencies of NEOs; cf. Note 9.1.]

The asteroid 1991 BA came within ~0.0011 AU (~165,000 km) of the Earth, a distance less than halfway (~0.43 × 384,400 km) to the Moon. The "20-kiloton airburst" represents an event that is roughly 0.2% of the energy of the Tunguska event of 30 June, 1908 (Chapman and Morrison, 1989, p. 278). Chapman and Morrison cite a frequency of about once every 300 years or so for a Tunguska-size event. Little imagination is required to realize what an explosion even "as small" as the equivalent of 20,000 tons of TNT would do to a populated region of the Earth, and a Tunguska-like event — itself a very "small" event on the scale of other impact phenomena discussed in this book — would be a disaster of unprecedented proportions if it occurred in the eastern U.S., Europe, or any other densely populated area of the globe (these remarks take on added significance when it is realized that the populated regions mentioned are in close proximity to the cratering nodes and swaths described in the present volume; cf. Figs. 4, 34, and 37; and Appendix 3). The significance of close approaches of near-Earth objects (NEOs) is given new perspective in the following discussion of spin-orbit resonances.

The theory of Allan (1967a, b) concerns the perturbing effects of mass heterogeneities in a primary on solutions of the classical equations for satellite orbits (Table 1, Sec. A). The circumstances of his analysis arose from the need for precise corrections for artificial-satellite orbits, and Allan worked out some general solutions for perturbing accelerations of circular orbits induced by the longitudinal anomalies of mass distributions in the Earth. Allan's results are shown in **Figure 20**, where α is the rotation frequency of the primary (Earth) and β is the orbital frequency of a satellite (e.g., $\alpha/\beta = T_S/T_E$, where T_S is the sidereal period of a geocentric satellite, and T_E is the sidereal period of rotation of the Earth). There are two general and important results. First, the predicted distribution of resonances corresponds to a "staircase" of spin-orbit ratios analogous to those of Figures 13 and 14 (cf. Appendix 9). Second, interior resonances ($\alpha < \beta$) can evolve to either low-inclination or high-inclination orbits, whereas exterior resonances ($\alpha > \beta$) can evolve only to high-inclination orbits (see Allan, 1967a, Figs. 1 and 2).

The parameter D_{max} in Figure 20 is a measure of relative stability in the "simple" case of weak nonlinearity and atmospheric drag. The central gap is dominated by the synchronous resonance *(1,1)*, using Allan's convention for the indexing of the resonances. Other resonances in the vicinity of synchronous resonance, including *(2,1)*, are weak and complicated (cf. Allan, 1965). Interior resonances that have the frequency $\alpha = 1$, however, are remarkably strong and long-lived. Although the disturbing function relates to short-period effects, secular stabilities are possible, barring large perturbations (see discussion of Saturn's rings by Allan, 1967a, pp. 72ff, and Table 2; cf. Jewitt, 1982; Greenberg and Brahic, 1984; Esposito and Colwell, 1989; Kolvoord et al., 1990; Esposito, 1991, 1992; Horn and Russell, 1991; Porco, 1991; Colwell and Esposito, 1993). Commensurabilities of some natural satellites were also examined by Allan (1967a, p. 69), but on the basis of discrepancies between orbital parameters and those predicted, he discounted a spin-orbit mechanism as the dominant effect. Many of the satellites of the planets are of the exterior type, and according to the theory should have evolved to high inclinations not observed. However, the diversity of resonance types in the Solar System, involving composite phenomena, are much too nonlinear and complex to be predicted by a simple two-body spin-orbit resonance model. This complexity is underscored by Allan's (1969) later reexamination of the commensurabilities of satellite orbits in the Solar System, as well as by my discussion of the Jupiter-Io-Europa-Ganymede interactions in the Prologue (cf. Note 6.1), and by the discussion of Neptune's satellites and ring arcs by Porco (1991); cf. B. A. Smith et al. (1989); Kolvoord et al. (1990), Ross and Schubert (1990), Esposito (1991, 1992), Colwell and Esposito (1993).

As pointed out by Alfvén and Arrhenius (1969, 1975, 1976), the Earth-Moon system could have experienced several types of resonances during its tidal evolution. The "spin" of the Moon about its own axis is locked with its period of revolution around the Earth, an effect that is taken so much for granted by every Moon-gazer that few persons ever think about this remarkable demonstration of spin-orbit resonance. Kaula (1968, pp. 203ff) explains this resonance in terms of the orbital-rotational balance of forces acting on the Moon, caused by the combined effects of its orbital eccentricity and the torque acting on the Moon's tidal bulge by the Earth. Because of the eccentricity of the lunar orbit, the torque acting on the Moon's rotation consists of two parts, one tending to slow the rotation, the other tending to speed it up. The retarding torque dominated until the Moon's rotation rate was slowed sufficiently that the two effects balanced, a condition in which the symmetry axis of the Moon's tidal bulge became oriented along the line of centers between the Earth and Moon, resulting in zero net torque averaged over the orbital period. At that stage the tidal bulge was trapped within a narrow range of lunar longitudes, a position that from then on was always facing toward the Earth, except for small oscillations (rocking motions called *librations*). Analogous effects have operated on the rotational periods of the larger satellites of the other planets, such that many of them are locked into the same type of synchronous resonance as our Moon (cf. Cassen et al., 1982, pp. 95ff).

Such a mechanism of tidal dissipation in the Moon is analogous to the inverse mechanism of tidal dissipation in the Earth related to the torque produced by the gravitational attraction of the Moon on Earth's tidal bulge, except that in the Earth, the tidal bulge is consistently carried forward by the Earth's greater rotation frequency relative to the Moon's orbital frequency about the Earth (cf. Fig. 21), so that the Earth is subjected to a persistent decelerating torque or braking mechanism. The net result of tidal friction in the Earth-Moon system is that the average period of rotation of the Moon exactly equals its average orbital period (**Note 6.2**), causing one hemisphere of the Moon's surface always to face the Earth, while Earth's rotation period (*length-of-day*) has gradually increased, together with an increase in the Moon's orbital distance (*length-of-month*), the energy of which is pumped up by the loss of rotational kinetic energy by the Earth to the Earth-Moon orbital motion as a whole (cf. Note 6.1). The point made by Alfvén and Arrhenius (1969), following Allan (1967a), is that in the evolution of the Moon's *rotational* synchronous resonance condition, it is quite likely that the Earth-Moon system passed through a more general sequence of resonances that were manifested by episodic changes in both the lunar rotation rate and its orbital period around the Earth. Besides the implication for conditions of lunar capture (representing a family of hypotheses that has, at least for now, been abandoned by nearly all geophysicists and astrophysicists in favor of a family of collision models; see caption of Fig. 19), the Alfvén-Arrhenius resonance model also predicted that the Earth's period of rotation (l.o.d. = length-of-day) would have changed episodically as the Earth-Moon system passed from resonance to resonance, so that the l.o.d. would have remained nearly constant during those intervals of time that corresponded to the main resonances. The record of the l.o.d. vs. age for the Earth, therefore, should exhibit plateaus during these times, an expectation that is supported in a general way by the fossil studies of the l.o.d. by Pannella (1972; cf. Williams, 1972, 1989a, b; Walker and Zahnle, 1986).

The spin-orbit relationship of the Earth and Moon, expressed in terms of the ratio α/β, gives a complicated fraction near $\alpha/\beta \cong 137/5 = 27.4$. (A more accurate ratio of sidereal periods of the Moon and Earth is $T_M/T_E = f_E/f_M = 27.3964652$.) The orbital evolution of the Moon illustrates a case where spin-orbit resonances relative to the Earth, which might have taken simpler integer ratios, were ultimately dominated by the combined effects of the size of the satellite and the dynamics of its orbital and internal gravitational anomalies acted on by the greater mass of the Earth (cf. Note 6.2; and Cassen et al., 1982). The Earth-Moon system could have experienced many complex resonance states since its origin, expressed by: (1) rotational resonance states of the Earth, (2) orbital resonance states of the Moon, and (3) complex resonance states of the Earth-Moon-Sun system, mediated by resonance states involving the inner planets and/or the planetary system in general. Planetary resonance states, which are considered in the next section of this chapter (see Fig. 21), address questions of the dynamical states of the Solar System as a whole in terms analogous to the types of interactions that are involved, for example, in the local dynamics of the Asteroid Belt. In view of recent developments showing that the Solar System itself is chaotic (cf. Wisdom, 1987a, b, c, 1990; Laskar, 1988, 1989, 1990; Milani and Nobili, 1992; Sussman and Wisdom, 1992), and that there are systematic resonance relations between several of the planets (Laskar, 1989; Laskar et al., 1992), then both the Earth and the Moon have been subjected to chaotic resonances of both aperiodic and nonlinearly periodic types (*chaotic bursts* in the orbital and rotational parameters of the Earth and the Moon would be expressed by sudden small changes in the length-of-day and the length-of-month).

In view of such large-scale resonance phenomena, direct effects of spin-orbit resonances of the Allan type cannot be discounted for objects that are "large" from the standpoint of Phanerozoic impact cratering (i.e., the order of ~10 km in radius). These objects most likely would be found in the exterior resonances at the right in Figure 20, hence would be in high-inclination orbits beyond ~40,000 km between the Earth and Moon. From the observational standpoint, these and possible objects in low-altitude interior resonances, at either high or low inclinations, could be moving fast either forward or backward relative to the point of observation. Thus, fast-moving objects (FMOs) and other undocumented objects mentioned in the discussion of asteroid discoveries are candidates. The discovery circumstances of asteroids (e.g., Tatum, 1988, 1991; Binzel, 1991; Binzel et al., 1989, 1991; cf. Yeomans, 1991b; Steel, 1991; Matthews, 1992; Ahrens and Harris, 1992) indicate that *successful exploration has in the past depended critically on knowing where to look* in terms of orbital parameters (see the discussion of "search strategies" in Note 6.5).

As nearly as I can tell from the literature, searches for natural near-Earth objects (i.e., those objects that differ from known artificial objects, including "space garbage" composed of defunct artificial satellites and fragments thereof; see Taff, 1985, Chap. 5) have been designed to scan mainly those regions of the celestial sphere where the experiences of successful asteroid and comet discoveries suggest *they might be expected*. Such regions have generally been restricted to the vicinity of the ecliptic plane at inclinations generally well below 20° (e.g., Tombaugh et al., 1959; Tombaugh, 1961a, b; Tombaugh and Moore, 1980; Shoemaker et al., 1990; Yeates, 1989; Frank et al., 1990); hence the probabilities of discoveries of objects in high-inclination orbits are very low, for a variety of reasons, among them the fact that the probability of searching in such regions is, itself, very low (see Note 6.5). In discussing cometary breakup and meteoroid debris in the inner Solar System, Olsson-Steel (1989) has stated: *"It is also suggested, from the evidence of the available meteoroid orbit data, that there is a substantial population of planet-crossing asteroids and short-period comets in high-inclination/retrograde orbits which wait discovery: such bodies would be of enormous importance with regard to the rate of impacts of large objects on Earth."*

In so saying, Olsson-Steel apparently did not have geocentric orbits in mind, but his scenario is analogous to the one I envision as most likely to lead to resonant capture of small satellitic objects by the Earth and Earth-Moon system (cf. Fig. 19, and Notes 6.1 and 6.5). Resonances analogous to Figure 20 theoretically exist outside the Earth-Moon distance — i.e., relative to the Earth-Moon center of mass (see Note 6.4) — where the Earth and Moon represent longitudinal mass anomalies of Earth-Moon "spin" of the combined system (cf. Fig. 21 and Appendix 11; the dynamics of the more distant orbital resonances would be caught up with the phenomena of *Lagrangian points, fractal spheres of influence, orbital transfer, and "fuzzy" orbital capture scenarios* outlined qualitatively in Note 6.1).

Up to this point in the present chapter I have given evidence — viz., consisting of (1) the prevalence of nonlinear periodic and chaotic resonance phenomena in the Solar System, (2) the similarities between spin-orbit and system-wide, relative-orbit-rotation ("orbit-orbit") resonances, as demonstrated by (a) the theoretical ordering of rotational vs. orbital frequencies plotted in Figure 20, (b) the observed Solar System reference-state orbital frequency vs. orbital frequencies (e.g., as plotted in Figure 21 and Appendix 11), and (c) the known resonance structures (Kirkwood Gaps) of the Asteroid Belt illustrated in Figures 12 and 13, and (3) the similarity between the known Solar System orbital- and rotational-frequency ratios and the numerical resonance structures of the Devil's Staircase in Figure 14 (and Appendix 9) — that the structure of a proximal geocentric reservoir of small natural satellites will have been self-organized relative to an invariant hierarchy of potentially stable resonance states, but that any one, or all, of these resonance states would have been "filled" and "emptied" intermittently and selectively as functions of the satellite mass distribution and time. In other words, the actual occupancies of potential resonance states are subject to the competing rates at which they can be filled and emptied, reflecting the relative strengths of different resonance conditions — i.e., in a manner analogous to that by which the Kirkwood Gaps in the Asteroid Belt have evolved.

Judging from the evidence just cited, the actual or potential evolution of a system of very small natural Earth satellites — small compared to the sizes of the named satellites of the planets — is self-similar in precisely the same way, dynamically speaking, that the generalized hierarchy of nonlinear-resonance structures (objects and/or reservoirs of objects) in the Solar System at large is self-similar, as schematically diagrammed in Figures 11 and 21 (and Appendix 11). In such a geometric context, there is some resemblance between this picture of Solar

System resonances and the model proposed by Rawal (1991, Fig. 5; cf. Rawal, 1992) of Oort Cloud-like cometary reservoirs around each of the planets at a very early stage in the orbital evolution of the planetary system. [According to the viewpoint espoused in the present volume, of course, any model of protosatellitic reservoirs, of any kind, would have been subject to the rigors of those phenomena that I have suggested — on several grounds, including the kinetic energy distributions of Fig. 10 and the astronomical arguments cited in Chap. 3 (cf. Notes 3.2 and 9.2) — were correlative with the chemical and dynamical changes, on a grand scale, that attended the cataclysmic transition between the stellar-binary and Sun-centered stages of the evolution of the planetary nebula.]

A reservoir of small natural satellites orbiting the Earth (and/or Earth + Moon) could, in principle (as in the "butterfly effect"; see Notes P.2 and P.6), eventually be perturbed and/or disrupted by objects penetrating the Earth's *sphere of influence* (within about 10^6 km from the Earth; cf. Note 6.2) or by the transient near-Earth passage and/or capture of larger satellitic objects — where the largest of such transient events might empty the reservoir completely for some period of time. Such disturbances are dynamically analogous to others proposed in theoretical models whereby passing stars at distances of the order of 10^5 km from the Sun (**Note 6.3**) perturb the outer Oort Cloud of comets (cf. Hills, 1975, 1981, 1984a, b, 1985, 1986; Tremaine, 1986; Stern, 1986, 1987, 1988, 1991; Duncan et al., 1987; Stern and Shull, 1988; Heisler, 1990; Weissman, 1990a, b, 1991b). Consequently, the intermittency of Earth's satellitic reservoir is expected to be related to the intermittency of comet and/or asteroid showers, hence related to the intermittency of the self-organized critical cascade of small objects throughout the Solar System (cf. Fig. 11). If the scenario of the breakup of a "giant comet" roughly 10^4–10^5 years ago, as advocated by several workers (cf. Clube, 1989b, 1992; Olsson-Steel, 1989; Clube and Napier, 1990), reflects reality, then it is possible that it disturbed Earth's small-object satellitic reservoir enough to empty it, so that it only began to fill again during the latest 10,000 years, a period during which there have been several low-altitude bolide events and/or impacts (see Appendix 3). Chaotic crises, as already discussed in Chapters 4 and 5, also would be expected to disrupt the satellitic reservoir — affecting some resonances more than others — in the same manner that I have argued should obtain within the asteroid and comet reservoirs (see Figs. 12, 13, and 15), and therefore should have been instrumental in controlling the populations of objects and the relative rates of filling and emptying of different resonance states (e.g., as in the Kirkwood Gaps of Fig. 13).

Many critical frequencies are possible in the processes described above (cf. Coffey et al., 1986, 1990, 1991), and tidal frequencies are likely to be important (there are many tidal components, e.g., relative to the Earth, the Moon, and the Earth-Moon system; cf. Munk and MacDonald, 1960; Kaula, 1968), especially for orbits near to conditions of synchronous resonance and/or other "nearly balanced" states, in a manner analogous to the secular variation of the lunar orbit (cf. Note 6.2) — for reasons analogous to those that are involved in *fractal basin boundaries* and *fuzzy transfer orbits*, as discussed in Note 6.1. The net flux (*delivery*) of objects to the proximal Earth reservoir is proportional to the intermittent total flux into the region of the inner planets, according to the model of critical self-organization, which calls for dynamical crises and collisional scattering, and feedback between the two, according to the respective influences of these processes on stable and unstable attractor structures in their vicinities (cf. Figs. 11–18). For example, I think of the K/T boundary interval as representing a peak occupancy of the proximal reservoir that had built up during the Cretaceous — or perhaps over a longer period of time — and was "dumped" at ~65 Ma by a cascade of chaotic crises involving both the Solar System dynamics and terrestrial dynamics (e.g., involving the self-organized motions of plate tectonics, core dynamics, and magma dynamics, etc., that modulate the geoid and the mass structure in the Earth that affects satellite orbital stabilities; cf. King-Hele, 1976, 1992; Hager et al., 1985; Anderson and Casenave, 1986; J. Davis et al., 1989; Dixon, 1991; Bowin, 1992; Engebretson et al., 1992; Richards and Engebretson, 1992), and that had, by terminal Cretaceous time, evolved to a net state that amounted to a "conspiracy of action," a form of feedback-systems effect representing — often counterintuitively so — *cybernetic self-tuning* (e.g., Forrester, 1961, 1973; Shaw, 1969, 1978, 1981; Shaw et al., 1980b; Richardson, 1991; cf. Chap. 5, and Notes P.4, I.2, 3.3, 5.2). This cascade of activity may have started with a chaotic crisis in the cometary regions, probably initiated by the Galactic tide (± stellar disruption), that then propagated through all the reservoirs of objects in Figure 11. My inferences, of course, are hypothetical, yet they relate to — and remain to be compared in any comprehensive way with — the records of geodynamical feedback phenomena (geomagnetism, magmatism, tectonism, etc.) that will be described in later chapters vis-à-vis their coupling with the integral record of impact-cratering phenomena.

In the remainder of this chapter I shall expand my examination of the concept of generalized self-similar resonances in the Solar System as a whole, and discuss the manner in which Earth's proximal reservoir of intermittent satellites may have controlled, and continues to control even now, within a given time frame — past, present, and future — the evolving patterns of terrestrial cratering that have already been impressed, and that *are yet to be impressed* (cf. Chap. 9; Note 9.1; and Appendixes 1–4) upon Earth's biosphere and tectonosphere (where I use the latter term loosely to include all geodynamic feedback processes). And in this way, by the means discussed above, the celestial reference frame (CRF) of Earth's geological-biological history has evolved in the codified manner that I have attempted to describe, both in the previous chapters and in the chapters to come.

Ordering of the Planets by Resonance with the "Spin" of the Solar System

The theory of Allan (1967a, b) is based on the simple idea that longitudinal mass anomalies in the gravitational field of a system give rise to periodic enhancements of the orbital frequencies of satellites, depending on their numbers and positions, hence frequencies, in the same reference frame. For Earth, this is the geocentric reference frame, and in this frame both the Earth and its satellites (the Moon and the geocentric reservoir of relatively "small" objects postulated to exist in the present volume) are corotating relative to the *center of mass* of the Earth-Moon system **(Note 6.4)**. The dynamical distinction between rotation and revolution disappears, except as a kinematic description, when we are discussing the orbital motions of a system relative to the same "primary" body (e.g., in terms of the ratios α/β and a/R specified relative to a given primary body; cf. Note 6.1 and Fig. 20). The two terms, rotation and revolution, are needed only when the motions are referred to different centers of mass (for example, the variation of the angular velocities of stars, spiral arms, and molecular clouds, relative to their distances from the center of the Galaxy, is called the "rotation curve" of the Galaxy; cf. Schmidt-Kaler, 1975; Rubin, 1983, 1984). The planetary system as a whole "rotates" about the center of mass (CM_{SS}) of the Solar System, which is near to but not identical with the center of the Sun. [The "orbital distance," and that is precisely what it is, between the Sun and the CM_{SS} is small compared to its balancing counterpart, the interplanetary distances, but the chaotic trajectory of the Sun relative to the CM_{SS} (see Jose, 1965) strays in and out, as though it is tied to the CM_{SS} by a tangled tether, by up to one solar diameter (~ 0.01 AU $\cong 1.5 \times 10^6$ km) or more within a time frame amounting to but a few tens of years, a distance that is roughly three times our distance from the Moon and is comparable to the radius of Earth's celestial sphere of influence.] For most purposes, this refinement does not change the usual measures of perihelia, q, or semimajor axes, a, with respect to distance from the Sun, but it *does* introduce another source of characteristic frequencies, identified by the average period or periods — nonlinearly periodic or chaotic — of the Sun's orbit around the CM_{SS}, that permeate the Solar System. [The Sun's orbit in the CM_{SS} frame of reference is especially important to the manifestation of torque episodes that influence the internal dynamics of the Sun (see Landscheidt, 1983, 1988) — incidentally, also representing a largely unevaluated nonlinear mechanism of orbital coupling with the planets, especially the inner planets (e.g., Wood, 1972) — and hence also influence solar sunspot activity and solar-terrestrial phenomena that, in turn, influence Earth's chaotic patterns of weather and climate (see Notes 6.4 and 8.4).]

If the mass distribution of a spin-orbit system is considered to consist of point masses with some approximately fixed mean period of relative rotation, then the form of the system is irrelevant, and resonances of the Allan type can occur in the same manner that they do relative to mass anomalies in the Earth (cf. Table 1 and Fig. 20). That is, if the rotational properties of the system of point masses are not uniformly distributed over all possible frequencies in phase space (i.e., if the system is not *ergodic*; cf. Note 4.2), then there is some sort of mass anomaly that will act on each of the individual objects in the sense of a system of spin-orbit resonances — depending on what rotational property of the system is chosen as the "primary" motion. Such a system of "spin-orbit resonances" would approach an invariant state if there were a feedback between the evolution of resonance states caused by nonuniform mass distributions and a particular set of mean-motions of the system as a whole that represented an "optimal" longitudinal mass anomaly that mediated a "complete" set of such resonances (e.g., analogous to the complete Devil's Staircase of Fig. 14).

In the absence of other perturbing influences (such as chaotic interactions between the power of the solar dynamo and the Sun's motion about the Solar System center of mass, as outlined in Note 6.4, or Galactic inputs of mass and kinetic energy), such a system would eventually conform to a *redundant state* analogous to the state of "least interaction action" of Ovenden (1972, 1973, 1975, 1976; cf. Hills, 1970; Ovenden et al., 1974; Ovenden and Byl, 1978); see also the tribute to Michael Ovenden early in the Introduction of the present volume. If a tendency toward stability existed, it would be fulfilled by feedback between sets of mean commensurabilities of mass that support compatible sets of spin-orbit resonances, which, in turn, would affect the mean commensurabilities of mass, and so on, until a quasisteady balance is achieved (e.g., representing a form of "stable chaos," in the terminology of Milani and Nobili (1992; cf. Murray, 1992). For the sake of illustration, and as a test of such a generalized model, I assume that the planetary system corresponds to a distribution of point masses rotating in the same reference frame, a distribution that is dynamically equivalent to the system of spin-orbit resonances of Figure 20. This reasoning treats the planets as a system of satellitic objects that are revolving in a Solar System that is "rotating" about its center of mass, just as Earth satellites revolve about the rotating Earth, or as satellites in orbits revolving about the Earth-Moon system are corotating with both the Earth and Moon about their common center of mass. These ideas are illustrated in **Figure 21**, where I compare the system of spin-orbit resonances of Figure 20, as computed for the Earth by Allan (1967a), with the distribution of planets in the Solar System expressed in the same way — i.e., as though the planets represent a "satellitic" system of spin-orbit resonances relative to the Solar System as a whole. In the case of the Solar System, any reference datum could be used for the position of the "primary body." In Figure 21, I have arbitrarily chosen the position of Pluto as the primary datum, so that all of the planets correspond to "interior" resonances. Other choices are illustrated in **Appendix 11**, where a datum based on the position of the *solar limb* ("surface" of the Sun), or one based on the "limb" of the Sun's orbit around the Solar System center of mass (cf. Note 6.4), would be analogous to the terrestrial case.

The same idea can be generalized to any number of objects in any number of subsystems of different scales that are ordered in the same way. Thus, a hierarchical system of such spin-orbit resonances will exhibit universality and fractal self-similarity, just as it is envisioned to operate in Figures 9–11. Each subsystem, in simplified terms, resembles the sine-circle function (Table 1, Eq. 5) in that the "circle" can be taken to represent the reference state of rotation of the "primary" object, and the iterated points can be taken to represent the motions of the satellites, either in the direct (*interior*, $\alpha < \beta$) or inverted (*exterior*, $\alpha > \beta$) senses of Figure 20. With zero nonlinear coupling ($K = 0$ in Eq. 5, Table 1), the satellites move independently and are ordered by phase differences (Ω) that can take any or "all" real values (i.e., every real number, rational and irrational, corresponds to a possible orbital frequency). In the paradoxical sense of order and disorder discussed in the Introduction, *such a system might be called a "perfectly ordered, random, two-dimensionally uniform, universe of concentric orbits."* With any degree of coupling ($K > 0$), however, the orbits are not independent, and their orbital frequencies are defined, on average, though precisely, by the winding number (W), which is equivalent to the frequency ratio of the primary to the satellite (i.e., $W \equiv \alpha/\beta$ of Figs. 14 and 20; cf. Notes 3.1 and 4.1). Now, instead of the continuous distribution of frequencies that characterized the system without coupling (i.e., the situation characterized by $K = 0$, which implied that *every orbit is independent of every other orbit*), the distribution of orbital frequencies has become a step function or "staircase function" of special resonant frequencies (the condition of *phase-locking* or *mode-locking* illustrated in Fig. 14; and as discussed in Notes 3.1 and 4.1), meaning that constant values of different orbital frequencies (equivalent to the winding numbers, W) exist over finite ranges of distances and conditions of forcing (corresponding to given combinations of K and Ω). Sets of these frequency steps (each corresponding to a constant value of W) make up fractal hierarchies of different resonance strengths, or depths of metastable gravitational potential wells (see Note 6.1), in a manner analogous to the sets of resonance ratios α/β in Figures 14 and 20.

The staircase of frequency states is completed when the coupling constant K reaches the critical value $K = 1$. At that point, discrete frequency steps exist for every integer ratio of $W \equiv \alpha/\beta$, without limit. [At higher values of K, the values of W vary irregularly, and they can be multivalued at fixed values of Ω and K (cf. Ding and Hemmer, 1988; Shaw, 1991, Fig. 8.36); in addition, trajectories in phase space are not invertible at $K > 1$. Thus, the condition $K > 1$ is regarded as the beginning of the chaotic regime, and $K = 1$ is called the *chaotic transition*, or *the edge of chaos*; but the "edge," of course, is a misnomer if it is thought of in the traditional Euclidean sense, for the reasons discussed in Note 6.2.] The total system of phase-locked resonances of the sine-circle function has an invariant fractal exponent of $\gamma \cong 0.87$, instead of the integer-value exponent $\gamma = 1$ that would apply to the case $K = 0$ describing sets of independent orbits, as discussed above (cf. Jensen et al., 1984; Bohr et al., 1984). The exponent γ is, in a sense, a measure of deviations from an idealized system of all possible real-number-valued simple harmonic motions. Kepler's Third Law, of course, holds true, as expressed by the 3/2 proportionality between $\log \alpha/\beta$ and $\log a/R$ shown in Figure 21 (i.e., the slopes of the heavy solid lines = 3/2) — meaning that the centers of the potential wells of the planets, satellites, etc., "line up," as it were, according to Kepler's laws — but the respective energy levels at different scales of motion are "quantized" in a fractal sense (the old analogy between the planets and electron levels in an atom holds true in the sense of nonlinear fractal similarities; cf. Richards, 1988; Jensen, 1992). These "quantized" properties of the planetary system are illustrated by the approximated Solar System resonances of Figure 21.

Within this structure there are several subsystems actually or potentially organized in the same way, such as the satellitic system of resonances for each planet, the resonances within the Asteroid Belt, within the Kuiper Belt, and within the Oort Cloud (including the inner part called the Hills Cloud), or resonances involving combinations of portions of these systems, or resonances within a system involving all comets and asteroids, etc.

If the Solar System as a whole is chaotic — as is indicated by recent studies (see below) and is reinforced by the discussion in Note 6.4 concerning the orbit of the Sun around the Solar System center of mass — it has entered a general region of orbital "quantum uncertainty," in the sense that the ordered resonances have become "detuned," or "broken up," or have disintegrated into orbital trajectories that are analogous to *quantum chaos* (cf. Berry, 1987; Richards, 1988; Jensen, 1992; Parmenter and Yu, 1992) in having both potentially stable orbital energy levels and excluded intervals (resonance gaps) that lack predictability in the trajectories of individual objects (i.e., especially if they were rescaled to resemble the orbital recurrence frequencies, sizes, and difficulties of instrumental resolution characteristic of an atomic structure). And if the chaotic state were fully developed at all scales of motion (i.e., in regions of phase space lacking resolvable periodic windows), then there would be no simple relationships among the orbital frequencies, and there would be no domains of stable periodic phase-locked frequencies (i.e., unstable periodic manifolds exist, but they only add to the uncertainty by inducing chaotic crises and bursting phenomena).

Within such a generally chaotic regime, the winding numbers (W) can: (1) "drift" in value, (2) become scattered in values over systematically varying ranges of forcing parameters, (3) change drastically in value with an infinitesimal change in a forcing parameter, or (4) take on two or more values at constant values of the forcing parameters (cf. Ding and Hemmer, 1988; Shaw, 1991, Fig. 8.36). The last of these cases is somewhat analogous to the quantum-dynamical paradoxes of "Schrödinger's cat" and/or "Wigner's friend," in the sense that a satellite or planet might be indistinguishably in either of two orbital states "at the same time" (cf. Schlegel, 1980, pp. 175ff). [Such analogies are not strict, for reasons discussed in the Prologue and in Notes P.4 and 4.2, having to do with fundamental disparities between the formalisms of quantum dynamics and "classical" nonlinear dynamics; see discussions by Berry (1987) and Ford et al. (1990).] The "quantum" paradox of alternative states also is reminiscent of the uncertainties discussed in Chapter 4 concerning "empty" and "filled" resonances in the Asteroid Belt (cf. Figs. 12 and 13), explained there by the phenomenon of chaotic bursts (a concept that applies to subatomic as well as to macroscopic dynamical phenomena; cf. Fig. 18; and Richards, 1988). "Empty" and "filled" resonance states scale self-similarly, because the bursting phenomena (Figs. 17 and 18) reflect fractal uncertainties in the boundaries of their basins of attraction (cf. McDonald et al., 1985; Moon and Li, 1985a, b; Moon, 1987, pp. 242ff; Shaw, 1991, Figs. 8.19, 8.20, and 8.36).

In general, if some part of an orbital system — or any general system of coupled oscillators — can be shown to be chaotic, and the system as a whole can be shown to be interdependent, then, formally, the system as a whole is chaotic (within characteristic relaxation times, and excepting the existence of periodic windows and sets of unstable periodic attractors that characterize system-wide chaotic crises). Such a maxim is implied by my argument that the universe as a whole, and in all of its parts, is a dissipative system, and seems to be confirmed by recent conclusions that all of the planetary orbits are chaotic at the time scales of 10^6 to 10^7 years (see Laskar, 1988, 1989, 1990; Wisdom, 1990; Milani and Nobili, 1992; Murray, 1992; Sussman and Wisdom, 1992). This does not mean, however, that there can be no periodic structures in the Solar System, because the chaotic regime contains periodic windows. It *does* imply, however, that a systematic regime of phase-locked resonances — of the sort discussed above and illustrated in Figures 20 and 21 — would be unstable. Such a general instability condition for nonlinear-phase-locked resonances in the Solar System has a number of possible implications and/or interpretations: (1) it might mean that, in the past, the Solar System was in a general state of stable orbital resonances and has only recently become "unlocked" from that condition, (2) alternatively, the Solar System might have been more chaotic in the past and is only now relaxing toward conditions where phase-locked orbital states can be expressed, or (3) since an early stage of nebular organization, the Solar System has persisted in a condition of marginal instability, where both phase-locked and chaotic orbital states have coexisted as an expression of chaotic intermittency operating near the critical transition between phase-locked and chaotic motions. The last of these cases is consistent with the prevalence of chaotic crises as an integral mechanism for orbital changes of state, and is consistent with the hypothesis that self-organized criticality is characteristic of the dynamics of the Solar System at large, as schematically illustrated in Figure 11. According to this interpretation, the Solar System is hierarchically poised at the "edge" of chaos — or, really, *within* the boundary that separates the stable nonlinear-periodic regime from the chaotic regime — so that it oscillates intermittently between ordered and chaotic resonances, both spatially and temporally, depending on perturbations involving the orbital motions of the Solar System's center of mass within the Galaxy, other stellar-Galactic phenomena, and so on (see Notes 4.2 and 6.4; cf. Ruthen, 1993). From this point of view, the Solar System as a whole is *frustrated* in the sense used to describe *spin-glass systems* (cf. Carroll et al., 1987; Dewdney, 1987; Stein 1989; Suarez et al., 1990; Abraham et al., 1991b).

Figure 14 portrays a metaphorical example of a system poised in the above manner, as generated from the sine-circle function at the transition (chaotic border) between phase-locked (+ quasiperiodic) behavior and chaotic behavior (the latter does not actually show up in the diagram because we are standing, as it were, "on the edge"). If the forcing, represented by the parameter K, is decreased, we fall into the realm of periodic regularity, and if it is increased, we are swept up into the more excited, and more exciting, realm of chaos, a situation perhaps resembling somewhat the paradigmatic transitions described by works such as *Alice in Wonderland* and *The Wizard of Oz* (except that — counter to conventional interpretations — the more exciting realm, the chaotic one, of nonlinear algorithms is actually the more realistic one). At a quasiperiodic (irrational) value of W equal to the inverse of the golden mean (an invisible point, at $W = 0.61803398..$, in the space of Fig. 14; see Note P.4; and Shaw, 1987b, Fig. 2), orbital ratios scale with power-law distributions of frequencies in a *multifractal singularity spectrum* (Fein et al., 1985; Halsey et al., 1986; Shaw and Chouet, 1989, 1991). Because this is a characteristic transition-state condition in a typical "route to chaos" in a number of complex coupled oscillators with clustered recurrence patterns (see Note P.4), it could occur in what others might call "random orbital patterns" of coupled celestial systems involving large numbers of objects (an example might involve the coupling between the Oort Cloud and the Galaxy, and perhaps the structure of the Galaxy itself; cf. Shaw, 1983b). According to this paradigm, the Solar System, viewed in its entirety, is also a multifractal singularity spectrum operating at the edge of chaos **(Note 6.5)**, and Figure 21 becomes a "quantum" planetary system with distinct resonance levels bordering on "quantum chaos" in any vicinity. [The history of discoveries of the less regular Solar System objects, such as Pluto and Charon — and the difficulties in making such discoveries (e.g., Tombaugh, 1961a, b, 1991; Tombaugh and Moore, 1980; Stern, 1991) — testify to the uncertainties revealed by such a viewpoint.]

I chose Pluto as the "primary" reference datum in Figure 21 ($\alpha = \beta = 1$) both for simplicity and because it is at the extant outer limit of the planetary sphere, at least until Stern's (1991) objects are documented. The choice, however, can be shifted arbitrarily and could just as well be taken, for example, at Jupiter, or anywhere else if what is desired is to expand the frequency intervals about a different frame of reference. The longitudinal mass anomalies are, of course, the instantaneous net distributions of planetary masses relative to the motions of the objects chosen as the perturbed "satellites."

To a first approximation, the dominant mass anomaly relative to the motions of all of the planets (and balanced by them) is identified by the gyration of the Sun about the center of mass of the Solar System, CM_{ss} (see Note 6.4; cf. Appendix 7A, Inset). At the next level of approximation, the asymmetric dumbbell-like or twirling-baton-like configuration of the Sun-Jupiter two-body motion around the CM_{ss} — with an average sidereal period equal to Jupiter's orbital period of 11.862 years, in the same sense that the period of rotation of the Earth-Moon system about its center of mass is equal to the sidereal period of the Moon's orbital motion — identifies the longitudinal mass anomaly, and disturbing function, for the relative motions of the remaining planets (cf. Table 1, Sec. A). The Sun, Jupiter, and Saturn, taken together, and considered to constitute a resonant three-body system, identify yet another form of "anomaly," perhaps an important one because Jupiter and Saturn are the largest and closest together ($\Delta = 4.336$ AU) of the giant planets, and the ratio of their sidereal periods is close to 2/5, equivalent to the frequency ratio $\alpha/\beta = 5/2$, when Jupiter is taken to be the "primary" object (cf. Fig. 21; and Appendix 11) — meaning that Saturn is in an exterior resonance state relative to Jupiter (i.e., Saturn's period is about 2.5 times that of Jupiter, so that Jupiter revolves five times around the Sun, relative to the fixed stars, for every two revolutions of Saturn). Thus, the planetary motions act as their own nonlinear "orrery" and map out the nonlinear resonances consistent with their mutual gravitational effects **(Note 6.6)**. An analogous problem is represented by the resonance effects in a primary that has strong differential rotations as a function of radius as well as heterogeneities of mass at depth, as is evidently true, for example, of the Sun and Jupiter (cf. Allan, 1967a, p. 69).

The slopes of the reference lines in Figure 21, given for the Solar System orbital resonances by $(\alpha/\beta) = (a/R)^{3/2}$ (therefore with a slope of 3/2 in the logarithmic plot) are defined by Kepler's Third Law. The planetary spacings are not exact, however, because of possible higher-order resonances, quasiperiodic states, and chaotic states. Values of relative orbital frequencies approximating simple ratios suggest that perturbations may arise from chaotic crises in the vicinities of exact, but unstable, resonant periods, perhaps representing past stable states that have become unstable, or a long-term condition of so-called *stable chaos*, as discussed by Milani and Nobili (1992) and Murray (1992). In contrast with other examples of chaotic crises in diverse types of physical systems (cf. Figs. 13, 15, 17, and 18), the amplitudes of chaotic transients for the planets would be small (examples might be some part of the microsecond variations in the Earth's period of rotation, the so-called *length-of-day,* or some part of the change in Earth's period of revolution around the Sun, as reflected in a nonconstant year relative to other standard measures of time, such as the "atomic clock"). The differing geometric progressions (resonance families) of Figure 21 again point to a major dynamical distinction between the planets of the outer and inner Solar System (cf. Figs. 9 and 10; and Note 4.2).

Transformations between different scales and "primaries" in Figure 21 are shown, for example, by the relation between the planetary sequence and the resonances in the Asteroid Belt (as indicated by the shaded segment marked MBA, and arrows connected by a dotted line along the Solar System resonance curve in Fig. 21). The position labeled ~1/4 Kirkwood corresponds to the 4:1 Kirkwood Gap of Figure 13 and defines the approximate inner limit of the main-belt asteroids **(Note 6.7)**, relative to Jupiter as the perturbing source (cf. Wisdom's 1987a, b, c, demonstration of chaotic behavior at the position of the 3:1 Kirkwood Gap, illustrated in Fig. 12 here).

The approximate value of the resonance ratio for Jupiter in Figure 21, relative to Pluto as the "primary" reference state, is $(\alpha/\beta) = (1/21)$. Noting that the asteroid resonance states are referred to Jupiter as the primary body, they can be transformed to equivalent resonance states referred to Pluto simply by taking the product of their resonance ratios and Jupiter's resonance ratio relative to Pluto. Taking the asteroid (A) resonance ratio relative to Jupiter (J) of 1/4 as an example, the transformed resonance ratio is defined by

$$(\alpha/\beta)_{J,A} \times (\alpha/\beta)_{P,J} = (\alpha/\beta)_{P,A}$$

giving (as shown by the horizontal dashed line with arrow in Fig. 21)

$$(\alpha/\beta)_{P,A} = (1/4) \times (1/21) = (1/84),$$

where the first of the binomial subscripts refers to the primary body and the second subscript refers to the object in question [e.g., $(\alpha/\beta)_{P,A}$ signifies the resonance ratio of the asteroid, A, referred to Pluto, P, as the primary body]. Therefore, the resonance ratio $\alpha/\beta = 1/84$ is the redefinition of the position of the 4:1 Kirkwood Gap according to the planetary sequence of resonance ratios referred to Pluto as the primary (**Note 6.8**). Obviously, then, in order to transform all of the Pluto-based resonance ratios to, for example, Jupiter-based resonance ratios, we simply multiply all of the Pluto-based resonance ratios by 21. In the *Jupiter-based reference frame*, therefore, we have the sequence of *exterior resonances* in the order: Pluto, $P \cong (21/1)$, Neptune, $N \cong (14/1)$, Uranus, $U \cong (7/1)$, and Saturn, $S \cong (5/2)$, and the sequence of *interior resonances* in the order: main-belt asteroids (evaluated for a mean semimajor axis near that of Ceres), $MBA \cong (7/18)$, Mars, $M \cong (7/44)$, Earth, $E \cong (7/83)$, Venus, $V \cong (1/19)$, and Mercury, $Me \cong (1/49)$. In this case, Jupiter, of course, represents the synchronous resonance $J = (1/1)$.

By choosing different planets (or other objects) as the primary body, local regions of the Solar System can be tested for subsets of resonance ratios. For example, choosing Uranus as the primary body, we find that the three outermost planets correspond roughly to a simple linear progression of ratios: $U = (1/1)$, $N = (2/1)$, $P = (3/1)$. Similarly, choosing Earth as the primary, we find that the resonance ratios of the terrestrial planets form the sequence: $M \cong (47/25) \cong 2$, $E = (1/1) = 1$, $V \cong (8/13) \cong 1/2$, $Me \cong (6/25) \cong 1/4$, where the actual ratio for Venus ($\sim 2/3$) is the largest departure from a simple progression given by the *period-halving sequence* 2^n, with $n = 1, 0, -1, -2$ (see discussions of the Titius-Bode relationship in Notes 4.2 and 6.8; cf. Appendix 6). The unmodified ratios in Figure 21 usually involve some combination of the roots *2, 3, 5,* and *7*, even though all primes are possible integers. [In Fig. 20, for example, *2/31* is an exception; in Fig. 21, ratios involving *1/11, 1/31,* and *1/103* are exceptions, although the ratio *1/103* is approximated by $3/(2 \times 5 \times 31)$.] Higher primes may occur when there are departures from the dominant resonances and/or when *quasiperiodic* orbital frequencies exist (*irrational frequency ratios*). Resonance ratios involving very large prime numbers, however, are suspect, because irrational numbers can always be approximated by some multiplicative combination of small integer terms (e.g., Hardy and Wright, 1985, Chap. 11).

To the degree of approximation represented by the resonance ratios of Figure 21, I do not feel confident that primes even as small as *31* differ significantly from their nearest rationals (e.g., $30 = 2 \times 3 \times 5$, or $32 = 2^5$). In prior attempts to reconcile geologic periodicities with chaotic dynamics, I found that the four roots *2, 3, 5,* and *7* accounted for most of the maxima in histograms of peak-to-peak intervals made up from diverse geologic time series (see Fig. 25 and Table 1 in Shaw, 1988a; cf. Shaw, 1987b; Rampino and Caldeira, 1993). I took this approach because combinations of the roots *2, 3, 5,* and *7* form conspicuous periods among the $2^n k$ periodic windows ($k = odd$ integers) of the chaotic regimes of quadratic and cubic equations beyond the stable period-doubling sequences [e.g., of the *logistic map*, a mapping of one form of quadratic equation, discussed in detail by May (1976) and Feigenbaum (1979a, b, 1980), and in the sine-circle map of Table 1, Sec. D]. Ratios involving these four roots also account for the most stable winding numbers (W) in the stable phase-locked resonances (nonlinear periods) of the sine-circle map (e.g., in Fig. 14 they account for all ratios but those shown in the magnified Inset), and for the most stable resonances of coupled physical oscillators in general (e.g., see the stable periodic points documented by Miles, 1984, pp. 310ff, in his elegant study of the nonlinear regimes of the spherical pendulum).

In the chaotic regimes of simple dissipative algorithms — such as the quadratic equation — simple geometric progressions exist because periodic windows are numerically ordered according to Sarkowskii's theorem (cf. May, 1976; Kaplan, 1987; Tufillaro et al., 1992, Chap. 2, Sec. 2.9). The widths and stabilities of these periodic windows are such that only the simple prime-number roots are likely to be encountered in all but the most precisely tuned experiments or computations. In the first sequence of periods beyond 2^n in quadratic maps, periodic windows in the generally aperiodic regime are in the order of increasing values of the ratios of primes up to *1/3* (thus perhaps contrary to what intuition would conclude on the basis of the period-doubling progression, 2^n). This is a crude statement of Sarkowskii's theorem, and is preferred by some as a definition for the "real beginning of chaos," as proposed in the early paper by Li and Yorke (1975) titled "Period Three Implies Chaos." In practice, periodic windows in the chaotic regime involving primes larger than 7 (periods such as 11×2^n, 13×2^n, etc.) are relatively narrow and unstable, hence they are often inconspicuous. What is more, simple periodic windows, as well as intermittent intervals of simple prime-number periodic roots (short intervals in chaotic time series involving intermittent repetitions of short sequences of various periods interspersed with other intervals of aperiodic and/or transient characteristics), continue to occur throughout the chaotic regimes of nonlinear-dissipative algorithms, even where measures of (dis)order asymptotically approach that of a coin toss (cf. R. Shaw, 1981; Ott, 1981).

Rather than appealing to the concept of relative stabilities or instabilities of periodic windows in the generally aperiodic regime as the criterion of chaos, as just discussed, a more general approach — and one that is easier to understand on physical grounds — concerns the practical question of *divergence* vs. *convergence* of trajectories in phase space. This concept, developed initially by Liapunov (1947; often spelled Lyapunov), is analogous to the meanings of the derivatives of a function, or of the relative "directions" of tangents to trajectories in phase space, evaluated over characteristic regions of phase space, or over characteristic durations in time. There are numerous — and often complicated — methods for evaluating Liapunov functions that depend on the topologic and fractal dimensions, respectively, of the phase space and of the attractor in question. The significance of the derived *Liapunov numbers,* or *Liapunov exponents,* as criteria for distinctions between periodic and chaotic regimes of nonlinear-dissipative systems is discussed, for example, by R. Shaw (1981), Wolf (1986), Farmer and Sidorowich (1988), and Nese (1989). *Relative Liapunov functions* are important discriminants for the characterization of both the periodic/chaotic nature of attractors and for their *degrees of complexity,* or *memory capacities.* These are qualitative terms, coined here (cf. Note I.2), that refer to the number of iterations, or the duration in time, over which a complex attractor keeps pace with or "remembers" the position on one trajectory relative to a nearby trajectory, or that refer, equivalently, to the faithfulness with which a nonlinear function computes the "same" relative state, or position in phase space relative to repeated starts from exactly the same initial parameter values of the function — as measured by an identical number of computational steps of the same generating algorithm.

Most of the research strategies used today in the development of prediction methods in chaotic dynamics are variations on the theme of Liapunov functions (cf. Note P.3; and Wolf, 1986; Farmer and Sidorowich, 1988; Nese, 1989; Sommerer and Ott, 1993). An alternative approach, exemplified by Sugihara and May (1990) — and analogous to the above discussion of Sarkowskii's theorem and characteristic geometric progressions — uses information on the numerical structures that are innate to the chaotic regimes of nonlinear functions to differentiate between the statistical properties of chaos and the statistical properties of "random," "Brownian," "uniform," or other patterns that often are assumed to characterize the numerical attributes of natural phenomena (cf. Tsonis and Elsner, 1992).

The analogous method of *fractal singularity analysis* (e.g., Jensen et al., 1985; Halsey et al., 1986; Gwinn and Westervelt, 1987; Shaw, 1987b; Shaw and Chouet, 1988, 1989) uses probabilistic evaluations of the recurrence frequencies at given locations in the characteristic fractal phase space of a chaotic attractor to evaluate a spectrum of likelihoods of next events. This method shares with the Sugihara-May approach the idea that attractors, and nonlinear-dynamical recursion algorithms in general, periodic and chaotic, have characteristic — i.e., nonuniform and nonrandom — multifractal structures that provide significant and realistic numerical information relative to the types of idealized lattice models (regularized or randomized) that have been applied in the past, for example to problems in statistical mechanics, celestial dynamics, and the dynamics of other orbital phenomena (subatomic structures, etc.). Such methods should prove useful in characterizing the structures of fractal basin boundaries between attractors, or the fuzzy routes of orbital transfers between celestial-dynamical spheres of influence (as described in Note 6.1 relative to interplanetary orbital transfer phenomena). This approach has revealed important properties of systems of coupled oscillators operating near the transition between phase-locked and chaotic regimes (the "edge" of chaos) — as seems to be true for the Solar System (cf. Notes 6.1 and 6.5) — because in that vicinity there are robust criteria for *universal multifractal spectra with ordered power-law scaling of frequencies,* representing forms of very rich, or orchestral, resonance phenomena (e.g., Fein et al., 1985, Fig. 2; Olinger and Sreenivasan, 1988, Figs 5 and 6; Shaw and Chouet, 1988, Fig. 16; 1989, Fig. 2; 1991, p. 10,205).

The line representing geocentric spin-orbit resonances relative to the Earth in Figure 21 is offset from the line representing analogous spin-orbit resonances relative to the Solar System, because the energy equation for rigid-body rotation is not continuous with the orbital energy distributions, as discussed in Note 6.4. That is, the resonance ratios have the same meaning, but the synchronous resonance *(1/1)* in the geocentric system of satellites occurs at a *synchronous distance* where the orbital velocity is equal to the rotational velocity of a point on the Earth's surface—a distance that occurs at $a = 6.61$ Earth radii (~42,000 km from the center of the Earth, or ~36,000 km altitude; see Fig. 20; cf. Allan, 1967a; Friesen et al., 1992)—in contrast with the planetary synchronous distance, where the equivalent "rotational" velocity of the reference primary (e.g., Pluto, or Jupiter, or whatever) is, by definition, identical with the orbital velocity at that distance. The actual velocity of the primary, Earth, at $a = R$ in the geocentric system is far smaller than is the orbital velocity at that distance (analogous to the fact that the Sun's orbital velocity around the Solar System center of mass is much smaller than the Kepler-Third-Law velocity at that distance). Furthermore, at distances $a \leq R$ (i.e., theoretically at and below the surface of the Earth, or inside of *Earth's limb*), satellite orbits are terminated. Obviously, then, as discussed in Note 6.4, there are major differences between the virtual orbital kinetic energies and the actual kinetic energies of particles at depth inside the Earth. This difference is related to the net energies of aggregation of Earth materials (reflecting the contrast between the energies of condensed-state rheologies and the rheologies of "free" motions in relatively evacuated, tenuous-gas systems).

The relativity of resonance states in diagrams such as Figure 21 (cf. Appendix 11) is perfectly general and could be applied, for example, to resonance ratios of satellites in other planetary systems, to planetary nebulae of other stellar objects, or perhaps to planetary and/or satellite spins relative to their primaries and/or their position in the Solar System. [The last of these examples concerns a general idea that was first explored by Daniel Kirkwood during the mid-nineteenth century in relation to possible coordinations among the rotation states of the planets; cf. Kirkwood (1849) and Walker (1850). Analogously, I noticed some years ago that there are certain commensurabilities among the planets between their patterns of orbital kinetic energy vs. distance and rotational kinetic energy vs. distance that are independent of mass (Shaw, 1983b, 1988a, Fig. 23A). The existence of such commensurabilities would suggest that planetary rotations also are ordered according to resonance hierarchies analogous to those of Figure 21 and Appendix 11, a conjecture that is illustrated in Appendix 6, Inset C).] Alternatively, ratios of rotational spins to "spin-orbit ratios" of a given set of "satellitic objects" relative to their "primary" can be compared as additional tests for organizational behavior of planets, satellites, and other objects in planetary nebulae (e.g., representing the dynamical states of a given planet and its satellites, the dynamical states of the major planets and the Sun, or relationships among sets of *minor planets*, the latter consisting of the asteroids and any undiscovered objects of planetary dimensions, such as the "*plutons*" predicted by Stern, 1991; cf. Weidenschilling, 1991; and Note 6.5 of the present volume). The generality of such comparisons essentially rests on the validity of self-organization, power-law relationships, and systematic geometric progressions in the dynamics of celestial phenomena (see Note P.4).

Exterior resonances beyond Pluto in Figure 21 are figurative in the sense that no known objects, other than comets and the Pluto-Charon binary, are identified at those distances. This is the realm of the Kuiper Belt, "Planet X" (see Note 6.5), "plutons," and the Hills and Oort comet reservoirs **(Note 6.9)**. The position labeled Inner Oort in Figure 21, representing the source vicinity of shower-type comets, is roughly the reciprocal of Mars' position (the "target vicinity" of comets inside 2 AU), as expressed in terms of orbital frequency ratios relative to Pluto. In this same reference frame, the orbital inclinations predicted by Allan (1967a, b), when referred to the planetary scale of Figure 21, correspond fairly well with those observed. That is, the interior resonances in the Pluto-as-primary convention (those among orbital distances inside of Pluto's orbit) evolved to low inclinations (except for the asteroids and "short-period comets"), while exterior resonances can persist at high inclinations (i.e., for values of i that are both smaller and larger than 90°, meaning for both high-inclination prograde and retrograde orbits), as expressed by orbital parameters of the Oort Cloud in general (cf. Duncan et al., 1987). Even relatively large objects in high-inclination orbits, as previously observed, are less likely to be discovered than objects in orbits near the ecliptic plane (cf. Notes 4.2 and 6.5).

Similarities and differences between the preceding discussions of resonances in the present volume and analyses of *secular resonances* (e.g., Scholl et al., 1989) and *mean-motion resonances* (e.g., Froeschlé and Greenberg, 1989) should be mentioned here. These phenomena are simply other forms of synchronization effects (cf. Blekhman, 1988; Hunt, 1991; Pecora and Carroll, 1991; Carroll and Pecora, 1992; Moon 1992) in which the criteria of commensurabilities are based on different kinematic reference states than those of the sidereal resonance states of Figures 20 and 21. The "secular method" of analysis separates orbital parameters into two parts: (1) short-period terms and (2) slowly varying terms. Secular periods are those assumed to be independent of short-period effects, as though there is no interdependence between short-term and long-term processes. This is an analytical convenience, but, as I have emphasized from the nonlinear-dynamical viewpoint, independence cannot be assumed — regardless of the separation in dimensional scales — because the types and degrees of "self-excitations" between short and long periods are not predictable (see discussions of *intermittency* and *crises* in Notes 5.1 and 5.2), and they have not been determined by observation in many instances (cf. Figs. 12, 15a, and 22 here; and Wisdom, 1987a, b, c; 1990). To the degree that they are representative of a fully coupled system of motions (e.g., if they act as unstable periodic manifolds relative to a general system of chaotic core attractors; see Chap. 5), secular resonances are important to considerations of bursting phenomena in a variety of contexts, such as: (1) a mechanism for the delivery of asteroids into the inner Solar System by secular planetary interactions, (2) a mechanism for the delivery of comets into the inner Solar System by secular interactions between the planetary system and the Galactic system, (3) a mechanism influencing instabilities of the putative system of geocentric Earth satellites relative to intermittencies in the differential rotational motions of the Earth's core, mantle, hydrosphere, and atmosphere, relative to secular cometary-planetary-Galactic resonances, and (4) mechanisms controlling the intermittencies of other geophysical phenomena, such as geomagnetic reversals, volcanism, climatic and biologic change, and so on, relative to the secular resonances of all of the respective interactions that influence Earth's dynamics. In respect to the first of these examples, some of the results discussed by Scholl et al. (1989) are analogous to previous discussions of resonances in the Asteroid Belt and delivery of asteroids into Earth-crossing orbits by means of chaotic crises (cf. Figs. 12 and 13, and Note 6.1; Scholl et al., 1989, pp. 858ff; Wisdom, 1987a, b, c).

Froeschlé and Greenberg (1989) have discussed mean-motion resonances (orbit-orbit resonances) for the Asteroid Belt. These resonances resemble those of Figure 21 in the sense that they are expressed in terms of the period of revolution of an asteroid relative to that of Jupiter, as measured by the ratios $(p + q)/p$, where p and q are small integers. The term $(p + q)$ means that the orbital frequency of the faster-moving object is given by the frequency of the slower-moving object plus a small integer. Such *relative* orbit-orbit resonances can be numerically the same as ratios of the effective "spin-orbit" frequencies among the planets, as expressed in Figure 21, if the orbital frequencies are referred to the same primary reference frame (e.g., Jupiter's orbital frequency relative to those of the asteroids). The concept of mean-motion resonances has some of the same implications for the *celestial reference frame hypothesis* (CRFH) as do the guidance mechanisms of Figures 20 and 21, where the frequencies were measured relative to the celestial configuration of fixed stars (*sidereal frequencies*).

In other words, mean-motion resonances, and the ratios $(p + q)/p$, deal with conditions of alignments between sets of orbit-orbit configurations in the same *numerical* sense as in Figures 20 and 21. A major difference, however, is that whereas the resonances ratios discussed so far have been expressed in terms of the celestial reference frame (meaning that repeat periods are measured relative to positions in the coordinate system of fixed stars), repeat periods in general — and their corresponding resonance ratios — can be measured relative to any chosen point of reference. For example, alignment between two planets, two asteroids, or a planet and an asteroid with *any* identifiable position in the celestial sphere (constituting a *fiducial mark*, meaning "taken as a — perhaps even temporary — standard of reference"), which heretofore has been a fixed star, represents such a condition. This condition of alignment is called a *conjunction*, simply because it satisfies a rule for an explicit *configuration*, in this case represented by the fact that three points define a straight line that is common to the three lines that join each pair of points. From the standpoint of recoverable configurations, however, "conjunctions" need not always be along straight-line intersections. They could just as well refer to the recovery of identical points on some invariant curve or surface (cf. conditions for "fuzzy routes" described in Note 6.1; and Belbruno, 1992, Fig. 2). Any general point of reference on one celestial object relative to another that satisfies a recoverable configuration between them is equivalent to a general conjunction between three celestial objects. For instance, a line or curve connecting the centers of the two bodies and passing through a particular point on the surface of one of them (a point of fixed latitude and longitude, such as a rocket launching site) represents a type of conjunction if the same configuration recurs. Windows of opportunity for launch of a spacecraft into a particular trajectory and orbital rendezvous with another planet represent conjunctions of this type, one in which the synodic periods (relative recurrence periods) of the planets, rather than their sidereal periods, are of crucial relevance (i.e., these are the types of cycles that would form the integers of any mean-motion resonances that may exist; cf. Froeschlé and Greenberg, 1989).

The so-called *synodic periods* of the Moon, the planets, and other celestial objects represent an important class of conjunctions, especially from the point of view of orbital transfer maneuvers. In this case, the position (fiducial mark) relative to which alignment or "conjunction" is referred is a position *on the Earth*. That is, it represents those positions of another celestial object and the Earth for which a given configuration, measured from the same point on the Earth, recurs. Because of the relative motions of the planets, the synodic period can differ significantly from the sidereal period. It is therefore the period of interest relative to opportune conditions of launch of a spacecraft from the Earth to another planet (e.g., Bate et al., 1971, p. 368; Taff, 1985, pp. 147 and 166; Roy, 1988a, p. 354).

If, instead of being interested in opportune conditions of launch, we are interested in the conditions of return of a space vehicle to a particular position on the Earth, then we are faced with the inverse question of alignment between the position of the Earth, a geographic location on the Earth, and the spacecraft. If the spacecraft is in geocentric orbit, then we are concerned with relative frequencies such as those of Figure 20 (neglecting the integrated effects of atmospheric drag during reentry). If the spacecraft is returning from heliocentric orbit, then we are concerned with conditions of collision or recapture into a geocentric orbit. That is, "conjunction" would be satisfied by a coincidence between the spacecraft, the Earth, and the geographic location, implying collision. But if we want the vehicle back in one piece, then we are confronted with the problem of either rocket-controlled or resonant capture scenarios, as discussed in Note 6.1. In other words, this type of "conjunction," and corresponding mean-motion and/or spin-orbit resonance phenomena, represent the class of approach scenarios of geographically ordered patterns of meteoritic impacts on Earth.

Two important types of alignments, from the standpoint of mean-motion resonances and the CRF, involve Sun-Jupiter-asteroid conjunctions and Sun-Earth-asteroid conjunctions. If we are dealing with an Earth-crossing asteroid, for example, then the existence and nature of possible mean-motion resonances of, and between, these two systems are important to conditions of capture of the asteroid by, and/or collision of the asteroid with, the Earth. Relative to the mean-motion resonance of a longitude, L, of a fiducial mark in the celestial coordinate frame — e.g., a conjunction of the Sun-asteroid-Jupiter type — the same configuration recurs after a certain integer number of revolutions of the asteroid and Jupiter. In the convention of the present volume, a 2/5 resonance means that the same alignment repeats after five revolutions of the asteroid and two revolutions of Jupiter (about every 23.7 years).

Alternatively, if the mean-motion resonance is of the synodic type, and we are concerned with the motion of an asteroid relative to a fixed latitude and longitude on the surface of the Earth — such as a cratering node of the CRF — the situation is harder to visualize, and the resonance conditions are more stringent. At conjunction, not only must there be an integer number of revolutions of the Earth about the Sun relative to an integer number of revolutions of the asteroid about the Sun, but this must occur identically at the same point on the celestial sphere (i.e., at a point representing the projection of the geographic coordinates of the cratering node on the coordinate system of fixed stars). In other words, there must be an alignment between the position of the Earth, the asteroid, and the cratering node, as mapped on the celestial sphere. From the geocentric point of view, the coordinates of the celestial sphere are the Earth's geographic coordinates. This means that there is not only an integer ratio between the number of revolutions of the Earth and asteroid around the Sun, at conjunction, but there is also an integer number of *rotations* of the Earth relative to the position of the cratering node (i.e., this is the number of rotations of the Earth needed to bring the geocentric coordinate frame, as mapped on the celestial sphere, into an identical configuration with the positions of the Earth and asteroid as seen from Earth; on a map of the celestial sphere, therefore, this would correspond to an identity between the positions of the cratering node and the asteroid). Such a configuration would imply, depending on the precision of location, one of the following phenomena: (1) capture of the asteroid by the Earth into a geocentric orbit synchronized with the rotational motion of the cratering node, (2) collision of the asteroid with the Earth, or (3) a near miss analogous to the hyperbolic flyby trajectory of Figure 19.

Sidereal resonance between the Earth and an asteroid is identified in Figure 21 by taking Earth as the "primary" reference instead of Pluto. A *2/5 planetary resonance* of an asteroid with Earth means that the same configuration relative to the fixed stars occurs every two years (and five sidereal orbital cycles of the asteroid around the Sun). Such a ratio can occur only if the orbit of the asteroid is effectively interior to that of the Earth (i.e., it is either totally interior to the Earth, or it is in an eccentric Earth-crossing orbit with an average orbital distance < 1 AU). If, on the other hand, we take a larger semimajor axis, corresponding, say, to the Ceres position in Figure 21 (cf. Fig. 13), the asteroid is in an exterior resonance relative to the Earth, and the sidereal resonance ratio is $\sim$*(248/54) = (124/27)* and repeats to the precision of Figure 21 every 124 Earth years, during which time the asteroid has revolved 27 times around the Sun. In this sidereal example, strict resonance means that the asteroid would never collide with the Earth.

Thus, the families of approach scenarios that would satisfy requirements of synchronization with an invariant system of geographic positions on the Earth, such as the CRF cratering nodes and/or cratering swaths of Figures 1–5 (and other cratering patterns shown in Appendixes 1–4), require forms of orbital evolution whereby the motions of asteroids or comets experience transformations from sidereal types of resonances to mean-motion resonances, and then probably to geocentric spin-orbit resonances analogous to those of Figure 20. This third stage would seem to be required because of the persistence of the patterns over geologic time and the requirement that a consistent system of mean-motion resonances would have to track the changing configurations of the geographic coordinates of the geocentric celestial sphere (i.e., this might occur, for example, if the ~26,000-year precessional motion of the Earth's Equator — hence the Equator of the geocentric celestial sphere — governed the configuration of the mean-motion resonance, a dynamical evolution even more stringent than orbital capture).

Figure 19 illustrates this more difficult problem. Precession of the equinoxes (the celestial equator) adds the requirement that an encounter at a particular geocentric latitude and longitude must either be synchronized with the period of precession (~26,000 years) or occur so quickly that the Earth's orientation is nearly constant (see Roy and Clarke, 1988, pp. 112ff). Since the orbit of an asteroid also typically precesses, the long-term problem (relative to the orbital time scales of the asteroids and inner planets) is unpredictable. In the shorter term, collision or capture might occur at mean-motion resonance if the asteroid is Earth-crossing and the precession of its orbit is also in resonance and is advancing fast enough to bring a locked configuration into coincidence with a given latitude and longitude on Earth. Such a configuration is easiest to see in a coplanar but Earth-crossing geometry (not likely, because eccentricity is coupled with inclination; cf. Fig. 22), wherein the perihelion of the asteroid circulates about the Sun like the asymmetric motion of the phase-space trajectory of a nonlinearly forced and damped spherical pendulum (cf. Miles, 1962, 1984; Tritton, 1986, 1988, 1989). Starting from a fiducial mark at the desired intersection with Earth's nearly circular orbit, the orbit of the asteroid would sweep through all possible intersections until the asteroid itself coincided with the resonant period at an integer multiple of orbital cycles and Earth years. If we allow the eccentricity e and inclination i of the asteroid's orbit to evolve rapidly through a sequence of resonances, the same effect could occur more quickly as the asteroid's orbit moves toward convergence with Earth's orbit. Similar scenarios probably occur, but if they are to represent a generalized guidance mechanism consistent with the CRF, there would have to be an evolving set of such resonances that have persisted for different objects over very long times (i.e., $>> 10^8$ years).

Satellitic Orbits and the Celestial Reference Frame

In view of the many possibilities suggested by Figure 21 and the above discussion, it is not possible, in the absence of observational evidence (direct or indirect; see Note 6.9), to restrict the source of CRF impacts to particular orbital or satellitic patterns specific to Earth. Ascertaining the full spectrum of possibilities will require explorations for diverse types of regular and chaotic orbits within ~300 R_E (Earth radii) — roughly two million kilometers — from the Earth so as to include the sorts of orbital transfer routes considered by Belbruno (1992) for artificial satellites in the Earth-Moon vicinity, with special emphasis on routes at high inclinations to the ecliptic plane. Such searches must be exhaustive — especially if the conclusion of Tombaugh et al. (1959) is indicative that no satellites, in *either* regular or irregular orbits, are likely to be found immediately — because even one object could confirm the principle of *episodic natural Earth satellites (NESs)* in spin-orbit resonances.

Whether or not these vicinities are barren of $\leq$ ~10-km objects at the present time, a larger system of high-inclination mean-motion resonances should be sought in the vicinity of "fuzzy routes" across all regions of low and/or reversed gravitational potential gradients involving the Earth ($\pm$ Moon) and the nearest-neighbor planets, relative to subsets of Sun-Earth-Jupiter conjunctions with asteroids and comets (i.e., regions in which approach scenarios relative to the Earth, as shown schematically in Fig. 19, may involve recurring states — transiently or in quasisteady evolving patterns — with similar Euclidean and/or fractal configurations, where *fractal configurations* refer to the fractally complex boundaries of interacting gravitational spheres of influence of the Sun, Earth-Moon system, and one or more other objects; cf. Fig. 12b; and Belbruno, 1992). Such broad regions of fractally uncertain potential gradients are vicinities in which *fuzzy surfaces* (zones of finite radial thickness) demark the nonlinear-dynamical "contacts" between the spheres of influence of celestial objects, where I interpret these contact zones as fractal basin boundaries between celestial attractors (see Note 6.1). I anticipate that among such configurations will be orbital subsets of simpler mean-motion resonances, including — and not necessarily in a mutually exclusive way — some sampling of the asteroids ($\pm$ defunct comets), short-period comets, and "new" comets (first appearances of long-period comets in the inner Solar System), as suggested, for example, by the meteor and debris streams discussed by Olsson-Steel (1989) and McIntosh (1991). There are also the more general configurations implied by the Mirror Theorem of Roy and Ovenden (1955) and all of the numerous periodic families of many-body mean-motion resonances discussed by Roy (1988a, pp. 144ff; cf. Chap. 4 and Note 6.1 of the present volume).

Ideally, the different depths of searching for *near-Earth objects (NEOs)* and/or NESs in the extraterrestrial environment should be conducted simultaneously by cooperating astronomical research teams, with emphasis on transitional configurations with known mean-motion resonance families (e.g., Froeschlé and Greenberg, 1989, p. 841) and/or with NES-resonance families (Figs. 19–21; and Allan, 1967a, b, 1971; Coffey et al., 1986, 1991; Belbruno, 1992; Friesen et al., 1992). Precedents for NES searches, and/or for the relevant conditions of their discovery, are few, and those that have been attempted have been of relatively short duration compared to traditional astronomical explorations beyond Earth's sphere of influence (i.e., beyond $\sim 2 \times 10^6$ km from Earth). None have been pursued with the commitment and longevity, for example, of the trans-Neptunian planet searches pioneered by Clyde W. Tombaugh, who also "holds the record," so far, for a dedicated search for NESs (Tombaugh et al., 1959). The history and conceptual strategies of the trans-Neptunian program are described in Tombaugh (1961a, b, 1991) and Tombaugh and Moore (1980).

By contrast with these Herculean efforts, the searches, to date, for NESs, including the one led by Tombaugh himself, have been of shorter duration, less certain search parameters, and less dedicated conviction of success — and are therefore far from complete. The study by Tombaugh et al. (1959) was supported by a U.S. Army Ordnance contract only during the period 12/53–10/58, a period in history during which Tombaugh — with his characteristic insight and determination — had anticipated the deployment of long-range ballistic missiles and artificial satellites (e.g., the first successful artificial satellite, Sputnik — freely translated as "traveling companion" — was launched into orbit by the Russian space program on 4 October, 1957) and had recommended, logically, that it would be prudent to know the distribution of any natural objects in the near-Earth vicinity before that vicinity became populated with artificial objects. His recommendation imparted a need to carefully monitor the ensuing history of space-age debris in Earth's vicinity, but it also marked a time after which almost no attention was given — with the exception of the isolated study — to searches for natural satellites of the Earth (at least until now). Geologists are familiar with this syndrome; the advancement of technology exploits knowledge, and forces the existence of new and complex archives that compound that knowledge (e.g., the program of monitoring artificial satellite debris; see Taff, 1985; Murr and Kinard, 1993), but — barring massive public demonstrations of disapproval — it seldom waits for it. A systematic program of searches for NEOs and NESs in the light of nonlinear-dynamical guidelines would be much further advanced if searches such as those by Bagby (1956, 1966, 1967, 1969), Baker (1958), and Tombaugh (1959) had been encouraged and supported more effectively during the developmental years of artificial-satellite and space-mission technology **(Note 6.10)**.

Depending on orbital distance ("altitude"), eccentricity, and inclination (a, e, and i), impacts on Earth originating from a reservoir of satellitic objects (NESs) should be clustered along arcs at intermediate to high angles to Earth's Equator (cf. Fig. 23), with separations along the arcs that depend on the mechanisms that dislodge the objects from their resonant states. Allan (1967a) studied only the resonance states of circular orbits, where $e = 0$ (Fig. 20), but Allan (1967b, 1971) extended the study to elliptical orbits (cf. Cook and Scott, 1967), including the so-called critical-inclination problem (cf. Coffey et al., 1986). Orbital eccentricity is clearly a very important parameter in the nonlinear case, as was confirmed by Wisdom's (1987a, b, c) studies of asteroid orbital instabilities (cf. Figs. 12 and 22). **Figure 22** illustrates schematically the interdependence of e and i in Wisdom's computations of asteroid instabilities (Wisdom, 1987a, Fig. 13). The chaotic irregularity of the recursive evolution of e-i values in Figure 22 introduces many possibilities for "strange orbits." The dissipative analog of such orbital trajectories in the inner Solar System might look like a braided swath with the fractal properties of a strange attractor (cf. discussion of Figs. 12b and 13b). In animation, the strange orbits might look something like the spin of an elliptical bicycle wheel that has been twisted and bent out of its original circular, dynamically balanced shape. Thus, the possible entry paths to Earth's sphere of influence (the natural analogs of artificial-satellite reentry paths) are unpredictably diverse, even though they are highly ordered by contrast with "random" collisions. The resulting system of geocentric satellites could be subsequently dislodged from spin-orbit resonance states, or from orbits satisfying conditions of "stable chaos" (Milani and Nobili, 1992), either by new objects (near-misses and/or captures of larger objects; see Fig. 19) or by chaotic crises relative to a variety of orbital frequencies (e.g., involving the Earth, Moon, Sun, other inner planets, and Jupiter). Chaotic crises should yield patterns scaled according to power-law distributions of bursting times seen as multiple nodes along trends (swaths) of cratering with different spacings and cratering magnitudes.

Patterns of crater sizes will in some way relate to the patterns of object sizes, because the strengths of the resonances "filter" the satellitic distribution in a manner generally but unpredictably similar to that of the Solar System (cf. Figs. 9, 10, and 21). These resonances would in turn correlate with impact energies limited by $\sim m_s(v_{es})^2$, where m_s is the satellite mass and v_{es} is the escape speed at the Earth's surface ($\cong$ *11.2* km sec^{-1}). By contrast, objects from elsewhere in low-eccentricity orbits have speeds comparable to the mean speed of the Earth in its orbit, which is given by $2\pi R_o/T_E \cong 29.8$ km sec^{-1}, where R_o is the mean orbital radius (km) and T_E is the orbital period (sec). In Earth's vicinity, exotic objects with large eccentricities can have higher velocities, but because the mass of an object increases as the cube of its radius while its kinetic energy increases as the square of its velocity, object size has the more important effect on impact energy (see Note 6.1 for discussion of orbital energies).

Thus a crater of given diameter formed by an object traveling at about 10 km sec^{-1} requires an object roughly double the diameter of one traveling at 30 km sec^{-1} if all other factors are equal. This scale relationship is offset by uncertainties in our knowledge of the dynamics of the crater-forming process (crater size vs. object diameter for given densities, cohesive strengths, friabilities, etc., of target and projectile materials). The average cohesive strength of a rock constrains the potential for fracture but not the fragmentation spectrum. A rock's friability is its ease of progressive comminution (reduction to fragments that conform to some characteristic size-frequency relation and limiting fineness), which is quantified by the fractal fragmentation-energy spectra of an impacting object relative to like properties of the substrate. Calculations of crater sizes, however, are generally based on gross scaling laws extrapolated from energetic events that are small compared to impacts of 1–10-km meteoroids, and their validity for large impacts is open to question (cf. Fujiwara et al., 1989; Melosh et al., 1992). Documented metamorphic zoning is probably a better basis for evaluating impact energy, but case histories are rarely adequate to tell a complete story. The composite rheology of the Earth's crust (+ atmosphere ± ocean) near ground zero is poorly known. Near ground zero, rocks can simultaneously, in differing proportions of course, melt, splash, fluidize, and vaporize — in addition to more distant, regionally more pervasive, fragmentation and fracture — with wide ranges of uncertainties in the dissipated energies of the respective processes (cf. Kieffer, 1971, 1975; Kieffer et al., 1976a, b; Kieffer and Simonds, 1980; Melosh and Vickery, 1991). The simple form of dimensional analysis performed by Hubbert (1937, pp. 1505–07, 1511–15) remains a valid approach to assessing the behavioral and rheological uncertainties involved in scaling meteoroid impacts and large-body collisions in the Solar System, but our detailed rheological knowledge of impact events with equivalent explosive energies ("TNT equivalent"; cf. Shoemaker, 1983, Fig. 1; Chapman and Morrison, 1989, p. 278) of more than a million megatons (globally catastrophic to us, but small in the spectrum of interplanetary collisional processes), and of the consequences of the propagating deformation waves in the Earth, is virtually nil in any experiential sense (cf. Fink et al., 1981, 1984; Gault and Sonett, 1982; McKinnon, 1982; Melosh, 1982, 1989; O'Keefe and Ahrens, 1982, 1989; Schmidt and Holsapple, 1982; Ahrens and O'Keefe, 1983; Schultz and Gault, 1990a, b; Melosh et al., 1992). Certain seismic analogies hold (e.g., the equivalent "earthquake magnitude" of an impact event calculated from explosive energy, discussed later), but the scaling is unknown for events analogous to seismic ruptures exceeding the circumference of the Earth, or explosions equivalent to detonation of the world arsenal in one place at an uncertain depth; see Kanamori (1986) for the scaling properties of great earthquakes.

Some constraints on CRF trajectories are indicated by the great-circle and generalized small-circle paths in **Figure 23**, which is based on the nodal points and cratering swaths of Figures 1–5 **(Note 6.11)**. Three great-circle sections, called *nodal great circles* (NGCs), each of which passes through two of the *Phanerozoic cratering nodes* (PCNs), are shown in Figure 23a (comparisons of the three NGCs with cratering patterns are shown in Appendixes 1–4). These sections can be compared with the "equatorial" great-circle section in Figure 23b that contains all three PCNs, as shown by the shaded small circle in the transformed projection of Figure 23c. At first glance, the implications of these relationships are deceptively simple. The three NGCs — I refer here to the traces on the Earth's surface of the mean orientations of geocentric "paleo-orbits," because the cratering patterns tell us nothing directly about orbital ellipticity, precession, etc. (see Note 6.11) — with inclinations ranging from about 45°–135° N (i.e., roughly ± 45° from the geographic poles), and with azimuths within the approximate range 75°–155° E (i.e., ± 40° of 115° E), provided, according to the *celestial reference frame hypothesis* (CRFH), ballistic patterns in some proximity to both the Phanerozoic and K/T circles. [I am using the term "ballistic" here in a loose sense, referring to that stage of orbital decay in which geocentric satellitic objects enter the Earth's atmosphere and impact Earth's surface (i.e., analogous to "reentry" of artificial Earth orbiters); orbital decay can occur by the cumulative effects of dissipation, leading to the loss of resonance conditions, or by perturbations by new satellitic objects or objects in near-Earth flyby heliocentric orbits; cf. Figs. 19 and 20; and Note 6.11).] Thus, the three orbital configurations suggested by Figure 23a are amazingly consistent with the high-inclination approach scenarios of Figure 19, where the ~26,000-year precession of the equinoxes doubles the incoming angles to the values indicated above relative to Earth's Equator (the initial inclinations, according to Fig. 19, would have been between roughly 65° and 115° *relative to the ecliptic*).

Combinations of heliocentric inclination, precessional position, and initial direction of approach determine whether the captured object is in a prograde or retrograde orbit relative to Earth's spin. For the above inclinations, it would seem that equal numbers of both prograde and retrograde orbit types could have occurred, either uniformly in time or varying with the precessional period of about 26,000 years. Some of the retrograde orbits, however, probably would have evolved to prograde orbits during capture (cf. Fig 19, upper right; and Alfvén and Arrhenius, 1969). Asymmetries in orbital decay and impact-cratering frequencies relative to the North and South poles would have depended on the relative frequencies of the several types of approach scenarios and the relative resonance relations derived from prograde vs. retrograde motions. For high-inclination approaches, relationships governing likely phase relations of orbital decay relative to the Northern and Southern hemispheres are ambiguous.

Comets, however, would have inherited a bias from the angular difference of ~60° between the Galactic and ecliptic planes (see Duncan et al., 1987, p. 1335; and Fig.15b herein), probably resulting in a preferred distribution of approaches for cometary sources over a particular range of relatively high inclinations (cf. Olsson-Steel, 1989; McIntosh, 1991; and Note 6.11 herein). Conceivably, the net result of all these relationships may have favored a dominance of the Northern Hemisphere over the Southern Hemisphere as the general locus of impacts, but these relationships are highly model-dependent and uncertain. Another possible influence on the apparent Northern Hemisphere dominance in the cratering patterns of Figures 1–4 (cf. Appendixes 2 and 4) might arise from a highly eccentric geocentric spin-orbit resonance near the critical-inclination condition (cf. Allan, 1971; Coffey et al., 1986). If the resonance favors a Northern Hemisphere apogee, a satellite would spend most of its time above the Northern Hemisphere (such a high-inclination elliptical satellite orbit relative to Earth would resemble a high-inclination cometary orbit relative to the inner Solar System; cf. Note 6.11).

Complexities enter the simple picture of Figure 23a from (1) the nonlinear effects of interactions between eccentricities and inclinations of the resonant orbits relative to their ground traces on Earth, due to tides, etc. (Fig. 22), (2) the relative stabilities of resonances, and (3) the possibilities of intermittent emptying and refilling of resonances with differing numbers and sizes of objects during the putative lifetime of > 600 m.y. The effect of (1) would be to widen and blur the ballistic swaths by the angular ranges of irregularities in oscillating elliptical shapes and inclinations, but by amounts that are not readily estimated. The net effect, however, would be consistent with the deviations required to skew the areal distributions as functions of time (Figs. 1–3). Evaluation of factors (2) and (3) depends on correlations between a more complete cratering history and the eventual astronomical documentation of the existences of and stabilities of satellitic objects.

In contrast with this description, Figures 23b and c illustrate problems of direct collisions without prior capture. For reference, the point marked "T. Epi." in Figure 23a designates the epicenter and inferred trajectory of the Tunguska event (Clube and Napier, 1982b; Rocchia et al., 1990). It could be consistent with collisional, satellitic, or transitional trajectories (e.g., the near-parabolic trajectory in Fig. 19). In order for direct collisions to have produced the arcuate swaths and/or nodes, the collisions had to be either roughly normal to the planes of the Phanerozoic nodes and K/T swath, or along paths similar to those of satellitic resonances (see the Epilogue). A more stringent test is that the collisions would have to have been invariant in time, something like a highly choked shotgun blast of grouped objects fired by a mechanism that gave essentially the same pattern every time. Otherwise, random impacts of single or grouped low-inclination objects would look like skewed zonal streaks spread over at least twice the angle of precession (say, within ± 45° of the Equator).

Hut et al. (1991) describe several possible scenarios for multiple impacts near the time of the K/T boundary. The breakup of a large comet during a first encounter with the Sun apparently could yield an instantaneous small-circle shot pattern similar to the K/T and/or Phanerozoic swaths. This assault would be limited to a one-shot event of a few hours duration, however, because repeated events of such "random" cometary disruptions would not be expected to produce nearly identical trajectories. That objection does not apply to multiple one-shot events from resonant geocentric orbits (Earth and/or Moon orbiters), because orbits of like orientations are likely to be refilled if they are emptied by any instantaneous but multiple event. If the K/T interval represents even a few events separated by intermissions lasting years (some workers have estimated at least 10^5 years for the K/T duration; e.g., Herman, 1990), then the requirement of repetitive coherence is severe. Otherwise, the discussion by Hut et al. (1991) emphasizes my earlier remarks concerning the abundances and trajectories of motion of debris near Earth-crossing orbits (cf. Olsson-Steel, 1989; McIntosh, 1991).

My assertions that cratering patterns are ordered rather than randomly distributed cannot be entirely proved until the global exploration of crater patterns is more complete, until more complete astronomical searches for satellitic reservoirs have been undertaken, and until our understanding of the nonlinear dynamics of small-object orbital evolution in the near-Earth vicinity is more advanced. The numbers of new craters and reinterpretations of known craters (see Appendix 1) since the inception of the impact-extinction debates suggest that the catalog of craters will continue to expand in proportion to incentive, just as the discoveries of asteroids and comets have accelerated intermittently (up to now) according to feedbacks between incentives and technology. Objections to the generality of the nodal circles displayed here can be made on the basis of deficient global coverage and special geologic settings, but such objections would apply equally, if not more stringently, to a thesis of crater origins by random impacts, because there are so many regions (some of which are remarkably close to the cratering nodes) in which craters originating from random impacts would be expected but which are, so far as has been discovered, barren of finds. Much the same can be said of the distributions of zodiacal dust bands, meteor streams, and radiants (e.g., Olsson-Steel, 1989; McIntosh, 1991), and the observation that near-chaotic resonances are typical of the Solar System suggests that all such structures should be ordered (cf. Alfvén, 1971a, b; Baxter and Thompson, 1971; Danielsson, 1971; Lindblad and Southworth, 1971; Roosen, 1971; Carusi and Valsecchi, 1979, 1985; Humes, 1980; Drummond, 1981, 1991; Olsson-Steel, loc. cit.; Stone and Miner, 1989; Sykes et al., 1989; McIntosh, loc. cit.). In lieu of direct astronomical documentations, the prevalence of these resonances and the potentials of all planets to form reservoirs of satellitic objects in a manner self-similar to that of the Solar System are themselves the strongest arguments for the CRF.

Spin-orbit resonances that have evolved because of the longitudinal (and other) mass anomalies in the Earth ($\pm$ Moon) also will have evolved with any changes that may have occurred in the distribution of these anomalies relative to the orientation of the Equator and rotation axis (i.e., relative to any changes in the orientations of the Earth's principal axial moments of inertia mapped on the celestial reference frame prior to those changes). Thus, the CRF is something like a "rainfall pattern" governed by celestial cycles interacting with long-term changes in the asymmetries of mass of the Earth-Moon system. In this sense the resultant geological and biological effects are analogous to those induced by patterns of weather and climate. The latter, however, are sensitive to many short-term effects of coupling between Earth's orbital parameters and the advective and/or convective effects of the glacier-ocean-atmosphere-magnetosphere (GOAM) system. Even so, patterns of weather and climate are globally distinctive, and are increasingly being seen as products of nonlinear dynamics and self-organized systems of attractors constrained by GOAM interactions. In addition to the "normal" effects associated with GOAM-orbital coupling (cf. Jose, 1965; Wood, 1972; Landscheidt, 1983, 1988; Berger et al., 1992), the geologically long-term (secular) changes in the GOAM system are, by the present hypothesis, also influenced by the integral magmatic-tectonic-geomagnetic effects of the intermittent impact-energy distribution (the meteoroid "shower activity" of the CRF reservoir), and by any one of the globally catastrophic impacts (and/or major pulses of rapid-fire multiple impacts) that can cause "impact winters" and greenhouse effects (e.g., Toon et al., 1982; O'Keefe and Ahrens, 1989; Clube, 1989a; Clube and Napier, 1990; Wolfe, 1991; Kerr, 1992a) such as at the K/T boundary and elsewhere.

Some workers have argued that volcanic eruptions of catastrophic scale can induce the same types of "winters," hence may be responsible for some — or all — mass extinctions (see McLean, 1978, 1988; Rampino and Volk, 1988; Rampino et al. 1988; Vogt, 1989; Courtillot, 1990; I. H. Campbell et al., 1992; Czamanske et al., 1992; cf. Note 8.11; and the historical context of the "volcanist" vis-à-vis "impactist" viewpoints, as reported by Glen, 1990, 1994). According to the CRFH, however, delayed-feedback coupling between integrated impact-dynamic effects and induced long-term magmatic-tectonic effects has become so intimately interrelated over geologic time that uniquely endogenous phenomena—such as the massive flood-basalt events that "volcanists" would argue are the immediate "cause" of some or all mass extinctions—are impossible to isolate from the secular consequences of the cratering process itself. Although we thereby lose the logic of the "smoking gun" as the "judicially definitive" forensic evidence of cause and effect between specific catastrophic events and specific "extinctions" — whether argued from the "volcanist," the "impactist," or any other special viewpoint—we gain the power of comparative nonlinear-dynamical scaling (universality) between impact-related phenomena and ordinary geologic processes (cf. Notes 8.1 and 8.11). The frequency signatures of diversified delayed-feedback phenomena are sometimes in phase and sometimes out of phase, yet in both instances they reflect the coupling of periodic/chaotic systems of attractors that — given sufficient data — eventually will be found to systematically describe all of the interrelated processes. In this respect, the consequences of time-delayed cosmological coupling with geological feedback processes display many of the characteristics shown by neurobiological coupled oscillators with delayed feedback, for example as described by Freeman (1992, pp. 467ff).

CHAPTER 7

Correlations with and Consequences of the CRF: In Situ Astrogeology

This chapter and the next offer a few examples of the geophysical and biological correlations, and/or implications, suggested by or required by the idea of an invariant celestial reference frame (CRF), especially with regard to (1) the distributions of iridium anomalies, (2) magmatic-tectonic relationships [e.g., the global patterns of flood-basalt fields and volcanic "hot spots" relative to the distributions of impact craters and Phanerozoic cratering nodes (PCNs) and swaths], (3) geomagnetic patterns [e.g., apparent polar-wander paths (APWPs) and "true" polar-wander paths (TPWPs), and intermittencies of paleomagnetic-reversal phenomena], (4) nonlinear-dynamical convection in the core dynamo, (5) concepts of internal vis-à-vis external dynamical-resonance phenomena, (6) paleogeography, (7) biostratigraphic codification, and (8) biological crises and concepts of background extinctions vis-à-vis mass extinctions. The examples have been chosen to illustrate the idea that principles of universality imply, if not require, that geophysical and biological processes be nonlinearly coupled with the celestial cavalcade symbolized by the discussions in the previous chapters (e.g., relative to the phenomena illustrated in Figs. 11 and 21; cf. Appendix 11). Fractal self-similarity, especially the power-law type of temporal and spatial frequencies that are characteristic of chaotic crises, is prevalent in Earth processes, perhaps most conspicuously in the time-delayed effects of the cratering process on the geomagnetic phenomena illustrated in Figures 6–8 (cf. Appendix 5), but also in the implicit coupling of both of these processes with the coupling between mantle convection and tectonic-magmatic cycles. [The illustrations in the Appendixes are designed to supplement discussions in the text concerning (a) impact-crater distributions and the concepts of orbital evolution considered in Chapters 1–6 (Appendixes 1–11), (b) distributions of "hot spots" on the Earth, and their antipodes (Appendixes 12–15), (c) "hot spots" and their antipodal distributions relative to impact-crater distributions and invariances (Appendixes 16–18), (d) polar-wander paths (Appendixes 19–21), (e) polar-wander paths relative to crater distributions and invariances (Appendixes 22 and 23), (f) comparisons of the "hot-spot" patterns of Jupiter's satellite Io with Earth's cratering and "hot-spot" patterns vis-à-vis cratering on the Moon and Mercury (Appendixes 24 and 25), (g) continental reconstructions relative to CRF-PCN invariances (Appendix 26), and (h) global cratering relative to the distribution of dominantly Precambrian basement rocks (Appendix 27).]

Again, the going may seem difficult, especially in the parts that deal with concepts of chaotic crises, but the difficulty — as it often turns out to be in the case of celestial processes — is largely induced by model-dependent preconceptions and the paucity of data that are specifically relevant to such models. Therefore, it may help to review the synopsis given in the Introduction at this point, and to recall that the nature of nonlinear-dynamical coupling phenomena are as relevant to the imperceptible motions of geological processes as they are to the more visually dramatic extraterrestrial phenomena that have been emphasized in the preceding chapters. The meaning of attractor dynamics (Note P.3) and its terminology (fractal basin boundaries, nonlinear crises, self-organization, universality, and so on) is precisely the same in the geophysical context as it is in the celestial context, hence the ancillary information given in the Notes to Chapters 1–6 should be consulted wherever the application seems unclear. In doing so, however, it may help to transfer attention from the previous focus on the motions of discrete objects (*particle* motions), a focus that suited the astronomical emphasis, to an analogous visualization of trajectories of motion of spatiotemporal *states* in phase space (the graph space displaying the properties of the system — e.g., according to the *control parameters* of Notes P.1 and P.3; cf. Note 5.1). In this way, we can describe even the slow pace of geological processes within a perspective that is self-similar to the one we habitually think of when we look outward from the Earth's surface toward the heavens. In either case we are looking back in time, sometimes as a function of depth in the Earth or as a function of distance in space, but often in the immediate field of view of a rock or meteorite specimen in the hand or under the microscope. Whatever the observational viewpoint, and at whatever the depth of penetration, we are continually confronted by issues of distinction and reconciliation — distinction between the proximal and the distal, reconciliation between the contemporaneous and the ancient. Phase portraits are but another way of bringing perspective to our processes of discrimination, hence to our capacities for recognition (cf. Notes P.1–P.4, 3.3, and 5.1).

Cratering Patterns, Iridium Anomalies (and Shocked Quartz), Flood-Basalt Provinces, and "Plant Pathology"

The notion of "plant pathology," to take up the last of these topics first, derives from a paper by Jack Wolfe (1991) that undertakes a Sherlock Holmesian analysis of the conditions that accompanied an at-least-double impact event at the

K/T boundary (cf. Kerr, 1992a, 1993b). By Wolfe's account, one of the two events was smaller than the other. The smaller event occurred in the greater vicinity of Teapot Dome, Wyoming, three to four months following the impact of a *"large, distant bolide."* This "bolide," a meteoroid (or meteoroids) of unknown type, unknown provenance, and unknown orbital-impact parameters, caused an "impact winter" — followed almost immediately by greenhouse warming (Wolfe, loc. cit., p. 422; cf. Toon et al., 1982; Rampino et al., 1988; Schneider, 1988; O'Keefe and Ahrens, 1989; Clube and Napier, 1990) — that Wolfe correlated with the death-by-freezing-and-disruption of aquatic plants deposited in a lily pond. Wolfe determined further that this larger impact occurred during the growing season after some plants had bloomed and produced seeds but while others still bore immature flowers and had not fruited, which by his reckoning placed the time of death in early June! This sudden killing freeze preserved fossil leaves and other plant material that became incorporated — after the ice thawed from post-freeze rainfall — into a chaotic bottom sediment containing only fine-grained impact fallout debris interpreted to have come from the putative large, distant impact event (Wolfe, loc. cit., Table 1, Bed 1). Microtektites and relatively coarse-grained shock-metamorphosed minerals from the second, nearby, smaller impact were superimposed (Bed 3) on the plant material and first-impact debris of Bed 1 following a depositional interval that recorded temperatures as high as 25–30° Celsius (Bed 2). Wolfe estimated the time interval between the first and second events from his interpretation of the palynomorphic sequence between Beds 1 and 5 (ibid., Fig. 1), the latter horizon containing the finer particles from the second impact event. In doing so, he evaluated the likely freezing-duration effects on the state of preservation of the abundant suddenly-killed lotus leaves — including a laboratory simulation of the effects of freezing on modern lotus leaves of equivalent type — and the relative flowering-seeding stages of the preserved pond lilies (completed) vs. the flowering-fruiting stages of the preserved lotus plants (immature flowers, no fruit). Wolfe used these indications in the manner of an experienced forensic examiner to judge that the evidence placed the freezing event (first impact) in early June and the burial of the remains by coarse debris from the second impact roughly two to four months later, placing the time of the nearby impact event no later than early October.

An independent description of stratigraphic successions across the K/T boundary at several dozen sites in western North America by Shoemaker and Izett (1992) would appear to correpond well with the above scenario, including the interpretation that "at least part of a growing season" intervened between a lower and an upper impact horizon. The single-impact, dual-phase model of the K/T stratigraphic succession by Bohor (1992) would seem to be invalidated by the growing-season observations, unless the "fireball phase" (upper layer) in Bohor's model is interpreted as being suspended in the upper atmosphere for months and then deposited as a *discrete* stratigraphic horizon. My own multi-event interpretation (see below; and the Epilogue; and Notes 1.1 and 8.1) implies that the upper strata are related to an actual fireball procession — and a series of separate small impacts — subsequent to, and possibly as a direct aftermath of, one or more large impacts (cf. Appendix 3 and Note A.1). It is similar to the interpretation by Shoemaker and Izett (1992), possibly excepting the nature of the source impactors and the proximal stages of their orbital evolution (see the Epilogue).

The ironies are many — as are the questions raised. A nameless and insignificant lily pond gives up what, 65 m.y. later, may be critically important astronomical data — data that are based simply on an examination and reasonable interpretation of a figurative handful of lily pads and lotus blossoms. Other data, concerning sedimentation sequences and times of decay vs. inferred temperatures, furnish helpful constraints on the nature of these events (cf. Kerr, 1992a, 1993b). But, in short, the inferences by Wolfe seem to be consistent with other indications that the K/T pattern of multiple events was imposed largely in the Northern Hemisphere (he does point out, however, that southern latitudes were at midwinter, and the effects there would accordingly have been quite different). Any of the large near-Arctic impacts (e.g., Popigai; cf. Fig. 3 and Note 1.1), and/or Chicxulub in the Caribbean, could have represented the "distant bolide," and the Manson, Iowa, or some other midcontinent event could have been the subsequent "nearby" event (e.g., Kerr, 1992a, 1993b; Shoemaker and Izett, 1992). If the impact site (or sites) of the large object (or objects) also was in the Northern Hemisphere — as is the case for the examples just mentioned — then the positions in the yearly cycle furnished by Wolfe's early June and early September to early October dates constrain some of the relationships for approach scenarios of the impact trajectories discussed in Chapter 6 (see Fig. 19). But if the K/T sites are as portrayed herein (Figs. 1–5; and Appendixes 1–4), these events probably were neither as limited in number nor as uniquely located as Wolfe's interpretation, or others like it, would suggest (cf. Kerr, 1992a, 1993b). For example, a satellitic reservoir may have been dumped progressively and almost completely over a period of months following a major perturbation by (1) a newly captured meteoroid, (2) a massive close encounter (atmosphere-grazing flyby; cf. Fig. 19), or (3) an internal chaotic crisis — e.g., by tidal modulation of sensitive resonance states (cf. Alfvén and Arrhenius, 1969) — within the satellitic reservoir, accompanied by catastrophic bursts (cf. Figs. 17 and 18, Chap. 5). According to these, or other, CRF scenarios, Wolfe's fossil site would not constrain the locality of any single smaller impact, or distinguish between a single, distant large event and several such events. Multiple impacts widely distributed over the globe, and/or widely separated in time, relative to Wolfe's "large, distant bolide" may be difficult to reconcile with his scenario — at least until the specific locations, times of impact, and scales of ecologic consequences of each K/T impact event, or rapid-fire succession of events, are known with much greater accuracy and precision than they are known now (cf. Kerr, 1992a, 1993b).

An alternative interpretation of Wolfe's data that could satisfy the same sequence of ecological effects might go somewhat as follows: a freight-train-like procession of intermittently clustered low-angle impacts (see the Epilogue; and Notes 1.2 and 6.1) — analogous to but with objects much larger than the "great fireball procession" (GFBP) illustrated in Appendix 3—could have traversed the region of Wolfe's fossil site but straddled it, and others like it, in leapfrog manner, with globally catastrophic effects of a multiple sequence of impacts releasing an immense cumulative impact energy. Such a globe-encircling multiple event would have created several large craters and many smaller ones, where any one of the latter might be interpreted as a "single small impact" but was really but one crater among many produced from the "tail-end" of the procession of those objects that were retained for some period of time in geocentric orbits. The procession as a whole, then, represented a geologically instantaneous global geospheric-biospheric crisis with distributed ecologic effects (where "geospheric" refers to the complex of core-mantle-surface tectonic, magmatic, and geomagnetic feedback phenomena discussed in this chapter and the next; cf. Note 8.1). A general implication of the celestial reference frame hypothesis (CRFH), as I am attempting to reconcile it with the panoply of observations concerning the K/T and other major boundary events of the geologic record, is that low-angle procession-like event sequences of all sizes—e.g., from that of the K/T event to that of the GFBP—have occurred intermittently throughout geologic history. This concept is favored because it would also satisfy many features of a scale-invariant hierarchy of biostratigraphic catastrophes in the sense discussed in Chapter 8 (cf. Ager, 1984; Raup, 1990, 1991).

But there remain some nagging questions concerning either of the scenarios outlined above, mine or Wolfe's, that are analogous to the questions raised by many authors concerning generally multiple, or systematically stepwise, extinction events across the K/T and other major biostratigraphic boundaries (cf. Notes 1.1 and 8.1; and Kauffman, 1984, 1985, 1988; Padian et al., 1984; Hut et al., 1987, 1991; DeRenzi, 1988; Hallam, 1988, 1989; McLean, 1988; Shaw, 1988a). An essential question concerns the "selectivity" of environmental and ecologic effects. As I have discussed elsewhere (Shaw, 1988a, Figs. 14–16), I have difficulty reconciling the devastation envisioned in many end-Cretaceous scenarios with the fact that there is so little *discontinuous* sign of it in the evolution of such delicate land fauna as, for example, the birds (cf. Fig. 36b herein). Those paleontologists I have queried on this issue have passed it off by saying that the fossil record of the birds is not good enough to make an evaluation of diversity changes at the K/T boundary, neglecting to answer such questions as: What actually *did* happen to the birds during the debacle? And why—even in a poor record—is there so little indication of offsets, of systematic terminations and originations of avian lineages, specific to the K/T boundary?

Analogous questions arise concerning the evidence of "wildfires" at the time of the K/T boundary (cf. Tschudy et al., 1984; Tschudy and Tschudy, 1986; Lerbekmo et al., 1987; Gilmour et al., 1989; Nichols and Fleming, 1990; Sweet et al., 1990; Wolbach et al., 1990). Wolfe (loc. cit., p. 420) points out that the "fern spike," known to be indicative of mass-kill of land vegetation by pervasive wildfires, "occurs below the impact layer" (referring to the Teapot Dome biostratigraphic sequence outlined above). This observation does not contradict

Wolfe's biostratigraphic interpretation, because the "distant bolide" left little evidence there in the way of recognizable impact-related deposits. But if there was a *regionally pervasive wildfire* prior to the events Wolfe describes, *why* were lily ponds and flowering plants still dotting the western interior of North America, *the type region* for the documentation of wildfires? If he were queried on this point, I suspect that Wolfe might answer in the same vein discussed by Gilmour et al. (1989, pp. 208ff; see especially their Table 2 that shows a list of twelve essential "environmental stresses," and their time durations, that are thought to occur following a major impact event), that wildfires must have been patchy and selective in their distribution. [Lily ponds and flowering plants certainly could have recovered quickly after a global wildfire, but not within the time frame deduced by Wolfe (loc. cit.) at Teapot Dome.] If such an argument holds water for lotus blossoms, then I guess it might do for the birds. But I am intuitively skeptical of the "convenient reasonableness" of selective effects appealed to by some workers as an explanation for survival of some biologic forms across a global extinction boundary — except as those effects can be demonstrated to have applied to organisms in which dormancy played a major role in the reproductive cycle (cf. Wolbach et al., 1990, p. 398). How much, even, of human culture would survive under similar circumstances (cf. Schneider, 1988)?

The crucial question in the above respects is, how pervasive is pervasive? Raup and Jablonski (1993), for example, present evidence that in the marine realm the end-Cretaceous extinction was uniformly intense in the world oceans, at least for the bivalves [see Note 8.14; cf. Chap. 8 concerning the "paired evolution" of clams and brachiopods, as described by Gould and Calloway (1980) and as plotted in Fig. 36a of the present volume]. The processional train or trains of multiple impacts envisioned in the present work would, if anything, have increased the spatiotemporal synchroneity, hence global uniformity, of wildfires by increasing the number of nucleation centers (whatever the actual ignition mechanisms might have been). Questions concerning the limitations of or ubiquity of biological extinctions represent areas of multidisciplinary study of global feedback phenomena. Such problems are notoriously difficult to solve, and the answers — when they emerge from feedback experiments, or when they are eventually manifested in Nature's patterns themselves — are often counterintuitive. This is not really surprising when one stops to realize that a question such as the selectivity of global catastrophes involves the chaotic dynamics of a broad spectrum of coupled phenomenologies, invoking, for instance, the principles of (1) population logistics, (2) epidemics (i.e., spatiotemporal propagation of such things as viral infections and chemical-reaction fronts — the latter applying to phenomena such as natural combustion processes, propagating star formation, and so on), (3) immunological evolution (e.g., see the section of the Introduction titled "The First Transformation of Thought"), (4) genetic/linguistic evolution, (5) neural-network and neurophysiological evolution (e.g., applying to economic chaos, "brain chaos," and "dynamical diseases"), and (6) tectonic-galactic self-organization. [Selected aspects of such phenomena, as cited in the present volume, are described by Forrester (1961, 1973), Hamilton (1964), Shaw (1970, 1978, 1983a, b, c, 1984, 1986, 1987a,b, c, 1988a, b, c, 1991, 1993), Prigogine et al. (1972), Aarseth (1973), Cole et al. (1973), Crick and Orgel (1973), Meadows and Meadows (1973), Lovelock and Margulis (1974a, b), Alfvén and Arrhenius (1975, 1976), May (1976, 1990), Shaw et al. (1980b), Alfvén (1981), Axelrod and Hamilton (1981), Margulis (1981, 1990, 1992), Platt (1981), Ortoleva et al. (1982), Seiden and Gerola (1982), Toon et al. (1982), Shaw and Doherty (1983), Edelman (1985, 1987), Lasota and Mackey (1985), Olsen and Degn (1985), Schaffer (1985), Margulis and Lovelock (1986), Mees (1986), Schaffer and Kot (1986), Vermeij (1986), Crutchfield and Kaneko (1987), Dewdney (1987), Mackey and Milton (1987), Skarda and Freeman (1987), Cavalli-Sforza et al. (1988), Chen (1988), Diamond (1988), Hewitt et al. (1988), Lewin (1988a), Lovelock (1988b), Martin and Simon (1988), Rapp et al. (1988), Shaw and Chouet (1988, 1989, 1991), Vogel et al. (1988), Chen and Bak (1989), Hunter and Gallagher (1989), Vogt (1989), Arthur (1990), Barbujani and Sokal (1990), Clube and Napier (1990), Goldberger et al. (1990), Sokal et al. (1990), Suarez et al. (1990), Bak and Chen (1991), Chouet and Shaw (1991), Freeman (1991, 1992), Hassell et al. (1991), Lauterborn and Holzfuss (1991), Ives (1991), Lewis and Glass (1991), Richardson (1991), Chyba and Sagan (1992), Greenberg and Ruhlen (1992), Levinton (1992), Nowak and May (1992, 1993), Podsiadlowski and Price (1992), Ruthen (1993).]

But here I must relegate such questions to our ignorance concerning how they might be applied to the K/T — or any other — boundary event, leaving them with the following queries concerning the end-Cretaceous extinctions: (1) Where were the bird sanctuaries? (2) What sorts of *refuges* (cf. MacArthur and Wilson, 1963; Eldredge and Gould, 1972; Simberloff, 1974, 1986; Gould and Eldredge, 1977; Fischer, 1981; Gould, 1984b, 1989b, 1992; Padian et al., 1984; Raup, 1984a, b, 1986a, b, 1988a, b; Boecklen and Simberloff, 1986; Parrish et al., 1986; Vermeij, 1986; Eldredge, 1989; Gallagher, 1991; McKinney and Frederick, 1992; Raup and Jablonski, 1993) might have been available to ensure a bottleneck-surviving number of mating pairs (cf. Note 8.1; and Raup, 1979; O'Brien et al., 1986) of an airborne and feathered fauna against the combined effects of a global impact winter, wildfires, and climatic disruption of the sorts envisioned by the proponents of the impact-extinction hypothesis? (3) How many refuges would have been required to ensure survival of *all* of the principal avian lineages (cf. Fig. 36, Chap. 8; and Shaw, 1988a, Figs. 14–16)? and (4) Where (what kind, how many, etc.) were the analogous plant and animal refuges that would have been required to preserve the North American land biota that is known to have survived the K/T "debacle" (see Fig. 36 and the discussion of Phanerozoic clam-brachiopod diversity patterns in Chap. 8 and Note 8.14)? Lacking answers, I reiterate the analogous reaction by Gilmour et al. (1989, pp. 209f) concerning the question of ecological effects on the pervasiveness of biologic extinctions, emphatically seconding their proposition: *"A first-order scientific task awaiting an imaginative ecologist is to rationalize the observed extinction patterns in terms of the above stresses"* (referring to ibid., Table 2).

A crucial question that may in due course be uniquely answered by paleobotanical studies is whether just one "impact winter" or several could have occurred during the time of the worldwide K/T biostratigraphic successions (cf. Lerbekmo et al., 1987; Herman, 1990; Johnson and Hickey, 1990; Sweet et al., 1990). If there was only one such event of global extent, then scenarios of multiple impacts widely separated in time (e.g., in the range 10^3–10^6 years) might be eliminated. One of the scenarios described by Hut et al. (1991), which was based on solar disruption of a comet, fits some of the constraints provided by Wolfe (1990, 1991), including compatibility with a Northern Hemisphere summer. Ironically, a CRF model in which craters were produced by objects from a generalized reservoir of geocentric natural satellites loses this discrimination, except for (1) a mechanism by which a high-inclination grazing encounter by a large bolide results in its breakup and the transient capture of its fragmentation products into decaying geocentric orbits (e.g., over a time frame of months for the larger fragments and years for the smaller fragments; see Fig. 19, Note 1.1, and Appendix 3; cf. Schultz and Gault, 1990a, b; Schultz and Lianza, 1992), and (2) spin-orbit mechanisms invoking mass exchanges between reservoirs orbiting both the Earth and the Moon (see the Epilogue; cf. Gault and Schultz, 1990). According to the analysis of Hut et al. (1991), however, one multiple event of cometary disruption could not by itself account for two impacts separated in time by months. But the disruption of a geocentric and/or selenocentric (Moon-centered) satellitic reservoir could explain coordinated multiple impacts. Wolfe's (1991) deductions raise hopes that there are other candidate paleobotanical sites somewhere (cf. Kerr, 1992a, 1993b) that might focus even more tightly on the sequence of happenings at the K/T and other biostratigraphic event boundaries. If such sites exist, they might even offer up evidence for more than a single processional impact trajectory [e.g., two or more trajectories such as those of the low-angle impact sites Campo del Cielo and Rio Cuarto, Argentina, described, respectively, by Cassidy et al. (1965) and Schultz and Lianza (1992), that form small angles with a great circle through the Australian and Eurasian PCNs (see Appendix 4), and/or trajectories subparallel to the other NGCs (nodal great circles) illustrated in Appendixes 1–4], which would in turn constitute important tests for the proposed invariances suggested by the CRF hypothesis. On present evidence, Wolfe's results neither confirm nor deny the existence of a CRF reservoir.

Iridium anomalies, shocked-quartz localities, and major flood-basalt provinces are compared in **Figure 24** with the impact loci and nodal positions illustrated in Figures 1–5 (see Alvarez et al., 1982; cf. Hsü et al., 1982; Hsü, 1984; Doehne and Margolis, 1990; Hsü and McKenzie, 1990; Orth et al., 1990). Occurrences of diamonds in K/T-boundary deposits, a subject of some interest with regard to terrestrial, extraterrestrial (Solar System), or even interstellar chemical sources, are described by Carlisle (1992), Gilmour et al. (1992), and Russell et al. (1992); see Wright (1992) for a discussion of interstellar diamonds [occurrences of diamonds in meteorites are discussed by Anders (1988) and Anders and

Kerridge (1988); cf. Note I.1 in the present volume]. The patterns of Ir anomalies, shocked-quartz localities, and tektite-strewn fields are consistent with the CRF loci, but they do not further constrain the locations of cratering nodes and swaths. Depending on future discoveries and interpretations of possible oceanic impact sites in the Pacific, and other sites at southern latitudes higher than that of the Australian node, geochemical-mineralogical anomalies might provide evidence concerning the possible presence or absence of additional cratering nodes (e.g., a second Southern Hemisphere node) symmetrically distributed with respect to a similar set of Northern Hemisphere nodes (cf. discussion of Figs. 27 and 28 below). As mentioned before, the present-day positions of flood-basalt provinces show patterns generally close to the principal impact loci, as illustrated in **Figure 25**, but the original positions of these provinces have not been worked out in detail, and the arrows shown in Figure 25 indicate only the general directions of backtracking without regard to the total magnitudes of rotations and/or translations — where "backtracking" refers to kinematic reconstructions by which the motions along plate-tectonic trajectories are reversed in time. In most cases the uncertainties associated with locations obtained by backtracking methods suggest that the proximities of flood-basalt localities to the cratering swaths, with arc widths of $10°-20°$, will be either similar or improved relative to the general proximities shown in Figure 25 (the flood-basalt distribution is shown in Mercator projection in **Figure 26**). However, the paleogeographic implications of the CRF, to be discussed later, may change some conventional interpretations of absolute plate motions, continental drift, and true polar-wander paths (see the synopsis of these relationships given in the Introduction). Accordingly, the indicated directions of rotation about Euler poles (see Gordon et al., 1984; Cox and Hart, 1986; Butler, 1992) placed at each of the three CRF nodes in Figure 25 are merely indicative of the senses of motion in kinematic reconstructions.

Kinematic Correlations with Impact Origins of Flood-Basalt Provinces

Many workers have suggested that inceptions of volcanism, continental rifting, and seafloor spreading were caused by or at least triggered by large meteoroid impacts. In a farsighted synthesis, Seyfert and Sirkin (1979, Table 12.2) compared tectonic and magmatic events of the Mesozoic and Cenozoic eras with what they call *impact epochs*. Although their classification of events cannot be specifically resolved in terms of the cratering patterns discussed in the present volume, there is considerable synchrony between the reasoning behind their classification scheme and my own reasoning concerning the nonlinear dynamics of coupled impact-magmatic-tectonic-geomagnetic processes. Stothers (1992) draws similar conclusions concerning long-term correlations between the lunar cratering record and orogenic episodes in Earth history. Stothers and Rampino (1990) discuss correlations between mass impacts, flood-basalt volcanism, and mass extinctions that are also generally consistent with my own conclusions (see, in addition, Burek et al., 1983; Burek, 1984; Burek and Wänke, 1988; Schmitt et al., 1991).

Clearly, the literature describing patterns of mass impacts on Earth has a much longer history than do the debates of the past decade on the impact-extinction hypothesis, although the latter really began among paleontologists and stratigraphers following the suggestion by McLaren (1970) that the Frasnian-Famennian (late Devonian) extinction was of such catastrophic proportions that an event of instantaneous and nearly global proportions would be required to explain it. In words difficult to paraphrase, words that some workers took to be tongue-in-cheek, McLaren (1970, p. 812) posed the idea of an impact as follows: *"I shall, therefore, land a large or very large meteorite in the Pacific at the close of the Frasnian. Presumably on impact with the ocean surface or at a certain depth below the surface, the missile will explode with an enormous release of energy. In the copious literature on meteorites, impact craters and astroblemes, I have been unable to find a calculation of the energy effect in terms of tidal waves from such an explosion. Dietz (1961) suggests that a giant meteorite falling in the middle of the Atlantic Ocean today would generate a wave twenty thousand feet high. This will do."*

The earlier literature cited by McLaren (1970) correlates with the successes of "astrogeology" and their implications for geology, a field pioneered by and still inspired by the work of E. M. Shoemaker (e.g., Shoemaker, 1960a, b; 1961; 1962; 1963; Shoemaker and Chao, 1961; Dietz, 1961, 1963; Gallant, 1964; Hartmann, 1972; cf. French, 1990) — and, of course, we should not forget the earlier pioneering studies of cratering phenomena by G. K. Gilbert (1893) and R. A. Daly (1946); cf. Hartmann and Davis (1975), Brush (1986), Baldwin and Wilhelms (1992), Spera and Stark (1993). Geologic evidence of major Precambrian cratering events and impact deposits are discussed, for example, by Green (1972), Frey (1977, 1980), and Grieve (1980); see also the recent notion that ancient "tillite" deposits may, in part, represent impactites, as proposed by Oberbeck et al. (1992), Marshall and Oberbeck (1992), Aggarwal and Oberbeck (1992), and Rampino (1992). Postplate-tectonic conceptions of relationships between magmatic and tectonic effects are discussed by Seyfert and Sirkin (1979, pp. 96f, and Table 12.2), Burek et al. (1983), Burek (1984), Alt et al. (1988), Burek and Wänke (1988), Rampino and Stothers (1988), Stothers and Rampino (1990), Hagstrum and Turrin (1991a, b), and Stothers (1992). Shoemaker (1983) remains probably the best discussion of impact phenomena on the continents, as do Melosh (1982) and Ahrens and O'Keefe (1983) for the effects of large impacts in the oceans (cf. Gault and Sonett, 1982; McKinnon, 1982; Ahrens and O'Keefe, 1990; Melosh et al., 1992). The effects of tidal waves, especially with regard to a Caribbean impact site, are treated, for example, in Hildebrand and Boynton (1990a), Alvarez et al. (1991), Roddy et al. (1991), Smit and Alvarez (1991), and Smit et al. (1991). An extensive literature exists on diverse aspects of impact phenomena, and the references cited in the present volume, though numerous, do not adequately express its scope. A few of the book-length surveys, symposium volumes, and textbooks that I have found to be helpful concerning the general context of Solar System impact phenomena include Gallant (1964), Seyfert and Sirkin (1979), Wood (1979), Clube and Napier (1982b, 1990), Silver and Schultz (1982), Horton et al. (1983), Berry et al. (1987), Wilhelms (1987), Kerridge and Matthews (1988), Binzel et al. (1989a), Chapman and Morrison (1989), Clube (1989a), Melosh (1989), Bailey et al. (1990), Beatty and Chaikin (1990), Kippenhahn (1990), Newsom and Jones (1990), Sharpton and Ward (1990), Kaufmann (1991), Pasachoff (1991), Yeomans (1991a), and Teisseyre et al. (1992).

There are also reports in the literature of large but hypothetical impact sites that have not been documented, hence those references — which are of considerable interest in their own right — are given only cursory mention in the present work (cf. esp. Gallant, 1964; Alt et al., 1988; Hildebrand and Boynton, 1991; Hagstrum and Turrin, 1991a, b). Several other sites of possible impact craters, and/or of impact-metamorphic features, beyond the impact sites already discussed, have been reported (many of these appear in a written communication from Grieve, 1986, the G-R Catalog mentioned in Chap. 1; cf. Grieve and Robertson, 1979; Grieve and Dence, 1979; Grolier, 1985; Grieve, 1987, 1991; Grieve and Robertson, 1987; Grieve et al., 1988). Pesonen and Henkel (1992a) present a collection of papers from a symposium with special focus on cratering in Fennoscandia; see especially Grieve and Pesonen (1992) and Henkel and Pesonen (1992). In addition to the established, probable, and possible impact sites documented in the compendiums by Grolier, Grieve and coworkers, and Pesonen and Henkel (cf. Pesonen and Henkel, 1992b), there are several other recent descriptions of cratering events, and/or evidence related to cratering events, in the literature, such as Ganapathy (1982), Hartnady (1986), Keller et al. (1987), Hargraves et al. (1990), Higgins and Tait (1990), Wang et al. (1991), Bice et al. (1992), Claeys et al. (1992a, b), Kerr (1992b, d), Lindström et al. (1992), Poag et al. (1992), Rajlich (1992), Rampino (1992), Schnetzler and Garvin (1992), Schultz and Lianza (1992), Wang (1992), and Gudlaugsson (1993). The locations of these impact sites (proven or putative) appear to be consistent with the patterns of Figures 1–5 and Appendixes 1–4.

Most of the models that have been proposed for tectonic and/or magmatic effects of impacts, such as the production of flood-basalt volcanism, refer to direct consequences at ground zero. Hagstrum and Turrin (1991a, b) and Hagstrum (1992), on the other hand, have proposed an interesting alternative model in which flood basalts occur at positions antipodal to causative impact sites, adding magmatism as a logical extension to more general models of antipodal tectonic effects on Mercury and on the Moon, as proposed by Schultz and Gault (1975); cf. Hughes et al. (1977) and Watts et al. (1991). They test their model against the positions of the antipodes to the K/T Deccan Traps, India, and the Miocene Columbia River flood basalts (CRB) of western North America. [Relationships between global patterns of volcanic "hot spots" and their antipodes are illustrated in **Appendixes 12–15**, and are compared with the distribution of impact craters, cratering swaths, Phanerozoic cratering nodes (PCNs), and nodal great circles (NGCs) in **Appendixes**

16–18 (see the recapitulation given in the Introduction).] By assuming that the Deccan flood-basalt province reflects the major K/T impact, they predict an antipodal impact site with present coordinates in the Pacific at 36° N, 205° E (155° W). This falls west of the K/T swath in Figure 24 (heavy short dashes), northeast of GPC-3 (the "giant piston core" cited in the studies of iridium anomalies and shocked quartz; see Alvarez et al., 1982, Table 1), and roughly midway between the K/T and Phanerozoic nodal swaths. Their reconstructed position would place the 65-Ma impact site within several degrees of 8° N and 224° E, or roughly antipodal to the Reunion "hot spot" (i.e., the present, zero-age, position of the "hot-spot track" that passes approximately through both the Deccan and Reunion localities; cf. Appendix 12).

In the framework of CRF invariance, the above reckoning places the Hagstrum-Turrin K/T impact site virtually on the Phanerozoic swath (long heavy dashes in Fig. 24), roughly 40° to the southwest (along the swath) from the North American node (large asterisk), and roughly 1000 km southeast of, and parallel to, the trend of the Hawaiian mantle-melting anomaly (HMMA), assuming it, too (like Reunion), to be roughly invariant (a "fixed hot spot"). The Hagstrum-Turrin hypothesis predicts that their putative K/T impact site was located on ocean crust that was subsequently subducted to the east, which would have destroyed both the crater and the nearby impact products. Except for the great difference in age, the suggestion of an impact site in the Pacific brings to mind once again the event envisioned by McLaren (1970) in the above quotation.

The proximity to the HMMA ($\sim 10^3$ km) and an uncertainty of similar magnitude in the relative position of that anomaly (Shaw et al., 1980a), pose the possibility of a near coincidence, especially if the putative impact was double or multiple. A K/T impact site obviously is not a candidate for initiating the HMMA, because the latter extends back to at least 73 Ma in the Emperor chain and is lost into subduction zones near the convergence of the Aleutian and Kamchatka trenches. But this constraint does not hold for prior events, or for the possibility of K/T and younger impact sites in the general vicinity of the Hawaiian-Emperor chain. These possibilities heighten an interest in certain arcuate structures recently discovered on the seafloor near the Hess Rise (see the caption of Fig. 3). If the Hagstrum-Turrin hypothesis proves viable, we may have to conclude that traces of other impact events near to or even somewhat prior to the K/T interval may still survive on the deep Pacific Ocean floor.

Such observations underscore the need for more intensive exploration of the ocean basins for potential impact localities. An important step in this direction would be provided by detailed side-scan sonar (GLORIA) bathymetric surveys currently documenting the U.S. Exclusive Economic Zones (EEZ), as illustrated with reference to submarine volcanism by Shaw and Moore (1988, cover illustration by R. T. Holcomb, W. R. Normark, and R. C. Searle). The arcuate scarps of Figure 3 were in fact detected during a recent GLORIA cruise (R. I. Tilling and R. T. Holcomb, written comm., 30 June, 1991). Unfortunately, these features were detected just inside the northern limit of the EEZ, and there is no information on

Consequences of the Celestial Reference Frame

their configurations or dimensions to the north. A detailed GLORIA survey in the vicinity of the Hess Rise and in the general vicinities of the CRF nodal swaths might greatly increase our knowledge of potential oceanic impact sites. Notably, the present position (36° N, 155° W) of the K/T impact site predicted by Hagstrum and Turrin (1991a, b) is only about twenty degrees or so east of the "mystery scarps" near the Hess Rise.

The Hagstrum-Turrin model applied to the CRB predicts an impact site at ~48° S, ~68° E, in the vicinity of Kerguelen Island (south Indian Ocean; cf. the discussion of "hot-spot" patterns and their relationships to cratering patterns in Appendixes 12–18). In Figure 24, this site falls within the small circle (short light dashes) of an uncertain swath in the Southern Hemisphere at latitudes above 25° to 30° S (cf. Fig. 26; and Appendix 18). The uncertainties mentioned earlier concerning the lack of evidence for impact cratering at high southern latitudes, now compounded by the Hagstrum-Turrin proposal, increase the incentives for more intensive searches in vicinities near mirror images of the Northern Hemisphere nodal swaths, especially in Antarctica. That is, although the existing evidence concerning orbital approach scenarios discussed in Chapter 6, and the evidence of the cratering patterns of Figures 1–5, appear to favor the Northern Hemisphere over the Southern Hemisphere as the dominant expression of Phanerozoic cratering, there is no conclusive proof of a Northern Hemisphere bias in the unmodified orbital properties of asteroids and comets (other than the possible effects, already mentioned, of mean-motion resonances and the Galactic-ecliptic angular difference on the inclinations of new comets). The mechanism of spin-orbit resonance with the Earth or Earth-Moon systems, however, does imply that—even in the absence of preferred Solar System orbits—a bias would likely be induced in a geocentric system of satellites by zonal and sectorial asymmetries in the mass anomalies of the Earth, the Moon, or both (e.g., see the hemispheric bias of cratering on the Moon, as illustrated in Fig. 37; cf. Wilhelms, 1987). Such a bias would influence precessional effects of eccentricity e and inclination i in the geocentric orbits that control the times spent near perigee and apogee near critical inclination, hence inducing some form of hemispheric asymmetries in the patterns of eventual orbital decay.

Another possible reason for a hemispheric bias in impact sites might be a preferred distribution of satellitic orbit-orbit resonances that became locked to the Earth's mass anomalies in such a way that collisions or chaotic crises preferentially triggered "reentries" (i.e., by analogy with the terminology of artificial satellites) at polar localities. It has been suggested, for example, that Antarctic meteorites may have been derived preferentially from high-inclination orbits (see Lipschutz et al., 1989, p. 760). If so, they may reflect the Southern Hemisphere analogues of orbital configurations that produced Northern Hemisphere impacts. And here the plot again thickens.

Although the age of the Antarctic ice sheet is thought to be greater than 10 Ma, the "stranding surfaces" containing meteorites are no greater than 1 Ma (the maximum "terrestrial age" of Antarctic meteorites; see Cassidy et al., 1992, p.

490; cf. Lipschutz et al., 1989, p. 760). Lipschutz et al. (ibid.) also point to an open question concerning how meteoroids — i.e., the presumed asteroidal fragments that became meteorites upon striking the Earth — retained their orbital distinctness during passage from their original orbits in the Asteroid Belt to become Earth-crossing or Earth-approaching meteorite parent bodies. It was long suspected by students of meteorites that the so-called HED meteorites (basaltic achondrites consisting of howardites, eucrites, and diogenites; cf. Lipschutz et al., 1989, Table V) were somehow derived from the asteroid 4 Vesta (~500 km in diameter, semimajor axis ~2.36 AU). Vesta's approximate orbital position lies just below the tip of the arrow that points to the 3/1 Kirkwood Gap in Figure 13b (cf. Appendix 9). [The number 4 in front of the name indicates that Vesta was the fourth asteroid discovered — even though it ranks third in size relative to the largest asteroid, Ceres, with a diameter of 913 km (Tedesco, 1989, p. 1093) — all of the first four having been discovered prior to the mid-nineteenth century (see Pilcher, 1979; cf. Considine, 1983, p. 242).]

Heretofore, Vesta was the main candidate for the basaltic achondrites essentially by default, because it was the only known asteroid with spectral properties consistent with a surface of basaltic composition (cf. Gaffey et al., 1989; Lipschutz et al., loc. cit.). Now, however, Binzel and Xu (1993) have identified 20 small main-belt asteroids (i.e., within the MBA of Fig. 13; cf. Fig. 12) that appear to have been spalled off of Vesta by impacts. All of these objects have diameters of 10 km or less. Twelve of them have orbits similar to Vesta's, and eight have orbits that span the orbital space between Vesta and the 3/1 Kirkwood Gap. Binzel and Xu (loc. cit., p. 186) state, conservatively, that these observations "establish Vesta as a dynamically viable source" for the HED meteorites. Gaffey (1993, p. 168), however, goes further, stating that "it is now clear that the basaltic achondrite meteorites in our museum collections are natural samples of the main-belt asteroid 4 Vesta." In this interpretation, the 20 small objects are viewed as impact fragments (cf. Fujiwara et al., 1989; Binzel and Xu, loc. cit.) whose orbits have retained some semblance of that of their parent body or have been "caught in the act," so to speak, of escaping from their parental MBA locality, perhaps to become a source of terrestrial meteorites at some chaotically unpredictable future date. It would seem to follow, therefore, that the basaltic achondrites bear testimony to a chaotic mechanism for the delivery of meteoroids to the Earth from MBA localities (cf. Wisdom, 1985, 1987a, b, c; and Fig. 12 herein), for example by the mechanism of chaotic crises discussed in Chapters 4 and 5, thus suggesting — as discussed in Notes 2.1 and 6.7 — that asteroids anywhere in the MBA might also represent meteorite parent bodies, not just the Earth-crossing and near-Earth asteroids of Figure 13 (i.e., the Apollos, Atens, and Amors), in keeping with the general notion of Solar System critical self-organization illustrated schematically by the "hourglass model" of Figure 11. And if a significant fraction of these source meteoroids ("daughters of Vesta," so to speak, and the like) have approached at high inclinations (see above), then the case for the highly nonlinear orbital focusing (natural selection) mechanisms discussed in Chapters 4–6 is greatly

Consequences of the Celestial Reference Frame

strengthened. Consequently, the celestial reference frame hypothesis (CRFH) might be tested, and perhaps significantly improved, by a detailed examination of the geographic patterns of meteorite falls. But this is an immense task — and one that, as yet, I have not undertaken.

It is evident, both on general grounds and from the above discussion of the Hagstrum-Turrin hypothesis, that interactions of impacts with the Earth's mantle could perturb the magmatic cycle, either directly or antipodally (cf. Appendixes 12–18; and Seyfert and Sirkin, 1979, Table 12.2). Either the direct or the antipodal interactions with the mantle are consistent with the proposed invariant patterns of impact cratering over time. I shall return to the relationship between impacts and magmatic processes following a discussion of geomagnetic patterns and reversal frequencies.

The Geomagnetic Signature of a Celestially Coupled Earth

Perhaps I should pause a moment and explain a bit more why I place so much emphasis on geomagnetic phenomena. The most obvious reason is that geomagnetic data are relatively abundant, and they come closest — among the geophysical data I have examined during the preparation of this volume — to illustrating the forms of phase-space phenomena already discussed (e.g., the critical-state phenomena of Chap. 5). The examples I have chosen for purposes of illustration also have been much in the scientific news during this period of time, and they form a bridge between some of the phenomenologies of Earth's orbital-rotational motions in space (e.g., relative to the nonlinear dynamics of the Sun and the properties of the solar wind) and phenomenologies that are characteristic of the Earth's inner-outer core, core-mantle, and lithosphere-crust-hydrosphere-atmosphere coupling with the CRF system of impact phenomena that correlate the system as a whole with its celestial setting. All of these phenomena are relevant to the convective motions of the Earth's mantle and to the mass distributions which, in turn, are coupled with Earth's plate-tectonic and magmatic history (cf. Note 8.3).

What is more, geomagnetic phenomena display some of the same global structures and partial symmetries that appear to characterize the nodal cratering patterns already discussed. These in turn form the basis for my proposal that Earth's cratering patterns, and its geomagnetic and magmatic signatures (e.g., as manifested in the patterns of volcanism referred to as the "hot spot reference frame"), have evolved systematically over Earth's history as a closely coordinated system of coupled processes. As such, for example, the role of chaotic crises can be seen as a common factor that reflects Earth's cosmological setting and, at the same time, describes the previously puzzling and controversial character of intermittencies in these several processes. Given such a general correlation of whole-Earth mechanisms, then, it is easier to see how other processes — processes that are related both to Earth's biostratigraphic history and to its impact-dynamic-geomagnetic-magmatic history — relate to the properties of the Earth as a planet.

The core dynamo, especially as it can be modeled in some respects as a very simple coupled oscillator (e.g., the two-disk dynamo models that have evolved from the simpler disk-dynamo model of Bullard, 1955, 1978; cf. Allan, 1958), offers an intriguing analogy with extraterrestrial orbital motion as well as with the numerical-dynamic experiments epitomized by the sine-circle function (Table 1, Sec. D) partially illustrated in Figure 14 and applied by Shaw (1987b) to concepts of geologic periodicities and by Shaw and Chouet (1988, 1989, 1991) to the properties of magmatic transport from the mantle to the surface of the Earth (cf. Note 8.3). The sine-circle map, in turn, bears distinct resemblances, for example, to the spatiotemporal structure of the Asteroid Belt (cf. Fig. 14 and Appendix 9), as well as comprising the simplest low-dimensional analog of self-organized criticality, a process that is directly relevant to hierarchical convective motions in the Earth's outer core (e.g., relative to the vortex structures that are satellitic to the inner core in a manner somewhat analogous to the motions of the asteroids relative to the inner Solar System), to magma percolation, and to the diverse scales of mantle convection and tectonic motions of the Earth's crust.

In what follows, I shall be considering three types of global geomagnetic patterns that show similarities to the cratering patterns and nonlinear-dynamical mechanisms already discussed. These are (1) the pattern of lobes in maps of the radial component of the magnetic field at the surface of the Earth's core for the years 1715 to 1980, as shown by Gubbins and Bloxham (1987); (2) the longitudinal distribution of transitional VGP (virtual geomagnetic pole) paths, as discussed and compared with the core flux by Clement (1991); and (3) the 150-m.y. record of geomagnetic-reversal frequencies (Fig. 6) illustrated and discussed by, among many others, Creer and Pal (1989), Gaffin (1989), Merrill and McFadden (1990), and Jacobs (1981, 1984, 1987, 1992a, b). Additional background concerning VGP patterns (see Note I.7) is given by K. A. Hoffman (1988, 1991, 1992, 1993), Bogue (1991), Laj et al. (1991, 1992), Bogue and Merrill (1992), Clement (1992), Constable (1992), Jackson (1992), Langereis et al. (1992), and Valet et al. (1992). Excellent reviews of geomagnetic-reversal phenomena are given by Gubbins (1987), K. A. Hoffman (1988), Merrill and McFadden (1988, 1990), Bloxham and Gubbins (1989), McFadden et al. (1991), and in the textbooks by Merrill and McElhinny (1983), Jacobs (1987, 1992b, pp. 99ff) and Butler (1992). Most of these data and discussions — from the present nonlinear-dynamical perspective — point to mechanisms of nonlinear criticality (phenomena influenced by relationships and interactions near to or crossing periodic-chaotic transition regions of phase space). These are the phase-space regions characterized by and susceptible to chaotic crises (bursting phenomena), as discussed in Chapter 5. Hence, I will be examining the geomagnetic records for possible relationships to similar behavior in the spatiotemporal signatures of the cratering record and in orbital phenomena governing the dynamics of impacting objects (the CRF and/or other near-Earth reservoirs of potential impactors). Crisis-induced intermittency appears to be evident in phase relations (illustrated below) between geomagnetic-reversal frequencies and impact-cratering frequencies (cf. Fig. 6). Similar chaotic crises are

Consequences of the Celestial Reference Frame

evident (Fig. 32, and Inset in Fig. 31) in numerical studies of the Rikitake (1958) two-disk dynamo by Ito (1980) and Hoshi and Kono (1988).

Contours near the Northern Hemisphere maxima of the radial component of the magnetic flux into the core are shown in North Polar projection in **Figure 27**. These are lobes 1 and 2 of Gubbins and Bloxham (1987, Fig. 2). The heavy dashed circle marks the profile of the inner core. The numbered flux lobes and the Northern Hemisphere CRF cratering nodes are roughly tangential to the inner core, and lobe 2 is nearly coincident with the North American cratering node. Lobe 1 is located about 90° E of the Eurasian node, which places it near the same meridian as the Australian cratering node but in the Northern Hemisphere. The flux lobes that represent the radial component of magnetic flux out of the core in the Southern Hemisphere are not shown in Figure 27. They are designated lobes 3 and 4, paired approximately with lobes 1 and 2. The flux pair (1, 2) fits with the broader Phanerozoic cratering maxima in the Northern Hemisphere, as shown by the light shading in Figure 27 (from Fig. 2), hence with the northern limit of the K/T swath (Fig. 3).

These relative relationships are clarified by the Mercator projection shown in **Figure 28**. Lobes 3 and 4 are not as sharply defined as 1 and 2, but the symmetry between (1, 3) and (2, 4) is evident, where flux lobe 3 correlates approximately with the Australian cratering node. Flux lobes 1 and 3 fall on a meridian that passes just west of the Australian cratering node and somewhat to the east of the Eurasian cratering node but within the general Eurasian cratering pattern shown by the light shading in Figure 27 (the western limit of the Eurasian magnetic flux contours in Fig. 28 extend parallel to the K/T swath almost to the Eurasian cratering node). If the North American cratering node in Figure 28 were paired with a virtual cratering node — a cratering node that is not evident in the patterns of Figures 1–5 either because it does not exist or because it represents an oceanic locus that remains undiscovered and/or has been partially or completely destroyed by submarine volcanism and/or subduction — west of the southern apex of the K/T swath (just west of South America), then that pair of cratering nodes would, in a crude way, be meridionally aligned with the (2, 4) pair of flux lobes.

Figure 28 (dashed curve in the lower graph) also shows the longitudinal distribution of transitional virtual geomagnetic pole positions (VGPs; see Note I.7) of Clement (1991, 1992; cf. K. A. Hoffman, 1988, 1991, 1992, 1993; Bogue, 1991; Laj et al., 1991, 1992; Butler, 1992; Constable, 1992; Jackson, 1992; Langereis et al., 1992; Valet et al., 1992). The longitudinal distribution of VGPs peaks at roughly 90°–110° E, 70°–80° W; thus it falls along a meridian that intersects the upper graph of Figure 28 roughly ten degrees or so east of the 90-degree W and 90-degree E meridians. The VGP distribution thus forms an approximate, though very broad, great-circle meridional band — centered roughly on the Americas and Eurasia — that is nearly aligned with the southern limits, respectively, of the K/T and Phanerozoic nodal swaths (Fig. 28, top), hence is also crudely aligned with the magnetic flux lobes and cratering patterns just described.

In addition to the geomagnetic and cratering patterns, Figure 28 (solid wedge-shaped symbols in bottom graph) also schematically portrays a correlation between the geomagnetic patterns and the approximate longitudinal positions of deep-mantle density anomalies determined by seismic imaging techniques (see Dziewonski and Woodhouse, 1987; cf. Morelli and Dziewonski, 1987). Laj et al. (1991) demonstrated that some patterns of transitional VGP paths show a remarkable correspondence with the almost meridional pattern of deep-mantle seismic P-wave anomalies determined by Dziewonski and Woodhouse (1987), a correlation that also corresponds in a crude way to topographic lows on the core-mantle boundary (Morelli and Dziewonski, 1987; Vogel, 1989, Fig. 3) and on the 660-km discontinuity (Fukao, 1992). Because the magnetic flux lobes, VGPs, cratering patterns, and mantle density anomalies, at first glance, would appear to span vastly different geologic time scales — as well as vastly different geodynamic mechanisms — these geographical comparisons are of considerable interest. Clement (1991) discusses the correlations between magnetic flux patterns and VGP patterns, and I explore their possible relationships to the cratering pattern in the following discussion.

The approximate matching of magnetic flux lobes and cratering patterns (Figs. 27 and 28) raises once again a question mentioned in the Introduction (and above), a question that concerns the possible existence of cratering nodes in addition to the three Phanerozoic nodes identified in Figures 1–5. Gubbins and Bloxham (1987) raised an analogous question concerning another pair of flux lobes that might complete an apparent 120° (trigonal) symmetry of a global pattern. Their "missing pair" of flux lobes would lie somewhere in the vicinity of the Greenwich meridian. If this were also the case for cratering nodes, as mentioned in the Introduction, then we would need three more nodes: one in the eastern South Pacific somewhere along a meridian offshore of Chile and intersecting the Amundsen Sea, Antarctica; another along a meridian between South Africa and Queen Maud Land, Antarctica (i.e., just east of the Greenwich meridian and aligning more or less with the Eurasian node); and one in easternmost Asia along a meridian between the Sea of Japan and the Laptev Sea, Arctic Ocean (interestingly, not far from the convergence of the Hawaiian-Emperor chain with the Aleutian and Kamchatka trenches). But, as in the case of the flux lobes, it may be that asymmetries and/or different symmetries are artifacts of the global system of dynamical effects. Any one or more of the symmetry positions of a given set may be poorly developed, masked, or latent, even though the spatial frequencies are related to the same general process (see below).

It does not seem fortuitous to me that the issue of lattice-like symmetries and "vacant sites" applies similarly to the pattern of magnetic flux lobes and the pattern of cratering nodes. That is, in both instances it would appear that there is an ambiguity between a possible fourfold or sixfold pattern which, whatever the truth of the matter may be, would seem to represent the *same* geographic distribution of nodal "lattice sites." Interest in the symmetry question is heightened by the additional observation that these alternatives would appear to be compatible either

Consequences of the Celestial Reference Frame

with the longitudinal phase dfferences (see Figs. 27 and 28) between the flux lobes (separated by about 120° and 240°) or with the longitudinal phase difference between the VGP bands (separated by about 180°). The cratering pattern has resemblances to both and might be viewed as an incomplete sixfold pattern, a longitudinally biased fourfold pattern, or a skewed twofold pattern (i.e., a pattern in which there is one very broad Northern Hemisphere swath-like PCN and another similarly smeared distribution in the Southern Hemisphere; cf. Figs. 1–5). The most important observation at the present juncture, however, is that the general orientations and uncertainties of these symmetry questions seem to be shared in common by both the geomagnetic and cratering patterns!

Kinematic similarities between the cratering and core-flux patterns raise seemingly paradoxical questions concerning the characteristic time scales of these phenomena. By hypothesis, the CRF invariance has persisted since at least the latest Precambrian (on the order of 10^9 years). During that time, the present westward drift of the nondipolar magnetic field ($\sim 0.18°$ per year = 2000 years per cycle; cf. Merrill and McFadden, 1990) would be equivalent to the motion of a marker that has circled the globe 300,000 times. Thus, we might infer either that the geographic similarities are fortuitous, or that there is a structural invariance in the core motions that controls the rotation of the flux lobes at the core surface in a manner analogous to that of the cratering nodes at the Earth's surface **(Note 7.1)**. At least some degree of invariance of the flux-lobe pattern is required to explain the ~ 200-year stationarity described by Gubbins and Bloxham (1987).

The relationship of the magnetic flux lobes mapped on the surface of the core (Figs. 27 and 28) to the geomagnetic-drift/planetary-rotation problem presents an interesting analogy with "spin-orbit" coupling (Fig. 21), in the sense that it concerns relative rotations in a relatively decoupled portion of a fluid interval in the geocentric system (cf. Notes 7.1, 7.3, and 7.4). In this case, the relatively inviscid outer core and/or interaction at the core-mantle boundary (cf. Merrill and McFadden, 1990, p. 346) — assuming that the flux-lobe pattern in Figure 27 is indeed rigorously stationary — somehow develops magnetic flux domains that do not rotate with the same average angular velocity of the magnetic field as a whole and are magnetohydrodynamically locked into a 1:1 resonance with the spin of the solid Earth. In contrast with the geocentric satellites, which resonate with the gravitational attractions of Earth's geoidal mass anomalies, the dominant effects on internal fluid motions in the core involve variations of magnetohydrodynamic moments of inertia and conservation of angular momenta within different parcels of fluid. The invariances of longitudinal velocities in the core therefore must be consequences of coupling between radial, zonal, and longitudinal momentum balances (this is also another way of describing the analogous force balances that govern the orbital motions of satellites, especially from the point of view of plasma dynamics, as discussed by Alfvén and Arrhenius, 1975, 1976; and Alfvén, 1980, 1981, 1984). In other words, the putative stationarity of flux lobes mapped on the core would be analogous to the stationarity of Jupiter's Great Red Spot (GRS; see Ingersoll, 1983, 1988; Marcus, 1987, 1988, 1990; Sommeria et al.,

1988; Gierasch, 1991; cf. Note 7.1 in the present volume; and color illustrations of the GRS compared with computer simulations by Marcus, in Gleick, 1987, Plate facing p. 115).

The experiments by Sommeria et al. (1988) show that in addition to a single stable vortex, ordered sets of vortices will form, depending on the "pumping rate" of fluid into a given rotational flow system. Their experiments employed a cylindrical rather than spherical geometry; that geometry would be appropriate to considerations of annular flow between the inner and outer core if the flux lobes were thought to reflect crudely the positions of convective rolls more or less aligned with the spin axis (cf. Busse, 1978; Childress, 1985; Gubbins and Bloxham, 1987). The term "convective roll," however, is figurative for such a supercritical, low-viscosity, high-density fluid in an effectively infinite annulus, and I assume that vortical structures in the fluid of the core will in some respects resemble those seen in Jupiter's atmosphere and in the experiments of Sommeria et al. (1988). In that case, as in Jupiter's gaseous atmosphere — and even though the viscosity of a gas tends to increase rather than decrease with increasing temperature, hence with increasing shear rate (cf. Note P.5) — we must be dealing with a chaotically driven system of complex flows capable of sustaining a variety of quasi-stable vortical singularities that depend on rates of fluid transport that are analogous to experimental pumping rates (cf. Gubbins, 1977; Loper, 1978; Olson and Hagee, 1990; Hagee and Olson, 1991; McFadden et al., 1991; Gilbert, 1991; Bogue and Merrill, 1992; Zhang and Gubbins, 1992; Galloway and Proctor, 1992; Weiss, 1993).

Sommeria et al. (1988) showed that even though the general vortical motion was chaotic (turbulent in the classical sense), ordered sets of vortices were generated in a tunable manner that is somewhat reminiscent of the numerical resonances of the sine-circle map previously discussed (cf. discussion of Figs. 14, 20, and 21). Thus, vortical singularities (resonances) in stratified flow within the fluid layers internal to a planet figuratively mimic satellitic resonances external to the planet. Hence, they may be ordered according to similar sets of winding numbers (see Table 1, Fig. 14, and Notes 3.1, 4.1, and 7.1). The (1/1) resonance implies an approximately constant longitudinal position of a single vortex, or perhaps paired sets of vortices (?), whereas higher or lower resonances may imply multiple vortices moving either backward or forward relative to the main flow. If the Great Red Spot (GRS) of Jupiter actually rotates at the same angular velocity as the planet as a whole, it would apparently correspond to the (1/1) resonance state, leaving the GRS seemingly motionless as one gigantic vortex (diameter $\cong$ Earth's circumference) in stroboscobic images taken at the period of Jupiter's rotation. There are, however, adjacent systems of numerous large secondary vortices interacting dynamically with the GRS that may represent more complex resonances (cf. Ingersoll, 1983, p. 65; Gierasch, 1991).

Small departures from either an equal-angular velocity of a marker or an equal-and-opposite-angular velocity of a marker are ambiguous, because very long times are required to determine actual deviations from idealized internal (viewed from inside the planet) or external (viewed from outside the planet) synchronizations (cf. Note 7.1). For example, if the westward drift of Earth's magnetic field ($\sim 0.18°$ yr^{-1}) represented a resonance state, the resonance ratio would be very small (or large, depending on perspective). From the point of view of an observer inside the Earth, it takes about 2000 years (730,000 rotation cycles of the Earth) to recover the same configuration between a marker on the Earth's surface and a marker moving with the mean angular velocity of the flow. Analogously, if we were to view the westward drift from a "fixed" position outside the Earth (i.e., as though we were "sidereal" observers) but only plotted the position of the marker moving with the flow every 2000 years, then that marker would appear to be nearly motionless to us. In nonlinear dynamics such a stationarity is sometimes called a "stroboscobic" portrait, representing a form of Poincaré section of the phase portrait as a whole (cf. Note P.3).

Singular dissipative systems of this kind are sometimes called *pumped systems far from equilibrum*, and are epitomized by the *optical laser* and/or the *gaseous maser* (see Note I.5; cf. Notes P.1 and P.5). Interactions between the pumping rate and the slaving of internal energy levels govern the states of nonlinear periodicities and/or chaotic bands (e.g., Haken, 1978). The most obvious source of pumping for the core flow (lacking either strong radiogenic or magnetohydrodynamic dissipative heating) is the radial mass flux from the presumably solid inner core, which is governed by its solidification rate and release of latent heat coupled with the thermochemical buoyancy of the fractionated outer core fluid, balanced against the rate of heat removal across the core-mantle boundary (cf. Gubbins, 1977; Loper, 1978). The radial geometry would also resemble the experimental conditions of Sommeria et al. (1988), wherein fluid was both introduced and removed through holes in the bottom of the rotating cylinder at different radial positions (in the outer core of the Earth, these flows would be analogous to the inward and outward mass exchanges owing to fractional crystallization mechanisms in the inner core as it responds to heat losses to the mantle; discussions of heat-transfer mechanisms in the core vis-à-vis core-mantle processes are given, from several different perspectives, by Vogt, 1975; Gubbins, 1977; Loper, 1978; Walzer and Maaz, 1983; Nicolaysen, 1985; Loper and McCartney, 1986; Gaffin, 1987, 1989; Jacobs, 1987, 1992a, b; Morelli and Dziewonski, 1987; Jeanloz, 1990; Olson and Hagee, 1990; Hagee and Olson, 1991; Knittle and Jeanloz, 1991; Larson and Olson, 1991; Zhang and Gubbins, 1992).

Since the geometries of magnetic flux tubes and fluid flow can differ, the resultant of the radially oriented mass pumping rate could be magnetically biased in either the northerly or southerly direction (e.g., by helical flows; cf. Clement, 1987), wherein the polarity structure will be determined by the net symmetries — stable or unstable — of the distribution of normal and/or reversed polarity domains (I refer here to positive/negative transport domains in spin-glass models of electromagnetic phenomena, and in analogous periodic-chaotic mechanical and reversing-flow systems; e.g., Figs. 18 and 29; cf. Moore and Spiegel, 1966; Robbins, 1977; Gorman et al., 1984; Rollins and Hunt, 1984; Carroll et al., 1987;

Stein, 1989; Tritton, 1989; Widmann et al., 1989; Ott et al., 1990; Gilbert, 1991; Singer et al., 1991; Bogue and Merrill, 1992; Galloway and Proctor, 1992; Weiss, 1993). In general, the magnetic flux is constrained by core flow and the conductivity distribution in the mantle, but not by the core-mantle interface per se, whereas the mass flow in the core is constrained by heat and mass transport across the core-mantle interface and by conservation of mass and angular momenta in the core, resulting in a chaotic distribution of vortical spins with locally diverse orientations (and components of electromagnetic transport) but with a characteristic mean angular velocity relative to the core flow as a whole (cf. Buffett, 1992; Buffett et al., 1992). Studies of the magnetic geometry of sunspots, for example, demonstrate the chaotic heterogeneity of magnetic orientations vis-à-vis hydrodynamic flow (cf. Weiss, 1993) — at the same time illustrating the coherence (e.g., the pupil-iris-like, umbra-penumbra, morphology) that can result from chaotic patterning in complex coupled-oscillator systems, such as occur in magnetohydrodynamic and chemical-reaction flow processes (cf. Swinney and Roux, 1984; Shaw, 1986; Sommeria et al., 1988; Scott, 1991). These sorts of chaotic biasing thus reflect, and affect, the oscillatory hemispheric vorticity balances, somewhat like the biasing of chaotically reversing flows in thermosyphon loops, where the spatiotemporal signs of the net rotations both intermittently oscillate and intermittently reverse over similarly short time scales (e.g., Fig. 29; cf. Moore and Spiegel, 1966; Robbins, 1977; Gorman et al., 1984; Tritton, 1989; Widmann et al., 1989; Ott et al., 1990; Shaw, 1991, Figs. 8.21–8.27; Singer et al., 1991).

An example of a chaotic bistable fluid oscillator is shown in **Figure 29**, which is modified from Singer et al. (1991). This experiment illustrates an example of chaotic fluid oscillations observed in simple convection loops (cf. Creveling et al., 1975; Shaw, 1991, pp. 122ff) and in their artificial suppression by negative feedback (cf. Ott et al., 1990; Ditto et al., 1990b; Hunt, 1991; Shaw, 1991, Figs. 8.33–8.36; Moon, 1992). Self-induced negative feedback analogous to that of Figure 29 (bottom) occurs in nature as well as in the laboratory, as will be discussed later (feedback phenomena in general are discussed in Notes P.1, P.3, P.5, P.6, I.4, I.5, 3.3, 4.2, 5.2, 6.4, and 6.8). The fluid (water) used in the experiment of Singer et al. (1991) is only a little less viscous than that usually assumed for the core fluid (e.g., Merrill and McFadden, 1990, p. 346), although a very large range of "viscosities" of the outer core has been cited in the literature **(Note 7.2)**.

If it is imagined that the fluid in the experiment of Figure 29 is a plasma subjected to a magnetic field (a much more difficult experiment to control and to visualize; cf. the solid-state spin-wave experiments of Carroll et al., 1987), then the oscillations in the sign of the flow would be analogous to oscillations in the electrical current (and magnetic polarities) — appropriately rescaled in time — generated by the feedback between the electromagnetic field and the mass flow rate of the plasma **(Note 7.3)**. There is no single characteristic "delay time," or simple set of characteristic delay times — or, it might be said that there is an infinity (a spectrum) of delay times — in such a system (by contrast with a finite system of relaxation oscillators; cf. Shaw and Chouet, 1991, pp. 10,203ff), because the flow is driven to be perpetually poised, as it were, at the unstable "brink" of a distribution of transiently active *separatrices* (see Note 5.1) between multitudes of different *attractor basins* (e.g., Shaw, 1987a, Fig. 51.16d). That is, the system as a whole corresponds to a condition of self-organized criticality such that the direction of flow, and/or the magnetic polarity of the flow, can change instantaneously in magnitude and direction according to the constantly changing ensemble distribution of chaotic motions (vortical "spin states").

In other words, I am describing a form of *critically fast dynamo* (CFD) that is analogous to the spatiotemporal power-law distribution of avalanches in the sandpile model of critical self-organization (Bak et al., 1988; Bak and Chen, 1989, 1991; Bak and Tang, 1989; Chen and Bak, 1989; Kadanoff et al., 1989; cf. Bohr et al., 1984; Jensen et al., 1984; Gwinn and Westervelt, 1987; Shaw, 1987b, 1991; Olinger and Sreenivasan, 1988; Shaw and Chouet, 1988, 1989, 1991; Gilbert, 1991; Galloway and Proctor, 1992; Feudel et al., 1993; Weiss, 1993; and Notes P.1, P.2, I.2, I.4, 4.1, 4.2, 5.1, and 6.1 in the present volume). Thus, the "speed" of partial reversals of flow velocities and/or polarities (cf. Fig. 29) corresponding to excursions and/or transient reversals of the geomagnetic field — or of complete reversals of net convection-related polarity orientations corresponding to "stable" paleomagnetic field reversals and *polarity chrons* (Harland et al., 1990, pp. 140ff) — is constrained only by the critical-state balances among the "infinite" sets of local attractor states **(Note 7.4)**. On the time scale of geological events, therefore, the CFD model implies that there is no constraint on the speed at which the geomagnetic field can change configurations, including stable reversals. The CFD model thus offers a simple explanation for the "astonishing properties" of polarity transitions, and the very brief time scales (of the order of a year or less) over which large magnetic-field excursions apparently can occur within the Earth (doubts concerning such high rates of change are expressed by Merrill and McFadden, 1990, p. 347; but see Courtillot and Le Mouël, 1984, 1985, 1988; Mankinen et al., 1985; Prévot et al., 1985; Champion et al., 1988; Coe and Prévot, 1989; Courtillot et al., 1992; Thouveny and Creer, 1992). [Geographic trajectories of stable paleomagnetic pole positions (i.e., the mean positions of assumed *geocentric axial dipoles*, the "GAD hypothesis"; see the Introduction) — averaged over time spans $> 10^5$ years (see Butler, 1992, p. 161) — reflecting continental drift, therefore called *polar wander*, apparent or "true" (APW or TPW), are illustrated in **Appendixes 19–21**. Comparisons between APWPs and cratering patterns are explored in **Appendixes 22 and 23**; see the recapitulation given in the Introduction.]

If the net magnetic flux direction can switch between the Northern and Southern hemispheres in such a manner, then it probably is operating across the nonlinear-periodic/chaotic transition (see Notes 5.1, 7.3, and 7.4). The nonlinear-periodic/chaotic transition states, in turn, are precisely where interior (and other) chaotic crises, as illustrated in Figures 17 and 18, are likely to be particularly conspicuous. Therefore, resonances and instabilities in the vortical domain structures of magnetohydrodynamics would be expected to resemble those of nonlinear periodicities and spin-wave instabilities demonstrated in solid-state experiments. For instance, the experimental time series for YIG (yttrium iron garnet) in Figure 18b resembles that of the fluid in Figure 29, right down to local and global oscillations, reversals in current direction, and suppression by negative feedback (cf. Carroll et al., 1987; Shaw, 1987a, Fig. 51.18; Ott et al., 1990; Singer et al., 1991). The stabilization shown in Figure 18 followed upon the eventual locking of transient chaotic trajectories of electronic states to a stable quasi-periodic attractor (i.e., the shaded central, toroidal, region of Fig. 18a correlates with the relatively regular low-amplitude oscillations of Fig. 18b).

The stabilization of the chaotically oscillating and reversing flow in Figure 29b was artificially induced by a servo-device that countered the rate of positive feedback (Singer et al., 1991; cf. Carroll et al., 1987; Shaw, 1987a, Fig. 51.18; 1991, Figs. 8.36 and 8.37; Ditto et al., 1990b; Ott et al., 1990; Hunt, 1991; Garfinkel, 1992; Moon, 1992). Analogous inhibiting effects would be induced naturally in the fluid cores of the Earth and the other planets, and in the Sun and other stars (e.g., pulsars; see Note I.3), by the "destruction" of chaotic attractor basins (e.g., by *boundary crises*, or perhaps by the *blue sky catastrophes* of Abraham and Stewart, 1986) — or, in the Earth and planets, by negative-feedback control of radial mass flow responding to heterogeneities induced by inner-core crystallization, core-mantle heat losses, and magma transfer. Such inhibitions might correlate with aborted excursions or reversals. However, the "wilder" types of field excursions and "sudden" reversals may be responses to typical chaotic bursting actions analogous to those illustrated in Figures 17 and 18 (cf. Fig. 32), which in turn are analogous to the orbital chaotic crises that I have argued are responsible for delivery of some asteroids and/or comets to Earth-crossing orbits (see Figs. 12, 13 and 15, and Appendix 9; cf. Wisdom, 1985, 1987a, b, c; Olsson-Steel, 1987, 1989; Greenberg and Nolan, 1989; Nobili, 1989; Bailey, 1990; Torbett and Smoluchowski, 1990; McIntosh, 1991; Yoshikawa, 1991; Ipatov, 1992), and that may eventually lead to capture events by the Earth-Moon system (cf. Note 6.1). The cyclic similarities between the self-organized chaotic crises implicit in this comparative description come together (cf. Fig. 11) in the implications of the interface expressing impact-dynamic/tectonic-geomagnetic coupling in the Earth, as discussed in this and the following chapter (and reviewed in the Introduction).

Clement (1991, p. 56; cf. Bogue, 1991; Laj et al., 1991, 1992; Clement, 1992; Constable, 1992; Jackson, 1992; Langereis et al., 1992; Valet et al., 1992) discusses possible relationships between virtual geomagnetic pole (VGP) patterns and the invariance of the core flux postulated by Gubbins and Bloxham (1987), suggesting that both may be stable for longer than one field reversal and possibly over the time scale of mantle convection (a range of $\sim 10^6$ to $> 10^8$ years). If so, then — maintaining the general notion of critical self-organization — the VGP patterns, as well as the core-flux patterns, would correspond to characteristic sets of "vortical" configurations, for example analogous to the subsidiary configurations of vortices in Jupiter relative to the Great Red Spot (GRS). The GRS, by this

Consequences of the Celestial Reference Frame

comparison, is in turn analogous to the "stable" dipole modes of Earth's magnetic field (or higher-order modes equivalent to more than one "GRS") that may or may not be approximately aligned with the planetary spin axis (i.e., just as the "lighthouse" sources — corresponding to magnetic axes — of pulsars may or may not be aligned with the spin axes of the stellar objects within which they are generated; see Note I.3). In other words, the core-flux patterns, VGP paths, VGP patterns, and paleomagnetic polarity events (polarity intervals, subchrons, chrons, and superchrons; cf. Harland et al., 1990, pp. 140ff; Butler, 1992, pp. 210ff) all map a spectrum of self-organized structural configurations of the core dynamo — each, perhaps, with its own characteristic spectrum of lifetimes, within a general power-law progression of spatiotemporal structures, ranging in durations from what would be called unstable (transient) to metastable (quasi-steady) types of geomagnetic/paleomagnetic patterns and/or events (cf. Note I.7).

If this general type of structural invariance can be documented from the distributions of characteristic VGP patterns over many polarity intervals, the result hopefully can provide the needed tests of resonant relationships between (1) the core dynamo, (2) mantle convection (including plate-tectonic motions evidenced by polar-wander paths), and (3) the cratering process (CRF patterns), especially as these processes are reflected in, and/or mediated by, the role of magma generation and transport (including "hot-spot" patterns and their antipodes; see Appendixes 12–15), where global magmatism also represents a universal system of critical self-organization (Shaw and Chouet, 1991) that is coupled with all three (cf. Note 8.3). VGPs will play a crucial role in this research because they can be measured relative to diverse geologic settings and ages around the world with good spatiotemporal resolution, at least in some instances (e.g., besides the references cited above, VGP studies have been in progress during the past several years by Jonathan Glen at the University of California, Santa Cruz, and he has joined with me, beginning during the fall of 1992, in a collaborative study of whether or not patterns of CRF-geomagnetic coupling are demonstrable in the VGP patterns). Until such information is forthcoming, a partial test of this idea is given by comparisons of (1) temporal correlations between reversal frequencies and impact frequencies, as shown in Figures 6–8 and 30–32, (2) temporal correlations between reversal frequencies and volume rates of ocean-crust production, as shown in Figure 8 (and other correlations shown in Appendix 5), and (3) patterns of APWPs vis-à-vis cratering patterns (see Appendixes 22 and 23; cf. the recapitulation given in the Introduction).

Geomagnetic Reversals and Coupling Between Interior Crises of the CRF and the Core Dynamo

Relationships between frequencies of cratering (data of Fig. 7) and geomagnetic reversals (data of Merrill and McFadden, 1990, Fig. 1) over the latest 165 Ma are shown in **Figure 30**. Figure 30a gives the ratios of reversal frequencies to cratering frequencies as a function of age. Maxima represent high reversal frequencies, and minima represent high cratering frequencies, relative to the average trends of both records. Because it reflects the composite record, the approximate 30-m.y. peak-to-peak average period (the "winding number") is interesting. More interesting is the fact that this periodicity *in the ratios* implies alternation and/or "syncopation" of the dominating effects, whatever the source of their coordinations. Between about 165 and 120 Ma, the reversals dominate, whereas from about 115 to 65 Ma, the cratering dominates. Between 65 and 5 Ma, the alternating effects are relatively balanced (representing an approximate out-of-phase synchronization; cf. the neurophysiological chaotic synchronization phenomena described by Freeman, 1992, pp. 467ff).

The respective frequencies are plotted against each other in Figure 30b, where the open circles connected by dotted lines are readings taken at the same ages. The mean trend of the open circles might imply that reversal frequencies increase roughly as the square of the impact frequencies. In detail, however, this trend (dotted lines) oscillates. At ~5 Ma it is moving toward equality, and at ~65 to ~115 Ma the impact frequencies nearly equal or dominate the reversal frequencies. An indication of how the average *ratios* change during episodes of increasing rates in either frequency is given by readings taken relative to the minima in the respective records in long-term trends (Fig. 6). The average trend is shown by the filled squares in Figure 30b (the oldest pair of readings represents 95 Ma on the geomagnetic curve and 165 Ma on the impact curve, etc.), which yield a straight line with slope of about 1.7, about that of the mean trend of the unlagged readings. Between about 15 and 55 Ma, however, the unlagged readings nearly coincide with the lagged readings. Thus, during this interval the unlagged behavior tracks the lagged behavior, suggesting that the phase responses of the shorter-term oscillations ($\leq$ 30 m.y.) are roughly synchronized with (in nonlinear resonance with) the long-term oscillations (> 200 m.y.). However, Figure 30a indicates that the oscillations during this resonant interval are directly out of phase with each other (reversal frequencies and impact frequencies alternately dominate), with about 15 m.y. between relative dominances. Yet there are also intervals of time in which the frequencies vary together, as if they were mutually reinforcing.

Only the impact record in Figure 6 shows the longer cycle (> 200 m.y.), but Merrill and McFadden (1990, p. 347) point out that there are similarly long variations between superchrons, and that the preceding superchron occurred about 200 m.y. before the Cretaceous superchron (e.g., the Permo-Carboniferous reversed polarity bias superchron of Harland et al., 1990, Fig. 6.10, spanning the age interval ~320–250 Ma; cf. McElhinny, 1971). This places it, like the Cretaceous normal-polarity-bias superchron, at a time of relatively high cratering rates (i.e., compared to adjacent intervals of mean rates; see Fig. 6). The different polarity senses in these two superchrons is not relevant to the dynamical interpretation if the long-term intermittency reflects a general forcing function common to the cratering mechanism and the threshold conditions for reversals of either sign in the core dynamo, as discussed above (i.e., the condition of *neutral instability* — the "poised" state mentioned above and implicit in Notes 7.3 and 7.4—is equivalent in either the normal or the reversed polarity mode).

Apparently this curious out-of-phase synchronization behavior between impact frequencies and geomagnetic reversal frequencies in the temporal record would apply to both short-term and long-term oscillations (cf. Fig. 6 and Fig. 30; and McElhinny, 1971; Harland et al., 1990, Fig. 6.10). Such an ambivalent (or multivalent) yet self-similar form of coupled intermittency would seem to make sense only in the context of a more inclusive system of intermittencies in which other processes take part (e.g., Freeman, 1992, pp. 467ff). Internal to the Earth, a key process involves what I call the *global magma-tectonic oscillator.* According to the previously described evidence that the global cratering mechanisms and the geodynamo mechanisms constitute a generally coupled terrestrial oscillator system, it must follow that coupling between the global cratering mechanisms and the global magmatic mechanisms—which I have illustrated in terms of the global distribution of flood-basalt provinces relative to the patterns of cratering nodes and swaths (e.g., Figs. 25 and 26; see Appendixes 16–18, which give analogous data for "hot-spot" volcanism; cf. Duncan, 1991; Duncan and Richards, 1991; D. L. Anderson, 1990, 1991a, b; Anderson et al., 1992; Fukao, 1992; Ray and Anderson, 1992; Shearer and Masters, 1992)—also implies coupling between the magma-tectonic oscillator and the core dynamo. *Accordingly, the terrestrial system of cratering mechanisms, geomagnetic mechanisms, and magmatic-tectonic mechanisms (which include the spectrum of mantle-convection mechanisms) comprises a global system of multiple time-delayed coupled oscillators* (cf. Note 8.3; and Appendixes 12–26). There is a remarkable conceptual parallel between this system of oscillators and *neurophysiological coupled oscillators with delayed feedback* (cf. Skarda and Freeman, 1987; Freeman, 1991, 1992, pp. 467ff).

What is more, the intermittencies of the cratering record, the cratering nodes and swaths of Figures 1–5 (and Appendixes 1–4), and the celestial reference frame (CRF) in general are coupled with the Solar System hierarchical flux of near-Earth objects (cf. Fig. 11) that are delivered in an intermittent but quasisteady manner to the geocentric system of natural Earth satellites. The net involvement of the terrestrial processes with extraterrestrial processes, therefore, also constitutes a multiple coupled-oscillator system wherein the intermittencies of cratering, geomagnetism, magmatism, tectonism, and mantle convection all are extensions of the general cascade of self-organized intermittencies portrayed in Figure 11 relative to the nonlinear-dynamical periodic/chaotic resonance phenomena of Figures 20 and 21 (and Appendix 11). The corresponding general-systems behaviors obviously must also affect the intermittent modes of all other systems coupled with it, including cyclic variations of solar-terrestrial heat balances (hence glaciation, etc.), geochemical cycles, and the net influences of all these phenomena on biological processes. There is an obvious kinship in these ideas with other "whole-Earth" viewpoints **(Note 7.5)**, and—without debating the merits and/or meanings of the term "living system"—there are strong parallels between the self-organized, and coupled, cosmological-galactic-solar-terrestrial feedback processes described in this book and analogous biological feedback processes at both the micro- and macroevolutionary levels (cf. Notes 8.14 and 8.15). Actually, I can think of no better qualitative introduction to feedback-systems behavior and the implications of roundabout, time-delayed information flow in nature [analogous to the semantic and strange loops of Hofstadter (1979), the neurophysiological loops of Skarda and Freeman (1987) and Freeman (1991, 1992), and those of the present volume] than the popular essays by Lewis Thomas (1974, 1979).

Indications of multiple and even self-opposing periodicities are not mutually contradicting, but are typical of nonlinear intermittencies that are both products of and susceptible to interior crises and bursting phenomena among several coupled oscillators, as discussed earlier and illustrated in Figures 17, 18, and 29. The putative existence of crisis-induced intermittency in the relative records of cratering and geomagnetic reversal frequencies is tested in **Figure 31** by assuming that the 165-Ma and 95-Ma minima, respectively, reflect proximity to an unstable quasiperiodic or phase-locked region in the general forcing function of each process. The unstable periodic manifold of the attractor phase space of this forcing function (see Chap. 5), relative to a common driving mechanism for both the cratering process and the core dynamo, is phase-shifted with respect to impact frequencies and reversal frequencies by the coupling of the latter processes with other nonindependent geodynamic processes (such as the core-mantle influence on reversal frequencies coupled with mantle convection and heat transfer coupled with magmatism — where magmatism responds directly to the impact frequencies which, in turn, are coupled with the asteroid-cometary flux coupled with the stellar and Galactic influences on transport crises in the planetary hierarchy, as outlined in Figs. 11 and 21). [Alternatively, it might be assumed that each process is coupled with different mechanisms that *happen* to approach equivalent unstable periodic manifolds independently within an age interval of 70 m.y., and in such a way that their frequency signatures achieve similarity. But such a happenstance would imply a form of "neutral drift" that is precluded by the existence of tendencies for nonlinear coupling between processes, because any coupling implies organization, and organization precludes random drift.]

Such a complex chain of time-delayed coupled effects does not necessarily rule out direct coupling phenomena, such as a direct perturbation of the core dynamo by a large impact, even though the general pattern of Figure 30 indicates that when impact frequencies are high, reversal frequencies are low (e.g., such a direct effect might be folded in — meaning that the immediate effect of a large impact might participate in the instantaneous dynamics of the core by influencing the mean distributions of its configurational states in such a way as to suppress polarity reversals for some interval of time — or it might represent a form of "self-induced noise" to the net time-series patterns, or both). Suppression of reversal frequencies by direct mechanical perturbation of the core dynamo is analogous to the suppression of reversals by high thermal perturbation in the model of core-mantle interaction advocated by Sheridan (1983), Olson and Hagee (1990), Hagee and Olson (1991), and Larson and Olson (1991).

Ambiguities in existing models for the mechanism or mechanisms of perturbation of the core flow by which reversal frequencies can be suppressed is an important issue to which I will return in Chapter 8 (cf. Loper and McCartney, 1986; Bloxham and Gubbins, 1987; Bloxham and Jackson, 1991, 1992; Bloxham, 1992; Jacobs, 1992a, p. 91). Unfortunately, such ambiguities cannot be uniquely resolved in multiple oscillator systems subject to interior crises (see Chap. 5). A case in point is the spin-wave system studied by Carroll et al. (1987). If there are multiple chaotic attractor basins and multiple periodic windows, then there are also multiple sets of conditions whereby interior crises can be activated. In the vicinity of any of the hierarchies of unstable periodic attractors, the recurrence times between chaotic transients (bursting times) may be indefinitely protracted. Therefore, a suppression of the chaotic motion might be incurred under conditions of either increased or decreased forcing (tuning) of the system—a situation that corresponds to suppression of reversal frequencies by either increasing or decreasing the excitation. Such an effect can be readily visualized in the case of a critically self-organized flow, such as the sandpile model. If one taps an hourglass, the unperturbed spatiotemporal scaling is disrupted, and the disruption can be sustained by repeated taps. That is, the normal distribution of small to large avalanches, with power-law scaling of recurrence times, can be modified by light tapping in such a way that the larger avalanches do not recur at all; conversely, if the sandpile is near its maximum amplitude, a tap can prematurely induce an avalanche of large size. Thus, the timing of avalanches becomes a function of the frequency of tapping relative to the growth rate of amplitude.

If I assume that the typical dynamical state of the core corresponds to the critically fast dynamo described above, then the unperturbed distribution of reversal frequencies is analogous to the distribution of sizes and recurrence times of avalanches in the hourglass. The effects of coupling with other processes, then, are analogous to what happens when I tap on the hourglass. In this regard, either the direct mechanical effects of impacts or the direct thermal effects of enhanced heat loss from the core could be analogous to the tapping effect. But, if I expand the self-organized critical-state concept to include the tapping mechanism as being induced by other critical-state mechanisms of both larger (lower-frequency) and smaller (higher-frequency) relative scaling, then the system as a whole performs in the spatiotemporal power-law mode, in which there are both synchronizations and phase delays at all scales of the frequency spectrum. This is not good news from the standpoint of sorting out specific causes and effects, or from the standpoint of assigning unique frequencies to specific mechanisms. But if the implication contained, a priori, in my original introduction of the concept of universal critical self-organization—in all the myriad forms caricaturized by Figure 11—has any general validity (cf. Shaw, 1987b; Chen and Bak, 1989), then the present interpretation of the coupling of impact dynamics with terrestrial dynamics—in all *its* myriad forms—is merely a self-fulfilling prophecy.

The trends of temporal intermittency values (the times, τ, between bursts) in Figure 31 correspond to the reciprocals of the average reversal and impact frequencies measured relative to their respective proximities to the long-term minima at 95 Ma (reversals) and 165 Ma (impacts). The power-law trends of τ in Figure 31 are of a form similar to the experimental data of Ditto et al. (1989; cf. Grebogi et al., 1986, 1987c, Savage et al., 1990; Sommerer et al., 1991a, b; Sommerer and Grebogi, 1992). The spread of values is also analogous to the lognormal distributions of experimental data at constant frequency (Ditto et al., 1989, Fig. 4; cf. Note 8.13). The resemblance between power-law exponents (slopes) is an equally important criterion, because it reflects similar (and universal) fractal densities of recurrences in phase space (cf. Ditto et al., 1989, Figs. 6 and 7; Savage et al., 1990, Figs. 10 and 11).

Interior Crises in Numerical Studies of the Rikitake Two-Disk Dynamo

The Inset in Figure 31 tests the power-law relationship against numerical simulations of the Rikitake two-disk dynamo by Ito (1980) and by Hoshi and Kono (1988), which are summarized in **Figure 32**. Measures of τ (Fig. 31, Inset) are estimated indirectly, and crudely, from the diagrams and descriptions in Hoshi and Kono (1988). To that degree of accuracy, the power-law trend appears to be of essentially the same form as that exhibited in the other curves of Figure 31. This consistency can be explained by comparing the effects shown in Figure 32 with the more explicit measurements of Figures 17 and 18. Portions of diagrams that illustrate key relationships are schematically redrawn to highlight the nature of the crisis mechanism in a reversing dynamo model. Figure 32a shows the bifurcation diagram of Ito (1980), which maps the relationships between nonlinear periodic and chaotic regimes as functions of the scaling parameter k and the coupling parameter μ for the dynamo model of Rikitake (1958; cf. Bullard, 1955, 1978; Allan, 1958; Robbins, 1977; Chillingworth and Holmes, 1980). Numerical experiments by Hoshi and Kono (1988) elucidate (1) relationships between μ and reversal frequencies as functions of k (Fig. 32b); (2) the form of the attractor at constant k and μ (Fig. 32c); and (3) characteristic forms of chaotic responses in the net current oscillations and polarity reversals (Fig. 32d). It may be noted that the minima of the functions in Figure 32b mimic the geomagnetic minimum (Cretaceous superchron of Figs. 6–8), where variations in μ symbolize the time-dependence of driving parameters in the core dynamo. The polarity sense is indifferent to the nature of an interior crisis, which can occur anywhere along a trajectory in Figure 32c relative to unstable periodic manifolds in the vicinities of N or R. Unless these manifolds reflect potential wells of identical level and depth, however, one or the other mode probably dominates once chaotic bursting begins.

The vicinities labeled MER in Figure 32 represent the *minimum entropy region* of Ito (1980). I interpret the MER to indicate proximity to a set of periodic windows and/or quasiperiodic trajectories that are interior to the chaotic field and unstable relative to neighboring chaotic attractors. The effect will be the same for approaches to this vicinity (shown by dashed lines and arrows in Figs. 32a and b) from stable chaotic-attractor basins of possibly more than one type (chaotic regions typically have banded sets of periodic and chaotic attractors). In this sense, there is a strong resemblance to phase relations in the YIG spin-wave system of Figure 18. The form of the attractor in Figure 32c makes these relationships more explicit. It resembles the Lorenz attractor, which possesses many periodic states, including two stable points such as N and R (cf. Lorenz, 1963, 1964, 1979; Sparrow, 1982; Shaw, 1991, Fig. 8.24). At other settings the stable form will be one of many possible chaotic attractors (e.g., one like the *butterfly strange attractor* shown in Gleick, 1987, Color Plate facing p. 114; cf. Shaw, 1991, Fig. 8.24), relative to which one or many of the periodic states will be unstable. It is when one or more such states are approached, as in the MER region, that interior crises of the types shown in Figures 17 and 18 occur (i.e., crises that give intermittent bursts with power-law response times, as in Fig. 31, Inset).

The frequency curves in Figure 32b show two higher but distinctive minima that would represent other weaker (higher-frequency) interior crises. The unstable periodic manifolds in phase space in those vicinities are more distant, relatively speaking, from the respective stable chaotic attractor basins. No specific periodic bands are directly correlated with these combinations of k and μ in Figure 32a, but the righthand minima are near the high-μ periodic field, and the lefthand minima are near the low-μ periodic field. Assuming that μ is also a measure of the ratios of mechanical to electromagnetic time scales in the core dynamo, then crisis-induced intermittency is also expected for conditions where either one or the other is of short term. This expectation fits with the variety of time scales involved in the ambivalent couplings between dynamo action, other coupled geophysical processes, and impact frequencies outlined above. Although I use the term descriptively above, *ambivalence* is sometimes applied in the sense of an actual measure of the complexity of chaotic intermittency. As a measure, ambivalence is related to the term *frustration,* which in current technical usage is a descriptor of the competitive mechanisms by which complexity is generated in oscillator networks, spin glasses, structural glasses, neural networks, and social strategies (cf. Palmer, 1986; Kinzel, 1987; Shaw, 1987b, p. 1658f; Dewdney, 1987; Carroll et al., 1987; Stein, 1989; Suarez et al., 1990; Mosekilde et al., 1991).

Formal and common usages of the term frustration merge in the context of the modeling of social/psychological interactions (e.g., Dewdney, 1987). There, the formal meaning of frustration refers to a measure of the net adjustments on a potential surface, or *social landscape*, owing to the attractions, repulsions, and psychological ambivalences of a system of protagonists. In this type of modeling, the social context would consist of a mixed social gathering of persons who know one another to varying degrees and who display a representative spectrum of normal likes and dislikes and psychological/sexual attractions and repulsions. In such a context, social interaction produces a nonlinear-dynamical system of responses wherein clustering phenomena of various types take place (conversational pairs, groups, and so on, as at a cocktail party). The time-dependent dynamics of the evolving interactions, assuming that the "party" goes on long enough, would represent a nonlinear-dynamical form of *evolutionary landscape* that is analogous, in some respects, to the statistical theories of biological populations developed extensively by Sewall Wright (see Provine, 1986; Wright, 1986). A caveat concerning the modeling of psychological phenomena should be added in that, so far, applications of nonlinear dynamics in psychology and sociology have been exploratory in nature (cf. Shafto, 1985a, b; Dewdney, 1987; Zeeman, 1987; Alper, 1989; F. D. Abraham, 1990; Suarez et al., 1990; Abraham et al., 1991b). By contrast, however, medical applications of nonlinear dynamics are much more advanced, and numerous laboratory and clinical studies in physiology and neurophysiology have been undertaken (cf. Edelman, 1978, 1985, 1987, 1989; Mackey and Milton, 1987; Rapp et al., 1987, 1988; Skarda and Freeman, 1987; Glass and Mackey, 1988; Holden, 1988; Pool, 1989; Goldberger et al., 1990; Lewis and Glass, 1991; Freeman, 1991, 1992; Peterson and Ezzell, 1992; Larnder et al., 1992; Garfinkel et al., 1992; Andreyev et al., 1992).

Figure 32d shows both normal chaotic intermittency, which may be either stable or due to transient bursting at high frequencies, and a sudden spike-like individual burst, which may be an isolated event or one among a series of transient reversals (not shown) with a very large recurrence time (τ). The reader who has at least scanned every chapter of the present volume might have noticed that these forms of behavior have recurred in nearly every type of data presented, starting with the chaotic bistabilities of asteroid orbits (Fig. 12). Interior chaotic crises are universally characteristic of bistability and multistability in systems of nonlinear fluid-state, solid-state, thermochemical, and thermomechanical oscillations (see review by Shaw, 1991). In short, chaotic crises of interior and boundary types are characteristic of all systems capable of critical self-organization (cf. Notes P.2, I.2, I.8), and this trait appears to be universal in astrophysical and geophysical processes.

Universality can be arbitrarily demonstrated by experiment, but its ubiquity in the solar-planetary hierarchy (e.g., Figs. 11 and 21; and Appendix 11) is founded in the fact that chaotic crises participate in coupled critical phenomena that operate near the periodic-chaotic transition at every scale from the Galactic perturbations of the Oort Cloud ($\leq \sim 10^{12}$ km, $\leq \sim 10^9$ years) down to the subtleties of fast geomagnetic variations ($\leq \sim 10^3$ km, $\leq \sim 10$ years) and the even faster phenomena of seismic tremor triggered by magma transport in the mantle (Shaw and Chouet, 1988, 1989, 1991). Given a common or homologous relationship, a constancy of scale ratios is implied by power-law behavior and universality, but for nonhomologous processes, the ratios may differ, because for those processes the scaling may involve different combinations of the same sets of universality parameters (see the discussion of scaling parameters for magma transport in Shaw and Chouet, 1991, p. 10,205; cf. Note P.4).

CHAPTER 8

Critical Self-Organization, Universality, Chaotic Crises, and the Celestial Reference Frame

The opening two paragraphs of Chapter 7 apply as well to the present chapter, wherein I will discuss some of the more general geological implications of, and questions raised by, the *celestial reference frame hypothesis* (CRFH) and concepts of critical self-organization, universality, and chaotic crises. The nonlinear-dynamical approach has some obvious implications for paleogeography, biostratigraphy, and paleontology, as well as other implications that are not so obvious. These latter topics will be taken up following a résumé of implications concerning consistent global dynamic models suggested by nonlinear coupling among processes simultaneously involving mass impacts, magmatism, fluid-core motions, and geomagnetism. Again, I suggest that rescanning the Prologue and Introduction from time to time should help in establishing the general plan of the book, as well as where I have come from in this survey, and where I am trying to go with it in the concluding discussions of Chapters 8 and 9 and the Epilogue.

Following the initial description of cratering patterns on Earth in Chapters 1 and 2, I undertook a survey of the phenomenologies and complexities of the Solar System as the dynamical context within which the sources of impacting objects must be understood (Chap. 3). This was followed by consideration of orbital evolution in the Solar System, which led to the concept of intermittent reservoirs of near-Earth objects (NEOs)—derived from the Solar System flux of asteroids and comets schematically portrayed in Figure 11—including a system (Chaps. 4 and 5) of natural Earth satellites (NESs) as a *proximal source*, or *holding pattern*, of Earth-impacting objects (a configuration analogous to the holding patterns in the vicinities of busy airports, which, like the reservoirs of NEOs, vary in their traffic patterns and in their congestion). Nonlinear-resonance phenomena within the Solar System, NEO reservoirs, and NES reservoirs were then evaluated (Chap. 6) from the standpoint of discovering an organizing principle that is sufficiently precise to provide geographic control of the pattern of Phanerozoic cratering nodes and swaths of Figures 1–5, a pattern that I have called the *celestial reference frame* (CRF). In Chapter 7, then, I began to consider the permutations of geophysical effects of the CRF pattern on Earth history, emphasizing correlations with flood-basalt magmatism and processes in the fluid core that relate to geomagnetic and paleomagnetic signatures in the Earth. In the present chapter, this progression is expanded to consider the ramifications of impact-related terrestrial effects within a broader spectrum of geological processes.

If one stands back a bit now to contemplate the cyclic dynamical influences on Earth history, a larger pattern may begin to come into focus. In the broadest sense—viewing the Earth as the locus of a system of coupled astrophysical-geophysical oscillators—perhaps it can be seen that the role of mass impacts represents a dynamical interface between a system of *convergent* astrophysical processes and a system of *divergent* geophysical processes. The astrophysical phenomena are convergent in the sense that they represent an extremely coherent focusing effect in their ability to deliver impacting objects from a vast, effectively infinite, spatial context (the Solar System, including the Oort Cloud, and perhaps even the Galaxy, as the sources of potential impactors) to a relatively infinitesimal landing site, the Earth. Conversely, once the focused effects of a mass impact, or of a shower of impacts, have occurred on the Earth's surface, the geophysical consequences are divergent in the sense that they propagate out in both their areal and radial influences to become engaged with the global system of geodynamic processes. Engagement with that system, however, does not put an end to the impact-dynamic influence. The affected geophysical processes themselves act subsequently in both divergent and convergent ways, in much the same manner as do the behaviors of attractors (*strange attractors* and *periodic attractors*, respectively; see Note P.3), both with respect to mutual interactions of their cyclic effects and with respect to their respective couplings with subsequent mass impacts. The result is a system of intricately *folded feedback* (Shaw, 1987a, Fig. 51.14; cf. Notes P.1, P.3, P.5, P.6, I.5, 1.2, 5.2, 7.3) in which the global system of mass impacts is *implicated* (literally) with the global Earth process. The *cyclic stretching and folding* of the generalized dynamical "mixing" phenomena constitute the essential kinematic nature of the nonlinear-dynamical process in general (e.g., Rabinovich, 1980; Ott, 1981; R. Shaw, 1981; Crutchfield et al., 1986; Abraham and Shaw, 1987; Shaw, 1987a, b, 1991; Ottino, 1989; Gilbert, 1991; Ottino et al., 1992; Weiss, 1993).

According to this overview, the Earth's surface (and near-surface; especially the envelope comprising the volcanogenic/tectonogenic "sphere," atmosphere, hydrosphere, and biosphere) represents a unique type of *cosmological interface*—that is to say, an interface between the nonlinear-dynamical evolution of a planet and the nonlinear-dynamical evolution of the planet's solar-galactic environment. This type of interface is generic and may exist within any spacetime locus of the universe wherein the analogously intricate windings—as in the recursive windings illustrated by circle maps (cf. Fig. 14 and Appendix 9), and by the intricate structures produced by irreversible chemical-feedback reactions (see Note P.7 on the candle as prototype example of *chemical chaos*; cf. Note P.5; also Swinney and Roux, 1984; Argoul et al., 1987; Tam, 1990; Scott, 1991) — of the *cosmodynamo* have reached an equivalent, but not necessarily homologous, degree of advancement (cf. Prigogine et al., 1972; Lima-de-Faria, 1988; Eigen, 1992). Earth's cosmological interface, as it were, is caught in a convergent/divergent *cosmic pincers* of cyclic, but intermittent, recursive singularities — where "singularities" refer to spectra of critically self-organized periodic/chaotic attractor states that are fed energetically and catalytically both from without and from within the membrane of evolving action.

This is the setting within which I view the instantaneous dynamics of impact-extinction phenomena vis-à-vis a more general evolutionary dynamics of geological-biological change. In this expanded context, the CRFH plays an integral role — rather than the role tacitly implied in most discussions of the impact-extinction hypothesis, where an impact represents only an instantaneous perturbation (event, or spike) as an independent cosmic device for catastrophic disruption of the status quo. According to the CRFH, the spatiotemporal record of all impact processes is doubly integrated, over geologic time and over the Earth's surface, so that the entire cratering history of the Earth is integrally involved with geodynamics, therefore with all other Earth processes. In other words, a single impact event is more than just a mechanism for the instantaneous and catastrophic tuning and/or resetting of what — up to then — would have been an unperturbed pattern of geological-biological evolution (i.e., many discussions of the impact-extinction phenomenon tacitly imply that there have been two dynamically different types of Earth history, the "normal one," and the normal one perturbed by offsets and restarts related to geologically instantaneous impact events). The CRFH is, by contrast, a persistent and pervasive *punctuational dynamics* operating at all levels and in all geologic processes, meaning that *all* environmental, ecological, and biological potentials have been coupled with the Earth's total record of impact-cratering events, regardless of event magnitudes and the repose times between impact events — and in this respect it is consistent with the general idea of *punctuated equilibria* in organic evolution (cf. Eldredge and Gould, 1972; Gould and Eldredge, 1977; Gould, 1982b, 1984a, b, 1987a, b, 1989b, c; Padian et al., 1984; Raup, 1984a, b, 1986a, b, 1988a, b, 1990; Raup and Sepkoski, 1984, 1986; Shaw, 1988a). I hope to build at least a circumstantial case for such a viewpoint in the following discussions of impact-dynamic coupling with several different types of geological processes.

In the following discussions of coupling between mass impacts and geological processes, loci of "direct effects" — unless specifically qualified — may refer both to the actual sites of impacts and to their antipodes (see Appendixes 16–18; and Schultz and Gault, 1975; Hughes et al., 1977; Hagstrum and Turrin, 1991a, b; Watts et al., 1991; Hagstrum, 1992).

Crisis-Network Resonances of External and Internal "Satellitic" Planetary Spaces

The general picture I have been developing describes a nonlinearly coordinated system of external objects (the CRF reservoir of satellites in spin-orbit coupling with the Earth, including, in an expanded context, other objects in nearby orbit-orbit resonances with the Earth and/or Earth-Moon system) that has been engaged — primarily via the agency of meteoroid impacts, but also through magnetospheric processes and radiation effects (e.g., solar and cosmic irradiances; cf. Milankovitch, 1969; Stacey, 1969, pp. 188f; Strahler, 1971, pp. 89f and 746ff; Woodard and Hudson, 1983; Ghil, 1985, pp. 348ff; Zirin, 1988, Chap. 2; Besançon, 1990, pp. 233ff and 1135ff; Cox et al., 1991; Sonett et al., 1991; Maran, 1992, pp. 136ff and 875ff) — in a systematic, if generalized, resonance relationship with Earth's magmatic, tectonic, and geomagnetic processes over most of the recognizable part of the geologic record. This system of resonances is universally coupled with that of the Solar System within the Galaxy, and, less directly, with that of the Galaxy within the context of supergalactic organization as well. The nonlinear dynamics of the Earth's core dynamo suggest that satellitic resonances, relative to the geodynamo as a whole, exist in forms analogous to those of the externalized resonances of the CRF system relative to the Solar System. Except for differences in the initial and boundary conditions of the motions, internalized resonances within the annular interval between the inner-core/outer-core boundary and the core-mantle boundary — involving the relatively inviscid fluid of the outer core (but see Note 7.2; and Smylie, 1992) — are analogous to the vortical resonances that are conspicuous, for example, in Jupiter's dense gaseous-aerosol atmosphere (e.g., Ingersoll, 1983, p. 66, 1988; Giovanelli, 1984, Plate III; Marcus, 1987, 1988; Atreya et al., 1989; Gierasch, 1991; Pasachoff, 1991, p. 212; cf. Toon et al., 1992), or that exist, by inference, in the internal dynamics of pulsars (see Note I.3; and Pasachoff, 1991, pp. 472ff; cf. van den Heuvel and Paradijs, 1988; Krolik, 1991; Blandford, 1992; Eichler and Silk, 1992; Lyne et al., 1992; Maran, 1992; Tavani and Brookshaw, 1992).

The similarities between the internal system of vortical motions "satellitic" to the inner core and the systems that exist in external reservoirs of impactors satellitic to the Earth, to the Earth-Moon system, and to the Sun-Earth system are remarkable. By hypothesis, these two generalized reservoirs of rotating mass singularities, one internal and the other external, represent systems of interacting resonances across the medium of intervening Earth layers (mantle, crust, and GOAM, the latter symbolizing the glacier-ocean-atmosphere-magnetosphere system of Earth's near-surface feedback interactions). The two reservoirs are analogous to two interacting systems of satellites coupled across a rheologically complex interval by processes of energy transfer involving impacts, magma generation, mantle convection, thermally pumped magnetohydrodynamic motions in the fluid core, and electromagnetically pumped thermal motions in the outer atmosphere. Thus, the Earth's outer (solid-Earth)/(fluid-Earth) interface — including (1) the GOAM, (2) the *magmasphere* (the upper-mantle and crustal layers that are tectonically modulated by magma distributions and transport phenomena), and (3) the coupled magma-hydrothermal interactions of oceanic and continental regions — is caught up in a *dynamical sandwich* between the mutually competing and/or mutually reinforcing external/internal satellitic reservoirs. In consequence, this interface is subjected to the effects of multiple coupled-oscillator systems, each with its own critically self-organized spectrum of intermittencies, as well as intermittencies related to interactions with all the other systems. At the nexus of these ambivalently competitive and cooperative effects is the biosphere.

The full ensemble process metaphorically resembles the chiming of a highly nonlinear hierachy of bells, each of which is coupled to all the others. Imagine, as well, that each of the bells is struck by a hammer suspended from a nonlinear pendulum that is also coupled, not only with the striking mechanism of each of the other bells, but also with another, more "primary," nonlinear pendulum (cf. Miles, 1962, 1984; Beckert et al., 1985; Gwinn and Westervelt, 1985, 1986; Tritton, 1986, 1988, 1989; Henderson et al., 1991; Matsuzaki and Furuta, 1991). Within this metaphor, the geocentric reservoir of impactors (the CRF source) is analogous to the primary pendulum for the impact ringing of the Earth. By the same system of couplings, however, the CRF source is resonantly coupled with the hierarchy of impacting objects in the Solar System that represent the ringing of the solar chimes by the Galaxy as the primary nonlinear pendulum. Neglecting geochemical and biological factors for the moment, we can likewise look upon the Earth and its reservoir of geocentric impactors as the primary nonlinear oscillator (pendulum) that activates a system of coupled magmatic and geomagnetic processes. In this hierarchical sense, these two processes taken together (magmatism/geomagnetism), as well as each of the processes considered individually, also represent systems of coupled nonlinear oscillators at successively lower levels of the hierarchy (e.g., in the magmatic branch of the hierarchy, the descending progression of pendulum-like oscillatory motions carries down to the scale of seismic tremor; see Aki and Koyanagi, 1981; Chouet, 1981, 1988, 1992; Chouet et al., 1987; Koyanagi et al., 1987; Chouet and Shaw, 1991; Shaw and Chouet, 1991).

Besides its role as the fluid portion of the pendulum-like mechanical oscillations of the mantle and crust of the Earth induced by mass impacts, magma also is the essential agency for heat transfer and convective coupling between the mantle and the outer parts of the solid Earth that are the loci of the impacts, just as the liquid-metal alloy in the outer core is the essential agency for heat transfer and coupling between the inner core, the vortex structures in the outer-core, the magnetic flux bundles at the core-mantle interface, the convective interactions between the outer core and the mantle, and perturbations at the core-mantle and/or inner-core/outer-core boundaries (cf. Jacobs, 1987, 1992a; Jeanloz, 1990). Besides the relatively slow perturbations of the core motions associated with heat-transfer phenomena, shorter-term perturbations might be incurred from a variety of mechanical instabilities, such as those reviewed by Young and Lay (1987) and

Jeanloz (1990); see also the discussion of hypothetical phase-change phenomena in Note 6.8 that might induce rapid density instabilities of the type proposed by Ramsey (1950). In addition to these effects, instantaneous perturbations of the core related to mass impacts themselves seem possible (see discussions of deformation waves and antipodal effects of impacts by Schultz and Gault, 1975; Hughes et al., 1977; and Watts et al., 1991). The phenomenon of impact-induced *spallation* of material at the core-mantle and/or inner-core/outer-core boundaries has been examined theoretically by Rice and Creer (1989), with the tentative conclusion that it is unlikely even for large impacts (a modification of their model, discussed later, however, suggests that spallation might occur if a more realistic mechanism of fluid-induced fracture were invoked in the theory).

Magma is coupled in both the short term and the long term with the geographic pattern of impacts as well as with the regional rates of impacting (e.g., the impact frequencies that are characteristic of — even though they are not as yet documented in sufficient detail to evaluate the local energy flux of — the respective cratering nodes and swaths of Figs. 1–5) through (1) the instantaneous effects of individual impacts and (2) the delayed effects of groups of impacts integrated over time within a given magmatic-tectonic setting. In some instances, such as in the examples of correlations between impacts and flood-basalt volcanism illustrated in Figures 24–26 (and in the correlations with "hot-spot" volcanism illustrated in Appendixes 16–18), individual impacts — or clusters of impacts that are closely coordinated in space and time — apparently have initiated the patterns of magmatic propagation that, in the literature, are sometimes called "hot-spot tracks" (cf. Hess, 1962; Wilson, 1963; Morgan, 1972, 1981, 1983; Molnar and Atwater, 1973; Shaw, 1973; Shaw and Jackson, 1973; Jackson et al., 1975; Jackson, 1976; Shaw et al., 1980a; Molnar and Stock, 1987; Hutchinson et al., 1990; Duncan, 1991; Duncan and Richards, 1991; Anderson et al., 1992; Fukao, 1992; Ray and Anderson, 1992; Shearer and Masters, 1992). [The history of the idea of "hot-spot tracks" (sometimes referred to as the Wilson-Morgan hypothesis; cf. Wilson, 1963; Morgan, 1971, 1972) relative to the evolution of the, now conventional, concepts of plate tectonics is admirably summarized by Glen (1982, Appendix C) — although it will be noted that some of the works just cited take exception to the simplistic interpretation of volcanic propagation as it was invoked by the Wilson-Morgan hypothesis; e.g., Shaw, 1973; Shaw and Jackson, 1973; Jackson et al., 1975; Shaw et al., 1980a; cf. Molnar and Atwater, 1973; Molnar and Stock, 1987).]

The mechanisms that are related to the chains of circumstances by which magmatic propagation in the Earth's mantle and/or crust has been induced or catalyzed by mass impacts has not been explained in any detail. Descriptive accounts that fit some aspects of my own viewpoints are given by Seyfert and Sirkin (1979, pp. 96f, 383ff, 464ff), Burek et al. (1983), Burek and Wänke (1988), Alt et al. (1988), Hagstrum and Turrin (1991a, b), and Stothers (1992). It seems evident to me, however, that whatever kinematic descriptions may be indicated by specific events of impact-induced magmatism, key steps in the dynamical effects that are subsequently manifested in magma generation and/or transport involve some combination of the following mechanisms: (1) dissipation of the kinetic energy of impact in the form of tectonic deformations and fracture in the Earth's crust and/or mantle, with or without significant partitioning of the impact energy into forms of magma generation, (2) activation of magma-induced fracture propagation in a manner analogous to the formation of ring-dike/cone-sheet complexes (cf. Anderson, 1936; Shaw, 1980) according to mechanisms of *fluid-injection extensional and/or extensional-shear failure cascades* (this being a general mechanism analogous to hydraulic fracture in which magma is the injecting fluid; see Shaw, 1980, Figs. 5 and 7; Shaw, 1987a, pp. 1368ff; Shaw and Chouet, 1991, Fig. 3 and pp. 10,202ff), and (3) the inception of thermo mechanical feedback cycles involving impact-induced tectonic deformations, gravitational energy releases associated with coupled magmatic-tectonic vertical motions (buoyancy instabilities), and shear melting.

Depending on the plate-tectonic setting of the impact event, magmatic propagation takes place according to a diversity of modes related to the local tectonic style of deformation and the intrusive/extrusive partitioning of the magmatic fraction (i.e., according to the balances of plutonic-volcanic igneous phenomena associated with magma transport in the vertical and horizontal directions; cf. Shaw and Swanson, 1970; Shaw et al., 1971; Shaw, 1973; Shaw, 1985a, 1987a, 1988b; Shaw and Chouet, 1988, 1991). In some instances, impacts may induce events of fracture-related magma transport from existing magma sources, while in other instances, the impact event is the initiating mechanism for the inception of magma generation in a locality where magma storage in the Earth's crust and/or mantle previously was not conspicuously developed. Short-term impact magmatism may be correlated with isolated igneous events, either plutonic (e.g., Sudbury, Canada; cf. Grieve et al. 1991) or volcanic (e.g., flood-basalt fields; cf. Alt et al., 1988; Hagstrum and Turrin, 1991a, b). The inception of tectonic rift systems—even plate-splitting ones—might be the result of single major-impact events (cf. discussions in Seyfert and Sirkin, loc. cit.; and back-cover illustration of Silver and Schultz, 1982), but it seems more likely that the rifting of continents and/or oceanic plates represents the culminations of prior magmatic inceptions that eventually gave rise to fully developed spreading centers according to the evolution of feedback cycles of the types outlined above.

The tectonic setting affects the longer-term magmatic consequences of impacts according to a variety of propagation styles, such as propagating linear island chains analogous to the Hawaiian-Emperor ridge system in the Pacific (e.g., Shaw, 1973; Shaw and Jackson, 1973; Jackson and Shaw, 1975; Jackson et al., 1975; Shaw et al., 1980a), and propagating nonspreading continental rift systems, such as the present-day East African rift or the Rio Grande rift of the western U.S., or the Precambrian Midcontinent Rift System of North America (cf. Hutchinson, 1990). Such evolving systems are reactivated in a repetitive manner over geologic time by coordinated systems of impact events, and only a select few of the evolving volcanic systems ever produces full-fledged continental rifting and the inception of new centers of truly oceanic sea-floor spreading. This conclusion is consistent with the remarkable similarities between patterns of impacts and patterns of "hot spots" on Earth and elsewhere in the Solar System [e.g., see the distributions of "hot spots" illustrated in Appendixes 12–18 relative to patterns of impacts on Earth (cf. Rampino and Caldeira, 1992; Yamaji, 1992; Stothers, 1993); this analogy is compared with the volcanic patterns on Io in **Appendix 24** (cf. Strom and Schneider, 1982; McEwen et al., 1985; Yamaji, 1991, 1992) and with impact patterns of other Solar System bodies in **Appendix 25** (cf. Carr, 1981; Head et al., 1981, 1992, Fig. 2f; Strom, 1987; Pike, 1988; Strom and Neukum, 1988; Kaula, 1990; Bindschadler et al., 1991; Greeley and Schneid, 1991; Trego, 1991; D. B. Campbell et al., 1992; Janle et al., 1992; Phillips et al., 1992; Schaber et al., 1992; Solomon, 1993)]. Stationary and/or propagating styles of magmatism all have nonlinear intermittencies with characteristic spatiotemporal correlations, as discussed in Shaw (1970, 1985a, 1987a, b; cf. R. L. Smith, 1979), Shaw et al. (1971, 1980a), Jackson et al. (1975), and Shaw and Chouet (1991).

Thus, I am proposing that coupled magmatic-tectonic-geomagnetic interactions are nonlinearly synchronized (resonant) with the cratering patterns, hence with the spatiotemporal characteristics of the CRF. Although I have not compiled a global database of temporal volcanic frequencies in the same form as the impact and geomagnetic reversal frequencies shown in Figures 6–8, 30, and 31, I anticipate — on the basis of correlations with the magma-production rates of Larson (1991a, b) shown in Figure 8 (cf. Appendix 5) — that they will show analogous relative variations of intermittencies (i.e., the long-term variations in global magma-production rates are analogous to the long-term variations in impact rates and geomagnetic-reversal frequencies).

In Figure 30, the ratios of geomagnetic-reversal frequencies and cratering frequencies show crude periodic oscillations that are sometimes qualitatively in phase with other geologic processes (e.g., subsidiary maxima of magma production, impact frequencies, and reversal frequencies occur at about 65–70 and 35–40 Ma, as shown by the data of Figs. 8 and 30a; note that the maxima and minima of the ratios in Fig. 30a depend on relative changes, and that the minima at 65–70 and 35–40 Ma reflect maxima in both terms, the relative increases in cratering frequencies exceeding those in reversal frequencies) and at other times are distinctly out of phase with each other, as well as with other geologic processes (e.g., during the Cretaceous normal superchron, when the reversal frequency was essentially zero, there were maxima both in global magma production and in the number of impacts). If the first of these two correlations had been considered generally valid, it might have been interpreted to suggest that near-surface magma production in the Earth is directly correlated with *both* impact frequency and reversal frequency. The same, however, is not true for the longer-term variation in global magma-production rate (with a maximum at $\sim 110 \pm 10$ Ma in Figure 8) that correlates with the minimum in the reversal frequency during the Cretaceous normal-polarity superchron and with two apparent short-term maxima in the impact rates [note that the long-term variation of impact rates (Fig. 6) has minima near ~ 165 and ~ 430 Ma].

Obviously, such limited correlations do not permit unambiguous conclusions to be drawn concerning the relationships, respectively, between (1) impact rates and the geomagnetic-reversal mechanism of the core dynamo, (2) impact rates and magma production in the mantle, and (3) magma production and delivery rates to the Earth's surface, or near-surface, in relation to heat-transfer mechanisms within the core dynamo and between the core dynamo and the deep mantle. It would be convenient, but simplistic and probably wrong, to conclude that short-term ($\leq \sim 30$ m.y.) peaks of geomagnetic-reversal frequencies occur when both impact rates and magmatic activity at the surface of the Earth are high, and that reversal frequencies are always low when long-term magmatic activity is high (e.g., Fig. 8 and Appendix 5; cf. Note 8.1). If the latter were true for times exceeding 100 million years (e.g., throughout the Phanerozoic), it would be tempting to conclude, as Larson (1991b) and Larson and Olson (1991) argue — on the basis of the theoretical work of Olson and Hagee (1990) and Hagee and Olson (1991) — that reversals of the geodynamo field are suppressed by high-energy transfer rates in the outer core (e.g., affected by major mechanical and thermal perturbations at the core-mantle boundary that signal, or are signaled by, what is happening at the Earth's surface). However, such a conclusion is not supported by the shorter-term variations mentioned above, where all three rates apparently can increase together over time intervals of $<$ 30 m.y.

The mechanisms of feedback coupling among impacts, magmatic cycles, and geomagnetic reversals clearly are multiple and complex **(Note 8.1)**. They begin with the direct effects of impact stress waves that radiate seismically to influence melting, radial displacements, and changes in moments of inertia that effect instantaneous changes of relative rotation speeds at the core-mantle boundary and the diffusion of vorticity pulses into the core flow. Such processes have been little explored, and in their inertial aspects resemble the mechanism proposed by Muller and Morris (1986; cf. Muller and Morris, 1989) for correlations between mass impacts and geomagnetic-reversal frequencies via coupling between the GOAM system and changes in rotation speeds. The resultant is directly analogous to the compaction effects discussed by Chao and Gross (1987) relative to the possible influence of earthquakes on changes in rotation speeds recorded at the Earth's surface from precise measurements of the length-of-day (l.o.d.); see the summary by Maddox (1988).

In the short term, therefore, impacts must be significant to moments of inertia, both because of the intense seismic-energy radiation of a moderately large impact event — which may be four orders of magnitude or so larger than that of the largest known earthquake, as discussed later — and because of the globally compacting effect produced by the upward displacements of magma that feed surface volcanism and are compensated by downward displacements of the unmelted solid portions of the Earth's crust, lithosphere, and deeper mantle. [The largest known earthquake has a moment magnitude of $M^* \cong 9.5$, as defined by Kanamori (1986), implying that the analogous moment magnitude of a large impact event is equivalent to an "earthquake" with $M^* \geq 13$.] Shaw (1980, p. 253) showed that the average present-day *volumetric* moment rate of global magma transport approximately compensates the present-day global *seismic* moment rate. This relationship appears to hold approximately both at the local scale in the Earth — as is the case in Hawaii (Shaw, 1980, pp. 239ff and Table 1) and even in the case of stope closure in deep mines (see McGarr, 1976) — as well as at the global scale, implying that the energy of tectonic motions is typically derived from the release of gravitational potential energy, and that the fraction of that energy dissipated at seismic frequencies is proportional to the buoyant energy released by magma rise. Earthquakes produced by mass impacts, on the other hand, represent the dissipation at seismic frequencies of some fraction of the kinetic energy of impact — reflecting the same inverse relationship between potential and kinetic energies that applies to the equations of orbital motions (see Note 6.1). The energy that drives the normal tectonic type of earthquake is analogous to the energy involved in changes of orbital distance, for an object of given mass, whereas an impact-related earthquake derives its energy from the velocity of the impacting object.

Impact-induced magmatism introduces large spikes or bursts into the normal tectonic-energy cycle, with a wide range of intermittencies, but with an $\sim$30-m.y. fluctuation in peak power, which is the mean value of one of the nonlinear frequency bandwidths of Hawaiian magma transport identified by Shaw and Chouet (1991, p. 10,205), as well as a characteristic nonlinear frequency maximum in a compilation representing many different types of geological processes prepared by Shaw (1988a, Table 1 and Fig. 25; see later discussion of Fig. 35 in the present volume). Correlations between impacts and flood-basalt magmatism have been discussed (Figs. 24–26), and possible correlations between "hot-spot" magmatism — represented by volcanic loci such as Hawaii, Yellowstone, and Iceland — and the direct and/or antipodal effects of impacts are illustrated in Appendixes 16–18. All of the above effects are transferred from outside in, so to speak, to influence core-mantle coupling and geomagnetic signatures **(Note 8.2)**.

Chaotic crises of many kinds are implicit in these geophysical-astrophysical resonances. In this sense, the correlation of the $\sim$110-Ma maximum in rates of ocean-crust production with the Cretaceous superchron (Fig. 8) represents a long intermission (τ) between intrinsic bursting times for coupled magmatism and global mantle convection. These intermissions correlate with the longer-term invariances of the core flow (flux lobes), VGPs, and intermittent cratering rates (the temporal CRF pattern of Fig. 6). Although the mantle flow cycle encompasses chaotic bursts of magmatism at higher mean nonlinear frequencies (e.g., corresponding to the mean nonlinear periodicity peaks near $\sim$65 Ma and $\sim$35 Ma seen in Fig. 8), the integrated magmatic distribution coupled with the delays of mantle convection control the cyclicity of geomagnetic superchrons, because they pace both the shorter-term and the longer-term cycles of integrated heat transfer from the core.

An integrated nonlinear function of this type would resemble in form that of the integrated curve of fictive biological diversity discussed below (Fig. 36a), where the ordinate would be replaced by nonlinear functions such as log {Σ

(volume of magma)}, *log* {Σ *(number of reversals)*}, and *log* {Σ *(volume of subducted lithosphere)*}, etc. The higher-frequency bandwidths would correspond to the shorter irregularities of such curves, while the lower-frequency bandwidths would correspond to the longer and smoother fluctuations. Recall that I am speaking of nonlinear frequencies analogous to the average winding number W of the circle map (Table 1, Eq. 5). In this context, the Phanerozoic record is too short for us to identify W for the lower frequencies of superchrons, global magma pulses, and impact frequencies (e.g., Fig. 8), hence mantle convection cycles. But by analogy with the theory of Galactic perturbations of average variations in the flux of objects from the Oort Cloud, W should be roughly equivalent to the frequency bandwidth of the Galactic circulation, which is identified with the rather poorly known and possibly irregular Galactic year of $\sim 225 \pm 25$ m.y. (cf. Innanen et al., 1978; Rubin, 1983; Duncan et al., 1987; Shaw, 1988a). Because even the age of the Earth is inadequate to establish such a low nonlinear frequency in any single process, we can rely only on the fluctuations of many different long-term processes to indicate the magnitude of this frequency bandwith, as outlined in Shaw (1988a, Table 1; cf. Williams, 1981), and/or on estimates made from extrapolations of self-similarities found in the same processes at higher frequencies, a procedure that would be made possible by the discovery of characteristic universality parameters (cf. Note P.4; and Shaw and Chouet, 1991). Within these uncertainties, the above correlations are consistent with both the direct and the integrated effects of coupling between (1) cratering magnitudes and frequencies and (2) geophysical processes induced initially at or near the Earth's surface. This reasoning is counter to a popular model in which mantle convection, core flow, and plate tectonics are all coordinated by thermal plumes or "superplumes" that, by hypothesis, emanate from the core-mantle boundary (e.g., Larson, 1991a, b; Larson and Olson, 1991; Sleep, 1992). The idea of discrete positive (upward-propagating) thermal plumes as the source of sharply defined trends of volcanic propagation at the Earth's surface (so-called hot-spot chains) is not supported by recent tomographic evidence on the thermal structure of the upper mantle, as illustrated by D. L. Anderson (1991a, b), Anderson et al. (1992), Fukao (1992), Ray and Anderson (1992), and Shearer and Masters (1992).

Energy Budgets and Straw Models: Self-Similar Cycles Coupled Between the Sun and Earth

What follows is an exercise designed partly to demonstrate the generality of extraterrestrial mechanisms and sources in Earth's energy budget, and partly to demonstrate possible redundancies and circularities in arguments concerning the primary driving mechanisms of global magmatism and tectonic motions. As the straw model, I will reiterate some of the factors governing the radial position and direction of propagation of magmatism in the Earth. I choose this focus because magmatism is, I assert, the sine qua non of all other geodynamical phenomena, including tectonism, mantle convection, glacier-ocean-atmosphere (GOA) dynamics, geomagnetism, and biogenesis (cf. Shaw, 1983a; and the Prologue and Notes P.5 and 8.3 of the present volume). Solar irradiance is necessary for life as we know it, but it is not necessarily essential to aphotic living processes. What *is* essential is the planetary circulatory system represented by magma and magmatic processes, as well as a diverse and heterogeneous bulk chemistry emphasizing diversity of representation in the periodic distribution of the chemical elements rather than particular chemical homologies, because it is the heterogeneity of subtle and recursive negentropic possibilities — for a given range of ambient temperatures and pressures — that is important, rather than a particular element or elements (cf. Prigogine et al., 1972; Cairns-Smith, 1985; Shaw, 1986; Lima-de-Faria, 1988; Eigen, 1992). Magma is both the symbol and the product of the dissipative process in the evolution of the planets and larger satellites (cf. Shaw, 1983a; and Notes P.5 and 8.3). The anthropomorphic analogy with the human circulatory system — wherein magma becomes the physiological medium and mechanism by which nutrients, thermodynamic states, tectonic states, and dynamical information are transported throughout the Earth's more solid body — would suggest that the magmatic process also represents the pulmonary system, the mechanism whereby both the ocean and the atmosphere were derived from earlier Mars-like and Venus-like states of the Earth and were subsequently modulated by coupled plate tectonics and volcanism, in synergetic conjunction with surficial geochemical and biological processes **(Note 8.3)**. In a protracted *Gaian* context, the magmatic system is the medium and the mechanism by which the evolution of the biosphere has maintained a communication with the Earth's interior via plate-tectonic processes (see Notes 7.5 and 8.3; cf. Margulis and Lovelock, 1986).

It is axiomatic that in any general system of critical processes such as those affecting global interactions in the Earth, in the Sun, and in the Solar System, a quasisteady state implies an approximate net balance among all relevant vectors in the phase space of the global attractor system, including the average power that drives the system (the pumping rate of the dissipative — hence irreversible — thermodynamic process; cf. Note 8.3). This simply says that the system is neither expanding nor contracting in phase space to a totally different global-attractor structure during the relevant spacetime frame of the power source (a principle of uniformitarianism relevant to globally interacting nonlinear systems). This means in turn that we are dealing with a sensibly steady and uniformitarian regime of a dissipatively pumped system far from equilibrium that "lives off of" its energy sources in a manner directly analogous to the life cycles of biological systems.

Dynamical models sometimes get stuck on what turn out to be irrelevant issues concerning dominant processes. The discussion of cosmological models by de Vaucouleurs (1970) is particularly instructive (see Note 4.2). Besides demonstrating that it would seem to be impossible to define an average density for the Cosmos (because the supergalactic clustering phenomenon has no apparent limit), he also showed that estimates of the age of the universe have been increasing exponentially, with a doubling time of about 16 years over the past 300 years! Note that this doubling time is about the same as I reported earlier (Chap. 4) for the discoveries of asteroids. This approximate doubling time apparently reflects a general growth property of astronomical studies. Alfvén's (1980, 1981, 1984) discussions of cosmic plasma and the role of matter-antimatter interactions is similarly informative with respect to the uncertainties of cosmological parameters. Nonetheless, textbooks in astronomy and astrophysics speak of the Big Bang, and of the age and mean density of the universe as matters of fact, subject only to uncertainties of measurements. There is a major difference between such a stance and one in which it is not possible, in principle, to make such measurements at all (astrophysicists, it would appear, are unwilling to concede that the universe cannot be put into a box so that we might measure its mean density — the way we can measure the mean density of a collection of particles in the laboratory — perhaps because then we would have also to concede that both its mass and its age are indeterminate). Analogously, terrestrial models, even before the advent of plate tectonics, have been immobilized by issues of existence, form, driving force, and dominant modes of circulation in mantle convection — as though, given the right data, and enough funding, the "right" model could be demonstrated. These are all — both the astrophysical and the geophysical types of measurements mentioned — important descriptive issues, but they are not unilaterally and once and for all decidable, as previous discussions of nonlinear dynamics in this volume have demonstrated.

Mantle convection, particularly as viewed through the lens of nonlinear dynamics, is far too complex for us to suppose we can identify one type of source or one dominant direction of circulation, just as the existences of relatively invariant sets of vortical structures in the outer core do not depend on only one direction of transport of heat and mass that must emanate either from (a) the inner core or (b) the core-mantle boundary — as though this were a problem for which only one of two possible answers could be right. The opening remarks by Larson (1991a) and the closing remarks by Shaw (1970) concerning Earth's heat engine are almost the same, from the standpoint of the general nature of global processes, yet the one infers the dominance of heat sources and heat-transfer mechanisms in the core, while the other makes the same inference for the outer few hundred kilometers of the mantle! From the standpoint of the processes emphasized in the present volume, neither inference can be valid — on general nonlinear-dynamical grounds — to the exclusion of the other. Furthermore, we now must include the additional role of mass impacts from sources external to the Earth, and this factor alone requires that we reconsider such issues. But, neither do impacts alone dominate other terrestrial processes, including the evolution of the biosphere.

Qualifications of the same general nature apply even to the Sun, where circulatory processes that correlate with sunspot cycles, solar prominences, and solar flares are not solely correlated with or spatiotemporally scaled by the clearly primary energy source of nuclear reactions in the Sun's deep interior. They apparently also depend to a significant degree on irregular — presumably chaotic — orbital motions of the Sun about the Solar System center of mass (CM), as demonstrated by Jose (1965) and Landscheidt (1983, 1988), and on the tides raised on the Sun's surface by the planets (Wood, 1972). Both phenomena have mean periods of ~11 years or so and are indistinguishable from the mean period of sunspot cycles (this period also resembles the orbital period of Jupiter, ~11.86 years, because of the balance of forces about the CM). This internal-external system of processes and periodic/chaotic resonances — involving what transpires on and in the Sun vis-à-vis what transpires within the Solar System as a whole — is directly analogous to the internal-external periodic/chaotic resonances affecting the Earth, where the forms of motion in the vicinity of Earth's fluid core are analogous to the processes in the vicinity of the Sun, and the orbital periodic/chaotic resonances of the Solar System as a whole are analogous to Earth's CRF system of satellitic objects. These two systems are of vastly different scales, but the interface of greatest interest to us from the standpoint of biological processes and survival — the Earth's surface and its interface with the rest of the Cosmos — is, in a qualitative sense, close to the locus of greatest action in both systems. That is, *the Earth's solid-fluid interface has taken the brunt of, therefore has reaped the benefit of, the most intense geophysical effects (both from the outside and from the inside) in the evolution of the Earth as a planet — and, likewise, the Earth, as principal member of the terrestrial planets of the Solar System, has taken the brunt of, therefore has reaped the benefit of, the most persistently focused astrophysical effects (both from the outside and from the inside) in the evolution of the solar nebula and planetary system (see Chap. 3).*

The issue of whether outside-in or inside-out (O-I or I-O) mechanisms drive the thermomechanical-thermochemical heat engine raises the long-standing and unresolved debate in geology concerning the relative dominance of endogenous or exogenous processes. I have not previously taken up that issue in this volume because neither one can be explicitly isolated in the hierarchical scaling of coupled processes in the Solar System. From the standpoint of an isolated cyclic process, however, a distinction can be made by arbitrarily specifying the source of power that is assumed to drive that portion of the system. I-O mechanisms can be made consistent with terrestrial circulations by assuming that there is a heat source within the core or at the core-mantle boundary that drives convective motions in both the core and mantle by magma generation and transport, and/or by simple diapiric upwellings of coupled buoyant mantle and fluid core. I-O mechanisms clearly exist in both geophysical and astrophysical problems and are important driving mechanisms for convection in planetary atmospheres and in the Sun. In such cases, it is possible to formulate nonlinear-dynamical models whereby chaotic intermittency — and therefore interior crises — persists without need of any other energy sources or interactions. For instance, given no proof that Galactic energy sources are involved, many of the nonlinear processes in the Earth and other planets that are discussed here might be labeled — to a certain degree of approximation — as endogenous to the Solar System (cf. Armbruster and Chossat, 1991). We have seen, however, that this assumption is not really valid relative to the Galaxy, nor is it really valid for internal motions within the Sun itself, even though such motions are driven primarily by hydrogen-fusion reactions.

The properties of the Earth's core are not known to an accuracy or precision that permits us to rule out significant radiogenic heating absolutely, or, for that matter, to rule out significant dissipation of mechanical energy due to cyclic tidal strains of low-Q materials (cf. Munk and MacDonald, 1960; MacDonald, 1964; Stacey, 1969; Shaw, 1970; Gruntfest and Shaw, 1974; Lambeck, 1980; Jacobs, 1987; Poirier, 1988). The usual assumption in recent discussions, however, appears to be that thermal states in the core have depended mainly on thermal storage of gravitational potential energy released during core formation, modulated by the thermal effects of fractional crystallization of the inner core coupled with the buoyant rise of less dense fractions from within the outer core to the core-mantle boundary (cf. Gubbins, 1977; Loper, 1978; Verhoogen, 1980; Jacobs, 1987, 1992a, b; O. L. Anderson, 1990; Jeanloz, 1990; Knittle and Jeanloz, 1991; Larson, 1991a, b; Larson and Olson, 1991; Smylie, 1992). But if we posit this as the main source that controls the core-mantle heat flux and the rise of mantle plumes, we eventually encounter a problem with regard to conservation of energy.

The radii of the Earth, total core, and inner core are, respectively, 6371, 3490, and 1221 km. These figures yield the following approximate volumes (km^3): (a) whole Earth $\cong 1.08 \times 10^{12}$, (b) mantle $\cong 0.91 \times 10^{12}$, (c) total core $\cong 0.18 \times 10^{12}$, (d) outer core $\cong 0.17 \times 10^{12}$, and (e) inner core $\cong 0.0076 \times 10^{12}$. Neglecting the comparatively small volume of the Earth's crust, the volumetrically significant volume ratios are: (a) (mantle)/(total core) = (0.91)/(0.18) $\cong$ 5.1:1, (b) (mantle)/(outer core) = (0.91)/(0.17) $\cong$ 5.4:1, and (c) (mantle)/(inner core) = (0.91)/(0.0076) $\cong$ 120:1. The ratios by mass would be reduced by one-third to one-half of these values for conventional density models (e.g., Jacobs, 1987; Jeanloz, 1990), but density models have large uncertainties (cf. O. L. Anderson, 1990; Duba, 1992; Smylie, 1992; Boehler, 1993).

Assuming an average rate for global magma production of 20 km^3 yr^{-1} for 4.5 $\times 10^9$ years yields $\sim 1 \times 10^{11}$ km^3 of magma (though that value of 20 km^3 yr^{-1} probably is too low by a significant factor, especially considering the speculative sequences of events and processes outlined in Fig. 33 attending and/or following the formation of the core of the Earth and the collisional formation of the Moon; cf. Drake, 1990; Ringwood, 1986, 1989, 1990; Ringwood et al., 1990; Harper and Jacobsen, 1992; Taylor, 1987, 1992). Assuming further that all of this magma was generated at or near the core-mantle boundary, and that all of that heat — given by the average heat capacities and latent heats integrated over the changes of temperature and pressure between the core-mantle boundary and ambient conditions at the Earth's surface — was removed from the core, each gram of magma represents $\sim 2 \times 10^{10}$ ergs of heat, or a total of $\sim 6 \times 10^{36}$ ergs of heat removed from the core. This is nearly equivalent to 10^9 years of heat flow at the Earth's present rate of $\sim 3 \times 10^{20}$ erg sec^{-1}, and is equivalent to $\sim 7 \times 10^9$ years of tidal dissipation at the present total rate of $\sim 3 \times 10^{19}$ erg sec^{-1}.

According to the usual model of Earth's thermal history — wherein at least half of the Earth's present-day heat flow comes from radiogenic heat production outside the core, and most of the rest of it represents heat stored in the mantle (e.g.,

Shaw, 1970; Verhoogen, 1980, p. 119) — the amount of heat given up by the core should be small compared to that given up by the mantle. If so, then the amount of heat loss from the core required to balance magma production over the age of the Earth is far too large (i.e., either the core would have been crystallized by now, or magma generation is a process that is characteristic of the normal heat-transfer processes of the mantle *without* significant contributions of heat from the core). The observation by Gubbins (1977, p. 463) that "Thermal convection is an inefficient way to generate (the) magnetic field and involves too high a heat," would appear to be consistent with the above conclusion. Similarly, Verhoogen (1980, p. 119) concluded that "A few years ago, the author was inclined to think that the core exerts a major influence on the mantle, that without core heat, convection in the mantle would be less vigorous than it is and possibly restricted to the upper part. This is not so clear anymore, for it is now certain, from the energy balance, that a large fraction of the Earth's heat must come from the mantle."

The model preferred by Gubbins (1977) is one in which the magnetic field is generated from the dissipation of the gravitational energy released by chemical differentiation in the core, a process that is more efficient and removes far less heat from the core — for a given field strength — than does a model based on strong thermal convection in the core. Analogously, Shaw (1969, 1970, 1973), Shaw et al. (1971), and Shaw and Jackson (1973) concluded that viscous dissipation in the upper mantle is a more reasonable and efficient means of generating magma than is the bulk transport of magma — with all of the attendant kinematic and chemical problems of intervening interactions throughout the mantle — from a source at the core-mantle boundary. The energy driving the dissipation in the models of Shaw and coworkers was the gravitational potential energy released by plate-tectonic motions, or the equivalent dissipation of energy by the solid-Earth tide. Stated differently, in the model of tidal dissipation proposed by Shaw (1970), Shaw et al. (1971), and Gruntfest and Shaw (1974), an overall thermal energy balance was achieved by considering the solid-state dissipation of tidal energy to be the heat source that generated the change of gravitational potential energy released by plate tectonics. In that sense, then, magma generation is simply the viscous loss term in a thermosyphon model of plate-tectonic circulations driven — at least in part — by tidal friction generated by the orbital interaction of the Earth with the Moon and Sun (cf. MacDonald, 1964; Shaw, 1970; Shaw et al., 1971). This model has the appeal of reconciling the endogenous and exogenous aspects of the thermal evolution of the Earth by integrating (1) the internal dynamics of magma generation in the Earth, (2) the dynamics of plate-tectonic motions, and (3) the dynamics of the orbital-rotational evolution of the Earth and Moon in the Solar System. The energy dissipated by mass impacts over the same lifetime amplifies the overall magmatic-tectonic energy balance, as discussed below (cf. the Prologue, and Notes P.5 and 8.3). Although the energy dissipation rates of mass impact *events* dominate all other sources, the impact-energy dissipation rate *averaged over the Phanerozoic* is much smaller that the tidal-tectonic dissipation rate just discussed. The thermal evolution of the early Earth was — in an analogous but grossly different proportional sense — determined by balances among (1) the energies of mass impacts (including the aftermath of the collisional formation of the Moon; see the later discussion of Fig. 33), (2) core formation (constraints on what might have already happened to and/or within the proto-Earth are not sufficient to rule out the possibility that core formation occurred, entirely or in part, before, more or less during, or/and after the Earth-disrupting collision; cf. Taylor, 1987, 1992; Drake, 1990; Ringwood, 1990; Ringwood et al., 1990; Harper and Jacobsen, 1992), (3) tidal dissipation related to the early evolution of the Moon's orbit (an unavoidable but uncertain process in view of the collision-related processes and uncertainties), and (4) early radiogenic heating by the shorter-lived isotopes. All four of these inputs of thermal energy would have been numerically much larger than the source terms in models of the Earth's later thermal evolution (e.g., during the late Precambrian and Phanerozoic eons), hence the uncertainties related to that stage of Earth's thermal evolution are — by the principle of sensitive dependence on initial conditions, if for no other reason — impossible to assess.

Unless there have been large contributions from a greatly superheated core and/or from radiogenic heating, the short-circuiting of magmatic heat to the Earth's surface by efficient mantle plumes and "superplumes" would seem to require the production of a prohibitive amount of heat from the volumetrically very small — in relation to the planet as a whole — core of the Earth. Core heat loss does represent some part of the net energy balance, but sole reliance on it as the dominant energy source driving global magmatism and plate motions seems to be both misleading and unnecessary. Caveats, of course, are many, in view of the large uncertainties in our knowledge of core properties. Current studies, though narrowing the range of some uncertainties (e.g., Jeanloz, 1990), have introduced additional ones related to the role of the D'' layer (read "D double-prime layer"), to the rheological properties of the inner and outer core, and to new or revised chemical compositions and equations of state of core materials (cf. Jacobs, 1987, 1992a, b; Poirier, 1988; O. L. Anderson, 1990; Jeanloz, 1990; Knittle and Jeanloz, 1991; Duba, 1992; Smylie, 1992; Boehler, 1993). In short, as things stand, the idea of controlling thermal states in the outer part of the Earth by "central heating" furnishes the wrong leverage relative to calculations of ordinary heat-transfer processes in the mantle and crust. Conduction-dominated cooling of the deep interior of the Earth over the latest 2 or 3 Ga of its history would be so shortened by the presently postulated mechanisms of thermal plumes and superplumes that these mechanisms contravene the longevity demanded by the thermodynamic states of the core. This seeming dilemma is analogous to the one represented by Kelvin's early attempt to establish the age of the Earth by means of heat-transfer calculations (see Dalrymple, 1991). In either case, the Earth's and/or the core's lifetimes are much too short.

The issue of I-O (inside-out) vs. O-I (outside-in) dominances in Earth's thermal history has been further confounded by recent models for the formation of the Moon that call for an early collision of a Mars-sized planetary object with the

Earth (e.g., an object somewhat larger than the residual planetesimals predicted by Stern, 1991; cf. Ip, 1989). Such a collision not only produced enough heat, by some models, to entirely melt the Earth but left a legacy of large satellitic objects in equatorial orbits that continued to bombard the early Earth with craters dwarfing those of its later history. Proponents of such a model presumably would argue that Earth's early magmatic history was dominantly an O-I process. [Related phenomena are discussed in later chapters here and in papers by Runcorn (1983, 1987), Wetherill (1985, 1990), Benz et al. (1986, 1987, 1989), Ringwood (1986, 1989, 1990), Stevenson (1987); Benz and Cameron (1990), Ringwood et al. (1990); cf. Daly (1946), Hartmann and Davis (1975), Baldwin and Wilhelms (1992), and Spera and Stark (1993).]

If we neglect the earliest stages of Earth history, the same considerations that were applied to the core in support of global magmatism can be applied to O-I mechanisms. These energy resources include all of those mentioned above plus: (1) radiogenic heating concentrated within the upper 100 km or so of crust and mantle, (2) the net dissipation of tidal processes, representing about 3×10^{19} erg sec^{-1} (Shaw, 1970) distributed within the glacier-ocean-atmosphere-magnetosphere (GOAM) system plus solid Earth (thus coupled with radial distributions of angular momentum by mechanisms analogous to those discussed in the model of Muller and Morris, 1986, 1989), and (3) the cumulative energy of all external sources of intermittently small and large magnitudes, including the energies of mass impacts, solar and cosmic radiation, tidal energy, etc.

An example of a surficial global effect that is usually, if unwisely, neglected in the solid-Earth energy budget (I shall also be discussing the cumulative energy of impacts) is the electromagnetic coupling between the Sun and the Earth (cf. Alfvén, 1984). This is not a trivial power source for near-surface processes in the Earth, and it is even more interesting with regard to coupled oscillations at different scales, as discussed above relative to coupling between the internal and external "satellitic" systems. For example, a chain reaction, or energy cascade, is established by the following sequence of cyclically fluctuating processes: sunspot cycles → solar flares → solar wind → magnetosphere → ionosphere → atmosphere → GOA → tectonosphere → magmasphere → core dynamo → magnetosphere. The coupling with the solar wind (and to the phenomenon of *coronal mass ejections*; cf. Appenzeller, 1992; Hoecksema, 1992) produces intermittent chaotic bursts of power as high as the order of 10^{20} erg sec^{-1} into the upper atmosphere (cf. Friedman, 1983) — and, geologically speaking, these are frequent events. [The model of Muller and Morris (1986, 1989) concerns portions of this circuit coupled with an input from the aftereffects (*impact winter* and *greenhouse warming*) of mass impacts (cf. O'Keefe and Ahrens, 1982; Toon et al., 1982; Gaffin, 1987, 1989; Schwarzschild, 1987; Schneider, 1988; Clube and Napier, 1990). The possibility of nonlinear-dynamical coupling between the external magnetic field and the core dynamo also was suggested by Chillingworth and Holmes (1980, p. 56).]

Sunspot activity fluctuates with an average period of ~11 years within a range of 7–17 years (Landscheidt, 1983), where the reciprocal of the average period corresponds to the *winding number* ($W \cong 0.09$ yr^{-1}), recalling that W is the frequency of an intermittent signal averaged over a large number of cycles (cf. discussion of Fig. 14). The average period of the sunspot cycle is similar to the average period of the chaotic circulatory motion of the Sun about the Solar System center of mass illustrated in Appendix 7A (see Jose, 1965; Landscheidt, 1983, 1988). The internal circulatory motions in the Sun at this frequency also appear to be modulated by planetary tides (Wood, 1972). In this and other ways, the possible mechanisms for temporal variations of sunspot activity are directly analogous to the phenomena of interior crises discussed earlier in this chapter and in Chapter 5, especially with regard to geomagnetic-reversal frequencies (see Fig. 31). For instance, the internal circulatory instabilities related to the Sun-CM motion — slowed by a factor of about 10^6 — are analogous to the Earth's magmatic and paleomagnetic instabilities (see discussion of Fig. 30 above) that, on average, fluctuate almost synchronously with crude average periods of 15-m.y. and 30-m.y. (cf. Mazaud et al., 1983; Negi and Tiwari, 1983; Raup, 1985; Stothers, 1986; Mazaud and Laj, 1991). There are also essentially contemporaneous fluctuations in Earth's weather patterns relative to the solar cycles and orbital motions, and even in correlations with volcanic instabilities (cf. Wood, 1972; White 1977; Sergin, 1980; Benzi et al., 1982; Landscheidt, 1983; McCormac, 1983; Ghil, 1985; Peltier, 1987; Rampino et al., 1988; Stothers, 1989, 1993; Friis-Christensen and Lassen, 1991; Berger et al., 1992; Winograd et al., 1992), where the scale differences in the timings of analogously related instabilities are functions of nonlinear-dynamical universality (e.g., the temporal range of nonlinear-dynamically correlated volcanic instabilities in Hawaii spans the incredible frequency range of roughly fifteen orders of magnitude; cf. Shaw, 1987a, Fig. 51.8; Shaw and Chouet, 1991).

In addition to these analogies between terrestrial and solar activities, a quiet interval of very low sunspot activity in the historical record of sunspot cycles — lasting about 70 years, between roughly 1645 and 1715, referred to as the *Maunder Minimum* after its discoverer E. W. Maunder in 1890 (Eddy, 1977, Figs. 2 and 4) — is analogous to the Cretaceous normal superchron in the history of cycles of geomagnetic-reversal frequencies (e.g., Figs. 8 and 31; and Appendix 5). The Maunder Minimum in sunspot activity, therefore, is analogous to the nearest unstable longer-term periodic mode of a chaotic solar attractor related to the sunspot process (i.e., a mode with a frequency near that of the null intervals, the analogous "sunspot superchrons"). According to this nonlinear-dynamical argument, variations in sunspot activity — as expressed by *sunspot numbers* (calculated from a normalization formula developed by A. R. Wolf in 1858 to correct for differences in spot counts and spot-group counts between different observatories; Zirin, 1988, pp. 303ff) — represent transient chaotic bursting phenomena manifested by an average sunspot recurrence time per sunspot cycle, τ (see Chap. 5

and Fig. 31), where τ is maximized near the critical frequency (the "sunspot-superchron" frequency), meaning that the number of sunspots and sunspot groups $\to 0$ in that vicinity (**Note 8.4**).

In the Sun, there are an unknown number of unstable periodic manifolds (**Note 8.5**). This means that there may be transient chaotic bursting in the vicinity of many different critical frequencies, and this multiplicity in and of itself influences the accessibility of more general chaotic attractor states to different critical frequencies (thus representing a form of crisis-induced self-organization analogous to infinite-dimensional critical self-organization of the sandpile type; see Chap. 5). For instance, the mechanism already mentioned, by which the effects of chaotic bursting affect the average recurrence time per sunspot cycle—which varies according to the number of spots per cycle (i.e., as τ increases without limit, the number of spots per cycle vanishes, and vice versa)—was referred to a critical frequency of Maunder Minimum type. By the same token, chaotic bursting phenomena would be expected to occur in the vicinity of the ~11-year sunspot cycle (i.e., at a frequency near 10^{-1} yr$^{-1} \cong 3 \times 10^{-9}$ Hz). The corresponding bursting phenomena at this critical frequency (~10^{-1} yr^{-1}) would therefore involve relationships between sunspots, solar flares, coronal mass ejections, and magnetospheric substorms (Appenzeller, 1992; Hoecksema, 1992)—reflecting a second form of chaotic crisis controlled by a periodic attractor state representing an ideally regular, but unstable, behavior of the sunspot cycle.

If the chaotic attractor characteristic of solar-flare activity and mass ejections approaches the vicinity of the unstable periodic sunspot frequency, then interior crises again occur with power-law distributions of bursting times analogous to those of the other types of processes already examined. Consequently, flares are expected to show great intermittent irregularity, with frequencies ranging from many per day to almost none per sunspot cycle (i.e., meaning that we might experience decade-length periods of time when there would be virtually no magnetic storms of the types that have recently disrupted terrestrial communications systems; cf. Giovanelli, 1984; Appenzeller, 1992; Hoecksema, 1992). In principle, therefore, it may be possible to understand solar-flare and coronal mass-ejection phenomena by studying their spatiotemporal variability (cf. Note 8.5) relative to that of the sunspot cycle (see the description of "butterfly diagrams" in Note 8.4), thereby providing a forecasting tool for magnetospheric substorms and terrestrial atmospheric disturbances (see Appenzeller, 1992; Hoecksema, 1992). Individual bursting events are unpredictable in character, but *patterns* of bursting activity are predictable in terms of their possible ranges of τ-values per critical nonlinear frequency bandwidth (cf. Fein et al., 1985; Carroll et al., 1987; Olinger and Sreenivasan, 1988; Shaw and Chouet, 1988, 1989, 1991).

This sequence of solar processes and time scales is analogous to the relationships between geomagnetic-reversal frequencies, impact frequencies, and frequencies of related magmatic events. Those aspects of global magmatism that are relatively periodic and roughly synchronous with impact frequencies are analogous to the sunspot regularity (the processes or events in the Sun that are analogous to the impact processes on Earth are the dissipative orbital phenomena mentioned previously). This terrestrial regime is represented, in part, by the post-95-Ma portion of the records of Figures 8 and 30, where the geomagnetic and impact frequencies are in phase during the ~30-m.y. peak-to-peak cycles of ratios shown in Figure 30a (the peaks in this plot are the *relative* maxima of the reversal frequencies, and the troughs are the *relative* maxima of the impact frequencies, meaning that minima can occur in Fig. 30a even when the impact and reversal frequencies both increase together). Figures 8 and 30a also show that in the post-95-Ma portion of the record, the maxima of magma production correlate almost directly with the maxima in impact rates (e.g., Fig. 8 shows subsidiary maxima of magma production at ~65 and ~35 Ma; Fig. 30a shows *relative* minima at these ages).

The ~30-m.y. cyclic regime in all three records—impact rates, magma-production rates, and paleomagnetic reversal rates—is demonstrable for only three cycles, because the signal is suppressed (i.e., by intermissions) in the longer-term chaotic record correlated with the geomagnetic superchrons and sustained intervals of low impact rates [e.g., I am referring to the normal superchron centered at ~100 Ma, and to the reversed superchron centered at ~290 Ma (Harland et al., 1990, Fig. 6.10); and likewise to the impact minima centered at ~170 Ma and at ~430 Ma (Fig. 6)]. The Frasnian-Famennian (F-F) and Permian/Triassic (P/T) boundaries, at 367 Ma and 245 Ma, respectively, should correspond to higher-frequency impact modes relative to the impact minimum at ~430 Ma—in the same sense that the Cretaceous/Tertiary impact episode occurred following the ~170 Ma impact minimum. In each case, an integer number of ~30-m.y. cycles of impact-rate fluctuations occurred between the longer-term impact minima and the shorter-term putative impact maxima (e.g., impact maxima would occur at two and six times the ~30-m.y. cycle following the ~430-Ma impact minimum, and at three and four times the ~30-m.y. cycle following the ~170-Ma impact minimum—the latter corresponding within half a cycle to the observed episodes of high impact rates at ~65 Ma and ~35 Ma). The impact record is not sufficiently complete at the older ages to test this progression. For example, there are no distinctive maxima in the existing impact record at either the P/T (245 Ma) or F-F (367 Ma) boundaries (but see Appendix 3a), and yet the regularity of the ~30-m.y. cycle is seen in global magmatic cycles throughout the Mesozoic and Cenozoic Eras (e.g., in Sierra Nevada plutonism and in averaged growth episodes of the Hawaiian-Emperor volcanic ridge; e.g., Kistler et al., 1971; Shaw et al., 1980a; cf. Johnson and Rich, 1986). The notion of an ~30-m.y. period in the variations of putative impact frequencies would appear to be as consistent with the timing of the 367-Ma (F-F) and 245-Ma (P/T) extinction events as it is for any others.

It should be recognized in such discussions of "periodic events" that I am speaking of mean values of nonlinear recurrence times in a manner analogous to the definition of the winding numbers, W, of sine-circle maps (see Table 1, Sec. D; cf. Notes P.4, 3.1, 4.1, 7.1, 8.5). In coupled nonlinear oscillations *self-tuned* to the vicinity of the *critical golden mean nonlinearity* (see Note P.4; and Shaw, 1987b; Shaw and Chouet, 1989), opportunities exist for period-doubling and/or period-halving bifurcations as well as phase-locked and quasiperiodic behaviors (cf. Notes P.1–P.3, P.7, I.4), hence half-cycle ambiguities in geological-cosmological coupled oscillations are inevitable in view of the few recurrences over which the mean values of W can be evaluated (cf. Shaw, 1988a, Table 1 and Fig. 25). The nonlinear character of natural periodicities would appear to be the essential reason why the "periodicity controversy" has raged in the earth sciences ever since radiometric methods of age determination became routine during the first half of the twentieth century (cf. Williams, 1981; Shaw, 1987b).

In view of the half-cycle period-doubling/period-halving uncertainties just mentioned, the coherence of a nonlinear-periodic ~30-m.y. signature in the geologic record (cf. Williams, 1981, Introduction, Table 1; Shaw, 1988a, Table 1 and Fig. 25) would appear to be remarkably well resolved in many different geologic processes. In the context of global tectonism, for instance, the comparative timings of the Mesozoic-Cenozoic episodes and "impact epochs" compiled by Seyfert and Sirkin (1979, Table 12.2) can be compared with the similarly timed orogenic-epeirogenic episodes documented by Damon and Mauger (1966), Evernden and Kistler (1970), and Kistler et al. (1971). Analogously, the longer-term commensurabilities of terrestrial-extraterrestrial correlations reviewed by Williams (loc. cit.) and by Shaw (loc. cit.) can be compared with the episodic mineralization history of the Earth described by Meyer (1988) and with the history of the Earth's major tectonic episodes relative to the Moon's major impact episodes, as reviewed by Stothers (1992). The half-cycle uncertainty (see above) has often confused attempts to reconcile, or refute, regional and global spatiotemporal correlations of these types—for example, correlations between plutonic and/or volcanic activity in western North America, regionally and globally (cf. Kistler et al., 1971; Smith et al., 1971; Sigurdsson, 1990, p. 103 and Table 2). Approximate correspondences between this generalized periodicity hierarchy and the paleontologic episodes are conspicuous in the data compiled by Shaw (loc. cit.; cf. later discussion of Figs. 35 and 36 in the present volume; and Newell, 1952, 1967; Raup, 1972, 1984a, 1986a; Fischer and Arthur, 1977; Fischer, 1981, 1984; Sepkoski, 1981, 1989, 1990; Ward, 1982; Benson et al., 1984; Rampino and Stothers, 1984; Raup and Sepkoski, 1984, 1986; Nance et al., 1986; Sepkoski and Raup, 1986; Walker and Zahnle, 1986; Knoll, 1991; Carroll, 1992; Kump, 1993; Rampino and Caldeira, 1993; Rogers et al., 1993; Simon et al., 1993; Widdel et al., 1993).

The general picture of scale-invariant self-similarity between solar and terrestrial processes is represented schematically by three distinct regimes of the same types, as follows: (1) a long-term regime of irregular intermittency (strong chaotic bursting) related to a more regular but unidentified long-term process, (2) an intermediate-term regime of normal intermittency (weak chaotic bursting) related to near-synchronizations of multiply coupled oscillations of several processes, and (3) a short-term regime of extremely irregular intermittency (strong chaotic bursting) related to the regularity of the second (intermediate-term) regime (note the resemblance to neurophysiological coupled oscillators with delayed feedback; e.g., Freeman, 1992, pp. 467ff; cf. Brown and Chua, 1991; Pecora and Carroll, 1991; Murali and Lakshmanan, 1991, 1992; Anishchenko et al., 1992; De Sousa Vieira et al., 1992; Carroll and Pecora, 1992; Rul'kov et al., 1992). The secondary stage in the Sun refers to the $\sim$11-year solar sunspot cycle driven by coupled magnetic/mass-transfer processes, and in the Earth by the nonlinearly coupled magmatic, geomagnetic, and impact frequencies with the relatively regular time scale of $\sim$30-m.y. The regularized regimes are analogous to the regime of controlled chaos in Figure 29, where the chaotic signal has been suppressed by subjecting the chaotic thermal oscillations to negative feedback (employing a servo device coupled to the thermal fluctuations of opposite signs). In natural systems of coupled oscillations, such negative feedback is self-organized by interactions with one or more additional oscillators. The proportionality factor separating the scale invariances between the solar and terrestrial processes is of the order of 10^6 (i.e., the three regimes in each set of processes represent two frequency bands, each spanning a factor of 10^3 in time, while the solar and terrestrial bands are separated by a factor of $\sim 10^6$ in time; cf. Note P.4; and Shaw and Chouet, 1991).

In the Sun, the irregular, and relative, "longer-term regime" refers to the variation of the sunspot record relative to intervals like the Maunder Minimum (Eddy, 1977), while the "shorter-term regime" in the Sun refers to the highly irregular solar-flare eruptions and coronal mass ejections (Appenzeller, 1992; Hoecksema, 1992). In the Earth, the analogous "longer-term regime" refers to (1) the geomagnetic superchrons, (2) the general fluctuation of impact-shower activity, and (3) the igneous-tectonic supercycles of global mantle convection (of either whole-mantle or upper-mantle type) that both effect and affect, because of the varieties of respective delayed-feedback modes, major fluctuations in ocean-crust production, subduction, orogenic-epeirogenic cycles, and episodic continental drift. The conspicuous forcing function for this latter time scale is the nonlinear Galactic year ($\sim$225 $\pm$ 25 m.y.), which interacts with the chaotic attractors of the Oort Cloud and those of the Solar System in general (cf. Duncan et al., 1987; Shaw, 1988a). The relative "shorter-term regime" in the Earth is represented by the irregularities of the geomagnetic-polarity time scale, the seeming randomness of meteor strikes and cometary apparitions during historical times, and the spatiotemporal unpredictability (on the historical time scale) of volcanic eruptions and earthquakes.

The approximate cumulative energy of Earth's cratering events, based on the events in Grieve (1982) and the energy scaling of Shoemaker (1983), is $\sim 10^{31}$ ergs (Shaw, unpubl. data, 1984). Events added or revised in this volume double or triple that figure, especially if the Chicxulub crater, Yucatan, is actually 200 km in diameter, making it the largest documented crater on Earth (see Sharpton et al., 1992). Because this estimate is mainly influenced by the known record of Phanerozoic impacts, an extrapolation that includes the late Precambrian and Phanerozoic portion of Earth's history would be at least 10^{32} erg. This total is increased another order of magnitude by the incompleteness of the cratering record, yielding in excess of 10^{33} erg. Judging from the cratering history of the Moon, it can be assumed that the cratering rate on Earth between about 4 and 3 Ga was conservatively an order of magnitude larger (cf. Shoemaker, 1983, p. 495), and prior to that there were incalculably large events associated with the earliest collisional history following lunar formation. Thus, we have, conservatively, more than 10^{34} erg as the minimum cumulative cratering energy during the latest 4 Ga, and the total cratering energy — including the energy of impacts directly associated with the collisional formation of the Moon — would greatly exceed this number (see Fig. 10 for orbital and rotational kinetic energies of the planets).

If I assume that the total impact energy was $\sim 10^{33}$ erg during the latest 1 Ga of Earth history, and that there were $\sim 10^3$ impacts during that time, then the average energy per event would have been $\sim 10^{30}$ erg. Energetically, this corresponds to a theoretical earthquake magnitude of $M^* \cong 12$, using the equation log $E \cong 11.8 + 1.5\ M$ (Richter, 1958; Kanamori, 1986), occurring at an average frequency of about one event per million years. The effect on the Earth of such an earthquake is seismologically inconceivable. In order to provide some perspective, however, ~ 1000 years of tidal dissipation at the present rate of $\sim 10^{27}$ erg yr^{-1} would be required to provide this amount of energy, and all of that energy would have to be deposited instantaneously at a single focus in the Earth! Employing another analogy, that amount of energy is roughly equivalent to the instantaneous generation of a volume of magma equivalent to the largest known crustal magma chambers (~ 20 km^3 per year for 10^3 years = 20,000 km^3), a buildup that normally would take many millions of years (cf. R. L. Smith, 1979; Shaw, 1985a).

Such estimates for the rates of deposition of impact energies into the Earth are not meaningful for impacts (or for resulting magmatic events) over times much shorter that $\sim 10^8$ years — because I have averaged a process that has a very wide range of intermittencies in time and size. But, thus qualified, the above energy estimates would be equivalent to ~ 100 impacts (each with an average energy of 10^{30} erg, and with accompanying large-scale magma production) with ages younger than about 100 Ma. For comparison, five of the basalt fields shown in Fig. 4 have ages ≤ 125 Ma. If each of these fields represented an "average" impact event, then there should have been more than ten times as many fields (assuming a completeness factor of 10% for the cratering record). However, as a rule-of-thumb, the ratio of the plutonic to the volcanic fraction of magmatic energy is approximately 10:1 (see Smith and Shaw, 1975, 1979; Smith et al., 1978; Crisp, 1984; Shaw, 1985a). This ratio would suggest that most of the estimated impact energy was deposited as the plutonic fraction of the Earth's magma production, and that only the very large and/or multiple impact events were directly associated with voluminous volcanic fields (e.g., the largest flood-basalt fields have volumes of the order of 10^6 km^3).

The number of "average" impacts estimated above should correspond roughly to the number of events with diameters > 20 km in Figure 6. In other words, there should be more than ten times the observed number of the largest events shown in Figure 6 to be equivalent to the estimated average rate during the Phanerozoic, but this comparison is only crudely comparable to the discussion of uncertainties in Chapter 2. Unfortunately, order-of-magnitude estimates based on crater scaling and cumulative energy are not very reliable. If the cumulative energy of cratering during the latest gigayear ($\sim 10^{33}$ erg) were taken literally, it would have been nearly equal to the rotational kinetic energy of the Earth (see Fig. 10). Therefore, if all of the impacts had been vectorially aligned to oppose the rotational angular momentum of the Earth with maximal efficiency, they could have slowed the rotation rate substantially (cf. Munk and MacDonald, 1960, p. 57; Lambeck, 1980). Such a comparison is absurd, of course, because most of the energy would have been dissipated internally, and because the polar scenario of approach envisioned in Figure 19, as well as the processes of capture, would have contributed components of both positive and negative angular accelerations (cf. Dones and Tremaine, 1993). But the point is well taken that, energetically speaking, the impact process during recent geologic time must have been significant relative to other mechanisms that excite transient variations in the length of day and wobble of the spin axis (e.g., Chao and Gross, 1987; Hyde and Dickey, 1991; Mathews and Shapiro, 1992; Dickey, 1993; cf. Note 8.5 and Appendixes 22 and 23 of the present volume).

The Celestial Reference Frame and Paleogeography

In attempting to reconcile the early history of impacts related to the formation of the Moon with the subsequent formation of the Earth's crust, we do well to suppose — if for no other reason, from the early history of the Moon (see Wilhelms, 1987, Plates 1–12) — that the events of that period would have formed large-scale geomorphic features that still persist (cf. Green, 1972). Remnants of these features are expected to align to some extent with mass anomalies associated with CRF satellitic resonances and with a meridional distribution of continents, for the reasons summarized in this chapter. Some of these features must have been so large as to leave little evidence of their impact loci because of subsequent magmatic and tectonic effects. Certainly, however, there are many topographic features of the continents and bathymetric features of the ocean floors existing today for which we can offer no other simple interpretation. The vicinities of the Gulf of Mexico–Caribbean Sea and Hudson's Bay are examples of what the scars might look like if they formed sometime around 4 Ga, but subsequent to the early stages of growth of thick sialic protocontinents. The ages and sizes of possible impact scars are indicated by the histories of craters and basining on the Moon (Wilhelms, loc. cit., Fig. 14.3 and Plate 3). Our immediate concern here, however, is how such early patterns might relate to the CRF system that formed subsequently, as proposed in the present work. [Dalrymple (1991, p. 400) cites an age of 3.96 Ga for the oldest demonstrably granitoid gneiss, found in northwestern

Canada, and ages of 4.0–4.3 Ga for zircons from sedimentary rocks, found in Western Australia, but controversial evidence (cf. Taylor, 1992) has been reported by Harper and Jacobsen (1992) for a 4.5-Ga event of geochemical fractionation of mantle sources of gneisses from western Greenland, which, if substantiated, would imply that the minimum age of sialic crustal material derived from the Earth's mantle is indistinguishable from the greatest known ages of the Earth and the Moon (see Dalrymple, loc. cit., Table 7.6 and pp. 355 and 401). As I remarked earlier in this chapter, however, there may be some ambiguities concerning pre- and post-collisional processes and products in the Earth, therefore concerning the early evolution of the Earth and Moon; see also the following, for example, which predate contemporary collision theories: Green (1972), Lowman (1976), Frey (1977, 1980), and Grieve (1980).]

Alt et al. (1988, p. 648) asked the reasonable question, "Where are the impact explosion craters more than 140 km in diameter?" The number of $\sim$100-km-diameter craters in the Phanerozoic record (Figs. 2 and 3) would suggest, from the prevalence of power-law statistics in crater counts on the Moon (cf. Baldwin, 1987; Wilhelms, loc. cit., Figs. 7.10–7.16) and elsewhere (cf. Hartmann, 1972; Hartmann and Davis, 1975; Carr, 1981; Head, 1981; Taylor, 1983; Pike, 1988; Schultz, 1988; Spudis and Guest, 1988; Strom and Neukum, 1988; Barlow, 1990; Trego, 1991; D. B. Campbell et al., 1992; Phillips et al., 1992; Schaber et al., 1992), and the power-law statistics of asteroid sizes (see Appendix 8), that there might be at least one Phanerozoic crater with a diameter of 300 km or more [recall that a crude rule of thumb is that the diameter of a crater is roughly an order of magnitude larger than the diameter of the impacting object (cf. Shoemaker, 1983, pp. 469ff)]. The largest asteroid (Ceres), for instance, has a diameter of about 1000 km, and there are three about half that size and more than twenty about one-fourth that size (Tedesco, 1989). The largest Phanerozoic crater thus far identified on Earth is Chicxulub, Yucatan (Hildebrand et al., 1991), with an age of 65 Ma (Swisher et al., 1992) and a diameter of $\sim$200 km (Sharpton et al., 1992). If we take the power-law statistics of cratering on other planetary bodies as well as the present-day counts of possible asteroids and comets in the inner Solar System to be indicative of the maximum size of impactors, then we might expect to see perhaps one or two more craters the size of Chicxulub, and one about twice this size, in the Phanerozoic record (cf. Shoemaker et al., 1990; Weissman, 1990a, b; Binzel et al., 1991, 1992). Alt et al. (1988) suggested that large flood-basalt plateaus are the terrestrial equivalents of lunar maria, and that the Deccan plateau is the site of an impact crater with a diameter of $\sim$600 km (see their Fig. 2; cf. Stothers and Rampino, 1990; Stothers, 1992). There is, however, no general agreement on their suggestion, and Hagstrum and Turrin (1991a, b) and Hagstrum (1992) argue that flood-basalt plateaus are in fact antipodal to the sites of the causative impacts [the issue of direct vs. antipodal correlations between impact events and magmatic events **(Note 8.6)** is explored in Appendixes 16–18]. Nevertheless, Alt et al. (1988) have identified one among several possible explanations for why very large craters, or basins initiated by craters, could have become obscured — they simply filled to overflowing with magma (cf. Oberbeck et al., 1992; Marshall and Oberbeck, 1992; Aggarwal and Oberbeck, 1992; Rampino, 1992). Analogous phenomena, on a much smaller scale, are commonplace in the filling of gravitational collapse craters — "pit craters" — formed during the intermittent growth of basaltic shield volcanoes, such as Kilauea, Hawaii (see the documentation of the growth stages of Hawaiian volcanoes by Holcomb, 1987; cf. Heliker and Wright, 1991).

Credibility diminishes with increasing size, of course, and the marshaling of evidence for any large-scale cratering event involves such a major mapping project that few geologists would have the wherewithal to attempt the documentation of such an event, even when sound indications of the event are suspected. The relatively small, but conceivably much larger, Avak structure, in Alaska, is a case in point (see Fig. 3; and Grantz and Mullen, 1991); it was discovered incidentally to a subsurface geological-geophysical survey (Kirschner et al., 1992). Similarly, an early suggestion that an arcuate structure on the Yucatan Peninsula, now identified as the Chicxulub structure, might be an impact crater was made in 1981 by G. T. Penfield and Z. A. Camargo as part of a gravity and aeromagnetics survey (see Penfield and Camargo, 1991; cf. the discovery circumstances in Hildebrand et al., 1991).

The number of person-years now taken up by the quest for, and documentation of, the — in retrospect — conspicuous Chicxulub structure illustrates how difficult, how costly, and how controversial it would be to make an analogous case for a larger and only slightly more obscure cratering event in a geologically less well-exposed setting. Chicxulub represents essentially the first concerted and interdisciplinary effort to find a large impact crater within a specific geologic setting and time frame (reflecting composite continental and oceanic ejecta materials and within the temporal uncertainty of the K/T boundary interval). Because nearly all crater finds have been incidental to other geological activities, it stands to reason that — should economic and/or other humanistic motivations justify it (such as a major program of long-term meteoroid-hazard mitigation at the funding level of the U.S. earthquake and radioactive-waste-disposal programs) — a high-priority research program dedicated to the documentation of impact structures, and evidence pertaining thereto, would undoubtedly bring to light many more examples of case histories just waiting to be documented. (Imagine what would happen if a new and technologically strategic mineral resource crucial to the continued economic survival of the major world powers were known to be associated only with meteoroid-impact localities.)

If we use the lunar history as a guide, events that produced terrestrial basin structures with diameters of the orders of $10^3 - 10^4$ km should exist on Earth but would be very old (see Wilhelms, 1987, Figs 7.15 and 7.16, and Plates 1–12). The largest and oldest of the extant lunar basins is Procellarum (pre-Nectarian basin group 1 in Fig. 14.3 of Wilhelms, 1987), with a reported diameter of 3200 km enclosing "Oceanus Procellarum," a complex of subsequent basining, tectonism, magmatism, and cratering events. The ratio of its diameter to that of the Moon, in equal-area projection, is about 0.7. If this same scale ratio is used to estimate the largest impact-triggered basining event on Earth, the basin diameter would be roughly 9000 km, or equivalent in size to the central Pacific basin ("Oceanus Pacificus"). Although this comparison is figurative, the possible size scales of the effects induced by early mass impacts must have been of global proportions! In view of the larger acceleration of gravity and the extensive tectonism on Earth that would have followed these relatively gigantic early impacts, the notion that the Pacific Ocean itself is an ancient impact scar is not so fanciful, especially after one has spent some time perusing Wilhelm's (1987) monograph *The Geologic History of the Moon*.

At such a phenomenal scale, questions concerning the chemical composition of the impactors (which would have been of the order of 100 km or more in diameter), and what became of that material, are significant. Smaller impact craters may contain associated meteoritic material, and/or they may have been accompanied by tektite and/or microtektite strewnfields (cf. Urey, 1957, 1962, 1963; Glass and Heezen, 1967; O'Keefe, 1976; Glass and Zwart, 1979; Glass et al., 1979; Ganapathy, 1982; Glass, 1982; Shoemaker, 1983; Keller et al., 1987; Barnes, 1990; Schneider and Kent, 1990; Izett et al., 1991a, b; Sigurdsson et al., 1991a, b; Smit et al., 1991; Wang et al., 1991; Blum et al., 1992; Claeys et al., 1992a, b; Schnetzler and Garvin, 1992; Swisher et al., 1992; Wang, 1992; Wasson and Heins, 1993), but the impactors responsible for the earliest basining events would have been incorporated into the crustal/upper-mantle compositional stratification of the Earth, hence would now represent some aspect of the chemical heterogeneity of the upper mantle (see below; and Chyba, 1991). If the impacting objects were of terrestrial affinities — in the sense that they belong to the suite of chemical compositions that characterize the rocky inner planets and asteroids of the Solar System — then the impact products would not necessarily be seen as conspicuous chemical, mineralogical, or mass anomalies (with the exception of iridium, shock-metamorphic minerals, and the like; see Alvarez et al., 1982; Orth et al., 1990; Huffman et al., 1990; cf. Oberbeck et al., 1992; Marshall and Oberbeck, 1992; Aggarwal and Oberbeck, 1992; Bice et al., 1992; Carlisle, 1992; Gilmour et al., 1992; Rampino, 1992). But such contemplations raise issues of compositional heterogeneity and mixing in the earliest post-lunar history of the Earth (cf. Kellogg, 1992), issues analogous to those currently being raised in the literature relative to the compositional heterogeneity of the impactors themselves (e.g., asteroids and comets; cf. Chapman, 1986; Sears and Dodd, 1988; Kerridge and Matthews, 1988; Wood, 1988; Bell, 1989; Chapman et al., 1989; Lipschutz et al., 1989; Carlisle, 1992; Fink, 1992; Rotaru et al., 1992; Russell et al., 1992; Sears et al., 1992).

Such speculations may never be satisfactorily answered, but some generalizations are suggested by comparative inferences based on the CRF hypothesis. To the extent that spin-orbit resonances have existed and/or have influenced the proximal trajectories of impacting objects, the patterns of geoidal mass anomalies in the Earth probably have been meridionally distributed more or less in opposite hemispheres, and with centers of mass that have maintained approximately constant phase relationships with the younger cratering nodes on sialic craton during at least the latest 1 to 2 Ga. Furthermore, such a cratering regime is envisioned to have been in a critically coupled nonlinear resonance relationship with the flux of impacting objects entering the inner Solar System during the same span of time. By contrast, the initial infall of debris from the formation of the Moon is assumed to have come from a massive orbiting equatorial ring of fractally distributed fragmentation products containing objects that ranged in size from dust up to the order of 10^2 to 10^3 km or more in diameter (cf. Bandermann, 1971; Brown et al., 1983; Donnison and Sugden, 1984; Turcotte, 1986; Stöffler et al., 1988; Fujiwara et al., 1989). Such a scenario is based on the collisional hypothesis for the origin of the Moon (e.g., Benz et al., 1986, 1987, 1989; Boss, 1986; Boss and Peale, 1986; Ringwood, 1986, 1990; Wood, 1986; Stevenson, 1987; Taylor, 1987; Chapman and Morrison, 1989, p. 165; Kerr, 1989; Melosh, 1989; Ringwood et al., 1990; Cameron and Benz, 1991; Kaufmann, 1991, Fig. 9–22; cf. Daly, 1946; Hartmann and Davis, 1975; Baldwin and Wilhelms, 1992; Spera and Stark, 1993) and computer simulations of other large-scale collisions in the early history of the planetary system (cf. Murray, 1983; Wetherill, 1985, 1988, 1990; Leake et al., 1987; Strom, 1987; Stewart, 1988; Cameron et al., 1988; Chapman, 1988; Fujiwara et al., 1989).

My general conclusions concerning the importance of the earliest and most massive period of bombardment of the Earth, however, do not necessarily depend on the above mechanism of the origin of the Moon. For example, Runcorn (1983, 1987) argues — without recourse to the collisional model — that the Moon was itself surrounded by an equatorial belt of large satellites that had evolved with it during earlier stages, but which became orbitally unstable and broke up into many fragments during orbital decay. By either view, the Earth was likely to have had an analogous early satellitic system, but the mechanism for the origin of that system, and for its impact record on the Earth, would have differed — in timing, sizes of objects, and sizes and geographic patterns of impact craters — from that of the CRF hypothesis. A system of the CRF type might nonetheless have existed as well, in the form of smaller and more distant satellitic objects with a variety of eccentricities and inclinations, depending on interactions with the equatorial satellites and mass anomalies existing in the Earth at that time. Such satellitic systems are, a priori, expectable (cf. Cruikshank et al., 1982; Jewitt, 1982; Greenberg and Brahic, 1984; Esposito and Colwell, 1989; Kolvoord et al., 1990; Binzel et al., 1991; Esposito, 1991, 1992; Horn and Russell, 1991; Porco, 1991; Colwell and Esposito, 1993).

By either model of initial states, the net effect of the equatorial in-fall regime would have been to establish long-lived mass redistributions in the Earth that eventually became the system of mass anomalies controlling the later CRF resonances. Accordingly, the nodal distribution of the CRF and its correlations with other geophysical observations would suggest that those early events are still reflected in the properties of a great-circle segment of the Earth — expressed at the surface as a meridional band with a width of roughly ± 30° of longitude. This meridional band coincides broadly with the global-scale distribution of the roots of continental masses and lithosphere subduction in a manner that is analogous to the deep-seated geophysical anomalies of Figures 27 and 28, and is more or less independent of the details of the relative positions of individual continents and continental fragments (see **Appendix 26**). The global map of "Lithospheric Graveyards" illustrated by Engebretson et al. (1992, Fig. 1; cf. Richards and Engebretson, 1992) provides visual drama to the location of the surface trace of this great-circle segment postulated by the CRF hypothesis — for example, by comparison with the VGP patterns, core-flux patterns, and mantle-density anomalies illustrated in Figures 27 and 28 of the present volume (also there are interesting correspondences with topographic anomalies on the core and on the 660-km discontinuity in the upper mantle; cf. Morelli and Dziewonski, 1987, cover illus.; Vogel, 1989, Fig. 3; Fukao, 1992, Fig. 4; Shearer and Masters, 1992, Fig. 4; Lay, 1992; Kellogg, 1992).

In other words, the mass distribution within the spherical segment at mantle depths putatively comprises the general guidance anomaly for the CRF satellitic reservoir over time, and it also comprises the general guidance anomaly (centering mechanism) for the distribution of the continents during the post-early-bombardment history of the Earth. The wide bandwidth, totaling about a sixth of the Earth's circumference, is consistent with geological evidence requiring that (1) large rotations of continental masses are permitted, (2) complex regenerations and/or redistributions of microplates occur within the band, (3) there has been intermittent and cyclical disassembly and reassembly of accretionary continents (e.g., Fischer, 1984; Nance et al., 1986, 1988; Murphy and Nance, 1992; cf. the accretionary maps of Eurasia and Africa, respectively, by Zonenshain et al., 1991, and de Wit et al., 1992), and (4) long-distance translation of accretionary-terrane slices and fragments is possible as a commonplace phenomenon in paleogeographic reconstructions (e.g., Irving, 1982; Howell, 1985, 1989; P. F. Hoffman, 1988, 1991; Stone, 1989; Stewart, 1990; Dalziel, 1991, 1992a, b; Hartnady, 1991; Moores, 1991; Zonenshain et al., 1991; de Wit et al., 1992; Elliott and Gray, 1992; Stump, 1992; Young, 1992; cf. Ernst, 1988). According to the CRF hypothesis, however, the portions of the continental cores that are locked to the Phanerozoic cratering nodes (PCNs) have remained globally invariant, in relation both to each other and to the principal axial moments of inertia (i.e., relative to the instantaneous spin axis and equatorial bulge). According to this interpretation, over geologic time the deep mass anomaly has acted literally like a *keel,* in the sense that it has guided both the evolution of the external geocentric satellitic system *and* the evolution of an internal system of "satellitic" motions reflected in the history of continental drift, plate tectonics, and paleomagnetism.

Descriptively, the means by which the mass redistribution could have taken place can be reduced to just two stages — stages that are exceedingly simple in general character but that are of admittedly mind-boggling proportions compared to the scale of Phanerozoic tectonic motions. The scaling, of course, is conjectural, and depends primarily on the sizes of the objects that were orbiting the Earth immediately following the formation of the Moon. I assume that these objects either equalled or exceeded the sizes of the lunar basin-forming objects that orbited the Moon during the same general astronomical epoch. The cratering process itself and the subsequent basin-forming tectonics on the Earth, even as it existed then, differed from those on the Moon because of Earth's higher gravity and greater range of rheological variability, probably favoring greater localization and deeper penetration for the same size impact (the Earth would in general be at higher temperatures and much more mobile than the Moon at the same age, relative to the formation of the Moon taken as the time of origin). Discussions of the protoplanet-Earth collisional hypothesis for the origin of the Moon are given in many references (e.g., Daly, 1946; Hartmann and Davis, 1975; Wetherill, 1985, 1990; Benz et al., 1986, 1987, 1989; Boss, 1986; Boss and Peale, 1986; Ringwood, 1986, 1990; Wood, 1986; Stevenson, 1987; Taylor, 1987; Chapman and Morrison, 1989; Garwin, 1989; Kerr, 1989; Melosh, 1989; Benz and Cameron, 1990; Ringwood et al., 1990; Cameron and Benz, 1991; Kaufmann, 1991; Baldwin and Wilhelms, 1992; Spera and Stark, 1993). Descriptions of cratering processes and basin formation on the Moon are given, for example, in Hartmann (1972), Howard et al. (1974), and Wilhelms (1987). The conjectural two-stage process is schematically outlined in **Figure 33** and described forthwith.

Stage One represented the formation of a Procellarum-like basin formed by orbital decay and impact of the next-to-largest objects in the belt of equatorial satellites. [The locus that became the Moon (e.g., Chapman and Morrison, 1989, p. 165; Kaufmann, 1991, Fig. 9–22), of course, was the center of attraction of the largest of these objects, but I assume that many satellites of both the Earth and the Moon with diameters $< 10^3$ km were present during at least the first 10^8 years or so following the main Moon-forming collision of a protoplanet with the Earth.] Scaled to Earth's diameter, the first major impact crater + basin was comparable in size to the central Pacific Basin. In fact, I will assume, for argument's sake, that it was in fact the ancestral Pacific and call it *Oceanus Pacificus* (an oxymoron for the most violent event of Earth's post-lunar history). One might be tempted to speculate that this central Pacific Basin was the birth-scar of lunar origin (e.g., Fisher, 1882, 1889, p. 338; Wood, 1985, pp. 54ff; cf. G. H. Darwin, 1879, 1907, 1962; Sridhar and Tremaine, 1992), but if that event involved collision of the Earth with anything like a Mars-sized object, as several different workers have inferred from computer simulations, the effect would have totally scrambled any prior configurations (**Note 8.7**).

Let us say that the Stage-One event occurred at about the same time as — although probably somewhat before because of the greater tidal effect of the Earth on the orbital degradation of large low-inclination satellites — the Procellarum Basin was formed on the Moon (Wilhelms, 1987, p. 278, assigned an age of 4.15 Ga to Procellarum for consistency with the chronology of his Fig. 14.3). This episode of cratering involved multiple impacts in an equatorial belt over a relatively brief interval of time (cf. Runcorn, 1983, 1987). The impacts had two effects on the equatorial mass: (1) a great deal of the ejecta was deposited at higher latitudes or totally escaped from the system, and (2) impact melting, lateral shear melting, and decompression melting accompanied and expedited an initially rapid isostatic in-filling of the "hole" (cf. Howard et al., 1974).

The net effect of these processes produced the great extent and complexity of the multi-basin structure that became Oceanus Pacificus. The volumetric in-filling rates of the initial crater or craters probably were greater than any others recorded in geologic history because of the viscous relaxations related to the complex of dissipative melting mechanisms (glacial oscillations, by comparison, involve time scales of the order of 10^3 years or less; cf. Walcott, 1972; Sergin, 1980; Hughes et al., 1981; Benzi et al., 1982; Held, 1982; Ghil, 1985; Peltier, 1987; Mitrovica and Peltier, 1989; Berger et al., 1992; Kerr, 1992f, 1993d; Winograd et al., 1992; Edwards et al., 1993). Even if most of the initial volume deficiency was entirely compensated on a time scale as short as thousands of years—which I assume was not the case because of its near-global scale—a mass deficiency of many percent would have been left by the net effects of melting and the mass ejected from Earth's sphere of influence (the depth of penetration is hard to define, but it certainly must have been great compared to Phanerozoic impacts, and therefore it undoubtedly affected the lower as well as the upper mantle). The most important immediate consequence of the impact was that Earth's axial moments of inertia were instantly changed, so that the previous maximum was now the minimum, and a very small one at that (cf. Fisher, 1889, pp. 380f; Gold, 1955; Goldreich and Toomre, 1969; Stacey, 1969, pp. 26f; Andrews, 1985; Schultz, 1985; Runcorn, 1987; D. L. Anderson, 1989, pp. 249ff). In less than 10^8 years—probably much less for such a stupendous change of mass—the great-circle swath that had been the equator became a meridian (see the hypothetical examples for the early cratering history of the Moon illustrated by Runcorn, 1987), with the greatest reoriented mass deficiency now at one of the rotation poles (say, the South Pole). During the subsequent 10^8 years or so, much of the magma ocean formed in this way became fractionally crystallized **(Note 8.8)**, leaving a stratiform deposit of residual iron-rich ultramafic minerals [i.e., with lower $Mg/(Mg + Fe)$ than the average mantle of the present day; cf. Shaw and Jackson (1973)] at what would now be intermediate to lower mantle depths. [The literature contains many models—and much confusion—concerning the early geochemical history of the Earth; e.g., compare the discussion in Notes 8.8 and 8.9 with Ringwood (1986, 1989, 1990), Ringwood et al. (1990), Garwin (1989), D. L. Anderson (1989, pp. 235ff, 1990, 1991a, b), Chyba (1991), Chyba and Sagan (1992), Ray and Anderson (1992), and Stevenson (1992); cf. Taylor (1987, 1992), Drake (1990), Harper and Jacobsen (1992), Ryder and Dalrymple (1992).]

Following Stage One, therefore, we had a very oblate Earth with a magmatically zoned axial-meridional anomaly that was very dense at mantle depths and perhaps of gabbroic densities near the surface at the poles (i.e., several deep-penetrating impact events were likely in the vicinity of the original equator, and the arc thus formed—after 90° of rotation during Stage One—was aligned in a meridional orientation; cf. Runcorn, 1987). Elsewhere there was a less deeply perturbed system of basaltic magmatism that had been moving petrogenetically during these 10^8 years or so toward the formation of sialic crust by the more normal processes of progressive zone refinement involving cyclical crystallization and remelting, frequently perturbed by smaller impacts. [As I mentioned earlier in this chapter, we cannot absolutely rule out pre-collisional stages of chemical segregation in — or even prior major collisions with — the proto-Earth that may have left their marks on the chemistries, degrees of advancement, and ages of the crust, mantle, and core, a subject that is likely to remain a source of controversy for some years; cf. Drake (1990), Ringwood et al. (1990), Harper and Jacobsen (1992), Taylor (1992); cf. the assessments of the ages of the Earth, the Moon, and meteorites by Dalrymple (1991). Chyba (1991) has estimated that $\sim 1.5 \times 10^{22}$ kg of meteoritic material has been accumulated by the Earth *subsequent* to the time of solidification of the lunar crust, which he places at $\sim$4.4 Ga. This mass would have been made up mainly from the largest and earliest meteoroids, hence it would have had its greatest influence on the heterogeneity of the Earth's post-lunar primordial mantle. This meteoritic contribution makes up only about 0.3% of the Earth's total mass, but it *exceeds* the total mass of the present-day continental crust down to a depth of about 25 km, and therefore it would constitute a major source of geochemical heterogeneity relative to plate-tectonics-related fractionation processes during the magmatic evolution of the Earth.]

Because of the major disturbance of geotherms in the axial-meridional vicinities of the Stage-One event, I assume that magmatism elsewhere in the Earth would have involved convective upwelling in the mantle, but now biased in favor of the equatorial vicinity. That is, the mantle flow and magmatic fractionation that compensated the Stage-One basining event decoupled that vicinity from further immediate involvement with large-scale mantle convection by major suppression of the geotherms in the axial direction owing to the "burst" of heat transfer from the then largely crystallized "magma ocean" (see Note 8.8). I assume also that the more typical magmatic processes elsewhere in the Earth continued until there was a substantial accumulation of sialic crust in the post-Stage-One equatorial vicinity — perhaps until about the age of the oldest demonstrable terrestrial granitoids at $\sim$4.0 Ga (cf. Galer, 1991; Schubert, 1991; Kröner and Layer, 1992; but see Harper and Jacobsen, 1992). On the other hand, if the oldest sedimentary zircons on Earth, found in Western Australia, with ages of 4.0–4.3 Ga (Dalrymple, 1991, p. 400) were derived from protocontinental masses — and/or if the 4.5-Ga event reported by Harper and Jacobsen (1992) is considered to apply to post-collisional stages of the Earth's mantle — then the timing of the above processes on Earth would be set back $\sim$0.3–0.5 Ga relative to the lunar chronology of Wilhelms (loc. cit.) that I adopted as an arbitrary reference. This difference in timing presumably would reflect an earlier orbital decay and infall of the satellitic objects responsible for basin formation on the Earth compared to basin formation on the Moon — or it might only reflect confusion over what aspects of the present-day mantle of the Earth have recorded both pre- and post-collisional fractionation processes. The formation of the ancestral liquid-water Pacific Ocean (Fig. 33, bottom left) would have coincided more or less with the end of Stage One because of the immense amount of degassing that accompanied the burst of magmatic crystallization (see Note 8.8). Catastrophic erosional conditions also must have attended the eventual precipitation and accumulation of the liquid-water ocean from the dense and hot atmosphere of complex gases that would have formed after the formation of the Moon and the degassing of the proto-Earth accompanying the magmatic events of Stage One (see Fig. 33, top).

In order to sustain a relative chronology, I will use the ages reported in Wilhelms (1987, Fig. 14.3), recognizing that they might have to be shifted backward by as much as a few hundred million years (see below). This chronology would have allowed at least $\sim$150 m.y. for the formation of protocontinental core complexes. If the production rate of sialic crust in these complexes had been at the average rate of magma generation given earlier ($\sim$20 km^3/yr), then by assuming that the refinement process had reached maximum efficiency, we might conclude that sufficient time had passed to produce $\sim$20 km of sialic crust over an area equal to the present area of the continents (by coincidence, the Earth's present land area is roughly 150 million km^2). Such a high rate of production of sialic crust, however, would not be expected on the basis of present-day igneous processes. Viewed in terms of the total magmatic production that can support the evolution of its most silicic fraction, our knowledge of present-day processes suggests that $\sim$10% or less of the magmatic source might have reached a sialic stage, giving of the order of 2 km^3/yr of sialic craton (cf. R. L. Smith, 1979; Shaw, 1985a). Across 150 m.y. this rate would have provided a reasonably thick protocontinental mass distributed in an equatorial belt perhaps 30° wide. During this interval, as well, isostatic compensation would have been largely completed in the polar regions, and the oblateness of the Earth would have approached an equilibrium value for the properties existing at that time.

At this juncture another major, but considerably smaller, barrage of equatorial impacts occurred, because the orbital decay time of the satellitic objects that still remained in an equatorial orbit happened to be similar to the above igneous cycle time of about 150 m.y. (a "coincidence" that could have been related, at least in part, to tidal coupling between the Earth and Moon, orbital decay of the belt of equatorial satellites, and tidal dissipation in the solid Earth). This event, **Stage Two**, produced a smaller mass deficiency over smaller depths, but one that was still large by present-day standards of geoidal anomalies. Now, the net result was to move this new equatorial mass deficiency toward the poles by the same mechanism as in Stage One, thus returning Oceanus Pacificus to lower latitudes again (cf. Schultz, 1985; Runcorn, 1987). This displacement took place at shallower depths, however, and over a longer time, because of the smaller impact-melting effects, mass deficiencies, and global perturbations of mantle convection. Assuming that the Stage Two cycle took somewhat longer than had Stage One to complete its reorientation, then it probably continued into the early Imbrian Period of lunar chronology (Wilhelms, 1987, Table 14.3; cf. Ryder, 1992; Ryder and

Dalrymple, 1992). During Stage Two, impacts from satellitic infall dwindled, and the subsequent impact history corresponded to the background rate characteristic of the Solar System at large (i.e., the CRF mechanisms began to be discernible relative to the massive initial bombardments of the satellitic infall stages). [Ryder (1992) and Ryder and Dalrymple (1992) have studied the distribution of ages in Apollo 15 KREEP basalts and the oldest known impact melts from lunar returned samples, finding a sharp cutoff in the ages of impact melts at ~3.9 Ga. They interpret their results to indicate that a "tightly constrained bombardment of the Moon" occurred at about that time, but that there is no evidence of any earlier bombardment events. According to the interpretation given herein (above), the ~3.9-Ga "event" may have marked the dynamical transition between, on the one hand, the latest inertial-inversion stage following the earliest regimes of massive-infall and impact-basining of the Moon (e.g., the scenarios of Fig. 33 and of Runcorn, 1987) and, on the other hand, the earliest record of extraterrestrial bombardments of the Moon from "normal" Solar System reservoirs of objects — recognizing that the general Solar System bombardment rates at that time may have been quite high relative to present-day rates (cf. Chap. 2; and Chyba, 1991; Chyba and Sagan, 1992). It may also be of interest to notice that such a transition interval on the Moon at ~4 Ga has some superficial resemblances to the putative resurfacing event on Venus at about 500 Ma, as discussed in Notes 8.8 and 8.9.]

At the end of Stage Two activity, Oceanus Pacificus (minus its ultramafic root) was again centered near the equator, while the band of sialic crust formed by normal igneous processes prior to, as well as by cratering at low latitudes during, Stage Two (perhaps representing the proto-core-complexes of the present-day continental masses) became reoriented in the meridional direction (cf. Fig. 33, and discussions of analogous reorientation sequences, or "true equatorial-wander" sequences, in Schultz, 1985, and Runcorn, 1987). Because of waning impact energy and magmatic activity, these configurations became effectively "frozen in," leaving a long-term (and probably extant) shallow mass deficiency at the poles, especially at the pole toward which the maximum of the Stage Two impacts migrated (Fig. 33, bottom). This deficiency was partially compensated by higher densities at depth and by a deep mass excess localized within a broad meridional band, especially beneath the Stage One (polar) position of Oceanus Pacificus (Fig. 33, top). The "frozen-in" mass configurations that resulted from the combined actions of Stages One and Two guided the subsequent convective motions in the mantle because of the anchoring and/or buttressing effects of the deep ultramafic anomaly (which had, and still has, a density, and ratios of Fe:Mg, higher than those of the average mantle) combined with the persistence of relatively shallow low-density and mass-deficient regions along the new meridional band and near the poles, induced by the effects of the reorientations of mass.

Unfortunately, information on radial distributions of mass within the Earth depends largely on interpretations of seismic velocity data (e.g., Dziewonski and Woodhouse, 1987; cf. Bowin, 1992). A more detailed picture of the radial and horizontal gradients of mass and inertia would depend on measurements of the

Earth's responses to torques exerted on the axis of rotation (e.g., Kaula, 1968, p. 75). Other than the dominating effects of the Sun and Moon on the Earth's axial precession, little is known with certainty about the fine structure of Earth's moments of inertia. The overall figure of the Earth (the *geoid*) represents an appropriately weighted average of many contributions, including, in addition to the average variations of mass with depth, the dynamical effects of all convective transfers of mass in the horizontal and radial directions. The famous "pear-shaped" figure of the Earth (Fig. 33, Inset 1; modified from King-Hele, 1976) is the resultant of this net effect, as measured by deviations of the surface from a reference shape assumed to be the Earth's equilibrium figure. These deviations could be referred to a sphere, but a dynamically more meaningful reference shape is that of a perfect liquid (or any condensed state of zero strength, or zero stress-relaxation time) rotating in a symmetric gravitational field. Because this assumption does not apply to the Earth, the reference surface for the geoid is taken as a liquid level (sea level) that conforms to an underlying spheroid with a flattening equal to that of the actual equatorial bulge due to the rotation of the Earth, *as measured by the torques acting on the orbits of artificial satellites crossing it at oblique angles* (King-Hele, 1976, Fig. 1). The preceding passage is emphasized not only because it reveals the sensitivity of artificial satellite orbits to the shape of the geoid, but because it also suggests that *relative* orbital perturbations within a system of artificial satellites may offer the most sensitive device for the detection of natural satellites, and/or other near-Earth objects passing near, or through, its vicinity (cf. Cook and Scott, 1967; Bagby, 1969; Wagner and Douglas, 1969; and Notes 6.5 and 6.10 of the present volume). The analysis of artificial-satellite orbit data gives an observed flattening of 1 part in 298.25 for the Earth (in other words, the polar diameter is less than the equatorial diameter by a roughly 0.3%).

The pear-shaped figure (Fig. 33, Inset 1) is based on a summation of spherical harmonic terms up to about $J_n = 20$ (King-Hele, 1976, Table 1 and Fig. 3; cf. Table 1, Eq. 2 in the present volume), where the odd harmonics beyond the dominant effect of flattening (J_2) are most significant to perturbations of satellite orbits. Of these, J_3 is the largest and therefore dominates the shape, as shown in Figure 33, hence giving Earth a somewhat toplike rotation ("upside-down," for the North Pole as "up") consistent with the propensity of the spin axis to wobble (like an unbalanced top, or out-of-round wheel (see Note 8.5). Top-like effects could be consistent with a South Polar bias in the axial part of the deep-mass anomaly (e.g., suggested by the flattened or indented South Polar end of the spinning pear-shaped figure), as suggested by the Stage-One scenario of Figure 33, but such interpretations are nonunique (cf. Bowin, 1992). Conceivably, the pear-shaped symmetry for the figure of the Earth relates in some way to the threefold symmetry of the Phanerozoic cratering nodes (PCNs) of the CRF, and to the averaged great-circle orbital trajectories in Figure 23. The bisector of these trajectories in the Northern Hemisphere (Fig. 23) defines a meridian that intersects the Equator at about 70° W and 110° E, and therefore the symmetry plane of the putative geocentric satellitic reservoir of the CRF system *almost coincides with the trend of the meridional*

Self-Organization and the Celestial Reference Frame

band of anomalies in Figure 28 (bottom) and Figure 33 (deep P-wave anomaly pattern shown in Inset 2).

I pointed out earlier (in Figs. 27 and 28) that the PCNs correlate in a general way with the patterns of principal geomagnetic flux lobes at the surface of the core (Gubbins and Bloxham, 1987). In the Northern Hemisphere, both the cratering and flux-lobe patterns are roughly tangential to the projected trace on the Earth's surface of the inner core (see Fig. 27). This correspondence does not appear to be true in the Southern Hemisphere, but we are hampered there by the lack of cratering data at latitudes higher (farther south) than about 30° S. The existence of the Acraman Crater in South Australia (Fig. 1) suggests that the Australian PCN may extend somewhat farther south than is shown in Figure 2, but the same sort of bias is also evident in the flux lobes of Gubbins and Bloxham (1987, Fig. 1). In Mercator projection (Fig. 28), flux lobes 3 and 4 are roughly 20° lower (farther north) in southern latitude than are lobes 1 and 2 in northern latitude. This suggests some skewing of PCN plus flux-lobe invariance in a sense equivalent to a rotation parallel to the plane of the meridional pattern of both phenomena (flux lobes and PCNs). Alternatively, the asymmetries of both patterns relate to the form of the deep-mass anomaly and its effects on spin-related phenomena both internal and external to the Earth. In that case, the PCN/flux-lobe asymmetric correspondence would be a form of external/internal (satellite-motion)/(core-motion) resonance mediated by the dynamics of the impact-cratering process.

Analogous effects can be seen in the structure of mass anomalies inferred from the pattern of maximum mantle P-wave velocities at a depth of ~2300 km (data of Dziewonski and Woodhouse, 1987, Fig. 9), as compared with VGP patterns and calculated trajectories of fluid motions in the core illustrated by Laj et al. (1991, 1992), where the 2300-km anomalies are only a few hundred km above the D″ (read "D double-prime") layer and core-mantle boundary. All of these patterns need to be compared with the implications of the two-stage model of mass inversions (Fig. 33) over longer intervals of geologic history than have so far been documented (cf. Figs. 27, 28, and 33). A detailed history of transitional VGP patterns — such as the systematic migrations, sudden jumps, and quasi-invariant clustering effects suggested by many of the studies cited above (cf. Clement, 1992; K. A. Hoffman, 1992, 1993) — over a time frame $\gg$ 10 m.y., compared with the details of an improved impact-cratering record, offers perhaps the most cogent test of (1) self-organized dynamo actions, (2) paleomagnetic-reversal mechanisms, (3) PCN-CRF invariances, and (4) their mutually coupled nonlinear resonances via magmatism and related tectonic-transport mechanisms (cf. Appendixes 5, 16–18, 22, and 23).

The picture I have laid out might appear at first glance to present a problem in terms of reconciliations between the deep-mass anomaly (the *meridional geodetic keel*, MGK, of anomalous mass frozen into the deep mantle by the processes outlined in Fig. 33) and mantle-wide convection (cf. Kellogg, 1992). However, inspection of the locations of all of the geophysical anomalies discussed above (e.g., Figs. 2–4, 23–28, and 33) shows that they also correlate crudely with the principal subduction zones, where dense lithosphere penetrates through the upper mantle, and possibly deeper (see the dotted trace of the approximate circumpacific rim in Fig. 3 and the location of the Mediterranean in Fig. 2; the complex Caribbean region coincides with K/T cratering and with the suggested locations of early giant cratering scars that align with the Americas). In other words, the original "frozen-in" configurations of the Stage-One and Stage-Two axial inversion processes that formed the MGK also provided a guide for subsequent subduction processes throughout Earth's history.

Such a long-term correlation is suggested by at least three sets of facts that support the notion of an Earth that is "partitioned" relative to the MGK: (1) the general meridional alignments of geophysical phenomena discussed above, (2) the generally compatible meridional distribution of the major continents (cf. Figs. 4, 28, 33, and 34; and Appendix 26), and (3) the coordination of the principal "lithospheric graveyards" of the most recent plate-tectonic cycle, dramatically illustrated by Engebretson et al. (1992). These authors show contours (ibid., Fig. 1) of the cumulative area of globally subducted lithosphere during the latest 180 m.y. of Earth's history, revealing a sharply defined meridional band that plots almost on top of the deep-mantle P-wave anomaly shown in the present volume (Fig. 33, Inset 2), *precisely along the meridional section of the putative MGK* and the other geophysical anomalies just discussed. And Africa only seems to be an exception to generalization (2) because, during the Mesozoic, it became "surrounded" and caught up in the upper-mantle ocean-floor production processes that attended the growth of the Mediterranean Sea and the Atlantic and Indian Oceans, thereby sufficiently decoupling it (and/or South America) from the deep mantle to partially override the centering effect of the MGK. [Compare the continental reconstruction given in Fig. 34 with that given in Appendix 26, and compare both with the distributions of (a) the lithospheric graveyards of Engebretson et al. (1992, Fig. 1), (b) the flood-basalt provinces of Figs. 26 and 28, (c) the "hot-spot" and "hot-spot"/cratering patterns of Appendixes 12–18, and (d) the volcanic patterns between the Cape-Fold-Belt-Karroo system of South Africa and the putative Permo-Triassic impact events on the Falkland Plateau between South America and South Africa, as proposed by Rampino (1992); cf. Oberbeck et al. (1992) and Note 8.11 of the present volume.]

Plate motions, subduction, and continental drift apparently represent mechanisms for orienting and "centering" the mass distributions of the upper mantle, lithosphere, and crust relative to the rotational figure of the Earth (cf. Monastersky, 1988). Upper-mantle magmatic and plate-tectonic processes, operating generally above the 660-km discontinuity, therefore "mimic," or "remember," the centering effects of the MGK, but in a relatively piecemeal and superficial way compared to the phenomena of Figure 33. For example, Bowin (1992) observes that the large anomalies of the Earth's satellite-determined geoid (of spherical harmonic degree 2–3) "must arise from the deep mantle or deeper." In other words, although the details of the gravity field cannot be correlated point for point between shallow and deep Earth structures, the general configuration and amplitude variations of the geoid would imply unreasonably large near-surface mass anomalies, and therefore the requisite masses must reside in the Earth's deep-mass distribution.

From a different perspective and set of observations — in this case related to a calculation of the net rotation of the outer "lithospheric" region of the Earth relative to the deeper mantle (cf. Shaw, 1970; Shaw et al., 1971; Bostrom, 1971, 1992; Gordon and Jurdy, 1986) — Ricard et al. (1991, Fig. 5) show a map of global "viscosity" variations that are consistent with net global plate motions and depth intervals beneath the continents and oceans. In other words, "viscosity," as portrayed in their illustration, is really just a shorthand for the comparative mean relative layer-thickness, thermal structure, and vertical gradient of horizontal velocity beneath the oceans and the continents necessitated by a net conservation of mass in the global flow system (cf. Note P.5; and Shaw, 1973). As expected, therefore, the contours of high "viscosity" mimic the distribution of the continents, and, accordingly, the map of the "viscosity-structure" of the Earth's lithospheric shell corresponds to the general pattern of the deep meridional anomaly (MGK) discussed in the present volume, thereby testifying to a marked coherence between the shallow and deep meridional structures in the Earth — a coherence that also is seen in the topographic configurations of the core-mantle boundary and of the 660-km discontinuity, which I shall describe in the next section.

The net westward drift of the lithospheric shell calculated by Ricard et al. (1991) is 2 centimeters per year — thus posing a problem, but only if that angular motion were assumed to be characteristic of the relative corotations of the Earth's outer layers over geologic time (e.g., the lithospheric structure as a whole would have been displaced westward by 2000 km during the latest 10^8 years, or 20,000 km since the late Precambrian). These distances are not necessarily prohibitive (the larger distance being only about one-half of a full cycle of relative rotation), but if the similarities between the meridional patterns with depth that I have pointed to reflect long-term properties of the Earth, as the model of Figure 33 and the MGK imply, then either the structural correspondences with depth are fortuitous or the relative rotation calculated by Ricard et al. (1991) has not been constant over geologic time (cf. Ricard et al., 1992). An alternative that would accommodate the simultaneous existence of lithospheric westward drift *and* the MGK would suggest a model of oscillatory relative rotations, a conclusion already tentatively reached for the general "confinement" of continental drift and cyclic continental reconfigurations within the meridional bandwidth mediated by the MGK. An advantage of an oscillatory model is that coupling between the relative rotations of different shells of the Earth and the tidal history of the Earth would be compensated indirectly through the dissipative processes of magma generation, the evolution of plutonism and volcanism, and the consequent magmatic-tectonic influences on gravitational potential gradients for lithospheric translations (cf. Shaw et al., 1971).

The role of the MGK over Earth history, in conjunction with plate tectonics and CRF impacts, becomes — in a dynamical sense — the metaphorical equivalent of the keel of an actual sailing vessel, where the mass distribution and waterline length of the hull plus the pressure distribution on the masts and sails act in counterpoise to the mass, length, and depth of the keel as the vessel cuts through the water, always seeking the configuration that minimizes the work required to overcome the net effect of the frictional resistance that both modulates and limits the maximum sailing speed. The feedbacks among the differently acting dynamical mechanisms (coupled oscillators) of an oceangoing sailing vessel, like the diverse mechanisms acting on and within the Earth, are complex. For instance, the keel not only stabilizes the furled-sail floating mode of the vessel, but its specifications provide the crucially important, continually varying angles of attack of the keel + hull in the water — according to the multiple-feedback system established with all of the other dynamical components of the vessel — thus representing a *tuning mechanism* by which the angular orientations of the masts, spars, and sails (relative to the area of sail, angle to the wind, and vector of resolved motion across the water) are sustained in nonlinear-dynamical periodic and/or chaotic resonance with the dynamics of the ocean and atmosphere.

The MGK performs a similar tuning function, relative to the internal dynamics of the Earth, as that of the sailboat keel, thereby sustaining analogous forms of resonances with the Earth's orbital parameters in space (representing "the interplanetary ocean") and with its buffeting by the solar wind and the showers of meteoroids that approach its vicinity (representing the "CRF-interplanetary weather and climate system"). In all of these analogous respects, the tuning mechanism provided by the MGK has been crucial to, among other things, the evolution of (1) the aggregate motions of the Earth's core, mantle, and tectonic plates, (2) the coupling between the Earth and the CRF system of satellitic resonances (e.g., Chap. 6 and Figs. 19–21), and (3) the postpartum orbital history of the Earth-Moon system. In this sense, Earth's paleogeographic history is really concerned with the reconciliation of feedback relationships between mass transfers in the outer shells of the solid Earth relative to the MGK and related feedback processes involving (a) the deep mantle, (b) the D″ layer just outside of the core-mantle boundary, and (c) the inner and outer core. As a figurative and schematic guide to such ideas, I expand a bit more in the next section on what may have happened following the Stage-One and Stage-Two inertial revolutions during Earth's primal history (as sketched in Figure 33, and as outlined above in terms of the formation and fate of Oceanus Pacificus) that attended the creation of the paleogeographic Pacific Basin of late Precambrian and Phanerozoic history.

Before going on with that, however, it is worth pointing out here that a paper by Pulliam and Stark (1993), just published, warns against the statistical artifacts that are often encountered in fitting smooth models to inhomogeneously distributed data on a sphere. Their discussion is particularly relevant to spherical harmonic analysis and the implications of structural correlations — especially those that would interpret any intricate type of small-scale structure as being characteristic of the global structure — on the basis of large-length-scale geophysical data arrays. Their remarks would appear to apply particularly to models that would view "thermal plumes," "bumps on the core-mantle boundary," etc., as being globally characteristic. With reference to seismic travel-time models, they have this to say: *"The usual model expressing travel time residuals as the sum of effects due to a finite-dimensional perturbation of Earth structure and independent, zero-mean Gaussian noise (plus station corrections, etc.) is almost surely false, since Earth is virtually certain to contain structure [sic] not expressible in the finite-dimensional set of models. The effect of that structure on the data is treated formally as 'noise,' which is then typically neither zero-mean nor independent"* (italics added for emphasis).

These points bring to mind remarks made elsewhere in the present volume concerning the implications of attractor theory for geophysics and astrophysics, viz. (1) coeval natural phenomena cannot be treated with any validity — even in the case of celestial dynamics — as outcomes of completely decoupled processes, and hence they cannot be equated with models of independent phenomena lacking dissipative interactions (e.g., Hamiltonian models of orbital trajectories; cf. Note P.3), and (2) because self-similarity and universality imply that spatiotemporal patterns are hierarchical over many scales of length and time (self-organized criticality), there is an implicit requirement that model correlations — such as those between volcanic "hot spots" and "thermal plumes" — be consistent with the appropriate hierarchical ranges of related processes documented for the Earth's mantle and crust, as well as with the patterns of volcanic propagation at the Earth's surface [i.e., they must be consistent with the many characteristic scales of volcanotectonic effects documented by Jackson et al. (1972, 1975), Shaw (1973, 1980, 1987a), Bargar and Jackson (1974), Jackson and Shaw (1975), Jackson (1976), Shaw et al. (1980a), Klein (1982, 1987), Dzurisin et al. (1984), Koyanagi et al. (1987), Ryan (1988), and Shaw and Chouet (1988, 1989, 1991); cf. other summaries cited herein of research conducted by the Hawaiian Volcano Observatory: Dalrymple et al. (1973), Wright et al. (1976, 1992), Decker et al. (1987), Holcomb (1987), Takahashi and Griggs (1987), Tilling et al. (1987), Heliker and Wright (1991), Tilling and Dvorak (1993)].

The mere demonstration, for example, that plumes (upward *or* downward) are *possible* in a hydrodynamic experiment in the laboratory does not constitute evidence that they do in fact exist in the Earth's mantle [e.g., compare the prototype model of an upwardly propagating, diapir-like, thermal plume, as proposed by Morgan (1971, 1972, 1981, 1983) — and the legions of its conceptual "descendants": cf. Vogt (1972, 1975, 1979, 1989); Loper and McCartney (1986), Dziewonski and Woodhouse (1987), Olson and Hagee (1990), D. L. Anderson (1991a, b), Hagee and Olson (1991), Larson (1991a, b), Larson and Olson (1991), D. L. Anderson et al. (1992), Sleep (1992) — with the downwardly penetrating residual-density-current model proposed as an alternative by Shaw and Jackson (1973)]. Now, however, simultaneous announcements by Tackley et al. (1993) and Honda et al. (1993) — describing new results of three-dimensional numerical modeling of mantle convection — have demonstrated the likelihood of downwelling cascades ("avalanches") of upper mantle material into the lower mantle that proceed in a manner that is qualitatively analogous to, if not dynamically equivalent to, the feedback-cascade model proposed by Shaw and Jackson (1973, pp. 8641ff). The issue of two-dimensionally localized positive and/or negative penetrative advective flows in the mantle, "plumes," "jets," "cascades," or "avalanches" would appear to be in as great a state of flux today as it was in 1971–73, when the quest for a fixed mantle reference frame for global plate motions (Goldreich and Toomre, 1969) was first being systematically pursued.

Schematic Paleogeographic History of the Earth Following Formation of the Moon by a Collisional Mechanism

If the *meridional geodetic keel* (MGK) that evolved from the original root system of the Stage-One magma ocean (Oceanus Pacificus) of Figure 33 had remained aligned with its later surface expression, the Pacific Basin, then it would now be misaligned with the meridional geophysical anomalies discussed above, which correlate generally with the Americas and the vicinity of Australia (e.g., Fig. 28; and Laj et al., 1991, 1992; Engebretson et al., 1992, Fig. 1). But I suggest that the transition from the regime of thin-skin crustal foundering that was characteristic of the Oceanus Pacificus magma ocean to the essentially "modern" regime of steady-state seafloor spreading and subduction involved a westward drift of the upper mantle relative to the deep mantle such that the early-formed protocontinents became centered above the deep-mass anomaly (the MGK) while the upper part of the crystallized magma ocean (i.e., what eventually became the present-day Pacific Basin) migrated to a position above "normal" preexisting mantle material (meaning that the deep and shallow fractions of the igneous differentiation process in the magma ocean became relatively displaced or "detached" relative to their original vertical continuity). This transitional westward drift coincided with the inception of the more familiar forms of seafloor spreading and subduction of the displaced upper mantle, which was generally less mafic than the "average" mantle (cf. Kellogg, 1992) because the most ultramafic fraction had been removed from it, and which, therefore, was subject to partial melting, upper-mantle magmatic upwelling, and the inception of lateral spreading episodes and subduction of lithosphere in the vicinity of the detached root zone. The familiar plate-tectonic cycles of today began therefore by involving the displaced upper mantle that originally had been directly above the MGK but that was now peripheral to it (cf. Monastersky, 1988). The MGK, acting as a sink for subducting lithosphere, thus generated a feedback circuit whereby the evolving subduction zones marginal to the protocontinental masses eventually "centered" themselves relative to the MGK **(Note 8.9)**. According to this scenario, the principal subduction zones of that time would have been on opposite sides of the laterally displaced, and lithologically fractionated, ancestral Pacific Ocean — a global asymmetry that persists today, except for enclaves such as the Mediterranean, Caribbean, and portions of the Atlantic and Indian Oceans that may also represent relicts of early Precambrian impact-basining phenomena.

In other words, during the early Precambrian, the primitive deep-mass anom-

Self-Organization and the Celestial Reference Frame

aly of Figure 33, which evolved to become the MGK, provided the primary gravitational gradient that guided the first increments of seafloor spreading and subduction that evolved from, and have interacted with, episodes of continent fragmentation and drift since the end of Stage Two in Figure 33 (i.e., representing a "thick-skinned" hence sluggish steady-state heat-transfer mechanism by comparison with the thin-skinned highly efficient heat-transfer crustal-foundering process in the primitive magma ocean; cf. Note 8.8; and Duffield, 1972; Galer, 1991; Schubert, 1991; Kröner and Layer, 1992). Paleogeographic patterns in the Precambrian, and perhaps even in the early Paleozoic, therefore, would have involved diverse trajectories of motions of continental fragments within an evolving, and westwardly migrating, meridional band near the Earth's surface. Later, during the Mesozoic and Cenozoic, large accretionary continents became more constrained in their movements, forming a more centered meridional band of near-surface processes — with exceptions and variations (e.g., the pattern of "lithospheric graveyards" illustrated by Engebretson et al., 1992, Fig. 1, shows a relatively weak north-south septum extending from the Arctic into the North Pacific, and an east-west septum extending from the vicinity of the Himalayas to the Mediterranean). In general, however, the MGK, and the mimicry of the longitudinally offset upper-mantle and continental meridional structures, have persisted as the principal guiding mechanism for plate tectonics, continental drift, and true vs. apparent polar-wander paths.

Because no paleogeographic models show the "closing" of the Pacific Ocean during these times (or at any other time, so far as I am aware), a consistent picture has it persisting over geologic time much as it is today (cf. Appendix 26). The Pacific Basin has been surrounded by a "ring of fire" and by subduction zones that have controlled both the depth of mantle convection and the cycles of upwelling to feed the mid-ocean spreading centers (cf. Walzer and Maaz, 1983), which have varied in positions and rates of upwelling during about the last 2 Ga or so (i.e., during about half the Earth's post-lunar lifetime, according to my interpretation of the collisional origin of the Moon; cf. Fig. 33, and Notes 8.8–8.11). Our present views concerning questions of mantle chemistry (especially models that are tied to any *single* mantle composition) are, I suggest, at least partly artifacts of the contrast between theories of homogeneous fluid convection and convection in a dissipatively coupled (magmatically coupled) heterogeneous mantle driven by self-organized rheological feedback phenomena of the sorts discussed in Notes P.5–P.7 (cf. the review of mantle chemistry by Kellogg, 1992). Earlier than this, and also during the time from about 2000 to 250 Ma, piecemeal processes of progressive continent evolution [e.g., by orogenic-epeirogenic cycles analogous to those discussed by Evernden and Kistler (1970), Kistler et al. (1971), and Shaw et al. (1971)], crustal rifting, and subduction-like sinking (foundering) of lithospheric slabs may have borne resemblances to several different, and chaotic, regimes of "lava lake tectonics" (Note 8.8; cf. Shaw et al., 1971, p. 878; Duffield, 1972). Sengör (1987, Fig. 3) refers to this oceanic hemisphere in pre-Mesozoic times as "Panthalassa," a later evolutionary stage of what I have called Oceanus

Pacificus in Figure 33 (certainly, by this interpretation, the birthplace of all terrestrial gods, whether of water, earth, air, or fire). The general processes outlined above, and in Figure 33, have constrained continental drift within longitudinal tolerances controlled by radial and horizontal compensations of mass, and hence have contributed to the maintenance of mass anomalies that have constrained CRF impacts within similar tolerances to feed the magmatism, mantle convection, and core motions that complete the perpetuation of this feedback cycle. [A subtle, and perhaps confusing, aspect of the above discussion of the relative symmetries and asymmetries of the Earth's deep and superficial mass distributions arises from the fact that the surface areas of the earliest postcollisional stages of the evolution of the Earth and Moon were each dominated by essentially one complex multi-ring impact basin, Procellarum on the Moon (cf. Wilhelms, 1987, Plates 3–8) and Oceanus Pacificus on the Earth (cf. Fig. 33 and Appendix 26).]

Major inversions of latitudes and longitudes in reconstructed subregions of what are now the major continents have been proposed for the late Precambrian and early Phanerozoic (cf. P. F. Hoffman, 1988, 1991; Dalziel, 1991, 1992a, b; Hartnady, 1991; Moores, 1991; Young, 1992). These and many other paleogeographic reconstructions for late Paleozoic and Mesozoic times rely heavily on combinations of geology and interpretations of true polar-wander paths (see Smith and Livermore, 1991). Other than inferences drawn from reconstructed geomagnetic-pole positions (and the assumption that they identify the approximate position of the Earth's pole of rotation), geophysical methods have provided no rigorous means for the determination of paleogeographic latitudes and longitudes (cf. Gordon, 1987). The CRF, however, offers an entirely new outlook for the evaluation of such constraints on kinematic manipulations, even if at this juncture the putative celestial reference frame is hypothetical and broadly defined.

According to the celestial reference frame hypothesis (CRFH), Phanerozoic subduction zones and continental paleogeography have been largely localized within the bandwidth of the net MGK anomaly described above—which is approximately delineated by the widths of the PCNs + CRF nodal paths, geomagnetic core-flux lobes, VGP paths, deep-mantle P-wave anomalies, and "lithospheric graveyards" (Engebretson et al., 1992, Fig. 1), all of which conform more or less to the meridional pattern shown in Figure 28. An additional point of major interest is that the existing data on the topography of the core-mantle boundary (Morelli and Dziewonski, 1987, Figs. 3–6; Vogel, 1989, Fig. 3) *and* on the topography of the 660-km discontinuity (Fukao, 1992, Fig. 4; Shearer and Masters, 1992, Fig. 4; Lay, 1992)—the latter usually being taken to represent the boundary between the lower and upper mantle (cf. D. L. Anderson, 1989, 1991a)—*also* have remarkable parallels with the MGK bandwidth. A global correspondence between the configurations of the core-mantle boundary (CMB) and the base of the upper mantle (BUM), amounting to a total relief of about 30 km or so on the "660-km" discontinuity (i.e., $660 \pm \sim 15$ km), would suggest that the MGK ($\pm$ components of relative westward drift between different shells) represents a long-

Self-Organization and the Celestial Reference Frame

term memory effect at the most fundamental levels of Earth structure — as would be predicted from the model of Figure 33. Lay (1992) reviews the long-standing controversy concerning mantle discontinuities as changes of bulk chemical composition or as isochemical changes in density (mineralogical phase transitions). By either point of view, or some combination of them (D. L. Anderson, 1991a), the influence of the deep-mantle mass anomaly of Figure 33 — which forms the basis of the MGK and CRFH global "centering" mechanisms — was likely to have been mimicked by the displacements of the equilibrium density contours at the CMB, and at any other transitional boundaries within the mantle, unless such mimicry was masked by the overriding dynamical effects of convective and/or advective transport mechanisms in regions of contrasting rheologies (e.g., as happens in the uppermost mantle relative to mass and heat transfers associated with the real or apparent boundaries of the asthenosphere, lithosphere, and crust). Shearer and Masters (1992, p. 795; cf. Fukao, 1992, p. 628) and Lay (1992, Fig. 2) have suggested that deflections of the BUM could have been caused by flattening of subduction slabs at that depth with attendant thermal depression of the density contours. For present purposes, however, the main thing is that the above topographic effects appear to be meridionally correlated with the MGK, with subduction, and with the other geophysical anomalies I have discussed.

Although I am not yet able to describe the motions of specific continents or continental fragments as being consistent or inconsistent with the CRFH, some general implications are summarized in **Figure 34** (cf. Appendix 26) on the basis of the reconstructions of Dietz and Holden (1970) for the breakup of Pangaea **(Note 8.10)**, which began, according to their model, at the end of the Permian, for which they gave an age of 225 Ma [the presently accepted age for the Permian/Triassic transition is 245 Ma, as given by Harland et al. (1990)]. Figure 34 was redrawn from the reconstruction by Dietz and Holden at "180 Ma" in their chronology (~200 Ma in the revised chronology), or ~45 m.y. after the breakup of Pangaea had begun (cf. Note 8.10; and Norton and Sclater, 1979; Bambach et al., 1980; Ziegler et al., 1982; Duncan and Hargraves, 1984; Royer and Chang, 1991; Smith and Livermore, 1991).

The modified drawing (Fig. 34) also shows the present positions of the continents (dotted outlines), the CRF nodes (heavy asterisks), and the Phanerozoic (double-dash-dot) and K/T (heavy short dash) nodal-cratering swaths. The three CRF nodes (PCNs) remain fixed in absolute relative positions — according to the CRFH — hence are fixed with respect to the immediately adjacent stable craton on which they are "imprinted" (i.e., CRF invariance does not constrain the deformations, rotations, and/or translations of continental fragments that do not either specifically include the PCNs or that can be shown to be decoupled from the continental structure that does include the PCNs). Otherwise, the continental masses that contain the PCNs can individually rotate (i.e., about an individual PCN) and/or they can rigidly rotate together as an invariant triangular set of points (i.e., "in lockstep") on the globe. In other words, if there were no fixed relationship between the PCNs and the Earth's spin axis, the meridional band (i.e., as projected on the Earth's surface by the MGK) could have experienced arbitrary shifts parallel to that meridian and still have retained the mutual invariance of the PCNs. If, however, the positions of the PCNs and of the orbital configurations of the system of geocentric CRF impactors (for a given astronomical epoch) *in concert with* the MGK represented the prevailing inertial configuration of the Earth plus the guidance system for CRF resonances and impacts originating from the geocentric system of natural Earth satellites — as is the fundamental principle of the CRFH — then the orbital configurations of the geocentric reservoir of objects, the PCNs, and the Earth's spin axis would have all been resonantly locked together, and they would have remained so until such time as the MGK, and all geodynamic processes coupled with it, experienced a revolution as profound as the one that established it in the first place. (An alternative paleogeographic reconstruction from that of Figure 34 — one in which I attempted to maintain an invariant spatial relationship between the three PCNs and the Earth's spin axis — is explored in Appendix 26).

The putative effect of impact cratering on the rifting process (e.g., Seyfert and Sirkin, 1979, p. 96; Alt et al., 1988) is imagined to act somewhat like the release mechanism of a zipper on oceanic or continental crust underlain by a magmatic or potential magmatic source. If an impact is sufficiently energetic to activate fracture-related magma percolation from the asthenosphere (cf. Shaw and Chouet, 1991), with or without contributing directly to the melting process itself, a self-perpetuating chain reaction or "thermal runaway" may ensue (cf. Shaw, 1969, 1973). Rifts can then "unzip" oceanic or continental crustal segments by lateral-extensional fracture propagation (cf. Shaw, 1973, 1980; Jackson and Shaw, 1975). The illustrations on the back and front covers of Silver and Schultz (1982) dramatize such effects at a time shortly before the stage shown in Figure 34 (back cover) and at a later time when intracontinental rifting occurred in northeastern North America (front cover). The relative positions of these dramatized impacts were at about the same location, just east of the North American PCN, a location that roughly bisected the distance between the North American and Eurasian PCNs at the time of the earlier impact. If these two nodes had maintained a constant relative distance, however, as proposed here, the configurations of the continents must have been somewhat different from the conventional configurations shown in Figure 34 or on the back cover of Silver and Schultz (1982). Such a reconfiguration for the opening of the North Atlantic would have required a greater counterclockwise rotation of Eurasia relative to North America and a more northerly compensating displacement of Africa (or segment thereof) and Greenland. If this were shown to be impossible, we would have to contemplate a revision wherein the North American and Eurasian PCNs would have reflected a more continuous cratering swath between them, but this might demand, as well, more evidence of cratering in contiguous portions of northwestern Africa — which, so far, does not exist.

It is evident that the 200-Ma reconstruction of Figure 34 does not depart greatly from a meridional $70°$ W–$110°$ E bilateral symmetry of hemispheres, although the continents have been compressed into one hemisphere by the traditionally assumed best fit of the continents. So as to conform approximately to this pattern, as mentioned above, different mutual rotations of northern North America and northern Eurasia would have been required, compared to those shown (the heavy circular arcs with arrows shown in Fig. 34 refer only to the relative sense of subsequent opening of the North Atlantic in the model of Dietz and Holden, 1970; cf. Ziegler et al., 1982; Duncan and Hargraves, 1984). Analogously, the relationship between Australia and Antarctica, interpreted according to the CRFH, would imply that the latter separated from the former in later times, rather than vice versa as is assumed in most models (cf. Royer and Chang, 1991). If evidence of glaciation requires that Australia be assigned a high southern latitude during this period (e.g., Crowley and Baum, 1991), that would have had to be accommodated by some amount of clockwise rotation of the three PCNs together, as discussed above **(Note 8.11)**. The main implication near the North Pole would be a revision of the opening of the Arctic Ocean, compensated in part by the suggested more northerly backtracking position of Greenland relative to the North American node. I have not yet undertaken detailed comparisons with published interpretations relevant to the history of the Arctic Ocean based on geologic and paleomagnetic data (cf. Hughes et al., 1981; Hillhouse and Grommé, 1988; Stone, 1989; Lawver et al., 1990; Harbert et al., 1990; Zonenshain et al., 1991).

My interpretation of a CRF that has been mapped out by the cratering history of the Earth, combined with the geomagnetic and plate-tectonic criteria of meridional invariances (the MGK effects discussed in the previous section), is consistent with a relatively persistent general pattern of continental aggregation. This idea, when viewed in light of the recent developments in accretionary terrane processes, may reconcile some of the discrepancies noted above (see Note 8.10; and the hypothetical reconstruction illustrated in Appendix 26). The motions of continental masses apparently have been somewhat pole-seeking at times rather than "pole-fleeing" (see interpretation of the Eötvös force by Goedecke and Ni, 1991), but *throughout the geological record there appears to have been a general oscillatory motion of the continents relative to the meridional distribution* (cf. Note 8.10). The position of Africa, roughly halfway between the opposite limbs of the meridional band, would seem anomalous by this interpretation, but it has been effectively surrounded by spreading centers since the late Jurassic and thereby held in counterbalance between the Atlantic Ocean, the Mediterranean, and the Tethys Sea. The degrees of freedom of local motions (continental drift ± piecemeal motions of accretionary terranes + vectors of seafloor spreading and subduction) are too great to be constrained by any existing paleogeographic models in the absence of a reference frame that is consistent both with the documented history of plate tectonics and with earlier episodes of analogous processes (cf. Gordon, 1987; and Note 8.10 of the present volume). To some extent this difficulty may be relieved by the use of the CRF as a test framework for the long-term evolution of the Earth's surface features that is largely consistent with a persistent meridional system of deep mantle and core structures (the MGK system).

The events in this sketch of the history of Earth's surface motions are analogous to the patterns of chaotic intermittency that I have described relative to other processes in the Earth and Solar System (cf. Note 8.10). In other words, there has been a hierarchy of event magnitudes in Earth history, of which the formation of the Moon and the two-stage mass distributions of the earliest post-lunar impacts on Earth are the largest. The same idea carries over with generally decreasing energetic magnitudes to the later sequences of smaller events associated with major continental drift and large-scale plate motions, the motions of continental fragments (exotic terranes), large intracontinental deformations and strike-slip faulting, large events associated with the GOAM system, and so on down to processes of smaller sizes and shorter time scales, ultimately reaching the local-event scales associated with more normal hydrologic processes of erosion, sedimentation, and so on (cf. Ager, 1984; Gould, 1987a). And this hierarchical outline is by no means a comprehensive description of the complexity of ordinary (uniformitarian) geological processes.

Clearly, the scales of processes in the ocean and atmosphere relative to the solid Earth have been analogously intermittent and hierarchical (the latter term refers only to levels of magnitudes, and does not discount the existence of parallel and cyclic feedback processes of great intricacy that involve cooperative and/or competitive mechanisms among all levels). Such observations lead to the view that the geological record is made up of hierarchies of variable event magnitudes, where each level of activity is intermittent and is characterized by abrupt changes relative to the operative time scale of that level. In a broad sense, an analogous principle applies to the biological record. For example, the biggest event was the origin of the first identifiable organism, and recent studies appear to be pushing the age of that event so far back in time that it may closely postdate my Stage Two of Figure 33 (cf. Knoll, 1991; Knoll and Walter, 1992). Similarly, the next-biggest identifiable event was the evolution of multicellular organisms during the late Precambrian, probably between about 1000 and 580 Ma, which led to the third-biggest event, the seemingly sudden appearance of macroscopic animal life (the Ediacaran fauna). And that was followed by the almost contemporaneous and analogously abrupt advent of the sophisticated global Cambrian fauna, which became the predecessor of the dominant Phanerozoic faunas (Harland et al., 1990; Morris, 1990, 1993; Knoll, 1991, p. 66; Foote and Gould, 1992). The evolution of Phanerozoic faunas represented a sequence of next-biggest steps, including many smaller ones of local scale and/or of brief duration.

In the biological context, magnitude is manifested only in the scaling of the events in the geological record, not in their importance from the standpoint of advancement or complexity (the largest event, by definition, in the context of biological evolution — the origin of the first reproducing organism — was, however, also the largest event in the geological context, since it has affected the entire planet and has had global consequences that have spanned the greatest length of time, etc.). But even in this regard there are some parallels with nonbiological events. For instance, the smaller hierarchies of nonbiological events exhibit a greater integral complexity (as in the trajectories of motion in phase portraits) than do the larger events. Thus, a picture emerges in which biological and nonbiological processes have analogous patterns of hierarchical intermittencies in space and time, ultimately converging on those properties that make up the biostratigraphical record itself, as a relative chronology of both phenomenologies in the Phanerozoic record. In all respects, the role of individual impacts and the CRF, in concert with other geodynamic events, has been to modulate and integrate these patterns hierarchically (cf. Note 8.1).

Nonlinear Biostratigraphic Codification and the Celestial Reference Frame

The essential commentary to be made on this vast and intricate field of research concerns the contextual implications of self-organization, nonlinear crises, power-law temporal behavior, and universality. As I interpret it, biostratigraphy is the attempt to come to terms with these nonlinear effects on the basis of readings of the evidence in the coupled stratigraphic-biologic recording of terrestrial processes — processes that are susceptible of preservation in lithologically and paleontologically recognizable temporal and spatial progressions that can be (ideally) correlated on a global basis. This is the goal sought, as well, in nonlinear dynamics, with respect to trajectories of variation in phase space, especially as regards interpretation of the spatiotemporal dynamics that can affect, as well as effect, the resulting patterns. The celestial reference frame hypothesis (CRFH) argues that critical self-organization and chaotic crises are keys to astrophysical, geophysical, and geological processes — thereby implicating both stratigraphy and biostratigraphy — where principles of universality offer us the best clues to the respective, actual or possible, self-similarities among the phenomena of those processes.

I suggest that the same principles have been at work in both the geological and biological processes that have made up the spatiotemporal signatures of the geologic record, regardless of the criteria by which that record is defined, so long as these criteria are consistent with respect to signs and proportionalities of changes in both space and time. Biostratigraphic, geomagnetic, and chronostratigraphic classifications of events (e.g., pulses of magmatic rise) in time and space may differ (e.g., Harland et al., 1990), but in the view of the CRFH they are all recording the same global system of coupled processes. This is what I have attempted to show with respect to spatiotemporal correlations of geomagnetic, magmatic, and impact processes (e.g., Figs. 8, 28, 30, and 31) that appear to be controlled and modulated by coupling, over time, between a geocentric CRF system of impactors and its counterparts in the Solar System (Figs. 11 and 19–21). These external mechanisms are effective only because a very active internal system of coupled dynamics is at work in the Earth's core, mantle, "magmasphere," glacier-ocean-atmosphere (GOA) system, and biosphere. The comparative dynamics of the Moon and the other inner planets demonstrate that no matter how great the input of external energy in the form of impacts, the global thermomechanical and thermochemical engine cannot be "jump-started" simply by O-I processes, without reciprocal interactions with I-O processes. The Moon showed signs of life in the form of magnetic and convective behavior only during the earliest and most massive phases of its bombardment (Runcorn, 1987), as is shown in graphic detail by the remarkable documentation of its history that Wilhelms (1987) summarizes.

The discussions by McLaren (1970, 1982, 1983, 1986), McLaren and Goodfellow (1990), and Ager (1980, 1984) indicate to me that there are strong affinities between the biostratigraphical problem and the principles of nonlinear dynamics I have addressed. Reduced to three essential properties—and I mean not to be reductionist about them—these cited publications identified the crucial importance of (1) the sharpness (meaning both temporal abruptness and brevity) of any stratigraphic event that can be of use in the derivation of a spatiotemporal framework applicable for purposes of correlations among different Earth processes, whether biological or geological, (2) the event's universality, i.e., its capacity to represent a stratigraphic process of global or potentially global extent that fits within a coordinated framework of dynamical processes that in turn can be understood within that same global spatiotemporal framework, and (3) the spatiotemporal frequencies of characteristic types of events (meaning only that such events have to occur often enough in time and with sufficient geographic representation to be identifiable in terms of the first two criteria). One type of event that might satisfy these attributes would be volcanic ash layers of global extent that came from single eruptions of well-understood cyclic volcanic processes. Although ash-layer stratigraphy has proved to be very useful in regional contexts, global correlations remain problematical. It is therefore not surprising that the possibility of mass impacts large enough to leave global signatures in the form of characteristic mineralogical, geochemical, *and* biological horizons in otherwise relatively continuous stratigraphic sequences became a very exciting prospect indeed (McLaren, 1970; Alvarez et al., 1980).

If we could add to this notion—i.e., to the notion that global biostratigraphic horizons of catastrophic character may exist—the additional notion that in the Phanerozoic history of the Earth there has been a concatenation of analogous horizons hierarchically scaled in magnitudes, therefore in time and space, and that the distribution of such horizons is organized, even if in a nonlinear way, then a very powerful framework would be available for global biostratigraphic correlations. A key factor in the idealized establishment of such a framework would be an ability to understand the nature of the biological consequences at each of the different event magnitudes. From the standpoint of nonlinear processes, this is tantamount to saying that it would be possible to understand the relationships between biological dynamics and impact dynamics, with the intervening geodynamic processes factored in. In other words, the processes addressed in this volume would be, ideally, interpretable at all levels of the biostratigraphic record (meaning, for instance, that we could distinguish between a biostratigraphic event that was directly related to a volcanic event that in turn was related to an impact event, as distinct from a biostratigraphic event that related to an impact event without involvement of volcanic phenomena at that particular biostratigraphic horizon).

The controversy that continues to rage (cf. Glen, 1990, 1994; Marvin, 1990; McLaren and Goodfellow, 1990) between advocates of mass-impacts-as-cause and volcanic-phenomena-as-cause (and/or advocates of other endogenous-processes-as-cause) of biological-extinction events (cf. Holser, 1977; McLean, 1978, 1988; L. Alvarez et al., 1979, 1980; Benson, 1984; Corliss et al., 1984; Gould, 1984a, 1987b; Kitchell and Pena, 1984; Mörner, 1984; Stanley, 1984; Webb, 1984; Officer and Drake, 1985; W. Alvarez, 1986; L. Alvarez, 1987; Hallam, 1987, 1988, 1989; Officer et al., 1987, 1992; Rampino and Volk, 1988; Rampino et al., 1988; Vogt, 1989; W. Alvarez and Asaro, 1990; Courtillot, 1990; McCartney et al., 1990; Montanari, 1990a, b; Rampino, 1992; I. H. Campbell et al., 1992; Czamanske et al., 1992) indicates that such a goal will not soon be achieved, if it is achievable at all at the smaller event magnitudes. However, if I have correctly read the biostratigraphical intentions of McLaren and of Ager, as cited earlier in this section, the above statements address some of the precise queries and goals set forth by them as essential to a correct "reading" of the biostratigraphic record. I shall take up the biological aspects of these diversely interrelated processes in the next section of this chapter, but let me here quote McLaren (1982, p. 478), who had the following to say about the relative nature of "background extinction" vis-à-vis "mass extinction":

> In spite of the convenience of dividing extinctions into two classes, however, it seems likely that in fact we are dealing with a single natural phenomenon. The essential major driving mechanism of evolution by natural selection is environmental change. This may be at any scale from micro-environmental variation affecting a few specialized species to a major catastrophe wiping out most living things on earth. That such events may be considered stochastically does not alter the fact that every extinction or evolutionary replacement must have a cause and, under fortunate circumstances, such a cause might be determinable. Doctrinally this should not offend uniformitarianism; this flexible principle may quite easily be extended to include new ideas on the history of the earth. It has already had to swallow some pretty big ones during the last twenty years. We may rest untroubled: Hutton and Darwin live.

I could not ask for a much better statement of the nonlinear-dynamical context as it operates in the coupled geosphere and biosphere. That context may change our views somewhat from the linear, harmonic, and stochastic ideas of normalcy we have been imbued with, hence also our sense of the meanings of "style," "pace," and "rhythm" of change, but they are alien neither to uniformitarianism nor to natural selection as guiding principles in our interpretations of natural history. For before any process can be identified as gradual or continuous it must be susceptible to inspection using methods of unlimited resolution. And when resolution is pushed to the limit, we find, typically, a hierarchy of intermittently discontinuous effects that is masked by the fabric of complexities seen in the larger and more general contexts (cf. Eldredge and Gould, 1972; Gould and Eldredge, 1977; Gould, 1982b, 1984a, b, 1987b, 1989b; Eldredge, 1989).

In an earlier work I attempted to illustrate the significance of relationships between biostratigraphic event horizons and their possible coincidences with mass impacts by using the metaphor of a nonlinear clock examined in terms of the numerical properties of the gambler's wheel of fortune (Shaw, 1987b). Without repeating that discussion here, I would point out that the properties discussed there are closely related to the properties of critical self-organization and power-law scaling of frequencies discussed here. These scaling factors are expressed in the nonlinear temporal structures of the "geologic column," such as those chronicled by Harland et al. (1990), and in the nonlinear biological life cycles of organic evolution, at various taxonomic levels, that, together, have been nonlinearly coordinated through the common action of those universality parameters that characterize complex networks of coupled oscillators. Such a parallelism between nonlinear-dynamical processes and the nonlinearly distributed records of those processes seems implicit in the above discussions of biostratigraphical phenomena.

[**A Linguistic Digression**: I should mention before going on that I have been using the concept of a "wheel of fortune" in the mathematical sense (e.g., Note 4.2), within the context of the seemingly oxymoronic phrase *deterministic chaos* that has become a sort of hallmark of nonlinear dynamics (cf. Shaw, 1987b; Schuster, 1988). In that context the symbolism implies a form of complex order that sometimes resembles and sometimes differs sharply from the usual connotation of gambler's luck — e.g., as it has been applied to the "Catch-22" of organismal fates during a geological history that has been filled with catastrophic surprises at almost every taxonomic and biostratigraphic level (cf. Ager, 1984; and see Shaw, 1987b vis-à-vis Gould, 1989b). And even then a residual resemblance is retained in the mutually inchoate attempts many of us share to capture the essential language that we hope will do justice to the fossil record as it was metaphorically expressed in Darwin's phrase, "endless forms most beautiful and most wonderful" (Gould, 1989b, p. 21). But this is only as it should be if language itself testifies to the nonlinear dynamics of the processes of genetic evolution — to processes of nonlinear recursion, of endlessly repeated stretchings and foldings of endlessly bifurcating vector fields of microscopic, subcellular, cellular, and variously packaged individually collective and collectively collective assemblages of these "microscopic" phenomena that do the same things simultaneously and all over again at the macroscopic scale — as many researchers are now finding to be self-evident (cf. Cavalli-Sforza et al., 1988; Diamond, 1988; Lewin, 1988a, b; Searls, 1988, 1992; Barton and Jones, 1990; Gould, 1989a; Shevoroshkin, 1990; Sokal, 1990; Deutsch, 1991, 1992a, b; Amato, 1992; Greenberg and Ruhlen, 1992; Nowak and May, 1992, 1993; Gibbons, 1993). Other aspects of such conundrums and paradoxes inscribed in the fossil record — graphite, chitin, shells, bones, amber, words, phrases, and conversations — of organic evolution are discussed in Notes 8.14 and 8.15).]

I understand that good stratigraphic practice has emphasized the importance of placing boundary markers within a continuous stratigraphic interval where "nothing happened." The apparent intention is a good one in the sense of avoiding possible discontinuities of unknown durations and other factors that cannot be readily correlated from place to place. But if an interval in question were really continuous in all respects, then such a marker would have the same flaw, a lack of resolution. And unless there is some feature that can be recognized as unique in time at a given place, then the marker, or "Golden Spike," is a floating datum relative to some other marker where something *did* happen, and thus allowed us to identify the continuous interval and its location in terms of a sharply defined global event (cf. Gallagher, 1991; Hanneman and Wideman, 1991; and Note 8.1). An attempt to establish correlations of natural phenomena in a manner that is consistent with nonlinear-dynamical principles is in essence an attempt to "crack the biostratigraphical code" (cf. House, 1985; Gaffin and Maasch, 1991; D. G. Smith, 1994; Kerr, 1992f; Winograd et al., 1992). For, if the interrelationships described here are valid — and in the general sense there is really no alternative because the universe is in fact nonlinear — it is precisely those arrangements of discontinuities in time and space (their intermittency patterns) that de facto *are* the geological record.

In technical terms, the "geologic column" or "time scale," including its spatial variations, constitutes a form of *singularity spectrum* (cf. Halsey et al., 1986; Shaw, 1987b; Olinger and Sreenivasan, 1988, Fig. 6; Shaw and Chouet, 1989, Figs. 1 and 2). A singularity spectrum is a representation of the continuous partitioning of recurring and recognizably discrete events in time and/or space, usually derived from a phase-space diagram. This qualitative definition implies only that selected variables of the process (position, rate, magnitude, time, population, etc.) are plotted and interpreted according to their recurrence properties or "densities" in the same parameter vicinity. Simplifications are possible, then, because in principle the essential nature of the pattern can be characterized by the time variation (absolute or relative) of a single parameter within a related system of parameters. In other words, the principles of construction of a nonlinear-dynamical singularity spectrum are not unlike the principles implicitly invoked in the construction of "The Geological Time Scale" (e.g., Harland et al., 1990), except that the singularity spectrum embodies no a priori assumptions concerning the linearity of time or the proper positioning of markers that would give it a characteristic and recognizable form. Event distributions in such a spatiotemporal nonlinear mapping are both discontinuous and intermittently clustered. The imposition of arbitrary markers and/or of imposed linear proportionalities (e.g., an assumption that a cyclic process should consist of a superposition of harmonic functions) can obscure the natural relationships unless they are appropriately calibrated relative to actual event discontinuities (hiatuses of diverse types that represent "crises," "catastrophes," etc.) over corresponding scales and ranges of time and space. The logic of this approach is familiar in the sense that it resembles the approach used to define "The Periodic Table of the Elements" (which may eventually require revisions based on concepts of "quantum chaos"; cf. Richards, 1988; Jensen, 1992; Parmenter and Yu, 1992) — or, more specifically, it resembles the approach used in the construction of Figure 21 as a means of exploring the patterns of nonlinear ordering among the planets. Admittedly, the problem confronting biostratigraphy, at least at the outset of the nonlinear approach, is much more formidable than are any of the examples mentioned above (cf. D. G. Smith, 1994; Winograd et al., 1992); but we cannot anticipate what simplifications may be found until "the code is cracked" (see the Introduction; and Note I.2). It is certainly true that, before the fact, no one had anticipated the order and symmetry that exists in the arrangements of the chemical elements, which, even now, may not be fully revealed (cf. Mazurs, 1974; Knight, 1978; W. B. Jensen, 1986; Hargittai, 1986; Larimer, 1988), or the simplicity of the biological genetic code (cf. Judson, 1979; Crick, 1981; Sampson, 1984; Shaw, 1986) within a grammatically ordered linguistic complexity that is just beginning to be explored (cf. Gatlin, 1972; Chaitin, 1979; Eigen et al., 1981; Shaw, 1986; Lima-de-Faria, 1988; Searls, 1988, 1992; Eigen, 1992).

Ager (1984) came close to defining a principle of self-similarity in the hierarchical classification of a system of sharp stratigraphic events ("catastrophes" in his terminology, and just so to the organisms affected). Two brief quotes from Ager accord with the above statements in describing both the stratigraphical theme and the theme of the present volume (Ager, 1984, pp. 93 and 95):

> To me, the whole record is catastrophic, not in the old-fashioned apocalyptic sense of Baron Cuvier and the others, but in the sense that only the episodic events — the occasional ones — are preserved for us.
>
> In other words, we are looking at evidence of a whole series of little catastrophes, followed by a bigger one, then by a still bigger one and then by a very big one indeed.

In the second passage **(Note 8.12)**, Ager (1984, pp. 94ff) was describing a late-Devonian biostratigraphic sequence (in France, of Frasnian age; "377.4–367 Ma," according to Harland et al., 1990, Fig. 1.7) in which the "very big one indeed" represented an abrupt transition to a seemingly new and different world (the Famennian Stage). This is, of course, the same transition that had been described on a global scale by McLaren (1970). [The age range is set in quotation marks to draw attention to the necessary elements of arbitrariness in the delineation of such a geologic time scale, wherein formal biostratigraphic Stage boundaries are stated with an implicit uncertainty, in the cited instance, of less than 50,000 years in a chronogram with an estimated absolute error of $\sim 2\%$ of the stated age. That is to say, such a scale of relative ages could not be written down at all without the excercise of arbitrary decisions, because the absolute error often exceeds the total duration of a Stage (see Harland et al., 1990, Fig. 1.7, Table 1.2, pp. 9ff, and p. 169).]

The above characterization does not imply any conventional sort of determinism but, rather, says that the style and rhythm of change had long since been set in the patterns of coupled oscillations among the cyclic interactions of the net system of nonlinear processes in the Earth and Solar System, as outlined in this volume. At the particular juncture chosen both by McLaren and by Ager to make their points concerning hierarchies of catastrophic happenings in stratigraphy — viz. the F-F catastrophe (or *debacle*, a term that nicely describes the nature of the dynamical conflicts that occur in the vicinity of a chaotic crisis, and that was used to describe the biological crisis at the Permian-Triassic transition by Gould and Calloway, 1980) — the global scale was "expectable" as a member of the hierarchical system of events, but I mean this only in the sense that the system of hierarchical catastrophes was part of a program of events that had already been set in motion within the self-organized Solar-planetary-CRF-terrestrial system of interactions. And by "program of events," I mean only that the nature of the intermittent fluctuations and the universality of the scaling parameters were consistent with the occurrence of the F-F event, not that anything had preordained such a fate for the "unfortunate" Frasnian organisms.

The memory of such a critical-state process is global, as is indicated by the notion of a *universality parameter,* but it is not necessarily local, because "local" is only meaningful relative to the scale of an observer (i.e., if, by observer, we are referring to scientists, engineers, etc., with human characteristics, then the observer does not have scale-invariant perceptions that would see the same dynamical proportionalities in the same way at different scales of time and size). This distinction concerns the meaning of "an event" in space and time relative to what is meant by a *local* vis-à-vis a *global* observer. For example, at the human scale of sensory perception, with or without instrumental assistance, one can observe the motions of small and large slumps of sand in an hourglass, and get a feeling for their timing, but it is something else again to have the same appreciation — *with* instrumental assistance — for global cascades of earthquake moments in the Earth, or comet showers from the Oort Cloud in the Solar System, even if the principles of self-organized criticality and relative scaling parameters *were identical* for the equivalent cascades of sand and earthquakes and comets. This makes the idea of "an event" scale-dependent and therefore "unpredictable" simply as a function of time and/or place in a space-time continuum, *even if every operable universality constant and every frequency of the universal spectrum of intermittencies were known exactly* (see Note P.4).

Bear in mind that I am referring to nonintegrable dissipative processes operating near the transition between nonlinear-periodic and chaotic modes. Events that are large and have long intermissions in such processes, relative to the observer, are predictable only in the sense that they represent a happening — after it has happened — that was consistent with the net cyclicity required to keep the critical-state process, the cyclic transfer of event sequences (such as in the figurative coupled-hourglass symbols of Figure 11), going. Alternative happenings might have taken place in the Earth, and/or elsewhere in the Solar System, that might have satisfied the same balance and the same general intermittency. In the Earth, such equivalencies might refer to alternative timings of geomagnetic, magmatic, tectonic, or other environmental processes that can be reflected in biostratigraphic processes, but such *endogenous* effects are meaningful *only* within the appropriately scaled magnitudes of the consequent phenomena. That is to say, in the universal spectrum of equivalently proportioned events, some of the events are consistent with intraplanetary phenomena, others are consistent with interplanetary phenomena, and the largest of all (in any reasonably meaningful framework of time and size for the survival of solar-terrestrial living systems) are consistent only with intra-Galactic phenomena. For perspective, however, if the universality of the process as a whole encompasses Galactic dynamics, as is implied by Figure 11, then it is also subject to intergalactic happenings. But the scale of time and size for intergalactic phenomena in our vicinity is so humongous (no other word will do) that the consequences of these phenomena are almost completely predictable from our point of view across almost any conceivable longevity of the human species *as we know it*.

When an intergalactic collision involving the Milky Way does one day occur, then the issue of what a local observer experiences will be irrelevant to us, as will be "prediction." (We must also allow the possibility that the present state of our Galaxy might someday be shown to be consistent with a scenario in which a past collision with another galaxy "has just happened," in somewhat the same sense that a collision with another protoplanet, creating the Moon and the Earth as we know them, or as I have imagined them to be from the schematic history of CRF-system history portrayed in Fig. 33, "has just happened.") To the "observers" at the time of the Frasnian-Famennian happening, the event — a seemingly disruptive, and still putative (but see the *McLaren Line* in Appendix 3a), impact of an object, or objects, from outside the ecologic bounds in question — though large in the above spectrum relative to endogenous happenings, was small in the context of commonplace interplanetary happenings. And throughout all of these diverse scenarios, the same principle of scaling and self-similarity has been constant, except for the observer-dependent criteria related to the subjective durations of intermissions and energy magnitudes of events, and, according to the CRF hypothesis, this has been true at all ecological levels in the Earth throughout the evolution of life on Earth.

Returning to the practical viewpoint — meaning to an observationally distorted, hence biased viewpoint — the search for and documentation of power-law behaviors in natural processes **(Note 8.13)**, with lognormal statistics in the instantaneous ranges of spatiotemporal data for the longer intermittencies (cf. Fig. 31, and Ditto et al., 1989, Fig. 4), are suggested as fruitful pursuits in numerical work where relationships are sought among the frequencies of related or potentially related processes of seemingly different periodic character. As discussed earlier and in Shaw (1988a), it appears that geological frequencies are numerically concatenated in terms of a net or critical balance among the geometric progressions of the prime numbers. A schematic summary of this effect, based on data from Shaw (1988a, Table 1, Fig. 25), is shown in **Figure 35**. The data represent measurements of peak-to-peak intervals of time in geological time series of diverse types (included were records of magmatic, geomagnetic, geochemical, and paleontological phenomena). I interpret this periodicity pattern to mean that there are characteristic sets of nonlinear (i.e., nonharmonic) bands that show maxima at certain multiples of the prime numbers (e.g., sets made up of geometric progressions to bases 2, 3, 5, 7, etc.). These are global and not necessarily individual properties of the data sets, the reason being that most geological time series with oscillatory intervals longer than a few million years have very few recurrences at that scale of time (Shaw, 1985b, 1988a). In Figure 35, for example, many different records, when added together, show a maximum near 32 m.y., and another below 8 m.y., where these two numbers represent natural breaks-in-slope of the logarithmic frequency histograms (natural and synthetic). In other words, these two numbers identify natural numerical boundaries between three power-law regimes among many other possible regimes at both smaller and larger scales of measurement (see Note 8.13).

The numerical regimes of Figure 35 have to do with the natural properties of the numbers as they appear in the nonlinear-dynamical context (I took this up earlier relative to the frequency structures of coupled oscillators; cf. Chaps. 4 and 5; Notes P.4, 4.1 and 5.3). The range from about 8 to 32 m.y. is a characteristic band in which several other average periods are evident (e.g., near 27, 18, and 14 m.y). There are many shorter average periods below a peak at about 6 m.y. that are not resolved in the longer-term records. The steepness of the fall-off above 32 m.y. reflects partly the paucity of data with longer average periods (48 m.y. is the limit for this particular data set; Shaw, 1988a, Fig. 25b). In larger data sets, additional frequency bands appear between about 50 and 70 m.y., between 70 and 90 m.y., and above 90 m.y. Their resolution becomes progressively poorer above about 50 m.y., again because of the incompleteness of the data. In the earlier study I tested the existence of these bands, as a reflection of the properties of nonlinear-periodicity arrays, constructing synthetic histograms of periods made up from best-fitting combinations of the integer periods (geometric bases) *2, 3, 5,* and *7* (Shaw, 1988a, Table 1 and Fig. 25a). There are strong similarities with the raw data even at the level of a comparison with a histogram made up from nearest values of periods 2^n. Those values composed of the combinations $2^n 3^m 5^p 7^q$ are nearly indistinguishable from the raw data. In a few instances, periods demanding the geometric bases *11* and *13* appeared to be required to satisfy particularly well-documented average periods (e.g., $2^1 \times 13^1$ is required to satisfy the average period of 26 m.y. for biological extinctions during the latest 250 m.y. derived by Raup and Sepkoski, 1984; cf. Rampino and Caldeira, 1993).

Patterns of this type (e.g., Shaw, 1985b) seem worth following up in studies of biostratigraphic and sedimentological cycles that might otherwise appear to be lacking periodic order (recall that a lack of harmonic regularity is not a sufficient criterion of disorder, nor is it evidence of random or null effects). In fact, any data that show hierarchical but possibly disproportionate cyclic sequences in space and/or time (absolute or relative), and at whatever scales may be characteristic of a systematic behavior, are candidates for such tests (with due regard to self-similar shifts in scaling; cf. Shaw and Chouet, 1991; D. G. Smith, 1994). Power-law behavior in distributions of biological diversity data, in either the incremental or the cumulative sense, also may provide important tests for similarities between long-term and short-term nonlinear effects, especially those in the vicinities of mass extinctions, or where there are sudden changes in diversity statistics. Two examples are given in the following section.

Intermittent Biological Crises, and Background Extinction vis-à-vis Mass Extinction

In the terminology of orbital dynamics, the geological-biological record is rich in complex nonlinear resonances. By the same token, evidence of the nonlinear coupling of geological and celestial processes would require that biological processes be part of the same system of cosmological phenomenologies. Such nonlinear commensurabilities among seemingly different dynamical processes are dramatically demonstrated at junctures such as the F-F, P/T, and K/T boundaries, but these are of course not the only boundaries at which crises have occurred. According to the preceding biostratigraphical discussion, all natural boundaries are phenomenologically analogous in representing some form of crisis, where the unmodified term crisis means simply that "something happened" and stands out sharply, hence discontinuously, relative to the scale of smaller happenings that may be construed (but only relatively) as continuous processes. Such a hierarchy has no a priori limit and depends only on the resolution of the description (hence it does not even exclude quantum "happenings"). In important respects, this viewpoint echoes ideas concerning the biostratigraphic principles advanced by McLaren (1970) in his landmark paper, "Time, life, and boundaries." Some of the central notions of biostratigraphy are elaborated by McLaren (1982, 1986), Ager (1980, 1984), and McLaren and Goodfellow (1990), among other sources cited in those papers.

In my view the dynamical consequences of mass impacts on biological diversity, in cause-effect scenarios, should be examined from the standpoint of energy partitioning between kinetic mechanisms of background originations as well as extinctions, to the extent this is possible from fossil data on biological diversity. Mass extinctions are, of course, more conspicuous than the more subtle processes of origination, but evidence of the latter before, during, and after extinction events may be informative when it is interpreted from the standpoint of interactive processes analogous to those of the coupled systems of environmental oscillators discussed here. The notion of coupled oscillations, or coupled nonlinear biological clocks, if you will, is an essential aspect of biological-feedback processes. This is especially true if the coupled geophysical processes described here are also assumed to be integrally coupled with them over geological time. Stated differently, the individual event in geological history is not the key to patterns of future geological or biological diversity; the key is the *integral pattern of prior events*, in the above biostratigraphical sense, by which the patterns of diversity have become mutually attuned during Earth's history, and by which they

Self-Organization and the Celestial Reference Frame

are being modified in the present by self-similar processes that accommodate even the rare catastrophic event. Self-similarity implies a likeness of possible regimes expressed in the language of *dynamical uniformitarianism,* a language that interprets past regimes according to present knowledge of nonlinear-dynamical processes. Retrospection of this sort is a principle of nonlinearly intermittent superposition, in both space and time, that provides an integrally continuous record (singularity spectrum) derived from a discontinuous system of environmental crises (recurrence states) in the Galactic-solar-terrestrial rhythm by which the nonlinear mechanisms of biological clocks have been calibrated (cf. Gould, 1965; McLean, 1988; Alvarez et al., 1989).

Nonlinear-dynamical uniformitarianism provides an inclusive context for an interpretation of Charles Darwin's metaphor of 10,000 wedges, which he used to describe the motivating role of natural selection in organic evolution, and which some subsequent workers have tended to take rather literally. In the context of the debates on the impact-extinction hypothesis, it has been tempting to infer that nonbiological wedges (such as mass impacts, or volcanic eruptions of catastrophic scale) have been hammered in so violently that many of the energetically smaller biological wedges fly out all-at-once in bunches, and that such coupled events are followed by a conservative rearrangement of the remaining "wedges" according to established principles of natural selection and survival of the fittest (e.g., previous lineages, and/or ones that were previously so latent that there was no prior fossil record of them, suddenly expand to refill the available ecologic energy space). This type of effect can appear to be so sudden and dramatic that it offers the singular attraction of providing a deterministic rule by which debacles imposed by "random" cosmic events might be terrestrially calibrated and explained. In its most literal form, such a cause-effect scenario would see the explosive energy deposited in the Earth by the action of a mass impact only as a fateful dice-throw-of-death. That is, on this basis large impacts become the random source of novelty in a theory that would use mass extinctions as happenings that influence the degrees and directions of change in macroevolution. In this sense the idea becomes a long-term, large-scale analogue of "random happenings" in the theory of neutral drift at the molecular level. The fateful dice-throw-of-death simply replaces the microevolutionary inverse, the fateful dice-throw-of-life that is represented by molecular genetic elements of randomly inheritable novelty (cf. Burnet, 1976; Shaw, 1987b; Gillespie, 1991). Both the large and the small are then exposed to the test of competitive survival.

Apart from the warlike implication that the universe operates on a basis of opportunities-for-dominance ("success"), measured by the body-counts of dead competitors, this view of the role of mass extinctions is a static one in which the origin of species proceeds against a desolate evolutionary landscape where the success of one group is measured against the degree of failure of one or more other groups (the *frustration* model of social landscapes and strategies is somewhat less stark in allowing self-organized forms of compromise; cf. Dewdney, 1987; Suarez et al., 1990; and Notes P.1, P.3, 3.1, 5.1–5.3, 8.1). It might seem that the alternative, the application of nonlinear dynamics to organic evolution as advocated here, does violence to the concept of conservation of energy. On the contrary, it says simply that with regard to interactions between the biosphere and the extended geosphere, we have so far been unable to identify all of the entries in the ledger, and that even if we could identify them all we might find that they severely modify the conventional wisdom of fixed-sum statistics suggested by Darwin's metaphor.

If Darwin's statements concerning natural selection are looked upon as if he were aware of nonlinear dynamics (his concept of evolution certainly implicated feedback and bifurcation, trademarks of nonlinear phenomena), they might be interpreted with some validity as a description of my own thesis. In a revolutionary paper — in the sense that it flew in the face of the concept that a species is in adversarial competition with other species for its piece of Earth's real estate, long held to be a truism in evolutionary dogma — Gould and Calloway (1980) cited a lesser-known quotation of Darwin's metaphor of the wedge to establish a context for discussion of two ecologically paired but genetically distinct lineages, the clams and the brachiopods (cf. Gould, 1989b). The lifespan of both lineages began in the Cambrian (i.e., at >510 Ma), and both continue to flourish today, although the brachiopods have had more than one brush with extinction. The passage from Darwin, as quoted by Gould and Calloway (1980, p. 383), goes as follows (**Note 8.14**), *"Nature may be compared to a surface covered with ten-thousand sharp wedges ... representing different species, all packed closely together and driven in by incessant blows, ... sometimes a wedge of one form and sometimes another being struck; the one driven deeply in forcing out others; with the jar and shock often transmitted very far to other wedges in many lines of direction."*

The driving in of the wedges by "incessant blows" reverberates as a paraphrasing of the intermittently continuous role of mass-impact events — and implicated mass-extinction events — as a major mediating factor in biological evolution. In a direct sense, the "incessant blows" symbolize the influence of mass impacts on physical disruption of the biosphere, as represented by the sharp events of McLaren (1970) and the self-similar hierarchy of little, big, bigger, bigger-yet, and biggest catastrophes of Ager (1984). In both the direct and indirect senses they symbolize those mutational, trophic, and other factors that mediate the broader criteria and conditions of biological originations within the context of geodynamic factors that are nonlinearly coupled with mass impacts through the magma-tectonic and geomagnetic implications of the cratering process. These factors act sometimes in one place and sometimes in another, in ways that both rearrange and yet coordinate the net geological-biological effects, with the "jar and shock" transmitted throughout the system "in many lines of direction." Besides fitting the imagery of impact-induced shock effects, these phrases are graphically reminiscent of the processes described here in terms of chaotic crises and bursting phenomena (**Note 8.15**). If he did not realize it then, I suspect that if Darwin had been working within the paradigm of mid-twentieth-century physics, he would have been among the first to recognize the role of nonlinear dynamics in the origin of species (cf. Notes 7.5, 8.14, and 8.15).

Self-Organization and the Celestial Reference Frame

In a recent paper, Raup (1990) explored the extent to which the background pattern of extinctions during the Phanerozoic, as well as the rare extinction events in the Phanerozoic, might be accounted for by mass impacts (cf. Raup, 1984a, b, 1991). For that purpose, he used a method of Monte Carlo simulations analogous to those I described earlier for the simulation of comet fluxes from the Oort Cloud (e.g., Fig. 16). He concluded conditionally that if the simulation were in accord with actual impact rates, then impact alone could account for the total record of biological extinctions. This idea opens the door to other conjectures, some of which are described below and in Shaw (1988a). If the portion of the energy input ascribed to impacts in the present volume were energetically sufficient to account for all death-dealing terrestrial effects, then it is by implication sufficient to account also for all life-generating effects, through its mechanisms of coupling with the geodynamic engine and the latter's incubational capacities to sustain life. This perception follows from the cyclic nonlinear interactions discussed here, which in league with solar insolation and photosynthesis determine the genetic and nutritive substrates for nearly all biological processes (with the exception of the chemosynthetic phenomena now known to be associated with submarine hydrothermal vent systems; cf. Shaw, 1986). In other words, if the energetic and functional role of impacts cannot be isolated from the "lifecycles" of geomagnetic, magmatic, and tectonic processes, then they cannot be isolated from the lifecycles of processes acting on the biosphere. I attempt to illustrate this point in **Figure 36**, using the diversity data of Gould and Calloway (1980) for the clam-brachiopod duo as one example and the largely ignored (but see Bock, 1979; Eldredge, 1989, Chap. 2) evolutionary history of the birds as another.

Fisher (1967) compiled the then-existing data for avian evolution in a family tree that illustrates patterns of radiation since the early Cretaceous. In recent years DNA-DNA hybridization techniques have been used to develop a DNA-based comparative phylogeny for birds (Sibley and Ahlquist, 1990) that revises, and sometimes differs in major respects from, the more conventional phylogeny by Fisher (1967). The data used here (from Shaw, 1988a) were compiled before the Sibley-Ahlquist taxonomy was completed. Because the conceptual basis for a molecular-genetic phylogeny differs from one based on conventional fossil cladistics, the two are not necessarily comparable, and the DNA-DNA hybridization results are in some respects controversial. The compendium by Sibley and Ahlquist (1990), however, is a monumental achievement, and the patterns of diversity shown in Figure 36 and in Shaw (1988a) should be recompiled from the DNA results in hopes of developing new perspectives, possibly with some major surprises, concerning rates of avian evolution. In the interim, I shall assume that the general numerical patterns shown by total counts at the level of families vs. age will not be drastically changed.

In Figure 36a, rather than attempting to interpret patterns of *standing diversity* for clam and brachiopod genera separately, relative to particular events of extinctions and originations, I have summed the standing diversities of each lineage to define a fictional *cumulative diversity* that is some measure of the total processes involved in both originations and extinctions in each lineage through the entire interval of geologic time. This is, of course, but an artifice used to show in a single diagram some of the effects discussed in detail by Gould and Calloway (1980), henceforth designated G & C, or G-C (the reading of which is mandatory as a check on the points discussed here). The curve of *fictive* cumulative diversity acts somewhat like a maximum thermometer, showing only the additions per stage in the relative chronology (i.e., the curve cannot reverse itself and has no extrema). G & C used the 72 divisions of the time scale of Harland et al. (1967) as a natural measure of characteristic rate phenomena, which seems consistent with present conclusions concerning possible differences between linear and nonlinear clocks. For the purpose of display, however, I assigned ages from Harland et al. (1982) at each of the Periods, Sub-periods, and Epochs in Table 1 of G-C. Each increment of numerical diversity in the compilation of clam-and-brachiopod genera represents the total count within a stage (the raw data). Notably, such data do not distinguish whether any specific genera persisted longer than one stage, nor do they indicate when within a given stage any genera may have originated and others may have become extinct (e.g., if two successive stages show the same value of diversity, we cannot tell from the numbers whether all genera of one stage were generated during that stage and became extinct during the same stage and repeated that process in the next stage, or whether all of the genera of the first stage persisted into the second stage without any new originations or extinctions; cf. McLaren, 1986).

The origination-extinction process obviously is intricate in detail. The gross approach shown in Figure 36a recommends itself only in showing some of the long-term trends, and all conclusions concerning detailed variations should be viewed with caution. Some real effects, however, *can* emerge, either earlier or later on the cumulative curve, as contrasted with variations on the incremental basis. For example, the standing diversity of clams (G-C, Table 1) was for the first time higher than that of brachiopods in the lower Triassic, immediately following the P/T "debacle" (this term appears in G-C and seems fitting because it can mean disruption, failure, or disaster without specifying how that came about, thus avoiding any stereotypical implication of the term "extinction"). The cumulative curve, however, swamps out these instantaneous reversals and shows only the time when the net generative potential of the bivalves has caught up with that of the brachiopods, as measured by the total numbers that had been generated or that had persisted up to a given time (the brachiopods had previously been "ahead" for a long time). As a result, the net crossover does not occur until a time (coincidentally) just prior to the K/T boundary. At that juncture, the rate of increase in "clam futures" is so large that the K/T debacle has a barely discernible effect on the clam curve even though it shows up as an almost 50% reduction in Paleocene standing diversity.

Other strange wrinkles occur in the cumulative data vis-à-vis the incremental data. The percentage reduction in brachiopod genera in the same interval is not much different from that for clam genera, yet the cumulative effect shows a tendency for brachiopod acceleration just prior to the K/T, followed by a sharp deflation (presumably because the debacle, combined with prior effects, hit the brachiopods harder than it did the clams). Another wrinkle concerns a conclusion based on G-C's careful statistical treatment of the standing diversity data. G & C observe that, on the basis of the incremental data, the P/T debacle was far and away the greatest crisis in comparative brachiopod vis-à-vis clam diversity (G-C, p. 393). In my Figure 36a, by contrast, the P/T event is distinctive in almost the same sense, and with about the same scale, as the K/T event. In the cumulative sense it would seem that a more profound effect for both lineages occurred following the F-F debacle, in the sense that the immediate effect at that juncture was about as large as the P/T and K/T effects for the brachiopods, was larger than the P/T and K/T effects for the clams, and apparently set the stage for the P/T effect on the brachiopods. McLaren has made similar observations for the magnitude of the F-F event in general (cf. McLaren, 1970, 1982, 1983; McLaren and Goodfellow, 1990). But perhaps even more surprising is the observation that the real beginning of the later relative crash for the brachiopods was evident as early as the mid-Ordovician (see the downward arrow labeled Mid-O in Fig. 36a). All of these effects seem to show up in both curves, but the growth impetus for the clams in each incident apparently was strong enough to overcome the damping effect of the crisis, whereas that of the brachiopods was insufficient for full recovery before the next crisis occurred.

The fluctuating behavior of the fictive diversity curves in Figure 36a bears a marked resemblance to the comparative behavior of two different systems of coupled oscillators in which the two systems have similar responses to forcing, but one has a lower integral damping factor than the other. In the integral sense, the clams were initially hit as hard by the F-F event as the brachiopods were, but between then and the P/T event the clams regained momentum sooner than the brachiopods (perhaps by a narrow margin, because the brachiopod curve also had begun to rise sharply before the P/T event). That is, the brachiopods had almost recovered a long-term robustness comparable to that of the clams when the P/T effect came "too soon" and knocked them for another loop. According to the oscillator analogy, such effects often have more to do with the timing of feedback between relative forcing and damping mechanisms than they do with the amplitudes of either oscillator (e.g., the *controlled chaos* illustrated in the bottom graph of Fig. 29 — which is analogous to steadily increasing rather than highly variable changes in diversity — would not have occurred if the timing of the feedback signal had been applied in a different part of the thermal convection loop, or if there had been a significantly different delay time; more complicated multiple oscillators of the types described in the present volume, however, bear a greater resemblance to the delayed-feedback-systems models discussed by Freeman, 1992, pp. 467ff).

The zero-age endpoints of the brachiopod-clam race, as it has been run so far, also are interesting. At the present time, the number of brachiopod genera is about one-eighth that of the clam genera, but the brachiopods have been on an exponential upswing since the K/T, with an average doubling time of roughly 30 m.y. (a fourfold increase in about 60 m.y.) and have exceeded the clams in percentage increase. These numbers do not appear to be an artifact of the "pull of the Recent" because similar episodes are evident at other times (cf. G-C). At the opposite ends, the "origins," of the cumulative curves, if we take as our criterion for the hypothetical *beginning* of the "race" the time when each lineage exceeded a few genera, then an extrapolation of the average power-law trends can be used to identify fictive starting times, as suggested by the intercepts of the light dashed lines in Figure 36a on the time axis (a situation somewhat analogous to the use of rolling starts and different starting times in a road race such that the rolling starts are of different durations but are equivalent to a fixed-time handicap). For the brachiopods, I take the summed statistics of both groups (open circles in Fig. 36a) as a fair guide, because they approximate the slope at a comparable stage of impetus for each group just prior to the mid-Ordovician, whereas the comparable curve for the clams (parallel heavy dashed curve) is almost tangent to the net curve at the present time and has been parallel to the average of the summed slope over the latest 400 m.y. or so. These arbitrary criteria, which make no attempt to account for "incubation" effects, etc., give respective starting ages of ~560 Ma for the brachiopods and ~550 Ma for the clams (i.e., as if a biotic handicapper had given the brachiopods a 10-m.y. headstart). There is no way to predict the longevity of these two lineages on the basis of these patterns or any theories of competition. The same relative robustness might continue to oscillate similarly for another 500 m.y., at which time it is possible that the brachiopods will have again recovered another temporary "lead in the race" — where "temporary" in such a scheme of things would refer to a race lasting more than a billion years.

The title of the paper by Gould and Calloway (1980) describes their main conclusion: "Clams and Brachiopods — Ships That Pass in the Night." The poetic metaphor reflects the fact that these two lineages show no indications of "paired replacement by competition," and that they have sailed along more or less independently, with little more than an occasional exchange of signals to indicate any biogenic "awareness" that they were pitted against each other in a competition (awareness is meant in the sense of roundabout ecologic-biogenic feedback signals of the sort described qualitatively in the essays of Lewis Thomas; e.g., Thomas, 1974, 1979; cf. Note 8.14 in the present volume). I would tamper with the metaphor of G & C only to the extent of suggesting that the curves of fictive diversity in Figure 36a might be likened to the ship-logs of two vessels that were sailing the same general course of circumnavigation, one starting a little later in the season so that they were always just over the horizon from each other but by a sufficient margin that the first ship (the brachiopods) got the worst of a series of storms that assaulted them both (perhaps only because of making slight course changes that led to different consequences, relative to factors of design, that may in due course turn out to be equally or more successful). In general, the two lineages responded roughly in tandem to the same oscillatory forces, but at any

Self-Organization and the Celestial Reference Frame

given time the chaotic effects and internal crises of the oscillations might have taken a form of resonant interaction that could be called (in traditional terms) "competitive replacement."

Incidentally, since we are already heavily engaged in anthropomorphisms — following the lead of G & C's "ships that pass in the night" — I venture to suggest that the slight indication of positive correlation mentioned by them (G & C, p. 393) might be construed, extending the tandem metaphor, to represent times when the two ships were within signaling distance of each other (e.g., by sending up flares), thereby revealing a sufficiently altruistic nature to warn "other shipping" of impending hurricane weather (see Note 8.14). By this same view, and by a form of "dangerous extrapolation" that G & C warn us against, a comparison with the types of coupled feedback cycles already discussed in the present volume leads me to suspect that the existing mutual pattern of clam-brachiopod interaction (or seeming lack thereof) represents a form of resonant stabilization provided by an additional interaction with one or more other systems of attractors (like those implicated in the magmatic-geomagnetic-impact interactions discussed earlier in this chapter, and/or like the neurobiological feedback cycles discussed by Skarda and Freeman, 1987, and Freeman, 1991, 1992). Within such a context, terms such as *ambivalence* and *frustration* — terms that are now used to describe the behaviors shown by complex networks of coupled oscillators (e.g., spin-glass systems; cf. Note 5.3) — may take on almost literal meanings.

The saga of the birds summarized in Figure 36b shows some similarities and some marked differences relative to the effects seen in the great adventure of the clams and brachiopods. Compared to the clam-brachiopod story, the avian lineage as a whole is, so far, relatively short-lived — extending back only 150 million years or so — and their story does not offer the same stake-race flavor that is so interesting in the relative diversity of the clam-brachiopod matchup during an odyssey that survived three of the Phanerozoic's greatest extinction events. But some effects are analogous to the above metaphors of tandem voyages and racing handicaps, an example being the contrasting styles of (1) an older group that can be traced back as far, at least, as the beginning of the Cretaceous (to > 140 Ma), and (2) a younger group that somehow split off from the older group near the end of the Cretaceous (at ~ 80 Ma), according to the graphical display of avian orders and families illustrated by Fisher (1967; cf. Note 8.1). The older group is epitomized in extant lineages, for example, by the Pelecaniformes and the Procellariiformes (which include many of the shorebirds, large seabirds, etc.), while the younger group is epitomized by the Passeriformes (which includes a wide diversity of perching birds). These two categories represent a crude indication of the traditional classification of lineages, which has been dramatically modified in some respects by DNA-hybridization studies (e.g., many lineages of the first group above fall within the new order Ciconiiformes of Sibley and Ahlquist, 1990, p. 492). Within each of the traditional groups there are subplots analogous to the clam-brachiopod story that involve differences in relative "starting times" and evolutionary rates (accelerations and decelerations), where the bimodal ("primitive/modern") character is revealed by the patterns of peak rates (maxima in plots of *standing diversity* at the level of avian families, as in the data of Fig. 36b, rather than cumulative diversity, as in the data of Fig. 36a; plots of standing diversity obviously can display maxima and minima). Shaw (1988a, pp. 22ff, Figs. 14–16) discusses the patterns shown by plots of peak rates vs. age at the level of families, using the data from Fisher (1967). Unfortunately, the fossil record is not adequate at lower taxonomic levels to follow details of such variations, and the age data are not adequate to resolve the relationship between the observed pattern of multiple peak events and the K/T transition with certainty.

The general character of the higher taxonomic relationship between the Pelecaniformes and the Passeriformes, for example, is revealed in Figure 36a by three effects: (1) The first is seen, in the general pattern of avian evolution, as the eventual attainment of a nearly stable number of families in both of these orders (for short, simply designated as being representatives of the "older" and "younger" lineages, respectively) following the K/T transition, numbers that have persisted as general diversity plateaus to the present time, as recorded by the data of Fisher (1967). [In view of the major taxonomic revisions implied by the work of Sibley and Ahlquist (1990), the age spans of avian evolution refer specifically only to Fisher's compilation (see Shaw, 1988a, pp. 22ff, Figs. 14–16), although the general dichotomy between an "older" and a "younger" line of descent is likely to be preserved in the revised taxonomy.] (2) Another effect is seen as a relatively slow response (by the older order) and a relatively fast response (by the younger order), both apparently affected by the K/T debacle but to different degrees, reflecting major changes in evolutionary rates at about that time (the older order rapidly decelerates, beginning some time before the K/T, then undergoes relative decline during the Tertiary, leveling out at about 10 Ma; the younger order abruptly accelerates, apparently beginning shortly before the K/T transition, and then abruptly levels out at constant diversity immediately afterward; cf. Shaw, 1988a, Fig. 15). (3) And the third effect is shown by the fact that the curve of summed diversity (filled circles), representing the standing diversity of all avian *families* at a given time, is subtended by intervals of invariant power-law behavior that appear to intersect at or near the K/T boundary—as if that event gave a positive genetic impetus to some taxons (at the level of orders, or below) and a negative impetus to other taxons, but where both influences had been "anticipated" before the fact. The ambiguity indicated by the phrase "taxons (at the level of orders, or below)" is occasioned by the fact that some avian lineages, at several different taxonomic levels, have been reassigned to different taxons in the classification of Sibley and Ahlquist (loc. cit.).

Both of the power-law trends identified in the third effect above appear to have been established prior to the K/T event, as is shown in Figure 36b by the contrast between the summed diversity of all families (solid circles and lines) and the diversities of the two orders Pelecaniformes and Passeriformes—i.e., representing the older group (pelicans, etc.) and the younger group (perching birds), respectively—which I chose to illustrate the range of relative rate effects in avian evolution. Other avian lineages have age spreads showing similar patterns, but each one is shifted somewhat relative to the K/T boundary (cf. Shaw 1988a, Figs. 14–16). It will be interesting to see how such trends will be modified by the DNA phylogeny of Sibley and Ahlquist (1990), a formidable task that I have not yet undertaken. [It seems worth pointing out that the relationship between avian evolution and that of the vascular plants (e.g. Knoll, 1984) should be of great interest. In fact the rates of evolution of the vascular plants (at the levels of species, families, and orders) as a function of age during the Mesozoic and Cenozoic Eras have some resemblances to the bipartite evolution of avian families shown in Figure 36b. For example, there is a rapid increase in angiosperms (flowering plants) and either a plateau or falloff in the diversities of the nonangiosperms, with peak rate changes — possibly of multiple-peaked form — occurring near the time of the Cretaceous-Tertiary transition (see Knoll, loc. cit., Figs. 1–5; cf. Shaw, 1988a, Fig. 16).]

The above phenomena, like those involved in clam-brachiopod evolution, are familiar in the context of the nonlinear dynamics of coupled oscillators involving delayed feedback (mutually and/or in relation to other oscillators). Oscillator-like advances and delays in genetics do not preclude synchronizations with major killing events — as occurs at the microlevel, for example, in the form of episodes of massive cell death during neurophysiological development in humans (cf. Cowan, 1973; Burgess and Nicola, 1983, pp. 253ff; Berns, 1983, pp. 239ff; Pons et al., 1991; Barinaga, 1993), and as occurred at the macrolevel in the transformations of global biodiversity patterns during and after the now well-established mass-impact events at the K/T boundary — which, as has been argued in the present study, were distributed in both space and time relative to loci of peak kill rates (cf. Figs. 3, 8, and 36; and Raup, 1990). The advances and delays do suggest, however, that the net effect of biogenetic-geologic feedback over time is far more involved and requires both cooperative and competitive phenomena, only part of which are represented by the killing effects of the most massive impact events. If we group the deadly effects of the more numerous small impacts with general factors of "environmental toxicity," then the general death-rate component of the kinetics may indeed work out somewhat as Raup (1990) has proposed.

Finally, we will do well to examine the naturalness of the power-law trends in the admittedly limited sets of diversity data available to us from the fossil record. Such trends are indeed what might be expected from the notion of self-similar scaling of catastrophes suggested by the nature of biostratigraphical hierarchies indicated by the work of McLaren (1970, 1982, 1983, 1986) and Ager (1980, 1984), as discussed in the preceding section of this chapter. Logarithmic diversity trends are in a sense fractal time plots in which the slopes are the dimensions, thus reflecting aggregate forms of fractal-kinetic rate constants for evolutionary progressions. These macroscopic kinetic rate constants would then be analogous to the microscopic rate constants of biological clocks, which in detail vary in average speeds and are not necessarily the same even within the same lineage (e.g., Sibley and Ahlquist, 1990). The comparative dynamics of micro-morphogenetic and macro-morphogenetic fields are of great interest, particularly when trends such as those displayed here involve contributions from both terrestrial and celestial nonlinear clocks — contributions that run the full gamut of environmental crises. Comparisons of the rates of evolution by different methods — e.g., by conventional paleontologic vis-à-vis DNA hybridization methods — hopefully will eventually bring some resolution to this difficult problem (cf. Note 8.15; and Shaw, 1986; Goodwin, 1991).

These remarks extend to the data of *allometry*, or the scaling of the comparative physiology of biological adaptation, where logarithmic proportionalities (power-law scale relationships) appear to be characteristic. Calder (1984), for example, has demonstrated that many types of allometric properties of animals are described by power-law relationships, such as (1) organ mass vs. body mass in the allometry of birds (ibid., Fig. 1–1), (2) metabolic rate vs. body mass for different animal lineages (ibid., Fig. 5–1), (3) physiological time scales — or what I would call the *invariance of recursive lifetimes* — vs. body mass for warm-blooded animals (ibid. Fig. 6–1), and (4) audible sonic frequency vs. interaural length scale in mammals (ibid., Fig. 8–12). Power-law scaling is characteristic of universal nonlinear-dynamical behavioral regimes such as those of the *critical golden mean nonlinearity* of Shaw (1987b) and Shaw and Chouet (1988, 1989, 1991; cf. Olinger and Sreenivasan, 1988, Figs. 5 and 6), those of *critical self-organization* of Bak and coworkers (e.g., Bak et al., 1988; Chen and Bak, 1989; Bak and Chen, 1989, 1991; Bak and Tang, 1989; cf. Chap. 5; and Notes 5.1–5.3), and those of *chaotic crises* (see Chap. 5, Figs. 17, 18, and 31; and Notes P.4, 5.1, and 5.2).

The living state is conditioned by phenomena such as allometric scaling and recursive lifespans, which — e.g., in the context of population dynamics (e.g., May, 1976; Edelman, 1987) — condition the phenomena of death and extinction (e.g., Calder, 1984, Figs. 6–1 and 6.2; cf. neural cell death during the development of a mature organism, as demonstrated by Cowan, 1973; Burgess and Nicola, 1983, pp. 253ff; Berns, 1983, pp. 239ff), but it is not defined by them. This maxim seems as valid for the celestial relationships between star-forming processes and supernova explosions, or for that matter between the "Big Bang" and whatever came before it in a transfinite cosmology (cf. Notes 3.3, 4.2, 8.15, E.1; and Alfvén, 1980, 1981, 1984; Lerner, 1991), as it is for the biosphere, or for biological lineages, or individual internal organs, or organismal cell populations. Perhaps the biological message and import of the point of view outlined in this volume is that it is impossible to describe birth and death as phenomena that are defined by separable summations of "positive" and "negative" energies (i.e., unlike a trophic energy budget that has a finite limit in a conservative ecosystem, and unlike the finite balance of energy and material resources that can be identified for a limited commons in social economics). If separable functions for lifegiving vs. deathdealing processes provided valid descriptions of the living state, someone in the medical community might decide that massive cell death (see above) during the ontogenetic stages of the human life cycle is a bad thing and set out to cure it by changing the rates of generation and decay of other cell cycles. I wonder if an analogous preoccupation with the negative connotations of "death" might not be distorting our interpretations of the role of mass extinctions in organic evolution? The analogous "cure" for that type of extinction is a global system for the detection and deflection of potential impactors on Earth that will be failproof for all time (cf. Matthews, 1992; Ahrens and Harris, 1992; Lindley, 1992b; Kerr, 1992g). In either of the cases mentioned (cell death and mass extinction), the "cure is worse than the disease," because it artificially overrides the existing states of universal self-organization and natural selection (ironically, although this would be true in the short term relative to the previous patterns of change, the ultimate effect of persistent artificial intervention — i.e., as a characteristic of the universe — simply becomes an aspect of the self-organization in the revised system).

Biologists and ecologists of today, however, are a lot more concerned about the decline in global biological diversity than they are about the possibility of an impact-induced extinction in the foreseeable future. And many conservation-minded people would say that the biological future of the world is in our hands, implying that the human species has both the power and the moral obligation to mitigate, if not to totally avoid, what some have dubbed "the last extinction" (cf. Kaufman and Mallory, 1986; Wilson, 1988; Raup, 1988a, b; Lovelock, 1988b; Grew, 1990; May, 1990). And yet, barring some long-term cooperative plan of international scope to deal with the inevitability of future bolide and meteoroid impacts — in the context of evolutionary cooperation, synergy, and even "altruism," as mentioned in Notes 5.3 and 8.14 (cf. Axelrod and Hamilton, 1981; Nowak and May, 1992, 1993) — the long-term prognosis for humanity even in "shallow time" (a term some ecologists have been using to distinguish between a time scale measured in tens of thousands to hundreds of thousands of years and the "deep" time of the Phanerozoic fossil record; cf. Raup, 1988a, b) is not any better than that of the redwood tree, or of the tropical rain forest. The fossil record of "shallow" time may yet provide a test of the concept of negative correlation for some new population of researchers — presumably more sentient ones than we have been — from a "fitter" world. In such a context, the human species cohabits the endangered-species list with all the rest of the "less fit" creatures that have been disappearing right before its eyes (cf. Asimov, 1981; Volk, 1992; Caldeira and Kasting, 1992; Kerr, 1993a).

On this pessimistic, or optimistic, note, depending on one's point of view, let me add that if an understanding of the nonlinear dynamics of organic evolution is a valid aim, we are missing something in the cycles of death-dealing mass impacts that are themselves the product of self-organized processes in the Cosmos. To either dramatize or bemoan the phenomena of mass extinction (whether induced by mass impacts, by volcanic crises, by global biological epidemics, or by the scourge of a rapacious biological species) is to emphasize only the negative feedback loops in a system that operates in oscillatory modes of both positive and negative feedback (cf. Fig. 29). Putting it bluntly, I doubt that a concept requiring us to see mass impacts as the most important cause of mass extinctions, and mass extinctions as one of the most important criteria of evolutionary change, is meaningful. And I say this notwithstanding the ironic fact that those extinctions which define the most distinctive events in Earth's biostratigraphic record (e.g., McLaren, 1970, 1982, 1983, 1986; L. Alvarez et al., 1979, 1980; W. Alvarez et al., 1982, 1989; Raup and Sepkoski, 1984; W. Alvarez, 1986, 1990; L. Alvarez, 1987; McLaren and Goodfellow, 1990; Raup, 1990) would not exist in such helpfully recognizable forms without the intermittent occurrences of the mass impacts that have established the characteristic nonlinear chronology that has calibrated Earth's tectonic history. Mass impacts might be more appropriately viewed as a cosmological device that plays the role of a nonlinear-dynamic escapement mechanism in the coupled geological-biological clock. According to this metaphor, Earth's history of mass impacts has influenced the variability of tempo and mode of evolution but has not necessarily determined its degree of advancement or "progress" — just as the nonlinear clock of the sine-circle map (see Table 1, Sec. D, and Fig. 14; cf. Shaw, 1987b, Fig. 1; 1991, Fig. 8.32) determines the instantaneous pace and irregularities of the clock's hands even though the clock may keep precisely the same time when the motion is averaged over many cycles.

After all, the fact that some phenomenon happens to provide a distinctive stratigraphic marker — even one in which its distinctiveness involves the disappearance of some large but selective portion of the global fauna and flora (i.e., relative to the full gamut of complexity ranging from microbial and hybrid organisms to certain groups of specialized lineages; cf. Fig. 36; and Notes 8.14–8.15) — does not at all mean that it has dominated the "balance of power" in the global system of bioenergetic feedback. The missing half of the global feedback loop — meaning those dynamical elements that are involved in the positive-feedback aspects of the global system — invokes the question: Is it possible that the injection of the exotic energies of mass impacts is essential to *both* the gradual and the punctuational processes that have controlled evolutionary growth, and hence has been responsible for both the populations and the circumstances that have permitted the mass extinctions to have occurred at their "proper" times in Earth history (cf. Eldredge and Gould, 1972; Gould and Eldredge, 1977; Gould, 1982b, 1984a, b, 1989b, 1992; Eldredge, 1989, pp. 197ff)?

In Shaw (1988a) I argued that the integral of intermittent impact energy supplied to Earth from environmentally catastrophic events may sometimes selectively (i.e., without replacement of, or obvious effects upon, other organisms) accelerate rates of evolution in some lineages, while yielding, at other times or in other lineages, either a conclusively terminal effect or no effect at all. In each case the effect realized depends on the net local dynamical interactions between the geosphere and the biosphere, and that involves many intermediate stages in the global complex of ecologic feedback loops (cf. Eldredge, 1989, pp. 197ff). *There is no fixed-sum energy budget in the global ecology that conditions the survival of species* — notwithstanding Darwin's wedges, or the influential concepts of supply-and-demand/prices-and-costs scenarios that (1) have conditioned our politico-economic beliefs concerning "balances" of world trade, and (2), in yet another kind of *Strange Loop* (see Hofstadter, 1979; and the Prologue to the present volume), were responsible for Darwin's metaphor of the wedge in the "economy of nature" through his study of Thomas Malthus's *Essay on the Principles of Population* (e.g., Gould, 1989b, p. 14; cf. Ridley, 1992).

From the dissipative nonlinear-dynamical point of view, there really are no simplistic economic balances (cf. Forrester, 1961, 1973; Arthur, 1990; Mosekilde et al., 1991; Ruthen, 1993) — and therefore a corollary must be that, in the conventional sense of compatibility between economics and political ideology, there really is no such thing as a political balance of power. That is to say, the politico-economic context of evolution is one sort of microcosm (multifractal attractor) nested within the macrocosm of biological evolution. And in that micro-macrocosm we are dealing with chaotic fluctuations at diverse spatiotemporal scalings — and the fact that we are ourselves embedded in the phase space of those fluctuations, *with them* so to speak, makes it impossible for us to perceive enough of the multifractal strange attractor to describe either its gross outlines or its fine structure. In the spirit of the original definition of the *strange* attractor (see Note P.3), then, perhaps we should call such complex structures *extraordinarily strange attractors*, or maybe even *weird attractors* (cf. Note 4.2).

The *evolutionary landscape* (cf. Provine, 1986; Wright, 1986) is deformable in kinetic phase space to an extent that exceeds prior expectation, simply because of the complexity of Earth's involvement within the larger context of the evolution of the Solar System within the Galaxy. This is just another way of saying that the Earth has been dynamically tuned to the evolution of complexity since early in its history. If the same cannot be said for other planets, it is because they have not experienced the same degree of coupling between their internal and external nonlinear-dynamical resonance states vis-à-vis the chaotic evolution of the Solar System. We have no need to invoke any special processes, such as "panspermia" (Arrhenius, 1908; Crick and Orgel, 1973; cf. Shaw, 1986; Marcus and Olsen, 1991; Chyba and Sagan, 1992), or "special creation," to explain complexity and the apparent uniqueness of living systems on Earth. This is the flip side of our seeing mass extinctions only as death blows dealt "from out of nowhere" to an Earth that, if only it had been unaffected by impacts, would have maintained a more "proper" gradualistic tempo and mode of evolution — hence demanding some sort of special event to get it all going in the first place (cf. Shaw, 1984, 1986, 1988a; Chyba and Sagan, 1992).

The nonbiological as well as biological processes discussed in this volume can be viewed as the resultants of nonlinear partitionings of phase space that have emerged according to the rates of recurrence at play within their own vicinities, thereby producing characteristic but complex and unpredictable structures. Some of these structures are comparatively invariant (e.g., stable and unstable fixed points, phase-locked resonances, and periodic windows in chaos). Some are comparatively mobile and changing, yet persistent (e.g., strange attractors and/or other chaotic attractors), while others — perhaps resembling chaotic attractors — oscillate with seeming violence in a characteristic vicinity between two or more attractor regimes, each with different temporal and spatial scalings (e.g., interior crises and chaotic bursts). Still others are seemingly destroyed by the passing of trajectories in their vicinity (e.g., boundary crises, "blue sky catastrophes," and extinction-like destructions of portions of attractor fields).

The coexistence of such a diversity of structures implies a kinetically self-tuned system (self-organized criticality), one that is "pumped" ultimately by a cosmological source or sources of energy. By contrast with cosmic-scale self-organization, the analogous laboratory description would invoke one or more attractor states that is/are invariant relative to an experimentally imposed set of tuning parameters (cf. Notes P.1, P.3, 3.1). The role of critical self-organization (cf. Chap. 5) is to run cyclically through the gamut of the possible structures that are made available, energetically, to the system by an enveloping hierarchy of source functions (e.g., in the Galaxy) that maintains energy levels in the system in states that are far from equilibrium (so-called *pumped states*, as in the states of lasers and masers; cf. Note I.5). In the the Galactic context, however, we can neither specify a priori nor pin down definitively the fluctuating pumping rates of the associated mass and energy fluxes. (Other variations on the themes of cosmological and biological evolution are discussed in Notes P.2–P.7, I.5, 8.1, 8.12, 8.14, and 8.15.)

CHAPTER 9

Summary of Conclusions

In January 1991, only months before I began writing this book, an event occurred that underscores the book's theme. An unnamed and orbitally undocumented asteroid (1991 BA) passed within 0.0011 AU (164,560 km) from Earth (Binzel et al., 1991, p. 91). It was not anticipated, and in fact it was discovered accidentally during routine observations only hours before it — fortunately for us — passed the Earth. Until its trajectory was established there could be no certainty that it would not collide with the Earth, or perhaps be captured by the Earth-Moon system (see Fig. 19, Chap. 6; cf. Steel, 1991; Yeomans, 1991a, b; Matthews, 1992; Kerr, 1992g; Ahrens and Harris, 1992; Lindley, 1992b). As it was, for a typical range of relative orbital speeds (roughly 10–20 km/sec), this near-Earth object (NEO) was near or inside the lunar distance when discovered and was at ~0.43 of the lunar distance when it passed the Earth. At a relative speed of 10–20 km/sec, an object would take roughly five to ten hours to traverse the distance between the Earth and the Moon along a direct course, and several to many times longer for a grazing trajectory — the shorter times being not nearly long enough to evacuate people and mitigate the devastation that would be wreaked in a metropolitan area, for example, by an object measuring even tens of meters in diameter.

Most of the NEOs identified in the references just cited are less than 100 meters in diameter **(Note 9.1)**, but this is large enough to do major damage to a populated area. For perspective, Shoemaker (1983, pp. 470ff) estimated that the object responsible for the approximately one-kilometer-diameter Meteor Crater, the famous tourist stop near Flagstaff, Arizona (see the geologic guidebook by Shoemaker and Kieffer, 1974), could have been no more than about 40 meters in diameter, but would have had an explosive energy of about 15 *megatons* TNT equivalent. By comparison, the energy of the Tunguska event — the bolide encounter of 1908 that devastated an immense but unpopulated area of Siberia — has been estimated at roughly 12 megatons TNT equivalent (e.g., Shoemaker, 1983, Fig. 1; Chapman and Morrison, 1989, p. 278). The near-equivalence of the energy scaling for Meteor Crater and the Tunguska event provides a cautionary tale for the comparative areas of direct-blast destruction of a relatively high-angle impact vs. a grazing encounter. That is, for the same energy, the areal devastation of Tunguska's grazing blast wave covered many times the area of analogous effects at Meteor Crater (cf. Krinov, 1963b; Shoemaker and Kieffer, 1974).

The best known energy "standard," by default, for the destructive potential of impacts is that established for large explosions set off by human technology — for which the physical destruction caused by the nuclear device tragically exploded over Hiroshima, Japan, on 6 August, 1945, is the most familiar documented event. This does not take account, obviously, of the great medical and psychological damage that was done by the radioactive aftermath of the nuclear explosion, as compared to the possible aftereffects of a meteoroid explosion of the same total energy (and neglecting possible long-term indirect effects on the Earth's atmosphere, etc., of low-altitude bolide encounters). The Hiroshima nuclear device had an energy yield of ~20 *kilotons* (not megatons) TNT equivalent (Besançon, 1990, p. 459; to find where this fits in a spectrum of other natural and technologic energy quantities see the perspective diagram given by Considine, 1983, p. 1099). Thus the Hiroshima blast was nearly a thousand times *smaller* than either of the above natural explosions (Tunguska or Meteor Crater), and approximately equivalent in energy to an "airburst event" — i.e., a "small" bolide-type explosion of the sort that occurs *annually* somewhere above the Earth's surface (cf. Shoemaker, ibid.; Chapman and Morrison, ibid.). Any of these annual events could conceivably be subjected to the same form of instability that occurred during the final stages of the Tunguska encounter — e.g., see Note 1.2 and the analysis of Chyba et al. (1993) — and it takes little imagination to realize the consequences of a low-altitude nuclear-scale explosion over any of the world's metropolitan areas (e.g., see the "fireball flyway" across eastern Canada and New York illustrated in Appendix 3).

Presumably, 1991 BA was discovered via the same circumstances by which other asteroids are discovered: it was noticed because it was on or near a set of orbital parameters known to be characteristic of asteroids (for a general description of astronomical observations of asteroids and the circumstances of their discovery, see Tatum, 1988; Bowell et al., 1989). [Veverka et al. (1993) report on a photograph of the asteroid 951 Gaspra taken from a distance of 5300 km by the Galileo spacecraft on 29 October, 1991, as Galileo made its way through the Flora region of the Asteroid Belt (see Fig. 13a) on its mission to Jupiter during the late 1990's (cf. Veverka et al., 1989). Compared with the objects discussed above, Gaspra is of a size equivalent to impactors that produced the largest known single-impact craters in the Earth's Phanerozoic record.]

If an unknown object the size of 1991 BA had approached at an inclination near 90°, on a trajectory *not* characteristic of asteroids or comets, what is the probability that it would have been seen before it passed, or at all? Zero, on both counts, according to the descriptions of discovery circumstances summarized in

Summary of Conclusions

Note 6.5 and elsewhere in this volume. I have discussed several types of resonance mechanisms that might account for the coordination of impactors with Earth's rotation period (or what is referred to, vis-à-vis Figs. 11 and 19–21, as *proximal flight control* of objects guided by mechanisms of the celestial reference frame hypothesis, CRFH). In the most obvious case, spin-orbit resonance has conspired with mean-motion resonances of NEOs, and/or with other focusing mechanisms affecting Earth-crossing objects, to create intermittent systems of natural Earth satellites (NESs) — where intermittent is meant both in the descriptive sense of seasonal and intermittent streams in an arid drainage basin and in the nonlinear-dynamical sense of chaotic intermittency, as discussed, for example, in Chapters 5 and 6 (e.g., Notes 5.1 and 6.1).

A seeming weakness in my thesis, one that I have been excruciatingly aware of since I began writing this book, and which was immediately pointed out by reviewers who had experience in astrophysics, is that it requires capture events that many investigators assume are unlikely because of the narrowness of the entry corridor for capture relative to the conditions for collision, ejection, or flyby (see Fig. 19). In response to such a criticism, however, I have (1) discussed new evidence concerning the dynamical circumstances that favor capture scenarios, especially with regard to fractal spheres of influence (basins of attraction) and "fuzzy" transfer orbits (see Note 6.1 and Belbruno, 1992), (2) pointed out that most planets have systems of satellites and/or particle rings, (3) cited putative examples of NESs, as well as candidate NEOs, described in the literature (Bagby, 1956, 1966, 1967, 1969; Baker, 1958; Kordylewski, 1961; Yeates, 1989; Frank et al., 1990), (4) illustrated examples of several nearly horizontal angle-of-incidence impacts, fireball processions, and subhorizontal bolide encounters cited in the literature (see Appendix 3; cf. Chant, 1913a, b; Mebane, 1955, 1956; LaPaz, 1956; Krinov, 1963a, b; Cassidy et al., 1965; Rocchia et al., 1990; Melosh, 1992; Schultz and Lianza, 1992), and (5) observed that even asteroids and comets sometimes appear to have satellites (cf. Van Flandern et al., 1979; Van Flandern, 1981; D. Davis et al., 1986, 1989; Weidenschilling et al., 1989; Chapman, 1990). As an example of how our concepts are changing with respect to satellitic systems, Porco (1991) has described six new satellites of Neptune and rings or arcs of smaller objects in resonance with them, and Esposito (1992) underscores the prevalence of such phenomena in the Solar System (cf. Cruikshank et al., 1982; Jewitt, 1982; Greenberg and Brahic, 1984; Esposito and Colwell, 1989; B. A. Smith et al., 1989; Kolvoord et al., 1990; Dones, 1991; Esposito, 1991; Colwell and Esposito, 1993). In another example, Binzel et al. (1991, p. 91) state that (cf. Burns, 1978), *"Phobos and Deimos, the two satellites of Mars, the eight outer moons of Jupiter, and the Saturnian satellite Phoebe appear to be captured asteroids."*

The question is, have we looked carefully for small satellitic objects around the Earth, the Moon, and the Earth-Moon system, especially those possibly in exterior polar orbits with resonant frequencies like those of Figure 20? An Earth-orbiter in that reference frame at a distance of 0.0011 AU (164,560 km) would correspond to a relative distance, a/R, of about 26 (the lunar distance of $\sim$384,000 km corresponds to a value of $a/R \cong 60.27$). This distance is off the scale of Figure 20 but can be located approximately in Figure 21 (righthand scale) just below the symbol for half the lunar distance. At half the lunar distance, $a/R \cong 30.1$, and the value of the orbital frequency ratio (α/β) is about (10/1). Therefore, near $a/R \cong 26$, orbital frequency ratios between (8/1) and (10/1) would be appropriate. The speed of such an object relative to an observer would depend on its inclination. At the value $i = 90°$, such an object would move with an angular velocity 1/8 to 1/10 that of an object circling the poles at the Earth's angular velocity. If an observer were looking in the polar direction, or if the object happened to be at a low declination and were big enough for direct observation, or if it happened to cross a star (occultation), then it *might* be detected. At lower inclinations, the relative speed could be much larger, depending on the orbital direction of the object, and some Earth-orbiters might be classed with FMOs (fast-moving objects of unknown orbital parameters and provenance) in sky explorations not dedicated to Earth-orbiting objects. According to the active observer Tatum (1988), *"Every month, astronomers observe additional very faint asteroids that they cannot identify. Many thousands of these 'unnumbered asteroids' exist, quite a few of them FMOs."*

Analogous remarks concerning the increasing awareness in the astronomical community of unexpectedly large numbers of near-Earth objects were made above (see Note 9.1) and in Chapter 6, an awareness that is underscored by the dramatically heightened attention to such objects in science-news reporting during 1991–92, by comparison with previous years (e.g., Binzel, 1991; Binzel et al., 1991, 1992; Steel, 1991; Tatum, 1991; Yeomans, 1991b; Ahrens and Harris, 1992; Kerr, 1992g; Matthews, 1992). An even stronger indication of concern over the potential multiplicity of near-Earth objects lies in the fact that distinguished scientists and large-budget, high-technology national research programs had begun to circle around these reports during 1992, including star-wars-like planning for rocket-launched deflection scenarios for small objects and nuclear-detonation schemes for objects larger than about 100 meters in diameter (e.g., Matthews, 1992; Ahrens and Harris, 1992). There has been talk — and even more rumors — about "anti-asteroid protection schemes" that would even invoke the development of the controversial neutron bomb — the "enhanced radiation weapon" that was so notorious during the late 1970's, and about which we have heard little since, until now — for use as a deflection mechanism (e.g., Matthews, 1992, p. 1205).

Even if searches demonstrated without exception that no objects with diameters ≥ 1 km currently exist in Earth and/or Earth-Moon orbital resonances, that would not falsify the CRF hypothesis. For if none were found at the present time, there remains the question of intermittent filling and emptying, as discussed earlier relative to the Asteroid Belt. It may be that spin-orbit lifetimes are short, or that they are long but are intermittently emptied by chaotic crises or by externally disruptive events (in the same sense that stars are thought to disrupt the Oort Cloud of comets). If that is the case, studies of other planetary systems may help to resolve the expected lifetimes of satellitic reservoirs (cf. Allan, 1969; Porco,

Summary of Conclusions

1991). For example, the ages of Phobos (~24 km in diameter) and Deimos (~13 km in diameter), the only presently recognized satellites of Mars, are not well determined, and further study of them, as well as evidence for any very small accompanying satellites, may be useful (cf. Burns, 1978; Veverka, 1978; Veverka and Thomas, 1979; D. Davis et al., 1981). Although Phobos and Deimos are small objects compared to many of the asteroids and satellites of other planets, they would correspond to "large-body" impactors from the standpoint of the impact-extinction hypothesis, for they are roughly comparable in size to the impactors that produced the largest of Earth's Phanerozoic craters.

In addition to spin-orbit resonances with Earth and/or the Earth-Moon system, there are several other possibilities for proximal control of CRF objects. These include mean-motion resonances with the Earth-Moon system and/or with one or more of the other inner planets. When the Sun is viewed as just another object revolving around the Solar System center of mass, other types of resonances are possible. They might pose trajectories analogous to the radiants of the meteor showers that are coordinated with Earth's orbital period. Many possibilities exist for resonance effects with the Sun. Some involve coupling between the Sun's motion about the center of mass, the orbital periods of the planets, and the Sun's internal mass anomalies, as manifested in the dynamics of sunspot cycles described earlier (cf. Wood, 1972; Landscheidt, 1983, 1988).

The mechanism or mechanisms by which the CRF became established represent, by the present hypothesis, both the greatest weakness and the greatest strength of that hypothesis. In the opinion of many workers in astronomy and astrophysics, the capture process, on which the geocentric system of spin-orbit resonances depends, is unlikely, relative to the prospects for direct collisions. My response is that many forms of nonlinear resonances exist, most of which are yet to be discovered, and that complex resonances apparently act in concert as sorting agents for the orbital trajectories of objects in the Solar System (see Note 6.1). It may be that were it not for such processes, we would be at *greater* risk from impacts than, to judge from the geological record, we have been. The postulated geocentric reservoir of NESs might well operate to modulate the intermittencies of impacts, hence to guide and mediate the incidences of multiple events, as well as to buffer much of the Earth's surface against high-frequency direct bombardments by small objects. The existing estimates of numbers of asteroids and comets in the near-Earth vicinity, as reviewed above and in Chapters 2 and 6, would seem to offer more opportunities for impacts than are found. Heightening this seeming discrepancy is the fact that discoveries of both comets and asteroids apparently are still increasing exponentially, and show no signs of tapering off to plateaus at any time in the near future. Several quantum leaps in the discovery of NEOs, as well as other objects in the Solar System, have been mentioned in this book *that occurred while I was writing it* (i.e., within a period of less than two years)!

A major strength of the CRFH, in my view, is the fact that having drawn the tentative conclusion that an invariant CRF reservoir has existed throughout the Phanerozoic, I discovered that the types of resonance mechanisms I have attributed to it are also applicable to many other types of celestial processes, as well as to terrestrial processes. The CRFH categorically predicts the existence of correlations with core motions, geomagnetic-reversal frequencies, and magmatic periodicities of types I previously did not have reason to suspect. These internal correlations are also coordinated through the CRFH with a biostratigraphic code that is globally synchronized by the same principles of nonlinear self-organization. Biostratigraphy is the terrestrial recorder of the global dynamic resonances mediated by the CRF impact signals, which in turn are mediated by the nonlinear dynamics of the Solar System, as outlined numerically in Figures 9–16 and 19–21 (cf. Appendixes 6–9, and 11). It may be that the orbital premise is flawed, but in that flaw is a universal truth concerning interrelationships among terrestrial and extraterrestrial processes.

Many issues have been considered in this volume; some that were stimulated by the CRFH and might not otherwise have been noticed include the following:

1. A group of craters, including those closely timed with the K/T boundary and others with ages roughly between 50 and 100 Ma, are clustered in a small-circle swath (Fig. 3) that would appear to be difficult or impossible to produce by any of the known orbital trajectories of asteroids or comets. The assumption that this pattern represented some form of local resonance is what stimulated the exploration of (a) spin-orbit and mean-motion resonances in the near-Earth vicinity and (b) geophysical correlations with systematically coordinated impacts over geologic time.

2. Given this assumption for the K/T interval, I plotted the known record of impact craters on Earth within five broad age classes (Fig. 2) and found it to show three local vicinities (which I have termed Phanerozoic cratering nodes, PCNs) in which events of all age classes coincided. These three PCNs fell on a small circle opposite to, and at a high angle to, the K/T swath, which may represent intersections between three great-circle swaths, or a swath similar to the K/T swath but mainly in the opposite hemisphere. Both possibilities suggest that Phanerozoic impacts have occurred dominantly in the Northern Hemisphere, or that one or two additional nodes at high southern latitudes are yet to be discovered (the key evidence may lie beneath the vast southern oceans, and/or conceivably beneath Antarctica's icecap).

3. Although the nodal swaths are drawn to represent the maxima of crater frequencies, some events have occurred between them, as for example Montagnais, in the North Atlantic (Fig. 3), the putative site of a major event of Permian age on the Falkland Plateau in the South Atlantic (Rampino, 1992), and possible sites in the Pacific Ocean (cf. Hagstrum and Turrin, 1991a, b; Hagstrum, 1992), one of which I suggested (Chap. 1, Fig. 3) might be looked for in the vicinity of arcuate scarps mapped on the sea floor by the GLORIA side-scan sonar survey of the Hawaiian Exclusive Economic Zone (Holcomb et al., 1991).

4. A systematic history of impacts on Earth suggests that the structure of the Solar System should show analogous evidence of ordering in the reservoirs of objects believed to be the ultimate sources of impactors. A review of the evidence

Summary of Conclusions

indeed shows the likelihood of a systematic hierarchy of fractal resonances — within and among the reservoirs of source objects and among the planets themselves — that reflects the evolution of a self-organized system of nonlinear processes beginning with the formation of the solar nebula from a pre-main-sequence binary star system **(Note 9.2)** that evolved through a T-Tauri star stage of the FU-Orionis type (see Herbert and Sonett, 1979; Herbert, 1989; Weissman and Wasson, 1990; Wasson, 1992). After its formation, the Solar System approached a quasi-steady state of nonlinear-dynamical intermittency that has since operated at the transition between chaotic resonances and phase-locked or quasiperiodic resonances. These effects are well displayed in the Asteroid Belt and probably exist in the Kuiper Belt and Oort Cloud as well. Viewed as chaotic attractor systems, these reservoirs of objects are subject to nonlinear crises, particularly of the type called chaotic bursts. Such bursting phenomena, with or without effects of stochastic triggering of the Oort Cloud by passing stars (see Note 6.3), can explain the shower-like intermittency in the delivery of impactors to the inner Solar System and to Earth.

5. Exploration of the mechanisms of spin-orbit coupling of satellitic objects in Earth's vicinity and in the planetary system itself relative to the Sun suggests that the CRF system of resonances has operated as a self-similar substructure of a general system of planetary and satellitic resonances in the Solar System (see Figs. 14, 20, and 21; and Appendixes 6, 9, and 11). The numerical ordering of objects includes phase-locking, period-doubling, and chaotic regimes, where any given substructure within the Solar System — such as the Asteroid Belt, satellitic rings and ring arcs, etc. — can show evidence of all three regimes. A generalized nonlinearly coupled orbital process thus explains the spacings of the planets (the Titius-Bode relation), the banded structure of the Asteroid Belt (the Kirkwood Gaps), and the existence of commensurable systems of satellites and satellitic rings around the planets. The reservoirs of comets (Kuiper Belt, Hills Cloud, and outer Oort Cloud) are less well known and appear to be largely chaotic, but there is a systematic hierarchical structure of cometary reservoirs that is partially "hybridized" or interwoven with the banded structure of the asteroids within the inner Solar System [the distribution of asteroid-like planetesimals and comets beyond the Asteroid Belt proper may be subject to revision over the coming years (cf. Fig. 13), judging from theories advanced and discoveries made during 1991–92 (e.g., Stern, 1991; Weidenschilling, 1991; Kerr, 1992e; Cowen, 1992]. Such a self-organized hierarchical system of resonances "straddling" the nonlinear-periodic/chaotic critical transition is not inconsistent with the collisional histories of objects within the Solar System, including the early fragmentation and rehealing of planet-size bodies now thought by many workers to apply to the histories of the terrestrial planets and the origin of the Moon.

6. The same principles of hierarchical self-organization, intermittency, and nonlinear crises also apply to the internal motions and transport mechanisms inside the planets and the Sun. The Moon, Mercury, and Mars mainly record impact dynamics because of their relatively sluggish and/or rigid behaviors,

compared with the Earth and Venus. Venus shows relatively little of its history of bombardment because of its rheological heterogeneities, mobilities, and volcanic activity (and/or catastrophic resurfacing prior to 500 Ma; see Note 8.8) — sharing this trait with the large and mobile satellites of the outer planets, where the mobility, in some cases, may be related to volcano-tectonic rheologies of volatile-rich silicate substrates, and, in other cases, may be explained by rheologies related to the melting and freezing of ice (H_2O "magmatism"), as on Jupiter's satellites Io and Europa, respectively. Earth is blessed with such a diversity of magmatic, tectonic, and surficial dynamics — including records of early sedimentation, metamorphism, and cratonic terrains — that partial records of some of its older impact structures have been preserved as evidence of the composite record of self-organized, coupled, impact-dynamic and geodynamic processes. The CRF system and the internal dynamics of the Earth emerged together as the evolutionary aftermath of the early history of major bombardments and impact-related basining events on Earth at scales proportional to, and dynamically and temporally commensurate with, those on the Moon following its formation. By hypothesis, the persistent Pacific Basin, and some of the large-scale geomorphic features preserved in the shapes of the continental margins, reflect the scars left by this seminal bombardment of the Earth's rehealed and/or rehealing surface.

The system of CRF invariances, which represents the proximal reservoir of subsequent external inputs of mass and energy to geodynamical processes, also coordinates and directly influences Earth's history of magmatism, tectonism, and core motions, through the intermediary of mantle convection. As a consequence of the nonuniform mass distribution in the Earth — with a long-term distribution and preferred orientation in the deep mantle that was also inherited from events and processes attending the collisional origin of the Moon (see Fig. 33) — the evolution of the CRF reservoir and intermittent impacts on Earth explains correlations between the timing of geomagnetic reversals, invariant magnetic flux lobes at the core-mantle boundary, metastable clustering of transitional VGPs and restriction of transitional VGP paths to dynamically selected meridional distributions, and correlations between all of these processes and the existence of the generally coextensive meridional pattern of deep P-wave anomalies. These structures and processes all align in part with a persistent meridional band through the Earth passing in the vicinity of 90° W and 90° E. This meridional band has acted as a stabilizing keel-like structure that, in conjunction with its influence on the CRF reservoir of impactors — by representing the longitudinal mass anomaly that has controlled the spin-orbit resonances of CRF objects — has also mediated the evolution of plate tectonics and continental drift throughout Earth's history by feedback coupling between geodynamic processes, thereby explaining the broadly meridional character of (1) the present-day continental distribution (and many of the published reconstructions of past continental distributions), (2) the patterns and periodicities of supercontinent cycles, and (3) the distribution of plate-tectonic subduction "graveyards." Lacking coordinating mechanisms of these types, one would have expected to see — instead of the above phenomenologies —

Summary of Conclusions

a tangle of globally distributed, and more "ergodic," trajectories of plate motions, continental drift, VGP paths, and linearly propagating volcanic trends. The self-organized nonlinear-dynamical features of Earth's history outlined here are discussed and illustrated in Chapter 8 (cf. the Appendixes in their entirety).

7. As a consequence of the processes summarized above, the geographic system of CRF coordinates is locked to the invariances of the cratering pattern (PCNs, etc.), hence with the deep-mass structure of the Earth that controls that pattern through its feedback with the dynamical evolution of spin-orbit resonances. According to the celestial reference frame hypothesis (CRFH), therefore, the cratering pattern represents a more or less fixed system of markers against which motions of other surface features of the Earth can be compared during the Phanerozoic Eon. This coordinate frame should correlate directly with the geographic coordinates of any crustal features that have remained kinematically fixed relative to the Phanerozoic cratering nodes (PCN's). The evaluation of such invariances, however, is complicated by processes of continental deformation and/or piecemeal transport of slivers, slices, and blocks of suspect (exotic) terranes relative to the loci of PCNs. That is, beyond some uncertain radius of influence about the PCNs, the cratering patterns have been partially disrupted by the phenomena of relative motions of microplates, etc. Conversely, the relative coherence of the global cratering pattern itself (PCNs plus cratering swaths) argues against major rearrangements of the cratering pattern as a whole, an observation that is consistent with my interpretation that even the motions of exotic terranes have been more or less constrained by the inferred meridional pattern of crustal kinematics. In other words, the coupling between the CRF, Earth's deep structure, and geodynamics has coordinated the motions of continental drift and plate tectonics within tolerances that mimic the bilateral keel-like meridional symmetry of the Earth. Kinematic departures from that general symmetry have been limited to such motions as (1) block rotations, (2) piecewise meridional translations of crustal slices, and (3) longitudinal kinematic oscillations within the bandwidth of the deep meridional anomaly, which, in turn, correlates crudely with the longitudinal bandwidth of the CRF system of impact craters (cf. Fig. 23).

8. I have described the dynamical history of the Earth as being the consequence of a system of time-delayed coupled oscillators operating within a larger system of time-delayed coupled oscillators (the Solar System) wherein the two systems are coordinated through the dynamics of processes common to both (the orbital evolution of, plus the impact-dynamic effects of, the CRF system of impactors). Furthermore, in addition to the dynamical effects of this hierarchical coupling, the common factor (the evolution of CRF impactors) both symbolizes, and acts as the printing mechanism for, a nonlinear clock that has recorded the history of the conjoined processes within both the Solar System and the Earth (e.g., by cross-correlating a planetary time scale — if we had one — with the Geologic Time Scale, as delineated by the intermittency structures of a nonlinear geologic code, which we call the Geologic Column). The nonlinear-dynamical

concepts by which these coupled processes have been described provide a new paradigm for the evaluation and/or reevaluation of the spatiotemporal histories of terrestrial processes, especially the stratigraphic and biostratigraphic processes by which the Geologic Column has been codified [the periodic chart of the singularity spectrum of discontinuities that mark the hierarchies of small to large chaotic crises (or catastrophes in the sense of Ager, 1984) that define its global structure]. The same principles of self-organized intermittency and chaotic crises apply to (1) the action of environmental processes that affect the rates of biological evolution, (2) the action of biological processes that affect the rates of biological evolution, (3) the environmental processes that generate the stratigraphic framework within which the biological processes are recorded as fossils, leading to (4) the nonlinear spatiotemporal biostratigraphical coordinate system that documents and maps the nonlinear-periodic chart of the Geologic Column. It should be noted that this progression comprises a system of feedback processes that *is* a feedback process, meaning that it is a self-referential system that operates by virtue of endless cycles of recursion, thereby forming what has been called an Endlessly Rising Canon, Strange Loop, or *Eternal Golden Braid* (see Hofstadter, 1979, p. 10). The formal implication of the *golden mean* in nonlinear dynamics plays an analogous role (see Note P.4; cf. Fein et al., 1985; Shaw, 1987b).

Other nonlinear spatiotemporal processes, such as magnetostratigraphy, ash-layer stratigraphy, dendrochronology (e.g., Fritts, 1991), etc. are folded into the generalized geologic code in analogous ways — according to the nonlinear-dynamical paradigm — because they involve processes that operate with similitudes like those of the fourfold list above. Such a system is in its essence a nonlinear periodic chart of the singularity spectrum of discontinuities (the endless musical score) that has recorded the hierarchies of small to large crises that define it. Mass extinctions, and the positive-negative feedbacks of originations and extinctions in general, are but another facet of the biological effects of Earth's chronicle of impacts. Of greater importance is the role of the CRF as a mediating factor in geodynamics and as a nonlinear clock that correlates the self-similar scaling of event magnitudes in time and space. By this view, background extinctions and mass extinctions are aspects of a self-similar and intermittent continuum of self-organized processes scaled according to the universality parameters of nonlinear dynamics. It was not simply impacts but rather a uniformitarian and nonlinearly intermittent terrestrial-celestial interaction that killed the dinosaurs. And it was the same process that gave them, along with the brachiopods, the clams, and the birds, their origination. Diversity patterns and crises (catastrophes) at smaller and smaller scales among lower and lower taxons of organisms make clearer how this process works.

In these ways, the astronomical clock of the Cosmos rules at all lesser scales as the giver of life, the essential correlation between the universe and the biological clock. In thus surviving, however, both the cosmological and the biological clocks reveal themselves to be nonlinear, and therefore do not tick with the

Summary of Conclusions

harmonic regularity that had been supposed to exist in the music of the spheres, the cycles of the planets, the processes of distant galaxies, and the calibration of the atomic clock. Betrayed by our harmonical predecessors, we are provoked to see a paradox of miraculous chaotic regularity, with a precision and accuracy made possible only through a synergy of self-organized critical states, universality parameters, and those divergent states of motion that dynamicists characterize by Liapunov numbers (e.g., Laskar, 1989, 1990; Sussman and Wisdom, 1992). Historians and philosophers might say that we have confounded ourselves, for an inaccuracy of only a millimeter in the position of a planet may be too great to allow us to postdict and predict, respectively, its past and future positions over even a few million years (cf. Sussman and Wisdom, 1992, p. 58), and therefore we have lost the certainty — that abstract sense of immortality that science has cherished for so long — that has been symbolized by the planetary orbits in the legends and lore of the common human experience ever since time, as we have defined it, began. Despite ourselves, however, a form of determinism has emerged in the universal behaviors of our coupled systems of nonlinear oscillators — the instruments of the new music of the spheres — a determinism capable of describing even the messier ingredients of a coupled Earth and Cosmic history, a history that many would abandon to the darkness of independent processes. I suspect that neither Hutton nor Darwin would find this state of affairs repugnant.

EPILOGUE

Copernican Craters and the Phases of the Moon

I have not previously considered the patterns of the craters on the Moon in this volume because knowledge of lunar age progressions comparable to those for the Earth during the Phanerozoic does not exist. Stothers (1992), in comparing the long-term history of cratering episodes on the Moon with episodes of orogenic tectonism on Earth, finds a general correspondence between six major episodes all younger than 3600 Ma. [This record should also be compared with the record of the major episodes of ore deposition in Earth history, especially as elucidated by Meyer (1988, Fig.1); because that record is far more intricate and complex than are the six generalized episodes of lunar cratering and orogeny, it broadens the theme of orogenic cycles and reveals other spectral properties of the phenomenological coupling between diverse Earth processes, including its cratering history, according to the feedback phenomena that are the central theme of this book (cf. Seyfert and Sirkin, 1979, Table 12.2).] In Figure 14.3 of Wilhelms' *The Geologic History of the Moon* (1987), the Copernican Period begins at about 1100 Ma, hence represents about twice the span of time of the principal data I have considered for the Earth. Still, the relative completeness of the young record on the Moon and the clarity of its spatial distribution may be of use in clarifying alternative interpretations of the CRF hypothesis for the Earth-Moon system. Therefore, in a final illustration, **Figure 37**, I have summarized the comparative histories of post-1100-Ma cratering for both the Earth and the Moon.

Copernicus, the largest of the multirayed impact structures visible on the near side of the Moon, has a central crater diameter of 93 km (Fig. 37; cf. Wilhelms, 1987, Plate 11A), which is roughly comparable to the diameters of the larger K/T craters of Earth discussed earlier (Fig. 3). The mapping of the Moon's younger craters is facilitated by features that cut across those of Copernicus, but since Copernicus itself is not the oldest crater of the Copernican Period, relative ages are somewhat equivocal (Wilhelms, 1987, Table 14.1, cites ages of ~810 Ma for Copernicus, ~109 Ma for the slightly smaller Tycho, and ≤ 50 Ma for "small craters"). Nonetheless, the spatial patterns of lunar cratering of the late Imbrian, Eratosthenian, and Copernican periods are relatively simple (see Plates 1–12 in Wilhelms, 1987). Rotations that resemble the effects illustrated in Figure 33 of the present volume (cf. Runcorn, 1987) may still have been active during part of this history, but presumably have not influenced the Copernican pattern.

Some of the geometric implications of Figure 37 are self-evident. Because not all of the lunar craters are identified on Wilhelms' maps, I have plotted those from the nearside and farside maps of his Plate 11 together with those listed in his Table 13.1 that have diameters of 9 km and larger. The diameters, shown by numbers in Figure 37, have a fairly uniform distribution between 9 and 93 km—presumably because the illustration is incomplete for the smaller craters (there are 58 craters shown in Fig. 37, with twice as many on the nearside as on the farside). The number of documented Copernican craters on the Moon in this size range is interesting relative to the number of documented Phanerozoic craters on Earth (which has a total of $\sim 10^2$ *known* craters of all sizes, about half of which have diameters $\geq$ 10 km). Thus, if we assume that Earth's cratering rate was roughly constant during the latest billion years, then, at the same preservation efficiency (see Chap. 2), there would be ~100 "discovered" craters with diameters $\geq$ 10 km on Earth during the time span of the Copernican Period on the Moon. This number will be taken as the extrapolated "catalog of discovered craters" on Earth in the following discussion.

The impact ratio between the Earth and the Moon should reflect in some way the relative orbital trajectories of impactors. If the number of Earth-crossing objects is considered to be equally available to both the Earth and Moon, then the numbers of collisions + captures would be given by the relative capture cross sections (Fig. 19). Taking the ratio of the surface areas of the Earth and Moon as a rough measure of the relative cross sections would imply a distribution proportional to the squares of the radii, or $(6341 \text{ km})^2 \div (1738 \text{ km})^2 = 13.31$. Multiplying by 58 (the number of Copernican craters $\geq$ 10 km on the Moon, from Fig. 37) gives ~770 craters as the areally proportional number on Earth back to 1100 Ma, assuming that the cratering process was unaffected by tidal or other orbital effects. The ratio of this hypothetical number of craters on Earth (~770), which is based on the assumption of areal proportionality with the number on the Moon, to the number of Earth's craters extrapolated from the Phanerozoic cratering record back to 1100 Ga (~100) is 7.7. If one assumes that this ratio identifies the number of undiscovered craters on Earth $\geq$ 10 km in diameter, then the recorded population is $1/(7.7) = 0.13$, or 13% of the expected population (this percentage agrees with the estimated completeness of ~10% for the Phanerozoic cratering record discussed in Chap. 2).

But if the ratio of craters on the Earth and Moon depends mainly on their mass ratio, then the predicted number of craters on Earth in the latest 1100 Ma would be about $(5.98 \times 10^{27} \text{ gm} \div 7.35 \times 10^{25} \text{ gm}) = 81.36$ multiplied by 58, which gives

Epilogue: Copernican Craters and Moon Phases

~4720 predicted craters ≥ 10 km in diameter on Earth during the latest 1100 m.y. On this basis, the Phanerozoic catalog would be only ~2% complete, or ~6% complete if normalized to the land area. On the same basis, the average cratering rate would have been $(4720)/(1100) \cong 4$ m.y.$^{-1}$, which is the approximate average rate of impacts *on the continents* predicted by the astronomical estimates by Shoemaker (1983) for craters ≥ 10 km in diameter, and by Shoemaker et al. (1990) for impactors ≥ 1 km in diameter.

Unfortunately, the above numbers — obtained from two different assumptions concerning cratering rates on the Earth relative to Copernican cratering on the Moon — leave us with the same range of uncertainties concerning the completeness of the catalog of documented Phanerozoic craters on the Earth as was discussed in Chapter 2. That is, the cratering catalog for the Earth is roughly 1–10% complete, depending on additional information for (1) the accuracy of the astronomically estimated number of impactors and/or (2) the geological factor to be applied to the Earth's cratering record for the numbers of craters that have been either hidden or destroyed by ordinary geological processes. Thus, the relative documentations of craters on the Earth and Moon are mutually consistent, but only within the order-of-magnitude range of uncertainties of the completeness of the terrestrial record. The above geometric and/or dynamical comparisons therefore do not resolve the problem of the relative cratering frequencies on the Earth and Moon, barring improved models for orbital delivery of impactors to the Earth-Moon vicinity and/or much better evidence for the distribution of the true cratering rates on Earth as functions of geologic setting and age (see **Appendix 27**; and Blatt and Jones, 1975; Derry, 1980; Kieffer and Simonds, 1980).

The same comparison can be expressed in another way by reversing our approach and using the astronomical estimate of ~4 m.y.$^{-1}$ for craters ≥ 10 km to calculate impact rates for both the Earth and Moon based on either the areal or mass proportionality defined above. By this approach, I assume that the total number of impacts on the Earth and Moon is given by the constant rate of ~4 m.y.$^{-1}$, giving a total of 4400 craters in 1100 m.y., and that there have been either $(4400)/(13.31) = 331$ or $(4400)/(81.36) = 54$ craters produced on the Moon according to the assumptions of, respectively, either area-dependent or mass-dependent partitioning of impacts between the Earth and Moon (the latter being consistent with the crater count in Fig. 37).

However, if I assume that the crater count on the Moon given in Figure 37 is accurate, I have an additional choice between two types of model — implying two different types of orbital-dynamic effects — for the numbers of craters on the Earth that are consistent with the observed number of craters on the Moon. In Model A, the proportional number of craters on Earth is assumed to be given by the ratio of surface areas, resulting in 772 craters on the Earth, for a total of $58 + 772 = 830$ craters on the Earth and Moon, a total cratering efficiency of $830/4400 \cong 0.2$ (20%), and a cratering ratio between the Earth and Moon of $772/58 \cong 13$. In Model B, the proportional number of craters on Earth is assumed to be given by the ratio of masses, resulting in 4719 craters on the Earth, for a total of $58 + 4719$

= 4777 craters on the Earth and Moon, a cratering efficiency of 4777/4400 ≅ 1.1 (110%), and a cratering ratio between the Earth and Moon of 4777/58 ≅ 82.

A cratering efficiency that exceeds 100% is not impossible and would imply that the flux of Earth-crossing impactors is somehow enhanced, or focused, in the near-Earth vicinity relative to the standard probabilities in calculations such as those of Shoemaker et al. (1990), whereas an efficiency of less than 100% implies that defocusing and/or selection phenomena result in fewer impacts than would be calculated from the unperturbed average flux of potential impactors. An efficiency of 20% results in only about one out of five of the potential impactors reaching the Earth, and about one out of 65 reaching the Moon. An explanation for such a sorting effect—one that is consistent with the CRFH—would be that the net effects of mean-motion resonances with the inner planets and Sun, plus the conditions of two-stage spin-orbit capture by the Earth-Moon system (and then by both the Earth and the Moon), "select out" a mutually proportional but relatively small percentage of the unperturbed flux of potential impactors. Although a yet smaller percentage of the objects selected out will be captured into stable, metastable, or transient spin-orbit resonances with the Earth and/or the Moon, the collision trajectories of the remaining objects will be biased toward lower angles of incidence, relative to the impacted surface, than would be the case for the unperturbed astronomical flux, where the most probable incident angle of impact for a "random" source of impactors is about 45° from the horizontal (cf. Gilbert, 1893; Shoemaker, 1962; Schultz and Lianza, 1992).

"Selection models" (i.e., models with significantly less than 100% cratering efficiency relative to the unperturbed astronomical estimate) would be consistent with a conspicuous number of multiple, low-angle, grazing, and/or near-horizontal impact trajectories—such as those pointed out in the present volume (see Appendix 3; cf. Baldwin and Sheaffer, 1971; Gault and Schultz, 1990; Schultz and Gault, 1990a, b, 1992; Trego, 1991; Schultz and Lianza, 1992). Craters depart significantly from a circular shape (in map view) only at extremely low angles of incidence (below ~15° from the horizontal); hence elongate and/or "skipping" craterforms are strong evidence of subhorizontal impact trajectories (Schultz and Gault, 1990b, 1992; Schultz and Lianza, 1992; cf. Melosh, 1992). [This general idea, with some new thoughts about the possible approach scenarios of the Tunguska event, is recapitulated in Note 1.2 (cf. Chyba et al., 1993; Melosh, 1993a).]

Although the mutual consistency of the above models seems reasonable, it is but a rationalization of the dilemma discussed in Chapter 2 concerning the completeness of the terrestrial record. The fact that the exposure of the larger craters on the Moon is almost 100% for the latest 1100 Ma (i.e., little has happened to the lunar surface during this time span other than reworking of the surface by relatively smaller impacts) would suggest that the lunar record is a good proportional indicator of impacts in the Earth-Moon system. On the other hand, the idea that on Earth there were about 43 craters ≥ 10 km for every one that has survived to be documented is hard to swallow, whether or not we normalize the data by land

Epilogue: Copernican Craters and Moon Phases

area. Actually, the assumed cratering rate of ~ 4 m.y.$^{-1}$ is the value given by Shoemaker (1983, p. 482) *for the continents*, making the total cratering rate for the Earth as a whole three times as large if the total number of impacts is simply a function of the surface area of the globe. Clustering in the oceans or in tectonically active land areas would help explain the discrepancy, but the argument that the areas of CRF nodal clustering represent higher than average impact probabilities would counter this explanation.

The regions underlying the Phanerozoic cratering nodes (PCNs) are relatively large, stable, and long-lived; hence an even denser clustering of craters should exist in those vicinities than is observed if there were, on average, forty times as many impacts even during the latest hundred million years or so. And if the cratering of the continents was not clustered in the way I have postulated it to be in the CRFH, then there are many relatively extensive areas of the continents that have been relatively undisturbed during the latest ten million years or so that should, in that case, yield conspicuous evidence of cratering. For example, at ~ 4 m.y.$^{-1}$, there would be forty craters ≥ 10 km in diameter on the continents with ages younger than 10 Ma to be accounted for — i.e., a number comparable to the entire Phanerozoic record for craters of that size range! Where are these putative craters in the relatively undisturbed areas of latest Tertiary age in North and South America, eastern and southern Africa, Siberia, and Asia? A detailed reinspection of crater distributions plotted specifically on geologic maps of the target rocks should be undertaken to explore the question of impact frequency vis-à-vis geologic preservability, especially for the Mesozoic and Cenozoic Eras (cf. Kieffer and Simonds, 1980, Table 1). I began such an undertaking following the first reviews of the present volume, but it proved to be far too formidable a task to accomplish single-handedly in the few months I had available to complete the revision of the text. In lieu of adequate geologic maps of actual and potential target areas, index maps showing the cratering distribution relative to the global distribution of Precambrian terranes are presented in Appendix 27. However, even these crude comparisons illustrate the problem discussed above and in Chapters 1 and 2.

In Chapter 9, I suggested that one effect of the tendency for capture of near-Earth objects (NEOs) into spin-orbit resonances is to effectively filter out some percentage of direct impacts, thus partly shielding the Earth from exposure to the full brunt of the unperturbed astronomical flux of Earth-crossing objects. If so, the difference between the astronomical flux and the number of craters observed on Earth would represent the net efficiency of the capture mechanism plus whatever geological factors affect the completeness of the impact record (i.e., as in Models A and B above). A low efficiency (as in Model A) implies that the net effect of resonance mechanisms during approach and capture is to reject the remaining objects. By the same token, the population of lunar impactors would be governed by the Earth and/or Earth-Moon mean-motion and/or spin-orbit mechanisms, suggesting that the latest lunar impact craters could have originated from an older stage of Earth-orbiting and/or Earth-Moon-orbiting objects. Such a two-stage or multiple-stage process also implies that some of these objects would be caught in local resonances with the Moon itself and therefore might be reinjected back to Earth. The transfer of meteoritic material from Moon to Earth is well documented (Lipschutz et al., 1989), and it is not unlikely that, at times in the past, more substantial populations of orbiting objects transiently captured by the Moon could also be returned to Earth, in the form of relatively focused meteoroid showers. [In general, it seems likely that material blasted off the surface of the Moon—and probably other planets, especially Mars—would in part reside in orbit about the parent body and in part be delivered to the Earth (cf. O'Keefe, 1980). The case for Mars as a source of meteoritic material—previously ruled out by some experts because the putative Martian igneous-rock fragments showed only weak shock effects—was given impetus recently by experimental demonstrations that high-velocity impacts can produce ejecta some portion of which displays only weak shock metamorphism (see Gratz et al., 1993; Melosh, 1993b).]

The features in Figure 37 that may bear upon this extension of the CRF hypothesis are the possible implications of clustering and alignments of craters on the Moon. It is clear that cratering has been dominantly preserved on the nearside of the Moon (twice as many craters occur on the nearside as on the farside, as shown in the upper-left equal-area stereographic projection of Fig. 37). Furthermore, the well-known asymmetry of the near and far hemispheres of the Moon evidently reflects the existence of perturbing longitudinal mass anomalies for spin-orbit resonances about the Moon, in a sense analogous to that illustrated in Figure 20 for the Earth (e.g., the nearside dominance of Copernican cratering on the Moon would—in the sense of some form of preferred spin-orbit-resonance effects—be analogous to the apparent Northern Hemisphere dominance of Phanerozoic cratering on the Earth). The heavy and dashed lines on the illustration of lunar craters (the same projection just mentioned in the upper left of Fig. 37) are arcs of great circles on the nearside and farside, respectively. These great-circle arcs, not necessarily complete, were drawn through conspicuous alignments of three or more craters on the equal-area stereonet. A few instances of small-circle patterns were found, but none is conspicuous enough or well-enough defined to warrant its being shown. There are few alignments along parallels that would suggest a zonal distribution (with the exception of the great circles aligned subparallel with the lunar equator).

At face value, then, the cratering pattern shown in Figure 37 for the Moon is consistent with one produced by spin-orbit resonances, because incidences of direct impacts are likely either to be unaligned or, if they involve multiple impacts, to be aligned along arcs of small circles. In other words, only approximately selenocentric (Moon-centered) initial trajectories and/or orbital trajectories of captured objects can fall along great-circle swaths of low to high inclinations—in the same sense that is illustrated for the Earth in Figure 20, following the theory of Allan (1967a, b). Furthermore, there is some suggestion that a partially resolved conjugate set of high-inclination and low-inclination trajectories is crudely focused at the center of the nearside of the Moon, a pattern that is similar to, but more centered than, the pattern of contemporary seismic activity and ancient

Epilogue: Copernican Craters and Moon Phases

volcanism illustrated by Wones and Shaw (1975, Fig. 1; cf. Wilhelms, 1987, p. 269). This pattern accords with predicted inclinations for a system of spin-orbit resonances of both the interior type (objects at relatively low altitudes) and the exterior type (objects at relatively high altitudes), where the interior resonances tend to migrate to low inclinations, and the exterior resonances tend to migrate to high inclinations (Allan, 1967a, Figs. 1 and 2). Because the Moon is itself locked into spin-orbit resonance with the Earth, the secondary resonances of selenocentric satellites are oriented more or less symmetrically to the direction of the — presently small — tide-generating potential of the Earth on the Moon, a symmetry that resembles the orientation of the volcanic hot-spot distribution on Io vs. the tides on Io caused by its own complex resonances with Jupiter and Jupiter's other satellites (see the Prologue, Appendixes 24 and 25, and Note 6.2; cf. Kaula, 1968, pp. 203ff; Peale et al., 1979; Cassen et al., 1982; Pearl and Sinton, 1982; Yamaji, 1991).

In a few instances, the great-circle trajectories on the Moon are consistent with continuation from the nearside to the farside, or vice versa (e.g., the paired craters Hayn and Bel'kovich K near the north pole). Without better knowledge of the relative timing of such craters, and study of the patterns of the more numerous craters with diameters < 10 km, of course, this effect could be fortuitous. Nonetheless, such patterns are expectable on the Moon in view of the discussion of spin-orbit resonances for Earth. Conceivably, the lunar pattern could be a by-product of an intermittent system of Earth satellites, and/or the Earth could have acted to assist capture of satellites by the Moon gravitationally. More likely, it would seem, there has been a paired system of intermittent satellitic reservoirs revolving about both the Earth and Moon since the mutual event whereby the Moon was created by the collision of a third body — a planetesimal possibly as large as Mars (cf. Daly, 1946; Hartmann and Davis, 1975; Wetherill, 1985, 1990; Stevenson, 1987; Chapman and Morrison, 1989, p. 165; Benz and Cameron, 1990; Cameron and Benz, 1991; Kaufmann, 1991, Fig. 9–22; Baldwin and Wilhelms, 1992; Spera and Stark, 1993) — with the proto-Earth, and where the dynamics of later satellitic processes have been mediated by resonant capture mechanisms dominated by the mass asymmetry of the Earth, or by the combined effects of the relative orbital motions *and* mass asymmetries of the Earth and Moon on capture of objects by both the Earth and the Moon. The present time would appear to be a minimum in satellitic occupancies for both the Earth and the Moon, which would be consistent with the idea that the statistics of large cratering events are also currently near a minimum, as discussed earlier (i.e., the cratering maximum evident during the latest 10 m.y. in Fig. 8 is past, and the cratering record during that time interval also shows an increase in the ratio of small to large objects, as was also discussed in Shaw, 1988a).

If a selenocentric satellitic reservoir has existed in the past, then — as just mentioned — the pattern of impacts on the Moon was influenced by the effects of spin-orbit resonances of the satellites with the lunar longitudinal mass asymmetry, as modified by the perturbations related to new captures and/or near misses of

Earth-crossing objects (i.e., this is the lunar analogue of Earth's CRF system). Accepting the possibility that a selenocentric reservoir of objects has existed intermittently in the past (in addition to the intermittent geocentric reservoir), a novel idea is suggested for multiple-impact events on Earth originating from the lunar reservoir — i.e., as a form of return mechanism to Earth from the Earth-Moon system of paired resonances. In effect, in such a model, the lunar reservoir becomes a staging platform for the launching of returned objects, constituting a sort of catapault effect whereby the Earth and the Moon have reciprocated in their respective cratering histories.

As mentioned in the Prologue, and above, the fact that the Moon keeps one side always facing toward Earth (a condition probably established long before the Copernican Period) is evidence of a rather complicated system of resonances involving the orbital-rotational interaction of both objects (cf. Note 6.2; and Kaula, 1968, pp. 203ff; Alfvén and Arrhenius, 1969). The net effect of these resonances on the lunar cratering mechanism apparently favored impacts on the nearside of the Moon. And since high-inclination great-circle orbits around the Moon are always potentially "aimed" toward Earth, any instability triggered by the near miss of a new incoming object might eject some portion of the lunar reservoir in Earth's direction. I take the liberty, therefore, to conclude this Epilogue — to a book in which conjectures concerning the evolution of chaotic, hence unpredictable, orbits are now familiar — with a somewhat more fanciful scenario, one that departs a bit from the relatively straightforward notion that impacts on Earth have been regulated, entirely or in part, by the existence of a single geocentric reservoir of natural Earth satellites, the NESs mentioned many times in the book. [Note that, as discussed in Chapter 6, if a "giant comet" broke up in the orbital vicinity of the terrestrial planets roughly 10^4–10^5 years ago — as advocated by Clube (1989b, 1992), Olsson-Steel (1989), and Clube and Napier (1990) — then it is possible that the larger objects from that event could have perturbed an already depleted Earth-Moon satellitic system enough to nearly empty it, and also possible, consequently, that it only recently began to fill again. As a matter of fact, numerous low-altitude bolide events and/or impacts have occurred during the Pleistocene and Holocene Epochs, events that are suspiciously coincidental with major episodes of continental glaciation on the Earth (see Appendix 3; and Clube and Napier, 1990, pp. 136ff and Table 5; cf. Walcott, 1972; Andrews, 1975; Sergin, 1980; Fischer, 1981; Hughes et al., 1981; Benzi et al., 1982; Glass, 1982; Held, 1982; Ghil, 1985; Landscheidt, 1983, 1988; Muller and Morris, 1986, 1989; Peltier, 1987; Crowley and North, 1988; Mitrovica and Peltier, 1989; Wasson, 1990; Gaffin and Maasch, 1991; Kerr, 1992f, 1993d; Winograd et al., 1992). An analogous event of even larger proportions could have occurred 65 million years ago (the K/T boundary event), when the Earth-Moon system of orbital resonances was maximally charged with satellites, thereby creating an unprecedented burst of multiple, systematically focused, impacts on both the Earth and the Moon.]

I accordingly consider what might have happened when a selenocentric reservoir "went critical" — either in the sense of a chaotic crisis or because it was

Epilogue: Copernican Craters and Moon Phases

disrupted by the capture and/or near miss of a new and relatively large-sized object in its vicinity — during the part of Earth's precessional cycle when the Earth was oriented as shown in the Inset of Figure 37. If this orientation represents a condition of northern summer (with the Sun to the right), and if we assume that the selenocentric reservoir was impulsively "emptied" toward Earth beginning at a specific time, then a train of orbital transfers would have been strung out toward Earth according to the positions and velocities of the objects in the selenocentric reservoir, as modified by the escape speeds induced by the perturbing chaotic crisis or newly introduced object. (This is obviously a highly complex phenomenon, and it would involve the sorts of "fuzzy routes" discussed in Note 6.1, but for the sake of concluding the story, I assume that the objects are strung out in a simple sequence such that they will reach the Earth like beads spilling from a broken necklace, or, more aptly, like moving freight cars spilling off a high railway trestle into a deep gorge.) The objects therefore reach the Earth in sequence, and in such a way that at least some of them make impact according to the spacings established when the objects left the Moon. The transit time to Earth depends on the Moon's position in its orbit at the "time of departure," etc. (i.e., like the launch of a space vehicle from the Moon back to Earth), a factor we cannot evaluate except to say that it was consistent with the ejection of some portion of the objects toward the Earth (see Bate et al., 1971, pp. 327ff). But let us say that things are optimal, and that the average speed during transfer was a few kilometers per second (the escape speed from the lunar surface is about one-fifth of Earth's escape speed of 11.2 km/sec, or a bit over 2 km/sec for low-altitude selenocentric satellites, and practically zero for those at high altitudes).

By direct analogy with Figure 20, a system of selenocentric spin-orbit resonances is divided into interior and exterior regimes relative to the condition of synchronous resonance, where the orbital period of the satellite is equal to the rotational speed of the Moon. Because the lunar period of rotation is locked with its period of revolution about the Earth, the rotation speed at the lunar surface is given by the Moon's circumference divided by the Moon's sidereal period of revolution about the Earth (i.e., it rotates just fast enough to keep the same point aimed toward the Earth (this speed is roughly 0.005 km/sec at the lunar surface, compared with almost 0.5 km/sec for the rotational speed of a fixed point on the Equator about the Earth's spin axis). Converting the synchronous selenocentric speed to the synchronous distance from the Moon, it can be demonstrated — i.e., by making calculations in parallel with those of Figures 20 and 21 for satellites around the Earth and planets around the Sun — that the regime of exterior resonances (where the satellite revolves more slowly around the Moon than the Moon rotates) begins at about 110,000 km from the Moon (nearly a third of the distance between the Moon and the Earth, which is 384,400 km). Objects in exterior selenocentric resonances would have two important characteristics: (1) they would tend to be at the critical inclination ($\sim 63°$) to the lunar equator (see discussion of Fig. 20; cf. Allan, 1969; Coffey et al., 1986), hence at inclinations relative to Earth's Equator of roughly $40°-86°$, and (2) they would be the first and easiest objects to be dislodged by a perturbing NEO (recall that many of the NEOs discussed in Chap. 6 and in Note 9.2 pass between the Earth and Moon, hence through the theoretical domains of the exterior selenocentric and geocentric satellites). For the sake of brevity, let us also assume that the trip from the lunar orbit to a collision with the Earth takes about the same amount of time as do our trips to the Moon (i.e., once started, the objects are essentially in free fall to Earth), or about a day (cf. Bate et al., 1971, Fig. 7.3–1). [Actually, I have neglected some of the "details" in this story, one of which is that the balance point for equal gravitational attraction between the Earth and the Moon is closer to the Moon than is the synchronous distance. We would also have to consider effects of proximity to the Lagrangian points in the vicinity of the Moon (cf. Note 6.1). Assuming, however, that there is considerable fuzziness in the dynamics at a distance of the order of 10^5 km from the Moon, I ignore these technicalities in order to conclude the story. It is evident, however, that any objects in exterior selenocentric resonances are likely to be there transiently, unless there is some nonlinear analogue of the Lagrangian points that has not yet been discovered.]

Because the Earth is, of course, rotating during this time, the first object hits the Earth at a position roughly 360° (about a day) from the position on the Earth that was facing the Moon when the object left its lunar orbit following the disruption of the lunar satellitic reservoir. The number and distribution of succeeding impacts obviously would depend on how many objects were in the lunar reservoir and what proportion were ejected by its disruption. If the ejection occurred over a span of at least a day—i.e., the entire lunar system became unstable at the same time (e.g., by the occurrence of chaotic bursts, perhaps influenced by passing NEOs)—then the impacts would strike the Earth along the latitudes of intersection with the Earth's rotation, therefore encircling the Earth during a day-long barrage. Furthermore, if the event occurred while the Moon was on the opposite side of Earth from the Sun, the sequence of impacts would produce—somewhat in the manner of a globe-encircling strafing run by a Star-Wars spaceship—a broad small-circle swath of craters at moderate to very high latitudes in the Northern Hemisphere. Now, because of the tilt of Earth's spin axis relative to the lunar equator in this summertime orientation, the highest latitudes of impact would be inside the Arctic Circle (i.e. above 66.5° N; also, the combined effect of the Moon's inclination and obliquity would add about one and a half degrees more tilt to the orientation of Earth's axis relative to a line drawn parallel to the plane of the lunar equator; see Fig. 37, Inset). Such latitudes are consistent with the fact that several terrestrial craters exist at latitudes above the Arctic Circle—e.g. Haughton, Northwest Territories, Canada, is at 75° N (diameter ~20 km, age ~22 Ma), Popigai Crater, Siberia, is at 71° 30' N (diameter ~100 km, age $\cong$ 65 Ma; but see Note 1.1), and the Avak structure (uncertain diameter and age), near Point Barrow, Alaska, is at about 71° N (see Fig. 3 and Appendix 3; cf. Grantz and Mullen, 1991; Kirschner et al., 1992). The numbers and locations of these craters suggest that northern high-latitude late-Phanerozoic cratering was com-

Epilogue: Copernican Craters and Moon Phases

monplace on the Earth (craters at near-polar latitudes, of course, are not likely to be discovered except by drilling and/or geophysical sensing techniques).

Oddly enough, with a little bit of poetic license this scenario would just about fit the requirements for the K/T cratering swath on the Earth — as depicted, from different perspectives, in Figure 3 and in Figure 37 (upper right). And because the paleobotanical reprise by Wolfe (1991; cf. Russell, 1982, p. 401; Kerr, 1992a) tells us that at least one of the major impacts occurred during the Northern Hemisphere summer, we can now add a new bit to the drama of the dying lily pads and the other hapless fauna and flora of the northern pre-K/T regimen. It could all have happened on a balmy summer eve while dinosaurs were foraging, primitive mammals were scurrying underfoot, and water birds were going to roost, just as each pair of eyes was drawn to the east, in that timeless awe experienced by all sensate organisms, as the horizon grew bright with the huge and numinous aura of the full Moon (**Note E.1**).

APPENDIXES

Appendixes:
The Supplementary Illustrations

These Appendixes consist of illustrations that elaborate on ideas presented in the text or present additional information that came to my attention too late in the revision process to incorporate without major reorganization of the text and/or text illustrations. The textual material and illustrations in the Appendix are organized more or less according to the subject matter of the chapters and figures in the body of the book — i.e., beginning with additional data pertaining to the cratering patterns and then proceeding to observations concerning orbital dynamics, geomagnetism, volcanism, and paleogeography in relation to the celestial reference frame hypothesis (CRFH). Each numbered Appendix refers to a new illustration, or set of illustrations, that supplements the figures already presented (*i.e., the principal illustrations are referred to in the text by Figure #; the supplementary illustrations are referred to by Appendix #*). Where expanded discussion is called for, in an Appendix, it is placed in the form of a note numbered serially and appended to the section of Chapter Notes (e.g., Note A.1, Note A.2, etc.). Primary citations in the text for the Appendix material, as for Figures and Chapter Notes, are indicated by boldface type, for instance (**Appendix 5**); notes to the Appendixes are indicated in boldface type only in the Appendix, for instance (**Note A.2**).

Where an Appendix supplements a specific chapter or chapters (including the Epilogue), the relevant chapters are indicated in parentheses following the appendix number (the Introduction is not indicated explicitly, because most of the Appendixes are cited there in outlining the contents of the book). For example, for Chapter 1, ancillary information is given in the Appendixes for new craters — and/or craters already in the literature — and/or new observations that came to my attention after the completion of Figures 1–4. The illustrations emphasize little-known aspects of cratering patterns not shown in Figures 1–4, especially low-angle, grazing, and/or Earth-orbiting trajectories, and inferred great-circle patterns of cratering trajectories not previously illustrated in the text — and their relationships to the Nodal Great Circles (NGCs) of Figure 23. Emphasis is placed on a description of a subhorizontal great-circle fireball trajectory that has apparently been lost in the literature since 1913, possibly because it was overshadowed by the drama of the earlier Tunguska event in Siberia on 30 June, 1908, an event that still commands great attention in the literature of terrestrial impact phenomena (e.g., Chyba et al., 1993; cf. Note 1.2). The

Appendix 1

1913 multiple-fireball event is impressive not because of its size but because of its clear exemplification of a subhorizontal, geocentric or quasi-geocentric — slowly decaying or newly captured — orbital trajectory precisely aligned with the NGC of Figure 23 that passes through the North American and Australian Phanerozoic cratering nodes (PCNs). This event is called the Great Fireball Procession (GFBP) — or the Great Canadian Fireball Procession — of 9 February, 1913, first described by Chant (1913a, b), and later by Mebane (1955, 1956) and LaPaz (1956); selected data from Chant (ibid.) are plotted in Appendix 3 and described in Note A.1.

Appendix 1 (Chapter 1)

North Polar equal-area projection showing new craters given by Grieve (1991) and other sources; new craters not listed in Grieve (1991) are indicated in parentheses at the bottom of the illustration and are numbered according to where they would appear in the G-R Catalog; literature sources are as follows: 3a, Avak, Alaska (Grantz and Mullen, 1991); 16a, Chicxulub, Yucatan (Hildebrand et al., 1991; Sharpton et al., 1992; Swisher et al., 1992); 20a, Cuarto, Argentina (Schultz and Lianza, 1992); 99a, Taihu Lake, China (Wang et al., 1991; Wang, 1992). The Beaverhead, Montana, structure of Hargraves et al. (1990) was not included in this or previous illustrations because of conflicting statements about its size and age. Hargraves et al. (ibid.) describe the age as "late Precambrian–early Paleozoic" and suggest that the diameter may have been as large as ~60 km; Grieve (1991) *lists* the age as ~600 Ma and gives a diameter of 15 km, while Grieve and Pesonen (1992) give ~600 Ma and a diameter of 60 km for the Beaverhead structure, presumably (?) after Hargraves et al. Notably, this structure is located near the intersection of the Phanerozoic and K/T cratering swaths (heavy dashed and dot-dashed lines) shown in text Figures 1–4 and Figure 26 (Mercator projection). The Taihu Lake structure is correlated by Wang et al. (1991) with the late Devonian Frasnian-Famennian (F-F) extinction event of McLaren (1970, 1982, 1986; cf. McLaren and Goodfellow, 1990). It does not fall on the Phanerozoic cratering swath, however, and the McLaren Line of Figure 4 (cf. Appendix 3a) would have to be rotated farther to the east to accommodate it. Claeys et al. (1992a, b) describe a stratigraphic section in Belgium that contains a microtektite horizon which they also correlate with an impact event at the F-F boundary. Analogously, Bice et al. (1992) describe evidence of shock-metamorphosed quartz grains in three shale beds of the uppermost Triassic in Italy, which they suggest correlate with three closely spaced impacts near, or at, the Triassic/Jurassic boundary. Rampino (1992) describes two possible impact structures on the Falkland Plateau of the eastern continental shelf of southernmost South America (cf. Oberbeck et al., 1992; and Note 8.11 of the present volume). The great circles shown in the illustration are the Nodal Great Circles (NGCs) of Figure 23. Phanerozoic cratering nodes (PCNs) are indicated by large asterisks (circled when they are in the lower hemisphere of the stereonet), as has been the custom in most of the figures of this volume. The data shown in the present stereoplot, and others like it, are easily visualized by reference to the standardized locations of the K/T and Phanerozoic cratering swaths in Figures 2–4; the same symbolic characterizations are included on nearly all North Polar stereographic projections for purposes of orientation.

Appendix 1

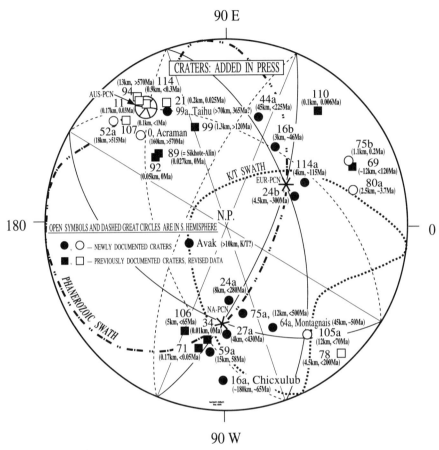

NEW CRATERS: Those listed in Grieve (1991, Table 1) are given a number and letter to indicate position in the G-R Catalog of Grieve and Robertson (1987); Craters not shown in either list are in parentheses.

0, Acraman, S. Australia	24b, Dobele, USSR	75a, Presqu'ile, Canada
(3a, Avak, Alaska)	27a, Glasford, USA	75b, Pretoria Salt Pan, S. Africa
(16a, Chicxulub, Yucatan)	44a, Kar-Kul, USSR	80a, Roter Kamm, Namibia
16b, Chiyli, USSR	52a, Lawn Hill, Australia	(99a, Taihu Lake, China)
(20a, Cuarto, Argentina)	59a, Marquez Dome, USA	105a, Vargeao Dome, Brazil
24a, Des Plaines, USA	64a, Montagnais, Canada	114a, Zapadnaya, USSR

Appendix 2

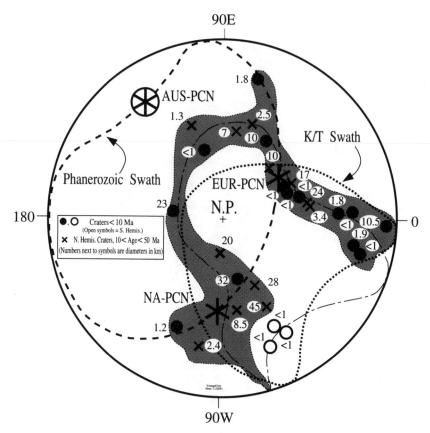

CRATERING PATTERN FOR AGES LESS THAN 50 Ma

Appendix 2 (Chapter 1)

The distribution pattern of all craters, regardless of size, with ages (1) less than 10 Ma, and (2) between 10 and 50 Ma, showing their approximate correspondence with the K/T cratering swath. The numbers on the figure are crater diameters. It is evident that the smallest and youngest craters — in other words, the subset of craters that should be the most indiscriminately distributed around the globe — retain nearly the same coherence as the largest K/T craters of Figure 3.

Appendix 3 (Chapters 1 and 6, and Epilogue)

Phanerozoic to modern impact trajectories based on diverse lines of ancillary evidence.

(a) Examples of organized impact trajectories based on cratering data that correlate approximately with the McLaren Line of Figure 4 (filled and open squares), South American "skipping impacts" (open circles) of the Holocene age (~10 Ka), from Cassidy et al. (1965) and Schultz and Lianza (1992), and the Great Fireball Procession (plane and circled

Appendix 3

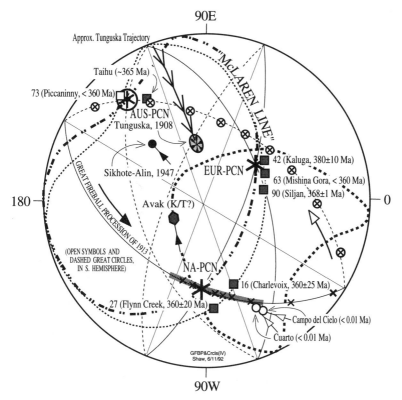

(a) EXAMPLES OF PALEOZOIC TO MODERN ORBITAL AND IMPACT SIGNATURES

(Numbered craters are from Grieve and Robertson, 1987)

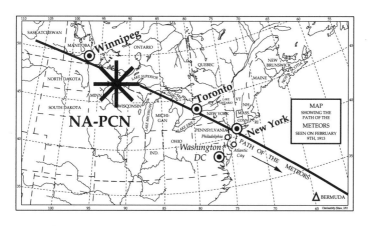

INSET: Index map of eastern North America (Modified from Chant, 1913a, Fig. 2)

337

Appendix 3

crosses) of 9 February, 1913, from Chant (1913a, Table 1). Additional information is given in **Note A.1**. The nodal great circle (NGC) that passes through the points from Chant (loc. cit.) is from Figure 23 and *is not*—contrary to the way it may appear—drawn through Chant's data points. Similarly, the McLaren Line of Figure 4 was drawn from descriptions of Frasnian-Famennian (F-F) biostratigraphic localities from McLaren (1970, 1982, 1986) and McLaren and Goodfellow (1990) before I plotted the craters. This illustration of possible F-F cratering localities was originally prepared at the suggestion of Digby McLaren (personal comm., 15 November, 1991) in criticizing—during his review of an early draft of this book—my designation of a "McLaren Line" based only on F-F extinction localities (it was only after his logical objection to my plotting such a line on the basis of fossil localities that we both plotted the late Devonian craters on Fig. 4, finding the correlation shown here). The recently described Taihu structure of possible F-F age near Shanghai, China (see Wang et al., 1991; Wang, 1992) departs significantly from the locus of the McLaren Line, but the deviation is within the tolerance of the swath-like patterns of Figure 2. The Avak structure, Alaska (Grantz and Mullen, 1991), and the Tunguska and Sikhote-Alin trajectories (Krinov, 1963a, b; Clube and Napier, 1982, pp. 140ff) are included as additional examples of intermediate-to-low-angle impacts (cf. Schultz and Gault, 1990a, b; Schultz and Lianza, 1992; Chyba et al., 1993). The orientation of the Tunguska trajectory is a compromise between descriptions given by Krinov (1963b) and Clube and Napier (1982, p. 140). I have subsequently concluded that, ironically, both descriptions could be valid (see Note 1.2). These three trajectories—and possibly the Taihu trajectory—are examples of patterns that roughly bisect the range of NGC orientations, suggesting the possibility, as discussed in the text (see Fig. 23), that the Phanerozoic, K/T, and other possible quasi-small-circle swaths may be the net effects of impact trajectories across the spectrum of NGC orientations that overlap to form a swath of preferred occurrences—e.g., as shown by the relation between the Avak trajectory deduced by Grantz and Mullen (1991) and the K/T swath. **Inset:** Index map of eastern North America, modified from Chant (1913a, Fig. 2), showing the path of the GFBP over Canada, the United States, and Bermuda. Embellishments include highlighting the locations of Winnipeg, Toronto, and New York with bold lettering, and adding the locations of Washington, DC, Philadelphia, and Atlantic City. The position of the North American Phanerozoic cratering node (NA-PCN), shown by the large asterisk, indicates the location and orientation of the map relative to the stereographic projection. The path (heavy solid line) is about 4000 km in length, starting near Mortlach, Saskatchewan, and ending just beyond Bermuda. The fireball procession at any one time was spread out over roughly a third of that length, judging from the time required to pass a given vantage point in the parade. For example, Chant (1913a, p. 161) gives an average duration of 3.3 minutes, which indicates a "train-length" of roughly 1600 kilometers.

(b) Comparison of the locations and trends of large North American and Asian lakes with cratering phenomena shown in (a). The North American lakes are aligned subparallel to the K/T swath and/or Avak-like impact trajectories. Although these lakes have obvious affinities with the pattern of continental glaciations (cf. Hughes et al., 1981; Mitrovica and Peltier, 1989), the source of such an aligned distribution of originating depressions would appear to be too systematic (e.g., relative to the oscillatory nature of glacial advances and retreats) to be explained by any simple mechanism of glacial erosion, unless there is a net memory of a single peripheral alignment of depressions in the phenomenon of glacio-isostatic rebound (cf. Andrews, 1975). Alternatively, there could be a correlation between glaciation and an existing arcuate pattern of depressions, such as those that might preexist from an alignment of small to intermediate-size impact craters. The analogous trend in Asia is not developed as coherently as the one in North America, at least at the same scale of existing lakes.

Appendix 3

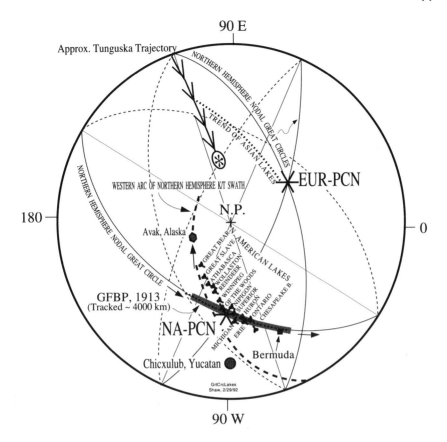

(b) YOUNG NORTHERN-HEMISPHERE GREAT-CIRCLE LAKES

Appendix 4

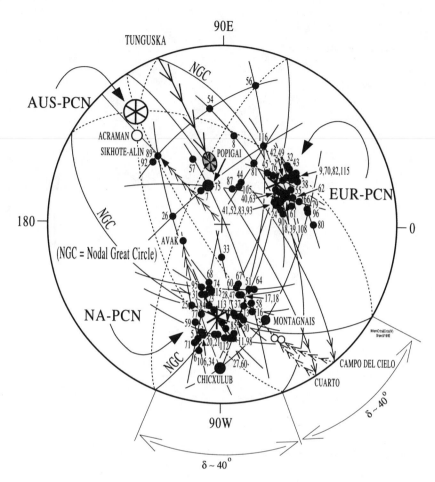

NORTHERN HEMISPHERE CRATERING PATTERN
(Craters are numbered according to Grieve and Robertson, 1987)

Appendix 4 (Chapters 1 and 6)

Composite diagram showing all Northern-Hemisphere craters, trajectories from Appendix 3a, and examples of possible great-circle trajectories through sets of three or more craters. Numbers are from the Catalog of Grieve and Robertson (1987). PCNs and NGCs are as given in the preceding Appendix illustrations. Note that the sets of great-circle arcs drawn through all Phanerozoic craters correspond roughly to the range of NGC orientations, within the uncertainties discussed in the previous figures and in the text (cf. Fig. 23). It is worthy of note that the Tunguska and Sikhote-Alin trajectories fall among a family of nearly polar great-circle arcs. Furthermore, a Tunguska trajectory originating from a somewhat more easterly direction would be essentially coplanar with — and with the same sense of motion as — the approach trajectory of the Rio Cuarto, Argentina, multiple ("skipping") crater field of Schultz and Lianza (1992), a trajectory that also passes through the Australian PCN.

Appendix 5

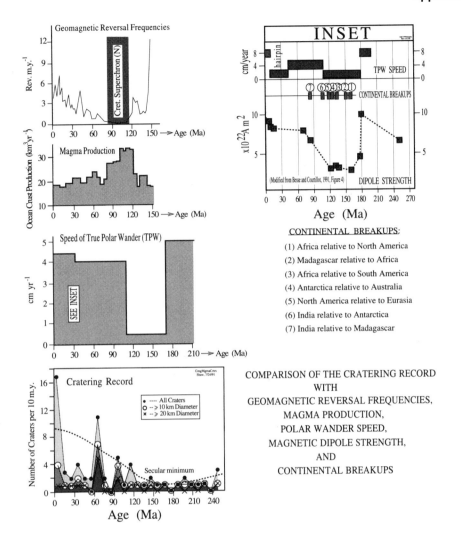

COMPARISON OF THE CRATERING RECORD
WITH
GEOMAGNETIC REVERSAL FREQUENCIES,
MAGMA PRODUCTION,
POLAR WANDER SPEED,
MAGNETIC DIPOLE STRENGTH,
AND
CONTINENTAL BREAKUPS

Appendix 5 (Chapters 2, 7, and 8)

Ancillary data pertaining to Chapter 2 referring to correlations between the time distribution of cratering and several types of geomagnetic, magmatic, and plate-tectonic processes, as discussed in Chapters 7 and 8 of the text (cf. Figs. 6–8). Detailed information on "true" polar-wander speeds, dipole field strengths, and events of continental breakup are shown in the Inset (modified from Besse and Courtillot, 1991, Fig. 4). Concise summaries of these relationships are given by K. Cox (1991) and Fuller and Weeks (1992). A cross section of the discussions of data, interpretations, and theories relevant to geomagnetic fields and field variations at many scales of time and size can be found—in addition to the sources just cited—in the following abbreviated list of references: Bullard (1955, 1978), Allan (1958), A. Cox (1968, 1969, 1981), McElhinny (1971), Vogt (1975), Gubbins (1977, 1987), Busse (1978), Chillingworth and Holmes (1980), Ito (1980), Verhoogen (1980), Jacobs (1981, 1984, 1987, 1992a, b), Mazaud et al. (1983), Merrill and McElhinny (1983),

Appendix 6

Burek (1984), Courtillot and Le Mouël (1984, 1985, 1988), Childress (1985), Mankinen et al. (1985), Prévot et al. (1985), Muller and Morris (1986, 1989), Bloxham and Gubbins (1987, 1989), Clement (1987, 1991, 1992), Courtillot and Besse (1987), Gubbins and Bloxham (1987), Schwarzschild (1987), Burek and Wänke (1988), Champion et al. (1988), Hoffman (1988, 1991, 1992), Hoshi and Kono (1988), Merrill and McFadden (1988, 1990), Poirier (1988), Coe and Prévot (1989), Creer and Pal (1989), Vogel (1989), O. L. Anderson (1990), Jeanloz (1990), Olson and Hagee (1990), Armbruster and Chossat (1991), Bloxham and Jackson (1991, 1992), Gilbert (1991), Hagee and Olson (1991), Knittle and Jeanloz (1991), Mazaud and Laj (1991), Laj et al. (1991, 1992), McFadden et al. (1991), Bloxham (1992), Bogue and Merrill (1992), Butler (1992), Constable (1992), Courtillot et al. (1992), Galloway and Proctor (1992), Valet et al. (1992), Smylie (1992), Thouveny and Creer (1992), Zhang and Gubbins (1992), and Gubbins and Coe (1993).

The more general ramifications of relationships such as those illustrated in this diagram are considered in the text of Chapters 7 and 8, especially in Notes 7.1–7.4 and 8.1–8.5 (see also the discussion of Ramsey's theory in Note 6.8). The diagram demonstrates that there is a systematic time-delayed signature in the illustrated suite of processes (e.g., scan the left column from bottom to top). The general interpretation of systematically related but nonsynchronous complex behavior is a form of nonlinear resonance—specifically a form that corresponds to *self-organized nonlinear coupled oscillations with delayed feedback*, as best exemplified in the chaotic dynamics of neurophysiological behavior in animals (e.g., Skarda and Freeman, 1987; Freeman, 1991; Freeman, 1992, pp. 467ff). This form of behavior can be thought of, alternatively, as *a system of multiple folded feedback*; compare the example of folded feedback described in Shaw (1987a, Fig. 51.14) with that illustrated on p. 83 of Freeman (1991) and/or Figure 14 in Freeman (1992). The implications of such interactions are sweeping, including the short-term and long-term behavior of continental motions, as shown by the timing of continental breakups shown in the Inset.

Appendix 6 (Chapter 3)

An alternative expression of the Titius-Bode relationship—modified from Shaw (1983b, Fig. 3)—as an ordered nonlinear-resonance diagram of the mass-energy space of a system of coupled orbital oscillators, analogous to the coupling of a set of nonlinear pendulums (cf. Table 1). The diagrams show the base-10 logarithms of the orbital velocities (top) and planetary distances (bottom) expressed in units of Earth's distance from the Sun (the Astronomical Unit, AU) and Earth's orbital period (year) vs. ordinal number, beginning with Mercury as the first order. The ordinal number is relative and can be shifted by taking another planet and/or position as marking the first order, and/or by deleting one or more other positions from the sequence, as is shown by the solid circles when Ceres is omitted from the planetary sequence. By hypothesis, the underlying structure—shown by the grids composed of two sets of intersecting light dotted lines—represents a nonlinear resonance net within which points of intersection constitute potentially stable or metastable orbital positions and velocities. The numbers of possible grids are unlimited, with the stipulation that the ratio of the slope of the gridline in velocity space (top) to the slope of the compatible gridline in distance space (bottom) must be equal to $-1/2$, as shown by the indicated values of slopes for one pair of lines. This is simply a requirement—or "artifact"—of Kepler's third law for the relationship between orbital distance and orbital period (e.g., Roy and Clarke, 1988, p. 145). It is evident that a single geometric progression, such as that of the Titius-Bode relation (see Inset B), does not work because there are shifts or offsets between at least two different compatible sets of gridlines, as shown in the figure. Pairs of gridlines comprise characteristic power-law invariances (spatiotemporal resonance families) in a multifractal phase space of orbital

Appendix 6

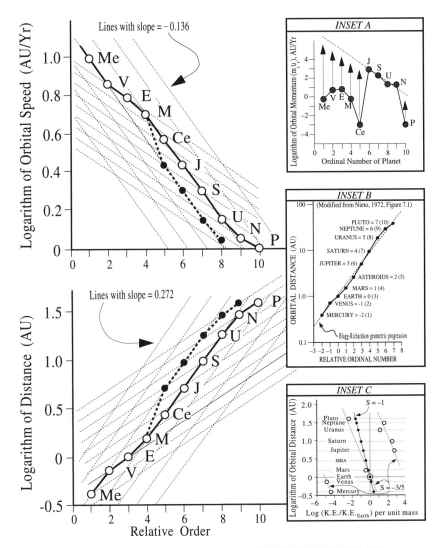

NONLINEAR RESONANCE GRID CORRESPONDING TO A
GENERALIZED TITIUS-BODE RELATIONSHIP FOR THE PLANETS

distributions. Unfortunately, there are "not enough planets" to give a reasonable estimate of the slopes of a second pair of gridlines. If such a secondary invariance (or sets of higher-order invariances) could be established, the slope ratios between sets of gridlines would constitute additional parameters that are characteristic of the Solar System—i.e., as a sort of orbital "fourth law"—in a manner somewhat analogous to Kepler's laws. The concept of universality would suggest that the same sort of relationship might apply to the orbital distributions of the satellites, asteroids, and comets, etc. (cf. Figs. 12–15, 20, and 21; Appendixes 7–9 and 11; and **Note A.2**). Elimination of the Asteroid Belt (symbolized by the position of Ceres, Ce)—thereby shifting the positions of Jupiter, Saturn, Uranus, Neptune, and Pluto one order to the left (solid circles)—complicates the pattern by requiring a third set of gridlines.

343

Appendix 6

A simple explanation for the origin of such a pattern may be offered by the hypothesis of the origin of the planetary nebula as a binary-star system, as discussed in Chapter 3 and Notes 3.2 and 9.2, combined with the notion that — going back in time — there were previously two principal resonance types, one for the "terrestrial planets" (at one time including Pluto as a member of the violently perturbed inner-planetesimal family) and another for the "giant" gaseous planets. Before that, perhaps, there was only one primordial resonance pattern (i.e., before the binary-star planetary nebula "exploded," as it were, to form the central-star planetesimal system that evolved to the resonance pattern that is characteristic of the Solar System as we know it today). Oddly enough, the progression from Mars through Neptune, including Ceres, fits reasonably well on one resonance line, suggesting the unorthodox notion that both Mars and the Asteroid Belt are locked in, or near, resonances that are more closely related to the outer planets than they are to the inner planets. The present-day trend of the planets Mercury, Venus, Earth, and Pluto (or Pluto-Charon) either would have been shuffled about from one that originally fitted the same trend as the outer planets, or — what appears to be more likely — they represent a second resonance family such as that shown by the second set of gridlines. Notice that the resonances, as shown, are independent of mass (hence orbital kinetic energy) — in the same sense that the period of a pendulum (in vacuo) depends only on the length of the ("massless") pendulum arm, not on the mass of the pendulum bob (e.g., Gamow, 1962, Fig. 1) — an observation that also fits with the complexity of the distance-mass-energy relationships in Figure 10 compared with the relative simplicities of orbital spacings and Kepler's laws. We are now treading, however, on the sacred ground of how the fundamental notions of force and mass were originally systematized by Newton on the basis of a simple subset of the relationships illustrated here and in Figure 10 (see the tribute to Michael Ovenden early in the Introduction of the present volume; cf. Shaw, 1983c, 1994; Milgrom, 1983; Schneider and Terzian, 1984; Milgrom and Sanders, 1993).

Inset A: Logarithm of orbital momentum $(m_r v_r)$ plotted against ordinal position, where m_r is the normalized planetary mass given by the mass of a planet divided by the mass of the Earth, and v_r is the orbital speed of a planet expressed in Astronomical Units (AU) per year. This plot shows the same structural dichotomy of the Solar System that is discussed in Chapter 3 and illustrated in Figures 9 and 10. The effect of mass "destroys" the simple spatiotemporal resonance pattern — or transforms it to one that is not as readily apparent — although the effect of mass is bimodal and peaked near the positions of the Earth and Jupiter. As in Chapter 3, I speculate that the deviations from a simple fractal distribution of momenta aligned with Jupiter, Saturn, and Uranus are somehow artifacts of the event, or sequence of events, that attended the disruption of the "Ghost Binary" of Figure 10 and stripped the inner Solar System of its more volatile elements (cf. Notes 3.2 and 9.2; and Appendix 7). The positions of Ceres and Pluto in this diagram are uncertain. Current estimates of mass for the Asteroid Belt and Pluto, respectively, would give somewhat greater negative values of momenta at these positions, but neither value of mass takes into account the possibility of larger masses related to "missing" — i.e., ejected, destroyed, or undiscovered — objects (cf. Ovenden and Byl, 1978; Van Flandern, 1978; Ip, 1989; Stern, 1991; Cowen, 1992; Kerr, 1992e; Weissman, 1993).

Inset B: General trend of the logarithm of orbital distance plotted against ordinal position, normalized to Earth as zeroth order (Nieto, 1972, Fig. 7.1), and the best fitting Blagg-Richardson regression curve (dashed line), as discussed in Note 4.2.

Inset C: Possible ordering relationships — analogous to the generalized resonance states of the main illustrations — between planetary orbital and rotational states expressed in terms of the common logarithm of the specific kinetic energies of the planets (K.E. per unit mass), normalized by that of the Earth, plotted against the logarithms of the planetary orbital distances from the Sun, in astronomical units [the small filled circles connected by the solid line of slope -1 are the specific orbital energy ratios of the planets (cf. Fish, 1967); open cir-

Appendix 6

cles and dashed line with slopes of $-3/5$ represent the specific rotational energy ratios of the planets, modified from Shaw (1983b, Fig. 5); cf. Shaw (1988a, Fig. 23A)]. The sets of points for planetary rotations (the open circles) show the deviations of the local condensed-state angular velocities of the planets — in which various short-range cohesive forces govern local states of rigidity and density relative to the long-range forces of orbital motion — from the Keplerian power-law relationships that describe the angular velocities of the planetary system (i.e., the solid line of slope -1 through the filled circles reflects the tautology of Kepler's third law, which states that the orbital period is proportional to the orbital distance raised to the power $3/2$, and therefore the specific kinetic energy — which is proportional to the square of the orbital speed hence is proportional to the square of the ratio of orbital distance to orbital period — is simply proportional to the reciprocal of orbital distance, as shown by Eqs. N6.1.1–N6.1.7 of Note 6.1). The above relationships govern the total masses and densities of local states of aggregation hence also the diameters of the planets [i.e., in order to reveal the possible resonances more accurately, I would have to show coordinates representing the planetary radii and densities, and/or other suitable condensed-state parameters, such as those parameters that describe the balances between the cohesional forces that maintain planetary integrity vs. the net actions of gravitational-rotational forces tending to separate the component parts constituting their states of aggregation (e.g., the "Roche" condition; cf. G. H. Darwin, 1879, 1962; Rubin, 1983; Landscheidt, 1988; Peale, 1988; Tassoul and Tassoul, 1989; Kaufmann, 1991, p. 287; Pasachoff, 1991, pp. 225f; Maran, 1992, pp. 606ff; Sridhar and Tremaine, 1992) — as in the formation of planetary rings, and in the early losses of volatiles from the terrestrial planets (see Appendix 7 and Chap. 3, Fig. 9, Inset; cf. Notes 3.2, 6.8, 8.3, 8.8, 9.2)]. The dashed lines with slopes of $-3/5$ represent one among an unlimited number of possible families of power-law exponents in the above distributions — i.e., in the same sense as the possible families of dashed lines that describe the ordering of planetary distances in the principal diagrams of this Appendix. The tripartite distribution of rotational states expresses the same relative groupings of the planets that are evident in Figures 9 and 10, in which Pluto appears to be affiliated with the terrestrial planets rather than with the giant gaseous planets, and where Venus and Mercury tend to stand apart. The Asteroid Belt (MBA of Fig. 13) would plot offscale to the left because of the small sizes of individual "minor planets" and rotation rates that vary from one to a few times those of the Earth and Mars (see Binzel et al., 1989b, Fig. 2; Pasachoff, 1991, Appendix 3a). Orbital/rotational resonance phenomena represent variations on the theme of spin-orbit resonances discussed in Chapter 6 and illustrated in Figures 20 and 21 and Appendix 11, thereby explaining qualitatively why Kirkwood (1849; cf. Walker, 1850) thought he could discern a systematic relationship between the rotation states of the planets and their spacings. [Kirkwood proposed a power-law relationship in which the square of the ratio of the sidereal orbital to rotation periods varied as the cube of the diameter of a planet's gravitational sphere of influence as determined from its spacings relative to the points of equal attraction between it and its neighboring planets, a relationship that perhaps fits an idealized nebular hypothesis (cf. Fig. 9 and Appendix 7) but fails in the light of modern data for specific planets). However, though the planets vary by many orders of magnitude in their ratios of specific orbital to rotational kinetic energies, certain pairs are nearly commensurable — such as Neptune and Uranus ($\sim 2:9$), Uranus and Saturn ($\sim 12:1$), Jupiter and Saturn ($\sim 1:1$), Mars and Earth ($\sim 2:1$), Venus and Mercury ($\sim 3:2$). It is also well known that Mercury's 59-day period of rotation is precisely 2/3 of its 88-day orbital period (see Pasachoff, 1991, Fig. 9-2 and pp. 162f), a spin-orbit relationship that is analogous to the synchronous relationship of the Moon's orbital and rotational periods relative to the Earth (cf. Note 6.2; and Alfvén and Arrhenius, 1969; Pannella, 1972), the synchronous relationship of Io relative to Jupiter, and diverse resonances of other satellite rotational periods relative to their primaries (see the Prologue).]

Appendix 7

Appendix 7 (Chapters 3, 4, 6, and 8)

Chapter 3 of the text is primarily concerned with the mass structure of the Solar System, especially as it can be characterized by fractal mass and/or energy distributions, as illustrated in Figures 9 and 10. Ancillary illustrations are given here that support and extend that general theme in relation to planetary spacings (Titius-Bode types of relationships), the fractal dimensions of the presently observed planetary distribution, and hypothetical distributions of mass related to different models of the solar-planetary nebula in the presence of a protostellar binary companion to the pre-main sequence Sun (the "Ghost Binary" of Fig. 10; see Notes 3.2 and 9.2).

(a) Actual (filled circles) and hypothetical (open squares with solid and dashed outlines) fractal distributions of the planets relative to various assumptions concerning the mass of the "Ghost Binary" as an end-member of the distribution. The hypothetical distribution with a slope approaching the value −3 (dashed line through the open squares with dashed outlines) would imply that virtually all of the mass resides in the "Ghost Binary" (point mass); this conclusion, however, is extremely sensitive both to the value of the slope and to the mass assumed for the "Ghost Binary." An example tabulation of calculated masses (expressed in multiples of present-day Earth mass) of the inner planets is shown for a slope between −2.75 (i.e., close to that for the present-day distribution of the outer planets) and a slope between that and the limiting slope of −3 (dashed line through the open squares with solid outlines). Such a solution implies that much of the present-day planetary mass (~445 Earth masses) would reside in the inner planets (e.g., ~335 Earth masses at the positions of Mercury through the Asteroid Belt, as shown) or that the total planetary mass was two or more times the present-day value. The effect of the putative

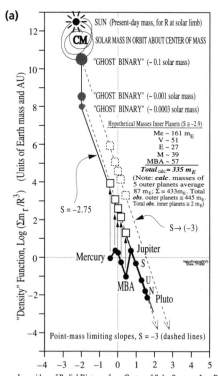

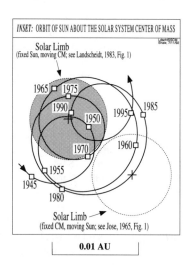

"DENSITY" OF FRACTAL PLANETARY NEBULA, RELATIVE TO ASSUMED NEBULAR MASS AND ASSUMED NUMBER OF PLANETS

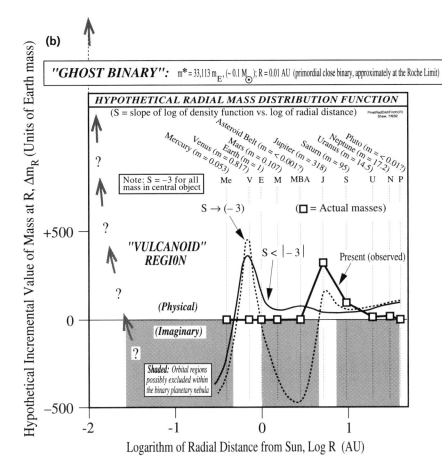

catastrophic disruption of the "Ghost Binary" discussed in Chapter 3 (see Fig. 9, Inset) and Notes 3.2 and 9.2 was to "collapse" the mass distribution of the inner planets to their present-day values, leaving the total planetary mass mainly in (or shifting it to) the positions of the outer planets. **Inset:** Relationship between the Sun and the center of mass (CM) of the Solar System according to Jose (1965) and Landscheidt (1983). The motion shown by the solid curve with arrows can be viewed either as the "orbit" of the Sun about the CM or the motion of the CM relative to a fixed position of the Sun. The range of variation in the plane of the ecliptic is approximately two solar diameters ($\sim 2.8 \times 10^6$ km) over the latest fifty years (i.e., the total potential variation over geologic time is not known).

(b) Schematic diagram of hypothetical mass distributions in the planetary nebula prior to catastrophic disruption of the "Ghost Binary." As the slope of the fractal distribution of planetary masses approaches the value -3 in (a) — i.e., where D, the fractal dimension, approaches zero as the mass distribution approaches the point-mass limit (see Fig. 9, Inset) — the total planetary mass becomes localized at positions corresponding to the inner and outer planets. Negative masses (shaded region of excluded states) are artifacts of the transition from continuous to discontinuous distributions (i.e., as planets coalesce, the mass excess at a given position is balanced by a mass deficiency in neighboring regions). If the "Ghost Binary" was as large as a tenth of the Sun's mass, the mass distribution (and positions of maxima and minima) would have been extremely sensitive to slight changes in fractal dimension (slope) relative to the point-mass state (i.e., nearly all of the mass is in the

Appendix 8

central binary star). The "vulcanoid region" refers to Leake et al. (1987), who speculate on planetesimals in the solar vicinity, perhaps states in which the mass-distribution curve was shifted to the left, as in (a). Note also that the position of the terrestrial planets is massive enough to have become a white drawf star (Sexl and Sexl, 1979, Fig. 4.10).

Appendix 8 (Chapters 3, 4, and 6)

Number vs. size distributions for the planets, satellites, and asteroids. The distributions include (1) the planets plus the Sun (large shaded circles), (2) the large satellites of Jupiter and Saturn (open squares and open circles, respectively), (3) the small satellites of Jupiter (small solid squares), and (4) the asteroids (shaded triangles, and Inset). Distributions (1) through (3) define fractal dimensions equal to or smaller than about 0.5 (i.e., approaching the point-mass-like distributions of Figs. 9 and 11, and Appendix 7), whereas the steep distribution for the asteroids corresponds to a fractal dimension between 2 and 3, typical of fragmentation spectra (Turcotte, 1986, Table 1). It seems clear that the asteroids, and the inner planets, do not correspond to the same genetic class as the outer planets and their satellites—an observation that would fit with the binary-star-catastrophe model of the planetary mass-energy spectrum discussed in Chapter 3 and in Notes 3.2, 6.8, and 9.2, and illustrated in Figures 9, 10, and 13 and Appendixes 6, 7, and 9.

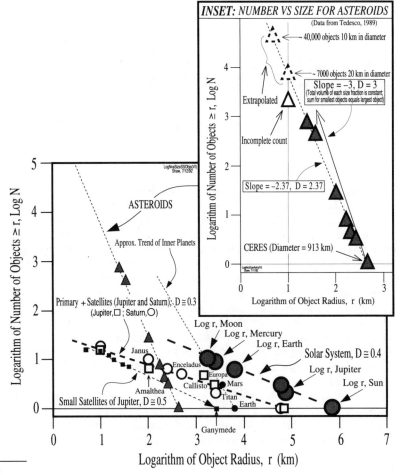

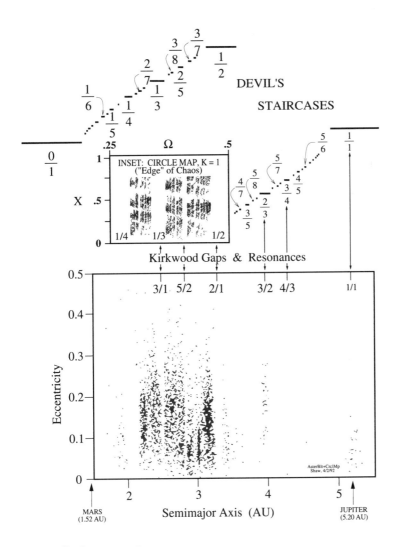

Appendix 9 (Chapter 4)

Elaboration of chaotic-resonance relations for the asteroids relative to the Kirkwood Gaps and the Devil's Staircase of Figure 14. The lower diagram is a schematic replica of the chaotic pattern of the asteroids, as illustrated by Nobili (1989, Fig. 1) in terms of their orbital eccentricities vs. semimajor axes (cf. Fig. 13 of the present volume). The **Inset** shows the analogous distribution of sine-circle-map quasiperiodic trajectories at the transition to chaos (computer experiment performed by the author; cf. Shaw, 1987b, 1991) and its general correspondence with the Kirkwood Gaps. Above and to the right of the Inset are portions of the completed Devil's Staircase of Figure 14. The steps in the staircase — viz., where the simple rational fractions correspond to the most stable mode-locked winding numbers; see Table 1 and Notes P.4, 3.1, 4.1, and 7.1 — are to be compared with the gaps in the trajectories of the sine-circle map and with the range of asteroid resonances between Mars and Jupiter. The righthand portion of the staircase required some distortion before it could be fit with the orbital distance relations of the lower diagram, but the general correspondences between a one-dimensional two-parameter dynamical map and the complexities of the orbital-parameter map of the Asteroid Belt is remarkable — if not amazing.

Appendix 10

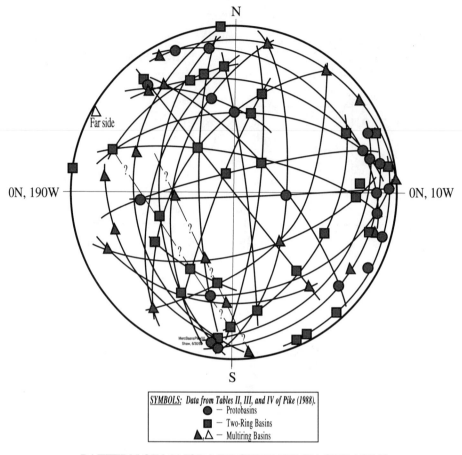

PATTERN OF MAJOR MERCURIAN IMPACT BASINS

(All great-circle arcs are in the front hemisphere; Queried dashed lines may represent artifactual swaths, as on Earth)

Appendix 10 (Chapter 6)

Equal-area stereographic projection of major impact basins on Mercury from data in Pike (1988, Tables II, III, and IV) derived from three Mariner 10 flyby encounters, the first (closest approach) on 29 March, 1974 (Considine, 1983, pp. 1858f; Chapman, 1988, p. 7; cf. Schultz, 1988; Spudis and Guest, 1988; Strom and Neukum, 1988; Wasson, 1988; Wetherill, 1988). Attempts were made, without success, to fit small-circle trajectories to sets of basins—a result that would appear to be at odds with models of low-inclination meteoritic encounters. On the other hand, nearly all of the large impact basins fell on sets of great-circle trajectories (four or more basins per trajectory) with inclinations ranging from 0–90° relative to the geographic coordinates. Mercury is in a state of spin-orbit resonance with the Sun, corresponding to a winding number, W, of 2/3, as discussed in Note 4.1 (cf. Table 1 and Notes P.4, 3.1, and 7.1), meaning that in the framework of the fixed stars Mercury rotates three times for every two trips around the Sun (cf. Peale, 1988; Pasachoff, 1991, Fig. 9-2).

Appendix 11

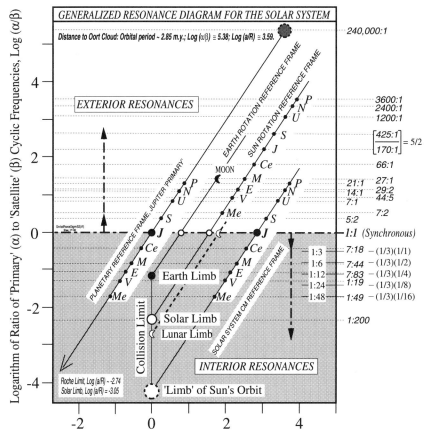

Appendix 11 (Chapter 6)

Elaboration and clarification of Figure 21. The region of interior resonances is shaded, and resonance ratios are indicated for systems defined according to several different reference states, ranging from the lower limits of the "limbs" of the respective Solar System objects (i.e., the surfaces of reference of the Sun, Earth, and Moon) out to the Oort Cloud, the effective "edge" of the Solar System (see discussions in Chap. 6 and Notes 4.1, 4.2, 6.1, 6.4, 6.8).

Appendix 12

Appendix 12 (Chapters 7 and 8)

North Polar equal-area stereographic projection (top) and Mercator projection (bottom) of the global distribution of present-day volcanic "hot spots," as illustrated by Duncan (1991). Numbers are assigned in the present volume to assist cross-referencing of names and locations with other Appendix illustrations (see also Crough and Jurdy, 1980; Duncan and Richards, 1991; Hagstrum and Turrin, 1991a, b; Hagstrum, 1992; Rampino and Caldeira, 1992; Yamaji, 1992; Stothers, 1993; cf. Shaw, 1969, 1970, 1973, 1980, 1983a, b, 1984, 1985a; Shaw and Swanson, 1970; Morgan, 1971, 1972, 1981, 1983; Shaw et al., 1971, 1980a; Jackson et al., 1972, 1975; Dalrymple et al., 1973; Molnar and Atwater, 1973; Shaw and Jackson, 1973; Bargar and Jackson, 1974; Gruntfest and Shaw, 1974; Jackson and Shaw, 1975; Jackson, 1976; Molnar and Francheteau, 1975; Crisp, 1984; Molnar and Stock, 1987; Sager and Bleil, 1987; Alt et al., 1988; Molnar et al., 1988; D. L. Anderson et al., 1992; Ray and Anderson, 1992; Sleep, 1992).

Upper Diagram: Northern Hemisphere "hot spots" are shown by filled circles with adjacent numbers and names; Southern Hemisphere "hot spots" are shown by encircled numbers and adjacent names; the Columbia River flood-basalt province (filled square labeled CRB) is shown for reference. Nodal great circles (NGCs) of Figure 23 are shown by the light (solid and dashed) great-circle arcs; Phanerozoic cratering nodes (PCNs) and cratering swaths of Figures 2–5 are also shown for reference (PCNs are shown by asterisks, circled in the lower hemisphere; the Phanerozoic swath is shown by the dot and double-dash trajectory, and the K/T swath is shown by the trajectory of closely spaced short dashes).

Lower Diagram: "Hot spots" are shown by encircled numbers (see upper diagram for names); the geographic coordinates of Balleny and Erebus fall outside the limits of the map and are indicated only schematically. The filled circles with smaller numbers 1 through 8 (highlighted but not circled) are flood-basalt provinces, as shown in Figure 26. The approximate small-circle trajectories shown by the heavy short and long dashes and the heavy short dashes are, respectively, the Phanerozoic and K/T cratering swaths of Figures 26 and 28 [the parallel (small circle) identified by the short dashes at a latitude of about 25° S gives the approximate position ($\pm$ ~15°) of a questionable Southern Hemisphere cratering swath, as shown schematically in Figs. 2–4]. The locations of Popigai (Siberia), Chicxulub (Yucatan), and Avak (Alaska) impact structures are shown for reference, as in Figures 3 and 26.

The predominance of "hot spots" in the Southern Hemispheres of these diagrams is consistent with a Northern Hemisphere predominance of impact craters if there is an antipodal bias in the relationship between cratering and associated volcanism (cf. Hagstrum and Turrin, 1991a, b; Hagstrum, 1992; Rampino and Caldeira, 1992).

Appendix 12

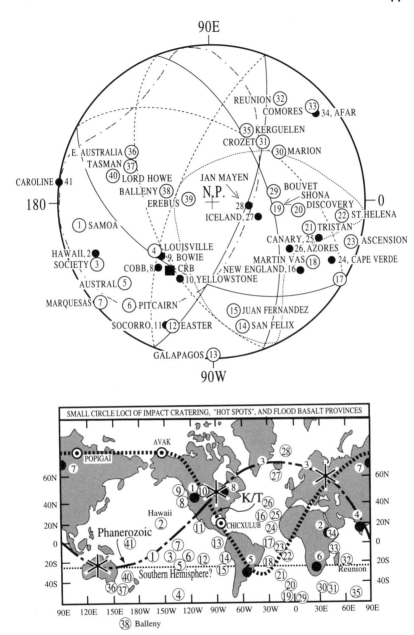

INDEX MAPS OF GLOBAL "HOT SPOTS", CRATERING NODES, SWATHS, AND NODAL GREAT CIRCLES

353

Appendix 13

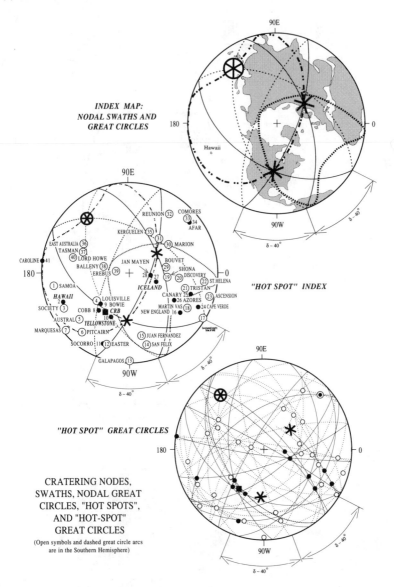

Appendix 13 (Chapters 7 and 8)

Orientation diagrams for North Polar equal-area stereographic projections of **(top)** the continents, the cratering swaths, and the nodal great circles (NGCs), **(middle)** the relative locations of the "hot spots" from Appendix 12, and **(bottom)** great circles drawn through the "hot spots," relative to the PCNs of Figures 2–4 and the intercepts of the NGCs on the Equator (i.e., for clarity the NGCs themselves are not shown; see middle diagram). "Hot spots" in the upper hemisphere are shown as filled circles, and those in the lower hemisphere by open circles (the filled circle enclosed by an open circle at the upper right is a near-coincidence of the positions of "hot spots" in the lower and upper hemispheres in this projection).

Appendix 14

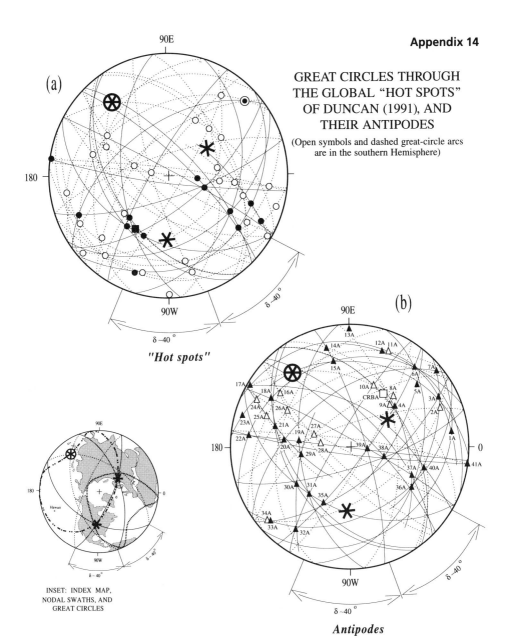

Appendix 14 (Chapters 7 and 8)

North Polar equal-area stereographic projections showing **(a)** great circles drawn through the locations of the "hot spots" of Appendix 12 (also shown in Appendix 13) and **(b)** great circles drawn through the locations of the antipodes of the "hot spots" (the antipodal positions are indicated by the "hot spot" numbers followed by the letter A; the Columbia River basalt province, indicated by the filled square in (a), is thus shown by an open square in (b) and labeled CRBA).

Inset: Index map of the Northern Hemisphere continents, and the PCNs, cratering swaths, and NGCs from Appendix 13 (top).

355

Appendix 15

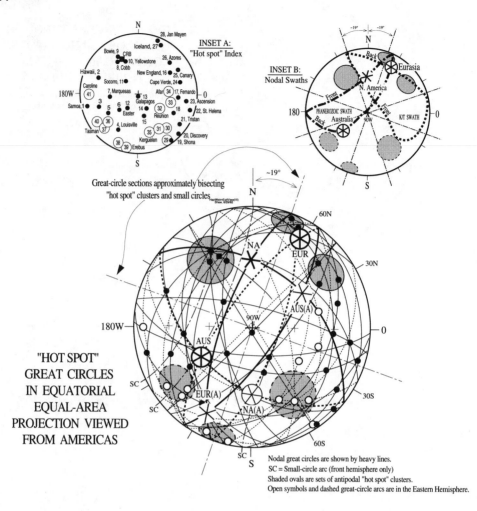

Appendix 15 (Chapters 7 and 8)

Equatorial equal-area stereographic projection of the global "hot spot" distribution of Appendix 12 viewed from longitude 90° W (the Americas), showing the NGCs (heavy lines, dashed in the far hemisphere) and great-circle arcs (lighter lines, dashed in the far hemisphere). Shaded ellipses show areas where groups of "hot spots" are situated in approximately antipodal locations (as distinguished from projected antipodes).

Inset A: Index map of "hot spots" as viewed in this projection (cf. Appendix 12).

Inset B: Schematic map showing similarities between symmetry elements of the "hot-spot" ellipses and the Phanerozoic cratering swaths and PCNs (see discussions in the Introduction and early in Chap. 8; cf. Rampino and Caldeira, 1992; Yamaji, 1992; Stothers, 1993).

Appendix 16

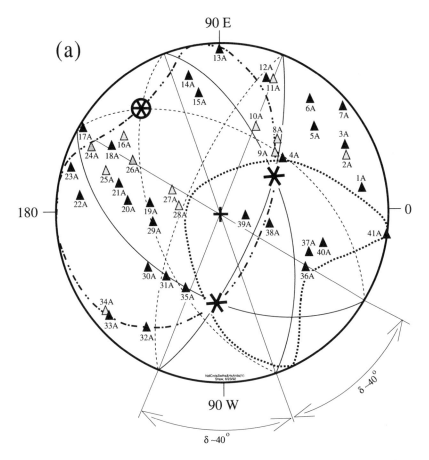

ANTIPODES TO THE GLOBAL "HOT SPOTS" OF DUNCAN (1991)
(Open symbols and dashed great-circle arcs are in Southern Hemisphere)

Appendix 16 (Chapters 7 and 8)

North Polar equal-area stereographic projections of the "hot spots" of Appendix 12 (Duncan, 1991) and their antipodes relative to Earth's impact-cratering patterns. "Hot spots" (circles; diagrams b–d) are identified numerically (see Appendix 12), and their antipodes (triangles; diagrams a–d) are identified by the same system of numbers, each followed by the letter A.

(a) "Hot spot" antipodes (triangles; dark in upper hemisphere, light in lower hemisphere) viewed against the Phanerozoic and K/T cratering swaths (heavy dash-dot and short-dash trajectories, respectively), PCNs (asterisks; circled in lower hemisphere), and nodal great circles (NGCs) of Figure 23 (medium-weight great-circle arcs, dashed in lower hemisphere). The δ-values plotted on the equatorial great circles of each projection show the approximate mean positions and angular deviations between the intercepts of the NGCs at Earth's Equator, as shown in Figure 23.

(cont.)

Appendix 16

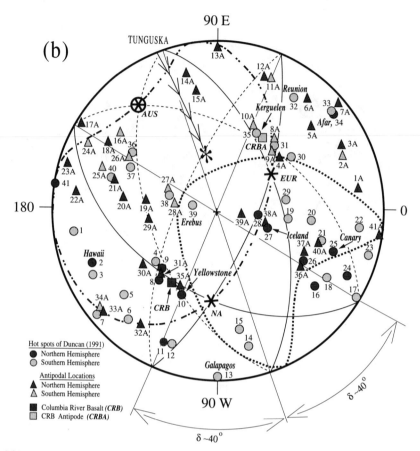

COMPARISON OF "HOT SPOTS" & ALL ANTIPODES WITH NODAL SWATHS AND GREAT CIRCLES
(Dashed great-circle arcs are in Southern Hemisphere)

(b) "Hot spots" (circles; dark in upper hemisphere, light in lower hemisphere) *and* "hot-spot" antipodes (triangles; dark in upper hemisphere, light in lower hemisphere) shown together as a test of direct and antipodal clustering of volcanism on the Earth (cf. Rampino and Caldeira, 1992; Yamaji, 1992; Stothers, 1993); see the discussions in the text (late in the Introduction and early in Chap. 8). Locations relevant to the discussions by Hagstrum and Turrin (1991a, b) and Hagstrum (1992) are labeled [e.g., Columbia River basalt (CRB) and its antipode (CRBA); Reunion ("hot spot" #32) and its antipode (#32A), Hawaii (#2) and its antipode (#2A), etc.]. The estimated trajectory of approach and epicenter of the Tunguska bolide (nested arrows with asterisk) is shown for perspective in the context of Note 1.2.

Appendix 16

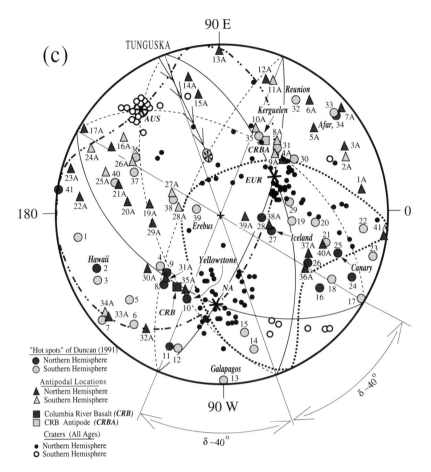

COMPARISON OF "HOT SPOTS" & ALL ANTIPODES WITH NODAL SWATHS AND GREAT CIRCLES WITH CRATERS ADDED

(c) Distribution of Phanerozoic impact craters (small black circles in upper hemisphere; small open circles in lower hemisphere) of Grieve and Robertson (1987) plotted on the projection in (b).

(cont.)

Appendix 16

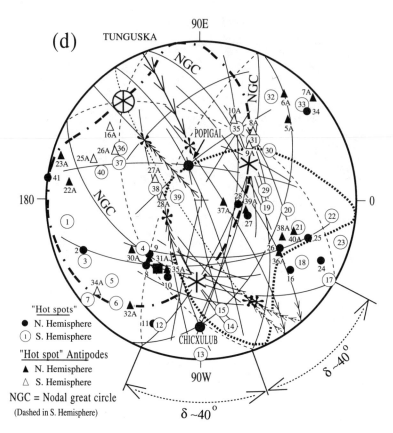

"HOT SPOTS" WITH PAIRED ANTIPODES,
AND ORBITAL/IMPACT SIGNATURES

(Twenty-three of forty-one "hot spots" have antipodes that correspond
to general locations of other "hot spots")

(d) Distribution of only those antipodes of the previous diagrams that plot generally in the same vicinities as "hot spots" (cf. Rampino and Caldeira, 1992; Yamaji, 1992; Stothers, 1993). Locations of (1) the largest K/T crater (Chicxulub, Yucatan) and possible K/T crater (Popigai, Siberia; cf. Note 1.1), (2) historic-to-Holocene bolide and/or grazing-impact trajectories from Appendix 3a, and (3) great-circle arcs through sets of impact craters (from Appendix 4) are shown for orientation relative to the likely terminal trajectories of proximal high-inclination impactors. The historic–Holocene trajectories (Tunguska, Sikhote-Alin, Campo del Cielo, and Cuarto) are shown by arcs with nested arrows and asterisks, as is the low-angle approach trajectory for the Avak structure, Alaska (of late-Cretaceous-to-Tertiary age), as deduced by Grantz and Mullen (1991); all of these trajectories are identified by name in Appendix 3a—i.e., in the identical stereographic projection.

Appendix 17

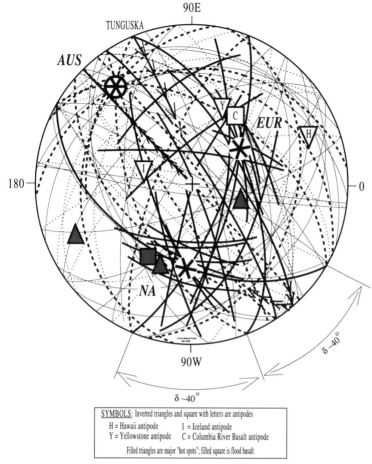

ORBITAL/CRATERING GREAT CIRCLES COMPARED WITH
"HOT SPOT" ANTIPODAL GREAT CIRCLES
(Heavy lines are cratering pattern, light lines antipodal "hot spot" pattern;
dashed arcs are in the Southern Hemisphere)

Appendix 17 (Chapters 7 and 8)

North Polar equal-area stereographic projections of sets of great-circle arcs drawn through sets of impact craters (arcs shown by heavy lines; solid in upper hemisphere and dashed in lower hemisphere) — as in Appendixes 4 and 16d — compared with great-circle arcs drawn through sets of "hot-spot" antipodes (arcs shown by light lines; solid in upper hemisphere and dashed in lower hemisphere). The set of all antipodes to the "hot spots" was used in order to test the model of Hagstrum and Turrin (1991a, b) and Hagstrum (1992). Other sets of great-circle patterns (not shown) drawn through the global "hot spots" of Appendix 12 (see Appendix 14a) have similar distributions relative to the impact pattern. Although the array of curves is quite "busy," it is evident — if one studies it awhile — that there are strong resemblances between the impact-derived and antipodal "hot-spot"-derived sets, an observation consistent with a relationship between the "hot spots" and impact patterns but not definitive as concerns direct or antipodal triggering of volcanism (cf. Appendix 12 and the discussion in the text of Chap. 7).

Appendix 18

Appendix 18 (Chapters 7 and 8)

North Polar equal-area stereographic projections showing a comparison between an artifactual small circle of "hot-spot" antipodes (a), the antipode of that small circle (Inset), and the conjectural small-circle pattern of Southern Hemisphere Phanerozoic impacts shown by the heavy dashed line in (b), a modified version of Figure 2 (see Chap. 1).

(a) A small circle drawn through a "coincidental" alignment of "hot-spot" antipodes, and conjunctions of corresponding antipodal great circles (cf. Appendix 14b), is shown by a heavy solid line skewed somewhat off center toward the upper right of the upper hemisphere (i.e., this is an antipodal "artifact" of some Southern Hemisphere alignment of "hot spots" and corresponding great-circle arcs skewed somewhat toward the lower left of the lower hemisphere).

(b) Reproduction of Figure 2 showing the previously drawn conjectural small circle of possible Southern Hemisphere impacts mentioned in Chapter 1 and its antipodal image.

Inset: Schematic comparison of an antipodal pair of small circles derived on the basis of the small circle in (a) — i.e., the lighter pair of small-circle trajectories — and another pair of small circles (heavy trajectories) representing the impact-derived swath shown in (b). The large uncertainties in latitudes of the small-circle patterns of either type would suggest that the antipodal "hot-spot" small circle in (a) could be consistent with a direct small-circle swath of impacts in the Southern Hemisphere, a conjectural conclusion that may lend some support to the idea of direct correlations between impacts and volcanic "hot spots" (cf. Appendix 12).

Appendix 18

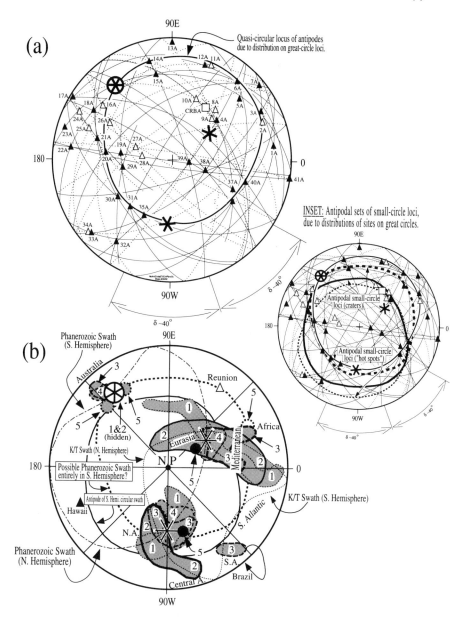

GREAT CIRCLES THROUGH ANTIPODES OF GLOBAL "HOT SPOTS", COMPARED WITH PHANEROZOIC CRATER PATTERN
(Open symbols and dashed great-circle arcs are in the Southern Hemisphere)

Appendix 19

Appendix 19 (Chapter 8)

Examples of apparent polar-wander paths (APWP) during the Phanerozoic, redrawn schematically from Butler (1992, Figs. 10.6 and 10.9).

(a) Apparent coincidences of major cusps in the long-term APWPs of Europe (open circles) and North America (solid circles) — where the European APWP has been rotated 38° about the Euler pole of Bullard et al. (1965) placed at 88.5° N and 27.7° E — at times that also appear to be coincident with Geologic Period and/or Stage boundaries within the Paleozoic and lower Mesozoic Eras (modified from Butler, 1992, Fig. 10.9b; ages added from Harland et al., 1990). Note especially the abrupt changes in the vicinities of 410 Ma (neighborhood of the Silurian-Devonian boundary), 370 Ma (neighborhood of the upper Devonian Frasnian-Famennian boundary), and 210 Ma (neighborhood of the Triassic-Jurassic boundary). The latter two events represent intervals of the Geologic Column for which correlations between global mass extinctions and large-body impacts have been proposed (see Chaps. 1, 2, and 8, Fig. 6, and Appendix 3a; cf. McLaren, 1970, 1982, 1983, 1986; McLaren and Goodfellow, 1990; Bice et al., 1992; Claeys et al., 1992a, b; Hodych and Dunning, 1992; Trieloff and Jessberger, 1992; Wang, 1992; Wang and Geldsetzer, 1992; Wang et al., 1992; Benton, 1993; Rogers et al., 1993). The exactitude of the temporal correlations is not the primary issue here, because reference is being made to a hierarchy of "catastrophes" in the sense of Ager (1984), or chaotic crises in the sense discussed in Chapter 5 (see Notes P.3, I.4, I.7, 1.1, 5.1, 6.1, 8.1, 8.10–8.15), that have evolved together in the context of a generalized *system* of chaotic resonances that has given expression to both the coarse and fine structures of the biostratigraphic event code (cf. Urey, 1973; Raup, 1990, 1991).

(b) Major cusps in APWPs of North America during the Mesozoic and Cenozoic Eras (modified from Butler, 1992, Fig. 10.6). Cusps apparently occurred near the times of the Triassic-Jurassic boundary (but see Hagstrum, 1993), sometime during the early Tertiary (significantly later than the K/T boundary), and somewhere within the Oligocene Epoch (35.4–23.3 Ma?). Data from Besse and Courtillot (1991) — e.g., as plotted in Appendix 20 — would suggest that significant directional changes may have occurred at ages (Harland et al., 1990) near 70 Ma (just prior to the K/T boundary), 50 Ma (just after the Paleocene-Eocene boundary), and 30 Ma (near the Rupelian-Chattian Stage boundary within the Oligocene Epoch); see **Note A.3**.

Appendix 19

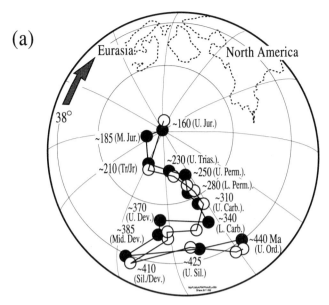

NORTH AMERICA/EUROPE PALEOZOIC-MESOZOIC APPARENT POLAR WANDER TRENDS & CUSPS
(Modified schematically after Butler, 1992, Figure 10.9)

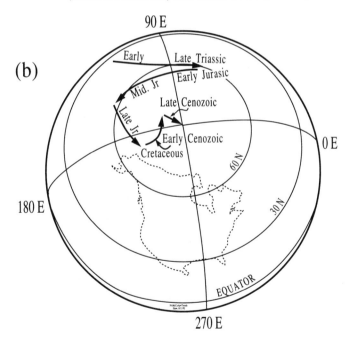

NORTH AMERICA POST-PALEOZOIC APW TRENDS AND CUSPS

Appendix 20

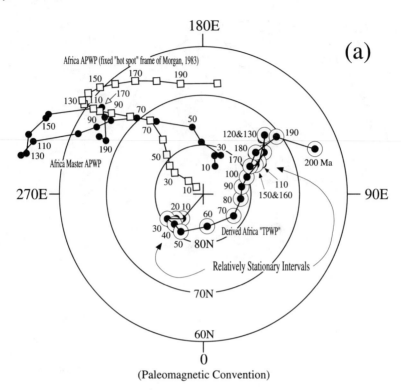

AFRICAN APPARENT AND "TRUE" POLAR WANDER PATHS

(Data from Besse and Courtillot, 1991, Table 7 and Figure 3a)

Appendix 20 (Chapter 8)

North-Polar equal-area projections illustrating African polar-wander paths relative to different assumptions concerning geographic reference frames, as derived by Courtillot and Besse (1987) and Besse and Courtillot (1991).

(a) A "true" polar-wander path ("TPWP") for Africa (solid circles within open circles) derived by Courtillot and Besse (loc. cit., Fig. 3) and Besse and Courtillot (loc. cit., Fig. 3a) on the basis of differences between the best paleomagnetic apparent polar-wander path (the "Africa Master APWP" shown by solid circles) and the apparent wander path of Africa relative to the "hotspot" reference frame (open squares) of Morgan (1983). The method of derivation is described in Besse and Courtillot (loc. cit., pp. 4042f, Table 7, and Fig. 3), but there is some apparent confusion between the labeling of Table 7 vs. Figure 3 [i.e., the data in Table 7 correspond to the points that define the "TPWP" in Figure 3a, rather than corresponding, as labeled, to the African APWP based on Morgan's (1983) "hot spot" model].

(b) Expanded plot showing the relationships between the positions of the Africa APWP (open squares) and "TPWP" (filled circles within open circles) relative to (1) the projection of the core-mantle boundary (heavy dashed circle forming the perimeter of the diagram), (2) the projection of the inner-core/outer-core boundary (light dashed circle), (3) the Phanerozoic cratering nodes (PCNs) of the present volume (large asterisks, circled

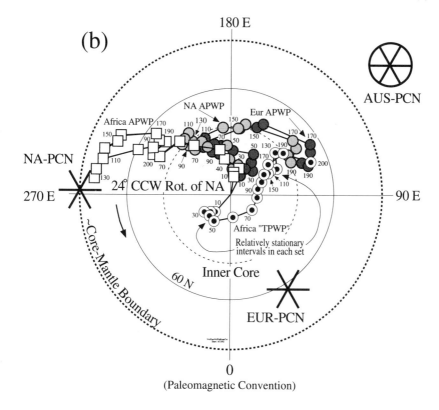

COMPARISON OF CONTINENTAL APPARENT AND "TRUE" POLAR WANDER PATHS
(Data from Besse and Courtillot, 1991)

in the lower hemisphere), (4) the Eurasian APWP (circles with dark shading) of Besse and Courtillot (loc. cit., Table 6), and (5) the North American APWP (circles with light shading) of Besse and Courtillot (ibid.) rotated 24° counterclockwise about the North Pole relative to the Eurasian APWP. The reconstructed set of APWPs bears an interesting bilateral symmetry relative to the African "TPWP," apparently reflecting (at face value) the difference between Northern Hemisphere apparent and "true" geomagnetic wander since the breakup of Pangaea (cf. Fig. 34 and Appendix 26; and Dietz and Holden, 1970; Bambach et al., 1980; Ziegler et al., 1982; Parrish et al., 1986; Smith and Livermore, 1991; Zonenshain et al., 1991).

Comparisons between the "true" polar-wander path in (a) and "TPWPs" computed by Livermore et al. (1984) and by Andrews (1985), as shown in Besse and Courtillot (loc. cit., Fig. 3c), demonstrate that the derivation of TPWPs is still far from an "exact science" (an observation compounded by the sorts of uncertainties in APWPs mentioned in Appendix 19). Andrews (loc. cit.), however, provides a concise discussion of some of the assumptions involved in evaluating the relative motions of different coordinate frames (cf. Gold, 1955; Goldreich and Toomre, 1969; Gordon, 1987).

Appendix 21

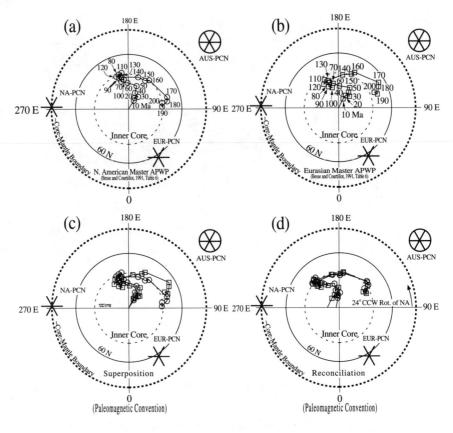

COMPARISON OF NORTH AMERICA-EURASIA APPARENT POLAR WANDER PATHS

Appendix 21 (Chapter 8)

Example of reconstruction of the North American and Eurasian APWPs (data from Besse and Courtillot, 1991, Table 6), using the North Polar equal-area coordinate frame of Appendix 20b; numbers are ages (Ma).

(a) North American master curve.
(b) Eurasian master curve.
(c) Superposition of North American and Eurasian master curves.
(d) 24° counterclockwise rotation of the North American APWP relative to the Eurasian APWP combined with a few degrees of subjective relative translation to give the "best overprint." It may be worth noting that this reconstruction requires 14° less rotation than that in the model of Bullard et al. (1965), as shown in Appendix 19 (cf. the arbitrary continental reconstruction of Appendix 26).

Appendix 22 (Chapter 8)

North Polar equal-area stereographic projections showing the Africa, India, North America, and Eurasia master APWPs of Besse and Courtillot (1991, Table 6) compared with essentially Paleozoic and Mesozoic impact craters (small black dots with crater numbers from Grieve and Robertson, 1987); younger craters were not plotted on these stereonets to avoid overprinting (see Appendix 23; cf. Figs. 1–4 and Appendixes 1–4). Notice that stereonets showing only the APWPs are oriented according to the paleomagnetic convention of Besse and Courtillot (loc. cit.)—i.e., with 0° E at the bottom (cf. Appendixes 20 and 21)—and stereonets showing both the APWPs and impact craters are oriented according to the usual convention of celestial mechanics, as employed elsewhere in the present volume—i.e., with 0° E at the right (cf. Appendixes 1–4). The projected outlines of the inner core (light dashed circle), outer core (heavy dashed circle), PCNs (the conventional asterisks of the present volume, circled in lower hemisphere), and the Tunguska trajectory and epicenter (nested arrows with asterisk) are shown on all diagrams for additional orientation. The Phanerozoic and K/T cratering swaths (labeled PCS and K/TCS, respectively) and NGCs of Figure 23 and Appendix 4, plus a few of the great-circle arcs from Appendix 4, are shown for additional perspective.

(a) Africa and India master APWPs. *(cont.)*

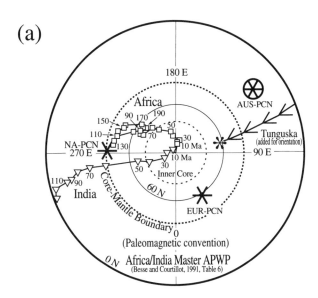

Appendix 22

(*Appendix 22 cont.*)
(b) Distribution of Paleozoic-Mesozoic craters in relation to the Africa and India APWPs.
(c) North America and Eurasia APWPs.
(d) Distribution of Paleozoic-Mesozoic craters in relation to the North America and Eurasia APWPs.

It may be of interest to notice that the relatively tightly constrained locations of the North America and Eurasia APWPs (i.e., at latitudes above 60° N, nearly superimposed on the projection of the inner-core/outer-core boundary) are somewhat centered relative to the PCNs and the clustering of pre-Cenozoic craters (cf. Appendix 23 and Note A.4).

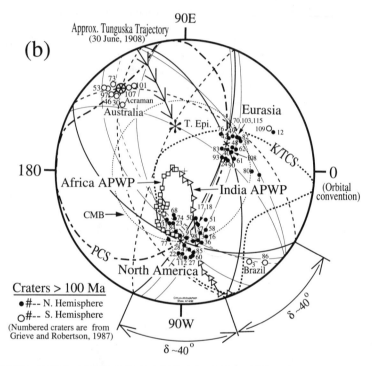

AFRICA-INDIA APWP COMPARED WITH CRATERING PATTERNS AND IMPACT TRAJECTORIES

Appendix 22

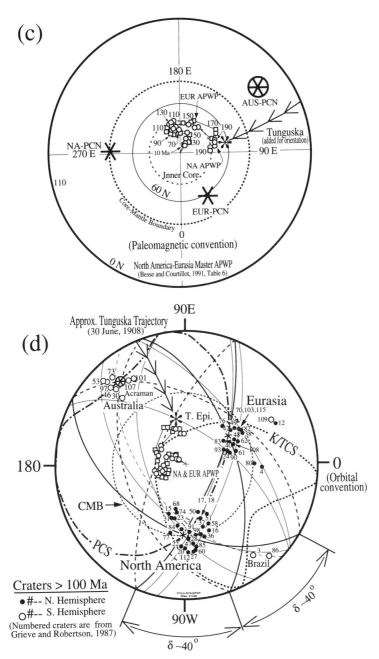

NORTH AMERICA-EURASIA APWP COMPARED WITH CRATERING PATTERNS AND IMPACT TRAJECTORIES

371

Appendix 23

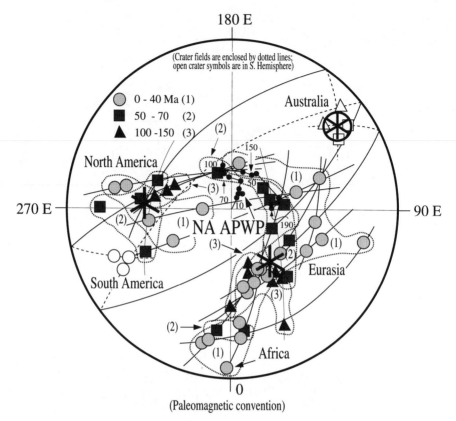

CRATER PATTERNS ≤ 150 Ma COMPARED
WITH NORTH AMERICA APWP

Appendix 23 (Chapter 8)

North Polar equal-area stereographic projection oriented according to the paleomagnetic convention of Appendix 22 (0° E at the bottom) showing the North America master APWP of Besse and Courtillot (1991, Table 6) compared with three selected age classes of craters younger than 150 Ma (cf. Figs. 2–4 and Appendixes 1–4): (1) middle Eocene to modern craters (circles with light shading in N. Hemisphere, open in S. Hemisphere; cf. Appendix 2), (2) craters spanning the K/T boundary interval, latest Cretaceous to early Eocene (squares with heavy shading in N. Hemisphere, open in S. Hemisphere), and (3) Cretaceous craters distinctly older than the K/T boundary interval (triangles, filled in N. Hemisphere, open in S. Hemisphere). Amoeboid areas (closely spaced dashes) enclose the respective age classes of craters. Possible implications of the cratering patterns vis-à-vis patterns of APWPs shown here and in Appendix 22 are considered qualitatively in **Note A.4**.

Appendix 24

Appendix 24 (Chapter 8 and Epilogue)

Patterns of volcanic "hot spots" on Io, a satellite of Jupiter, as discussed in the Prologue and Epilogue (cf. Johnson and Soderblom, 1982; Kieffer, 1982; Pearl and Sinton, 1982; Schaber, 1982; Strom and Schneider, 1982; McEwen et al., 1985; Yamaji, 1991; Rothery, 1992).

(a) Distribution of volcanic features on Io, modified from Strom and Schneider (1982, Table 16.1 and Fig. 16.21). Shaded circles and ellipses are in the front hemisphere of the coordinate system of Strom and Schneider, and open circles and ellipses are in the far hemisphere [the Voyager Imaging Science Team of the Galilean satellite cartographic program presumably assigned a prime meridian for Io on the basis of the side facing Jupiter at a specific astronomical epoch relative to the Voyager 1 observations, because Io had no obvious permanent features on which to base a topographic coordinate system; see Davies (1982, pp. 912f and Fig. A.1)]. The features portrayed by Strom and Schneider were (1) active volcanic "plumes" [i.e., nozzle-like eruptions that formed fountain-like or solar-prominence-like plume heads in the tenuous atmosphere of Io (e.g., Strom and Schneider, loc. cit., Fig. 16.1; Kieffer, 1982, 1989); see also the discussion of Kilauean "lava prominences" in Note P.3], (2) areas of diffuse deposits resembling those associated with active eruptions, and (3) small dark diffuse spots with diameters of 30 km or less. These three types of features are not differentiated in the present schematic illustration, except with regard to approximate areas of plumes or plume deposits. The small dark spots of Strom and Schneider presumably were quiescent eruptive sites (simple and compound vents, calderas, etc.; cf. McEwen et al., 1985; Rothery, 1992, Plates 2–5). Pele (named for the "Hawaiian volcano goddess"; see Davies, 1982, Table A2) was the largest of the active eruption plumes photographed by Voyager 1 (see Morrison, 1982b; Strom and Schneider, 1982, Figs. 16.1–16.4; Rothery, 1992, Plates 3 and 4). *(cont.)*

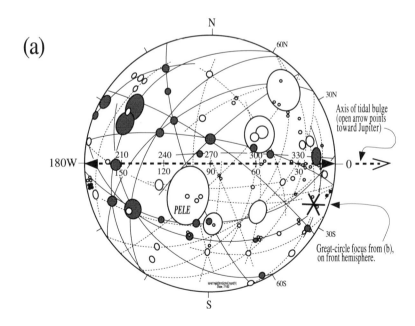

Appendix 24

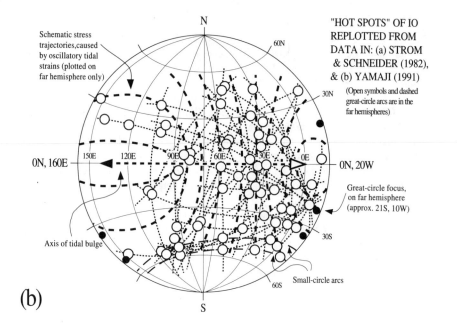

(b) "Hot spots" of Io as plotted by Yamaji (1991, Fig. 1) from data in McEwen et al. (1985, Table 1) based on interpretation of the distribution of "low-albedo calderas" in far hemisphere (open circles; solid circles are in front hemisphere); notice that the coordinate system is skewed 20° to the "west," so that the 180° limb in (a) is in the far hemisphere in (b). The heavy dashed lines are inferred tidal-stress trajectories (Yamaji, ibid.) plotted on the far hemisphere. I have drawn arcs of great circles, and a few small circles, through sets of "hot spots" (light solid and dashed lines), as in Appendix 13 for the Earth, instead of portraying the small-circle pattern deduced by Yamaji (loc. cit., Figs. 1–3) on the basis of the statistical "point-to-point self-correlation" method (cf. Yamaji, 1992). [The reason for this departure is to make the suggestion that the pattern of volcanic venting on Io may represent a combination of the following processes: (1) nucleation of transient magmatic loci by impact cratering (i.e., craters are quickly filled and obliterated so that the only record of their occurrence is the evolving pattern of active volcanic centers), (2) rapid convective migration, and/or propagating volcanic migration, of the surviving loci of active volcanism to regions of maximum tidal deformation and sustained dissipative mantle melting (see the discussion in the Prologue; and Peale et al., 1979; Cassen et al., 1982; Rothery, 1992, pp. 23f), and (3) maintenance of the "hot-spot" pattern by steady-state creation and destruction of magmatic sources according to cyclic feedback among the frequencies of impacts, tidal-dissipative melting and transport, and volcanic venting of the magmatic products (combined with mantle assimilation of the residual products left behind by the melting process). It is my firm conviction that the same general feedback process applies to the volcano-tectonic cycle on Earth, except that the time constants of the respective stages are longer, and evidence of the impact-crater-nucleation stage is sometimes preserved (see the Prologue and Notes P.5, 4.2, 6.2, 7.2, 8.1–8.3, 8.13; cf. Shaw, 1969, 1970, 1973, 1991; Shaw et al., 1971, 1980a; Gruntfest and Shaw, 1974; Wones and Shaw, 1975; Seyfert and Sirkin, 1979; Rampino, 1987; Alt et al., 1988; Rampino and Stothers, 1988; Stothers and Rampino, 1990; Hagstrum and Turrin, 1991a, b; Hagstrum, 1992; Oberbeck et al., 1992; Rampino and Caldeira, 1992; Stothers, 1992, 1993).]

Appendix 25

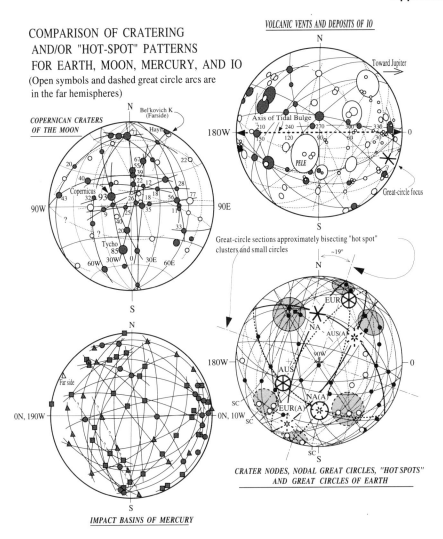

Appendix 25 (Chapter 8 and Epilogue)

Comparison of the cratering characteristics of the Earth, Moon, and Mercury relative to the patterns of volcanic "hot spots" on the Earth and Io. Each of the labeled stereographic projections has already been discussed in (1) Figure 37 (cratering pattern on the Moon), (2) Appendix 10 (cratering pattern on Mercury), (3) Appendix 24 ("hot-spot" pattern on Io), and (4) Appendix 15 ("hot-spot" pattern on Earth relative to the characteristic elements of the Phanerozoic cratering pattern).

Appendix 26 (Chapter 8)

Stereographic projection showing an arbitrary pre-Phanerozoic reconstruction of the continents that would approximately satisfy the theoretical invariance of the PCNs in the CRF system. The continents (which obviously did not then have the outlines shown by these modern configurations) were rotated and translated by trial and error, using the program Terra Mobilis, while keeping the PCNs (asterisks) as invariant as possible (cf. Fig. 34). Several aspects of this reconstruction are, of course, unorthodox, such as the placement of Greenland within what is now the Arctic Ocean, and the inverted orientations of Australia, India, and East Antarctica. But these rearrangements are no more "innovative" than some of the paleogeographic reconstructions proposed in various models of "suspect-terranes," "accretionary continents," "supercontinent cycles," "expanding-earth theory," the "revisionist" impactite-tillite hypothesis, etc. The following sources include examples of most of the newer models as well as their precedents: Dietz and Holden (1970), Crowell (1978), Norton and Sclater (1979), Seyfert and Sirkin (1979), Bambach et al. (1980), Irving (1982), Irving and Irving (1982), Ziegler et al. (1982), Carey (1983), Owen (1983a, b), Vogel (1983), Hughes (1986), Parrish et al. (1986), Piper (1987), Hillhouse and Grommé (1988), P. F. Hoffman (1988, 1991), Molnar et al. (1988), Nance et al. (1988), Howell (1989), Stone (1989), Harbert et al. (1990), Lawver et al. (1990), Stewart (1990), Crowley and Baum (1991), Dalziel (1991, 1992a, b), Hartnady (1991), Moores (1991), Murphy and Nance (1991, 1992), Royer and Chang (1991), Smith and Livermore (1991), Zonenshain et al. (1991), Aggarwal and Oberbeck (1992), Elliott and Gray (1992), Marshall and Oberbeck (1992), Oberbeck et al. (1992), Rampino (1992), Stump (1992).

Appendix 26

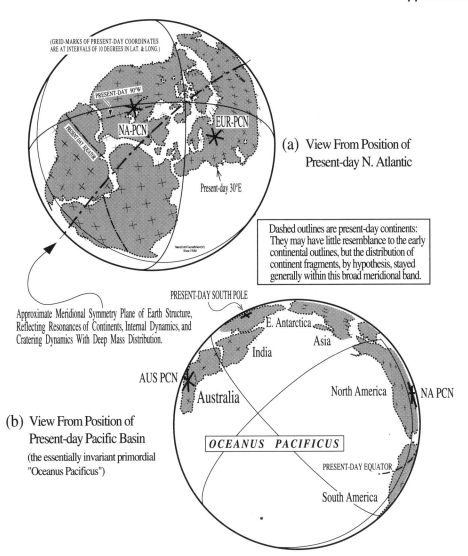

**SCHEMATIC PRE-PHANEROZOIC
CONTINENT DISTRIBUTION:**

(ARBITRARY EXAMPLE RECONSTRUCTED TO BE ROUGHLY CONSISTENT
WITH INVARIANCE OF CRATERING NODES; REDRAWN FROM A COMPUTER SIMULATION)

377

Appendix 27

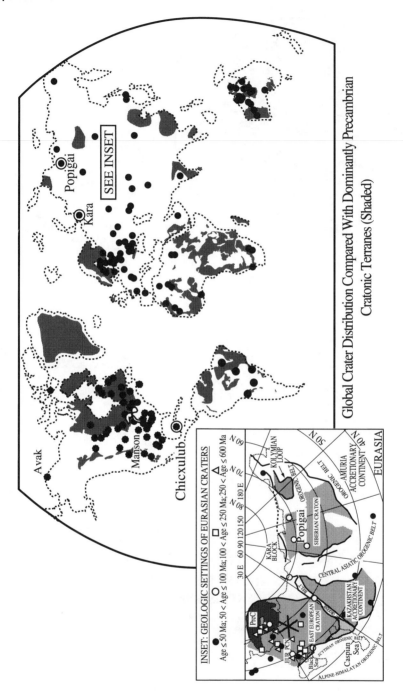

Global Crater Distribution Compared With Dominantly Precambrian Cratonic Terranes (Shaded)

Appendix 27 (Chapters 1, 2, 7, and 8 and Epilogue)

Schematic index map of the world showing the approximate distribution of Precambrian basement rocks (shaded areas) relative to the principal impact craters discussed in the present volume (filled circles). The main point to be made is that the distribution of craters is not obviously dominated by the geographic patterns of the oldest cratonic shield areas. The areas of Precambrian rocks were drawn on the basis of D. R. Derry's (1980) *World Atlas of Geology and Mineral Deposits*.

Inset: Schematic sketch map of the Eurasian "accretionary continent," modified from Zonenshain et al. (1991). A representative suite of impact craters from Grieve and Robertson (1987) is shown, subdivided according to four age groups: (1) 600–250 Ma (open triangles), (2) 250–100 Ma (open squares), (3) 100–50 Ma (open circles), and < 50 Ma (filled circles). The cratering pattern for ages < 100 Ma obviously extends beyond the pattern of the older craters because of the distribution of outcrop ages (not shown; cf. Blatt and Jones, 1975; Kieffer and Simonds, 1980), but this is at least partly an artifact of the swath-like character of the global pattern shown in Figure 3 and Appendix 2 for craters of late Cretaceous and younger ages (i.e., the youngest, smallest terrestrial craters are the ones most likely to be found *anywhere* on land, yet they are distributed in much the same pattern as the K/T swath, as shown in Appendix 2). [Additional information on the impact-cratering history of Europe and Fennoscandia is available in Lindström et al. (1992) and Pesonen and Henkel (1992a, b). Rajlich (1992) describes new evidence of breccias and recrystallized glass with shock-metamorphic mineralogies in an ~260-km-diameter circular structure in Czechoslovakia (Bohemian Massif) apparently of late Precambrian age — established on the basis of field evidence that an *"Upper Proterozoic sequence begins with pillow lavas and is terminated by the sedimentary sequence with shallow water fossils . . . , indicating the successive filling of the hole."* If this structure proves to be an impact crater, it would be an example of a very large Proterozoic (< 2.5 Ga) impact in the vicinity of the Eurasian PCN (cf. Figs. 2 and 4; and Appendixes 1–4). The putative infilling sequence beginning with basaltic lavas is consistent with many other descriptions of the magmatic aftermath of major impacts (cf. Figs. 24–26 and 28; Appendixes 12–18; and Rampino, 1987, 1992; Alt et al., 1988; Rampino and Stothers, 1988; Stothers and Rampino, 1990; Hagstrum and Turrin, 1991a, b; Hagstrum, 1992; Oberbeck et al., 1992; Stothers, 1992, 1993).]

Faraday, on the Fate of Hypotheses

> *Hypotheses, treated as mere poetic fancies in one age, scouted as scientific absurdities in the next — preparatory only to their being altogether forgotten — have often, when least expected, received confirmation from indirect channels, and, at length, become finally adopted as tenets, deducible from the sober excercise of induction.*
>
> — Michael Faraday, 1853
> from the frontispiece in Knight (1978)

Knight's book, *The Transcendental Part of Chemistry* — a phrase attributed to Sir Humphry Davy (1778–1829) in a lecture given in 1809 — addresses that period of the history of science before atoms, molecules, and such were known to have any "real existence." Davy's and Faraday's statements expressed the idea that the "atomic theory" — as posited, for example, by John Dalton (1766–1844) — must be of transcendental nature, because no such atomic entity was demonstrable at that time, hence the laws of chemistry could follow only deductively from its postulated, and/or imagined, properties (cf. Atkins, 1991, p. 6; Zewail, 1993). Knight (1978, p. ii of the Introduction) goes on to say that *"It was not until the beginning of the twentieth century, with the work particularly of Perrin and Einstein on Brownian movement, that independent lines of reasoning began to converge upon atoms, making it implausible to doubt their real existence."* Chemical technology would indeed be held in awe today if it could be seen from such an eighteenth-and-nineteenth-century perspective, for it has sealed the earlier expectation of plausibility by actually identifying, taking pictures of, and even manipulating the positions of *individual* atoms, arrayed pretty much as theory had expected them to be (e.g., Atkins, 1991, pp. 6 and 15; von Baeyer, 1992; Zewail, 1993). But as we near the end of the century, nonlinear dynamics and quantum chaos bid fair to bring the transcendental part of chemistry once again to the fore. To see this, one needs but compare the descriptions of attractors and nonlinear astronomical orbits described herein — for instance in the historical perspectives of Notes P.3, 4.1, and 4.2 — with the configurational concepts of the orbital electron shells of atoms, particularly as they can be viewed in the light of quantum-orbital-chaotic trajectories of motion [see for example the resonances in the diagram of the chaotic even-parity hydrogen atom illustrated by Richards (1988; data from Holle et al., 1988); see also Prigogine et al. (1991), Jensen

(1992), and Parmenter and Yu (1992)]. The *celestial reference frame hypothesis* (CRFH) — and its role in the systematization of the impact-extinction hypothesis (IEH) — can be viewed, analogously, as the transcendental part of *A New Theory of Earth*, the "real existence" of which remains to be demonstrated by the uniformitarian application of the principles of nonlinear dynamics, as advocated in the present volume. Faraday's remark, as elaborated by Knight, reverberates within the nonlinear-dynamical coupled-oscillation modes of scientific discovery that promise, almost imminently, revelations across a broad front (see the Prologue, and Note 8.15).

CHAPTER NOTES

Chapter Notes
(including notes to the Prologue,
Epilogue, and Appendix)

Few of the notes that follow are notes in the usual sense; rather, they are extended discussions of ideas introduced, tangentially or at length, in the text or the Appendixes. In the text, primary citations of notes are indicated by **boldface** type, and fall consistently in note-number order; cross-references to all notes in the text (Prologue, Introduction, Chapters, Epilogue, and Appendixes) are indicated in plain type. Notes to the Prologue are indicated (in text and below) by P plus a decimal number; notes to the Introduction are indicated by I plus a decimal number; notes to Chapters are indicated by a whole number plus a decimal number (e.g., 8.2); notes to the Epilogue and Appendixes follow the Chapter Notes and are indicated, respectively, by the letters E and A followed by decimal numbers. Primary citations of the notes for the Appendix (given in boldface type) are all located in the Appendixes themselves.

Prologue

Note P.1: The results of nonlinear feedback-systems computations — i.e., computations involving general systems of fully coupled nondegenerate partial differential equations (cf. Forrester, 1961, 1973; Gruntfest, 1963; Lorenz, 1963, 1964, 1989; Shaw, 1969, 1978, 1981; Franceschini and Tebaldi, 1979; Moore et al., 1983; Prüfer, 1985; Knobloch et al., 1986; Mees, 1986; Morrison, 1991; Richardson, 1991; Pachner, 1993) — may be of many types. For example, computations may (1) converge to one stable value, or to sets of stable values called *fixed points* and/or *limit cycles*, but only within certain characteristic ranges of the *tuning* parameters of the system (see below; cf. Notes P.3, P.4, and 3.1) that symbolize the "forces" acting on, and within, the system; e.g., May, 1976; Feigenbaum, 1980; Hofstadter, 1981; Fein et al., 1985; Mees, 1986); (2) be *numerically unstable* (a computation "blows up," giving an error message that, ironically, may in fact be an accuracy message, in the sense that the numerical instability manifested could be physically real; e.g., Gruntfest, 1963; Gruntfest et al., 1964; Shaw, 1969, 1973, 1978, 1981; Wones and Shaw, 1975; Nelson, 1981; Lorenz, 1989); (3) never converge to constant, single-valued answers (a computation may oscillate "wildly" in the neighborhood of fixed points and/or limit cycles, or it may wander indefinitely — "ergodically" — over the range of all possible

385

Chapter Notes: Prologue

finite-valued solutions; e.g., Shaw, 1988b, 1991), or (4) converge to certain types of seemingly fixed patterns — e.g., in terms of unique ranges of numerical values, or unique patterns of oscillating values — but repeated computations give different arrays of numerical values within those patterns when the algorithm is restarted at different times relative to the same initial values, and/or the iteration scheme is varied in the slightest; e.g., Lorenz, 1963, 1964, 1985, 1989; Shimizu and Morioka, 1978; Sparrow, 1982; Moore et al., 1983; Prüfer, 1985; Knobloch et al., 1986; Pachner, 1993).

Plots of the iterated values of feedback-systems computations, which are called *phase-space maps* (see Note P.3), can become extremely complex in systems represented by control functions as simple as quadratic equations, or by analog control curves in which there is at least one maximum (e.g., May, loc. cit.; Jensen, 1987; Devaney, 1989; Tufillaro et al., 1992, Chap. 2). Quadratic functions portray a simple form of coupled feedback in which intervals of positive feedback alternate with intervals of negative feedback. It is qualitatively evident, in such a system of dynamically compensated actions, that if the effect of positive feedback becomes too large, the system as a whole becomes unstable, and the solutions to control functions of quadratic form globally *diverge*, producing sets of random-looking solutions or escape from the parameter space of physically realizable solutions. Conversely, if negative feedback dominates, the solutions globally *converge* to some set of stable solutions (i.e., to one or more *fixed points* in phase space). It is also qualitatively evident that in the neighborhood of critical imbalance between convergence and divergence the behavior of systems of differential equations will be extremely sensitive to (1) values of nonconstant physical coefficients (especially kinetic ones, such as viscosity, diffusivity, etc.), and (2) the nature of the finite numerical schemes by which the computations are carried out (e.g., the dimensional scaling of computational increments in time and space, whether the scaling is fixed or "tunable," and so on). Issues concerning integration methods are brought to a focus by contrasting the respective emphases, for example, of Prüfer (1985) and Pachner (1993), especially relative to phenomena discussed by Lorenz (1989) and Nese (1989).

Balances or imbalances between growth intervals and decay intervals are controlled by the sharpness or curvature of maxima in "hump-shaped" control curves or surfaces. The process of adjusting the amplitude, hence curvature, of the control function is called *tuning* (also see the analogous meaning of *control function* in the context of *pumped systems far from equilibrium*, dynamical *slaving* mechanisms, etc., as developed by Haken, 1978). Coefficients in the system equations that govern the rates of change — and hence the curvatures and feedback balances near maxima and minima of a system's control function — are called the system's *tuning parameters* or *control parameters*. *Cyclic recursion* in such a system (i.e., its responses to reiterative computations carried out by the statements of an *algorithm*) can be thought of, in the above sense, as *"folded feedback"* (e.g., Shaw, 1987a, Fig. 51.14).

As the tuning parameter (system forcing) is increased from a negligible value to larger and larger values, fixed-point solutions progressively increase in their proportional degree of divergence relative to convergence, typically multiplying in sequences of *period-doubling bifurcations* up to a critical value of the tuning parameter (the *point of infinite collection of period-doubling bifurcations*; cf. May, 1976), where there is a global transition to steady aperiodic behavior (i.e., irregular behavior that is not simply an artifact of the transient solutions that occur during an approach to the condition of steady-state recursion). Such a sequence of solutions is often called the *period-doubling route to chaos*, or — because of the figurative analogy with experimental behavior in certain simple types of fluid convection (e.g., Gollub and Swinney, 1975; Franceschini and Tebaldi, 1979; Gollub and Benson, 1980; Iooss and Langford, 1980) — the period-doubling route to turbulence; cf.

Chapter Notes: Prologue

May, loc. cit.; Feigenbaum, 1980; Iooss and Langford, 1980; Lorenz, 1980; Croquette and Poitou, 1981; Hofstadter, 1981; Helleman, 1983; Moore et al., 1983; Beckert et al., 1985; Devaney, 1986, 1989, 1991; Shaw, 1986, 1988b; Kaplan, 1987; Boe and Chang, 1991; Komuro et al., 1991; Lauterborn and Holzfuss, 1991). [The period-doubling route to chaos should be contrasted with the so-called *quasiperiodic route to chaos* that is characteristic of systems of coupled oscillators; e.g., Gollub and Benson, 1980; Ostlund et al. (1983), Jensen et al. (1984, 1985), Bohr et al. (1984), Dowell and Pezeshki (1986), Haucke and Encke (1987), Shaw (1987b, 1991), Olinger and Sreenivasan (1988), Shaw and Chouet (1988, 1989), Ling et al. (1991); cf. Sreenivasan and Meneveau (1986), Thompson and Stewart (1986), Deissler (1987), Meneveau and Sreenivasan (1987), and Moon (1987). Period-doubling also occurs in coupled-oscillator systems for tuning conditions under which the control function develops a maximum; see Jensen et al. (1984, Fig. 1), Shaw (1987b, 1991, Figs. 8.31–8.35).]

Note P.2: Henri Poincaré (1854–1912) is credited with being the first person (cf. Note 8.15) to enunciate the principles of nonlinear dynamics in terms that can be traced directly to its current forms of expression. In the first English translation of his book *Science and Method*, Poincaré (1952, p. 68) described what has come to be known as the "butterfly effect" — a phrase originating with Edward Lorenz (1963, 1964) and his *re*discovery of the *principle of sensitive dependence on initial conditions* (see the opening chapter in Gleick, 1987). Poincaré put it thus: *"Small differences in the initial conditions produce very great ones in the final phenomena."* This statement anticipated, and epitomizes, the most avant-garde studies in nonlinear dynamics, an area of research now known — depending on emphasis — by several different names, such as "theory of nonlinear systems," "theory of chaotic systems," "theory of attractors," "theory of dissipative systems," "numerical dynamics," "deterministic chaos," "chaos science," or just "chaos." The first attempt at a comprehensive historical overview of the first flourish of research in nonlinear dynamics, a book titled simply *Chaos,* by James Gleick (1987), gives a perspective that is as laudable in its own way concerning the happenings that attended the emergence of a new way of seeing and doing science as the historical surveys in biology and geology, respectively, by Judson (1979) and Glen (1982). With reference to my opening remarks in the Prologue about time-delayed resonances in the history of science, Gleick's book was published about a decade after Robert M. May, a mathematically inclined biologist and student of *population dynamics*, published a paper in *Nature* titled "Simple Mathematical Models with Very Complicated Dynamics" (May, 1976). And, with reference to longer-term precursors, Gleick showed up about a quarter of a century after Edward Lorenz's demonstrations of the butterfly effect, and roughly a century after Poincaré's pronouncements concerning initial conditions and nonpredictability. May's (1976) paper indeed boosted nonlinear dynamics into a new orbital trajectory, because he put his finger on the central mystery of how and why mathematical formalisms, even ones as simple as the equation of a parabola, can disguise — and/or completely hide — the complexity (richness is a better word) of the phenomena they would represent.

The feedback among the simple precursory statements by Poincaré, Lorenz, and May — particularly if one includes the like implications for the material sciences stated in equally simple terms by Gruntfest (1963) — encompasses nearly all of the most significant studies to date within the general realm of nonlinear dynamics. For example, much is made of the chaotic consequences of deterministic indeterminism that stem simply from an unavoidable sensitivity to initial conditions, but it is rarely pointed out that the same uncertainty applies equally well to the potential existences of — and unpredictable future states of — simpler behaviors, many of which are commonplace, such as: (1) "stable chaos," representing cyclic phenomena or processes that are regular when they are described within certain tolerances but are irregular and unpredictable when they are examined with greater precision (the motions of the planets themselves, discussed later, has been described in this way; cf. Murray, 1992; Milani and Nobili, 1992; Sussman and Wisdom, 1992); (2) stable or unstable limit cycles of diverse kinds; and (3) stable or unstable periodic behavior of diverse kinds (nonlinear resonance phenomena and/or systems of *fixed-point attractors* of dissipative systems; see Notes P.1 and P.3). All of these nonlinear phenomena, as alternative manifestations of the butterfly effect, are important in the context of the evolution of planetary nebulas and of precisely, if chaotically, structured planetary progressions — a relevance especially important to discussions in the present volume concerning the history of the Solar System. May's article — when I finally saw it several years later — resonated with my own ambivalence toward the implications of applied mathematics for the natural sciences, and I think it did, too, for many others in the arena of interdisciplinary research. For if few saw the implications of nonlinear dynamics with regard to the fatal flaw implied, post hoc, by the word "initial," fewer still have seen the self-organizational implications of the deterministic-nondeterministic paradox in the phenomenology of the universe. Beyond its more concrete implications, nonlinear dynamics, for the first time in the history of science, would dare to promulgate a methodology of scientific research that would threaten to erase the lines of departure that identify our most cherished scientific totems — the scientific disciplines. Nonlinear dynamics, in its broadest connotations, is neither a "discipline" nor a "field" of research. It is, in a much larger sense, a paradigm by which to view *all* of the divisions of the sciences, physical and otherwise, within a common behavioral context that does not isolate science from humanity.

Note P.3: An *attractor*, in simplest terms, is exactly what it sounds like. It is a locus in (1) physical space, (2) the graphical space of abstract equations of motion, (3) the real graphical space of imaginary functions, and/or (4) the imaginary space of imaginary functions, toward which "things" are "attracted." These "things" may be: (a) symbolic points, representing variables of abstract and/or descriptive mathematical functions, plotted on a graph, (b) objects, such as asteroids and comets within and/or around, respectively, the Solar System, or organisms within the biosphere (e.g., moths around a light bulb, insects around a highway, wolves around a caribou migration), or (c) fictional creatures around fictional constructions (foxes around Watership Downs, halos around angels, etc.). It is only when we consider the diversity of phenomena that can represent the attractor and the attractee that the story gets complicated. The attractor itself — though formally a steady-state construction (e.g., in a computer experiment) — may exist only in some relatively transient and/or intermittent sense (e.g., geologically, or even in "real time"), or it may cease to exist at all (e.g., a light that is turned off and on vis-à-vis one that burns out, relative to a moth attractor), or it may for some reason become a *"repellor,"* which is the formal term for everything outside of the attractor's "sphere of influence" (e.g., see the analogy with the *celestial sphere of influence* discussed in Note 6.1).

A sometimes subtle point that distinguishes these usages of attractor and attraction from physical "laws" of attraction — as in the laws of gravitational and/or electromagnetic fields of attraction — is that the attractor represents both the field of attraction as well as the things by which it is recognized as an object [e.g., nonlinear-dynamical theories of information storage and retrieval are beginning to recognize that attractors can be treated as objects that correspond to sources of information, as in Andreyev et al. (1992, p. 503); cf. Nicolis (1991); see also the concept of *bifurcation synapses* in Shaw (1991, Fig. 8.35, p. 134)]. For example, a magnet attracts iron filings and arranges them in a characteristic pattern in a manner analogous to points on a computer screen, but neither the magnet nor the iron filings

Chapter Notes: Prologue

(as objects) are the attractor. They simply mimic the form of the attractor field to a certain degree of approximation. The attractor, as object, may contain or imply a great deal more information that is invisible until it either acts on another object or is acted upon by another field that is associated with another object. The attractors of nonlinear dynamics also identify a source mechanism, but that mechanism may have no material counterpart, or it may have many material counterparts that are interrelated by feedback in remote and roundabout ways that identify a *field* or *basin* of attraction.

In dynamics, attractors imply dissipation in some guise. This dissipation often is expressed only in the topological sense, simply as a contraction of the graphical space of representation—such as in the area visited by trajectories in a two-dimensional plot that describes the motion of an oscillator, such as the simple pendulum with forcing and damping (cf. Gwinn and Westervelt, 1986; Baker and Gollub, 1990). In the actual motion of physical objects, however, such as molecules, pendulums, vehicles, asteroids, planets, stars, etc., there is dissipation of energy in various forms. Without some form of dissipation, attractors theoretically do not exist. *This abstraction points to a fundamental assumption concerning my usage of the word attractor in the present work. In contrast to the usual theoretical assumption upon which the history of ideas in celestial mechanics is based, I assume that dissipation is essential to the nature of cosmological organization, and that it is ubiquitous.* In many of the textbooks and other literature of nonlinear dynamics that will be cited in this volume, formal developments are usually subdivided into *conservative systems* and *dissipative systems* (cf. Berry, 1978; Helleman, 1980; Bergé et al., 1984; Schmidt, 1987; Schuster, 1988; Nicolis and Prigogine, 1989; Baker and Gollub, 1990). The study of conservative Hamiltonian systems, for example, employs much of the same language as does the terminology of nonlinear dissipative systems in general (terms such as nonlinear resonance, nonlinear periodicity, chaos, etc., are common to both conservative-systems theories and dissipative-systems theories), except that attractors do not exist in conservative phase-space systems (Helleman, 1980, gives a revealing dissertation on these contrasts; cf. Meiss, 1987; Schmidt, 1987; Smith and Spiegel, 1987). Some texts, and the literature in general, intermingle the two different contexts, often without emphasizing their profoundly different physical implications [compare Helleman (1980) with texts such as those by Schuster (1988) and Jackson (1989–90)]. In my opinion, this duality is both confusing and misleading because, while Hamiltonian dynamics may be meaningful to the person who has been trained to study idealized systems, such as certain restricted kinematic classes of fluid flow, or the *three-body problem* of classical mechanics (as in traditional studies of the motions of the planets, stabilities of electron beams, and so on), to everyone else—and particularly to those persons interested in the more realistic realms of the periodic and chaotic behavior of dissipative systems—it is often both puzzling and unphysical (e.g., it is not always easy to see how the results of the conservative assumption connect up with forms of behaviors that are found to exist in simple dissipative algorithms (see especially the discussion by Helleman, 1980; cf. Jensen, 1987; Schmidt, 1987; Shaw, 1987b, 1991; Smith and Spiegel, 1987; Coffey et al., 1990; Wisdom, 1990).

The persistence of the conservative-Hamiltonian paradigm, or, more accurately, the conservative-Hamiltonian *gestalt*—because it represents a pattern of habitual response among its adherents that persists no matter what subsuming and/or transforming world view might be offered in its place (cf. Glen, 1991; Shaw, 1991, 1994)—in the post-Poincaréan practices of nonlinear dynamics rests on the so-called *Kolmogorov-Arnold-Moser (KAM) theorem* (see definitions and applications in Horton et al., 1983; Thompson and Stewart, 1986, pp. 324ff; Jensen, 1987, p. 174; Moon, 1987, p. 265; Percival, 1987; Schmidt, 1987), especially with regard to applications in the fields of (1) dynamical modeling (e.g., Morrison, 1991, pp. 199ff), (2) celestial dynamics (e.g., Roy, 1988a, pp.

Chapter Notes: Prologue

198ff; Wisdom, 1987a, b, c), (3) magnetohydrodynamics (e.g., Chandrasekhar, 1968; Busse, 1983; Childress, 1985; Roberts, 1987a, b; Smith and Spiegel, 1987), (4) quantum dynamics (e.g., Richards, 1988; Jensen, 1992), (5) theories of electron beams (e.g., Chirikov, 1979, 1991), (6) theories of mixing (e.g., Ottino, 1989; Ottino et al., 1992), etc. An immense literature exists on KAM dynamics, a dynamics which says — boiled down to its essence — that if you take an integrable Hamiltonian system and add small perturbations, then the evolution of the system will remain approximately regular, although somewhat distorted, relative to *most* initial conditions (i.e., maps of KAM systems have some recognizable similarity to maps produced by integrations of linear or weakly nonlinear integrable systems). However, for all of the mathematical effort, and all of the intricate map geometries produced by conservative Hamiltonian functions (see Fig. 12c, and almost any of the journals of nonlinear dynamics cited in the present volume), the KAM theorem leaves us in the dark about systems of attractors (meaning, among other things, systems that can evolve and/or have evolved — thus meaning all of the life sciences and all of the natural sciences, including geology and cosmology; cf. Prigogine et al., 1972, 1991; Prigogine, 1978; Horton et al., 1983; Krinsky, 1984; Prigogine and Stengers, 1984; Jensen, 1987, 1992; Shaw, 1987b, 1991, 1994). [In his introductory remarks, Robert H. G. Helleman (1980) briefly describes his educational experiences in classical mechanics and theoretical physics, emphasizing the absence of any insights into the relationships between conservative and dissipative systems — admitting that, in point of fact, there was hardly any mention of nonlinear phenomenologies at all — thereby providing us with revelations as poignant as those given by Robert M. May (1976) concerning the nonrepresentation of the nonlinear paradigm in science education. These two citations, among many — including other examples alluded to in the present volume, and embodied by it — are telling indictments of the classical system of scientific education, especially in the light of the discussions by Glen (1991) and Shaw (1991, Secs. 8.1 and 8.2; 1994).]

Much of the research effort in the above fields in recent years has been devoted to methods by which the frontiers of nonintegrability can be extended (e.g., see the tremendous mathematical feats represented by the integrations of the equations of motion of the Solar System by Laskar, 1988, 1989, 1990; cf. Milani and Nobili, 1992; Sussman and Wisdom, 1992). But the Hamiltonian-KAM paradigm tells us almost nothing about dissipative processses, except for the varieties of ways a fully *de*coupled system of *independent* oscillators would behave if they existed [e.g., see the evolution of *phase-locked resonances* displayed in the cover illustration of my review article (Shaw, 1987b); but see also Meiss (1987), Schmidt (1987), and Smith and Spiegel (1987)]. The fact that a belief in independent processes has held sway for so long explains the amazement of the general scientific community over the discoveries during the 1980's of the chaotic motions of the planets. But I had already been struck during the first flush of my newfound freedom in 1981, represented by my "discovery" of *numerical dynamics* (see the Prologue), by the realization that, in one sense or another, *the Solar System must likewise display the recursive properties of dissipative maps* (Shaw, 1983a). This was a freedom offered by the spatiotemporal paradigm displayed by computer algorithms of dissipative systems, algorithms that permitted me the personally unprecedented opportunity to explore for myself a systematically coordinated cosmos — even if it was a cosmos described by only the simplest of coupled quadratic maps and their cyclical counterparts, the circle maps of Table 1, Sec. D (see Shaw, 1983b; 1987b, cover illus.; 1988a, Figs. 17–20; 1988b, Figs. 9–11; 1991, Figs. 8.31–8.36; Shaw and Doherty, 1983; cf. Gu et al., 1984; Kaneko, 1986, 1990; Crutchfield and Kaneko, 1987; Willeboordse, 1992).

The tradition of duality in the classification of dynamical systems is only an artifact of archaic conventions, but it has conditioned some dynamicists to think of the conservative

Chapter Notes: Prologue

Hamiltonian formalisms as the "real" problems of physics and the computer demonstrations of attractors as representing only the schematics of the intractable oddities that might occur in the presence of dissipation. This attitude, usually unacknowledged in the literature, obscures the essential importance of convergences related to dissipative effects that, in concert with divergences owing to various chaos-inducing perturbations, generate the richness of actual structures in nature (cf. Lorenz, 1984, 1989; Nese, 1989; Schmidt, 1987; Smith and Spiegel, 1987; Nicolis and Prigogine, 1989). The term *attractor* thus represents to me the general case for the evolution of natural structures. The qualified terms *chaotic attractor* and *strange attractor* refer to the manner in which the trajectories of attraction are played out. *Chaotic*, in the simplest sense, just refers to behaviors that are so sensitive to small deviations in the starting conditions—i.e., those conditions that are assumed to initiate the occurrence of a characteristic phenomenon in nature, in an experiment, or in a computer model (or, analogously, conditions reassigned to restart an interrupted process in any of these contexts)—that the future paths of nearby trajectories tend to diverge exponentially "so fast" (cf. Lorenz, 1985) that future states are unpredictable, even though they may be absolutely determined by the equations and parameters assigned (this is the origin of the frequently used phrase *deterministic chaos*; cf. Lorenz, 1963, 1964, 1979, 1984).

A *strange attractor* is a chaotic attractor in which the patterns of the trajectories of motion—remembering that the notion of the attractor as a characteristic structure still holds—continue to wind recurrently about the space of the characteristic form in such a way that a section through the structure, if examined at finer and finer magnifications, will show the trajectories to be clustered, or nested, in finer and finer but geometrically *self-similar* sets (where self-similar just means that, in the absence of a characteristic standard of length, or of measurement in general, any two sets look indistinguishably alike however much they may differ in actual size; cf. Zaslavsky, 1978; Schmidt, 1987; Smith and Spiegel, 1987). The term "strange attractor" is usually attributed jointly to the mathematical physicist David Ruelle and the mathematician Floris Takens, who had together developed a new theory of turbulence almost 25 years ago (Ruelle and Takens, 1971a; cf. Mandelbrot, 1974). I recently encountered two enlightening (and heartfelt) asides by Ruelle concerning the difficulties of getting that paper published—a paper that, with Lorenz's work, is one of the cornerstones in contemporary theories of chaos (cf. Hofstadter, 1981; Baker and Gollub, 1990; Tufillaro et al., 1992)—in his graceful essay *Chance and Chaos* (Ruelle, 1991, Chap. 9, p. 56; Chap. 10, p. 63), that read, respectively, as follows:

> These strange attractors came from Steve Smale's paper, but the name was new, and nobody now remembers if Floris Takens invented it, or I, or someone else. We submitted our manuscript to an appropriate scientific journal, and it soon came back: rejected. The editor did not like our ideas, and referred us to his own papers so that we could learn what turbulence really was.

> Let me now come back to the paper "On the nature of turbulence" which I wrote with Floris Takens and which we abandoned in the last chapter. The paper was eventually published in a scientific journal. (Actually, I was an editor of the journal, and I accepted the paper myself for publication. This is not a recommended procedure in general, but I felt that it was justified in this particular case.) "On the nature of turbulence" has some of the same ideas that Poincaré and Lorenz developed earlier (we were unaware of this). But we were not interested in motions of the atmosphere and their relevance to weather prediction. Instead, we had something to say about the general problem of hydrodynamic turbulence. Our claim was that turbulent flow was *not* described by a superposition of many modes (as Landau and Hopf proposed) but by *strange attractors*.

Chapter Notes: Prologue

The first of Ruelle's remarks, about publishing (i.e., about the "peer" review process; cf. Glen, 1989), is echoed by the experiences of many others who can be recognized only in retrospect by realizations that their contributions were actually *without peer* [see, for example, (1) my comments in Note P.7 concerning the life of the now famous — in the age of fractals, thanks to Benoit Mandelbrot's (1977, 1983) books on *fractal geometry* — but once infamous, Georg Cantor (1845–1918), the inventor of *transfinite set theory*, and (2) a remark, cited in Note 4.2, made by the contemporary pioneer of *satellite geodesy* Desmond King-Hele, author of the book *A Tapestry of Orbits* (cf. Hughes, 1993), concerning the "refereeing" of manuscripts]. The reference by Ruelle to the then-unrecognized precedents of the Ruelle-Takens ideas in 1971 is reminiscent of what I later call the *uncertainty-of-precedence principle* in science (see the Prologue, and Note 8.15). "Steve Smale" in the first quotation from Ruelle is a mathematician who has published a number of seminal ideas on nonlinear-dynamical formalisms [cf. Abraham and Marsden (1987) and Abraham and C. D. Shaw (1982–88, 1987); the paper Ruelle was referring to is Smale (1967)]. The amusing apology in the second quotation, concerning the editor (himself) who finally accepted the paper on turbulence, is reminiscent of the television commentator who offers a gratuitous public-service message concerning a daredevil sporting event, to the effect that "The performers are highly trained professionals, and such a feat should not be attempted by the viewers."

In a subsequent note, in response to a reminder from the Russian mathematicians Y. G. Sinai and V. I. Arnold, it was acknowledged by Ruelle and Takens that a substantial Russian literature already existed on analogous bifurcation phenomena. [It is not clear in this written acknowledgment by Ruelle and Takens (1971b), however, whether or not the Russian workers had indeed documented the fractal "strangeness" of the attractor field.] Ruelle and Takens (1971a, p. 170), in a remarkably evocative sentence, described their attractor as follows: *"Going back to the vector field X, we have thus a 'strange' attractor which is locally the product of a Cantor set and a piece of two-dimensional manifold."* A Cantor set is the simplest prototype of the fractal object, being constructed from *recursive cutouts of a line* (e.g., Mandelbrot, 1977, pp. 97ff), although Mandelbrot (1983, p. 80) elevates the Cantor "set" to the more expressive term *Cantor dust*. Interpreting "manifold" to mean literally what it says, "many" and "fold," it is easy to visualize a strange attractor as being made up of recursive cutouts, or "shreddings," of a folded surface.

Ruelle (1980) illustrates this process by means of computer algorithms and pictures of natural flows that convey the multiply-folded-and-cross-shredded manifold idea [a form of structural geometry that is kinematically analogous to *petrographic fabrics* seen in some types of banded mineralization and in multiply folded and stretched high-grade metamorphic rocks and/or *migmatites* — the latter referring to the products of a form of high-grade metamorphism in which conditions of partial melting and/or supercritical dense-gas mass transfer (a form of *metasomatism*) have been partially attained, usually at deep crustal levels and/or in deformed and remobilized portions of batholithic terrains at higher crustal levels (cf. Strahler, 1971, cover illus., and Chap. 26; Ortoleva et al., 1982)]. The two examples Ruelle chose to represent natural analogues of strange attractors are the familiar images of (1) smoke rising from a cigarette and (2) the striate folded and vortical flow structures in the atmosphere of Jupiter (see Chap. 7, and Notes I.4 and 7.1). Excellent color photographs that show analogous images of altocumulus clouds on Earth and vortical structures on Jupiter — each of which would show finer self-similar detail at higher magnifications — are shown by Giovanelli (1984, Plates I–III), where the cortex-like lobate structures (see the discussion of Fig. 12b in Chap. 3 of the present volume) of the altocumulus clouds can be compared with the less well resolved black-and-white photographs of solar granulation (ibid., Figs. 6–10), which "burst" or erupt to form the pupil-iris-like umbra-

Chapter Notes: Prologue

penumbra structure of sunspots, in a manner somewhat like the expansion of gas-inflated hemispherical blisters of lava seen during the Mauna Ulu eruptive episode (1972–74) of Kilauea Volcano, Hawaii (Tilling et al., 1987).

The hemispherically shaped, bubble-like blisters of gas-inflated lava — each hemispherical blister measuring some tens of meters in diameter when fully formed — emerged from the surface of relatively degassed transiently ponded lava near the vent of the Mauna Ulu eruption, growing from the surfaces of the cooled but still-mobile lava in a manner somewhat resembling the growth of solar prominences on the Sun. [The lava surfaces through which such blisters emerged had cooled below the visible radiation colors, thus appearing dark, but the expanding blisters and "dome fountains" were of bright-red to near-incandescent hues: compare, for example, the photographs of a "lava jet" and a "lava prominence," respectively, on pp. 850 and 859 of Takahashi and Griggs (1987); the average temperatures were above 1000° Celsius, the temperature below which Kilauean lavas rapidly stiffen and become highly viscous or even "rigid" in appearance (cf. Shaw et al., 1968, 1977; Wright et al., 1976, 1992; Tilling et al., 1987).] The surface areas of the "lava prominences" grew so fast and demonstrated such extreme states of expansion that the partially incandescent films of magma that defined their perimeters consisted of volumetrically separated and (as expansion continued) increasingly tenuous arrays of islands that formed smaller and smaller fractions of the total area of the bubbles while bubble growth continued. As is the case with solar prominences, this growth-rate dependence would appear to be essential to the development of the characteristic spherical, or more distended and extended, geometric forms (cf. Giovanelli, 1984, Figs. 85 and 86; Zirin, 1988, Chap. 9, Figs. 9.5 and 9.6; Koutchmy et al., 1991, Figs. 9 and 10; Kaler, 1992, pp. 114f). As far as I am aware, no one has made any systematic study of this "classical" problem in *fractal geometry* — for either the case of lava prominences or that of solar prominences (cf. Mandelbrot, 1983, Chaps. 29, 33, and 34; Feder, 1988, Chaps. 12–14). For example, the dimension of the total surface area of the molten-lava fraction, relative to the area of a sphere, as expressed in terms of the exponent of the radius of the sphere — where the dimension of a spherical surface is $D = 2$, as expressed by that exponent — is clearly $D < 2$ [i.e., the area of the lava fraction of the surface "shell" progressively decreased with expansion and would have approached the point-like limit $D = 0$ at infinite expansion; cf. the discussion of the "blazing sky" effect in Note 4.2, and the discussion of the Dirac-Jordan-Carey theory of an expanding Earth in Note 6.8; see also Coniglio (1986), Lucchin (1986), and Pietronero and Kupers (1986)]. My impression, gathered from visual inspection of photographs of lava jets and prominences, is that the fractal dimension was between 1 and 2 (perhaps about 1.5). Chouet and Shaw (1991) discuss the *fractal dimension* of a strange attractor generated by gas-piston magmatic activity in an eruptive vent of Kilauea Volcano (as reconstructed from its seismic-tremor signature; cf. Aki and Koyanagi, 1981; Chouet, 1981, 1988, 1992; Chouet et al., 1987; Koyanagi et al., 1987), and Shaw and Chouet (1991) discuss the fractal dimensions and dynamics of magmatic clustering and percolation through the lithosphere (cf. the discussion of *dynamical dimensions* in Note 5.3). There are some remarkable parallels between terrestrial and solar eruptive behaviors, the study of which, in addition to the comparisons of prominences and so on, might be instructive to both disciplines [see, for example, the hypothetical substructures of solar flares and spots as illustrated by Giovanelli (1984, Fig. 40), Pecker (1991, Fig. 2), and Rabin et al. (1991, Figs. 17 and 18)]. Temporal aspects of solar phenomena are emphasized in Sonett et al. (1991); cf. Kurths and Herzel (1987), Chouet and Shaw (1991), and Shaw (1991).

An excellent, and more easily studied, physical example of strange attractors is provided by the motions of fine particles floating on the surface of a slowly swirling tidal pond. The best example I have seen occurred on the surface of a salt evaporation pond on

Chapter Notes: Prologue

the south side of the east end of the Dumbarton Bridge crossing the San Francisco Bay between Palo Alto and Newark, California (a nature preserve and trails provide easy access; the phenomenon is best viewed following windy periods in the late spring and summer, when abundant evaporite dust and/or pollen have formed an extensive surface film). A theoretical analysis of an analogous phenomenon, produced in a controlled (*tunable*) laboratory experiment, was just published in *Science* by Sommerer and Ott (1993), and I would suggest that one take a copy of that issue of *Science* along the next time you visit a salt pond, or Jupiter—especially the description of the flow setup (Fig. 1) and the vividly colored fluorescent flow trajectories shown in the *Science* cover illustration and Figure 2 of that article. All of the above examples fit well with the emphasis of the present volume on attractor systems (cf. Treve, 1978; Rabinovich, 1980), assuming that the treatment of the dynamics appropriately reflects the dynamics of dissipative systems (as distinct from the conservative Hamiltonian systems discussed above; cf. Note I.4; and Meiss, 1987; Jensen, 1987; Schmidt, 1987; Smith and Spiegel, 1987)—a necessary condition for them to be strange attractors—even if aspects of the *kinematics* of the flows can be modeled by Hamiltonian equations of motion (cf. Marcus, 1987, 1988, 1990; Ottino, 1989; Coffey et al., 1990; Ottino et al., 1992).

Ruelle (1980) could not have picked two better examples for the purpose of illustrating the comparative chaotic dynamics and structural geometry of diverse cosmological, meteorological, oceanographical, and geological flow systems—because those examples illustrate the coevolution of contrasting relative dynamical regimes displayed in single portraits ("snapshots") of the same system (cf. Sommerer and Ott, loc. cit.). To illustrate, the following descriptions can be applied to either of Ruelle's examples: (1) the flow structures are both *chaotic* and *fractal* (cf. Moon, 1980, 1987, pp. 250ff; Moon and Li, 1985a, b; Ditto et al., 1990b; Shaw, 1991, Figs. 8.19–8.23) in the nonlinear-dynamical context (cf. the discussion of "Liapunov" numbers by Sommerer and Ott, loc. cit.; and see Note 5.3, and Chap. 6 in the present volume), (2) they display the types of irregularities typically associated with percepts of fluid-dynamical turbulence, (3) both also show internal regimes of flow that are well-ordered and/or laminar (nonturbulent-looking) in appearance, and (4) both show singular (stationary, or quasisteady) types of structures that, from a different dynamical perspective (a different mapping of the flow), might correspond to fixed-point attractors. [Compare the formal mathematical definitions of chaotic attractors given by Devaney (1986, 1987) with the qualitative descriptions given above, and with the various discussions of mathematical formalities given in these Chapter Notes and in Shaw (1987b, 1988c). The *chaotic bursts* of Devaney (1987) are analogous to those discussed in Chap. 5 of the present volume; they differ rigorously in referring to (a) transcendental control functions, and (b) mappings in the complex plane—also noting the fact that the only way any of us, mathematicians, physicists, or geologists, is likely to discover the existence of such crisis-like conditions is through *experimentation*. This last point, made by a mathematician, reiterates the theme of the present work (cf. Ruelle, 1980), as well as that of Shaw (1987a, b, c, 1988a, b, c, 1991).]

The characteristics described above reveal that the unqualified terms *turbulence* and/or *chaos* are not sufficient to describe the motion (there are ordered structural regimes within both examples that not even the concept of the strange attractor adequately describes). In both instances, special structures can emerge from the general flow field, representing singular and/or bifurcating smoke-ring-like and/or whirlpool-like steady-state vortical phenomena. The smoke-ring type of regime is familiar to cigarette smokers, but better examples—ones that invoke period-doubling and/or higher-order geometric progressions of vortex-ring bifurcation sequences (vortex "trees")—are given by the streaming motions of density currents in viscous-fluid flow (forced and/or free advective and/or

Chapter Notes: Prologue

convective motions of positive and negative buoyancy currents, plumes, and jets). The most illustrative example I know of for this type of bifurcation sequence was described by W. H. Bradley (1965, 1969) in an experimental study designed to study rapid sedimentation — compared to single-particle settling rates — of finely divided particles in alpine lakes (deep freshwater lakes that are annually mixed by convective overturn caused by winter cooling of the near-surface water; cf. Hutchinson, 1957). Following Bradley's lead, I later investigated the phenomenon of ring-vortex bifurcations in other modes, discovering that there are dynamical families of bifurcation sequences, each corresponding to a different geometric progression (Shaw, 1986; cf. Chen and Chang, 1972; Lim and Nickels, 1992). Atmospheric or hydrologic vortex structures (tornadoes, whirlpools, salt fingers of certain types, atmospheric "spots" on the giant planets, analogous to Jupiter's Great Red Spot, etc.) also can display bifurcation phenomena of analogous types (e.g., Snow, 1978; Shaw, 1986; Ingersoll, 1983, 1988; Sommeria et al., 1988; Gierasch, 1991).

A passing observation by Ruelle (1980, p. 128) identifies yet another exceedingly important characteristic of chaotic systems. They are *tunable*, meaning that their behavioral regimes are subject to parametric adjustment by either artificial or natural means (see Notes 3.1 and 5.2). Under some parametric conditions, arbitrary or natural, variations in the tuning of the strange attractor may cause it to suddenly change in appearance, or to disappear — phenomena related to the chaotic bursts of Devaney (1987) called *chaotic crises* (see Chap. 5, and Notes I.4, 5.1, and 5.2). In the example mentioned by Ruelle — called a *boundary crisis* in the present volume (see Chap. 5, and the Notes just cited) — the strange attractor is replaced abruptly by a fixed-point attractor (a periodic attractor of period 7 in the example described by Ruelle, meaning that there was a repetitive recurrence of a seven-fold set of *fixed points* in phase space; see below). This is a geometric object with dimension zero, or, if viewed collectively, it may form part of a more extensive object with a fractal dimension between 0 and 1 (cf. the discussion of the ensemble dimension of the "Devil's Staircase" of phase-locked nonlinear periods in Jensen et al., 1984), representing some form of multiplicative sequence (i.e., the base-7 geometric progression is directly analogous to one of the ring-vortex bifurcation sequences mentioned above).

Periodic attractors — by default, or by process of elimination of chaotic attractors — include all other types of loci of attraction that display rigorous forms of numerical recurrences in *phase space* (the space of the graphical portrayal, usually called a *phase portrait,* variants of which are called *return maps*, representing the first, second, third, etc. differences between the value of a function and its subsequent iterates, and *Poincaré sections* — the latter being, topologically, sections through phase portraits or return maps that are chosen arbitrarily, or, in oscillatory systems, at *stroboscopic* intervals related usually to the natural frequency of the system or related to the natural frequency of an external forcing function).

Phase portraits may be abstract, or they may portray the recurring behaviors of a concrete physical space (e.g., trajectories connecting earthquake hypocenters in the Earth as a function of time constitute a phase portrait of the earthquake process characteristic of a specified region — or characteristic of the Earth as a whole — if it can be shown that the earthquake process displays attractors, or attractor-like behavior, as is the case in volcanological contexts such as that in Hawaii; cf. Shaw, 1987a, 1988b; Shaw and Chouet, 1988, 1989, 1991; Chouet and Shaw, 1991). Periodic attractors, therefore, represent one or more subsets of points that repeat "exactly" within the precision of measurement (meaning to roundoff error of a computer or measurement error of a standard oscillator, such as a tuning fork or its equivalent) for the same relative condition of a cyclic recurrence algorithm, which may be a mathematical function or a self-organized function of a natural process. An example of what I call a *natural algorithm* is given by the cyclic orbital motions

Chapter Notes: Prologue

of the planets, periodic or chaotic, relative to other types of orbital behaviors that are not typical of the Solar System [e.g., the behaviors of some asteroids and comets are not obviously related in a systematic way to the attractor-like behaviors of specific reservoirs of asteroids and comets (see Fig. 13, Insets; and Hoffmann et al., 1993); however, such "anomalous" behaviors mean that a larger context, or more encompassing algorithm, is needed to describe the system, often at a larger scale of coupling—e.g., as may be represented by Galactic phenomena that are coupled with the Solar System]. But even if measurements and computations could be made with vanishing error, nonlinear periodic repetitions need not be—and usually are not—spaced uniformly in space or time, as they would be in a harmonic series, a standard of repetition that most people—scientist or layperson—habitually, and erroneously, associate with the generic term *periodicity* (cf. Shaw, 1987b).

Note P.4: In nonlinear dynamics *metric universality* refers to sets of invariant numerical relationships, expressing the fact that certain properties of *phase space* (see Notes P.3 and 5.1) depend neither on the specific form of a phase portrait nor on the particular physical makeup of the system described by it (cf. May, 1976; Feigenbaum, 1979a, b, 1980; Shaw, 1987b, 1991, p. 143; Schuster, 1988, p. 167; Baker and Gollub, 1990, p. 81; Shaw and Chouet, 1991, p. 10,205). The implications of this statement cut across all disciplinary boundaries in science and are more fundamental than the most fundamental "laws" of physics. These numerical relationships are characterized by so-called *universality parameters*, which, in the simplest terms, are numerical ratios that express the immutability of certain characteristic dynamical proportionalities—representing a role somewhat similar to that of the geometric proportionality constant pi, the ratio of the circumference of a circle to its diameter, which is essential to the formulation of the law of gravity, wherein the properties of rectilinear and angular motions of massive objects are reconciled (cf. Gamow, 1962; and Note P.6). In contradistinction with pi, however, which is a purely geometrical ratio, the universality parameters of nonlinear dynamics subsume—or "wed"—both geometry and dynamics, in the sense that they express constancies of numerical ratios that are both geometrical and "dynamical" in the phase-space context (the quotes point to the fact that the numerical character of the equations of motion that describe universal properties of nonlinear-dynamical phase portraits are not dependent on definitions of mass, force, or physical constants—such as the gravitational "constant"). The bifurcational properties of the equations of the pendulum, of "kicked rotors," of the sine-circle map, or of the logistic map of population dynamics (cf., respectively, Gwinn and Westervelt, 1986; Jensen, 1987; Shaw, 1987b; May, 1976)—to name just a few types of nonlinear algorithms—are all characterized by the same relationships between the driving parameters of the motion and the topologic properties of the phase portraits produced by the motion (cf. Schuster, 1988, p. 167; Shaw, 1987b). In this context, the numerical ratio called the *golden mean* (see below; cf. Shaw, 1987b)—together with other proportionalities, such as the Feigenbaum ratios (Feigenbaum, 1979a, b, 1980; cf. May, 1976; Schuster, 1988), etc. (e.g., Helleman, 1983; MacKay, 1983)—plays an essential role in demarcating the sets of singularities that represent the transition between periodic and chaotic motions of coupled-oscillator systems (cf. Ling et al., 1991; Shaw and Chouet, 1989, 1991). In a real sense, universality parameters constitute the architectural girders of nonlinear dynamics, hence are the sine qua non of "deterministic chaos" and of its potentiality for the quantitative description of complex order in natural systems, an order that up to now could only be called random according to conventional methods of analysis and/or conventional measurements of complexity—whether traditional or "algorithmic" (see Ruelle, 1991, Chap. 22).

Universal Order, Geometry, and Number: The numbers *pi* (π) and the *golden mean*

Chapter Notes: Prologue

($), together with less commonly known universality parameters of nonlinear dynamics, undergird virtually all of human experience — from the practical to the theoretical, from the artistic to the occult. I employ the dollar sign (gold standard) to represent the golden mean in this note for several reasons, some of them given below. Vast libraries of both ancient and modern vintage are filled with writings about pi and the golden mean, yet they both express the order of the universe — the Cosmic proportions — as dimensional ratios of axiomatic simplicity. The number π is used so often, by rote, that we rarely think of it as anything other than a numerical factor needed to get the right answer when we are dealing with spherical measures and trigonometric relationships — and $, if it is known at all, conjures up images of alchemy and arcane practices of metaphysics, or the aura of magical incantations. Compare *A History of π (Pi)* by Petr Beckmann — noting the matter-of-fact tone of the Preface to the Third Edition (1974) — with the evocative mystery of number theory as the ominous theme, disguised as a history of chess (certainly a more provocative gambit than "the mapping of a square onto itself"), of the historical novel by Katherine Neville (1989) titled *The Eight*. (*The Eight* was a gift from my daughter Andrea, because she sensed a relationship — prior to any communications from me about the cosmological implications of the book I was writing — with the mystery of, and the age-old quest for, the "secret" of the power of the universe, as alluded to in Neville's story.) In the following I employ the number π as an entrée to the golden mean — both being dimensional constants that express the meaning of metric universality.

Geometrically, π *is the ratio of the circumference of a circle to its diameter*, but the formula for the circumference of a circle [the diameter times pi ($= 2r\pi$)] has been so etched in our memories since grammar school that we rarely think of it in those words. The number $, the golden mean, if it is recognized at all, is usually described as a vague notion having to do with the most pleasing proportions in works of art, architecture, anatomy, and nature (cf. James and James, 1992, p. 187). Few persons can describe its properties in words, and fewer still would recognize it as the numerical constant derived from the limit $ $= 2 \div (5^{1/2} - 1)$ $= 1.61803398\ldots$, where the dots signify "without end," because $ is an *irrational number*, meaning that it can only be approximated as a ratio of integers by employing, for example, the method of *continued fractions* (see Hardy and Wright, 1985, p. 163; cf. Shaw, 1987b, 1988c; Morrison, 1988). Yet, despite all the rumors of arcane knowledge and esoteric number theory associated with its noble name, $ is geometrically defined by the same sort of invariant geometric ratio as the number π, viz., in my own words, *the golden mean is that ratio of lengths between the longer and shorter virtual segments of a continuous two-part linear measure that preserves an identity with the ratio between the total length of the linear measure and the length of the longer segment* (cf. Ghyka, 1977, Chaps. 2 and 3).

As innocuous as this definition may seem, it is the nonlinear-dynamical key to universal order, because it is the key to (1) the proportionalities encountered in the *critical golden mean nonlinearity (CGMN)* of the sine-circle map, the essential metaphor of nonlinear coupled oscillations (e.g., Bohr et al., 1984; Jensen et al., 1984, 1985; Fein et al., 1985; Halsey et al., 1986; Gwinn and Westervelt, 1987; Shaw, 1987b; Olinger and Sreenivasan, 1988; Shaw and Chouet, 1988, 1989, 1991), *and* (2) the generalized proportionalities of *self-organized criticality* (see Chap. 5; cf. Bak et al., 1988; Kadanoff et al., 1989), the two proportionalities together symbolizing the essential metaphor of a dissipative, open-system, autopoietic universe (see Jantsch, 1980, p.7; Shaw and Chouet, 1989, p. 197). In other words, an infinite system (the universe) proportioned by such a geometric rule (with due regard for higher powers than one, the linear measure) will be self-similar in dynamical structures *without regard to size or conventional definitions of material states* — a statement of *universal homology* that underlies and surpasses all other concepts of

Chapter Notes: Prologue

homology and/or analogy based on physical, chemical, or biological principles. It is this property of the golden mean that turns up ubiquitously, one way or another, in all descriptions of Nature, whether the described object is "animal, vegetable, or mineral," an artform, a symphony hall, an acoustic chamber of a recording studio—or a volcano, a solar prominence, or a spiral galaxy (e.g., Huntley, 1970; Ghyka, 1977).

A curious thing about the relationship between the dimensional constants π and $\$$ is that twice their product happens to be approximately ten [i.e., $2\pi\$ = 10.166407\ldots$, where precise numerical values of π and $\$$, respectively, can be found, for example, in Beckmann (1974, p. 103) and Ghyka (1977, p. 7)—or the latter can be calculated from the formula given above using a high-precision calculator]. For the purposes of this note, and according to the reasoning set out below, I shall arbitrarily call this quantity—viz. *the quantity that is equal to two times the product of pi and the golden mean*—the *nonlinear cosmic proportion*, $© = 2\pi\$$. The product $2\pi\$$, symbolized by $©$—where the encircled letter c stands for the "universal constant of nonlinear hierarchical systems of circles in golden-mean cosmic proportions"—as I will explain shortly, has several interesting properties, viz. (1) it is directly analogous to π in representing the ratio of the circumference of a circle to a special line segment, but this special line segment is one that is irrationally proportional to the radius of the circle (see below), (2) it has the *mnemonic* property that it is a multiplicative combination of the fundamental scaling ratios of (a) the line and (b) the circle (the latter explaining the typologic choice of $\$$ for the golden mean, symbolizing both its noble-element and sine-circle properties), (3) it is approximately equal in numerical value to the base of the common logarithm (i.e., $© \cong 10$), and, together with its square and cube, provides a geometric rationale for the commonly encountered proportions tenfold, one-hundredfold, one-thousandfold in natural measures (as in decade, centigrade, century, percent, per mil, mill, millennium, etc.), and (4) it therefore "explains" (only semi-facetiously) why the "order of magnitude," is the most frequently encountered "scale factor" in science; i.e., the two factors $© \cong 10$ and $©^3 \cong 10^3$, for example, correspond to the loosely defined measure called the "order of magnitude" familiarly applied in terrestrial physics and cosmological physics, respectively (where the latter is sometimes called, pejoratively, the "astronomical order of magnitude," as by a parochial scientist speaking of cosmological measurements; cf. Notes 4.2 and I.3).

Examples of both the tenfold and the thousandfold scaling factors are numerous even in the context of terrestrial scale relationships. There are some curious ones—such as the fact that the *tenth* power of two, a member of the nonlinear bifurcation sequence called the *period doubling hierarchy* (see Note P.1) happens to be approximately 10^3 (i.e., $2^{10} = 1,024$). Besides such numerological "coincidences"—and despite the obvious importance of progressions involving other geometric bases, such as base-e, base-2, and, in general, base-n (with n the sequence of integers; cf. Note 8.15; and Shaw, 1988a, Table 1 and Fig. 25)—there is strong circumstantial evidence that hierarchies consisting of powers of ten and powers of ten cubed are not merely accidents of our arbitrary dimensional standards (cf. the Introduction). For instance, the proportion 10^3 is mimicked by the prefixes currently in vogue for the systems of units used in atomic physics and chemistry, computer technology, and astronomy, where dimensions are specified in units of 10^{3n} (where n represents, from small scale to large, the negative to positive integers)—as in the 10^3-unit hierarchy, from $n = -5$ to $n = +4$, *femto-pico-nano-micro-milli/kilo-mega-giga-tera* for the numerical orders-of-magnitude range 10^{-15} to 10^{12} (omitting the unit value $10^0 = 1$) in powers of 10^3. The prevalence of decimal scaling in the sciences is exemplified, for example, in the following books: *Powers of Ten*, published by Scientific American Books (1982); *Units and Conversion Charts* by Wildi (1991), *Astronomy* by Pasachoff (1991, Chap. 1 and Appendix 1), *Universe* by Kaufmann (1991, Chap. 1), and *Atoms, Electrons, and Change* by Atkins

Chapter Notes: Prologue

(1991, p. 126). For a mathematical definition of "order of magnitude" see James and James (1992, p. 259).

Thus © and its square and cube are reminders of the relationships between linear, areal, and volumetric scaling in ordinary space, as well as their fractal analogues in the dimensionally normalized "circle-line space" of nonlinear dynamics—where the latter space symbolizes, in turn, the little-explored cosmological interface between nonlinear-geometric mappings (e.g., analogous to *Riemannian mapping*; James and James, 1992, pp. 360f) and linear-geometric mappings (e.g., on the basis of ordinary Euclidean geometry). The implication should not be missed that the "circle-line space" of nonlinear dynamics, and the cosmological implications of ©, would appear to wed the various "unified theories," such as general relativity, quantum geometrodynamics, superstring theory, etc. into one universal class—perhaps with a name that evokes, not entirely with tongue in cheek, the nonlinear-cosmological-quantum-dynamical characteristics of deterministic chaos, such as *CQ-chaology* (where CQ refers to "cosmo-quantum," and "chaology" is an ancient term of theological derivation recast as a description of "semiclassical" behaviors—meaning behaviors that span the realms of classical chaos, in the modern dynamical sense, and quantum dynamics—by Berry, 1987; cf. Ford, 1987; Kaku and Trainer, 1987; Davies and Brown, 1988; Ford et al., 1990; Jensen, 1992; cf. also Note 4.2).

The geometrical meanings of the universal numerical "coincidences" related to pi and the golden mean can be visualized by returning to the primary definitions given above. First, express π as the length ratio $C/2r$, where C stands for the "length" (the continuous linear measurement per cycle) of the circumference of a circle. Next, express the golden mean, $\$$, as the ratio L_o/L, where L_o is the length of a standard measure (e.g., a "meter stick," or any length defined as a standard) and L is the golden-mean segment of the standard length, $L_o/\$$. If we now define a "standard circle" in like manner to be a circle with radius equal to the standard length L_o, then we can rewrite the definition of pi in the form $\pi = C_o/2L_o$, yielding $2\pi = C_o/L_o$. Multiplying by the golden mean, we then find that $2\pi\$ = (C_o/L_o) \times (L_o/L) = C_o/L$. We now have the identities $© = 2\pi\$ = C_o/L$. The geometric meaning of the *nonlinear cosmic proportion*, ©, is now evident: *it is the ratio of the circumference of the standard circle to the golden-mean segment (L) of the standard radius L_o of the standard circle*. In a sense, then, $L = C_o/2\pi\$$ can be thought of as a segment of *the golden circle*, which, multiplied by $© = 2\pi\$$, *the golden equivalent of pi*, yields the *circumference of the golden circle*. In other words, we have now defined a self-similar hierarchy of golden circles scaled relative to the golden-mean proportioning of a line of arbitrary length—and therefore, in like manner, we have also defined the third-power-law scaling of a self-similar hierarchy of golden spheres relative to golden cubes and the second-power-law scaling of golden circular sections relative to golden squares. It therefore seems intuitive to suppose that, if the oscillation frequencies of the linear elements of a sine-circle-like mechanism scale as powers of $\$$—as occurs reciprocally in the device described by Fein et al. (1985), and in other devices that yield the CGMN singularity spectra cited above—then the oscillation frequencies of a hierarchy of golden circles should scale as powers of $© \cong 10$, and the oscillation frequencies of a hierarchy of golden spheres should scale in some proportion to powers of $©^3 \cong 10^3$. Although this conjecture simply reflects the qualitative circumstances of the dimensional scaling of the sine-circle algorithm—as I have mentioned several times, and discuss further below—it seems worthy of experimental investigation.

The original motivation for my enquiry into the relationship between pi and the golden mean came from an empirical observation made by Bernard Chouet and myself in our joint studies of *volcanic tremor* (the long-period low-amplitude seismic, or acoustical, vibrations encountered in volcanoes and their substrates owing to the subterranean flow of magma; see Notes P.5, 4.2, 6.2, 8.1–8.3, 8.13; cf. Aki and Koyanagi, 1981; Chouet, 1981, 1988, 1992;

Chapter Notes: Prologue

Chouet et al., 1987; Koyanagi et al., 1987; Schick and Mugiono, 1991). In Shaw and Chouet (1988, 1989), we showed that the CGMN of the sine-circle function (Table 1, Sec. D) — as is the case in the diverse physical systems already cited, such as coupled fluid convection and electronic transport (cf. Ostlund et al., 1983; Bohr et al., 1984; Jensen et al., 1984, 1985; Fein et al., 1985; Gwinn and Westervelt, 1987; Shaw, 1987b) — provides at least an approximate numerical metaphor for the nonlinear frequency spectrum (*multifractal singularity spectrum*) of complex coupled oscillations in Nature, as exemplified by the coupling of magma transport and seismic tremor. The apparent reason for the success of the CGMN model is the combined *tuning* condition that involves *critical forcing* (which refers to the value of the forcing parameter at the transition between phase locking and chaos in the sine-circle function; cf. Shaw, 1987b) and a value of phase delay (*bias*) that is set in such a way — either by arbitrary tuning in the laboratory or by *cybernetic self-tuning* in Nature — that the resulting nonlinearly coupled *winding number*, W (Table 1, Sec. D), is *quasi-periodic* (irrational) and acquires a value equal to the inverse of the golden mean [i.e., $W = \$$ (mod 1), as computed from the sine-circle function of Table 1, Sec. D; cf. Fig. 14 and Appendix 9]. The experimental study of coupled fluid convection in one dimension by Fein et al. (1985) — i.e., reflecting the coupling of a simple convective loop with, effectively, one-degree-of-freedom periodic forcing at the sidewalls — was of great importance in demonstrating that the CGMN condition does indeed yield an infinite spectrum of frequency bundles that scale as powers that are integer multiples of $\$$ (i.e., the frequencies of the coupled oscillator scale logarithmically in cycles of $\$$; see Fein et al., loc. cit., Figs. 2 and 3).

Bernard Chouet and I noticed that the golden-mean frequency bundles of the CGMN — which are therefore internally ordered according to *Fibonacci sequences* (cf. Fein et al., loc. cit., p. 80; James and James, 1992, p. 164) — are similar in complexity to the frequency bundles of seismic tremor (see Chouet et al., 1987; Shaw and Chouet, 1988, 1989; Chouet and Shaw, 1991). Later we also found that when the tremor process was expressed as a hierarchy of coupled magmatic-tectonic *relaxation oscillations*, it yielded a logarithmic hierarchy of frequencies in self-similar bundles that spanned several decades (powers of ten) in the time domain — and we already knew that the time-dependent evolution of volcanic structures in the Earth tends to be self-similar in like manner (see Shaw, 1985a, 1987a; Shaw and Chouet, 1991; cf. Note 8.13). At first we were puzzled that the dominant volcanic-tectonic frequency bundles seemed to differ by powers of ten — rather than being more densely spaced according to powers of $\$$, as in Fein et al. (loc. cit., Fig. 2; cf. Olinger and Sreenivasan, 1988, Figs. 5 and 6) — and even more puzzling was the fact that the center-to-center distances between the volcanologically dominant frequency bundles appeared to be about three orders of magnitude (i.e., forming a hierarchy of 10^{3n}, as described above), and spanned a temporal range from about one second to the order of thirty million years (cf. Note 8.13)! The most obvious explanation that occurred to us for such a large difference in power-law scaling — volcanism (three-dimensional in the traditional Euclidean sense) vis-à-vis phenomena that can be scaled according to the essentially one-dimensional metaphor of the sine-circle algorithm — was the existence of some combination of universality parameters acting in concert with the higher spatial dimensions of volumetric transport that characterize magma percolation through systems of fractures. Becoming convinced of this general explanation, I eventually came to realize that the relationships decribed in this note — whether or not other factors are involved — provide the right kind of proportionalities [cubic scaling of © ($= \sim 10^{3n}$)] to reconcile the geodynamical analysis with the more precise measurements made in the laboratory, as cited here and in our papers on volcanic tremor.

In another type of natural system, mentioned in Note I.3, the same sort of 10^{3n} scaling seems relevant. That note concerns the astrophysical scale relationships between the oscillating stellar objects called *pulsars* relative to the quasi-stellar objects called *blazars*

Chapter Notes: Prologue

(see Kaufmann, 1991, pp. 548f). The possible analogy between pulsar-blazar scaling and the scaling of terrestrial volcanism, magma transport, and seismicity is historically interesting, because one of the inspirations for analyzing magma-activated seismic tremor in the way we did was derived from the seminal paper on cosmological fractal self-similarity by de Vaucouleurs (1970) titled "The Case for a Hierarchical Cosmology"!

The dimensional considerations given in this note elucidate two questions that up to now have never been adequately explained — viz., (1) What is the physical/geometric basis for the prevalence of the golden mean in Nature, in the broadest sense (i.e., one that includes art, architecture, and so on)? and (2) Why is the sine-circle algorithm so successful in nonlinear dynamics as the universal numerical scaling function for general systems of coupled oscillators? Besides the conspicuous rule of self-similarity exposed by the golden mean itself (i.e., the self-organized steady-state proportioning of a heterogeneously distributed hierarchy of length scales), the procedure of nonlinearly mapping the sine function onto the circle not only invokes both pi and the golden mean but also brings together the multiplicative properties of the rational, the irrational, and the transcendental numbers in their clearest and most elegant manifestations of *complexity in simplicity* (cf. Note P.6) — the other side of the Janus-like face of *simplicity in complexity* that is usually cited as the paradigmatic signature of deterministic chaos.

Note P.5: *Strain-energy feedback* refers to phenomena where the local amplitude of strain increases as it becomes more and more localized, accompanied by the accelerating growth of local "hot spots" and/or by the relatively isothermal but accelerating growth of endothermic solid-state phase changes, until the rate of mechanical energy dissipation "runs away" and endothermic reactions become dominant, eventually exceeding the thermal stabilities of minerals on the local scale. On the global scale, and especially in the case of bodily deformations, such as the tides of the Earth, the Moon, and Jupiter's satellite Io (cf. Shaw, 1970; Shaw et al., 1971; Gruntfest and Shaw, 1974; Wones and Shaw, 1975; Cassen et al., 1982; Greeley and Schneid, 1991), the runaway mode typically will be expressed in the form of melting reactions (as in the mantle-generated magma in the Earth and Io as sources of global volcanism) and/or volatilization reactions [as in the gaseous phases of terrestrial diatremes and kimberlites, and Ionian eruption plumes (e.g., Kieffer and Delany, 1979; Kieffer, 1982, 1989; cf. Notes I.5 and 8.3 in the present volume)]. There would be unlimited propagation of runaway modes if they were not stabilized in some way to yield kinematic stationary states, and/or other forms of singularities, of mass and thermal energy distributions (e.g., Shaw, 1973; Shaw and Jackson, 1973; Anderson and Perkins, 1975). The irreversible growth of endothermic reaction fronts typically is countered by the conductive and convective (and/or advective) transport of mass and heat from the locus of strain-energy feedback at rates that balance the rates of mechanical energy dissipation; cf. Gruntfest (1963), Gruntfest et al. (1964), Shaw (1969, 1970, 1973, 1987a, pp. 1368ff, 1991, Figs. 8.1–8.9), Pierce (1970), Shaw et al. (1971), Shaw and Jackson (1973), Gruntfest and Shaw (1974), Wones and Shaw (1975), Lachenbruch (1980), Nelson (1981), Nicolis and Prigogine (1989, Sec. 6.2). [Of historical interest is the resemblance of the above description to a theory of volcanic heat proposed by the nineteenth century Irish civil engineer Robert Mallet. Mechanical work done in the compression of rock caused by thermal and gravitational contraction of the Earth as a whole was converted by Joule heating to volcanic energy (Mallet, 1873; cf. Shaw, 1991, pp. 100ff). Fisher (1889, pp. 298ff), however, "showed" that simple thermal contraction of the Earth did not provide enough energy to drive volcanism, and Mallet's idea was essentially ignored by geologists and geophysicists. Rheologically, however, Mallet's proposal could easily have been transformed to other forms of cyclical dissipative phenomena analogous to the role of viscous heating in plate tectonics (Oxburgh and Turcotte, 1968; McKenzie and Sclater, 1968; Shaw, 1969, 1970, 1973, 1991; Shaw et al., 1971).]

As a general statement, the result of thermal-feedback deformation of solids and liquids in natural systems (i.e., systems lacking arbitrary control of initial and boundary conditions) is to reduce the effective flow field to a regime of, on average, roughly constant shear rate (e.g., as in the average motion rates of plate tectonics, the average velocities of sinking or rising masses in the Earth, or the average zonal and sectorial velocities of large air and water masses in the atmosphere and oceans). That is, the *mean* shear rate becomes relatively constant, but the *local* flow field oscillates intermittently with large fluctuations of shear rate and temperature; cf. Shaw, 1991, Figs. 8.1 and 8.2). This is because both the shear stress and shear rate, if not arbitrarily imposed, act to mediate one another according to the thermal history and the amounts of energy dissipated in phase transformations, including melting, vaporization, and fracture (cf. Shaw, 1969, and 1987a, pp. 1368ff). In other words, *if the initial shearing stress is high* (making the shear rates, hence viscosities, the dependent variables) within some region of the Earth (e.g., for reasons of motions on a larger scale combined with a high resistance to flow, as in the solid or solid-like aggregates of heterogeneous chemical phase states that exist at high temperatures and pressures in the Earth), then the *temperatures and shear rates* begin to increase at coalescing loci of maximum dissipation rates, at first exponentially relative to the initial ambient conditions, the rock "weakens," and—following the initial "runaway" phase, which may be fast or slow compared to a given standard of explosiveness, a relativity that depends on the diffusive dissipation of heat and/or mass from the locus of maximum deformation rate—the anomalous shear stresses and the shear rates eventually fall and thereafter oscillate about average levels governed by the total available energy of the flow field (see Shaw, 1991, Figs. 8.1 and 8.2). This is a time-dependent feedback process that is characterized by sets of relaxation times, such as those discussed in Shaw (1987a, pp. 1368ff). Conversely, *if the initial rate of shear is high*, then the *temperatures and shear stresses* increase rapidly at the loci of highest initial dissipation rates—in this case almost instantaneously, because if the material is viscoelastic or relatively rigid, the stress is driven "to infinity" superexponentially, or essentially as a step function between the initial ambient condition and a "failure" condition—until the rock weakens (perhaps by melting, perhaps by fracture, perhaps by vapor-induced fluidization), and the shear stress falls to an oscillatory level commensurate with the average shear rate and energy of the flow field.

The *viscosity* at any point in the flow field (ignoring the direction-dependent properties of the general three-dimensional case) is, *by definition*, the scalar *ratio of* the local—referring to a spatially infinitesimal locus of the flow field—*shear stress* (resolved shearing force at a point in space normalized to a given unit of area) *to shear rate* (the gradient of velocity normal to the plane of shear) resolved at the same point. Thus, viscosity—contrary to the tacit assumption implied by the usual practices in fluid dynamics, and often even in rheology—is a time-dependent, energy-dependent dynamical property of the flow field, not an assigned physical parameter (consider the profound contrast between the above statement and the exercises of college physics courses where one looks up the "value" of the viscosity of a fluid in a handbook, on the tacit assumption that it can be taken as a "physical constant" of the problem; cf. Shaw, 1991, pp. 100ff). As a consequence of such dissipative-dynamical balances, the viscosity is an oscillating measure of the relative shear-stress/shear-rate history of the local flow field, hence it is a spatiotemporal variable of the flow at a point. In effect, therefore, an approximately adiabatic flow system—meaning that the system as a whole is fairly well insulated against net gains and/or losses of heat and/or mass—is one in which the relaxation oscillations of the dissipative effects (e.g., the time series of viscosity variations at characteristic locations in the flow) are analogous to a multidimensional network of nonlinearly coupled oscillators (cf. Shaw and Doherty, 1983; Kaneko, 1986; Crutchfield and Kaneko, 1987; Ling et al., 1991; Shaw, 1991, Figs. 8.1 and 8.2; Shaw and Chouet, 1991, pp. 10,203ff). Such a description is consistent with the idea that natural flows are equivalent to systems of self-organized critical-state processes which, in turn, are equivalent to multifractal systems of periodic/chaotic attractors (cf. Chap. 5; and Notes P.1, I.2, I.4, 4.1, 4.2, 6.1, 6.5, 7.3). The analysis of fracture-related fractal percolation of magma from the mantle to the surface beneath Hawaii by Shaw and Chouet (1988, 1989, 1991) has confirmed the multiphase fluid-dynamical, mechanical, and thermomechanical characteristics of magma transport in the Earth (cf. Shaw, 1969, 1980, 1991), characteristics that bear out the dynamical universalities that the work by Shaw and Chouet share, alike, with nonlinear-dynamical studies of fluid turbulence and electronic forms of transport (charge-density waves, etc.) in the solid state (cf. Bohr et al., 1984; Jensen et al., 1984, 1985; Fein et al. 1985; Gwinn and Westervelt, 1987; Olinger and Sreenivasan, 1988).

The same reasoning applies, in an inverted sense, to gaseous systems, because the temperature dependence of viscosity is generally the inverse of that in condensed systems (i.e., the viscosity of gases tends to increase with increasing temperature, hence with increasing local shear rate, rather than decreasing, as it does in most liquids and solids). Shearing flow in a gaseous system at constant shear stress will increase the local viscosity, and therefore the rate of shear and rate of dissipation in a gaseous system will progressively decrease, rather than increase as they would normally do in thermal feedback shear of a liquid or solid. Thus, the shear of a gaseous material at constant stress stabilizes — rather than destabilizes — the flow, corresponding to negative rather than positive feedback. With constant shear rate, however, the dissipation rate will progressively increase in a gaseous system, because the viscosity, hence the shear stress, continues to increase as long as the shear rate is sustained. These relationships have relevance at galactic scales — where the gaseous state prevails — hence to the deformations of molecular clouds that result in star formation, just as volcanism and coupled mantle convection are natural consequences of thermal feedback melting in the Earth, Moon, and Io (satellite of Jupiter) where condensed states (solids and liquids) are the norm.

These phenomena suggest the novel idea that the shear of a dense gas in the laboratory might be performed in such a way as to cause it to condense to liquid-like and/or solid-like states by virtue of the viscosity increase (and the implicitly enhanced energies of atomic and/or molecular interactions) — an inversion of the case where thermal-feedback deformation of the solid state leads to melting and vaporization (Shaw, 1969, 1991). If this were to happen, however, the inverse feedback effects in the condensed-state modes eventually ensue, and the system would therefore be caused to oscillate in a quasisteady manner between the gaseous and condensed states of matter. In effect, the general responses of gaseous and condensed-state systems to the same shear deformation are, figuratively speaking, mirror images of each other. Significantly, in any natural flow, where feedback tends to oscillate between the constant-stress and constant-shear-rate modes of deformation, depending on the relevant spectrum of relaxation times (Zener, 1948; Shaw, 1987a), the fluctuations of dissipative states will be similar in kind, even though they are — in the language of vibrational dynamics — essentially out of phase with each other with regard to the algebraic signs of their stress and strain-rate amplitudes (e.g., Shaw, 1991, Fig. 8.2). Such observations evidently would have important implications for flow regimes in stars and in massive planets at the interface between condensed-core regions and gaseous envelopes (cf. Note 7.1).

One need but compare the above description of rheological processes with the analogous description of biological processes, as in the discussions by Prigogine et al. (1972), or papers in Krinsky (1984), to see that dissipative structures comprise the most fundamental

Chapter Notes: Prologue

elements in the architectural plan of a self-organized universe. This is a recurring idea in the present work (e.g., Notes I.2, I.4, 4.1, 4.2, 5.1, 6.1, 6.8, and 7.1).

Note P.6: Even though it is widely known that Laplace was misled by false confidence in his expectations of mathematics, in the everyday world of research programs and NSF grants the physical sciences are usually practiced as if the Laplacian ideal still rules. Research continues to be funded as if the answer to weather prediction, earthquake prediction, volcano-eruption prediction, and even meteoroid-impact prediction is essentially a matter of obtaining a large body of descriptive data and a matching repertoire of "predictive models." This belief would appear to be based on a conviction that natural processes can be simulated to arbitrary precision by systems of partial differential equations [compare the discussions, for example, of Pachner (1993) and Lorenz (1989); cf. Note P.1], and that, given a large enough array of computers, such equations can be solved to give unique answers for given initial and boundary conditions — an outlook that is essentially unchanged from the mathematical method "invented" by Isaac Newton (i.e., the differential and integral calculus), as published in *Philosophiae Naturalis Principia Mathematica* in 1687 (Gamow, 1962, Chap. 3), more than 300 years ago! Until the time of Poincaré (1854–1912), no one seemed to be concerned about what the word "initial" really meant in such a formulation (cf. Notes P.1–P.3 and I.8), and — despite Poincaré's influence on many theoretical aspects of celestial mechanics and mathematical physics — that pernicious mindset, in disguised forms, still dominates theoretical physics and cosmology (e.g., in cosmology, theoreticians speak of "the beginning," and "the age of the universe," as if they refer to a uniquely specifiable set of initial conditions; cf. Notes I.8, 4.2 and 6.1).

Belief in a paradigm of intrinsic predictability, in the form either of numerically exact outcomes or of numerically unique long-term average outcomes — as in the axiomatic foundations of probability theory and statistical thermodynamics, representing belief in the parity of alternative states in successive flips of a coin or throws of a die, defining a standard of prediction called "the chance event" (cf. Gould, 1989c) — is traceable only in part to old habits, or to fears of losing research funds. At the root of the belief that certainty is an achievable goal in problem-solving by "the scientific method" lies a fundamental gestalt that affects all scientists to varying degrees. This is the gestalt of historical conditioning, a form of naturally selected homeostasis in which the mechanisms of cognitive change are ruled by time-delayed feedback in the expression of our subconsciously reinforced beliefs (e.g., Glen, 1991; Shaw, 1991, Sec. 8.1; 1994; cf. Note 8.15 in the present volume). At some cognitive level, most scientists still partly "live" in the time of Newton (1642–1727) or Laplace (1749–1827), even as their mechanisms of conscious cognition extol a new paradigm founded, for example, in the discoveries epitomized by Poincaré, Einstein, or Schrödinger, symbolizing, respectively, nonlinear dynamics, relativity, and quantum mechanics. During their university training, students of the physical sciences, or any other discipline, are subjected to intense "academic stress," literally representing a form of conditioned response that differs only in relative circumstances — and our interpretation of the conscious intentions of the "professors" (an archaic title that itself hints at the nature of the academic process) — from "brainwashing," a process by which the anachronisms of scientific beliefs are built into the standard curriculum, and hence into the standard mindset of scientific research (cf. Note P.3 concerning the *"conservative-Hamiltonian gestalt"*).

If such a multivalent gestalt constitutes a general attribute of the nature of "evolving thought" in science and natural philosophy, as seems evident, then it must also have affected the acknowledged giants of scientific research. Scientifically, then, we are living in a past that was living in *its* past, a dizzying thought that resembles the process of audio-visual feedback (see Note 8.15; and Crutchfield et al., 1986) and that is not unlike the recursive

Chapter Notes: Prologue

nature of evolving thought concerning, for example, definitions of *inertial reference frames*—where the Newtonian paradigm is "safely" pinned to the fixed stars, and the Einsteinian paradigm recurs within local free-falling spacetime fixed to nothing (see Sexl and Sexl, 1979, Fig. 1.2; cf. Shaw, 1994; cf. Notes 4.2, 6.1, and 8.1 herein). The conceptual hierarchies inherent within the above description—both within the thought processes of individual scientists and within the evolution of thought processes in the history of science—resemble concepts of time-delayed feedback in nonlinear dynamics, as they are related, for example, to behaviors characteristic of biological *population dynamics* (e.g., May, 1976; Hassell et al., 1991; Rey and Mackey, 1992) and *systems of coupled oscillators* (e.g., Shaw, 1987b; Freeman, 1992, pp. 467ff). A bonanza of research funding probably awaits the development of a learning paradigm—rather than a "teaching paradigm"—that will express the natural principles of nonlinear dynamics and chaos science. Sadly, however, we are already trying to make of nonlinear dynamics what we made of Newtonian physics, a set of immutable laws within which there is also a "chain of command" to be followed as blindly as an army (humans or ants; cf. Hölldobler and Wilson, 1990) follows its hierarchy of leaders. The remarks by May (1976) concerning the implications of nonlinear dynamics for science education are of even greater import now, when the paradigm is closer to being academically accepted than ever before—hence closer to being academically *taught* than ever before—than they were, perhaps, originally intended to be (an analogous comparison with religious doctrines and hierarchies is explored in Shaw, 1994; cf. Rotman, 1993a, b).

Actually, much of what falls under the rubric of nonlinear dynamics and chaos science is fundamentally simple. It is simple because it is natural to the cognitive mechanisms of organic data processing already familiar to us at the unconscious, or at least untutored, level of recognition that everyone already practices without thinking about it [for enlightening discussions of the neurophysiology of such mechanisms, see Skarda and Freeman (1987; especially the Open Peer Commentary, pp. 173–83, and Author's Response, pp. 183–92) and Freeman (1991, 1992)]. That this is so may be seen more cogently by contemplating what we really know about the older, supposedly more easily understood and clearer, ideas of classical mechanics and dynamics as they are presented academically—e.g., within a standard physics curriculum.

The popular textbooks by Halliday and Resnick, *Physics*, Parts I and II (1966, p. 32), and *Fundamentals of Physics* (1970, p. 25), define *Mechanics* as "the study of the motion of objects." They include among their examples of "motion," in this definition, the motions of any and all "objects," whether they involve the calculated paths of artillery shells, or space probes, or the tracks formed by the decay and interactions of fundamental particles in a bubble chamber. Consider carefully the implications of such a statement. Generically, the study of motion would have to involve the recognition and/or description of geometry, time, and the nature of "interactions." The geometric description of motion in time is a subdivision of Mechanics called, literally, *Kinematics*. When the motion is formally related to the action of forces associated with it, the corresponding descriptive discipline is called, also literally, *Dynamics,* another formal subdivision of Mechanics.

It is important to realize that there is a great deal of confusion already embodied in such definitions. For instance, Mechanics also is described in the same textbooks as "the oldest of the physical sciences," hence one might think that it also would be, by simple stages of ideological bifurcation, the *most general* category, the taproot, as it were, of physical studies. By that reasoning, Mechanics should include information concerning the nature of the objects themselves as well as the processes and properties of the objects' interactions with other objects. Consequently, it would follow that Mechanics implicates knowledge of such things as chemical states of matter, chemical reactions, electronic

Chapter Notes: Prologue

phenomena, and even nuclear reactions. But such a view of Mechanics definitely is not in accord with its common usage in science, where only the motions of chemically nondissipative and nonreactive discrete objects are considered. [For example, in the monumental modern works in Mechanics, the *Foundations of Mechanics* by Abraham and Marsden (1987) and *Manifolds, Tensor Analysis, and Applications* by Abraham et al. (1988), there is little or no mention of terms such as temperature, chemical state, diffusion, etc. — and *"dissipation"* is described only as a constraint on the theory of orbital motions expressed by the mathematics of Hamiltonian systems (e.g., Abraham and Marsden, 1987, pp. 234f).]

Dynamics differs from Mechanics only in being more explicit about the specification of the forces acting on the objects, hence it is (nominally) more restrictive and precise concerning the descriptions of the properties of motion. By physical implication (and the introductory remarks of Halliday and Resnick, cited above), however, both Mechanics *and* Dynamics *should* include all phenomena involved with the dissipative, hence *thermal*, properties of interactions and reactions associated with the intrinsic nature of objects, hence with the internal motions of, and interactions between, those objects. The thermal properties of objects, however, are classified as yet another discipline called *Thermodynamics*, a discipline that *in practice* is carefully removed from the "essential" problems of Mechanics and Dynamics — despite the fact that none of these disciplines can function at all without at least tacit assumptions concerning the thermal states of objects, with or without simple interactions and reactions — that are traditionally restricted to such things as rigid-body motion, elastic collisions, and so forth.

In truth, all branches of Physics inevitably implicate the nature of dissipative processes — or, stated differently, none of the formal disciplinary classes of the Physics curriculum is ultimately meaningful without the involvement of dissipative processes (see my remarks in Note P.3 concerning the educational experiences of R. M. May and R. H. G. Helleman, respectively). Yet the types of motions associated with dissipative processes (diffusion, viscosity, etc.) are relegated to a lower, or peripheral, status having to do with the subject called *Kinetics* (oddly enough, the one term that refers to the *essential* nature of time-dependent phenomena), a subdivision of both Physics and Chemistry that serves to artificially patch together the "purer" disciplines of Dynamics and Kinematics.

From the standpoint of academic classifications of subject matter, of course, all of these remarks might be viewed as semantic nitpicking. I submit, however, that they identify what has become, subliminally and habitually, a strangely distorted view of Nature as something that can be sliced up, laid out, and patched together in much the same way we have done with the scientific curriculum. In geology and geophysics, for example, this attitude is manifested in the way the concept of plate tectonics evolved. Until recently, few students of plate tectonics would, or could, distinguish between the geometric formalities of plate motions (kinematics) and the more fundamental issues involved in plate-margin evolution and internal deformations (dynamics, thermodynamics, and *rheology*, the "subdiscipline" that deals with the general principles of flow). More tellingly, at the academic level, many students were taught (in effect) that plate kinematics *was* plate dynamics, in the sense that the phenomena normally considered under the above headings of mechanics, dynamics, kinetics, and thermodynamics — especially with respect to the rheology and scaling of "fast" or "slow" (even if geologically "fast") deformations — became secondary or peripheral to the ruling kinematic model. In plate-tectonic thinking there has thus evolved what I would call *the flat-world effect*, meaning an ingrained response to seeing the world in a particular way, such that the habit, or paradigm, can be broken only through the intervention of a greater revolution and/or transformation of thought — a transformation that comes only after a period of ideological stasis that, from the nonlinear-dynamical viewpoint, is described by a form of conceptual *chaotic intermittency* wherein the intermis-

Chapter Notes: Prologue

sions are of unpredictable durations, during which everyone reinforces, by rote, the same rites of initiation (cf. Note P.7; and Chap. 5, Note 5.1).

The concept of force requires the specification of units of mass and acceleration as well as their distributions and directions of action in space and time. In order to be described completely, therefore, the mass-acceleration components of a system require vectorial representation. And when interactions are involved, a more general accounting is required in the form of *tensor* mathematics, meaning a mathematics capable of expressing the diversity of vectorial propagations, as well as vectorial properties and vectorial components of the processes of interactions in as many "dimensions" as may be required (e.g., the scalar and vectorial consequences of collisions between objects, "frictional" tractions, and other effects of direct and indirect influences — including the phenomena of vibratory and oscillatory synchronizations and nonlinear-resonance phenomena that are the subject of this book). Any real interaction whatsoever within this tensorial description changes it to yet a different, and sometimes a drastically different, system of vector fields — metaphorically in the same sense that a slight touch can change the configuration of an artificial edifice of matchsticks, or a slight deflection of magma transport from a mantle source can change the configuration of volcanic edifices and volcanic propagation at the Earth's surface — by transformations that are traceable to the action of dissipative processes involving the most innocuous of phenomena, phenomena analogous to those by which the motion of a butterfly's wing in Bombay can eventually influence the weather in London (e.g., see the opening chapter in Gleick, 1987).

Ironically, however, the very complexity of such problems can produce very simple types of behavior (*attractors* of diverse types; cf. Note P.3) compared to what we would have been led to believe from the disciplinary studies described above. A central discovery in nonlinear dynamics is that both qualitative and quantitative statements often can be made about complex systems — systems that are intractable by any of the standard methods of physics — using certain characteristic and global properties of nonlinear regimes already referred to in the Prologue by the term *universality* (see Notes P.2–P.4). A token sampling from the theoretical and descriptive literature of attractors, hence referring to the emergence of simplicity from complexity — arbitrarily chosen from the hundreds of references cited in the present volume that describe the application of nonlinear dynamics to diverse disciplinary fields of study — and with emphasis on concepts of nonlinear-dynamical synchronization and resonance, is given by the following selected list of references (and by the precedents they cite): Treve (1978), Shaw and Doherty (1983), Gu et al. (1984), Kaneko (1986, 1990), Crutchfield and Kaneko (1987), Meiss (1987), Shaw (1987a, b; 1988a, Figs. 17–19; 1988b, Figs. 3–11; and 1991, exemplifying attractors in diverse contexts), Schmidt (1987), Smith and Spiegel (1987), Blekhman (1988), Brown and Chua (1991), Chouet and Shaw (1991), Freeman (1991, 1992, pp. 467ff), Kapitaniak (1991), Murali and Lakshmanan (1991, 1992), Pecora and Carroll (1991), Shaw and Chouet (1991), Anishchenko et al. (1992), Carroll and Pecora (1992), De Sousa Vieira et al. (1992), Garfinkel et al. (1992), Moon (1992), Rul'kov et al. (1992), and Willeboordse (1992).

According to the definitions of either Mechanics or Dynamics found in standard college textbooks, one would conclude that, in natural systems, we are faced with problems of transcendent complexity. In these natural systems, the number of dimensions alluded to above is conventionally assumed to be "infinite." Before nonlinear dynamics entered the picture, mechanics, dynamics, and thermodynamics had, by tradition, come to be associated with, or restricted to, problems that could be cast into mathematically exact and analytically soluble formats. It was therefore only a matter of habit, or gestalt (Glen, 1991), that our conceptions of real and tangible problems in physics, or "allowable" problems in physics, were restricted largely to the criteria of the *mathematically soluble formulations* of carefully

Chapter Notes: Prologue

selected problems in physics. Physics—as it was commonly perceived—dealt only in terms of "existence theorems" concerning classes of continuous functions (*differentiable functions*), and, among such functions, it dealt only with those histories of motion that could be described by functions that were amenable to solution by analytical or numerical methods (*integrable functions*). This meant that classical Mechanics, Dynamics, and Thermodynamics were restricted to a minuscule subset of real physical problems, most of which involve discontinuities, singularities, or other mathematical peculiarities of nondifferentiable and/or nonintegrable forms (cf. Helleman, 1980). In the past, the latter problems were often referred to as *pathological*, conveying strong subliminal if not overt messages of disapproval—e.g., by the reigning role models in science (often represented by graduate thesis advisors)—as valid subjects of research (cf. May, 1976; Mandelbrot, 1977, pp. 3ff; 1983, pp. 3ff).

The tacit separation of Dynamics into "classical dynamics" and "nonlinear dynamics" perpetuates a misconception concerning both Mechanics and Dynamics as branches of physics. "Nonlinear dynamics" is really just Dynamics—and actually, if we accept the Halliday-Resnick textbooks as a sample of the way physics has been taught in colleges and universities up to a decade or so ago, it is really just an aspect of Mechanics. Unfortunately, the frequencies of usages ("conditioned responses") in the nomenclature of science are dominating influences on the practices and gestalts of science, just as they are in every other area of human experience—including beer commercials. In the analog and/or digital algorithms prevalent in computer-experimental nonlinear dynamics—often just called *numerical dynamics*—the notion of "force" is not present at all, or is left to be implied, either by parallel reasoning or by analogy with mathematical functions used to describe the action of forces in classical dynamics. Thus, there is a lot of circular reasoning—and a "Catch-22" here and there—in the nature of dynamical studies. Like it or not, circular reasoning, dilemmas, and paradox (e.g., in the sense of multiple reentrant signaling, sine-circle functions, indeterminacy, and so on) constitute the nonlinear information-processing mechanisms of the brain-mind plexus (cf. Edelman, 1978, 1985, 1987, 1989; Skarda and Freeman, 1987; F. D. Abraham, 1990; Freeman, 1991, 1992; Abraham et al., 1991b; Shaw, 1987b, 1988c, 1994; and Note 4.2 of the present volume concerning the analogous properties of the universe at large). In numerical dynamics there seems to be a tacit agreement, however unspoken, that in the exploration of the nonconventional properties of mathematical forms and equations, the experimenter is investigating regions of mathematical "space" that, at the very least, may be relevant to regions of physically realizable space—regions of space that have never been explored because they were ruled out-of-bounds by the expediencies of historical mathematical-physical acceptance, and/or by academic authoritarianism.

During the late 1960's I was often told by reputable geophysicists that my descriptions of "thermal feedback" ("thermal runaway"; cf. Gruntfest, 1963; Shaw, 1969; Anderson and Perkins, 1975; Nelson, 1981) were not valid because the exponential equations of positive feedback were "numerically unstable," hence would "blow up," apparently implying, to their minds, that the physical instabilities ("blowups") of my experimental demonstrations were no better than computational artifacts (see the examples of experiments I performed in the 1960's, published in Shaw, 1991, Figs. 8.1–8.9; cf. Note P.1). I was not aware at that time of the work by Lorenz (1963, 1964), or that my demonstration, and Gruntfest's before me, of exponential dependence on local conditions—representing the feedback between temperature, viscosity (dissipation), and shearing stress in a material body (such as the Earth's mantle)—was essentially equivalent to Lorenz's demonstration of the now well-known principle of *sensitive dependence on initial conditions*. Furthermore, the applications of the method of computational scaling of feedback instabilities to physical problems

Chapter Notes: Prologue

by Gruntfest and by me, individually and in collaboration (e.g., Gruntfest, 1963; Gruntfest et al., 1964; Shaw, 1969, 1973, 1987a, 1991; Gruntfest and Shaw, 1974; Wones and Shaw, 1975), were directly analogous to the strategies of computer experimentation that have become a major part of the nonlinear-dynamical method, especially in the exploration of *metric universalities* (the scaling parameters of universal self-similarity; see Note P.4). Even then, however, I was convinced that at the most fundamental level of physical reality there was really no operative distinction between "blowups" in digital computers and "blowups" in physical phenomenologies, including those of economics, psychology, and social change ("wars" in one form or another).

Digital-computer, analog-computer, and physical experimentation truly represent "the wild blue yonder" of the research frontier in nonlinear dynamics, especially in the scope of what-there-is-to-be-discovered relative to what-we-have-been-taught-is-already-"nailed-down" by conventional mathematics and physics. The "mystery" and "incomprehensibility" of nonlinear dynamics that many of its detractors appeal to as criticisms of its applicability to "real" and/or "practical" problems (see, for example, the Open Peer Commentary, pp. 173–83, on the target article by Skarda and Freeman, 1987, "How Brains Make Chaos in Order to Make Sense of the World") are simply admissions that the realm of the unknown remains as fearsome to them as was the realm of the "pathological function" and the "mathematical monster" to traditional mathematical authorities. Mandelbrot (1977, 1983) put the lie to the "mathematical witchcraft" implied by such terms, and he has shown that the "pathological function," from the mathematical standpoint, is as normal and ubiquitous as the geometry of running, jumping, and breathing in our everyday world. Unfortunately, the phrase "fractal geometry," which Mandelbrot coined to describe unconventional geometric forms, has, to some, itself become the "mathematical monster" that it would expose to the light of familiarity. [To get a hint of the "real-time" nature of fractal complexity, and the geometry of *strange attractors* (see Note P.3), imagine a system of trajectories traced out by the motions of the players during a game of basketball. Within the complexity are "rest states" (tip-offs, inbound plays, time-outs, and times when free-throws are about to be attempted) that are analogous to static fixed-point attractors. The rest of the time there is a swirling, sometimes bursting (e.g., "fast breaks"), sometimes intermittently double-helical (e.g., alternate setups of the "half-court game" at both ends, mimicking, as it were, aspects of the organismal structure), back-and-forth oscillatory movement between the backboards that makes up the net pulsatory fabric woven by the intertwining paths of the players — e.g., as is displayed dramatically on a TV screen by a videotape of a game that is viewed at fast-forward — a motion that is analogous to the particle trajectories that describe the motions of a bistable oscillator (see Shaw, 1991, Figs. 8.13–8.24), or a pattern of coupled fluid convection analogous to a doubly inverted Lorenz attractor (cf. Sparrow, 1982, Fig. 1.1; Shaw, 1991, Fig. 8.24; see also Figs 15b, 17a, 22, 29, and 32, and Notes P.3 and 8.4 herein).]

The surprises and revelations of nonlinear dynamics are often but the recognition that the "pathological" at one spatiotemporal scale of physical behavior is synonymous with intrinsic, autonomic, or involuntary understanding at other levels of cognition. "Understanding" of this type is contextual, hence differs in nature from formal logic in which one proceeds according to a sequence of demonstrations (logical closures) to prove that a theorem is both complete within itself and "true," meaning that the hierarchy of closures at each step of the logic also closes "globally" to affirm the statement made by the theorem (see Nagel and Newman, 1958). The real theorems of life, the nonlinear-dynamical ones, do not fit a system of logical closures — except in the sense of belonging to a self-organized and endless-beginningless Escher-like Strange Hierarchy within a self-referential "voidism" (cf. Spencer-Brown, 1972; Hofstadter, 1979; Rotman, 1993a, b). Apparently there are still

Chapter Notes: Prologue

some mathematicians who hold to a tradition in which the cycles of formal logic are the only acceptable demonstrations of communication and "understanding." I do not deny anyone the right to such a conviction, nor do I deny a poet the right to view haiku as the ultimate form of "logic" (cf. Sewell, 1951; Shaw, 1994). In the mathematical sense of self-reference, however, such "definitions" can live only within a nonlinear body-mind context that itself ultimately "decides" what form of communication constitutes "understanding." In that sense we really have no choice about the mechanisms of meaning, even as we draw lines and boxes and laws about our perceptions — or inventions — of "truth" (see Spencer-Brown's, 1972, *Laws of Form*). Statements of preference in the forms of communication are but attempts to fragment and separate "truth" from contextual reality arbitrarily (contrast the example of "mathematical understanding" cited by Ruelle, 1991, p. 3, with the ability "to experience the world clearly" in the symbolic calculus of Spencer-Brown, 1972, p. 101).

Through the auspices of dissipative-systems theories and demonstrations of discretely ordered attractors and fractally ordered strange attractors, the realms of dynamics and thermodynamics, which heretofore have been artificially held as separate disciplines, can now be seen to merge within an encompassing view of mechanics (e.g., see Shaw, 1991, Secs. 8.1–8.3). Nonlinear dynamics is but a more inclusive member of the set that includes kinematics, classical dynamics, thermodynamics, and kinetics (hence also rheology). An attempt to impose a rigid terminology on such a transformed view of physics and mathematics is fruitless, because it will but reiterate the arbitrary constructions of the past that we are only now beginning to see for what they are. In my estimation, the history of science will show that the attempt to impose disciplinary boundaries in science ultimately defeats its own purpose, assuming that the intention of science is "to experience the world clearly" rather than to become its nemesis. Ironically, when the reverence for logical "rigor" is overcome — a tall order, I admit, for anyone trained in its observance — I have inevitably found that everyone already knows a great deal about nonlinear dynamics, hence about the role of chaos in expressing the evolution of simplicity *as* complexity within that context (cf. Note P.4).

Note P.7: Ignorance of mathematics certainly did not hinder Michael Faraday (1791–1867), who, some believe, made a greater impact on modern physics, chemistry, and technology — perhaps even mathematics itself — than any other nineteenth-century scientist. In the book *Faraday as Discoverer*, published a year after Faraday's death, "Professor Tyndall" — see Faraday (1920, p. 14); I infer, because no first name is mentioned, in the British tradition of assuming that everyone knows who is who, that this citation refers to John Tyndall (1820–1893), a British physicist noted for his study of the scattering of light, the *Tyndall effect*, as related to why the daytime sky appears to be blue; cf. Halliday and Resnick (1966, p. 1168f); Besançon (1990, p. 658f) — called Faraday "the greatest experimental philosopher the world has ever seen." I do not disagree, and I have made analogous remarks to colleagues many times during my career, concerning Faraday's precedent in establishing the value of the experimental-experiential method of enquiry as the essential paradigm of scientific research — this being independent of particular disciplines, such as mathematics, or of particular doctrines of science, such as Newton's laws, Gauss's laws, and so on.

It is widely acknowledged not only that Faraday was untutored in mathematics, but that he apparently had little aptitude for *mathematical manipulations* [i.e., referring to the methodology of mathematics that is rooted in the formal axiomatic development of closed logical loops of numerical-algebraic-symbolic proof as the sole arbiter of valid thought, a facility that most persons, even in the field of nonlinear dynamics, apparently assume is the

Chapter Notes: Prologue

rightful role played by mathematics as the absolute, and only, criterion available to us for the demonstration of "truth" — noting that, even within mathematics, there is no consensus on such matters; cf. Post (1965), M. Davis (1965), Spencer-Brown (1972), van Dalen (1981), Rössler (1986), Ruelle (1991)]. It is my belief, on the contrary, whether or not such an aptitude was absent in Faraday, or — what pragmatically amounts to the same thing — was not consciously cultivated by him, that Faraday's curiosities about, enquiries about, and general views about natural phenomena had early outstripped the realms in which the practices of mathematics (as they would have been known to him during his eighteenth- and nineteenth-century formative years) could offer any advantage in the ranging of his mind. Even had he been so inclined, the pursuit of an essentially natural or *intuitional* form of mathematical expression probably would have been — as it later became for L. E. J. Brouwer (1881–1967), the founder of modern *mathematical intuitionism* (cf. Kramer, 1970, pp. 630 and 692; van Dalen, 1981) — so totally absorbing that it would have denied us the fruits of his more practical philosophy (cf. the inverted world of "the brain-in-a-box" described by Shafto, 1985b). In other words, Faraday represented a paradigm, an experiential worldview that indicated a reality of "forms within the void" (Spencer-Brown, 1972) that went far beyond any of the mathematical practices of his day. The practical importance of such an expanded view of mathematics is only now, during the middle to late twentieth century, coming to the forefront of mathematical enquiries that have been forced upon us, as it were, by the — at first sight only technologic — challenges of computability and the related questions concerning *decidable* and/or *undecidable propositions* (in the preceding phrase, "and/or" indicates a self-referential relativity of logical closures that computer experimentation, especially now in nonlinear dynamics, demands we contemplate; cf. M. Davis, 1965; Garey and Johnson, 1979; Hodges, 1983, pp. 90ff, 107, 111ff; Bolter, 1984, pp. 70f; Casti, 1990, pp. 329ff and 367ff; Ruelle, 1991).

Karl Friedrich Gauss (1777–1855) would already have established himself as a preeminent authority in mathematics by the time Faraday came of age, and — judging from the fact that James Clerk Maxwell (1831–1879), roughly two generations Gauss's junior, subsumed both Faraday's experimental observations and Gauss's laws of electrical and magnetic fields into a universal mathematical description of electromagnetism now called "Maxwell's Equations" (e.g., Halliday and Resnick, 1966, pp. 647ff, and Table 38-3, p. 964), much as Newton had done with Galileo's and Kepler's observations in formulating his law of gravitation ("laws of motion") — it would appear that the more purely mathematical interest, as displayed by Gauss and the like, was to establish an artificial body of law ("rules") rather than to explore the mathematically inaccessible realms of natural phenomena which, according to the viewpoint expressed here, represent the context of scientific law (cf. Shaw, 1988c, 1994). A case in point is Dauben's (1983) biographical-mathematical description of the difficulties encountered by Georg Cantor (1845–1918) in pursuing *transfinite set theory*, as it came to be known, in the face of the mathematical authoritarians of his time (cf. ibid, p. 125, concerning a dictum by Gauss regarding the impossibility of "completed infinities"; and pp. 122 and 127 concerning personal attacks on Cantor's intentions by Leopold Kronecker, an influential mathematics professor, one-time teacher of Cantor's, and journal editor who called Cantor a charlatan — and worse, in at least one instance doing what he could to block publication of Cantor's work; cf. Kramer, 1970, p. 577). The idea of *completed infinities* is illustrated numerically by Dauben (1983, pp. 125ff), and demonstrations of that idea in terms of denumerable sets and hierarchies of infinities — the essential meaning of the word *transfinite* in the number-theoretic sense developed by Cantor — are given almost trivially by the simplest of nonlinear-dynamical bifurcation sequences (e.g., May, 1976; Feigenbaum, 1980; cf. Hofstadter, 1981; Shaw, 1987b, cover illus., and Fig. 2; Shaw, 1991, Figs. 8.24–8.35). That is to say, the *period-*

Chapter Notes: Prologue

doubling cascade is an infinite set, and therefore every odd period in the hierarchy of periodic windows of the chaotic regimes of both quadratic maps and circle maps represents a different period-doubling cascade, hence a different infinite set (e.g., Shaw, 1991, Figs. 8.33–8.35). Noting that a doubling cascade is a geometric progression with the numerical base 2 (i.e., 2^n, where n represents an infinite series of integers), then there are, in general, other progressions (infinite hierarchies) of *tripling cascades, quadrupling cascades, quintupling cascades*, etc., ad infinitum, as I have shown to be the likely case in *ring-vortex bifurcations of strange-attractor fields* (Shaw, 1986; cf. Note 8.15; and Bradley, 1965, 1969; Chen and Chang, 1972; Lim and Nickels, 1992).

Ironically, even Poincaré (Dauben, 1983, p. 122) is alleged to have condemned the theory of transfinite numbers as "a disease." In a sense, perhaps, he was right, in that Cantor lived at that "edge of mind" [referring to a sort of chaotic dichotomy of mental-numerical organization, following the ideas of Sewell (1950) on the "numerical structure" of language and the nature of poetical communication (cf. Hutchinson, 1953; Shaw, 1994)] where convention finds it easy to label mental states as "psychotic," because Cantor succumbed from time to time to what is now called the manic-depressive psychosis. Such a "diagnosis" is hardly surprising in view of the near-metaphysical quality of headiness in the intellectual concepts Cantor pursued, which stood in stark contrast with the negative criticism, character assassination, and often abusive attitudes ("props-knocked-out" states) displayed by his academic peers and even more so by the politico-scientific authorities who ruled over the vehicles of his self-expression (refereed journals; cf. Glen, 1989). Cantor worked under conditions of psychological stress that many a modern psychotherapist probably would consider to be untenable if he were their "client"—putting it mildly—a stress that he managed to withstand, even if with chaotic intermittency, over nearly three decades of besieged research (note the nonlinear implication for applied psychology; cf. F. D. Abraham, 1990; and Chap. 5 and Note 5.1 in the present volume). Ambivalent and/or ambiguous viewpoints are held even today concerning the significance of Cantor's work (cf. Mandelbrot, 1977, 1983; Morrison, 1988; Shaw, 1988c). The irony of Poincaré's statement is that he was himself, at least marginally, afflicted with the same "disease" in recognizing the implicitly uncomputable realities with which nonlinear dynamics is faced (cf. Note P.2). Cantor's research obviously intersected the notions of mathematical incompleteness and self-reference (see both the Prologue and Introduction to the present volume), issues that remain at the edge of the mathematical vis-à-vis the universal conceptions of nature, conceptions with which Faraday was apparently more familiar (and presumably more comfortable) than were the mathematicians of his day.

Atkins (1991) employs the heuristic technique of comparing Faraday's "Six Christmas Lectures for Young People"—later transcribed and published by Sir William Crookes under the title *Faraday's Lectures on a Candle*, and in a later edition as *The Chemical History of a Candle* (Faraday, 1920)—with our present-day technologic understanding of chemical reactions at the levels of molecular and atomic structures and transformations, down to the space-time scale of *femtochemical reactions* (reactions involving length scales measured over femtosecond periods of a laser pump/probe pulse, where 1 femtosecond = 10^{-15} sec, equivalent to a light path-length of ~ 0.0003 mm; ibid., p. 126f; cf. Wildi, 1991; and Note 6.1). Together, Atkins's and Faraday's expositions span a great range of physicochemical phenomena, including the modern arena of *coupled chemical oscillations* and *chemical chaos* (Atkins, 1991, Chap. 6; cf. Simoyi et al., 1982; Swinney and Roux, 1984; Argoul et al., 1987; Tam, 1990; Scott, 1991). Although Faraday's emphasis was on what the candle could teach his audience—composed of the young people, and "enquiring minds" of any age, who came to hear him talk, and for whom the descriptions and analogies were given in simple terms, and with simple graphics, not unlike his scientific writings—concerning the variety of phenomenologies attending combustion, and chemical reactions in general, there was implicit in his discussion all of the regimes of the nonlinear-dynamical motions of the candle flame itself.

If one were to devise an experimental study of the nonlinear-dynamical motions of a candle-flame—e.g., in a manner analogous to that by which Robert Shaw quantitatively described the complex intermittency of the common dripping faucet (R. Shaw, 1984; cf. Wu and Schelly, 1989; Cahalan et al., 1990)—then the equally common experience of the observation of candle flames would suggest, just as it did with regard to the dripping of faucets long before any quantitative data were collected (who has not been mesmerized by the motions of a candle flame, during power blackouts, or other circumstances, just as who has not been driven wacky by the dripping of a leaky faucet, or worse, leaky roof?), that the oscillatory motion of the flame could be *tuned* (Note 3.1) to reveal a *universal spectrum of dynamical regimes* (see Notes P.4, I.4, I.8). For instance, by controlling the ratio of the volumetric flow of air past the candle wick relative to the inflammability of the wax and the capillary-feed-rate of volatile liquid wax through the wick to the combustion zone (e.g., Faraday, 1920, Figs. 3 and 4), it seems likely that one could demonstrate virtually all of the types of behaviors characteristic of the so-called *period-doubling route to chaos* (e.g., May, 1976; Feigenbaum, 1980; Hofstadter, 1981; R. Shaw, 1981; Ott, 1981; cf. Note P.1), as well as the so-called *quasiperiodic route to chaos* that also demonstrates *transitions to chaos by interactions of resonances in dissipative systems* (cf. Jensen et al., 1984; Crutchfield, 1986, p. 57; Shaw, 1987b, cover illustration). The latter route to chaos is the one that plays such a prominent role in the present volume relative to the structure of the Solar System and the orbital evolution of asteroids, comets, and meteoritic objects (cf. Chap. 6, and Note 6.1). Different aspects of these nonlinear-dynamical behaviors comprise what are now referred to as systems of *fixed-point attractors*, *strange attractors*, and *chaotic attractors* in general (cf. Notes P.3, P.4, and 3.1).

Although I have made no search of Faraday's published works on this point, it is my guess that Faraday probably would have noted the variety of relative oscillations of the candle flame under different conditions—perhaps in unpublished laboratory notes, if not in published works—a guess that is supported by the observations he made (Faraday, 1920, pp. 42f) about the more general context of distributed sources and dendritic bifurcations of complex flame structures. Thus, Faraday's studies of the candle flame—had he taken the time to develop them with the experimental detail he applied to his studies of electricity and magnetism—could well have been the prototype for present-day studies in the fields of coupled mechanical oscillators and coupled chemical oscillators, as well as a prototype (among many other experimental studies, such as Rumford's experiments in the cannon-boring mill; cf. Sandfort, 1962; Shaw, 1991) for the study of dissipative structures in nature (e.g., Prigogine et al., 1972; Prigogine, 1978; Krinsky, 1984; Prigogine and Stengers, 1984). Faraday's study of the candle certainly stands as one of the prototypes of what I have called "folded feedback" (Shaw, 1987a, Fig. 51.14), invoking recurrently coupled positive-negative feedback cycles akin to iterations of a function with one or more maxima. Physically, the positive-feedback modes of such cyclic processes constitute what I call *thermal-feedback* or *thermal-runaway* phenomena representing, in general, *explosive chemical reactions* (cf. Notes P.5 and 1.2; and Gruntfest, 1963; Gruntfest et al., 1964; Gruntfest and Shaw, 1974; Shaw, 1969, 1991; Wones and Shaw, 1975; Anderson and Perkins, 1975; Nelson, 1981). In all of these respects, and others, Faraday's study of the humble candle continues to enlighten us, even with regard to the celestial-dynamical phenomena that are emphasized in the present volume.

Chapter Notes: Introduction

Introduction

Note I.1: A qualitative notion put forward by Rawal (1991, 1992) holds that at the time of their formation, each of the planets was surrounded by an "Oort Cloud" of comet-like material left over from the nebular stages of organization as a form of self-similar subset of the Solar System Oort Cloud (Oort, 1950; Oort and Schmidt, 1951). The fate of such "clouds" — in Rawal's qualitative model — was eventually caught up in all of those processes that gave rise to the present-day systems of satellites of the planets, during which time the original material was redistributed. It is difficult to devise a test for such a proposal, barring distinctive original chemistries and the ability to sample all present-day fractions of interplanetary dust-size objects. Such a possibility seems doomed from the outset, if for no other reason than the implication of *sensitive dependence on initial conditions,* and the consequent mixing effects of chaotic dynamics (cf. Crutchfield et al., 1986; Ottino et al., 1992). The implications of chaotic dynamics for the chemical heterogeneity of the cosmos — and the lack of any absolute cosmological provenance by which we can define a characteristic chemistry (e.g., the model of the early Solar System presented in Chap. 3 of the present volume) — is brought home by a recent study of the chemical histories of carbonaceous chondrites by Rotaru et al. (1992), in which our very standard of "average cosmic abundances" (Sears and Dodd, 1988, Table 1.1.2) seems to represent a heterogeneous mix of source compositions (cf. Chapman, 1986; Kerridge and Matthews, 1988; Wood, 1988; Chapman et al., 1989; Bell, 1989; Sears et al., 1992). Fink (1992) has reported the discovery of a new class of carbon-poor comets that differs from known comet compositions and adds weight to the evidence of chemical heterogeneity in Solar System and/or Galactic source materials (cf. Owen et al., 1991; Farley and Craig, 1992; Farley and Poreda, 1993). Adding to the confusion, there is now an interesting diversity of data and interpretations on occurrences of diamonds in meteorites, at the K/T boundary, and in the interstellar medium, some evidently formed by shock compression while others are apparently formed metastably (or even stably?; cf. Nuth, 1988, p. 986) by condensation, perhaps within stellar outflows (e.g., Carlisle, 1992; Gilmour et al., 1992; Russell et al., 1992; Wright, 1992; cf. Anders, 1988; Anders and Kerridge, 1988). If the hypothesis I advance in Chapter 3, and elaborate in Notes 3.2 and 9.2, is valid in principle — where a pre-main-sequence binary stage of the Solar nebula evolved through a T-Tauri star stage of the FU-Orionis type, with violent disruption of the binary companion accompanying and/or following planetesimal formation — a mixed signature of Solar and/or Galactic compositions might be expected in "random" samples of either condensation-type or shock-type diamonds, a possibility that would be in keeping with the other findings of heterogeneous sources for the chemistry of asteroids and comets cited above. (The complexity of coupled chemical and dynamical processes in the early Solar System is a recurring theme of many of the fifty chapters of the nearly 1300-page book *Meteorites of the Early Solar System* edited by John F. Kerridge and Mildred S. Matthews, 1988; an interesting historical account of meteorite studies is given by Burke, 1986.)

The hypothesis offered in the present volume concerning meteoritic source reservoirs in the Solar System differs from that of Rawal (1991, 1992) in several respects, and especially in terms of the dynamical processes of relative "delivery," or kinematic partitioning, of small cometary and asteroidal objects, and any other forms of objects that may be relics of earlier stages of the solar nebula or that may come from surrounding regions of the Galaxy (see Chaps. 3 and 4, and Figs. 11 and 13). A dynamical model of one form of cometary history that originates in the vicinity of the outermost planets, as developed by Duncan et al. (1987), has some resemblance to Rawal's (1991) proposal, as does Stern's

Chapter Notes: Introduction

(1991) concept of relict planetesimals (cf. Weidenschilling, 1991; Kerr, 1992e; Cowen, 1992). See also Notes 3.2 and 9.2.

Note I.2: The theory of *algorithmic complexity* represents a systematic mathematical basis for the description and quantification of the degrees of randomness that can be expressed in terms of numerical algorithms (e.g., Chaitin, 1979, 1982, 1987a, b, 1988; Ruelle, 1991). This approach to the characterization of the random state reveals at least two paradoxes that are intrinsic to our attempts to define the complexity of a set of data. One form of paradox concerns distinctions between the analytical vis-à-vis the synthetical points of view. A second form of paradox concerns the self-referential nature of the universe and of ourselves vis-à-vis the meaning of "effort."

Numerical sets that are constructed from measurements of natural data pose problems that are *analytical* in character, in the sense that we would attempt to "decode" the data, or "compress" the number of statements required to describe every detail of a data set. In this context, according to measures of algorithmic complexity, any set that cannot be described by fewer bits of information than it takes to describe the set in its entirety is a "random" set. This statement, however, does not preclude the possibility that an undiscovered algorithm might exist which would greatly "compress" that description in terms of some statement requiring fewer bits of information than the total. Such a view of randomness is useful, but, like the more traditional methods of statistical analysis, it leaves us with no option — in the absence of a new discovery of a simplifying algorithm — other than the conclusion that all *seemingly* incompressible data sets are maximally complex, or random. According to the *axiomatic principle of the ubiquity of nonlinear interactions and dissipative systems in the universe*, invoked in this volume (cf. Note P.3), however, all natural systems are inherently compressible according to the subsets of numerical structures intrinsic to the nature of, and universality parameters of, chaotic dynamics (cf. Note P.4). I will call this the *analytical* form of the algorithmic paradox in theories of complexity.

Taking a different tack, it is also instructive to look at concepts of complexity from the constructional or *synthetic* viewpoint. According to this alternative, it is evident that our perspective suddenly has changed, or *has appeared to change*, from one of "code breaking" to one of "code making." In the context of synthesis, then, complexity is seen to be a measure (or "test") of how *difficult* it is to make something that can be called "complex." Ruelle (1991, p. 136) states the situation in these words: *"An entity is complex if it embodies information that is hard to get."* For instance, it has taken a lot of effort to build and program rockets and space probes to do our bidding in the exploration of the Solar System, hence these are complex objects and systems of objects. But the problem has not been intractable — in hindsight — because such things now exist. We can conclude, therefore, simply by the expedient of making dynamical comparisons, that the motions of the planets and satellites of the Solar System must be more complex than our systems of space probes, because even our best computational programs have great difficulty with any but the simpler forms of orbital motions and orbital transfers (cf. Friesen et al., 1992; Belbruno, 1992). This realization is a far cry from the thinking of only a few decades ago, because back then we thought that the system of planets, and many of the satellites of the planets, were so perfectly cyclical and regular in their motions (in the sense of Kepler and Newton) that their orbits were almost perfectly redundant (minimally complex, or maximally simple). However, now that the methodologies of nonlinear dynamics are being actively applied to problems in celestial mechanics, the motions of the planets are being shown to be so complex that their orbits are, in fact, uncomputable (they are chaotic, but they are not random; see comments below, and Chap. 3) — and this means that they are of a different, and

Chapter Notes: Introduction

higher, order of complexity relative to the system of space probes that have been built to study them. The synthetic study of planetary motions now must be placed within those classes of complex problems carrying stigmatic labels such as "hard problems," "incomplete problems," and/or "intractable problems" (cf. Ruelle, 1991, p. 140).

Analogously, synthetic algorithms of living-systems theories are even more complex, where "more" now can be seen to mean something quite different and beyond the usual implication that an unspecified but finite additional effort will get the job done — even if the brainpower and computer power made available to the problem were to greatly exceed "star-wars" proportions (cf. Farr and Goodfellow, 1992). So, here is the *analytical vis-à-vis synthetical paradox* in a nutshell (or in a cell-shell): (1) our current understanding of the biogenetic code, and of the diversity in the replicative machinery of the living cell (e.g., Mandel et al., 1992), represent huge compressions of information relative to the descriptive and *analytic* capabilities of pre-1953 biology (1953 being the publication date of the Watson-Crick determination of the DNA structure; cf. Judson, 1979); but, (2) turning the problem around — replacing the analytical question with the synthetical question ("flipping the coin of comprehension," so to speak) — no amount of effort has been or ever will be sufficient to design and build a living system that works in exactly the same way that the organic cells of our bodies work (the question of whether or not it is possible to build a living system of *any* kind involves the deeper — though contextually analogous — question of what the word "living" means; cf. Schrödinger, 1992, a reprinting of his provocative book *What is Life?*, first printed in 1944; also see the discussions of living-systems theories from the standpoint of inorganic constructions, sometimes referred to as "the mineral origins of life," as exemplified by Cairns-Smith, 1985; cf. Miller, 1978; Shaw, 1986; Chyba and Sagan, 1992). The experiences that have accrued from the studies of the much less intricate, but also chaotic, planetary system would strongly, if not definitively, argue — by simple nonlinear-dynamical "comparative anatomy" — that such a goal of genetic "self-synthesis" must be impossible, meaning that it is even more intractably intractable than the mentioned problems of orbital dynamics (cf. Note 8.15). We have no way of knowing, much less controlling, how all of the requisite trajectories of motion of a living human cell are to be put together "starting from scratch." (Notions of "cloning" and "genetic engineering" are different ideas altogether, because they involve — when reduced to their bottom-line essentials — simply the redundant operations of copying that which has already been built, whatever rearrangements of the already-living parts, or arrangements and rearrangements of other constructs copied after some aspect or other of the already-living parts, might be invoked in the manufacturing context.)

From the synthetical point of view, truly intractable problems elude us faster and faster the harder and faster we pursue them, a statement that fits with the ambiguities encountered in studies of the "size" and "age" of the universe (e.g., de Vaucouleurs, 1970; van den Bergh, 1992; cf. Note 4.2). In that context, we place ourselves in the position of being the universe in pursuit of itself. And if the universe has arranged itself (cybernetically) in such a way that "the game is not up" just because it has "designed," in the sense of self-organization, a better computer (e.g., represented by "us" ± our engineering technologies), then it always must recede from itself faster than it can synthetically re-create itself. Otherwise the universe would be absolutely convergent, hence could ultimately define itself (in the mathematical sense), "end," or "halt" — and that is informationally no different than if it had never been. It should be noticed, in an analogous sense, that conventional demonstrations of random processes "in nature" involve some form of contrived demonstration. The kinetic theory of gases relies on *the existence and definition of a geometrically regular container*, hence on the specification of symmetrically ordered boundary conditions. The theory cannot be formulated in an *open* universe, because in that case self-organization and clustering of states is the expectable outcome (expectable because that is what we see, and what we are). Analogously, the rolling of dice and the flipping of coins depend on the specification of a special geometry that permits the relevant observations to be made—as does the operation of a computer in simulating any "random" process.

The fact that one can make a series of observations of a regular process arbitrarily more complex than a given initial state is not sufficient grounds to assert that the random state is representative of natural processes. In other words, *the only way that one can quantify "randomness" (maximum relative complexity) is first to impose order on the mechanisms and/or processes being observed.* In applying such ideas to such things as subatomic-particle dynamics (including atomic "clocks," etc.), it should be evident that in order to define the initial state of an ordered set of observations, we would immediately encounter the uncertainty principle—and therefore the accepted criterion of the random process is not testable, because there are no absolutely determined particle states (absolutely error-free specifications in time and space) equivalent, in exactly the same sense, to "coins" or "dice." Actual particle states, somewhat like the sets of numerical-dynamical states generated by a nonlinear algorithm, are statistical in the analogous sense that they reflect regimes of relatively simple to complex motions that are inherently ordered by the recursive self-organization of the process itself (i.e., they are inherently ordered in some sense, rather than "random," even though we cannot collect the proper numerical data to establish relative complexities in the sense of testing the data for "compression," as previously discussed).

In the synthetic context, then, complexity goes hand-in-hand with the question of whether a problem is tractable or intractable (or whether it is computable or uncomputable). At this juncture, however, another seemingly vengeful paradox interjects itself—blindsiding us in yet another way. Even though we cannot algorithmically reconstruct the Solar System, an organism, or an atomic nucleus exactly the way they are in nature, it is also true that their constructions have taken absolutely no effort on our part, and, to a significant extent—meaning relative to our interests in and uses for relative structural information, even allowing for chaotic properties—their essential structures are clearly describable in terms of far fewer bits of information than it took to build them. Indeed, at the most fundamental mathematical level, no amount of computing power will permit us to compute the properties of the numbers by which we measure and describe information (and computers) even if we were to limit such numbers to sequences of integers. In that sense, the next integer that "shows up" in our universe can only be a matter of "chance" (cf. Ruelle, 1991, p. 149). Conversely, however, through the action of chaotic self-organization, and dissipative interaction, the universe also "serves up" systems of numbers that are already "predigested," as it were, to variable degrees and that are recognizable to us by the simple fact that we are related to those numerical structures in the same self-referential context that mathematics invokes to prove that such numbers are produced by chance and must be random. Apparently we can have our cake and eat it too—just as the metaphorical snake can eat its tail—simply because, from the self-referential algorithmic point of view, we *are* the universe. And from that perspective none of these problems are intractable—or even "hard." In this paradoxical "floating crap game," it would appear to be possible—and to me preferable—to experience the world from a viewpoint of mutual numerical recognizability rather than from a viewpoint of demonstrably intractable complexity and chance. Evidently we can have it either way, and argue about it until "doomsday," as an intellectual variation on the theme of Turing's "halting problem" (cf. Casti, 1990, pp. 329ff and 367ff; Ruelle, 1991, p. 148).

Note I.3: The study of *pulsars* has a number of interesting parallels and analogies with the present study. A pulsar is an object in space, now believed to be a type of neutron

Chapter Notes: Introduction

star, that emits electromagnetic radiation in relatively sharp pulses ("blinks") which often, but not always, are spaced very regularly in time, ranging in period from thousandths of a second to a few seconds (see Strahler, 1971, pp. 85ff; Kaufmann, 1991, pp. 451–57; Pasachoff, 1991, pp. 472ff; Blandford, 1992; Lyne et al., 1992; Maran, 1992). The original discovery, in 1967, was made—again by "accident" (*synergetic serendipity*) while studying something else—during a study of *scintillation*, or "twinkling," in objects radiating energy at radio frequencies, an effect caused by variations of electron density in the solar wind that is analogous to atmospheric optical scintillation. However, the effect was fixed relative to the stars, rather than relative to the Earth's motion. It was soon demonstrated that the pulse period of the original object was quite precise (it had a period of about 1.3373011 sec) and that, once recognized, there also was a pulsating optical signal associated with it. By a process of elimination of possible dynamical mechanisms, it has been concluded by the astrophysics community that the most likely fundamental mechanism accounting for the pulsar is rapid rotation of a very dense neutron star. Many neutron stars exist in binary associations, giving rise to irregularities in the pulse frequency—suggesting to me that the process is, in general, one of nonlinear coupled oscillations in which a rich variety of periodic to chaotic resonance effects are likely (cf. Lyne et al., 1992). The generally accepted explanation for the sharpness of the pulsations, called "the lighthouse model," is based on the idea that the magnetic axis of the star is aligned in a different direction from the spin axis, hence giving rise to a directional beam that rotates in much the same manner as does a lighthouse on Earth, the beam coming to a sharp maximum in brightness as it sweeps past the position of an observer.

The studies of pulsars also have interesting parallels and analogies with studies of the class of astronomical objects first characterized during the period between 1960 and 1963 as quasi-stellar radio sources, *QSRs*, and quasi-stellar objects, *QSOs*. QSRs and QSOs are collectively called *quasars* (see Pasachoff, 1991, pp. 579f; cf. Kaufmann, 1991, p. 531), some of which are radio-loud (QSRs) while others are radio-quiet (QSOs). Parallels with the studies of pulsars include the fact that the discovery of quasars also came with the development of radio astronomy and with the reconciliation of the optical and radio characteristics of a suite of objects that differed from optically familiar stellar objects. A qualitative parallel—but one that identifies major differences in spatiotemporal scaling between pulsars and quasars—is that quasars also vary (oscillate) in the intensities of their electromagnetic radiation, but this oscillation takes place over time scales of weeks to months instead of less than seconds (i.e., quasars oscillate with periods $\geq \sim 10^6$ sec, while typical pulsars oscillate at frequencies that are more than 10^6–10^9 times faster), a fact requiring that quasars be small relative to the distance light can travel during the period of oscillation (cf. Pasachoff, 1991, p. 585).

Since light travels at a speed of about 299,792 km/sec (see Notes 6.1 and 6.3), quasars must be of the order of $\sim 3 \times 10^{11}$ km (roughly 2000 AU) or less in diameter, a size comparable to that of the Solar System out to the inner Oort Cloud (Hills Cloud)—or approximately the size of the solar nebula prior to formation of the planets (i.e., this crude criterion of size assumes that the classical Oort Cloud at $\sim$20,000 AU represents the distal margin of the protoplanetary nebula, and that most of the mass of the Solar System is at distances from the Sun less than the outer limit of the Hills Cloud; cf. Figs. 9–11 and 21; and Appendixes 7, 8, and 11). Quasars therefore have large diameters compared to typical main-sequence stars such as the Sun, but very small diameters compared to those of galaxies (the Milky Way Galaxy, for example is of the order of 10^5 parsecs, roughly 10^{10} AU, in diameter). Paradoxically, however, typical masses for quasars may range from millions to billions of solar masses, and, accordingly, they have the very peculiar property of resembling stellar systems in size and galactic systems in mass. Along with this property goes an

Chapter Notes: Introduction

interpretation that quasars are associated with black holes at the centers of a class of very old and highly energetic galaxies, the age representing the fact that quasars are receding from us at rates of up to 90% or more of the speed of light (see Pasachoff, 1991, pp. 584ff, and Note 4.2 of the present volume; for an alternative interpretation, see Jayawardhana, 1993).

The studies of pulsars and quasars provide interesting lessons in spatiotemporal scaling and cosmological self-similarity (cf. Strahler, 1971, pp. 85ff; van den Heuvel and van Paradijs, 1988; Krolik, 1991; Blandford, 1992; Eichler and Silk, 1992; Lyne et al., 1992; Tavani and Brookshaw, 1992; Wolszczan and Frail, 1992). For example, pulsars display some features in common with the *celestial reference frame (CRF) hypothesis* of the present volume: (1) it was noticed that the position of pulsars was fixed relative to the stars, not with respect to the Earth and solar wind, an observation analogous to my conclusion that orbital resonances in the Solar System govern the approach scenarios and cratering patterns of impacting objects on Earth in such a way that certain nodes of those patterns trace out paths on the celestial sphere that are invariant relative to the motions of the Earth's Equator and axis of rotation on the celestial sphere, (2) the temporal regularity of the oscillatory patterns of pulsars was inconsistent with the scintillation effect for which the study was originally designed, just as I conclude that the Earth's cratering pattern is inconsistent with the expected distribution of impacts computed from assumptions that impactors were stochastically distributed in time and space, (3) the sharpness of some pulsar signals, as well as the diversity of signal types, would appear to originate from the phenomenology of coupled nonlinear rotational dynamics, representing the same essential mechanism that is invoked in Chapter 6 for the fine-tuning of the cratering pattern on Earth, and (4) the preferred model for the existence of radio pulsations is related to a misalignment between the spin axes of the source objects and the magnetic axes of the dynamos that generate beam-like sources of electromagnetic radiation, a model that closely parallels my conclusion that Earth's geomagnetic dipole axis has migrated among a number of different unstable and metastable paleopositions relative to the spin axis, which — by comparsion — has remained relatively stationary, or at least meridionally constrained, during the last several hundred million years. [In contrast with the simplified dynamics of pulsars, the intermittent migratory (stuttering) motion of Earth's geomagnetic dipole axis involves several types of coupled phenomena that have operated over a variable and irregular (chaotic) range of low frequencies, frequencies that are measurable only by the geochronology of the comparative patterns of geomagnetic, magmatic, and other impact-dynamic changes over time scales from thousands to tens of millions of years (see Chap. 7; and Appendixes 12–23).]

In addition to the above properties of pulsars and quasars (and terrestrial-cosmological phenomena), other characteristics are reminiscent of systems that exhibit stimulated emission of radiation, such as *lasers* and *masers* (see Note I.5). For example, the beam-like characteristics of pulsar radio emissions may be qualitatively paralleled by quasars, because there appears to be a class of objects, called *blazars*, in which beam-like radio emissions are interpreted to emanate from special orientations of a quasar system with respect to relativistic jets and central "supermassive" black holes (e.g., Kaufmann, 1991, pp. 548f). In other words, somewhat like the lighthouse model of the pulsar, a blazar represents that state of a beam-emitting quasar that occurs when its electromagnetic axis is pointed in our direction. It is therefore also very interesting that the relative scaling of pulsation periods between quasars and pulsars (representing a range of ratios from $< 10^6$ to $> 10^9$) resembles the scaling of masses between them (i.e., between the respective stellar and galactic masses). This type of frequency/size relationship, in turn, resembles that of the seismovolcanic spectrum of terrestrial oscillations discovered by Shaw and Chouet (1991, p. 10,205), which

Chapter Notes: Introduction

varies in scale over frequency increments (bandwidths) that are measured in proportional cycles of 10^3 in the mass and energy of magma-transport rates (see Universal Order, Geometry, and Number, in Note P.4). Magma transport rates, in turn, represent the "stimulated emission" of a dissipative system (magma generation and transport from the Earth's mantle) relative to the orienting effects of the stress field that guide both the transport and the relaxation oscillations that govern the frequency distribution of the seismovolcanic signals (Shaw and Chouet, loc. cit.; cf. Shaw, 1988b).

Note I.4: The terminology of nonlinear dynamics (often called "deterministic chaos") is rather complicated and confusing — largely because it is a hybrid terminology that has evolved from many disciplines and subdisciplines of the physical sciences and mathematics (see Note P.6). Mastery of this terminology, however, is not really required to follow the ideas presented in this book, where emphasis is placed on the comparative recognizability of universal patterns rather than on the synthetic modeling and/or reconstruction of particular patterns. It may be useful nonetheless to scan a few of the more general references to nonlinear concepts and frequently encountered terms. A concise overview of phase-space portraiture (see Note P.3) is given by Abraham and Shaw (1987). Simplified introductions to chaotic dynamics given, for example, by Baker and Gollub (1990) and Tufillaro et al. (1992), can be used as guides to more complete textbooks, such as those of Lichtenberg and Lieberman (1983), Bergé et al. (1984), Thompson and Stewart (1986), Moon (1987), or Schuster (1988). Jackson (1989–90) offers a comprehensive two-volume exposition of much of this field of study, and Cvitanovic (1984), Hao Bai-Lin (1984, 1987, 1990), and Holden (1986) provide anthologies of many of the seminal papers in nonlinear dynamics. Gleick (1987) places some of these ideas into a highly readable historical perspective (cf. Ruthen, 1993). The journal articles by May (1976), Feigenbaum (1980), Hofstadter (1981), R. Shaw (1981), and Jensen (1987) are particularly recommended for general perspectives.

In brief, nonlinear dynamics is subdivded into two broad classes of numerical studies: (1) dissipative systems and (2) spatially conservative ("Hamiltonian") systems. According to the viewpoint expressed in the present work, conservative systems include the types of unrealizable systems discussed in Notes P.3 and P.6. Studies in Class 2 have proven useful in the demonstration of first-order departures of orbital phenomena from classical equations of motion, but conceptual emphasis in the present volume is placed on Class-1 behaviors (in contrast with this emphasis, most of the recent discussions and demonstrations of "Solar System chaos" employ variations of classical and Class-2 methodologies; cf. Wisdom, 1987a, b, c; Mallove, 1989; Laskar, 1988, 1989, 1990; Wisdom, 1990; Kerr, 1992c; Milani and Nobili, 1992; Sussman and Wisdom, 1992). Jensen (1987) gives a clear treatment of the underlying principles of phase-space maps of conservative Hamiltonian functions. Analogous phase-space portraits and/or return maps of dissipative functions displaying attractors, strange attractors, and/or combinations thereof, are explained more fully in the texts cited above (cf. Note P.3). The graphical methods of Class-1 systems often emphasize the bifurcational character of nonlinear functions and the rates of divergences of nearby trajectories in phase space, as measured by so-called Liapunov exponents (the study of chaotic evolution of the Solar System by Sussman and Wisdom, 1992, for example, illustrates divergences from nearby trajectories of "real-space," "real-time" planetary motions in which the difference in initial conditions between two adjacent trajectories differed by as little as *1 mm*!). The paper by R. Shaw (1981) provides a useful discussion of the meaning of Liapunov exponents, particularly with regard to concepts of dynamical "memory" and Shannon measures of dynamical information contents (cf. Liapunov, 1947; Wolf, 1986; Nese, 1989; and see Swinney and Roux, 1984; Argoul et al., 1987, vis-à-vis *chemical chaos*).

Chapter Notes: Introduction

A rather complex terminology exists relative to concepts of "periodicity" within special types of systems with varying degrees of algorithmic degeneracy [e.g., there are diverse kinds of "quasiperiodic," "nonlinearly periodic" ("mode-locked" or "phase-locked") resonance states, plus the periodic fixed-point bifurcation sequences of nonchaotic regimes, and the periodic bifurcation windows of chaotic regimes, depending on the nature of the problem and on the strengths of systematic forcing and interaction, and on the extent to which dissipation is taken into account in the nonlinear algorithm; cf. Note P.1; and Meiss, 1987; Jensen, 1987; Schmidt, 1987; Smith and Spiegel, 1987; Shaw, 1987b, 1991)]. Confusion over the meaning of "periodicity" in either the linear or nonlinear contexts is partly an artifact of our paradoxical views of disorder and order, or chance and determinism ("chance and necessity"), and partly an artifact of the types of algorithms that we appeal to as "standards" of periodic behavior. Usages in this book of the terms "period," "periodic," and "periodicity" are explained insofar as possible where they occur, but in general they simply refer to conditions and numerical properties of recurrences in time and/or in space without appeal, as has been customary in science, to the idealized notion of harmonic regularity. Systems of attractors that operate near the steady state but far from thermodynamic equilibrium correspond to what is often called *critical self-organization* (see Shaw, 1987b, 1991; Bak et al., 1988; Shaw and Chouet, 1988, 1989, 1991; Bak and Tang, 1989; Chen and Bak, 1989; Kadanoff et al., 1989). A related theme concerns a phenomenon called the *chaotic crisis* (cf. Grebogi et al., 1982, 1983, 1987a, b, c; Gwinn and Westervelt, 1986; Ditto et al., 1989; Savage et al., 1990; Sommerer et al., 1991a, b; Shaw, 1991; Sommerer and Grebogi, 1992), a theme that recurs frequently, beginning with Chapter 4.

Note I.5: Rheologies of feedback type, even in very "weak" or nearly "inviscid" materials, behave according to a mosaic of heterogeneously focused gradients of strain — as is seen in lava flows, in ponded magma bodies, in the flow of glaciers, and even in the cavitation-flow of water — where the mechanisms of locally coherent amplification of strain involve strain-rate feedback (see Notes P.1 and P.5) thereby "slaving" the many degrees of freedom of the generalized deformation field to a relatively few degrees of freedom of local and directionally focused motions (e.g., Shaw, 1988b, Figs. 9 and 10). Such behaviors resemble, from a qualitative dynamical viewpoint, optical slaving mechanisms epitomized by the *laser* (= *l*ight *a*mplification by *s*timulated *e*mission of *r*adiation; cf. Haken, 1978; Arrechi, 1991), or, analogously, mass-transfer slaving mechanisms epitomized by the *gaseous maser*, where the slaving of atomic and/or molecular energy levels leads to coherent *m*icrowave *a*mplification by *s*timulated *e*mission of *r*adiation (a phenomenon discovered in 1965 — during the "decade of discoveries" discussed in the Prologue — that occurs in interstellar space, often in star-forming regions of galactic molecular clouds; see Maran, 1992, pp. 414ff; Cohen, 1992; Elitzur, 1992; cf. Note P.5).

The general phenomenon of "stimulated emissions" would appear to be more common in Nature than is implied by its nominal restriction to lasers and masers. It exists, at least analogously, in the — compared to lasers and masers — low-frequency phenomena of magma transport in the Earth (cf. Shaw, 1988b, pp. 189ff), and perhaps in other types of high-frequency emissions, such as "earthquake lights" on Earth (cf. Lockner et al., 1983; Suslick and Flint, 1987; Maddox, 1993b) and the powerful beam-like cosmic jets called *blazars* that appear to emanate from quasi-stellar radio sources (*quasars*), and which can be seen only when the axial beam is pointed toward an observer (e.g., Kaufmann, 1991, pp. 548f; cf. Note 4.2). Low-frequency "stimulated emission" of mass-energy transport would have different implications with respect to mantle flow in the Earth, for instance, than do the many-mode mechanisms of homogeneous flow that invoke simplistic theories of natural, or "free," convection (cf. Shaw, 1969, 1988a, b, 1991; Shaw and Chouet, 1991). Theories of convection

Chapter Notes: Introduction

have represented the standard paradigm for descriptions of, and dynamical modeling of, mantle-flow phenomena, including the penetrative structures called "plumes"—whether they be "hot" and positively buoyant or "cold" and negatively buoyant (the latter, as distinct from "subducting slabs," are sometimes called "inverted plumes," "downwellings," "density currents," etc.). Convective models are simplistic, if not entirely specious, if they do not take into account the rheological consequences of the local thermal states that are implied by mechanisms of *thermal feedback* (cf. Notes P.5 and 8.3); recent computer models of partially coupled three-dimensional mantle convection are somewhat more realistic in their predictions of local convective instabilities. [The "avalanche" models of Tackley et al. (1993) and Honda et al. (1993) are cases in point, to be compared, for example, with the more qualitative feedback models of Shaw (1969, 1991) and Shaw and Jackson (1973); cf. King (1993).]

Note I.6: *Phase space* (also called *state space*, special cases of which are called *return maps*) is simply the parameter space of time and/or position—however topologically complex this may be—employed in both classical and nonlinear dynamics to plot ("map") the history of motion (the history of the states of a system) in a given context; e.g., Abraham and Shaw (1987), Shaw (1987a, b; 1988a, b; 1991), Baker and Gollub (1990), Shaw and Chouet (1988, 1989), Chouet and Shaw (1991); cf. Notes P.3 and 5.1.

Note I.7: A *virtual geomagnetic pole* (VGP) represents the position of a pole of a virtual dipole axis relative to the North geographic pole of the Earth based on a set of measurements of the observed geomagnetic field direction at *one specific location* on the Earth's surface and at *one specific time*. Thus, a single VGP, in principle, represents a unique measurement, or *point event*, with zero range in both time and space that— compared with like measurements at different localities on the globe but of the same age, and like *sets* of measurements at other ages—gives information on the spatiotemporal configurations of the geodynamo. For example, a given data set of the same age might be chaotically coherent in spatiotemporal phase contrasts — as in the trajectories of motion of a *low-dimensional strange attractor*—or incoherent—as in the trajectories of motion of *chaotic transients of transitional attractor-state crises* (see Note P.3; and Chap. 5, Fig. 18 and Note 5.1). In practice, averaging of various kinds—depending on the variety of techniques involved in sampling, in characterizing the locality in question, and in reducing the geomagnetic measurements—is exceedingly important to the statistical significance of a given set of VGPs. For example, a number of discussions in the literature have questioned the statistical significance of the longitudinal distributions of VGP determinations, as summarized in Figure 28 of the present volume (from compilations by Clement, 1991), and as shown in analogous global-geographic patterns by Laj et al. (1991); cf. prior discussions of global symmetries and *dynamo families* in K. A. Hoffman (1991; cf. K. A. Hoffman, 1988, 1992, 1993; McFadden et al., 1991; Courtillot et al., 1992; Gubbins and Coe, 1993). Although the bases for these criticisms are valid (e.g., relative to interpretations of VGP patterns for the latest ~1–10 million years of Earth's history; cf. Langereis et al., 1992; Valet et al., 1992), there seems to be growing evidence that certain spatial coherences exist in global VGP patterns during particular intervals of time (cf. Clement, 1992; Constable, 1992; Courtillot et al., 1992; Jackson, 1992; Laj et al., 1992; Hoffman, 1993; Gubbins and Coe, 1993).

In the context of paleomagnetic studies of field reversals, a VGP obviously represents only a single measurement of a past state of the geomagnetic field at a single sampling site over the shortest interval of time that can be resolved at that site (see Butler, 1992, Chaps. 1 and 7). In general, such a state may be, a priori, "infinite-dimensional" in terms of the degrees of freedom of the core motion, or "high-dimensional" in the context of characterization of chaotic attractors (cf. Tsonis and Elsner, 1990, 1992; Lorenz, 1991; Pierrehumbert, 1991; Chouet and Shaw, 1991). Clearly, in such a context, a VGP cannot faithfully represent an idealized point event because of the finite nature of the sampling problem and the characteristic temporal ranges of the relevant geologic processes (e.g., relative to sedimentary sequences in a lake deposit, emplacement sequences of lava flows, and so on). Much effort currently is being devoted to refining these problems of paleomagnetic VGPs, because they offer the most tangible means by which details of geomagnetic field configurations during the reversal process — as well as during field excursions relative to a normal or reversed *polarity chron* or *subchron* (cf. Harland et al., 1990, pp. 140ff) and/or during transient reversals (reversal events too short to be given the status of a subchron) — can be documented. The patterns of VGP distributions cited above would suggest that individual paleomagnetic reversal paths may have identifiable (hence ordered, from the standpoint of self-organized critical-state processes) geometric characteristics. If so, a totally new paradigm may be emerging for the spatiotemporal configurations of the core dynamo, a paradigm that would appear to call for the application of chaotic-systems theories rather than emphasis on the theories of stochastic processes that characterized most theoretical treatments of geomagnetic reversals since the pioneering work of Allan Cox (cf. Cox, 1968, 1969, 1981) — hence a paradigm that is consistent with the emphasis of the present volume on the nonlinear-periodic/chaotic nature of Earth processes within the cosmological context.

Note I.8: The Principle of Uniformitarianism, like the principle of natural selection (cf. Notes 8.12–8.15), is, I assert, really an expression of the nonlinear-dynamical view of geological history, and this can be seen most readily (cf. Note P.6) by reinterpreting its articulation as it was stated in the words of James Hutton (1726–1797) in his *Theory of the Earth* (1795), as quoted in Mather and Mason (1939, p. 92; cf. Dean 1992):

> *In examining things which actually exist, and which have proceeded in a certain order, it is natural to look for that which had been first; man desires to know what had been the beginning of those things which now appear. But when, in forming a theory of the earth, a geologist shall indulge his fancy in framing, without evidence, that which had preceded the present order of things, he then either misleads himself, or writes a fable for the amusement of his reader. A theory of the earth, which has for object truth, can have no retrospect to that which had preceded the present order of this world; for, this order alone is what we have to reason upon; and to reason without data is nothing but delusion. A theory, therefore, which is limited to the actual constitution of this earth, cannot be allowed to proceed one step beyond the present order of things.*

This famous quotation has been read in many different ways by succeeding generations of geologists since Hutton's time. Today, most geologists see it simply as an observation that the laws of Nature operated the same way in the past as they do now. In other words, one cannot appeal to some process or principle that operated long ago but which no longer applies to the processes of Nature — or, conversely, that certain principles are operative now that have no precedent. Examples of such discontinuities in geologic principles might be a belief that (1) the law of gravity did not exist in its present form prior to the Phanerozoic Eon, (2) tomorrow there may be an entirely new set of operational rules of physics and/or chemistry that invalidates prior applications of physicochemical argumentation in geology, or (3) the available types of mechanisms, processes, feedbacks, and organizing principles of evolution did not exist prior to the first appearance of organic forms, and/or that the "ground rules" of evolution changed abruptly one or more times in geologic history (e.g., as if someone or something suddenly reconfigured the alphabets and characters of all languages — including the genetic language — "overnight").

Chapter Notes: Introduction

In Hutton's view, such lines of argument applied to explanations of past events simply represented invention, for if all we needed to do to "explain" observations concerning a given phenomenon in the past — and/or to explain observations concerning a given phenomenon in the present — is to invent a characteristic principle that acts uniquely only in such an instance, and in other like instances, then all "explanations" of geological processes and geologic history are indistinguishable from works of fiction. In that case we might as well read Jules Verne as read any "serious" work that would interpret the phenomena of Nature. My graduate supervisor at UC Berkeley during the late 1950's, Charles (Chuck) Meyer (1915–1987) — who was for me a natural attractor to the field of geology and laboratory research, and to whom I owe my first two jobs in geology (one of which I still hold) — was fond of saying, with reference to special genetic interpretations of zoned ore deposits, that Waldemar Lindgren (1860–1939) called such approaches to geologic interpretation *the spigot method*; viz., that whenever a particular mineral, ore, gangue, or wallrock-alteration mineralogical assemblage either suddenly appeared or disappeared, or changed in character from ore deposit to ore deposit, all one had to do to explain such variations was to open or close the appropriate (expedient) spigot at an appropriate (expedient) time from an appropriate (expedient) source (cf. Mather, 1967, pp. 65ff). Although this attribution may be literally inaccurate or apocryphal, I have remembered it often over the years, and — since the advent of Plate Tectonics — I have often thought I was seeing it being practiced in the guise of specially created kinematic models that happened to be needed to "satisfy" this or that ramification of plate-tectonic theory.

Looking through the "lens" of nonlinear dynamics, I now see Hutton's statement as essentially equivalent to — though semantically inverted from — the *principle of sensitive dependence on initial conditions* (see Note P.2). That is to say, nonlinear-dynamical principles tell us that the general process of geologic evolution is noninvertible, that — in other words — there is no way to extrapolate, model, or linearly deduce the exact properties of an "initial" state in Nature, or a specific system of initial mechanisms characteristic of that state. The "beginning" is dynamically unknowable, and any theory that purports to describe it in specific terms can only be a work of fiction (see Notes 3.3, 4.2, 8.15; cf. Notes P.6, I.2, and E.1). I have come to this conclusion by making systematic observations of how nonlinear-dynamical systems operate in the present — both as the present is seen in the local context, and as it is seen as though it happened locally in the past, by making observations as functions of increasing geochronologic age or increasing cosmic distances. If one claims to see something different back then, out there, it is more likely to represent artifacts of viewing methods than it is to reflect reality — at least until it can be demonstrated that the observational methods used are also valid at much larger (more contextual) distances and ages, and that the claimed discontinuity persists within the phenomenologies of that more encompassing framework.

In another quoted passage (Mather and Mason, 1939, pp. 94ff), Hutton expounded further on what is wanted and needed in a geologic theory (parenthetical words and phrases are inserted in nonitalicized characters set off within the quote in order to "modernize" Hutton's statement in ways that bring it into the present by virtue of parallels and alignments with the principles of nonlinear dynamics):

"*Nothing can be admitted as a theory of the earth which does not, in a satisfactory manner, give the efficient causes* (i.e., effective causes, or, in my own terminology, effect-effect rather than cause-effect coupling of sequential phenomena) *for all these effects already enumerated* (seven categories of familiar geologic processes and/or phenomena Hutton listed as requiring explanation; see Mather and Mason, 1939, pp. 93 and 94). *For, as things are universally to be acknowledged in the earth, it is essential in a theory to explain those natural appearances* (here, Hutton was actually enunciating, or unwittingly anticipat-

Chapter Notes: Introduction

ing, the universality principles of nonlinear dynamics, which hold with equal relevance for all categories of geologic phenomena, such as the seven listed by him).

"*But this is not all. We live in a world where order everywhere prevails* (i.e., Nature is demonstrably nonlinear, but its manifestations are, de facto, what persons in Hutton's day called the 'natural order' of things, as in the arrangements of land and sea and sky, fauna and flora, and so on); *and where final causes are as well known, at least* (or, as little known, at best), *as those which are efficient. The muscles, for example, by which I move my fingers when I write, are no more the efficient cause of that motion, than this motion is the final cause for which the muscles had been made* (i.e., it is just as meaningless in any effect-effect sequence to ask the immediate cause of an effect as it is to ask the ultimate, or primal, cause of any immediate cause — which amounts to a two-hundred-year old description of a nonlinear-dynamical system of multiple coupled oscillators with delayed feedback, such as those discussed, for example, in the present volume and by Freeman, 1992, pp. 467ff). *Thus, the circulation of the blood is the efficient cause of life* (i.e., the circulatory system is contextually essential, along with other analogous systems, to what we call the living state, at least to the extent that we — the makers of the theories of the Earth — are capable of understanding what life is at the same time that we are its manifestation, meaning that we would attempt to define that by which we are defined); *but, life is the final cause, not only for the circulation of the blood, but for the revolution of the globe* (meaning that the principle of universality must hold for biological as well as cosmological processes): *Without a central luminary* (unifying principle) *and a revolution of the planetary body* (a contextually unifying system of motion), *there could not have been a living creature upon the face of the earth* (in other words, organic evolution could not operate in isolation from the cosmic principle); *and, while we see a living system on this earth* (namely, us), *we must acknowledge* (admit to be self-evident), *that in the solar system* (the Cosmos), *we see a final cause* (i.e., finally admitting that — de facto — because we judge everything relative to ourselves as the system of reference, we therefore can only see ourselves, along with all other creatures judged to be living according to the same criteria, as the central, hence essential, consequence of cosmic action)."

As I have paraphrased his words, Hutton was enunciating his *Theory of the Earth* in much the same way that I have attempted to do in the present volume from the point of view of nonlinear dynamics. Central to his, as well as to my own, statements are the related principles — whether accorded any other name — of self-organization and self-reference (see Notes P.1–P.4, P.6, I.2, I.4, 4.1, 4.2, 5.1, 6.1, 8.14, 8.15). In a paragraph immediately following the above statements, Hutton further stated his case that the Earth (Cosmos) and living systems in that realm must be explained by the same universal principles (cf. Note 7.5). Words used by Hutton that might have been construed to imply divine intervention and/or design are now easily seen — by nonlinear-dynamical intervention — to be synonymous with technical terms used in the present volume. For example, his conjoined usage of "Being" and "design" can now be seen to refer to the unifying principle of universal self-organization. And his disallowance of "chaos and confusion" as appropriate terms for the explanation of natural phenomena (cf. Shaw, 1994) is analogous to my enjoinder not to be sold a bill of goods by authoritative appeals to random models — except as such models may reflect qualifications that fall within the purview of the incompleteness theorem of mathematics, of algorithmic complexity theory, or, equivalently, of the principles of self-referential phenomena (cf. Note I.2). By the same token — which now, I hope, adds Hutton's posthumously implicit approval to that of Charles Darwin (1809–1882), which I invoke in Chapter 8 and Notes 8.12–8.15, for the application of nonlinear dynamics to Nature — neither should the terminology of nonlinear dynamics confound us, for, as I and many others have been attempting to show, it stands only for phenomena that are entirely

Chapter Notes: Chapter 1

consistent with the most tangled views of an Earth that has been subjected to evolving punctuations and catastrophes of many kinds, and at many scales of involvement.

Chapter 1

Note 1.1: While this volume was being revised following initial reviews (see the persons acknowledged at the end of the Prologue), Swisher et al. (1992) provided concrete documentation for the K/T age and size of Chicxulub crater, Yucatan Peninsula, Mexico. They performed $^{40}Ar/^{39}Ar$ laser incremental-heating analyses on samples of "glassy melt rock" as well as on samples of glasses from tektites that are more widely distributed in the Caribbean. Their age for the melt-rock glassy material, based on repeated determinations, is an amazingly precise 64.98 ± 0.06 Ma. Even more remarkable is the compatibility of tektite ages, both as determined by them and as recalculated by them from the results of Izett et al. (1991a, b) for tektites from Haiti. The two suites of ages differ within less than 100,000 years from an age of 65.0 Ma. Swisher et al. (1992, Fig. 1) also give a stratigraphic cross section of the Chicxulub structure based on petroleum exploration drill holes by Petróleos Mexicanos (PEMEX), showing that the andesitic glassy melt rock they analyzed came from a depth of nearly 1500 m near the center of the structure. The drill-hole information appears to support other evidence based on gravity and magnetic anomalies suggesting that the diameter of the crater is about 180 km. Kerr (1992b) reviews the history of the discovery of Chicxulub and the debate concerning its status as "the smoking gun" of the K/T extinction event.

Ironically, although the report by Deino et al. (1991) on a redetermination of the age of the Popigai impact structure, Siberia, was what had forcibly brought to my attention the pattern labeled the "K/T swath" in Figure 3 (see Chap. 1), I noticed subsequently that publications by Grieve and coworkers continued to cite ages of less than 40 Ma for Popigai (e.g., Grieve, 1991; Grieve and Pesonen, 1992). The K/T age for the Popigai structure given by Deino et al. (loc. cit., Fig. 2) was reported on the basis of what appears to be a quite stable $^{40}Ar/^{39}Ar$ laser step-heating profile from the more homogeneous fraction of an impact-melt glass—i.e., the profile is stable in the sense that it shows a fairly constant plateau of apparent ages between 55 and 75 Ma, with a mean value of ~66 Ma, and an error that is, conservatively, about ± 2 Ma [this is my own subjective estimate of the variation shown by Fig. 2 in Deino et al. (loc. cit.); that report gives two estimates of the apparent age, a plateau age of 65.4 ± 0.3 Ma and an integrated total-fusion age of 66.3 ± 0.4 Ma]. The sample was apparently a bona fide glass (not the partially or completely recrystallized material that is sometimes loosely called "impact glass"), and it should have yielded the most reliable age for Popigai (as in the analyses of Chicxulub impact glass cited above) vis-à-vis analyses of recrystallized, and therefore possibly altered, "glassy" material. The analyses by Deino et al., however, were performed on glass fragments from a single sample of material provided by V. L. Masaitis (1975) — then at the Ministry of Geology in Leningrad, USSR — evidently from a study completed nearly twenty years ago. And it would appear that no recent sampling by any research group has been accomplished because of the remote (71.5° N, 111° E) Siberian location of Popigai.

In the meantime, other age redeterminations on apparent impact-melt materials from Popigai (recrystallized glass?), presumably from the same collection, evidently have been carried out by Grieve and coworkers, also using the $^{40}Ar/^{39}Ar$ laser step-heating method, in their case giving calculated ages of ~34 Ma (as cited by Garvin and Deino, 1992). Ambiguities thus persist in the age of Popigai, related to problems of sampling and/or sample histories of impact materials, and until a definitive suite of demonstrably unaltered impact glass samples can be obtained and analyzed, the age for Popigai given in all of the Grieve catalogs cited here, as well as the K/T age of Deino et al. (loc. cit.), can only be

Chapter Notes: Chapter 1

viewed as nondefinitive — at least until all of the sample-provenance and alteration-history problems have been worked out [e.g., determinations by different workers have now yielded apparent ages for Popigai spanning an interval from the latest Cretaceous to the Oligocene (Garvin and Deino, loc. cit.; Deino, written comm., 20 April, 1993)].

Ironical as it is that the present study was partially motivated by a report that would now appear to be the subject of controversy, it seemed evident from the outset that the "K/T swath" of Figure 3 represents a spectrum of ages spanning the K/T boundary (see Chap. 1). Furthermore, it has been found in the present study that the same swath-like pattern has been characteristic of Earth's impact history during the latest 100 Ma, including the very youngest and smallest of impact structures (cf. Appendix 2), which supports the notion not only that impacts were spatiotemporally episodic near the time of the K/T transition but that the same general pattern persisted throughout the Cenozoic.

Regardless of the ambiguities concerning the age of Popigai, and/or other Russian impact structures, as I have mentioned elsewhere in the present volume, I do not subscribe to any of the proposed single-impact scenarios as the unique cause — or "smoking gun" — of the K/T extinction event, or events (cf. Hut et al., 1987, 1991; Shaw, 1988a). There is increasing evidence that the K/T boundary "event" involved more than a single impact — as the cratering pattern discussed in Chapter 1 strongly suggests (see Fig. 3). For example, Kerr (1993b) reports that a drill hole into the crater at Manson, Iowa, has recovered a sample of impact-melted rock, and that this material gives a K/T age that is analytically indistinguishable from the 65.0-Ma age for Chicxulub, as documented by Swisher et al. (loc. cit.). Rapid-fire multiple events that span major arcs of world-encircling orbits are indicated in the present work to be characteristic of at least the K/T extinctions and probably also the late Devonian (Frasnian-Famennian) extinctions described by McLaren (1970, 1982, 1983, 1986) and McLaren and Goodfellow (1990), as suggested by the impacts of approximately the same age shown in Appendix 3a (also see the evidence concerning the primary radiation of land vertebrates, first documented in the Famennian, reviewed by Carroll, 1992). Multiple sets of "self-similar catastrophes" that span the order of a million years or so are more in keeping with persisting evidence that extinctions are not limited to a single event horizon with a chronographic resolution as small as ± 0.1 Ma (see discussions in Chaps. 7 and 8, and Note 8.1; see the Epilogue for one sort of scenario that might give a very short-term — "Uzi-like" — burst of multiple impacts). Kerr (1992b) also cites studies by Peter Ward (University of Washington, Seattle), who continues to argue that — though he now agrees that the fossil record of the ammonites appears to have been truncated sharply right at the K/T boundary — the inoceramid clams disappeared 2 million years earlier, and that "something phenomenal" happened at this earlier time that was not the result of the impact responsible for the principal K/T extinctions (see the discussion of Fig. 36 in Chap. 8; cf. Note 8.1; and Shaw, 1988a, Figs. 14–16; Ward, 1990a, b; Shoemaker and Izett, 1992). Analogous debates also persist concerning late Triassic extinction events important to the origination and radiation of dinosaurs (e.g., Bice et al., 1992; Hodych and Dunning, 1992; Trieloff and Jessberger, 1992; Rogers et al., 1993; Benton, 1993; cf. Note 8.1).

Note 1.2: Following Schultz and Gault (1990a, b), a single large impactor conceivably could, in one pass, produce a closely timed composite sequence of craters consisting of (1) a single crater with a diameter of the order of 100 km, plus (2) several craters with diameters of the order of tens of kilometers, plus (3) many craters with diameters of the order of 1 km (cf. Baldwin and Sheaffer, 1971; Chyba et al., 1993). Depending in part on the initial direction of approach, the cratering line, or swath, could form either a small-circle or great-circle arc, the former consisting of incomplete arcs or complete but distorted small circles, while the latter could form either partial great-circle arcs or potentially recurring

Chapter Notes: Chapter 1

great-circle cyclic swaths. If the first "ricochet" transfers the trajectory of the object, or portions of the object, into a geocentric orbit (or orbits), subsequent impacts will fall along great-circle paths (great-circle routes for first-pass objects would be subject to the sorts of constraints illustrated in Fig. 19). The mechanism of low-angle to grazing impacts thus provides an effective feedback mechanism by which to charge (capture), recharge, and redistribute a system of natural Earth satellites (see Chap. 6; cf. Gault and Schultz, 1990) — e.g., by secondary, tertiary, etc., impacts, some of which will occur only after protracted histories of geocentric orbital decay, which, depending on orbital-period/orbital-distance relationships, object size, nonlinear interactions, enhanced atmospheric drag, etc. (cf. Notes 3.3 and 6.1; and Allan, 1967a, b, 1969, 1971; Greenberg, 1978, 1982; Pollack et al., 1979; Pollack and Fanale, 1982; Greenberg and Brahic, 1984; Atreya et al., 1989; Toon et al., 1992; Zahnle et al., 1992), may represent (1) a fraction of one orbital cycle, (2) an unstable or transient geocentric orbit, or (3) a semipermanent geocentric orbit. Such scenarios are similar to those that would appear — circumstantially at least — to be necessary to account for the high frequency (geologically speaking) of low-angle trajectories in the relatively young prehistoric and historic records of meteoroid encounters with the Earth (e.g., Appendix 3): (a) two events of multiple skipping impacts occurred in Argentina within the last ten thousand years; they differ significantly in approach trajectories even though they are in close proximity to each other in the global sense and correspond to quite similar great-circle orbital trajectories (see Cassidy et al., 1965; Schultz and Gault, 1992; Schultz and Lianza, 1992; cf. Grantz and Mullen, 1991), (b) the Tunguska event of 1908 that allegedly traveled over western China to Siberia at a fairly shallow angle, because it was reportedly seen by the naked eye for a distance of more than a thousand kilometers from the epicenter — i.e., the angle of incidence could not have been more than about 10°, otherwise the altitude would have exceeded several hundred kilometers when the bolide was first sighted (cf. Krinov, 1963b, Figs. 1 and 2; Clube and Napier, 1982b, p. 140), and (c) the Great Canadian Fireball Procession of 1913 that traveled more than 4000 km at nearly constant altitude (see Appendix 3a; and Chant, 1913a, b; Mebane, 1955, 1956).

The Tunguska event is still shrouded in the mystery of conflicting reports and analyses. If Clube and Napier (1982b, p. 140) were correct in saying that the bolide entered the atmosphere over western China, then it had to travel well over 2000 km in a direction several degrees east of north, yet Krinov (1963b, Fig. 1) shows a terminal flight azimuth that is nearly normal to the trend of Lake Baikal, or about forty degrees west of north. Although trajectories can rotate during reentry (cf. Arho, 1971), this angular difference is extreme, and in the wrong sense — unless there was a spiraling effect analogous to that of an unstable rocket near burnout. In my illustrations, I have compromised on the trajectory and have shown it to be only slightly west of north (see Figs. 23a and 37; cf. Appendixes 3, 4, 16, 17, and 22). Adding to this confusion — at first glance — Chyba et al. (1993) take the most probable angle of incidence (angle of impact measured from the horizontal) to be ~45°, basing their inference on model consistency with the treefall pattern at the epicenter of the blast (cf. Krinov, 1963b, Fig. 2; Burke, 1986, Fig. 56). Such a large angle would put the first alleged sightings at an altitude of fifteen hundred kilometers or more, which is five times "higher" than the usual limit of the ionosphere and three times that of the thermopause (cf. Roy, 1988a, pp. 307ff), assuming that the angle of descent was constant all the way — which I am about to suggest may not have been the case.

There appear to be two classes of observations of the Tunguska event that, at first blush, would appear to be totally contradictory to one another: (1) long-distance, low-angle, visual reports and (2) site-related observations consistent with a high-incidence-angle blast wave. The catastrophic fragmentation model developed by Chyba et al. (1993, p. 41; cf. Melosh, 1992, 1993a) is essentially one of explosive runaway followed by enhanced ablation and aerobraking effects (i.e., fast spreading of the fragmentation front followed by fast deceleration of the fragmented aggregate, something like the spreading of a fragmentation shell fired into a shock-absorbing medium). This model puts an interesting twist (perhaps literally) to the story, because such a model of positive-negative feedback is a description of a fast-reacting route to chaos (like rapidly turning up the gain — the "tuning knob" — on the logistic map; cf. Note P.1; and May, 1976; Hofstadter, 1981). If the reaction grew exponentially, say during the last thousand kilometers of travel, as occurs in positive-feedback processes — i.e., at first slowly, during the latent stage (see Note 8.1), while the distant observers were watching — and then culminated in an even faster negative-feedback exponential decay, then the most chaotic, vortex-generating stage would have occurred just prior to impact (hypothetically with a counterclockwise vortical twist of forty to sixty degrees during the final hundred kilometers or so). By hypothesis, the reaction front quickly lost altitude during the final stage of the climactic transition from the positive to the negative feedback mode — i.e., as it "spun in," or screwed itself into the ground, somewhat like a crippled helicopter, or the out-of-control rocket mentioned above, or like the touching down of an atmospheric vortex to form a "twister." No wonder the fellow sitting on his porch in the factory town of Vanavara sixty kilometers south of the point of touchdown was "thrown several meters, felt a sensation of heat, and lost consciousness" (Krinov, 1963b, p. 210); he may have just been knocked for a loop by the edge of a hot tornado. Given a more protracted history, and a more fully developed vorticity, such a chaotic progression might have had a chance to go through an actual bifurcational splitting of the blast wave, something like the multiple-vortex phenomena encountered in some tornadoes, and in other rotationally induced vortex processes (cf. Snow, 1978; Sommeria et al., 1988).

Chapter 2

Note 2.1: The argument presented in Chapters 1 and 2 that most terrestrial impacts are concentrated at geographic nodes, because of orbital focusing effects in the inner Solar System and spin-orbit resonances with mass anomalies in the Earth, implies that the flux of impactors conventionally calculated from astronomical estimates is not representative of the delivery rate of meteoroids to the Earth-Moon vicinity (see Chap. 9, and Notes 6.1 and 9.1; cf. Steel, 1991; Kerr, 1992g). This is because effects of nonlinear organization are not taken into account in search and computational strategies (see Note 6.5), hence beam-like effects in the delivery of meteoroids usually are not considered (but see Alfvén, 1971a, b; Baxter and Thompson, 1971; Danielsson, 1971; Lindblad and Southworth, 1971; Drummond, 1981, 1991; Olsson-Steel, 1989; McIntosh, 1991). Consequently, an assumption of uniform partitioning over the continents and oceans is not valid. Beam-like effects can significantly modify estimates of delivery rates, particularly for high-inclination approach scenarios (cf. Fig. 19, and the Epilogue). An analogous assumption, relative to the usual one that the flux of impactors is uniform and/or random, would be to consider that the flux inferred from the astronomical record applies mainly to the continents, increasing the discrepancies discussed in Chapter 2 by a factor of three.

Chapter 3

Note 3.1: Dynamical resonance of any kind can be defined in a nontechnical sense as an interaction of two or more coupled oscillating systems in such a way that there is *synchronization* in some form, often with reinforcement, of numerically unique and reproducible phase-space patterns (cf. Note 4.1; and Shaw and Doherty, 1983; Gu et al., 1984; Kaneko, 1986, 1990; Thompson and Stewart, 1986; Crutchfield and Kaneko, 1987; Meiss, 1987; Moon, 1987; Schmidt, 1987; Smith and Spiegel, 1987; Shaw, 1988a, Fig. 19; 1988b, Figs. 9 and 10; Blekhman, 1988; Brown and Chua, 1991; Ling et al., 1991; Pecora and

Carroll, 1991; Moon, 1992; Garfinkel et al., 1992; Murali and Lakshmanan, 1991, 1992; Anishchenko et al., 1992; Carroll and Pecora, 1992; De Sousa Vieira et al., 1992; Freeman, 1992, pp. 467ff; Rul'kov et al., 1992; Willeboordse, 1992). Phase space is the "canvas" on which the portraiture of patterns in space and time is plotted or "painted" as in a diagram of the celestial position of an object vs. its orbital velocity (cf. Note P.3). *Linear resonance* implies that temporal periods of the spatial motion within each of two or more interacting oscillators are very nearly identical in form (e.g., sinusoids or other waveforms of constant shape and repeat period, and theoretically of "infinite" length), and that these periods combine to produce a simple harmonic curve of spatial amplitudes in the time or frequency domains that is a linear superposition of the amplitudes of the interacting oscillators (*period* and *frequency* represent the same measure, one being the reciprocal of the other). The combination of linear periodic cycles results in an oscillation of the same period, with an amplitude that is the summation of the individual contributions, an amplitude that theoretically increases in direct proportion to the energy supplied by the two oscillators.

Linear systems are exemplified by *idealized* springs, pendulums, vibrating reeds, acoustical speakers, surface waves on ideal fluids (fluids "without viscosity," called *inviscid fluids*, which exist only in the abstract), celestial orbital motion, and so on, which are characterized in their simplest forms by single-valued *natural periods* and *natural frequencies* of oscillation (cf. Table 1). To the extent that no two objects can be exactly identical in the absence of artificial and arbitrary manipulations of properties, natural frequencies typically are unique (this is a sometimes subtle and nitpicking point, but it is the essential reason why there are needs for standards of time as measured by the periods of standardized phenomena, such as the *sidereal day*—defined as Earth's period of rotation relative to a fixed position on the celestial sphere, such as one of the "fixed" stars—or the natural period of a tuning fork, or quartz crystal, or atomic "clock"). When an external oscillating source, represented by a periodic impulse or periodic force field of arbitrary frequency, is imposed on such an idealized linear oscillator, the system is driven to respond in ways that are proportional to the imposed force, depending on the timing of the applied force relative to the *phase* of the *driven system* (where phase refers to the position in a cycle of oscillation, as in the position of a sweep second hand on a circular clock face). If the frequency of the external forcing function is caused to vary until it is equal to the natural frequency of the driven linear system, at constant phase, a condition of *linear resonance* is encountered in which the motion of the driven system is suddenly amplified. The adjustment of the imposed periodic force is called *tuning* (cf. Notes P.1, P.3, and P.4). The suddenness and amount of *amplification* of the linear oscillator depends in part on the sharpnesses of the natural and imposed frequencies, as measured by their *bandwidths,* and on the duration of the applied forcing frequency. Since no frequency is exactly single-valued, there is no system with zero bandwidth (even the bandwidth of a laser is finite, and the bandwidths of frequencies assigned by a computer are functions of roundoff errors).

The general idea of "tuning" applies to both artificial and natural systems as well as to both linear and nonlinear systems. In linear systems, coincidence of the externally applied and internally characteristic oscillation frequency reinforces the amplitude of response of the driven oscillator at its constant natural frequency and bandwidth (theoretically, the idealized linear system is free of distortions, or deformation-related changes of properties). This occurs because the power of the source is assumed to be unlimited and of constant characteristics relative to the driven system—meaning that it is "unrelentingly" imparted to the driven system at its natural frequency (the term "continuously applied" is inaccurate if the source is an intermittently applied discrete, hence discontinuous, pulse of energy, such as a precisely timed *"kick"*; cf. discussion of the *kicked rotor* by Jensen, 1987). The amplitude of linear resonant motion therefore theoretically grows without limit as long as energy from the periodic source continues to be supplied, and the response remains linear at the constant characteristic frequency.

It is evident from these definitions that *perfectly linear resonance is intrinsically unstable*. No matter how small the initial response may be, the amplitude of the driven motion at resonance eventually exceeds the range of response appropriate to the assumption of linear and constant properties (elastic moduli and other properties eventually change, as in a suspension bridge in danger of collapse due to wind stresses that oscillate near its characteristic frequency — or, as in a playground swing that eventually goes erratically out of control from repeated shoves at its characteristic period, and so on). In other words, even if a perfectly linear resonance condition were possible, some form of *nonlinear effect* eventually is incurred with persistent forcing. The time interval from the onset to the culmination of a nonlinear effect may be so short, and so disastrous to a system, that sustained, controlled nonlinear interactions often are not easily tracked and documented, especially in manmade objects and experiments (unless such effects are anticipated and monitored accordingly). In long-term natural coupled oscillations driven by a relatively constant forcing frequency (e.g., in terrestrial, solar, galactic, and celestial tidal phenomena), on the other hand, the onset of nonlinear instabilities may be of long duration — assuming it were possible to identify and follow their evolution from some initial linear mode of coupled oscillation. As I have attempted to point out elsewhere in the present volume, however, even if such a state ever existed, it would soon have become unstable as described above. Natural systems, therefore, are intrinsically nonlinear and tend to be operating in the vicinity of resonance breakdown — which is another way of saying that natural systems are typically operating near the transition between nonlinear resonances and chaotic self-organization. For example, near-resonance phenomena, discussed in Chapter 4, are almost ubiquitous in the organizational structure of the Solar System (cf. Note 4.1). The advent of *numerical dynamics* (computer experimentation with functions resembling those of physical dynamics), however, has made it possible to simulate this sensitive transition region of dynamical interactions by means of numerical algorithms (e.g., Gwinn and Westervelt, 1986; Shaw, 1987b).

When the responses of driven systems become nonlinear, they are no longer directly proportional to the magnitudes of the imposed forces, hence irregularities are induced in the cycles of the resulting motions. In other words, a periodic force imposed on a nonlinear (disproportionately reacting) oscillator results in a condition where the cyclic motion is neither invariant nor single-valued in characteristic amplitudes and frequencies. Nonetheless, the complicated natural modes of the nonlinear oscillator are activated and absorb energy just as in the linear case. Now, however, the heterogeneity creates internal conflicts and *frustrations* that distribute the energy in complex ways which, in material experiments, must be represented by response-dependent elastic moduli and *damping coefficients* (properties associated with *detuning* of electronic oscillators, with *attenuation* of vibrating reeds, with variable *quality factors* of ringing bells, with *internal friction* in fluttering airplane wings, and so on). Usually a finite range of forcing frequencies exists for which nonlinear responses can remain exactly repetitive in their net cyclicities. Because the driven response keeps pace on average with cycles of the periodic source, a steady-state balance of *modulated amplification* is enforced for special subsets of the driven states. These are the nonlinear-resonance states of the system, already discussed, expressed as a form of generalized synchronization (see the references cited above), just as in the linear case.

A major distinction between linear resonance and nonlinear resonance is that, in the latter, energy partitioning leads to nonuniform incremental oscillatory responses that are more *orchestral* in that they are of unequal magnitudes and durations from place to place but interact in concert to produce a form of controlled synchronization that does not increase without limit at a constant power and/or amplitude of forcing. Disastrous instabilities do not necessarily occur in nonlinear resonances, below a *critical amplitude* or *threshold of chaotic oscillations*, because, in the steady state, the nonrecoverable (or irreversible) part of the energy supplied is balanced by consequent *dissipation of energy* (the part involved in the diffusion of heat and mass induced by the nonlinear deformations of the driven system). Below the critical threshold, the net motion is exactly repetitive over some number of cycles of forcing, but the *wave form* is not a simple sinusoid, nor is it necessarily any simple harmonic function (if transformed to the audible frequency range, it may or may not sound pleasing, an exception being that certain kinds of self-organized nonlinear resonances result in orchestral rhythms that seem to be at the heart of our senses of musical and visual esthetics; see discussion of the *golden mean* in Note P.4; cf. Shaw, 1987b). The net motion is sometimes expressed as a combination, or mixture, of harmonic modes, or as the ratio of the number of cyclic oscillations of the forcing function relative to the number of oscillations of the driven system, a fraction called the *winding number* (cf. Notes P.4 and 4.1; Fein et al., 1985; Shaw, 1987b, Fig. 1; Ding and Hemmer, 1988; Baker and Gollub, 1990).

A resemblance should be evident between the above description of nonlinear resonances and *commensurability ratios of orbital coupling*, for example, as described in Chapter 4 for the Asteroid Belt (cf. Note 4.1, and Figs. 12–14). There the nonlinearities of the driven system (the system of asteroids) are represented by the varieties of responses of the orbital parameters of different objects in the system to an imposed source of periodically forced oscillation. In the case of the asteroids, this forcing is associated with the periodic gravitational perturbations produced by the cyclic variations of the orbital distance between each asteroid and Jupiter (i.e., Jupiter's orbital frequency is one of the dominant forcing frequencies by virtue of its mass and proximity). More accurately, the resonance states of the asteroids are resultants of the net cyclic perturbations associated with the Sun and *all* of the planets relative to the center-of-mass (CM) of the Solar System (cf. Fig. 21).

A simpler example of nonlinear resonance is the motion of points around a circle map, as in the motion of the hands of a clock (cf. Table 1, Sec. D). Circle maps elucidate distinctions between *quasiperiodic* resonances, *phase-locked* resonances, and *chaotic* resonances. The following remarks refer to phase-locked resonances. The other two regimes are elaborated in Shaw (1987b, 1991). In the *nonlinear phase-locked regime* (sometimes called the *mode-locked regime*, or just the *nonlinear periodic regime*), the hands of the clock do not move in equal increments, even though the clock still keeps perfect time over a complete cycle (tick-marks on the clock-face are unevenly spaced, as if the clock-face were subjected to a Dali-like distortion). The relation between the nonlinear motion of the hands relative to a resonantly coupled periodic source is expressed in terms of ratios of the numbers of iterations (ticks) of each oscillator during a complete cycle. If the number of ticks of the driven nonlinear clock is an integer multiple of the number of ticks of a coupled periodic source function, then the two oscillators are in nonlinear resonance. However, it might take, say, 27 cycles (or any other integer number of cycles) of the hand in the nonlinear clock to arrive at an identical position in the cycle (the position of *conjunction*) relative to the hand of the linear clock coupled with it, where the latter takes some other integer number of cycles (the question of whether the ratio of these integers is less than 1 or greater than 1 depends simply on convention as to whether the forcing frequency or the driven frequency is in the numerator; see discussion in Chap. 4 of the *Kirkwood Gaps* in Figs. 12 and 13 vis-à-vis Fig. 14). Between the times of identity (conjunction), the hands of the two clocks do not move synchronously (mode-locked resonances of a nonlinear clock are illustrated by Shaw, 1987b, Figure 1, using the *sine-circle function* of Table 1, Sec. D).

It should be evident that the meaning of nonlinear resonance in complex mechanical systems has a close correspondence with the more commonly encountered meaning of resonance in ordinary experience, such as in music and poetry. The analogy is appropriate because it emphasizes the inherent property of nonlinear dynamics and chaos as the source of the complex organization and richness that we associate with esthetic beauty. The related implications, numerical and metaphorical, of the golden mean provide one of the common denominators (universality parameters) for such analogies (see Note P.4). A remarkable treatise on the relation between the structures of numbers and the structures of poetic metaphor is given by Elizabeth Sewell (1951) in her book *The Structure of Poetry* [its relevance to patterns in Nature is pointed out by Shaw (1994), and in a marvelous essay by G. Evelyn Hutchinson, based on an address given upon receiving the Leidy Medal of the Academy of Natural Sciences of Philadelphia (Hutchinson, 1953)]. Stewart (1989, pp. 252–61) gives an explanation in analogous terms for nonlinear resonances of orbital phenomena, basing his discussion on the dynamics of Hamiltonian systems (cf. Note P.3).

Note 3.2: My suspicion that the Solar System had experienced a binary-star stage in its evolution was formulated many years ago (see Shaw, 1983b, p. 16 and Fig. 7; 1988a, p. 38 and Fig. 23; cf. Fig. 9, Inset). No theoretical basis for this conjecture (which had occurred to me as an explanation for the curiously bimodal mass and energy distributions of diagrams like Figs. 9 and 10 of the present volume) was known to me at that time from searches of the available literature in astronomy (cf. Note 9.2). The idea had been planted in my mind, perhaps, by a passing exclamation I had noticed in Sexl and Sexl (1979, p. 72, Fig. 4.10) that "Jupiter is almost a star!" And the hypothesis advanced by Williams (1972) that the planets and satellites condensed from a "tornado-like prominence or jet of gases ejected from the primeval Sun" had some of the implications of the "stripping mechanism" that seemed to be needed to leave the terrestrial planets as a "naked core" of lithophilic materials from a small pre-hydrogen-burning star (perhaps one of the *contracting star* systems described by Appenzeller, 1985), conceivably including a white-dwarf-like condensate from the protoplanetary disk of a ternary solar nebula (Appendix 76; cf. Stevenson, 1989; Pasachoff, 1991, Fig. 26-5). I was thus primed when several years later a title caught my eye: "Naked Stars and Hot Meteorites" (Weissman and Wasson, 1990). This paper described stellar systems called T-Tauri stars of FU-Orionis type that matched some of the above characteristics (cf. Note 9.2; and Hartmann and Kenyon, 1985; Sargent, 1989).

A different, but in some respects kindred, hypothesis of later-stage collisional processes among planetesimals, offered by Cameron et al. (1988, Figs. 5–6) as a possible explanation for "the strange density of Mercury," illustrates one aspect of what I envision could have happened during the dynamical stripping and differentiation of the prototerrestrial planets some time between the cataclysmic destruction of the binary protostellar or massive-disk stage of the Solar System and a planetary stage that would have had some resemblance to the planetary system as it is conventionally modeled (possibly including a planet, now represented by the "minor planets" of the Asteroid Belt, that did not survive intact the intensity of the cataclysmic transition processes; cf. Meyer-Vernet and Sicardy, 1987; Boss, 1988). One of the immediate consequences of such a "nebulous" scenario is that the early planetary history in the vicinities of the terrestrial planets was rife with (relatively) large-scale collisions, but the history of the relationship between the protostellar stage and the protoplanetary stage as a consequence of the cataclysmic inversion process is obscure, and must differ in major respects from previous models of planet formation (cf. Note 6.8 and Note 9.2; see also Petersons, 1984; Appenzeller, 1985; Kerridge and Matthews, 1988; Herbig, 1989; Sagan and Druyan, 1989; Lada and Shu, 1990). Such a sequence of happenings, perhaps with other intermediate stages of self-organization prior to the cataclysmic inversion process — should the close-binary model turn out to be a realistic

Chapter Notes: Chapter 3

picture of Solar-System origin (see Fig. 9, Inset; and Note 9.2) — would have profound consequences for the science of "planet-building" and the concept of a "standard model" (e.g., Wetherill, 1989c, 1990, 1991b; cf. Herbig, 1989, p. 299).

Just how the collisional events, such as the one that was imagined to have turned Mercury inside out (Stewart, 1988; Cameron et al., 1988; cf. Wasson and Wetherill, 1979; Murray, 1983; Leake et al., 1987; Strom, 1987; Chapman, 1988; Strom and Neukum, 1988; Wasson, 1988; Wetherill, 1988), and the one that many planetologists think was responsible for the origin of the Moon (see Note 6.8), fit into this picture is not clear. But those stories, and those of the so far incompletely explained origins of the asteroids, and especially their relationships to the origins of the chondritic/achondritic meteorites, or vice versa (perhaps a chicken-and-egg sort of question) — where I use the plural form because sources from both the innermost and outermost regions of the Solar System, at least, are implicated in the current makeup of the Asteroid Belt (cf. references below) — must represent part of an overprint on a history that was far richer in its structural self-organization than most theories of planet formation have, so far, contemplated (e.g., Sonett, 1971; Chapman and Davis, 1975; Weidenschilling, 1977, 1983, 1988; Chapman, 1986; Caffee and Macdougall, 1988; Cassen and Boss, 1988; Grossman, 1988; Hewins and Newsom, 1988; Levy, 1988; MacPherson et al., 1988; McSween et al., 1988; Steele, 1988; Wood, 1988; Wood and Morfill, 1988; Zolensky and McSween, 1988; Chapman et al., 1989; Gaffey et al., 1989; McSween, 1989; Scott et al., 1989; Weidenschilling et al., 1989; Wetherill, 1989b; Binzel et al., 1991, 1992; Rotaru et al., 1992; Russell et al., 1992; Sears et al., 1992; Binzel and Xu, 1993; Gaffey, 1993; Hoffmann et al., 1993). For example, Herbig (1989), in summarizing his views on major issues in planetary system formation, had the following to say about binary systems:

> I want now to talk about binaries from the point of view of the conventional disk model. Bodenheimer mentioned that about 80% of the main sequence stars are in binary systems, and that a representative separation is about 10 AU. Yet curiously, in discussions of the properties and behavior of pre-main sequence stars, one hears very little about the possible effects of a close companion. The realization that symbiotic stars and cataclysmic variables are binaries revolutionized the interpretation of those objects. One wonders if there are close-binary phenomena taking place in the pre-main sequence that we have not yet appreciated.

And Herbig continued, emphasizing an observation that first struck me in contemplating the kinetic energy distributions of Figure 10, *"Very obviously, the presence of another star 10 or 100 AU away* (or even much closer) *would grossly modify the structure of a conventional accretion disk, for which dimensions of the order of hundreds of AU are often mentioned"* (parenthetical phrase added, befitting a close binary companion to the proto-Sun).

Herbig was here contrasting the types of behaviors predicted by conventional models of accretion disks around pre-main sequence single stars, as presented in the same symposium (Weaver and Danly, 1989), with what might happen if, instead, the stellar object consisted of two stars in fairly close proximity to one another. I have essentially asked the same question vis-à-vis my interpretation of the mass-energy distribution in the Solar System shown in Figure 10 of the present volume (*see the update on pre-main sequence binary star systems*, Note 9.2). And yet the situation may be even more complex. In view of Appenzeller's (1985) observations on contracting stars (pre-main sequence, "embryonic," star systems that have not yet reached the hydrogen-burning stage), it is conceivable — at least from the standpoint of expectable nonlinear-dynamical phenomena (cf. Chapman et al., 1992; Cannizzo and Kaitchuck, 1992) — that the evolution of the so-called "circumstellar disk" is more or less continuous with the evolution of binary and multiple-star loci that may condense within a common contraction process as forms of bifurcational and/or multifurcational fixed-point phenomena of types that occur in chemically chaotic vortex processes (cf. Swinney and Roux, 1984; Shaw, 1986; Argoul et al., 1987; Meyer-Vernet and Sicardy, 1987; Scott, 1991).

Although I was convinced that the nebular binary-star idea could at least partially account for some of the confusion in the literature concerning "primordial" chemical signatures (or at least my confusion concerning what I saw to be a deficient rational basis for the definition of chemical "standards" for cosmological abundances, as typically discussed in the geochemical literature), no definitive data were available to test this idea. Cosmic abundances typically are discussed in terms of the compositions of meteorite parent bodies that have been assumed to represent the standards of the most primitive chemical states in the Solar System (cf. Sears and Dodd, 1988; Larimer, 1988; and other chapters in Kerridge and Matthews, 1988) — and I have never understood upon what basis it could be assumed that any meteoritic composition could be representative of a homogeneous state in what must have been a chaotically evolving, hence heterogeneous, system (e.g., Shaw, 1983a, b). After Chapter 3 was written, however, at least two reports have been published concerning reinterpretations of "cosmic chemistry": (1) Fink (1992) has described a new chemical type of carbon-poor comet that, whether it represents an in situ source or an interstellar source, demonstrates nonuniformity of primordial materials, and (2) Rotaru et al. (1992) published a study that appears to demonstrate the existence of major heterogeneities in the types of stellar source materials found in carbonaceous chondrites — carbonaceous chondrites being the very meteoritic materials long thought to be most representative of the solar nebula, and indeed even of the universe as a whole (e.g., Class CI.1 chondrites have been assumed to reflect the "cosmic abundance" of the elements; cf. Anders and Kerridge, 1988; Larimer, 1988; Sears and Dodd, 1988, Tables 1.1.1 and 1.1.2; Carlisle, 1992; Russell et al., 1992; Sears et al., 1992; Wright, 1992).

The chemical anomalies reported by Rotaru et al. (1992) appear to be largest for the neutron-rich isotopes of the iron-group refractory materials. According to the binary-star model of solar evolution proposed in the present volume, such refractory materials would be those that would most likely record: (a) the history of stellar sources, (b) the dynamical disintegration of a companion star, and (c) the chaotically "multi-folded" distribution of relict chemistries throughout the Solar System (cf. Fig. 11). Because the process of stellar evolution, as envisioned in a generally chaotic model of cosmological evolution, is a form of self-similar cyclic process within the Galaxy, the dynamical heterogeneity evidently should record a palimpsest of chemical signatures — in a manner somewhat analogous to the way the record of impacts on Earth records the palimpsests of smaller-scale processes in the planetary nebula, and in the subsequent planetary evolution (see Middlehurst and Kuiper, 1963; Morrison, 1982a; Carusi and Valsecchi, 1985; Birmingham and Dessler, 1988; Kerridge and Matthews, 1988; Vilas et al., 1988; Atreya et al., 1989; Binzel et al., 1989a; Ringwood, 1989; Bailey et al., 1990; Beatty and Chaikin, 1990; Kippenhahn, 1990; Lagerkvist et al., 1990; Newsome and Jones, 1990; Cox et al., 1991; Newburn et al., 1991; Sonett et al., 1991; Yeomans, 1991a; Teisseyre et al., 1992; cf. the concise summary of the diversity of theories for the origin of the Earth, the Moon, and their chemistries by Drake, 1990). Accordingly, the structural implications of chaotic models — with regard to chaotic vs. random numerical signatures (see Note I. 2) — would affect our thinking not only about dynamical ordering in the universe, but about chemical ordering as well. In this context, the cosmic abundances (see Sears and Dodd, 1988, Table 1.1.2; Larimer, 1988) — and even the "periodicities" of the chemical states in the periodic table of the elements (e.g., Jensen, 1986, Figs. 5–9; cf. Mazurs, 1974) — demonstrate the universal characteristics of the chaotic ordering process, thus providing the "glue" that holds together or "integrates" the

435

Chapter Notes: Chapter 3

fundamental phenomenologies of nature, such as: (1) the spatiotemporal (geometrodynamic) structures of cosmology, (2) the nature of macroscopic dynamical states vis-à-vis laws of mass-energy scaling derived from those structures (the "fundamental forces" and spacetime geometries of action at a distance, as in gravity and electromagnetism), and (3) the analogous microscopic dynamical states and laws of mass action (the "fundamental forces" acting at chemical and nuclear distances) reflected in the chemical states of matter. [After this note was written, Farley and Poreda (1993) published a study of the abundance of neon in the mantle (cf. Owen et al., 1991; Farley and Craig, 1992), and two of their conclusions may be relevant to the above description of the early planetary nebula: (1) the Earth seems to have "accreted with gases nearly solar in composition," and (2) "Importantly, the inferred mantle Ne/Ar ratios are much higher than atmospheric, which is consistent with simultaneous fractionation of both the atmospheric neon isotope ratio and the Ne/Ar ratio by massive hydrodynamic escape." The relationship, or relationships, between "hydrodynamic escape" and early Solar System hydrolytic alteration phenomena is a question that may be of great relevance to the model proposed herein. As expressed by Zolensky and McSween (1988, p. 114), *"of even greater importance* (than heat) *in the evolution of some of the most compositionally primitive meteorites* (chondrites) *was the chemical reactivity of water at relatively low temperatures"* (parenthetical phrases added). It is virtually self-evident that in the putative nebular inversion process that I have described, the heterogeneity of chemical potential gradients in the "volatility hierarchy" of cosmochemistry (cf. Larimer, 1988; Lipschutz and Woolum, 1988) must have been great— e.g., compression-rarefaction waves could readily explain the intimate juxtapositions of low-temperature hydrous and carbonate phases with high-temperature early-formed and/or relict phases in chondrites (e.g., Zolensky and McSween, loc. cit., Table 3.4.1, and pp. 127 and 139; cf. Grossman, 1988; Grossman et al., 1988; Kerridge and Matthews, 1988, Part 7, Chaps. 7.1–7.10). See also Appendix 7 herein.]

Note 3.3: The orbital structure of the Solar System itself, as is seen conspicuously in the discussion of Chapter 3, is nonlinear dynamics in action. The orbital trajectories are directly analogous to those of dissipative phase portraits. This viewpoint departs from the conventional approach to celestial mechanics, but it is now supported by the results of orbital calculations cited in the text (Peterson, 1992, highlights some of these results). It stands in sharp contrast (if not shocking contrast, by traditional beliefs) with the usual assumption that structures due to the universal action of dissipative coupling and feedback do not exist in celestial mechanics (the term "universal" is used here in the formal sense of universality parameters in nonlinear dynamics; e.g., Shaw, 1987b; Baker and Gollub, 1990; cf. Notes P.1, P.4, I.2, and 3.1). In textbooks on astrophysics and planetology, this stance is qualified relative to only the most conspicuous occurrences of dissipation in the guise of collisions, tidal friction, atmospheric and plasma drag, and so on (but see Allan, 1967a, b, 1969, 1971; Alfvén and Arrhenius, 1975, 1976; Greenberg, 1978, 1982; Pollack et al., 1979; Alfvén, 1980, 1981, 1984; Jewitt, 1982; Pollack and Fanale, 1982; Greenberg and Brahic, 1984; Roy, 1988a, Chap. 10; Atreya et al., 1989; Dormand and Woolfson, 1989; Esposito and Colwell, 1989; Kolvoord et al., 1990; Cox et al., 1991; Dones, 1991; Esposito, 1991, 1992; Toon et al., 1992; Zahnle et al., 1992; Colwell and Esposito, 1993). Such effects, however, usually are reduced to linearized computations by assuming linear viscoelasticity, interactions of point masses, and the virtual absence of accelerations other than Newtonian gravity, even in view of the ubiquity of cosmic plasmas and the seeming necessity of antimatter structures in our vicinity of the Galaxy (see, Alfvén, 1980, 1981, 1984; Lerner, 1991; Papaelias, 1991).

Chapter Notes: Chapter 3

Linear cosmology carries over to the probabilistic computations of stochastic models and Monte Carlo simulations (cf. Duncan et al., 1987; Heisler, 1990) where nonlinear effects are viewed as though they can be mimicked by reiterating complex combinations of linear interactions dominated by chance events (but see the concept of *stochastic resonance* discussed by McNamara and Wiesenfeld, 1989; cf. Ford, 1978, 1983; Benzi et al., 1982; Meiss, 1987; Schleich and McClintock, 1989; Irwin et al., 1990; Kloeden et al., 1991). The assumption of conservation of length, area, volume, or hyperdimension in *phase space* (see Notes P.3 and 5.1) is probably the most profound misconception of celestial mechanics (cf. Schmidt, 1987; Smith and Spiegel, 1987). That assumption constrains the world view to limited dynamical outcomes, as did the geographical world view of the Flatlanders vis-à-vis that of the post-Columbian geodesists. It is now impossible to look at the Solar System as though it is an independent dynamical structure set in motion as a relativistic ballistic aftermath of the "Big Bang." Traditionally, each substructure of the universally evolving structure is dynamically described as if it were independent of other substructures composing the Milky Way Galaxy, which itself is now known to represent only one form of organization within a diversity of other galactic phenomena that rivals biologic diversity in its complexity (cf. Lerner, 1991; Flam, 1992a, b). Orbital systems can no longer be viewed as if they were gyroscopes spun up by an unkown hand that granted a guaranteed future consisting of a largely frictionless and exchangeless independence of action.

In the numerical structures of sine-circle maps and other dissipative algorithms, there are no physical forces at work except those of the computer, subject only to assurances of its digital integrity. The assignment of coupling constants (see Table 1, Sec. D) — though they may be intended to guide computer outcomes relative to analog experiments, thought by some workers to be more intuitively reasonable or appealing — says nothing more than that numerical interactions exist within the context of a generalized recursion algorithm. When it is recognized that a state of zero interaction in any phase space is as unattainable as is the absolute zero in thermodynamics, then it is also recognized that in a universe where there are no arbitrary zero or nonzero limitations (no existential physical boundaries), there must be natural interactions that are numerically analogous in possible types and strengths to the results of numerical computer experiments. When it is also stipulated that there is no limitation on the numbers of recursion steps in a natural computation (steps in which the output of one iteration of the computation is reentered into the computation as the input for the next iteration, as in the natural "algorithms" of biological processes), then self-organized, hence naturally *cybernetic*, patterns of numerical evolution must emerge from the computational process. Such patterns of strictly numerical character not only look like many kinds of natural structures, they often are found to dimensionally resemble the scaling properties of complex natural phenomena, celestial to biological, and even poetical (cf. Sewell, 1951, for a discussion of numerical realms of order and disorder vis-à-vis experiential realms of "language," "dream," and "nightmare" in the structure of poetry; and Shaw, 1987b, 1988c, 1994 for discussion of computational realms of chaotic natural structures).

There is a predictable wonder and pleasure in seeing that nonlinearly resonant structures exist in the Solar System and Cosmos (cf. Figs. 12–16). I am not surprised to find, for example, that when the orbital eccentricities and inclinations of the asteroids in the space between Mars and Jupiter (actually, it would now appear, between the Sun and neighboring stars; see Note 6.3; cf. Figs. 11, 13, 15, and 21) are plotted against their semimajor axes, they are organized in essentially the same numerical fashion as are the points in a computer plot of the sine-circle map tuned to the vicinity of the golden mean (see Fig. 14 and Appendix 9). I am not surprised only because I had already experienced my surprise in seeing such structures emerge from purely numerical ("nonphysical") computer algorithms

Chapter Notes: Chapter 4

seemingly devoid of any resemblances to the laws of celestial mechanics. Surprise is an essential ingredient of nonlinear dynamics, and future surprises will be limited only by the systems of disciplinary beliefs that are imposed on it.

The viewpoint outlined here concerning orbital attractor structures in the Solar System was in important respects anticipated long ago by the work of A. E. Roy and M. W. Ovenden, originally in the implications of the Mirror Theorem, and later in the work by Ovenden on what he called the *Principle of Least Interaction Action* (cf. Roy and Ovenden, 1954, 1955; Ovenden, 1975, 1976; and Roy, 1988a, pp. 119f, 141ff, and 269ff).

Chapter 4

Note 4.1: Technical definitions of the terms *commensurability* and *nonlinear resonance*, like the terminology discussed in earlier notes, are more confusing in appearance than in substance. Commensurability just means that two or more periodic phenomena are divisible by a common factor — which may be any natural number — in such a way that the result can be expressed as a system of integers (cf. Figs. 20 and 21). Thus, as is the custom in nonlinear dynamics and celestial mechanics, commensurabilities are expressed in the form of hierarchical sets of rational fractions, for example in the descriptions and comparisons of orbital mean motions (e.g., Öpik, 1976, pp. 118ff; Roy, 1988a, Chap. 1; Froeschlé and Greenberg, 1989; cf. discussions of nonlinear *synchronization* in other contexts discussed by Shaw and Doherty, 1983; Gu et al., 1984; Kaneko, 1986, 1990; Crutchfield and Kaneko, 1987; Blekhman, 1988; Shaw, 1988a, Fig. 19; 1988b, Figs. 9 and 10; Brown and Chua, 1991; Murali and Lakshmanan, 1991, 1992; Pecora and Carroll, 1991; Anishchenko et al., 1992; Carroll and Pecora, 1992; De Sousa Vieira et al., 1992; Freeman, 1992, pp. 467ff; Garfinkel et al., 1992; Moon, 1992; Rul'kov et al., 1992; Willeboordse, 1992).

I would suggest that, if one is in doubt about the subject matter and meaning of nonlinear dynamics, some trust can be placed in the "literary" sense of nonlinear phenomena — i.e., as long as that sense is not superficially truncated by contrived literalness. The verbal sense of what is considered to be nonlinear is a better "measure" even than the musical and the visual (acoustical and optical) senses — because the "literary sense" implicates (folds together), in addition to sight and sound, the incredibly involuted and convoluted realms of touch, taste, and smell. [Compare, for example, the discussion of "How the Brain Makes Chaos in Order to Make Sense of the World" by Skarda and Freeman (1987) with Freeman's (1991, 1992) tutorials on neurophysiology, especially as he applies the principles of nonlinear dynamics to the coupled oscillatory phenomena of the olfactory system, an application that is qualitatively parallel with those of the present work relative to the coupled oscillations of the Earth, Solar System, Galaxy, and Cosmos.] The figurative "common denominator" that begins to make numerical sense of all this is *linguistics*, taken, in its broadest sense, to embrace everything from poetry to evolutionary genetics and cosmic structure (e.g., Sewell, 1951; Searls, 1988, 1992; Shaw, 1994; see also the anthropological implications cited in A Linguistic Digression, Chap. 8; cf. Notes 8.14 and 8.15). In other words, linguistics comes closer to describing the meaning of nonlinear dynamics — especially with regard to the potential richness of nonlinear phenomena — than do the stilted, and sometimes misleading, formal definitions and proofs one finds in the specialized textbooks and literature of nonlinear dynamics. The word *resonance* is a case in point. Technical papers and textbooks in nonlinear dynamics rarely discuss the physical meaning of resonance. For example, when you look for the word "resonance" in the textbooks on nonlinear dynamics cited in the present work, you will often find that its meaning is introduced only in the context of mathematical solutions of the equations of motion of a pendulum with forcing and damping — or in the context of the equations of motion of some analogous oscillator.

Chapter Notes: Chapter 4

To me, it is symptomatic of the impasse that mathematical physics had gotten itself into prior to the advent of revelations made by Poincaré, Lorenz, May, Feigenbaum, and others previously mentioned—as discussed in the Prologue, and by Shaw (1991, 1994)—that "the physics" of a problem became confused with the mathematics of a system of equations that was tacitly alleged to *be* the problem. That is, systems of equations came to be thought of as *the physics itself* (contrast this stance with the experimental method epitomized by Michael Faraday, Note P.7). Textbooks reveal this evolutionary dead end by defining a phenomenon such as resonance only in terms of the solutions of an idealized system of equations. One of the profound results of nonlinear dynamics has been the demonstration that the equations of a physical system *are not* the physical system, and that a physical system—even when it is "acted out" in the form of numerical structures—is something much richer and more complex than is indicated by restricted sets of solutions of any system of equations that usurps the place of the actual system (cf. May, 1976; Morrison, 1991).

Resonance, in my view, and in its nonlinear-dynamical context, is much more an experiential phenomenon than it is an analytic phenomenon. For example, the condition of *linear resonance* of an "ideal" spring or pendulum is the simplest of experiential notions. If energy, in the form of a "push," a "kick," or other imposed cyclic force is contributed to any idealized (dissipation-free) motion at intervals of time equal to the natural cycle time of the motion (the *natural period*, or inverse of the *natural frequency*, of an *oscillator*), the amplitude of the motion grows "without limit" in proportion to the energy contributed by that properly timed impetus. No equations are needed to see this, and nearly all of us have experimented with it in the motions of playground swings. But when an equation is written down to represent the essence of such an oscillator—an easy matter in this case—it immediately obscures the more interesting aspects of the general phenomena of physical resonance. For one thing—despite dramatizations by resonant instabilities of suspension bridges, or of automobiles with bad shock absorbers on an appropriately undulating road—the implication of unlimited reinforcement is absurd, and consequently the "pendulum equation" had to be labeled with the "warning" that it is valid only in the limit of small displacements (linear states). This is about as far as the elucidation of the physics of resonant phenomena went for most of us in our first courses in physics. And, until recently, even in the field of basic research in physics, most of the really interesting dynamical properties of the pendulum (and oscillators in general) had been largely ignored (cf. Miles, 1984; Gwinn and Westervelt, 1986; Tritton, 1986, 1988, 1989).

Nonlinear resonance, by comparison with linear resonance, represents a rich array of interactive phenomena that cannot be captured in a simple definition illustrated by one unequivocal and explicit system of equations—although, to the extent that this can be accomplished in a metaphorical sense with the help of equations, the sine-circle equation of Table 1 (Sec. D), in my view, offers the clearest prototype (cf. Shaw, 1987b, Figs. 1 and 2). Yet, in the experiential sense, nonlinear resonance has some simple parallels with linear resonance. It implies that (1) there is reinforcement of a natural frequency (or array of natural frequencies), but the response of the nonlinear oscillator is modulated and transformed by dissipation in such a way that (a) the natural frequency of the oscillator in question, or (b) the frequency of forcing, or (c) both the "natural" and "forcing" frequencies, are modified by the resonance condition and can persist over a range of frequencies that may be either simple or complex (a condition called *mode-locking* or *phase-locking*; cf. Fig. 14), and (2) if the energy of the forcing frequency (amplitude) persists, or is increased, over and above that which is balanced by dissipation associated with a given frequency of response, the system undergoes a sequence of phase shifts that may cause it to pass into and through other mode-locked resonance states (frequency arrays or bundles; cf. Note P.4)—

Chapter Notes: Chapter 4

and if the forcing continues unabated, the system eventually passes through an unlimited series of increasingly complex states until it encounters conditions of irreversible responses (noninvertable trajectories in phase space) which induce diverse behaviors consisting of multiplicative combinations of aperiodic and periodic motions of the types that are characteristic of the chaotic regimes of nonlinear systems in general.

The idea of a forcing frequency, while convenient for the purpose of making laboratory demonstrations and mathematical representations of the phenomena of coupled oscillations, does not apply in the same idealized sense in nature, because, there, it is often impossible to clearly differentiate between the "natural frequency" and the "forcing frequency." In that case, a system evolves interactively, by *self-organization*, through the analogous systems of frequency structures that were just described (cf. Shaw, 1987b; Shaw and Chouet, 1988, 1989, 1991; and Note I.4).

In the nonlinear dynamics of coupled oscillators, resonances usually are expressed as fractions, rational or irrational — relative to a given system of one or more absolute normalizing frequencies — where the fractions are called *winding numbers* (cf. Notes P.4 and 3.1, and Fig. 14; cf. Shaw, 1987b). The significance of these fractions concerns the relative frequencies of the two coupled oscillators (cf. Fein et al., 1985; Ding and Hemmer, 1988) — such as the relative frequencies of two coupled celestial phenomena. The meaning of a given fraction depends on how a given cyclical phenomenon is described; e.g., the resonance fraction may refer to (a) the orbital frequency of a satellite relative to the rotational frequency of its primary (cf. Fig. 20), (b) the mean orbital frequency of one object in the Solar System relative to another (Jupiter vs. Saturn, an asteroid vs. Jupiter, etc.; cf. Fig. 21 and Appendix 11), sometimes called *mean-motion resonances* (see Froeschlé and Greenberg, 1989), or (c) in the general case, the motion of one "counter" of a nonlinear clock relative to the motion of another counter (cf. Fig. 14; and Shaw, 1987b, Fig. 2) — noting that, in a nonlinear clock, the counts (the increments of discrete "ticking" motion around a clock face), though they sum to 2π per cycle and repeat exactly in subsequent cycles, neither are equal to each other nor are they in integer proportions to each other within a cycle (it is as though the clockwork is nonuniformly warped and distorted, where some increments are stretched and others are shortened, relative to the linear case; cf. Shaw, 1987b, Fig. 1). The meaning of a particular fraction depends on the convention chosen for the two objects. For example, the fraction 1/2 means that one object cycles once while the other object cycles twice, relative to a given fundamental frequency. If it is preferable, for some reason, to refer to the denominator as the slower of the two motions, the same ratio would be expressed as 2/1, and so on. The latter is the convention for commensurabilities in the Asteroid Belt, first established by Daniel Kirkwood in 1857 as the locations of numerous gaps, now called Kirkwood Gaps (see Kirkwood, 1876, p. 365). In Kirkwood's convention, the numerator refers to the orbital frequency of an asteroid relative to the orbital frequency of Jupiter (the ratio 4/1 means that an asteroid, if present at that position, would revolve four times around the Sun during the time that Jupiter revolves once around the Sun, relative to the fixed stars; in Fig. 13, for example, this is shown to occur for asteroids at a distance of just over 2 AU from the Sun, a distance beyond Mars that is about the same as the distance of Mars beyond the Earth, whereas asteroids near the orbital distance of Jupiter, a distance of 5.2 AU from the Sun, correspond to a resonance ratio of 1/1).

Note 4.2: The circumstances of the discovery of Ceres, the first documented asteroid (cf. Kippenhahn, 1990, Chap. 8; Yeomans, 1991a, pp. 148ff), is one of those stories that edifies and enlightens us concerning the methods of, history of, and nature of astronomical research, and of research in general. Except for the time, place, scope, and methods of observation, and the phenomenological scale and state of advancement of knowledge

Chapter Notes: Chapter 4

and/or theories in astronomy, physics, and mathematics, the story is remarkably similar to the circumstances of the documentation of the first-discovered pulsating stellar object, now called a *pulsar*, an event that occurred 166 years later (see Note I.3; cf. Pasachoff, 1991, pp. 472ff; Maran, 1992). Although the discovery of Ceres (with a diameter of 913 kilometers, according to the catalog of Tedesco, 1989) in the year 1801 is credited to Giuseppi Piazzi — then the new director of the astronomical observatory in Palermo, Sicily — it might be argued (admittedly in an officious and nitpicking manner) that according to the present-day rules of documentation for the numbered asteroids (Bowell et al., 1989) the *documented* discovery of Ceres as an object with an established ephemeris, one sufficiently precise to qualify it for inclusion in the list of numbered asteroids, was not made until almost a year later by Franz Xaver von Zach at an observatory in Gotha, Germany. It was almost simultaneously discovered by Heinrich Wilhelm Olbers, an ophthalmologist and amateur astronomer — an amateur who, according to Yeomans (1991a, p. 144), had built one of the best contemporary observing facilities in Germany, above his own house — testifying to the nature of the eighteenth- and nineteenth-century paradigm for personal funding of research — though whether Olbers's affluence in building a state-of-the-art astronomical observatory came from inherited wealth or from an, even then, lucrative medical practice, is not clear (Olbers also reappears farther along in this Note in relation to the so-called "blazing sky effect").

In a sense, the discovery of Ceres might even be credited to Karl Friedrich Gauss, now held up as one of the great icons in mathematical physics, who even then, in his early twenties, had a distinguished reputation that was further enhanced by this incident. Gauss had taken up the challenge of calculating the orbital parameters of Piazzi's object with an aim to establishing the necessary circumstances for its recovery. His success in accomplishing that feat from only 41 days of observations by Piazzi was confirmed by Zach and by Olbers, who searched in the location predicted by Gauss. The established recovery of Ceres by Zach and by Olbers is dated 1 January 1802, almost exactly one year after Piazzi's first sighting. Gauss's prediction was indeed a good one, in that the orbital period of Ceres is about 4.6 years (semimajor axis, $a \cong 2.767$ AU; see Fig. 13a), so that in one year it had moved a significant fraction of a revolution about the Sun; also, its orbit is more eccentric than Earth's ($e_c \cong 0.08$, compared with $e_E \cong 0.017$) and is inclined to the ecliptic by $i_c \cong$ *10.6°*, compared with $i_E \cong 0.0°$ for Earth. But, in the same sense that Tombaugh's real discovery of Pluto was based on its apparition in a photographic plate — noticed by the presence of a new spot on the plate relative to the positions of known stars — almost a month before the accepted date of discovery (see Tombaugh and Moore, 1980, pp. 125ff), so, too, Zach had recovered Ceres on 7 December, 1801 (Yeomans, 1991a, p. 149) by noticing that, on reexamination of the same location on 18 December, one of the "stars" present on the previous date was no longer there (though Zach only became convinced of this after another two weeks of observations and deliberations).

The scientists, historical events, and cosmological phenomena outlined in the above sketch have a significance much greater than might be assumed from their being recorded as incidental happenstances in the history of astronomy. Piazzi actually was engaged in compiling a catalog of the fixed stars when — a few hours after midnight on the first day of the first year of the nineteenth century! — he first noticed a "star" that did not appear on his star chart. He then noticed that this "star" (which he soon realized was an object that resembled a star only in its twinkling star-like aspect) moved relative to the other stars on successive nights of observation, via movements which he carefully recorded and which were later used by Gauss to ascertain that the object followed an elliptical orbit resembling those of the planets. Ironically, the first discovery of a pulsar, by Jocelyn Bell in 1967 (see Pasachoff, 1991, p. 474), likewise occurred while she was studying something else —

which, in her case, was the study of *scintillation* in astronomical radio sources. This "twinkling" of radio sources was thought to be caused by the solar wind, in much the same way that the optical twinkling of stars and asteroids is caused by variations in the density of the Earth's atmosphere [i.e., owing to fluctuations in temperature, water-vapor concentrations, particle concentrations, etc., that, in turn, cause both the optical paths and optical intensities of light rays to fluctuate so much that the objects themselves appear to be flashing on and off, or *scintillating*; planets twinkle less — except when they are low on the horizon, or are viewed through especially turbulent air — because they are close enough to be seen as areally distributed sources rather than as easily perturbed infinitesimal "rays" of light; e.g., Pasachoff, 1991, p. 90; cf. SEEING (Astronomy) in Considine, 1983, pp. 2539f]. Bell noticed, however, that not all twinkles of pulsars were alike, and that one of the aberrant types did not move relative to the fixed stars, nor was the character of its "scintillation" affected by variations in the electron density of the solar wind. After a month of observations, Bell was able to show that this object was a new kind of star that pulsed radio waves at regular intervals of a little more than one second (see Note I.3). Again, the history of astronomy was changed simply by someone's noticing how an object moved relative to the fixed stars, a criterion that the present volume attempts to show — in combination with evidence concerning the nature of nonlinear orbital phenomena — is also relevant to the interpretation of the cratering record of the planets.

This note could be expanded to address virtually the full scope of nonlinear-dynamical phenomena in the Solar System and in the Cosmos, where the latter term is capitalized because I am using it here in the sense of the nonlinearly ordered structure of the universe, a usage that expands on the literal definition of "cosmos" to encompass chaotic as well as periodic concepts of order — another irony, in view of the definition of "chaos" as essentially the antonym of "cosmos" in the dictionary. The achievement by Gauss, as impressive as it was to the astronomers of that time, could have been greater if he had taken Kepler's "obsession" with numerical periodicities more seriously, and had integrated Kepler's notions with those of Johann Daniel Titius, who, in 1766 — according to the description by Tombaugh and Moore (1980, p. 46; cf. Nieto, 1972, p. 1), which rings most true to me among the many accounts I have read — noted in the margin of a book by the French philosopher Charles Bonnet the following observation:

> For once, pay attention to the widths of the planets from each other and notice that they are distant from each other almost in proportion as their bodily heights increase. Given the distance from the Sun to Saturn as 100 units, then Mercury is distant 4 such units from the Sun; Venus, $4 + 3 = 7$ of the same; the Earth $4 + 6 = 10$; Mars $4 + 12 = 16$; etc. But see, from Mars to Jupiter there comes forth a departure from this so exact progression. From Mars follows a place of $4 + 24 = 28$ such units, where at present neither a chief nor a neighboring planet is to be seen. And shall the Builder have left this place empty? Never!

The numbers listed by Titius can be recognized as the first few terms of an infinite series in which the ratios of successive distances asymptotically approach *a period-doubling progression* (e.g., for the tenth term — which corresponds to Pluto if we assume that the Asteroid Belt marks the position of a "missing" planet — the ratio of that orbital distance to the preceding one is, according to Titius's prescription, $772 \div 388 \cong 1.99$, differing by only 1 part in 200 from the factor 2; with two more iterations this deviation is only about 1 part in 1000. As it later turned out, the rapid divergence from the actual spacings of the planets beyond Uranus — none of which, including Uranus, was known in 1766, Uranus being discovered "accidentally" by William Herschel (1738–1822) in 1781 (see Nieto, 1972, Chap. 4; Tombaugh and Moore, 1980, Chap. 3; Considine, 1983, p.

Chapter Notes: Chapter 4

2899) — was the beginning of what all the fuss has been about in Solar System observational astronomy during this century, for instance in the design and execution of the *trans-Neptunian planet search* (see Tombaugh, 1961a, b, 1991; Tombaugh and Moore, 1980). This search has been pursued in the U.S. by intrepid astronomers such as Percival Lowell (1855–1916) and Clyde Tombaugh (b. 1906, discovered Pluto in 1930, and was still active at the time of this writing, judging from his lively discussion of "Planets X" in *Sky and Telescope* in 1991; see Tombaugh, 1991) ever since Lowell and William H. Pickering (1858–1938) founded the Lowell Observatory at Flagstaff, Arizona, in 1894. [The trans-Neptunian planet search continues today in a variety of guises (see Notes 6.5 and 6.8; cf. Matthews, 1991; Stern, 1991, 1993; Kerr, 1992e; Cowen, 1992).]

The orbit of Uranus happened to fit Titius's formula quite well (perhaps partly because the distance to Saturn falls short of the original idea of using its position as an orbital "meter stick" with 100 divisions, analogous to centimeters; i.e., on this scale, Saturn actually is at 95.39 divisions), and the post-facto reactions to the discovery of Uranus gave Titius's formula a reputation that eventually helped to stimulate searches for such objects as Planets X and O (e.g., Tombaugh and Moore, 1980, p. 68), and was, to that extent, responsible — by dint of the determined trans-Neptunian planet search by astronomers at the Lowell Observatory at Flagstaff — for the discovery of Pluto. Ironically, however, Titius's formula played little part in the discovery of the next planet to be discovered after Uranus, Neptune — although the basis for Neptune's discovery fits with the theme of nonlinear orbital phenomena. Neptune was discovered, to cut the story short, because celestial mechanicists were having great difficulty in computing the orbit of Uranus, and eventually it was realized that a proper orbit could be calculated only if the gravitational perturbation by another large planet beyond Uranus was taken into account. [The persons and events that led to the eventual telescopic documentation of such an object, named Neptune (another name, like Pluto, suggesting godly but earth-like *depths* rather than ethereal distance), and finally credited to the German astronomer Johann Gottfried Galle in 1846, are described in some detail by Tombaugh and Moore (1980, Chap. 5; cf. Considine, 1983, p. 1966). Subsequently, it was realized that a complex system of nonlinear planetary resonances was responsible for the origin and evolution of the three outermost planets (see Stern, 1991; 1992, pp. 228ff) — and eventually, indeed, all of the planets (cf. Wisdom, 1980, 1987a, b, c, 1990; Laskar, 1988, 1989, 1990; Milani, 1989; Milani et al., 1989; Laskar et al., 1992; Milani and Nobili, 1992; Sussman and Wisdom, 1992; Laskar and Robutel, 1993).]

As it happens, there is some confusion, at least in translation, about the exact wording of Titius's marginal note, as well as about when he wrote it — a phenomenon of repetitive distortion in the literature of science that is analogous to the parlor game wherein a whispered statement is passed from ear to ear, and then the several-times reiterated statement is compared with the original statement, and is inevitably found to be different, sometimes to the extent of changing its meaning entirely, much to the hilarity of the assembled group who, even if they are scientists, do not seem to notice the implication for the historical foundation of "our state of knowledge." Nieto (1972, p. 1) confirms the year (1766) and the source (presumably the same one cited by Tombaugh and Moore, loc. cit.) as being found in a German translation of *Contemplation de la Nature* by Charles Bonnet (1720–1793), but Nieto states the rule as being normalized to an arbitrary distance of 10 units for the Earth, giving the general form, starting with Mercury, as $r_n = 4 + 3 \times 2^n$, for $n = -\infty, 0, 1, 2, 3, 4, 5$, where $n = 5$ corresponds to Saturn — which gives the values $r_1 = 10$ and $r_5 = 100$ (agreeing numerically with the above quotation of Titius's statement, in which the actual distance, measured in AU, is obtained by dividing the result by ten; there is also a hint here that the success of the simple relation has something to do with the fact that there is roughly one common-log order of magnitude difference between the distances of Earth and

Chapter Notes: Chapter 4

Saturn from the Sun; cf. Note P.4 on "powers of ten"). Roth (1962, pp. 19f) gives a statement similar to the above quotation but dates it as being made in 1772, apparently from a different source of German authorship, and in this source Titius allegedly went on to say that Jupiter fitted the position $4 + 48 = 52$ and Saturn the position $4 + 96 = 100$, speculating that the "missing" position could be accounted for by "moons" (sic) of Mars and/or of Jupiter (an odd statement; perhaps implying that the missing matter was captured by its neighboring planets?). Johann Elert Bode, director of the Berlin Observatory in 1772, enters the picture — as the story goes — by being the person who allegedly first published Titius's observation in the literature of astronomy and actively promoted the idea that there was a "missing planet" between Mars and Jupiter, as well as additional planets beyond Saturn (see the literature sources cited by Nieto, 1972). Thus, although the label "Bode's Law" for the relationship authored by Titius would appear to be a miscarriage of scientific justice, its impact on the history of planetary exploration is in fact owed significantly to Bode's advocacy.

There are almost as many mathematical expressions for the Titius-Bode relationship as there are authors of astronomical texts. Many of them are empirical fits to exponential functions, based on semi-log plots of planetary distances vs. ordinal numbers. For example, Nieto (1972, Fig. 7.1; cf. Barut, 1986, Fig. 1) shows that a reasonably good fit is obtained in this way for all of the planets out to Pluto (cf. Appendix 6, Inset B), where the ordinal number is referred to the Earth as $n = 1$ (making Mercury $n = -2$, and Pluto $n = 7$). The most precise work of this nature is due to M. A. Blagg (1913) and D. E. Richardson (1945), who showed, independently, as discussed in Nieto (1972, Chap. 7), that the best general fit corresponds to a geometric progression to the base ~ 1.728, rather than 2, the latter representing the essential progression in the relationship originally enunciated by Titius (cf. Roy, 1953). [I find it interesting, in view of my discussion of the relationships between geometry and the numbers pi and the golden mean in Note P.4, that 1.728 differs from the golden mean by the factor $(1.728/1.618) = 1.068$, or less than 7%.] The Blagg-Richardson relationship was shown by these authors to be equally applicable to the satellites of the planets, and they developed correction terms on the basis of oscillatory error functions to better fit the deviations of planets or satellites from the simple geometric order vs. distance relationship (cf. Appendix 6, Inset B herein; and Nieto, 1972, Fig. 7.2, and Eq. 7.4). (In Appendix 6, such "deviations" are shown to be more reasonably interpreted as selective occupancies on a grid or lattice of nonlinear resonance nodes.) Kaula (1968, Chap. 5) almost dismisses the "Titius-Bode law" in his discussion of resonance phenomena in the Solar System, mentioning only that *"The ratio of planetary semimajor axes is rather constant: if we count the asteroids as a planet and disregard Pluto, we get for this ratio about $(a_{n+1})/a_n \cong 1.75 \pm 0.20$."* Kaula's proportionality is simply an alternative way of saying that the planetary spacing conforms generally to a geometric progression with roughly the same base that was identified above in the Blagg-Richardson relationship.

Despite discussions and refinements, the original statement by Titius retains a simplicity, elegance, and precision (out to Saturn) that is arresting. A formula composed of only the integers 2 and 3 appears to describe a dynamically complex orbital evolution! Analogously, I have shown (Shaw, 1988a, Fig. 25) that geometric progressions to bases 2 and 3 account for many of the dominant periodicities in the geological record, and that nearly all geological periodicities can be expressed in terms of multiplicative geometric series to the bases 2, 3, 5, and 7 (with exceptions that appear to require higher prime-number bases in multiplicative combinations with 2, 3, 5, and 7). [In view of my discussion of number theory in Note P.4, I should point out that such progressions might be approximated equally well by a logarithmic distribution with base equal to the golden mean and periods arranged according to Fibonacci series (cf. Fein et al., 1985, Fig. 2; Shaw, 1987b). The ambiguities between such alternatives — i.e., between a universal period-doubling type of distribution and a universal quasiperiodic type of distribution — are conspicuous in the periodic/chaotic resonance structures of the Solar System, as discussed below and in Chap. 6 (see Fig. 21 and Appendix 11).] Just as the circumstances and date that attended Piazzi's discovery of Ceres might seem to be almost eerily fortuitous, so, too, were the historical circumstances of astronomical observations that led Titius to write his now-famous marginal note in 1766. If Neptune had already been discovered, for example, he might never have posed such a relationship — because, if he had normalized the planetary distances relative to Neptune, he would not have got the same simple progression that is identified with his name.

Most scientists and historians probably would pass off such circumstances as being fortuitous events, events that have no greater implication than the turn of the cards in a game of chance. But such odd discovery circumstances, I submit, have profound implications with regard to the ordering of scientic knowledge and theories. These implications stem from the fact that "bizarre" or "fortuitous" discovery events are folded into the "recurrence algorithm" of scientific experience and advancement, hence represent a form of self-organization, a self-organization that, over a sufficiently great number of recursions, corresponds to a critical-state process. A terrestrial, rather than celestial, example of such "fateful circumstances," related to the theme of the present work, is revealed in a review by David W. Hughes (1993) of *A Tapestry of Orbits* by Desmond King-Hele (1992). If the "right" sequence of events had not occurred, the famed work by King-Hele on artificial satellite orbits — and its application to the modern field of artificial-satellite geodesy (e.g., King-Hele, 1976), as well as to the terrestrial resonance concepts emphasized in the present volume (cf. Fig. 33) — would not have proceeded in just the way it did, and the rate of development of artificial-satellite observations and applications might have been significantly delayed. Comments of this sort are encountered quite often in the history of science, especially in the personal reflections of scientists on their own work (see, for example, my quote of Clyde W. Tombaugh in Note 6.5 concerning the circumstances preceding and surrounding his discovery of the planet Pluto on 18 February, 1930). Hughes (loc. cit.) also points to another aspect of such serendipitous happenings, as found in a remark by King-Hele concerning the contemporary peer-review system, to wit, "fortunately, the refereeing system had not reached its present level of oppressiveness" (see the analogous evaluation by Glen, 1989). The progress of science, as I suggested in the Prologue, apparently is not independent of the nonlinear dynamics that it has set itself both to analyze and to simulate.

The Titius-Bode relation was widely known at the time of the discovery of Ceres in 1801, and it was therefore presumably known to Piazzi, Zach, and Olbers, as well as to Gauss, but — by the luck of the cards, or, as I interpret it, by the chaotic dynamics of scientifc research — it played little part in the discovery circumstances of Ceres because Piazzi's attention was on the stars rather than on the planets, nor did it play any overt role (as far as I can tell from the fragmentary excerpts on the history of astronomy I have read) in the computations by Gauss. Even now, little notice seems to be taken of the fact that the original form of the statement by Titius is rather closely related to the forms of geometric progressions that have emerged as the common numerical structure of the chaotic regime of nonlinear algorithms — and that modified statements of the Titius-Bode relation are analogously close to resembling the *critical golden-mean nonlinearity* discussed in Note P.4 (cf. Fein et al., 1985; Shaw, 1987b; Shaw and Chouet, 1988, 1989, 1991). Except for the algebraic complication represented by the additive constant 4 — which could be removed by a transformation of coordinates, making Venus the limit of the stable progression, and Mercury one among an infinity of possible stable and unstable states — the geometric progression at positions outside of Venus is identical to a period-doubling cascade of period three, the progression made famous by Li and Yorke (1975) in the statement "Period Three

Implies Chaos" (arguing that the interval between the end of the first period-doubling cascade and the period-three window of, for example, a quadratic equation does not contain all possible periodic and related aperiodic numerical sequences and should not be called fully "chaotic"; cf. May, 1976; Kaplan, 1987). Positions inside of Venus, then, would represent either a chaotic interval or a different periodic band. Alternatively, if the planetary distances were normalized to the position of Saturn, in the transformed coordinate system, we would have the inward sequence $S = 96/96 = 1$, $J = 48/96 = 1/2 = 0.5$, $Ce = 24/96 = 1/4 = 0.25$, $M = 12/96 = 1/8 = 0.125$, $E = 6/96 = 1/16 = 0.0625$, $V = 3/96 = 1/32 = 0.03125$, $Me = 1.5/96 = 1/64 = 0.015625$. If we now redefine the progression to be consistent with the normalization $E = 1$, by adding 0.0375 to each term and multiplying by 10, we obtain the values $Me = 0.5$ $(0.4; 0.387)$, $V = 0.7$ $(0.7; 0.723)$, $E = 1$ $(1; 1.00)$, $M = 1.6$ $(1.6; 1.524)$, $Ce = 2.875$ $(2.8; 2.767)$, $J = 5.4$ $(5.2; 5.203)$, $S = 10.4$ $(10.0; 9.539)$, $U = 20.4$ $(19.6; 19.19)$, $N = 40.4$ $(38.8; 30.06)$, and $P = 80.4$ $(77.2; 39.53)$. [The numbers in parentheses are the Titius-Bode distances followed by the actual distances in AU; i.e., *(Titius-Bode; Actual)*, where the actual planetary distances are from Pasachoff (1991, Appendix 3), except for Ce, Ceres, which is from Considine (1983, p. 242).]

This description does not fit quite as well as Titius's original relationship, but it is now seen to be just the inverted (and reversed) period-doubling bifurcation sequence $1/2^n$, where n is the set of integers *1, 2, 3, 4,* etc. Oscillatory correction functions — or shifts in the numerical base of the geometric progression (e.g., Appendix 6) — are needed in either case to make the correspondences with the actual planetary distances precise (cf. Nieto, 1972, Chap. 7). The greatest departures from this simpler geometrical progression occur at distances near to and beyond the orbit of Uranus, and near to and inside Earth's orbit (cf. Appendixes 6 and 11). Variations on the theme of geometric sequences of resonance ratios in the Solar System are discussed in Chapter 6 (cf. Note 6.8) and are illustrated in Figure 21 and Appendix 11. It can be seen by comparing Appendixes 6 and 11 that, while a period-doubling progression gives a reasonable first-order fit to the planetary distances, more accurate relationships involve transformations of the Titius-Bode relationship — e.g., as expressed by a family of nonlinear-periodic resonances that form an array of geometric progressions embedded within a system of generally chaotic nature. Nonlinear-periodic resonance states, approximated by idealized geometric progressions, are analogous, for example, to the unstable periodic attractor states of Figure 17a.

Although such discussions of numerical progressions sometimes appear to be mathematically arcane, I have argued in the Prologue, as well as in the Introduction and elsewhere (e.g., Shaw, 1987b, 1988a), that the periodic structure of nature is analogous to that of the chaotic regime of simple equations, such as the quadratic equation — an example being the so-called *logistic function*, iterations on which produce the much-cited *logistic map* made famous by the work of May (1976) and Feigenbaum (1979a, b; 1980) — and the sine-circle equation of Table 1, Section D (cf. Shaw, 1987b). The qualitative basis for this assertion is very simple: if the complex numerical fabric of the periodic-aperiodic structure of such chaotic phenomena can be displayed by computer experiments, then it can certainly be displayed by natural processes that operate in computer-like modes, as exemplified by biological phenomena and, as I have attempted to illustrate (Shaw, 1987a; 1988a, b), by volcanological and other geological processes.

Another irony of astrophysical nature raised by the subject of this note — and one of profound import relative to questions of order vs. disorder in the universe — concerns a celebrated observation made by the amateur astronomer Heinrich Olbers. Even if Olbers had been credited with the discovery of Ceres, it is likely that he would still be remembered in science for quite a different sort of observation, but still an observation that impacts questions concerning the periodic structure of natural phenomena. This observation is referred to either as "Olbers's paradox" or as the "blazing sky effect." Probably no one knows (cf. Note 8.15) who first noticed this seeming contradiction of astronomical observations [according to Mandelbrot (1983, pp. 91ff) it goes back at least to the time of Galileo and Kepler; according to Harrison (1984, 1986) it can be traced to a man named Thomas Digges in 1576, when Galileo was 12 years old, and so on], but—just as Bode is often credited in the literature as the author of the planetary relationship previously jotted down by Titius in the margin of a book he was reading—Olbers is credited with enunciating the conundrum, or riddle, of why the night sky is dark.

Astronomers had long noted that no matter in what direction they pointed their telescopes they encountered a star shining against a dark background, an observation that has continued to hold true even as the technology of astronomical observations has grown, and as the depth of penetration of the universe has increased. Olbers pointed out that, assuming it to be true that one should see a star wherever, and however closely, they looked, then the night sky—instead of being dark—should be of uniform brightness, even of a "blazing" luminosity (cf. Pasachoff, 1991, pp. 599ff). Like the Titius-Bode relationship, Olbers's paradox seems to have nearly as many interpretations as there are authors. The dominant one at the present time would appear to attribute the darkness of the sky to a finite age for the universe, and that the farther we look, the greater is the age of the objects that we "sample." If we are looking at a background that represents what existed at a finite time in the past before there were any stars, we would be seeing "beyond" the effective age of the universe, hence eventually we can see nothing (or the primal darkness before God said, "Let there be light"; cf. Note E.1). Some such explanation would follow from the point of view of the Big Bang hypothesis, or—depending on a variety of physical assumptions—from the point of view of the finite durations of stellar luminosities relative to an infinite static universe, or relative to the proper sorts of relative growth rates in an expanding universe. This kind of explanation, to me, has the simplistic ring of the hypothesis-dependent (or icon-dependent) statement, one that logically "explains away" a paradoxical dynamical phenomenon that stands at the very heart of our understanding of reality (see challenges to the Big Bang hypothesis chronicled by Lerner, 1991)—noting that a "paradox" is nothing more than a *seemingly* self-contradictory observation or statement, hence it always refers to a system of logic, and systems of logic inevitably are suspect (cf. discussion in the Prologue; and Spencer-Brown, 1972; Shaw, 1994).

Mandelbrot (1983, pp. 91ff), while not contradicting the relativistic sorts of interpretations made by astrophysicists, points out that such a phenomenon could also be a simple artifact of fractal clustering of stars and galaxies, clustering in which the effective fractal dimension is less than two [i.e., the sources of light are not plane-filling, hence represent a dimension less than that of a filled area, the rest being devoid of light—which could mean primal darkness, cool "dark matter" (cf. Aumann and Good, 1990; Mateo, 1992), or what have you (cf. Milgrom, 1983; Shaw, 1983b, c; Milgrom and Sanders, 1993; Shaw, 1994]. Such a purely geometric interpretation has a natural, even unavoidable, dynamical counterpart, because an essential characteristic of nonlinear-dynamical systems—especially as they are discussed in this book—is that they are fractal-generating phenomena (meaning that the dynamics is characterized in many respects by power-law statistics, whether one considers the dimensions of period-doubling sets, phase-locked sets, systems of strange attractors, or systems of self-organized critical phenomena). Perhaps the simplest nonlinear-dynamical analogue of Mandelbrot's observation is a model in which the dynamics of stellar-galactic structures is so clustered (in the long-range, long-time sense) that it approximates the geometry of a system of phase-locked attractors. Such a system is nonergodic, meaning that it does not uniformly sample all of the potentially available energy space (or equivalently, the phase-space trajectories do not visit every conceivable state of the system with equal probability; or, equivalently, any sequence of trajectories, or sample of the system, is not representative of the system as a whole). Such a system cannot be averaged to give a property that is uniquely representative of the whole, because every average is path-dependent and hence is a function of the scale of observation. Therefore no matter how many point sources of light one allows in the system (e.g., each one being analogous to a point in a computer printout of a nonlinear-dynamical algorithm, where the size of the graph can be arbitrarily large relative to the print font that represents the light source; in other words, depending on the physics of the light sources and sinks in the system relative to its size), the result will be predominantly dark at every scale of magnification — or predominantly light in the case of a black-and-white printer plot).

The problem of scale-invariant (*fractally self-similar*) clustering in cosmology was pointed out more than twenty years ago by de Vaucouleurs (1970) and was recently recapitulated and modified by Luo and Schramm (1992). The analogous problem of a, loosely speaking, phase-locked universe — or, better yet, one that oscillates near the nonlinear-periodic/chaotic transition, the critical-golden-mean nonlinearity (CGMN) of Shaw (1987b) and Shaw and Chouet (1988, 1989, 1991), representing the "edge" of chaos (cf. Chen and Bak, 1989; Ruthen, 1993) — offers a simple model by which we can view the nonuniformity of cyclic space/time processes. As I pointed out in Shaw (1987b), the phase-locked regime of the sine-circle function is directly analogous to a wheel of fortune (e.g., Dauben, 1983) on which only the rational fractions are marked. Thus, by making the divisions arbitrarily small, or the wheel arbitrarily large, one can always "see" a rational fraction by adding more subdivisions (note that the same reasoning applies to harmonic series, and to any clock that measures time in rational increments, linear or nonlinear; cf. Shaw, 1987b, Fig. 1). If the distribution of stars and galaxies in the universe is significantly ordered in time and/or space (i.e., if it is not absolutely ergodic at every scale of observation; cf. Luo and Schramm, 1992; Barnes and Hernquist, 1992), then the universe can be viewed as being analogous to a wheel of fortune (cf. Dauben, 1983). The important property of such a system, one that fits dynamically with Mandelbrot's (1983) geometrical observation concerning the blazing sky effect, and with de Vaucouleurs' (1970) observations concerning self-similar clustering in a nonuniform cosmos, is that no matter how tightly we pack the rational markings of the wheel, or the rational clusterings of cosmological objects, there is always an uncountable infinity of real but irrational numbers (or structures) that exist between any two rational numbers (or structures), however small the rational numbers may be. In other words, just as Mandelbrot observed, such a systematically clustered rational geometry is of "measure zero" (meaning that in the limit of an infinitely packed, infinitely expanding system, there is zero probability of hitting a rational number on the wheel — each representing a star and/or a galaxy — or, equivalently, there is zero probability of "seeing" stars and galaxies at infinite distances even though every pixel of a digitally designed camera of infinite resolving power contains at least one star; see Note 9.1). Such a numerical-dynamical implication is independent of specific physical models or concepts of age, as I have discussed elsewhere in the present volume (cf. Note P.4).

Harrison (1984, 1986) reviews the history of Olbers's paradox from the standpoint of physical models. He attributes the discovery of the "correct" solution to the riddle of the blazing sky effect — preferring this term because he would expunge from the record any credit to Olbers, and perhaps would expunge even the word paradox — to an unlikely pair of observers, one being (expectably) Lord Kelvin (William Thomson, 1824–1907), the other being Edgar Allan Poe (1809–1849)! Harrison's descriptions of the respective arguments advanced by these authors is an intriguing study in contrasts between poetic and scientific metaphors (cf. Shaw, 1994). However, the physical interpretation by Harrison — the "correct one," though there may be more than one resolution to this riddle (a possibility with ontological overtones) — implicates the rate-dependent relationships between star-forming/galaxy-forming processes and the phenomena of ordered clustering that I have emphasized. It should be evident that the two viewpoints — one representing the time-dependence of physical processes, and the other the geometrodynamics of scale-invariant fractal clustering — are not independent and must fit together from the standpoint of cyclic feedback phenomena.

Star formation and galaxy formation are density-dependent processes [e.g., among many treatments of star formation, one by Seiden and Gerola (1982) is particularly simple and instructive; cf. Note 9.2 in the present volume; and Vogel et al. (1988), Shaw (1988a, p. 30), Hunter and Gallagher (1989), Pasachoff (1991, pp. 533ff), Maran (1992, pp. 696ff)]. Accordingly, the fractal scale-invariant nonuniform clustering of density distributions in the universe must somehow correspond to a form of *folded feedback* (cf. Shaw, 1987a, Fig. 51.14) consisting of *positive feedback* (representing the formation of countable stars and galaxies) opposed by *negative feedback* (representing regions where stars are either destroyed by *explosions*, thereby evacuating the locus of a given star while transferring mass and dynamical shocks to other regions of active star formation, or by *implosions*, thereby evacuating the locus of a given star and/or galaxy by "consuming" mass in black-hole-like phenomena, hence also removing mass from the local star-forming process). In such a "folded" feedback system, the more heterogeneous is the clustering of matter, the greater is the local rate of star formation and destruction, hence the greater is the rate of growth of *tunable* density heterogeneity. Does the above description of the heavens constitute a vicious circle, a Catch-22, or a genetic cycle of perpetual rebirth? In the steady state, this type of feedback system is self-similar over many scales of time and size, and — in the context of open-system dynamics — it conforms to, or is *self-tuned* to, the state of *self-organized criticality* (cf. Shaw, 1987b). But if the process is steady and cyclic, rather than static — and is of the open-system type — the return loop by which the densest regions recycle to feed the material balance of star formation in the more tenuous regions also must satisfy the dynamical balance needed to preserve the fractal dimension (or multifractal dimensions) of the overall process within the universal limitation, whatever that may be, or whether, in fact, there is one (i.e., we do not know as yet what is implied by the *faint ripples in the cosmological microwave background* recently announced by Silk, 1992; cf. de Vaucouleurs, 1970; Aarseth, 1973; Alfvén, 1980, 1981, 1984; Lucido, 1985; Lerner, 1991; Shaw and Chouet, 1991; Luo and Schramm, 1992; van den Bergh, 1992; Flam, 1993).

In other words, generically speaking, the description just given fits many kinds of *regenerative feedback* systems, and the types it most nearly resembles are those that describe *biological regeneration*. Both supernova explosions and black holes of various types (e.g., Pasachoff, 1991, pp. 475ff and 585f; Maran, 1992) would appear to represent "return" mechanisms in this cyclic process, a process that is analogous at the galactic and supergalactic scales to other cyclic feedback processes — such as the cycling of meteoritic objects in the Solar System illustrated in Figure 11, and/or the cycling of magma in evolving cratonic volcanic systems that can vary in size and longevity while maintaining approximately the same invariant spectrum of eruptive frequencies and volume rates of magma supply (cf. Smith and Shaw, 1975; Smith et al., 1978; R. L. Smith, 1979; Shaw, 1985, 1988b). Ironically, therefore, cratonic volcanic systems (cf. Smith and Luedke, 1984), such as the evolving system of Idaho-Montana-Wyoming that culminates in the caldera complex of Yellowstone National Park, and the evolving system of Arizona-New Mexico that culminates in the caldera complex of the Jemez Mountains, New Mexico (encompassing the Valles Caldera and flanking cultural landmarks known as the Los Alamos National Laboratory and the Bandelier National Monument, juxtapositions that have always struck me as horribly funny in their mixed metaphors of anthropological-volcanological-cosmological

Chapter Notes: Chapter 4

black humor, a cosmic joke of myriad dimensions that humanity is yet to get, otherwise we would have long since changed our ways), are one form of terrestrial analogue of the expanding universe.

There is nothing contextually different about terrestrial and cosmological processes that require — as most researchers would presume, and as university curricula and federally funded research programs reinforce — separate theoretical descriptions for them. For example, theoretical cosmology is now discussed largely in the languages of special and general relativity, quantum mechanics, and superstring theory, while volcanology and biology are normally restricted to the older and — in practice at least — conceptually "decoupled" languages of Newtonian mechanics and thermodynamics. The only approach that would place all of these studies on a par is epitomized by the methodology of nonlinear dynamics as applied to studies of the phenomenologies of nonlinear systems (*folded feedback*). If a unified physical theory of the universe could be formulated within the rules that have represented the tradition in physics as it has evolved from Newton through Einstein and Schrödinger to contemporaries such as Geoffrey Chew of UC Berkeley, Yoichiro Nambu of the University of Chicago, John Schwartz of CalTech, and Michael Green of Queen Mary College, London — some of the key players in the development of "superstring theory" (Kaku and Trainer, 1987; Davies and Brown, 1988) — it must be capable of explaining the relationship between, on the one hand, structures at every scale that are produced by a given system of axioms and postulated forces (e.g., the fourfold system of the gravitational, the electromagnetic, and the weak and strong nuclear forces that most physicists, and textbooks, say are characteristic of real systems and must be explained in any unified theory) and, on the other hand, structures that are simply generated algorithmically by nonlinear, recursive, critical-state processes that are independent of any physical concepts of mass and force (hence any mass-length-time constructs).

The "theoretical physics" approach (e.g., superstring theories) must eventually — of necessity — invoke the nonlinear dynamics approach (e.g., CGMN theories, as outlined in Note P.4) if it is supposed to be capable of explaining the reality of ordered complexity at all scales of physical phenomenology, whereas nonlinear dynamics does not require *any* traditional assumptions in its descriptions of, and/or syntheses of, ordered complexity, hence is in every respect the more universal, hence the more fundamental theory — i.e., by the usual criteria of science, such as simplicity, elegance, universality, independence from physical assumptions, and even "craziness" (the latter being a criterion that some theoretical physicists seem to cultivate with a traditional pride that smacks of the fraternity brother during hazing week; e.g., see the comment attributed to Bohr, as cited in Kaku and Trainer, 1987, p. 12, concerning a lecture by Pauli in 1958, *"We all agree that your theory is crazy. The question which divides us is whether it is crazy enough."*). Actually, some of the fundamental characteristics of superstring theory — such as those concerning numerical structures, interaction, resonance, symmetry, etc. — are intrinsic to the nonlinear dynamics of systems of coupled oscillators [e.g., compare the discussions in Chaps. 5 and 6 of Kaku and Trainer (1987) with studies in the field of nonlinear oscillations, such as those reviewed by Shaw (1987b, 1991) and in the present volume]. Within characteristic limits — referring to the quasisteady-state regimes of relatively local critical-state processes in systems of much larger or smaller scales (generically representing pumped dissipative structures far from equilibrium; cf. Notes P.2–P.5, I.2–I.6; and Prigogine et al., 1972, 1991; Prigogine, 1978, Krinsky, 1984; Prigogine and Stengers, 1984; Nicolis and Prigogine, 1989) — the biogenetic, magmatic, terrestrial, solar-galactic, and celestial feedback systems described in the present volume are of analogous self-similarity. Furthermore, this self-similarity — if it can be shown to correspond to the same set of metric universality parameters, such as the set that includes the CGMN set that Shaw (1987b) and Shaw and Chouet (1988, 1989,

1991) proposed to be characteristic of many, if not all, natural systems — can only be described as *that which is essential to the living state.*

The geometric/nonlinear-dynamic argument above would refute any finite-age hypotheses, hence the "Big-Bang hypothesis," simply on the grounds that they are indeterminate, or are analogous to undecidable propositions — as discussed in Notes I.2 and I.8 — because we are confronted by functions that involve recursively complex and interdependent variables, such as "distance" and its derivative with respect to "time," as in *Hubble's law* [see the discussion by de Vaucouleurs (1970, p. 1211) vis-à-vis Pasachoff (1991, pp. 559ff) and van den Bergh (1992)]. The same argument suggests, however, that all such problems are dynamically interrelated through the recursive ordering process, and that even if the universe were of infinite age and of unlimited energy, and the production rates and durations of stars and galaxies were infinite, a nonlinear-dynamically *ordered* universe would not display the blazing sky effect. Comparisons and tests exploring nonlinear-dynamical alternatives of this sort would seem to be required for all of the so-called explanations of cosmological phenomena that are based on special physical theories (not excluding those based on Einstein's "general" theory of relativity).

Consideration of the time-distance relation in cosmology — e.g., as expressed by Hubble's law (which, in simplistic form, states that the velocity of recession of a galaxy is proportional to its distance; therefore also explaining the intense current interest in *quasars*, objects with dimensions resembling the stellar range, masses in the galactic range, black holes at their centers, receding at up to 90% or more of the speed of light, and originating from an energy burst that occurred at a very early stage of the "Big-Bang universe"; cf. Pasachoff, 1991, Chap. 32; Kaufmann, 1991, Chap. 27) — suggests that the limits of universal fractal invariance are functions of the size-age indeterminacy, because the closer we get to such limits, the more nebulous and elusive they become (cf. Note I.2 concerning limitations in the concept of algorithmic complexity). Here is another example where fame distorts the nature of an original contribution in science, because, in his 1936 book *The Realm of the Nebulae*, Edwin P. Hubble himself, as quoted by van den Bergh (1992, p. 421), revealed the nature of the cosmological indeterminacy in this poetic passage, *"With increasing distance, our knowlege fades, and fades rapidly. Eventually, we reach the dim boundary — the utmost limits of our telescopes. There, we measure shadows, and we search among ghostly errors of measurement for landmarks that are scarcely more substantial."* It is within the "shadows" and "ghostly errors" of this "dim boundary" that we lose track of our measures of fractal invariance, or any measures of finite limitation (e.g., of the sort discussed by Luo and Schramm, 1992; cf. Coniglio, 1986; Lucchin, 1986; Pietronero and Kupers, 1986). And all physical law must fail us, if on no other grounds than uncertainty, a point that de Vaucouleurs (1970, p. 1211) expressed in terms of stochastic density fluctuations and their effects on gravity and the integral path lengths of light rays of ever increasing length. A fractal or multifractal model might permit one to account for this effect, but limiting uncertainties in the relative periodic-chaotic domains within the "dim boundary" — i.e., oscillations near the "edge" of the nonlinear-periodic/chaotic transition, as well as of the universe — would suggest that if a boundary could be defined for such a universe, it would refer to the *algorithmical'y knowable universe* in a context somewhat analogous to that of fractal basin boundaries (i.e., conceptually not unlike the "fuzzy boundaries" discussed in Note 6.1). According to this general notion, the universe, instead of resembling a spacetime sphere, topological swiss cheese, ensemble of infinitesimal vibrating "strings" (in the sense of an ensemble of "tiny violin strings," etc.), or a wriggling mass of cosmic vermicelli (cf. Kaku and Trainer, 1987, p. 5; Davies and Brown, 1988, pp. 124ff), would thus resemble — as expressed in the only language that is capable of expressing it, as Coleridge has demonstrated in both poetry and essay, the language of metaphor — *a nebu-*

lous puff of vibrant down cast upon the virgin air, sounding as of a symphony of bells and luted strings, a halo, a glory, a coronal orchestration emanating from the ethereal, and yet the recognized, the spectral resonance known but in the eternal mind, the unthinking quality of peace.

Thus is described a cosmic afferent-efferent, reentrant-signaling, *weirdly strange attractor* (a maximum-dimensional, multifractal system of attractors that includes all possible bifurcation, n-furcation, and *strange attractor* sequences) suspended within the phase space that sustains it, its dynamical inverse, the unknowable envelope of transfinite nonrandom possibility (cf. Shaw, 1988c). In this respect, the universe as a "whole" — meaning at both the most local and the most global scales — represents a sort of *cosmosynaptic interface (CSI)* between the algorithmically computable realm and the algorithmically uncomputable realm; i.e., between the knowable and the unknowable, at least in any form that can be numerically quantified by nonlinear dynamics and theories of algorithmic complexity, thus implicating mathematics and physics. The CSI can therefore only be envisioned as a universally ordered open-system *numerometric* (contraction of number + geometry; see Note P.4) structure, a name chosen for want of anything better for a sort of nonlinear euclidean-noneuclidean-Pythagorean paradigm, named by analogy with the relatively low-dimensional structures called *bifurcation bubbles* by Moore et al. (1983) and Knobloch et al. (1986), and *bifurcation synapses* by Shaw (1991, Figs. 8.33–8.35, pp. 131ff). With reference to the properties of metric universality and tunability of nonlinear-dynamical coupled-oscillator systems, it would seem that the CSI model is the nonlinear context within which superstring theory — or any form of GUT ("grand unified theory") — must operate, where the metaphor for its fundamental elements is the musical string in ensemble open (strings with ends) and closed (strings in circles) configurations (cf. Kaku and Trainer, 1987, pp. 3ff and 95ff), which also describes the tunable nonlinear-dynamical coupled-oscillator system called a violin (i.e., involving the tunable frequency spectrum made possible by the vibration modes of each string, the array of strings, feedback, and the resonator box).

If we describe the universe by the same analogy, the "music of the spheres" so avidly sought by Pythagoras, by Kepler, and by uncounted persons before and after them (see Notes P.4, 8.15; and Neville, 1989, pp. 295ff) is played by a metaphorical superstringed instrument wherein the resonator box is the self-referential boundary of the instrument itself. It should be noticed that the modes of the two types of string elements in superstring theory (Kaku and Trainer, 1987, pp. 95ff) are analogous, in nonlinear dynamics, to two of the simplest types of algorithms, (1) the quadratic-map algorithm (*analogous to an open-ended string with free ends*; cf. May, 1976; Shaw and Doherty, 1983; Jensen, 1987; and Note P.1 of the present volume) and (2) the sine-circle-map algorithm (*analogous to a closed, hence cyclical, string without free ends*; cf. Shaw, 1987b; and Notes P.1, P.4, 7.4, and Table 1, Sec. D, of the present volume). By analogy with superstring theory, then, the simplest nonlinear-dynamical model of a stringed-oscillator universe would be a quadratic function coupled with a sine-circle function (see Note P.4). But the quadratic part is already satisfied, in the coupled sense, by the properties of the sine-circle function, and so — voilà — *the sine-circle map is the simplest nonlinear-dynamical model of a superstringed universe*, and the color-coded cover illustration presented with Shaw (1987b), or the same type of construction presented on p. 57 of Crutchfield et al. (1986), is a recursively scanned pictograph (bifurcation diagram) of the music of the spheres, qualitatively including the thermal implications of the universal "optical" spectrum [the uniformly cold, unrippled and unrealizable, microwave background is at the bottom in Shaw (1987b, cover illus.) and at the left in Crutchfield et al. (1986, p. 57)].

It is interesting, and once more ironical, that the digression concerning the fame of the amateur astronomer H. W. Olbers has led to a tour of classical and modern physics, and ends where it started, with the question of growth rates and limits of astronomical observations. Now, however, the growth rates of interest concern the "age" (and/or "size") of the universe (cf. Note 8.15; and van den Bergh, 1992; Freedman, 1992). As de Vaucouleurs (1970) also pointed out in his, lamentably almost ignored, paper on the density-radius relation of clustering and superclustering in the universe—a paper that reveals so much about our ignorance concerning cosmological phenomena that it should be the first thing read, along with May (1976), by every high-school student, and certainly by every first-year college student, regardless of curriculum—the estimated age of the universe has grown exponentially during the past three centuries with a doubling time of 16 years! Obviously, if the Big Bang never happened (cf. Alfvén, 1980, 1981, 1984; Lerner, 1991, Chap. 6), then there is no restriction on the "age" of the universe, and the concept of age is relevant only to the identifiable event markers (calibration points of nonidentical spacetime mechanisms) in a multifractal continuum. This viewpoint is held by a few cosmologists, and most texts give the "age of the universe" as between about 10 and 20 billion years (e.g., Pasachoff, 1991, p. 617; Kaufmann, 1991, p. 556; Dalrymple, 1991, pp. 393ff; cf. van den Bergh, 1992; Freedman, 1992), a number that seems unlikely even on intuitive grounds, because the Earth—an infinitesimal part of a galaxy that is an infinitesimal part of the system of galaxies—has an age that is of nearly the same magnitude (roughly one-fourth to one-half of the age of the universe; cf. Dalrymple, 1991, p. 401). This "compressed" age hierarchy forces the conclusion that, on such a time scale, the galaxies all formed almost simultaneously, because the Milky Way Galaxy, of which we are a part, is statistically indistinguishable in age from the estimates of the age of the universe (cf. Dalrymple, 1991, p. 393; van den Bergh, 1992). Such a conclusion does not fit well with the fact that the density states of galactic clusters and superclusters are heterogeneous and hierarchical, implying a complex and protracted evolutionary process, as discussed above (cf. Barnes and Hernquist, 1992). [If we imagine the universe, as *we* know it, to be analogous to a *chaotic burst* within a continuously intermittent chaotic process (see Chap. 5; esp. Note 5.1), then the "origin" of *our* particular aspect of the universe is analogous to an *intermission* of indeterminate duration.]

The fact that the "age" of the universe has been an ever-changing concept, a fact rarely mentioned in university courses in physics and astronomy, should transform one's entire outlook on "knowledge," yet it seems to make not a dent in the erudition of some scientists who, for almost as long, have written about the "absolute" age of this or that phenomenon and the "completeness" of discoveries in astronomy and in science in general. We can see at once that the near-equality between the growth rates of cosmological age and of asteroid discoveries (Chap. 4) may mean that attention to studies of the asteroids is simply keeping pace with discovery rates in astronomy in general—consistent with the additional observation by de Vaucouleurs (1970, p. 1204) that a doubling time of 16 years just about matches "the growth rate of astronomical progress in general." Furthermore, the doubling time of the discovery rate of comets, which lags that of the asteroids by nearly a factor of two (Chap. 4), would imply that our attention to and understanding of the problems of cometary origin and evolution are "retarded" relative to other aspects of astronomy, a possibility supported by comments in the present volume concerning the speculative state of knowledge of cometary phenomena (e.g., Fig. 13, Insets). If this is true, then we may be in for a dramatic upswing in the observational information on comets and cometary processes in the near future (or even as I write; cf. Fink, 1992), representing a form of critical-state intermittency (see Chap. 5, and Note 5.1) that also describes the nature of episodic advances in all other areas of science. The impact-extinction debates may, in fact, be a sign that a critical-state cascade of comet studies to a new level somewhat closer to parity with other advancements in astronomical knowledge has already begun.

Note 4.3: After searching many books and research reports regarding our state of knowledge concerning the sources of the comets, I feel that a general comment is in order. Several different ideas concerning the original source or sources of cometary materials in the inner Solar System have been mentioned in the text and other Chapter Notes of the present volume. The books alone, not to mention the research papers, that I have cited for background material on the comets — e.g., in the context of relationships with the planets, the satellites, the planetary rings, the asteroids, the Galaxy, and so on — include Kirkwood (1867, 1888); Roth (1962); Struve et al. (1962); Gallant (1964); Berlage (1968); Kaula (1968); Gehrels (1971a, 1979); Nieto (1972); Gingerich (1975); Alfvén and Arrhenius (1975, 1976); Öpik (1976); Szebehely and Tapley (1976); Béland et al. (1977); Carusi and Valsecchi (1979, 1985); Duncombe (1979), Wood (1979); Alfvén (1981); Skinner (1981); Williams (1981); Clube and Napier (1982b, 1990); Morrison (1982a, b); Russell and Rice (1982); Silver and Schultz (1982); Spitzer (1982); Considine (1983); Greenberg and Brahic (1984); Hodge (1984); Holland and Trendall (1984); Roberts (1985); Burns and Matthews (1986); D. Davis et al. (1986, 1989); Raup (1986b); Smoluchowski et al. (1986); Suess (1987); Birmingham and Dessler (1988); Carey (1988); Lissauer et al. (1988); Muller (1988); Roy (1988a, b); Vilas et al. (1988); Atreya et al. (1989); Binzel et al. (1989a); Chapman and Morrison (1989); Clube (1989a); Dormand and Woolfson (1989); Melosh (1989); Weaver and Danly (1989); Xu et al. (1989); Bailey et al. (1990); Beatty and Chaikin (1990); Bertotti and Farinella (1990); Kippenhahn (1990); Lagerkvist et al. (1990); Newsom and Jones (1990); Sharpton and Ward (1990); Kaufmann (1991); Newburn et al. (1991); Pasachoff (1991); Yeomans (1991a); Kaler (1992); Maran (1992); Teisseyre et al. (1992). The fact that several of these literature sources are primarily devoted to asteroids, satellites, etc., provides a major clue to the difficulties involved in formulating genetic theories of the comets. For example, a discussion of the orbital states of, and/or the orbital evolution of, the asteroids inevitably involves consideration of the evolution of the comets (e.g., Weissman et al., 1989). In other words, the origin of the asteroids is inextricably entangled with the origin of the comets, and vice versa (see Fig. 13, Insets, and Note 6.4; cf. Duncan et al., 1987; Stern, 1991; Hoffmann et al., 1993; McFadden et al., 1993).

Although the "majority vote" would indicate that the interim source of "new" comets is the Hills Cloud (including the Kuiper Belt or Disk) — where the Hills Cloud (Hills, 1981, 1986) and Kuiper Belt together comprise that part of the cometary reservoir surrounding the Solar System inside of $\sim 10^4$ AU, the inner cutoff of the comet "cloud" originally delineated by Oort (1950); cf. Figures 9, 11, 13 (Insets), 15, and 21; and Oort and Schmidt (1951), Kaula (1968, pp. 231–38), Bailey (1990), Weissman (1990b, 1991b), Weidenschilling (1991) — there are several different ideas about how objects got into these reservoirs in the first place (see below). In a sense, the question of cometary "paleogeography" has conceptual uncertainties and diversities analogous to paleogeographic reconstructions of Earth's continents. No one knows the original distribution, and the present distribution is not free from theoretical assumptions and justifications. To draw the geological parallel, I contrast five different ideas below — some of them drastically different — for the provenances of comets during the Phanerozoic that might be compared with analogous ideas about the motions of the continents over the same time span (differences in paleogeographic reconstructions for the early Paleozoic seen in recent literature involving interpretations of exotic and/or accretionary terranes should amply demonstrate the uncertainty and diversity of

Chapter Notes: Chapter 4

opinion on the paths of the continents and/or their fragments; cf. Fig. 34 and Appendix 26; and discussion in Chap. 8).

The hypothetical source-concepts for the comets that populate the outer reservoirs are outlined as follows: (1) the perturbed planetesimal "residues" of planet formation (e.g., Kuiper, 1951; Wetherill and Shoemaker, 1982; Duncan et al., 1987; Safronov, 1990; Wetherill, 1991a); (2) intragalactic and/or interstellar sources (e.g., Clube, 1967, 1978, 1982, 1985, 1989a, b, 1992; Clube and Napier, 1982a, b, 1984a, b, 1986, 1990; Napier, 1985, 1989; Stern, 1986, 1987; cf. Stern, 1988, 1991, 1992, 1993; Stern and Shull, 1988); (3) an "exploded" asteroidal planet source (e.g., Ovenden, 1972, 1973, 1975, 1976; Ovenden et al., 1974; Ovenden and Byl, 1978; Van Flandern, 1978, 1981; Van Flandern et al., 1979); (4) protosatellitic cometary halos around all of the protoplanets as well as the protoplanetary system as a whole, i.e., in the terminology of the present volume, a hierarchical fractal distribution of cometary clouds throughout the evolving protoplanetary system (Rawal, 1991; 1992; cf. Bailey, 1987; Stern, 1987; Bailey et al., 1990); and (5) the model presented in Chapter 3 of the present volume (see Figs. 9–11, Appendixes 6 and 7, and Notes 3.2 and 9.2).

Source-concept (5) proposes that the evolution from a binary protostellar stage of the protoplanetary system to the present fractal bimodal-multimodal planetary system — bimodal in gross structure of the planetary mass-distance organization, and multimodal in the properties of the planets + "minor planets" + satellites + comets + particle rings and arcs + dust streams and clouds, etc. (e.g., Appendix 7) — involved a violent event of self-organized destruction of a small close-binary-companion star that produced the terrestrial-planet/giant-planet dichotomy by cataclysmic ejection of mass from the inner protoplanets that were originally formed in a more conventional way (e.g., Petersons, 1984; Sagan and Druyen, 1989), as well as the chaotically self-organized asteroidal-cometary system of interacting, at least partially interwoven, small-object reservoirs (cf. Figs. 11, 13, and 15). This cataclysmic event occurred at a time roughly coincident with, or shortly after, that at which the pre-main-sequence protosolar binary system reached the stellar main sequence (cf. Kaufmann, 1991; Pasachoff, 1991; Maran, 1992), thus representing the birthing event whereby the protosolar nebula became a distinctive central-star planetary system resembling the present-day Solar System. These five concepts do not exhaust published ideas concerning comets or comet-like objects (cf. antimatter "meteors" of Papaelias, 1991), but they serve to highlight some of the major dynamical distinctions between several different interpretations of the history of the Solar System.

So far as I can tell, there is no exclusivity in these five concepts of cometary sources and phenomenologies, and the differences seem to involve mainly issues of locations, scales of effects, and timings. Perhaps the major import is that different workers, using essentially the same sets of contextual observations and principles of celestial dynamics, have argued for drastically different material sources, orbital locations, and trajectories of approach to the inner Solar System in the course of cometary evolution. Debates on this subject are far from closed — rather, in fact, they have just begun. According to my own hypothesis (concept 5 above), the comets might be viewed as the chaotic commingling of the icy breath from the cataclysmic origin of the planets, satellites, and minor planets arising from the partial destruction of a protostellar-protoplanetary stage of a close-binary pre-main-sequence stage of stellar evolution. Such a chaotic evolution encompasses elements of the other four concepts of cometary hypotheses cited above, including possible Oort-Cloud-like condensations around the protoplanets (4), the existence, at some intermediate stage of this evolution, of a residual Kuiper-Disk-like band from which some objects evolved to Hills-Oort-Cloud distances (1), the possibility of residual fragments derived from fragmentation of early-formed and partially icy protoplanets and/or protosatellites (3), and the exchange of cometary objects with the Galactic medium, which contains numerous sources of episodic star formation as sources of active pre-main-sequence binary systems that evolve in similar ways and therefore interact in kind with other protostellar-stellar main-sequence systems (2). My hypothesis concerning the evolution of the Solar System (5) is discussed at greater length in Chapter 3, and in Notes 3.2 and 9.2.

Chapter 5

Note 5.1: The phrase "Big Picture" does not refer to size. It is meant in the sense of the usage by Abraham and C. D. Shaw (1987, Fig. 25). They define a *state space*, S, and a function $D(S)$ that represents the set of all *smooth dynamical systems* mapped on (or within) that state space. Each dynamical system is represented by a point in the state space, hence $D(S)$ is a sort of index of dynamical systems. They call the function $D(S)$ of state space S the *big picture*. Emphasis is placed by Abraham and Shaw on the behavior of trajectories drawn upon the topologies of phase-space surfaces or manifolds, and on sections through these structures — or on time series drawn from the amplitude record of a single coordinate of a trajectory on such structures — which can be classified as to types of attractors, repellors, attractor basins, and transitional relationships between attractor basins. In the present context, there is no restriction on smoothness or types of attractor relationships. The state space represents the dynamical space (*phase space*) of the Cosmos at any and all scales in both continuous and discontinuous modes (note, however, that a discontinuous process — such as the dripping faucet — can be modeled as a continuous and smooth dynamical system, or as a set of such systems; cf. R. Shaw, 1984; Smith and Spiegel, 1987; Shaw, 1991, Figs. 8.28–8.31). Here, I make the assertion that, within this cosmological state space, attractors, and systems of attractors, are related to one another by sets of *universality parameters* (see Notes P.1, P.3, and P.4). These parameters constitute the numerical scaling structure of cosmological phenomena, from the largest supergalactic scale to the smallest subatomic scale.

Crises in phase space can be understood, categorically, as the sets of transient trajectories that describe the changes of state parameters (motions of the system components) that attend transitions between one attractor state and another, or that attend the reconfiguration of a complex attractor state to a modified attractor state or set of multiple attractor states. And whenever there are changes in attractor states — whether they occur as the result of smoothly varying changes in the tuning parameters (see Note 5.2) of either periodic or chaotic attractors, or as a result of abrupt shifts in tuning parameters and/or attractor basins (as in the chaotic transients outlined below) — they are accompanied by changes in the patterns of *spatiotemporal intermittency* in phase space (as represented by time series, wave numbers, and so on). Intermittency as a formal concept has come into vogue in nonlinear dynamics, and is sometimes given specialized mathematical interpretations. For example, reference in the literature is often made to *"types I, II, and III intermittency,"* as though this is a complete classification of transitional spatiotemporal states. These classes of intermittency are descriptively important, but they refer to a particular mathematical criterion for the stability of a trajectory in phase space, relative to a set of partial differential coefficients in the vicinity of a point in that space, called a *Floquet matrix*, which admits of only three types of solutions (see Bergé et al., 1984, Chap. 9; Gwinn and Westervelt, 1986; Schuster, 1988, Chap. 4). This approach offers a graphic way to visualize transition-state effects for analytical systems with specific equations of motion, but it only hints at the types of variations that might be involved in experimental and natural attractor systems. In the present work, *the term intermittency is used to imply that there is a break in a pattern of recurring phenomena — algorithmic, experimental, or natural — in time and/or in space,*

Chapter Notes: Chapter 5

such that any prior criteria used for interpolation, extrapolation, or prediction at particular scales of the motion are interrupted, disrupted, or generally invalidated, at least for some indeterminate sequence of data points or number of iterations of the subsequent recurrences.

The concept of intermittency is the nub of the so-called "periodicity problem" in the natural sciences (cf. Shaw, 1987b; 1994). Whereas any transition between attractor states involves some interval of transient trajectories in phase space, intermittency — as described above — is inevitable in some form and to a greater or lesser degree. For example, transitions between periodic attractors can occur in so few iterations of a computer algorithm, or over so short a time in a physical experiment, that the break in continuity associated with transition-state trajectories (*transients*) is "over with" so quickly that the new periodic attractor state is almost immediately recognizable. More often, especially near the transition to chaos and "beyond," a transient trajectory may persist for a long time, and the distinction between transient trajectories and stable trajectories of a complex attractor may be impossible to determine on the basis of limited data. For example, in computer experiments, thousands of iterations — or even millions of iterations in an attractor with the complexity of biochemical recursion mechanisms — may be required to establish and reproduce a recognizable form, and/or to digitally quantify its characteristics (cf. Shaw, 1988b, Figs. 3 and 4).

In some cases, involving marginal instabilities oscillating in the neighborhood of two or more attractor states, intermittency can take the form of intervals of highly regular and seemingly recognizable patterns interspersed with chaotic breaks and/or intervals of different periodic patterns (cf. Shaw and Chouet, 1988, p. 11 and Fig. 8). *The "intermittencies of nature" are analogous to this artificial example in at least one important respect: short intervals of rigorously repeated and seemingly periodic data do not establish the nature of a recurrence pattern, nor even that a pattern is periodic rather than formally chaotic in character.* This effect essentially sums up the "periodicity controversy" in literature debates concerning periodic extinctions in the fossil record relative to the impact-extinction hypothesis (IEH). [The history of discoveries and ideas underlying the pros and cons of the IEH is reviewed by Glen (1990). Sepkoski (1989) and A. Hoffman (1989) summarize two diametrically opposing viewpoints, in the context of the IEH, concerning whether or not mass impacts and extinctions are directly coupled as a fundamental periodic cause-effect mechanism in biologic evolution. Shaw (1987b) presented a nonlinear-dynamical argument suggesting that neither the statistical criteria used by the advocates of a periodic IEH nor those of opponents who would discredit such analyses are appropriate to the nature of nonlinear-feedback phenomena implied by the IEH, as further developed in the present work.]

Chaotic crises are those crises where a chaotic attractor is the initially stable state, and where the crisis constitutes the intersection — or "collision" — between some part of the chaotic attractor and another attractor state that — initially at least — is unstable relative to the chaotic attractor (cf. Gwinn and Westervelt, 1986; Sommerer and Grebogi, 1992). In many dynamical systems, the boundaries of the stability fields of attractors (their *basins of attraction*) are geometrically complex, and they are often *fractal* in character, meaning that they have scale-invariant but noninteger "widths" that invoke varying degrees of nondeterminism and uncertainty (see Note 6.1) in the specification of attractor divides — analogous to topographic drainage divides of a landscape — called *separatrices* (e.g., McDonald et al., 1985; Abraham and C. D. Shaw, 1987; Moon, 1987, pp. 242ff). In such systems, the conditions that permit the occurrence of chaotic crises, and the accompanying types of chaotic intermittency — or the occurrence of crises and phase-space intermittency in general (see Grebogi et al., 1982, 1983, 1986, 1987a, b, c; Gwinn and Westervelt, 1985, 1986; Sommerer et al., 1991a, b) — may be diverse. The types of temporal and/or spatial intermit-

457

tency can be ambiguous whenever there are small differences in the relative stabilities of two or more attractors (cf. Crutchfield and Kaneko, 1987; Shaw, 1987a, Figs. 51.15–51.18; 1988b, Figs. 6–8; 1991, Figs. 8.17–8.23). In certain cases, a chaotic attractor may "collapse" abruptly to a simple periodic attractor, or to a degenerate state (e.g., a single fixed-point attractor, such as the stable equilibrium position of a pendulum with friction). In such a case, the chaotic attractor is said to be *destroyed* by the crisis, and this type of crisis is called a *boundary crisis* (cf. Grebogi et al., 1982, 1983; Jensen and Myers, 1985; Bucher, 1986; Gwinn and Westervelt, 1986). More complicated chaotic crises, in which simple to complex unstable periodic attractor states are embedded within ("hidden within") the stable basin of attraction, are referred to generically as *interior crises* (cf. McDonald et al., 1985; Gwinn and Westervelt, 1986; Ditto et al., 1989; Savage et al., 1990; Sommerer and Grebogi, 1992).

Sommerer and Grebogi (1992) discuss methods of classification and mediation of interior crises. In contrast to the stated aim of Sommerer and Grebogi, however, that "it is usually desirable to operate nonlinear systems away (in parameter space) from crisis," the present volume takes the almost diametrical position that *natural systems typically tend to operate in the neighborhood of crises*. In fact, I would go so far as to say that the natural state *is* the state of universal crisis. The reason for this statement is that natural systems are in a general state of change (crisis) involving two or more (usually many more) attractors. This is a universal state of flux that might be described as a "competition" between different fractal attractor basins (a phenomenon important to the celestial-dynamic concept of orbital *spheres of influence*, which are equivalent — from the perspective of the present volume — to attractor basins of the planets and their satellites in the Solar System; cf. Note 6.1). In this sense, however, competition just means that many alternative choices, or switching mechanisms, are available to trajectories of the system, giving rise to an indeterminate number of evolutionary paths through an uncountable number of attractor states — a form of self-selection process ("natural" selection) wherein all states are interrelated according to a common numerical structure, a structure ordered by the sets of all possible nonlinear-dynamical universality parameters. Intersections or collisions between attractor basins are occurring perpetually and at all scales according to a *universal principle of critical self-organization*. Within such a universal system, discrimination between attractor states sometimes may be possible only in terms of comparative sets of universality parameters (cf. Note P.4; and Shaw and Chouet, 1991), or by the conceptually equivalent — hence characteristic — scaling in time or in space of families of chaotic transients, the *chaotic bursts* of the sets of interior crises encountered along the way. This general theme is the subject of Chapter 5.

Note 5.2: In nonlinear dynamics, the concept of *tuning* is analogous to the tuning of a musical instrument, but it usually lacks the implication of correspondence with a standard of perfect pitch (an exception occurs where a particular nonlinear periodicity with vanishing bandwidth is sought in an experiment; cf. discussion of the golden-mean tuning of a system of coupled oscillators by Fein et al., 1985; see also Shaw 1987b; Olinger and Sreenivasan, 1988, Figs. 5 and 6; Shaw and Chouet, 1988, 1989, 1991). The tuning of a nonlinear system is a more general idea than the tuning of a piano, but it retains the common implication that the system response is arbitrarily adjusted, either for the purpose of general exploration of the parameter space of a nonlinear algorithm (or experiment), or for the purpose of imposing a particular type of behavior previously demonstrated and/or expected on the basis of other experience or theory. The idea also can be applied to *natural tuning*, or *self-organized tuning*, where the system itself seeks out — in the *cybernetic* sense — a particular mode or modes of response of a subsystem to a more general system (e.g., the dynamics of the cometary reservoirs of the Solar System are tuned to the Solar-Galactic interaction).

Chapter Notes: Chapter 5

The tuning parameter is usually represented by a coefficient that defines the strength of forcing of a system, or the strength of coupling between two interacting systems. In the general case, however, at least two adjustable parameters are implied, one that defines the forcing term, and one that defines resistive forces intrinsic to the system. In other words, tuning depends on terms that define the nature of the energy *added to* and/or *subtracted from* the system and on terms that define the nature of the energy *dissipated by* the system. Although the response of nonlinear systems — except for certain types of static fixed-point attractors and limit cycles, such as the rest position of a pendulum or the homeostatic limit of a feedback response in population dynamics — generally involves some form of oscillation, in the sense of a cyclic phenomenon, the terms *oscillator* and *coupled oscillator* sometimes are restricted to the types of phenomena traditionally called classical oscillators, in the sense of the theory of mechanical vibration (cf. any text in physics). I do not follow such a tradition in the present volume, nor is it followed in the terminology of nonlinear flow, population dynamics, and/or chemical reaction phenomena. My own bias is to view all nonlinear systems as oscillators (actual or potential), because *oscillatory behavior is the norm in feedback systems operating far from mechanical and/or thermodynamic equilibrium*.

Generally speaking, however, in the literature of nonlinear dynamics, terms used to describe the tuning parameters of a system tend to follow conventional usage. For example, the tuning of nonlinear vibratory phenomena (cf. Moon, 1987) is usually expressed in terms of the frequency and amplitude of a forcing function, relative to specified values of damping parameters. Thus, the tuning of the magnetoelastic experiment described by Ditto et al. (1989) is defined in terms of an imposed frequency measured in cycles per second (Hz). Damping factors, in this case, are inherent in the material response function for the magnetoelastic ribbon (Ditto et al., 1989, Fig. 1; cf. Savage et al., 1990) and in the amplitude of the response measured by the movement sensor (see Figs. 17a, b). The critical frequency thus reflects a time-dependent balance between the cycles of forcing and the amplitude of the response that occurs during that cycle time. This net response function is what is mapped onto the phase space of the problem (Fig. 17a) to define its phase portrait and/or Poincaré section (as described in Note P.3) depending on how the response function is sampled. In oscillatory systems, especially at high frequencies, the sampling is accomplished by taking readings at finite intervals of time, usually at some multiple of the forcing period, and the resulting map is a Poincaré section, or what is now often called a *stroboscopic section* of the phase portrait — although, in its original sense, the Poincaré section referred to a map in which the sampling period was equal to the forcing period (cf. Thompson and Stewart, 1986; Moon, 1987).

It is evident that the specification of a tuning frequency does not fully define the response of a nonlinear oscillator, hence the critical frequency of the core attractor of a chaotic crisis is a function of other parameters as well (such as the way in which a sample is clamped, its time-dependent rheology, etc.). Thus the critical condition for a chaotic crisis is a complicated function of experimental conditions (or natural conditions in cases of interest in the present volume). Rather than specifying a critical frequency, it is sometimes more meaningful to describe the critical condition in terms of a critical stress, a critical ratio of stress to strain rate, a critical dissipation rate vs. the power of the forcing function, or some other expression of the net rate-dependent process (e.g., Rollins and Hunt, 1984; Carroll et al., 1987). A general discussion of the critical state in chaotic crises is given by Sommerer and Grebogi (1992). *More at Note 5.1.*

Note 5.3: A major controversy exists in the literature of theoretical and applied nonlinear dynamics concerning the issue of the *dynamical dimensions* of (1) phase portraits (see Note P.3), (2) numerical data sets that record the history of a dynamical experiment

Chapter Notes: Chapter 5

(analog and/or digital computer experiments, physical and/or chemical experiments in the laboratory, feedback systems models, etc.), and (3) data sets that record the spatiotemporal characteristics of natural phenomena (for example, time-series data representing some natural phenomenon, such as fluid turbulence, weather patterns, climate patterns, volcanic eruption patterns, paleomagnetic reversal patterns, solar-flare activity, etc. (e.g., Ruelle and Takens, 1971a, b; Farmer et al., 1980; Packard et al., 1980; Ruelle, 1980, 1989; R. Shaw, 1981, 1984; Takens, 1981; Grassberger and Procaccia, 1983; Nicolis and Nicolis, 1984; Grassberger, 1986; Essex et al., 1987; Kurths and Herzel, 1987; Shaw, 1987a, b, c, 1988a, b, c, 1991, p. 143; Farmer and Sidorowich, 1988; Procaccia, 1988; Shaw and Chouet, 1988, 1989, 1991; Casdagli, 1989; Gaffin, 1989; Sugihara and May, 1990; Tsonis and Elsner, 1990, 1992; Chouet and Shaw, 1991; Gaffin and Maasch, 1991; Lorenz, 1991; Pierrehumbert, 1991; Sornette et al., 1991).

The controversy appears to revolve around a belief that the *dynamical dimension* of the phase space that describes the attractor state or states of a system should also identify the *degrees of freedom* in the dynamics of the process or processes that generated the attractor or attractors (e.g., Nicolis and Nicolis, 1984; Grassberger, 1986; Nicolis and Prigogine, 1989, Appendix 4; Lorenz, 1991). It is my experience — heavily weighted by empirical observations of experimental computer dynamics, resonating, as it were, with my previous studies of natural and experimental systems — that, among many special attributes, two generalizations can be made about nonlinear-dynamical systems: (I) a system of equations that underlies a given *recursion algorithm* (i.e., the computer algorithm used to generate a given set of dynamical data, as expressed by a time series, by a spatial pattern, or by a spatiotemporal distribution of phase-state parameters) *tells us little or nothing about the actual dimensional complexity of a phase portrait generated by that algorithm* (e.g., the simplest recursive scheme, numerical, analogue, or natural, is capable of producing data with dimensional complexities ranging from perfectly redundant fixed-point periodicities to highly uncertain pseudo-random, space-filling, maximally complex, aperiodic arrays; cf. Lorenz, 1963, 1964, 1979, 1982, 1984, 1989; May, 1976; R. Shaw, 1981, 1984; Shaw, 1983b, 1986, 1987b, 1988a, b, 1991; Shaw and Doherty, 1983; Gu et al., 1984; Kaneko, 1986, 1990; Crutchfield and Kaneko, 1987; Wu and Schelly, 1989; Cahalan et al., 1990; Henderson et al., 1991; Ling et al., 1991; Willeboordse, 1992), and (II) the rigorous analytical determination of the dimension of the phase-space geometry of an attractor, whether periodic or chaotic, strange or nonfractal, *tells us little or nothing about the degrees of freedom of the generating algorithm*, whether numerical or natural, beyond the fact that it is recursive in nature (i.e., feedback interaction is indicated, demanding a function with at least one maximum, such as a quadratic equation; cf. Note P.1).

In order to tell more about the equivalent algorithmic complexity of the generating function — in the operationally literal and minimal sense of the word *algorithm* — we can do so only in terms of the comparative anatomies of *systems* of attractors (cf. Lorenz, 1969). It may help me to make this point by rephrasing it in the form of a question. Can some unknown phase portrait (e.g., one created from a set of natural data) be mapped one to one onto a numerically quantified phase space — e.g., one produced by a numerical algorithm analogous to those I have been discussing — generated by a minimal *type function*, where the term "type function" is used in the same sense that we would use it in identifying a *type specimen* of a biologic species? Obviously, this is not a trivial question, and the phrase "mapped one to one" is a loaded one (cf. Lorenz, 1969, 1984, 1985, 1989). If the system in question contains chaotic elements, the general answer to the preceding question is "no." *But* this does not mean that we cannot learn something fundamental about the system that produced the data set. Here an *operational* qualification becomes important. The operational context suggests to me that the first thing I want to know about the characterization of a *blind sample* of phase space (the "unknown" data set from a natural system) is the extent to which it can be compared to phase portraits generated by algorithms that I know something about: (A) Could it have been produced by a one-independent-variable quadratic equation, and the like? (B) Does a "one-humped control curve," or equivalent control surface, adequately describe the system of attractors? or (C) Does the unknown pattern require, at the very least, a cyclic control function analogous to a sine-circle map, etc. (cf. Shaw, 1983b, 1986, 1987b; Jensen et al., 1984, 1985; Bohr et al., 1984; Halsey et al., 1986; Olinger and Sreenivasan, 1988; Shaw and Chouet, 1988, 1989, 1991)? And again, even if the answers happened to be "yes" to question (B) and "no" to question (C), for example, that *still* does not tell me that such an equation had anything to do with the process or mechanism that produced the data. What it *does* tell me is that the natural process includes at least some elements of *scaling* that are analogous to the scaling coefficients of a numerical algorithm. And *that* information *is* important, because it begins to give me clues about the possible universality parameters that may be characteristic of the natural system (cf. Notes 8.13–8.15).

The reason that we need information on *systems* of attractors, and why the dimensional specification of the blind sample *as a unique attractor* is not sufficient to characterize the dynamical context of a natural phenomenon, is that a specific attractor state can be a member of the sets of all attractors produced by a large and *nondeterminable* number of different algorithms (see the discussion of self-referential systems in the Prologue; cf. Note P.7). For example, the period-doubling sets of a quadratic map and a sine-circle map — or the period-doubling sets from different regimes of either one of them — are topologically indistinguishable from the standpoint of scale invariance. That is to say, given a picture of a period-doubling numerical cascade — meaning one that represents some range of *tuning parameters* (see Note 5.2) in any one of an indeterminate number of different possible generating functions (e.g., May, 1976; Feigenbaum, 1979a, b, 1980; Jensen, 1987; Shaw, 1987b, cover illus.) — it is impossible, out of context, to identify the algorithm that generated it. This is a qualitative, and alternative, statement of the concept of *metric universality* (see Notes P.1 and P.4) that was forcibly brought to the attention of the nonlinear-dynamics community by Mitchell Feigenbaum (loc. cit.; cf. Gleick, 1987) — and by predecessors that included Stanislaw Ulam and a distinguished list of early researchers at the Los Alamos National Laboratory (cf. Anon., 1987). In other words, certain multiplicative regimes of maps produced by different nonlinear-dynamical algorithms *possess common sets of universality parameters*. Therefore, what we first really want to know in an analysis (really meaning "a quest for recognition") involving any blind sampling of dynamical data, natural or experimental, is what universality class or classes, and what set or sets of universality parameters are compatible with it as a representative sample of a unique system of attractors (cf. Horton et al., 1983; and the discussions of universalities and/or *pattern periodicities* in Notes 8.4 and 8.14; and Shaw, 1987a, b, 1988b, 1991; Shaw and Chouet, 1991). If we cannot do at least this much, then we have essentially no clue concerning the type of control function that *might* be required to describe the system in any deterministic mathematical sense, or that *might* be expressed by a specific set of partial differential equations of the motion — not to mention the equations needed for a prescription of all the forces and flows involved in the dynamics that are relevant to a determination of how the system came into being relative to any other "neighboring" dynamical systems in the first place (see Note P.6).

Given information about the universality class, of course, we are still a long way from describing the physical conditions that generated our blind sample. But we do know, in that case, what types of experimental systems, numerical or physical, behave similarly — at least to that level of approximation. To my way of thinking, as a person who has spent a great deal of time making qualitative and semiquantitative observations on natural systems, this is a

Chapter Notes: Chapter 5

huge conceptual advance (e.g., Note 8.13). Even if we can tell only the qualitative similarities and differences between samples of natural data — e.g., a behavior that looks like an attractor sequence from a quadratic map vs. one that looks like an attractor sequence from a sine-circle map — then we have taken a major step in guiding decisions concerning further discriminatory tests. The former pattern could represent any simple oscillatory feedback system, but the latter pattern suggests a new order of complexity, such as that seen in a cyclic system involving coupled oscillators. As a case in point, I recently noticed (Shaw, 1991, Fig. 8.31) an interesting parallel between the behavior of bifurcation sequences of the sine-circle map and a set of experimental data that described the tuned intermittent modes of a *dripping-faucet system* plotted as a bifurcation sequence (Wu and Schelly, 1989, Fig. 4). In other words, the comparison showed that some regimes of a dripping-faucet experiment are directly analogous to *phase-bias tuning* of the sine-circle function of Table 1, Section D. And *this* observation underscored the fact that the characteristics of the dripping faucet are universal in nature, a fact that had not been emphasized in previous work where the dripping-faucet model had been used only as an aid to the visualization of an intermittent natural process that otherwise had no similarity to a faucet (cf. Hones, 1979; R. Shaw, 1984; Cahalan et al., 1990). And the recognition that many classes of intermittent behavior in natural systems resemble both the dripping-faucet analogy *and* the coupled-oscillator analogy — rather, that the coupled-oscillator analogy, as a characterization of an *actual* dripping faucet, applies to complex natural systems in general — is, in my research experience with volcanic systems and seismic phenomena, a landmark discovery (cf. Shaw, 1987b; Shaw and Chouet, 1988, 1989, 1991), just as it has been in other research fields that have addressed such physical phenomena as (to name just a few applications) solid-state electronic devices, mechanical oscillators (such as pendulums, etc.), convecting flows, magnetospheric dynamics, and acoustic cavitation (e.g., Bohr et al., 1984; Jensen et al., 1984, 1985; Gwinn and Westervelt, 1986, 1987; Lauterborn and Holzfuss, 1986, 1991; Sreenivasan and Meneveau, 1986; Meneveau and Sreenivasan, 1987; Olinger and Sreenivasan, 1988; D. N. Baker et al., 1990; Sreenivasan, 1991).

The parallelisms, and even self-similarities, between different physical systems are far more informative to me than the premature attempt to characterize, out of context, the dimension of an attractor constructed from a time series (cf. Lorenz, 1969, 1985, 1991). Given a context, however, the more conventional attempt to measure a characteristic *correlation dimension*, or other scaling dimensions, of a set of data (e.g., Grassberger and Procaccia, 1983; Nicolis and Nicolis, 1984; Grassberger, 1986; Essex et al., 1987; Procaccia, 1988; Gaffin, 1989; Tsonis and Elsner, 1990; Gaffin and Maasch, 1991; Lorenz, 1991; Pierrehumbert, 1991) is useful as ancillary information on the variety of possible attractor states a given system may display (e.g., Chouet and Shaw, 1991), and as information on possible resemblances the system may have to other systems that are physically analogous. Interesting cases that come to mind in the terrestrial-extraterrestrial context (cf. Shaw, 1991), in addition to the examples from celestial dynamics discussed in Chapters 3–6 of the present volume, are the treatment of solar oscillations by Kurths and Herzel (1987) and the analyses of the nonlinear dynamics of the Earth's magnetosphere by Hones (1979) and Baker et al. (1990).

We are so conditioned by mathematical descriptions of simple physical laws, theories, and models of idealized mechanisms and processes, that we have transferred that experience to our beliefs about our "understanding" of complex phenomena (cf. Note P.6) — in the present context, meaning natural systems. Lacking a one-to-one correspondence between natural systems and these artificial models, laws, and theories, it is assumed a priori that natural systems are "infinite-dimensional" — an assumption that is related to another assumption, or worldview, of many scientists that the universe is "random" unless or until special cases are proven otherwise. I cannot help but notice an analogy between the (random)/(self-organized) duality of beliefs about the dynamical states of the universe and the (guilty)/(innocent) duality of beliefs about systems of social justice (dynamics). This point may be worthy of note vis-à-vis nonlinear-dynamical studies of social phenomena — such as solutions to the so-called *Prisoner's Dilemma,* for example as studied in the context of *cellular automata, evolutionary games,* and the *evolution of cooperative behavior in social dynamics* (see Note 8.14; and Hamilton, 1964; Axelrod and Hamilton, 1981; Hassell et al., 1991; Ives, 1991; Nowak and May, 1992, 1993; Sigmund, 1992; cf. Mosekilde et al., 1991). Another approach to analogous questions of social guilt and/or innocence stems from concepts of *ambivalence* and/or *frustration* as applied in the context of *spin-glass theories* and their applications to social dynamics (see the tribute to Michael Ovenden early in the Introduction; cf. Chap. 7; and Palmer, 1986; Kinzel, 1987; Shaw, 1987b, pp. 1658f; Dewdney, 1987; Carroll et al., 1987; Stein, 1989; Suarez et al., 1990).

Another point that concerns the *tuning* of complex systems is the observation that in networks of coupled oscillators an increase in the degrees of freedom does not necessarily increase the dynamical dimensions of the resulting attractors. For example, fixed-point behaviors were observed in systems of two to ten coupled quadratic equations — the *logistic equations* of May (1976) and Feigenbaum (1980) — by Shaw and Doherty (1983), and analogous structures have been demonstrated in other studies, such as those by Gu et al. (1984), Kaneko (1986, 1990), Crutchfield and Kaneko (1987), and Willeboordse (1992). Employing the method of Shaw and Doherty (1983), Shaw (1988a, Fig. 19; 1988b, Figs. 9 and 10) demonstrated that a regime of chaotic attractors, previously established in one of the equations, could be systematically slaved to lower dimensions by coupling it with a second equation, even when the tuning amplitude was being increased starting from an initially chaotic state. A fixed point has dimension zero, implying zero degrees of freedom by the above arguments concerning dynamical dimensions. In reality, however, the cases described involved many degrees of freedom in the equations of the system, and the fixed-point and/or simple periodic behaviors were self-determined, in the sense that they were not predictable from those equations. That is, they occurred within, or starting from, a chaotic field of motion. In that sense, the experiments that I just described constitute examples of *controlled chaos,* and/or *synchronization of chaos,* subjects that are currently receiving much attention in the literature of nonlinear dynamics (see Fig. 29; cf. Rapp et al., 1987, 1988; Ditto et al., 1990b; Kaneko, 1990; Brown and Chua, 1991; Hunt, 1991; Pecora and Carroll, 1991; Singer et al., 1991; Garfinkel et al., 1992; Moon, 1992; Murali and Lakshmanan, 1991, 1992; Anishchenko et al., 1992; Carroll and Pecora, 1992; De Sousa Vieira et al., 1992; Rul'kov et al., 1992; Willeboordse, 1992; Shinbrot et al., 1993).

To sum up, the (redundant) dimension of a system tells us nothing whatsoever about the generating functions — and, in nature, evidence of low-dimensional systems, as determined from correlation dimensions, singularity spectra, and so on, tells us nothing in and of itself about the dynamical mechanisms and processes that run the system. The fractal dimension or dimensions of a system are useless to us for the purpose of describing the specific physical makeup of a natural process or processes, but they *are* useful in telling us that there is something universal in the dimensional scaling produced by those physical processes. Often we can infer, therefore, that deterministic but nondeterminable (undecidably complex) factors are involved in the physical phenomena that are operating in a system, but that, by the same token, these phenomena have become mutually coupled and/or interactive in such a way that nonrandom patterns of motion have evolved that have recursively "organized themselves" so as to inculcate ("breed"), within the system, dynamical characteristics that *mimic the dynamical signatures of much simpler systems* (e.g., Chouet and Shaw, 1991).

Chapter Notes: Chapter 6

It does not take a great conceptual leap to realize that the human mind is an epitome of such self-organized dynamical systems. Is the mind infinite-dimensional or low-dimensional? Clearly, it is both, and much more, depending on what modes are induced within it by self-tuning, and/or seem to be imposed on it by "external" tuning. In the social context, therefore, these contrasts may appear to represent non sequiturs, and then the mind is capable of all of the ambivalences and frustrations of the Prisoner's Dilemma, of social automata—and stigmata—and other analogies vis-à-vis spin-glass behaviors. In tandem with these "externally" interactive behaviors, the mind also induces shifts and/or inversions of physiological redundancy structures throughout the globally coupled mind-body "universe" (e.g., involving self-tuned nonlinear-periodic/chaotic behaviors of the heart, circulatory and respiratory systems, endocrine system, reentrant neural feedback system, and immune system—to name a few key response systems and/or mechanisms).

From the standpoint of nonlinear dynamics, the universal mind-body system is arbitrarily and artificially split by disciplinary rules concerning the external/internal vis-à-vis mind/body dualities, and in this we can begin to see the therapeutic violence that resides in the traditional division between the practices of—relative to the applications introduced above—physiological and psychological medicine. At the present juncture in the evolution of humanistic applications of nonlinear dynamics, this artificial separation survives in the tendency to distinguish between different disciplinary types of *dynamical diseases*, even when their common basis in dynamical universality is implicitly acknowledged (for a sampling of these burgeoning fields of therapeutic applications of nonlinear dynamics, cf. Edelman, 1978, 1985, 1987, 1989; Bramble and Carrier, 1983; Mackey and Milton, 1987; Rapp et al., 1987, 1988; Skarda and Freeman, 1987; Glass and Mackey, 1988; Holden, 1988; Alper, 1989; Pool, 1989; F. D. Abraham, 1990; Goldberger et al., 1990; Freeman, 1991, 1992; Lewis and Glass, 1991; Garfinkel et al., 1992; Peterson and Ezzell, 1992). Parallels between the universal nonlinear-dynamical cosmology of the mind-body syndrome and the universal astrophysical nonlinear-dynamical cosmology are rather transparent, including the gamut of psychoses that, at the limit of autism, mimics the behavior of an infinite-dimensional fixed-point attractor with zero degrees of freedom.

Chapter 6

Note 6.1: Conditions of capture or escape of an object relative to the vicinity of a specified primary body depend on a number of factors, such as the *escape speed*, the *sphere of influence* of the primary body, the importance of third-body (fourth-body, etc.) interactions, and the energy and timing of *transfer orbits*. If capture is considered from a strictly linear Newtonian point of view—for example, viewed as the inverse of the problem of placing an artificial object into orbit around the Moon or another planet (called orbital transfer)—it is almost impossible (with certain exceptions), because the conditions for a precise transfer are highly stringent (i.e., in the classical problem, there is little tolerance, or "fuzziness," between different states of the system, so that, as in the alternative states of a perfect two-well potential where there is an exact balance point of unstable equilibrium at the maximum of the separating potential field, an object immediately "falls" one way or another before other factors that may be operating in that vicinity of energy space, analogous to the function of a navigation-rocket "burn," can be brought to bear on the outcome). The technology of space navigation avoids this problem by imposing guidance accelerations that are strategically timed to correct for orbital deviations and that provide the necessary adjustment of velocity for insertion of a space vehicle into a new sphere of influence—meaning within sufficient proximity to another body that its gravitational field dominates that of the original primary (cf. Dubyago, 1961; Bate et al., 1971; Taff, 1985; Roy, 1988a; Roy and Clarke, 1988). In this sense, technology provides the flexibility, or orbital "fuzziness," needed to implement orbital transfers that are inversely analogous to the capture scenarios of natural objects. Nonlinear dynamics, the central emphasis of the present volume, significantly modifies this situation, so that instead of being almost impossible, capture may be essentially inevitable for certain classes of orbital configurations. Only recently has it been realized that the same implication exists for the planning of *optimally efficient transfer orbits* of space probes designed to rendezvous with particular primary objects in the Solar System.

A pioneering breakthrough in the technology of energy-efficient travel to the Moon is described by Edward Belbruno (1992) in terms that are directly relevant to the present discussion. He refers to his theory for an energy-saving route to the Moon as going "through the fuzzy boundary," a name that resonates with the discussions of the present volume concerning fractal (fuzzy) basins of attraction and chaotic crises of diverse types (cf. Chaps. 4 and 5; and Note 5.1). The problem Belbruno set out to solve in 1985 was an attempt to demonstrate the existence of a new and more energy-efficient route to the Moon, one that all the experts in the space program emphatically assured him did not exist because, if it did, someone among the rocket and space scientists working on such problems since the 1920's surely would have found it already. After working on the problem for more than five years, and eventually running out of research support, Belbruno was forced to throw in the towel in 1990 and, temporarily at least, take a teaching job.

At that juncture, one of those "quirks of fate" that I have argued are the socioscientific analogues of chaotic crises (cf. Note 4.2), occurred, and Belbruno had an opportunity to see his theory tested on an almost aborted lunar mission by the Japanese mother ship *Muses-A* that had been launched into Earth orbit in January, 1990, for the purpose of subsequently launching a daughter spacecraft into lunar orbit. That intention had failed for technical reasons, and an adviser to that mission, James K. Miller of the Jet Propulsion Laboratory, who remembered Belbruno's work, asked if his fuzzy-boundary theory might enable *Muses-A* itself to complete the trip to the Moon. This possibility was confirmed, and *Muses-A*, renamed *Hiten* for a Buddhist angel, was diverted to the "fuzzy route" in late April, 1991, arriving in lunar orbit on 15 February, 1992, after performing experiments on the Earth-Moon system since 2 October, 1991, its first arrival near the Moon made possible by the fuel saved in taking the new — if longer — route (Belbruno, 1992, Fig. 2). The energy saved by the new route was about 25 percent — an academic comparison because the previously considered route had already been ruled out of the question for purposes of salvaging the mission of *Muses-A*. [The previously considered route, called the *Hohmann transfer orbit*, was named after the German astrophysicist W. Hohmann who, in 1925, was among the first to foresee the need for and to work out some possible routes of interplanetary missions (cf. Dubyago, 1961; Bate et al., 1971, pp. 163ff; Roy, 1988a, p. 388; Roy and Clarke, 1988, pp. 157ff); Hohmann's method is still the standard one used for computing transfer orbits of interplanetary probes.] And if the energy saved were figured on a "cost per mile" basis it would be far greater. The serendipitous success of *Hiten* made Japan the third nation in history to reach the Moon — and one would think that this achievement would qualify Belbruno as one of Japan's national heroes!

I have described this story at some length because it intersects what some observers and reviewers of the present volume have claimed to be a weak point in my thesis. That is, while some critics have based their chief criticism on a belief that the pattern of impact craters on Earth is "random" (cf. Note I.2), others have based it on my inability to disprove — or my inability to "prove" the alternative to — another long-held belief in geophysics and planetology that orbital capture is a rare and almost nonexistent phenomenon in the Solar System (I have seen it mentioned as an alternative to other mechanisms for the origin of satellites of the planets in only a handful of the nearly one-thousand references

465

Chapter Notes: Chapter 6

I have searched on this general subject). I fully acknowledge the truth of this criticism, in the sense that I cannot give any unique solution to the problem of satellite capture by the Earth. My reply, however, is that this inability is not a deficiency of my hypothesis but an inevitability of nonlinear dynamics. It is now widely known that the planetary system is in every real sense a chaotic dynamical system (Wisdom, 1987a, b, c, 1990; Laskar, 1988, 1989, 1990; Laskar et al., 1992; Sussman and Wisdom, 1992; Milani and Nobili, 1992; Berger et al., 1992). It is therefore impossible to carry out unique solutions to orbital-evolution models. By the same token, however, broad classes of nonlinear-resonance and chaotic-boundary states must be recognized as intrinsic structural attributes in problems of transfer orbits and of orbital motions in general. I return to the significance of the fuzzy-boundary method of Belbruno (and colleagues Jaume Llibre and James K. Miller) below, especially as an entrée to a related form of orbital fuzziness in notions of nonlinear-dynamical orbits and fractal attractor-basin boundaries, after setting forth a few relevant terms and parameters of the problem.

One of the most important parameters of orbital-transfer problems, superficially at least, is the *escape speed* of an object relative to a given primary body. The escape speed is the minimum speed to which an object must be accelerated in order for its trajectory of motion to "escape" from the gravitational influence of its *primary body*, meaning that its trajectory must transcend all ballistic trajectories that would return it to the primary, as well as all trajectories that would leave it locked in a primary-centered orbit (the terms *heliocentric*, *geocentric*, and *selenocentric* are used in reference to the orbital motions about the primary bodies when these bodies are, respectively, the Sun, Earth, and Moon). This definition of escape speed is simplistic, especially for the Earth and other primary objects that have an atmosphere, because it does not take into account the effect on the projectile of frictional resistance, for example, as it passes through the Earth's atmosphere — and many other effects related to the location of launch, to launch configurations, etc. By contrast with the projectile, a space vehicle launched from Earth is gradually accelerated to the escape speed as it passes through the atmosphere, thus avoiding damaging physical effects and requiring a relatively small final acceleration to achieve a velocity exceeding the escape speed as a function of altitude above the surface of the primary.

Restricting the discussion to the Earth, a *capture trajectory* of approach of an external, nongeocentric, orbit is one in which the velocity difference between the approaching object and the Earth's velocity must be less than the escape velocity by the time the object has passed the point of closest approach, or the point at which all braking mechanisms have become negligible (cf. Figs. 19 and 20), noting that — somewhat like Zeno's paradox of the hare and the tortoise — if the approach velocity is very small, whatever the speed, the object will never reach the Earth, unless that velocity difference arises from frictional drag phenomena following entry into the sphere of influence (cf. Notes 1.2 and 3.3), and, in that case, an instantaneous velocity difference of zero means that the object has intersected (crashed into) the Earth's surface (the mean velocity difference between the Earth and a geocentric satellite, of course, is zero relative to Earth's orbital velocity). The escape velocity at the surface of the Earth is about 11.2 km/sec (by contrast, the velocity of the Earth in its orbit, and any other objects orbiting the Sun at this distance, is about 29.8 km/sec), decreasing effectively to zero at a distance equal to the *gravitational sphere of influence* (cf. Bate et al., 1971, p. 40; Roy and Clarke, 1988, p. 162, Table 13.2). Geocentric escape velocities of orbiting objects vary from nearly zero at the limit of the sphere of influence, which for Earth is at a distance of approximately 10^6 km (about two and a half times the lunar distance and about 0.7 percent of the distance to the Sun), to nearly the maximum escape velocity at very low altitudes.

These relationships can be seen in Figure 20, which shows the ratio of Earth's rotation frequency α to a satellite's orbital frequency β vs. the ratio of the satellite's orbital distance a from the Earth to the Earth's Equatorial radius R_E (actually, a is the semimajor axis of an elliptical orbit, but Fig. 20 was calculated for a circular orbit, where the mean orbital distance r and semimajor axis a are equal). An object "falling" directly into the Earth from the limit of the sphere of influence would have an impact velocity (neglecting drag) that is approximately equal to the escape velocity (11.2 km/sec). By contrast, an object caught in the types of resonance states shown in Figure 20 eventually evolves through a sequence of different frequency ratios and distances until it approaches one of the near-surface interior resonances at the far left of the diagram (unless it is slowed sufficiently while it is in an exterior resonance state to cause rapid terminal decay) — although some of the high-inclination exterior resonances can be relatively stable for very long times (cf. Allan, 1971; Coffey et al., 1986; 1991, Fig. 1). In either case — i.e., either a low-altitude interior resonance or a quasistable exterior resonance that is perturbed enough to cause direct terminal decay — the object is moving at a relatively high velocity, at least compared with the rotational velocity at the Earth's surface (~0.464 km/sec at the Equator), at the moment of impact. But in the former case (low-altitude orbit) the object is traveling at a high relative speed along a path *nearly parallel to the Earth's surface* (e.g., see the discussion of the Great Fireball Procession given in Appendix 3a and Note A.1).

Let us say, for example, that an object approaches a distance ratio of $a/R_E \cong 1$; then the frequency ratio will be less than the smallest value shown in Figure 20 ($\alpha/\beta = 1/16$, the strongest resonance calculated by Allan, 1967a). In that particular case ($\alpha/\beta = 1/16$), the horizontal velocity is $16 \times 0.464 \cong 7.4$ km/sec, and the altitude is roughly 200 to 300 km. In other words, whatever evolutionary route a captured object may follow, it will eventually reach the Earth's surface with a relative velocity that is of the same order of magnitude as the escape velocity (11.2 km/sec). But if the object has passed through a close-in resonance state, it will be moving in a trajectory that is almost parallel to the Earth's surface on impact (the geographic location and direction of motion relative to the Earth's surface depend on the particular resonance and on complicated factors involving the object's orbital eccentricity and inclination; cf. Allan, 1967b, 1971). A close-in orbit in which the velocity is very nearly equal to the escape velocity of 11.2 km/sec would correspond to the resonance state $\alpha/\beta \cong 1/24$, if it were attainable. Because of frictional energy losses, however, the geocentric orbital kinetic energy of the object is diminished from its maximum value, and, at this frequency ratio, the apparent value of a is given by $a/R_E < 1$ — meaning that the object has already crashed (cf. Figs. 20 and 21; and Appendix 11). In other words, a geocentric orbital velocity of 11.2 km/sec is not attainable at very low frequency ratios because of the asymptotically increasing effects of friction (cf. Baldwin and Sheaffer, 1971) — in the limit corresponding to the grazing and low-angle "ricochet" trajectories of Schultz and Gault (1990a, b, 1992) and Schultz and Lianza (1992). Orbital decay from such low-altitude resonance states (altitudes of the order 10^2 km or less), with terminal velocities of 7 to 8 km/sec, may explain the dramatic subhorizontal approach trajectories of very-low-altitude bolides — e.g., events such as Tunguska and other grazing meteor encounters and fireballs, as illustrated in Appendix 3a (cf. Notes 1.2 and A.1; and Chant, 1913a, b; Mebane, 1955, 1956; LaPaz, 1956; Cassidy et al., 1965; Baldwin and Sheaffer, 1971; Schultz and Lianza, 1992; cf. Melosh, 1992) — because an object in a hyperbolic orbit would not be expected to travel at a nearly horizontal angle of approach over thousands of kilometers (barring direct insertion on the first pass, and at extremely low altitude, into the *entry corridor* of Fig. 19, representing the very narrow parabolic transition interval between hyperbolic flyby orbits and elliptical capture orbits, and/or collisions).

An efficient capture orbit is equivalent to an efficient transfer orbit in artificial systems control. "Brute-force capture," like brute-force orbital transfer, requires a great expenditure of energy, hence is analogous to the least efficient transfer orbit. By contrast with the case where we are concerned with how much fuel energy must be expended in an artificial satellite orbit to achieve the proper state for transfer to a different primary body, a capture orbit has no equivalent artificial source ("rocket power"), and we are therefore concerned instead with how to dissipate enough orbital kinetic energy in the capture process to compensate the drop in potential energy that a natural object experiences in changing from a larger to a smaller orbit. In both cases, we are concerned with finding a route that requires the least exchange with some external energy source or sink. And this is where Belbruno's (1992) concept of fuzzy routes again enters the picture.

It is my guess that no one had ever tried to use routes analogous to the Belbruno route to the Moon because of a tacit belief that is somewhat analogous to a misinterpretation of the Second Law of Thermodynamics, a principle which can be shown, in one context, to mean simply that there is no such thing as a perpetual motion machine (the inverse statement, that any conceivable form of motion in the universe involves some form of dissipation of energy, is the premise of the present study). In short, the Second Law reduces to the adage "one cannot get something for nothing." Applied to the problems of space navigation — and despite the use of "dissipationless" conservative Hamiltonian equations of motion — it is widely believed that rocket engines (artificial energy expenditures) of some sort are essential to space navigation. Without rockets, navigating a space vehicle is analogous to driving a car down a hilly slope without engine or brakes. This is a risky business, to be sure, but it is possible, by "reading the topography," to negotiate successively smaller valleys, or even to stop, by strategic lateral upgrade "transfers" (detours) — assuming, of course, that the surface is negotiable by automobile, and there is no traffic. The role of dissipation in this case is to make the route taken by the car "fuzzy" relative to what it would be if the contact between the tires and the ground were devoid of friction and there was no frictional drag whatsoever in the wheel bearings and other moving parts of the car (to visualize the role of even a nominal amount of road friction, imagine such a trip on a surface of wet ice; i.e., without friction, the problem of stopping becomes delicate indeed). Analogously, Belbruno (1992, Fig. 2) has shown that paths in space ("fuzzy routes") exist for which, theoretically, no expenditure of energy (acceleration or braking) is necessary, other than that required to get up and out of Earth's gravitational potential well in the first place — just as the car had to get to at least one relative highpoint on the topographic slope before we could let it coast — to enter into orbit about a smaller primary body, such as an orbit about the Moon relative to an initial orbit about the Earth. In effect, Belbruno's fuzzy route to the Moon makes use of the fact that *there is* dissipation of energy in space travel, just as there is for an automobile coasting down a hill on Earth. Otherwise, computation of the exact orbit (steering the vehicle) — like the car coasting on wet ice — would have to be accomplished with vanishing uncertainty, and that is impossible on several grounds (such as roundoff errors, and uncertainties of location in phase space, where the latter uncertainties are subject to any form of nonlinear aberration, not to mention ultimate quantum indeterminism).

Two major effects are important here. One concerns the total change of energy between the alternative orbits, and the other concerns the location of, and time spent in the vicinity of, a "fuzzy boundary" between two or more orbital systems. At this juncture, I will refer to these boundaries — both by analogy, and in principle — as *fractal attractor-basin boundaries*.

In the literature of celestial dynamics and astrophysics, usages of energy terms tend to be lax, an artifact of the tacit assumption that we are dealing with nothing but mechanical changes of kinetic and potential energies of a well-defined system. Although in this note I do not stray far from that assumption (cf. Note P.3), knowledge of the energy differences between alternative paths is important. In many instances, "orbital energy" is simply equated with the reciprocal $x = 1/a$, where a is the semimajor axis of an elliptical orbit (e.g., Oort, 1950; Duncan et al., 1987). This shorthand, where $x \equiv$ *"energy,"* can be understood from the equation for the "total" energy of a two-body system (the total available mechanical energy of the system). Although two-body motion identifies the nature of the energy equation, it should be understood that approximations (sometimes left unstated) are implicit even in textbooks of astronomy and astrophysics. I have found, therefore, that where equations are not explicitly derived it is wise to examine several sources. For one thing, as described in Note 6.4, few treatments of orbital problems pay much heed to the motion of the dominant mass in 2-body to n-body systems, such as the Sun–planet and planet–satellite systems, relative to the center of mass (CM) of the system, even though the role of the CM is implicit, for example, in the equations of motion. For example, Roy (1988a, pp. 100f) shows that the partitioning of velocities of each body in a two-body system depends on the mass ratio of each body relative to the total mass of the system. When one mass, such as that of the Sun, greatly exceeds the sum of all other masses, it is tempting to neglect the effect of the smaller masses on the total energy of the system, or, analogously, the distance between the center of the dominant mass and the CM. Therefore, energy equations — and even Kepler's third law (Roy, 1988a, p. 78; Roy and Clarke, 1988, p. 145) — are usually written in approximate forms, such as the following for two-body motion:

$$E = (mV^2)/2 - (m\mu)/r, \qquad (N6.1.1)$$

in which

$$\mu = G(M_{Sun} + m), \qquad (N6.1.2)$$

and where E is the "total" energy (sum of the kinetic and potential energies, respectively), V is orbital velocity, r is orbital distance from the Sun, μ is the so-called *gravitational parameter*, G is the *universal constant of gravitation*, M_{Sun} is the mass of the Sun, and m is the mass of the planet (cf. Bate et al., 1971, pp. 14ff; Roy and Clarke, 1988, p. 142). Although V is typically called the orbital "velocity," it is really meant in the sense of orbital speed. For example, Montenbruck (1989, p. 54) gives equations of the Cartesian components $(dV/dx, dV/dy, dV/dz)$ of the velocity vector $\mathbf{v} = d\mathbf{r}/dt$, where $\mathbf{r}$ is the spatial position vector and t is time, which may be useful, for example, relative to approach scenarios of Figure 19. Because $m \ll M_{Sun}$ for the planets, the gravitational parameter is usually just written $\mu = GM$. Thus, with

$$G = 6.672 \times 10^{-8}\ cm^3\ gm^{-1}\ sec^{-2}$$

and

$$M_{Sun} = 1.9891 \times 10^{33}\ gm,$$

we obtain for the gravitational parameter

$$\mu = 1.32713 \times 10^{11}\ km^3\ sec^{-2}.$$

I have intentionally used the *cgs* (centimeter-gram-second) system of units, because it is traditional in physics. Celestial dynamics uses a mixture of *cgs* units, *mks* (meter-kilogram-second) units, *English units*, and a variety of other special units, including so-called *SI units (Système international d'unités)*. Taff (1985, p. 94) gives a perceptive and heartfelt commentary on the question of units and conversion factors in celestial mechanics, pointing out that the situation is "a mess," adding "I don't even want to bring up SI units" (a viewpoint with which I emphatically concur). Bate et al. (1971, Appendixes A and B) give tables of some of the commonly used units, and commonly encountered constants, in *canonical units*

Chapter Notes: Chapter 6

(the normalized units of astronomy, such as the astronomical unit, AU), *English units*, and *metric units*. [See standard physics texts — such as Halliday and Resnick (1966, pp. 90ff) — for distinctions between units of mass and units of weight.]

Unfortunately, the accuracy and precision of many of the parameters of astronomical measurements reported in standard textbooks and reference works have changed by small, but sometimes significant, amounts even over the past several years, and one commonly encounters somewhat different values in different literature sources. [Where possible, I have used *The Astronomical Almanac for the Year 1992* (Anon., 1991). The above values of G and M_{Sun}, respectively, are from that work (ibid., p. K6), which gives the value of the canonical astronomical unit (AU) as $1.495\ 978\ 706\ 6 \times 10^8$ km.] If it is necessary, or desired, to convert from or to SI units, an extensive, but simple, system of conversions is provided by Wildi (1991). For example, conversion factors for units of length, ranging from *one parsec* ($= 3.2616$ *light year* $= 2.0626 \times 10^5$ *AU* $= 3.0857 \times 10^{13}$ *km* $= 3.0857 \times 10^{16}$ *m*) down to *one femtometer* ($=$ *one fermi* $= 10^{-15}$ *m*), are displayed in hierarchical format, so that intermediate conversions (to five or six significant figures) can be seen at a glance (e.g., Wildi, 1991, p. 25). Analogous diagrams are available for units of time, energy, angular measurement, electromagnetic measurement, luminous intensity, and so on [except for relative magnitude scales for stars, etc., which can be found in standard texts in astronomy, such as Pasachoff (1991), Kaufmann (1991), or Maran (1992)].

The term

$$(mV^2)/2 \qquad (N6.1.3)$$

is the kinetic energy (K.E.) of the body in its orbit, and

$$-(m\mu)/r \qquad (N6.1.4)$$

is its potential energy (P.E.) relative to the center of mass. The equivalent *specific energies* (energies per unit mass) are obtained by dividing the above terms by m, giving

$$\mathscr{E} = E/m = V^2/2 - \mu/r. \qquad (N6.1.5)$$

In elliptical orbits, specific energies are expressed equivalently, where r is replaced by the semimajor axis a, and the energy equation is written in two parts (because of the eccentricity e), one for the velocity state at *periapsis* (closest point on the orbit relative to the primary) and the other for the velocity state at *apoapsis* (farthest point on the orbit). In the limit of a circular orbit, however, where $e = 0$, and $r = a$, we have

$$\mathscr{E} \to 0,$$

for

$$r = a \to \infty,$$

representing an evaluation of the energy integral in the limit of an infinite circular orbit,

$$V^2/2 - \mu/a \to 0, \text{ ("infinite" circular orbit)} \qquad (N6.1.6)$$

(cf. Bate et al., 1971, p. 16; Roy, 1988a, p. 144f; Roy and Clarke, 1988, p. 144). Every textbook seems to describe the energy relationships somewhat differently. The seeming confusion relates to the fact that the energy function $\mathscr{E}$ is a relative quantity. There is no absolute datum of energy, except by convention, which is equivalent to saying that the evaluation of energy can be given only in terms of some definite integral relative to a chosen zero datum and limit of integration. The "infinite" orbit is clearly the simplest choice in celestial dynamics, where the range of distance scales of interest is very large. Relative to this limit, Eq. (N6.1.6) is equivalent to saying that changes of kinetic and potential energies with change of orbital distance, and/or time, are identified by the proportional relation $\Delta V^2/2 = -\Delta\mu/a$, meaning that as the distance increases, P.E. increases (algebraically) and K.E. decreases. Halliday and Resnick (1966, p. 413) graphically illustrate the relationship between the kinetic, potential, and "total" energies of a mechanical system (i.e., in the absence of other energy sources and sinks: thermal, chemical, electromagnetic, etc.), and Roy (1988a, pp. 113ff) gives the derivation of energy integrals for n-body systems. These equations are the "dissipation-free" idealizations of celestial mechanics (the "conservative-Hamiltonian" energy equations). Because energy is related to positions and velocities in space, and in the equations of motion, Hamiltonian dynamical systems conserve areas and volumes in phase space.

Thus, it is seen that the shorthand

$$x = 1/a \qquad (N6.1.7)$$

for "orbital energy" is a measure proportional to the orbital specific potential and/or kinetic energy of the object (space probe, planet, asteroid, comet, etc.) relative to its distance from the Sun, or artificial or natural satellite relative to its distance from its primary, expressed in terms of the circular reference state [the sign of the potential energy may be confusing; the value $x = 0$ corresponds to the maximum potential energy (least negative value), where the orbital kinetic energy approaches zero (corresponding to an "infinite" orbital period and distance).] Thus, a difference Δx between two orbits identifies the potential energy difference that is compensated by an equivalent change in orbital kinetic energy and velocity. The condition of orbital capture, as identified by these terms, is the *energy of compensation* needed to reconcile the orbital energy balance. In many of the natural circumstances of interest, this compensation takes the form of braking, or dissipation of kinetic energy. Assuming this to be the case, we can now examine, hypothetically, how the nature of fractal basin boundaries may modify this picture.

In the present context, the simplest way to envision the meaning of an *attractor-basin boundary* in celestial mechanics is by analogy with basin boundaries of a watershed. The *sphere of influence* of an object in the Solar System is analogous to a basin in a landscape where water collects, the radius of the sphere being a measure of the "basin size." If we further imagine that the *drainage divides* between the basins are knife-sharp, then we see that each basin has an area of rainfall collection (energy collection in the celestial sense) equal to its size. These knife-sharp divides that separate the basins, in this idealized case, are analogous to linear demarcations between attractor basins in phase space, called *separatrices* (cf. Abraham and C. D. Shaw, 1987, Fig. 9, p. 549). Nonlinear conservative-Hamiltonian systems have analogous structures that define various regions of characteristic "periodicities" of the system — such as parameter values of the equations of orbital motion expressed in terms of eccentricity e, inclination i, and semimajor axis a, etc. (e.g., Table 1, and Fig. 12) — but these structures *are not* describable as loci of convergent flows (attractor basins). Little modification is needed, however, to see that any dissipation whatsoever transforms such systems to evolving *attractor landscapes* of the types described here, wherein the details of the trajectories of that evolution form the essence of the dissipative structures generated within and/or among the attractor basins of the relevant phase space (cf. Jensen, 1987; Schmidt, 1987; Smith and Spiegel, 1987; Lee, 1989).

So far, the analogy of a sharply defined basin of attraction represents the way in which orbital stabilities, energy changes, and transfers from "basin to basin" are treated in traditional methods of orbital navigation. However, if the proper "knife-sharp ridge" and corresponding energy are not located exactly, an infinitesimal error can mean that the sought-for basin is missed relative to adjacent ones. Further, because the spheres of influence of the Sun, planets, satellites, asteroids, comets, etc., differ in size by many orders of magnitude, and because they are moving relative to each other in complex, revolving and rotating reference frames — by contrast with the comparatively static analogy of a drainage landscape — then locating oneself in the space of such an oscillating landscape is difficult indeed. So, too, is the determination of any path by which a relatively small object might be captured by a particular sphere of influence (attractor basin) of interest.

Actually, still using the landscape analogy, the contrast between conservative and dissipative systems is more clearly differentiated by the very concept of an attractor basin and its boundaries. In the first place, the attractor landscape is a kinetic one (consistent with a thermodynamically *open-system* viewpoint) rather than a static one (conservative, hence limited, *closed-system* viewpoint), and the problem is more nearly analogous to following (and/or controlling) the motion of a rolling ball as it plunges, climbs, and jumps from basin to basin — and where, in an analogous *three-dimensional celestial roulette game*, the "jumps" may entirely miss small gravitational potential wells (representing local basins of attraction). Thinking of the problem in this way points to its crux, the real questions that concern assessments of realistic transfer (capture) orbits and their energy exchanges. The central question is: What is the nature of the more realistic boundaries between the various spheres of influence, as attractor-basin boundaries, that represent the potential-energy divides that were characterized in the above idealized case as knife-sharp separatrices? As a first refinement of ideas, we can recognize that these boundaries are not really perfectly sharp, but have finite widths with finite geometric irregularities. In short, they have a certain *fuzziness* of definition, which, as Belbruno (1992) has shown, can be exploited. But, rather than attempting to calculate exactly how "fuzzy" they are, let us first imagine how a system analogous to a watershed would behave if the drainage divides were extremely broad and crenulated instead of sharp.

It is evident that the trajectories of motion describing the collection process in particular basins now will be spread out over an area of uncertainty. Within an increment of area, the trajectories of motion of individual particles of the flow may flow toward more than one basin, depending on the "topography" of the boundary and the gradients of the superimposed irregularities, just as the flow of water in any real watershed depends on the details of the dendritic structures at the distal ends of its tributary systems. These dendritic structures are metaphorically symbolic of the fractal structures of the complex and/or fractal attractor-basin boundaries of nonlinear-dynamical flows, especially those of chaotic systems. We can now rephrase the landscape model of orbital evolution as follows: What are the characteristics of chaotic attractor-basin boundaries vis-à-vis trajectories of motion within and between attractor basins?

Some insights can now be obtained from the actual dynamical behavior of simple chaotic attractor basins. The simplest type of behavior relevant to the hydrologic landscape analogy is the two-well potential, or bistable oscillator, a form of which we have already encountered in the discussion of chaotic crises in Chapters 4 and 5 (e.g., Fig. 17; cf. Shaw, 1991, Figs. 8.10–8.20). In brief, whereas a symmetrical two-well potential field acts effectively as an either/or switch when it is tuned to operate in the periodic mode of bistable oscillation — because trajectories of motion in the neighborhood of the unique maximum of the potential field (separatrix) are equally distributed, with vanishing uncertainty, between the two potential minima (in contrast with the asymmetrical example of Fig. 17) — the same field becomes a fractally self-similar "fuzzy switch" with quantum-like uncertainty when it is tuned to operate at transitional frequencies between periodic and chaotic modes and/or in the chaotic regime. Moon and Li (1985a, b) performed an elegant experiment, using the equation of motion of a two-well potential oscillator, that illustrates the transition between smooth and fractal basin boundaries (e.g., Moon, 1987, pp. 244ff, Figs. 6-23 through 6-28; cf. Shaw, 1991, pp. 112ff, Figs. 8.10–8.20). In the limit of a fully developed fractal

Chapter Notes: Chapter 6

boundary, there is a region of broad but variable width in phase space (and in the physical space of the celestial problem) for which the trajectories of motion will seek one or the other potential minima according to a fractal distribution of probabilities. That is, within a broad range of proximities to the maximum of the potential barrier, individual orbits of the motion will move toward one or the other potential well according to a probability density that has a fractal distribution. In other words, the fuzziness of the attractor-basin boundary may be so great that trajectories beginning within any small increment of area — and at any distance from the potential-energy maximum within that fuzzy boundary — may move, with fractal weightings, toward *either* of the potential wells (or, toward *both* of the potential wells, in a sense somewhat like the paradox of Schrödinger's cat that is both alive and dead at the same time). That is, with repeated tries, a certain fractally weighted fraction of trajectories will move one way, while a complementary fraction moves the other way (the quantum-dynamical overtones of this phenomenology should not be missed, and are raised in Chap. 6 relative to interpretations of Fig. 21 and concepts of coexisting nonlinear-resonant and chaotic attractors in the Solar System). The concept of fractal-basin-boundary widths is treated quantitatively by McDonald et al. (1985), whose Figure 5 shows a fractal attractor basin that is analogous to a *celestial fractal-sphere of influence*, and where details of the fractal structure of the basin boundary are shown at scales of about one-tenth and about one-ten-millionth (!) of the attractor basin as a whole. Analogous properties of attractor-basin boundaries of the driven and damped pendulum have been worked out by Gwinn and Westervelt (1986, Figs. 1–3). This latter case is particularly relevant to problems in celestial dynamics, because there are formal similarities between the orbital equations of motion and the pendulum equation (e.g., Table 1, Sec. B; Kaula, 1968, pp. 207f and 226; Wisdom, 1987a, p. 111; cf. Miles, 1962, 1984; Tritton, 1989; Baker and Gollub, 1990, pp. 4ff).

The similarity between fractal attractor-basin boundaries and Belbruno's (1992) "fuzzy boundaries" is striking. Belbruno's Earth-Sun fuzzy boundary is at a distance of about 1.5 million kilometers from Earth, or roughly half again more distant than the limit of the conventional radius of influence. Other forms of fuzzy boundaries evidently must exist for particular pairs or groups of objects. Assuming, for the sake of illustration, that the fuzzy boundary is the principal fractal basin boundary between the Earth and Sun — within which are the fractal-attractor basins of the Earth and Moon (much like the illustration of nested attractors in Abraham and C. D. Shaw, 1987, Fig. 9) — with a width of hundreds of thousands of kilometers, we can see how the boundary structure can influence the outcomes of particular orbital trajectories. At any particular position within the boundary, a trajectory may move either toward the Earth's basin of attraction or toward the Sun's basin of attraction, with an uncertainty that depends on the fractal structure of the boundary. The nature of the outcome depends on the inverse nature of the two types of problems previously considered in this note: (1) navigational control of orbital transfer, or (2) natural control of orbital capture. It is easy to see that in case (1), a vehicle can be injected into the boundary in such a way that little or no energy need be expended to nudge the outcome back toward the Earth-Moon attractor space (Belbruno, 1992, Fig. 2). Conversely, if a natural object (Earth-crossing asteroid or comet), by reason of other mean-motion resonances or by reason of the probability distribution of encounters within the boundary bandwidth centered about 0.01 AU from Earth (a "huge" target relative to the probability of an orbital encounter with Earth itself), it may become "stuck" in that vicinity as an outcome of one form of *boundary crisis* (cf. Jensen and Myers, 1985; Bucher, 1986; Gwinn and Westervelt, 1986, Fig. 5; Meiss, 1987; Schmidt, 1987; Smith and Spiegel, 1987; Lee, 1989; Cahalan et al., 1990, Fig. 15). Such a crisis, therefore, provides a certain amount of *selection pressure* for the eventual decay of some trajectories toward capture by the attractor basins of the Earth-Moon system. The function of a "crisis" is to retain such objects within that vicinity for a protracted, though formally transient, interval of time, during which time there is an enhanced opportunity for small dissipation effects (from the net actions of the solar plasma, cosmic "dust," gravitational perturbations by other objects, etc.; cf. Alfvén, 1971a, b; Baxter and Thompson, 1971; Danielsson, 1971; Lindblad and Southworth, 1971; Trulsen, 1971; Drummond, 1981, 1991; Wisdom, 1985; Greenberg and Nolan, 1989; Olsson-Steel, 1989; McIntosh, 1991) to accumulate sufficiently to nudge some fraction of these objects toward the Earth-Moon system (*chaotic collisions and scattering* phenomena also play a role in such mechanisms; cf. Eckhardt, 1988; Hénon, 1988; Bleher et al., 1989; Whelan et al., 1990; Korsch and Wagner, 1991). By analogy with the Belbruno problem, some fraction of these chaotically evolving orbits are "selected out" for capture by the Earth, according to the particular sets of routes that involve the least amount of energy dissipation to be captured within Earth's sphere of influence.

Attractor-basin boundaries, in general, are the sites of dynamical crises, meaning simply that such boundaries are the loci of change in the evolution of attractor structures (cf. Grebogi et al., 1982, 1983, 1986, 1987a, b, c; McDonald et al., 1985; Gwinn and Westervelt, 1986; Sommerer and Grebogi, 1992). Within both periodic and chaotic regimes, long-term intermittencies within narrow regions of phase space represent one form of expectable outcomes during transitions between attractor basins (so-called *Pomeau-Manneville intermittencies of Types I and III*; Gwinn and Westervelt, 1986, Fig. 5; cf. Shaw, 1987a, Figs. 56.16 and 56.18, 1988b, Figs. 6 and 9). It is also likely that unstable periodic manifolds exist in the vicinities of chaotic attractors, and, if so, *interior crises* and *chaotic bursts* — discussed in Chapters 4 and 5, and at various places throughout the text — would be variants of the behavior in the vicinities of fractal basin boundaries analogous to Belbruno's (1992) fuzzy boundary. Under those conditions, systematic power-law distributions of bursting times (see Figs. 17, 18, and 31) would contribute to local shower-like bursts of near-Earth objects (cf. Binzel, 1991; Yeomans, 1991b; Anderson and Friedman, 1991; Binzel et al., 1991, 1992; Steel, 1991; Kerr, 1992g).

Candidates for possible unstable periodic manifolds in the vicinity of Earth are numerous. Since the dynamical structure of the Solar System as a whole is chaotic (see Chap. 3), it can be viewed as a nested chaotic attractor within which numerous local attractor states of both locally chaotic and periodic types exist, and the same can be said for the vicinities of each of the planet-satellite attractor systems — and for any satellite-subsatellite attractor systems, etc. Such descriptions of self-similar quasi-scale-invariant behavior (e.g., at the several different scales of orbital motion in the Solar System) also are implied by the concept presented in Chapter 4, and illustrated in Figure 11, that the motions of small objects within the Solar System correspond to a general state of *critical self-organization*, a condition that is closely related to the concept of chaotic crises, as described in Chapter 5. Besides the unstable periodic manifolds of the planets themselves (cf. Note 4.2), a number of other types of periodic structures exist. An obvious type is the family of all possible Lagrangian balance points of three-body interactions (plus, of course, analogous balance points of higher-order interactions).

The *Lagrangian points* — first derived by Joseph Louis Lagrange in 1772 as a prize-winning solution of the "three-body problem" (Kippenhahn, 1990, pp. 39ff) — represent the set of orbital points in a two-body orbital interaction at which the gravitational and rotational accelerations acting on a third body are in balance, where the third body is of negligible mass compared with the primary and its main satellitic body, such as the Earth and the Moon. There are five such points in the simplified three-body problem of Lagrange, three of which are on line with the primary and its satellite (Lagrange points L_1–L_3), and two others (L_4 and L_5) are at the apexes of two opposing equilateral triangles with a common base defined by the line connecting the primary and the satellite (cf. Roy, 1988a,

Fig. 5.2, p. 126; Kippenhahn, 1990, Fig. 18, p. 40). In the Earth-Moon system, L_1 and L_2 straddle the position of the Moon at distances of roughly 58,000 km from its center (i.e., these two points are at about 326,000 km and 442,000 km from the Earth, where the distance to the Moon is about 384,400 km). Points L_4 and L_5 fall on the Moon's orbit (assumed to be circular) at distances ahead of and behind the position of the Moon that are given by chords with lengths equal to the lunar distance. Point L_3 is at a distance of about 379,000 km from the Earth on a line connecting the Earth and Moon, but on the side of the Earth that is opposite to the position of the Moon.

Additional analyses of this problem (Roy, 1988a, pp. 135ff) have shown that the Lagrangian points L_1–L_3 on the Earth-Moon line are unstable to small perturbations, while L_4 and L_5 are conditionally stable *libration points* (meaning that objects in these two vicinities *oscillate* in the vicinities of the theoretical Lagrangian points). In view of nonlinear effects and uncertainties in the scaling of "small" objects, it may be preferable to view all of such points of Lagrangian types as being *metastable points, relative to the stable two-body problem*, and as unstable periodic points of the actual chaotic planetary motions. Metastability in this orbital context can be readily visualized in terms of the attractor-landscape metaphor employed above. Instead of a two-well potential, we have a multi-well potential field with small satellitic potential wells (small local minima of the field) on the flanks of the two relatively large potential wells. Thus, chaotic third-body orbits passing near such locally metastable but globally unstable periodic points may experience chaotic crises and transient bursting phenomena analogous to the transient trajectories displayed in Figures 17 and 18. As mentioned above (cf. Fig. 31), such transient bursts then occur with power-law intermittencies in time. When they are viewed in this way, the Lagrangian points may be sites of significant orbital perturbations of near-Earth objects—and they may act, perhaps, more like localized "popcorn machines" rather than as stable sites of librating objects. This possibility may explain why the Earth-Moon Lagrangian points have been the subject of considerable controversy over the years as sites of small satellitic objects and/or "cosmic" dust (cf. Kordylewski, 1961; Peale, 1968). The same observation applies to controversies concerning small satellitic objects in general, because the prevalence of chaotic crises in the Earth-Moon vicinity implies that the occupancies of the satellitic resonances of Figure 20 are intermittent over a wide range of time scales. Similar conclusions may hold for the Lagrangian points of other planetary bodies. The complex resonances and strong tidal interactions between Jupiter and its satellites, for example, greatly modify the local dynamics, making possible a rich variety of periodic-chaotic phenomena (see the discussion of the Jupiter-Io-Europa-Ganymede interactions in the Prologue compared with the discussion of satellitic resonances and rings around Neptune by Porco, 1991; cf. Cruikshank et al., 1982; Jewitt, 1982; Greenberg and Brahic, 1984; Esposito and Colwell, 1989; B. A. Smith et al., 1989; Kolvoord et al., 1990; Dones, 1991; Esposito, 1991, 1992; Horn and Russell, 1991; Colwell and Esposito, 1993).

When the ideas of multi-well gravitational potential landscapes, and multifractal basins of attraction, are generalized to the scale of the Solar System as a whole, they are found to have an interesting relationship to Ovenden's (1975, 1976) *principle of least interaction action*. This principle was described by Ovenden (1975, p. 486) in the following words, *"viz. that a planetary system of N point masses spends most of its time close to a configuration for which the time-mean of the action associated with the mutual interactions of the planets is a (local) minimum."* If we combine Ovenden's concept with the existence of *multifractal basins of attraction* (i.e., attractor basins involving all fractal spheres of influence of the planets, satellites, etc., and subsets thereof), it would appear that the general concept of a global minimum of the mutual interactions is analogous to the notion of a global basin of attraction of the planetary system as a whole. In other words, the most stable

Chapter Notes: Chapter 6

condition of the planetary system — where the system as a whole may be both multiply periodic and chaotic — is one that corresponds to a very complex "local potential well" relative to the scale of the Galaxy, meaning a potential well in which the local topography consists of a complex system of nested fractal attractor basins (it might also be noted that such a concept may have added implications for path-dependent relativistic phenomena, "gravity waves," and so on).

According to the drainage-basin metaphor, then, the Solar-System composite attractor basin is not only a figurative basin of interior drainage, but it also represents the time-dependent complex topology of that basin that minimizes the potential energy of the landscape as a whole, relative to hierarchies of local potential minima of characteristic amplitudes (planetary attractor basins, etc.) and relative to a persisting system of time-dependent rejuvenating effects (e.g., analogous to the time-dependent tectonic forcing effects that continually modify terrestrial drainage basins). There are many possible variations on this theme relative to the most stable configuration of a single, smoothly varying, continuous basin. Such a configuration is not allowed in the case of a planetary system, because the planets introduce irregularities in the potential surface with varying characteristic amplitudes (analogous to topographic hills and valleys of characteristic sizes in the terrestrial landscape). The local relief, or roughness, of the internal "terrain," in the orbital case, is represented by the variety of possible conjunctions and other configurations of the orbital motions (e.g., the *mirror configurations* and other periodic mean-motion conjunctions of the planets; cf. Roy, 1988a, pp. 119f, and pp. 144ff) relative to the long-term tendencies of the local systems, and the system as a whole, to be always moving toward the most stable limiting state — a state that will never be attained as long as planets exist, because dissipation always will persist in some guise. The system as a whole never achieves a single-valued absolute potential minimum because it is a *driven (pumped) system operating far from equilibrium* (cf. Notes P.2–P.7, and 4.2). The driving energy comes, for example, from spin-orbit coupling between the internal energy of the Sun and the orbital motions of the Sun-planet system about the Solar System center of mass (e.g., Landscheidt, 1988, pp. 270f), and from external coupling of the Solar System motion as a whole relative to more general motions within the Galaxy (e.g., Galactic tidal energy is deposited in the Solar System, coupled with changes in the Solar-System orbit in the Galaxy — a scaled-up analogue of the Earth-Moon tidal interactions relative to their rotational and orbital parameters). In essence, this type of nonlinearly coupled and time-dependent hierarchical configuration is the general notion upon which the construction of Figure 21 is based (cf. Appendixes 7 and 11).

Note 6.2: The present-day synchronous lunar spin-orbit resonance condition (cf. Fig. 19; and Alfvén and Arrhenius, 1969) appears to have been enhanced by a net mass anomaly inherited from the early stages of lunar evolution, an anomaly corresponding to a significant difference between the principal moments of inertia acting about axes normal to the Moon's axis of rotation (cf. Kaula, 1968, Eq. 4.5.38, p. 205). This anomaly would appear to be correlated with the well-known asymmetry of morphological features between the near and far hemispheres of the Moon (cf. Wones and Shaw, 1975; Wilhelms, 1987, p. 103). I infer that the mass asymmetry in the Moon is analogous to the type of longitudinal mass anomaly in the Earth proposed here to have formed during the earliest stages of Earth's history (see the discussion of Fig. 33, Chap. 8; cf. Notes 8.8–8.11; and Appendix 26). According to the *celestial reference frame hypothesis* (CRFH), both of these anomalies are consequences of the earliest and most massive bombardments of both the Moon and Earth accompanying and following formation of the Moon (cf. Runcorn, 1983, 1987) — and both were later modulated by tidal dissipation, with magmatic processes as a contributing factor (e.g.,

Shaw, 1970; Shaw et al., 1971; Gruntfest and Shaw, 1974; Wones and Shaw, 1975; Greeley and Schneid, 1991, p. 998). This interpretation heightens the importance of the tidal evolution of the Earth-Moon system, and of planetary systems in general (cf. Jeffreys, 1920, 1970; Gerstenkorn, 1955, 1967, 1969; Munk and MacDonald, 1960; MacDonald, 1964; Goldreich, 1965; Olson, 1968; Shaw, 1970; Bostrom, 1971, 1992; Shaw et al., 1971; Pannella, 1972; Gruntfest and Shaw, 1974; Klein, 1976; Brosche and Sündermann, 1978, 1982; Melosh and Dzurisin, 1978; Peale et al., 1979; Dzurisin, 1980; Lambeck, 1980; Williams, 1981, 1989a, b; Yoder et al., 1983; Walker and Zahnle, 1986; Sridhar and Tremaine, 1992) by placing a premium on models in which the orbital and rotational parameters of both the Earth and Moon have evolved through an episodic sequence of coupled resonance states since their formation (Alfvén and Arrhenius, 1969, 1975, 1976; cf. Figs. 19 and 37 in the present volume).

Note 6.3: The nearest star at the present epoch is Proxima Centauri, at a distance of about 1.31 parsec $\cong$ 4.27 light years $\cong$ 2.7 $\times$ 10^5 AU [cf. Gliese and Jahreiss (1986), Pasachoff (1991, Appendix 7), or Kaufmann (1991, p. 371), where 1 parsec $\cong$ 3.262 light years, 1 light year $\cong$ 6.324 $\times$ 10^4 AU, and 1 AU $\cong$ 1.496 $\times$ 10^8 km; see Wildi, 1991, p. 25; and Note 6.1 of the present volume]. In addition to the long-time advocacy by S. V. M. Clube and W. M. Napier for consideration of Galactic sources of comet encounters with the inner Solar System (Clube, 1967, 1978, 1982, 1985, 1989b, 1992; Clube and Napier, 1982a, b, 1984a, b, 1986, 1990; Napier, 1985, 1989; Bailey et al., 1990), studies by S. A. Stern (1986, 1987, 1988) have brought yet another perspective to questions of extra-Solar System sources of comets (cf. Smoluchowski et al., 1986). Stern and Shull (1988) also point to thermal processing of Oort Cloud comets (heating events that are analogous to thermal metamorphism in the Earth) by passing stars and supernovae during the lifetime of the Solar System, an effect that adds to the conclusions of the present volume that there is virtually no such thing as a "pristine" meteoritic object or representative sample of the "average cosmic abundance of the elements" (cf. Rotaru et al., 1992; Fink, 1992). Following de Vaucouleurs (1970), I go further to conclude that an average cosmic sample cannot be defined (cf. Notes 3.2 and 4.2; and Maddox, 1993a).

Weissman (1991b) criticizes the Clube-Napier type of hypothesis on the grounds that there is no clear evidence of an interstellar trajectory in the existing record of long-period or "new" comets that have entered the inner Solar System (cf. Oort, 1950; Kaula, 1968, pp. 231ff; Bahcall, 1986; Bash, 1986; Bailey et al., 1990, p. 347). Unless most of the putative interstellar comets were captured by the Solar System, there should be evidence of hyperbolic "flyby" comet trajectories analogous to the diagram at the bottom of Figure 19, where the central object now represents the Sun rather than the Earth (cf. Fig. 13, Insets). However, there could be several reasons why this phenomenon has not been observed over the history of astronomical observations. For one thing, two or three hundred years of observations is a very small sample compared with the intermittency time scales of comet-forming processes in the Galaxy. From this same viewpoint, if interstellar comets emanated from different foci of comet formation in the Galaxy at different times, just as the radiants of meteor showers do, as seen from the Earth (e.g., Kippenhahn, 1990, Fig. 58, p. 168), they would approach the Sun, relatively speaking, on essentially parallel paths (e.g., as if the Sun were overtaking past nuclei of comet formation that were moving with the Galaxy at the average rotation rate of the spiral arms (cf. Schmidt-Kaler, 1975; Shaw, 1988a, pp. 29ff). Accordingly, their trajectories would be near-parabolic ones, where the parabola is the transition between the elliptic and hyperbolic cases of the conic section — or the limit of maximum eccentricity, $e = 1$, of the ellipse. Neither parabolic nor hyperpolic orbits are periodic, but, since the period of any high-eccentricity elliptic orbit with semimajor axis

Chapter Notes: Chapter 6

greater than 100 AU is longer than the recorded history of astronomical observations, it is virtually impossible to say whether orbits with $e \cong 1$ are elliptic, parabolic, or hyperbolic. Such a description would fit with the fact that near-parabolic orbits ($e \cong 1$, $1/a \cong 0$) are in the preponderance (cf. Kaula, 1968, Table 5.7, p. 233), and also with the inference that all such cometary orbits are "new" (first-pass objects) relative to the inner Solar System (inside of Jupiter's orbit). In this same scenario, a capture process might not be as improbable as has been assumed. In addition to the Sun, some number of other stars would already have interacted with each of the "comet factories" in the path of the Sun's Galactic orbit (cf. Clube, 1985). These interactions would have perturbed the relative trajectories of the comets in such a way that all trajectories but those near the "radiant" would have been deflected hyperbolically from the path of the Sun. In effect, therefore, this "scattering" mechanism would act as a collimating device to give a spread of near-parabolic approach trajectories. If there are many such stars in the path of the Sun's (or any star's) approach to a Galactic source of comets, then the collimating mechanism would act as a sequence of third-body capture mechanisms relative to each of the stars (cf. Note 6.1). If such a process has gone on for a long time, relative to the number of different nuclei of comet formation in the Galaxy, then all stars in the Galaxy might be expected to have "Oort Clouds."

According to Stern (1987, p. 187), "Extra-Solar Oort clouds" (ESOCs) are not expected to exist within $\sim 10^4$ AU from the Solar System at the present time, but there would have been many such interactions in the past — i.e., whenever a neighboring star passed within about a tenth of the present nearest-neighbor distance, mentioned above. Stern (1987) discounts mechanisms of direct orbital capture of comets by the Sun from such stellar systems, following the conclusion of Valtonen and Innanen (1982). Thus, we would not expect to see hyperbolic comets at the present epoch by either argument just advanced (mine or Stern's). Weissman's criticism concerning interstellar sources is therefore weakened, either in the sense that hyperbolic orbits are not expected — from either general trajectories in the Galaxy, or relative to an ESOC — or in the sense that extra-Solar comets may represent secular parabolic capture mechanisms and formation of "Oort Clouds" by multiple stellar interactions (orbital-Galactic "agglomeration" mechanisms; cf. Bailey et al., 1990, Chap. 15). Other discussions by Stern (1986, 1987, 1988), Bailey (1987), and Stern and Shull (1988) are informative with regard to the general nature of stellar "Oort Clouds" and their "metamorphic" histories.

The above discussion does not exhaust the possibilities for nonlinear-chaotic ("diffusive") orbital exchanges between the outer regions of interacting stellar comet clouds, especially in view of possible "third-object" effects — e.g., the effect of "fuzzy" orbital resonances involving two other stars and the Sun, analogous to the fuzzy-boundary-transfer mechanism of Belbruno (1992) involving the Sun, Moon, and Earth to achieve orbital capture of an artificial satellite by the Moon. Such exchanges would concern the chaotic histories of chemical source compositions of Oort Cloud comets relative to those from other stellar objects. An attempt to distinguish chemically between bona fide Oort Cloud comets and those from ESOCs, according to the chaotic mixing argument, is probably doomed to failure. The compositions of ESOC comets, like those of solar-system Oort Cloud comets, would be compositionally clustered and heterogeneous on all scales — and even if ESOCs are compositionally zoned in a statistically self-similar manner, chaotic mixing would imply that such zones would not correlate with the orbital parameters of individual comets, a problem analogous to the compositional heterogeneity of the asteroids (cf. Fink, 1992; Rotaru, 1992). Identification of comets according to different Galactic sources would not be possible in such models, because, as concluded above, criteria of chemical provenance in a chaotic cosmos are suspect. Galactic sources of comets may have added to, mixed with, and energized the Oort Cloud population of comets at periodic intervals over the age of the

Solar System—for example, most recently from Gould's Belt sources (see Clube, 1967, 1982, 1985, 1989b, 1992; Clube and Napier, 1982a, 1984a, 1986, 1990; Bailey et al., 1990) — thereby disguising all criteria for distinguishing between Solar System comets and interstellar comets (cf. Weissman, 1986, 1990b, 1991b). Chaotic capture processes involving neigboring stars might also "kick out" Solar System Oort Cloud comets into closer comet reservoirs in a manner analogous to the mechanisms of chaotic crises already discussed for the Asteroid Belt, Kuiper Belt, and the inner Oort Cloud (Hills Cloud) of comets (cf. discussion of Figs. 12, 13, and 15; and Wisdom, 1987a, b, c; Duncan et al., 1987; Torbett and Smoluchowski, 1990).

Note 6.4: The *center of mass* (CM) of a rotating system of objects, or of a system of objects that appear to be revolving about one "central" object, or a system of objects that are rotating and revolving in complex ways without any obvious central object, is the mass-weighted mean of the positions of the objects in some coordinate frame. That is, if one viewed the geometry of a system as being described by a mass-weighted or gravitationally weighted coordinate system, the "distances" in such a system would be related to distances in a linear coordinate system by weighting (normalizing) factors determined by the amount of mass associated with each point in the system (e.g., Roy, 1988a, pp.98ff and 113ff). For a system of point masses, therefore, the normalization simply involves a set of coefficients given by the "point masses," or equivalently by the object masses in a celestial system of separate bodies discretely distributed in space. In celestial mechanics, the CM is called the *barycenter* of the system.

Accordingly, reduced to the case of a two-body system, the position of the CM, designated x_{CM}, is given by the simple proportions

$$x_{CM} = (m_1 x_1 + m_2 x_2) \div (m_1 + m_2), \qquad (N6.4.1)$$

where m_1 and m_2 are the masses of objects 1 and 2 at positions x_1 and x_2, and the respective normalizing coefficients are given by $m_1/(m_1 + m_2)$ and $m_2/(m_1 + m_2)$. This is simply the *lever law* for the relative position of the fulcrum of a gravimetric balance, seesaw, or dumbbell, neglecting the masses of the connecting arms. Expressed in vector notation for a noncollinear *n*-body system in spherical coordinates, the position of the CM is given by

$$\boldsymbol{r}_{CM} = \Sigma(m_i \boldsymbol{r}_i)/M, \qquad (N6.4.2)$$

where the bold $\boldsymbol{r}$'s are the radius vectors of the CM and point masses m_i in the spherical coordinate system, and M is the total mass of the system.

The properties of celestial systems relative to their CMs are implicit in the equations of motion of celestial dynamics, but, oddly enough — and with exceptions — discussion of barycentric motions may be mentioned only in passing, or may not be mentioned at all, in textbooks dealing with the orbital dynamics of the Solar System (cf. Dubyago, 1961; Bate et al., 1971; Kaula, 1968; Taff, 1985; Roy, 1988a; Roy and Clarke, 1988). It is obvious that — by a straightforward transformation of coordinates — the CM should be the natural origin of the coordinate frame for a given orbital system, such as the Earth-Moon system or the Solar System. One reason that the CM is not used to define the origin — other than the traditional view that the primary body is the "center" of a system of satellitic objects — is that the CM is a *virtual position*. There is no object at the CM to identify it as a describable point in the heavens. Yet it has — rather, the orbiting system relative to it has — very interesting properties.

Although we ordinarily speak of the planets as revolving around the Sun, or the Moon as revolving around the Earth, etc., these statements are incorrect. All of these objects really revolve in elliptical orbits around their CMs (e.g., Roy, 1988a, p. 100). And, in the same context, the primary bodies themselves also revolve in elliptical orbits around their CMs. Technically speaking, then, the orbits of the planets represent a family of generally elliptical motions that includes the orbit of the Sun as part of the family (this statement is reminiscent of the concept of self-reference discussed in the Prologue and Introduction; i.e., rather than being distinct from the planets, the Sun is simply another member of a *self-referential set* of orbital motions — and this concept could be carried hierarchically from the smallest scale of orbital motions (e.g., in atoms) to the scale of the universe; see Note 4.2; cf. Breuer, 1991; Zimmermann, 1991).

Ordinarily, little meaning is lost in neglecting the motions of satellites and their "primary" relative to the CM of a system when the "central" mass — the "primary" mass — greatly exceeds the sum of all the other masses. There are exceptions, however, where neglect of orbital motions relative to the CM either complicates the physics of orbital phenomena, especially in the vicinity of a "primary" (e.g., the inner Solar System) or completely obscures important aspects of coupling between the orbital and internal dynamics of the primary body (e.g., the Sun; cf. Landscheidt, 1983, 1988).

One of the simpler cases in point concerns an understanding of the details of the tide-producing forces in the Earth. The center of mass of the Earth-Moon system, CM_{EM}, lies at a point along the line of centers between the Earth and the Moon at a depth of about 1718 km beneath the Earth's surface (e.g., Wood, 1986, p. 498). Considered as a two-body system, the CM_{EM} defines the point relative to which the centrifugal force of the Earth-Moon rotation is defined. This force is constant relative to the centers of mass of the Earth and Moon taken separately. The gravitational/centrifugal forces in the Earth, however, vary as a function of distance between the Moon and points in the Earth. As a consequence, the *net* tide-raising force varies with equal and opposite signs at radial positions in the Earth nearest and farthest from the Moon at any given instant, resulting in the symmetric tidal deformation of the Earth. Rotation of the Earth about its own center of mass is essentially irrelevant to the magnitude and orientation of this force balance, because it does not affect the centrifugal force relative to the CM_{EM}. However, the Earth's rotation *is* relevant to what happens to the tidal bulge raised by the tide-generating force, because, as a tangible distortion of the Earth's shape, the affected mass anomalies (plural because of the variation of the Earth's physical properties, hence the amplitudes of the tidal deformations, with depth) are subjected to another force balance represented by the opposing effects of the forward rotation of the tidal bulge by the greater rotation frequency about the Earth's center of mass, relative to the rotation of the Earth-Moon system about the CM_{EM}, balanced by the gravitational attraction of the Moon on the mass *anomalies* of the tidally deformed Earth. The result, at the current epoch, and throughout much of Earth's history, has been a braking torque (*tidal friction*) on the Earth's rotation (cf. earlier discussion in Chap. 6, and Note 6.2). This simplified account is condensed from a detailed development of tidal phenomena in the Earth-Moon system by Fergus J. Wood (1986; cf. Munk and MacDonald, 1960).

When other bodies are taken into account, the dynamical significance of the CM is even more interesting, as exemplified by the Solar System as a whole. This problem has been studied by relatively few authors, and it seems to be an aspect of orbital dynamics of the Solar System that has been, in certain respects, "swept under the rug." The motion of the Sun relative to the Solar System center of mass, CM_{SS}, is illustrated in Appendix 7A (Inset) on the basis of the data of Jose (1965) and the dynamical phenomena discussed by Landscheidt (1983, 1988). The distance between the center of the Sun and the CM_{SS} varies by up to about 2.2 solar radii, where one solar radius $\cong$ 696,000 km, or by about 1,530,000 km (~0.01 AU). The average period of the Sun's revolution about the CM_{SS}, during the recorded history, varies from about 9 to 14 years, with an average of about 11 years (Landscheidt, 1983, p. 294). The average period is similar to that of solar sunspot cycles, and to the orbital period of Jupiter (the latter is explained by noting that, if the Solar System is treated as a highly asymmetric two-body system, analogous to the Earth-Moon system, the period of "rotation" about the CM is approximately equal to the orbital period of the satellitic body). Clearly, however, the rather large excursions of the Sun relative to the CM_{SS} are of considerable dynamical significance, certainly for the internal dynamics of the Sun, particularly with regard to interactions with the tide-raising effects of the inner planets (cf. K. D. Wood, 1972).

The similarity of the average frequency of the Solar sunspot cycle (cf. Jose, 1965; Eddy, 1977; Landscheidt, 1983, 1988; Friis-Christensen and Lassen, 1991) and the orbital frequencies of the Sun and Jupiter about the CM_{SS} are suggestive of nonlinear, and probably chaotic, resonance phenomena. That is, now that it is known that the orbital motions of the planets are, in detail, chaotic (cf. Laskar, 1990; Laskar et al., 1992; Milani and Nobili, 1992; Sussman and Wisdom, 1992) it is not surprising to see analogous behavior in the orbital motion of the Sun relative to the CM_{SS}. In fact, I would suggest that the documented motion of the Sun relative to the CM_{SS} had already indicated that the Solar System as a whole is chaotic (see Appendix 7A, Inset; and Jose, 1965, Fig. 1; Landscheidt, 1983, Fig. 1). Although the recorded history is only about three hundred years, the orbital motion would appear to be exponentially divergent, at least within the range of about 0.01 AU (i.e., within the historic time span, about three hundred years, the Sun's orbit has not "closed" or converged to an exact period, suggesting a form of "stable chaos," in the sense of Milani and Nobili, 1992). It is possible, therefore, that nonlinear-periodic and/or chaotic motions of the Sun's orbit — analogous to the vibrations of an out-of-balance flywheel — are important influences on the resonance states of the inner planets and of small-object trajectories within the inner Solar System, if not within the planetary system as a whole.

The motion of the Sun about the CM_{SS} is a curious "hybrid" of the essentially "rigid-body" rotation of a two-body system and the Keplerian motion of a satellite relative to its CM. For example, if the Sun, rather than a cohesive object, consisted of a self-gravitating cluster of relatively frictionless particles, the particles would move about the CM_{SS} according to Kepler's laws of orbital motion. The orbital motion of the Sun does describe an (open) ellipse, but its orbital period is far greater than the Keplerian period for the equivalent orbital distance in the Solar System. A particle at an orbital distance from the CM_{SS}, for example, equal to the Sun's radius would have a Keplerian frequency of the order of 10^4 times Earth's orbital frequency, or a period of ~ 1 hour, yet the measured orbital motion of the Sun about the CM_{SS} has a mean period of about 11 years. The difference obviously must be accounted for in terms of the energies associated with material cohesive forces, plus the forces associated with internal motions within the Sun (i.e., associated with the power of the solar dynamo relative to mass transfers within the Sun and frictional dissipation) and in its general vicinity (including frictional dissipation in the nearby planets Mercury and Venus, and, to a lesser extent, in the Earth). The amount of such energy apparently could be computed from the amount of orbital kinetic energy — integrated over the orbital sphere of the Sun's motion relative to the CM_{SS} — that would have to be dissipated to "slow" the Keplerian velocities to those observed. By analogy with tidal friction in the Earth — which, as we have seen, results in the slowing of its speed of rotation (increase in the length-of-day) as well as feedback of the dissipated rotational energy into the orbital motion of the Moon (actually, both the Earth and the Moon relative to the Earth-Moon CM), which leads also to an increase in the length of the month — the dissipation of the "rotational" energy of the Sun-CM_{SS} system feeds back into the orbital motions of the planetary system in general, implying a complex system of tidal phenomena involving the Solar plasma flux, the flux of small-body motions, and even the motions of the planets themselves (see Landscheidt, 1983, 1988, for a discussion of *impulses of torque* in the Solar System; cf. Munk and

MacDonald, 1960; Challinor, 1971; Brosche and Sündermann, 1978, 1982; Hide et al., 1980; Lambeck, 1980; Ward, 1982; Yoder et al., 1983; Carter et al., 1984; Chao and Gross, 1987; Maddox, 1988; Jin and Jin, 1989; Hinderer et al., 1990; Hyde and Dickey, 1991; Dones and Tremaine, 1993; Laskar and Robutel, 1993; Laskar et al., 1993; Murray, 1993).

Note 6.5: It is evident from the searches of the literature cited here, even though these searches are by no means comprehensive, that the dynamical relationships between objects in the Solar System are far more complex than would be indicated by readings of most of the textbooks of astronomy and planetology. In the Third Edition of his treatise on celestial dynamics, titled simply *Orbital Motion*, however, Archie E. Roy (1988a, Chap. 8, p. 240) — in one of those flashes of candor that reveals our common experience, and illuminates the student mind by liberating it from a self-conserving but stupefying awe — remarks, concerning geological evidence on the orbital history of the Earth, "It is humiliating to acknowledge that, even today, celestial mechanics is not capable of making such confident statements on the age, stability, and evolution of the Solar System." According to the theme of the present volume, it is not in *that* regard that celestial mechanics is wanting — and, as Roy's treatise testifies, celestial dynamics, as he has practiced it, has made remarkable sense out of what some would call a "random" universe, or one ruled by "chance" — but only in its tacit reluctance, as in other scientific disciplines, to acknowledge the evidence before its eyes that Poincaré came to emphasize, and that both Michael Ovenden and Archie Roy have abundantly documented, concerning the *necessary ambiguity*, or "symbiosis," between order and disorder in the universe. But this statement is "true" only in the dichotomous logic of a science that has been schizophrenically divided — and remains largely so even today when the paradigm of *nonlinear dynamics and deterministic chaos*, itself loosed among us by Poincaré's experiences of celestial dynamics, has come to the fore — between "the harmonic" and "the random," that some pay vacillating homage to as the realities of shadows cast on cave walls of scientific rites that have denied us the homologies of mind that are so obvious in our language (cf. Shaw, 1994).

Given some reflection on the implications of nonlinear dynamics, the reason for this state of affairs is fairly obvious. For instance, with respect to the identification and enumeration of asteroids, comets, satellites of the planets, and possible undiscovered distant planets or planetoids, the principal requisite for "discovery" is the existence of orbital parameters that are both categorically known ahead of time and are sufficiently regular that the object can be recovered over a predicted sequence of subsequent apparitions. Witness the arduous and still ongoing search for the trans-Neptunian "Planet X," long predicted to exist not only on the basis of the Titius-Bode relation but also from anomalies in the orbital motions of Uranus and Neptune (see Tombaugh and Moore, 1980, Chap. 7).

A dedicated program of optical searches for Planet X was instituted in 1905 by Percival Lowell (1855–1916) of the "Boston Lowells," a man who spent his early adult years in the diplomatic service before surrendering to his fascination with astronomy (see Tombaugh and Moore, 1980, Chap. 7). He founded the Lowell Observatory at Flagstaff, Arizona, ostensibly to further a cause that argued for a Martian civilization on the basis of his observations and interpretations of that planet's "canal system." Eventually discredited, if not defamed, in astronomical circles for that advocacy, he apparently decided to establish prestige for the Lowell Observatory by having it be the first to discover a new planet beyond Neptune (Tombaugh and Moore, loc. cit., p. 83). His dedication to that cause was rewarded on 18 February, 1930, a quarter of a century later, with the discovery of Pluto by Clyde W. Tombaugh (ibid., Chap. 11). Tombaugh did not, as popularly imagined, discover Pluto by staring through a telescope, but by using a so-called "blink comparator" to scan photographic plates taken of certain "cultivated" regions of the night sky (see Tombaugh,

1991). [A blink comparator is an optical device that rapidly alternates between two photographic plates of the same celestial field of view — the background of "fixed" stars — taken at different times. Because relatively fixed celestial objects are superimposed in the time-shifted fields of view, nonstationary objects appear to jump out of the images ("blink"), depending on the frequency of alternation and the time differential between the two photographic exposures. Use of this technique for discovering new objects obviously depends on the proper timing of photographic exposures with respect to the field of view, the relative distance, and the apparent orbital speed of the new objects (see Gehrels, 1971a, b, 1979; Morrison and Niehoff, 1979, regarding "future" discoveries of asteroids; cf. Note 9.1). An analogy might be noticed between this photographic technique and the natural blinking phenomena that led to the discovery of pulsars (see Note I.3).]

It was during Tombaugh's exhaustive scanning of such carefully planned sets of photographic plates that he *confirmed*, on 18 February, 1930 (the historically assigned "discovery date"), the *apparition* — astronomy's somewhat theatrical term for the "appearance," as it were, as if on stage either for the first time or during subsequent orbital *recursions* (the nonlinear-dynamical term for "repeat performances") — of a new object that had apparently first shown up on a plate taken on 21 January, 1930, and was again present in a plate taken on 23 January, 1930 (the so-called *discovery plate*; see Tombaugh and Moore, 1980, p. 127; cf. Tombaugh, 1991). Even though the general regions in which to search were known, it took decades of dogged determination on the part of several different astronomers to accomplish what Lowell had set out to do in 1905 (see the Preface by Tombaugh in Tombaugh and Moore, 1980). Tombaugh himself was both lucky and talented in carrying out the photographic scans that led to the discovery of Pluto, because he accomplished this within thirteen months of joining the Lowell Observatory staff, lacking any university degrees and with only a midwestern farm background of interest in amateur astronomy to recommend him (Tombaugh and Moore, 1980, Chap. 1).

No sooner had the discovery of Pluto been accomplished than doubts arose that it was *the* "Planet X" they had been looking for (cf. Tombaugh and Moore, 1980, Chap. 13). Incidentally, a question that personally interested me, answered on p. 136 of this reference, was how Pluto got its name. Answer, essentially by lottery: the vote went to a suggestion by Venetia Burney, age 11, of Oxford, England (so much for the wisdom of venerable judgment in the naming of the planets). The debate concerning Planet X — as well as interest in the search for it — has flared anew since 1980, stimulated by its possible relevance to the impact-extinction hypothesis (cf. Anderson and Standish, 1986; Matese and Whitmire, 1986; Tombaugh, 1991; Matthews, 1991; Kerr, 1992e; Cowen, 1992). But even before that happened, many astronomers were convinced that something was still missing from the planetary picture. Pluto, including Charon, is much too small to produce the apparent anomalies in the orbits of Uranus and Neptune that some astronomers still insist must be attributed to an object more massive than the Earth. Tombaugh and the staff of the Lowell Observatory were so influenced by this possibility that Tombaugh continued his search for another 13 years after the discovery of Pluto (see Tombaugh, 1961a, b, 1991). Contemplating the idea that he might have missed another object in the scanning of his photographic plates, he wrote (Tombaugh and Moore, 1980, p. 151), "The thought of this possibility makes me shudder yet, after fifty years." The controversy over Planet X persists today, and the basis for it is still the issue of observational coverage versus orbital predictions (cf. Stern, 1991; Tombaugh, 1991; Matthews, 1991, 1992; Kerr, 1992e; Cowen, 1992). Now that we know chaotic orbits are the rule rather than the exception in the Solar System (cf. Wisdom, 1987a, b, c, 1990; Laskar, 1989, 1990; Milani, 1989; Mallove, 1989; Kerr, 1992c; Milani and Nobili, 1992; Sussman and Wisdom, 1992), we are no longer justified in assuming — concerning Planet X or any other planet-size objects at distances greater than

Chapter Notes: Chapter 6

about 30 AU (see Stern, 1991; Weidenschilling, 1991) — that "if they were there, someone would have found them by now" (cf. Tombaugh, 1991).

Such considerations naturally raise questions of scale in the definition of planet-size objects. Here, as in many other descriptive aspects of the Solar System of dynamical importance, criteria often are vague and/or ambiguous. For example, the asteroids, comprising objects ranging from less than a meter to nearly a thousand kilometers in diameter, are generally referred to as *the minor planets* (cf. Bowell et al., 1989). The objects predicted by Stern (1991) likewise would be classed as minor planets, and perhaps there is even one or more "major planets" among them, if any of them approach the size of Pluto. [The diameter of Pluto is ~ 2000–3000 km, depending on the data source: *The Astronomical Almanac for the Year 1992* (Anon, 1991, p. E88) gives the mean equatorial radius of Pluto as 1500 km, while Pasachoff (1991, Appendix 3) cites an equatorial radius of 1150 km, agreeing with a value of 1151 ± 6 km given by Stern (1992, Table 1).] Stern (1991, p. 271) gives a generalized set of criteria for any object that might be called a "planet": (1) it must directly orbit the Sun, or another star, (2) it must be massive enough that gravity overcomes its material strength — so that its rotational figure approximates that of hydrostatic equilibrium, and (3) it must not be so massive that it generates energy through nuclear fusion. As I have observed elsewhere, taxonomic criteria often are not dynamically meaningful. By these criteria, objects in solar orbit down to the order of 200 km in diameter are "planets," whereas the satellites, perhaps only by definition, do not qualify as planetary objects even though several of them (including our Moon) exceed the size of Pluto. The largest satellites of our Solar System planets — Ganymede of Jupiter, and Titan of Saturn — exceed 5000 km in diameter, hence are about double Pluto's diameter and thirty times its volume! At the other end of the "planetary scale," Jupiter plots at the lower mass limit of white-dwarf stars. As Sexl and Sexl (1979, Fig. 4.10) exclaim: *"Jupiter is almost a star!"* Even the Earth shows evidence that runaway nuclear reactions (though not nuclear fusion) have occurred in the past (see Cowan, 1976; Brookins, 1976; Shaw, 1978, p. 90; Jakubick and Church, 1986). It is evident, therefore, that our knowledge of Solar System phenomena is less complete than stereotypical descriptions would imply, and that taxonomic classifications in astrophysics may be meaningless unless they are placed within a context of self-consistent nonlinear-dynamical evolution.

Admittedly, Pluto was discovered despite the fact that its orbit is more irregular than those of the other planets, but the above remarks indicate not only that the task was difficult, but that the circumstances of Pluto's discovery were marked by a number of unusual events in the history of the search, not the least of which was the availability of an observer with tenacity and a "nose for discovery." Tombaugh himself said of the discovery of Pluto (Tombaugh and Moore, 1980, p. 14): "Reflecting upon this whole episode in later years, I came to realize that the discovery of Pluto was due to a remarkable chain of events spanning several decades, decreed by fate." What he called fate, I would say, was the self-organizational effect of pattern recognition that evolved from a diligently pursued but flexible ("learnable") research opportunity (Tombaugh certainly was the right person at the right place and time to satisfy the research opportunity that was particularly suited to his learning abilities). Examples of more "unusual" orbits exist among the asteroids and comets (see Fig. 13, Insets), and they are even more likely to exist for undiscovered "vulcanoids" (relict planetesimals inside the orbit of Mercury; cf. Leake et al., 1987) and "plutons" (relict planetesimals in the outer Solar System; see Weidenschilling, 1991; Stern, 1991).

It seems evident that standard practices in astronomy almost guarantee that *those objects that depart greatly from stable Keplerian, or more generally LaPlacian, LaGrangian, or other simple Hamiltonian forms of orbital motions, will be excluded automatically from astronomical catalogs.* This applies especially to objects of variable *a-e-i* relationships, meaning objects with irregular and even transiently chaotic orbits (crises), hence nonrecoverable orbital parameters. This statement is underscored by descriptions of astronomical *search strategies*, such as the one by Taff (1985, pp. 148–56; cf. Shoemaker et al., 1989, 1990) with special reference to asteroid discoveries. Discovery is predicated on the validity of a methodological "blueprint for success," which is a systematic sequence of operations designed to identify and delineate: (1) a two-dimensional *search space*, (2) a so-called *a priori target distribution* within that space, (3) a *cost function*, reflecting — in practice — the time spent per field of view, and (4) a *search plan*, analogous to an algorithm for sequencing the fields of view. This general outline makes clear two major caveats: (a) the "blueprint for success," though it may appear to be meticulously organized, is heavily conditioned by a belief in precisely recoverable planetary orbital elements, hence excludes the important class of short-period chaotic phenomena that — in the present volume, and in the published work of many others cited in the text — is postulated to exist, and (b) the so-called *cost function* and *search plan* are but fancy ways to equivocate about fundamental commitments to exploration.

If the Lowell Observatory had required Tombaugh to work according to such a camouflage for administrative "*cost*-effectiveness," in all probability "Pluto" — by some other name — would have remained undiscovered until the era of nonlinear celestial dynamics, meaning *about now*! Notice, for example, that Neptune was discovered *sixty-five years* after the discovery of Uranus (the discovery years being, respectively, 1781 and 1846; see Note 4.2) and eighty years after Titius made his remark about the spacing of the planets (Note 4.2), a period of time during which astronomy struggled to digest the possible implications of the Titius-Bode relation and deviations from it, and to come to terms with computing gravitationally perturbed distant orbits (the "n-body problem"; see Note 6.1 herein and Roy, 1988a, pp. 113ff; cf. Tombaugh and Moore, 1980, Chap. 5; and Roy, 1988b). That amount of time seems remarkably long in view of the fact that (1) Neptune is optically "immense" compared to Pluto and large compared to asteroids, many of which were discovered earlier than Neptune — beginning *forty-five years* earlier with the discovery of Ceres in 1801 (cf. Note 4.2; and Pilcher, 1979) — and (2) Neptune's orbit is a "better circle" than the orbit of any other planet except Venus (cf. Pasachoff, 1991, Appendix 3b). Consider, in this light, the *"blueprint for success"* behind the search for "Planet X" cited above, a strategy that began at the turn of the century, resulted in the discovery of Pluto in 1930 after a quarter-century of trial-and-error exploration, but is still not a "success" by the above rules of the game, because — for a variety of reasons related to the theme of this book — the search goes on (cf. Stern, 1991; Weidenschilling, 1991; Tombaugh, 1991; Kerr, 1992e; Cowen, 1992). It is also evident — with regard to the fractal spheres of influence discussed in Note 6.1 — that in the *era of the chaotic Solar System,* which has just begun, a realistic "a priori target distribution" in point (2) of the above four-point strategy for success is likely to be described by a multifractal singularity spectrum of the sort discussed in Chapter 6 (cf. Note P.4).

The problem of nonrecoverable orbits is particularly crucial to the documentation of near-Earth objects (NEOs), natural Earth satellites (NESs), fast-moving objects (FMOs), and comets in general. Near-Sun objects (the "vulcanoids," or other objects entering the Sun's deep gravitational potential well) will be especially subject to chaotic orbits, highly energetic bursts, and other undecipherable transients. Methods for dealing with problems of nonlinear and chaotic orbits will be the primary challenge of observational astronomy in the twenty-first century. On the basis of present experience, one major need is some form of running documentation (analogous to a phase-space plot) of all nonrecovered orbital trajectories maintained in graphical as well as digital format in such a way that evolving patterns may emerge over time. In this respect, the problem is analogous to that of artificial satellite tracking, which in the U.S. is the mission of NORAD's SPACETRACK system (e.g., Taff, 1985, Chap. 5). From the standpoint of discovering natural Earth orbiters, the artificial-satellite database provides a major source of orbital information from which perturbations by other objects can be inferred (e.g., Bagby, 1969; Wagner and Douglas, 1969; cf. Cook and Scott, 1967; King-Hele, 1976, 1992; Hughes, 1993). By the same token, the existing catalogs for orbital elements of the more regular objects in a given vicinity of the Solar System (e.g., the best-documented NEOs) can provide information for global inferences concerning objects that may be introducing local irregularities. This will be something of a bootstrap operation (perhaps "Project SPACESTRAP"), in the sense that there will be little to go on, initially, concerning expectable patterns until something begins to emerge. Given that eventuality, however, further discoveries should grow rapidly in the same sense that phase-space portraits of numerical experiments become increasingly recognizable beyond some finite number of iterations, a phenomenon I have elsewhere called *pattern periodicity* (cf. Note 8.4; and Shaw, 1987b, 1988b).

Murr and Kinard (1993) have reported on preliminary examinations of the Long-Duration Exposure Facility (LDEF) that was originally placed in low Earth orbit in 1984 (initial altitude = 650 km) and was recovered in 1990 (final altitude = 180 km) after almost six years in space and ~34,000 orbits of the Earth. During that time the orbit remained within ± 28.5° of the Equator. The LDEF was designed to test near-Earth space for potential deleterious effects on spacecraft, but evidently only those spacecraft in low equatorial orbits [by contrast, a diagram given by Coffey et al. (1991, Fig. 1) would indicate that most of the older artificial satellites were in high-inclination orbits — i.e., at inclinations resembling those shown in Fig. 19 of the present volume]. Among the first analyses of the surfaces of the test objects were attempts to determine the compositions of debris particles that struck them (photographic studies of the LDEF's entire surface revealed $\sim 10^9$ "craters" with diameters larger that 0.01 micrometer (ibid., p. 159). Initial results of the analyses of residues in the cratered objects indicates that *two-thirds of the projectiles that struck them were of natural origin*, as distinguished from artificial debris residues, on the basis of the relative abundances of the chemical elements (ibid., p. 157). These preliminary chemical results were consistent with "chondritic" meteoroidal materials that — judging from uncertain estimates of impact velocities (ibid., p. 161) — were derived from asteroidal sources (rather than from sources in geocentric orbits). It is evident, however, that the calculations of impact velocities were weighted by models of asteroid and/or cometary debris trajectories — the very factors under consideration in Chapter 6 of the present volume, and in this Note — and therefore they do not represent adequate tests for the percentage of particles that may have been in geocentric orbits.

Note 6.6: Neglecting the satellites of the planets, Jupiter constitutes 71.2% of the planetary mass, while (Jupiter + Saturn) constitutes 92.5% of the planetary mass (Pasachoff, 1991, Appendix 3). Thus, the "spin-frequencies" of the two-body and three-body "rotors," Sun-Jupiter and Sun-Jupiter-Saturn, constitute major, though not exclusive, gravitational disturbing functions for the resonance states of the other planets, their satellites, and the asteroids and comets. That is, there are also subsidiary resonances of 2-body to n-body interactions among the other planets, the asteroids, and the systems of satellites relative to these major systems. It may be significant that the symmetries of these major systems, involving the dominant mass distributions among the planets, are not unlike the symmetries of the mass distributions in the Earth, described later in the present volume (see Chap. 8, and Fig. 33), which, putatively, constitute the principal gravitational disturbing functions for the systems of geocentric satellitic resonance states schematically illustrated in Figure 20 (cf. Table 1, Sec. A).

Chapter Notes: Chapter 6

Note 6.7: The most densely populated region of the Asteroid Belt (see Fig. 13 and Appendix 9), which I have called the region of the main-belt asteroids (MBA), with semimajor axes between about 2 and 3.3 AU, is closely bracketed by the *1/4* (4:1 Kirkwood) and *1/2* (2:1 Kirkwood) resonance ratios, although I have included the range between *1/2* and *1/1* in Figure 21 to emphasize the resonance relationships with Jupiter (cf. Appendix 11). Several other taxonomic groups of asteroids exist inside of the *1/4* resonance position—i.e., toward the Earth—including, in the order of decreasing *mean* distances from the Sun, the Hungarias, Amors, Apollos, and Atens of Figure 13a (cf. Gradie et al., 1989). The Hungaria group flanks the inner limit of the MBA in a manner similar to the way the Cybele group flanks its outer limit. Taxonomically, the Aten group consists of asteroids with orbits entirely inside of Earth's orbit (i.e., they comprise the innermost asteroids whose orbits are not Earth-crossing). Analogously, the Amors taxonomically include those asteroids with orbits just outside of Earth's orbit. In taxonomic contrast with the Atens and Amors, the Apollo asteroids have orbits that "straddle" Earth's orbit, because their semimajor axes and orbital eccentricities constrain their *perihelion* distances (points on their orbits coming closest to the Sun) to lie inside of Earth's orbit and their *aphelion* distances (points farthest from the Sun) to lie outside of Earth's orbit (cf. Inset 1, Fig. 13). A better name for these objects than "Apollo asteroids"—which is an assigned taxonomic classification that does not necessarily indicate an origin and dynamical history the same as, or even similar to, the asteroid Apollo (see Tholen, 1989, and Tholen and Barucci, 1989; and index headings: "Asteroid taxonomy" and "Asteroids, individual" in Binzel et al., 1989a; cf. Kaula, 1968, Fig. 5.9; Nobili, 1989; Weissman et al., 1989)—is simply "Earth-crossing asteroids" (cf. Fig. 11 and Insets of Fig. 13).

It seems likely, in view of discussions in the present volume concerning resonance phenomena involving the inner planets and the Sun, as well as the existence of chaotic orbits and crises (cf. Notes 4.2 and 6.1), that the asteroids in general, and the Aten-Apollo-Amor asteroids in particular, constitute a dynamically evolving system that involves both resonant and chaotic orbital states relative to some subset of the inner planets, Jupiter ($\pm$ the outer planets), and the Sun. This categorical viewpoint would seem to be supported by the discussion of Greenberg and Nolan (1989, pp. 783ff) concerning the *orbital delivery* of asteroids to the inner Solar System. In fact, these authors (ibid., Fig. 2) have shown that the distribution of the orbital parameters *a*, *e*, and *i* (*semimajor axes*, *eccentricities*, and *inclinations to the ecliptic*, respectively) of asteroids in the vicinity of the Earth's orbit display wide variations within the same taxonomic classes, as would be expected from the studies by Wisdom (1987a, b, c) illustrated in Figures 12 and 22 of the present volume (cf. Wisdom, 1985; Weissman et al., 1989, Fig. 1). Values of *a* vary within the approximate range *0.75–4.2* AU; values of *e* vary within the range *0.06–0.9* (where *e = 1 is the parabolic transition between elliptical and hyperbolic orbits*); and the values of *i* are grouped according to whether they are *< 25°* or *> 25°* (roughly one-fifth of this general population has inclinations *i > 25°*). [New observations by Binzel and Xu (1993) concerning likely parent bodies for basaltic achondrite meteorites would appear to underscore these remarks (cf. Gaffey et al., 1989; Lipschutz et al., 1989, p. 760; Gaffey, 1993).]

The wide range of ellipticities, *e*, recorded among the asteroids in the data of Greenberg and Nolan (loc. cit., Fig. 2) emphasize the possibility that the orbits of near-Earth asteroids could have evolved chaotically from widely different initial states—including the cometary reservoirs (compare the analogous diagram in Weissman et al., 1989, Fig. 1; cf. Hoffmann et al., 1993; McFadden et al., 1993)—an idea that can be appreciated better after studying the relationships illustrated schematically in Figure 13 (Insets). The following citations, beginning with the work of Daniel Kirkwood in the nineteenth century, provide an entreé—in one way or another, but by no means a complete one even in the References

Chapter Notes: Chapter 6

Cited in this volume — into a literature that deals with some of the indeterminacies in the dynamics of asteroid-comet orbital evolution: Kirkwood (1871, 1872, 1888, 1891), Oort and Schmidt (1951), Herrick (1961), Roemer (1961, 1971), Jacchia (1963), Brady (1965), Kaula (1968), Van Flandern (1978), Kresák (1979, 1991), Wasson and Wetherill (1979), Drummond (1981, 1991), Valtonen and Innanen (1982), Weissman (1982, 1990a, b, 1991b), Wetherill and Shoemaker (1982), Matese and Whitmire (1986), Duncan et al. (1987, 1988), Olsson-Steel (1987, 1989), Stern (1987, 1991), Stern and Shull (1988), Greenberg and Nolan (1989), Kresák and Klacka (1989), McFadden et al. (1989), Weissman et al. (1989), Wetherill (1989b, 1991a), Yeates (1989), Froeschlé (1990), Safronov (1990), Shoemaker et al. (1990), Torbett and Smoluchowski (1990), Binzel (1991), Fernández and Ip (1991), Hajduk (1991), Ip and Fernández (1991), McIntosh (1991), Meech (1991), Clube (1992), Ipatov (1992), Hoffmann et al. (1993), McFadden et al. (1993).

Note 6.8: In considering simple geometric progressions analogous to the Titius-Bode relationship (cf. Note 4.2), consisting of the inverse period-doubling progression $1/2^n$, for $n = 0{-}10$, it is interesting that, on the basis of the orbital period of Pluto as the synchronous resonance, $1/1$, there are several close alignments with other planetary resonance ratios, some gross misalignments, and yet another "missing planet," in addition to the usual assumption that there is a correspondence between the position of the Asteroid Belt and the potential (or previous) existence of a major planet. Also note that this particular hypothetical progression is expressed in terms of orbital frequencies rather than distances; the corresponding distance relationship is obtained from the proportional logarithmic transformation given by the 3/2 slope of the resonance line in Figure 21, as required by Kepler's Third Law (also see Appendixes 6 and 11). The period-halving sequence corresponding to $1/2^n$ is compared below with resonance ratios from Figure 21:

Progression $1/2^n$			Planets (resonances from Fig. 21)	
(1/1)	P	———	*(1/1)*	Pluto
			(2/3)	Neptune
(1/2)	N*			
			(1/3)	Uranus
(1/4)	U*			
(1/8)	S	———	*(5/42)*	Saturn
(1/16)	J*1			
			(1/21)	Jupiter
(1/32)	J*2			
(1/64)	A	———	*(1/54)*	MBA (within range)
(1/128)	M	———	*(1/132)*	Mars
(1/256)	E	———	*(1/248)*	Earth
			(1/400)	Venus
(1/512)	V*			
(1/1024)	Me	———	*(1/1030)*	Mercury

In contrast with the Titius-Bode relationship, as presented in Note 4.2, this progression would "predict" that, relative to the progression $1/2^n$, the planets Neptune, Uranus, and Venus have shifted to a staircase-type of sequence (cf. Fig. 14) — perhaps owing to secondary resonances — and that *two hypothetical planets bracketing the general position of Jupiter* are needed to complete the period-halving progression (i.e., there are *eleven* planetary positions in this progression — where revised positions are marked by asterisks — rather than the ten known positions, including the Asteroid Belt as a "major planet" in both cases). Dynamically, such a sequence would imply a more "gradualistic" transition between the compositions of the terrestrial planets (Mercury, Venus, Earth, Mars, and some portion of the Asteroid Belt) and the present-day distribution of the giant planets (Jupiter, Saturn, Uranus, and Neptune). The Pluto-Charon pair, some portions of the Asteroid Belt, Stern's "plutons" (cf. Weidenschilling, 1991; Kerr, 1992e), and the comets would appear to have compositional affinities that are transitional between those of the terrestrial planets and the giant volatile-rich planets (see Chap. 3 and Figs. 9 and 10 of the present volume; cf. Atreya et al., 1989). The simplicity of the period-halving sequence, and the existence of a Jupiter "doublet," would suggest that if such a distribution ever existed, it was probably at an early stage of protoplanetary organization, a stage for which there is little structural evidence other than the evolved planetary distribution (cf. Stern, 1991; Kerr, 1992e). Unfortunately, collisional models of planet formation, and/or modification, involving planet-size objects cannot be very explicit about where such impactors came from, or how their orbital interactions evolved (see Notes 3.2 and 9.2; and Appendixes 7 and 8). Two important examples of such models, from the standpoint of the evolution of the terrestrial planets, are (1) the collisional origin of the Moon (e.g., Daly, 1946; Hartmann and Davis, 1975; Wetherill, 1985, 1990; Benz et al., 1986, 1987, 1989; Boss, 1986; Boss and Peale, 1986; Ringwood, 1986, 1989, 1990; Wood, 1986; Stevenson, 1987; Taylor, 1987; Ringwood et al., 1990; Kerr, 1989; Cameron and Benz, 1991; Baldwin and Wilhelms, 1992; Spera and Stark, 1993), and (2) the collisional model for the evolution of Mercury (e.g., Murray, 1983; Leake et al., 1987; Strom, 1987; Chapman, 1988; Cameron et al., 1988; Stewart, 1988; Wetherill, 1988).

In a sense, the period-halving sequence would have been a better predictive tool than the Titius-Bode sequence for the *trans-Neptunian planet search* (Tombaugh, 1961a, b), because it does not show the extreme divergence beyond Uranus that seemingly invalidated the Titius-Bode "predictions." Furthermore, the dynamical circumstances in the Uranus-Neptune and Jupiter-Asteroid-Belt vicinities might be "easier to explain" — e.g., relative to initial states of the planetary nebula (e.g., Appendixes 7 and 8) — than the observed relationship in which there is a major mass discontinuity (cf. Fig. 9) between Jupiter and the Asteroid Belt. This discontinuity has been the basis for proposals by various workers over the years that a major planet once existed in this region, an idea that other workers have not confirmed in computer models of planetesimal evolution (cf. Wetherill, 1985, 1988, 1989b, 1990, 1991b). However, the specific orbital-evolution models originally invoked to support the idea of a "missing planet" apparently have not been tested since the work by Ovenden (1972, 1973, 1976; cf. Hills, 1970), Ovenden et al. (1974), Ovenden and Byl (1978), and Van Flandern (1978). These earlier studies suggested that there is great uncertainty in the stability conditions of the Mars–Jupiter mass distribution (cf. Fig. 9, Inset), possibly even involving a significant change in the mass of Jupiter at a relatively recent epoch (using epoch in the astronomical sense, but in the context of geologic time) associated with the disruption of a "missing planet" (see Figs. 9–13 and 21; cf. Appendixes 6–9, and 11). Present-day concepts of Solar System chaos may resurrect and/or place a new light on these older ideas (cf. Wisdom, 1987a, b, c, 1990; Laskar, 1988, 1989, 1990; Mallove, 1989; Milani, 1989; Milani et al., 1989; Laskar et al., 1992; Milani and Nobili, 1992; Sussman and Wisdom, 1992).

Ovenden (1973, 1976) coined the name Aztex for his "missing planet," a name he chose in order to acknowledge a period of research at the University of Texas, Austin, during which he was influenced by the hospitality he received during that period of time from the Universidad Nacional Autónoma de México (see Ovenden, 1973, p. 332). Van Flandern (1978) apparently was the only one among the proponents of a missing planet to suggest a mechanism of disruption that did not involve a collision of planet-size objects, a

Chapter Notes: Chapter 6

cataclysmic phenomenon subscribed to by many for early stages of planet formation—and for the formation of the Moon (see Chap. 8; cf. Chapman and Davis, 1975; Leake et al., 1987)—but for which there is no concrete evidence within the inner Solar System during the Phanerozoic Eon (the latest 570 million years of Earth's history). As I have pointed out in the text (see the Introduction and Chap. 8), Phanerozoic impact phenomena on Earth, and Copernican impact phenomena on the Moon, are dwarfed by the scale of events associated with planet-size collisions during the early stages of the evolution of the planetary system as we know it today. Van Flandern (1978), however, proposed a novel mechanism for the disruption of a planet-size object at the position of the Asteroid Belt on the basis of a theoretical model by Ramsey (1950) for sudden and potentially violent phase-change instabilities in certain types of planetary cores. Other than the meteoritic evidence that the materials making up the asteroids (rocks, minerals, etc.) are very old (see Tilton, 1988; Caffee and Macdougall, 1988; and other contributions in Kerridge and Matthews, 1988)—and the general resemblance of their size distribution to fractal fragmentation spectra (see Appendix 8)—there is no clear understanding of the original mechanism by which the asteroids formed in the first place or exactly how their present heterogeneous states of aggregation and/or disaggregation came about (e.g., Chapman and Davis, 1975; Weidenschilling, 1977, 1988; D. Davis et al., 1986, 1989; Cassen and Boss, 1988; Wood and Morfill, 1988; Lipschutz et al., 1989; Wetherill, 1989b). Those candidate meteorite parent bodies—generally *assumed* to come from source objects in the Asteroid Belt (cf. Wetherill and Chapman, 1988)—for which optical information exists, such as from *occultations* (e.g., Millis and Dunham, 1989), rotation *lightcurves* (e.g., Magnusson et al., 1989; Binzel et al., 1989b), *photopolarimetry* (e.g., Dollfus et al., 1989), and so on, are usually of irregular shapes, some apparently representing rubble-like aggregates (cf. Fujiwara et al., 1989; McKay et al., 1989; Chapman, 1990; Binzel et al., 1991; Belton et al., 1992; Veverka et al., 1993). Obviously, those fractions of Solar System objects that are available for inspection on Earth as meteorites (cf. Shoemaker, 1960a, b, 1963, 1983; Anders, 1963; Beals et al., 1963; Krinov, 1963a, b; Hewins and Newsom, 1988; Kerridge and Matthews, 1988; Lipschutz and Woolum, 1988; McSween et al., 1988; Sears and Dodd, 1988; McSween, 1989; Huss, 1990; Cassidy et al., 1992), especially those with metallic phases, are fragments of much larger objects (cf. Appendix 8; and Bandermann, 1971; Brown et al., 1983; Donnison and Sugden, 1984; Turcotte, 1986; MacPherson et al., 1988; Steele, 1988; Stöffler et al., 1988; Fujiwara et al., 1989). It is clear that the asteroids, individually and collectively, have had diverse and complex histories in which the role of core-forming processes—caught at various stages of completion—are conspicuous (see many of the contributions in Kerridge and Matthews, 1988; and Scott et al., 1989, pp. 726ff; Taylor, 1992).

It is interesting that state-of-the-art studies of the Earth's core and core-mantle interface (e.g., Young and Lay, 1987; Jeanloz, 1990) are now emphasizing the importance of solid-solid and solid-liquid phase changes in their analyses of heat transfer from the core and fluid-dynamical regimes of the geodynamo. It therefore remains conceivable that seismically "slow" transformations (i.e., slow relative only to the lack of any present-day evidence of "normal" earthquake events, requiring instability time scales of the order of seconds, at core-mantle depths in the Earth) could exist in deep planetary interiors, transformations that would be catastrophic over geologically short time intervals for planetary masses of a certain critical size and composition, and yet be of sufficiently local influence and sufficiently low kinetic energy at the scale of the planetary nebula as a whole that orbital mass distributions would be rearranged only within "small" intervals of distance (e.g., > 0.1 AU, < 10 AU; cf. Appendixes 6–9). If the main result was to reduce transiently the cohesive strength of a host planet or planets (e.g., by the reverberant

Chapter Notes: Chapter 6

collapse, or block-caving instabilities, of a sudden evacuation at great depth owing to a negative volume change), then those planetary objects that became sufficiently fluidized may also have become susceptible to combinations of rotational-collisional instabilities (cf. G. H. Darwin, 1879, 1907, 1962; Sridhar and Tremaine, 1992) in the form of following waves that would have permitted progressive disruptions to occur in successive stages over protracted, but geologically short, intervals of time — e.g., reminiscent of the nature of chondritic meteorites, rubble-pile asteroids, and the like (see chapters dealing with textures and structures of early Solar System materials in Kerridge and Matthews, 1988; cf. Chapman, 1990; Binzel et al., 1991).

The Ramsey-implosion mechanism is not unlike, figuratively speaking, instabilities thought to occur in neutron stars just prior to the supernova explosion stage — the end stage in the thermonuclear life cycle of massive stars — which, in an analysis featured on the cover of *Science*, 16 October, 1992, are preceded by transient convective instabilities of violent character (Burrows and Fryxell, 1992). The supernova mechanism, in general, is a variation on the theme of positive/negative feedback (see Note P.5), in which — because of the thermonuclear mechanisms of heating and the nucleus-crunching pressures of high mass (characterized by the so-called *critical Chandrasekhar mass*, ~ 1.4 times the mass of the Sun) — there are potentially two *runaway modes* (cf. ibid., p. 430). One is the usual form of explosive runaway which, when fully developed, becomes the well-known — but incompletely understood — supernova explosion (compare this form of explosive runaway with ordinary chemical explosions, and the analogous *shear-instability thermal runaway* demonstrated by Shaw, 1969; 1991, Fig. 8.1; cf. Anderson and Perkins, 1975). The other runaway mode is the nucleus-crunching density instability that leads to the dynamical regime called a *black hole* (cf. Note 4.2). Between these alternative forms of instability are modulated states (i.e., states wherein the tendencies for one or the other form of runaway are mediated by the opposing types of reaction rates, e.g., involving thermal expansion vs. densification, mass-energy loss vs. transformational instabilities and gravitational collapse, and so on) such as those described by Burrows and Fryxell (loc. cit.). The spectrum of scales over which the analogous mechanisms can operate should not be missed, from the scale of the supernova mechanism to that of deep-focus earthquakes and avalanche-like mantle instabilities in the Earth (e.g., Kirby et al., 1991; cf. Notes 4.2 and 8.2 herein). These are all forms of *dissipative mechanisms*, but the mediated states are capable of producing quasi-steady entropy-producing structures of great intricacy (see cover illustration of Burrows and Fryxell, loc. cit.; cf. Prigogine et al., 1972, 1991; Prigogine, 1978; Krinsky, 1984; Prigogine and Stengers, 1984; and Note P.7).

The stellar instability modeled by Burrows and Fryxell (1992) represents *the first 30 ms after core bounce and shock generation*. Rescaled to a regime of core-mantle densification reactions in a planet, the same mechanism might indeed be disruptive. Even if "core bounce" were subseismic (subsonic), geologically instantaneous convective overturns of catastrophic character could ensue (cf. Note 8.8). On the planetary scale, such an event presumably would leave an imprint on the nature of the orbital evolution of the planetary system as a whole (e.g., the *principle of least interaction action*, discussed in Notes 3.3 and 6.1, states that the orbital interactions of all of the planets will be transformed by a change in the energy state of one of the planets, e.g., as effected by a change of rotational angular momentum and tidal interactions with satellites and/or nearby planets), even if there were no actual disruption of the unstable planet. Because of the implicit rebound effects, and other resurgence phenomena, of such a global instability, some of the residual structural features of planet-wide scope might resemble the effects of dilatation, or planetary expansion. Perhaps some of the seeming extensional characteristics of long-lived surface features of the inner planets might record events of this type, surviving — depending on the varying

Chapter Notes: Chapter 6

intensities and timings of the original instabilities—from early stages of crustal evolution (see Chap 8). Ramsey's (1950) hypothesis—which would seem to be relegated to a category of either "outrageous" or "wrong" ideas in geophysics (cf. Jacobs, 1987, pp. 308ff, 1992b), and which has been both forgotten and ignored in most recent studies of the Earth's deep interior—thus may have an important, even intriguing, bearing on the nature of relevant observational evidence concerning the more general hypothesis of changing cosmological constants and the expansion of astronomical bodies (cf. Dirac, 1937; Jordan, 1971; Van Flandern, 1976; Carey, 1988). This form of expansionist viewpoint (cf. Note 4.2) is expressed specifically—and even more "outrageously"—by S. W. Carey (1976) in his theoretical treatise *The Expanding Earth*; cf. Carey 1983, 1988; Bailey and Stewart, 1983; Dachille, 1983; Hora, 1983; Owen, 1983a, b; Taylor, 1983; Tryon, 1983; Vogel, 1983; Walzer and Maaz, 1983; Hughes, 1986). Perhaps the overriding message is analogous to that of de Vaucouleurs (1970) concerning *the* density of the universe. A universe of positive/negative mass/density fluctuations cannot be characterized as being of either an expanding or contracting type (it cannot be characterized by an average density, or by any one generalized dynamical state). Both expansion and contraction, in such a universe, are occurring "simultaneously" at all spatiotemporal scales of nonlinear-dynamical fractal interaction (cf. Note 4.2; and de Vaucouleurs, 1970; Alfvén, 1980, 1981, 1984; Coniglio, 1986; Lucchin, 1986; Pietronero and Kupers, 1986; Lerner, 1991; Luo and Schramm, 1992).

Note 6.9: A disconcerting variety of research reports of new and exciting discoveries, analyses, and theories in astronomy, geology, and geophysics have appeared during 1992–93 (the period during which this book was being revised following the reviews by persons acknowledged in the Prologue, and during editing by Stanford University Press). I have tried to keep the book current during this process, but, alas, I had to abandon that effort by the second quarter of 1993. It is clear that 1991–93 has been a period in which reports on observations of the types discussed in this volume were accelerating. Without attempting to list all of the classes of the new research (most of the 1992–93 entries in References Cited are of this nature), there is creative ferment, if not foment, in the arenas of, for example: (1) core-dynamo observations and theories, (2) documentation of transitional geomagnetic pole paths (VGPs, etc.), polar-wander paths, etc., (3) large-scale computing of Solar System chaos and nonlinear resonance phenomena (esp. Laskar et al., 1992; Milani and Nobili, 1992; Sussman and Wisdom, 1992), (4) observational astronomy and astrophysics, and (5) geological observations and documentations relative to the impact-extinction hypothesis, paleogeographic reconstructions of the continents, etc.

I have sometimes been so startled by similarities between many of these new reports and the theme of the present volume that every new issue of *Science* or *Nature* is anticipated with mixed feelings—including some chagrin that the pace has gotten too fast for me to keep up with the literature while the book is going to press. One of the latest examples of an exciting astronomical observation (as of 26 September 1992) was the announcement of the discovery of an approximately 240-km diameter planetary object at a distance of about 41 to 42 AU from the Sun (see Kerr, 1992e; Cowen, 1992), the product of a joint study by David Jewitt of the University of Hawaii, Honolulu, and Jane X. Luu of the University of California, Berkeley, using the 2.2-meter telescope on Mauna Kea, Hawaii. In a sense, this discovery falls within the historical class of discoveries—along with Pluto and Charon—that are long-term products of the original *trans-Neptunian planet search* for "Planet X" begun by Percival Lowell in 1905 (see Tombaugh, 1961a, b, 1991; Tombaugh and Moore, 1980; Matese and Whitmire, 1986; Anderson and Standish, 1986; Matthews, 1991; cf. Note 6.5).

Chapter Notes: Chapter 6

The distance of the new object indicates that it is in a Pluto-crossing orbit (cf. Stern, 1992, 1993). Pluto, itself, with a mean orbital distance of ~39.5 AU, is in a Neptune-crossing orbit—because its perihelion distance of ~29.7 AU is inside of Neptune's mean distance of ~30.1 AU from the Sun, and its aphelion distance is ~49.3 AU, which is well beyond the present distance of the new object (for perspective on the orbital scaling of the Solar System, see the Insets in Fig. 13), temporarily designated 1992 QB_1 (see the initial reports by Kerr and by Cowen, loc. cit.; and Kaufmann, 1991, pp. 309 and 321). The duration of the observations of 1992 QB_1 is not sufficient, as yet, to define an orbit—but it may be before the end of 1993 (keeping in mind that if the orbit is chaotic a limited number of observations is inadequate to establish large variations in its orbital semimajor axis, eccentricity, and inclination; cf. Figs. 12, 15, and 22 of the present volume; and Wisdom, 1987 a, b, c). At the moment, it is not known whether the orbit is nearly circular or highly eccentric (for that matter, whether it is in the elliptical class or the parabolic-hyperbolic class of orbital trajectories), nor is its inclination to the ecliptic plane of the Solar System known as yet. Frankly, from the point of view of nonlinear-resonance phenomena illustrated in Figures 20 and 21 of the present volume, even an accurate determination of the orbital parameters at the present epoch, though ultimately related to the original source of 1992 QB_1, is not sufficient to distinguish between—or even to identify—the classes of Solar System objects favored by this or that person who has expressed their excitement about its discovery (see Kerr, 1992e; Cowen, 1992). But I *do* agree with the quote of Alan Stern by Cowen (1992), that *"Now there's a whole new game in planetary science."* [See the update by Paul Weissman (1993); cf. Hoffmann et al. (1993).]

The orbital distance of 1992 QB_1 excites advocates of at least two, perhaps three, undoubtedly overlapping, types of chaotic Solar System structures: (1) the Kuiper Belt (or "disk") that postulates the existence of a distinct structure of low-inclination/moderate eccentricity, but otherwise nebulously defined, relict "icy" planetesimals (Kuiper, 1951) at distances near Uranus and Neptune, and beyond, that may be the source of at least some portion of the Solar System's relatively "short-period" comets, meaning periods of hundreds of years and less (cf. Figs. 11 and 15; and Duncan et al., 1987, 1988; Bailey, 1990; Torbett and Smoluchowski, 1990; Rawal, 1991, 1992), (2) a system of cometary objects that originally nucleated from the planetary nebula near the Uranus-Neptune distance but thereafter evolved to form the much more distant inner Oort Cloud (Hills Cloud) and outer Oort Cloud of comets at distances of the orders of 10^3 and 10^4 AU, respectively [Duncan et al. (1987, 1988); see Figs. 13 (Insets) and 15 in the present volume and Bailey (1990) and Torbett and Smoluchowski (1990) concerning chaotic perturbations within such regions; see also the alternative, but not necessarily mutually exclusive, Galactic models of comet formation, as enunciated in studies by, among others, Clube, Napier, M. E. Bailey, and Stern, as cited in the present volume], and (3) the conceptually related but startling prediction by Stern (1991) that numerous relict, ~10^3-km-diameter "planets" should exist throughout the same general range of distances (cf. Weidenschilling, 1991; Rawal, 1991, 1992; Weissman, 1993).

The main distinctions between Kuiper's concept and Stern's concept concern (a) the total mass, (b) the range of object sizes, (c) the mean densities, (d) the bulk chemical compositions, and (e) the states of orbital stability and/or transience of the putative objects that originally resided within these overlapping spatial realms near and beyond the orbits of Uranus and Neptune. Most discussions of the Kuiper disk appear to view it as being a rather flat (ring-like) region that is more or less restricted to the ecliptic plane and is made up of icy materials of comet-like characteristics, whereas Stern's (1991) objects are alleged to be of "planetary" characteristics. But such distinctions would appear to be poorly understood—if they are indeed real distinctions (cf. Weissman, 1993).

493

Chapter Notes: Chapter 6

It is becoming more and more obvious that we do not fundamentally understand what the difference between the taxonomic classes "asteroids" and "comets" really is in terms of genetic models — or even in terms of their orbital locations relative to the planets (see Fig. 13, Insets; cf. Hoffmann et al., 1993; McFadden et al., 1993). The Asteroid Belt evidently contains objects from both types of sources, whatever that may imply concerning distinguishing characteristics once the coma-producing stage of decaying comets has run its course (cf. Oort and Schmidt, 1951; Ovenden and Byl, 1978; Van Flandern, 1978; Kresák, 1979; Carusi and Valsecchi, 1979, 1985; Duncan et al., 1988; Weissman et al., 1989; Froeschlé, 1990; Safronov, 1990; Fernández and Ip, 1991; Hajduk, 1991; Ip and Fernández, 1991; Kresák, 1991; Meech, 1991; Wetherill, 1991a). Stern's objects, dubbed "plutons" by Weidenschilling (1991), have uncertain (read completely unknown) compositional affinities with Pluto, and/or Charon — the "satellite" of Pluto — or even with the asteroids (cf. Ip, 1989). Because of the relative parity in sizes between Pluto and Charon, as compared to the radius ratios between satellites and primary planets in general, together with the uncertainty of the dynamical histories of either object, the Pluto-Charon pair is sometimes referred to as a "binary planet." But, so little is known about the chemical compositions of Pluto and Charon, or even their positions in the mass/radius spectrum of planetary objects (cf. Fig. 10; and Teisseyre et al., 1992, Fig. 4.1.1; Anon., 1991, p. K7), that the similarities or differences between them and Stern's planetesimals is a completely open question (Stern's criteria of planetary characteristics are outlined in Note 6.5; the diameter of 1992 QB_1 is near the lower limit of "planet-size" objects).

It could be that none of the three — Pluto, Charon, or "pluton" (see Chap. 3; cf. Note 6.5) — differs significantly from the compositions and densities represented by objects currently residing in the Asteroid Belt (cf. discussion of Figs. 9 and 10 in Chap. 3; cf. Stern, 1992, 1993). Ceres, an example of a 10^3-km "minor planet," has a mass that is 5 to 6 $\times$ 10^{-10} solar mass (cf. Millis and Dunham, 1989, Table 2; Hoffmann, 1989a, p. 230; Anon., 1991, p. K7), giving it a density of ~ 2.5–2.7 gm cm^{-3}, near the median of the planetary range of densities (see Teisseyre et al., 1992, Fig. 4.1.1). Some of the larger asteroids (Millis and Dunham, 1989, Table 2) appear to have densities similar to those of Earth's Moon and Jupiter's Io and Europa (cf. Pasachoff, 1991, Appendix 4; Anon., 1991, p. K7). Small asteroids may have much lower or higher densities. A density of about 2 gm cm^{-3}, or even less, may be characteristic of many of them (e.g., the "rubble-pile" asteroids described by Binzel et al., 1991; cf. Chapman, 1990). Small asteroids that represent fragments of terrestrial mantle-like mineralogies, or metal-rich fragments of once much larger, possibly core-forming, planetesimals obviously may have densities exceeding 3 gm cm^{-3} (see the general index of individual asteroids, and their properties, in Binzel et al., 1989a; cf. Belton et al., 1992; Veverka et al., 1989, 1993).

Judging from the new studies that have been published in 1992 concerning issues of asteroid and comet compositions (e.g., Rotaru et al., 1992; Fink, 1992; Belton et al., 1992), it seems likely that a revision of the chemical classifications of comets, asteroids, and other objects of actual or potential meteoritic affinities are in the offing.

Note 6.10: The criteria of "discovery" of astronomical objects in the Solar System invoke regular recurrence as a necessary property of recoverable orbits. Orbits in nonlinear resonances, and chaotic orbits, therefore, have lowered probabilities of being found a second time by routine types of search strategies based on prior knowledge of orbital characteristics (Note 6.5). The story of the search for "Planet X," a program established by Percival Lowell in 1905, which provided the setting, precedent, and persistent attitude that later permitted Clyde W. Tombaugh to make his famous discovery of Pluto in 1930, reveals the difficulty of any quest to find "unusual" objects in astronomy, objects for which there is

Chapter Notes: Chapter 6

no precedent (see Note 6.5). The discovery of pulsars tells an analogous story (see Note I.3). An even greater challenge faces us today in our attempts to discover nonlinear-dynamical orbits of *near-Earth objects (NEOs)* and *natural Earth satellites (NESs)*. The difficulties of designing search strategies for possible NEOs and NESs have been exacerbated by the fact that the artificial satellite program gained priority during a time when some workers were beginning to report examples of NESs (e.g., Bagby, 1956, 1966, 1967, 1969; Baker, 1958). One might think that the programs of surveillance that have accompanied the space program would have provided many opportunities to discover NEOs and NESs if they were present. This, in fact, was true in an exceptional case described by Bagby (1969). He used the body of orbital data for artificial Earth satellites (AESs) that existed even at that early date in the space program to identify perturbations that might be caused by NESs. But the focus of interest and the ranges of orbital parameters in the AES program have been too specialized, up to now, to provide further evidence of that kind; and any objects that may have entered the space of that surveillance, but differed from the orbital parameters of artificial objects, apparently were not recorded.

Taff (1985, Chap. 5) gives an excellent overview of tracking systems, numbers of satellites, and amount of "space garbage" that had accrued up to the year 1985. Such objects often are at relatively low orbital altitudes, altitudes analogous to, or corresponding to, the *interior* resonances of Figure 20. Some special resonance conditions are now well known (e.g., Allan, 1967b, 1971; Coffey et al., 1986, 1991; Meiss, 1987; Friesen et al., 1992), but—insofar as I have been able to discover by talking to personnel at NASA and NORAD data centers (cf. Taff, 1985, pp. 135ff)—no comprehensive classification of geocentric resonance states of artificial satellites has been published in the nonclassified literature. I recommend in the present volume (see *Project SPACESTRAP*, Note 6.5), and have proposed elsewhere (H. R. Shaw, U.S.G.S. Gilbert Fellowship Program Proposal, submitted 30 March, 1992; see the caption to Fig. 19, Inset), that the present proliferation of AES orbital data (e.g., Taff, 1985), as well as the computational expertise developed in the space program since the 1960's (e.g., Coffey et al., 1986, 1990, 1991), be used to develop an AES-NES *nonlinear-dynamical comparator program (NDCP)* in order to test, modify, and/or transform, in the light of nonlinear-dynamical algorithms, the analysis of AES orbital perturbations by NEOs and/or NESs that were the subject of preliminary examination by Bagby (1969; cf. Wagner and Douglas, 1969). Such a nonlinear-dynamical comparator program (NDCP) would enhance our knowledge of AES behavior, improve the precision and ranges of application of artificial-satellite geodetic analysis (e.g., King-Hele, 1976), and provide a long-term and evolving data base and system of algorithms to screen near-Earth space for natural and/or artificial objects newly entering Earth's sphere of influence (the environment within $\sim 2 \times 10^6$ km around the Earth, which includes the Moon, the Earth-Moon Lagrangian points, and "fuzzy" orbital-transfer routes of the type described by Belbruno, 1992; cf. Note 6.1 of the present volume). Our present state of knowledge of the size-distance-frequency spectrum of NEOs entering this region—enhanced by several very recent examples of unanticipated passages of NEOs at distances between the Earth and Moon—is discussed, for example, by Chapman and Morrison (1989), Steel (1991), Binzel (1991), Matthews (1992), and Ahrens and Harris (1992).

Without the analytical strategy and data sources of the NDCP, the task of mounting a systematic global search for NESs in exterior as well as interior resonances is enormous (cf. Tombaugh et al., 1959; Tombaugh and Moore, 1980; Tombaugh, 1991). The scope of such a global study is underscored by the difficulties that attended the analogous, but more focused, searches by Yeates (1989) and Frank et al. (1990) for near-Earth objects of comet and/or asteroid affinities under circumstances such that these workers knew where to look for the objects they sought (the study by Frank et al., 1990, was at the limit of optical

495

Chapter Notes: Chapter 6

resolution of ~10-m-diameter relatively dark objects, and the results apparently could not definitely establish whether the objects discovered were in near-Earth heliocentric orbits or were captured in geocentric synchronous orbits; cf. Aksnes, 1971; Friesen et al., 1992).

The most persistent search for NESs to date was made by Tombaugh et al. (1959) during the earliest phases of the space program. In the approximately five years of systematic searching by Tombaugh et al. (ibid., p. 61), the search team finished with the terse statement, *"No satellites were found."* This statement has sometimes been pointed out to me as a firm "conclusion" by Tombaugh — rather than as an *observation* (there is a major difference, which, I am sure, Tombaugh would be the first to point out, from personal experience) — in objecting to the theme that I have espoused in the present work. I point out, in rejoinder, that the only meaningful statement that can be made about *any* astronomical search that gives negative results, in the context of the present narrative, is that no satellites in regular orbits were *recovered.* Such a result is consistent with the conclusion of the present volume that objects in nonlinear-resonant orbits with variable *a, e,* and *i* relationships, and/or in chaotic orbits, are not likely to be recovered by search routines based on traditional methods of discovery of near-Earth objects, objects that are, a priori, inferred to be in heliocentric orbits (cf. Bagby 1969; Aksnes, 1971). A sampling of the closing remarks by Tombaugh et al. (1959, pp. 61f) renders the situation more graphic (italics added):

> It should be emphasized that these are the coverages attained for *circular orbits.* Orbits with some eccentricity would not have been covered as thoroughly, and in general, the higher the eccentricity one wishes to consider, the less meaningful these coverages become.
>
> The search was confined almost exclusively to two specific planes, but they are the two planes considered most likely to contain natural satellite bodies. A large portion of these planes was covered, and no satellite bodies were found. Although recovery attempts were made for a few of the more promising image suspects, results were always negative. *If any of these were actual satellites in orbits of considerable eccentricity, they could have escaped recovery.*
>
> A more comprehensive statistical probing, photographically, of *all* satellite space about the Earth would not be impossible. However, the time and expense required for such an undertaking are formidable, and far beyond the scope of this project.

The "two specific planes" were the Earth's Equatorial plane and the ecliptic plane, neither of which are the preferred loci of the NESs proposed to exist in the present study. The above search also made some explorations at high geocentric inclination (85.4°, presumably to the Equator) at a distance of 6010 miles (9672 km). This distance is at a ratio of $a/R_E = 1.5165$ in Figure 20, a region in which high-inclination orbits probably are not stable (i.e., stable high-inclination orbits likely would be in exterior orbits at distances greater than 40,000 km; cf. Allan, 1967a, Figs. 1 and 2).

The comments about eccentricity are perhaps most revealing. Although the study by Allan (1967a) considered only circular orbits, Allan (1967b, 1971) studied resonant eccentric orbits of low to high inclination. Nearly all of these conditions would be outside of the planes covered by traditional search routines such as that of Tombaugh et al. (1959). Allan (1971) studied highly eccentric *($e_o \cong 0.74$)* orbits near the so-called "critical inclination" of about 63.4° (cf. Coffey et al., 1986) and period of about 12 hours *($\alpha/\beta \cong 1/2$)*, simulating Russian *Molniya I class communication satellites*. At this initial eccentricity, and with ratios a/R_E of 4 or so, these objects have closest approaches (perigees) of a few hundred kilometers and therefore reach greatest altitudes (apogees) of about $2a$, or roughly 50,000 km. Near apogee they would be moving slowly, but at perigee they would be a comparative blur (see the description of "fast-moving objects," called FMOs, by Tatum, 1988), perhaps even unnoticeable in telescopic searches (depending on the photographic resolution and frequency) unless they were anticipated.

It is evident that realistic orbits for nonlinearly resonant or chaotic NESs would be difficult to track by the methods of Tombaugh et al. (1959; cf. Tombaugh, 1991; Kerr, 1992e). It also seems evident that any new search strategies must be designed to take account of resonance phenomena of the types studied by Allan (1967a, b, 1971) and Coffey et al. (1986, 1991).

Note 6.11: The nodal great circles of Figure 23 — like the resonance ratios of Figure 20 — refer to the repeat periods of like configurations relative to the Earth's surface. During the time between the commensurable (integer) values of rotations and revolutions, however, the satellite is in some noncommensurable state. In general, it is precessing in longitude relative to the Earth's surface until it again comes into conjunction with the particular fixed configuration that corresponds to a given recurrence ratio (cf. Bagby, 1956; Wagner and Douglas, 1969). Because the geocentric-resonance orbits of Figure 20 are "locked" to the Earth, the Earth's precession relative to the celestial sphere (cf. Roy and Clarke, 1988, pp. 112ff) is not involved in determining the ballistic trajectories that determine the latitudes and longitudes of the final impact sites related to the geocentric reservoir (the celestial reference frame, CRF). The geocentric system of satellites of Figure 20, hence the CRF, precesses with the Earth, meaning that its pattern remains invariant relative to the spin axis and Equator (as do the consequent patterns of impact cratering), while the spin axis, Equator, and geographic system of latitudes and longitudes precess relative to the ecliptic plane and system of fixed stars (the ~26,000-year precession of the equinoxes).

Allan (1967a, b), on whose work Figure 20 and the general hypothesis of spin-orbit resonances described in the present volume rests, considered essentially two limiting types of orbital configurations: (1) so-called i-type resonances, in which a non-zero inclination (i) is essential and the eccentricity e is sufficiently small to be taken as zero (this is the case of Fig. 20), and (2) so-called e-type resonances, in which a non-zero eccentricity is essential and the inclination is sufficiently small to be taken as zero (inclination refers to the ecliptic plane for heliocentric orbits and to the Earth's Equator for geocentric orbits; cf. Figs. 19 and 37). The second case is analogous to the description of the orbital motion of asteroids in Figure 12 (after Wisdom, 1987a, b, c), where the eccentricity is the critical parameter of orbital stability or instability. In Figure 12, however, the motion is chaotic, whereas in the spin-orbit resonance situation we are considering here, the motion is described by a set of essentially phase-locked resonance conditions analogous to those of Figure 14. The complexity of (heliocentric) asteroid orbital evolution is underscored by the chaotic relationships between eccentricities (e) and inclinations (i) shown in Figure 22. That is, orbital evolution is unlikely to occur at either constant e or constant i. The same inference holds for the nonlinear evolution of the geocentric system of natural Earth satellites (NESs) in either commensurable near-resonance conditions or noncommensurable near-chaotic conditions. Allan (1971), for example, shows that an important geocentric resonance condition involves both a high inclination and a large eccentricity (the so-called critical-inclination condition at $i \cong 63.4°$ to the Equator; cf. Coffey et al., 1986). In the interacting e-i mode, the complexity of spin-orbit resonances is potentially much greater than is indicated by Figure 20, and is not susceptible — in the numerical sense shown — to a general solution.

If, as I infer to be the general situation in terrestrial and Solar System dynamics, nonlinear interactions occur categorically near the "edge of chaos" (near the golden-mean transition between the phase-locked and chaotic regimes; cf. Jensen et al., 1984; Bohr et al., 1984; Fein et al., 1985; Shaw, 1987b; Olinger and Sreenivasan, 1988; Shaw and Chouet,

Chapter Notes: Chapter 7

1989, 1991), then there is likely to be a fuzzy (fractal) range of orbital trajectories, and consequently a fuzzy range of cratering trajectories. This general nonlinear-dynamical uncertainty principle would predict that cratering trajectories should have swath widths that mimic the range of uncertainties of these orbital-dynamic effects. This "fuzziness" is imposed on the general range of heliocentric orbital inclinations that *do* reflect the variations of the ~26,000-year precession of the equinoxes (see Fig. 19). Secular variations (meaning over times $\gg$ 26,000 years — which, in the celestial context, is generally a much longer time frame than, for example, the accepted meaning of the word "secular" as it is applied in paleomagnetism; cf. Butler, 1992, pp. 161ff), however, both smear out and effectively average the two types of e-i effects (the respective heliocentric and geocentric effects), so that the net result is to produce the very broad, but relatively invariant, swaths of cratering trajectories and Phanerozoic cratering nodes observed (cf. Figs. 1–4 and 23–26; and Appendixes 1–4).

It is evident from the discussion in the text that a system of NESs may be subject to a variety of nonlinear forcing effects. I have used the formal theory of Allan (1967a, b) to illustrate the categorically expected types of effects because it is physically grounded in the existence of longitudinal variations in the Earth's mass distribution, as well as being mathematically grounded in the numerical affinities with the simple nonlinear-dynamic resonance phenomena mentioned above. The actual patterns of impacts obviously may vary greatly within the general bounds of the respective resonance conditions and the phase angles for which orbital decay is incurred.

It is interesting that the ecliptic-plane/Galactic-plane angle of ~60° (cf. Duncan et al., 1987, p. 1335; and Fig.15b here) is similar to the critical inclination (63.4°) of geocentric spin-orbit resonances (Allan, 1971; Coffey et al., 1986). Averaged over long times ($\gg$ 26,000 years), therefore, the incidence of cometary objects would be almost uniformly distributed relative to Earth's critical-inclination angle of 63.4°, because the long-time average orientation of Earth's spin axis is perpendicular to the ecliptic plane (see Fig. 37, Inset). That is, the orientation of the geocentric critical-inclination resonance, relative to a cometary source in the Galactic disk, would oscillate relative to the ecliptic plane in fixed relationship to the Earth's spin axis and Equator. The secular spread of geocentric inclinations would therefore range between 63.4 ± 23.5° and 116.6 ± 23.5°, or between about 40–87° and 93–140°, which is nearly identical with the range of ± 45° from the Poles inferred from Fig. 23, as discussed in Chapter 6 [i.e., this distribution represents a "cone" of i-values about the North Pole — e.g., similar to the configuration shown in Fig. 19, allowing for the precessional cycle — that is maximized in the vicinity of the critical-inclination resonance condition; the inset in Figure 19 (Coffey et al., 1991, Fig. 1) illustrates the analogous distribution for artificial satellites].

Chapter 7

Note 7.1: I use the terms *stationarity* and/or *geographic invariance* with regard to geophysical patterns in much the same sense that the term *conjunction* is used with reference to celestial configurations (see Notes 3.1, 4.1, and 6.1). A stationarity could refer to an invariant position on the Earth's surface (or on the celestial sphere), or it could refer to a fixed relationship that repeats the same *configuration* at specified intervals (periods) of time in such a way that it always shows the same geometric and/or geographic aspect when it is examined at the same relative cyclical phase difference, as measured in time, or in cycles of the Earth's rotation, or in phases of the Moon, or in cycles of the precession of the equinoxes, etc. (i.e., relative to any recoverable criterion for the measurement of a specified configuration). In this respect, a stationarity, as I employ the term, corresponds to a state in which a given geographic locus remains unchanged after some characteristic number of

Chapter Notes: Chapter 7

cycles of the Earth's rotation. For instance, if a particular configuration (such as the center of one of the geomagnetic flux lobes of Fig. 27) is rigidly locked to a particular latitude and longitude, it repeats that same configuration — as seen from the viewpoint of the fixed stars — once per day (i.e., it moves with the same mean angular velocity as the Earth itself, in exactly the same sense that the Phanerozoic cratering nodes and invariant CRF system, according to the present work, are postulated to move). By contrast, if that same locus were to drift westward relative to the Earth at a relative angular velocity numerically equal to the Earth's rotational velocity (i.e., equal in magnitude but of opposite sign), it would appear to be fixed, like a star, relative to the celestial sphere, because it would be locked to one position in the coordinate frame of the fixed stars (equivalently, its sidereal period of rotation would effectively become infinite). These criteria have essentially the same implication as the term conjunction in astronomy, where a conjunction refers to the recovery of a like configuration among a specified system of objects (see Chap. 6; and Notes 6.1, 6.11). The accuracy of this sort of description, of course, depends on other components of motion not considered in the simplified case of corotation. Ironically, the relative motions of different parts of the Earth are even more complex than the forms of celestial-dynamic orbital motions previously discussed, largely because there are no analogous criteria for the relative coupling of all of the possible three-dimensional modes of motions within and between those different parts (consider, for example, the relative zonal and longitudinal motions in the atmosphere and oceans relative to volcanogenic ocean-floor heat sources that may be implicated in the El Niño cycle; see discussion by Shaw and Moore, 1988).

The stationarity of the magnetic flux lobes, as mapped on the surface of the Earth's core by Gubbins and Bloxham (1987; cf. Figs. 27 and 28 here), refers to the apparent absence of significant westward drift of the steady-state field over a 200-year interval of time, an interval during which the known westward drift of the magnetic field (cf. Note 7.1A) would have displaced the centers of each of the lobes by nearly 40° in longitude. More detailed studies of the time-dependent and steady-state parts of the geomagnetic field, respectively, by Bloxham and Jackson (1991, 1992) and Bloxham (1992) appear to support these earlier conclusions (cf. Gubbins and Coe, 1993). Such a stationarity is *stroboscobic* if photographs taken at the appropriate intervals of time corresponding to like configurations are identical. This is the approximate situation for the flux lobes of Figure 27 if, in fact, they are not being progressively smeared out by westward drift (cf. Bloxham, 1992). For example, Jupiter's Great Red Spot (GRS) appears to be fixed in space when photographs are taken at the proper relative period of rotation — allowing for the mean angular velocity of rotation of Jupiter and the relative angular velocity (drift) of the GRS measured in a coordinate frame rotating with Jupiter's mean angular velocity (see discussion in the text, Chap. 7). Although the GRS changes its internal appearance and its relationships to surrounding vortices in successive cycles (cf. Gierasch, 1991), its net angular velocity of drift is sufficiently slow that it is nearly stroboscopic at Jupiter's mean rotation period of about 9.925 hours (i.e., ~0.41 of Earth's rotation period; Maran, 1992, p. 540); see, for example, a series of twelve images of the GRS made every fourth rotation of Jupiter ($= 44$ cycles) by the Voyager 1 spacecraft (Ingersoll, 1983, p. 65).

In the context of nonlinear dynamics, such relative motions are equivalent to *nonlinear resonances* (nonlinear *phase locking* or *mode locking*). In that context, recovery of like configurations may involve multiple cycles of motion of the geometric object in question as well as of the coordinate system of reference (see discussion of Fig. 14; and Shaw, 1987b, Fig. 1). In the gaseous outer planets, as in the Sun, the mean period of rotation is difficult to determine because of the variations of angular velocity with latitude and depth. The mean rotation periods of the outer planets Jupiter, Saturn, Uranus, and Neptune are based on radio emissions modulated by the magnetic fields, hence are analogous to the rotation of the

Chapter Notes: Chapter 7

Earth's magnetic field, which differs slightly — about 0.18 degree per year, or $\sim 5 \times 10^{-4}$ degree per rotation cycle (day) — from the mean sidereal rotation period. Analogously, the astronomically accepted rotation period of the Sun is pinned to observations of the sunspots (see the *Astronomical Almanac for the Year 1992*, Anon., 1991, p. C3), hence the cited sidereal period of 25.38 standard days (86,400 SI seconds/day) is subject to our understanding of the nonlinear dynamics of the Sun's electromagnetic circulation (e.g., Tassoul and Tassoul, 1989; Maran, 1992, p. 848; cf. Foukal, 1990, p. 36, with regard to differential rotation in the Sun). The same conclusion applies to the outer gaseous planets, as can be seen, in principle, by the discrepancy between the Earth's "rigid-body" rotation rate and its electromagnetic rotation rate (i.e., the giant planets have analogous stratifications and differential rotations; cf. Atreya et al., 1989). Thus, unless the Great Red Spot of Jupiter (GRS) is rigidly coupled with the mean electromagnetic rotation rate, it should have some net drift relative to the astronomically accepted rotation period of 9.925 hours.

If we were to assume, for the sake of illustration, that the stroboscobic pictures of the GRS shown by Ingersoll (loc. cit.) were taken at intervals of exactly 40 hours, then there would be a discrepancy of $440 - (44 \times 9.925) = 3.3$ hours relative to the mean electromagnetic rotation period, during the sequence of 12 images. This amounts to a "slippage" or drift in the direction opposite to that of the mean rotation (i.e., it takes longer than 9.925 hours to recover the same configuration), hence the GRS hypothetically rotates slower than the mean rate by about one cycle in every 133 cycles of Jupiter's rotation. Compared to Earth's magnetic drift rate, this is an exceedingly fast relative motion (e.g., the ratio 1/133 is equivalent to a drift of $\sim 2.7°$ per cycle; and if the same *cyclic* rate were applied to the Earth, the equivalent westward drift would be 2.7° per day, or nearly three full cycles per year! Although this is only a figurative example (I have not sought out the literature on measurements of Jupiter's vortical drift rates), it illustrates how large a seemingly insignificant relative motion — as measured telescopically, or from photographs taken by space probes in repeated flybys — can be, especially when it is compared to the magnitudes of motions that are relevant to interpretations of geophysical patterns, such as the geomagnetic flux lobes of Figures 27 and 28, and the invariance of the Phanerozoic cratering nodes (and CRF) proposed in the present volume. Unfortunately, most reference works in astronomy do not adequately describe the criteria by which rotation periods of the planets, asteroids, and satellites are determined (cf. Anon., 1991; Kaufmann, 1991; Pasachoff, 1991, Appendix 3; Maran, 1992). I suspect that, given a systematic compilation of planet-satellite-asteroid rotation data (cf. Binzel et al., 1989b), the nonlinear dynamics of the coupled orbital/rotational structure of the Solar System can be taken to a new level of understanding in terms of its resonance-state/chaotic-state interactions, along the lines of Figures 14, 20, and 21, and Appendix 11 (cf. Chap. 6; and Notes 4.1, 4.2, 5.1, 6.1, 6.5–6.8, 6.10, 6.11), a goal that would culminate a proposal apparently first put forward by Kirkwood (1849; cf. Walker, 1850; see also Inset C of Appendix 6 in the present volume). Work by Touma and Wisdom (1993), Laskar and Robutel (1993), and Laskar et al. (1993) would appear to substantiate the above statement (cf. Murray, 1993).

It should be evident from the discussion of "westward drift" just how difficult it is to ascertain and/or to measure nonlinear resonances in tectonics, plate tectonics, drift of the "hot spot reference frame," true polar wander (TPW), or other forms of relative rotations of different "shells" or depth regions of the Earth [e.g., the radial interval encompassing the net motions of the global atmosphere; the radial interval encompassing the net motions of the global lithospheric envelope; the radial interval of the deep mantle; of the D" (read "D double-prime") layer; of the outer core; of the inner core; etc.]. For example, the secular westward drift of the geomagnetic field — presumed to be related to the net fluid motions in the outer core — has a resonance ratio, mentioned in the text, of 1 cycle every 730,000 cycles of the Earth's rotation, which translates to a frequency ratio $W = \alpha/\beta = 1/730,000$, or its reciprocal, depending on which frequency is taken to be the primary state of reference (cf. Figs. 14 and 20), where W is the *winding number* that expresses the (average) relative number of rotations required to wind up (or wind down) two corotating (or co-oscillating) systems to the same configuration as the initial state of reference (see Notes P.4, 3.1, and 4.1). The precision of measurement required in this example, in order to identify the ratio 1/730,000 as a specific nonlinear-resonance condition, would have to be considerably better than one part per million — and the westward drift of the geomagnetic field is a *very high* value of relative angular velocity compared to other possible examples of relative global-tectonic rotations (cf. Munk and MacDonald, 1960; Lambeck, 1980).

When the rotations are not precisely coaxial, the notion of the winding number still holds, but it now refers to configurations involving added degrees of freedom. In the case of approximately cylindrical rotations about axes that are inclined to one another, W can be expressed in two ways: an approximate one, W', say, that describes the cyclic ratio in terms of the motion of markers as projected on the plane that is normal to the rotation axis of the primary rotator (i.e., the projected "longitudes" of the pair of markers), and a more accurate one, W'', that describes the cyclic ratio read in terms of the varying positions measured *along* the *two* trajectories formed by plotting the relative motions of the initially coincident markers of the two rotators in cylindrical space, where the cylindrical axis is the direction of the primary rotation axis. In other words, one of the trajectories describes a helical motion about that rotation axis, while the reference trajectory describes simple circular motion. The condition of synchronization (the resonance configuration) occurs when the two markers recover the identical values of their coordinates that they had at the "start," as measured in this cylindrical coordinate frame.

In a spherical coordinate frame, of course, the winding number is even more complex, involving two pseudo-values (by projection) and a "true" value that occurs when the two reference markers again recover the same identical values of latitude and longitude. If there are other components of "wobble" or "nutation" — generally meaning other forms of oscillations in the trajectories of the motions — then the necessary precision of measurement of relative rotations, and the criteria of resonant frequencies, are even more stringent. There is little wonder that the issue of plate-tectonic "reference frames" in the Earth has been so controversial and ambiguous (e.g., Molnar and Stock, 1987; Molnar et al., 1988), especially when there are significant uncertainties in the mean axes of corotations between different systems of tectonic motions (cf. Shaw et al., 1971; Bostrom, 1971, 1992; Shaw and Jackson, 1973; Gordon and Jurdy, 1986). In such instances, criteria concerning the nonlinear-dynamical likelihood of periodic or chaotic resonance states can be of crucial importance to the meaning of a given form of relative rotation, and to the validity of any given set of "measured" apparent motions. I have in mind, especially, the possible complexities and ambiguities in defining the Earth's "hot-spot reference frame" relative to the mean rotational states of plate-tectonic motions, with or without the contributing effects of tidal-rotational deformation states of the Earth as a whole, as discussed, for example, by Bostrom (1971, 1992), Gordon and Jurdy (1986), Ricard et al. (1991, 1992), Cadek and Ricard (1992), Spada et al. (1992), Dickman (1992), and in the Prologue of the present volume (cf. Note P.5 and Appendixes 12–18; and Shaw, 1969, 1970, 1973; Shaw et al., 1971, 1980a; Shaw and Jackson, 1973; Gruntfest and Shaw, 1974; Jackson et al., 1975).

Note 7.2: Viscosity is *not* a physical constant of materials — especially not for heterogeneous mixtures of phase states — a reality that is belied by the usual assumptions of geophysical fluid dynamics and magnetohydrodynamics (see discussion in Note P.5). Viscosity is a *dynamical-thermophysical* relationship, one that can be resolved only by the

Chapter Notes: Chapter 7

actual acting out of the dissipative relationships between the fields of forces and fields of flow in a given system (Shaw, 1969, 1991). Therefore, indirect measurements of the "viscosities" of natural systems, such as from particular forms of flowing lava (e.g., Shaw et al., 1968), from models of mantle flow (e.g., Shaw, 1973), or from any geophysical measurement that is characteristic of a particular dynamical process, are special cases that, lacking dimensional justification from the scale-dependent (e.g., Hubbert, 1937) and thermomechanical (e.g., Gruntfest, 1963; Shaw, 1969; Gruntfest and Shaw, 1974) viewpoints, are meaningful only for the processes from which the measurements were derived.

The "viscosity" of the Earth's outer core is a case in point. Estimates and reported values vary over the incredible range of roughly fourteen orders of magnitude (e.g., Jacobs, 1987, p. 55; Poirier, 1988), a range that spans nearly the full spectrum of zero-shear-rate viscosities (Newtonian limits) of mixed-phase (solid-liquid-gas) magmatic systems (e.g., Shaw, 1965, Fig. 2; 1969, Fig. 4; 1972, Fig. 6; 1973, Fig. 9). For example, Merrill and McFadden (1990, p. 346) refer to a value of viscosity for the outer core that is not much greater than that of water at standard temperature and pressure. By contrast, Smylie (1992, p. 1682) — from an analysis of translational oscillations of the inner core (diameter $\sim$1221 km) with periods of the order of 4 hours — estimates a *kinematic viscosity* for the outer core of $\sim 7.7 \times 10^7$ cm^2 sec^{-1}. Since the kinematic viscosity is given by the dynamic viscosity divided by the density, and the density of the fluid outer core is ~ 12 gm cm^{-3} (e.g., Jacobs, 1987, p. 49; Smylie, 1992, Fig. 5), the corresponding dynamic viscosity of the outer core would be of the order of 10^9 poise (i.e., more than 10^{10} times the value cited by Merrill and McFadden, loc. cit.). This entire range of "viscosities" could conceivably reflect variations of shear rates at constant shear stress, or some disproportionate variation of both shear rate and shear stress, as in "power-law fluids," which in turn must reflect — in one way or another — feedback coupling of the local thermal and mechanical fields in the fluid (cf. Shaw et al., 1968; Shaw, 1969, 1973).

The egregious uncertainty shown by the range of outer-core viscosity estimates points up a fatal flaw in the modeling of large-scale geophysical flow systems. From the standpoint of classical hydrodynamic instabilities (e.g., Chandrasekhar, 1968), the length scale of the flow is so large that the Rayleigh-number criterion for turbulent flow is exceeded for almost any choice of viscosity within the cited range (cf. Shaw, 1974, Table 5). It is evident, however, from the discussion in Note P.5, that the classical assumptions that were applied to calculations of homogeneous turbulence are not valid for thermal-feedback instabilities in adiabatic systems (e.g., Gruntfest, 1963; Lorenz, 1963, 1984, 1989; Gruntfest et al., 1964; Shaw, 1969, 1991; cf. Ruelle and Takens, 1971a; Mandelbrot, 1974; Gollub and Swinney, 1975; Gubbins, 1977; Robbins, 1977; Feigenbaum, 1979b; Franceschini and Tebaldi, 1979; Gollub and Benson, 1980; Iooss and Langford, 1980; Siggia and Aref, 1980; Takens, 1981; Lauterborn and Holzfuss, 1986; Sreenivasan and Meneveau, 1986; Clement, 1987; Meneveau and Sreenivasan, 1987; Aranson et al., 1988; Widmann et al., 1989; Bloxham and Jackson, 1991; Sreenivasan, 1991). In large-scale flows — as determined by large conductive and/or convective heat-transfer path lengths relative to the path lengths for transfer of momentum by viscosity — the diffusion of heat is slow relative to the diffusion of momentum, and the system is effectively "adiabatic," hence unstable, from the standpoint of thermomechanical feedback (noting, for perspective, that momentum diffuses "instantaneously" through a theoretically rigid body, as an unrealizable limit, and at the speed of the appropriate stress waves through elasticoviscous bodies, meaning in some proportion to the speed of sound). In other words, the nature of geophysical flows requires nonlinear-dynamical modeling of flow regimes, both according to feedback models of convecting systems *and* according to models of chaotic dynamics (e.g., Fein et al., 1985; Jensen et al. 1985; Halsey et al., 1986; Shaw, 1986, 1991; Shaw and Chouet, 1988, 1989). Demonstration

Chapter Notes: Chapter 7

of this principle was my rationale in performing the experiments that I used to illustrate thermal-feedback-coupled convection of a viscous fluid, as published in Shaw (1991, Figs. 8.1–8.9) — even though at the time of those experiments (1968), the formalisms of chaotic dynamics existed only in the computer studies of weather patterns by Lorenz (1963, 1964), and in his analogous deductions concerning the role of variability and instabilities in atmospheric convection (e.g., Lorenz, 1969, 1984).

Note 7.3: The current literature of astrophysical dynamics distinguishes "fast" dynamo action from "ordinary" dynamo action (or diffusion-limited dynamo action), depending, respectively, on the presence or absence of rapid, chaotically organized, magnetohydrodynamic convection (e.g., Gilbert, 1991; Galloway and Proctor, 1992; Feudel et al., 1993). Gilbert (1991, p. 483) observed *"intense stretching and folding of the magnetic field in the chaotic regions of the flow. The folding brings together field that is largely aligned in the same direction, and the average field in a chaotic region grows exponentially with time."* He also observed that the main role of weak diffusion was to average the field *locally* (notice here a certain conceptual analogy with the significance of the local thermal field in the thermal-feedback instabilities of conventional fluid convection). Thus, although concepts of critical self-organization are not invoked by the above authors, their general description of the fast dynamo seems to be consistent with an extension of their ideas to the sorts of attractor dynamics discussed in the present volume. Modeling of astrophysical dynamos (e.g., in the Sun; see the review of the properties of the Sun's magnetic field by Foukal, 1990; cf. Cox et al., 1991; Sonett et al., 1991; Weiss, 1993) usually is *kinematic*, in the sense that viscosity and diffusion — the hydrodynamic parameters that are needed to characterize the dissipation of energy in a fully coupled dynamical analysis — are assumed to be zero or very small. These are the same assumptions that are usually made in celestial dynamics and often even in geophysical fluid dynamics (cf. Note P.3). The usual rationale for taking this tack, or course, derives from an avoidance of the vicious circle that stems from mathematical intractability, an avoidance that haunted the scientific conception of realistic nonlinear models in dynamics before the time of Poincaré and — in practice — has continued to do so even since the time of Poincaré and the advent of contemporary chaotic-systems theories (cf. Notes P.1–P.7, I.4, 3.1, 6.1, 6.5). In other words, the nonlinear-dynamical metaphor and methods of calculation used in most studies of astrophysical dynamo problems follow the tradition of conservative-Hamiltonian analyses wherein dissipative structures and attractors are presumed, purely by "virtue" of the mathematical formalism itself, not to exist, meaning that the more realistic phenomena represented by the self-organized critical state — a prophetic property of dissipative systems — cannot be adequately addressed (e.g., Schmidt, 1987; Smith and Spiegel, 1987). Nonetheless, the recursive stretching-folding mechanisms are characteristic of nonlinear-dynamical systems in general (cf. *folded feedback* of Shaw, 1987a, Fig. 51.14), and the alignment mechanism described by Gilbert (loc. cit.) clearly can "flip" at exponential rates from one dipolarity to its opposite, and/or to higher-order polarity-switching modes.

Note 7.4: Concepts of self-organized criticality can be described from the standpoint of (1) *cellular automata* (e.g., Bak et al., 1988, p. 366; Kadanoff et al., 1989; Chen and Bak, 1989), (2) the *critical golden-mean nonlinearities* (CGMNs) of Shaw and Chouet (1988, 1989, 1991) — these exemplified, in simplest form, by the periodic/chaotic transition states of the *quasiperiodic route to chaos* in sine-circle maps (see Note P.4; cf. Ostlund et al., 1983; Bohr et al., 1984; Jensen et al., 1984, 1985; Fein et al., 1985; Shaw, 1987b; Olinger and Sreenivasan, 1988) — or (3) chaotic crises (e.g., Grebogi et al., 1982, 1983, 1986, 1987a, b, c; Gwinn and Westervelt, 1986; Sommerer et al., 1991a, b; Sommerer and

503

Chapter Notes: Chapter 7

Grebogi, 1992; cf. Chapter 5 of the present volume). The one-dimensional sine-circle function (Table 1, Eq. 5) — one-dimensional because, in the simple case, it describes the nonlinear periodicities of point events on a line (circle), as in the nonlinear clock mechanism described by Shaw (1987b, Fig.1; 1991, Fig. 8.32, pp. 131ff), which are somewhat analogous to the nonlinear periodicities of earthquakes and/or volcanic events around the Circumpacific "Ring of Fire" — is the simplest numerical generator of critically self-organized power-law recurrence statistics (see Note P.4 and discussions in Chap. 5 vis-à-vis the *CGMN singularity spectrum* illustrated by Ostlund et al., 1983; Bohr et al., 1984; Fein et al., 1985; Jensen et al., 1985; Halsey et al., 1986; Gwinn and Westervelt, 1987; Shaw, 1987b, 1991; Olinger and Sreenivasan, 1988; Shaw and Chouet, 1988, 1989, 1991). To reemphasize this point, I quote from Chapter 5: *"A self-organized critical cascade in a natural system effectively has an infinite number of degrees of freedom (e.g., an 'infinite' hourglass, sandpile, or magnetic-domain structure in which there are events of all possible sizes and durations), yet the sizes of events and their durations and intermissions are neither random nor Poissonian, but, rather, accord with power-law scaling — just like the power-law spectra of chaotic crises — as functions of proximity to a critical condition in finite-dimensional attractor systems. The singularity spectrum of the one-dimensional sine-circle map at $K = 1$, $W =$ inverse of the golden mean (Table 1, Eq. 5), is directly analogous to such a cascade (cf. Shaw, 1987b, 1991; Shaw and Chouet, 1988, 1989, 1991). The spectrum is one-dimensional in the sense that it is restricted to motion on the circle, but for the above values of K and W the system is 'frustrated' by attempting to generate simultaneously all of the possible phase-locked frequencies (a few of which are shown in Fig. 14) while also trying to focus on an exact quasifrequency (the irrational number given by the inverse of the golden mean; see Shaw, 1987b), all the while balancing itself exactly at the onset of noninvertibility and chaos (thus representing a condition 'at the edge of chaos')."* (See also Universal Order, Geometry, and Number, in Note P.4.)

Note 7.5: If *Gaian-symbiogenetic* concepts of terrestrial-organic evolution (cf. Lovelock, 1972, 1988a, b, 1992; Lovelock and Margulis, 1974a, b; Margulis, 1981, 1990, 1992; Margulis and Lovelock, 1986; Margulis and Sagan, 1986a, b) have any validity at all — and in my opinion there is no question but what they do, despite attempts to isolate and separate the biochemical-as-effect from the geochemical-as-cause in a physical/chemical, periodic/chaotic coupled-oscillator system that *is* the dynamical nature of Earth as we live it and know it (cf. Des Marais et al., 1992; Veizer, 1992) — then the Gaian paradigm is a property of the galactic-solar-terrestrial phenomenological cascade (cf. Note 8.3; and Shaw, 1983a). Unless we are separable gods deposited on a discrete substrate — like parameciums on a petri dish — by a greater god, using nonuniversal robotics, then one can no more isolate an *originating-chemical-cause* from a *chemical-effect-as-an-originating-chemical-cause* than one can isolate the originating causes of the Moon or Mars from the originating cause of Earth (see discussion of the origin of asteroids in Chap. 6; cf. Appendixes 6–9). This statement is but a variation on the theme of *sensitive dependence on initial conditions* of nonlinear dynamics (see Note P.2; cf. Notes I.1 and I.8). The logical closure, which says that if there were no Galaxy, then there would be no Sun, and if there were no Sun, then there would be no Earth, and if there were no Earth, then there would be no biosphere, is also true — and if there were no biosphere, then there would be no logic. But the real question is not answered by the ground truth of logical proof, the inevitable downfall of adherence to "reductionist" science to the bitter end (cf. Spencer-Brown, 1972, p. 101; and the Prologue of the present volume).

Perhaps, in view of the rate of sociological/technological evolution (cf. Platt, 1981; Shaw, 1983a) and the human propensity for escalation of the dramatic, *crescendo* would be

Chapter Notes: Chapter 8

a better orchestral term for the contemporaneous stage of the cosmically coupled *strange loop* of life in which we are implicated (see Hofstadter, 1979). And life *is* characteristic of the Cosmos, because the Cosmos is the self-organized, self-pumping, one-and-only context there is by which any material form of life can be defined at all. What other material context could there be in a self-referential universe (cf. the Prologue of the present volume)? In the broadest sense, still within the operating rules of scientific reductionism, geological-evolution-as-life/life-as-geological-evolution must be referred to the sustaining chemical potential gradients of the hierarchical systems of far-reaching entropic "loopings" among universal configurations that are essential, in one way or another, to the very nature of *vis viva* — not to be confused with the truncated meaning of the "vis-viva law" in astronomy (see Montenbruck, 1989, pp. 54 and 106). In such an open-system context, no universal standard state for the absolute measurement of life energy can be defined, therefore vis viva — as I have just used the term — stands for a real, but nonquantifiable, measure of an indefinable source of mass-length-time energy that is manifested in the recursive states of the physico-chemical forms of living matter (cf. Prigogine et al., 1972; J. G. Miller, 1978; Chaitin, 1979; Krinsky, 1984; Lima-de-Faria, 1988; Eigen, 1992; Schrödinger, 1992).

Chapter 8

Note 8.1: The correlation between the paleomagnetic-reversal time scale and the Vine-Matthews-Morley hypothesis of a geomagnetically coded sequence of sea-floor spreading episodes (essentially a tape recording of plate-tectonic motions) was established convincingly in 1966 with the independent, and almost simultaneous, discoveries — by (1) the team of Allan Cox, Dick Doell, and Brent Dalrymple (UC Berkeley and U.S. Geological Survey), and (2) a team led by Neil Opdyke of the Lamont-Doherty Geological Observatory (see Glen, 1982, Fig. 6.1 and pp. 310f) — that an ~ 0.9-Ma normal polarity event (the *Jaramillo event*) was present in (a) the polarity-reversal time scale of Cox et al. (1964), subsequently published by Doell and Dalrymple (1966), refining the paleomagnetic record of a post-caldera rhyolite dome sequence in the continental volcanic system of the Valles Caldera, New Mexico, (b) magnetic profiles on the sea floor (see Vine and Matthews, 1963; Glen, 1982), and (c) deep-sea sediment drill cores (e.g., Opdyke, 1969). During the following quarter of a century, a host of research papers have been written on the subject of correlations between recurrence patterns involving, in one way or another, (1) geomagnetic phenomena in time and/or space, such as (a) paleomagnetic reversal (PMR) frequencies vs. age, (b) average polarity-sense percentage (per specified time interval) vs. age, (c) magnetic dipole strength vs. age, (d) present-day magnetic-flux patterns at the core-mantle boundary, (e) virtual geomagnetic pole (VGP) patterns during the latest 10 m.y., (f) Phanerozoic polar-wander paths, etc., and (2) geomagnetic phenomena and other Earth processes (including, both directly and indirectly, correlations with biological processes), such as (a) PMR frequencies vs. magma production rates; (b) PMR frequencies vs. eustatic sea level, (c) PMR boundaries vs. microtektite horizons and/or other criteria of impacts, (d) long-term PMR frequency variation and heat flow, (e) PMR frequencies vs. tectonic events, (f) PMR frequencies vs. geomagnetic field intensities during the latest 150 m.y., (g) PMR frequencies vs. polar-wander rates, (h) PMR correlations with impact events, and (i) direct and/or indirect PMR correlations with diverse geological processes (e.g., tectonic pulsations and biologic extinction events). Some of these correlations are discussed in the text of Chapter 8. [A new refinement of the Late Cretaceous and Cenozoic geomagnetic polarity time scale by Cande and Kent (1992), demonstrates that the spatiotemporal variability seen in many of the self-organized critical-state processes discussed in the present volume would also appear to describe the variability of seafloor spreading rates, in accord with the prediction by Shaw et al. (1971, Fig. 5) that rates of seafloor spreading, rather than being

Chapter Notes: Chapter 8

steady and constant over a given characteristic plate-tectonic length scale, must reflect averages of widely fluctuating rates over many scales of length and time among transiently accelerating — both positively and negatively in order to maintain the global balance of no net plate acceleration — local systems of self-similar oscillatory instabilities (compare spreading-rate profile CK92 in Fig. 41 of Cande and Kent, 1992, with Figs. 8.37–8.40 of Shaw, 1991).]

The numbers of papers written on the above types of correlations are now legion. An even greater literature exists on the subject of statistical tests and/or criticisms of periodicity — or putative demonstrations of numerically independent, hence nonperiodic, "renewal processes" (processes that are devoid of self-organized structures and have no memory; cf. Phillips et al. 1975) — in the paleomagnetic record. I have cited only a few of the latter types of papers because I want mainly to draw attention here to the types of self-organized coupling between the core dynamo and other geological phenomena that many workers have proposed during the past quarter of a century. Several examples of broadly commensurable geological phenomena are illustrated in Shaw (1988a). These were based on data I compiled during 1984, the year I was introduced to William Glen (both of us were at UC Berkeley during the 1950's, and both were acquainted with the paleomagnetic work by Allan Cox and Dick Doell, but apparently the fact that Bill was in paleontology while I was in geology, then in separate buildings, precluded more than a passing acquaintance; cf. the Foreword and Prologue of the present volume; and Glen, 1982, 1990, 1994).

The following is a sketchy sampling of the literature cited in the present volume that should provide an entreé to the diversity of geologic phenomena that may be correlated with the paleomagnetic record:

Temporal variation of heat flow and the paleomagnetic-reversal (PMR) record: Gubbins (1977), Loper (1978), Sprague and Pollack (1980), Jacobs (1981, 1992a), Olson and Hagee (1990), Hagee and Olson (1991), Larson (1991a, b), Larson and Olson (1991).

Tectonics, hot-spot reference frame, and true polar wander (TPW): Vogt (1972, 1975, 1979), Seyfert and Sirkin (1979), Harrison and Lindh (1982), Sheridan (1983), Burek (1984), Rampino and Stothers (1984), Courtillot and Besse (1987), Molnar and Stock (1987), Sager and Bleil (1987), Burek and Wänke (1988), Besse and Courtillot (1991), Cox (1991), Duncan (1991), Duncan and Richards (1991), Anderson et al. (1992).

TPW rate, magnetic field strength, and the PMR record: Loper and McCartney (1986), Olson and Hagee (1990), Besse and Courtillot (1991), Duncan (1991), Duncan and Richards (1991), Hagee and Olson (1991), Larson (1991b), Larson and Olson (1991), Fuller and Weeks (1992).

Sea level, climate change, and the PMR record: Muller and Morris (1986, 1989), Gaffin (1987), Schwarzschild (1987), Burek and Wänke (1988), Gaffin and Maasch (1991), Larson (1991b), Marzocchi et al. (1992).

Mass impacts and the PMR record: Glass and Heezen (1967), Glass and Zwart (1979), Glass et al. (1979), Glass (1982), Burek (1984), Muller and Morris (1986, 1989), Burek and Wänke (1988), Shaw (1988a), Burns (1989), Rice and Creer (1989), Schneider and Kent (1990), Blum et al. (1992), Schnetzler and Garvin (1992).

Mass impacts, volcanism, and the PMR record: Seyfert and Sirkin (1979), Muller and Morris (1986, 1989), Rampino (1987), Rampino and Stothers (1988), Rampino et al. (1988), Shaw (1988a), Stothers and Rampino (1990).

Hot-spot volcanism and the PMR record: Vogt (1972, 1975, 1979), Kennett et al. (1977), Moberly and Campbell (1984), Johnson and Rich (1986), Shaw (1988a), Stothers and Rampino (1990), Duncan (1991), Duncan and Richards (1991), Larson (1991a, b), I. H. Campbell et al. (1992), Czamanske et al. (1992), Renne et al. (1992).

Seafloor-spreading phenomena and the PMR record: Vogt (1971, 1972, 1975,

Chapter Notes: Chapter 8

1979), Seyfert and Sirkin (1979), Le Douaran et al. (1982), Rich et al. (1986), Larson (1991a, b), Cande and Kent (1992).

Periodicity of PMR frequencies: Mazaud et al. (1983), Negi and Tiwari (1983), Courtillot and Le Mouël (1984, 1985, 1988), Mankinen et al. (1985), Prévot et al. (1985), Stothers (1986), Raup (1986b), Champion et al. (1988), Coe and Prévot (1989), Gaffin (1989), Marzocchi and Mulargia (1990), Stothers and Rampino (1990), Mazaud and Laj (1991), McFadden et al. (1991).

Multiple correlations, including biological phenomena, with the PMR record: McLean (1978, 1988), Clube (1982, 1989a), Clube and Napier (1982a, 1984b), Burek et al. (1983), Rich et al. (1986), Champion et al. (1988), Rampino and Volk (1988), Shaw (1988a), Hallam (1989), Vogt (1989), Larson (1991b), Rampino and Caldeira (1993).

The investigation of coupled periodicity phenomena in the geological record that I began in 1984 was the extension of an idea that I had formulated in 1981 with my awakening to the literature of nonlinear dynamics as it was being practiced by mathematicians, physicists, chemists, meteorologists, and biologists. This interest in global geological periodicity phenomena originated from an earlier and mutually shared idea — in separate collaborations with, respectively, R. W. Kistler, R. L. Smith, and E. D. Jackson — that magma, in effect, represents the circulatory medium by which the pulse of the Earth can be read (cf. Joly, 1930; Umbgrove, 1939a, b, 1947), given comparable attention to data gathering, and research-budget support, that exists in the analogous studies of both atmospheric and oceanic dynamics. The rejuvenation of interest on my part, however, was specifically stimulated by the dialogue with Bill Glen, who dramatized for me what I have come to think of as the second wave of the debates of the *impact-extinction hypothesis* (IEH) — second wave because a strong precedent had already been established by two happenings: (1) the international program of astrogeological studies during the 1960's, following the launching of Sputnik on 4 October, 1957, which galvanized the first wave of U.S. and international enthusiasm for space exploration, and the first step made by humankind — or, so far as is known, contact by any living organism — on the face of the Moon, 20 July, 1969 (for background, see Shoemaker, 1961, 1962, 1963; Dietz, 1961, 1963; Gallant, 1964; Glass and Heezen, 1967; Pasachoff, 1991, pp. 147ff), and (2) McLaren's (1970) invocation of the world's attention on space-oriented studies in support of an idea of extraterrestrial intervention, in the form of a massive meteoroid strike, as an explanation for an abrupt and implicitly catastrophic biogenetic transition recorded in his biostratigraphic research (cf. Ager, 1980, 1984).

The renewed impetus for the IEH debates, of course, came in 1979–80 with the discoveries of iridium anomalies at the K/T boundary (the work of UC Berkeley researchers led by the father/son team of Luis and Walter Alvarez, with Frank Asaro and Helen Michel; cf. W. Alvarez, 1986; L. Alvarez, 1987; Alvarez and Asaro, 1990; Glen, 1990, 1994; Marvin, 1990; McLaren and Goodfellow, 1990), and it quickly received another major boost from the serendipitous coincidence of J. J. Sepkoski's (1982) publication of his compendium of fossil marine organisms that provided the data for the "first," and explosively controversial, statistical analysis of extinction periodicities by Raup and Sepkoski (1984). [The Raup-Sepkoski analysis was "first" mainly with regard to its timing relative to the IEH debates and the fact that it galvanized strong advocacies for or against not only the concept of periodicity but also the particular analytical methods by which periodicities of any kind in the natural record might be demonstrated (cf. Raup, 1985, 1986a, b; Raup and Sepkoski, 1986; Sepkoski and Raup, 1986; Sepkoski, 1989, 1990).] The IEH periodicity debate began a few years after the inception of an even broader debate of concepts of periodicity that was taking place at multidisciplinary levels among nonlinear

Chapter Notes: Chapter 8

dynamicists in the more universal context of spatiotemporal order and chaos — an "applied-chaos" debate, as it were, especially as it was sparked, not surprisingly, by the UC Santa Cruz "Chaos Cabal" described in the Prologue (e.g., Farmer et al., 1980; Packard et al., 1980; Crutchfield et al., 1986; Farmer and Sidorowich, 1988). A paper written jointly by Norman Packard, James Crutchfield, Doyne Farmer, and Robert Shaw — prime movers in that Cabal — titled "Geometry from Time Series" (Packard et al., 1980), made a great impact on concepts of nonlinear periodicity, judging from the frequency of its citation and the following wave of papers that even now continue to express its central idea, the idea that, in essence, a *complete* description of a dynamical system is contained, at least latently, in its temporal record (see the references cited, for example, in Shaw, 1987b; and Shaw and Chouet, 1988, 1989, 1991).

The geometry-from-time-series idea was given great impetus by the mathematical reasoning of Floris Takens (1981; cf. Ruelle, 1989, Chaps. 6 and 13) and the far-sighted attempt by Nicolis and Nicolis (1984) to apply it to the problem of climate analysis, an endeavor that — like the attempt to defend a specific extinction periodicity by Raup and Sepkoski (loc. cit.) — spawned a furor of dissent among analysts and modelers, especially those with strong traditional biases. [A different caveat, one that has received relatively little attention in the heat of the debates of extinction periodicity and nonlinear periodicity alike, is expressed in Shaw (1987b) and in Note 5.3.] Few persons seem to have noticed that the same types of goals and motivations — hopes, apologies, despairs, or vehement disclaimers — are implied by the work of those who have pursued the issue of nonlinear-time-series analysis of natural phenomena, namely the work of those who have contributed to the building of the Geologic Column (cf. Williams, 1981; Harland et al., 1990), and even by the work of those who have seen glimpses of periodic order in patently poor geologic records (see the commentary in Shaw, 1987b). Concepts of geologic periodicities, even "exact" ones, had been hotly debated long before the issue was rejuvenated by Raup and Sepkoski (1984), as those authors were keenly aware in their enthusiasm for what seemed to be a particularly clear-cut example. In fact, the periodic concept in geology has an old and distinguished literature, some of which is recapitulated in Williams (1981) and in Shaw (1987b, 1988a). Papers such as those by Damon and Mauger (1966), Newell (1967), Evernden and Kistler (1970), Damon (1971), Kistler et al. (1971), Fischer and Arthur (1977), and Fischer (1981, 1984), among many others, have contributed to earlier ideas that periodic components exist in the record of terrestrial evolution, whether that evolution refers to the fossil record or to magmatic phenomena (both plutonism and volcanism), tidal phenomena, and geomagnetism.

Although, at first, I viewed my involvement in the IEH and periodicity debates only as an opportunity to extend the studies of igneous (magmatic) plutonic and volcanic recurrence patterns that have been a persistent subject of my own research (especially as influenced by collaborations with R. W. Kistler, R. L. Smith, T. L. Wright, D. A. Swanson, E. D. Jackson, and I. J. Grundfest), I quickly became fascinated by the range of phenomenological interplays that had been suggested and/or documented between diverse types of Earth processes, and between them and cosmological processes. The self-similarities in geological recurrence patterns with which I was familiar from the magmatic record appeared to exist in many other types of records as well. I was greatly aided in the focus of that effort by the appearance in 1981 of the annotated compilation of the literature on geologic recurrence phenomena by G. E. Williams, titled *Megacycles*. His introductory chapter on concepts of geological cyclicity, which includes examples of geological, geochemical, paleontological, oceanographic, climatographic, and astronomical time-series data, and some of their *commensurabilities* (e.g., Williams, 1981, p. 6; cf. Note 4.1 of the present volume concerning usages of the word *commensurability*), was both a major help and an inspiration in support of my growing conviction that there is a common and systematic recursive, architecturally self-organized, and codifiable pattern — a taxonomic scheme of *dynamical bauplans*, as it were (cf. Hennig, 1966, p. 10), based on *universality-parameter systematics* (see Note P.4) — hidden within the nonlinearities and seeming discordances among geological-cosmological phenomena, a pattern that I predict will ultimately reveal itself as a form of integrating factor for terrestrial-extraterrestrial history (e.g., Shaw, 1988a, Table 1 and Fig. 25).

The precursor of the second wave of the IEH debates — the second wave being the one-two punch of the Alvarez-Alvarez/Raup-Sepkoski resonance — was comparatively subtle: the 1970's took little notice of Digby McLaren's seminal paper (1970), which was not readily available to a broad spectrum of interdisciplinary researchers (i.e., outside the controversies it stirred within the fields of biostratigraphy and paleontology); that experience stands in marked contrast with the later, seemingly sudden, emergence (rather re-emergence) of interest that involved many different disciplines, many of which were now primed for action. Historically, therefore, I would identify the emergence of the IEH debates with McLaren's (1970) presidential address to The Society of Economic Paleontologists and Mineralogists and The Paleontological Society, titled "Time, Life, and Boundaries." This acknowledgment of McLaren's insight and foresight seems accurate to me from the standpoint that the spatiotemporal realities and significances of the putative meteoritically sourced geochemical anomalies at major boundaries of the Geologic Column, and the evidence for discontinuities in the successions of fossil biodiversity data across those boundaries — issues that are at the very heart and soul of the IEH debates — have had to be evaluated, initially and subsequently, on the basis of the biostratigraphic principles that McLaren addressed in 1970. Lacking that context, the IEH debates would have been severely threatened (with much more havoc than has been the case) by *chaos of the first kind* (the Biblical kind), which has made fleeting many an interesting hypothesis of Earth history. Even so, published criticisms of stratigraphic setting, geographic scope, temporal coherence, and geochemical provenance have not been trivial, as two large symposium volumes edited by Silver and Schultz (1982) and by Sharpton and Ward (1990), and critiques by Officer and Drake (1985), Officer et al. (1987), and Courtillot (1990), among many others, have testified (see Glen, 1990, 1992, 1994).

The earlier period of *latency* in the evolution of the IEH debates is a lesson in the nature of evolutionary rate phenomena. Ironically, the importance of latency is sometimes forgotten in paleontology and biogenetics, even when it is invoked to dramatize the minuscule duration and technological explosion in the social history of *Homo sapiens* vis-à-vis geological history. The prototypical geological example, from my own perspective, is given by the initial stages of exponential growth of coupled magmatic-tectonic evolutionary progressions, as illustrated by Shaw (1969, 1973, 1987a, 1991; cf. Shaw et al., 1971, 1980a; Gruntfest and Shaw, 1974). The evolution of magmatic instabilities in the Earth clearly demonstrates the nature of a progressive feedback process that is at first imperceptible (even though accelerating), then perceptively creeps, then marches, then gallops, then explodes through history (much as human technologic "progress" has done since the Stone Age) — and wherein the first stage, the latent stage, *represents nearly all* of that history.

In a Strange Loop (Hofstadter, 1979) of connectivity to the IEH debates, the early and almost imperceptible stages of latent evolutionary change in organic evolution (as represented by data on biologic diversity, evolutionary rates, etc.) are necessarily devalued in characterizing "explosive biologic radiations" in the fossil record, simply because — like the surface manifestations of latent magmatism — there is no record of them. I am speaking generically, here, with regard to progressive instabilities of many kinds and durations, not with reference to the issue of single, multiple, and/or sequential bursts of impact events

509

relative to any particular biostratigraphic extinction horizon. Analogously, however, the effect of a relatively instantaneous impact-extinction event also influences the following rates of accelerating evolution, as well as the relative shifts in genetic makeup of the resulting multispecies populations, according to at least two mechanisms that operate at low diversity levels, one being *genetic drift*, and the other being the *founder effect* (cf. Raup and Stanley, 1978, p. 106). Genetic drift is enhanced at low populations, and so, too, are the species shifts that originate from the enhanced genetic representation of atypical individuals (the founder effect). I cannot help but note the (unintended?) double entendre in the word "founder"; i.e., a major fraction of the previous population *founders* during the event, while a possibly significant number of originations of new species is *founded* from the atypical survivors of the preexisting population, depending on the abruptness of the event and its relationships to *genetic bottleneck effects* (cf. Raup, 1979). All of these phenomena are examples of feedback phenomena that are subject to accelerative as well as decelerative regimes of behavior, hence to latency effects of both positive and negative types.

Bottleneck phenomena are complex and, in some respects, resemble the types of nonlinear-dynamical phenomena called *chaotic intermittencies* (see Chap. 5, and Note 5.1). In some cases, intermittency is identified with the iterative "stagnation" of phase-space recursion cycles that eventually become attracted to other regions of phase space (e.g., to new and/or expanded attractor states of the system, or to states beyond its bounds that are equivalent to the destruction of the preexisting dynamics), which is analogous to a biological population that gets "stuck" for a protracted period of time at a small, and oscillating, number of reproducing individuals before it proceeds either to extinction or to a new stage marked by an exponentially increasing diversity of surviving forms (cf. Gwinn and Westervelt, 1986, Fig. 5; Shaw, 1987a, Figs. 51.15–51.18; Shaw, 1988b, Fig.6). An interesting modern example of a complicated bottleneck effect has been identified from studies of the African cheetah (O'Brien et al., 1986), which is alleged — in order to explain the extraordinary genetic uniformity, and lack of immunological rejection reactions, between different individuals — to have gone through one or more near-extinction events over the past ten thousand years or more (the geologic age at which the first among a possible series of bottleneck effects was expressed in this lineage apparently is not known). The cheetah may yet become extinct, or it may evolve to a more viable form (O'Brien et al., 1986, p. 92; cf. Gibbons, 1993).

Lacking such definitive tests to aid the interpretation of the fossil record, it may be difficult to distinguish between genetic radiation effects related to (1) catastrophic environmental events (e.g., the IEH), (2) bottlenecks of other types, and (3) the effects of long-protracted threshold stages of exponential growth that were set in motion long before a specific extinction horizon and were just reaching the "climactic" part of the growth curve (relative to other species, preservation statistics, etc.) near the time of a putative extinction horizon. Any of these modes might produce dynamical bifurcation patterns that are numerically similar in form (see Fig. 36 in the present volume; and examples discussed in Silver and Schultz, 1982, and Sharpton and Ward, 1990; cf. Fisher, 1967; Russell, 1977, 1982; Russell and Rice, 1982; Schopf, 1982; Shaw, 1988a, Figs. 14–18; Carroll, 1992).

Biostratigraphic evidence of the early stages of the *latent-acceleration effect* is subject to rarefaction phenomena that are somewhat analogous to the reverse of the *Signor-Lipps effect* (Signor and Lipps, 1982; cf. Raup, 1972, 1976; Lipps, 1986; Gould, 1992). The latter effect represents a *rolloff* (an abrupt, perhaps unanticipated and even seemingly unjustified, decline in the value of a function — not unfamiliar to seismologists and stock-market investors) in the probability of finding fossil evidence of a given biologic population of constant diversity level as a function of progressively smaller and smaller stratigraphic exposures — further accentuated by discontinuous habitats and population clustering — with

Chapter Notes: Chapter 8

increasing proximity to an extinction horizon (cf. Béland, 1977; Russell, 1977, 1982; Clemens, 1982; Russell and Rice, 1982; Schopf, 1982; Smit, 1982; Fastovsky, 1986, 1990; Bourgeois, 1990; Flessa, 1990; Herman, 1990; Pospichal et al., 1990; Gallagher, 1991; Hanneman and Wideman, 1991). In other words, the position of a putative extinction horizon in a "complete" biostratigraphic section that extends both below and above it — based on last occurrences in the section of diagnostic fossils of the preextinction biota and first occurrences in the section of diagnostic fossils of the postextinction biota (a variation on time-honored field methods for bracketing geologic contacts that may be either gradational or sharp) — is uncertain within a variable bandwidth that depends on factors related to, among other things (cf. Eldredge and Gould, 1972; Raup et al., 1973; Simberloff, 1974, 1986; Gould and Eldredge, 1977; Raup and Stanley, 1978; Raup, 1979, 1984a, b, 1986a, 1988b; Gould, 1984b; Kauffman, 1984, 1985, 1988; Boecklen and Simberloff, 1986; Colbert, 1986; Vermeij, 1986; Wilson, 1988; Eldredge, 1989; Raup and Jablonski, 1993; and papers in Russell and Rice, 1982; Silver and Schultz, 1982; Berggren and Van Couvering, 1984; Nitecki, 1984; Kaufman and Mallory, 1986; Sharpton and Ward, 1990; MacLeod and Keller, 1991), (1) any changes in actual species diversity levels, (2) any changes in the biogeography of a given species at constant diversity, relative to the location of the section, (3) any changes in the sedimentation processes by which biotic remains are incorporated into their characteristic lithic matrices, and/or the geochemical processes by which they are preserved as fossils, (4) any processes of destruction of the relevant fossil-bearing horizons (fossil hiatus effects), and (5) the stratigraphic intermittency of a normally complete biostratigraphic section for a given fossil type.

The last of these effects represents the probability of finding a given fossil within a specified length of normal section in which there is no extinction horizon — meaning that unless the section is totally made up of fossils (as is sometimes approximated in the sedimentary accumulation of pelagic organisms), any given "horizon" always represents some finite, and often unknown, range in time. In the case of extinction horizons, the potentially barren range can extend well below a putative event horizon (theoretically representing an instant in time). In the case of origination, however, the barren range may extend far above an extinction horizon, or even farther above an origination horizon that occurred prior to the extinction event. This ambiguity is a function of the latent-acceleration effect plus all of the other factors mentioned above, because in that case the population grows from essentially zero over some protracted interval of time, and during that time all of the nonoccurrence probability factors are multiplied by the rarity of preservation of the first fossils of a new species.

In Shaw (1988a, pp. 22ff), I used the record of fossil birds (at the levels of orders and families) to illustrate the role of prior latent acceleration phenomena vis-à-vis the K/T extinction event as the cause of adaptive radiations. This is an excellent example for present purposes because, as many paleontologists have advised me in no uncertain terms, the fossil record of avian evolution is one of the worst in the geologic column (but see Bock, 1979; Eldredge, 1989, Chap. 2) — therefore the intermittencies of fossil preservation should be large, and the *barren-section uncertainty length* should be maximized. Ignoring questions of the proper taxonomic relationships — because, for purposes of illustration, I used the compilation of Fisher (1967), aware that the entire avian phylogenetic classification scheme was being revised by Charles Sibley and Jon Ahlquist on the basis of DNA hybridization techniques (see Sibley and Ahlquist, 1990, for the results of that monumental work) — it is evident that the record of any identifiable lineage and its branches, rightly or wrongly identified as to genetic relationships, gives only a rough thumbnail sketch of the evolutionary tree (bush, thicket, or what have you) that has expanded to the present-day number of avian families. Fisher (1967) plotted the avian family tree in a form that is directly

Chapter Notes: Chapter 8

analogous to the bifurcation sequences of a sine-circle function plotted on a circle, hence is equivalent to a circle map (cf. Shaw, 1987b, cover illus. and Figs. 1 and 2; Shaw, 1991, Figs. 8.32 and 8.33), where the radius is proportional to geologic age, thereby accommodating the increased packing density with time. In that form (replotted as Fig. 14 in Shaw, 1988a), the diversity appears to "explode" in the numbers of families at an age near the K/T boundary (making a liberal allowance for the effects of time-scale corrections of biostratigraphic stages and possible errors in diversity estimates per stage).

At this juncture, one should know that in the bifurcations of circle maps, as in quadratic maps, the number of states of the system (analogous to number of lineages) increases geometrically within a given sequence (e.g., within a period-doubling "family") in proportion to the *tuning* order (Shaw, 1988a, Figs. 16 and 17; cf. Shaw 1987b; cf. Notes P.1 and 3.1, and Table 1, Sec. D, of the present volume). If the tuning order is proportional to the time since origination (a branching point of the family tree), then the diversity (number of lineages at a given time) is approximately an exponential function of the time since the beginning of a branch (e.g., number of orders within the class Aves; number of families within an order; total number of families within the class, etc.). On the other hand, if the evolution of diversity is a critical cascade process, then the number of lineages grows as a power-law function of time (logarithmic growth, giving linear log-log plots).

Shaw (1988a, Figs. 15–18) demonstrated that the diagrammatic rate of avian evolution, measured in terms of family diversity, is not an exponential function of time but grows in quasi-logarithmic decaying pulses (cf. Fig. 36b), each of which — representing the (nominal) families within the (nominal) orders of Fisher (1967) — reaches a peak rate at different characteristic diversity levels and ages (originations $\gg$ extinctions), and then either remains at constant diversity (originations = extinctions) or retrogresses somewhat (originations < extinctions) to the present diversity level. These effects are shown in Figure 36b of the present volume in the form of logarithmic plots of (1) the standing diversity of all avian families of Fisher (1967), and (2) the standing diversities within two different avian orders [e.g., cormorant-type birds vis-à-vis perching birds; the former is longer-lived in the fossil record, perhaps befitting the relatively unspecialized fish-eating marine lifestyle, whereas the perching birds evolved after — or cooperatively with (assuming that birds contributed significantly to seed dispersal) — the evolution of flowering and fruit-bearing plants (the angiosperms; e.g., Knoll, 1984)]. The effects mentioned above are shown by the abrupt falloffs or reversals in logarithmic rates within the latest 40 Ma for each order. The logarithmic rate for total family diversity in Figure 36b shows (1) a constant power-law slope of the older evolutionary regime, (2) some rolloff relative to the initial power-law slope just prior to the K/T boundary, (3) a steeper (larger exponent) power-law slope during a portion of the Tertiary Period, followed by (4) a rolloff (failure of the power-law trend) and decay or stabilization at the present-day level of avian diversity. The end-Cretaceous transition in power-law regimes would appear to be an artifact of the balances between the negative evolutionary rates of the pre-K/T lineages and the accelerating rates of evolution of the post-K/T lineages (in a manner somewhat analogous to the pre- and post-K/T patterns of change in the vascular plants; see Knoll, 1984, Figs. 1–5). Times of inception and decline, however, bracket the K/T boundary over a span of about ± 10 m.y.

According to the preceding discussion of the latent-acceleration and Signor-Lipps effects, it is possible that the bimodal rate effect in avian evolution (e.g., Fig. 36b here; and Shaw, 1988a, Fig. 16) was a consequence of the rapid radiation of post-K/T originations superimposed on the effect of the K/T extinctions on the decline of pre-K/T lineages (which, however, have persisted with fairly stable diversity to the present time). [A number of revisions in the traditional avian taxonomy, for example, as compiled and illustrated by Fisher (1967) at the levels of orders and families, have been made by Sibley and Ahlquist

(1990), including many taxonomic reassignments among existing orders and the creation of some new orders. Although there are some big surprises in the reassignments of some of the traditional affiliations of birds that were previously thought to be well-known in their genetic relationships, the general distinction between the more primitive and the more modern lines of descent seems to be preserved (see the discussion of Fig. 36 in Chap. 8; cf. Shaw, 1988a, Figs. 14–16).] But this does not explain why the general forms of the growth curves are similar within all orders and are partitioned intermittently in peak rates over an age span between about 120 Ma and 40 Ma, with origination ages partitioned over an age span from ~150 Ma—representing the direct ancestors of the penguins, and common ancestors of all but the kiwis, ostriches, rheas, emus, cassowaries, and tinamous, which Fisher shows, questionably, to branch off of the ancestral penguin lineage at 90 Ma or thereabout (cf. Sibley and Ahlquist, 1990, pp. 528ff)—to about 60 Ma, representing the ancestors of the woodpeckers, toucans, etc. It seems at least possible that, because of the latency effect, *all* of the lineages of birds could have originated prior to the K/T boundary and be relatively little affected by the K/T extinction event—except with regard to relative rates of evolution between different lineages. From the perspective of nonlinear dynamics, such rate effects might be viewed as the consequences of *implicitly deterministic* but not *determinable* functions (cf. Note P.7) that control the intermittency of fossil preservation, rock-matrix preservation, and all related biogenetic-geodynamic functions that influence both the origination and the extinction of species. [The importance of such effects apparently is being seen in recent studies of the relationships between dinosaurs and primitive birds (cf. Altangerel et al., 1993; Milner, 1993) and also in controversies over possibilities of multiple Triassic extinction events relative to the initial radiation of the dinosaurs (see below; cf. Bice et al., 1992; Hodych and Dunning, 1992; Rogers et al., 1993; Benton, 1993). Analogous discussions exist concerning multiple end-Cretaceous extinctions (cf. Note 1.1; and Kauffman, 1984, 1985, 1988; Padian et al., 1984; Hut et al., 1987, 1991; DeRenzi, 1988; Hallam, 1988, 1989; Shaw, 1988a). In the context of the present Note, proponents of the IEH place undue emphasis on finding *the* "smoking gun" for each putative impact-related extinction event, an emphasis, perhaps, that reflects our contemporary social concerns with actual smoking guns and issues of proof in courts of law.]

Returning to the concepts of coupling and/or correlations between geomagnetic phenomena and other geodynamic phenomena, the discussion of latency and the Signor-Lipps effect points to an *ambivalence* and/or *frustration* in our abilities to discriminate between and/or to definitely identify specific coupled processes and/or cause-effect sequences. If we were to assume, for example, that a massive impact is not definitely established, from other evidence, as the causative factor in the K/T and other mass-extinction events in the geologic record, then the uncertainties of the above rate-dependent processes would be too great to prove a one-to-one correlation between the exact timing of an impact mechanism and a biostratigraphic boundary crisis. However, there is another criterion that does not formally depend on the *exact* timing of cause and effect. This is the concept of multidimensional-systems resonances that I introduced in 1987 (Shaw, 1987b) by arguing that if there is general evidence for synchroneities between two complex dynamical systems, then concepts of nonlinear-dynamical coupling imply that the two systems are unlikely to represent independent processes (i.e., corresponding to the wheel-of-fortune argument that I have applied in the present work to a universe in which strict independence is impossible). This is both good news and bad news. The bad news is that instances of "incontrovertible" one-to-one correlations of various kinds that have been cited either for or against cause-effect phenomenologies are not definitive one way or the other [e.g., see the diametrically opposed conclusions of Schneider and Kent (1990) and Glass and Zwart (1979, pp. 341f) concerning the correlation of the Ivory Coast microtektite horizons with the Jaramillo geomagnetic polarity-event horizons in oceanic sedimentary drill-core sections; cf. Burns (1989)]. The good news is that universal spatiotemporal scaling of generally correlated stratigraphic and/or biostratigraphic intermittency patterns of different phenomena argues for a generally coupled system of oscillatory genetic mechanisms. This is a reflection of the general principle of *nonlinear-dynamical stratigraphic correlation* to which I first drew attention in Shaw (1987b).

In the same sense in which fossils represent the preserved remains of living organisms (dynamical systems), the preserved remains of the past dynamical states of the geomagnetic fields (fossil fields) are represented by paleomagnetic measurements. The complexities of geomagnetic phenomena, and their interactions with other geological processes, are therefore analogous to the geological-biological processes described above. In this sense, direct cause-effect coupling not only is impossibly difficult to prove for geomagnetic correlations with other processes, terrestrial and extraterrestrial, but it is not essential (and may be antithetical) to the actuality of multidimensional systems resonances among the variety of coupled processes that affect and/or are affected by the dynamics of the core dynamo. The apparent synchronizations identified in the text are better viewed as nonlinear resonances within a general system of time-delayed interactions, rather than as one-to-one correlations. I would expect, then, that the frequency signatures of such diversified feedback phenomena sometimes would be in phase and sometimes would be out of phase, yet in both cases they reflect the coupling of the periodic/chaotic systems of attractors that—given unlimited data—would describe all of the respective processes. In the context of the coupled processes discussed above—and presumably in many other contexts as well that I have had neither the data nor the time to consider (cf. Note 8.5)—the geological-cosmological coupling of feedback phenomena possesses many of the characteristics shown by neurobiological coupled oscillators with delayed feedback (Freeman, 1992, pp. 467ff).

Note 8.2: The possible mechanisms by which impact-induced tectonic and/or magmatic phenomena could influence the core dynamo are diverse (cf. Note 8.1; and Stothers and Rampino, 1990, Fig. 5). I briefly consider two modes: (1) the influence of inward-propagating effects initially produced at the Earth's surface, and (2) the influence of outward-propagating effects initially produced within the core and/or at the core-mantle boundary by the effects in (1), as follows:

(1) The most obvious effect is the impact-induced initiation of propagating magmatism, as manifested in so-called hot-spot volcanism (cf. Appendixes 12–18), at either the direct or antipodal sites of sufficiently large and/or strategically located impacts (e.g., Hagstrum and Turrin, 1991a, b). In some cases, a single impact, or cluster of impacts, may induce only an incipient melting instability in the mantle (e.g., according to the dissipative feedback mechanism of Shaw, 1969; cf. Shaw, 1991; Hagstrum, 1992), whereas, in other cases, the same impact energy may activate a rapidly propagating system of rifts and the progressive fission of continental masses (e.g., by initiating rifting events that participate in continental drift and the growth of ocean basins; cf. Seyfert and Sirkin, 1979, Table 12.2). In this way, the initial magmatic instability is amplified to wholesale participation with plate-tectonic motions and deep-mantle convection, thereby influencing the longer-term thermal balances at the core-mantle boundary (cf. Walzer and Maaz, 1983). Such effects presumably will contribute to the cyclicity of the $>$ 100-m.y. scale of polarity superchrons (cf. McElhinny, 1971; Harland et al., 1990, Fig. 6.10). Indirect outside-in phenomena also are possible, as in the model of Muller and Morris (1986, 1989; cf. Schwarzschild, 1987), wherein there are exchanges of rotational angular momentum between the atmosphere, mantle, and core through the agency of glacier-ocean-atmosphere dynamics responding to impact-winter phenomena.

(2) The direct influence of a mass impact on the core and/or core-mantle boundary concerns the propagation of stress waves through the Earth—hence on path-dependent focusing effects, refraction effects, etc.—a subject of controversial and/or ambiguous modeling studies (cf. Schultz and Gault, 1975; Hughes et al., 1977; Rice and Creer, 1989; Watts et al., 1991). I restrict this commentary to the rather easily visualized phenomenon of *spallation*, whereby slabs of relatively "brittle" material at the D″ layer (read as "the D double-prime layer"; see Young and Lay, 1987; Knittle and Jeanloz, 1991) of the core-mantle interface, or at the solidifying interface between the inner (solid) and outer (liquid) core might split off, thereby perturbing the fluid motion within the outer core. For example, even though density stratification presumably would constrain the effects of spallation to these interfaces (cf. Jacobs, 1987, 1992a, b; Jeanloz, 1990; Smylie, 1992; Duba, 1992; Boehler, 1993), decoupled slabs of solid material might tumble along solid-liquid interfaces and/or effectively roughen and perturb the preexisting boundary-layer flows and vorticity states of the outer core. Rice and Creer (1989) outlined the conditions under which spallation might occur at either interface, given a suitably weak but brittle material. In their estimation, the strengths of even partially molten states at either interface probably would be too great to permit spallation to occur. Apparently, however, they did not consider the possible effects of fluid-injection fracture.

Shaw (1980) pointed out that under conditions where the fluid pressure in the Earth is approximately equal to the vertical (lithostatic) load, there is no limit to the depth at which extensional fracture (analogous to hydraulic fracture) can occur under conditions of non-hydrostatic stress, given a sufficiently small tensile strength (cf. Anderson, 1936). Tensile strengths, even in solid materials, tend to be small. More important, however, partially molten materials—even with very high liquid fractions, hence very low strengths—can behave in a brittle manner. In this context, the term "brittle" refers to the sharpness and angularity of fracture surfaces and the relatively rigid-body continuity of the unfractured material. And in *that* sense, crystal-liquid portions of magma bodies often display evidence of brittle fracture even at very high melt percentages (cf. Shaw, 1980, Fig. 8). It seems likely, therefore, that crystal-liquid assemblages at either the inner or outer core temperatures and pressures would be subject to some form of brittle spallation even at very small stress differences. If so, the rotational tractions at those interfaces could be significantly affected by the stress-wave transients of even moderately large impacts, perhaps causing rotational impulses or "jerks" analogous to those seen in modern records of variations in the Earth's rotation (e.g., Lambeck, 1980, pp. 84f) and in secular accelerations of the geomagnetic field (e.g., Courtillot and Le Mouël, 1984, 1985; Alldredge, 1984; McLeod, 1985).

In the same speculative vein, it also seems conceivable that transient effects of shock compression and rarefaction waves related to large impacts might induce phase-change instabilities in those parts of the D″ layer where complex mineral reactions are thought to occur (e.g., Young and Lay, 1987; Jeanloz, 1990; Knittle and Jeanloz, 1991). Such effects would be analogous to the reconstructive phase transformations thought by some workers to explain the source mechanism of deep-focus earthquakes at depths of about 350 to 700 km in the mantle (cf. Kirby et al., 1991; Iidaka and Suetsugu, 1992; Green, 1993). In Note 6.8, I suggested that under some circumstances phase-change instabilities might be sufficiently rapid—even if at subseismic frequencies—and extensive to resemble Ramsey's (1950) concept of violent instabilities in certain types of planetary cores. Such a concept is a deep-mantle analogue of episodic—perhaps even "catastrophically" abrupt, though not so compared with the implication of the Ramsey mechanism—upper-mantle composition-dependent phase-change instabilities that have recently come into vogue (cf. Anderson, 1987, 1989, 1991a; Ito and Takahashi, 1989; Machetel and Yuen, 1989; Ito and Sato, 1991; Liu et al., 1991; Machetel and Weber, 1991; Peltier and Solheim, 1992; Honda et al., 1993;

Tackley et al., 1993; Weinstein, 1993). Such instabilities, in turn, are analogous to the magma-related composition-dependent phase-change instability proposed by Shaw and Jackson (1973) as an anchoring mechanism of a relatively stationary mantle reference frame for plate kinematics that is based on the "ground truth" of volcanologic patterns rather than on the remotely hypothesized mechanism of a system of stationary core-mantle thermal plumes. Any of such episodic and radially inward modes of penetrative convection — especially one as potentially violent as the Ramsey mechanism — could, in conjunction with large cratering events at the Earth's surface, perturb the outer core for geologically significant periods of time, thereby providing impact-related coupling between the Solar-System-CRF-system and multiple mantle instabilities that in turn contribute to persistent wobble-like effects in the Earth's rotation (cf. Munk and MacDonald, 1960, Chap. 10; Lambeck, 1980, Chap. 8; Hyde and Dickey, 1991; Mathews and Shapiro, 1992; Dickey, 1993).

Note 8.3: *Magma* is the generic name for multiphase partially or wholly *molten* material states of a planetary or satellitic body (disregarding analogous states of stellar bodies in the present work; cf. Zirin, 1988; Cox et al., 1991; Maran, 1992). The word *molten*, in this usage, refers to the thermal instability of crystalline states (with or without dissolved and/or occluded volatile components, including such things as carbon dioxide, noble gases, etc.), hence it has a very broad, and very vague, meaning in the general context of heterogeneous equilibria. The brine of a saline lake is "molten" in this generic sense — as is, in general, the solvent phase of any (liquid $\pm$ gas)-solution/solute heterogeneous equilibrium. These are important realizations to keep in mind when dealing with the possible meanings of the word *magma* in the icy satellites of the outer planets, or in the sulfurous magmas of the Jovian satellite Io, or in the carbonatitic and other "odd" magmas in the Earth, etc. (e.g., see the Prologue and Note P.5). Although one is usually referring to an essentially condensed bulk phase state when talking about magma, this inference, too, is ambiguous. Most magmas that can be directly sampled in the form of near-surface and/or volcanic materials contain gaseous components in both the dissolved (liquid and/or solid) and exsolved (bubbles in liquid and/or occluded gas in solid) states. The term *magma* also must apply to phase states in which the solvent state, and/or the solution state, and/or the solvent, solution, and solid-phase states are partly or completely supercritical states (gas-like *plasma* states of widely varying compositions and densities). The word *plasma* often is seen only in reference to stellar and/or interstellar conditions, but the concept is dynamically relevant to planetary and interplanetary processes (cf. Alfvén, 1980, 1981, 1984; Alfvén and Arrhenius, 1975, 1976) and to the dynamical meaning of the word *magma* applied to the core materials of planets and large satellites (e.g., in the D″ layer at the Earth's core-mantle boundary where pressures are high enough to be supercritical relative to phase states of similar bulk chemistries in the upper mantle and at the Earth's surface; cf. Jacobs, 1987, 1992b; Jeanloz, 1990; O. L. Anderson, 1990; Knittle and Jeanloz, 1991; Smylie, 1992).

The vagueness and ambiguity of the terms *magma* and *molten* are examples of the wisdom often contained in the roots of scientific terms. The word *magma* is derived from an ancient etymological lineage — one that was used in many other contexts before it was adopted by geology — stemming from a Greek root meaning *to knead* (cf. The Compact Edition of the *Oxford English Dictionary*, 1981 Printing). Thus, the word *magma* conveys the essential nature of the magmatic process in a way that usually is not mentioned in the textbooks of the earth and planetary sciences. Beyond the implication that partial or complete melting has taken place is the implication that the rate of deformation (the dissipative *kneading rate*) of the Earth — or other parent body — is intimately involved in how these phase states have come into being (Shaw, 1969, p. 533; cf. Shaw, 1983a). In other words, beyond the mere existence of a particular type of fusible material substance, magma testifies to the dissipative history of the parent body as the product of an *irreversible process operating far from equilibrium* (e.g., Shaw, 1970, 1983a, 1991; Gruntfest and Shaw, 1974; Wones and Shaw, 1975; Peale et al., 1979; Cassen et al., 1982; Greeley and Schneid, 1991; cf. Notes P.5 and I.5), even when there is evidence that—at any intermediate or end stage of the process—metastable thermodynamic equilibrium may have been approximated.

Note 8.4: Solar activity is a fascinating subject in its own right. Its relevance to Earth processes, once largely ignored, has become the focus of intensive study, perhaps because of the contemporary concern with problems of weather and climate, "ozone holes," and other factors involved in environmental degradation. Justice cannot be done in this brief note to the relevant literature, but the overviews given by Eddy (1976, 1977), McCormac (1983), Landscheidt (1983, 1988), and Zirin (1988, Chap. 10) illustrate the indisputably powerful involvement—far beyond the direct effects of total irradiance—of processes in the Sun with processes in the Earth. An aspect of sunspot activity not discussed in the text is the spatiotemporal patterning of sunspot distributions. These patterns—like much of what is known about sunspots—were first recognized a century ago in the research of E. W. Maunder and others during the late nineteenth century. Of particular interest from the standpoint of nonlinear-dynamical recursion is the fact that the sunspots, individually and in groups, evolve systematically during each $\sim$11-year cycle within a bilaterally symmetric latitude zone of $\pm \sim 45°$. This pattern—first demonstrated by Maunder and later developed by J. A. Eddy (loc. cit.)—is referred to as the *butterfly diagram* (Zirin, 1988, Fig. 10.2), because the bilateral symmetry gives the appearance of the open wings of a butterfly mirrored across the Solar Equator as the spatial distribution of sunspot activity is plotted as a function of time.

At the beginning of each sunspot cycle, activity begins at latitudes of roughly $10°$–$30°$, expands to both higher and lower latitudes, and then converges toward the Solar Equator at the end of the cycle. The descriptive resemblance to Lorenz's *butterfly attractor* (Lorenz, 1963; Sparrow, 1982; Gleick, 1987; cf. Shaw, 1991, Fig. 8.24) is striking. I have referred to the recursive regeneration of such attractors—whether fractal (i.e., "strange"; see Note P.3) or nonfractal, and whether chaotic or nonchaotic (cf. Ditto et al., 1990a)—as *pattern periodicity* (e.g., Shaw and Doherty, 1983; Shaw, 1988b, Figs. 3 and 4) for the reason that the periodic repetition refers to the complex *form* of the attractor rather than to the individual points (e.g., plotted points of a computer algorithm; sunspots on the Sun; basaltic volcanic vents on the Earth) or clusters of points (clusters of plotted points of a computer algorithm; sunspot groups on the Sun; basaltic vent fields on the Earth) that go to make it up. In other words, the periodic structure of a complex pattern refers to the *recurrence of recognizability* rather than the periodic recurrence of identical temporal or spatial phase states of the process. This criterion is the same one by which we identify two different specimens of *actual* butterflies of the same species—meaning that it refers to recurrence relationships of organic complexity (cf. Shaw, 1987b). No two points or sequences of points in the phase space of the two different attractor systems (specimens) are identical, even though the pattern is—within the uncertainty parameters of the human sensory response (the nonlinear-dynamical analogue of an all-purpose power-spectral analyzer)—"identical." This just means that a rose is a rose, etc., that a Monarch butterfly is a Monarch butterfly, etc., and that, from cycle to cycle, a sunspot field is a sunspot field—and that volcanoes and volcanic fields are of sufficient organic complexity that they can be identified as to individuals and species by the same criteria we use to identify birds, butterflies, or people (e.g., R. L. Smith, 1979; Shaw, 1985a, 1987a, 1988b).

Chapter Notes: Chapter 8

Within the longer variations of the sunspot cycle (see Note 8.5), the character of the individual butterfly portraits changes only quantitatively — i.e., with regard to the numbers of spots and spot groups as a function of latitude and time — but otherwise remains recognizably the same, just as individuals within an organic species can vary in size and other features but still be, unmistakably, members of the same species (compare Figs. 3 and 4 in Shaw, 1988b; cf. Calder, 1984). This does not mean, however, that one or the other of such "species" cannot evolve, either gradually or abruptly, as is evident in the cited examples from Shaw (1988b; cf. Shaw and Doherty, 1983), to a distinctly different form. Conceivably such a transformation might occur in phase portraits of the solar butterfly diagram over the "supersecular" (see Note 8.5) variations of solar activity for which detailed information on spot distributions is not available. Such transformations would imply shifts — gradual or abrupt — in the *tuning* (cf. Notes P.1, P.3, 3.1) of the dominant attractor or system of attractors that generates the recognizable characteristics of the species patterns.

Note 8.5: The argument given in the text concerning coupling between solar activity and Earth processes is qualitative. There are a number of different pseudo-periodic cycles in solar activity, each of which may display associated chaotic bursting effects analogous to the many-level chaotic intermittencies of spin-wave phenomena (e.g., Fig. 18; and Carroll et al., 1987). Landscheidt (1983, Figs. 5 and 6) illustrates a "secular" variation in sunspot activity with an average period of ~83 years, and a "supersecular" variation with a peak-to-peak repeat period of roughly 200–400 years, the latter being represented by only three or four cycles in the historical record (cf. Eddy, 1977, Fig. 10). Neither of these "periods" is established well enough to define a *winding number* in the sense described in the text for the sunspot cycles (see Notes P.4, 3.1, and 4.1; and the discussion of Fig. 14 in the text; cf. Appendix 9). A yet longer variation in solar activity is indicated by relative carbon-14 (^{14}C) variations from dendrochronologically dated tree rings over the past 7000 years or so (Damon, 1977, Fig. 1; and Eddy, 1977, Fig. 9; Landscheidt, 1983, Fig. 2). Only about one-half cycle of the ^{14}C variation is revealed by these data, suggesting a poorly defined cycle time of the order of 10^4 years (cf. Merrill and McElhinny, 1983, Fig. 4.9). According to the cited authors, this ^{14}C variation reflects an approximately "sinusoidal" variation in the Earth's magnetic dipole moment, underscoring the importance of coupling between the Earth's magnetic field and the solar output (e.g., White, 1977; McCormac, 1983; Sonett et al., 1991; Cox et al., 1991). Data cited by Merrill and McElhinny (1983, Figs. 4.5–4.9), however, would indicate that the variations of field intensity and dipole moment during the latest 10^4 years are probably chaotic rather than harmonic, a complexity that would appear to extend to the order of 10^5 years (ibid., Fig. 4.9).

In the same context, it is interesting to note that the approximate frequency of the sunspot cycle ($W \cong 0.09$ yr^{-1}; see the discussion in the text) is about an order of magnitude lower than the mean frequency of the *Chandler wobble*, or *free nutation* of the Earth (named after S. C. Chandler for his late-nineteenth-century studies of anomalous variations in latitude; see Munk and MacDonald, 1960, Chap. 10; Lambeck, 1980, Chap. 8). I mention this because the excitation of the wobble of the Earth's spin axis, with a crudely bimodal power spectrum comprising an annual peak and another at about 14 months, corresponding to the frequencies ~1.0 and ~0.9 yr^{-1} (e.g., Munk and MacDonald, 1960, Fig. 10.2) — and even though the mechanisms of its excitation are still not understood after a century of study — is thought to be coupled with some factor or factors that induce relative rotational accelerations of the fluid layers in the Earth. These factors are thought to involve mainly the atmosphere and fluid core (e.g., Wahr, 1983; Naito and Kikuchi, 1990; Hinderer et al., 1990) — but with possible instantaneous excitations related to earthquakes (e.g., Mansinha and Smylie, 1967; Kanamori, 1976; O'Connell and Dziewonski, 1976; Chao and Gross, 1987; Maddox, 1988).

The proposed earthquake-excitation source would appear to be too small by an order of magnitude or so to be a significant factor in the analysis of the Chandler wobble (cf. Kanamori, 1976; Chao and Gross, 1987), but I mention it because in Appendixes 22 and 23 I refer to the analogous effects of mass impacts—many of which exceed the largest known earthquakes by many orders of magnitude in kinetic energy—on potential rotational instabilities of the core dynamo (hence perhaps indirectly affecting the Chandler wobble) as an unevaluated and chaotically intermittent forcing function relative to the general spectrum of variations in the Earth's rotation rate (length-of-day, usually abbreviated l.o.d.; cf. Munk and MacDonald, 1960; Hide et al., 1980; Lambeck, 1980; Hyde and Dickey, 1991; Dickey, 1993) and free nutations (wobble; cf. Mathews and Shapiro, 1992). Furthermore, it seems evident that the external and internal geomagnetic fields are mutually involved in complex ways with both the orbital and rotational motions of the Sun in relation to the Solar System center of mass and the internal dynamics of the Sun's plasma ejections (solar wind, etc.), hence—in reference to the diversity of geophysical correlations discussed in Note 8.1—they must both be generally coupled with the glacier-ocean-atmosphere (GOA) interactions and with variations in the Earth's rotational motions and wobble. It therefore seems necessary—from the above evidence, together with other evidence given in the present volume (see ancillary illustrations given in Appendixes 1–4, 15–18, 22, and 23)—that mass impacts are energetically significant relative both to the direct perturbations *and* the indirect perturbations of (1) global magmatism, (2) the geodynamo, (3) long-term sea-level, atmospheric, and climate perturbations, (4) variations in the l.o.d. and "free" (intermittently kicked but not regularly forced) nutations of the Earth's spin axis, and (5) variations in the Earth's magnetic axis relative to virtual geomagnetic pole (VGP) and apparent polar-wander (APW) patterns. Circumstantial evidence (see Notes 8.1–8.3) argues strongly for the involvement of impact dynamics with a generalized nonlinear-periodic/chaotic multiple-oscillator spectrum representing *all* Earth processes.

Note 8.6: In Appendixes 12–18 of the present volume I have examined the patterns of both the direct and antipodal locations of volcanic "hot spots," as illustrated by Duncan (1991), relative to the distribution of impact craters. [In order to explore the possible relationship between cratering antipodes and volcanism, one can plot either (1) the antipodes to the distribution of craters and compare them directly with the locations of "hot spots" or (2) the antipodes of the "hot spots" and compare them with the locations of craters. Either result illustrates the extent to which antipodal correlations may be significant, within whatever constraints exist for comparing the ages of the two phenomena. I decided to use the second scheme because I wished to examine the extent to which there might be direct *and/or* antipodal correlations between the global *patterns* of cratering and the global *patterns* of volcanic "hot spots" (see Appendixes 16–18).] The general patterns of either direct or antipodal impact sites are remarkably similar to either direct or antipodal "hot-spot" patterns—suggesting that analogous mechanisms of self-organization were involved in the generation of both patterns, and that there was some indifference to either the direct or antipodal manifestations of those mechanisms. The simplest interpretation would seem to be that the "hot-spot" patterns were initiated either directly or antipodally by impacts, whether or not a crater is evident at either the direct or antipodal sites. For example, this could be true for Jupiter's satellite Io, but there would be no way to observe the pattern of impacts (cf. Appendix 24). In that case, however, there is also evidence for a tidal influence on the pattern as a whole—and I would argue that there is some form of spin-orbit influence on the organization of volcanic propagation on the Earth as well (e.g., Appendix 15). Both cases would appear to be systematic in the sense of Yamaji's (1991, 1992) criteria, and I interpret these correlations to mean that the global distribution of "hot-spot" patterns on Solar System objects are influenced by three principal factors, in order of systemic importance (i.e., in individual cases, one or another of the factors may have dominated the magmatic evolution): (1) by the pattern of impact-crater formation on a planet or large satellite, (2) by tidal deformation, both as a dissipative mechanism and as a major long-term influence on the global three-dimensional stress-field patterns that have guided magmatic propagation, at depth and to the surface, and (3) by plate tectonics, which acts to enhance dissipative effects and to modify the kinematics of volcanic propagation relative to plate motions and changes in the Earth's principal axial moments of inertia (the result being to skew the relative coordinate frames on the basis of, alternatively or in combination, plate-tectonic motions or the fixity and coherence of a "hot-spot" reference frame; cf. the Introduction to the present volume; and Molnar and Stock, 1987; Molnar et al., 1988). In my judgment, magma generation in the Earth's crust and/or mantle may accompany the direct and/or the antipodal loci of cratering events. I have not been able to identify correlations that are exclusively specific to one or the other mode—i.e., to a direct impact site or its antipode (but see the comparisons of *patterns* in Appendixes 15–18)—nor have I been able to identify any particular critical magnitude for magma-generating impact events, apparently because the critical magma-generating impact energy depends on the magmatic potential of the crust-mantle section at the impact site relative to that at its antipode, as well as on the path-dependent focusing/refracting properties of the core (cf. Note 8.2) relative to the surface location of impact-induced stress waves (cf. Schultz and Gault, 1975; Hughes et al., 1977; Rice and Creer, 1989; Watts et al., 1991; Hagstrum, 1992).

Rampino and Caldeira (1992) have shown that a significant number of volcanic "hot spots" on Earth correspond, statistically, to antipodal pairs. Their work should be compared with similar demonstrations in the present volume (e.g., Appendix 16d), and with the conclusion that there is a striking resemblance between the global distributions of volcanic "hot spots" and the great-circle patterns of impacts on Earth—a type of global pattern that may hold as well for the Jovian satellite Io, and for Mercury, Venus, and Mars (cf. Appendixes 15–18, 24, and 25; and Head et al., 1981, 1992; Schaber, 1982; Strom and Schneider, 1982; Pike, 1988; Strom and Neukum, 1988; Kaula, 1990; Bindschadler et al., 1991; Greeley and Schneid, 1991; D. B. Campbell et al., 1992; Janle et al., 1992; Phillips et al., 1992; Schaber et al., 1992; Solomon, 1993).

Note 8.7: There are many possible models and scenarios for collisions between the proto-Earth and the smaller protoplanets (e.g., Wetherill, 1985; cf. Ip, 1989). The results differ in detail, depending on whether the collision was barely grazing, low-angle, or high-angle, and whether the orbital inclination was polar, intermediate, or equatorial (see Fig. 19, and references cited in the preceding paragraph; cf. Melosh, 1989; Chapman and Morrison, 1989, p. 165; Kaufmann, 1991, Fig. 9–22). Kerr (1989) shows an artist's conception of orbiting objects some time after the formation of the Moon, taken from the work of the astronomer-painter W. Hartmann. Stewart (1988) portrays time-lapse stages of a direct-hit scenario for the collisional history of the planet Mercury (cf. Cameron et al., 1988).

Note 8.8: The mechanism that I propose by which an early-formed magma ocean (cf. Garwin, 1989; Spohn and Schubert, 1990; Stevenson, 1992) was cooled efficiently is a form of thin-skin heat transfer, resembling, on a gigantic scale, the phenomenon of wholesale crustal foundering in Hawaiian lava lakes (see Shaw et al., 1971, footnote 9, p. 878; Duffield, 1972; cf. Fisher, 1889, pp. 47ff and pp. 379ff; Wood, 1985, pp. 198ff). If post-cratering basin formation and tectonic-magmatic infilling processes proceeded as rapidly as

Chapter Notes: Chapter 8

imagined (Fig. 33), the surface layers of the resulting magma ocean would have been maintained in a perpetually seething motion, owing to the crustal foundering process, until a significant fraction of the magma was crystallized. The foundering process, as it has been revealed to operate in lava lakes, is driven by degassing of the magma and trapping of gases under the developing slag-like crust, which then episodically founders into the underlying low-density bubble-rich layer of the magma immediately below the crust, where gases are transiently collected between foundering episodes (Shaw et al., loc. cit.; Wright et al., 1992, p. 106; cf. Mangan and Helz, 1985; Helz, 1991). The vesiculation gradient, especially for carbon dioxide and relatively inert gases, extends to great depths in the magma, hence, in the case of the magma ocean, to great depths in the Earth. In the case of Hawaiian lava lakes (Shaw et al., loc. cit.; cf. Shaw et al., 1968; Wright et al., 1976), the volumes of magma affected by near-surface vesiculation and foundering processes are progressively displaced downward with every new foundering episode — i.e., as they are replaced by new subcrustal vesiculating magma — aided by tendencies for the eventual densification of foundered crust by thermal metamorphism, metasomatism, and partial melting at depth. Analogous mechanisms must have operated more efficiently and at vastly larger scales in the early magma ocean (or oceans) — where the importance of densification was increased by the feedback between the sinking rate and the much greater increase in pressure with depth — thereby influencing chemical gradients over great depths in the Earth, and influencing lateral chemical heterogeneity over length scales related to the local penetrative depths of the foundering process.

Wright et al. (1992, Fig. 116) present color photographs of an episode of piecemeal crustal foundering in Makaopuhi lava lake that occurred on March 8, 1965 — photographs taken from the south rim of Makaopuhi crater, more than 200 meters above the lake surface, by Dallas L. Peck, then a co-worker with Thomas L. Wright, Reginald Okamura, and me on the emplacement history and rheology of that magma body (cf. Shaw et al., 1968; Wright et al., 1976), and later Director of the U.S. Geological Survey. In the episodic-foundering mode, in addition to bursts of radiative heat loss, latent heats of crystallization and vaporization are catastrophically released from great depths in the magma body, as determined by the length scales of the mixing process between quenched crust and fresh magma — length scales that are constrained only by the average cycle time of crustal-foundering episodes, a cycle time that determines, and is determined by (see below), the thickness of insulating crust that can grow between any two foundering episodes. The average thickness of successive layers of cyclically foundered crust in Makaopuhi lava lake, Hawaii, was of the order of 10 cm, representing growth during an average cycle time of a few hours (estimated from unpublished measurements I made in 1965; cf. Wright et al., 1976, p. 380).

Other characteristics of the foundering process in the 1965 Makaopuhi lava lake that are applicable to the early stages of mixing in a magma ocean are indicated by additional observations made during 1965 (many of these features I observed myself, and others are described from copies of the notes taken by numerous other observers, a set of which I have kept since that time). The durations of piecemeal to wholesale foundering episodes in Makaopuhi lava lake ranged from almost continuous disruption of the surface layer (there was no maximum duration because episodic-foundering activity merged with more or less continuous crustal mixing at early stages of the post-filling history of 1965 Makaopuhi) to a minimum of about 4 minutes (the time required to sweep across the $\sim$350-m diameter of the lake and replace the $\sim 10^5$ m^2 surface area) to many hours. The cycle times (intermissions), which varied inversely with the foundering durations, ranged from essentially zero (continual disruption of the surface) to about 12 hours (corresponding to the intermission just prior to the most catastrophic — 4-minute — episode of wholesale overturn of the lake

521

Chapter Notes: Chapter 8

surface). These numbers are, of course, characteristic only of the diameter, magma volume, and gas concentration of the 1965 Makaopuhi lava lake. They do indicate, however, that the corresponding rates of crustal replacement for a given crustal thickness would have been much greater in a magma ocean. The equivalent "most catastrophic" episode of crustal foundering in the magma ocean would, by analogy, have occurred only after a much protracted intermission — perhaps only when the crust had eventually grown to many kilometers in thickness. The corresponding cycle time, however, is not readily evaluated because, at that stage of solidification, the simplified "quenching" analogy of the lava-lake model would no longer have applied (a conductive model — which would predict millions of years for such a thickness — is not directly applicable; see below).

As crystallization proceeded in the magma ocean, the average crustal thickness, and average foundering periodicity, must have increased from nearly vanishing values to maxima determined by the initial content of volatiles, the total magma depth, and the rate of volatile rise (representing the forcing function for the average rate and total duration of crustal foundering) vis-à-vis the rate of crustal growth and changing rheological properties (representing the "damping function" that would have resisted crustal foundering by increasing the crustal strength and therefore, in feedback with the rate of volatile-driven tumescence, mediated the lengths of the intermissions between foundering episodes). Not surprisingly, catastrophic crustal foundering is a typical example of a folded-feedback process (Shaw, 1987a, Fig. 51.14), which, when strongly forced, is characterized by chaotic intermittency, crises, and bursting phenomena (see Chap. 5).

The catastrophic-foundering mechanism described above, and by Shaw et al. (loc. cit.), is a more efficient heat-transfer mechanism than is the model of steady-state crustal spreading and subduction described by Duffield (1972), because, in the former model, several mechanisms of heat transfer are maximized over the entire surface area of the magma body, whereas they are restricted to the vicinities of linear spreading centers in the steady-state regime (compare Figs. 115 and 116 in Wright et al., 1992). The vesicular lava crust in the steady-state regime (cf. Duffield, 1972; Wright et al., loc. cit., Fig. 115) becomes a good insulator as it thickens with distance away from the centers of active spreading — i.e., regions marked by zigzaggy, but roughly linear, zones of incandescent magma — whereas in wholesale, or even in piecemeal crustal foundering (e.g., Wright et al., loc. cit., Fig. 116), large areas of partially quenched magma are instantaneously replaced by new incandescent magma (great bursts of radiative heat loss accompany each episode of surface renewal because radiative heat transfer is proportional to the fourth power of the surface temperature; cf. Shaw and Swanson, 1970, p. 289 and Table 2).

Besides rapid cooling, thin-skin convective overturn by the crustal foundering mechanism has other interesting properties. The efficiency of heat transfer at the magma-ocean surface would, to some extent, have been compensated at great depth by enhanced accumulation of the most refractory minerals. For the most part, however, the chemical implications would have been much different from simple crystal-settling fractionation as long as the almost unfractionated (except for volatiles) quenched crust continued to be replaced by fresh magma and the replaced crust was progressively mixed into the underlying magma body. This process would have continued until the net change in crystal content and compositional drift of the liquid fraction resulted in a rapidly increasing bulk viscosity of the magma (e.g., Shaw et al., 1968, Fig. 15; Shaw, 1969, Figs. 2–4; 1972, Fig. 6; 1974, Fig. 10). In ordinary crustal magma chambers, the transition interval between boundary-layer flow and wholesale convective motion of the chamber occurs when the bulk crystal content has exceeded about 50% (Shaw, 1974). In a "magma ocean," on the other hand, the thin-skinned (quenched boundary-layer) crustal regime would have persisted to higher bulk-crystal contents because of the magma's great depth and lateral extent (i.e., the crustal-

Chapter Notes: Chapter 8

foundering mechanism is highly unstable and would have persisted to much higher bulk-crystal contents and viscosities until, according to linear theory, the thin-skinned regime would have merged with what most geophysicists would call whole-mantle convection, plus or minus plate-tectonic regimes — where the later stages of crystallization processes in such regimes would have become very complex and rheologically unpredictable; see Note P.5; cf. Shaw, 1969, 1991; Turcotte, 1991, pp. 165ff).

A geochemical consequence of the thin-skinned (quenched) solidification modes is that trace-element fractionation trends are different from, and presumably less pronounced than, standard fractional crystallization models. Accordingly, compositional trends in the crystallized products of the magma ocean (which represented future heterogeneous mantle, as outlined in Fig. 33) would not have departed dramatically from the initial bulk composition until very late in the solidification process (and, perhaps, in the deepest, most refractory, mineral assemblages). Some geochemists and petrologists have objected to the general idea of a magma ocean on the grounds that predicted magmatic fractionation trends do not fit with standard models for the composition of the upper mantle (see the summary of a recent meeting on "Magma Oceans" by Stevenson, 1992; cf. Ringwood, 1986, 1989, 1990; Garwin, 1989; Ringwood et al., 1990; Anderson, 1989, pp. 235ff, 1990, 1991a, b; Taylor, 1992; Tonks and Melosh, 1993). Besides eliminating that particular objection, and objections based on models of protracted cooling histories, a thin-skinned quenched crustal-foundering model favors a laterally and vertically heterogeneous mantle (one that is iron-enriched in some places and iron-depleted in others, depending on points of view concerning what is considered to have been the "normal" and/or "average" composition of the primitive Earth; cf. Notes I.1, 3.2, 4.2, 6.3, 6.9, 8.9). This view of mantle heterogeneity, derived from the above description of crustal foundering, differs in many respects from standard models of crystal-liquid fractionation, which lead either to a stagnant mantle or to a closed-system mantle with either wholesale or stratified convection (cf. Note 8.2; and Anderson, 1989, pp. 235ff; Garwin, 1989; Turcotte, 1991; Stevenson, 1992).

Until a large fraction of the magma ocean was crystallized, the heat-loss rate governed by catastrophic crustal foundering would have been defined by a nearly constant surface thermal gradient of the order of 100° Kelvin *per centimeter* (for comparison, a geothermal gradient of 100° *per kilometer* is several times the present-day average gradient at the Earth's surface). Thus, the thermal gradient at the surface of a crystallizing and repetitively foundering magma ocean may have approached the order of 10^6 times the present-day average geothermal gradient in the Earth, and such a gradient would have shortened the cooling time of the magma ocean by about the same order of magnitude. A million-fold decrease in a purely conductive cooling time as great, for example, as the age of the Earth would predict an effective solidification time of the magma ocean of the order of only 10^4 years (!). As incredibly brief as this estimate would appear to be, it is consistent with the transience of the early basining phenomena discussed in the text (see Fig. 33, Chap. 8). Two other factors would have contributed to the shortening of the total cooling history of a magma ocean: (1) the radiative component of heat loss that occurred during each cycle of crustal foundering, compensating each of the intervals of exposure of new incandescent magma to the sky, is additive to the short-circuited conduction-path-length effect (this "sky" would have been a relatively dense gaseous atmosphere that — if it had not already existed — would have developed rapidly after formation of the magma ocean), and (2) when the time-averaged temperature at the surface of the magma ocean fell below the boiling point of the saline H_2O solution that eventually precipitated to begin forming a water ocean, an additional quenching effect would have been provided by cycles of contact vaporization of water at the magma-ocean/water-ocean interface during some part of the foundering cycle, coupled with condensation of the vaporized water in the atmosphere and initiation of a rainfall cycle analogous to that of modern-day lava lakes (i.e., such a vaporization cycle removes the latent heat of vaporization from the magma surface and deposits it in the atmosphere, where expansion, cooling, condensation, and rainfall would have completed the efficient transfer of latent heats from the magma ocean; cf. Peck et al., 1977; Shaw et al., 1977).

Factor (2) above is problematical, because of the high temperatures in the early-formed atmosphere, which may have maintained the forming atmosphere at supercritical temperatures. If so, revaporization of rainfall may not have been important until very late in the history of the magma ocean, perhaps occurring for the first time only after the early regime of catastrophic crustal foundering had ended (?). If the vaporization-condensation cycle operated while crustal foundering was still going on (at a stage when the average crustal thickness was large and the cycle time was protracted), it could have enhanced the cooling rate by a significant but unknown factor. As an example, the solidification time of a thin, sheet-like "static" Hawaiian lava lake — meaning a ponded lava body that is not influenced by significant magmatic convection and/or crustal foundering — is reduced by a factor of about five by the effect of vaporization of rainfall relative to a model of simple conductive cooling (Shaw et al., 1977, p. 401; cf. Peck et al., 1977).

The phrase "catastrophic crustal foundering" for the above process is not an exaggeration. As an observer in Hawaii during the crustal-foundering episodes described above — in a magma body roughly one-third kilometer in diameter and about 80 meters deep (Shaw et al., 1968; Wright et al., 1976, Table 1) — I can testify to the remarkable, even awesome, violence of the process. At night, observing the lava-lake surface from 200 meters or so above it on the crater rim, the first episode of wholesale crustal foundering seemed virtually to "explode" before our eyes, replacing the entire lake surface within minutes and lighting up the foggy sky like an artillery barrage, instantly igniting trees many tens of meters above the lava surface on the crater wall (the ignition process presumably was by thermal radiation, because there was virtually no time for even a blast of hot gas to reach that high). In a magma ocean, the analogous foundering may have qualitatively resembled — though scaled to the diameter of the Earth — the eruptive flares and prominences displayed by the Sun (e.g., Giovanelli, 1984; Zirin, 1988; Foukal, 1990; cf. description of Mauna Ulu "lava prominences" in Note P.3 of the present volume).

In a summary just published by Solomon (1993), describing the results of recent studies of impact cratering on Venus, using techniques of high-resolution radar imaging (e.g., Phillips et al., 1992; Schaber et al., 1992; cf. Saunders, 1990), a general conclusion seems to have been reached that the "average age" of Venus's surface is about 500 Ma. And Solomon (loc. cit.) points out that some workers have interpreted the data to indicate that there was an event of catastrophic resurfacing of the planet at that time. Although this idea is undoubtedly simplistic (cf. Phillips et al., 1992), it is reminiscent of the wholesale-foundering mechanism described above. If so, Venus at 500 Ma conceivably resembled a much earlier stage of Earth's history (i.e., $\gg 1$ Ga), perhaps representing the time interval between the end of wholesale foundering of crustal layers and the beginning of systematic piecemeal crustal spreading and subduction processes. But why should there be a "common age" for a conspicuous change in the geologic record of the two planets at roughly 500–600 Ma? Partial, or even substantially "complete" (e.g., $> 70\%$ or so), "catastrophic overturn" of the lithospheres of both planets at that time might be a possibility. Analogous models have been proposed as a mechanism for catastrophic downwelling ("avalanches") in the Earth's present-day mantle — the downwelling representing an internal phase-change instability at and/or above the 670-km seismic discontinuity, analogous to the asthenospheric Fe/Mg-composition-dependent, negatively buoyant density instability of Shaw and Jackson (1973; cf. Note I.5 of the present volume) — but no information is available on the

Chapter Notes: Chapter 8

timing of such events, other than may be given by the possible relationships of the analogous plate-tectonic processes during the latest 200 Ma or so of Earth history.

Variations on the theme of avalanche-like transition phenomena in the mantle have been proposed by Honda et al. (1993), Tackley et al. (1993), and Weinstein (1993); cf. Anderson (1987, 1989), Ito and Takahashi (1989), Machetel and Yuen (1989), Ito and Sato (1991), Kirby et al. (1991), Liu et al. (1991), Machetel and Weber (1991), Iidaka and Suetsugu (1992), Peltier and Solheim (1992), Green (1993), and King (1993). [It seems worth noting here that there are some formal resemblances between these ideas and the "Ramsey-implosion mechanism" (Ramsey, 1950) discussed in Note 6.8.] I have not had an opportunity to compare the cratering patterns of the present work (including or excluding the patterns of volcanic "hot spots") — and their qualitative comparisons with patterns of cratering on the Moon and Mercury, and patterns of volcanism on Io (see the Prologue and Appendixes 24 and 25) — with those of the putative 500-Ma history on Venus, but when that is done, I suspect that it will demonstrate, if scaled appropriately, an impact-related volcano-tectonic history for Venus homologous with those of the Earth and Io.

Note 8.9: The two-stage model for the evolution of the protomantle of the Earth following a collisional origin of the Moon, as outlined in Figure 33, clearly is simplistic from the geochemical point of view. The process must have been considerably more heterogeneous than I have envisioned it — and it may have involved several stages of inertial inversions (episodes of "true *equatorial* wander"), as Runcorn (1983, 1987) proposed for the Moon, and as Schultz (1985) proposed for Mars [this might also explain why there is little or no evidence of the impact products of the infall stages of massive bombardment on the Moon, a point emphasized by the seeming cutoff of the oldest impact-melt ages — e.g., as determined by Ryder and Dalrymple (1992) — at about 3.9 Ga (i.e., a simplistic interpretation might be that there was an effective "resurfacing event" on the Moon at roughly 4 Ga analogous to the putative resurfacing event on Venus at 500 Ma, as discussed in Note 8.8)]. The notion of the rapid crystallization and differentiation of a *single* "magma ocean" (Oceanus Pacificus), as outlined in Note 8.8, probably should be revised somewhat to involve a more irregular distribution of otherwise areally equivalent "magma seas" — as is more consistent with the geographical irregularities of the deep ocean basins and seas of today. If so, the chemical and petrographic properties of the upper mantle — both beneath the protocontinental masses and in the regions of impact-related basining, as on the Moon — would have been as heterogeneous in horizontal gradients of chemistry and mineralogy as they were in the analogous gradients with depth. This may explain some of the differences in plate-tectonic models of mantle melting (i.e., models based on end-member types of melting sources, and/or on idealized chemistries postulated to exist in this or that plate-tectonic setting, or in the deep mantle vs. the upper mantle; cf. Hart et al., 1992; Farley and Craig, 1992).

Such uncertainties of provenance in a heterogeneous mantle have been underestimated in petrology because of its emphasis on an idealized plate tectonics which, even if it were valid now, is not representative of the chemical sources in the pre-Mesozoic Earth — even if the general character of the plate-tectonic mechanism has persisted since the early Precambrian. As I see it, the question of chemical heterogeneity and hybridization in the mantles of the terrestrial planets is directly related to the issue of chemical heterogeneity in the "cosmic abundances" of the elements, which — as I have pointed out repeatedly in the present volume — probably is much greater than has heretofore been suspected, or assumed, in geochemistry (see Chap. 8 on the earliest bombardment history of the Earth and Moon; cf. Notes I.1, 3.2, 6.3, 6.9; and Rotaru et al., 1992; Fink, 1992; Ray and Anderson, 1992).

Chapter Notes: Chapter 8

Note 8.10: Although the conception of Pangaea as a more or less continuous "supercontinent" appears to survive in the current geological literature, the newly developing concepts of "accretionary continents" would seem to introduce many degrees of kinematic freedom—especially with regard to in situ block rotations and long-distance microcontinental translations—not previously evaluated in paleogeography relative to continental coherence at the scale of "supercontinents" (cf. Norton and Sclater, 1979; Bambach et al., 1980; Irving, 1982; Fischer, 1984; Howell, 1985, 1989; Nance et al., 1986, 1988; P. F. Hoffman, 1988, 1991; Stone, 1989; Stewart, 1990; Dalziel, 1991, 1992a, b; Hartnady, 1991; Moores, 1991; Murphy and Nance, 1991, 1992; Smith and Livermore, 1991; Zonenshain et al., 1991; Elliott and Gray, 1992; de Wit et al., 1992; Stump, 1992; Young, 1992). In fact, it would appear that many of the simple geometric "rules" by which late-Mesozoic and Cenozoic continental drift was made plausible in relation to the motions of oceanic plates have been, if not abandoned, chaotically scrambled for geological periods prior to the Triassic. This is not necessarily implausible, either in relation to the principles of nonlinear dynamics emulated here, or in relation to the specific concepts embodied in the CRFH. The added degrees of freedom imply, however, that paleogeographic states are analogous to those of the chaotic regime of nonlinear dynamcs, wherein phase-space transitions that are represented geographically by the locations and orientations of coherent pieces and/or slivers and fragments of continents—in place of the trajectories of points on a graph or computer screen—may be subject to *chaotic crises* and *transient bursts* of continental drift at many scales of time and size relative to steady-state and/or periodic fixed-point behaviors of stable continental configurations, such as that existing today, or that portrayed by Figure 34 as having existed in the form of the single "supercontinent" Pangaea during the latest Paleozoic and earliest Mesozoic times.

Pangaea itself putatively came together at a somewhat earlier, but not well-determined, time by the "collision" of "Laurasia" to the north and "Gondwana" to the south (e.g., Hartnady, 1992; cf. Strahler, 1971, pp. 470ff; Bambach et al., 1980; Nance et al., 1988; Murphy and Nance, 1992, p. 85)—and Laurasia and Gondwana had before that been joined together in the "original" supercontinent that Wegener reasonably had named Pangaea but which some others apparently referred to as Gondwanaland, a name coined (tautologically) before Wegener by E. Suess for some part of what is now called Pangaea, or Pangea (cf. Strahler, 1971, pp. 470ff; Bambach et al., 1980; Wood, 1985, Chap. 3). The confusing terminology of this "forward and reverse" bifurcation and/or multifurcation of the supercontinent Pangaea prior to and following the end of the Paleozoic is exceeded only by the variety of scenarios (in the literature cited above) by which such an oscillatory inside-out/outside-in/inside-out phenomenon could have taken place and still be consistent with the modern notions of plate tectonics (cf. Dalziel, 1992b, Fig. 2).

If the diversely scaled notions of continental drift (ranging from small slivers of "suspect terranes" through "microcontinents" to "continental nuclei" to "accretionary continents," and, ultimately, to "supercontinents") all have merit, then the notion of a "supercontinent cycle" — which has been around in one form or another essentially as long as the idea of continental drift (e.g., Glen, 1982, pp. 4ff; Wood, 1985, Chaps. 3 and 4; cf. Wilson, 1966; Fischer, 1984; Nance et al., 1986, 1988; Dalziel, 1992b)—would have to be described by "continent-fragment-motion trajectories" that resemble the *chaotic crisis* of Figure 18 in the seeming wildness of their behaviors, where the dashed trajectories in that figure would represent "explosive" transient motions of continental fragments, while the most coherent states of continental cohesion would be represented by the *core attractor*. According to the concept of meridionally mediated geodynamic processes, therefore, the periodic and chaotic continental motions would represent coupled oscillatory deviations from a stable and continuous "meridional continent" in equilibrium with the long-term

Chapter Notes: Chapter 8

"gyroscopic" motions of the MGK in space (e.g., something like the illustration by Murphy and Nance, 1992, p. 85; cf. the alternative "meridional continent" that I arbitrarily pieced together to illustrate this idea, as shown in Appendix 26).

Note 8.11: The accepted record of glaciation in Earth history has sometimes been puzzling, contradictory, and/or controversial in relation to continental drift and paleomagnetic apparent polar-wander paths (APWPs). If the usual criteria of continent-scale glaciations (massive tillite deposits, etc.) are taken to indicate polar latitudes, or even latitudes higher than ~45°, they severely limit possible paleogeographic reconstructions of the affected continents. Contemporary notions of accretionary continents (see Chap. 8) have sometimes helped and sometimes exacerbated this situation to varying, and often highly subjective, degrees — because the motions of continental slivers, slices, microplates, and so on ("suspect terranes"; see Howell, 1985, 1989; cf. Ernst, 1988) increase the degrees of freedom for selective interpretation of paleolatitudes. Even so, the issue remains unresolved in terms of the paleogeographic history of continental nuclei relative to climatological evidence and correlations with latitude (e.g., Crowell, 1978; Parrish et al., 1986; Crowley and North, 1988; Crowley and Baum, 1991).

An arbitrary continental reconstruction is given in Appendix 26 to illustrate an example of intact-continent rotations and translations that more or less satisfies the assumption of celestial-reference-frame and Phanerozoic-cratering-node (PCN-CRF) invariance throughout Earth's history — relative to the added constraint that the continents tend to be positioned according to some permutation of crudely meridional clusters, as postulated by the tectonic-centering concept of Chapter 8, involving the ballast-like stabilization effects of an ancient *meridional geodetic keel* (MGK hypothesis of Fig. 33) on the Earth's orbital-rotational interactions. In making that reconstruction, I ignored all of the usual criteria of geologic-tectonic contiguity, especially evidence on paleolatitudes based on correlations of major glacial deposits with past polar climates, or evidence on tropical (fossil) fauna and flora indicating past equatorial climates. There are a number of reasons — sometimes not so obvious ones — why the usual geological assumptions concerning correlations between climate and latitude are simplistic. One that obviates all correlations between latitude and climate in one fell swoop is the possibility that the obliquity of the ecliptic has varied significantly in the geologic past. Williams (1972) argues, for example, that the Earth's spin axis was once aligned within the orbital plane, a relationship that instantly turns such tacit correlations on their heads (cf. Ward, 1982; Walker and Zahnle, 1986). And if we add to this early Precambrian, hence safely remote, possibility the fact that the orbital history of the Solar System has been shown to be chaotic (see Chaps. 3–6), then the amplitude and timing of the past variations of the obliquity of the Earth's spin axis to the ecliptic plane are formally unknown for ages prior to the order of ten million years ago (cf. Wisdom, 1987a, b, c; Laskar, 1988, 1989, 1990; Sussman and Wisdom, 1992; Laskar and Robutel, 1993; Laskar et al., 1993).

What I mean by this is that the *divergence time* for significant deviations of orbital parameters from some set of initial values (which, in this case, are totally unknown) — representing an alternative statement of the *principle of sensitive dependence on initial conditions*, as described in Note P.2 (see also the Prologue and Notes I.1 and 7.5) — is something like 10^7 years (e.g., see the graphical demonstrations given by Sussman and Wisdom, 1992). And because we actually are dealing with *systems of attractors,* all bets are off after that time (or perhaps long before that time), concerning future states of the system. [This observation gives some insight into why modelers in celestial mechanics habitually adhere to — or at least do not depart far from — the traditional *gestalt* of conservative-Hamiltonian systems theories (see Note P.3; and Shaw, 1994). *The modelers* might say that

Chapter Notes: Chapter 8

they do so for want of the "infinite" computer power needed to explore more complex models (e.g., see the proposed use of massively parallel-processing computers to bring to fruition claims implied by headlines such as "Ocean-in-a-Machine Starts Looking Like the Real Thing"; see Kerr, 1993c; cf. Lorenz, 1982) — and might point, for instance, to the incredible computational feats of numerical integration of the orbital equations of motion that were carried out by Laskar (1988, 1989, 1990) before most of us would accept the fact that the Solar System is operating according to chaotic-dynamical rather than classical principles. But in that adherence the modelers continue to miss the implications of sensitive dependence on initial conditions and its role in the *chaotic crises* of attractor dynamics that, at least qualitatively, describe the behaviors of complex coupled-oscillator systems in the laboratory and in Nature (see Chap. 5, and Figs. 17 and 18).] Nonetheless, a team with a strong background in geophysics, climatology, and celestial mechanics, which includes Laskar (see Berger et al., 1992), has retained sufficient faith in their gestalt that they have assured us that the changes in Earth's orbital eccentricity and obliquity have been, for climatological purposes, negligible during the past 500 m.y. or so (cf. Ward, 1982; Walker and Zahnle, 1986).

A possibility even more startling than are the uncertainties in the variations of Earth's orbital parameters in the geologic past is the suggestion that the geological criteria for identifying a particular type of deposit with a particular type of climate have been mistaken in at least one major respect. Such a dramatic turn of events may materialize from recent work by Oberbeck and coworkers (see Oberbeck et al., 1992; Marshall and Oberbeck, 1992; Aggarwal and Oberbeck, 1992; cf. Rampino, 1992), which suggests that the criteria habitually used to identify particular sedimentary deposits as glacial in character (tillites, diamictites, etc.) are essentially the same as those of ejecta deposits left by large impact-cratering events (except for the presence or absence of shocked minerals, shock-deformation fabrics, large-scale impact-structural patterns, etc.)! If this allegation is valid, and if the impact-related examples of such deposits can be distinguished from glacial examples (e.g., by the distribution of shock-metamorphic mineral assemblages, etc.), then not only do we have a new tool for expanding the catalog of cratering events, but we also have another criterion for testing the CRF patterns during geologic intervals when large craters have been expectable but have not yet been documented. This possibility would be particularly important for Paleozoic and Precambrian impact events [e.g., at the Permian/Triassic (P/T) boundary, at the late-Devonian Frasnian-Famennian (F-F) boundary (cf. the *McLaren Line* in Appendix 3a), and at the Precambrian/Cambrian, or other early Paleozoic and/or Precambrian boundaries; cf. Urey, 1973; Xu et al., 1988, 1989; Wang et al., 1991; Claeys et al., 1992a, b; Kerr, 1992d; Rampino, 1992; Wang, 1992; Wang and Geldsetzer, 1992)]. Oberbeck et al. (1992) further point out that tillites often are spatially associated with, and are postdated by, flood basalts, a correlation that would seem to parallel the conclusion of the present volume that the CRF pattern of cratering swaths correlates closely with Phanerozoic flood-basalt provinces (see Figs. 1–4, 24–26, 28; cf. Appendix 12).

Rampino (1992) has specifically identified two circular basins (~200 and ~350 km in diameter) in a major region of South America, and contiguous pre-rifting areas of South Africa, on the basis of the new conception of tillite deposits as possible impactites containing suggestive evidence of shock-metamorphic minerals within them, surrounding the present-day Falkland Plateau in rocks that show evidence of a major deformational event with an age of ~248 Ma (not distinguishably different from the age of 245 Ma for the P/T boundary given by Harland et al., 1990). Should these circular basins be documented as impact structures, then the seeming lack of major impacts of this age (e.g., Fig. 6) will have been dramatically revised — just as the present record of impacts at the K/T boundary has been dramatically revised relative to the record of only a few years ago — to reveal a major

Chapter Notes: Chapter 8

pulse of impact events at a strategic time and location in Earth history. To wit, that a Falkland location and P/T age for such a pulse of impact events is unique in representing perhaps the most massive extinction event in geologic history (cf. Fig. 36), an event occurring at a time of major continental rifting of the Gondwana portion of the Pangaea supercontinent (i.e., the Southern Hemisphere portion of the shaded reconstruction of Pangaea shown in Fig. 34, at ~50 m.y. earlier in its history) — a time that ushered in the present-day style and related regimes of plate tectonics. [Arguments were advanced recently by Czamanske et al. (1992) and I. H. Campbell et al. (1992) that the P/T debacle was the result of catastrophic Siberian volcanism (but see Baksi and Farrar, 1991). Such a correlation, however, would appear to be equally consistent with the concept of impactite- "tillite"/flood-basalt associations cited above. Furthermore, according to the CRFH, cratering phenomena and volcanic phenomena are so intimately correlated that the so-called "volcanist" interpretation of extinction events no longer constitutes an "alternative" hypothesis relative to the IEH (see Note 8.1).]

The reconstructed location of the putative Falkland Plateau impact structures places them at the juncture of continental rifting between Africa and South America and the opening of the South Atlantic, as explicitly suggested by Rampino (1992), and also as suggested independently in the studies of Oberbeck et al. (1992). From my own perspective, and that of the CRF hypothesis of the present volume, the Falkland location would place a major impact episode *just south of the trace of the K/T cratering swath* [see Fig. 34 for comparisons between the present-day and early Mesozoic geographic configurations of that region; cf. the sketch map of tillite deposits shown in Strahler (1971, Fig. 29.13, p. 523) relative to a different reconstruction] and *roughly antipodal* to the Phanerozoic cratering swath (cf. Figs. 1–5) — a location that was only a matter of speculation when Chapter 1 (and the cited illustrations in the present volume) were first drafted, more than a year before the above work became known to me. Furthermore, should the P/T age be substantiated for a major pulse of impacts confined to a position on, or along, the southern extension of the K/T cratering swath in the vicinity of the Falkland Plateau, the possibility suggests itself that the pattern of cratering that existed during the Paleozoic predominantly along the Phanerozoic swath experienced a transition that somehow brought it into alignment with the late Mesozoic-Cenozoic (and hence K/T) cratering pattern (cf. Appendixes 1–4 and 23). A global shift in the pattern of impact cratering may have occurred following the late Devonian Frasnian-Famennian (F-F) extinction at 367 Ma (Harland et al., 1990), or — as seems more likely from the evidence that the orbits of young Earth-orbiting meteoroids appear to be consistent with impact localities along *both* the K/T and the Phanerozoic swaths (cf. Appendixes 2, 3, and 23) — impacts have oscillated between the two swaths intermittently throughout the Phanerozoic (perhaps emphasizing one or the other swath for longer or shorter durations at different times). That is, the global pattern of impacts could have been intermittently continuous along both swaths at all time scales, or it could have oscillated between them at periodic intervals of time — thus alternating between dominantly "Phanerozoic-type" geometries and dominantly "K/T-type" geometries over periods of perhaps tens of millions of years (cf. Figs. 5 and 23; and Appendixes 1–4).

Note 8.12: Ager (1984, pp. 94ff) chose for the illustration of his concept of *scale-invariant stratigraphic catastrophes* (named here according to my nonlinear-dynamical interpretation of his description, so as not to burden him with words or meaning beyond his intentions) an event horizon of global scale, the late Devonian, Frasnian-Famennian (F-F) boundary at 367 Ma (Harland et al., 1990, Fig. 1.7). Whether by design or good fortune, Ager was reiterating a description of the very horizon that McLaren (1970) had earlier described as representing such a globe-encircling biostratigraphic change (cf. the *McLaren*

Line in Fig. 4 and Appendix 3a of the present volume). The only physical mechanism McLaren could think of that might bring it about was a meteoroid impact in the ocean large enough that the tsunami and other immediate consequences would have spread over the globe so catastrophically as to seemingly wash the biostratigraphic slate clean. McLaren was referring to bottom-dwelling shallow-water marine organisms, therefore necessitating the almost complete devastation of epicontinental-sea and shallow continental-shelf communities. It would also seem to be evident that such a complete devastation of epicontinental seas could not have happened without major changes in subaerial and deep-sea environments as well. By the present hypothesis, this event, its smaller predecessors, and the cumulative history of all predecessors of these smaller predecessors represented the direct recording of a nonlinear-dynamical cascade phenomenon that has generated the biostratigraphic codification that we know as the Geologic Column. In other words, the above sequence of events (1) characterized the immediate vicinity of the late Devonian, Frasnian-Famennian (F-F) boundary, (2) reflected its prehistory, and (3) both effected and affected the history that was to come (see the discussion of Fig. 36a in Chap. 8).

It is sometimes said that time is the great arbiter of geological change and the seeming miracles and mysteries of biological evolution. But, in a sense, *time has nothing to do with it*, because it is not time, in and of itself, that makes all this possible. It is the inconceivably *re*recursive, multiply folded, feedback-feedforward loops upon and within loops of self-evident complexity, such as that displayed before our eyes in the biostratigraphic record, that accomplishes all this (cf. Note 8.1). There were direct geological and biological forms of feedback in the rates of environmental and genetic change that attended and immediately followed the F-F extinction, and there were complex and roundabout forms of feedforward in the geological and biological futures of phenomena that would one day happen only because *all* of the smaller Frasnian "catastrophes" *as well as the biggest* of all those catastrophes, the F-F event, had happened. In other words, the same sorts of cyclic reentrant signaling were recurring in the temporally scale-invariant sense (meaning without regard to the absolute "size" of the time dimension) at every spatiotemporal scale of the above hierarchy of environmental catastrophes, even down to the scales of mutations and physicochemical mechanisms of natural selection, which are largely inferential because we have so far been unable to "de-scribe" the biostratigraphic record with sufficiently discriminating instruments — i.e., in terms of the detail necessary to see the real inscription well enough that the task of deciphering it is given some chance of success. This argument would fail, just as the one below concerning the structure of the universe would fail, if some *absolute* dimensional datum in time and/or space intervened. For example, in the preceding discussion, such a datum would exist if every trace of living organisms down to the most primitive, including the viruses, had been obliterated from the face of the Earth at or before (see Note 8.1 concerning *latency*) some particular time. Everything living after that time would then represent an entirely unrelated beginning — *assuming* that the continuity of the living state *in the universe* were absolutely unique to the Earth and was not a general property of the universe at large (in this regard, all previous theories of the universe that have taken Earth to be the origin have inevitably failed miserably — leaving only their fossil remains in the anthropological museums of scientific thought).

Such a description, of course, poses the fundamental question that is common to all recordings, whether those recordings are in the form of (1) digital (bit-map) phase portraits from a nonlinear-dynamical computer algorithm, (2) lithological and fossil "bit maps" displayed through the actions of geological processes, or (3) stellar, galactic, and supergalactic bit maps displayed through the actions of cosmological processes. A question, then: Is the fundamental symbol size, or array of symbol sizes, as measured in a common system of bits, and according to which the record was transcribed, sufficiently fine-grained to convey the full range of information contained within the "living" system? The word "living" is in quotes to emphasize that I am talking about the distinction between phenomena that are expressed by bits of information as they are happening vis-à-vis the symbolic fossilized entities that represent frozen and relatively discontinuous bits, pieces, and clumps of the phenomenological information that is stored for later inspection, and that may be subject to changes, ranging from subtle to wholesale, while the data are in storage. A simple example is the distinction between printed and spoken languages. The "fossil" symbols in that case are the printed letters of the alphabet and the words made from them that are typed out according to rules of grammar, syntax, and semantics. By contrast, spoken language has much greater flexibility as regards grammar and syntax, and it contains elements and nuances of inflection, pace and rhythm, sound amplitude, semantic reiteration and feedback, contextual tone and/or feeling, etc., that give it a greater range and depth of meaning. And beyond all of those supplementary multiplicative qualities, spoken language usually involves feedback from another speaker (sometimes with the same result as in electronic audio feedback), as well as feedback to and from the internal sensory apparatuses of each speaker that are relatively quiet and uninformative, except by indirect evocation, in printed language. Simplifying further, I propose that *"living languages"* (by which I mean any communicative body of information that is in the process of happening as we speak, be it in the form of patterns of acoustic vibrations between two people, or in other forms of coupled oscillatory phenomena — such as patterns of radio waves and light received from deep space) can be expressed by fundamental measures of meaning that I shall call *cognons* (for cognitive signaling), and that *"fossil languages"* (analogous to printed languages) are expressed in fundamental units called *logons*, or *codons* (for symbolic signaling).

The above question, then, boils down to the old chestnut concerning the difference between language and meaning — except that, in the way I just defined them, both are expressed, hypothetically at least, in measurable units. All we now need to know is the conversion factor between the *cognon* and the *logon* to complete a full numerical transliteration between a living language and a fossil language, and vice versa — in other words, a conversion factor by which we can transform any printed language statement into *a universal measure called meaning* (cf. Sewell, 1951; Hutchinson, 1953; Shaw, 1994). It hardly needs saying that such a transformation is physically impossible, for a variety of reasons, one of which is that a truly living language, as I defined it, does not exist. For one thing, even without appealing to a quantum uncertainty principle, the spoken word is fossilized within its acoustic voice print before it gets out of the mouth of the speaker (as in the cartoon of a person trying to talk in subzero weather, where each word freezes and falls onto a pile of broken verbal stalactites).

Looking at the situation mathematically, my assertion that the hypothesized meaningful conversion — cognon-to-logon, logon-to-cognon — is physically impossible is but another way, among uncountable other ways, of saying that language, written and spoken, is a self-referential system and is incompatible with completeness and/or computability (cf. the Prologue, Introduction, and Notes P.1, P.6, and I.2). On the other hand, it is possible to compare relative degrees of linguistic complexity between symbolic languages, hence it is possible to compare relative degrees of meaning on the assumption of like cognon/logon complexity ratios. This assumption just means that meaning tends to decrease with decreasing symbolic complexity. However, while this can be shown to be true for certain categories of simple language statements, it is neither "provable" nor even intuitively plausible for sophisticated language statements that are capable of vastly different evocational transformations (such as is possible in poetry). Nonetheless, with this caveat in mind, perhaps it can be seen that it is possible in principle to categorize different types of natural records according to relative complexities. Perhaps it can also be seen that every type of language-like record also will have a characteristic dynamical frequency range, meaning that there is some limit of fineness and some limit of coarseness outside of which a given language loses its structural coherence (e.g., a simple printed language may retain its characteristic structure only within a certain range of alphabetic variability and word size). But this criterion does not necessarily rule out *composite languages* that may or may not be easy to discover.

The most famous example of a composite language is the genetic code, which is composed of relationships between two alphabets, the 4-letter alphabet of the nucleic-acid code (referring to the four bases of DNA, etc.) and the 20-letter alphabet of the amino-acid code (other amino acids exist, but this is the number that, in some combination, specifies, or codes for, the characteristic structures of the proteins, which are therefore analogous to complex words and word groups; cf. Sampson, 1984, pp. 45ff). The amino acid alphabet, in turn, was discovered to be made up from a subset of 20 combinations of the 64 possible triplet codes (cf. Judson, 1979; Crick, 1981), which represent all of the possible combinations of the four-letter alphabet taken three letters at a time, where each of the three-letter words is called a *codon* (i.e., these particular three-letter combinations comprise the unit *logon*, as outlined above, needed to convey genetic information in the living cell). In this way, astronomically large composite alphabets, dictionaries of words, and grammars can be made up from primitive alphabets (cf. Gatlin, 1972, pp.122ff; Searls, 1988, 1992). The real problem, and effort (cf. Note I.2), resides in what it takes to recognize that such alphabets and word structures exist within a given context — because once the multiplicative folding actions of nonlinear-dynamical recurrence and recombination have been set in motion within a suitably rich system of numerical structures, it takes no time at all to produce a dictionary the size of the genetic code (cf. Shaw, 1986, 1988c). If one uses "the brute force method" in such an endeavor, by which I mean basically a bootstrap operation of trial and error with a few guiding principles and a lot of determination — the way that, in effect, both the genetic code (Judson, 1979) and the geomagnetic time scale (Glen, 1982) were discovered (see the Prologue to the present volume) — then an exceedingly large sample of the resulting language is required, such as exists both in the case of living systems (cf. Miller, 1978) and in the case of the geomagnetic complexity of the Earth. In the latter instance, we have not progressed much beyond the recognition that something systematic and complex has been written in the natural record — a stage of recognition analogous to that by which complex long-chain molecules were first identified in biology. I do not say this in a disparaging way, because data arrays in geophysics — as large as the compilations of geomagnetic patterning on the sea floor would appear to be — are minuscule by comparison with the data arrays that have long been available in biology for the dedicated purposes of recognizing and systematically classifying the structures of organic systems.

Without straying farther into the esoterica of linguistics, suffice it to say that there are certain parallels between (1) nonlinear-dynamical codes (e.g., codified structures such as strange attractors that can be expressed, within limits, in terms of fractal dimensions, singularity spectra, etc.), (2) natural codes or languages, such as the genetic code, and (3) written codes and languages, such as English. Each of these categories of codification — not just the first — is expressible, within limits, in terms analogous to those of strange attractors, fractal dimensions, singularity spectra, and the like. Such studies applied to natural systems, other than in molecular genetics, are in their infancy (cf. Shaw, 1987a, b, c, 1988a, b, c; Shaw and Chouet, 1988, 1989, 1991; Chouet and Shaw, 1991), but thinking in these terms can help to clarify distinctions between regimes of complexity in different types of natural systems, and/or between homologous and/or analogous systems at vastly different scales of behavior. An example that may seem trivial, but is not so trivial — seemingly trivial with respect to the large-scale geometric complexity of the observable universe but profound relative to the fundamental question of its origin — specifically concerns the fractal dimension of the distribution of clustered astronomical objects and phenomena in the heavens. This subject transcends the question of the origin of particular astronomical objects, or their background, in tackling the question of the origin of the recursive nonlinear-dynamical phenomena that have brought us to this ability to contemplate such an origin.

Within limits, the relationship between density and radius in the universe is scale-invariant, a power-law (fractal) function (cf. de Vaucouleurs, 1970; Aarseth, 1973; Mandelbrot, 1977, 1983; Lucido, 1985; Shaw and Chouet, 1991). As is true of all other simple fractal invariances, there are limits to the validity of a concept of *the* fractal dimension of the universe — or of anything else treated as an object or set of objects. But this does not necessarily mean that there is anything unique about such demarcations — e.g., the system may be continuously multifractal but with data points distributed in such a way that there is insufficient information to characterize the "ends" of the spectrum and/or certain intervals of the spectrum (cf. Shaw and Chouet, 1989). Luo and Schramm (1992), for example, point out that a fractal description of the density distribution in the universe must fail in the limit of the *cosmological microwave background* (cf. Coniglio, 1986; Lucchin, 1986; Pietronero and Kupers, 1986). This is certainly true in terms of what is presently known about the microwave background structure (e.g., Silk, 1992), but this tells us only that our discriminatory techniques have failed beyond certain maximum and minimum limits of resolution, and says nothing about the informational relationships between the different realms as they might be expressed in terms of a common language or code (or in terms of a *universal system of cognons*; cf. Note 4.2). In other words, there is a lot going on out there — between the largest and/or most distant galactic structures and the so-called "cosmological background" — that we are just now barely beginning to glimpse [e.g., see the quotation from Hubble's (1936) book *The Realm of the Nebulae* cited in Note 4.2 (from van den Bergh, 1992, p. 421); cf. Finkbeiner (1992); Flam (1992a, b, 1993)].

Note 8.13: One of the implications of the nonlinear-dynamical regimes discussed in the present work is that geometric progressions and power-law behavior — or, generally speaking, logarithmic scaling — is usually, but not always (see discussion of Fig. 31, Chap. 7), typical of spatiotemporal relationships, hence is also an expectable relationship for the magnitudes of extensive quantities (mass, volume, energy, etc.) as functions of the spatial and/or temporal scales of behavior (e.g., Shaw, 1988a; Chouet and Shaw, 1991; Shaw and Chouet, 1991). The scaling of earthquake magnitudes as functions of tectonic length-scale and time (e.g., paleoseismicity vis-à-vis the spatiotemporal scaling of fault lengths; cf. Shaw and Gartner, 1984, 1986, 1987; Shaw et al., 1981; Shaw, 1987c) is a conspicuous example of such logarithmic distributions. Magmatic phenomena, too, tend to follow logarithmic relationships between the scaling of volumetric magnitude, length-scale, and time (cf. Smith and Shaw, 1975; Smith et al., 1978; R. L. Smith, 1979; Smith and Luedke, 1984; Shaw, 1985a, 1987a; Shaw and Chouet, 1991), but there are conspicuous exceptions of exponential form related to local, relatively short-term phenomena (e.g., Shaw, 1973, 1987a, 1991; Jackson et al., 1975; Shaw et al., 1971, 1980a). Furthermore, the distinction between the logarithmic and exponential regimes appears to be characteristic of both magmatic and tectonic phenomena, as well as of chaotic crises, because the same distinction is seen in earthquake distributions (see Chap. 5, Figs. 17–18; Chap. 7, Fig. 31; and Nishenko and Buland, 1987; Ditto et al., 1989; Savage et al., 1990; Sommerer et al., 1991a, b). That is to say, the distribution of recurrence times for an earthquake fault of characteristic length tends to be lognormal, whereas the relationship between earthquake frequency, fault length, and magnitude tends to be logarithmic.

Chapter Notes: Chapter 8

At a constant value of the average intermittency ($1/\tau$), in chaotic crises, the frequency distribution of recurrence times, τ, is semilogarithmic (Ditto et al., 1989, Fig. 4), a relationship that is analogous to the distributions of earthquake recurrence time intervals analyzed by Nishenko and Buland (1987, Table 1), which are defined according to *characteristic segments* of plate boundaries, and are normalized by the average recurrence time of earthquakes within each segment. They call the earthquake events within each of these characteristic segments a *characteristic earthquake,* in the sense that it refers to a specific plate segment in each case, hence their definition is analogous to or identical with the characteristic earthquake defined relative to a particular fault of given length, as in Schwartz and Coppersmith (1984), or as analyzed by Shaw and Gartner (1986) and Shaw (1987c). Interpreted according to the concept of interior crises and self-organization, therefore, this class of characteristic earthquakes represents events scaled to a particular recurrence frequency relative to a characteristic critical frequency for earthquake instabilities involving a more inclusive system of faults. In the general case, then, this more inclusive system is the global system.

The critical frequency of the earthquake mechanism could represent a form of external forcing, but in the context of self-organized criticality it is an intrinsic property of the aggregate behavior. Thus, a power-law distribution of faults, plate segments, and earthquake magnitudes for the global system implies a spectrum of recurrence frequencies related to greater or lesser proximities to system-wide critical frequencies. The longer recurrence intervals therefore refer to the longer segments or faults of a generally interrelated system with a fractal distribution of lengths (Shaw and Gartner, 1986; Shaw, 1987c). The maximum characteristic event of such a system, relative to a globally characteristic critical frequency, would be of effectively infinite length compared to plate dimensions (i.e., the spectrum is truncated at the longest fault lengths; and also at the shortest fault lengths, where fracturing vs. intergranular-intragranular deformation phenomena interfere). The actual maximum event, however, is associated with the largest rupture length (or moment) that can exist on a finite Earth of given rheological properties and dynamical dissipation rate. As in laboratory studies of crises and self-organized criticality, these forms of geometry-related frequency truncations are simply artifacts of the finite sample size and energy sources driving the system.

Within the time frame of historical observations and earthquake hazard assessments for human activities, the closest approach to the minimum critical frequency (largest event with longest recurrence time) would correspond to a catalog of the greatest earthquake events (e.g., Kanamori, 1986). Over shorter times, smaller geographic regions, and smaller magnitudes, spatiotemporal scaling of events corresponds to fractal fault-length budgets analogous to those described by Shaw and Gartner (1986, 1987). According to models of *critical self-organization* and/or chaotic crises, these fractal spatial distributions will be correlated with recurrence frequency spectra analogous to those described by Shaw and Chouet (1989; cf. Shaw and Gartner, 1987). In that case, normalized frequency bundles (meaning spatiotemporal patterns of event frequencies the amplitudes of which are normalized by the size of the maximum event of a given characteristic scale) are identical, and are logarithmically spaced at equal self-similar intervals in the frequency domain (i.e., they are evenly distributed in a plot of the logarithm of the normalized amplitude vs. the logarithm of the characteristic frequency, as illustrated by Fein et al., 1985, Fig. 2; and Olinger and Sreenivasan, 1988, Fig. 5; cf. Note P.4). In this sense, the scaling of the seismotectonic frequency spectrum is essentially identical with that of the magmatic energy spectrum (cf. Shaw, 1980, p. 253; Shaw and Chouet, 1991, p. 10,205).

At the resolution of magmatic-tectonic processes provided by the history of Hawaiian volcanism, a history that—including its extension backward in time and northward in

Chapter Notes: Chapter 8

direction along the Emperor Ridge — has spanned more than 70 m.y. in time and 6000 km in length (almost 60° of arc on the globe), the logarithmic spacing of frequency bundles is ~3 (base-10 logarithms), or about a thousandfold variation between characteristic frequency bundles (cf. Note P.4). This approximate scaling factor was shown by Shaw and Chouet (1991) to hold with remarkable consistency over an amazing range of frequencies from about one cycle per second (the ~1-Hz frequency of volcanic tremor) to about one cycle per thirty million years! (Here, ~30 m.y. is probably the dominant average period in the record of terrestrial magmatism and geomagnetism; cf. Fig. 35; and Kistler et al., 1971; Shaw et al., 1971; Negi and Tiwari, 1983; Johnson and Rich, 1986; Rampino and Stothers, 1986; Stothers, 1986; Shaw, 1988a, Figs. 10 and 25; Stothers and Rampino, 1990; Rampino and Caldeira, 1993.) The distribution of possible frequencies within each of the characteristic frequency bundles (cf. Fein et al., 1985, Fig. 2), or between pairs of frequency bundles, however, is quite broad (e.g., representing the lognormal spread of frequencies at constant critical frequency, as discussed above for chaotic crises, and as given by the recurrence times of characteristic global earthquakes, from the analysis by Nishenko and Buland, 1987). From a natural-hazards point of view it is obvious, of course, that the coarseness of this hierarchical scaling relationship does not provide a predictive tool, especially at the time scales of historic activity and longer — at least not until ancillary information is forthcoming to resolve the finer intrabundle frequency structures (cf. Fein et al., 1985, Fig. 2; Olinger and Sreenivasan, 1988, Fig 5; Shaw, 1987a, 1988a). The relationship between this sort of intermittency pattern and humanistic concepts of prediction is discussed in Chapter 8.

The same three-decade logarithmic scaling factor shows up in other ways in the records of Hawaiian volcanism. Before Bernard Chouet and I had tumbled to the scale invariances and great range of the seismomagmatic frequency spectrum (cf. Note P.4) — which we discovered as an outgrowth of Bernard's career-long fascination with, and multifaceted study of, volcanic tremor, the seemingly most innocuous of seismic phenomena [see Aki and Koyanagi (1981), Chouet (1981, 1988, 1992), Chouet et al. (1987), Koyanagi et al. (1987), Shaw and Chouet (1988, 1989, 1991), and Chouet and Shaw (1991); cf. Schick and Mugiono (1991)] — I had collaborated with my earlier coworkers Dale Jackson and Keith Bargar in delineating some of the larger-scale features of the Hawaiian and Hawaiian-Emperor magmatic-tectonic scaling phenomena (see Jackson et al., 1972; Shaw, 1973; Shaw and Jackson, 1973; Bargar and Jackson, 1974; Jackson and Shaw, 1975; Jackson et al., 1975; Jackson, 1976; Shaw, 1980; Shaw et al., 1980a; Shaw, 1987a). (Tragically, Dale Jackson died in 1978, long before his time, having contracted complications to a simple respiratory infection during an arduous tour of duty on a deep-sea drilling expedition that he had been instrumental in designing as a test of our model of Hawaiian-Emperor episodic volcanic propagation. The characteristic intensity of his day-and-night vigil on that expedition severely lowered his resistance to pneumonia, and that, in turn, apparently accelerated a latent and previously unsuspected disease that proved to be fatal.) In Shaw (1987a, Figs. 51.1–51.8), for example, I summarized some of the self-similarities of Hawaiian and Hawaiian-Emperor volcanism at several different spatiotemporal scales, relying heavily on the compilations by Dzurisin et al. (1984) and Klein (1982) for the records of twentieth-century actvity, respectively, of Kilauea and Mauna Loa volcanoes. In Figure 51.8 of Shaw (1987a; cf. Fig. 51.1) I illustrated self-similarity in the long-term volumetric record of Hawaiian and Hawaiian-Emperor (H-E) volcanism by overlaying the 70-m.y. H-E record on the most recent portion of the volumetric record of eruptions at Kilauea (the latter providing a relatively continuous, though intermittent and oscillatory, record of erupted volume since ca. 1920). In order to fit both records on the same diagram, of course, I had to plot the H-E record at one-millionth of the scale of Kilauea's eruptive

535

Chapter Notes: Chapter 8

record. After this reduction, the two graphs could be nearly superposed, except for the greater irregularity of the Kilauean curve caused by the greater effect of short-term intermittency. In other words, a scale factor of 10^6 — at the coarseness of the above composite graph, representing two cycles of the thousandfold scale factor — describes both the variation in time and the variation in volume of Hawaiian eruptive activity over a span of the order of 10^8 years. The "missing" frequency band, with a mean period of the order of 30,000 years, is not resolvable at the scale of Figure 51.8 in Shaw (1987a); in the seismomagmatic phenomena described by Shaw and Chouet (1991), it may correspond to the mean period of volcano-edifice growth. A 30,000-year period may be more characteristic of continental glaciation (cf. Walcott, 1972; Hughes et al., 1981; Benzi et al., 1982; Held, 1982; Ghil, 1985; Peltier, 1987; Berger et al., 1992), suggesting that the place to look for its volcanotectonic signature is at high latitudes (e.g., in the Aleutian chain and Alaska).

The rather coarse 10^3-fold bandwidth, or frequency range per band, does not preclude a more finely divided spectrum with the same relative logarithmic spacings (cf. Note P.4). It is a truism that, in Hawaii, and probably in many other magmatic and/or tectonic circumstances, the thousandfold bandwidth is accentuated by the nature of the phenomena that are available to be recorded, either instrumentally or by the observation of geological phenomena. In other words, that frequency and bandwidth are characteristic of the processes of mid-ocean volcanism — as recorded by the Hawaiian-Emperor system — where the temporal scaling is represented by (1) instrumental seismovolcanic records, (2) historical records of eruptive events, (3) topographically distinctive volcanic shield-building processes, and (4) bathymetrically mappable oceanic ridge-building processes. That is, these are the phenomena that are conspicuously susceptible to reasonably precise measurement. They stand out as demarcations of the frequency bundles described above — i.e., as tick marks, as it were, on the "logarithmic slide rule" of Hawaiian volcanism that registers, more faithfully than anywhere else in the world, the progress of global volcanism — because they *are* the maximal event frequencies in the volcanotectonic process and the globally coupled magmatic-tectonic process (where the latter includes contemporaneous global volcanic eruptibility and seismicity as characteristic manifestations that are equated, in the human equation, with "volcanic hazards" and "earthquake hazards"). In other processes — such as processes of global glaciation (Hughes et al., 1981), sea-level variation, geomagnetic striping of the sea floor, erosional uplift vs. denudation of the Himalayas, or whatever — different frequencies may be more conspicuous. But all of the processes mentioned are, in one way or another, linked to the same dissipative global geodynamic engine that is expressed by the seismic-paleoseismic-tectonic-volcanic-magmatic-paleomagnetic rhythms (cf. Shaw, 1980; Lindholm et al., 1991; Shaw and Chouet, 1991). *It is expectable, therefore, that the frequency structure that is characteristic of Hawaiian volcanism and seismicity is also characteristic of global dynamic processes of virtually every kind!*

Other supportive documentation for the above phenomena is provided by the scaling of continental silicic volcanism, as described by Smith and Shaw (1975), Smith et al. (1978), Smith (1979), and Smith and Luedke (1984). Notably, the logarithms of the volumes of high-level crustal silicic magma chambers and ash-flow eruptions are proportional to the logarithm of geologic age over a wide range of spatiotemporal scales. Shaw (1985a, Fig. 1; following Smith, 1979, Fig. 12) compared the scaling of continental silicic volcanism with Hawaiian basaltic volcanism, finding that the volumetric ratios are scale-invariant over approximately eight decades in the ages and volumes of the silicic systems (i.e., a constant proportionality factor holds over approximately the same span of time as H-E volcanism). Although specific frequencies are not given directly by the above data on silicic volcanism, the conceptual relationship between continental volcanism and plutonism (Shaw, 1985a)

Chapter Notes: Chapter 8

would suggest that they fit the same broad frequency structure that was identified above for Hawaiian volcanism (cf. Shaw, 1988a, Fig. 10). The study by Kennett et al. (1977) of Cenozoic Circumpacific volcanism, mainly from continental and/or island-arc sources of mafic to silicic magmas, also supports the idea that there is a common spectrum for global volcanism, for these authors also found strong resemblances to the periodic variations of H-E volcanism (cf. Vogt, 1971, 1972, 1975, 1979; Jackson et al., 1975; Jackson, 1976; Shaw et al., 1980a).

Note 8.14: The revolutionary aspect of the study by Gould and Calloway (1980) of clam and brachiopod diversity discussed in the text — where this citation is indicated by G-C or G & C — concerns the concept of biological competition. It had become effectively a *gestalt* (cf. Glen, 1991) in evolutionary biology to equate *natural selection* with *competition for survival*, in the sense of the "survival of the fittest" as *negative* interaction. This would be a literal interpretation of Charles Darwin's metaphor of the wedge (see the discussion of Fig. 36 in the text), at least for two biological groups that functioned somewhat similarly and lived in the same habitat over a substantial interval of geologic time (cf. G & C, pp. 383 and 393). According to such an interpretation, one group can flourish (expand in diversity) only at the expense of the other. Therefore, the diversity patterns of the two groups, in the statistical sense, should be negatively (almost reciprocally) correlated — and this should be evident over enough twists and turns of geological fate that the issue of false or adventitious correlation is effectively eliminated (cf. G & C, p. 384). The clams and brachiopods are almost ideally suited to such a study because of their great longevities at both local and global scales (> 500 m.y.) through the thick and thin of geological circumstances, including all of the greatest global mass extinctions since the advent of the Phanerozoic fossil record (with some ambiguity concerning what was going on near the beginning of that record; cf. Gould, 1983; Morris, 1990, 1993; Knoll, 1991; Knoll and Walter, 1992; Levinton, 1992; Kerr, 1992d; Foote and Gould, 1992).

The study by Gould and Calloway (1980) is one of the few that has sought to directly test this long-held but tacit belief. In brief, what they discovered was, if anything, the opposite of what would have been expected if one group tended to be "wedged out" by the other. Over the course of the greater than 500-million-year history of the paired groups, the clams and the brachiopods, a description of their global interaction was that they behaved more like the phrases from Longfellow, "ships that pass in the night" — "only a signal shown and a distant voice in the darkness" (Gould and Calloway, 1980, p. 393). But here I would emphasize another point made in that study. Quoting from the same page, "The only recorded interaction is, in fact, a positive one"!

While it could be argued that the clam-brachiopod story does not prove anything one way or another — or that the study demonstrated only that these two groups were not really in competition over the "same piece of real estate" after all — its conceptual impact was liberating, if for no other reason than that it debunked a tenet long held as gospel concerning those two specific lineages. Furthermore, it opened the door to an unsettling possibility — viz. that the cohabitation of two major biological groups, rather than resembling a shaky arm wrestle at best and a battle of gladiators at worst, even involved elements of what some might call cooperative behavior (at least in the sense of co-operation), if not actual *synergy* (see the roles of cooperative behavior and even "altruism" in "spatial game" theories; e.g., Hamilton, 1964; Axelrod and Hamilton, 1981; Nowak and May, 1992, 1993; cf. Hölldobler and Wilson, 1990, Chap. 4; and Note 5.3 in the present volume). Taken to its logical conclusion — unwarranted in the case at hand — the old adage "survival of the fittest" might have to be changed in some situations to read "survival of the most synergistic," or at least

537

modified to sometimes equate "fittest" with "most cooperative." The latter term certainly is no more anthropomorphic than the former, and is equally ambiguous in the time-honored tradition of Darwinian metaphors (cf. Gould, 1989b).

In the context of nonlinear dynamics, the analogously anthropomorphic terms *ambivalence* and *frustration* are sometimes used to describe situations where complex, subtle, or even delicate nuances of orchestral self-tuning are involved in systems of social intercourse (see the discussion in tribute to Michael Ovenden early in the Introduction; cf. Notes 3.1, 5.3). Whereas these ideas invoke a major theme that I cannot go into in the present volume, I shall leave the subject of natural selection to the few hints I offer in this Note (and Note 8.15, and here and there in the text) with one more — perhaps unwelcome, or even disturbing — observation. The whole idea of *natural selection*, in the sense of *survival of the fittest*, has been flawed all along — in exactly the same sense that celestial mechanics has been flawed — *if* those terms have been meant to imply that individuals, societies, or species are pitted against each other in a struggle to the death over some part of a *conservative Hamiltonian world* (the dynamical analogue of Charles Darwin's conservative system of "wedges"). The nature of the flaw, of course — which has been pointed out many times throughout this book — concerns the implication of hierarchical dynamical independence at many levels of many different categories of natural processes. *Sensitive dependence on initial conditions* demonstrates that not even the most subtle interaction can be discounted in nonlinearly interacting systems. This issue looms large for the Darwinian and neo-Darwinian views of evolution, which are based on an edifice of divided conceptions of what sorts of interactions are and are not allowed (mutational phenomena are treated by one set of rules, and the phenomenology of natural selection by another). Nonlinear dynamics does not deny the possibility that different levels and rates of dynamical interaction can exist, but it does point to the dangers of assuming that any part of an even weakly and/or remotely coupled sequence, set, or system of processes can be treated as though it has no influence on any of the other parts (see analogous remarks by Gillespie, 1991; cf. Lenski and Mittler, 1993; Karlin and Brendel, 1993). It may look that way for a while in the seemingly "normal" behavior of the system (e.g., as is asserted by the authors of "neutral models"; see the contrasting observations by Gillespie, 1991, Chaps. 6 and 7), and suddenly the consequences of a weak and/or remote interaction become definitive of the system behavior as a whole. The processes by which mass impacts — as an outgrowth of long-term self-organization of solar-planetary phenomena (e.g., Fig. 11) — have affected the evolution of the Earth, both geologically and biologically, is a case in point that should be taken to heart in our conceptions of mutational phenomena, as well as in our testing of the literal implications of selective wedging.

It should be obvious in this discussion that nonlinear dynamics cannot, as yet, deal quantitatively with semantics — but neither can biology, paleontology, genetics, or "information theory." I have always been puzzled about the meaning attached to "natural selection" in the literature of different disciplines, such as in paleontology, molecular biology, immunology, and so on, because that meaning always seems to represent a *natural camouflage* for the fact that the *natural ambiguity* of the words conveys a provocation to argue about their meaning — to an end that, ironically (or, "of course," depending on one's stance vis-à-vis "competition"), is dedicated to the survival of one's own point of view (cf. Edelman, 1992; Rose, 1992; Partridge and Harvey, 1992). [I am reminded, in this regard, of the title of the award-winning book on the "genesis of modern evolutionary thought" by Ernst Mayr (1991), *One Long Argument*, which, in the context of the preceding sentence, should be read in the vernacular — as in remarking on a common analogy between (a) Darwinism, (b) the Hundred Years War (i.e., 1338–1453 vis-à-vis 1859–1993, so far), and (c) the nature of nineteenth- and twentieth-century global political conflicts — "Man, that was one *long* argument!"] Yet, notwithstanding such ambiguities, a subliminal message is conveyed that natural selection is about conflict, even though that conflict may be reduced to the scale of semantic debate. And we seem unable to interpret conflict in any sense other than the jugular issue of our own individual mortality identified with (confused with) the mortalities of cells and/or species. It seems rarely to be emphasized in such debates — even when confronted with the facts — that interaction often *enhances* rather than diminishes the viability of analogous traits at all levels, in biological cells, in individuals, in corporate enterprises, and in species (cf. Miller, 1978; L. Thomas, 1974, 1979; Margulis, 1981, 1990, 1992; Sonea and Panisset, 1983; Margulis and Sagan, 1986a, b; and Notes 7.5, 8.15). Synergy, most would believe — as a self-fulfilling prophecy — must inevitably violate some "law" or other, and if none other, then at least the Second Law of Thermodynamics, and therefore must invoke a "cost" (e.g., Pyke, 1991; Partridge and Harvey, 1992). In closed-system thermodynamics, dissipation invokes a cost in the form of a loss of useful energy, but the word "useful" is not questioned. In numerical dynamics — and in theories of open-system dissipative structures (e.g., Prigogine et al., 1972, 1991; Prigogine, 1978; Krinsky, 1984; Prigogine and Stengers, 1984; cf. Notes P.1, P.5, P.6) — dissipation is the source of self-organized complexity, which, ironically, is the source of the biological mechanism (us) that invented the notion of "cost." In other words, the derivative concept of a *cost* is owed directly to a self-organized structure that can only be described as a *reward* if the living system is the criterion of selective advantage. Clearly, we are dealing here with a prototypical *Strange Loop* (Hofstadter, 1979), which returns the theme of the present volume to its origin (see the Prologue and Introduction; cf. Notes P.6, 4.2, 7.5, 8.15).

Note 8.15: Paleontologists and/or molecular geneticists, in attempting to retrace the history of organic evolution by interpreting the fossil record (shells, bones, amber, DNA, etc.; cf. Crichton, 1990; Gillespie, 1991; Morell, 1992; Poinar, 1992; Poinar et al., 1993), are confronted by many obstacles, one of which is analogous to what I think of as the *uncertainty-of-precedence principle* in the history of science. This "principle" has been thrust upon me unsought on a number of occasions — viz. every time I have published something that attributed priority for a scientific discovery or idea to a particular person and/or date in history. Inevitably, I have later found other sources for the same concept or phenomenon — perhaps disguised in different words, or expressed in a discipline that differed in name from mine (one that "obviously had no business invading" the area of my vested interest) — sources belonging unmistakably to the same conceptual lineage (examples of how this principle works are indigenous to the present work, and those of which I became consciously aware are pointed out in these Notes).

Rediscovery of already-published observations, experiments, hypotheses, and principles is an aspect of this uncertainty that has a progressively shrinking time constant, or "halving time." The time has come when one can experience the obsolescence and manifold rediscovery of one's "own" published ideas even while their career is still unfolding. The "scientific generation gap" is now as short as — or shorter than, in some disciplines — the Ph.D. cycle time, meaning about four years, or less. Such an inverse progression fits with the nonlinear-dynamical concept of the history of science explored in the Prologue. Numerically, the "retention time," or rediscovery rate, of four years is roughly the square root of the "astronomical discovery rate" of about 16 years discussed in Chapters 4 and 8 — or shorter by two cycles in the base-2 geometrical progression 16, 8, 4, etc. (this is, of course, the inverse of the period-doubling bifurcation sequence expressed by quadratic maps, such as the logistic equation of population dynamics; cf. Notes P.1 and 4.2; and May, 1976; Feigenbaum, 1980; Hofstadter, 1981). My own career in the earth sciences has spanned about two astronomical doubling times — which, in the example of the inverse progression just mentioned, is three or more period-halving cycles — meaning that I am now seeing my own publications forgotten in less than their generation time, and the ideas in them, which to me are still "new and fresh," have been "independently" rediscovered and forgotten again several times over (cf. "Faraday, on the Fate of Hypotheses," which appears at the end of the Appendixes). The metaphorical parallel with audio-visual feedback phenomena has become literal — a parallel that is literally deafening and blinding to John Q. Scientist. [An example of video recursion is given by a full-page color illustration in the *Scientific American* article "Chaos" by James Crutchfield, Doyne Farmer, Norman Packard, and Robert Shaw (Crutchfield et al., 1986, p. 46). Notice especially the property of intermittent recurrence of an original image — and then contemplate the amplification of that process in the spoken, written, and diagrammatic processing of scientific information. I rest my case.]

I have mentioned several examples of scientific rediscovery in the present volume (e.g., as elaborated especially in Notes P.1–P.7, 4.2, 6.1, and 8.1) wherein nonlinear dynamics, theories of chaos, the celestial reference frame hypothesis (CRFH), and the impact-extinction hypothesis (IEH) are themselves prime examples. Another interesting and timely example — especially in view of the emphases on nonlinear resonances, intermittency, chaotic crises, controlled chaos, and so on of the present volume (e.g., see Chap. 5, and Notes cited there) — concerns the history of the study of "greenhouse warming," positive-negative feedback controls of climatic oscillations, and a turn-of-the-century, yet essentially modern, conception of that subject by T. C. Chamberlin, as succinctly summarized by J. R. Fleming (1992; cf. Williams, 1981, pp. 1–14; Shaw, 1987b). Yet another example concerns Reginald Daly's early-twentieth-century conceptions of a collisional model for the origin of the Moon vis-à-vis the "new" models (evidently unaware of any precedent) that have been developed only during the past decade (see recapitulations given by Baldwin and Wilhelms, 1992; Spera and Stark, 1993).

The complexity of nonlinear-dynamical evolution, wherein feedback and recursion are prevalent, meaning repeated folding and stretching of the phase-space trajectories of the motion, hence kneading, and fractal implication of information (e.g., Crutchfield et al., 1986, p. 46) — thereby implicating evolution in general — is such that evidence of a unique initial state in the fabric of the evolved states no longer exists. Or, equivalently, there is evidence of infinitely many initial states. Like the Gordian Knot, or the edge of the universe (cf. Note I.2; and de Vaucouleurs, 1970; van den Bergh, 1992; Freedman, 1992), there is no trace of a beginning or of an ending, an observation that, in geology, is attributed — together with the *principle of uniformitarianism* — to James Hutton during the late eighteenth century (see Mather and Mason, 1939, pp. 92ff; Dean, 1992; Taylor, 1992; cf. Note I.8). Another way of saying the same thing is that *the initial state (states) of a nonlinear-dynamical evolutionary process is not (are not) knowable by any future complex state, or set of complex states, that evolves from it (them)* — cf. Spencer-Brown (1972, p. 105) — an adage that also applies to the notion of the "origin" of the universe (e.g., the "Big Bang"; cf. Alfvén, 1980, 1981; Lerner, 1991) as well as to the "origin" of a paradigm.

These observations pose questions for Darwinian and neo-Darwinian concepts of evolution — if those concepts are predicated on the idea that "family trees" can be traced back to unique sets of common ancestors. For example, in his book *Macroevolutionary Dynamics*, Niles Eldredge (1989, p. 1) opens with these words, *"Evolution is the proposition that all organisms on earth, past, present, and future, are descended from a common ancestor that lived at least 3.5 billion years ago, the age of the oldest fossil bacteria yet reliably identified."* According to the uncertainty-of-precedence principle, the idea of either (1) a unique ancestral asexual individual, or (2) a unique ancestral sexual pair of individuals is, at best, ambiguous, and, at worst, fallacious (cf. Margulis, 1981, 1990, 1992; Margulis and Sagan, 1986a, b; Grant and Grant, 1992; Farr and Goodfellow, 1992). However, a qualified view of organic evolution — constructed in parallel with the above remarks on recursive complexity (nonlinear-periodic/chaotic complexity) — would see the dendritic structure of macroscopic biologic diversity as "surficial" in character, analogous to dendritic streams traced upon an infinitely complex geomorphological landscape (cf. Bock, 1979; Eldredge, 1989, Chap. 2). In this sense, macroscopic organic evolution is analogous to dendritic bifurcations of *systems* of chaotic attractor fields, where any particular fractal dendrite, if geometrically abstracted from its chaotic microstructure (recognizing the dynamical absurdity of such an abstraction), can be exceedingly simple (algorithmically reduced, or compressed; cf. Note I.2) relative to the fractal geometry of the particle trajectories of that same form within the context ("dynamical matrix") of the interdependent system as a whole. In other words, there is a fundamental distinction between tracing a dendritic pattern to a point of convergence (as in walking downstream along tributaries, or upstream along distributaries, to the trunk stream of a river system) and tracing (inverting) the trajectories of particle motions that give the system as a whole its nonlinear-dynamical reality. Chaotic systems are *noninvertible* in the latter context, and there can be no doubt that the feedback processes of organic evolution, as well as the turbulent flows of alpine streams, are chaotic in such a context. Nonetheless, because simple dendrites are readily generated by chaotic dynamics, the real question apparently comes down to what one means by a "common ancestor." If it is meant to imply that one could trace even a single identical gene structure back to its source in a unique pair of sexually reproducing individuals — or that one specific word in a language could be traced back to the first individual who uttered it — then the idea is dynamically untenable (cf. Lewin, 1988a, b; Gillespie, 1991).

Striking nonbiological examples of this figurative double standard are given by *bifurcations of ring-vortex fields*, where an initial vortex ring bifurcates, or "*n*-furcates" into some integer multiple of rings (subdivides into two or more unique entities that appear to be macroscopic replicas of the original structure) — passing through stages analogous to *gastrulation* (cf. Turing, 1952; Sampson, 1984, Sec. 8.3.1, Fig. 8.1) and fission — following which each new ring does the same, and so on (abstractly) to infinity (e.g., Bradley, 1965; Chen and Chang, 1972; Shaw, 1986; cf. Lim and Nickels, 1992). The integer number, n, of rings per new generation may be even or odd and tends to be constant for a relatively invariant set of flow parameters, but it is highly sensitive to variations (*variable tuning*) of these parameters and would appear to have no upper numerical limit (i.e., the logarithmic base of the possible geometric progressions appears to be unconstrained except by factors that are extremely sensitive to the diversity of local accelerations in the matrix fluid relative to the trajectories of marker particles that trace out the forms of the vortices; see Shaw, 1986). In the terminology of nonlinear dynamics, the fundamental bifurcation sequence is called a *period-doubling cascade*, or *period-doubling route to chaos* (i.e., where chaos, in this case, is a form of turbulence that is characteristic of doubling progressions; see Note P.1; cf. Gollub and Swinney, 1975; May, 1976; Feigenbaum, 1980; Hofstadter, 1981), and therefore the ring-vortex cascades just described are *period-tripling* (*-quadrupling*, *-quintupling*, etc.) *routes* to forms of turbulence that differ subtly from each other in the vicinity of the periodic-chaotic transition (testifying to the exquisite intricacy of nonlinear-dynamical regimes that tend to be lumped under the unqualified rubric "chaos"; see examples discussed by Shaw, 1987b; cf. Gollub and Benson, 1980). The *microscopic* flow is both locally and globally chaotic (e.g., each ring resembles a *strange attractor* of the type illustrated by Ruelle, 1980, as discussed in Note P.3 of the present volume), while the *macroscopic* structure consists of a system of fixed-point attractors (the sets of rings). Such macroscopic attractors are subject to annihilations (global catastrophes), particularly given a punctuational change in tuning parameters, whereas the many-dimensional *chaotic at-*

tractor field — being a *pumped dissipative system operating far from equilibrium* — consists of globally persistent, though intermittent, regions of *transient chaos* (cf. Chap. 5; and Note 5.1).

The above description is analogous to the idea that macroscopic organic evolution is superimposed on a *microevolutionary fabric* in which there is a fundamental and essentially immutable genetic structure analogous to the chaotic attractor field. And, in fact, there are a number of interesting parallels between such a nonlinear-dynamical paradigm and the ideas of Theodosius Dobzhansky, Ernst Mayr, George Gaylord Simpson, Sewell Wright, W. J. Bock, and others in paleontology (see the comprehensive overview of paleontological concepts by Eldredge, 1989; cf. Mayr, 1982, pp. 616ff; 1991, pp. 137ff), given some interpretive sensitivity and "selectivity" concerning the respective contexts. If one gets too literal, of course, the parallels fail — thus hinting that the same sort of dichotomy (at least) occurs in our perceptions of evolutionary dynamics. For instance, concerning micro- and macro-states, Eldredge (1989, p. 5) quotes a statement made by Dobzhansky half a century earlier that, *"We are compelled at the present level of knowledge reluctantly to put a sign of equality between the mechanisms of micro- and macro-evolution, and, proceeding on this assumption, to push our investigations as far ahead as this working hypothesis will permit."* Obviously, if taken literally and out of context, that "equality" could not be pushed ahead at all. And yet, when one looks at the related notions developed, for example, by Simpson, Wright, and Bock (e.g., Eldredge, 1989, Figs. 2.2–2.7), they bear a strong resemblance to the bifurcational structures of vortex fields as described in Shaw (1986), an idea I presented more schematically in Shaw (1988a, Figs. 17–19). Even so, there is a sort of *quantum uncertainty* that goes beyond that implied by Simpson's provocative ideas of *"quantum evolution"* developed in his 1944 book, *Tempo and Mode in Evolution* — an idea he apparently later backed away from, perhaps prematurely, even if some of the criticisms leveled at specific applications of that metaphor were valid (see Eldredge, 1989, pp. 23ff).

A priori, and by analogy with the concept of nonlinear recursion and vortex fields, there is no reason why there cannot be genetic languages other than the one built up from the 4-nucleic-acid-based (letter), 64-trigram-based (syllable), and 20-amino-acid-based (super-letter/word) genetic code (cf. Note 8.12; and Gatlin, 1972, pp. 122ff; Crick and Orgel, 1973; Eigen et al., 1981; Sampson, 1984, pp. 45ff; Cairns-Smith, 1985; Shaw, 1986, pp. 96ff, 122ff; Lima-de-Faria, 1988, Chap. 19; Eigen, 1992). If so, then the fundamental genetic structure of organic evolution is not so immutable after all, and we need to explain why the "matrix" of evolution has remained as stable as it has for billions of years despite catastrophes of the sorts discussed in the present volume. The simple answer lies in the concept of the *bandwidth of nonlinear-dynamical tuning and universality* of the system (cf. Notes P.1, P.4, I.2, 3.1, 3.3, 5.1, 5.2). Just as I found was the case in the, by comparison, delicate structures of vortex fields, each structural regime has a characteristic dynamical stability range in terms of *dissipative pumping rates* (cf. Chap. 7; and Notes P.1 and P.3). It would appear that the universally coupled oscillators of the geodynamic engine — including both the "endogenous" system of coupled processes and the "exogenous" processes coupled with them, in relation to Earth's participation with the dynamical evolution of the Solar System — have modulated, mediated, and "buffered" the dynamical range of the Earth as a whole, despite mass impacts and thermochemical evolution, to fall within the appropriate range for complex organic syntheses. Is this just a convenient cop-out? Perhaps, but it fits all of the qualities one would expect to find in a complex feedback-feedforward system of multiply-coupled oscillators capable of simultaneously generating structures at both the micro- and macro-levels of self-organized complexity.

The nonlinear-dynamical paradigm just described immediately reconciles some of the age-old controversies in geology and paleontology, as well as in the the impact-extinction debates, concerning *uniformitarianism* vs. *catastrophism* as alternative dynamical regimes in nature (cf. Raup, 1984b) — as well as reconciling controversies concerning the independence of micro-states vis-à-vis macro-states involved in biological change and *morphogenetic fields*, where the latter term is applied in its broadest sense (cf. Shaw, 1986; Goodwin, 1991). It is obvious, in the general context of natural processes, that these are not dynamically separable concepts (the "destruction" of a ring vortex is catastrophic only to a particular geometric form, not to the system of dynamical states that is essential to the existence of the form; cf. Lim and Nickels, 1992). Of greater relevance than the destruction of a particular macroscopic form, in the above dynamics, is the event, process, or processes that can change the tuning of the flow regime. Episodic or punctuational changes in tuning parameters are likely to correlate with major changes in macroscopic forms of fixed-point systems of vortex trees (attractors), but abrupt changes and/or "destructions" of fixed-point attractors also can occur with arbitrarily gradual changes in tuning parameters — hence the nonlinear-dynamical paradigm offers no definite resolution to the chronic debate of *gradual vs. punctuational* change in macroscopic Darwinian evolution, nor does it support a common belief that wholesale macroscopic extinctions are necessarily correlated with wholesale physical cataclysms. What it *does* support is a viewpoint in which *patterns* of punctuational changes in the *environment* of macroscopic forms (as expressed in the underlying dynamics at all scales from the microscopic to the global) *do* go together with *patterns* of punctuational changes in the *diversities* of those macroscopic forms (barring the more contrived meanings of negative feedback and "controlled chaos"; see, for example, the discussion of Fig. 29).

This is not good news to the advocates of the impact-extinction hypothesis (IEH) who would argue their case *only* on the basis of cause-effect correlations, because, in that case, the before/after, true/false nature of the evidence on the sequential timing and precision of events (cf. Note 8.1) — which has been of such paramount importance to the gaining of impetus by the IEH — loses its power of absolute falsification, hence also loses its power of implicit circumstantial verification (as in scientific hearsay, with legislative overtones). The importance attached to such evidence can be seen from the size of the literature that argues endlessly about the before-or-after issue in boundary-event correlations (cf. papers in Sharpton and Ward, 1990). But by the same token argued above, whereby the nonlinear-dynamical paradigm ironically resolves the paradox of the dichotomous view of uniformity vs. catastrophe in nature (i.e., by replacing a *logic* paradigm of mutual exclusion with a *coupled-dynamical-action* paradigm of holistic inclusion), the abandonment of absolutism may turn out to be the salvation of the IEH in a future where the sharpness of the tools of stratigraphic correlation — and the instances of false correlations — are likely to increase with time. For example, I recently saw a report that gave a convincing demonstration that a microtektite horizon in a sedimentary sequence definitely occurred within, rather than at the base of, a geomagnetic polarity chron, thereby "proving that the related impact event or events could not be causally related to the dynamical mechanism or mechanisms of polarity reversals." According to the revised paradigm just described, however, such evidence loses its power of absolute falsification — but *it retains* the stratigraphic property of relative-pattern correlation (self-similarity and/or universality) between comparative sequences.

A *universal principle of evolutionary change* is implied by the above discussion — where universality refers to the existence of invariances in numerical scaling that are independent of material states, size, and time (cf. the Prologue, and Chap. 5). The role of universality parameters in global (holistic) systems helps to explain some of the seemingly dichotomous aspects of evolutionary change as evidenced by the fossil record, such as: (1) the hierarchical classification of evolutionary principles (e.g., microevolution vis-à-vis macroevolution; cf. Burnet, 1976; Edelman, 1978, 1985, 1987, 1989; Gould, 1982a, b,

1989c; Shaw, 1986, 1987b, 1988a; Eldredge, 1989; Goodwin, 1991), (2) distinctions between mass extinctions and background extinctions (cf. Jablonski, 1986a, b; Raup, 1986a, 1990, 1991), (3) the relativity of multiple — and seemingly independent or even antithetical — contemporaneous outcomes ("parallel universes"), for instance as they might be compared in terms of Sewell Wright's concept of an *adaptive landscape* (cf. Provine, 1986; Eldredge, 1989, Chap. 2), and (4) reconciliation of the "tree of life" (taxonomically as well as dynamically) with the emerging biologic syntheses that deal with the implications of *symbiotic cohabitations of microbial fields* (cf. Margulis, 1981, 1990, 1992; Sonea and Panisset, 1983; Margulis and Sagan, 1986a, b). The type of systematics that would seem to be implied by category (4) is remarkably analogous to the hierarchical *vortex-tree model* by which I attempted, above, to illustrate the nature of macroscopic heterogeneity in terms of structural dynamics and the emerging diversity of coexisting, transitional, and/or saltational forms (cf. Eldredge and Gould, 1972; Gould and Eldredge, 1977; Bock, 1979; Gould, 1982b, 1984a, b, 1989b) — forms that are not just geometrically, or taxonomically, possible but that are dynamically compatible with differing relative simplicities within the same general range of nonlinear-dynamical dissipative complexity (cf. Notes 4.2, 8.1, and E.1).

A recent report by Lenski and Mittler (1993) reviews the nature of the *assumed independence* — or schism — between "random mutation" and natural selection in a clear and concise discussion of *"directed mutation."* In the context of this Note, directed mutation implies nonindependence between mutation and certain kinds of selective mechanisms [in protein evolution, for example, enhanced rates of amino-acid substitutions that occur in response to a changing environment are called *environmentally challenged* substitutions (Gillespie, 1991, p. 41)]. Natural selection is, in a fundamental sense, a form of coupling, hence feedback, between genetic and environmental phenomena — meant in the sense that no natural mechanism can be truly independent of other mechanisms, hence there can be no instances of perfectly one-sided dynamical coupling of natural mechanisms, to be read in the sense that there is no such thing as perfectly unidirectional information flow between natural systems (see Note P.3, and the Introduction) — whether that occurs at the microscopic or macroscopic scales of biological processes. I avoid the nebulous, ambiguous, and/or anthropomorphically biased, meanings of terms such as "cost," "genetic advantage," "utility," "value," and so on, as representing preconceived and arbitrarily imposed criteria, or "measures," of the nonlinear-dynamical consequences of natural selection. Therefore, directed mutation represents the idea — however weak it may be in many instances (as in the nonlinear-dynamical phenomenon called the *butterfly effect*; see Notes P.2 and P.6), as is the case for "mutational events" in celestial dynamics — that there is feedback between mutation and natural selection. Although such a conclusion may be anathema to many, if not most, evolutionary paleontologists and neo-Darwinian geneticists (but see Cairns et al., 1988; Gillespie, 1991, Chaps. 1 and 7; Thompson and Burdon, 1992; Cairns, 1993; Karlin and Brendel, 1993), it seems unavoidable in principle [cf. Shaw (1987b); and the Introduction and Note 8.14 of the present volume].

The concept of independence between the concepts of random mutation and selection was profoundly challenged in 1988 in a paper published in *Nature* titled "The Origin of Mutants," by John Cairns, Julie Overbaugh, and Stephen Miller. That paper created a comparative furor in evolutionary biology that continues today, as exemplified by the Lenski-Mittler retrospective (above; cf. Cairns, 1993), because Cairns, Overbaugh, and Miller made a statement in their Abstract that challenged the central dogma of neo-Darwinian evolution — e.g., as expressed by Burnet's (1976) dictum (cf. Shaw, 1987b, p. 1659) that "Ever since the publication of the *Origin of Species* in 1857 [sic] it has become more and more evident that evolution has found no way to introduce novelty other than to produce a wide diversity of inheritable patterns in some essentially random fashion, and then to expose those patterns to the test of competitive survival." Contrast Burnet's statement with the "inflammatory" one by Cairns et al. (1988), to wit: *"As the result of studies of bacterial variation, it is now widely believed that mutations arise continuously and without any consideration for their utility. In this paper, we briefly review the source of this idea and then describe some experiments **suggesting that cells may have mechanisms for choosing which mutations will occur**."* [Italics in the quotation from Cairns et al. (loc. cit.) are retained as in the original *Nature* article; boldface type was added for emphasis.] The key concept in such definitions is usually thought to revolve around the notion of "randomness" — which, as I attempted to explain in the Introduction, can be defined only algorithmically (i.e., by default; cf. Chaitin, 1979, 1982, 1987a). But, as all good magicians know, misdirection is the key to selling the trick, and by placing the emphasis on the nature of the "random state" — an idea that most of us are conditioned to accept without question — the real issue is hidden, and hardly anyone thinks to ask what is meant by such seemingly benign words as "utility" and "value." Ironically, however, the latter are self-referential terms that have no absolute meaning, and the debate boils down to one that is analogous to the issue of completeness in mathematics. The mutational-selectional relationship resides within a cosmos that is defined by the outcomes of that relationship, hence any attempt to subdivide that cosmos into mutational and selectional parts is as doomed as an attempt to isolate processes of terrestrial evolution from processes of cosmological evolution. One could say that both reside within a context of algorithmic randomness, in the sense that the ultimate source of any kind of genetic information — cosmic or biologic (cf. Shaw, 1988c, p. 670) — is uncomputable, hence incompressible (see Note I.2). In that sense, however, the evolutionary problem is just placed on a par with the mathematical definition of the "edge" of the universe, or the meaning of the microwave background (cf. Notes 4.2 and 8.12). [The year of publication of the first edition of *On the Origin of Species* was 1859 (see Darwin, 1859; Stauffer, 1975, p. 1; Mayr, 1991, Chap. 1). Burnet (1976, p. 158) may have been referring to the year in which Darwin first announced a portion of his work under the title *Natural Selection* in a letter to Asa Gray (see Stauffer, ibid.). Stauffer's book, *Charles Darwin's Natural Selection*, in turn, is an edited version of the second part of what Darwin referred to as "my big book" (Stauffer, loc. cit., p. 5) — *On the Origin of Species* being but an "abstract" hurried to completion in his ironical "struggle for survival" under the pressure of the famous letter received from Alfred Russel Wallace in June 1858 (cf. Mayr, 1991, p. 7). In a not-so-rare, but rarely acknowledged, example of cooperation and synergy in the evolution of scientific thought (cf. Note 8.14), Wallace's manuscript together with excerpts from Darwin's manuscripts and letters were jointly presented for the first time in public by Charles Lyell and Joseph Hooker, associates of Darwin's, at a meeting of the Linnean Society of London on July 1, 1858).]

The purpose of Lenski and Mittler (1993) in writing their paper apparently was to critique the putatively documented instances of directed mutation, especially the experiments by Cairns et al. (1988; cf. Cairns, 1993), but they did their job too well and thereby exposed the essential fallacy of "random mutation" as the central dogma of neo-Darwinian theories. The very data cited by Lenski and Mittler (loc. cit., Table 1) in their questioning of published allegations of directed mutation reveal — even if their specific criticisms of the experimental data are valid — some of the types of feedback effects that act at the level of mutational mechanisms, as well as in general at all levels of natural processes. Whatever the outcome of that debate may be in the immediate future (cf. Cairns, 1993), the paper is a model of clear reporting and enunciation of fundamental issues in evolutionary biology, an enunciation that reveals "random" mutation to be a nonlinear-dynamical phenomenon that is genetically kindred to those of other types that have been addressed in the present volume. *Nonlinear mutation*, and/or *chaotic mutation*, would appear to offer more accurate

Chapter Notes: Chapter 8

descriptions of the underlying principles of microevolution, just as it/they would appear to do for the meanings of both mutation and natural selection as they have been used to describe macroevolution ever since Darwin enunciated his metaphor of the wedge (see the discussion of Fig. 36 in Chap. 8; cf. Note 8.14).

Use of the word "selection" in descriptions of the principles of change has "infected" the conceptual development of evolutionary biology since the time of Darwin in a manner analogous to the way a computer virus scrambles the machine logic and/or semantics of a computational algorithm—especially now in the era of neo-Darwinian molecular genetics and neuroimmunology (e.g., Edelman, 1978, 1985, 1987, 1989; Gillespie, 1991; cf. Rose, 1992; Morgan, 1993; and Notes 5.3 and 8.14 of the present volume). Because "selection" can refer either to "that which selects," or "that which is selected," it leads to a semantic stalemate in the untutored mind [i.e., one not yet dynamically *slaved* (cf. Notes P.1, P.5–P.7, I.5, 5.3; and Haken, 1978; Skarda and Freeman, 1987; Glass and Mackey, 1988; Alper, 1989; F. D. Abraham, 1990; Freeman, 1991, 1992; etc.) to an automatic response when terms such as "natural selection," "Darwinism," "gradualism," "punctuation," "mutation," "neutral drift," etc., are seen or mentioned], hence—like the proverbial deer (squirrel, person, fish, or fowl) caught dead center in the headlights of an oncoming vehicle (subaerial, submarine, or aerial)—to utter confusion and frustration [see, for example, Michael Morgan's (1993) analogous commentary on a recent book that is alleged to profess a "selectionist's" views on human thought, emotions, sexuality, language, and intelligence; cf. Rose (1992)]. From a nonlinear-dynamical-feedback-systems point of view, however, "stalemate," "confusion," and *frustration* imply a null balance in the *rates* of information flow that would act to change the character of the balances between alternative complex states, relative to those "functions" by which the global behavior of the system is *tuned* (see Notes P.1, P.3, 3.1, 5.3, 8.1, 8.14). Therefore—as a descriptive term for self-organized feedback in Nature (biological and otherwise)—*natural selection* is appropriate to a *mutually* self-tuned system in a state of *universal critical self-organization* (see Notes I.4, 5.1, 6.1, 8.13; cf. Shaw, 1987b; Shaw and Chouet, 1988, 1989, 1991). In that context nonlinear-dynamical natural selection refers alike to the neo-Darwinian phenomenologies that are stereotypically classified under the dichotomous labels "mutation" *and* "natural selection" (see, for example, the in-depth theoretical and empirical discussions of the ambiguities and overlaps between these two types of genetic models by Gillespie, 1991, Chaps. 6 and 7; see also "A Linguistic Digression" in Chap. 8 of the present volume).

The nonlinear-dynamical view of natural selection would imply that biological evolution is subject to phenomena analogous to chaotic intermittency, nonlinear resonance, chaotic crises (bursting phenomena), and controlled chaos (a form of chaotic resonance). According to this concept, then, it is conceivable that a resonance could occur between frequencies of mutation and environmental change (selection "pressures") such that an organism that might otherwise be expected to be subject to rapid and variable genetic drift (highly unstable biological clock mechanisms) would become synchronized—by a form of feedback control, as it were (see Chap. 5, and notes cited therein)—and thus persist indefinitely in the resonant state as a form of "controlled chaos," or "stable chaos" (cf. Milani and Nobili, 1992; Murray, 1992; Shinbrot et al., 1993). Such "stabilities," of course—as is the case with long intermissions in chaotic bistable oscillators (cf. Figs. 17b, 18b, 29b, 32d)—also are subject to unpredictable and sudden change (punctuation). Accordingly, genetic states that are manifestations of stable chaos could be phenomenologically indistinguishable from ones assumed to be nearly constant because of very slow neutral drift (cf. Gillespie, 1991, pp. 139ff, 289f)—a situation analogous to the distinction between a complex system that displays a multifractal chaotic singularity spectrum and one that corresponds to statistically "random" fluctuations (e.g., thermodynamic equilibrium;

Chapter Notes: Chapter 9

see the discussion of a "structured gas" in the Introduction; cf. Hoover and Moran, 1992). In principle, however, it is possible that a state of stable chaos could persist indefinitely in an invariant genetic state if the tuning is sufficiently global — whereas, according to the theory of random mutation, there must be some genetic "drift," no matter how slow.

One might think that a rapidly reproducing organism such as the fly — famous in laboratory studies of genetic variation [e.g., see the index heading *Drosophila* in Sampson (1984) and Gillespie (1991)] — would diversify rapidly over geologic time. That this is not necessarily the case is shown by the fossil record of Drosophilinae (e.g., Beverley and Wilson, 1985) — *and* the records of everything else from single-celled plants (and, in general, the protist-protoctist "kingdoms"; see Margulis and Sagan, 1986b) to elephants (cf. Eldredge, 1989; Poinar et al., 1993). Recently, in reading an account of insects preserved in amber (Whalley, 1992), I was suddenly struck by the incongruity of the near-identity of a 120-million-year-old fly relative to its modern equivalent, an observation that also apparently applies to bacteria, protozoa, ants, bees, beetles, termites, and fictional dinosaurs (cf. Crichton, 1990; Hölldobler and Wilson, 1990, Plate 1, facing p. 178; Morell, 1992; Poinar, 1992; Poinar et al., 1993). Suddenly, molecular-genetic neo-Darwinian explanations for mechanisms of speciation and rates of evolution struck me as peculiarly absurd (see the much more carefully reasoned and substantiated, but analogous, remarks by Gillespie, 1991, Chaps. 1 and 7) — especially in view of the myriad potentialities for "coevolution" and punctuational change in the realms of (1) protist-protoctist evolution (traditionally, unicellular and acellular organisms, "plant" and/or "animal"; see Margulis and Sagan, 1986a, p. 168; 1986b, pp. 72ff and Figs. 20–22; Graham, 1992; cf. Poinar et al., 1993), (2) hybridization, microbial and plant-parasite symbioses, etc. (e.g., Margulis, 1981, 1990, 1992; Sonea and Panisset, 1983; Margulis and Sagan, 1986a, b; Graham, 1992; Grant and Grant, 1992; Thompson and Burdon, 1992), (3) colonial insect communities (e.g., Hölldobler and Wilson, 1990), and so on. At the same time, however, I was also struck by the notion that all of these observations — geologically and in the laboratory — could be reconciled by the point of view espoused in the present volume if they represented, in effect, a complex of mutually coupled mutational-environmental systems of globally interacting oscillators *subject to the universalities of spatiotemporally tunable dynamical resonances* (a concept not unlike that of the relationship between *superstring theory* and nonlinear dynamics, as discussed in Note 4.2). For only in that way — i.e., a dynamics that is fundamentally indifferent to the material and kinetic makeup of its component parts (as in the Cosmos, Galaxy, Solar System, Earth, *and* Biosphere) — does the nature of a biologic record that simultaneously accommodates robust invariance and ephemeral transience among otherwise similar groups of microscopic/macroscopic organisms make sense to me in the context of a CRF-coordinated Earth [e.g., consider the concept of coevolution described by Thompson and Burdon (1992) and the microbial symbioses described by Margulis (1981), Sonea and Panisset (1983), and Margulis and Sagan (1986a, b) as complex forms of nonlinear-dynamical geobiogenetic synchronizations, or stable-chaotic "bioresonances"].

Chapter 9

Note 9.1: No sooner had I revised Chapter 9 than an announcement appeared in *Science*, Research News for 16 October, 1992, titled "Earth Gains a Retinue of Mini-Asteroids" (Kerr, 1992g; cf. Pike, 1991; Yeomans, 1991b; Matthews, 1992; Ahrens and Harris, 1992; Lindley, 1992b). Kerr's report indicates that about 50 NEOs of approximately 5–100 meters in diameter pass between the Earth and the Moon *each day*. This rate is *fifty times greater* than the already startling rate highlighted in Chapter 6 of one NEO per day (Steel, 1991)! The transformed estimate was made possible by the use of a charge-coupled-device (CCD) camera attached to a telescope at the Kitt Peak Observatory, Arizona (Project

"Spacewatch"), operated by a team of observers led by T. Gehrels of the University of Arizona. These new observations imply that this higher rate of near-misses has been going on in Earth's vicinity for some unknown period of time and is only now being documented, by means of the CCD technique. The significance of these observations, to me, concerns (1) my earlier remarks about technology-related growth rates in astronomy (e.g., growth rates of the discoveries of asteroids and comets discussed in Chaps. 4 and 6), (2) the implication that, if this many NEOs exist in the Earth-Moon *"resonance space,"* so to speak (i.e., applying the term by analogy with the inapplicable aviation term "airspace"), then the negative results of searches for NESs (natural Earth satellites), such as those of Tombaugh et al. (1959), tell us little or nothing about the number that may exist, and (3) the existing reports of NESs in the literature — especially reports such as those of Bagby (1969), Yeates (1989), and Frank et al. (1990) — may be indicating a much greater population of NESs than anyone has ever before imagined possible (i.e., if anyone had claimed a few years ago that objects large enough to produce craters 1 km in diameter, or larger, are passing close to the Earth and Moon *daily*, their claim might well have been discredited or ignored).

A CCD — the device that has given us this enhanced ability to monitor near-Earth space — is a solid-state detector consisting of a two-dimensional array of pixels, each of which is composed of a photoelectrically sensitive material that interacts with incoming photons to produce electrons that are trapped within the charge-potential well of each pixel. The CCD greatly enhances sensitivity over standard photometry and provides *direct computer readout pixel by pixel*, thereby greatly improving our ability to generate "light-curves" of varying intensity that give information on asteroid sizes, rotation (spin) parameters, and so on (see Mackay, 1986; Binzel, 1989, p. 5; French and Binzel, 1989, pp. 56ff; Harris and Lupishko, 1989; cf. Note 7.1 of the present volume). The significance for NES search strategies — especially for those strategies that have not been considered because conventional interpretations of orbital probabilities automatically exclude them (see Note 6.5) — is incalculable.

Note 9.2: Wasson (1992) invokes the notion of an FU-Orionis type event (cf. Hartmann and Kenyon, 1985; Kaufmann, 1991; Pasachoff, 1991; Maran, 1992) to provide a mechanism for what may have happened to the parent bodies of the "differentiated meteorites" to explain their histories of melting and magmatic differentiation, events presumed to have taken place in the vicinity of the Asteroid Belt (cf. Anders, 1963; Jacchia, 1963; Millman and McKinley, 1963; Nininger, 1963; Sonett, 1971; Herbert and Sonett, 1979; Wasson and Wetherill, 1979; Wisdom, 1985; Chapman, 1986; Taylor et al., 1987; McSween et al., 1988; Greenberg and Nolan, 1989; Bell, 1989; Bell et al., 1989; Chapman et al., 1989; Herbert, 1989; McSween, 1989; Huss, 1990; Weissman and Wasson, 1990; Grimm and McSween, 1993). Wasson's hypothesis (loc. cit.; cf. Sonett, 1971; Herbert and Sonett, 1979; Herbert, 1989) — like the alternative hypothesis of radiogenic heating advanced by Grimm and McSween (1993) — is concerned with transient heating events, while I am inferring (Note 3.2) a cataclysmic stellar transition event involving the destruction of a close-binary companion star or massive protostellar disk, but the implications may be similar for the chemical history of the inner Solar System (cf. Weidenschilling, 1977; Möhlmann, 1985; Meyer-Vernet and Sicardy, 1987; Kerridge and Matthews, 1988).

Understanding of the T-Tauri, FU-Orionis type of stellar evolution, as well as research on the evolution of young, pre-main-sequence binary star systems, has been evolving rapidly, even as the present volume is going to press. Prior to 1983 no spectroscopic binaries had ever been discovered (or really looked for?) among the young pre-main-sequence stars, many of which are thought to be less than a million years old, according to a summary by Cathie Clarke (1992) of research reported at the IAU Colloquium "Complementary Ap-

Chapter Notes: Chapter 9

proaches to Double and Multiple Star Research" held at Pine Mountain, Georgia, during April, 1992. Since 1983, at least twenty pre-main-sequence binary stars have been discovered! The following *italicized* passages reiterate some of the key points made in Clarke's summary report, a report that provides background for my conjecture in Chapter 3 that the Sun once had a binary companion — the "Ghost Binary" shown in Figure 10 — that was destroyed, or that self-destructed (perhaps by a form of chaotic crisis in the nonlinear dynamics of binary-star convective-advective oscillations), leaving the curious bimodal dynamical and chemical discontinuity (see Figs. 9 and 10; cf. Appendixes 6 and 7) between the inner (terrestrial) planets and the giant outer (gaseous) planets as its legacy:

(1) The new results indicate that binaries may be even more common among pre-main-sequence stars than they are among their older main-sequence counterparts. This implies that the formation of stellar pairs is inextricably linked with the initial processes of star formation, and that theories of star formation must be able to account for this preference in the star-forming process for the paired state. **Remarks:** Star formation is a cyclic process (at least within spiral galaxies, such as our own Milky Way Galaxy) that begins, in essence, with "condensation" from molecular cloud complexes owing to some form of nucleating factor, such as shock-wave compressions emanating from loci of supernova explosions (which, in themselves, represent an end stage of the stellar birth-death cycle), or from the self-organized generation and propagation of spiral density waves, radiative shocks, etc. (cf. Schmidt-Kaler, 1975; Vogel et al., 1988; Hunter and Gallagher, 1989; Chapman et al., 1992; Podsiadlowski and Price, 1992). In a highly simplified context, star formation can be viewed as a form of propagating autocatalytic "ignition" process, a process that, in more familiar form, is analogous to the propagation of a forest fire burning in a very large forest (e.g., Seiden and Gerola, 1982). Such cyclic processes maintain themselves as long as there are mechanisms to initiate the ignition and runaway of the combustion process and fuel to sustain it. Star formation, therefore, is the fulfillment of a self-perpetuating feedback cycle that continues until the star-forming potential is destroyed by material losses to, or exchanges with, the surrounding environment — which may involve collision with another galaxy, or some form of interference that results in the breakup and/or transformation of star-forming structures into other galactic processes (e.g., destruction of the spiral-wave structure that correlates with active star-forming regions, such as in the Orion arm of our Galaxy, where periodic star-formation is in progress at the present time; cf. Spitzer, 1982; Rubin, 1983; Shaw, 1988a, Fig. 21B). The "main sequence," in the traditional classification of stellar structures (which must itself evolve to take account of the dynamical implications of the bifurcational and multifurcational aspects of the stellar birth-death cycle, or cycles, that are rapidly being documented by modern observational astronomy), represents a particular trend in mass-luminosity-age phase space that is typical of at least the most obvious type of stellar aging process, and within which the Sun is an example of a mature stage (cf. Rubin, 1984; Shaw, 1988a, Fig. 21A). Some aspects of this description have distinct resemblances to the processes of self-organized criticality discussed in Chapter 5, and to the vortical bifurcation processes discussed in Chapter 7 of the present volume (cf. Shaw, 1986). General information on stellar and galactic processes, ca. 1991, can be found in the remarkably up-to-date and beautifully illustrated textbooks of astronomy by Kaufmann (1991) and Pasachoff (1991), and in *The Astronomy and Astrophysics Encyclopedia* (Maran, 1992); cf. Fig. 9 (Inset), and Notes 3.2 and 4.2 of the present volume.

(2) The apparent excess of pre-main-sequence binaries implies that even single stars on the main sequence (such as the Sun) might have originated in binaries! If so, what caused some binary pairs to part company? (Exclamation added.) **Remarks:** This is the first definite statement I have encountered in the literature of astronomy and astrophysics that specifically mentions a binary origin for the Sun as a likely scenario (see item 3; cf.

Chapter Notes: Chapter 9

Chapman et al., 1992). I hasten to add that a variant of the binary-star idea was put forward in the first heat of the impact-extinction debates. The motivation in that case, however, had to do with a perceived need for a timing mechanism that could account for the pseudo-periodic nature of the impact-extinction process, a motivation that also signaled a resurgence in the search for "Planet X," which already had a venerable history (see Notes 6.5 and 6.10). Some workers held that a "companion star" to the Sun, "Nemesis," representing a distant, "loosely bound," wide binary star (see item 3 below) would provide a suitable forcing function for the activation of crudely periodic comet showers, large-body impacts on the Earth, and catastrophic extinction events spaced at intervals of 30 million years or so, thereby giving substance to the menacing name "Death Star" in the popular media (cf. Whitmire and Jackson, 1984; Hut, 1986; Hills, 1984a, b, 1985, 1986; Muller, 1986, 1988; Raup, 1986a, b; Tremaine, 1986; Gould, 1984a, 1987b; Perlmutter et al., 1990). According to the present state of knowledge, as summarized in this note, Nemesis is neither confirmed nor denied by my postulated close-binary origin of the Solar System. The disruption of the close-binary companion to the primordial Sun—putatively associated with the inception of the planetary order more or less as we know it (see Chap. 3, Figures 9 and 10; cf. Notes 3.2 and 4.2)—conceivably could have left a residual trace in the existence of a small and weakly bound, highly eccentric, low-luminosity, distant Solar companion that has survived to this day. Should such an object still exist, it would be the direct descendant of the original companion star to the Sun during the pre-main-sequence stage of the close-binary-star system and the earliest stages of the Solar System, for which I have suggested both "Genesis" and "Siva"—for its early and late phases, respectively—the latter adopted from Gould (1984a); see the Section titled "The Legend of the Beginning Written in the Earth" in the Introduction.

(3) Current estimates suggest that the fraction of binaries among solar-type stars is at least 65 percent! Such binaries span six orders of magnitude in separation, from systems that are almost in contact to those that are so wide that they are destined to be dissolved eventually by encounters with field stars! (Exclamations added.) **Remarks:** This point refers to stars on the main sequence (cf. Chapman et al., 1992). In other words, the odds are much better than fifty-fifty, at the very least, that the Sun began as part of a binary system. The binary model deduced from Figure 10 of the present volume suggests that the companion was of the close-binary type (I have guessed that the separation was ~ 0.01 AU, roughly the diameter of the Sun, making the infant solar system almost a contact binary). Nemesis (or Siva), if it exists, would be among the examples of evolved wide binary systems among the sample of all Sun-like stars (not just young stars) to which the proportion of 65 percent refers. Such wide binaries have highly eccentric orbits, which, combined with the age of the Solar System and the frequency of close stellar encounters during that time (cf. Perlmutter, 1990; and Stern, 1986, 1987, 1988; Stern and Shull, 1988), might argue against the survival of a twin birth, even without a cataclysmic version of the FU-Orionis type of event.

(4) Among the pre-main-sequence binary systems with separations of from ten to hundreds of astronomical units (the Earth is at 1 AU), there would appear to be an excess of binaries compared to the main-sequence counts. **Remarks:** The lower limit of 10 AU in this observation apparently is an artifact of an interferometric technique that is sensitive to binaries of intermediate separation, suggesting the possibility that binaries with smaller separations may be as abundant (or even more abundant?). In other words, the already large percentage of main-sequence binaries is apparently amplified in the youngest systems, an observation that leads to the final item I wish to emphasize.

(5) Either a large number of systems are being missed in main-sequence surveys, or else many binaries in this separation range are somehow destroyed during their main-sequence lifetimes! (Exclamation added.) **Remarks:** In this concluding observation, Clarke

Chapter Notes: Epilogue

(1992) comes full circle back to the premise of the present work that the Solar System once represented one of those young binary systems that did not survive, as such, to maturity. If the rate of progress in the study of binary star systems continues apace, the answer to the fate of young stellar twins may be forthcoming before the present volume has gone to the printers. Another interesting angle on this problem, and on the nature of the physical states of matter in pre-main-sequence stellar systems, is offered by the work of Appenzeller (1985) on "contracting stars." These are stars that are "almost not quite stars," in the sense that they are so young that core hydrogen burning has not yet begun and their luminosity is generated simply by quasi-static contraction (cf. Boss, 1988). According to Appenzeller (1985, p. 76), such systems have "cool and dusty circumstellar disks where the birth of new solar-system-like planetary systems can probably be witnessed" (cf. Cannizzo and Kaitchuck, 1992). By inference, it was at a dynamical stage somewhere between the evolution of a contracting-star type of system and the robust (i.e., surviving) binary-star type of system that the phenomena described in Chapter 3 of the present volume, attending the formation of the Solar System in a form more or less as we know it today, may have taken place (cf. Fig. 9, Inset). It would appear that, where star-forming processes in general are concerned, this is a nonlinear-dynamical regime capable of generating periodic/chaotic structures with a richness rivaling that found in studies of so-called chemical chaos (cf. Swinney and Roux, 1984; Shaw, 1986; Argoul et al., 1987; Scott, 1991; Winston et al., 1991).

Epilogue

Note E.1: The cosmic world order and *sensitive dependence on initial conditions*: A scientist's conversation about *IT* with some kind of Bear, or a bear of a conversation with some kind of *IT* (see concluding note). A story of the Beginning for all students of nature, fuzzy or otherwise — with apologies to L. Tzu, Pooh-Tao, and other masters (cf. Capra, 1977, Chaps. 2 and 8; Hoff, 1982).

Bear: Sir, what was here before the world?

Scientist: (*Clearing throat*) Well, that is a very difficult question, Mr. B. Let me see if I can simplify. I guess the one thing that would best describe the beginning is that there was an immense "explosion" — a word so inadequate that many scientists just call it the "Big Bang" — a flash of light so intense that it instantly provided all the energy needed to run the universe as we know it today.

B: Boy, that must have been some kind of thunder and lightning. Did you see it or hear it, Sir?

Sci: Well, no, of course not. You see, that was a very long time ago, much before I was born. In fact, long before anyone was born, before there were any planets or Sun.

B: Then who told you about the bang and the flash of light?

Sci: Well, *no* one told me about it. In fact, no one *could* have seen it *or* heard it, because there wasn't any place from which to do either one. It was more like an intense vibration made up of all the frequencies that we now call by names such as light, sound, and so on.

B: (*With a shake of the head*) But, Sir, that makes me dizzy! I don't understand.

Sci: Well, as I said, it's very complicated. We have done a great deal of study, and I will have to go into it very slowly, step by step.

B: (*Settling back*) That's OK with me, Sir. I will listen very carefully. I have lots of time.

Sci: (*Frowning slightly*) Well, I can't take *too* much time to talk about it. I have a

Chapter Notes: Epilogue

	very important appointment with the President's Science Advisor at two o'clock.
B:	(*Looking impressed*) Dear me, that sounds very important. I hope, Sir, that our conversation will help you decide what you want to say.
Sci:	(*Frowning more*) Well, really; I already know what I want to say. It is a very serious meeting about our research budget.
B:	You mean it isn't about what was here before the world?
Sci:	Well, yes, in a way it is. You see there are a lot of things to investigate before we will feel very confident in advising the President about *that*. You know, he has some very pressing economic problems on his mind, and we don't want to create the impression that scientists just think about the origin of the universe and aren't interested in the immediate practical matters.
B:	(*Looking puzzled again*) You mean, Sir, that explaining what was here before the world is not practical?
Sci:	(*Looking a bit nonplussed*) No, no, that's not at all what I mean. I mean we don't want to be disrespectful of the important problems of state and the serious social dilemmas the President has on his mind.
B:	(*Looking even more confused*) You mean there are two different kinds of scientific questions, and one kind is practical and the other isn't?
Sci:	(*Pulling at one of his ears*) No, no, no! We're getting off the track. Forget about all that for now. Let me explain.
B:	(*Brightening a bit*) OK, Sir, I'm ready.
Sci:	Well, now, let's take stock. Here we are on planet Earth and we look up from the ground and what do we see?
B:	Clouds. If it's a nice day, lots of blue sky and sun.
Sci:	Exactly. We observe that there is an atmosphere that we breathe, and that there is weather and climate — and beyond that is the Sun, and at night the Moon, and beyond all that lots of points of light we call the planets and stars. With a very powerful looking glass we can tell the difference, and we can see that the points of light come from many objects of many sizes, and that the objects are at vast distances from us and from each other. We can see all this because of light. And when we look long enough, we can also see that these objects all move relative to us and to each other, and the planets alternate between closer and farther and the stars keep getting farther away from each other and from us too.
B:	Wow! You can tell all that from looking through a spyglass? Even an eagle can't see that well.
Sci:	Well, it has taken a long time, and lots of people looking to see it. But it's a fact, and you can see for yourself if you want to go to one of the Astronomical Observatories and spend lots of cold nights looking (*muttering to himself—"and if you look long enough maybe a comet or an asteroid will conk you on the head"!*).
B:	(*Shivering slightly*) No, thanks. I'll take your word for it — but what was that about conking someone on the head?
Sci:	Oh, hmmmph. . . . nothing, nothing at all. (*Then, thinking fast*) I was referring to a colleague of mine out in Flagstaff, Arizona, who spends so much time looking through any telescope he can get his hands on that you would think he might get hit by a shooting star (*chuckles to himself*). Hrrrmmmph. . . . well, then, after we learned all this, some brilliant thinkers came along and asked what would happen if you reversed what we see happening and extrapolated all of it backwards in time.

Chapter Notes: Epilogue

B: Excuse me, Sir. Sounds to me like those observatories are sorta dangerous — and what does that word "extrapolate" mean? I think I understand "backwards," though I've usually found that I can't see very well when I back up, and I tend to stumble over things.

Sci: (*A little impatiently*) No, no, sorry about that, I was just making a little joke — and "backwards" is just a figurative expression for some rather complicated mathematics. What I mean is that they sort of mathematically *imagined* what it would be like. You know, as if you pretended to reverse the gears of a clock. You can understand that. Extrapolate means to reach in our imaginations beyond the limits of the part of history that we have seen directly.

B: Ummm. . . . You mean it's like daydreaming?

Sci: (*Searching for words*) No — well, a little like that, except that it is done with very precise computers and it is very quantitative and reproducible.

B: You mean you can have the same daydream over and over exactly the same way?

Sci: No — Yes. What I mean is, we construct a thought model that we can test over and over and see how it matches our observations, so we can change it.

B: (*Sitting up straight and cocking his head*) I see now. You've found a way to use computers to daydream better!

Sci: (*Resignedly*) Well, OK, if that's the way you want to look at it. Let's take a different tack, and maybe you will see why this is the way to understand what happened in the past before there was anyone to see it.

B: OK. I'm with you all the way.

Sci: Well, let's see. You are familiar with the Earth, with the forests and the layers of soil and the layers of rock underneath. You've seen the streams at work and how they erode the mountains and deposit the sand and the silt in the lakes, along the lower stretches of rivers, and in the oceans. You can see that this takes time, and that each year the layers are a little different, and that they eventually pile up on top of each other.

B: Yes, I've seen things like that, and I know that old stumps eventually get buried in the marshy country.

Sci: Exactly. Well, there have been lots of other scientists — some of those who weren't *always* looking at the sky — who studied those layers for years and years and years, and they found out how long it takes to make this or that kind of soil, this or that kind of rock. And they also became very quantitative, using many advances of modern technology, so they can say exactly how long it has taken to make mountains and to erode them away, and even how long it has taken to make all the continents and the oceans. And, beyond all that, they even know how often chunks of material from nearby rocky objects, called asteroids, or faraway icy objects, called comets, come down and hit the Earth — you know, the shooting stars I mentioned, and I bet you've seen lots of *them* — sometimes making holes as big as all of Washington, D.C., and part of Virginia and Maryland too.

B: (*With wide eyes*) My goodness me! How long will it be before I have to look for another mountain — or worry about it getting blown up?

Sci: (*With reassuring smile*) Now, now, not to worry — it takes millions and millions of years for mountains to wear out, and really very few of them ever get blown up these days — (*then amending his remark to make it sound more innocuous*) — Well, at least it's no more to worry about than an airplane crashing into your favorite blackberry patch. And we've found out with more and

553

Chapter Notes: Epilogue

B: more sophisticated techniques that the Earth and Moon have been around for billions and billions of years.

B: (*Relaxing, but with a mild doubletake*) Oh, I feel better — hmmmm, except that now that you mention it, a friend of mine said that a friend of a friend of his who lives out near Dulles Airport had a near thing the other day when a suitcase almost hit him on the head while he was rummaging around in the bushes. But, anyway, I see what you mean. There's been lots of water under the bridge.

Sci: (*Ignoring the suitcase . . . but looking a little deflated*) Ahhh . . . , yes, I guess you could say that.

B: (*Wrinkling his nose*) But what's that got to do with that word — the one that means daydreaming backwards about starlight?

Sci: (*Muttering under his breath*) The word is *extrapolate*, and it doesn't mean just thinking about the past. It's a technique for going in our minds beyond the limits of the observed, in any direction of space or time. And that's the point. You see, everything I have explained about reading the history of the rocks and the Earth, and meteors and comets and so on connects up with reading the history of the stars. And it's because of light.

B: (*Perking his ears*) Boy howdy! How does that work?

Sci: Well, I'm sure you've noticed that plants and trees — in fact, all of the cycles of nature, including us — grow and thrive because of sunlight and the cycles of seasonal weather and climate.

B: (*With a droll look*) Yes, as a bear, I think I have an intimate acquaintance with the seasons. Hibernating is very pleasant. Have you tried it? But, I think I like the spring and summer best, when there are lots of flowers, berries, and honey (*starts licking his lips and drifting into a daydream*).

Sci: (*Abruptly*) Hold it, hold it! You're getting off the track again. I'm talking about the practical question of how we relate the cycles of the seasons with the history of the Earth, and of the planets, and the stars.

B: (*Puzzled again*) Isn't eating and sleeping a practical matter?

Sci: (*Looking slightly disheveled*) Of course. But right now we're talking about practical ways to think about how all these things relate to each other.

B: OK, I understand that. Yes, Sir — as long as it won't postpone lunch too long.

Sci: (*Seizing the opportunity*) Indeed. I think we can wind this up pretty soon. You seem to get *that* point very well. Obviously, the factors that provide our food and our pleasures also relate to the way light has influenced both the Earth and the stars. That means that we can understand something about how long the Sun and stars have been shining.

B: I see what you are getting at. But I still don't see how that helps answer my question about what was here before the world.

Sci: Well, just stop and think. If we know how long the stars have been moving away from each other, we begin to see how all this light was once much closer together, and that where we are right now in the universe isn't where we used to be. And I hasten to add before your next question that by "we" I mean the Earth, planets, and Sun, not you and me individually. That makes sense, doesn't it?

B: (*Candidly*) Sure. You mean once upon a time everything was more together — and when things are more together everything looks brighter?

Sci: (*Mollified*) Yes. I see you grasp ideas very quickly.

B: No, Sir. Not really. I just see that the world used to be different and that it has always been different and will keep getting different. But what I want to know is what being different *means*. I don't see how it changes the question about what

Chapter Notes: Epilogue

there was before there was a world. If we weren't where we are when there was more light, then light wasn't where it is when there was more light. Isn't that what you have been saying, Sir?

Sci: (*Taking a deep breath*) Well, wait a minute. Let's be careful here. You're jumping to conclusions. Besides, if you don't watch out, someone will start saying that the whole investigation is hopeless, and that there's no reason to be doing scientific research at all.

B: (*Looking penitent*) Oh, Sir, I don't want to give that impression. I have a very high regard for scientists, and I'm happy you are doing all that thinking so I can enjoy talking with you and asking all these questions. But if the purpose of science is to give practical answers to practical questions, then, unless that is an impractical question, there must be an answer. Doesn't that follow, Sir? In fact, if *I* were going to talk to the President about *my* budget, I would want to be sure I knew exactly why I was asking him for money. That's almost as hard to think about as what it was like before there was light, but I imagine it would be because I wanted to be sure there was going to be plenty of berries and honey in the meadows and forests. Is that why you want to ask him for money to do research on light?

Sci: (*Muttering some more and feeling rather damp around the collar*) Ah (*clears throat noisily*). I'm not meeting with the President, it's the Science Advisor.

B: (*Innocently*) But Sir, doesn't it amount to the same thing? Don't you have to be able to explain why it's important so that whatever it is *you* like — if it isn't berries and honey, though that's hard for *me* to imagine — you will be paid so you can have it by thinking about where light has been for billions and billions of years?

Sci: (*Now a bit flustered but still game*) When you boil it down that way, I guess it amounts to the same thing. And my answer is the same. Our whole way of life depends on how well we think. All the things we enjoy that are equivalent to your berries and honey have come from the minds that have been investigating, one way or another, the history and source of light. We wouldn't have electricity, or transportation — we wouldn't have the energy to run cities without that capacity for thoughtful research, for investigating and using the properties of life that come from light. Without that resource, and therefore without our ability to think about its functioning in the universe, there would be no life as we know it.

B: (*Scratching an itch and wrinkling his nose again*) I hear you loud and clear, Sir. Let's see if I understand this. If what you say is true, then are you also saying that if we don't keep up with the speed that light is moving away from itself we will lose it totally?

Sci: (*Vehemently*) No! No! I didn't say that at all. There is always light in some form and there always will be light. When I say the Sun and stars and so on weren't here in the beginning and have been moving away from each other ever since, I just mean that it would only appear that way if we were looking at the universe as it was then with the point of view we have now. But we can't really do that because we're also moving with the light. So it's not relevant to say that light can ever get away from us, or that there was no light in the part of space where we are now when we, or the stars, weren't here. But that gets even more complicated and involves the Theory of Relativity, Quantum Field Theory, Superstrings and such, and we don't have time to go into such esoteric questions today (*looking at his watch*).

555

Chapter Notes: Epilogue

B: (*Holding his head and looking properly humbled*) Yes, Sir, I see what you mean about how complicated an answer can get. My head hurts already, and I don't even know *how* to ask you a question about those theories — Relativity and Quant-something with strings. I guess I always knew that such questions are truly awesome and beyond me. I sure hope the Science Advisor can understand better than I can.

Sci: (*Relaxing, and nodding in a comforting manner*) There, there, you've done very well and asked some very penetrating questions. You could go far in science if you were trained properly and learned about all the knowledge and theories I have studied for the past thirty years. You have nothing to be ashamed of, and I'm glad we've had a chance to chat, so I could explain a little about the scientific method.

B: Thank you, Sir. I'm grateful for your patience with me. There are just one or two things I need to clear up so I can go and enjoy my lunch. If my question isn't about what it would be like without light, then does it mean it's really about what the world would be like if there was no life, or is it about what it would be like if there wasn't anyone around smart enough to think about it?

Sci: Well, if there had been no life, the question couldn't be asked, could it? And if no one was smart enough to think about it, then I guess no one would ever have thought about it, would they? So there's no way to think about what that would be like.

B: (*Sitting very still*) Yes, Sir, I guess that's right. So it sounds like you're saying that my question is the same as asking what was here before there was any thinking, and it can't be answered because science is about thinking. Is that it?

Sci: (*Edging forward, getting ready to head for the door*) From a layman's viewpoint, I guess you could say that that's what our dialogue has been about, for all practical purposes. Thank goodness we have advanced far enough to be able to think about the past and learn from it so we can provide for the future.

B: (*Glancing out the window, squinting his eyes to see the nearest berry patch far in the distance*) Thank you again, Sir. I see it is time for you to go to your meeting. Please give my respects to the President. But before you go, let me check one more time to see if I have learned my lesson:

When I ask a scientist — meaning a person who is a serious investigator of the way things are — a question about the world, what I'm really going to find out — if the scientist is as honest as you've been — is that there really isn't any answer at all, just more complicated questions. And that would seem to imply, by simple logic, that the only possibility for real answers is if there aren't any questions. But to my mind, it's also simple to ask questions. To a Bear this is a very puzzling conundrum. Therefore, I can only conclude that anyone who is as simpleminded as I am should realize that if he asks a question the answer will be too complicated for him to understand it. So it would follow that the only kind of research such a person can do is to ask questions and *not think* about trying to understand the answers. Isn't that what you have taught me today, Sir? What do you think? Would you please ask the Science Advisor if I could have a research grant to provide berries and honey so I can have the time to go around asking a lot more questions not to think about?

Sci: (*Finding himself grinning foolishly as he walks unsteadily toward the door*) Good day, Mr. B. I'm not sure who has taught whom what today. I think I don't want to think about it. I'll be glad to deliver your request. But just in case, would

Chapter Notes: Appendixes

you mind if I stop by tomorrow and get your opinion on where to find a good berry patch?

B: (*Good-naturedly*) I would be delighted, Sir. I know just the place to chat about such matters while we sample a few. I should have some more questions to ask by then.

Endnote: IT = (the) *Ineffable Transfinite* (see the Prologue), or, perhaps, the Intransitive Tremolant.

Appendixes: Supplementary Illustrations

Note A.1: Curiously, other than the papers by Chant (1913a, b), Mebane (1955, 1956), and LaPaz (1956), there is no mention of the Great Fireball Procession (GFBP) in the substantial literature of other meteoritic events cited in the present volume. To judge from the many Canadian descriptions recorded by Chant immediately after the event — which was first observed in western Ontario at about 9:05 P.M. local time on 9 February, 1913 (i.e., well after dark on a winter evening, but still early enough that many people were still out and about) — the GFBP was made up of a stately procession of three clusters consisting of several (perhaps 2 to 4) fireballs, each fireball estimated by Chant (1913a, p. 162) to measure (allowing for exaggerations of their luminosity) roughly 100 feet ($\sim$30 m) in diameter, coincidentally a size typical of many recently observed near-Earth asteroids (see Chap. 9 and Note 9.1). The speed of individual objects along the path of the procession — estimated by Chant both from observations along the path and from the theoretical orbital speed of an object near the Earth's surface — was roughly 5 miles/sec, or a bit over 8 km/sec (an appropriate fraction, at that altitude, of the escape speed at the Earth's surface, $\sim$11.2 km/sec; cf. Fig. 20). The appearance of the procession in an artist's sketch — drawn from a northern vantage point at sufficient distance for the artist to see it passing on the horizon across the nighttime profile of the city of Toronto — reminded me of the legendary "Ride of the Valkyries." Estimates of the duration of the procession relative to a fixed observer ranged from one to seven minutes, with an average of 3.3 minutes (Chant, 1913a, p. 161), implying an arc length of the train (relative to the observer) somewhere in the range of 480–3400 km, the average being $\sim$1600 km. Even if the duration was somewhat exaggerated, there is no doubt about the stately procession-like nature of the phenomenon. Most observers reported that the procession was accompanied by a rumbling sound "like distant thunder," and one person likened the sound to "the crashing together of two railway trains at a distance." The estimated speed of the front of the procession greatly exceeded the speed of sound — which is about 0.33 km/sec at low altitude and, of course, zero in vacuum — and the descriptions therefore accord with the notion of a repetitive train of low-frequency sonic booms emanating from a region of rarefied density (if the altitude had been decreasing significantly during the passage of the procession, such a "hypervelocity" encounter would have produced increasingly sharp sonic booms, or even an airburst explosion, as the atmospheric density increased). The "rumbling" nature of the sound undoubtedly was mediated by the fact that the groups of fireballs were intermittently breaking up as they progressed [suggesting — in view of the path length and the persistence of the parade-like aspect of the groups of objects — that the altitude of observable effects was but the "keel" of a more general progression of objects just beginning to plow the top of the atmosphere enough to produce a quasi-steady stream of replenished fireballs (i.e., something like the ablation that would stream off of the lower part of an artificial space capsule that was reentering the atmosphere at a very shallow angle)]. Chant (ibid.) pointed out that the great length of the procession was supported to some extent by the reports of some observers that the sound reached them when the last objects were just disappearing, while other observers said that it did not arrive until up to three minutes after the objects had gone. The range of differences between the sonic first arrivals and the visual last departures would appear to be consistent with my inference that the numbers of fireballs and the lengths of the train were variable from place to place and time to time along the trajectory — befitting the chaotic nature of a frictional feedback process in which an object, or group of objects, was skipping along tangentially to the atmospheric envelope like a flat (but crumbly) stone skipping across water.

In some respects, the GFBP is kindred to the more explosive Tunguska event of 1908 discussed in Note 1.2 (cf. Krinov, 1963a, b; Clube and Napier, 1982b; Rocchia et al., 1990; Chyba et al., 1993). And even though the GFBP was not low enough to create an atmospheric shock wave (explosion) when it was observed over southeastern Canada and the northeastern United States, it was on a highly dangerous, and dangerously low, flight path that took it near densely populated areas of the western Great Lakes region and then almost directly over Toronto and Manhattan. Furthermore, from the standpoint of the celestial reference frame hypothesis (CRFH), this flight path would appear to be one of the more probable ones for natural Earth satellites (NESs), as shown by the GFBP path plotted in Appendix 3a. Chant (1913a) was intrigued by the horizontal trajectory of the GFBP, asking himself — if he were to extrapolate from that trajectory — what other parts of the world might have observed it after it was last seen over Bermuda, more than 4000 km (one-tenth of the Earth's circumference) from where it was first observed while still maintaining essentially the same general visual aspect and roughly the same altitude ($\sim$100 km, within a factor of two; see Chant 1913a, Tables II and III; Chant 1913b, pp. 439ff). Assuming that it was on a great-circle path, Chant (1913a, p. 153) found that — after a flight entirely over the South Atlantic and Indian Oceans — its next "landfall" would have been Western Australia! It apparently never reached there, or was not noticed in the unpopulated outback, or conceivably — if it was in orbital transfer and was evolving by atmospheric drag from a parabolic to an elliptical orbit (see Figs. 19 and 20; and Notes 1.2, 3.3, 6.1) — had regained enough altitude to be invisible to the naked eye even at night (cf. Chant, 1913b, pp. 446f). At 8 km/sec, however, it would have arrived over Australia in less than an hour from the time when it was first observed, and hence would have been inconspicuous in the daylight even if it had maintained the same altitude (an altitude essentially within the lower ionosphere; cf. Roy, 1988a, Fig, 10.1, p. 307).

When I noticed that Chant (1913a, Table I) had tabulated his great-circle path, I plotted its coordinates on one of my stereonets, finding — with an eerie sense of astonishment that raised hairs on the back of my neck — that his points plotted *exactly* along my North-American-Australian nodal great circle (NGC) (see the open and inscribed crosses in Appendix 3a). In fact, Chant's points, *written down almost eighty years before I had given any thought to such matters*, plotted so precisely on the NGC that I would have had to fudge them *to avoid giving the impression that I had drawn the NGC through his points.* In point of fact, however, the locations of the Phanerozoic cratering nodes (PCNs), and the orientations of the NGCs, were established more than a year before I knew of the existence of the GFBP, which I discovered "by accident" only because I happened across Mebane's (1955, 1956) reports in which he attempted to recreate the nature of the event from observations made in the U.S. (Mebane's conclusions confirmed Chant's original reports in every essential respect.) And some might say that my discovery of Mebane's reports was itself a "chance event." While waiting for the first reviews of my manuscript to come back during the winter of 1991–92, I spent some time scanning old issues of *Meteoritics* and *Icarus* because I was intentionally looking for pre-Sputnik literature on meteors, bolides, and fireballs. I was doing so because I suspected that the Space Program had changed the outlook of observers and/or searchers for grazing and near-grazing near-Earth objects.

Chapter Notes: Appendixes

Actually, even then, few observers seem to have had thoughts of looking for NESs, with the notable exceptions of Clyde W. Tombaugh (see Note 6.10; and Tombaugh et al., 1959), John Bagby (1956, 1966, 1967, 1969), and R. M. L. Baker, Jr. (1958). My suspicions were supported by the fact that nearly all of the alleged and partially documented NES reports that I found date from the 1960's or earlier (e.g., Bagby, 1969).

Mebane's (loc. cit.) reports whetted my interest in the GFBP of 1913, and, after some weeks of waiting for an interlibrary loan of an 80-year-old volume of the *Journal of the Royal Astronomical Society of Canada*, I was rewarded by the observations set forth above. Reading the numerous eyewitness accounts recorded by Chant, I was struck by their consistency, and by Chant's conservative but insightful remarks. His article struck me forcibly as recording a class of hazardous — and potentially devastating — small impactors, because the GFBP seemed to represent the breakup of an object into fragments ranging in size up to a few tens of meters in diameter (cf. Bandermann, 1971; Baldwin and Sheaffer, 1971; Brown et al., 1983; Turcotte, 1986), fragments that would be capable of explosive energies ranging from the "~ 20-*kiloton* annual airburst event" to the ~ 15-*megaton*-equivalent Tunguska event or Meteor Crater event (cf. Shoemaker, 1983, Fig. 1; Chapman and Morrison, 1989, p. 278). The potential for catastrophe struck me, not because of any immediate threat of an event of biologically catastrophic character, such as those I had been contemplating, but because this great-circle trajectory recorded by Chant, and its alignment with the North-American-Australian NGC, is one that is uniquely oriented to intersect, in addition to the Canadian cities of Winnipeg and Toronto, the greater New York Metropolitan Area. And a many-kiloton to megaton burst in *that* vicinity would constitute a human disaster beyond anything experienced in recorded history.

New York City is not the only major city of the world that is uniquely, if hypothetically, at risk from such events — as judged on the basis of the CRFH and NGCs. To be sure, once an object has passed North America on the North-American-Australian NGC, it stays clear of any other major populated region around the globe, with the possible exceptions of Adelaide and Perth, Australia. But the North-American-Eurasian NGC, by contrast, poses hypothetical "threats" to a global swath that includes Moscow, northern Fennoscandia, middle-American cities from Winnipeg to Mexico City, and parts of the Middle East. And the Australian-Eurasian NGC crosses the vicinities of Melbourne and Sydney, parts of the densely populated regions of Burma, Thailand, and China, much of continental Europe and the British Isles, and possibly Buenos Aires and coastal cities in Chile (e.g., note the grazing and/or ricocheting impacts at Cuarto and Campo del Cielo, Argentina, within the latest 10,000 years or so, as shown in Appendix 3a). These incidents — the 1913 GFBP, Tunguska, Meteor Crater, Cuarto, and Campo del Cielo, among others like them that I have overlooked or have been forgotten — happened "just yesterday and the day before," and they are harbingers of meteoritic hazards we cannot afford to ignore. These are not random artillery shells fired from howitzers by Lady Luck without aim. They represent focused barrages that, in view of the CRFH and the events mentioned, have certain parts of the world "bracketed," and it is only a matter of time before there is a direct hit of a Tunguska-like event — or worse — somewhere other than in a remote and unpopulated region of Siberia (see Appendixes 1–4).

Note A.2: It could be said that the Solar System, together with the various orbital subsystems within it, represents a form of *nonlinear pendulum-percussion clockwork modulated by sine-circle-map escapement mechanisms* (e.g., Figs. 11 and 14 and Appendix 9; cf. Shaw, 1987b, Fig. 1; Shaw, 1991, Fig. 8.32; and Allen, 1983; Morrison, 1991, Fig. 8.1, pp. 92ff). A "clockwork" is any mechanism or combination of mechanisms (process) that metes out "time" — rather, metes out any measurable cyclic quantity that can be expressed

Chapter Notes: Appendixes

as a function that "progresses" in a quantitative way and can be calibrated in an explicit way against a standard of progression or history of advancement from a definable "past" to a definable "future." Two end-member mechanisms exemplify this meaning: (1) the linear mechanical clock, as epitomized by a spring-driven or gravity-driven gear system in which the increments of "time" (progression) are meted out according to a rhythm determined by a linearized escapement mechanism coupled directly with the gear action (the escapement mechanism limits the rate of advancement of the gear drive to equal linear increments by the action of a proportional oscillator, such as a perfect pendulum), and (2) the linear randomized clock epitomized by the "atomic clock" in which the average rate of radioactive decay determines the rate of advancement of "time" as measured from a standard datum, and as calibrated against units of time based on a standard concept of progression. In both instances, the calibration is nominally related to astronomical time, as expressed by the standard second, day, year, etc. Notably, however, subjective evaluations are required at some stage of calibration in terms of what unit or units can be taken as constant or as standing in constant invariance to one another. The astronomical "day" and "year" are not constant, nor are they invariant relative to each other, a problem made especially conspicuous by the presently agreed upon chaotic state of the Solar System (cf. Laskar, 1989; Milani and Nobili, 1992; Murray, 1992; Sussman and Wisdom, 1992).

Clockwork (1) is easily visualized because it is abundantly on display in museums and other public places, with its inner workings graphically exposed to view. Clockwork (2) is harder to envision in detail, because no one has counted each and every atomic emission involved in radioactive decay — at least not in the same sense that one can count each tick of a mechanical clock. The atomic clock is made equivalently simple, however, by imagining that it is directly analogous to the mechanical clockwork, but differs in involving an escapement mechanism in which the gear wheel advances in increments that vary "randomly" within a range of magnitudes that give a constant average rate relative to a standard time cycle. In having such a high incremental frequency and such a small scale of incremental amplitudes that the rate of progression averages to a constant value to high precision, the atomic scale of measurement offers great advantage. By analogy, we might build a mechanical clock of pendulum type in which the "timing gear" is rather large (consisting, say, of a billion teeth), but where the distances between the gear teeth varies in a random way over a range of a thousand parts, from, say, one micron to one millimeter (but note that if the mean value is as small as a micron, the perimeter of the timing gear measures a kilometer in length, making this a rather awkward device to build and deploy). The construction of the gear, of course, would be based on a precise program for machining the tooth-spacing between the limits of one micron and one millimeter, according to a good random number generator, or according to successive flips of a coin toss. (But a question arises: Considering a clock as thus described, would you feel confident in planning a space mission for which the mission chronometer was to be run by a coin-flipping machine?)

Given a good clock-maker, both of the above clockworks can "tell time" to great accuracy and precision, even though neither one is perfectly linear and invariant. The important point, however, is that they are end-member constructions of a generic type of nonlinear clock, the prototype for which is the sine-circle map (Shaw, 1987b). The number of clock types between these limits is uncountable. Each tuning state of the sine-circle function (see Table A.6.1) is possible, and each one differs in an explicitly deterministic way from all the others. Thus, there are clock types that range from linearly periodic to nonlinearly mode-locked, quasiperiodic, chaotic, and n-period-tupling multiplicative variants of quasiperiodic fundamental modes. Within either the mode-locked or chaotic regimes there are characteristic numerical structures that distinguish one class of clock from the other, as well as from the linearly periodic or randomized clockworks. The generic sine-

Chapter Notes: Appendixes

circle nonlinear clockworks are "natural" clockworks, in contrast with clockworks (1) and (2), which are "artificial." Processes in nature are more likely to resemble the mixed modes of the nonlinear clockworks than either of the artificial clockworks. This is why so much emphasis is being placed on the "chaotic time-series analysis" of natural processes (cf. Shaw, 1987b, 1988a; Farmer and Sidorowich, 1988; Shaw and Chouet, 1988, 1989; Sugihara and May, 1990; Tsonis and Elsner, 1992). Ironically, the system of planetary motions, upon which our most fundamental notions of time are based, turns out to be exactly such a natural chaotic clockwork—just as it must be according to the above reasoning.

Note A.3: Several caveats should be offered relative to the apparent polar-wander paths (APWPs) illustrated in Appendix 19. It is evident by inspection that there are significant differences between the APWPs of diagrams (a) and (b) during the Triassic-Jurassic interval. Nominally, the APWP in (b) should be the more accurate, because it is based on the "most reliable" paleomagnetic poles available—i.e., without applying the sorts of statistical methods used for data of varying time intervals and/or precision (see Butler, 1992, p. 251)—whereas the APWPs in (a) are plotted on the basis of mean paleomagnetic poles within 25-m.y. intervals of time (Butler, loc. cit., p. 258). Even without looking into the literature on this point, it should be evident from the Cenozoic APWP in (b) that great variation in APWP trajectories can occur within geologically short spans of time. I would also point out that—partly as a consequence of the sort of uncertainty just mentioned—the apparent "excellence" of the match between the rotated European and North American APWPs in (a) may depend significantly on the location chosen for the paleomagnetic Euler pole, and hence is subject to several types of interpretative uncertainties. [That is, the choice of the Euler pole is itself a subjective function of the general configurations of the APWPs to be compared; cf. Appendixes 19 and 21—the latter plotted on the basis of the Mesozoic-Cenozoic paleomagnetic data given in Besse and Courtillot (1991), where only 24° (rather than 38°) of relative rotation of the North American and European APWPs is needed to achieve nearly identical superposition of the two data sets.] The interpretation of the Triassic-Jurassic APWP for North America—hence implicitly for Europe, too, if reconstructions such as (a) are to be believed—is currently the subject of a major controversy, one that remains unresolved at the time of writing. Some of the issues involved are mentioned in Butler (1992, Chap. 10) and are recapitulated in more detail by Hagstrum (1993). Evidently, the extent of longitudinal motions and the sharpness of particular cusps in the APWPs must be considered questionable. Even so, existing evidence still supports the notion that relatively abrupt changes in the directions of apparent polar wander might be correlated with geologic boundaries (i.e., between certain Periods, Epochs, or Stages), hence with geodynamic events that have occurred in the geologic record. If such correlations exist, they are probably not precise, nor need they be (cf. Note 8.1). For example, the data from Besse and Courtillot (1991) plotted in Appendix 20 would suggest that little directional change in APWPs occurred between about 70 Ma and 50 Ma, conceivably for reasons suggested by symmetries between K/T impacts, the dynamics of core motions during that period of time, and the actual wander of the mean axis of an inclined geomagnetic dipole (see the Introduction and Chap. 7, and the patterns illustrated in Appendixes 22 and 23).

Note A.4: It seems odd to me that little attention has been given in the literature of geophysics to relationships between mass impacts and mechanisms for excitation of impulsive changes in the Earth's parameters of rotation and nutation. Exceptions—which I will expand on somewhat further below—include (1) some passing remarks by Munk and

Chapter Notes: Appendixes

MacDonald (1960, pp. 55ff) on impulsive displacements of the spin axis, (2) an attempt by Gallant (1964, Appendix I) to calculate the angle of shift of the spin axis caused by the formation of an impact crater, and (3) a qualitative analysis by Muller and Morris (1986) attempting to demonstrate how the indirect effects of impacts can induce polarity reversals in the core dynamo (cf. Note 8.2; and Schwarzschild, 1987; Muller and Morris, 1989; Rice and Creer, 1989). The comparative neglect of theories of the Earth as a dissipative *kicked rotor* (excluding the crop of large-scale computations carried out during the past decade concerning the collisional origin of the Moon; cf. the Introduction, Chap. 8, and Notes 3.2 and 6.8) — i.e., where the *non*dissipative "kicked rotor" is one of the standard algorithms of conservative-Hamiltonian dynamics, hence is often called the *standard map* (cf. Table 1, Sec. C; Notes P.3, P.4, P.6, I.4, 3.1; and Helleman, 1983; Bohr et al., 1984; Jensen, 1987; Percival, 1987; Schmidt, 1987; Prigogine et al., 1991) — is underscored by the large geophysical literature devoted to more mundane relationships between the Chandler wobble and (a) earthquakes (Mansinha and Smylie, 1967; Kanamori, 1976; O'Connell and Dziewonski, 1976; Chao and Gross, 1987; Maddox, 1988), (b) the dynamics of core motions (Alldredge, 1984; Courtillot and Le Mouël, 1984, 1985, 1988; Nicolaysen, 1985; Courtillot and Besse, 1987; Hinderer et al., 1990), (c) coupling of ocean-atmosphere torques with the "solid" Earth (Hide et al., 1980; Wahr, 1983; Naito and Kikuchi, 1990), and (d) other factors that may influence impulsive changes in the Earth's variable rotation (cf. Munk and MacDonald, 1960, Chap. 10; Stacey, 1969, Chaps. 2 and 7; Challinor, 1971; Lambeck, 1980, Chap. 8; Yoder et al., 1983; Carter et al., 1984; Landscheidt, 1983, 1988; Jin and Jin, 1989; Muller and Morris, 1989; Friis-Christensen and Lassen, 1991; Hyde and Dickey, 1991; Jacobs, 1992a; Smylie, 1992).

The subject of relationships between impacts and displacements of the spin axis began to interest me more as I watched the comparative patterns between polar-wander paths and impact-crater distributions emerge while I was preparing Appendixes 22 and 23. The illustrations struck me as rather provocative vis-à-vis the issue of short-term to geomagnetically secular (i.e., from zero to the order of 10^6 years) excitations of rotational effects in the Earth — even aside from the intriguing question of possible direct effects on the spin axis — such as the question of the orientation of an inclined-magnetic-dipole axis in the core. These provocations resonate with the discussions in Chapters 7 and 8 concerning feedback relationships between impacts and a variety of spatiotemporal scales of nonlinear-dynamical rotational-tectonic coupling phenomena in the Earth. Thus I suggest that new perspectives are in the offing with regard to dynamical mechanisms of polar wander.

The essential question to be considered can be stated in two parts: Do impacts represent significant factors affecting the Earth's rotational parameters (1) in the sense of instantaneous perturbations and/or (2) in the sense of long-term integrated effects? Attempts to evaluate these questions quantitatively are beyond the scope of this book, but if the answer to (1) is in the affirmative, then, according to the theory of dissipative systems, so must be the answer to (2) — with due regard for vectorial compensations, amplifications, and synchronizations (resonances of the sorts discussed in Chaps. 7 and 8).

Although studies of the effects of earthquakes as a mechanism for exciting the Chandler wobble have had a checkered history (see the literature cited above), some investigators apparently still take the idea seriously — at least in the case of great earthquakes and their indirect relationships to net changes in the rotational-moment distribution in the Earth (e.g., Chao and Gross, 1987). Great earthquakes have moment magnitudes of the order of $M^* \sim 9$ (see Kanamori, 1986). But the moment magnitude equivalent to the energy of the *average* impact event is of the order of $M^* \sim 12$ or more (see the discussion of Earth's cumulative impact energy in Chap. 8)! Therefore, if seismic excitations are considered to be even marginally significant as a perturbing factor in the rotational dynamics of

Chapter Notes: Appendixes

the Earth, then the mass impacts in Earth's history have implicitly induced major instantaneous effects, and, in the long term, these effects have been cumulative. The frequency of the M* ~12 "average impact" is of the order of one event every 10^5–10^6 years, and is therefore much too low to be the *primary* forcing function for short-term nutational effects — such as the Chandler wobble with a period of about 14 months (cf. Note 8.5) — but this does not mean that it may not represent the forcing function for a system of nonlinear resonances that have consequent short-term modes (e.g., in a manner analogous to the dynamics of the solar sunspot cycle; cf. Landscheidt, 1988; Cox et al., 1991; Sonett et al., 1991; Stothers, 1993). [Curiously enough, a time span of the order of 10^5 years resembles or encompasses the time scales of several types of characteristic geomagnetic phenomena, such as (a) the so-called "secular" variations of the magnetic field, (b) magnetic-polarity "excursions," and (c) "short polarity reversals"; see the discussions, for example, in Champion et al. (1988), Courtillot and Le Mouël (1988), Coe and Prévot (1989), Butler (1992, Figs. 1.7–1.10, and pp. 9ff, 161ff), Courtillot et al. (1992), and Thouveny and Creer (1992).]

Returning to my opening remarks in this Note, let us now consider the length scale of the perturbation of the Earth's spin axis by a single impact — i.e., the problem considered by Munk and MacDonald (1960, pp. 56f) and Gallant (1964, Appendix I). Gallant apparently calculated the shift of the spin axis by equating the kinetic energy of an impact (its cratering energy) with the mechanical torque required to rotate the spin axis a given amount about another axis normal to it, suggesting that rather large displacements (tens of meters) might be induced by relatively small impacts. In a more complete analysis — apparently provoked by a fictional event that had disproportionately great historical reverberations — Munk and MacDonald (1960, p. 57) commented on an attempt by Jules Verne (1828–1905), as described in his 1889 book *Les Voyages Extraordinaires: Sens Dessus Dessous*, "to fire a projectile of 180,000 tons in order to displace the pole 23° and so to remove the obliquity of the ecliptic." Verne, however, apparently neglected to take account of the Earth's equatorial bulge, as a French engineer later pointed out, the effect of which would have reduced the displacement produced by Verne's projectile — according to Munk and MacDonald (ibid.) — to about a tenth of a micron! In the same historical footnote, however, Munk and MacDonald calculated that a projectile placed at a latitude of 45° and propelled to escape velocity by an energy of 10^{24} ergs (or an equivalent exchange of mass) would produce a polar displacement of about 1 micron. Accordingly, because the energy required to fire a rocket to escape velocity is equivalent to the energy of a free-fall impact (cf. Note 6.1), we can make rule-of-thumb estimates for the polar displacements of the spin axis by impacts either from their masses or from their cratering energies. Taking a value of 10^{30} ergs, as estimated in the discussion of energy budgets in Chapter 8 for the energy of the "average impact" — i.e., the energy of an M* ~12 event — during the latest billion years of Earth's history, the resulting polar displacement would be increased a million-fold over the Munk-MacDonald estimate, corresponding to an impulsive displacement of about one meter every million years, or about one kilometer every billion years. This result clearly represents a negligible effect *if it is viewed, for example, in the context of discussions of "true" polar wander*, or in terms of the discussion of Figure 33 in Chapter 8 (cf. Appendix 5). [Estimates of the cumulative energy of impact cratering on the Earth vs. the completeness of the cratering catalog are discussed in Chaps. 2 and 8, and in the Epilogue; cf. Shoemaker (1983, Fig. 1).]

But even if a one-meter polar displacement *is* negligible in terms of the characteristic tectonic length scales in the Earth, *believe me* when I say that as a *torque impulse* applied to the Earth as a whole it would truly send all of our cultural artifacts topsy-turvy (e.g., think about the erratic motion that occurs when one tries to rotate a toy gyroscope in space arbitrarily — or even when one tries to move a small room-cooling electric fan, forgetting to turn it off first). The indubitable violence of the effect is underscored by the notion of *an equivalent $M^* \sim 12$ to $M^* \sim 13$ earthquake*, as discussed in the section of Chapter 8 cited above. Such energy calculations, of course, tell us nothing about how the impulse would actually be distributed locally in the Earth. [The actual (i.e., structurally damaging) magnitude scales of the catastrophic consequences would be larger in some places and smaller in others, just as they are in ordinary earthquakes — e.g., as was epitomized by the effects of the Loma Prieta earthquake of 18 October, 1989 (Greenwich mean time) in the San Francisco Bay Area (local residents, such as myself, remember the event as happening during the afternoon "rush hour" of Tuesday, October 17, but the time difference between San Francisco, California, and Greenwich, England, just nudged it into the next day, as recorded in standard seismic catalogs).] However, if a significant amount of the energy were dissipated in the core — for example according to some variant of the spallation-traction model of Note 8.2 — then the torque impulse might be equivalent to an exceedingly large acceleration of core fluid at the core-mantle boundary (e.g., producing propagating wave fronts with velocities of the order of a meter divided by the pulse time, which theoretically approaches zero). Such an event in the core might well induce a "paleomagnetic memory" in its effect on the overall vorticity-helicity distributions in the core — hence on the polarity states of the magnetic field (cf. Note 7.3) — as well as a geodynamic memory in the variety of coupled-feedback phenomena associated with it (e.g., Note 8.1).

Thus it would seem that Jules Verne came very close to the contextual truth of chaotic dynamics in choosing for his title *Les Voyages Extraordinaires: Sens Dessus Dessous*. In actuality, that title is perfectly suited to the extraordinary — even "topsy-turvy" (*sens dessus dessous*) — implications of the present volume, wherein I have attempted to apply the emerging principles of nonlinear dynamics to the development of a new theory of the Earth.

References Cited

References Cited

All works cited in the text or elsewhere in this volume are given in full here. Where a work is by two authors, both are given in the text citation. Works by three or more authors, however, are all cited *in text* as "et al." (that is, the authors other than the first-named are not given in the text citation). Where two or more such multiauthor works have the same first author, they are entered here *by year* (since normal alphabetization, by second, third, and further authors, would be of no help in locating them here); such works fall *after* all works (if any) by either the first-named author or that author and one other.

Aarseth, S. J., Computer simulations of star cluster dynamics, *Vistas in Astron.*, v. 13, p. 13–37, 1973.

Abraham, F. D. (with R. H. Abraham and C. D. Shaw), *A Visual Introduction to Dynamical Systems Theory for Psychology*, Santa Cruz, CA, Aerial Press, Inc., ca. 290 pp., 1990.

Abraham, R. H., and J. E. Marsden, *Foundations of Mechanics*, New York, Addison-Wesley Pub. Co., 806 pp., 1987.

——, and C. D. Shaw, Dynamics: A visual introduction, in F. E. Yates, ed., *Self-Organizing Systems*, p. 543–597, NY, Plenum Publishing Corporation, 1987.

——, *Dynamics – The Geometry of Behavior, VISMATH, v. 1–4*, Santa Cruz, CA, Aerial Press, Inc., 678 pp. (4-volume set), 1982–88.

Abraham, R. H., and H. B. Stewart, A chaotic blue sky catastrophe in forced relaxation oscillations, *Physica, v. 21D*, p. 394–400, 1986.

Abraham, R. H., J. E. Marsden, and T. Ratiu, *Manifolds, Tensor Analysis, and Applications*, New York, Springer-Verlag, 654 pp., 1988.

Abraham, R. H., J. B. Corliss, and J. E. Dorband, Order and disorder in the toral logistic lattice, *Internat. Jour. Bifurcation and Chaos, v. 1*, p. 227–234, 1991a.

Abraham, R. H., A. Keith, M. Koebbe, and G. Mayer-Kress, Computational unfolding of double-cusp models of opinion formation, *Internat. Jour. Bifurcation and Chaos, v. 1*, p. 417–430, 1991b.

Ager, D. V., *The Nature of the Stratigraphical Record, Second Edition*, London, Macmillan/Halstead Press, 117 pp., 1980.

——, The stratigraphic code and what it implies, in W. A. Berggren and J. A. Van Couvering, eds., *Catastrophes and Earth History: The New Uniformitarianism*, p. 91–100, Princeton, NJ, Princeton University Press, 1984.

References Cited

Aggarwal, H. R., and V. R. Oberbeck, Mathematical modeling of impact deposits and the origin of tillites (Abstract), *Eos, Trans. Am. Geophys. Union, v. 73* (Supplement, 27 Oct.), p. 325, 1992.

Ahrens, T. J., and A. W. Harris, Deflection and fragmentation of near-Earth asteroids, *Nature, v. 360*, p. 429–433, 1992.

Ahrens, T. J., and J. D. O'Keefe, Impact of an asteroid or comet in the ocean and extinction of terrestrial life, *Proc. 13th Lunar and Planetary Sci. Conf., Part 2, Jour. Geophys. Res., v. 88*, Suppl., p. A799–A806, 1983.

——, Large impact craters (Abstract), *Eos, Trans. Am. Geophys. Union, v. 71*, p. 1429, 1990.

Aki, K., and R. Koyanagi, Deep volcanic tremor and magma ascent mechanism under Kilauea, Hawaii, *Jour. Geophys. Res., v. 86*, p. 7095–7109, 1981.

Aksnes, K., Manmade objects — A source of confusion to asteroid hunters?, in T. Gehrels, ed., *Physical Studies of Minor Planets*, p. 649–652, Washington, DC, National Aeronautics and Space Administration, NASA SP-267, 1971.

Alfvén, H., Motion of small particles in the Solar System, in T. Gehrels, ed., *Physical Studies of Minor Planets*, p. 315–317, Washington, DC, National Aeronautics and Space Administration, NASA SP-267, 1971a.

——, Apples in a spacecraft, *Science, v. 173*, p. 522–525, 1971b.

——, Cosmology and recent developments in plasma physics: The Dirac Lecture in Theoretical Physics for 1979, *The Australian Physicist*, p. 161–165, Nov., 1980.

——, *Cosmic Plasma*, Boston, MA, D. Reidel Publ. Co., 164 pp., 1981.

——, Magnetospheric research and the history of the Solar System, *Eos, Trans. Am. Geophys. Union, v. 65*, p. 769–770, 1984.

Alfvén, H., and G. Arrhenius, Two alternatives for the history of the Moon, *Science, v. 165*, p. 11–17, 1969.

——, *Structure and Evolutionary History of the Solar System*, Boston, MA, D. Reidel Publ. Co., 276 pp., 1975.

——, *Evolution of the Solar System*, Washington, DC, National Aeronautics and Space Administration (NASA SP-345), 599 pp, 1976.

Allan, D. W., Reversals of the Earth's magnetic field, *Nature, v. 182*, p. 469–470, 1958.

Allan, R. R., On the motion of nearly synchronous satellites, *Proc., Roy. Soc. London, Ser. A., v. 288*, p. 60–68, 1965.

——, Resonance effects due to the longitude dependence of the gravitational field of a rotating primary, *Planet. Space Sci., v. 15*, p. 53–76., 1967a.

——, Satellite resonance with longitude-dependent gravity — II: Effects involving the eccentricity, *Planet. Space Sci., v. 15*, p. 1829–1845, 1967b.

——, Evolution of Mimas-Tethys commensurability, *Astronomical Jour., v. 74*, p. 497–506, 1969.

——, Commensurable eccentric orbits near critical inclination, *Celestial Mech., v. 3*, p. 320–330, 1971.

Alldredge, L. R., A discussion of impulses and jerks in the geomagnetic field, *Jour. Geophys. Res., v. 89*, p. 4403–4412, 1984.

Allen, T., On the arithmetic of phase locking: Coupled neurons as a lattice on R^2, *Physica D, v. 6*, p. 305–320, 1983.

Alper, J., The chaotic brain: New models of behavior (Mind and Brain), *Psychology Today*, May, p. 21, 1989.

Alt, D., J. M. Sears, and D. W. Hyndman, Terrestrial maria: the origins of large basalt plateaus, hotspot tracks, and spreading ridges, *Jour. Geol., v. 96*, p. 647–662, 1988.

Altangerel, P., M. A. Norell, L. M. Chiappe, and J. M. Clark, Flightless bird from the Cretaceous of Mongolia, *Nature, v. 362*, p. 623–626, 1993.

Alvarez, L. W., Mass extinctions caused by large bolide impacts, *Phys. Today, v. 40*, p. 24–33 (July), 1987.

Alvarez, L. W., W. Alvarez, F. Asaro, and H. V. Michel, Extraterrestrial cause for the Cretaceous-Tertiary extinction: Experiment and theory, *University of California, Berkeley; Lawrence Berkeley Report, LBL-9666*, 86 pp., 1979.

References Cited

———, Extraterrestrial causes for the Cretaceous-Tertiary extinction, *Science, v. 208*, p. 1095–1108, 1980.

Alvarez, W., Toward a theory of impact crises, *Eos, Trans. Am. Geophys. Union, v. 67*, p. 649–658, 1986.

———, Interdisciplinary aspects of research on impacts and mass extinctions: A personal view, in V. L. Sharpton and P. D. Ward, eds., *Global Catastrophes in Earth History; An Interdisciplinary Conference on Impacts, Volcanism, and Mass Mortality*, p. 93–97, Boulder, CO, The Geological Society of America, Special Paper 247, 1990.

Alvarez, W., and F. Asaro, Debate: What caused the mass extinction? An extraterrestrial impact, *Sci. Am., v. 263*, p. 78–84, 1990.

Alvarez, W., L. W. Alvarez, F. Asaro, and H. V. Michel, in L. T. Silver and P. H. Schultz, eds., *Geological Implications of Impacts of Large Asteroids and Comets on the Earth*, p. 305–315, Boulder, CO, The Geological Society of America, Special Paper 190, 1982.

Alvarez, W., T. Hansen, P. Hut, E. G. Kauffman, and E. M. Shoemaker, Uniformitarianism and the response of Earth scientists to the theory of impact crises, in S. V. M. Clube, ed., *Catastrophes and Evolution: Astronomical Foundations*, p. 13–24, NY, Cambridge University Press, 1989.

Alvarez, W., J. Smit, M. H. Anders, F. Asaro, F. J-M. R. Maurrasse, M. Kastner, and W. Lowrie, Impact-wave effects at the Cretaceous-Tertiary boundary in Gulf of Mexico DSDP cores, *Lunar and Planetary Sci. Conf. XXII, Abstracts of Papers*, p. 17–18, Lunar and Planetary Science Institute, Houston, TX, March 18–22, 1991.

Amato, I., DNA shows unexplained patterns writ large, *Science, v. 257*, p. 747, 1992.

Anders, E., Meteorite ages, in B. M. Middlehurst and G. P. Kuiper, eds., *The Moon, Meteorites, and Comets*, p. 402–495, Chicago, The University of Chicago Press, 1963.

———, Circumstellar material in meteorites: Noble gases, carbon and nitrogen, in J. F. Kerridge and M. S. Matthews, eds., *Meteorites and the Early Solar System*, p. 927–955, Tucson, AZ, The University of Arizona Press, 1988.

Anders, E., and J. F. Kerridge, Future directions in meteorite research, in J. F. Kerridge and M. S. Matthews, eds., *Meteorites and the Early Solar System*, p. 1155–1186, Tucson, AZ, The University of Arizona Press, 1988.

Anderson, A. J., and A. Casenave, eds., *Space Geodesy and Geodynamics*, New York, Academic Press, 490 pp., 1986.

Anderson, C. M., and L. D. Friedman, Prospecting the future: The Planetary Society Asteroid Program, *The Planetary Report, v. 11*, p. 20–22 (Nov./Dec.), 1991.

Anderson, D. L., Thermally induced phase changes, lateral heterogeneity of the mantle, continental roots and deep slab anomalies, *Jour. Geophys. Res., v. 92*, p. 13,968–13, 980, 1987.

———, *Theory of the Earth*, Boston, MA, Blackwell Scientific Publications, 366pp., 1989.

———, Planet Earth, in J. K. Beatty and A. Chaikin, eds, *The New Solar System, Third Edition*, p. 65–76, Cambridge, MA, Sky Publishing Corporation, 1990.

———, Chemical boundaries in the mantle, in R. Sabadini, K. Lambeck, and E. Boschi, *Glacial Isostasy, Sea-Level and Mantle Rheology*, p. 379–401, Boston, MA, Kluwer Academic Publishers, 1991a.

———, Plumes, plates, and deep Earth structure (Bowie Lecture, Abstract), *Eos, Trans. Am. Geophys. Union, v. 72, Supplement (Oct. 29, 1991)*, p. 64, 1991b.

Anderson, D. L., T. Tanimoto, and Y.-S. Zhang, Plate tectonics and hotspots: The third dimension, *Science, v. 256*, p. 1645–1651, 1992.

Anderson, E. M., The dynamics of the formation of cone-sheets, ring-dykes, and cauldron-subsidences, *Roy. Soc. Edinburgh Proc., v. 56*, p. 128–163, 1936.

Anderson, J. D., and E. M. Standish, Jr., Dynamical evidence for Planet X, in R. Smoluchowski, J. N. Bahcall, and M. S. Matthews, eds., *The Galaxy and the Solar System*, p. 286–296, Tucson, AZ, The University of Arizona Press, 1986.

Anderson, O. L., The high-pressure triple points of iron and their effects on the heat flow from the Earth's core, *Jour. Geophys. Res., v. 95*, p. 21,697–21,707, 1990.

Anderson, O. L., and P. C. Perkins, A plate tectonics model involving nonlaminar asthenospheric

References Cited

flow to account for irregular patterns of magmatism in the southwestern United States, *Phys. Chem. Earth, v. 9*, p. 113–122, 1975.

Andrews, J. A., True polar wander: An analysis of Cenozoic and Mesozoic paleomagnetic poles, *Jour. Geophys. Res., v. 90*, p. 7737–7750, 1985.

Andrews, J. T., *Glacial Systems: An Approach to Glaciers and Their Environments*, North Scituate, MA, 191 pp., 1975.

Andreyev, Y. V., A. S. Dmitriev, L. O. Chua, and C. W. Wu, Associative and random access memory using one-dimensional maps, *Internat. Jour. Bifurcation and Chaos, v. 2*, p. 483–504, 1992.

Anishchenko, V. S., T. E. Vadivasova, D. E. Postnov, and M. A. Safonova, Synchronization of chaos, *Internat. Jour. Bifurcation and Chaos, v. 2*, p. 633–644, 1992.

Anonymous, *Stanislaw Ulam 1909–1984*, Los Alamos Science, Number 15, Special Issue, 318 pp., 1987.

———, *The Astronomical Almanac for the Year 1992*, Washington, DC, U. S. Government Printing Office, 546 pp., 1991.

———, Radar spies second dumbbell asteroid, *Science, v. 259*, p. 314, 1993.

Appenzeller, I., Contracting stars, *Physica Scripta, v. T11*, p. 76–81, 1985.

Appenzeller, T., Hope for magnetic storm warnings, *Science, v. 255*, p. 922–924, 1992.

Aranson, I. S., A. V. Gaponov-Grekhov, and M. I. Rabinovich, The onset and spatial development of turbulence in flow systems, *Physica D, v. 33*, p. 1–20, 1988.

Arecchi, F. T., Rate processes in nonlinear optical dynamics with many attractors, *Chaos, v. 1*, p. 357–372, 1991.

Argoul, F., A. Arneodo, P. Richetti, J. C. Roux, and H. L. Swinney, Chemical chaos: From hints to confirmation, *Accounts Chem. Res. (Am. Chem. Soc.), v. 20*, p. 436–442, 1987.

Arho, R., On the rotation of the orbital plane during ballistic re-entry, *Planet. Space Sci., v. 19*, p. 1215–1224, 1971.

Armbruster, D., and P. Chossat, Heteroclinic orbits in a spherically invariant system, *Physica D, v. 50*, p. 155–176, 1991.

Arrhenius, S., *Worlds in the Making*, New York, Harper and Brothers, Publishers, 230 pp., 1908.

Arthur, W. B., Positive feedbacks in the economy, *Sci. Am., v. 262*, p. 92–99 (February), 1990.

Asimov, I., *A Choice of Catastrophes: The Disasters That Threaten Our World*, New York, Fawcett Columbine, 377 pp., 1981.

Atkins, P. W., *Atoms, Electrons, and Change*, New York, Scientific American Library, A Division of HPHLP, 242 pp., 1991.

Atreya, S. K., J. B. Pollock, and M. S. Matthews, eds., *Origin and Evolution of Planetary and Satellite Atmospheres*, Tucson, AZ, The University of Arizona Press, 881 pp., 1989.

Aumann, H. H., and J. C. Good, IRAS constraints on a Cold Cloud around the Solar System, *Astrophys. Jour., v. 350*, p. 408–412, 1990.

Axelrod, R., and W. D. Hamilton, The evolution of cooperation, *Science, v. 211*, p. 1390–1396, 1981.

Babcock, K. L., and R. M. Westervelt, Avalanches and self-organization in cellular magnetic-domain patterns, *Phys. Rev. Lett., v. 64*, p. 2168–2171, 1990.

Badjukov, D. D., M. A. Nazarov, and L. D. Barsukova, Chemistry of the Kara impact structure, *Lunar and Planetary Sci. Conf. XXII, Abstracts of Papers*, p. 43–44, Lunar and Planetary Science Institute, Houston, TX, March 18–22, 1991.

Bagby, J. P., The Holpuch satellite ring, *Twentieth Century Observations, v. 1*, p. 49–56 (March), 1956.

———, Evidence of an ephemeral Earth satellite, *Nature, v. 211*, p. 285, 1966.

———, Radio anomalies associated with an ephemeral satellite still in orbit, *Nature, v. 215*, p. 1050–1052, 1967.

———, Terrestrial satellites: Some direct and indirect evidence, *Icarus, v. 10*, p. 1–10, 1969.

Bahcall, J. N., The galactic environment of the Solar System, in R. Smoluchowski, J. N. Bahcall, and M. S. Matthews, eds., *The Galaxy and the Solar System*, p. 3–12, Tucson, AZ, The University of Arizona Press, 1986.

Bailey, D. K., and A. D. Stewart, Problems of ocean water accumulation on a rapidly expanding

References Cited

Earth, in S. W. Carey, ed., *The Expanding Earth: A Symposium, Sydney, 1981*, p. 67–69, Hobart, Tasmania, University of Tasmania, 1983.
Bailey, M. E., The formation of comets in wind-driven shells around protostars, *Icarus, v. 69*, p. 70–82, 1987.
——, Comet orbits and chaos, *Nature, v. 345*, p. 21–22, 1990.
Bailey, M. E., S. V. M. Clube, and W. M. Napier, *The Origin of Comets*, New York, Pergamon Press, 577 pp., 1990.
Bai-Lin, H., *SEE:* Hao, B.-L.
Bak, P., The Devil's staircase, *Phys. Today, v. 39*, p. 38–45, 1986.
Bak, P., and K. Chen, The physics of fractals, *Physica D, v. 38*, p. 5–12, 1989.
——, Self-organized criticality, *Sci. Am., v. 264*, p. 46–53, 1991.
Bak, P., and C. Tang, Earthquakes as self-organized phenomena, *Jour. Geophys. Res., 94*, p. 15635–15637, 1989
Bak, P., C. Tang, and K. Wiesenfeld, Self-organized criticality, *Phys. Rev. A, v. 38*, p. 364–374, 1988.
Baker, D. N., A. J. Klimas, R. L. McPherron, and J. Büchner, The evolution from weak to strong geomagnetic activity: an interpretation in terms of deterministic chaos, *Geophys. Res. Lett., v. 17*, p. 41–44, 1990.
Baker, G. L., and J. P. Gollub, *Chaotic Dynamics: An Introduction*, New York, Cambridge University Press, 182 pp. (computer software available by prepaid order), 1990.
Baker, R. M. L., Jr., Ephemeral natural satellites of the Earth, *Science, v. 128*, p. 1211–1213, 1958.
Baksi, A. K., Estimation of lava extrusion and magma production rates for two flood basalt provinces, *Jour. Geophys. Res., v. 93*, p. 11,809–11,815, 1988.
——, Timing and duration of Mesozoic-Tertiary flood-basalt volcanism, *Eos, Trans. Am. Geophys. Union, v. 71*, p. 1836–1836, 1840, 1990a.
——, The search for periodicity in global events in the geologic record: Quo vadimus?, *Geology, v. 18*, p. 983–986, 1990b.
Baksi, A. K., and E. Farrar, $^{40}Ar/^{39}Ar$ dating of the Siberian Traps, USSR: Evaluation of the ages of the two major extinction events relative to episodes of flood-basalt volcanism in the USSR and the Deccan Traps, India, *Geology, v. 19*, p. 461–464, 1991.
Baldwin, B., and Y. Sheaffer, Ablation and breakup of large meteoroids during atmospheric entry, *Jour. Geophys. Res., 76*, p. 4653–4668, 1971.
Baldwin, R. B., On the history of lunar impact cratering: The absolute time scale and the origin of planetesimals, *Icarus, v. 14*, p. 36–52, 1971.
——, Relative and absolute ages of individual craters and the rate of infalls on the Moon in the post-Imbrium period, *Icarus, v. 61*, p. 63–91, 1985.
——, On the current rate of formation of impact craters of varying sizes on the Earth and Moon, *Geophys. Res. Lett., v. 14*, p. 216–219, 1987.
Baldwin, R. B., and D. E. Wilhelms, Historical review of a long-overlooked paper by R. A. Daly concerning the origin and early history of the Moon, *Jour. Geophys. Res., v. 97*, p. 3837–3843, 1992.
Bambach, R. K., C. R. Scotese, and A. M. Ziegler, Before Pangea: The geographies of the Paleozoic world, *Am. Scientist, v. 68*, p. 26–38, 1980.
Bandermann, L. W., Remarks on the size distribution of colliding and fragmenting particles, in T. Gehrels, ed., *Physical Studies of Minor Planets*, p. 297–303, Washington, DC, National Aeronautics and Space Administration, NASA SP-267, 1971.
Barbujani, G., and R. R. Sokal, The zones of sharp genetic change in Europe are also linguistic boundaries, *Proc. Natl. Acad. Sci., v. 87*, p. 1816–1819, 1990.
Bargar, K. E., and E. D. Jackson, Calculated volumes of individual shield volcanoes along the Hawaiian-Emperor chain, *U. S. Geol. Survey, Jour. Res., v. 2*, p. 545–550, 1974.
Barinaga, M., Death gives birth to the nervous system. But how?, *Science, v. 259*, p. 762–763, 1993.
Barlow, N. G., Application of the inner Solar System cratering record to the Earth, in V. L. Sharpton and P. D. Ward, eds., *Global Catastrophes in Earth History; An Interdisciplinary*

References Cited

Conference on Impacts, Volcanism, and Mass Mortality, p. 181–187, Boulder, CO, The Geological Society of America, Special Paper 247, 1990.

Barnes, J. E., and L. Hernquist, Formation of dwarf galaxies in tidal tails, *Nature, v. 360*, p. 715–717, 1992.

Barnes, V. E., Tektite research 1936–1990, *Meteoritics, v. 25*, p. 149–159, 1990.

Barton, N. H., and J. S. Jones, The language of the genes, *Nature, v. 346*, p. 415–416, 1990.

Barut, A. O., Symmetry and dynamics: Two distinct methodologies from Kepler to supersymmetry, in B. Gruber and R. Lenczewski, eds., *Symmetries in Science II*, p. 37–50, New York, Plenum Press, 1986.

Bash, F., The present, past and future velocity of nearby stars: The path of the Sun in 10^8 years, in R. Smoluchowski, J. N. Bahcall, and M. S. Matthews, eds., *The Galaxy and the Solar System*, p. 35–46, Tucson, AZ, The University of Arizona Press, 1986.

Bass, T. A., *The Eudaemonic Pie*, New York, Vintage Books, 324 pp., 1985.

Bate, R. R., D. D. Mueller, and J. E., White, *Fundamentals of Astrodynamics*, New York, Dover Publications, Inc., 1971.

Baxter, D. C., and W. B. Thompson, Jetstream formation through inelastic collisions, in T. Gehrels, ed., *Physical Studies of Minor Planets*, p. 319–326, Washington, DC, National Aeronautics and Space Administration, NASA SP-267, 1971.

Beals, C. S., M. J. S. Innes, and J. A. Rottenberg, Fossil meteorite craters, in B. M. Middlehurst and G. P. Kuiper, eds., *The Moon, Meteorites, and Comets*, p. 235–284, Chicago, The University of Chicago Press, 1963.

Beatty, J. K., and A. Chaikin, eds., *The New Solar System, Third Edition*, New York, Cambridge University Press, 326 pp., 1990.

Beckert, S., U. Schock, C.-D. Schulz, T. Weidlich, and F. Kaiser, Experiments on the bifurcation behavior of a forced nonlinear pendulum, *Phys. Lett., v. 107A*, p. 347–350, 1985.

Beckmann, P., *A History of π (Pi), Third Edition*, New York, St. Martin's Press, 200 pp., 1974.

Béland, P., Models for the collapse of terrestrial communities of large vertebrates, in P. Béland et al. (The K-TEC Group), eds., *Cretaceous-Tertiary Extinctions and Possible Terrestrial and Extraterrestrial Causes*, p. 25–37, Ottawa, Canada, The National Museums of Canada, Syllogeus No. 12, 1977.

Béland, P., J.-R. Roy, and D. Russell, Chains of events leading to mass extinctions: Two synopses, in P. Béland et al. (The K-TEC Group), eds., *Cretaceous-Tertiary Extinctions and Possible Terrestrial and Extraterrestrial Causes*, p. 155–158, Ottawa, Canada, The National Museums of Canada, Syllogeus No. 12, 1977.

Belbruno, E., Through the fuzzy boundary: A new route to the Moon, *The Planetary Report, v. 12*, p. 8–10 (May/June), 1992.

Bell, J. F., Mineralogical clues to the origins of asteroid dynamical families, *Icarus, v. 78*, p. 426–440, 1989.

Bell, J. F., D. R. Davis, W. K. Hartmann, and M. J. Gaffey, Asteroids: The big picture, in R. P. Binzel, T. Gehrels, and M. S. Matthews, eds., *Asteroids II*, p. 921–945, Tucson, AZ, The University of Arizona Press, 1989.

Belton, M. J. S., J. Veverka, P. Thomas, P. Helfenstein, D. Simonelli, C. Chapman, M. E. Davies, R. Greeley, R. Greenberg, J. Head, S. Murchie, K. Klaasen, T. V. Johnson, A. McEwen, D. Morrison, G. Neukum, F. Fanale, C. Anger, M. Carr, and C. Pilxher, Galileo encounter with 951 Gaspra: First pictures of an asteroid, *Science, v. 257*, p. 1647, 1992.

Benson, R. H., The Phanerozoic "crisis" as viewed from the Miocene, in W. A. Berggren and J. A. Van Couvering, eds., *Catastrophes and Earth History: The New Uniformitarianism*, p. 437–446, Princeton NJ, Princeton University Press, 1984.

Benson, R. H., R. E. Chapman, and L. T. Deck, Paleoceanographic events and deep-sea ostracodes, *Science, v. 224*, p. 1334–1336, 1984.

Benton, M. J., Late Triassic extinctions and the origin of the dinosaurs, *Science, v. 260*, p. 769–770, 1993.

Benz, W., and A. G. W. Cameron, Terrestrial effects of the Giant Impact, in H. E. Newsom and J. H. Jones, *Origin of the Earth*, p. 61–67, New York, Oxford University Press, 1990.

References Cited

Benz, W., W. L. Slattery, and A. G. W. Cameron, The origin of the Moon and the single impact hypothesis I, *Icarus, v. 66*, p. 515–535, 1986.

———, The origin of the Moon and the single impact hypothesis II, *Icarus, v. 71*, p. 30–45, 1987.

Benz, W., A. G. W. Cameron, and H. J. Melosh, The origin of the Moon and the single-impact hypothesis III, *Icarus, v. 81*, p. 113–131, 1989.

Benzi, R., G. Parisi, A. Sutera, and A. Vulpiani, Stochastic resonance in climatic change, *Tellus, v. 34*, p. 10–16, 1982.

Bergé, P., Y. Pomeau, and C. Vidal, *Order Within Chaos: Toward a Deterministic Approach To Turbulence*, New York, John Wiley & Sons, Inc., 329 pp., 1984.

Berger, A., M. F. Loutre, and J. Laskar, Stability of the astronomical frequencies over the Earth's history for paleoclimate studies, *Science, v. 255*, p. 560–566, 1992.

Berggren, W. A., and J. A. Van Couvering, eds., *Catastrophes and Earth History: The New Uniformitarianism*, Princeton, NJ, Princeton University Press, 464 pp., 1984.

Berlage, H. P., *The Origin of the Solar System*, New York, Pergamon, 130 pp., 1968.

Berns, M. W., *Cells, Second Edition*, New York, Saunders College Publishing, 256 pp., 1983.

Berry, M. V., Regular and irregular motion, in S. Jorna, ed., *Topics in Nonlinear Dynamics: A Tribute to Sir Edward Bullard*, p. 16–120, New York, American Institute of Physics, 1978.

———, Quantum chaology: The Bakerian Lecture, 1987, in M. V. Berry, I. C. Percival, and N. O. Weiss, eds., *Dynamical Chaos*, p. 183–198, Princeton, NJ, Princeton University Press, 1987.

Berry, M. V., I. C. Percival, and N. O. Weiss, eds., *Dynamical Chaos*, Princeton, NJ, Princeton University Press, 199 pp., 1987.

Bertotti, B., and P. Farinella, *Physics of the Earth and the Solar System*, Boston, MA, Kluwer Academic Publishers, 479 pp., 1990.

Besançon, R. M., ed., *The Encyclopedia of Physics, Third Edition*, New York, Van Nostrand Reinhold, 1378 pp., 1990.

Besse, J., and V. Courtillot, Revised and synthetic apparent polar wander paths of the African, Eurasian, North American and Indian plates, and true polar wander since 200 Ma, *Jour. Geophys. Res., v. 96*, p. 4029–4050, 1991.

Beverley, S. M., and A. C. Wilson, Ancient origin for Hawaiian Drosophilinae inferred from protein comparisons, *Proc. Natl. Acad. Sci., v. 82*, p. 4753–4757, 1985.

Bice, D. M., C. R. Newton, S. McCauley, P. W. Reiners, and C. A. McRoberts, Shocked quartz at the Triassic-Jurassic boundary in Italy, *Science, v. 255*, p. 443–446, 1992.

Bindschadler, D. L., G. Schubert, W. M. Kaula, and A. Lenardic, Venus hotspots vs. cold-spots: Magellan images, mantle dynamics, and global tectonics (Abstract), *Eos, Trans. Am. Geophys. Union, v. 72, Supplement* (29 October), p. 286, 1991.

Binzel, R. P., An overview of the asteroids, in R. P. Binzel, T. Gehrels, and M. S. Matthews, eds., *Asteroids II*, p. 3–18, Tucson, AZ, The University of Arizona Press, 1989.

———, Asteroids and comets in near-Earth space, *Planetary Report, v. 11*, p. 8–11 (Nov./Dec.), 1991.

Binzel, R. P., and S. Xu, Chips off of asteroid 4 Vesta: Evidence for the parent body of basaltic achondrite meteorites, *Science, v. 260*, p. 186–191, 1993.

Binzel, R. P., T. Gehrels, and M. S. Matthews, eds., *Asteroids II*, Tucson, AZ, The University of Arizona Press, 1258 pp., 1989a.

Binzel, R. P., P. Farinella, V. Zappalà, and A. Cellino, Asteroid rotation rates: Distributions and statistics, in R. P. Binzel, T. Gehrels, and M. S. Matthews, eds., *Asteroids II*, p. 416–441, Tucson, AZ, The University of Arizona Press, 1989b.

Binzel, R. P., M. A. Barucci, and M. Fulchignoni, The origins of the asteroids, *Sci. Am., v. 265*, p. 88–94, 1991.

Binzel, R. P., S. Xu, S. J. Bus, and E. Bowell, Origins for the near-Earth asteroids, *Science, v. 257*, p. 779–782, 1992.

Birmingham, T. J., and A. J. Dessler, eds., *Comet Encounters*, Washington, DC, American Geophysical Union, 352 pp., 1988.

Blagg, M. A., On a suggested substitute for Bode's Law, *Month. Not. Roy. Astron. Soc. v. 73*, p. 414–422, 1913.

References Cited

Blandford, R. D., A brief history of pulsar time, *Nature, v. 359,* p. 675, 1992.

Blatt, H., and R. L. Jones, Proportions of exposed igneous, metamorphic, and sedimentary rocks, *Bull. Geol. Soc. Am., v. 86,* p. 1085–1088, 1975.

Bleher, S., E. Ott, and C. Grebogi, Routes to chaotic scattering, *Phys. Rev. Lett., v. 63,* p. 919–922, 1989.

Blekhman, I. I., *Synchronization in Science and Technology* (Translated from the Russian by E. I. Riven), New York, ASME Press, 255 pp., 1988.

Bloxham, J., The steady part of the secular variation of the Earth's magnetic field, *Jour. Geophys. Res., v. 97,* p. 19,565–19,579, 1992.

Bloxham, J., and D. Gubbins, Thermal-mantle interactions, *Nature, v. 325,* p. 511–513, 1987.

———, The evolution of the Earth's magnetic field, *Sci. Am., v. 261,* p. 68–75 (December), 1989.

Bloxham, J., and A. Jackson, Fluid flow near the surface of Earth's outer core, *Rev. Geophys., v. 29,* p. 97–120, 1991.

———, Time-dependent mapping of the magnetic field at the core-mantle boundary, *Jour. Geophys. Res., v. 97,* p. 19,537–19,563, 1992.

Blum, J. D., D. A. Papanastassiou, C. Koeberl, and G. J. Wasserburg, Neodymium and strontium isotopic study of Australasian tektites: New constraints on the provenance and age of target materials, *Geochimica et Cosmochimica Acta, v. 56,* p. 483–492, 1992.

Bock, W. J., The synthetic explanation of macroevolutionary change—a reductionistic approach, in J. H. Schwartz and H. B. Rollins, eds., *Models and Methodologies in Evolutionary Theory,* p. 20–69, Bulletin of the Carnegie Museum of Natural History, No. 13, 1979.

Boe, E., and H-C. Chang, Transition to chaos from a two-torus in a delayed feedback system, *Internat. Jour. Bifurcation and Chaos, v. 1,* p. 67–81, 1991.

Boecklen, W. J., and D. Simberloff, Area-based extinction models in conservation, in D. K. Elliott, ed., *Dynamics of Extinction,* p. 247–276, New York, John Wiley & Sons, Inc., 1986.

Boehler, R., Temperatures in the Earth's core from melting-point measurements of iron at high static pressures, *Nature, v. 363,* p. 534–536, 1993.

Bogue, S., Reversals of opinion, *Nature, v. 351,* p. 445–446, 1991.

Bogue, S. W., and R. T. Merrill, The character of the field during geomagnetic reversals, *Ann. Rev. Earth Planet. Sci., v. 20,* p. 181–219, 1992.

Bohm, D., and F. D. Peat, *Science, Order, and Creativity,* New York, Bantam Books, 280 pp., 1987.

Bohor, B. F., Shocked quartz and more: Impact signatures in Cretaceous/Tertiary boundary clays, in V. L. Sharpton and P. D. Ward, eds., *Global Catastrophes in Earth History; An Interdisciplinary Conference on Impacts, Volcanism, and Mass Mortality,* p. 335–342, Boulder, CO, The Geological Society of America, Special Paper 247, 1990.

———, Large meteorite impacts: The K/T model, *Papers Presented at the International Conference on Large Meteorite Impacts and Planetary Evolution, 31 August–2 September, 1992, Sudbury, Ontario, Canada,* p. 8–9, Lunar and Planetary Institute, Houston, TX (LPI Contribution No. 790), 1992.

Bohr, T., P. Bak, and M. H. Jensen, Transition to chaos by interacting resonances in dissipative systems: II. Josephson junctions, charge-density waves, and standard maps, *Phys. Rev. A, v. 30,* p. 1970–1981, 1984.

Bolter, J. D., *Turing's Man: Western Culture in the Computer Age,* Chapel Hill, NC, The University of North Carolina Press, 264 pp., 1984.

Boss, A. P., The origin of the Moon, *Science, v. 231,* p. 341–345, 1986.

———, High temperatures in the early solar nebula, *Science, v. 241,* p. 565–567, 1988.

Boss, A. P., and S. J. Peale, Dynamical constraints on the origin of the Moon, in W. K. Hartmann, R. J. Phillips, and G. J. Taylor, eds., *Origin of the Moon,* p. 59–102, Houston, TX, Lunar and Planetary Inst., 1986.

Bostrom, R. C., Westward displacement of the lithosphere, *Nature, v. 234,* p. 536–538, 1971.

———, TPW and lithosphere-rotation component of plate motion (Abstract), *Eos, Trans. Am. Geophys. Union, v. 73* (Supplement, 27 Oct.), p. 60, 1992.

Bottomley, R. J., D. York, and R. A. F. Grieve, $^{40}Ar-^{39}Ar$ dating of Scandinavian impact craters, *Meteoritics, v. 12,* p. 182–183, 1977.

References Cited

Bourgeois, J., Boundaries: A stratigraphic and sedimentologic perspective, in V. L. Sharpton and P. D. Ward, eds., *Global Catastrophes in Earth History; An Interdisciplinary Conference on Impacts, Volcanism, and Mass Mortality*, p. 411–416, Boulder, CO, The Geological Society of America, Special Paper 247, 1990.

Bowell, E., T. Gehrels, and B. Zellner, Magnitudes, colors, types, and adopted diameters of the asteroids, in T. Gehrels, ed., *Asteroids*, p. 1108–1129, Tucson, AZ, The University of Arizona Press, 1979.

Bowell, E., N. S. Chernykh, and B. G. Marsden, Discovery and followup of asteroids, in R. P. Binzel, T. Gehrels, and M. S. Matthews, eds., *Asteroids II*, p. 21–38, Tucson, AZ, The University of Arizona Press, 1989.

Bowin, C. O., Mass anomalies and deep Earth structure (Abstract), *Eos, Trans. Am. Geophys. Union, v. 73* (Supplement, 27 Oct.), p. 61, 1992.

Bradley, W. H., Vertical density currents, *Science, v. 150*, p. 1423–1428, 1965.

———, Vertical density currents, *Limnology and Oceanography, v. 14*, p. 1–3, 1969.

Brady, J. L., Effect of the planetary system on the nearly-parabolic comets, *Astron. Jour., v. 70*, p. 279–282, 1965.

Bramble, D. M., and D. R. Carrier, Running and breathing in mammals, *Science, v. 219*, p. 251–256, 1983.

Breuer, R., *The Anthropic Principle: Man As the Focal Point of Nature*, Boston, MA, Birkhäuser, 261 pp., 1991.

Brookins, D. G., Shale as a repository for radioactive waste: The evidence from Oklo, *Environmental Geol., v. 1*, p. 255–259, 1976.

Brosche, P., and J. Sündermann, *Tidal Friction and the Earth's Rotation*, New York, Springer-Verlag, 241 pp., 1978.

———, *Tidal Friction and the Earth's Rotation II*, New York, Springer-Verlag, 345 pp., 1982.

Brown, R., and L. Chua, Generalizing the twist-and-flip paradigm, *Internat. Jour. Bifurcation and Chaos, v. 1*, p. 385–416, 1991.

Brown, W. K., R. R. Karp, and D. E. Grady, Fragmentation of the universe, *Astrophysics and Space Sci., v. 94*, p. 401–412, 1983.

Brush, S. G., The early history of selenogony, in W. K. Hartmann, R. J. Phillips, and G. J. Taylor, eds., *Origin of the Moon*, p. 3–15, Houston, TX, Lunar and Planetary Inst., 1986.

Bucher, M., Universal scaling of windows from one-dimensional maps, *Phys. Rev. A, v. 33*, p. 3544–3546, 1986.

Buffett, B. A., Constraints on magnetic energy and mantle conductivity from the forced nutations of the Earth, *Jour. Geophys. Res., v. 97*, p. 19,581–19,597, 1992.

Buffett, B. A., H. E. Huppert, J. R. Lister, and A. W. Woods, Analytical model for solidification of the Earth's core, *Nature, v. 356*, p. 329–331, 1992.

Bullard, E. C., The stability of a homopolar dynamo, *Proc. Cambridge Philos. Soc., v. 51*, p. 744–760, 1955.

———, The disk dynamo, in S. Jorna, ed., *Topics in Nonlinear Dynamics: A Tribute to Sir Edward Bullard*, p. 373–389, New York, American Institute of Physics, 1978.

Bullard, E. C., J. E. Everett, and A. G. Smith, A symposium on continental drift. IV. The fit of the continents around the Atlantic, *Phil. Trans. Roy. Soc. London, v. A258*, p. 41–51, 1965.

Burek, P. J., Impacts, magnetic field reversals and tectonic episodes, *Lunar and Planetary Sci. Conf. XV Abstracts of Papers*, p. 102–103, Lunar and Planetary Science Institute, Houston, TX, March 12–16, 1984.

Burek, P. J., and H. Wänke, Impacts and glacio-eustasy, plate-tectonic episodes, geomagnetic reversals: A concept to facilitate detection of impact events, *Phys. Earth Planetary Interiors, v. 50*, p. 183–194, 1988.

Burek, P. J., H. Wänke, and J. D. Arneth, K/T environmental changes: Lattengebirge/SE-Germany (Abstract), *Meteoritics, v. 18*, p. 275, 1983.

Burgess, A., and N. Nicola, *Growth Factors and Stem Cells*, New York, Academic Press, 355 pp., 1983.

Burke, J. G., *Cosmic Debris: Meteorites in History*, Berkeley, CA, University of Califonia Press, 445 pp., 1986.

References Cited

Burnet, M., A homeostatic and self-monitoring immune system, in F. M. Burnet, ed., *Immunology: Readings From Scientific American*, p. 158–161, San Francisco, W. H. Freeman & Co., 1976.

Burns, C. A., Timing between a large impact and a geomagnetic reversal and the depth of NRM acquisition in deep-sea sediments, in F. J. Lowes, D. W. Collinson, J. H. Parry, S. K. Runcorn, D. C. Tozer, and A. Soward, eds., *Geomagnetism and Palaeomagnetism*, p. 253–261, Boston, MA, Kluwer Academic Publishers, 1989.

Burns, J. A., The dynamical evolution and origin of the Martian moons, *Vistas in Astronomy*, v. 22, p. 193–210, 1978.

Burns, J. A., and M. S. Matthews, eds., *Satellites*, Tucson, AZ, The University of Arizona Press, 1021 pp., 1986.

Burrows, A., and B. A. Fryxell, An instability in neutron stars at birth, *Science, v. 258*, p. 430–434, 1992.

Busse, F. H., Magnetohydrodynamics of the Earth's dynamo, *Ann. Rev. Fluid Mech., v. 10*, p. 435–462, 1978.

Butler, R. F., *Paleomagnetism: Magnetic Domains to Geologic Terranes*, Boston, MA, Blackwell Scientific Publications, 319 pp., 1992.

Byl, J., and M. W. Ovenden, On the satellite capture problem, *Month. Not. Roy. Astron. Soc. v. 173*, p. 579–584, 1975.

Cadek, O., and Y. Ricard, Toroidal/poloidal energy partitioning and global lithospheric rotation during Cenozoic time, *Earth Planet. Sci. Lett., v. 109*, p. 621–632, 1992.

Caffee, M. W., and J. D. Macdougall, Compaction ages, in J. F. Kerridge and M. S. Matthews, eds., *Meteorites and the Early Solar System*, p. 289–298, Tucson, AZ, The University of Arizona Press, 1988.

Cahalan, R. F., H. Leidecker, and G. D. Cahalan, Chaotic rhythms of a dripping faucet, *Computers in Physics, JUL/AUG*, p. 368–382, 1990.

Cairns, J., Letter: Directed mutation (and response by Lenski and Mittler), *Science, v. 260*, p. 1221–1224, 1993.

Cairns, J., J. Overbaugh, and S. Miller, The origin of mutants, *Nature, v. 335*, p. 142–145, 1988.

Cairns-Smith, A. G., *Seven Clues to the Origin of Life*, New York, Cambridge University Press, 131 pp., 1985.

Caldeira, K., and J. F. Kasting, The life span of the biosphere revisited, *Nature, v. 360*, p. 721–723, 1992.

Calder, W. A., III, *Size, Function, and Life History*, Cambridge, MA, Harvard University Press, 431 pp., 1984.

Cameron, A. G. W., and W. Benz, The origin of the Moon and the single impact hypothesis IV, *Icarus, v. 92*, p. 204–216, 1991.

Cameron, A. G. W., B. Fegley, Jr., W. Benz, and W. L. Slattery, The strange density of Mercury: Theoretical considerations, in F. Vilas, C. R. Chapman, and M. S. Matthews, eds., *Mercury*, p. 692–708, Tucson, AZ, The University of Arizona Press, 1988.

Campbell, D., and H. Rose, *Order in Chaos*, New York, North-Holland, 362 pp., 1983.

Campbell, D. B., N. J. S. Stacy, W. I. Newman, R. E. Arvidson, E. M. Jones, G. S. Musser, A. Y. Roper, and C. Schaller, Magellan observations of extended impact crater related features on the surface of Venus, *Jour. Geophys. Res., v. 97*, p. 16,249–16,277, 1992.

Campbell, D. K., ed., *CHAOS/XAOC: Soviet-American Perspectives on Nonlinear Science*, New York, American Institute of Physics, 496 pp., 1990.

Campbell, I. H., G. K. Czamanske, V. A. Fedorenko, R.I. Hill, and V. Stepanov, Synchronism of the Siberian Traps and the Permian-Triassic boundary, *Science, v. 258*, p. 1760–1763, 1992.

Cande, S. C., and D. V. Kent, A new geomagnetic polarity time scale for the Late Cretaceous and Cenozoic, *Jour. Geophys. Res., v. 97*, p. 13,917–13,951, 1992.

Cannizzo, J. K., and R. H. Kaitchuck, Accretion disks in interacting binary stars, *Sci. Am., v. 266*, p. 92–99, 1992.

Capra, F., *The Tao of Physics: An Exploration of the Parallels Between Modern Physics and Eastern Mysticism*, New York, Bantam Books, 332 pp., 1977.

Carey, S. W., *The Expanding Earth*, Amsterdam, Holland, Elsevier, 488 pp., 1976.

References Cited

———, The necessity for Earth expansion, in S. W. Carey, ed., *The Expanding Earth: A Symposium, Sydney, 1981*, p. 375–393, Hobart, Tasmania, University of Tasmania, 1983.

———, *Theories of the Earth and Universe: A History of Dogma in the Earth Sciences*, Stanford, CA, Stanford University Press, 413 pp., 1988.

Carlisle, D. B., Diamonds at the K/T boundary, *Nature, v. 357*, p. 119–120, 1992.

Carnevale, G. F., Y. Pomeau, and W. R. Young, Statistics of ballistic agglomeration, *Phys. Rev. Lett., v. 64*, p. 2913–2916, 1990.

Carr, M. H., The geology of Mars, in B. J. Skinner, ed., *The Solar System and its Strange Objects: Readings from American Scientist*, p. 100–109, Los Altos, CA, William Kaufmann, Inc., 1981.

Carroll, R. L., The primary radiation of terrestrial vertebrates, *Ann. Rev. Earth Planet. Sci., v. 20*, p. 45–84, 1992.

Carroll, T. L., and L. M. Pecora, A circuit for studying the synchronization of chaotic systems, *Internat. Jour. Bifurcation and Chaos, v. 2*, p. 659–667, 1992.

Carroll, T. L., L. M. Pecora, and F. J. Rachford, Chaotic transients and multiple attractors in spin-wave experiments, *Phys. Rev. Lett., v. 59*, p. 2891–2894, 1987.

Carter, W. E., and D. S. Robertson, Studying the Earth by very-long baseline interferometry, *Sci. Am., v. 254*, p. 46–54, 1986.

Carter, W. E., D. S. Robertson, J. E. Pettey, B. D. Tapley, B. E. Schutz, R. J. Eanes, and M. Lufeng, Variations in the rotation of the Earth, *Science, v. 224*, p. 957–961, 1984.

Carusi, A., and G. B. Valsecchi, Numerical simulations of close encounters between Jupiter and minor bodies, in T. Gehrels, ed., *Asteroids*, p. 391–416, Tucson, AZ, The University of Arizona Press, 1979.

Carusi, A., and G. B. Valsecchi, eds., *Dynamics of Comets: Their Origin and Evolution*, Boston, MA, D. Reidel Publishing Company, 439 pp., 1985.

Casdagli, M., Nonlinear prediction of chaotic time series, *Physica D, v. 35*, p. 335–356, 1989.

Cassen, P., and A. P. Boss, Protostellar collapse, dust grains and solar-system formation, in J. F. Kerridge and M. S. Matthews, eds., *Meteorites and the Early Solar System*, p. 304–328, Tucson, AZ, The University of Arizona Press, 1988.

Cassen, P. M., S. J. Peale, and R. T. Reynolds, Structure and thermal evolution of the Galilean satellites, in D. Morrison, ed., *Satellites of Jupiter*, p. 93–128, Tucson, AZ, The University of Arizona Press, 1982.

Cassidy, W. A., L. M. Villar, T. E. Bunch, T. P. Kohman, and D. J. Milton, Meteorites and craters of Campo del Cielo, Argentina, *Science, v. 149*, p. 1055–1064, 1965.

Cassidy, W., R. Harvey, J. Schutt, G. Delisle, and K. Yanai, The meteorite collection sites of Antarctica, *Meteoritics, v. 27*, p. 490–525, 1992.

Castellan, G. W., *Physical Chemistry, Second Edition*, Menlo Park, CA, Addison-Wesley Publishing Co., 866 pp., 1971.

Casti, J. L., *Searching for Certainty: What Scientists Can Know About the Future*, New York, William Morrow and Co., Inc., 496 pp., 1990.

———, Chaos, Gödel, and Truth, in J. L. Casti and A. Karlqvist, eds., *Beyond Belief: Randomness, Prediction and Explanation in Science*, p. 280–327, Boston, MA, CRC Press, 1991.

Cavalli-Sforza, L. L., A. Piazza, P. Menozzi, and J. Mountain, Reconstruction of human evolution: Bringing together genetic, archaeological, and linguistic data, *Proc. Natl. Acad. Sci., v. 85*, p. 6002–6006, 1988.

Chaitin, G. J., Toward a mathematical definition of life, in R. D. Levine and M. Tribus, eds., *The Maximum Entropy Formalism*, pp. 477–498, Cambridge, MA, MIT Press, 1979.

———, Gödel's theorem and information, *Internat. Jour. Theoretical Phys., v. 22*, p. 941–954, 1982.

———, *Information, Randomness and Incompleteness: Papers on Algorithmic Information Theory*, Singapore, World Scientific, 272 pp., 1987a.

———, *Algorithmic Information Theory*, New York, Cambridge University Press, 175 pp., 1987b.

———, Randomness in arithmetic, *Sci. Am., v. 259*, p. 80–85 (July), 1988.

Challinor, R. A., Variations in the rate of rotation of the Earth, *Science, v. 172*, p. 1022–1025, 1971.

References Cited

Champion, D. E., M. A. Lamphere, and M. A. Kuntz, Evidence for a new geomagnetic reversal from lava flows in Idaho: Discussion of short polarity reversals in the Brunhes and late Matuyama polarity chrons, *Jour. Geophys. Res., v. 93*, p. 11,667–11,680, 1988.

Chandrasekhar, S., *Hydrodynamic and Hydromagnetic Stability*, New York, Oxford University Press (corrected reprinting of 1961 Edition), 654 pp., 1968.

Chant, C. A., An extraordinary meteoric display, *Jour. Roy. Astron. Soc. Canada, v. 7*, p. 145–215, 1913a.

——, Further information regarding the Meteoric display of February 9, 1913, *Jour. Roy. Astron. Soc. Canada, v. 7*, p. 438–447, 1913b.

Chao, B. F., and R. S. Gross, Changes in the Earth's rotation and low-degree gravitational field induced by earthquakes, *Geophys. Jour. Roy. Astron. Soc., v. 91*, p. 569–596, 1987.

Chapman, C. R., Implications of the inferred compositions of asteroids for their collisional evolution, in D. R. Davis, P. Farinella, P. Paolicchi, and V. Zappalà, eds, *Catastrophic Disruption of Asteroids and Satellites* (Internat. Workshop, Pisa, Italy, July 30 – Aug. 2, 1985), p. 103–114, Jour. Italian Astronomical Soc., v. 57, No. 1, 1986.

——, Introduction to an end-member planet, in F. Vilas, C. R. Chapman, and M. S. Matthews, eds., *Mercury*, p. 1–23, Tucson, AZ, The University of Arizona Press, 1988.

——, Rubble-pile parent bodies, asteroids, and satellites (Abstract), *Meteoritics, v. 25*, p. 353–354, 1990.

Chapman, C. R., and D. R. Davis, Asteroid collisional evolution: Evidence for a much larger early population, *Science, v. 190*, p. 553–556, 1975.

Chapman, C. R., and D. Morrison, *Cosmic Catastrophes*, New York, Plenum Press, 302 pp., 1989.

Chapman, C. R., P. Paolicchi, V. Zappalà, R. P. Binzel, and J. F. Bell, Asteroid families: Physical properties and evolution, in R. P. Binzel, T. Gehrels, and M. S. Matthews, eds., *Asteroids II*, p. 386–415, Tucson, AZ, The University of Arizona Press, 1989.

Chapman, S., H. Pongracic, M. Disney, A. Nelson, J. Turner, and A. Whitworth, The formation of binary and multiple star systems, *Nature, v. 359*, p. 207–210, 1992.

Chen, C. H., and L.-M. Chang, Flow patterns of a circular vortex ring with density differences under gravity, *Jour. Appl. Mech., v. 39*, p. 869–872, 1972.

Chen, K., and P. Bak, Is the universe operating at a self-organized critical state?, *Phys. Lett., A, v. 140*, p. 299–302, 1989.

Chen, P., Empirical and theoretical evidence of economic chaos, *System Dynamics Review, v. 4*, p. 81–108, 1988.

Childress, S., An introduction to dynamo theory, in M. Ghil, R. Benzi, and G. Parisi, eds., *Turbulence and Predictability in Geophysical Fluid Dynamics and Climate Dynamics*, p. 200–225, New York, North-Holland, 1985.

Chillingworth, D. R. J., and P. J. Holmes, Dynamical systems and models for reversals of the Earth's magnetic field, *Math. Geol., v. 12*, p. 41–59, 1980.

Chirikov, B. V., *A Universal Instability of Many-Dimensional Oscillator Systems*, Physics Reports (Review Section of Physics Letters), v. 52, p. 263–379, Amsterdam, North-Holland Publ. Co., 1979.

——, Patterns in chaos, *Chaos, Solitons and Fractals, v. 1*, p. 79–103, 1991.

Chouet, B., Ground motion in the near field of a fluid-driven crack and its interpretation in the study of shallow volcanic tremor, *Jour. Geophys. Res., v. 86*, p. 5985–6016, 1981.

——, Resonance of a fluid-driven crack: Radiation properties and implications for the source of long-period events and harmonic tremor, *Jour. Geophys. Res., v. 93*, p. 4375–4400, 1988.

——, A seismic model for the source of long-period events and harmonic tremor, in R. W. Johnson, G. Mahood, and R. Scarpa, eds., *IAVCEI Proceedings in Volcanology, Volume 3*, (P. Gasparini, R. Scarpa, and K. Aki, eds., Volcanic Seismology), p. 133–156, New York, Springer-Verlag, 1992.

Chouet, B., and H. R. Shaw, Fractal properties of tremor and gas piston events observed at Kilauea Volcano, Hawaii, *Jour. Geophys. Res., v. 96*, p. 10,177–10,189, 1991.

Chouet, B., R. Y. Koyanagi, and K. Aki, The origin of volcanic tremor in Hawaii, II, Theory and discussion, in R. W. Decker, T. L. Wright, and P. H. Stauffer, eds., *Volcanism in Hawaii*, Vol. 2,

References Cited

Chap. 45, p. 1259–1280, U.S. Geol. Survey, Prof. Paper 1350, Washington, DC, U.S. Gov't Printing Office, 1987.

Chyba, C. F., Terrestrial mantle siderophiles and the lunar impact record, *Icarus, v. 92*, p. 217–233, 1991.

Chyba, C. F., and C. Sagan, Endogenous production, exogenous delivery, and impact-shock synthesis of organic molecules: An inventory for the origins of life, *Nature, v. 355*, p. 125–132, 1992.

Chyba, C. F., P. J. Thomas, and K. J. Zahnle, The Tunguska explosion: Atmospheric disruption of a stony asteroid, *Nature, v. 361*, p. 40–44, 1993.

Claeys, P., J.-G. Casier, and S. V. Margolis, Microtektites and mass extinctions: Evidence for a late Devonian asteroid impact, *Science, v. 257*, p. 1102–1104, 1992a.

Claeys, P., J.-G. Casier, and S. V. Margolis, A link between microtektites and late Devonian mass extinctions (Abstract), *Eos, Trans. Am. Geophys. Union, v. 73* (Supplement, 27 Oct.), p. 328, 1992b.

Clarke, C., Stars arriving two by two, *Nature, v. 357*, p. 197–198, 1992.

Clemens, W. A., Patterns of extinction and survival of the terrestrial biota during the Cretaceous/Tertiary transition, in L. T. Silver and P. H. Schultz, eds., *Geological Implications of Impacts of Large Asteroids and Comets on the Earth*, p. 407–413, Boulder, CO, The Geological Society of America, Special Paper 190, 1982.

Clement, B. M., Paleomagnetic evidence of reversals resulting from helicity fluctuations in a turbulent core, *Jour. Geophys. Res., v. 92*, p. 10,629–10,638, 1987.

———, Geographical distribution of transitional VGPs: Evidence for non-zonal equatorial symmetry during the Matuyama-Brunhes geomagnetic reversal, *Earth Planet. Sci. Lett., v. 104*, p. 48–58, 1991.

———, Evidence for dipolar fields during the Cobb Mountain geomagnetic polarity reversals, *Nature, v. 358*, p. 405–407, 1992.

Clube, S. V. M., The kinematics of Gould's Belt, *Month. Not. Roy. Astron. Soc., v. 137*, p. 189–203, 1967.

———, Does our Galaxy have a violent history?, *Vistas in Astronomy, v. 22*, p. 77–118, 1978.

———, On episodic catastrophism, in D. A. Russell and G. Rice, eds, *K-TEC II: Cretaceous-Tertiary Extinctions and Possible Terrestrial and Extraterrestrial Causes*, p. 72–76, 136–137, Syllogeus No. 39 (Proceedings of workshop, Ottawa, 19–20 May, 1981), Ottawa, National Museums of Canada, 1982.

———, Molecular clouds: Comet factories?, in A. Carusi and G. B. Valsecchi, eds., *Dynamics of Comets: Their Origin and Evolution*, p. 19–30, Boston, MA, D. Reidel Publishing Company, 1985.

———, ed., *Catastrophes and Evolution: Astronomical Foundations*, New York, Cambridge University Press, 239 pp., 1989a

———, The catastrophic role of giant comets, in S. V. M. Clube, ed., *Catastrophes and Evolution: Astronomical Foundations*, p. 81–112, New York, Cambridge University Press, 1989b.

———, The role of giant comets in Earth History, *Celest. Mech. and Dynamical Astron., v. 54*, p. 179–193, 1992.

Clube, S. V. M., and W. M. Napier, Spiral arms, comets and terrestrial catastrophism, *Quart. Jour. Roy. Astron. Soc., v 23*, p. 45–66, 1982a.

———, *The Cosmic Serpent: A Catastrophist View of Earth History*, New York, Universe Books, 299 pp. 1982b.

———, Comet capture from molecular clouds: A dynamical constraint on star and planet formation, *Month. Not. Roy. Astron. Soc., v. 208*, p. 575–588, 1984a.

———, The microstructure of terrestrial catastrophism, *Month. Not. Roy. Astron. Soc., v. 211*, p. 953–968, 1984b.

———, Giant comets and the galaxy: Implications of the terrestrial record, in R. Smoluchowski, J. N. Bahcall, and M. S. Matthews, eds., *The Galaxy and the Solar System*, p. 260–285, Tucson, AZ, The University of Arizona Press, 1986.

———, *The Cosmic Winter*, Cambridge, MA, Basil Blackwell, Inc., 307 pp., 1990.

References Cited

Coe, R. S., and M. Prévot, Evidence suggesting extremely rapid field variation during a geomagnetic reversal, *Earth Planet. Sci. Lett., v. 92*, p. 292–298, 1989.

Coffey, S. L., A. Deprit, and B. R. Miller, The critical inclination in artificial satellite theory, *Celest. Mech., v. 39*, p. 365–406, 1986.

Coffey, S. L., A. Deprit, E. Deprit, and L. Healy, Painting the phase space portrait of an integrable dynamical system, *Science, v. 247*, p. 833–836, 1990.

Coffey, S. L., A. Deprit, and E. Deprit, Painting phase spaces to put frozen orbits in context, *Paper AAS 91-427*, 27 pp., AAS/AIAA Astrodynamics Specialist Conference, Durango, CO, 19–22 August, 1991.

Cohen, R. J., Celestial beams, *Nature, v. 357*, p. 204–205, 1992.

Colbert, E. H., Mesozoic tetrapod extinctions: A review, in D. K. Elliott, ed., *Dynamics of Extinction*, p. 49–62, New York, John Wiley & Sons, Inc., 1986.

Cole, H. S. D., C. Freeman, M. Jahoda, and K. L. R. Pavitt, *Models of Doom*, New York, Universe Books, 244 pp., 1973.

Colwell, J. E., and L. W. Esposito, Origins of the rings of Uranus and Neptune: 2. Initial conditions and ring moon populations, *Jour. Geophys. Res., v. 98*, p. 7387–7401, 1993.

Coniglio, A., An infinite hierarchy of exponents to describe growth phenomena, in L. Pietronero and E. Tosatti, eds, *Fractals in Physics*, p. 165–168, New York, North-Holland Physics Publishing, a division of Elsevier Science Publishers, 1986.

Considine, D. M., ed., *Scientific Encyclopedia, Sixth Edition*, New York, Van Nostrand Reinhold Co., 3067 pp., 1983.

Constable, C., Link between geomagnetic reversal paths and secular variation of the field over the past 5 Myr, *Nature, v. 358*, p. 230–233, 1992.

Cook, G. E., and D. W. Scott, Lifetimes of satellites in large-eccentric orbits, *Planet. Space Sci., v. 15*, p. 1549–1556, 1967.

Corliss, B. H., M.-P. Aubry, W. A. Berggren, J. M. Fenner, L. D. Keigwin, Jr., and G. Keller, The Eocene/Oligocene boundary event in the deep sea, *Science, v. 226*, p. 806–810, 1984.

Courtillot, V. E., Debate: What caused the mass extinction? A volcanic eruption, *Sci. Am., v. 263*, p. 85–92, 1990.

Courtillot, V., and J. Besse, Magnetic field reversals, polar wander, and core-mantle coupling, *Science, v. 237*, p. 1140–1147, 1987.

Courtillot, V., and J.-L. Le Mouël, Geomagnetic secular variation impulses, *Nature, v. 311*, p. 709–716, 1984.

———, Comment on "A discussion of impulses and jerks in the geomagnetic field" by L. R. Alldredge, *Jour. Geophys. Res., v. 90*, p. 6897–6898, 1985.

———, Time variations of the Earth's magnetic field: From daily to secular, *Ann. Rev. Earth Planet. Sci., v. 16*, p. 389–476, 1988.

Courtillot, V., J.-P. Valet, G. Hulot, and J.-L. Le Mouël, The Earth's magnetic field: Which geometry?, *Eos, Trans. Am. Geophys. Union, v. 73*, p. 337–342, 1992.

Cowan, G. A., A natural fission reactor, *Sci. Am., v. 235*, p. 36–47, 1976.

Cowan, W. M., Neuronal death as a regulative mechanism in the control of cell number in the nervous system, in M. Rockstein and M. L. Sussman, eds., *Development and Aging in the Nervous System*, p. 19–41, New York, Academic Press, Inc., 1973.

Cowen, R., Distant object hints at the Kuiper Belt, *Science News, v. 142*, p. 196 (26 September), 1992.

Cox, A., Lengths of geomagnetic polarity intervals, *J. Geophys. Res., v. 73*, p. 3247–3260, 1968.

———, Geomagnetic reversals, *Science, v. 163*, p. 237–245, 1969.

———, A stochastic approach towards understanding the frequency and polarity bias of geomagnetic reversals, *Phys. Earth Planet. Interiors, v. 24*, p. 178–190, 1981.

Cox, A., and R. B. Hart, *Plate Tectonics: How it Works*, Palo Alto, CA, Blackwell Scientific Publications, Inc., 392 pp, 1986.

Cox, A., R. R. Doell, and G. B. Dalrymple, Reversals of the Earth's magnetic field, *Science, v. 144*, p. 1537–1543, 1964.

Cox, A. N., W. C. Livingston, and M. S. Matthews, eds., *Solar Interior and Atmosphere*, Tucson, AZ, The University of Arizona Press, 1416 pp., 1991.

References Cited

Cox, K. G., A superplume in the mantle, *Nature, v. 352*, p. 564–565, 1991.
Creer, K. M., and P. C. Pal, On the frequency of reversals of the geomagnetic dipole, in S. V. M. Clube, ed., *Catastrophes and Evolution: Astronomical Foundations*, p. 113–132, New York, Cambridge University Press, 1989.
Creveling, H. F., J. F. De Paz, J. Y. Baladi, and R. J. Schoenhals, Stability characteristics of a single-phase free convection loop, *Jour. Fl. Mech., v. 67*, p. 65–84, 1975.
Crichton, M., *Jurassic Park*, New York, Ballantine Books, 399 pp., 1990.
Crick, F. H. C., The Genetic code: III, in C. I. Davern, ed., *Genetics (Readings from the Scientific American)*, p. 151–157, San Francisco, W. H. Freeman & Co., 1981.
Crick, F. H. C., and L. E. Orgel, Directed panspermia, *Icarus, v. 19*, p. 341–346, 1973.
Crisp, J. A., Rates of magma emplacement and volcanic output, *Jour. Volc. Geoth. Res., v. 20*, p. 177–211, 1984.
Croquette, V., and C. Poitou, Cascade of period doubling bifurcations and large stochasticity in the motions of a compass, *Jour. Physique — Lettres, v. 42*, p. L537–L539, 1981.
Crough, S. T., and D. M. Jurdy, Subducted lithosphere, hotspots, and the geoid, *Earth Planet. Sci. Lett., v. 48*, p. 15–22, 1980.
Crowell, J. C., Gondwanan glaciation, cyclothems, continental positioning, and climate change, *Am. Jour. Sci., v. 278*, p. 1345–1372, 1978.
Crowley, T. J., and S. K. Baum, Estimating Carboniferous sea-level fluctuations from Gondwanan ice extent, *Geology, v. 19*, p. 975–977, 1991.
Crowley, T. J., and G. R. North, Abrupt climate change and extinction events in Earth history, *Science, v. 240*, p. 996–1002, 1988.
Cruikshank, D. P., J. Degewij, and B. H. Zellner, The outer satellites of Jupiter, in D. Morrison, ed., *Satellites of Jupiter*, p. 129–146, Tucson, AZ, The University of Arizona Press, 1982.
Crutchfield, J. P., and K. Kaneko, Phenomenology of spatio-temporal chaos, in Hao Bai-lin, *Directions in Chaos, Volume 1*, p. 272–353, Singapore, World Scientific Publ. Co., 1987.
Crutchfield, J. P., J. D. Farmer, N. H. Packard, and R. S. Shaw, Chaos, *Sci. Am., v. 254*, p. 46–57, 1986.
Cvitanovic, P., ed., *Universality in Chaos*, Bristol, U.K., Adam Hilger Ltd., 513 pp., 1984.
Czamanske, G. K., V. A. Fedorenko, B. G. Dalrymple, I. H. Campbell, J. L. Wooden, and N. T. Arndt, The Siberian Traps: A flip of the dynamo and 600,000 years of Hell on Earth (Abstract), *Am. Geophys. Union, Abstracts of Meetings (Fall, 1992) Preprint*, 1992.
Dachille, F., Great meteorite impacts and global geological responses, in S. W. Carey, ed., *The Expanding Earth: A Symposium, Sydney, 1981*, p. 267–276, Hobart, Tasmania, University of Tasmania, 1983.
Dalrymple, G. B., *The Age of the Earth*, Stanford, CA, Stanford University Press, 474 pp., 1991.
Dalrymple, G. B., E. A. Silver, and E. D. Jackson, Origin of the Hawaiian Islands, *Am. Scientist, v. 61*, p. 294–308, 1973.
Daly, R. A., Origin of the Moon and its topography, *Proc. Am. Philos. Soc., v. 90*, p. 104–119, 1946.
Dalziel, I. W. D., Pacific margins of Laurentia and East Antarctica-Australia as a conjugate rift pair: Evidence and implications for an Eocambrian supercontinent, *Geology, v. 19*, p. 598–601, 1991.
———, On the organization of American plates in the Neoproterozoic and the breakout of Laurentia, *GSA Today, v. 2*, p. 237–241 (November), 1992a.
———, Antarctica: A tale of two supercontinents?, *Ann. Rev. Earth Planet. Sci., v. 20*, p. 501–526, 1992b.
Damon, P. E., The relationship between late Cenozoic volcanism and tectonism and orogenic-epeirogenic periodicity, in K. K. Turekian, ed., *The Late Cenozoic Glacial Ages*, p. 15–33, New Haven, CT, Yale University Press, 1971.
———, Solar induced variations of energetic particles at one AU, in O. R. White, ed., *The Solar Output and Its Variation*, p. 429–448, Boulder, CO, Colorado Associated University Press, 1977.
Damon, P. E., and R. L. Mauger, Epeirogeny-orogeny reviewed from the Basin and Range province, *Soc. Mining Eng. (AIME), Trans., v. 235*, p. 99–112, 1966.

References Cited

Danielsson, L., The profile of a jet stream, in T. Gehrels, ed., *Physical Studies of Minor Planets*, p. 353–362, Washington, DC, National Aeronautics and Space Administration, NASA SP-267, 1971.

Darwin, C. *On the Origin of Species by Means of Natural Selection, or, The Preservation of Favored Races in the Struggle for Life*, London, John Murray, 502 pp., 1859.

Darwin, G. H., On the precession of a viscous spheroid and on the remote history of the earth, *Phil. Trans. Roy. Soc., Part II*, v. *170*, p. 447–530, 1879.

———, *Scientific Papers*, Cambridge, MA, Cambridge Univ. Press, 5 Volumes, 1907.

———, *The Tides*, San Francisco, W. H. Freeman & Co., 379 pp., 1962.

Dauben, J. W., Georg Cantor and the origins of transfinite set theory, *Sci. Am.*, v. *248*, p. 122–131, and 134 (June), 1983.

Davies, M. E., Appendix: Cartography and nomenclature for the Galilean satellites, in D. Morrison, ed., *Satellites of Jupiter*, p. 911–933, Tucson, AZ, The University of Arizona Press, 1982.

Davies, M. E., V. K. Abalakin, M. Bursa, G. E. Hunt, J. H. Lieske, B. Morando, R. H. Rapp, P. K. Seidelmann, A. T. Sinclair, and Y. S. Tjuflin, Report of the IAU/IAG/COSPAR Working Group on cartographic coordinates and rotational elements of the planets and satellites: 1988, *Celest. Mech. and Dynamical Astron.*, v. *46*, p. 187–204, 1989.

Davies, P., *The Mind of God: The Scientific Basis for a Rational World*, New York, Simon & Schuster, 254 pp., 1992.

Davies, P. C. W., and J. Brown, eds., *Superstrings: A Theory of Everything?*, New York, Cambridge University Press, 234 pp., 1988.

Davis, D. R., K. R. Housen, and R. Greenberg, The unusual dynamical environment of Phobos and Deimos, *Icarus*, v. *47*, p. 220–233, 1981.

Davis, D. R., P. Farinella, P. Paolicchi, and V. Zappalà, eds., *Catastrophic Disruption of Asteroids and Satellites* (Internat. Workshop, Pisa, Italy, July 30–Aug. 2, 1985), Jour. Italian Astronomical Soc., v. 57, No. 1, 114 pp., 1986.

Davis, D. R., S. J. Weidenschilling, P. Farinella, P. Paolicchi, and R. P. Binzel, Asteroid collisional history: Effects on sizes and spins, in R. P. Binzel, T. Gehrels, and M. S. Matthews, eds., *Asteroids II*, p. 805–826, Tucson, AZ, The University of Arizona Press, 1989.

Davis, J. L., W. H. Prescott, J. L. Svarc, and K. J. Wendt, Assessment of global positioning system measurements for studies of crustal deformation, *Jour. Geophys. Res.*, v. *94*, p. 13,635–13,650, 1989.

Davis, M., ed., *The Undecidable: Basic Papers on Undecidable Propositions, Unsolvable Problems, and Computable Functions*, Hewlett, NY, Raven Press, 438 pp., 1965.

Dean, D. R., *James Hutton and the History of Geology*, Ithaca, NY, Cornell University Press, 303 pp., 1992.

Decker, R. W., T. L. Wright, and P. H. Stauffer, eds., *Volcanism in Hawaii (Volumes 1 and 2)*, U.S. Geol. Survey, Prof. Paper 1350, Washington, DC, U.S. Gov't Printing Office, 1667 pp., 1987.

Deino, A. L., J. B. Garvin, and S. Montanari, K/T age for the Popigai impact event, *Lunar and Planetary Sci. Conf. XXII, Abstracts of Papers*, p. 297, Lunar and Planetary Science Institute, Houston, TX, March 18–22, 1991.

Deissler, R. J., Spatially growing waves, intermittency, and convective chaos in an open-flow system, *Physica D*, v. *25*, p. 223–260, 1987.

DeMets, C., Earthquake slip vectors and estimates of present-day plate motions, *Jour. Geophys. Res.*, v. *98*, p. 6703–6714, 1993.

DeMets, C., R. G. Gordon, D. F. Argus, and S. Stein, Current plate motions, *Geophys. Jour. International*, v. *101*, p. 425–478, 1990.

De Renzi, M., What happens after extinctions, in M. A. Lamolda, E. G. Kauffman, and O. H. Walliser, eds., *Palaeontology and Evolution: Extinction Events. 2nd International Conference on Global Bioevents (Bilbao Conf., 20–23 October, 1987)*, Revista Española de Palaeontologia, N°. Extraordinario, p. 107–112, Madrid, Sociedad Española de Paleontologia, 1988.

Dermott, S. F., The Mimas-Tethys resonance formation problem, *Month. Not. Roy. Astron. Soc.*, v. *153*, p. 83–96, 1971.

References Cited

Derry, D. R., *World Atlas of Geology and Mineral Deposits*, London, Mining Journal Books, 110 pp. 1980.
Des Marais, D. J., H. Strauss, R. E. Summons, and J. M. Hayes, Carbon isotope evidence for the stepwise oxidation of the Proterozoic environment, *Nature, v. 359*, p. 605–609, 1992.
De Sousa Vieira, M., P. Khoury, A. J. Lichtenberg, M. A. Lieberman, W. Wonchoba, J. Gullicksen, J. Y. Huang, R. Sherman, and M. Steinberg, Numerical and experimental studies of self-synchronization and synchronized chaos, *Internat. Jour. Bifurcation and Chaos, v. 2*, p. 645–657, 1992.
Deutsch, D., The tritone paradox: An influence of language on music perception, *Music Perception, v. 8*, p. 335–347 (Summer), 1991.
——, Some new pitch paradoxes and their implications, *Phil. Trans. Roy. Soc. London, B, v. 336*, p. 391–397, 1992a.
——, Paradoxes of musical pitch, *Sci. Am., v. 267*, p. 88–95 (August), 1992b.
Devaney, R. L., An *Introduction to Chaotic Dynamical Systems*, Menlo Park, CA, The Benjamin/Cummings Publ. Co. Inc., 320 pp., 1986.
——, Chaotic bursts in nonlinear dynamical systems, *Science, v. 235*, p. 342–345, 1987.
——, *Transition to Chaos: The Orbit Diagram and the Mandelbrot Set*, New York, Science Television, Am. Inst. Phys., Videotape (65 min.), 1989.
——, e^z: Dynamics and bifurcations, *Internat. Jour. Bifurcation and Chaos, v. 1*, p. 287–308, 1991.
de Vaucouleurs, G., The case for a hierarchical cosmology, *Science, v. 167*, p. 1203–1213, 1970.
Dewdney, A. K., Computer recreations: Diverse personalities search for social equilibrium at a computer party, *Sci. Am., v. 257*, p. 112–115, 1987.
de Wit, M. J., C. Roering, R. J. Hart, R. A. Armstrong, C. E. J. de Ronde, R. W. E. Green, M. Tredoux, E. Peberdy, and R. A. Hart, Formation of an Archean continent, *Nature, v. 357*, p. 553–562, 1992.
Diamond, J. M., Linguistics: Genes and the tower of Babel, *Nature, v. 336*, p. 622–623, 1988.
Dickey, J. O., Angular momentum exchange between atmosphere and Earth, *EOS, Trans. Am. Geophys. Union, v. 74*, pp. 17 and 22, 1993.
Dickman, S. R., Preserving a sense of direction, *Nature, v. 360*, p. 421–422, 1992.
Dietz, R. S., Astroblemes, *Sci. Am., v. 205*, p. 50–58, 1961.
——, Astroblemes: Ancient meteorite-impact structures on the Earth, in B. M. Middlehurst and G. P. Kuiper, eds., *The Moon, Meteorites, and Comets*, p. 285–300, Chicago, The University of Chicago Press, 1963.
Dietz, R. S., and J. C. Holden, Reconstruction of Pangaea: Breakup and dispersion of continents, Permian to present, *Jour. Geophys. Res., v., 75*, p. 4939–4956, 1970.
Ding, E. J., and P. C. Hemmer, Winding numbers for the supercritical sine circle map, *Physica D, v. 32*, p. 153–160, 1988.
Dirac, P. A. M., The cosmological constants, *Nature, v. 139*, p. 323, 1937.
Ditto, W. L., S. Rauseo, R. Cawley, C. Grebogi, G-H. Hsu, E. Kostelich, E. Ott, H. T. Savage, R. Segnan, M. L. Spano, and J. A. Yorke, Experimental demonstration of crisis-induced intermittency and its critical exponent, *Phy. Rev. Lett., v. 63*, p. 923–926, 1989.
Ditto, W. L., M. L. Spano, H. T. Savage, S. N. Rauseo, J. Heagy, and E. Ott, Experimental observation of a strange nonchaotic attractor, *Phys. Rev. Lett., v. 65*, p. 533–536, 1990a.
Ditto, W. L., S. N. Rauseo, and M. L. Spano, Experimental control of chaos, *Phys. Rev. Lett., v. 65*, p. 3211–3214, 1990b.
Dixon, T. H., An introduction to the global positioning system and some geological implications, *Rev. Geophys., v. 29*, p. 249–276, 1991.
Doehne, E., and S. V. Margolis, Trace-element geochemistry and mineralogy of the Cretaceous/Tertiary boundary: Identification of extraterrestrial components, in V. L. Sharpton and P. D. Ward, eds., *Global Catastrophes in Earth History; An Interdisciplinary Conference on Impacts, Volcanism, and Mass Mortality*, p. 367–382, Boulder, CO, The Geological Society of America, Special Paper 247, 1990.
Doell, R. R., and G. B. Dalrymple, Geomagnetic polarity epochs—a new polarity event and the age of the Brunhes-Matuyama boundary, *Science, v. 152*, p. 1060–1061, 1966.

References Cited

Dohnanyi, J. S., Fragmentation and distribution of asteroids, in T. Gehrels, ed., *Physical Studies of Minor Planets*, p. 263–295, Washington, DC, National Aeronautics and Space Administration, NASA SP-267, 1971.

Dollfus, A., M. Wolff, J. E. Geake, D. F. Lupishko, and L. M. Dougherty, Photopolarimetry of asteroids, in R. P. Binzel, T. Gehrels, and M. S. Matthews, eds., *Asteroids II*, p. 594–616, Tucson, AZ, The University of Arizona Press, 1989.

Dones, L., A recent cometary origin for Saturn's rings?, *Icarus, v. 92*, p. 194–203, 1991.

Dones, L., and S. Tremaine, Why does the Earth spin forward? *Science, v. 259*, p. 350–354, 1993.

Donnison, J. R., and R. A. Sugden, The distribution of asteroid diameters, *Month. Not. Roy. Astron. Soc., v. 210*, p. 673–682, 1984.

Dormand, J. R., and M. M. Woolfson, *The Origin of the Solar System: The Capture Theory*, New York, Halsted Press, Div. of John Wiley & Sons, Inc., 230 pp., 1989.

Dowell, E. H., and C. Pezeshki, On the understanding of chaos in Duffings equation including a comparison with experiment, *Jour. Appl. Mech. (Trans. ASME), v. 53*, p. 5–9 (Mar.), 1986.

Drake, M. J., Experiment confronts theory, *Nature, v. 347*, p. 128, 1990.

Drummond, J. D., Earth-orbit-approaching comets and their theoretical meteor radiants, *Icarus, v. 47*, p. 500–517, 1981.

——, Earth-approaching asteroid streams, *Icarus. v. 89*, p. 14–25, 1991.

Duba, A., Earth's core not so hot, *Nature, v. 359*, p. 197–198, 1992.

Dubyago, A. D., *The Determination of Orbits* (Translated from Russian by R. D. Burke, G. Gordon, L. N. Rowell, and F. T. Smith, The Rand Corp.), New York, The Macmillan Co., 431 pp., 1961.

Duffield, W. A., A naturally occurring model of global plate tectonics, *Jour. Geophys. Res., v. 77*, p. 2543–2555, 1972.

Duncan, M., T. Quinn, and S. Tremaine, The formation and extent of the Solar System comet cloud, *Astronomical Jour., v. 94*, p. 1330–1338, 1987.

——, The origin of short-period comets, *Astrophys. Jour., v. 328*, p. L69–L73, 1988.

Duncan, R. A., Ocean drilling and the volcanic record of hotspots, *GSA Today, v. 1*, p. 213–219 (October), 1991.

Duncan, R. A., and R. B. Hargraves, Plate tectonic evolution of the Caribbean region in the mantle reference frame, *Geol. Soc. Am., Memoir 162*, p. 81–93, 1984.

Duncan, R. A., and M. A. Richards, Hotspots, mantle plumes, flood basalts, and true polar wander, *Rev. Geophys., v. 29*, p. 31–50, 1991.

Duncombe, R. L., ed., *Dynamics of the Solar System*, Boston, MA, D. Reidel Publishing Company, 330 pp., 1979.

Dziewonski, A. M., and J. H. Woodhouse, Global images of the Earth's interior, *Science, v. 236*, p. 37–48, 1987.

Dzurisin, D., Influence of fortnightly tides at Kilauea Volcano, Hawaii, *Geophys. Res. Lett., v. 7*, p. 925–928, 1980.

Dzurisin, D., R. Y. Koyanagi, and T. T. English, Magma supply and storage at Kilauea Volcano, Hawaii, 1956–1983, *Jour. Volc. Geoth. Res., v. 21*, p. 177–206, 1984.

Eckhardt, B., Irregular scattering, *Physica D, v. 33*, p. 89–98, 1988.

Eddy, J. A., The Maunder Minimum, *Science, v. 192*, p. 1189–1202, 1976.

——, Historical evidence for the existence of the solar cycle, in O. R. White, ed., *The Solar Output and Its Variation*, p. 51–71, Boulder, CO, Colorado Associated University Press, 1977.

Edelman, G. M., Group selection and phasic reentrant signaling: A theory of higher brain function, in G. M. Edelman and V. B. Mountcastle, *The Mindful Brain*, p. 51–100, Cambridge, MA, M.I.T. Press, 1978.

——, Neural Darwinism: Population thinking and higher brain function, in Shafto, M., ed., *How We Know* (Nobel Conference XX), p. 1–30, San Francisco, Harper and Row, Publishers, 1985.

——, *Neural Darwinism: The Theory of Neuronal Group Selection*, New York, Basic Books Inc., Publishers, 371 pp., 1987.

——, *The Remembered Present: A Biological Theory of Consciousness*, New York, Basic Books Inc., Publishers, 346 pp., 1989.

References Cited

———, *Bright Air, Brilliant Fire: On the Matter of the Mind*, New York, Basic Books/Allen Lane, 280 pp., 1992.
Edwards, R. L., J. W. Beck, G. S. Burr, D. J. Donahue, J. M. A. Chappell, A. L. Bloom, E. R. M. Druffel, and F. W. Taylor, A large drop in atmospheric $^{14}C/^{12}C$ and reduced melting in the Younger Dryas, documented with ^{230}Th ages of corals, *Science, v. 260*, p. 962–968, 1993.
Eichler, D., and J. Silk, High-velocity pulsars in the Galactic halo, *Science, v. 257*, p. 937–942, 1992.
Eigen, M., *Steps Toward Life, A Perspective on Evolution*, New York, Oxford University Press, 173 pp., 1992.
Eigen, M., P. Schuster, and R. Winkler-Oswatitsch, The origin of genetic information, *Sci. Am., v. 244*, p. 88–118 (April), 1981.
Eldredge, N., *Macroevolutionary Dynamics*, New York, McGraw-Hill, 226 pp., 1989.
Eldredge, N., and S. J. Gould, Punctuated equilibria: An alternative to phyletic gradualism, in T. J. M. Schopf, ed., *Models in Paleobiology*, p. 82–115, San Francisco, Freeman, Cooper and Co., 1972.
Elitzur, M., *Astronomical Masers*, Boston, MA, Kluwer Academic Publishers, 351 pp., 1992.
Elliott, C. G., and D. R. Gray, Correlations between Tasmania and the Tasman-Transantarctic orogen: Evidence for easterly derivation of Tasmania relative to mainland Australia, *Geology, v. 20*, p. 621–624, 1992.
Elliott, D. K., ed., *Dynamics of Extinction*, New York, John Wiley & Sons, Inc., 294 pp., 1986.
Engebretson, D. C., K. P. Kelley, H. J. Cashman, and M. A. Richards, 180 million years of subduction, *GSA Today, v. 2*, p. 93–100, 1992.
Ernst, W. G., Metamorphic terranes, isotopic provinces, and implications for crustal growth of the western United States, *Jour. Geophys. Res., v. 93*, p. 7634–7642, 1988.
Erwin, D. H., and T. A. Vogel, Testing for causal relationships between large pyroclastic volcanic eruptions and mass extinctions, *Geophys. Res. Lett., v. 19*, p. 893–896, 1992.
Esposito, L. W., Ever decreasing circles, *Nature, v. 354*, p. 107, 1991.
———, Running rings around modellers, *Nature, v. 360*, p. 531–532, 1992.
Esposito, L. W., and J. E. Colwell, Creation of the Uranus rings and dust bands, *Nature, v. 339*, p. 605–607, 1989.
Essex, C., T. Lookman, and M. A. H. Nerenberg, The climate attractor over short time scales, *Nature, v. 326*, p. 64–66, 1987.
Everhart, E., Chaotic orbits in the Solar System, in T. Gehrels, ed., *Asteroids*, p. 283–288, Tucson, AZ, The University of Arizona Press, 1979.
Evernden, J. F., and R. W. Kistler, Chronology of emplacement of Mesozoic batholithic complexes in California and western Nevada, *U. S. Geol. Survey, Prof. Pap. 623*, 42 pp., 1970.
Faraday, M., *The Chemical History of a Candle* (edited by W. R. Fielding), New York, E. P. Dutton and Co., 158 pp., 1920.
Farley, K. A., and H. Craig, Mantle plumes and mantle sources, *Science, v. 258*, p. 821, 1992.
Farley, K. A., and R. J. Poreda, Mantle neon and atmospheric contamination, *Earth Planet. Sci. Lett., v. 114*, p. 325–339, 1993.
Farmer, D., J. Crutchfield, H. Froehling, N. Packard, and R. Shaw, Power spectra and mixing properties of strange attractors, in R. H. G. Helleman, ed., *Nonlinear Dynamics*, p. 453–472, New York, Annals of the New York Acad. Sciences, v. 357, 1980.
Farmer, J. D., and J. J. Sidorowich, *Exploiting Chaos to Predict the Future and Reduce Noise*, Los Alamos, NM, Los Alamos National Laboratory Report LA-UR-88-901, 1988.
Farr, C. J., and P. N. Goodfellow, Hidden messages in genetic maps, *Science, v. 258*, p. 49, 1992.
Fastovsky, D. E., Sedimentology, stratigraphy, and extinctions during the Cretaceous-Paleogene transition at Bug Creek, Montana, *Geology, v. 14*, p. 279–282, 1986.
———, Rocks, resolution, and the record; a review of depositional constraints on fossil vertebrate assemblages at the terrestrial Cretaceous/Paleogene boundary, eastern Montana and western North Dakota, in V. L. Sharpton and P. D. Ward, eds., *Global Catastrophes in Earth History; An Interdisciplinary Conference on Impacts, Volcanism, and Mass Mortality*, p. 541–548, Boulder, CO, The Geological Society of America, Special Paper 247, 1990.
Feder, J., *Fractals*, New York, Plenum Press, 283 pp., 1988.

References Cited

Feigenbaum, M. J., The universal metric properties of nonlinear transformations, *Jour. Stat. Phys.*, v. 21, p. 669–706, 1979a.

———, The onset spectrum of turbulence, *Phys. Lett.*, v. 74A, p. 375–378, 1979b.

———, Universal behavior in nonlinear systems, *Los Alamos Science*, v. 1, p. 4–27, 1980.

Fein, A. P., M. S. Heutmaker, and J. P. Gollub, Scaling at the transition from quasiperiodicity to chaos, *Physica Scripta*, v. T9, p. 79–84, 1985.

Feit, S. D., Characteristic exponents and strange attractors, *Commun. Math. Phys.*, v. 61, p. 249–260, 1978.

Fernández, J. A., and W.-H. Ip, Statistical and evolutionary aspects of cometary orbits, in R. L. Newburn, Jr., M. Neugebauer, and J. Rahe, eds., *Comets in the Post-Halley Era, Volume 1*, p. 487–535, Boston, MA, Kluwer Academic Publishers, 1991.

Feudel, U., W. Jansen, and J. Kurths, Tori and chaos in a nonlinear dynamo model for solar activity, *Internat. Jour. Bifurcation and Chaos*, v. 3, p. 131–138, 1993.

Fink, J. H., R. Greeley, and D. E. Gault, Impact cratering experiments in Bingham materials and the morphology of craters on Mars and Ganymede, *Proc. Lunar Planet. Sci.*, v. 12B, p. 1649–1666, 1981.

Fink, J. H., D. Gault, and R. Greeley, The effect of viscosity on impact cratering and possible application to the icy satellites of Saturn and Jupiter, *Jour. Geophys. Res.*, v. 89, p. 417–423, 1984.

Fink, U., Comet Yanaka (1988r): A new class of carbon-poor comet, *Science*, v. 257, p. 1926–1929, 1992.

Finkbeiner, A., Mapping the river in the sky, *Science*, v. 257, p. 1208–1210, 1992.

Fischer, A. G., Climatic oscillations in the biosphere, in M. H. Nitecki, ed., *Biotic Crises in Ecological and Evolutionary Time*, p. 103–131, New York, Academic Press, 1981.

———, The two Phanerozoic supercycles, in W. A. Berggren and J. A. Van Couvering, eds., *Catastrophes and Earth History: The New Uniformitarianism*, p. 129–150, Princeton, NJ, Princeton University Press, 1984.

Fischer, A. G., and M. A. Arthur, Secular variations in the pelagic realm, in H. E. Cook and P. Enos, eds., *Deep Water Carbonate Environments*, p. 18–50, Soc. Econ. Paleontologists and Mineralogists, Special Public. 25, 1977.

Fish, F. F., Jr., Angular momenta of the planets, *Icarus*, v. 7, p. 251–256, 1967.

Fisher, J., Fossil birds and their adaptive radiation, in W. B. Harland, C. H. Holland, M. R. House, N. F. Hughes, A. B. Reynolds, M. J. S. Rudwick, G. E. Satherthwaite, L. B. H. Tarlo, and E. C. Willey, eds., *The Fossil Record*, p. 133–154, London, Geol. Soc. London, 1967.

Fisher, Rev. O., On the physical causes of the ocean basins, *Nature*, v. 25, p. 243–244, 1882.

———, *Physics of the Earth's Crust, Second Edition*, New York, Macmillan and Co., 391 pp., 1889.

Flam, F., Giving the galaxies a history, *Science*, v. 255, p. 1067–1068, 1992a.

———, Hubble sees a zoo of ancient galaxies, *Science*, v. 258, p. 1733, 1992b.

———, Microwave ripples have a reprise, *Science*, v. 259, p. 31, 1993.

Flaschka, H., and B. Chirikov, eds, *Progress in Chaotic Dynamics: Essays in Honor of Joseph Ford's 60th Birthday* (Physica D, Volume 33), Amsterdam, North-Holland, 323 pp., 1988.

Fleming, J. R., T. C. Chamberlin and H_2O climate feedbacks: A voice from the past, *Eos, Trans. Am. Geophys. Union*, v. 73, pp. 505 and 509, 1992.

Flessa, K. W., The "facts" of mass extinction, in V. L. Sharpton and P. D. Ward, eds., *Global Catastrophes in Earth History; An Interdisciplinary Conference on Impacts, Volcanism, and Mass Mortality*, p. 1–7, Boulder, CO, The Geological Society of America, Special Paper 247, 1990.

Foote, M., and S. J. Gould, Cambrian and Recent morphological disparity, *Science*, v. 258, p. 1818, 1992.

Ford, J., A picture book of stochasticity, in S. Jorna, ed., *Topics in Nonlinear Dynamics: A Tribute to Sir Edward Bullard*, p. 121–146, New York, American Institute of Physics, 1978.

———, How random is a coin toss?, in C. W. Horton, Jr., L. E. Reichl, and V. G. Szebehely, eds, *Long-Time Prediction in Dynamics*, p. 79–92, New York, John Wiley & Sons, Inc, 1983.

References Cited

——, Directions in classical chaos, in Hao Bai-Lin, ed., *Directions in Chaos, Volume 1*, p. 1–16, Singapore, World Scientific, 1987.

Ford, J., G. Mantica, and G. H. Ristow, The Arnol'd cat: Failure of the correspondence principle, in D. K. Campbell, ed., *CHAOS/XAOC: Soviet-American Perspectives on Nonlinear Science*, p. 477–493, New York, American Institute of Physics, 1990.

Forrester, J. W., *Industrial Dynamics*, Cambridge MA, M.I.T. Press, 464 pp., 1961.

——, *World Dynamics, Second Edition*, Cambridge, MA, Wright-Allen Press, Inc., 144 pp., 1973.

Foukal, P. V., The variable Sun, *Sci. Am., v. 262*, p. 34–41 (February), 1990.

Franceschini, V., and C. Tebaldi, Sequences of infinite bifurcations and turbulence in a five-mode truncation of the Navier-Stokes equations, *Jour. Stat. Phys., v. 21*, p. 707–726, 1979.

Frank, L. A., J. B. Sigwarth, and C. M. Yeates, A search for small Solar-System bodies near the Earth using a ground-based telescope technique and observations, *Astron. Astrophys. v. 228*, p. 522–530, 1990.

Freedman, W. L., The expansion rate and size of the universe, *Sci. Am., v. 267*, p. 54–60, 1992.

Freeman, W. J., The physiology of perception, *Sci. Am., v. 264*, p. 78–85 (February), 1991.

——, Tutorial on neurobiology: From single neurons to brain chaos, *Internat. Jour. Bifurcation and Chaos, v. 2*, p. 451–482, 1992.

French, B. M., 25 years of the impact-volcanic controversy; Is there anything new under the Sun or inside the Earth?, *Eos, Trans. Am. Geophys. Union, v. 71*, p. 411–414, 1990.

French, L. M., and R. P. Binzel, CCD photometry of asteroids, in R. P. Binzel, T. Gehrels, and M. S. Matthews, eds., *Asteroids II*, p. 54–65, Tucson, AZ, The University of Arizona Press, 1989.

French, L. M., F. Vilas, W. K. Hartmann, and D. J. Tholen, Distant asteroids and Chiron, in R. P. Binzel, T. Gehrels, and M. S. Matthews, eds., *Asteroids II*, p. 468–486, Tucson, AZ, The University of Arizona Press, 1989.

Frey, H., Origin of the Earth's ocean basins, *Icarus, v. 32*, p. 235–250, 1977.

——, Crustal evolution of the early Earth: The role of major impacts, *Precambrian Res., v. 10*, p. 195–216, 1980.

Friedman, H., The legacy of the IGY, *Eos, Trans. Am. Geophys. Union, v. 64*, p. 497–499, 1983.

Friesen, L. J., A. A. Jackson, IV, H. A. Zook, and D. J. Kessler, Analysis of orbital perturbations acting on objects in orbits near geosynchronous Earth orbit, *Jour. Geophys. Res., v. 97*, p. 3845–3863, 1992.

Friis-Christensen, E., and K. Lassen, Length of the solar cycle: An indicator of solar activity closely associated with climate, *Science, v. 254*, p. 698–700, 1991.

Fritts, H. C., *Reconstructing Large-Scale Climatic Patterns from Tree-Ring Data: A Diagnostic Analysis*, Tucson, AZ, The University of Arizona Press, 286 pp., 1991.

Froeschlé, C., Chaotic behavior of asteroidal and cometary orbits, in C.-I. Lagerkvist, H. Rickman, B. A. Lindblad, and M. Lindgren, eds., *Asteroids, Comets, Meteors III*, p. 63–76, Proceedings of Meeting at Astronomical Observatory of Uppsala University, June 12–16, 1989, Uppsala, Sweden, Uppsala universitet Reprocentralen HSC, 1990.

Froeschlé, C., and R. Greenberg, Mean motion resonances, in R. P. Binzel, T. Gehrels, and M. S. Matthews, eds., *Asteroids II*, p. 827–844, Tucson, AZ, University of Arizona Press, 1989.

Frouzakis, C. E., R. A. Adomaitis, and I. G. Kevrekidis, Resonance phenomena in an adaptively-controlled system, *Internat. Jour. Bifurcation and Chaos, v. 1*, p. 83–106, 1991.

Fujiwara, A., P. Cerroni, D. Davis, E. Ryan, M. Di Martino, K. Holsapple, and K. Housen, Experiments and scaling laws for catastrophic collisions, in R. P. Binzel, T. Gehrels, and M. S. Matthews, eds., *Asteroids II*, p. 240–265, Tucson, AZ, The University of Arizona Press, 1989.

Fukao, Y., Seismic tomogram of the Earth's mantle: Geodynamic implications, *Science, v. 258*, p. 625–630, 1992.

Fuller, M., and R. Weeks, Superplumes and superchrons, *Nature, v. 356*, p. 16–17, 1992.

Gaffey, M. J., Forging an asteroid-meteorite link, *Science, v. 260*, p. 167–168, 1993.

Gaffey, M. J., J. F. Bell, and D. P. Cruikshank, Reflectance spectroscopy and asteroid surface mineralogy, in R. P. Binzel, T. Gehrels, and M. S. Matthews, eds., *Asteroids II*, p. 98–127, Tucson, AZ, The University of Arizona Press, 1989.

References Cited

Gaffin, S., Phase difference between sea level and magnetic reversal rate, *Nature, v. 329*, p. 816–819, 1987.

———, Analysis of scaling in the geomagnetic polarity reversal record, *Phys. Earth Planet. Int., v. 57*, p. 284–290, 1989.

Gaffin, S. R., and K. A. Maasch, Anomalous cyclicity in climate and stratigraphy and modeling nonlinear oscillations, *Jour. Geophys. Res., v. 96*, p. 6701–6711, 1991.

Galer, S. J. G., Interrelationships between continental freeboard, tectonics and mantle temperature, *Earth Planet. Sci. Lett., v. 105*, p. 214–228, 1991.

Gallagher, W. B., Selective extinction and survival across the Cretaceous/Tertiary boundary in the northern Atlantic Coastal plain, *Geology, v. 19*, p. 967–970, 1991.

Gallant, R., *Bombarded Earth*, London, John Baker, 256 pp., 1964.

Galloway, D. J., and M. R. E. Proctor, Numerical calculations of fast dynamos in smooth velocity fields with realistic diffusion, *Nature, v. 356*, p. 691–693, 1992.

Gamow, G., *Gravity*, Garden City, NY, Anchor Books, Doubleday and Company, Inc., 157 pp., 1962.

Ganapathy, R. Evidence for a major meteorite impact on the Earth 34 million years ago: Implication on the origin of North American tektites and Eocene extinction, in L. T. Silver and P. H. Schultz, eds., *Geological Implications of Impacts of Large Asteroids and Comets on the Earth*, p. 513–516, Boulder, CO, The Geological Society of America, Special Paper 190, 1982.

Garey, M. R., and D. S. Johnson, *Computers and Intractability: A Guide to the Theory of NP-Completeness*, New York, W. H. Freeman & Co., 340 pp.(1983 Reprinting), 1979.

Garfinkel, A., M. L. Spano, W. L. Ditto, and J. N. Weiss, Controlling cardiac chaos, *Science, v. 257*, p. 1230–1235, 1992.

Garvin, J. B., and A. L. Deino, New perspectives on the Popigai impact structure, *Papers Presented at the International Conference on Large Meteorite Impacts and Planetary Evolution, 31 August–2 September, 1992, Sudbury, Ontario, Canada*, p. 27–28, Lunar and Planetary Institute, Houston, TX (LPI Contribution No. 790), 1992.

Garwin, L., Tales of a lost magma ocean, *Nature, v. 338*, p. 1920, 1989.

Gatlin, L. L., *Information Theory and the Living System*, New York, Columbia University Press, 210 pp., 1972.

Gault, D. E., and P. H. Schultz, Earth-orbiting debris from lunar impact ejecta: Environmental effects? (Abstract), *Eos, Trans. Am. Geophys. Union, v. 71*, p. 1429, 1990.

Gault, D. E., and C. P. Sonett, Laboratory simulation of pelagic asteroidal impact: Atmospheric injection, benthic topography, and the surface wave radiation field, in L. T. Silver and P. H. Schultz, eds., *Geological Implications of Impacts of Large Asteroids and Comets on the Earth*, p. 69–92, Boulder, CO, The Geological Society of America, Special Paper 190, 1982.

Gehrels, T., ed., *Physical Studies of Minor Planets*, Washington, DC, National Aeronautics and Space Administration, NASA SP-267, 687 pp., 1971a.

Gehrels, T., Future work, in T. Gehrels, ed., *Physical Studies of Minor Planets*, p. 653–659, Washington, DC, National Aeronautics and Space Administration, NASA SP-267, 1971b.

Gehrels, T., ed., *Asteroids*, Tucson, AZ, The University of Arizona Press, 1182 pp., 1979.

Gerstenkorn, H., Über gezeiten Reibung beim zwei Körber-problem, *Z. Astrophys., v. 36*, p. 245–274, 1955.

———, On the controversy over the effect of tidal friction upon the history of the Earth-Moon system, *Icarus, v. 7*, p. 60–167, 1967.

———, The earliest past of the Earth-Moon system, *Icarus, v. 11*, p. 189–207, 1969.

Ghil, M., Theoretical climate dynamics: An introduction, in M. Ghil, R. Benzi, and G. Parisi, eds., *Turbulence and Predictability in Geophysical Fluid Dynamics and Climate Dynamics*, p. 347–402, New York, North-Holland, 1985.

Ghyka, M., *The Geometry of Art and Life*, New York, Dover Publications, Inc., 174 pp., 1977.

Gibbons, A., Geneticists trace the DNA trail of the first Americans, *Science, v. 259*, p. 312–313, 1993.

Gierasch, P. J., Jupiter's stratosphere mapped, *Nature, v. 351*, p. 103–104, 1991.

Gilbert, A. D., Fast dynamo action in a steady chaotic flow, *Nature, v. 350*, p. 483–485, 1991.

References Cited

Gilbert, G. K., The Moon's face (Presidential Address), *Phil. Soc. Washington, Bull., v. 12*, p. 241–292, 1893.

Gillespie, J. H., *The Causes of Molecular Evolution*, New York, Oxford University Press, 336 pp., 1991.

Gilmour, I., W. S. Wolbach, and E. Anders, Major wildfires at the Cretaceous-Tertiary boundary, in S. V. M. Clube, ed., *Catastrophes and Evolution: Astronomical Foundations*, p. 195–213, New York, Cambridge University Press, 1989.

Gilmour, I., S. S. Russell, J. W. Arden, M. R. Lee, I. A. Franchi, and C. T. Pillinger, Terrestrial carbon and nitrogen isotopic ratios from Cretaceous-Tertiary boundary diamonds, *Science, v. 258*, p. 1624–1626, 1992.

Gingerich, O., ed., *New Frontiers in Astronomy: Readings From Scientific American*, New York, W. H. Freeman & Co., 369 pp., 1975.

Giovanelli, R., *Secrets of the Sun*, New York, Cambridge University Press, 116 pp., 1984.

Glass, B. P., Possible correlations between tektite events and climate change, in L. T. Silver and P. H. Schultz, eds., *Geological Implications of Impacts of Large Asteroids and Comets on the Earth*, p. 251–256, Boulder, CO, The Geological Society of America, Special Paper 190, 1982.

Glass, B. P., and B. C. Heezen, Tektites and geomagnetic reversals, *Sci. Am., v. 217*, p. 32–38, 134, 1967.

Glass, B. P., and P. A. Zwart, The Ivory Coast microtektite strewnfield: New data, *Earth Planet. Sci. Lett., v. 43*, p. 336–342, 1979.

Glass, B. P., M. B. Swincki, and P. A. Zwart, Australasian, Ivory Coast, and North American tektite strewnfields: Size, mass and correlation with geomagnetic reversals and other Earth events, *Proc. Lunar Planet. Sci. Conf., 10th*, p. 2535–2545, 1979.

Glass, L., and M. C. Mackey, *From Clocks to Chaos*, Princeton, NJ, Princeton University Press, 248 pp., 1988.

Gleick, J., *Chaos: Making A New Science*, New York, Viking Penguin, Inc., 352 pp., 1987.

Glen, W., *Continental Drift and Plate Tectonics*, Columbus, OH, Charles E. Merrill Publ. Co., 188 pp., 1975.

———, *The Road to Jaramillo*, Stanford, CA, Stanford University Press, 459 pp., 1982.

———, Musings on the review process, *PALAIOS, v. 4*, p. 397–399, 1989.

———, What killed the dinosaurs?, *Am. Scientist, v. 78*, p. 354–370, 1990.

———, The power of gestalt, standards of appraisal, and the polemic mode, Biannual Meeting, *Int. Soc. History, Phil., and Social Studies in Biol.*, Abstracts Volume, p. 25, 12 July, 1991.

———, Mindset, Standards, and Style in the Mass-Extinction Debates, *Abstracts Volume, 29th International Geological Congress*, 24 August–3 September, 1992.

Glen, W., ed., *The Mass-Extinction Debates: How Science Works in a Crisis*, Proceedings of a Symposium held at Northwestern University, 12 July, 1991, by the Institute for Historical, Philosophical, and Sociological Studies in Biology, Stanford, CA, Stanford University Press, 370 pp., 1994.

Gliese, W., and H. Jahreiss, Stars within 25 parsecs of the Sun, in R. Smoluchowski, J. N. Bahcall, and M. S. Matthews, eds., *The Galaxy and the Solar System*, p. 13–34, Tucson, AZ, The University of Arizona Press, 1986.

Gödel, K., Über formal unentscheidbare Sätze der Principia Mathematica und verwandter Systeme I, *Monatshefte für Mathematik und Physik, v. 38*, p. 173–198, 1931.

Goedecke, G. H., and J. F. Ni, Eötvös force on the lithosphere, in T. W. C. Hilde and R. L. Carlson, eds., *Silver Anniversary of Plate Tectonics*, p. 251–257, Amsterdam, Elsevier Science Publishers *(Tectonophysics, v. 187)*, 1991.

Gold, T., Instability of the Earth's axis of rotation, *Nature, v. 175*, p. 526–529, 1955.

Goldberger, A. L., D. R. Rigney, and B. J. West, Chaos and fractals in human physiology, *Sci. Am., v. 262*, p. 42–49 (February), 1990.

Goldreich, P. An explanation of the frequent occurrence of commensurable mean motions in the Solar System, *Month. Not. Roy. Astron. Soc., v. 130*, p. 159–181, 1965.

Goldreich, P., and A. Toomre, Some remarks on polar wandering, *Jour. Geophys. Res., v. 74*, p. 2555–2567, 1969.

References Cited

Gollub, J. P., and S. V. Benson, Many routes to turbulent convection, *Jour. Fluid Mech.*, v. *100*, p. 449–470, 1980.

Gollub, J. P., and H. L. Swinney, Onset of turbulence in a rotating fluid, *Phys. Rev. Lett.*, v. *35*, p. 927–930, 1975.

Goodwin, B. C., The generic properties of morphogenetic fields, in J. L. Casti and A. Karlqvist, eds., *Beyond Belief: Randomness, Prediction and Explanation in Science*, p. 181–198, Boston, MA, CRC Press, 1991.

Gordon, R. G., Polar wandering and paleomagnetism, *Ann. Rev. Earth Planet. Sci.*, v. *15*, p. 567–593, 1987.

Gordon, R. G., and D. M. Jurdy, Cenozoic global plate motions, *Journ. Geophys. Res.*, v. *91*, p. 12,389–12,406, 1986.

Gordon, R. G., A. Cox, and S. O'Hare, Paleomagnetic Euler poles and the apparent polar wander and absolute motion of North America since the Carboniferous, *Tectonics*, v. *3*, p. 499–537, 1984.

Gorman, M., P. J. Widman, and K. A. Robbins, Chaotic flow regimes in a convection loop, *Phys. Rev. Lett.*, v. *52*, p. 2241–2244, 1984.

Gostin, V. A., P. W. Haines, R. J. F. Jenkins, W. Compston, and I. S. Williams, Impact ejecta horizon within late Precambrian shales, Adelaide geosyncline, South Australia, *Science*, v. *233*, p. 198–200, 1986.

Gostin, V. A., R. R. Keays, and M. W. Wallace, Iridium anomaly from the Acraman impact ejecta horizon: Impacts can produce sedimentary iridium peaks, *Nature*, v. *340*, p. 542–544, 1989.

Gould, S. J., Is uniformitarianism necessary?, *Am. Jour. Sci.*, v. *263*, p. 223–228, 1965.

———, Darwinism and the expansion of evolutionary theory, *Science*, v. *216*, p. 380–386, 1982a.

———, The meaning of punctuated equilibrium and its role in validating a hierarchical approach to macroevolution, in R. Milkman, ed., *Perspectives in Evolution*, p. 83–104, Sunderland, MA, Sinauer Associates Inc. Publishers, 1982b.

———, Nature's great era of experiments, *Natural History*, v. *92*, p. 12–21, 1983.

———, The cosmic dance of Siva, *Natural History*, v. *93*, p. 14–19, 1984a.

———, Toward the vindication of punctuational change, in W. A. Berggren and J. A. Van Couvering, eds., *Catastrophes and Earth History: The New Uniformitarianism*, p. 9–34, Princeton, NJ, Princeton University Press, 1984b.

———, *Time's Arrow/Time's Cycle*, Cambridge, MA, Harvard University Press, 222 pp., 1987a.

———, This View of Life: The godfather of disaster, *Natural History*, v. *99*, p. 20–29 (September), 1987b.

———, This View of Life: Grimm's greatest tale, *Natural History, February*, p. 20–28, 1989a.

———, This View of Life: The Wheel of Fortune and the wedge of progress, *Natural History, March*, p. 14–21, 1989b.

———, This View of Life: Through a lens, darkly, *Natural History, September*, p. 16–24, 1989c.

———, This View of Life: Dinosaurs in the haystack, *Natural History, March*, p. 2–13, 1992.

Gould, S. J., and N. Eldredge, Punctuated equilibria: The tempo and mode of evolution reconsidered, *Paleobiology*, v. *3*, p. 115–151, 1977.

Gould, S. J., and C. B. Calloway, Clams and brachiopods — ships that pass in the night, *Paleobiology*, v. *6*, p. 383–396, 1980.

Gradie, J. C., C. R. Chapman, and E. F. Tedesco, Distributions of taxonomic classes and the compositional structure of the Asteroid Belt, in R. P. Binzel, T. Gehrels, and M. S. Matthews, eds., *Asteroids II*, p. 316–335, Tucson, AZ, The University of Arizona Press, 1989.

Graham, J., *Cancer Selection: The New Theory of Evolution*, Lexington, VA, Aculeus Press, Inc., 213 pp., 1992.

Graham, R. L., and J. H. Spencer, Ramsey theory, *Sci. Am.*, v. *263*, p. 112–117 (July), 1990.

Grant, P. R., and B. R. Grant, Hybridization of bird species, *Science*, v. *256*, p. 193–197, 1992.

Grantz, A., and M. W. Mullen, Possible implications of the low-incidence angle Avak astrobleme, Arctic Alaska, for the K/T event, *preprint*, 1991.

Grassberger, P., Do climate attractors exist?, *Nature*, v. *323*, p. 609–612, 1986.

References Cited

Grassberger, P., and I. Procaccia, Characterization of strange attractors, *Phys. Rev. Lett., v. 50*, p. 346–349, 1983.
Gratz, A. J., W. J. Nellis, and N. A. Hinsey, Observations of high-velocity, weakly shocked ejecta from experimental impacts, *Nature, v. 363*, p. 522–524, 1993.
Grebogi, C., E. Ott, and J. A. Yorke, Chaotic attractors in crisis, *Phys. Rev. Lett., v. 48*, p. 1507–1510, 1982.
———, Crises, sudden changes in chaotic attractors, and transient chaos, *Physica D, v. 7*, p. 181–200, 1983.
———, Critical exponent of chaotic transients in nonlinear dynamical systems, *Phys. Rev. Lett., v. 57*, p. 1284–1287, 1986.
———, Chaos, strange attractors, and fractal basin boundaries in nonlinear dynamics, *Science, v. 238*, p. 632–638, 1987a.
Grebogi, C., E. Ott, J. A. Yorke, and H. E. Nusse, Fractal basin boundaries with unique dimension, in J. R. Buchler and H. Eichhorn, eds., *Chaotic Phenomena in Astrophysics*, p. 117–126, New York, Annals of the New York Academy of Sciences, Volume 497, 1987b.
Grebogi, C., E, Ott, F. Romeiras, and J. A. Yorke, Critical exponents for crisis-induced intermittency, *Phys. Rev. A, v. 36*, p. 5365–5380, 1987c.
Greeley, R., and B. D. Schneid, Magma generation on Mars: Amounts, rates, and comparisons with Earth, Moon, and Venus, *Science, v. 254*, p. 996–998, 1991.
Green, D. H., Archaean greenstone belts may include terrestrial equivalents of lunar maria, *Earth Planet. Sci. Lett, v. 15*, p. 263–270, 1972.
Green, H. W., II, The mechanism of deep earthquakes, *Eos, Trans. Am. Geophys. Union*, v. 74, p. 23, 1993.
Greenberg, J. H., and M. Ruhlen, Linguistic origins of Native Americans, *Sci. Am., v. 267*, p. 94–99, 1992.
Greenberg, R., Orbital resonance in a dissipative medium, *Icarus, v. 33*, p. 62–73, 1978.
———, Orbital evolution of the Galilean satellites, in D. Morrison, ed., *Satellites of Jupiter*, p. 65–92, Tucson, AZ, The University of Arizona Press, 1982.
Greenberg, R., and A. Brahic, eds., *Planetary Rings*, Tucson, AZ, The University of Arizona Press, 784 pp., 1984.
Greenberg, R., and M. C. Nolan, Delivery of asteroids and meteorites to the inner Solar System, in R. P. Binzel, T. Gehrels, and M. S. Matthews, eds., *Asteroids II*, p. 778–804, Tucson, AZ, The University of Arizona Press, 1989.
Grew, P. C., An atmosphere of thought (Book Reviews), *Nature, v. 346*, p. 230, 1990.
Grieve, R. A. F., Impact bombardment and its role in proto-continental growth on the early Earth, *Precambrian Res., v. 10*, p. 217–247, 1980.
———, The record of impact on Earth: Implications for a major Cretaceous/Tertiary impact event, in L. T. Silver and P. H. Schultz, eds., *Geological Implications of Impacts of Large Asteroids and Comets on the Earth*, p. 25–37, Boulder, CO, The Geological Society of America, Special Paper 190, 1982.
———, Terrestrial impact craters, *Ann., Rev. Earth Planet. Sci., v. 15*, p. 245–270, 1987.
———, Terrestrial impact: The record in the rocks, *Meteoritics, v. 26*, p. 175–194, 1991.
Grieve, R. A. F., and M. R. Dence, The terrestrial cratering record: II. The crater production rate, *Icarus, v. 38*, p. 230–242, 1979.
Grieve, R. A. F., and P. B. Robertson, The terrestrial cratering record: I. Current status of observations, *Icarus, v. 38*, p. 212–229, 1979.
———, Terrestrial impact structures, *Geol. Survey of Canada, Map 1658A* (scale = 1:63,000,000), 1987.
Grieve, R. A. F., and L. J. Pesonen, The terrestrial impact cratering record, *Tectonophys., v. 216*, p. 1–30, 1992.
Grieve, R. A. F., C. A. Wood, J. B. Garvin, G. McLaughlin, and J. F. McHone, *Astronaut's Guide to Terrestrial Impact Craters*, Houston, TX, Lunar and Planetary Inst. Tech. Report No. 88–03, 89 pp., 1988.
Grieve, R. A. F., D. Stöffler, and A. Deutsch, The Sudbury structure: Controversial or misunderstood?, *Jour. Geophys. Res., v. 96*, p. 22,753–22,764, 1991.

References Cited

Grimm, R. E., and H.Y. McSween, Jr., Heliocentric zoning of the Asteroid Belt by aluminum-26 heating, *Science, v. 259*, p. 653–655, 1993.

Grolier, M. J., Bibliography of terrestrial impact structures, *NASA Report, TM-87567*, 539 pp., 1985.

Grossman, J. N., Formation of chondrules, in J. F. Kerridge and M. S. Matthews, eds., *Meteorites and the Early Solar System*, p. 680–696, Tucson, AZ, The University of Arizona Press, 1988.

Grossman, J. N., A. E. Rubin, H. Nagahara, and E. A. King, Properties of chondrules, in J. F. Kerridge and M. S. Matthews, eds., *Meteorites and the Early Solar System*, p. 619–659, Tucson, AZ, The University of Arizona Press, 1988.

Gruntfest, I. J., Thermal feedback in liquid flow: Plane shear at constant stress, *Trans. Soc. Rheology, v. 7*, p. 195–207, 1963.

Gruntfest, I. J., and H. R. Shaw, Scale effects in the study of earth tides, *Trans. Soc. Rheol., v. 18:2*, p. 287–297, 1974.

Gruntfest, I. J., J. P. Young, and N. L. Johnson, Temperatures generated by the flow of liquids in pipes, *Jour. Appl. Phys., v. 35*, p. 18–22, 1964.

Gu, Y., M. Tung, J.-M. Yuan, D. H. Feng, and L. Narducci, Crises and hysteresis in coupled logistic maps, *Phys. Rev. Lett., v. 52*, p. 701–704, 1984.

Gubbins, D., Energetics of the Earth's core, *Jour. Geophys. Res., v. 43*, p. 453–464, 1977.

——, Mechanism for geomagnetic polarity reversals, *Nature, v. 326*, p. 167–169, 1987.

Gubbins, D., and J. Bloxham, Morphology of the geomagnetic field and implications for the geodynamo, *Nature, v. 325*, p. 509–511, 1987.

Gubbins, D., and R. S. Coe, Longitudinally confined geomagnetic reversal paths from non-dipolar transition fields, *Nature, v. 362*, p. 51–53, 1993.

Guckenheimer, J., Computational environments for exploring dynamical systems, *Internat. Jour. Bifurcation and Chaos, v. 1*, p. 269–276, 1991.

Guckenheimer, J., and P. Holmes, *Nonlinear Oscillations, Dynamical Systems, and Bifurcations of Vector Fields*, New York, Springer-Verlag, 453 pp., 1983.

Gudlaugsson, S. T., Large impact crater in the Barents Sea, *Geology, v. 21*, p. 291–294, 1993.

Gwinn, E. G., and R. M. Westervelt, Intermittent chaos and low-frequency noise in the driven damped pendulum, *Phys. Rev. Lett., v. 54*, p. 1613–1616, 1985.

——, Fractal basin boundaries and intermittency in the driven damped pendulum, *Phys. Rev. A, v. 33*, p. 4143–4155, 1986.

——, Scaling structure of attractors at the transition from quasiperiodicity to chaos in electronic transport in Ge, *Phys. Rev. Lett., v. 59*, p. 157–160, 1987.

Haase, R, Kepler's Harmonies, between Pansophia and Mathesis Universalis, in A. Beer and P. Beer, eds, *Kepler—Four Hundred Years (Vistas in Astronomy, v. 18)*, p. 519–533, New York, Pergamon Press, 1975.

Hagee, V. L., and P. Olson, Dynamo models with permanent dipole fields and secular variation, *Jour. Geophys. Res., v. 96*, p. 11,673–11,687, 1991.

Hager, B. H., R. W. Clayton, M. A. Richards, R. P. Comer, and A. M. Dziewonski, Lower mantle heterogeneity, dynamic topography and the geoid, *Nature, v. 313*, p. 541–545, 1985.

Hagstrum, J. T., Is flood basalt volcanism a seismically-induced response to large antipodal bolide impacts?, *preprint*, 1992.

——, North American Jurassic APW: The current dilemma, *Eos, Trans. Am. Geophys. Union, v. 74*, p. 65–69, 1993.

Hagstrum, J. T., and B. D. Turrin, Is flood basalt volcanism a seismically-induced response to large antipodal bolide impacts? (Abstract), *Eos, Trans. Am. Geophys. Union, v. 72*, p. 516 (29 October, Supplement), 1991a.

——, Flood basalt volcanism as a seismically-induced response to large antipodal bolide impacts, *preprint*, 1991b.

Hajduk, A., Evolution of cometary debris: Physical aspects, in R. L. Newburn, Jr., M. Neugebauer, and J. Rahe, eds., *Comets in the Post-Halley Era, Volume 1*, p. 593–606, Boston, MA, Kluwer Academic Publishers, 1991.

Haken, H., *Synergetics, An Introduction*, New York, Springer-Verlag, 355 pp., 1978.

References Cited

Hallam, A., End-Cretaceous mass extinction event: Argument for terrestrial causation, *Science, v. 238*, p. 1237–1242, 1987.

——, A compound scenario for the end-Cretaceous mass extinctions, in M. A. Lamolda, E. G. Kauffman, and O. H. Walliser, eds., *Palaeontology and Evolution: Extinction Events. 2nd International Conference on Global Bioevents (Bilbao Conf., 20–23 October, 1987), Revista Española de Palaeontologia, Nº. Extraordinario*, p. 7–20, Madrid, Sociedad Española de Paleontologia, 1988.

——, Catastrophism in geology, in S. V. M. Clube, ed., *Catastrophes and Evolution: Astronomical Foundations*, p. 25–55, New York, Cambridge University Press, 1989.

Halliday, D., and R. Resnick, *Physics: Parts I and II*, New York, John Wiley & Sons, Inc., 1324 pp., 1966.

——, *Fundamentals of Physics*, New York, John Wiley & Sons, Inc., 837 pp., 1970.

Halsey, T. C., M. H. Jensen, L. P. Kadanoff, I. Procaccia, and B. I. Shraiman, Fractal measures and their singularities, *Phys. Rev. A, v. 33*, p. 1141–1151, 1986.

Hamilton, W. D., The genetical evolution of social behavior, *Jour. Theoretical Biol., v. 7*, p. 1–52, 1964.

Hanneman, D. L., and C. J. Wideman, Sequence stratigraphy of Cenozoic continental rocks, southwestern Montana, *Bull. Geol. Soc. Am., v. 103*, p. 1335–1345, 1991.

Hao, Bai-Lin, *Chaos*, Singapore, World Scientific Publ. Co., 576 pp., 1984.

——, *Directions in Chaos, Volume 1*, Singapore, World Scientific Publ. Co., 353 pp., 1987.

——, *Chaos II*, Singapore, World Scientific Publ. Co., 737 pp., 1990.

Harbert, W., L. Frei, R. Jarrard, S. Halgedahl, and D. Engebretson, Paleomagnetic and plate-tectonic constraints on the evolution of the Alaskan-eastern Siberian Arctic, in A. Grantz, L. Johnson, and J. F. Sweeney, eds., *The Arctic Ocean Region*, p. 567–592, Boulder, CO, Geol. Soc. Am., The Geology of North America, Vol. L, 1990.

Hardy, G. H., and E. M. Wright, *An Introduction to the Theory of Numbers, Fifth Edition*, Oxford, Clarendon Press, 426 pp., 1985.

Hargittai, I., *Symmetry: Unifying Human Understanding*, New York, Pergamon Press, 1045 pp., 1986.

Hargraves, R. B., and R. A. Duncan, Does the mantle roll?, *Nature, v. 245*, p. 361–363, 1973.

Hargraves, R. B., C. E. Cullicott, K. S. Deffeyes, S. Hougen, P. P. Christiansen, and P. S. Fiske, Shatter cones and shocked rocks in southwestern Montana: The Beaverhead impact structure, *Geology, v. 18*, p. 832–834, 1990.

Harland, W. B., C. H. Holland, M. R. House, N. F. Hughes, A. B. Reynolds, M. J. S. Rudwick, G. E. Satterthwaite, L. B. H. Tarlo, and E. C. Willey, eds., *The Fossil Record*, London, Geological Society of London, 827 pp., 1967.

Harland, W. B., A. V. Cox, P. G. Llewellyn, C. A. G. Pickton, A. G. Smith, and R. Walters, *A Geologic Time Scale*, New York, Cambridge University Press, 131 pp., 1982.

Harland, W. B., R. L. Armstrong, A. V. Cox, L. E. Craig, A. G. Smith, and D. G. Smith, *A Geologic Time Scale 1989*, New York, Cambridge University Press, 263 pp., 1990.

Harper, C. L., Jr., and S. B. Jacobsen, Evidence from coupled ^{147}Sm-^{143}Nd and ^{146}Sm-^{142}Nd systematics for very early (4.5-Gyr) differentiation of the Earth's mantle, *Nature, v. 360*, p. 728–732, 1992.

Harris, A. W., and D. F. Lupishko, Photometric lightcurve observations and reduction techniques, in R. P. Binzel, T. Gehrels, and M. S. Matthews, eds., *Asteroids II*, p. 39–53, Tucson, AZ, The University of Arizona Press, 1989.

Harrison, C. G. A., and T. Lindh, Comparison between the hot spot and geomagnetic field reference frames, *Nature, v. 300*, p. 251–252, 1982.

Harrison, E. R., The dark night-sky riddle: A "paradox" that resisted solution, *Science, v. 226*, p. 941–945, 1984.

——, Kelvin on an old, celebrated hypothesis, *Nature, v. 322*, p. 417–418, 1986.

Harry, D. L., and D. S. Sawyer, Basaltic volcanism, mantle plumes, and the mechanics of rifting: The Paraná flood basalt province of South America, *Geology, v. 20*, p. 207–210, 1992.

Hart, S. R., E. H. Hauri, L. A. Oschmann, and J. A. Whitehead, Mantle plumes and entrainment: Isotopic evidence, *Science, v. 256*, p. 517–520, 1992.

References Cited

Hartmann, L., and S. J. Kenyon, On the nature of FU Orionis objects, *Astrophys. Jour., v. 299*, p. 462–478, 1985.

Hartmann, W. K., Interplanet variations in scale of crater morphology — Earth, Mars, Moon, *Icarus, v. 17*, p. 707–713, 1972.

Hartmann, W. K., and D. R. Davis, Satellite-sized planetesimals and lunar origin, *Icarus, v. 24*, p. 504–512, 1975.

Hartnady, C. J. H., Amirante Basin, western Indian Ocean: Possible impact site of the Cretaceous/Tertiary extinction bolide?, *Geology, v. 14*, p. 423–426, 1986.

——, About turn for continents, *Nature, v. 352*, p. 476–478, 1991.

Hartung, J. B., and R. R. Anderson, The geology of the Manson impact structure: Sample studies reveal a well preserved complex impact crater, *Lunar and Planetary Sci. Conf. XXII, Abstracts of Papers*, p. 525–526, Lunar and Planetary Science Institute, Houston, TX, March 18–22, 1991.

Hartung, J. B., M. J. Kunk, and R. R. Anderson, Geology, geophysics, and geochronology of the Manson impact structure, in V. L. Sharpton and P. D. Ward, eds., *Global Catastrophes in Earth History; An Interdisciplinary Conference on Impacts, Volcanism, and Mass Mortality*, p. 207–221, Boulder, CO, The Geological Society of America, Special Paper 247, 1990.

Hassell, M. P., H. N. Comins, and R. M. May, Spatial structure and chaos in insect population dynamics, *Nature, v. 353*, p. 255–258, 1991.

Haucke, H., and R. Encke, Mode-locking and chaos in Rayleigh-Bénard convection, *Physica D, v. 25*, p. 307–329, 1987.

Head, J. W., C. A. Wood, and T. A. Mutch, Geologic evolution of the terrestrial planets, in B. J. Skinner, ed., *The Solar System and its Strange Objects: Readings from American Scientist*, p. 71–79, Los Altos, CA, William Kaufmann, Inc., 1981.

Head, J. W., L. S. Crumpler, J. C. Aubele, J. E. Guest, and R. S. Saunders, Venus volcanism: Classification of volcanic features and structures, associations, and global distribution from Magellan data, *Jour. Geophys. Res., v. 97*, p. 13,153–13,197, 1992.

Heisler, J., Monte Carlo simulations of the Oort comet cloud, *Icarus, v. 88*, p. 104–121, 1990.

Held, G. A., D. H. Solina II, D. T. Keane, W. J. Haag, P. M. Horn, and G. Grinstein, Experimental study of critical-mass fluctuations in an evolving sandpile, *Phys. Rev. Lett., v. 65*, p. 1120–1123, 1990.

Held, I. M., Climate models and the astronomical theory of the Ice Ages, *Icarus, v. 50*, p. 449–461, 1982.

Heliker, C., and T. L. Wright, The Pu'u 'O'o-Kupaianaha eruption of Kilauea, *Eos, Trans. Am. Geophys. Union, v. 72*, p. 521–526, 1991.

Helleman, R. H. G., Dynamics revisited: A glossary, in S. Jorna, ed., *Topics in Nonlinear Dynamics: A Tribute to Sir Edward Bullard*, p. 400–403, New York, American Institute of Physics, 1978.

——, Self-generated chaotic behavior in nonlinear mechanics, in E. G. D. Cohen, ed., *Fundamental Problems in Statistical Mechanics, Vol. 5*, p. 165–233, 1980.

——, One mechanism for the onsets of large-scale chaos in conservative and dissipative systems (with an Appendix by R. S. MacKay), in C. W. Horton, Jr., L. E. Reichl, and V. G. Szebehely, eds, *Long-Time Prediction in Dynamics*, p. 95–126, New York, John Wiley & Sons, Inc, 1983.

Helz, R. T., Kilauea Iki, a model magma chamber, *Eos, Trans. Am. Geophys. Union, v. 72*, p. 315, 1991.

Henbest, N., *Mysteries of the Universe*, New York, Van Nostrand Reinhold Co., 184 pp., 1981.

Henderson, M. E., M. Levi, and F. Odeh, The geometry and computation of the dynamics of coupled pendula, *Internat. Jour. Bifurcation and Chaos, v. 1*, p. 27–50, 1991.

Henkel, H., and I. J. Pesonen, Impact craters and craterform structures in Fennoscandia, *Tectonophys., v. 216*, p. 31–40, 1992.

Hennig, W., *Phylogenetic Systematics*, Urbana, IL, University of Illinois Press, 263 pp., 1966.

Hénon, M., Chaotic scattering modelled by an inclined billiard, *Physica D, v. 33*, p. 132–156, 1988.

References Cited

Herbert, F., Primordial electrical induction heating of asteroids, *Icarus, v. 78*, p. 402–410, 1989.

Herbert, F., and C. P. Sonett, Electromagnetic heating of minor planets in the early Solar System, *Icarus, v. 40*, p. 484–496, 1979.

Herbig, G. H., Major issues in planetary system formation: 5. Summarizing remarks on the astronomical evidence for circumstellar disks, in H. A. Weaver and L. Danly, eds., *The Formation and Evolution of Planetary Systems*, p. 296–304, New York, Cambridge University Press, 1989.

Herman, Y., Selective extinction of marine plankton in the Paratethys at the end of the Mesozoic Era; a multiple interaction hypothesis, in V. L. Sharpton and P. D. Ward, eds., *Global Catastrophes in Earth History; An Interdisciplinary Conference on Impacts, Volcanism, and Mass Mortality*, p. 531–540, Boulder, CO, The Geological Society of America, Special Paper 247, 1990.

Herrick, S., The orbits of asteroids and meteors, *Proc. Lunar Planet. Explor. Colloq., v. 2*, p. 12–14, 1961.

Hess, H. H., History of the ocean basins, in A. E. J. Engel, H. L. James, and B. F. Leonard, eds., *Petrologic Studies: A Volume in Honor of A. F. Buddington*, p. 599–620, New York, The Geological Society of America, 1962.

Hewins, R. H., and H. E. Newsom, Igneous activity in the early Solar System, in J. F. Kerridge and M. S. Matthews, eds., *Meteorites and the Early Solar System*, p. 73–101, Tucson, AZ, The University of Arizona Press, 1988.

Hewitt, G. M., R. A. Nichols, and M. G. Ritchie, Ecological genetics: 1868 and all that for Magicicada, *Nature, v. 336*, p. 206–207, 1988.

Hide, R., N. T. Birch, L. V. Morrison, D. J. Shea, and A. A. White, Atmospheric angular momentum fluctuations and changes in the length of the day, *Nature, v. 286*, p. 114–117, 1980.

Higgins, M., and L. Tait, A possible new impact structure near Lac de la Presqu'ile, Québec, Canada, *Meteoritics, v. 25*, p. 235–236, 1990.

Hildebrand, A. R., and W. V. Boynton, Proximal Cretaceous-Tertiary boundary impact deposits in the Caribbean, *Science, v. 248*, p. 843–847, 1990a.

———, Locating the Cretaceous/Tertiary boundary impact crater(s) (Abstract), *Eos, Trans. Am. Geophys. Union, v. 71*, p. 1424, 1990b.

———, Cretaceous ground zero, *Natural History*, June, p. 46–53, 1991.

Hildebrand, A. R., and G. T. Penfield, A buried 180-km-diameter probable impact crater on the Yucatan Peninsula, Mexico (Abstract), *Eos, Trans. Am. Geophys. Union, v. 71*, p. 1425, 1990.

Hildebrand, A. R., G. T. Penfield, D. A. Kring, M. Pilkington, A. Camargo Z., S. B. Jacobsen, and W. V. Boynton, Chicxulub Crater: A possible Cretaceous/Tertiary boundary impact crater on the Yucatán Peninsula, Mexico, *Geology, v. 19*, p. 867–871, 1991.

Hillhouse, J. W., and C. S. Grommé, Early Cretaceous paleolatitude of the Yukon-Koyukuk Province, Alaska, *Jour. Geophys. Res., v. 93*, p. 11,735–11,752, 1988.

Hills, J. G., Dynamic relaxation of planetary systems and Bode's Law, *Nature, v. 225*, p. 840–842, 1970.

———, Encounters between binary and single stars and their effect on the dynamical evolution of stellar star systems, *Astron. Jour., v. 80*, p. 809–825, 1975.

———, Comet showers and the steady-state infall of comets from the Oort cloud, *Astron. Jour., v. 86*, p. 1730–1740, 1981.

———, Dynamical constraints on the mass and perihelion distance of Nemesis and the stability of its orbit, *Nature, v. 311*, p. 636–638, 1984a.

———, Close encounters betweeen a star-planet system and a stellar intruder, *Astron. Jour., v. 89*, p. 1559–1564, 1984b.

———, The passage of a "Nemesis-like" object through the planetary system, *Astron. Jour., v. 90*, p. 1876–1882, 1985.

———, Deflection of comets and other long-period solar companions into the planetary system by passing stars, in R. Smoluchowski, J. N. Bahcall, and M. S. Matthews, eds., *The Galaxy and the Solar System*, p. 397–408, Tucson, AZ, The University of Arizona Press, 1986.

Hinderer, J., H. Legros, D. Jault, and J.-L. Le Mouël, Core-mantle topographic torque: A

References Cited

spherical harmonic approach and implications for the excitation of the Earth's rotation by core motions, *Phys. Earth Planet. Int.*, v. 59, p. 329–341, 1990.

Hodge, P. W., ed., *The Universe of Galaxies: Readings From Scientific American*, New York, W. H. Freeman & Co., 113 pp., 1984.

Hodges, A., *Alan Turing: The Enigma*, New York, Simon and Schuster, 587 pp., 1983.

Hodych, J. P., and G. R. Dunning, Did the Manicouagan impact trigger end-of-Triassic mass extinction?, *Geology*, v. 20, p. 51–54, 1992.

Hoecksema, J. T., Solar sources of geomagnetic storms, *Eos, Trans. Am. Geophys. Union*, v. 73, p. 34 (21 January), 1992.

Hoff, B., *The Tao of Pooh*, New York, Penguin Books, 158 pp., 1982.

Hoffman, A., Mass extinctions: The view of a sceptic, *Jour. Geol. Soc. London*, v. 146, p. 21–35, 1989.

Hoffman, K. A., Ancient magnetic reversals: Clues to the geodynamo, *Sci. Am.*, v. 258, p. 76–83 (May), 1988.

——, Long-lived transitional states of the geomagnetic field and the dynamo families, *Nature*, v. 354, p. 273–277, 1991.

——, Dipolar reversal states of the geomagnetic field and core-mantle dynamics, *Nature*, v. 359, p. 789–794, 1992.

——, Do flipping magnetic poles follow preferred paths?, *Eos, Trans. Am. Geophys. Union*, v. 74, p. 97, 1993.

Hoffman, P. F., United Plates of America, the birth of a craton: Early Proterozoic assembly and growth of Laurentia, *Ann. Rev. Earth Planet. Sci.*, v. 16, p. 543–603, 1988.

——, Did the breakout of Laurentia turn Gondwanaland inside out?, *Science*, v. 252, p. 1409–1412, 1991.

Hoffmann, M., Asteroid mass determination: Present situation and perspectives, in R. P. Binzel, T. Gehrels, and M. S. Matthews, eds., *Asteroids II*, p. 228–239, Tucson, AZ, The University of Arizona Press, 1989a.

——, Impactless asteroid collisions: Opportunities for mass determinations and implications from actual close encounters, *Icarus*, v. 78, p. 280–286, 1989b.

Hoffmann, M., U. Fink, W. M. Grundy, and M. Hicks, Photometric and spectroscopic observations of 5145 Pholus, *Jour. Geophys. Res.*, v. 98, p. 7403–7407, 1993.

Hofstadter, D. R., *Gödel, Escher, Bach: An Eternal Golden Braid*, New York, Vintage Books, A Division of Random House, 777 pp., 1979.

——, D. R., *Metamagical themas:* Strange attractors: Mathematical models delicately poised between order and chaos, *Sci. Am.*, v. 245 (Nov.), p. 22–43, 1981.

Holcomb, R. T., Eruptive history and long-term behavior of Kilauea Volcano, in R. W. Decker, T. L. Wright, and P. H. Stauffer, eds., *Volcanism in Hawaii*, Vol. 1, Chap. 12, p. 261–350, U.S. Geol. Survey, Prof. Paper 1350, Washington, DC, U.S. Gov't Printing Office, 1987.

Holcomb, R. T., and U. S. Hawaiian EEZ-SCAN Scientific Staff, Completed GLORIA side-scan sonar survey of the Hawaiian Exclusive Economic Zone (Abstract), *Eos, Trans. Am. Geophys. Union*, v. 72, Supplement (Oct. 29, 1991), p. 249, 1991.

Holden, A. V., ed., *Chaos*, Princeton, NJ, Princeton University Press, 324 pp., 1986.

——, What makes them tick?, *Nature*, v. 336, p. 119, 1988.

Holland, H. D., and A. F. Trendall, *Patterns of Change in Earth Evolution*, New York, Springer-Verlag, 431 pp., 1984.

Hölldobler, B., and E. O. Wilson, *The Ants*, Cambridge, MA, The Belknap Press of Harvard University Press, 732 pp., 1990.

Holle, A., J. Main, G. Wiebusch, H. Rottke, and K. H. Welge, Quasi-Landau spectrum of the chaotic diamagnetic hydrogen atom, *Phys. Rev. Lett.*, v. 61, p. 161–164, 1988.

Holser, W. T., Catastrophic chemical events in the history of the ocean, *Nature*, v. 267, p. 403–408, 1977.

Honda, S., D. A. Yuen, S. Balachandar, and D. Reuteler, Three-dimensional instabilities of mantle convection with multiple phase transitions, *Science*, v. 259, p. 1308–1311, 1993.

Hones, E. W., Jr., Transient phenomena in the magnetotail and their relation to substorms, *Space Sci. Rev.*, v. 23, p. 393–410, 1979.

Hoover, W. G., and B. Moran, Viscous attractor for the Galton board, *Chaos, v. 2*, p. 599–602, 1992.

Hora, H., Degenerate plasma phases to explain the expansion of the Earth while releasing energy, in S. W. Carey, ed., *The Expanding Earth: A Symposium, Sydney, 1981*, p. 363–364, Hobart, Tasmania, University of Tasmania, 1983.

Horn, L. J., and C. T. Russell, Unusual behavior of spiral density waves in Saturn's A ring (Abstract), *Eos, Trans. Am. Geophys. Union, v. 72, Supplement* (29 October), p. 283, 1991.

Horton, C. W., Jr., L. E. Reichl, and V. G. Szebehely, eds., *Long-Time Prediction in Dynamics*, New York, John Wiley & Sons, Inc., 496 pp., 1983.

Hoshi, M., and M. Kono, Rikitake two-disk dynamo system: Statistical properties and growth of instability, *Jour. Geophys. Res., v. 93*, p. 11,643–11,654, 1988.

House, M. R., A new approach to an absolute timescale from measurements of orbital cycles and sedimentary microrhythms, *Nature, v. 315*, p. 721–725, 1985.

Howard, K. A., D. E. Wilhelms, and D. H. Scott, Lunar basin formation and highland stratigraphy, *Rev. Geophys. Space Phys., v. 12*, p. 309–327, 1974.

Howell, D. G., Terranes, *Sci. Am., v. 253*, p. 116–125 (Nov.), 1985.

———, *Tectonics of Suspect Terranes: Mountain Building and Continental Growth*, New York, Chapman and Hall, 232 pp., 1989.

Hsü, K. J., Geochemical markers of impacts and of their effects on environments, in H. D. Holland and A. F. Trendall, eds., *Patterns of Change in Earth Evolution*, p. 63–74, New York, Springer-Verlag, 1984.

Hsü, K. J., and J. A. McKenzie, Carbon-isotope anomalies at era boundaries: Global catastrophes and their ultimate cause, in V. L. Sharpton and P. D. Ward, eds., *Global Catastrophes in Earth History; An Interdisciplinary Conference on Impacts, Volcanism, and Mass Mortality*, p. 61–70, Boulder, CO, The Geological Society of America, Special Paper 247, 1990.

Hsü, K. J., J. A. McKenzie, and Q. X. He, Terminal Cretaceous environmental and evolutionary changes, in L. T. Silver and P. H. Schultz, eds., *Geological Implications of Impacts of Large Asteroids and Comets on the Earth*, p. 317–328, Boulder, CO, The Geological Society of America, Special Paper 190, 1982.

Hubbert, M. K., Theory of scale models as applied to the study of geologic structures, *Bull. Geol. Soc. Am., v. 48*, p. 1459–1520, 1937.

Huffman, A. R., J. H. Crocket, N. L. Carter, P. E. Borella, and C. B. Officer, Chemistry and mineralogy across the Cretaceous/Tertiary boundary at DSDP Site 527, Walvis Ridge, South Atlantic Ocean, in V. L. Sharpton and P. D. Ward, eds., *Global Catastrophes in Earth History; An Interdisciplinary Conference on Impacts, Volcanism, and Mass Mortality*, p. 319–334, Boulder, CO, The Geological Society of America, Special Paper 247, 1990.

Hughes, D. W., Satellites in decay, *Nature, v. 361*, p. 509, 1993.

Hughes, H. G., F. N. App, and T. R. McGetchin, Global seismic effects of basin forming impacts, *Phys. Earth Planet. Interiors, v. 15*, p. 251–263, 1977.

Hughes, T. J., Book Review: Atlas of Continental Displacement, 200 Million Years to the Present, by H. G. Owen, *Palaeogeog., Palaeoclim., Palaeoecol., v. 55*, p. 95–99, 1986.

Hughes, T. J., G. H. Denton, B. G. Andersen, D. H. Schilling, J. L. Fastook, and C. S. Lingle, The last great ice sheets: A global view, in G. H. Denton and T. J. Hughes, eds., *The Last Great Ice Sheets*, p. 275–318, New York, John Wiley & Sons, Inc., 1981.

Humes, D. H., Results of Pioneer 10 and 11 meteoroid experiments: Interplanetary and near-Saturn, *Jour. Geophys. Res., v. 85*, p. 5841–5852, 1980.

Hunt, E. R., Stabilizing high-period orbits in a chaotic system: The diode resonator, *Phys. Rev. Lett, v. 67*, p. 1953–1955, 1991.

Hunter, D. A., and J. S. Gallagher, III, Star formation in irregular galaxies, *Science, v. 243*, p. 1557–1563, 1989.

Huntley, H. E., *The Divine Proportion: A Study in Mathematical Beauty*, New York, Dover Publications, 186 pp., 1970.

Huss, G. R., Meteorite infall as a function of mass: Implications for the accumulation of meteorites on Antarctic ice, *Meteoritics, v. 25*, p. 41–56, 1990.

Hut, P., Evolution of the Solar System in the presence of a solar companion star, in R.

References Cited

Smoluchowski, J. N. Bahcall, and M. S. Matthews, eds., *The Galaxy and the Solar System*, p. 313–337, Tucson, AZ, The University of Arizona Press, 1986.

Hut, P., W. Alvarez, W. P. Elder, T. Hansen, E. G. Kauffman, G. Keller, E. M. Shoemaker, and P. R. Weissman, Comet showers as a cause of mass extinctions, *Nature, v. 329*, p. 118–126, 1987.

Hut, P., E. M. Shoemaker, W. Alvarez, and A. Montanari, Astronomical mechanisms and geologic evidence for multiple impacts on Earth, *Lunar and Planetary Sci. Conf. XXII, Abstracts of Papers*, p. 603–604, Lunar and Planetary Science Institute, Houston, TX, March 18–22, 1991.

Hutchinson, D. R., R. S. White, W. F. Cannon, and K. J. Schulz, Keweenaw Hot Spot: Geophysical evidence for a 1.1 Ga mantle plume beneath the Midcontinent Rift System, *Jour. Geophys. Res., v. 95*, p. 10,869–10,884, 1990.

Hutchinson, G. E., The concept of pattern in ecology, *Proc. Acad. Nat. Sci. Philadelphia, v. 105*, p. 1–12, 1953.

———, *A Treatise on Limnology: Vol. 1. Geography, Physics, and Chemistry*, New York, John Wiley & Sons, Inc., 1015 pp., 1957.

Hyde, R., and J. O. Dickey, Earth's variable rotation, *Science, v. 253*, p. 629637, 1991.

Iidaka, T., and D. Suetsugu, Seismological evidence for metastable olivine inside a subducting slab, *Nature, v. 356*, p. 593–595, 1992.

Ingersoll, A. P., Jupiter and Saturn, in B. Murray, ed., *The Planets: Readings From Scientific American*, p. 60–70, San Francisco, W. H. Freeman & Co., 1983.

———, Models of Jovian vortices, *Nature, v. 331*, p. 654–655, 1988.

Innanen, K. A., A. T. Patrick, and W. W. Duley, The interaction of the spiral density wave and the Sun's galactic orbit, *Astrophysics and Space Sci., v. 57*, p. 511–515, 1978.

Iooss, G., and W. F. Langford, Conjectures on the routes to turbulence via bifurcations, in R. H. G. Helleman, ed., *Nonlinear Dynamics*, p. 489–505, New York, Annals of the New York Acad. Sciences, v. 357, 1980.

Ip, W.-H., Dynamical injection of Mars-sized planetoids into the asteroidal belt from the terrestrial planetary accretion zone, *Icarus, v. 78*, p. 270–279, 1989.

Ip, W.-H., and J. A. Fernández, Steady-state injection of short-period comets from the trans-Neptunian cometary belt, *Icarus, v. 92*, p. 185–193, 1991.

Ipatov, S. I., Evolution of asteroidal orbits at the 5:2 resonance, *Icarus, v. 95*, p. 100–114, 1992.

Irving, E., Fragmentation and assembly of the continents, mid-Carboniferous to present, *Geophysical Surveys, v. 5*, p. 299–333, 1982.

Irving, E., and G. A. Irving, Apparent polar wander paths Carboniferous through Cenozoic and the assembly of Gondwana, *Geophysical Surveys, v. 5*, p. 141–188, 1982.

Irwin, A. J., S. J. Fraser, and R. Kapral, Stochastically induced coherence in bistable systems, *Phys. Rev. Lett., v. 64*, p. 2343–2346, 1990.

Isomäki, H. M., J. von Boehm, and R. Räty, Devil's attractors and chaos of a driven impact oscillator, *Phys. Lett., v. 107A*, p. 343–346, 1985.

Ito, E., and H. Sato, Aseismicity in the lower mantle by superplasticity of the descending slab, *Nature, v. 351*, p. 140–141, 1991.

Ito, E., and E. Takahashi, Postspinel transformations in the system Mg_2SiO_4–Fe_2SiO_4 and some geophysical implications, *Jour. Geophys. Res., v. 94*, p. 10,637–10,646, 1989.

Ito, K., Chaos in the Rikitake two-disk dynamo system, *Earth Planet. Sci. Lett., v. 51*, p. 451–456, 1980.

Ives, A. R., Chaos in time and space, *Nature, v. 353*, p. 214–215, 1991.

Izett, G. A., G. B. Dalrymple, L. W. Snee, and M. S. Pringle, $^{40}Ar/^{39}Ar$ age (66–64 Ma) of K-T boundary tektites, *Lunar and Planetary Sci. Conf. XXII, Abstracts of Papers*, p. 627–628, Lunar and Planetary Science Institute, Houston, TX, March 18–22, 1991a.

Izett, G. A., G. B. Dalrymple, and L. W. Snee, $^{40}Ar/^{39}Ar$ age of Cretaceous-Tertiary boundary tektites from Haiti, *Science, v. 252*, p. 1539–1542, 1991b.

Jablonski, D., Background and mass extinctions: The alternation of macroevolutionary regimes, *Science, v. 231*, p. 129–133, 1986a.

———, Causes and consequences of mass extinctions: A comparative approach, in D. K. Elliott, ed., *Dynamics of Extinction*, p. 183–229, New York, John Wiley & Sons, Inc., 1986b.

References Cited

Jacchia, L. G., Meteors, meteorites, and comets: Interrelations, in B. M. Middlehurst and G. P. Kuiper, eds., *The Moon, Meteorites, and Comets*, p. 774–798, Chicago, The University of Chicago Press, 1963.

Jackson, A., Still poles apart on reversals?, *Nature, v. 358*, p. 194–195, 1992.

Jackson, E. A., *Perspectives of Nonlinear Dynamics, Volume 1*, New York, Cambridge University Press, 495 pp., 1989.

———, *Perspectives of Nonlinear Dynamics, Volume 2*, New York, Cambridge University Press, 632 pp., 1990.

Jackson, E. D., Linear volcanic chains on the Pacific plate, in *The Geophysics of the Pacific Ocean Basin and its Margins*, p. 319–335, Washington, DC, Am. Geophys. Union, Geophys. Monograph 19, 1976.

Jackson, E. D., and H. R. Shaw, Stress fields in central portions of the Pacific plate: Delineated in time by linear volcanic chains, *Jour. Geophys. Res., v. 80*, p. 1861–1874, 1975.

Jackson, E. D., E. A. Silver, and G. B. Dalrymple, Hawaiian-Emperor chain and its relation to Cenozoic circum-Pacific tectonics, *Geol. Soc. Am. Bull., v. 83*, p. 601–618, 1972.

Jackson, E. D., H. R. Shaw, and K. E. Bargar, Calculated geochronology and stress field orientations along the Hawaiian chain, *Earth Planet. Sci. Lett., v. 26*, p. 145–155, 1975.

Jacob, F., and E. L. Wollman, Viruses and genes, in D. Kennedy, ed., *Cellular and Organismal Biology: Readings From Scientific American*, p. 33–46, San Francisco, W. H. Freeman & Co., 1974.

Jacobs, J. A., Heat flow and reversals of the Earth's magnetic field, *Jour. Geomag. Geoelectr., v. 33*, p. 527–529, 1981.

———, *Reversals of the Earth's Magnetic Field*, Bristol, U.K., Adam Hilger, Ltd., 230 pp., 1984.

———, *The Earth's Core*, New York, Academic Press, 413 pp., 1987.

———, Causes of changes in reversals of the Earth's magnetic field: Inside or outside the core?, *Eos, Trans. Am. Geophys. Union, v. 73*, p. 89–91, 1992a.

———, *Deep Interior of the Earth*, New York, Chapman and Hall, 167 pp., 1992b.

Jakubick, A. T., and W. Church, *Oklo Natural Reactors: Geological and Geochemical Conditions – A Review*, Ottawa, Canada, Atomic Energy Control Board, Res. Report INFO-0179, 53 pp., 1986.

James, G., and R. C. James, *Mathematics Dictionary, Fifth Edition*, New York, Van Nostrand Reinhold, 548 pp., 1992.

Janle, P., A. T. Basilevsky, M. A. Kreslavsky, and E. N. Slyuta, Heat loss and tectonic style of Venus, *Earth, Moon, and Planets, v. 58*, p. 1–29, 1992.

Jansa, L. F., M.-P. Aubry, F. M. Gradstein, Comets and extinctions: Cause and effect?, in V. L. Sharpton and P. D. Ward, eds., *Global Catastrophes in Earth History; An Interdisciplinary Conference on Impacts, Volcanism, and Mass Mortality*, p. 223–232, Boulder, CO, The Geological Society of America, Special Paper 247, 1990.

Jantsch, E., *The Self-Organizing Universe: Scientific and Human Implications of the Emerging Paradigm of Evolution*, New York, Pergamon Press, 343 pp., 1980.

Jayawardhana, R., Could quasars get their shine from stars?, *Science, v. 259*, p. 1692–1693, 1993.

Jeanloz, R., The nature of the Earth's core, *Ann. Rev. Earth Planet. Sci., v. 18*, p. 357–386, 1990.

Jeffreys, H., Tidal friction in shallow seas, *Phil. Trans. Roy. Soc. London, v. A221*, p. 239–264, 1920.

———, *The Earth (Fifth Edition)*, New York, Cambridge University Press, 525 pp., 1970.

Jensen, M. H., P. Bak, and T. Bohr, Transition to chaos by interaction of resonances in dissipative systems: I. Circle maps, *Phys. Rev. A, v. 30*, p. 1960–1969, 1984.

Jensen, M. H., L. P. Kadanoff, A. Libchaber, I. Procaccia, and J. Stavans, Global universality at the onset of chaos: Results of a forced Rayleigh-Bénard experiment, *Phys. Rev. Lett., v. 55*, p. 2798–2801, 1985.

Jensen, R. V., Classical Chaos, *Am. Scientist, v. 75*, p. 168–181, 1987.

———, Quantum chaos, *Nature, v. 355*, p. 311–318, 1992.

Jensen, R. V., and C. R. Myers, Images of the critical points of nonlinear maps, *Phys. Rev. A, v. 32*, p. 1222–1224, 1985.

References Cited

Jensen, W. B., Classification, symmetry and the periodic table, *Computers & Mathematics with Applications, v. 12B*, p. 487–510, 1986.

Jewitt, D. C., The rings of Jupiter, in D. Morrison, ed., *Satellites of Jupiter*, p. 44–64, Tucson, AZ, The University of Arizona Press, 1982.

Jin, R-S., and S. Jin, The ~60-year power spectral peak of the magnetic variations around London and the Earth's rotation rate fluctuations, *Jour. Geophys. Res., v. 94*, p. 13,673–13,679, 1989.

Johnson, G. L., and J. E. Rich, A 30 million year cycle in Arctic volcanism, *Jour. Geodynamics, v. 6*, p. 111–116, 1986.

Johnson, K. R., and L. J. Hickey, Megafloral change across the Cretaceous/Tertiary boundary in the northern Great Plains and Rocky Mountains, U.S.A., in V. L. Sharpton and P. D. Ward, eds., *Global Catastrophes in Earth History; An Interdisciplinary Conference on Impacts, Volcanism, and Mass Mortality*, p. 433–444, Boulder, CO, The Geological Society of America, Special Paper 247, 1990.

Johnson, T. V., and L. A. Soderblom, Volcanic eruptions on Io: Implications for surface evolution and mass loss, in D. Morrison, ed., *Satellites of Jupiter*, p. 634–646, Tucson, AZ, The University of Arizona Press, 1982.

Joly, J., *The Surface-History of the Earth, Second Edition*, Oxford, Clarendon Press, 211 pp., 1930.

Jordan, P., *The Expanding Earth: Some Consequences of Dirac's Gravitation Hypothesis* (Translated and revised from the German edition: Pascual Jordan, *Die Expansion der Erde*, Friedr. Vieweg, Braunschweig, 1966), New York, Pergamon, 202 pp., 1971.

Jorna, S., ed., *Topics in Nonlinear Dynamics: A Tribute to Sir Edward Bullard*, New York, American Institute of Physics, 404 pp., 1978.

Jose, P. D., Sun's motion and Sunspots, *Astron. Jour., v. 70*, p. 193–200, 1965.

Judson, H. F., *The Eighth Day of Creation: Makers of the Revolution in Biology*, New York, A Touchstone Book, Simon and Schuster, Inc., 686 pp., 1979.

Kadanoff, L. P., S. R. Nagel, L. Wu, and S-M. Zhou, Scaling and universality in avalanches, *Phys. Rev. A, v. 39*, p. 6524–6537, 1989.

Kaku, M., and J. Trainer, *Beyond Einstein: The Cosmic Quest for the Theory of the Universe*, New York, Bantam Books, 226 pp., 1987.

Kaler, J. B., *Stars*, New York, Scientific American Library, A Division of HPHLP, 273 pp., 1992.

Kanamori, H., Are earthquakes a major cause of the Chandler wobble?, *Nature, v. 262*, p. 254–255, 1976.

——, Rupture process of subduction-zone earthquakes, *Ann. Rev. Earth Planet. Sci., v. 14*, p. 293–322, 1986.

Kaneko, K., *Collapse of Tori and Genesis of Chaos in Dissipative Systems*, Singapore, World Scientific Publishing Co., 264 pp., 1986.

——, Clustering, coding, switching, hierarchical ordering, and control in a network of chaotic elements, *Physica D, v. 41*, p. 137–172, 1990.

Kapitaniak, T., The loss of chaos in a quasiperiodically-forced nonlinear oscillator, *Internat. Jour. Bifurcation and Chaos, v. 1*, p. 357–362, 1991.

Kaplan, H., A cartoon-assisted proof of Sarkowskii's theorem, *Am. Jour. Phys., v. 55*, p. 1023–1032, 1987.

Karlin, S., and V. Brendel, Patchiness and correlations in DNA sequences, *Science, v. 259*, p. 677–680, 1993.

Kauffman, E. G., The fabric of Cretaceous marine extinctions, in W. A. Berggren and J. A. Van Couvering, eds., *Catastrophes and Earth History: The New Uniformitarianism*, p. 151–246, Princeton, NJ, Princeton University Press, 1984.

——, *Dynamics of Cretaceous Epicontinental Seas (A 1985 Distinguished Lecture)*, Tulsa, OK, American Association of Petroleum Geologists, 2 Audio Tapes, 66 35-mm Slides, 1985.

——, The dynamics of marine stepwise mass extinction, in M. A. Lamolda, E. G. Kauffman, and O. H. Walliser, eds., *Palaeontology and Evolution: Extinction Events. 2nd International Conference on Global Bioevents (Bilbao Conf., 20–23 October, 1987)*, Revista Española de

References Cited

Palaeontologia, N°. Extraordinario, p. 57–71, Madrid, Sociedad Española de Paleontologia, 1988.

Kaufman, L., and K. Mallory, eds., *The Last Extinction*, Cambridge, MA, The MIT Press, 208 pp., 1986.

Kaufmann, W. J., *Universe, Third Edition*, New York, W. H. Freeman & Co., 648 pp. (with 12 Simplified Star Charts), 1991.

Kaula, W. M., *An Introduction to Planetary Physics*, New York, John Wiley & Sons, Inc., 490 pp., 1968.

———, Venus: A contrast in evolution to Earth, *Science, v. 247*, p. 1191–1196, 1990.

Keller, G., S. L. D'Hondt, C. J. Orth, J. S. Gilmore, P. Q. Oliver, E. M. Shoemaker, and E. Molina, Late Eocene impact microspherules: Stratigraphy, age, and geochemistry, *Meteoritics, v. 22*, p. 25–60, 1987.

Kellogg, L. H., Mixing in the mantle, *Ann. Rev. Earth Planet. Sci., v. 20*, p. 365–388, 1992.

Kennett, J. P., A. R. McBirney, and R. C. Thunell, Episodes of Cenozoic volcanism in the Circum-Pacific region, *Jour. Volc. Geoth. Res., v. 2*, p. 145–163, 1977.

Kerr, R. A., Making the Moon, remaking the Earth, *Science, v. 243*, p. 1433–1435, 1989.

———, Extinction by a one-two comet punch?, *Science, v. 255*, p. 160–161, 1992a.

———, Huge impact tied to mass extinction, *Science, v. 257*, p. 878–880, 1992b.

———, From Mercury to Pluto, chaos pervades the Solar System, *Science, v. 257*, p. 33, 1992c.

———, The earliest mass extinction?, *Science, v. 257*, p. 612, 1992d.

———, Planetesimal found beyond Neptune, *Science, v. 257*, p. 1865, 1992e.

———, A revisionist timetable for the Ice Ages, *Science, v. 258*, p. 220–221, 1992f.

———, Earth gains a retinue of mini-asteroids, *Science, v. 258*, p. 403, 1992g.

———, Catastrophes of every ilk at the geophysics fest, *Science, v. 259*, p. 28–29, 1993a.

———, Second crater points to killer comets, *Science, v. 259*, p. 1543, 1993b.

———, Ocean-in-a-machine starts looking like the real thing (Sidebar: The parallel route to an ocean model), *Science, v. 260*, p. 32–33, 1993c.

———, How Ice Age climate got the shakes, *Science, v. 260*, p. 890–892, 1993d.

Kerridge, J. F., and M. S. Matthews, *Meteorites and the Early Solar System*, Tucson, AZ, The University of Arizona Press, 1269 pp., 1988.

Kiang, T., The distribution of asteroids in the direction perpendicular to the ecliptic plane, in T. Gehrels, ed., *Physical Studies of Minor Planets*, p. 187–195, Washington, DC, National Aeronautics and Space Administration, NASA SP-267, 1971.

Kieffer, S. W., Shock metamorphism of the Coconino sandstone at Meteor Crater, Arizona, *Jour. Geophys. Res., v. 76*, p. 5449–5473, 1971.

———, From regolith to rock by shock, *The Moon, v. 13*, p. 301–320, 1975.

———, Dynamics and thermodynamics of volcanic eruptions: Implications for the plumes on Io, in D. Morrison, ed., *Satellites of Jupiter*, p. 647–723, Tucson, AZ, The University of Arizona Press, 1982.

———, Geologic nozzles, *Rev. Geophys., v. 27*, p. 3–38, 1989.

Kieffer, S. W., and J. M. Delany, Isentropic decompression of fluids from crustal and mantle pressures, *Jour. Geophys. Res., v. 84*, p. 1611–1620, 1979.

Kieffer, S. W., and C. H. Simonds, The role of volatiles and lithology in the impact cratering process, *Rev. Geophys. Space Phys., v. 18*, p. 143–181, 1980.

Kieffer, S. W., P. P. Phakey, and J. M. Christie, Shock processes in porous quartzite: Transmission electron microscope observations and theory, *Contrib. Mineral. Petrol., v. 59*, p. 41–93, 1976a.

Kieffer, S. W., R. B. Schaal, R. Gibbons, F. Hörz, D. J. Milton, and A. Dube, Shocked basalt from Lonar Crater, India, and experimental analogues, *Proc. Lunar Sci. Conf., 7th*, p. 1391–1412, 1976b.

King, S. D., Seeing the mantle in the round, *Nature, v. 361*, p. 688–689, 1993.

King-Hele, D., The shape of the Earth, *Science, v. 192*, p. 1293–1300, 1976.

———, *A Tapestry of Orbits*, New York, Cambridge University Press, 244 pp., 1992.

Kinzel, W., Spin glasses and memory, *Physica Scripta, v. 35*, p. 398–401, 1987.

Kippenhahn, R., *Bound to the Sun: The Story of Planets, Moons, and Comets*, New York, W. H. Freeman & Co., 282 pp., 1990.

References Cited

Kirby, S. H., W. B. Durham, and L. A. Stern, Mantle phase changes and deep-earthquake faulting in subducting lithosphere, *Science, v. 252*, p. 216–225, 1991.

Kirkwood, D., A new analogy in the periods of rotation of the primary planets (Communicated by S. C. Walker), *Jour. Franklin Institute, v. 18, 3rd Ser.*, p. 324–328, 1849.

———, Instances of nearly commensurable periods in the Solar System, *The Mathematical Monthly, v. 2*, p. 126–127, 1860.

———, *Meteoric Astronomy: A Treatise on Shooting-Stars, Fire-balls, and Aerolites*, Philadelphia, PA, J. B. Lippincott & Co., 129 pp., 1867.

———, On the formation and primitive structure of the Solar System, *Proc. Am. Phil. Soc., v. 12*, p. 163–167, 1871.

———, On the disintegration of comets, *Nature, v. 6*, p. 148–149, 1872.

———, The asteroids between Mars and Jupiter, *Smithsonian Report*, p. 358–371, 1876.

———, *Asteroids, or Minor Planets Between Mars and Jupiter*, Philadelphia, J. B. Lippincott Company, 60 pp., 1888.

———, On the origin of the gaps in the zone of asteroids, *Sidereal Messenger, v. 10*, p. 194–196, 1891.

Kirschner, C. E., A. Grantz, and M. W. Mullen, Impact origin of the Avak structure, Arctic Alaska, and genesis of the Barrow gas fields, *Am. Assoc. Petrol. Geologists, v. 76*, p. 651–679, 1992.

Kistler, R. W., J. F. Evernden, and H. R. Shaw, Sierra Nevada plutonic cycle: Part I, Origin of composite granite batholiths, *Geol. Soc. Am. Bull., v. 82*, p. 853–868, 1971.

Kitchell, J. A., and D. Pena, Periodicity of extinctions in the geologic past: Deterministic versus stochastic explanations, *Science, v. 226*, p. 689–691, 1984.

Klein, F. W., Earthquake swarms and the semidiurnal solid earth tide, *Geophys. Jour. Roy. Astron. Soc., v. 45*, p. 245–295, 1976.

———, Patterns of historical eruptions at Hawaiian volcanoes, *Jour. Volc. Geoth. Res., v. 12*, p. 1–35, 1982.

Klein, F. W., R. Y. Koyanagi, J. S. Nagata, and W. R. Tanigawa, The seismicity of Kilauea's magma system, in R. W. Decker, T. L. Wright, and P. H. Stauffer, eds., *Volcanism in Hawaii*, Vol. 2, Chap. 43, p. 1019–1185, U.S. Geol. Survey, Prof. Paper 1350, Washington, DC, U.S. Gov't Printing Office, 1987.

Kloeden, P. E., E. Platen, and H. Schurz, The numerical solution of nonlinear stochastic dynamical systems: A brief introduction, *Internat. Jour. Bifurcation and Chaos, v. 1*, p. 277–286, 1991.

Knight, D. M., *The Transcendental Part of Chemistry*, Folkestone, Kent, UK, Wm. Dawson & Son Ltd., 289 pp., 1978.

Knittle, E., and R. Jeanloz, Earth's core-mantle boundary: Results of experiments at high pressures and temperatures, *Science, v. 251*, p. 1438–1443, 1991.

Knobloch, E., D. R. Moore, J. Toomre, and N. O. Weiss, Transitions to chaos in two-dimensional double-diffusive convection, *Jour. of Fluid Mechanics, v. 166*, p. 409–448, 1986.

Knoll, A. H., Patterns of extinction in the fossil record of vascular plants, in M. H. Nitecki, ed., *Extinctions*, p. 21–68, Chicago, IL, The University of Chicago Press, 1984.

———, End of the Proterozoic Eon, *Sci. Am., v. 265*, p. 64–73, 1991.

Knoll, A. H., and M. R. Walter, Latest Proterozoic stratigraphy and Earth history, *Nature, v. 356*, p. 673–678, 1992.

Koeberl, C., V. L. Sharpton, T. M. Harrison, D. Sandwell, A. V. Murali, and K. Burke, The Kara/Ust-Kara twin impact structure; a large-scale impact event in the late Cretaceous, in V. L. Sharpton and P. D. Ward, eds., *Global Catastrophes in Earth History; An Interdisciplinary Conference on Impacts, Volcanism, and Mass Mortality*, p. 233–238, Boulder, CO, The Geological Society of America, Special Paper 247, 1990.

Kolvoord, R. A., J. A. Burns, and M. R. Showalter, Periodic features in Saturn's F ring: Evidence for nearby moonlets, *Nature, v. 345*, p. 695–697, 1990.

Komuro, M., R. Tokunaga, T. Matsumoto, L. O. Chua, and A. Hotta, Global bifurcation analysis of the double scroll circuit, *Internat. Jour. Bifurcation and Chaos, v. 1*, p. 139–182, 1991.

Kordylewski, K., Photographische Untersuchungen des Librationspunktes L_5 im System Erde-Mond, *Acta Astronomica, v. 11*, p. 165–169, 1961.

References Cited

Korsch, H. J., and A. Wagner, Fractal mirror images and chaotic scattering, *Computers in Physics, Sept/Oct*, p. 497–504, 1991.

Koutchmy, S., J. B. Zirker, R. S. Steinolfson, and J. D. Zhugzda, Coronal activity, in A. N. Cox, W. C. Livingston, and M. S. Matthews, eds., *Solar Interior and Atmosphere*, p. 1044–1086, Tucson, AZ, The University of Arizona Press, 1991.

Koyanagi, R. Y., B. Chouet, and K. Aki, Origin of volcanic tremor in Hawaii, I. Data from the Hawaiian Volcano Observatory, in R. W. Decker, T. L. Wright, and P. H. Stauffer, eds., *Volcanism in Hawaii*, Vol. 2, Chap. 45, p. 1221–1257, U.S. Geol. Survey, Prof. Paper 1350, Washington, DC, U.S. Gov't Printing Office, 1987.

Kramer, E. E., *The Nature and Growth of Modern Mathematics*, New York, Hawthorn Books, Inc. Publishers, 758 pp., 1970.

Kresák, L., Dynamical interrelationships among comets and asteroids, in T. Gehrels, ed., *Asteroids*, p. 289–309, Tucson, AZ, The University of Arizona Press, 1979.

———, Evidence for physical aging of periodic comets, in R. L. Newburn, Jr., M. Neugebauer, and J. Rahe, eds., *Comets in the Post-Halley Era, Volume 1*, p. 607–628, Boston, MA, Kluwer Academic Publishers, 1991.

Kresák, L., and J. Klacka, Selection effects of asteroid discoveries and their consequences, *Icarus, v. 78*, p. 287–297, 1989.

Kring, D. A., A. R. Hildebrand, and W. V. Boynton, The petrology of an andesitic melt rock and a polymict breccia from the interior of the Chicxulub structure, Yucatán, Mexico, *Lunar and Planetary Sci. Conf. XXII, Abstracts of Papers*, p. 755–756, Lunar and Planetary Science Institute, Houston, TX, March 18–22, 1991.

Krinov, E. L., Meteorite craters on the Earth's surface, in B. M. Middlehurst and G. P. Kuiper, eds., *The Moon, Meteorites, and Comets*, p. 183–207, Chicago, The University of Chicago Press, 1963a.

———, The Tunguska and Sikhote-Alin meteorites, in B. M. Middlehurst and G. P. Kuiper, eds., *The Moon, Meteorites, and Comets*, p. 208–234, Chicago, The University of Chicago Press, 1963b.

Krinsky, V. I., ed., *Self-Organization: Autowaves and Structures Far from Equilibrium*, New York, Springer-Verlag, 263 pp., 1984.

Kristian, J., Gravitational lenses, *Carnegie Institution of Washington Year Book 91* (The President's Report, July 1991–June, 1992), p. 24–34, 1992.

Kroeber, T., *Ishii in Two Worlds: A Biography of the Last Wild Indian in North America*, Berkeley, CA, University of California Press, 258 pp., 1961.

———, *The Inland Whale: Nine Stories Retold from California Indian Legends*, Berkeley, CA, University of California Press, 205 pp., 1971.

Krolik, J. H., Creation by stellar ablation of the low-mass companion to pulsar 1829–10, *Nature, v. 353*, p. 829–831, 1991.

Kröner, A., and P. W. Layer, Crust formation and plate motion in the early Archean, *Science, v. 256*, p. 1405–1411, 1992.

Kuiper, G. P., On the origin of the Solar System, in J. A. Hynek, *Astrophysics: A Topical Symposium*, p. 357–424, New York, McGraw-Hill, 1951.

Kump, L., Bacteria forge a new link, *Nature, v. 362*, p. 790–791, 1993.

Kunk, M. J., G. A. Izett, R. A. Haugerud, and J. F. Sutter, ^{40}Ar/^{39}Ar dating of the Manson Impact Structure; A Cretaceous-Tertiary boundary crater candidate, *Science, v. 244*, p. 1565–1568, 1989.

Kurths, J., and H. Herzel, An attractor in a solar time series, *Physica D, v. 25*, p. 165–172, 1987.

Lachenbruch, A. H., Frictional heating, fluid pressure, and the resistance to fault motion, *Jour. Geophys. Res., v. 85*, p. 6097–6112, 1980.

Lada, C. J., and F. H. Shu, The formation of sunlike stars, *Science, v. 248*, p. 564–572, 1990.

Lagerkvist, C.-I., H. Rickman, B. A. Lindblad, and M. Lindgren, eds., *Asteroids, Comets, Meteors III*, Proceedings of Meeting at the Astronomical Observatory of Uppsala University, June 12–16, 1989, Uppsala, Sweden, Uppsala universitet Reprocentralen HSC, 620 pp., 1990.

Laj, C., A. Mazaud, R. Weeks, M. Fuller, and E. Herrero-Bervera, Geomagnetic reversal paths, *Nature, v. 351*, p. 447, 1991.

References Cited

———, Geomagnetic reversal paths, *Nature, v. 359,* p. 111–112, 1992.

Lambeck, K., *The Earth's Variable Rotation,* New York, Cambridge University Press, 449 pp., 1980.

Landscheidt, T., Solar oscillations, sunspot cycles, and climate change, in B. M. McCormac, ed., *Weather and Climate Responses to Solar Variations,* p. 293–308, Boulder, CO, Colorado Associated University Press, 1983.

———, Solar rotation, impulses of the torque in the Sun's motion, and climatic variation, *Climatic Change, v. 12,* p. 265–295, 1988.

Langereis, C. G., A. A. M. van Hoof, and P. Rochette, Longitudinal confinement of geomagnetic reversal paths as a possible sedimentary artefact, *Nature, v. 358,* p. 226–230, 1992.

LaPaz, L., The Canadian Fireball Procession of 1913 February 9, *Meteoritics, v. 1 (4),* p. 402–405, 1956.

Larimer, J. W., The cosmochemical classification of the elements, in J. F. Kerridge and M. S. Matthews, eds., *Meteorites and the Early Solar System,* p. 375–389, Tucson, AZ, The University of Arizona Press, 1988.

Larnder, C., N. Desaulniers-Soucy, S. Lovejoy, D. Schertzer, C. Braun, and D. Lavallée, Universal multifractal characterization and simulation of speech, *Internat. Jour. Bifurcation and Chaos, v. 2,* p. 715–719, 1992.

Larson, R. L., Latest pulse of Earth: Evidence for a mid-Cretaceous superplume, *Geology, v. 19,* p. 547–550, 1991a.

———, Geological consequences of superplumes, *Geology, v. 19,* p. 963–966, 1991b.

Larson, R. L., and P. Olson, Mantle plumes control magnetic reversal frequency, *Earth Planet. Sci. Lett., v. 107,* p. 437–447, 1991.

Laskar, J., Secular evolution of the Solar System over 10 million years, *Astron. Astrophys., v. 198,* p. 341–362, 1988.

———, A numerical experiment on the chaotic behaviour of the Solar System, *Nature, v. 338,* p. 237–238, 1989.

———, The chaotic motion of the Solar System: A numerical estimate of the size of the chaotic zones, *Icarus, v. 88,* p. 266–291, 1990.

Laskar, J., and P. Robutel, The chaotic obliquity of the planets, *Nature, v. 361,* p. 608–612, 1993.

Laskar, J., T. Quinn, and S. Tremaine, Confirmation of resonant structure in the Solar System, *Icarus, v. 95,* p. 148–152, 1992.

Laskar, J., F. Joutel, and P. Robutel, Stabilization of the Earth's obliquity by the Moon, *Nature, v. 361,* p. 615–617, 1993.

Lasota, A., and M. C. Mackey, *Probabilistic Properties of Deterministic Systems,* New York, Cambridge University Press, 358 pp., 1985.

Lauterborn, W., and J. Holzfuss, Evidence of a low-dimensional strange attractor in acoustic turbulence, *Phys. Lett., v. 115A,* p. 368–372, 1986.

———, Acoustic chaos, *Internat. Jour. Bifurcation and Chaos, v. 1,* p. 13–26, 1991.

Lawver, L. A., R. D. Müller, S. P. Srivastava, and W. Roest, The opening of the Arctic Ocean, in U. Bleil and J. Thiede, eds., *Geological History of the Polar Oceans: Arctic Versus Antarctic,* p. 29–62, Boston, MA, Kluwer Academic Publishers, 1990.

Lay, T., Wrinkles on the inside, *Nature, v. 355,* p. 768–769, 1992.

Leake, M. A., C. R. Chapman, S. J. Weidenschilling, D. R. Davis, and R. Greenberg, The chronology of Mercury's geological and geophysical evolution: The Vulcanoid hypothesis, *Icarus, v. 71,* p. 350–375, 1987.

Le Douaran, S., H. D. Needham, and J. Francheteau, Pattern of opening rates along the axis of the Mid-Atlantic Ridge, *Nature, v. 300,* p. 254–257, 1982.

Lee, K.-C., What makes chaos border sticky?, *Physica D, v. 35,* p. 186–202, 1989.

Lenski, R. E., and J. E. Mittler, The directed mutation controversy and neo-Darwinism, *Science, v. 259,* p. 188–194, 1993.

Lerbekmo, J. F., A. R. Sweet, and R. M. St. Louis, The relationship between the iridium anomaly and palynological floral events at three Cretaceous-Tertiary boundary localities in western Canada, *Bull. Geol. Soc. Am., v. 99,* p. 325–330, 1987.

References Cited

Lerner, E. J., *The Big Bang Never Happened*, New York, Times Books, A Division of Random House, 466 pp., 1991.

Levine, A. J., *Viruses*, New York, Scientific American Library, A Division of HPHLP, 240 pp., 1992.

Levine, D. S., Survival of the synapses, *The Sciences, November/December*, p. 46–52, New York, The New York Academy of Sciences, 1988.

Levinton, J. S., The Big Bang of animal evolution, *Sci. Am., v. 267*, p. 84–91 (November), 1992.

Levy, E. H., Energetics of chondrule formation, in J. F. Kerridge and M. S. Matthews, eds., *Meteorites and the Early Solar System*, p. 697–711, Tucson, AZ, The University of Arizona Press, 1988.

Lewin, R., Trees from genes and tongues, *Science, v. 242*, p. 514, 1988a.

———, Linguists search for the Mother Tongue, *Science, v. 242*, p. 1128–1129, 1988b.

Lewis, J. E., and L. Glass, Steady states, limit cycles, and chaos in models of complex biological networks, *Internat. Jour. Bifurcation and Chaos, v. 1*, p. 477–483, 1991.

Li, T-Y., and J. A. Yorke, Period three implies chaos, *American Mathematical Monthly, v. 82*, p. 985–992, 1975.

Liapunov, A. M., Problème général de la stabilité du mouvement (Transl. from Russian by E. Davaux; Reprinted from Annales de la Faculté de Sciences de Toulouse, 2nd Ser., v. 9, 1907), *Annals of Mathematics Studies, No. 17*, p. 203–474, Princeton, NJ, Princeton University Press, 1947.

Lichtenberg, A. J., and M. A. Lieberman, *Regular and Stochastic Motion*, New York, Springer-Verlag, 499 pp., 1983.

Lim, T. T., and T. B. Nickels, Instability and reconnection in the head-on collision of two vortex rings, *Nature, v. 357*, p. 225–227, 1992.

Lima-de-Faria, A., *Evolution Without Selection: Form and Function by Autoevolution*, New York, Elsevier, 372 pp., 1988.

Lin, E. C. C., R. Goldstein, and M. Syvanen, *Bacteria, Plasmids, and Phages: An Introduction to Molecular Biology*, Cambridge, MA, Harvard University Press, 316 pp., 1984.

Lindblad, B. A., and R. B. Southworth, A study of asteroid families and streams by computer techniques, in T. Gehrels, ed., *Physical Studies of Minor Planets*, p. 337–352, Washington, DC, National Aeronautics and Space Administration, NASA SP-267, 1971.

Lindholm, C. D., J. Havskov, and M. A. Sellevoll, Periodicity in seismicity: Examination of four catalogs, *Tectonophys., v. 191*, p. 155–164, 1991.

Lindley, D., Messing around with gravity, *Nature, v. 359*, p. 583, 1992a.

———, Earth saved from disaster!, *Nature, v. 360*, p. 623, 1992b.

Lindström, M., V. Puura, T. Flodén, and Å. Bruun, Ordovician impacts at sea in Baltoscandia, *Papers Presented at the International Conference on Large Meteorite Impacts and Planetary Evolution, 31 August–2 September, 1992, Sudbury, Ontario, Canada*, p. 47, Lunar and Planetary Institute, Houston, TX (LPI Contribution No. 790), 1992.

Ling, F. H., G. Schmidt, and H. Kook, Universal behavior of coupled nonlinear oscillators, *Internat. Jour., Bifurcation and Chaos, v. 1*, p. 363–368, 1991.

Lipps, J. H., Extinction dynamics in pelagic ecosystems, in D. K. Elliott, ed., *Dynamics of Extinction*, p. 87–104, New York, John Wiley & Sons, Inc., 1986.

Lipschutz, M. E., and D. S. Woolum, Highly labile elements, in J. F. Kerridge and M. S. Matthews, eds., *Meteorites and the Early Solar System*, p. 462–487, Tucson, AZ, The University of Arizona Press, 1988.

Lipschutz, M. E., M. J. Gaffey, and P. Pellas, Meteoritic parent bodies: Nature, number, size and relation to present-day asteroids, in R. P. Binzel, T. Gehrels, and M. S. Matthews, eds., *Asteroids II*, p. 740–777, Tucson, AZ, University of Arizona Press, 1989.

Lissauer, J. J., S. W. Squyres, and W. K. Hartmann, Bombardment history of the Saturn system, *Jour. Geophys. Res., v. 93*, p. 13,776–13,804, 1988.

Liu, M., D. A. Yuen, W. Zhao, and S. Honda, Development of diapiric structures in the upper mantle due to phase transitions, *Science, v. 252*, p. 1836–1839, 1991.

References Cited

Livermore, R. A., F. J. Vine, and A. G. Smith, Plate motions and the geomagnetic field, II. Jurassic to Tertiary, *Geophys. Jour. Roy. Astron. Soc., v. 79*, p. 939–961, 1984.

Lockner, D. A., M. J. S. Johnston, and J. D. Byerlee, A mechanism to explain the generation of earthquake lights, *Nature, v. 302*, p. 28–33, 1983.

Loper, D. E., The gravitationally powered dynamo, *Geophys. Jour. Roy. Astron. Soc., v. 54*, p. 389–404, 1978.

Loper, D. E., and K. McCartney, Mantle plumes and the periodicity of magnetic field reversals, *Geophys. Res. Lett., v. 13*, p. 1525–1528, 1986.

Lorenz, E. N., Deterministic nonperiodic flow, *Jour. Atmospheric Sciences, v. 20*, p. 130–141, 1963.

——, The problem of deducing the climate from the governing equations, *Tellus, v. 16*, p. 1–11, 1964.

——, Atmospheric predictability as revealed by naturally occurring analogues, *Jour. Atmos. Sci., v. 26*, p. 636–646, 1969.

——, On the prevalence of aperiodicity in simple systems, *Lecture Notes in Math., v. 755*, p. 53–75, 1979.

——, Noisy periodicity and reverse bifurcation, in R. H. G. Helleman, ed., *Nonlinear Dynamics*, p. 282–291, New York (*Annals of the New York Acad. Sciences, v. 357*), 1980.

——, Atmospheric predictability experiments with a large numerical model, *Tellus, v. 34*, p. 505–513, 1982.

——, Irregularity: A fundamental property of the atmosphere, *Tellus, v. 36A*, p. 98–110, 1984.

——, The growth of errors in prediction, in M. Ghil, R. Benzi, and G. Parisi, eds., *Turbulence and Predictability in Geophysical Fluid Dynamics and Climate Dynamics*, p. 243–265, New York, North-Holland, 1985.

——, Computational chaos — A prelude to computational instability, *Physica D, v. 35*, p. 299–317, 1989.

——, Dimension of weather and climate attractors, *Nature, v. 353*, p. 241–244, 1991.

Lovelock, J. E., Gaia as seen through the atmosphere, *Atmospheric Environment, v. 6*, p. 579–580, 1972.

——, *The Ages of GAIA: A Biography of Our Living Earth*, New York, W. W. Norton & Co., 252 pp., 1988a.

——, The Earth as a living organism, in E. O. Wilson and F. M. Peter, eds., *Biodiversity*, p. 486–489, Washington, DC, National Academy Press, 1988b.

——, Rethinking life on Earth. The sum: Gaia takes flight, *Earthwatch, September/October*, p. 22–24, 1992.

Lovelock, J. E., and L. Margulis, Atmospheric homeostasis by and for the atmosphere: The Gaia hypothesis, *Tellus, v. 26*, p. 1–10, 1974a.

——, Homeostatic tendencies of the Earth's atmosphere, *Origins of Life, v. 1*, p. 12–22, 1974b.

Lowman, P. D., Crustal evolution in silicate planets: Implications for the origin of continents, *Jour. Geol., v. 84*, p. 1–26, 1976.

Lucchin, F., Clustering in the universe, in L. Pietronero and E. Tosatti, eds, *Fractals in Physics*, p. 313–318, New York, North-Holland Physics Publishing, a division of Elsevier Science Publishers, 1986.

Lucido, G., Magma-clustering and galaxy-clustering: A comparison, *Phys. Earth Planet. Int., v. 39*, p. 134–140, 1985.

Lull, R. S., *Organic Evolution*, New York, The Macmillan Company, 744 pp., 1949.

Luo, X., and D. N. Schramm, Fractals and cosmological large-scale structure, *Science, v. 256*, p. 513–515, 1992.

Lyne, A. G., F. G. Smith, and R. S. Pritchard, Spin-up and recovery in the 1989 glitch of the Crab pulsar, *Nature, v. 359*, p. 706–707, 1992.

MacArthur, R. H., and E. O. Wilson, An equilibrium theory of insular zoogeography, *EVOLUTION, (Internat. Jour. Organic Evol.), v. 17*, p. 373–387, 1963.

MacDonald, G. J. F., Tidal friction, *Rev. Geophys. Space Phys., v. 2*, p. 467–541, 1964.

Machetel, P., and P. Weber, Intermittent layered convection in a model mantle with an endothermic phase change at 670 km, *Nature, v. 350*, p. 55–57, 1991.

References Cited

Machetel, P., and D. A. Yuen, Penetrative convective flows induced by internal heating and mantle compressibility, *Jour. Geophys. Res., v. 94*, p. 10,609–10,626, 1989.

Mackay, C., Charge-coupled devices in astronomy, *Ann. Rev. Astron. Astrophys., v. 24*, p. 255–275, 1986.

MacKay, R. S., Period doubling as a universal route to stochasticity, in C. W. Horton, Jr., L. E. Reichl, and V. G. Szebehely, eds., *Long-Time Prediction in Dynamics*, p. 127–134, New York, John Wiley & Sons, Inc, 1983.

Mackey, M. C., and J. G. Milton, Dynamical diseases, *Annals, N. Y. Acad. Sci., v. 504*, p. 16–32, 1987.

MacLeod, N., and G. Keller, Hiatus distributions and mass extinctions at the Cretaceous/Tertiary boundary, *Geology, v. 19*, p. 497–501, 1991.

MacPherson, G. J., D. A. Wark, and J. T. Armstrong, Primitive material surviving in chondrites: Refractory inclusions, in J. F. Kerridge and M. S. Matthews, eds., *Meteorites and the Early Solar System*, p. 746–807, Tucson, AZ, The University of Arizona Press, 1988.

Maddox, J., Earthquakes and the Earth's rotation, *Nature, v. 332*, p. 11, 1988.

———, New ways with aggregation (News and Views), *Nature, v. 345*, p. 661, 1990.

———, Galactic origin of cosmic rays (News and Views), *Nature, v. 361*, p. 201, 1993a.

———, Sonoluminescence in from the dark (News and Views), *Nature, v. 361*, p. 397, 1993b.

Magnusson, P., M. A. Barucci, J. D. Drummond, K. Lumme, S. J. Ostro, J. Surdej, R. C. Taylor, and V. Zappalà, Determination of pole orientations and shapes of asteroids, in R. P. Binzel, T. Gehrels, and M. S. Matthews, eds., *Asteroids II*, p. 66–97, Tucson, AZ, The University of Arizona Press, 1989.

Mallet, R., Volcanic energy: An attempt to show its true origin and cosmical relations, *Phil. Trans. Roy. Soc. London, v. 163*, p. 147–227, 1873.

Mallove, E. F., The Solar System in chaos, *The Planetary Report, v. 9*, p. 4–7, 1989.

Mandel, J.-L., A. P. Monaco, D. L. Nelson, D. Schlessinger, and H. Willard, Genome analysis and the human X chromosome, *Science, v. 258*, p. 103–109, 1992.

Mandelbrot, B. B., Intermittent turbulence in self-similar cascades: Divergence of high moments and dimension of the carrier, *Jour. Fluid Mech., v. 62*, p. 331–358, 1974.

———, *Fractals: Form, Chance, and Dimension*, San Francisco, W. H. Freeman & Co., 365 pp., 1977.

———, *The Fractal Geometry of Nature* (Updated and Augmented), New York, W. H. Freeman & Co., 468 pp., 1983.

Mangan, M. T., and R. T. Helz, Vesicle and phenocryst distribution in Kilauea Iki lava lake, *Eos, Trans. Am. Geophys. Union, v. 66*, p. 1133, 1985.

Mankinen, E. A., M. Prévot, C. S. Grommé, and R. S. Coe, The Steens Mountain (Oregon) geomagnetic polarity transition: 1. Directional history, duration of episodes, and rock magnetism, *Jour. Geophys. Res., v. 90*, p. 10,393–10,416, 1985.

Mansinha, L., and D. E. Smylie, Effect of earthquakes on the Chandler wobble and the secular polar shift, *Jour. Geophys. Res., v. 72*, p. 4731–4743, 1967.

Maran, S. P., *The Astronomy and Astrophysics Encyclopedia*, New York, Van Nostrand Reinhold, 1002 pp., 1992.

Marcus, P., Spatial self-organization of vorticity in chaotic shearing flows, *Nuclear Physics (Proc. Suppl.), v. 2*, p. 127–138, 1987.

———, Numerical simulation of Jupiter's Great Red Spot, *Nature, v. 331*, p. 693–696, 1988.

———, Vortex dynamics in a shearing zonal flow, *Jour. Fluid Mech., v. 215*, p. 393–430, 1990.

Marcus, J. N., and M. A. Olsen, Biological implications of organic compounds in comets, in R. L. Newburn, Jr., M. Neugebauer, and J. Rahe, eds., *Comets in the Post-Halley Era, Volume 1*, p. 439–462, Boston, MA, Kluwer Academic Publishers, 1991.

Margulis, L., *Symbiosis in Cell Evolution: Life and Its Environment on the Early Earth*, W. H. Freeman & Co., 419 pp., 1981.

———, Speculation on speculation, in J. Brockman, ed., *Speculations: The Reality Club*, p. 157–167, New York, Prentice Hall Press, 1990.

———, Rethinking life on Earth. The parts: Power to the protoctists, *Earthwatch, September/October*, p. 25–29, 1992.

References Cited

Margulis, L., and J. E. Lovelock, The atmosphere as circulatory system of the biosphere — the Gaia hypothesis, in A. Kleiner and S. Brand, eds., *Ten Years of CoEvolution Quarterly: News that Stayed News, 1974–1984*, p. 15–25, San Francisco, North Point Press, 1986.

Margulis, L., and D. Sagan, *Microcosmos: Four Billion Years of Evolution from Our Microbial Ancestors*, New York, Summit Books, 301 pp., 1986a.

———, *Origins of Sex: Three Billion Years of Genetic Recombination*, New Haven, CT, Yale University Press, 258 pp., 1986b.

Marshall, J. R., and V. R. Oberbeck, Textures of impact deposits and the origin of tillite (Abstract), *Eos, Trans. Am. Geophys. Union*, v. 73 (Supplement, 27 Oct.), p. 324, 1992.

Martin, A. P., and C. Simon, Anomalous distribution of nuclear and mitochondrial DNA markers in periodical cicadas, *Nature*, v. 336, p. 237–239, 1988.

Marvin, U. B., Impact and its revolutionary implications for geology, in V. L. Sharpton and P. D. Ward, eds., *Global Catastrophes in Earth History; An Interdisciplinary Conference on Impacts, Volcanism, and Mass Mortality*, p. 147–154, Boulder, CO, The Geological Society of America, Special Paper 247, 1990.

Marzocchi, W., and F. Mulargia, Statistical analysis of the geomagnetic reversal sequences, *Phys. Earth Planet. Int.*, v. 61, p. 149–164, 1990.

Marzocchi, W., F. Mulargia, and P. Paruolo, The correlation of geomagnetic reversals and mean sea level in the last 150 m.y., *Earth Planet. Sci. Lett.*, v. 111, p. 383–393, 1992.

Masaitis, V. L., et al., *The Popigai Meteorite Crater*, Moscow, Nauka Press, 123 pp., 1975.

Mateo, M., Hunting for dark matter, *Carnegie Institution of Washington Year Book 91* (The President's Report, July 1991–June, 1992), p. 34–48, 1992.

Matese, J. J., and D. P. Whitmire, Planet X as the source of the periodic and steady-state flux of short period comets, in R. Smoluchowski, J. N. Bahcall, and M. S. Matthews, eds., *The Galaxy and the Solar System*, p. 297–309, Tucson, AZ, The University of Arizona Press, 1986.

Mather, K. F., *Source Book in Geology 1900–1950*, Cambridge, MA, Harvard University Press, 435 pp., 1967.

Mather, K. F., and S. L. Mason, *A Source Book in Geology*, New York, McGraw-Hill Book Company, Inc., 702 pp., 1939.

Mathews, P. M., and I. I. Shapiro, Nutations of the Earth, *Ann. Rev. Earth Planet. Sci.*, v. 20, p. 469–500, 1992.

Matsuzaki, Y., and S. Furuta, Bifurcation analysis of the motion of an asymmetric double pendulum subjected to a follower force: Codimension three problem, *Nonlinear Dynamics*, v. 2, p. 199–214, 1991.

Matthews, R., Planet X: Going, going . . . but not quite gone, *Science*, v. 254, p. 1454–1455, 1991.

———, A rocky watch for Earthbound asteroids, *Science*, v. 255, p. 1204–1205, 1992.

May, R. M., Simple mathematical models with very complicated dynamics, *Nature*, v. 261, p. 459–467, 1976.

———, Taxonomy as destiny, *Nature*, v. 347, p. 129–130, 1990.

Mayr, E., *The Growth of Biological Thought: Diversity, Evolution, and Inheritance*, Cambridge, MA, The Belknap Press of Harvard University Press, 974 pp., 1982.

———, *One Long Argument: Charles Darwin and the Genesis of Modern Evolutionary Thought*, Cambridge, MA, Harvard University Press, 195 pp., 1991.

Mazaud, A., and C. Laj, The 15 m.y. geomagnetic reversal periodicity: A quantitative test, *Earth Planet. Sci. Lett.*, v. 107, p. 689–696, 1991.

Mazaud, A., C. Laj, L. de Seze, and K. L. Verosub, 15-Myr periodicity in the frequency of geomagnetic reversals since 100 Myr, *Nature*, v. 304, p. 328–330, 1983.

Mazurs, E. G., *Graphic Representations of the Periodic System During One Hundred Years*, University, AL, The University of Alabama Press, 251 pp., 1974.

McCartney, K., A. R. Huffman, and M. Tredoux, A paradigm for endogenous causation of mass extinctions, in V. L. Sharpton and P. D. Ward, eds., *Global Catastrophes in Earth History; An Interdisciplinary Conference on Impacts, Volcanism, and Mass Mortality*, p. 125–138, Boulder, CO, The Geological Society of America, Special Paper 247, 1990.

References Cited

McCormac, B. M., ed., *Weather and Climate Responses to Solar Variations*, Boulder, CO, Colorado Associated University Press, 623 pp., 1983.

McDonald, C. Grebogi, E. Ott, and J. A. Yorke, Fractal basin boundaries, *Physica D, v. 17*, p. 125–153, 1985.

McElhinny, M. W., Geomagnetic reversals during the Phanerozoic, *Science, v. 172*, p. 157–159, 1971.

McEwen, A. S., D. L. Matson, T. V. Johnson, and L. A. Soderblom, Volcanic hot spots on Io: Correlation with low-albedo calderas, *Jour. Geophys. Res., v. 90*, p. 12,345–12,379, 1985.

McFadden, L. A., D. J. Tholen, and G. J. Veeder, Physical properties of Aten, Apollo, and Amor asteroids, in R. P. Binzel, T. Gehrels, and M. S. Matthews, eds., *Asteroids II*, p. 442–467, Tucson, AZ, The University of Arizona Press, 1989.

McFadden, L. A., A. L. Cochran, E. S. Barker, D. P. Cruikshank, and W. K. Hartmann, The enigmatic object 2201 Oljato: Is it an asteroid or an evolved comet?, *Jour. Geophys. Res., v. 98*, p. 3031–3041, 1993.

McFadden, P. L., R. T. Merrill, M. W. McElhinny, and S. Lee, Reversals of the Earth's magnetic field and temporal variations of the dynamo families, *Jour. Geophys. Res., v. 96*, p. 3923–3933, 1991.

McGarr, A., Seismic moments and volume changes, *Jour. Geophys. Res., v. 81*, p. 1487–1494, 1976.

McIntosh, B. A., Debris from comets: The evolution of meteor streams, in R. L. Newburn, Jr., M. Neugebauer, and J. Rahe, eds., *Comets in the Post-Halley Era, Volume 1*, p. 557–591, Boston, MA, Kluwer Academic Publishers, 1991.

McKay, D. S., T. D. Swindle, and R. Greenberg, Asteroidal regoliths: What we do not know, in R. P. Binzel, T. Gehrels, and M. S. Matthews, eds., *Asteroids II*, p. 617–642, Tucson, AZ, The University of Arizona Press, 1989.

McKenzie, D. P., and J. G. Sclater, Heat flow inside the island arcs of the northwestern Pacific, *Jour. Geophys. Res., v. 73*, p. 3173–3179, 1968.

McKinney, M. J., and D. Frederick, Extinction and population dynamics: New methods and evidence from Paleogene foraminifera, *Geology, v. 20*, p. 343–346, 1992.

McKinnon, W. B., Impact into the Earth's ocean floor: Preliminary experiments, a planetary model, and possibilities for detection, in L. T. Silver and P. H. Schultz, eds., *Geological Implications of Impacts of Large Asteroids and Comets on the Earth*, p. 129–142, Boulder, CO, The Geological Society of America, Special Paper 190, 1982.

McLaren, D. J., Presidential address: Time, life, and boundaries, *Jour. Paleontology, v. 44*, p. 801–815, 1970.

———, Frasnian-Famennian extinctions, in L. T. Silver and P. H. Schultz, eds., *Geological Implications of Impacts of Large Asteroids and Comets on the Earth*, p. 477–484, Boulder, CO, The Geological Society of America, Special Paper 190, 1982.

———, Bolides and biostratigraphy, *Geol. Soc. Am. Bull., v. 94*, p. 313–324, 1983.

———, Abrupt extinctions, in D. K. Elliott, ed., *Dynamics of Extinction*, p. 37–46, New York, John Wiley & Sons, Inc., 1986.

McLaren, D. J., and W. D. Goodfellow, Geological and biological consequences of giant impacts, *Ann. Rev. Earth Planet. Sci., v. 18*, p. 123–171, 1990.

McLean, D. M., A terminal Mesozoic "greenhouse": Lessons from the past, *Science, v. 201*, p. 401–406, 1978.

———, K-T transition into chaos, *Jour. Geol. Educ., v. 36*, p. 237–243, 1988.

McLeod, M. G., On the geomagnetic jerk of 1969, *Jour. Geophys. Res., v. 90*, p. 4597–4610, 1985.

McNamara, B., and K. Wiesenfeld, Theory of stochastic resonance, *Phys. Rev. A, v. 39*, p. 4854–4869, 1989.

McSween, H. Y., Jr., Chondritic meteorites and the formation of planets, *Am. Scientist, v. 77*, p. 146–153, 1989.

McSween, H. Y., Jr., D. W. G. Sears, and R. T. Dodd, Thermal metamorphism, in J. F. Kerridge and M. S. Matthews, eds., *Meteorites and the Early Solar System*, p. 102–113, Tucson, AZ, The University of Arizona Press, 1988.

References Cited

Meadows, D. L., and D. H. Meadows, *Toward Global Equilibrium: Collected Papers*, Cambridge, MA, Wright-Allen Press, Inc., 358 pp., 1973.

Mebane, A. D., A preliminary report on U. S. observations of the Great Fireball Procession of 1913 February 9, *Meteoritics, v. 1 (3)*, p. 360–361, 1955.

———, Observations of The Great Fireball Procession of 1913 February, 9, made in the United States, *Meteoritics, v. 1 (4)*, p. 405–421, 1956.

Meech, K. J., Physical aging in comets, in R. L. Newburn, Jr., M. Neugebauer, and J. Rahe, eds., *Comets in the Post-Halley Era, Volume 1*, p. 629–669, Boston, MA, Kluwer Academic Publishers, 1991.

Mees, A. Chaos in feedback systems, in A. V. Holden, ed., *Chaos*, p. 99–110, Princeton, NJ, Princeton University Press, 1986.

Meiss, J. D., Resonances fill stochastic phase-space, in J. R. Buchler and H. Eichhorn, eds., *Chaotic Phenomena in Astrophysics*, p. 83–96, New York, The New York Academy of Sciences (Annals Volume 497), 1987.

Melosh, H. J., The mechanics of large meteoroid impacts in the Earth's oceans, in L. T. Silver and P. H. Schultz, eds., *Geological Implications of Impacts of Large Asteroids and Comets on the Earth*, p. 121–127, Boulder, CO, The Geological Society of America, Special Paper 190, 1982.

———, *Impact Cratering*, New York, Oxford University Press, 245 pp., 1989.

———, Airblast scars on Venus, *Nature, v. 358*, p. 622–623, 1992.

———, Tunguska comes down to Earth, *Nature, v. 361*, p. 14–15, 1993a.

———, Meteorite origins: Blasting rocks off planets, *Nature, v. 363*, p. 498–499, 1993b.

Melosh, H. J., and D. Dzurisin, Mercurian global tectonics: A consequence of tidal despinning?, *Icarus, v. 35*, p. 227–236, 1978.

Melosh, H. J., and A. M. Vickery, Melt droplet formation in energetic impact events, *Nature, v. 350*, p. 494497, 1991.

Melosh, H. J., E. V. Ryan, and E. Asphaug, Dynamic fragmentation in impacts: Hydrocode simulation of laboratory impacts, *Jour. Geophys. Res., v. 97*, p. 14,735–14,759, 1992.

Meneveau, C., and K. R. Sreenivasan, Simple multifractal cascade model for fully developed turbulence, *Phys. Rev. Lett., v. 59*, p. 1424–1427, 1987.

Merrill, R. T., and M. W. McElhinny, *The Earth's Magnetic Field: Its History, Origin and Planetary Perspective*, New York, Academic Press, 401 pp., 1983.

Merrill, R. T., and P. L. McFadden, Secular variation and the origin of geomagnetic field reversals, *Jour. Geophys. Res., v. 93*, p. 11,589–11,597, 1988.

———, Paleomagnetism and the nature of the geodynamo, *Science, v. 248*, p. 345–350, 1990.

Meyer, C., Ore deposits as guides to geologic history of the Earth, *Ann. Rev. Earth Planet. Sci., v. 16*, p. 147–171, 1988.

Meyer-Vernet, N., and B. Sicardy, On the physics of resonant disk-satellite interaction, *Icarus, v. 69*, p. 157–175, 1987.

Middlehurst, B. M., and G. P. Kuiper, *The Moon, Meteorites, and Comets*, Chicago, IL, The University of Chicago Press, 810 pp., 1963.

Middleton, G. V., ed., *Nonlinear Dynamics, Chaos, and Fractals: With Applications to Geological Systems*, Toronto, Ontario, Geol. Assoc. Canada, Short Course Notes Volume 9, 235 pp. (includes software disk), 1991.

Milani, A., Emerging stability and chaos, *Nature, v. 338*, p. 207–208, 1989.

Milani, A., and A. M. Nobili, An example of stable chaos in the Solar System, *Nature, v. 357*, p. 569–571, 1992.

Milani, A., M. Carpino, G. Hahn, and A. M. Nobili, Project SPACEGUARD: Dynamics of planet-crossing asteroids. Classes of orbital behavior, *Icarus, v. 78*, p. 212–269, 1989.

Milankovich, M., *Canon of Insolation and the Ice-Age Problem*, Jerusalem, Israel Program for Scientific Translations, 484 pp., 1969.

Miles, J., Stability of forced oscillations of a spherical pendulum, *Quart. Appl. Math., v. 20*, p. 21–32, 1962.

———, Resonant motion of a spherical pendulum, *Physica D, v. 11*, p. 309–323, 1984.

Milgrom, M., A modification of the Newtonian dynamics as a possible alternative to the hidden mass hypothesis, *Astrophys. Jour., v. 270*, p. 365–370, 1983.

References Cited

Milgrom, M., and R. H. Sanders, More on modified dynamics, *Nature, v. 362*, p. 25, 1993.
Miller, J. G., *Living Systems*, New York, McGraw-Hill, 1102 pp., 1978.
Millis, R. L., and D. W. Dunham, Precise measurement of asteroid sizes and shapes from occultations, in R. P. Binzel, T. Gehrels, and M. S. Matthews, eds., *Asteroids II*, p. 148–170, Tucson, AZ, The University of Arizona Press, 1989.
Millman, P. M., and D. W. R. McKinley, Meteors, in B. M. Middlehurst and G. P. Kuiper, eds., *The Moon, Meteorites, and Comets*, p. 674–773, Chicago, The University of Chicago Press, 1963.
Milner, A. C., Ground rules for early birds, *Nature, v. 362*, p. 589, 1993.
Mitrovica, J. X., and W. R. Peltier, Pleistocene deglaciation and the global gravity field, *Jour. Geophys. Res., v. 94*, p. 13,651–13,671, 1989.
Moberly, R., and J. F. Campbell, Hawaiian hotspot volcanism mainly during geomagnetic normal intervals, *Geology, v. 12*, p. 459–463, 1984.
Möhlmann, D., Origin and early evolution of the planetary system, *Earth, Moon, and Planets, v. 33*, p. 201–214, 1985.
Molnar, P., and T. Atwater, Relative motions of hotspots in the mantle, *Nature, v. 246*, p. 288–291, 1973.
Molnar, P., and J. Francheteau, The relative motion of hotspots in the Atlantic and Indian Oceans during the Cenozoic, *Geophys. Jour. Roy. Astron. Soc., v. 43*, p. 763–774, 1975.
Molnar, P., and J. M. Stock, Relative motions of hotspots in the Pacific, Atlantic, and Indian Oceans since the Late Cretaceous time, *Nature, v. 327*, p. 587–591, 1987.
Molnar, P., T. Atwater, J. Mammerickx, and S. M. Smith, Magnetic anomalies, bathymetry and the tectonic evolution of the South Pacific since the Late Cretaceous, *Geophys. Jour. Roy. Astron. Soc., v. 40*, p. 383–420, 1975.
Molnar, P., F. Pardo-Casas, and J. Stock, The Cenozoic and Late Cretaceous evolution of the Indian Ocean Basin: Uncertainties in the reconstructed positions of the Indian, African, and Antarctic plates, *Basin Res., v. 1*, p. 23–40, 1988.
Monastersky, R., The whole-Earth syndrome, *Sci. News, v. 133*, p. 378–380, 1988.
Montanari, A., Two Ir anomalies near the *semiinvoluta/cerroazulensis* boundary in the E/O boundary stratotype of Massignano (Italy): New evidence for multiple late Eocene events (Abstract), *Eos, Trans. Am. Geophys. Union, v. 71*, p. 1425, 1990a.
———, Geochronology of the terminal Eocene impacts: An update, in V. L. Sharpton and P. D. Ward, eds., *Global Catastrophes in Earth History; An Interdisciplinary Conference on Impacts, Volcanism, and Mass Mortality*, p. 607–616, Boulder, CO, The Geological Society of America, Special Paper 247, 1990b.
Montenbruck, O., *Practical Ephemeris Calculations*, New York, Springer-Verlag, 146 pp., 1989.
Moon, F. C., Experiments on chaotic motions of a forced nonlinear oscillator: Strange attractors, *Jour. Applied Mech. (Trans. ASME), v. 47*, p. 638–644, 1980.
———, *Chaotic Vibrations*, New York, John Wiley & Sons, Inc., 309 pp., 1987.
———, Coming to grips with chaos, *Nature, v. 355*, p. 675–676, 1992.
Moon, F. C., and G.-X. Li, Fractal basin boundaries and homoclinic orbits for periodic motion in a two-well potential, *Phys. Rev. Lett., v. 55*, p. 1439–1442, 1985a.
———, The fractal dimension of the two-well potential strange attractor, *Physica D, v. 17*, p. 99–108, 1985b.
Moore, D. R., J. Toomre, E. Knobloch, and N. O. Weiss, Period doubling and chaos in partial differential equations for thermosolutal convection, *Nature, v. 303*, p. 663–667, 1983.
Moore, D. W., and E. A. Spiegel, A thermally excited non-linear oscillator, *Astrophysical Jour., v. 143*, p. 871–887, 1966.
Moores, E. M., Southwest U. S.–East Antarctic (SWEAT) connection: A hypothesis, *Geology, v. 19*, p. 425–428, 1991.
Morell, V., 30-million-year-old DNA boosts an emerging field, *Science, v. 257*, p. 1860–1862, 1992.
Morelli, A., and A. M. Dziewonski, Topography of the core-mantle boundary and lateral homogeneity of the liquid core, *Nature, v. 325*, p. 678–683, 1987.
Morgan, M., All in the mind (Book Review), *Nature, v. 362*, p. 124, 1993.

References Cited

Morgan, W. J., Convection plumes in the lower mantle, *Nature*, v. 230, p. 42–43, 1971.

———, Deep mantle convection plumes and plate motions, *Am. Assoc. Pet. Geol. Bull.*, v. 56, p. 203–213, 1972.

———, Hotspot tracks and the opening of the Atlantic and Indian Oceans, in C. Emiliani, ed., *The Sea, Volume 7*, p. 443–475, New York, Wiley Interscience, 1981.

———, Hotspot tracks and the early rifting of the Atlantic, *Tectonophys.*, v. 94, p. 123–139, 1983.

Mörner, N.-A., Eustasy, geoid changes, and multiple geophysical interaction, in W. A. Berggren and J. A. Van Couvering, eds., *Catastrophes and Earth History: The New Uniformitarianism*, p. 395–415, Princeton, NJ, Princeton University Press, 1984.

Morris, S. C., Late Precambrian and Cambrian soft-bodied faunas, *Ann. Rev. Earth Planet. Sci.*, v. 18, p. 101–122, 1990.

———, The fossil record and the early evolution of the Metazoa, *Nature*, v. 361, p. 219–225, 1993.

Morrison, D., ed., *Satellites of Jupiter*, Tucson, AZ, The University of Arizona Press, 972 pp., 1982a.

———, Introduction to the satellites of Jupiter, in D. Morrison, ed., *Satellites of Jupiter*, p. 3–43, Tucson, AZ, The University of Arizona Press, 1982b.

Morrison, D., and J. Niehoff, Future exploration of the asteroids, in T. Gehrels, ed., *Asteroids*, p. 227–250, Tucson, AZ, The University of Arizona Press, 1979.

Morrison, F., On chaos, *Eos, Trans. Am. Geophys. Union*, v. 69, p. 668–669, 1988.

———, *The Art of Modeling Dynamic Systems: Forecasting for Chaos, Randomness, and Determinism*, New York, John Wiley & Sons, Inc., 387 pp., 1991.

Mosekilde, E., E. Larsen, and J. Sterman, Coping with complexity: Deterministic chaos in human decisionmaking behavior, in J. L. Casti and A. Karlqvist, eds., *Beyond Belief: Randomness, Prediction and Explanation in Science*, p. 199–229, Boston, MA, CRC Press, 1991.

Moser, J., Is the Solar System stable?, *Mathematical Intelligencer*, v. 1, p. 65–71, 1978.

Müller, N., J. B. Hartung, E. K. Jessberger, and W. U. Reimold, ^{40}Ar-^{39}Ar ages of Dellen, Jänisjärvi, and Sääksjärvi impact craters, *Meteoritics*, v. 25, p. 1–10, 1990.

Muller, R. A., Evidence for Nemesis: A solar companion star, in R. Smoluchowski, J. N. Bahcall, and M. S. Matthews, eds., *The Galaxy and the Solar System*, p. 387–396, Tucson, AZ, The University of Arizona Press, 1986.

———, *Nemesis: The Death Star*, New York, Weidenfeld and Nicolson, 193 pp., 1988.

Muller, R. A., and D. E. Morris, Geomagnetic reversals from impacts on Earth, *Geophys. Res. Lett.*, v. 13, p. 1177–1180, 1986.

———, Geomagnetic reversals driven by sudden climate changes (Abstract), *Eos, Trans. Am. Geophys. Union*, v. 70, p. 276, 1989.

Munk, W. H., and G. J. F. MacDonald, *The Rotation of the Earth*, New York, Cambridge University Press, 323 pp., 1960.

Murali, K., and M. Lakshmanan, Bifurcation and chaos of the sinusoidally-driven Chua's circuit, *Internat. Jour. Bifurcation and Chaos*, v. 1, p. 369–384, 1991.

———, Transition from quasiperiodicity to chaos and Devil's staircase structures of the driven Chua's circuit, *Internat. Jour. Bifurcation and Chaos*, v. 2, p. 621–632, 1992.

Murphy, J. B., and R. D. Nance, Supercontinent model for the contrasting character of Late Proterozoic orogenic belts, *Geology*, v. 19, p. 469–472, 1991.

———, Mountain belts and the supercontinent cycle, *Sci. Am.*, v. 266, p. 84–91, 1992.

Murr, L. E., and W. H. Kinard, Effects of Low Earth Orbit, *Am. Scientist*, v. 81, p. 152–165, 1993.

Murray, B. C., Mercury, in B. Murray, ed., *The Planets: Readings From Scientific American*, p. 5–15, San Francisco, W. H. Freeman & Co., 1983.

Murray, C. D., Wandering on a leash, *Nature*, v. 357, p. 542–543, 1992.

———, Planetary rotation: Seasoned travellers, *Nature*, v. 361, p. 586–587, 1993.

Nagel, E., and J. R. Newman, *Gödel's Proof*, New York, New York University Press, 118 pp., 1958.

Naito, I., and N. Kikuchi, A seasonal budget of the Earth's axial angular momentum, *Geophys. Res. Lett.*, v. 17, p. 631–634, 1990.

References Cited

Nance, R. D., T. R. Worsley, and J. B. Moody, Post-Archean biogeochemical cycles and long-term episodicity in tectonic processes, *Geology, v. 14*, p. 514–518, 1986.

——, The supercontinent cycle, *Sci. Am., v. 259*, p. 72–79, 1988.

Napier, W. M., Dynamical interactions of the Solar System with massive nebulae, in A. Carusi and G. B. Valsecchi, eds., *Dynamics of Comets: Their Origin and Evolution*, p. 31–41, Boston, MA, D. Reidel Publishing Company, 1985.

——, Terrestrial catastrophism and galactic cycles, in S. V. M. Clube, ed., *Catastrophes and Evolution: Astronomical Foundations*, p. 133–167, New York, Cambridge University Press, 1989.

Nazarov, M. A., D. D. Badjukov, L. D. Barsukova, and A. S. Alekseev, Reconstruction of original morphology of the Kara impact structure and its relevance to the K/T boundary event, *Lunar and Planetary Sci. Conf. XXII, Abstracts of Papers*, p. 959–960, Lunar and Planetary Science Institute, Houston, TX, March 18–22, 1991a.

Nazarov, M. A., A. L. Devirts, E. P. Lagutina, A. S. Alekseev, D. D. Badjukov, and Yu. A. Shukolyukov, The Kara impact structure: Hydrogen isotopic composition in the impact melts and constraints on the impact age, *Lunar and Planetary Sci. Conf. XXII, Abstracts of Papers*, p. 961, Lunar and Planetary Science Institute, Houston, TX, March 18–22, 1991b.

Negi, J. G., and R. K. Tiwari, Matching long term periodicities of geomagnetic reversals and galactic motions of the Solar System, *Geophys. Res. Lett., v. 10*, p. 713–716, 1983.

Nelson, S. A., The possible role of thermal feedback in the eruption of siliceous magma, *Jour. Volc. Geoth. Res., v. 11*, p. 127–137, 1981.

Nese, J. M., Quantifying local predictability in phase space, *Physica D, v. 35*, p. 237–250, 1989.

Neville, K., *The Eight*, New York, Ballantine Books, 598 pp., 1989.

Newburn, R. L., Jr., M. Neugebauer, and J. Rahe, eds., *Comets in the Post-Halley Era*, Boston, MA, Kluwer Academic Publishers, 1360 pp., 1991.

Newell, N. D., Periodicity in invertebrate evolution, *Jour. Paleontology, v. 26*, p. 371–385, 1952.

——, Revolutions in the history of life, *Sci. Am., v. 208*, p. 76–92, 1967.

Newell, A. C., D. A. Rand, and D. Russell, Turbulent transport and the random occurrence of coherent events, *Physica D, v. 33*, p. 281–303, 1988.

Newsom, H. E., and J. H. Jones, eds., *Origin of the Earth*, New York, Oxford University Press, 378 pp., 1990.

Nichols, D. J., and R. F. Fleming, Plant microfossil record of the terminal Cretaceous event in the western United States and Canada, in V. L. Sharpton and P. D. Ward, eds., *Global Catastrophes in Earth History; An Interdisciplinary Conference on Impacts, Volcanism, and Mass Mortality*, p. 445–455, Boulder, CO, The Geological Society of America, Special Paper 247, 1990.

Nicolaysen, L. O., Renewed ferment in the Earth sciences—especially about power supplies to the core, for the mantle and for crises in the faunal record, *South African Jour. Sci., v. 81*, p. 120–132, 1985.

Nicolis, C., and G. Nicolis, Is there a climatic attractor, *Nature, v. 311*, p. 529–532, 1984.

Nicolis, G., and I. Prigogine, *Exploring Complexity: An Introduction*, New York, W. H. Freeman & Co., 313 pp., 1989.

Nicolis, J. S., *Chaos and Information Processing: A Heuristic Outline*, Singapore, World Scientific, 285 pp., 1991.

Nieto, M. M., *The Titius-Bode Law of Planetary Distances: Its History and Theory*, New York, Pergamon Press, 161 pp., 1972.

Nininger, H. H., Meteorite distributions on the Earth, in B. M. Middlehurst and G. P. Kuiper, eds., *The Moon, Meteorites, and Comets*, p. 162–182, Chicago, The University of Chicago Press, 1963.

Nishenko, S. P., and R. Buland, A generic recurrence interval distribution for earthquake forecasting, *Bull. Seis. Soc. Am., 77*, p. 1382–1399, 1987.

Nitecki, M. H., ed., *Extinctions*, Chicago, The University of Chicago Press, 354 pp., 1984.

Nobili, A. M., Dynamics of the outer Asteroid Belt, in R. P. Binzel, T. Gehrels, and M. S. Matthews, eds., *Asteroids II*, p. 862–879, Tucson, AZ, The University of Arizona Press, 1989.

Norton, I. O., and J. G. Sclater, A model for the evolution of the Indian Ocean and the breakup of Gondwanaland, *Jour. Geophys. Res., v. 84*, p. 6803–6830, 1979.

References Cited

Nowak, M. A., and R. M. May, Evolutionary games and spatial chaos, *Nature, v. 359*, p. 826–829, 1992.

——, The spatial dilemmas of evolution, *Internat. Jour. Bifurcation and Chaos, v. 3*, p. 35–78, 1993.

Nuth, J. A., III, Astrophysical implications of presolar grains, in J. F. Kerridge and M. S. Matthews, eds., *Meteorites and the Early Solar System*, p. 984–991, Tucson, AZ, The University of Arizona Press, 1988.

Oberbeck, V. R., J. R. Marshall, and H. R. Aggarwal, Did impacts initiate breakup of Gondwanaland? (Abstract), *Eos, Trans. Am. Geophys. Union, v. 73* (Supplement, 27 Oct.), p. 324, 1992.

O'Brien, S. J., D. E. Wildt, and M. Bush, The Cheetah in genetic peril, *Sci. Am., v. 254*, p. 84–92 (May), 1986.

O'Connell, R. J., and A. M. Dziewonski, Excitation of the Chandler wobble by large earthquakes, *Nature, v. 262*, p. 259–262, 1976.

Officer, C. B., and C. L. Drake, Terminal Cretaceous environmental events, *Science, v. 227*, p. 1161–1167, 1985.

Officer, C. B., A. Hallam, C. L. Drake, and J. D. Devine, Late Cretaceous and paroxysmal Cretaceous/Tertiary extinctions, *Nature, v. 326*, p. 143–149, 1987.

Officer, C. B., C. L. Drake, J. L. Pindell, and A. A. Meyerhoff, Cretaceous-Tertiary events and the Caribbean caper, *GSA Today, v. 2*, p. 69–75, 1992.

O'Keefe, J. A., *Tektites and Their Origin*, New York, Elsevier Scientific Publishing Co., 254 pp., 1976.

——, The terminal Eocene event: Formation of a ring system around the Earth, *Nature, v. 285*, p. 309–311, 1980.

O'Keefe, J. D., and T. J. Ahrens, The interaction of the Cretaceous/Tertiary extinction bolide with the atmosphere, ocean, and solid Earth, in L. T. Silver and P. H. Schultz, eds., *Geological Implications of Impacts of Large Asteroids and Comets on the Earth*, p. 103–120, Boulder, CO, The Geological Society of America, Special Paper 190, 1982.

——, Impact production of CO_2 by the Cretaceous/Tertiary extinction bolide and the resultant heating of the Earth, *Nature, v. 338*, p. 247–249, 1989.

Olinger, D. J., and K. R. Sreenivasan, Nonlinear dynamics of the wake of an oscillating cylinder, *Phys. Rev. Lett., v. 60*, p. 797–800, 1988.

Olsen, L. F., and H. Degn, Chaos in biological systems, *Quart. Rev. Biophys., v. 18*, p. 165–225, 1985.

Olson, P., and V. L. Hagee, Geomagnetic polarity reversals, transition field structure, and convection in the outer core, *Jour. Geophys. Res., v. 95*, p. 4609–4620, 1990.

Olson, W. S., The Moon: Time of appearance and nearest approach to Earth, *Science, 161*, p. 1364, 1968.

Olsson-Steel, D., Planetary close encounters: Probability distributions of resultant orbital elements and application to Hidalgo and Chiron, *Icarus, v. 69*, p. 51–69, 1987.

——, Meteoroid streams and the Zodiacal dust cloud, in S. V. M. Clube, ed., *Catastrophes and Evolution: Astronomical Foundations*, p. 169–193, New York, Cambridge University Press, 1989.

Oort, J. H., The structure of the cloud of comets surrounding the Solar System, and a hypothesis concerning its origin, *Bull., Astron. Inst. Netherlands, v. 11*, p. 91–110, 1950.

Oort, J. H., and M. Schmidt, Differences between old and new comets, *Bull. Astron. Inst. Netherlands, v. 11*, p. 259–270, 1951.

Opdyke, N. D., The Jaramillo event as detected in oceanic cores, in S. K. Runcorn, ed., *The Application of Modern Physics to Earth and Planetary Interiors*, p. 549–552, New York, John Wiley & Sons, Inc., 1969.

Öpik, E. J., *Interplanetary Encounters: Close-Range Gravitational Interactions*, New York, Elsevier Scientific Publishing Co., 155 pp., 1976.

Orth, C. J., M. Attrep, Jr., and L. R. Quintana, Iridium abundance patterns across bio-event horizons in the fossil record, in V. L. Sharpton and P. D. Ward, eds., *Global Catastrophes in*

References Cited

Earth History; An Interdisciplinary Conference on Impacts, Volcanism, and Mass Mortality, p. 45–59, Boulder, CO, The Geological Society of America, Special Paper 247, 1990.

Ortoleva, P., J. Chadam, M. El-Badewi, R. Feeney, D. Feinn, S. Haase, R. Larter, E. Merino, A. Strickholm, and S. Schmidt, Mechanism of bio- and geo-pattern formation and chemical signal propagation, in L. E. Reichl and W. C. Schieve, *Instabilities, Bifurcations, and Fluctuations in Chemical Systems*, p. 125–195, Austin, TX, University of Texas Press, 1982.

Ostlund, S., D. Rand, J. Sethna, and E. Siggia, Universal properties of the transition from quasiperiodicity to chaos in dissipative systems, *Physica D, v. 8*, p. 303–342, 1983.

Ott, E., Strange attractors and chaotic motions of dynamical systems, *Rev. Mod. Phys., v. 53*, p. 655–671, 1981.

Ott, E., C. Grebogi, and J. A. Yorke, Controlling chaos, *Phys. Rev. Lett., v. 64*, p. 1196–1199, 1990.

Ottino, J. M., *The Kinematics of Mixing: Stretching, Chaos, and Transport*, New York, Cambridge University Press, 364 pp., 1989.

Ottino, J. M., F. J. Muzzio, M. Tjahjadi, J. G. Franjione, S. C. Jana, and H. A. Kusch, Chaos, symmetry, and self-similarity: Exploiting order and disorder in mixing processes, *Science, v. 257*, p. 754–760, 1992.

Ouspensky, P. D., *In Search of the Miraculous: Fragments of an Unknown Teaching*, New York, Harcourt, Brace & World, Inc., 399 pp., 1949.

Ovenden, M. W., Bode's Law and the missing planet, *Nature, v. 239*, p. 508–509, 1972.

———, Planetary distances and the missing planet, in B. D. Tapley and V. Szebehely, eds., *Recent Advances in Dynamical Astronomy*, p. 319–332, Dordrecht, Holland, D. Reidel Publ. Co., 1973.

———, Bode's Law—truth or consequences, *Vistas in Astronomy, v. 18*, p. 473–496, 1975.

———, The principle of least interaction action, in V. Szebehely and B. D. Tapley, eds., *Long-Time Predictions in Dynamics*, p. 295–305, Boston, MA, D. Reidel Publ. Co., 1976.

Ovenden, M. W., and J. Byl, Comets and the missing planet, in V. Szebehely, ed., *Dynamics of Planets and Satellites and Theories of Their Motion*, p. 101–107, Dordrecht, Holland, D. Reidel Publ. Co., 1978.

Ovenden, H. G., T. Feagin, and O. Graf, On the principle of least interaction action and the Laplacean satellites of Jupiter and Uranus, *Celestial Mech., v. 8*, p. 455–471, 1974.

Overstreet, H. A., *The Mature Mind*, New York, W. W. Norton, 295 pp., 1949.

Owen, H. G., Ocean-floor spreading evidence of global expansion, in S. W. Carey, ed., *The Expanding Earth: A Symposium, Sydney, 1981*, p. 31–58, Hobart Tasmania, University of Tasmania, 1983a.

———, *Atlas of Continental Displacement: 200 Million Years to the Present*, New York, Cambridge University Press, 159 pp., 1983b.

Owen, T., A. Bar-Nun, and I. Kleinfeld, Noble gases in terrestrial planets: Evidence for cometary impacts?, in R. L. Newburn, Jr., M. Neugebauer, and J. Rahe, eds., *Comets in the Post-Halley Era, Volume 1*, p. 429–437, Boston, MA, Kluwer Academic Publishers, 1991.

Oxburgh, E. R., and D. L. Turcotte, Problem of high heat flow and volcanism associated with zones of descending mantle convective flow, *Nature, v. 218*, p. 1041–1043, 1968.

Pachner, J., The nature of chaos in two systems of ordinary nonlinear differential equations, *Computers in Physics, v. 7*, p. 226–247, 1993.

Packard, N. H., J. P. Crutchfield, J. D. Farmer, and R. S. Shaw, Geometry from time series, *Phys. Rev. Lett., v. 45*, p. 712–716, 1980.

Padian, K., W. Alvarez, T. Birkelund, D. K. Fütterer, K. J. Hsu, J. H. Lipps, D. J. McLaren, D. M. Raup, E. M. Shoemaker, J. Smit, O. B. Toon, and A. Wetzel, The possible influences of sudden events on biological radiations and extinctions (Group Report), in H. D. Holland and A. F. Trendall, eds., *Patterns of Change in Earth Evolution*, p. 77–102, New York, Springer-Verlag, 1984.

Palmer, R. G., Parallels and contrasts between glass and spin glass, in C. A. Angell and M. Goldstein, eds., *Dynamic Aspects of Structural Change in Liquids and Glasses*, p. 109–120, New York, The New York Academy of Sciences (Annals Volume 484), 1986.

References Cited

Pannella, G., Paleontological evidence on the Earth's rotational history since the early Precambrian, *Astrophysics and Space Sci.*, v. 16, p. 212–237, 1972.

Papaelias, P. M., Energy deposition in antimatter meteors, *Earth, Moon, and Planets*, v. 55, p. 215–222, 1991.

Parmenter, R. H., and L. Y. Yu, The periodically kicked quantum spin, *Chaos*, v. 2, p. 589–598, 1992.

Parrish, J. M., J. T. Parrish, and A. M. Ziegler, Permian-Triassic paleogeography and paleoclimatology and implications for Therapsid distribution, in N. H. Hotton, II, P. D. McLean, J. J. Roth, and E. C. Roth, eds., *The Ecology and Biology of Mammal-Like Reptiles*, p. 109–131, Washington, DC, Smithsonian Institution Press, 1986.

Partridge, L., and P. H. Harvey, Evolutionary Biology: What the sperm count costs, *Nature*, v. 360, p. 415, 1992.

Pasachoff, J. M., *Astronomy: From the Earth to the Universe, Fourth Edition*, Philadelphia, PA, Saunders College Publishing, 682 pp. (Incl. 11 Appendices, Selected Readings, Glossary, Index; + 8 Sky Maps), 1991.

Peale, S. J., Evidence against a geocentric contribution to the Zodiacal light, *Jour. Geophys. Res.*, v. 73, p. 3025–3041, 1968.

——, The rotational dynamics of Mercury and the state of its core, in F. Vilas, C. R. Chapman, and M. S. Matthews, eds., *Mercury*, p. 461–493, Tucson, AZ, The University of Arizona Press, 1988.

Peale, S. J., P. Cassen, and R. T. Reynolds, Melting of Io by tidal dissipation, *Science*, v. 203, p. 892–894, 1979.

Pearl, J. C., and W. M. Sinton, Hot spots of Io, in D. Morrison, ed., *Satellites of Jupiter*, p. 724–755, Tucson, AZ, The University of Arizona Press, 1982.

Peck, D. L., M. S. Hamilton, and H. R. Shaw, Numerical analysis of lava lake cooling models: Part II. Application to Alae lava lake, Hawaii, *Am. Jour. Sci.*, v. 277, p. 415–437, 1977.

Pecker, C.-P., The global Sun, in A. N. Cox, W. C. Livingston, and M. S. Matthews, eds., *Solar Interior and Atmosphere*, p. 1–30, Tucson, AZ, The University of Arizona Press, 1991.

Pecora, L. M., and T. L. Carroll, Driving systems with chaotic signals, *Phys. Rev. A*, v. 44, p. 2374–2383, 1991.

Peltier, W. R., A relaxation oscillator model of the Ice Age cycle, in C. Nicolis and G. Nicolis, eds., *Irreversible Phenomena and Dynamical Systems Analysis in Geosciences*, p. 399–416, Boston, MA, D. Reidel Publishing Company, 1987.

Peltier, W. R., and L. P. Solheim, Mantle phase transitions and layered chaotic convection, *Geophys. Res. Lett.*, v. 19, p. 321–324, 1992.

Penfield, G. T., and A. Camargo Z., Interpretation of geophysical cross sections on the north flank of the Chicxulub impact structure, *Lunar and Planetary Sci. Conf. XXII, Abstracts of Papers*, p. 1051, Lunar and Planetary Science Institute, Houston, TX, March 18–22, 1991.

Percival, I. C., Chaos in Hamiltonian systems, in M. V. Berry, I. C. Percival, and N. O. Weiss, eds., *Dynamical Chaos*, p. 131–143, Princeton, NJ, Princeton University Press, 1987.

Perlmutter, S., R. A. Muller, C. R. Pennypacker, C. K. Smith, L. P. Wang, S. White, and H. S. Yang, A search for Nemesis: Current status and review of history, in V. L. Sharpton and P. D. Ward, eds., *Global Catastrophes in Earth History; An Interdisciplinary Conference on Impacts, Volcanism, and Mass Mortality*, p. 87–91, Boulder, CO, The Geological Society of America, Special Paper 247, 1990.

Pesonen, L. J., and H. Henkel, eds., *Terrestrial Impact Craters and Craterform Structures with a Special Focus on Fennoscandia*, (Tectonophysics, v. 216, Nos. 1/2), Amsterdam, Elsevier Science Publishers, 241 pp., 1992a.

Pesonen, L. J., and H. Henkel, Impact cratering record of Fennoscandia, *Papers Presented at the International Conference on Large Meteorite Impacts and Planetary Evolution, 31 August–2 September, 1992, Sudbury, Ontario, Canada*, p. 57, Lunar and Planetary Institute, Houston, TX (LPI Contribution No. 790), 1992b.

Peters, T., *Thriving on Chaos: Handbook for a Management Revolution*, New York, Alfred A. Knopf, 561 pp., 1987.

Peterson, I., Chaos in the clockwork, *Science News*, v. 141, p. 120–121, 1992.

References Cited

Peterson, I., and C. Ezzell, Crazy rhythms: Confronting the complexity of chaos in biological systems, *Sci. News, v. 142*, p. 156–159 (September), 1992.

Petersons, H. F., Possible conditions for formation of satellite from protoearth by ejection of material, *Earth, Moon, and Planets, v. 31*, p. 15–24, 1984.

Phillips, J. D., R. J. Blakely, and A. Cox, Independence of geomagnetic polarity intervals, *Geophys. Jour. Roy. Astron. Soc., v. 43*, p. 747–754, 1975.

Phillips, R. J., R. F. Raubertas, R. E. Arvidson, I. C. Sarkar, R. R. Herrick, N. Izenberg, and R. E. Grimm, Impact craters and Venus resurfacing history, *Jour. Geophys. Res., v. 97*, p. 15,923–15,948, 1992.

Pickover, C. A., The world of chaos, *Computers in Physics, SEPT/OCT*, p. 460–467, 1990.

Pierce, J. R., *An Introduction to Information Theory: Symbols, Signals, and Noise, Second Edition*, New York, Dover Publications, Inc., 305 pp., 1980.

Pierce, W. H., A thermal speedometer for overthrust faults, *Geol. Soc. Am. Bull., v. 81*, p. 227–231, 1970.

Pierrehumbert, R. T., Dimensions of atmospheric variability, in J. L. Casti and A. Karlqvist, eds., *Beyond Belief: Randomness, Prediction and Explanation in Science*, p. 110–142, Boston, MA, CRC Press, 1991.

Pietronero, L., Hierarchically constrained thermodynamics in metastable systems and glasses, in L. Pietronero and E. Tosatti, eds., *Fractals in Physics*, p. 417–420, New York, North-Holland Physics Publishing, a division of Elsevier Science Publishers, 1986.

Pietronero, L., and R. Kupers, Stochastic approach to large scale clustering of matter in the universe, in L. Pietronero and E. Tosatti, eds., *Fractals in Physics*, p. 319–324, New York, North-Holland Physics Publishing, a division of Elsevier Science Publishers, 1986.

Pike, J., The sky *is* falling: The hazard of near-Earth asteroids, *The Planetary Report, v. 11*, p. 16–19 (Nov./Dec.), 1991.

Pike, R. J., Geomorphology of impact craters on Mercury, in F. Vilas, C. R. Chapman, and M. S. Matthews, eds., *Mercury*, p. 165–273, Tucson, AZ, The University of Arizona Press, 1988.

Pilcher, F., Circumstances of minor planet discovery, in T. Gehrels, ed., *Asteroids*, p. 1130–1154, Tucson, AZ, The University of Arizona Press, 1979.

——, The circumstances of minor planet discovery, in R. P. Binzel, T. Gehrels, and M. S. Matthews, eds., *Asteroids II*, p. 1002–1033, Tucson, AZ, The University of Arizona Press, 1989.

Piper, J. D. A., *Palaeomagnetism and the Continental Crust*, New York, Halsted Press, Division of John Wiley & Sons, Inc., 434 pp., 1987.

Platt, J., The acceleration of evolution, *The Futurist, v. 15*, p. 14–23, 1981.

Poag, C. W., D. S. Powars, L. J. Poppe. R. B. Mixon, L. E. Edwards, D. W. Folger, and S. Bruce, Deep Sea Drilling Project Site 612 bolide event: New evidence of a late Eocene impact-wave deposit and a possible impact site, U.S. east coast, *Geology, v. 20*, p. 771–774, 1992.

Podsiadlowski, P., and N. M. Price, Star formation and the origin of stellar masses, *Nature, v. 359*, p. 305–307, 1992.

Poinar, G. O., Jr., *Life In Amber*, Stanford, CA, Stanford University Press, 350 pp., 1992.

Poinar, G. O., Jr., B. M. Waggoner, and U.-C. Bauer, Terrestrial soft-bodied protists and other microorganisms in Trassic amber, *Science, v. 259*, p. 222–224, 1993.

Poincaré, H., *Science and Method* (First English Translation), New York, Dover Publications, Inc., 288 pp., 1952.

Poirier, J. P., Transport properties of liquid metals and viscosity of the Earth's core, *Geophys. Jour. Roy. Astron. Soc., v. 92*, p. 99–105, 1988.

Pollack, J. B., and F. Fanale, Origin and evolution of the Jupiter satellite system, in D. Morrison, ed., *Satellites of Jupiter*, p. 872–910, Tucson, AZ, The University of Arizona Press, 1982.

Pollack, J. B., J. A. Burns, and M. E. Tauber, Gas drag in primordial circumplanetary envelopes: A mechanism for satellite capture, *Icarus, v. 37*, p. 587–611, 1979.

Pons, T. P., P. E. Garraghty, A. K. Ommaya, J. H. Kaas, E. Taub, M. Mishkin, Massive cortical reorganization after sensory deafferentation in adult macaques, *Science, v. 252*, p. 1857–1860, 1991.

References Cited

Pool, R., Is it healthy to be chaotic? (sidebar, The footprints of chaos, p. 605), *Science, v. 243*, p. 604–607, 1989.

Pope, K. O., A. C. Ocampo, and C. E. Duller, Hydrogeological evidence for a possible 200 km diameter K/T impact crater in Yucatan, Mexico, *Lunar and Planetary Sci. Conf. XXII, Abstracts of Papers*, p. 1083–1084, Lunar and Planetary Science Institute, Houston, TX, March 18–22, 1991a.

———, Mexican site for K/T impact crater?, *Nature, v. 351*, p. 105, 1991b.

Porco, C. C., An explanation for Neptune's ring arcs, *Science, v. 253*, p. 995–1001, 1991.

Pospichal, J. J., S. W. Wise, Jr., F. Asaro, and N. Hamilton, The effects of bioturbation across a biostratigraphically complete high southern latitude Cretaceous/Tertiary boundary, in V. L. Sharpton and P. D. Ward, eds., *Global Catastrophes in Earth History; An Interdisciplinary Conference on Impacts, Volcanism, and Mass Mortality*, p. 497–507, Boulder, CO, The Geological Society of America, Special Paper 247, 1990.

Post, E. L., Absolutely unsolvable problems and relatively undecidable propositions — Account of an anticipation, in M. Davis, ed., *The Undecidable: Basic Papers on Undecidable Propositions, Unsolvable Problems, and Computable Functions*, p. 338–433, Hewlett, NY, Raven Press, 1965.

Prévot, M., E. A. Mankinen, R. S. Coe, and C. S. Grommé, The Steens Mountain (Oregon) geomagnetic polarity transition: 2. Field intensity variations and discussion of reversal models, *Jour. Geophys. Res., v. 90*, p. 10,417–10,448, 1985.

Prigogine, I., Time, structure, and fluctuations, *Science, v. 201*, p. 777–785, 1978.

Prigogine, I., and I. Stengers, *Order Out of Chaos: Man's New Dialogue with Nature*, New York, Bantam Books, 349 pp., 1984.

Prigogine, I., G. Nicolis, and A. Babloyantz, Thermodynamics of evolution (in two parts), *Physics Today, v. 25*, p. 23–28 (November), and p. 38–44 (December), 1972.

Prigogine, I., T. Y. Petrosky, H. H. Hasegawa, and S. Tasaki, Integrability and chaos in classical and quantum mechanics, *Chaos, Solitons, and Fractals, v. 1*, p. 3–24, 1991.

Procaccia, I., Universal properties of dynamically complex systems: The organization of Chaos, *Nature, v. 333*, p. 618–623, 1988.

Provine, W. B., *Sewall Wright and Evolutionary Biology*, Chicago, The University of Chicago Press, 545 pp., 1986.

Prüfer, M., Turbulence in multistep methods for initial value problems, *SIAM, Jour. Appl. Math., v. 45*, p. 32–69, 1985.

Pulliam, R. J., and P. B. Stark, Bumps on the core-mantle boundary: Are they facts or artifacts?, *Jour. Geophys. Res., v. 98*, p. 1943–1955, 1993.

Pyke, G. H., What does it cost a plant to produce floral nectar?, *Nature, v. 350*, p. 58–59, 1991.

Rabin, D. M., C. R. DeVore, N. R. Sheeley, Jr., K. L. Harvey, and J. T. Hoeksema, The solar activity cycle, in A. N. Cox, W. C. Livingston, and M. S. Matthews, eds., *Solar Interior and Atmosphere*, p. 781–843, Tucson, AZ, The University of Arizona Press, 1991.

Rabinovich, M. I., Strange attractors in modern physics, in R. H. G. Helleman, ed., *Nonlinear Dynamics*, p. 435–452, New York, Annals of the New York Acad. Sciences, v. 357, 1980.

Rajlich, P., Bohemian circular structure, Czechoslovakia: Search for the impact evidence, *Papers Presented at the International Conference on Large Meteorite Impacts and Planetary Evolution, 31 August–2 September, 1992, Sudbury, Ontario, Canada*, p. 57–59, Lunar and Planetary Institute, Houston, TX (LPI Contribution No. 790), 1992.

Rampino, M. R., Impact cratering and flood basalt volcanism, *Nature, v. 327*, p. 468, 1987.

———, A major late Permian impact event on the Falkland Plateau (Abstract), *Eos, Trans. Am. Geophys. Union, v. 73* (Supplement, 27 Oct.), p. 336, 1992.

Rampino, M. R., and K. Caldeira, Antipodal hotspot pairs on the Earth, *Geophys. Res. Lett., v. 19*, p. 2011–2014, 1992.

———, Major episodes of geologic change: Correlations, time structure and possible causes, *Earth, Planet. Sci. Lett., v. 114*, p. 215–227, 1993.

Rampino, M. R., and R. B. Stothers, Geological rhythms and cometary impacts, *Science, v. 226*, p. 1427–1431, 1984.

———, Geologic periodicities and the galaxy, in R. Smoluchowski, J. N. Bahcall, and M. S.

Matthews, eds., *The Galaxy and the Solar System*, p. 241–259, Tucson, AZ, The University of Arizona Press, 1986.

———, Flood basalt volcanism during the past 250 million years, *Science, v. 241*, p. 663–668, 1988.

Rampino, M. R., and T. Volk, Mass extinctions, atmospheric sulphur, and climatic warming at the K/T boundary, *Nature, v. 332*, p. 63–65, 1988.

Rampino, M. R., S. Self, and R. B. Stothers, Volcanic winters, *Ann. Rev. Earth Planet. Sci, v. 16*, p. 73–99, 1988.

Ramsey, W. H., On the instability of small planetary cores, *Month. Not. Roy. Astron. Soc., v. 110*, p. 325–338, 1950.

Rapp, P. E., I. D. Zimmerman, A. M. Albano, G. C. deGuzman, N. N. Greenbaun, and T. R. Bashore, Experimental studies of chaotic neural behavior: Cellular activity and electroencephalographic signals, in H. G. Othmer, ed., *Nonlinear Oscillations in Biology and Chemistry*, p. 175–205, Berlin, Springer-Verlag, 1987.

Rapp, P. E., R. A. Latta, and A. I. Mees, Parameter-dependent transitions and the optimal control of dynamical diseases, *Bull. Math. Biol., v. 50*, p. 227–253, 1988.

Raup, D. M., Taxonomic diversity during the Phanerozoic, *Science, v. 177*, p. 1065–1071, 1972.

———, Species diversity in the Phanerozoic: An interpretation, *Paleobiol., v. 2*, p. 289–297, 1976.

———, Size of the Permo-Triassic bottleneck and its evolutionary implications, *Science, v. 206*, p. 217–218, 1979.

———, Evolutionary radiations and extinctions, in H. D. Holland and A. F. Trendall, eds., *Patterns of Change in Earth Evolution*, p. 5–14, New York, Springer-Verlag, 1984a.

———, Death of species, in M. H. Nitecki, ed., *Extinctions*, p. 1–19, Chicago, IL, The University of Chicago Press, 1984b.

———, Magnetic reversals and mass extinctions, *Nature, v. 314*, p. 341–343, 1985.

———, Biological extinction in Earth history, *Science, v. 231*, p. 1528–1533, 1986a.

———, *The Nemesis Affair: A Story of the Death of Dinosaurs and the Way of Science*, New York, W. W. Norton and Co., 220 pp., 1986b.

———, The role of extraterrestrial phenomena in extinction, in M. A. Lamolda, E. G. Kauffman, and O. H. Walliser, eds., *Palaeontology and Evolution: Extinction Events. 2nd International Conference on Global Bioevents (Bilbao Conf., 20–23 October, 1987), Revista Española de Palaeontologia, N°. Extraordinario*, p. 99–106, Madrid, Sociedad Española de Paleontologia, 1988a.

———, Diversity crises in the geological past, in E. O. Wilson and F. M. Peter, eds., *Biodiversity*, p. 51–57, Washington, DC, National Academy Press, 1988b.

———, Impact as a general cause of extinction: A feasibility test, in V. L. Sharpton and P. D. Ward, eds., *Global Catastrophes in Earth History; An Interdisciplinary Conference on Impacts, Volcanism, and Mass Mortality*, p. 27–32, Boulder, CO, The Geological Society of America, Special Paper 247, 1990.

———, *Extinction: Bad Genes or Bad Luck?*, New York, W. W. Norton & Co., 210 pp., 1991.

Raup, D. M., and D. Jablonski, Geography of end-Cretaceous marine bivalve extinctions, *Science, v. 260*, p. 971–976, 1993.

Raup, D. M., and J. J. Sepkoski, Jr., Periodicity of extinctions in the geologic past, *Proc. Natl. Acad. Sci., v. 81*, p. 801–805, 1984.

———, Periodic extinctions of families and genera, *Science, v. 231*, p. 833–836, 1986.

Raup, D. M., and S. M. Stanley, *Principles of Paleontology*, San Francisco, W. H. Freeman & Co., 481 pp., 1978.

Raup, D. M., S. J. Gould, T. J. M. Schopf, and D. S. Simberloff, Stochastic models of phylogeny and the evolution of diversity, *Jour. Geol., v. 81*, p. 525–542, 1973.

Rawal, J. J., A hypothesis on the Oort clouds of planets, *Earth, Moon, and Planets, v. 54*, p. 89–102, 1991.

———, Physical significance and role of Lagrangian points and the Oort Clouds of planets in the Solar System, *Earth, Moon, and Planets, v. 58*, p. 153–161, 1992.

Ray, T. W., and D. L. Anderson, Hotspots, ridges, slabs, and mantle geochemistry: Correlation

References Cited

with deep mantle tomography (Abstract), *Eos, Trans. Am. Geophys. Union*, v. 73 (Supplement, 27 Oct.), p. 61, 1992.

Rega, G., F. Benedettini, and A. Salvatori, Periodic and chaotic motions of an unsymmetrical oscillator in nonlinear structural dynamics, *Chaos, Solitons and Fractals*, v. 1, p. 39–54, 1991.

Renne, P. R., M. Ernesto, I. G. Pacca, R. S. Coe, J. M. Glen, M. Prévot, and M. Perrin, The age of Paraná flood volcanism, rifting of Gondwanaland, and the Jurassic-Cretaceous boundary, *Science*, v. 258, p. 975–979, 1992.

Rey, A. D., and M. C. Mackey, Bifurcations and traveling waves in a delayed partial differential equation, *Chaos*, v. 2, p. 231–244, 1992.

Ricard, Y., C. Doglioni, and R. Sabadini, Differential rotation between lithosphere and mantle: A consequence of lateral mantle viscosity variations, *Jour. Geophys. Res.*, v. 96, p. 8407–8415, 1991.

Ricard, Y., R. Sabadini, and G. Spada, Isostatic deformations and polar wander induced by redistribution of mass within the Earth, *Jour. Geophys. Res.*, v. 97, p. 14,223–14,236, 1992.

Rice, A., and K. M. Creer, Geomagnetic polarity reversals: Can meteor impacts cause spall disruption into the outer core?, in F. J. Lowes, D. W. Collinson, J. H. Parry, S. K. Runcorn, D. C. Tozer, and A. Soward, eds., *Geomagnetism and Palaeomagnetism*, p. 227–230, Boston, MA, Kluwer Academic Publishers, 1989.

Rich, J. E., G. L. Johnson, J. E. Jones, and J. Campsie, A significant correlation between fluctuations in seafloor spreading rates and evolutionary pulsations, *Paleoceanography*, v. 1, p. 85–95, 1986.

Richards, D., Order and chaos in strong fields, *Nature*, v. 336, p. 318–319, 1988.

Richards, M. A., and D. C. Engebretson, Large-scale mantle convection and the history of subduction, *Nature*, v. 355, p. 437–440, 1992.

Richardson, D. E., Distances of the planets from the Sun and of satellites from their primaries in the satellite systems of Jupiter, Saturn, and Uranus, *Pop. Astron.*, v. 53, p. 14–26, 1945.

Richardson, G. P., *Feedback Thought in Social Science and Systems Theory*, Philadelphia, PA, University of Pennsylvania Press, 374 pp., 1991.

Richter, C. F., *Elementary Seismology*, San Francisco, W. H. Freeman & Co., 768 pp., 1958.

Ridley, M., Triumph of the embryo (Book Review), *Nature*, v. 357, p. 203–204, 1992.

Rikitake, T., Oscillations of a system of disk dynamos, *Proc. Cambridge. Philos. Soc.*, v. 54, p. 89–105, 1958.

Ringwood, A. E., Terrestrial origin of the Moon, *Nature*, v. 322, p. 323–328, 1986.

———, Significance of the terrestrial Mg/Si ratio, *Earth Planet. Sci. Lett.*, v. 95, p. 1–7, 1989.

———, Earliest history of the Earth-Moon system, in H. E. Newsom and J. H. Jones, *Origin of the Earth*, p. 101–134, New York, Oxford University Press, 1990.

Ringwood, A. E., T. Kato, W. Hibberson, and N. Ware, High-pressure geochemistry of Cr, V, and Mn and implications for the origin of the Moon, *Nature*, v. 347, p. 174–176, 1990.

Robbins, K. A., A new approach to subcritical instability and turbulent transitions in a simple dynamo, *Math. Proc. Camb. Phil. Soc.*, v. 82, p. 309–325, 1977.

Roberts, M. S., *Astronomy and Astrophysics*, Washington, DC, American Association for the Advancement of Science, 383 pp., 1985.

Roberts, P. H., Convection in spherical systems, in C. Nicolis and G. Nicolis, eds., *Irreversible Phenomena and Dynamical Systems Analysis in Geosciences*, p. 53–71, Boston, MA, D. Reidel Publishing Company, 1987a.

———, Dynamo theory, in C. Nicolis and G. Nicolis, eds., *Irreversible Phenomena and Dynamical Systems Analysis in Geosciences*, p. 73–133, Boston, MA, D. Reidel Publishing Company, 1987b.

Roberts, W. W., Jr., Theoretical aspects of galactic research, *Vistas in Astronomy*, v. 19, Pt. 1, p. 91–109, 1975.

Robin, E., D. Boclet, Ph. Bonté, L. Froget, C. Jéhanno, and R. Rocchia, The significance of Ni-rich magnetites for the study of the K-T boundary event, *Lunar and Planetary Sci. Conf. XXII, Abstracts of Papers*, p. 1125–1126, Lunar and Planetary Science Institute, Houston, TX, March 18–22, 1991.

Rocchia, R., P. Bonté, C. Jéhanno, E. Robin, M. de Angelis, and D. Boclet, Search for the

Tunguska event relics in the Antarctic snow and new estimation of the cosmic iridium accretion rate, in V. L. Sharpton and P. D. Ward, eds., *Global Catastrophes in Earth History; An Interdisciplinary Conference on Impacts, Volcanism, and Mass Mortality*, p. 189–193, Boulder, CO, The Geological Society of America, Special Paper 247, 1990.

Roddy, D. J., R. A. Schmitt, and S. H. Schuster, Asteroid and comet impacts on continental and oceanic sites: Computer simulations of cratering and inferred Fe/Ir ratios in ejecta vapor compared with Fe/Ir ratios measured at the K/T boundary from the Shatsky Rise (Pacific Ocean), *Lunar and Planetary Sci. Conf. XXII, Abstracts of Papers*, p. 1129–1130, Lunar and Planetary Science Institute, Houston, TX, March 18–22, 1991.

Roemer, E., A program of comet observing, *Proc. Lunar Planet. Explor. Colloq., v. 2*, p. 75–84, 1961.

———, Discovery and observation of close-approach asteroids, in T. Gehrels, ed., *Physical Studies of Minor Planets*, p. 643–648, Washington, DC, National Aeronautics and Space Administration, NASA SP-267, 1971.

Rogers, R. R., C. C. Swisher, III, P. C. Sereno, A. M. Monetta, C. A. Forster, and R. N. Martinez, The Ischigualasto tetrapod assemblage (late Triassic, Argentina) and ^{40}Ar/^{39}Ar dating of dinosaur origins, *Science, v. 260*, p. 794–797, 1993.

Rollins, R. W., and E. R. Hunt, Intermittent transient chaos at interior crises in the diode resonator, *Phys. Rev. A, v. 29*, p. 3327–3334, 1984.

Roosen, R. G., Spatial distribution of interplanetary dust, in T. Gehrels, ed., *Physical Studies of Minor Planets*, p. 363–375, Washington, DC, National Aeronautics and Space Administration, NASA SP-267, 1971.

Rose, S., Selective attention (Book Review), *Nature, v. 360*, p. 426–427, 1992.

Ross, M. N., and G. Schubert, The coupled orbital and thermal evolution of Triton, *Geophys. Res. Lett., v. 17*, p. 1749–1752, 1990.

Rössler, O. E., How chaotic is the universe, in A. V. Holden, ed., *Chaos*, p. 315–320, Princeton, NJ, Princeton University Press, 1986.

Rotaru, M., J. L. Birck, and C. J. Allègre, Clues to early Solar System history from chromium isotopes in carbonaceous chondrites, *Nature, v. 358*, p. 465–470, 1992.

Roth, G. D., *The System of Minor Planets*, New York, D. Van Nostrand Co., Inc., 128 pp., 1962.

Rothery, D. A., *Satellites of the Outer Planets: Worlds in Their Own Right*, Oxford, Clarendon Press, 208 pp., 1992.

Rotman, B., *Signifying Nothing: The Semiotics of Zero*, Stanford, CA, Stanford University Press, 111 pp., 1993a.

———, *Ad Infinitum . . . The Ghost in Turing's Machine: Taking God Out of Mathematics and Putting the Body Back In*, Stanford, CA, Stanford University Press, 203 pp., 1993b.

Roy, A. E., Miss Blagg's formula, *Jour. British Astron. Assoc., v. 63*, p. 212–215, 1953.

———, *Orbital Motion, Third Edition*, New York, Adam Hilger, 532 pp., 1988a.

Roy, A. E., ed., *Long-term Dynamical Behaviour of Natural and Artificial N-Body Systems*, Boston, MA, Kluwer Academic Publishers, 526 pp., 1988b.

Roy, A. E., and D. Clarke, *Astronomy: Principles and Practice, Third Edition*, Philadelphia, PA, Adam Hilger, 357 pp., 1988.

Roy, A. E., and M. W. Ovenden, On the occurrence of commensurable mean motions in the Solar System, *Month. Not. Roy. Astron. Soc., v. 114*, p. 232–241, 1954.

Roy, A. E., and M. W. Ovenden, On the occurrence of commensurable mean motions in the Solar System: II. The mirror theorem, *Month. Not. Roy. Astron. Soc., v. 115*, p. 296–309, 1955.

Royer, J.-Y., and T. Chang, Evidence for relative motions between the Indian and Australian plates during the last 20 m.y. from plate tectonic reconstructions: Implications for the deformation of the Indo-Australian plate, *Jour. Geophys. Res., v. 96*, p. 11,779–11,802, 1991.

Rubin, V. C., The rotation of spiral galaxies, *Science, v. 220*, p. 1339–1344, 1983.

———, Dark matter in spiral galaxies, in P. W. Hodge, ed., *The Universe of Galaxies: Readings From Scientific American*, p. 31–43, New York, W. H. Freeman & Co., 1984.

Ruelle, D., Strange attractors, *The Mathematical Intelligencer, v. 2*, p. 126–137, 1980.

———, *Chaotic Evolution and Strange Attractors*, New York, Cambridge University Press, 96 pp., 1989.

References Cited

———, *Chance and Chaos*, Princeton, NJ, Princeton University Press, 195 pp., 1991.

Ruelle, D., and F. Takens, On the nature of turbulence, *Commun. math. Phys.*, v. 20, p. 167–192, 1971a.

———, Note concerning our paper "On the nature of turbulence," *Commun. math. Phys.*, v. 23, p. 343–344, 1971b.

Rul'kov, N. F., A. R. Volkovskii, A. Rodriguez-Lozano, E. Del Rio, and M. G. Velarde, Mutual synchronization of chaotic self-oscillators with dissipative coupling, *Internat. Jour. Bifurcation and Chaos*, v. 2, p. 669–676, 1992.

Runcorn, S. K., Lunar palaeomagnetism, polar wandering and the existence of primaeval lunar satellites (Abstract), *Meteoritics*, v. 18, p. 389–390, 1983.

———, The Moon's ancient magnetism, *Sci. Am.*, v. 257, p. 60–68, 1987.

Russell, D. A., The biotic crisis at the end of the Cretaceous Period, in P. Béland et al. (The K-TEC Group), eds., *Cretaceous-Tertiary Extinctions and Possible Terrestrial and Extraterrestrial Causes*, p. 11–23, Ottawa, Canada, The National Museums of Canada, Syllogeus No. 12, 1977.

———, A paleontological consensus on the extinction of the dinosaurs?, in L. T. Silver and P. H. Schultz, eds., *Geological Implications of Impacts of Large Asteroids and Comets on the Earth*, p. 401–405, Boulder, CO, The Geological Society of America, Special Paper 190, 1982.

Russell, D. A., and G. Rice, eds., *K-TEC II: Cretaceous-Tertiary Extinctions and Possible Terrestrial and Extraterrestrial Causes*, Syllogeus No. 39 (Proceedings of workshop, Ottawa, 19–20 May, 1981), Ottawa, National Museums of Canada, 151 pp., 1982.

Russell, S. S., C. T. Pillinger, J. W. Arden, M. R. Lee, and U. Ott, A new type of meteoritic diamond in the enstatite chondrite Abee, *Science*, v. 256, p. 206–209, 1992.

Ruthen, R., Trends in nonlinear dynamics: Adapting to complexity, *Sci. Am.*, v. 268, p. 130–140 (January), 1993.

Ruzmaikina, T. V., V. S. Safronov, and S. J. Weidenschilling, Radial mixing of materials in the asteroidal zone, in R. P. Binzel, T. Gehrels, and M. S. Matthews, eds., *Asteroids II*, p. 681–700, Tucson, AZ, The University of Arizona Press, 1989.

Ryan, M. P., The mechanics and three-dimensional internal structure of active magmatic systems: Kilauea Volcano, Hawaii, *Jour. Geophys. Res.*, v. 93, p. 4213–4248, 1988.

Ryder, G., Coincidence in time of the Imbrium Basin impact and Apollo 15 KREEP volcanic series: Impact-induced melting?, *Papers Presented at the International Conference on Large Meteorite Impacts and Planetary Evolution, 31 August–2 September, 1992, Sudbury, Ontario, Canada*, p. 61–62, Lunar and Planetary Institute, Houston, TX (LPI Contribution No. 790), 1992.

Ryder, G., and G. B. Dalrymple, Apollo 15 impact melts, the age of Imbrium, and the Earth-Moon impact cataclysm, *Papers Presented at the International Conference on Large Meteorite Impacts and Planetary Evolution, 31 August–2 September, 1992, Sudbury, Ontario, Canada*, p. 62–63, Lunar and Planetary Institute, Houston, TX (LPI Contribution No. 790), 1992.

Safronov, V. S., Problems of the origin of asteroids and comets, *Meteoritics*, v. 25, p. 243–248, 1990.

Safronov, V. S., and E. V. Zvjagina, Relative sizes of the largest bodies during the accumulation of planets, *Icarus*, v. 10, p. 109–115, 1969.

Sagan, C., and A. Druyan, Comets: Mementos of creation, *The Planetary Report*, v. 9, p. 4–9, 1989.

Sager, W. W., and U. Bleil, Latitudinal shift of Pacific hotspots during the late Creatceous and early Tertiary, *Nature*, v. 326, p. 488–490, 1987.

Sager, W. W., and M. S. Pringle, Mid-Cretaceous to early Tertiary apparent polar wander path of the Pacific plate, *Jour. Geophys. Res.*, v. 93, p. 11,753–11,771, 1988.

Sampson, J. R., *Biological Information Processing: Current Theory and Computer Simulation*, New York, John Wiley & Sons, Inc., 310 pp., 1984.

Sandfort, J. F., *Heat Engines*, New York, Doubleday and Co., 292 pp., 1962.

Sargent, A. I., Molecular disks and their link to planetary systems, in H. A. Weaver and L. Danly,

References Cited

eds., *The Formation and Evolution of Planetary Systems*, p. 111–129, New York, Cambridge University Press, 1989.

Saunders, R. S., The surface of Venus, *Sci. Am., v. 263*, p. 60–64, 1990.

Savage, H. T., W. L. Ditto, P. A. Braza, M. L. Spano, S. N. Rauseo, and W. C. Spring, III, Crisis-induced intermittency in a parametrically driven, gravitationally buckled, magnetoelastic amorphous ribbon experiment, *Jour. Appl. Phys., v. 67*, p. 5619–5623, 1990.

Schaber, G. G., The geology of Io, in D. Morrison, ed., *Satellites of Jupiter*, p. 556–597, Tucson, AZ, The University of Arizona Press, 1982.

Schaber, G. G., R. G. Strom, H. J. Moore, L. A. Soderblom, R. L. Kirk, D. J. Chadwick, D. D. Dawson, L. R. Gaddis, J. M. Boyce, and J. Russell, Geology and distribution of impact craters on Venus: What are they telling us?, *Jour. Geophys. Res., v. 97*, p. 13,257–13,301, 1992.

Schaffer, W. M., Can nonlinear dynamics elucidate mechanisms in ecology and epidemiology?, *IMA Journal of Mathematics Applied in Medicine and Biology, v. 2*, p. 221–252, 1985.

Schaffer, W. M., and M. Kot, Differential systems in ecology and epidemiology, in A. V. Holden, ed., *Chaos*, p. 158–178, Princeton, NJ, Princeton University Press, 1986.

Schick, R., and R. Mugiono, eds., *Volcanic Tremor and Magma Flow*, German-Indonesian-Cooperation in Scientific Research and Technological Development, Jülich, ZENTRALBIBLIOTHEK, 200 pp., 1991.

Schlegel, R., *Superposition & Interaction: Coherence in Physics*, Chicago, The University of Chicago Press, 302 pp., 1980.

Schleich, W., and P. V. E. McClintock, Humpty Dumpty to Moslem art, *Nature, v. 339*, p. 257–258, 1989.

Schmidt, G., Hamiltonian and dissipative chaos, in J. R. Buchler and H. Eichhorn, eds., *Chaotic Phenomena in Astrophysics*, p. 97–109, New York, The New York Academy of Sciences (Annals Volume 497), 1987.

Schmidt, R. M., and K. A. Holsapple, Estimates of crater size for large-body impact: Gravity-scaling results, in L. T. Silver and P. H. Schultz, eds., *Geological Implications of Impacts of Large Asteroids and Comets on the Earth*, p. 93–102, Boulder, CO, The Geological Society of America, Special Paper 190, 1982.

Schmidt-Kaler, T., The spiral structure of our galaxy: A review of current studies, *Vistas in Astronomy, v. 19, Pt.1*, p. 69–89, 1975.

Schmitt, R. A., Y.-G. Liu, and R. J. Walker, Shatsky Rise evidences support hypothesis that both a bolide (asteroid or comet) impact (BI) and Deccan trap floodings (DT) caused Cretaceous/Tertiary (K/T) extinctions and not hypothesis of either BI or DT alone, I and II, *Lunar and Planetary Sci. Conf. XXII, Abstracts of Papers*, p. 1187–1190, Lunar and Planetary Science Institute, Houston, TX, March 18–22, 1991.

Schneider, D. A., and D. V. Kent, Ivory Coast microtektites and geomagnetic reversals, *Geophys. Res. Lett., v. 17*, p. 163–166, 1990.

Schneider, S. E., and Y. Terzian, Between the galaxies, *Am. Scientist, v. 72*, p. 574–581, 1984.

Schneider, S. H., Whatever happened to nuclear winter — an editorial, *Climatic Change, v. 12*, p. 216–219, 1988.

Schnetzler, C. C., and J. B. Garvin, Search for the 700,000-year-old source crater of the Australasian tektite strewn field, *Papers Presented at the International Conference on Large Meteorite Impacts and Planetary Evolution, 31 August–2 September, 1992, Sudbury, Ontario, Canada*, p. 63–64, Lunar and Planetary Institute, Houston, TX (LPI Contribution No. 790), 1992.

Scholl, H., Ch. Froeschlé, H. Kinoshita, M. Yoshikawa, and J. G. Williams, Secular resonances, in R. P. Binzel, T. Gehrels, and M. S. Matthews, eds., *Asteroids II*, p. 845–861, Tucson, AZ, The University of Arizona Press, 1989.

Schopf, J. W., and B. M. Packer, Early Archean (3.3-billion to 3.5-billion-year-old) microfossils from Warrawoona Group, Australia, *Science, v. 237*, p. 70–73, 1987.

Schopf, T. J. M., Extinction of the dinosaurs: A 1982 perspective, in L. T. Silver and P. H. Schultz, eds., *Geological Implications of Impacts of Large Asteroids and Comets on the Earth*, p. 415–422, Boulder, CO, The Geological Society of America, Special Paper 190, 1982.

References Cited

Schrödinger, E., *What is Life?*, with *Mind and Matter*, and *Autobiographical Sketches*, New York, Cambridge University Press (Canto Edition), 184 pp., 1992.

Schroeder, M., *Fractals, Chaos, Power Laws: Minutes from an Infinite Paradise*, New York, W. H. Freeman & Co., 429 pp., 1991.

Schubert, G., The lost continents, *Nature, v. 354*, p. 358–359, 1991.

Schultz, P. H., Polar wandering on Mars, *Sci. Am., v. 253*, p. 94–102 (Dec.), 1985.

———, Cratering on Mercury: A relook, in F. Vilas, C. R. Chapman, and M. S. Matthews, eds., *Mercury*, p. 274–335, Tucson, AZ, The University of Arizona Press, 1988.

Schultz, P. H., and D. E. Gault, Seismic effects from major basin formation on the Moon and Mercury, *The Moon, v. 12*, p. 159–177, 1975.

———, Environmental consequences from oblique impacts (Abstract), *Eos, Trans. Am. Geophys. Union, v. 71*, p. 1429, 1990a.

———, Prolonged global catastrophes from oblique impacts, in V. L. Sharpton and P. D. Ward, eds., *Global Catastrophes in Earth History; An Interdisciplinary Conference on Impacts, Volcanism, and Mass Mortality*, p. 239–261, Boulder, CO, The Geological Society of America, Special Paper 247, 1990b.

———, Recognizing impactor signatures in the planetary record, *Papers Presented at the International Conference on Large Meteorite Impacts and Planetary Evolution, 31 August–2 September, 1992, Sudbury, Ontario, Canada*, p. 64–65, Lunar and Planetary Institute, Houston, TX (LPI Contribution No. 790), 1992.

Schultz, P. H., and R. E. Lianza, Recent grazing impacts on the Earth recorded in the Rio Cuarto crater field, Argentina, *Nature, v. 355*, p. 234–237, 1992.

Schuster, H. G., *Deterministic Chaos, Second Edition*, Weinheim, Federal Republic of Germany, VCH Verlagsgesellschaft, 270 pp., 1988.

Schwartz, D. P., and K. J. Coppersmith, Fault behavior and characteristic earthquakes: Examples from the Wasatch and San Andreas fault zones, *Jour. Geophys. Res., v. 89*, p. 5681–5698, 1984.

Schwarzschild, B., Do asteroid impacts trigger geomagnetic reversals?, *Phys. Today, v. 40*, p. 17–20 (February), 1987.

Scott, S. K., *Chemical Chaos*, Oxford, Clarendon Press, 454 pp., 1991.

Scott, E. R. D., G. J. Taylor, H. E. Newsom, F. Herbert, M. Zolensky, and J. F. Kerridge, Chemical, thermal and impact processing of asteroids, in R. P. Binzel, T. Gehrels, and M. S. Matthews, eds., *Asteroids II*, p. 701–739, Tucson, AZ, The University of Arizona Press, 1989.

Scotti, J. V., D. L. Rabinowitz, and B. G. Marsden, Near miss of the Earth by a small asteroid, *Nature, v. 354*, p. 287–289, 1991.

Searls, D. B., Representing genetic information with formal grammars, in Anonymous, *Proceedings AAAI-88: Seventh National Conference on Artificial Intelligence*, p. 386–391, San Mateo, CA, Morgan Kaufmann Publishers, Inc., 1988.

———, The linguistics of DNA, *Am. Scientist, v. 80*, p. 579–591, 1992.

Sears, D. W. G., and R. T. Dodd, Overview and classification of meteorites, in J. F. Kerridge and M. S. Matthews, eds., *Meteorites and the Early Solar System*, p. 3–31, Tucson, AZ, The University of Arizona Press, 1988.

Sears, D. W. G., L. Jie, P. H. Benoit, J. M. DeHart, and G. E. Lofgren, A compositional classification scheme for meteoritic chondrules, *Nature, v. 357*, p. 207–210, 1992.

Seiden, P. E., and H. Gerola, Propagating star formation and the structure and evolution of galaxies, *Fundamentals Cosmic Phys., v. 7*, p. 241–311, 1982.

Sengör, A. M. C., Tectonics of the Tethysides: Orogenic collage development in a collisional setting, *Ann. Rev. Earth Planet. Sci., v. 15*, p. 213–244, 1987.

Sepkoski, J. J., Jr., A factor analytic description of the Phanerozoic marine fossil record, *Paleobiology, v. 7*, p. 36–53, 1981.

———, A compendium of fossil marine families, *Milwaukee Public Museum, Contrib. Biol. Geol., No. 51*, 125 pp., 1982.

———, Periodicity in extinction and the problem of catastrophism in the history of life, *Jour. Geol. Soc. London, v. 146*, p. 7–19, 1989.

———, The taxonomic structure of periodic extinction, in V. L. Sharpton and P. D. Ward, eds., *Global Catastrophes in Earth History; An Interdisciplinary Conference on Impacts, Volca-*

nism, and Mass Mortality, p. 33–44, Boulder, CO, The Geological Society of America, Special Paper 247, 1990.

Sepkoski, J. J., Jr., and D. M. Raup, Periodicity in marine extinction events, in D. K. Elliott, ed., *Dynamics of Extinction*, p. 3–36, New York, John Wiley & Sons, Inc., 1986.

Sergin, V. Ya., Origin and mechanism of large-scale climatic oscillations, *Science, 209*, p. 1477–1482, 1980.

Sewell, E., *The Structure of Poetry*, London, Routledge & Kegan Paul Ltd. (Reprinted by Richard West, 1977), 196 pp., 1951.

Sexl, R., and H. Sexl, *White Dwarfs-Black Holes: An Introduction to Relativistic Astrophysics*, New York, Academic Press, 203 pp., 1979.

Seyfert, C. K., and L. A. Sirkin, *Earth History and Plate Tectonics, Second Edition*, New York, Harper and Row, Publishers, 600 pp., 1979.

Shafto, M., ed., *How We Know* (Nobel Conference XX), San Francisco, Harper and Row, Publishers, 171 pp., 1985a.

———, Afterword: The brain-in-a-box, in M. Shafto, ed., *How We Know* (Nobel Conference XX), p. 169–171, San Francisco, Harper and Row, Publishers, 1985b.

Shannon, C. E., and W. Weaver, *The Mathematical Theory of Communication*, Urbana, IL, University of Illinois, 125 pp., 1949.

Sharpton, V. L., and K. Burke, Cretaceous-Tertiary impacts, *Meteoritics, v. 22*, p. 499–500, 1987.

Sharpton, V. L., and R. A. F. Grieve, Meteorite impact, cryptoexplosion, and shock metamorphism: A perspective on the evidence at the K/T boundary, in V. L. Sharpton and P. D. Ward, eds., *Global Catastrophes in Earth History; An Interdisciplinary Conference on Impacts, Volcanism, and Mass Mortality*, p. 301–318, Boulder, CO, The Geological Society of America, Special Paper 247, 1990.

Sharpton, V. L., and P. D. Ward, eds., *Global Catastrophes in Earth History; An Interdisciplinary Conference on Impacts, Volcanism, and Mass Mortality*, Boulder, CO, The Geological Society of America, Special Paper 247, 631 pp., 1990.

Sharpton, V. L., B. C. Schuraytz, K. Burke, A. V. Murali, and G. Ryder, Detritus in K/T boundary clays of western North America: Evidence against a single oceanic impact, in V. L. Sharpton and P. D. Ward, eds., *Global Catastrophes in Earth History; An Interdisciplinary Conference on Impacts, Volcanism, and Mass Mortality*, p. 349–357, Boulder, CO, The Geological Society of America, Special Paper 247, 1990.

Sharpton, V. L., B. C. Schuraytz, D. W. Ming, J. H. Jones, E. Rosencrantz, and A. E. Weidie, Is the Chicxulub structure in N. Yucatan a 200 km diameter impacts crater at the K/T boundary? Analysis of drill core samples, geophysics, and regional geology, *Lunar and Planetary Sci. Conf. XXII, Abstracts of Papers*, p. 1223–1224, Lunar and Planetary Science Institute, Houston, TX, March 18–22, 1991.

Sharpton, V. L., G. B. Dalrymple, L. E. Marin, G. Ryder, B. C. Schuraytz, and J. Urrutia-Fucugauchi, New links between Chicxulub impact structure and the Cretaceous/Tertiary boundary, *Nature, v. 359*, p. 819–821, 1992.

Shaw, H. R., Obsidian-H_2O: Viscosities at 1000 and 2000 bars in the temperature range 700°–900° C, *Jour. Geophys. Res., v. 68*, p. 6337–6343, 1963.

———, Comments on viscosity, crystal settling and convection in granitic magmas, *Am. Jour. Sci., v. 263*, p. 120–152, 1965.

———, Rheology of basalt in the melting range, *Jour. Petrology, v. 10*, p. 510–535, 1969.

———, Earth tides, global heat flow, and tectonics, *Science, v. 168*, p. 1084–1087, 1970.

———, Viscosities of magmatic silicate liquids: An empirical method of prediction, *Am. Jour. Sci., v. 272*, p. 870–893, 1972.

———, Mantle convection and volcanic periodicity in the Pacific: Evidence from Hawaii, *Bull. Geol. Soc. Am., v. 84*, p. 1505–1526, 1973.

———, Diffusion of H_2O in granitic liquids: Part 1, Experimental data; Part 2, Mass transfer in magma chambers, in A. W. Hofmann, B. J. Giletti, and H. S. Yoder, Jr., eds., *Geochemical Transport and Kinetics*, p. 139–170, Washington, DC, Carnegie Institution of Washington Publication No. 634, 1974.

———, Methods of simulation analysis applied to questions of geological stability of the refer-

References Cited

———, ence system, in J. E. Campbell et al., *Risk Methodology for Geologic Disposal of Radioactive Waste: Interim Report, Appendix 2A*, p. 87–160, Albuquerque, NM, Sandia National Laboratory Report SAND78–0029 (NUREG/CR-0458), 1978.

———, The fracture mechanisms of magma transport from the mantle to the surface, in R. B. Hargraves, ed., *Physics of Magmatic Processes*, p. 201–264, Princeton, NJ, Princeton University Press, 1980.

———, Analysis of salt dissolution, in G. L. Benson, ed., *WISAP Release Scenario Analysis Workshop 1979*, p. 25–41, Richland, WA, Battelle Memorial Institute Pacific Northwest Laboratory Report PNL-SA-8257, 1981.

———, Sociotectonics and the new geology, U.S. Geological Survey, Unpub. Res. Report, 38 pp., 16 Figs., 1982.

———, Magmatic processes and orbital evolution, *Geol. Soc. Am., Abstracts with Programs, v. 15*, p. 684, 1983a.

———, Mathematical attractor theory and geological patterns: The distribution of the planets as a resonant mapping of chaos, *U. S. Geological Survey, Unpub. Res. Report*, 32 pp., 9 Figs., 1983b.

———, Conjectures on cosmogenic attractors, *U. S. Geological Survey, Unpub. Res. Report*, 23 pp., 1 Fig., 1983c.

———, Nonlinear evolution related to catastrophic impacts: Alternatives to fixed periodicities, *U. S. Geological Survey, Unpub. Res. Report*, 48 pp., 10 Figs., 1984.

———, Links between magma-tectonic rate balances, plutonism, and volcanism, *Jour. Geophys. Res., v. 90*, p. 11,275–11,288, 1985a.

———, Phase portraits of terrestrial evolution, U.S. Geological Survey, Unpub. Res. Report, 30 pp., 24 Illus., 1985b.

———, Ring vortex bifurcations, information cascades, and volcanogenic sources of prebiotic evolution, *U. S. Geological Survey, Unpub. Res. Report*, 163 pp., 30 Figs., 1986.

———, Uniqueness of volcanic systems, in R. W. Decker, T. L. Wright, and P. H. Stauffer, eds., *Volcanism in Hawaii*, Vol. 2, Chap. 51, p. 1357–1394, U.S. Geol. Survey, Prof. Paper 1350, Washington, DC, U.S. Gov't Printing Office, 1987a.

———, The periodic structure of the natural record, and nonlinear dynamics, *Eos, Trans. Am. Geophys. Union, v. 68*, p. 1651–1665, 1987b.

———, *A Linguistic Model of Earthquake Frequencies, Applied to the Seismic History of California*, U. S. Geol. Survey, Open-File Report 87–296, 129 pp., 1987c.

———, *Terrestrial-Cosmological Correlations in Evolutionary Processes*, U. S. Geol. Survey, Open-File Report 88–43, 85 pp., 1988a.

———, Mathematical attractor theory and plutonic-volcanic episodicity, in C.-Y. King and R. Scarpa, eds., *Modeling of Volcanic Processes*, p. 162–206, Braunschweig/Wiesbaden, Friedr. Vieweg & Sohn, 1988b.

———, Reply: On Chaos, *Eos, Trans. Am. Geophys. Union, v. 69*, p. 669–670, 1988c.

———, Magmatic phenomenology as nonlinear dynamics: Anthology of some relevant experiments and portraits, in G. V. Middleton, ed., *Nonlinear Dynamics, Chaos and Fractals; With Applications to Geological Systems: Chapt. 8*, p. 97–149, Toronto, Geol. Assoc. Canada, Short Course Notes, v. 9, 1991.

———, The liturgy of science: Chaos, number, and the meaning of evolution, in W. Glen, ed., *The Mass Extinction Debates: How Science Works in a Crisis*, p. 170–199, Proceedings of Symposium held at Northwestern University, 12 July, 1991, by the Institute for Historical, Philosophical, and Sociological Studies in Biology, Stanford, CA, Stanford University Press, 1994.

Shaw, H. R., and B. Chouet, *Application of Nonlinear Dynamics to the History of Seismic Tremor at Kilauea Volcano, Hawaii*, U. S. Geol. Survey, Open-File Report 88–539, 78 pp., 1988.

———, Singularity spectrum of intermittent seismic tremor at Kilauea Volcano, Hawaii, *Geophys. Res. Lett., v. 16*, p. 195–198, 1989.

———, Fractal hierarchies of magma transport in Hawaii and critical self-organization of tremor, *Jour. Geophys. Res., v. 96*, p. 10,191–10,207, 1991.

References Cited

Shaw, H. R., and P. Doherty, Strangely familiar attractors: Maps of recursions on hyperparabolic functions, *U. S. Geological Survey, Unpub. Res. Report*, 15 pp., 3 Figs., 1983.

Shaw, H. R., and A. E. Gartner, *Empirical Laws of Order Among Rivers, Faults, and Earthquakes*, U. S. Geol. Survey, Open-File Report 84-356, 51 pp., 1984.

——, *On the Graphical Interpretation of Paleoseismic Data*, U. S. Geol. Survey, Open-File Report 86-394, 92 pp., 1986.

——, Earthquake distributions as multifractal singularity spectra, *U. S. Geological Survey, Unpub. Res. Report*, 37 pp., 8 Figs., 1987.

Shaw, H. R., and E. D. Jackson, Linear island chains in the Pacific: Result of thermal plumes or gravitational anchors?, *Jour. Geophys. Res., v. 78*, p. 8634-8652, 1973.

Shaw, H. R., and J. G. Moore, Magmatic heat and the El Niño cycle, *Eos, Trans. Am. Geophys. Union, v. 69*, p. 1553, 1564-1565, 1988.

Shaw, H. R., and D. A. Swanson, Eruption and flow rates of flood basalts, in E. H. Gilmour and D. Stradling, *Proceedings of the Second Columbia River Basalt Symposium*, p. 271-299, Cheney, WA, Eastern Washington State College Press, 1970.

Shaw, H. R., T. L. Wright, D. L. Peck, and R. Okamura, The viscosity of basaltic magma: An analysis of field measurements in Makaopuhi lava lake, Hawaii, *Am. Jour. Sci., v. 266*, p. 225-264, 1968.

Shaw, H. R., R. W. Kistler, and J. F. Evernden, Sierra Nevada plutonic cycle: Part II., Tidal energy and a hypothesis for orogenic-epeirogenic periodicities, *Bull. Geol. Soc. Am., v. 82*, p. 869-896, 1971.

Shaw, H. R., M. S. Hamilton, and D. L. Peck, Numerical analysis of lava lake cooling models: Part I. Description of the method, *Am. Jour. Sci., v. 277*, p. 384-414, 1977.

Shaw, H. R., E. D. Jackson, and K. E. Bargar, Volcanic periodicity along the Hawaiian-Emperor chain, in A. J. Irving and M. A. Dungan, eds., *The Jackson Volume: A Special Volume of the American Journal of Science in Memory of Everett Dale Jackson, 1925-1978*, p. 667-708, New Haven, CT, Kline Geology Laboratory, Yale University (Am. Jour. Sci., v. 280-A), 1980a.

Shaw, H. R., A. E. Gartner, C. Sontag-Eng, and L. Casey, *Methods of Simulation Analysis Applied to Questions of Geological Stability of a Radioactive Waste Depository in Bedded Salt*, U. S. Geol. Survey, Open-File Report 80-705, 160 pp., 1980b.

Shaw, H. R., A. E. Gartner, and F. Lusso, *Statistical Data for Movements on Young Faults of the Conterminous United States: Paleoseismic Implications and Regional Earthquake Forecasting*, U. S. Geol. Survey, Open-File Report 81-946, 353 pp., 1981.

Shaw, R. S., Strange attractors, chaotic behavior, and information flow, *Z. Naturforsch., v. 36a*, p. 80-112, 1981.

——, *The Dripping Faucet as a Model Chaotic System*, Santa Cruz, CA, Aerial Press, 113 pp., 1984.

Shearer, P. M., and T. G. Masters, Global mapping of topography on the 660-km discontinuity, *Nature, v. 355*, p. 791-796, 1992.

Shepard, R. N., Circularity in judgments of relative pitch, *Jour. Acoustical Soc. Am., v. 36*, p. 2346-2353, 1964.

Sheridan, R. E., Phenomena of pulsation tectonics related to the breakup of the eastern North American margin, *Tectonophys., v. 94*, p. 169-185, 1983.

Shevoroshkin, V., The Mother Tongue, *The Sciences, May/June*, p. 20-27, 1990.

Shimizu, T., and N. Morioka, Transient behavior in periodic regions of the Lorenz model, *Phys. Lett., v. 69A*, p. 148-150, 1978.

Shinbrot, T., C. Grebogi, E. Ott, and J. A. Yorke, Using small perturbations to control chaos, *Nature, v. 363*, p. 411-417, 1993.

Shoemaker, E. M., Penetration mechanics of high velocity meteorites, illustrated by Meteor Crater, Arizona, *21st Internat. Geol. Congress., Copenhagen*, p. 418-434, 1960a.

——, Brecciation and mixing of rock by strong shock, *U.S. Geol. Survey, Prof. Paper 400-B*, p. B423-425, Washington, DC, U.S. Gov't Printing Office, 1960b.

——, Interplanetary correlation of geologic time, *Abstract, Bull. Amer. Assoc. Petroleum Geologists, v. 45*, p. 130, 1961.

References Cited

———, Interpretation of lunar craters, in Z. Kopal, ed., *Physics and Astronomy of the Moon*, p. 283–359, New York, Academic Press, 1962.

———, Impact mechanics at Meteor Crater, Arizona, in B. M. Middlehurst and G. P. Kuiper, eds., *The Moon, Meteorites, and Comets*, p. 301–336, Chicago, The University of Chicago Press, 1963.

———, Asteroid and comet bombardment of the Earth, *Ann. Rev. Earth Planet. Sci.*, v. 11, p. 461–494, 1983.

———, Large body impacts through geologic time, in H. D. Holland and A. F. Trendall, eds., *Patterns of Change in Earth Evolution*, p. 15–40, New York, Springer-Verlag, 1984.

Shoemaker, E. M., and E. C. T. Chao, New evidence for the impact origin of the Ries basin, Bavaria, Germany, *Jour. Geophys. Res.*, v. 66, p. 3371–3378, 1961.

Shoemaker, E. M., and R. J. Hackman, Stratigraphic basis for a lunar time scale, in Z. Kopal and Z. K. Mikhailov, eds., *The Moon*, p. 289–300, New York, Academic Press, 1962.

Shoemaker, E. M., and G. A. Izett, K/T boundary stratigraphy: Evidence for multiple impacts and a possible comet stream, *Papers Presented at the International Conference on Large Meteorite Impacts and Planetary Evolution, 31 August–2 September, 1992, Sudbury, Ontario, Canada*, p. 66–68, Lunar and Planetary Institute, Houston, TX (LPI Contribution No. 790), 1992.

Shoemaker, E. M., and S. W. Kieffer, *Guidebook to the Geology of Meteor Crater, Arizona*, Tempe, AZ, Center for Meteorite Studies, Arizona State University, 66 pp., 1974.

Shoemaker, E. M., and R. F. Wolfe, Cratering time scales for the Galilean satellites, in D. Morrison, ed., *Satellites of Jupiter*, p. 277–339, Tucson, AZ, The University of Arizona Press, 1982.

———, Evolution of the Uranus-Neptune planetesimal swarm, *Lunar and Planetary Sci. Conf. XV, Abstracts of Papers*, p. 780–781, Lunar and Planetary Science Institute, Houston, TX, 1984.

———, Mass extinctions, crater ages and comet showers, in R. Smoluchowski, J. N. Bahcall, and M. S. Matthews, eds., *The Galaxy and the Solar System*, p. 338–386, Tucson, AZ, The University of Arizona Press, 1986.

Shoemaker, E. M., J. G. Williams, E. F. Helin, and R. F. Wolfe, Earth-crossing asteroids: Orbital classes, collision rates with Earth, and origin, in T. Gehrels, ed., *Asteroids*, p. 253–282, Tucson, AZ, The University of Arizona Press, 1979.

Shoemaker, E. M., B. K. Lucchitta, J. B. Plescia, S. W. Squyres, and D. E. Wilhelms, The geology of Ganymede, in D. Morrison, ed., *Satellites of Jupiter*, p. 435–520, Tucson, AZ, The University of Arizona Press, 1982.

Shoemaker, E. M., C. S. Shoemaker, and R. F. Wolfe, Trojan asteroids: Populations, dynamical structure and origin of the L4 and L5 swarms, in R. P. Binzel, T. Gehrels, and M. S. Matthews, eds., *Asteroids II*, p. 487–523, Tucson, AZ, The University of Arizona Press, 1989.

Shoemaker, E. M., R. F. Wolfe, and C. S. Shoemaker, Asteroid and comet flux in the neighborhood of Earth, in V. L. Sharpton and P. D. Ward, eds., *Global Catastrophes in Earth History; An Interdisciplinary Conference on Impacts, Volcanism, and Mass Mortality*, p. 155–170, Boulder, CO, The Geological Society of America, Special Paper 247, 1990.

Sibley, C. G., and J. E. Ahlquist, *Phylogeny and Classification of Birds: A Study in Molecular Evolution*, New Haven, CT, Yale University Press, 976 pp., 1990.

Siggia, E. D., and H. Aref, Scaling and structures in fully turbulent flows, in R. H. G. Helleman, ed., *Nonlinear Dynamics*, p. 368–376, New York, Annals of the New York Acad. Sciences, v. 357, 1980.

Sigmund, K., On prisoners and cells, *Nature*, v. 359, p. 774, 1992.

Signor, P. W., III, and J. H. Lipps, Sampling bias, gradual extinction patterns and catastrophes in the geologic record, in L. T. Silver and P. H. Schultz, eds., *Geological Implications of Impacts of Large Asteroids and Comets on the Earth*, p. 291–296, Boulder, CO, The Geological Society of America, Special Paper 190, 1982.

Sigurdsson, H., Assessment of the atmospheric impact of volcanic eruptions, in V. L. Sharpton and P. D. Ward, eds., *Global Catastrophes in Earth History; An Interdisciplinary Conference*

References Cited

on Impacts, Volcanism, and Mass Mortality, p. 99–110, Boulder, CO, The Geological Society of America, Special Paper 247, 1990.

Sigurdsson, H., S. D'Hondt, M. A. Arthur, T. J. Bralower, J. C. Zachos, M. van Fossen, and J. E. T. Channell, Tektite glass from the Cretaceous-Tertiary boundary in Haiti, *Lunar and Planetary Sci. Conf. XXII, Abstracts of Papers*, p. 1259–1260, Lunar and Planetary Science Institute, Houston, TX, March 18–22, 1991a.

Sigurdsson, H., Ph. Bonté, L. Turpin, M. Chaussidon, N. Metrich, M. Steinberg, Ph. Pradel, and S. D'Hondt, Geochemical constraints on source region of Cretaceous/Tertiary impact glasses, *Nature, v. 353*, p. 839–842, 1991b.

Silk, J., Cosmology back to the beginning, *Nature, v. 356*, p. 741–742, 1992.

Silver, L. T., and P. H. Schultz, eds., *Geological Implications of Impacts of Large Asteroids and Comets on the Earth*, Boulder, CO, The Geological Society of America, Special Paper 190, 528 pp., 1982.

Simberloff, D. S., Permo-Triassic extinctions: Effects of area on biotic equilibrium, *Jour. Geol., v. 82*, p. 267–274, 1974.

——, Are we on the verge of a mass extinction in tropical rain forests?, in D. K. Elliott, ed., *Dynamics of Extinction*, p. 165–180, New York, John Wiley & Sons, Inc., 1986.

Simon, L., J. Bousquet, R. C. Lévesque, and M. Lalonde, Origin and diversification of endomycorrhizal fungi and coincidence with vascular land plants, *Nature, v. 363*, p. 67–69, 1993.

Simoyi, R. H., A. Wolf, and H. L. Swinney, One-dimensional dynamics in a multicomponent chemical reaction, *Phys. Rev. Lett., v. 49*, p. 245–248, 1982.

Singer, J., Y.-Z. Wang, and H. H. Bau, Controlling a chaotic system, *Phys. Rev. Lett., v. 66*, p. 1123–1125, 1991.

Skarda, C. A., and W. J. Freeman, How brains make chaos in order to make sense of the world, *Behavioral and Brain Sciences, v. 10*, p. 161–195, 1987.

Skinner, B. J., ed., *The Solar System and its Strange Objects: Readings from American Scientist*, Los Altos, CA, William Kaufmann, Inc., 193 pp., 1981.

Sleep, N. H., Hotspot volcanism and mantle plumes, *Ann. Rev. Earth Planet. Sci., v. 20*, p. 19–43, 1992.

Smale, S., Differentiable dynamical systems, *Bull. Am. Math. Soc., v. 73*, p. 747–817, 1967.

Smit, J., Extinction and evolution of planktonic foraminifera at the Cretaceous/Tertiary boundary after a major impact, in L. T. Silver and P. H. Schultz, eds., *Geological Implications of Impacts of Large Asteroids and Comets on the Earth*, p. 329–352, Boulder, CO, The Geological Society of America, Special Paper 190, 1982.

Smit, J., and W. Alvarez, Is the "mid-Cretaceous unconformity" in the Gulf of Mexico a Cretaceous-Tertiary boundary impact-wave erosion surface?, *Lunar and Planetary Sci. Conf. XXII, Abstracts of Papers*, p. 1275–1276, Lunar and Planetary Science Institute, Houston, TX, March 18–22, 1991.

Smit, J., A. Montanari, and W. Alvarez, Microkrystites and (micro)tektites at the K/T boundary: Two different sources or one?, *Lunar and Planetary Sci. Conf. XXII, Abstracts of Papers*, p. 1277–1278, Lunar and Planetary Science Institute, Houston, TX, March 18–22, 1991.

Smith, A. G., and R. A. Livermore, Pangea in Permian to Jurassic time, in T. W. C. Hilde and R. L. Carlson, eds., *Silver Anniversary of Plate Tectonics*, p. 135–179, Amsterdam, Elsevier Science Publishers *(Tectonophysics, v. 187)*, 1991.

Smith, B. A., L. A. Soderblom, D. Banfield, C. Barnet, A. T. Basilevsky *et al.* (60 additional coauthors), Voyager 2 at Neptune: Imaging science results, *Science, v. 246*, p. 1422–1449, 1989.

Smith, D. E., R. Kolenkiewicz, P. J. Dunn, J. W. Robbins, M. H. Torrence, S. M. Klosko, R. G. Williamson, E. C. Pavlis, N. B. Douglas, and S. K. Fricke, Tectonic motion and deformation from satellite laser ranging to LAGEOS, *Jour. Geophys. Res., v. 95*, p. 22,013–22,041, 1990.

Smith, D. G., Cyclicity or chaos? Orbital forcing versus nonlinear dynamics, in P. L. de Boer and D. G. Smith, eds., *Orbital Forcing and Cyclic Sequences*, p. 531–544, Special Publication 19, International Association of Sedimentologists, 1994.

Smith, J. G., E. H. McKee, D. B. Tatlock, and R. F. Marvin, Mesozoic granitic rocks in northwestern Nevada: A link between the Sierra Nevada and Idaho batholiths, *Geol. Soc. Am., Bull., v. 82*, p. 2933–2944, 1971.

References Cited

Smith, L. A., and E. A. Spiegel, Strange accumulators, in J. R. Buchler and H. Eichhorn, eds., *Chaotic Phenomena in Astrophysics*, p. 61–65, New York, Annals of the New York Academy of Sciences, Volume 497, 1987.

Smith, R. L., Ash-flow magmatism, in C. E. Chapin and W. E. Elston, eds., *Ash-Flow Tuffs*, p. 5–27, Boulder, CO, The Geological Society of America Special Paper 180, 1979.

Smith, R. L., and R. G. Luedke, Potentially active volcanic lineaments and loci in western conterminous United States, in Geophysics Study Committee, eds., *Explosive Volcanism: Inception, Evolution, and Hazards*, p. 47–66, Washington, DC, National Academy Press, 1984.

Smith, R. L., and H. R. Shaw, Igneous-related geothermal systems, in D. E. White and D. L. Williams, eds., *Assessment of the Geothermal Resources of the United States*, p. 58–83, U. S. Geological Survey, Circ. 726, 1975.

——, Igneous-related geothermal systems, in L. J. P. Muffler, ed., *Assessment of Geothermal Resources of the United States – 1978*, p. 2–17, U.S. Geological Survey, Circ. 790, 1979.

Smith, R. L., H. R. Shaw, R. G. Luedke, and S. L. Russell, *Comprehensive Tables Giving Physical Data and Energy Estimates for Young Igneous Systems of the United States*, U. S. Geological Survey Open-File Report 78–925, 18 pp., 2 Tables, 1978.

Smoluchowski, R., J. N. Bahcall, and M. S. Matthews, eds., *The Galaxy and the Solar System*, Tucson, AZ, The University of Arizona Press, 483 pp., 1986.

Smylie, D. E., The inner core translational triplet and the density near Earth's center, *Science, v. 255*, p. 1678–1682, 1992.

Snow, J. T., On inertial instability as related to the multiple-vortex phenomenon, *Jour. Atmos. Sci., v. 35*, p. 1660–1677, 1978.

Sokal, R. R., N. L. Oden, P. Legendre, M.-J. Fortin, J. Kim, B. A. Thomson, A. Vaudor, R. M. Harding, and G. Barbujani, Genetics and language in European populations, *Am. Naturalist, v. 135*, p. 157–175, 1990.

Solomon, S. C., Keeping that youthful look, *Nature, v. 361*, p. 114–115, 1993.

Sommerer, J. C., and C. Grebogi, Determination of crisis parameter values by direct observation of manifold tangencies, *Internat. Jour. Bifurcation and Chaos, v. 2*, p. 383–396, 1992.

Sommerer, J. C., and E. Ott, Particles floating on a moving fluid: A dynamically comprehensible physical fractal, *Science, v. 259*, p. 335–339, 1993.

Sommerer, J. C., E. Ott, and C. Grebogi, Scaling law for characteristic times of noise-induced crises, *Phys. Rev. A, v. 433*, p. 1754–1769, 1991a.

Sommerer, J. C., W. L. Ditto, C. Grebogi, E. Ott, and M. L. Spano, Experimental confirmation of the theory for critical exponents of crises, *Phys. Lett. A, v. 153*, p. 105–109, 1991b.

Sommeria, J., S. T. Meyers, and H. L. Swinney, Laboratory simulation of Jupiter's Great Red Spot, *Nature, v. 331*, p. 689–693, 1988.

Sonea, S., and M. Panisset, *A New Bacteriology*, Portola Valley, CA, Jones and Bartlett Publishers, Inc., 140 pp., 1983.

Sonett, C. P., The relationship of meteoritic parent body thermal histories and electromagnetic heating by a pre-main sequence T-Tauri Sun, in T. Gehrels, ed., *Physical Studies of Minor Planets*, p. 239–245, Washington, DC, National Aeronautics and Space Administration, NASA SP-267, 1971.

Sonett, C. P., M. S. Giampapa, and M. S. Matthews, eds., *The Sun in Time*, Tucson, AZ, The University of Arizona Press, 990 pp., 1991.

Sornette, A., J. Dubois, J. L. Cheminée, and D. Sornette, Are sequences of volcanic eruptions deterministically chaotic?, *Jour. Geophys. Res., v. 96*, p. 11,931–11,945, 1991.

Spada, G., Y. Ricard, and R. Sabadini, Excitation of true polar wander by subduction, *Science, v. 360*, p. 452–454, 1992.

Sparrow, C., *The Lorenz Equations: Bifurcations, Chaos, and Strange Attractors*, Berlin, Springer-Verlag, 269 pp., 1982.

Spencer-Brown, G., *Laws of Form*, New York, The Julian Press, Inc., 141 pp., 1972.

Spera, F. J., and L. E. Stark, Comment on "Historical review of a long-overlooked paper by R. A. Daly concerning the origin and the early history of the Moon" by Ralph B. Baldwin and Don Wilhelms, *Jour. Geophys. Res., v. 98*, p. 3087, 1993.

References Cited

Spiegel, E. A., A class of ordinary differential equations with strange attractors, in R. H. G. Helleman, ed., *Nonlinear Dynamics*, p. 305–312, New York, Annals of the New York Academy of Sciences, Volume 357, 1980.

Spiegel, E. A., and A. Wolf, Chaos and the solar cycle, in J. R. Buchler and H. Eichhorn, eds., *Chaotic Phenomena in Astrophysics*, p. 55–60, New York, Annals of the New York Academy of Sciences, Volume 497, 1987.

Spitzer, L., Jr., *Searching Between the Stars*, New Haven, CT, Yale University Press, 179pp., 1982.

Spohn, T., and G. Schubert, Cooling and solidification of the Earth following a giant impact (Abstract), *Eos, Trans. Am. Geophys. Union, v. 71*, p. 1430, 1990.

Sprague, D., and H. N. Pollack, Heat flow in the Mesozoic and Cenozoic, *Nature, v. 285*, p. 393–395, 1980.

Spudis, P. D., and J. E. Guest, Stratigraphy and geologic history of Mercury, in F. Vilas, C. R. Chapman, and M. S. Matthews, eds., *Mercury*, p. 118–164, Tucson, AZ, The University of Arizona Press, 1988.

Sreenivasan, K. R., Fractals and multifractals in fluid turbulence, *Ann. Rev. Fluid Mech., v. 23*, p. 539–600, 1991.

Sreenivasan, K. R., and C. Meneveau, The fractal facets of turbulence, *Jour. Fluid Mech., v. 173*, p. 357–386, 1986.

Sridhar, S., and S. Tremaine, Tidal disruption of viscous bodies, *Icarus, v. 95*, p. 86–99, 1992.

Stacey, F. D., *Physics of the Earth*, New York, John Wiley & Sons, Inc., 324 pp., 1969.

Stanley, S. M., Marine mass extinctions: A dominant role for temperature, in M. H. Nitecki, ed., *Extinctions*, p. 69–117, Chicago, IL, The University of Chicago Press, 1984.

Stauffer, R. C., ed., *Charles Darwin's Natural Selection: Being the Second Part of His Big Species Book Written from 1856 to 1858*, New York, Cambridge University Press, 692 pp., 1975.

Steel, D., Our asteroid-pelted planet, *Nature, v. 354*, p. 265–267, 1991.

Steele, I. M., Primitive material surviving in chondrites: Mineral grains, in J. F. Kerridge and M. S. Matthews, eds., *Meteorites and the Early Solar System*, p. 808–818, Tucson, AZ, The University of Arizona Press, 1988.

Stein, D. L., Spin glasses, *Sci. Am., v. 261*, p. 52–59 (July), 1989.

Stern, S. A., The effects of mechanical interaction between the interstellar medium and comets, *Icarus, v. 68*, p. 276–283, 1986.

———, Extra-solar Oort cloud encounters and planetary impact rates, *Icarus, v. 69*, p. 185–188, 1987.

———, Collisions in the Oort cloud, *Icarus, v. 73*, p. 499–507, 1988.

———, On the number of planets in the outer Solar System: Evidence of a substantial population of 1000-km bodies, *Icarus, v. 90*, p. 271–281, 1991.

———, The Pluto-Charon system, *Ann. Rev. Astron. Astrophys, v. 30*, p. 185–233, 1992.

———, The Pluto reconnaissance flyby mission, *Eos, Trans. Am. Geophys. Union, v. 74*, p. 73–78, 1993.

Stern, S. A., and J. M. Shull, The thermal evolution of comets in the Oort cloud by stars and supernovae, *Nature, v. 332*, p. 407–411, 1988.

Stevenson, D. J., Origin of the Moon — The collision hypothesis, *Ann. Rev. Earth Planet. Sci., v. 15*, p. 271–315, 1987.

———, Implications of the giant planets for the formation and evolution of planetary systems, in H. A. Weaver and L. Danly, eds., *The Formation and Evolution of Planetary Systems*, p. 75–90, New York, Cambridge University Press, 1989.

———, Stalking the magma ocean, *Nature, v. 355*, p. 301, 1992.

Stewart, G. R., A violent birth for Mercury, *Nature, v. 335*, p. 496–497, 1988.

Stewart, I., *Does God Play Dice: The Mathematics of Chaos*, Cambridge, MA, Basil Blackwell, 348 pp., 1989.

Stewart, J. A., *Drifting Continents and Colliding Paradigms*, Bloomington, IN, Indiana University Press, 285 pp., 1990.

Stöffler, D., A. Bischoff, V. Buchwald, and A. E. Rubin, Shock effects in meteorites, in J. F.

References Cited

Kerridge and M. S. Matthews, eds., *Meteorites and the Early Solar System*, p. 165–202, Tucson, AZ, The University of Arizona Press, 1988.

Stone, D. B., Paleogeography and rotations of arctic Alaska—an unresolved problem, in C. Kissel and C. Laj, eds., *Paleomagnetic Rotations and Continental Deformation*, p. 343–364, Boston, MA, Kluwer Academic Publishers, 1989.

Stone, E. C., and E. D. Miner, The Voyager 2 encounter with the Neptunian System, *Science, v. 246*, p. 1417–1421, 1989.

Stothers, R. B., Periodicity of the Earth's magnetic reversals, *Nature, v. 322*, p. 444–446, 1986.

———, Volcanic eruptions and solar activity, *Jour. Geophys. Res., v. 94*, p. 17,371–17,381, 1989.

———, Impacts and tectonism in Earth and Moon history of the past 3800 million years, *Earth, Moon and Planets, v. 58*, p. 145–152, 1992.

———, Hotspots and sunspots: Surface tracers of deep mantle convection in the Earth and Sun, *Earth Planet. Sci. Lett., v. 116*, p. 1–8, 1993.

Stothers, R. B., and M. R. Rampino, Periodicity in flood basalts, mass extinctions, and impacts; a statistical view and a model, in V. L. Sharpton and P. D. Ward, eds., *Global Catastrophes in Earth History; An Interdisciplinary Conference on Impacts, Volcanism, and Mass Mortality*, p. 9–18, Boulder, CO, The Geological Society of America, Special Paper 247, 1990.

Strahler, A. N., *The Earth Sciences, Second Edition*, New York, Harper & Row, Publishers, 824 pp., 1971.

Strom, R. G., Mercury: The Elusive Planet, Washington, DC, Smithsonian Institution Press, 197 pp., 1987.

Strom, R. G., and G. Neukum, The cratering record on Mercury and the origin of impacting objects, in F. Vilas, C. R. Chapman, and M. S. Matthews, eds., *Mercury*, p. 336–373, Tucson, AZ, The University of Arizona Press, 1988.

Strom, R. G., and N. M. Schneider, Volcanic eruption plumes on Io, in D. Morrison, ed., *Satellites of Jupiter*, p. 598–633, Tucson, AZ, The University of Arizona Press, 1982.

Struve, O., B. Lynds, and H. Pillans, *Astronomie: Einführung in ihre Grundlagen*, Berlin, de Gruyter, 468 pp., 1962.

Stump, E., The Ross Orogen of the Transantarctic Mountains in light of the Laurentia-Gondwana split, *GSA Today, v. 2*, p. 25–27, 30–31 (February), 1992.

Suarez, I. M., T. J. Hoffman, L. R. Smith IV, J. C. Lam, E. N. La Joie, and C. Boekema, Effects of frustration: A computational study, in L. Lam and H. C. Morris, eds., *Nonlinear Structures in Physical Systems: Pattern Formation, Chaos, and Waves*, p. 103–110, New York, Springer-Verlag, 1990.

Suess, H. E., *Chemistry of the Solar System*, New York, John Wiley & Sons, Inc., 145 pp., 1987.

Sugihara, G., and R. M. May, Nonlinear forecasting as a way of distinguishing chaos from measurement error in time series, *Nature, v. 344*, p. 734–741, 1990.

Suslick, K. S., and E. B. Flint, Sonoluminescence from non-aqueous lquids, *Nature, v. 330*, p. 553–555, 1987.

Sussman, G. J., and J. Wisdom, Chaotic evolution of the Solar System, *Science, v. 257*, p. 56–62, 1992.

Sweet, A. R., D. R. Braman, and J. F. Lerbekmo, Palynofloral response to K/T boundary events; a transitory interruption within a dynamic system, in V. L. Sharpton and P. D. Ward, eds., *Global Catastrophes in Earth History; An Interdisciplinary Conference on Impacts, Volcanism, and Mass Mortality*, p. 457–469, Boulder, CO, The Geological Society of America, Special Paper 247, 1990.

Swindle, T. D., J. S. Lewis, and L-A. A. McFadden, Near-Earth asteroids and the history of planetary formation, *Eos, Trans. Am. Geophys. Union, v. 72*, p. 473, 479–480, 1991.

Swinney, H. L., and J. C. Roux, Chemical chaos, in C. Vidal and A. Pacault, *Non-Equilibrium Dynamics in Chemical Systems*, p. 124–140, New York, Springer-Verlag, 1984.

Swisher, C. C., III, J. M. Grajales-Nishimura, A. Montanri, S. V. Margolis, P. Claeys, W. Alvarez, P. Renne, E. Cedillo-Pardo, F. J.-M. R. Maurrasse, G. H. Curtis, J. Smit, and M. O. Mc-Williams, Coeval $^{40}Ar/^{39}Ar$ ages of 65.0 million years ago from Chicxulub crater melt rock and Cretaceous-Tertiary boundary tektites, *Science, v. 257*, p. 954–958, 1992.

Sykes, M. V., R. Greenberg, S. F. Dermott, P. D. Nicholson, J. A. Burns, and T. N. Gautier, III,

Dust bands in the Asteroid Belt, in R. P. Binzel, T. Gehrels, and M. S. Matthews, eds., *Asteroids II*, p. 336–367, Tucson, AZ, The University of Arizona Press, 1989.

Szebehely, V., and B. D. Tapley, *Long-Time Predictions in Dynamics*, Boston, MA, D. Reidel Publishing Company, 358 pp., 1976.

Tackley, P. J., D. J. Stevenson, G. A. Glatzmaier, and G. Schubert, Effects of an endothermic phase transition at 670 km depth in a spherical model of convection in the Earth's mantle, *Nature, v. 361*, p. 699–704, 1993.

Taff, L. G., *Celestial Mechanics: A Computational Guide for the Practitioner*, New York, John Wiley & Sons, Inc., 520 pp., 1985.

Takahashi, T. J., and J. D. Griggs, Hawaiian volcanic features: A photoglossary, in R. W. Decker, T. L. Wright, and P. H. Stauffer, eds., *Volcanism in Hawaii*, Vol. 2, Chap. 36, p. 845–902, U.S. Geol. Survey, Prof. Paper 1350, Washington, DC, U.S. Gov't Printing Office, 1987.

Takens, F., Detecting strange attractors in turbulence, in D. Rand and L.-S. Young, eds., *Dynamical Systems and Turbulence*, p. 366–381, New York, Springer-Verlag, 1981.

Tam, W. Y., Pattern formation in chemical systems, in L. Lam and H. C. Morris, eds., *Nonlinear Structures in Physical Systems: Pattern Formation, Chaos, and Waves*, p. 87–102, New York, Springer-Verlag, 1990.

Tassoul, J. L., and M. Tassoul, The internal rotation of the Sun, *Astron. Astrophys., v. 213*, p. 397–401, 1989.

Tatum, J. B., Tracking asteroids, *The Planetary Report, v. 8*, p. 9–11 (May/June), 1988.

——, Tracking asteroids: Why do it?, *The Planetary Report, v. 11*, p. 12–15 (Sept./Oct.), 1991.

Tavani, M., and L. Brookshaw, The origin of planets orbiting millisecond pulsars, *Nature, v. 356*, p. 320–322, 1992.

Taylor, G. I., Tidal friction in the Irish Sea, *Phil. Trans. Roy. Soc. London, v. A220*, p. 1–93, 1919.

Taylor, G. J., P. Maggiore, E. R. D. Scott, A. E. Rubin, and K. Keil, Original structures, and fragmentation and reassembly histories of asteroids: Evidence from meteorites, *Icarus, v. 69*, p. 1–13, 1987.

Taylor, S. R., Limits to Earth expansion from the surface features of the Moon, Mercury, Mars, and Ganymede, in S. W. Carey, ed., *The Expanding Earth: A Symposium, Sydney, 1981*, p. 343–347, Hobart, Tasmania, University of Tasmania, 1983.

——, The origin of the Moon, *Am. Scientist, v. 75*, p. 469–477, 1987.

——, Vestiges of a beginning, *Nature, v. 360*, p. 710–711, 1992.

Tedesco, E. F., Asteroid magnitudes, UBV colors, and IRAS albedos and diameters, in R. P. Binzel, T. Gehrels, and M. S. Matthews, eds., *Asteroids II*, p. 1090–1138, Tucson, AZ, The University of Arizona Press, 1989.

Tedesco, E. F., J. G. Williams, D. L. Matson, G. J. Veeder, J. C. Gradie, and L. A. Lebofsky, Three-parameter asteroid taxonomy classifications, in R. P. Binzel, T. Gehrels, and M. S. Matthews, eds., *Asteroids II*, p. 1151–1161, Tucson, AZ, The University of Arizona Press, 1989.

Teisseyre, R., J. Leliwa-Kopystynski, and B. Lang, eds., *Evolution of the Earth and Other Planetary Bodies*, New York, Elsevier Science Publ. Co., Inc., 583 pp., 1992.

Tholen, D. J., Asteroid taxonomic classifications, in R. P. Binzel, T. Gehrels, and M. S. Matthews, eds., *Asteroids II*, p. 1139–1153, Tucson, AZ, The University of Arizona Press, 1989.

Tholen, D. J., and M. A. Barucci, Asteroid taxonomy, in R. P. Binzel, T. Gehrels, and M. S. Matthews, eds., *Asteroids II*, p. 298–315, Tucson, AZ, The University of Arizona Press, 1989.

Thomas, E., Late Cretaceous-early Eocene mass extinctions in the deep sea, in V. L. Sharpton and P. D. Ward, eds., *Global Catastrophes in Earth History; An Interdisciplinary Conference on Impacts, Volcanism, and Mass Mortality*, p. 481–495, Boulder, CO, The Geological Society of America, Special Paper 247, 1990.

Thomas, L., *The Lives of a Cell*, New York, Bantam Books, 180 pp., 1974.

——, *The Medusa and the Snail: More Notes of a Biology Watcher*, New York, The Viking Press, 175 pp., 1979.

Thomas, P., and J. Veverka, Amalthea, in D. Morrison, ed., *Satellites of Jupiter*, p. 147–173, Tucson, AZ, The University of Arizona Press, 1982.

References Cited

Thompson, J. M. T., and H. B. Stewart, *Nonlinear Dynamics and Chaos*, New York, John Wiley & Sons, Inc., 376 pp., 1986.

Thompson, J. N., and J. J. Burdon, Gene-for-gene coevolution between plants and parasites, *Nature, v. 360*, p. 121–125, 1992.

Thompson, R., Atomic clocks: Limits to improvements?, *Nature, v. 362*, p. 789–790, 1993.

Thouveny, N., and K. M. Creer, Geomagnetic excursions in the past 60 ka: Ephemeral secular variation features, *Geology, v. 20*, p. 399–402, 1992.

Tilling, R. I., and J. J. Dvorak, Anatomy of a basaltic volcano, *Nature, v. 363*, p. 125–133, 1993.

Tilling, R. I., R. L. Christiansen, W. A. Duffield, E. T. Endo, R. T. Holcomb, R. Y. Koyanagi, D. W. Peterson, and J. D. Unger, The 1972–1974 Mauna Ulu eruption, Kilauea volcano: An example of quasi-steady-state magma transfer, in R. W. Decker, T. L. Wright, and P. H. Stauffer, eds., *Volcanism in Hawaii*, Vol. 1, Chap. 16, p. 405–469, U.S. Geol. Survey, Prof. Paper 1350, Washington, DC, U.S. Gov't Printing Office, 1987.

Tilton, G. R., Age of the Solar System, in J. F. Kerridge and M. S. Matthews, eds., *Meteorites and the Early Solar System*, p. 259–275, Tucson, AZ, The University of Arizona Press, 1988.

Tittemore, W. C., Chaotic motion of Europa and Ganymede and the Ganymede-Callisto dichotomy, *Science, v. 250*, p. 263–267, 1990.

Tombaugh, C. W., Asteroids and the trans-Neptunian planet search, *Proc. Lunar Planet. Explor. Colloq., v. 2*, p. 15–17, 1961a.

——, The trans-Neptunian planet search, in G. P. Kuiper and B. M. Middlehurst, eds., *Planets and Satellites*, p. 12–30, Chicago, The University of Chicago Press, 1961b.

——, Plates, Pluto, and Planets X, *Sky and Telescope, v. 81*, p. 360–361, 1991.

Tombaugh, C. W., and P. Moore, *Out of the Darkness: The Planet Pluto*, Harrisburg, PA, Stackpole Books, 221 pp., 1980.

Tombaugh, C. W., J. C. Robinson, B. A. Smith, and A. S. Murrell, *The Search for Small Natural Earth Satellites* (Final Technical Report, U. S. Army, Office Ord. Res.), University Park, NM, Publ. of New Mexico State University Physical Science Laboratory, 110 pp., 1959.

Tonks, W. B., and H. J. Melosh, Magma ocean formation due to giant impacts, *Jour. Geophys. Res., v. 98*, p. 5319–5333, 1993.

Toon, O. B., J. B. Pollack, T. P. Ackerman, R. P. Turco, C. P. McKay, and M. S. Liu, Evolution of an impact-generated dust cloud and its effects on the atmosphere, in L. T. Silver and P. H. Schultz, eds., *Geological Implications of Impacts of Large Asteroids and Comets on the Earth*, p. 187–200, Boulder, CO, The Geological Society of America, Special Paper 190, 1982.

Toon, O. B., C. P. McKay, C. A. Griffith, and R. P. Turco, A physical model of Titan's aerosols, *Icarus, v. 95*, p. 24–53, 1992.

Torbett, M. V., and R. Smoluchowski, Chaotic motion in a primordial comet disk beyond Neptune and comet influx to the Solar System, *Nature, v. 345*, p. 49–51, 1990.

Touma, J., and J. Wisdom, The chaotic obliquity of Mars, *Science, v. 259*, p. 1294–1297, 1993.

Trego, K. D., Multiple-impact mechanisms on Venus and other solid planets and satellites, *Earth, Moon and Planets, v. 55*, p. 69–72, 1991.

Tremaine, S., Is there evidence for a solar companion, in R. Smoluchowski, J. N. Bahcall, and M. S. Matthews, eds.,*The Galaxy and the Solar System*, p. 409–416, Tucson, AZ, The University of Arizona Press, 1986.

Treve, Y. M., Theory of chaotic motion with application to controlled fusion research, in S. Jorna, ed., *Topics in Nonlinear Dynamics: A Tribute to Sir Edward Bullard*, p. 147–220, New York, American Institute of Physics, 1978.

Trieloff, M., and E. K. Jessberger, $^{40}Ar/^{39}Ar$ ages of the large impact structures Kara and Manicouagan and their relevance to the Cretaceous-Tertiary and Triassic-Jurassic boundary, *Papers Presented at the International Conference on Large Meteorite Impacts and Planetary Evolution, 31 August–2 September, 1992, Sudbury, Ontario, Canada*, p. 74–75, Lunar and Planetary Institute, Houston, TX (LPI Contribution No. 790), 1992.

Tritton, D. J., Ordered and chaotic motion of a forced spherical pendulum, *Eur. Jour. Phys., v. 7*, p. 162–169, 1986.

——, *Physical Fluid Dynamics, Second Edition*, Oxford, Clarendon Press, 519 pp., 1988.

——, Deterministic chaos, geomagnetic reversals, and the spherical pendulum, in F. J. Lowes,

D. W. Collinson, J. H. Parry, S. K. Runcorn, D. C. Tozer, and A. Soward, eds., *Geomagnetism and Palaeomagnetism*, p. 215–226, Boston, MA, Kluwer Academic Publishers, 1989.

Trulsen, J., Collisional focusing of particles in space causing jetstreams, in T. Gehrels, ed., *Physical Studies of Minor Planets*, p. 327–335, Washington, DC, National Aeronautics and Space Administration, NASA SP-267, 1971.

Tryon, E. P., Cosmology and the expanding Earth hypothesis, in S. W. Carey, ed., *The Expanding Earth: A Symposium, Sydney, 1981*, p. 349–358, Hobart, Tasmania, University of Tasmania, 1983.

Tschudy, R. H., and B. D. Tschudy, Extinction and survival of plant life following the Cretaceous/Tertiary boundary event, western interior, North America, *Geology, v. 14*, p. 667–670, 1986.

Tschudy, R. H., C. L. Pillmore, C. J. Orth, J. S. Gilmour, and J. D. Knight, Disruption of the terrestrial plant ecosystem at the Cretaceous-Tertiary boundary, Western Interior, *Science, v. 225*, p. 1030–1032, 1984.

Tsonis, A. A., and J. B. Elsner, Comments on "Dimension analysis of climatic data," *Jour. Climate, v. 3*, p. 1502–1505, 1990.

———, Nonlinear prediction as a way of distinguishing chaos from random fractal sequences, *Nature, v. 358*, p. 217–220, 1992.

Tufillaro, N. B., T. Abbott, and J. Reilly, *An Experimental Approach to Nonlinear Dynamics and Chaos*, New York, Addison-Wesley Publishing Company, 420 pp. (includes software disk), 1992.

Turcotte, D. L., Fractals and fragmentation, *Jour. Geophys. Res., v. 91*, p. 1921–1926, 1986.

———, Nonlinear dynamics of crust and mantle, in G. V. Middleton, ed., *Nonlinear Dynamics, Chaos and Fractals; With Applications to Geological Systems: Chapt. 9*, p. 151–184, Toronto, Geol. Assoc. Canada, Short Course Notes, v. 9, 1991.

Turing, A. M., The chemical basis of morphogenesis, *Phil. Trans. Roy. Soc. London, Ser. B, v. 237*, p. 37–72, 1952.

Umbgrove, J. H. F., The relation between magmatic cycles and orogenic epochs, *Geol. Mag., v. 76*, p. 444–450, 1939a.

———, On rhythms in the history of the Earth, *Geol. Mag., v. 76*, p. 116–129, 1939b.

———, *The Pulse of the Earth, Second Edition*, The Hague, Martinus Nijhoff, 358 pp., 1947.

Urey, H. C., Origin of tektites, *Nature, v. 179*, p. 556–557, 1957.

———, Origin of tektites, *Science, v. 137*, p. 746, 748, 1962.

———, Cometary collisions and tektites, *Nature, v. 197*, p. 228–230, 1963.

———, Cometary collisions and geological periods, *Nature, v. 242*, p. 32–33, 1973.

Valet, J.-P., P. Tucholka, V. Courtillot, and L. Meynadier, Paleomagnetic constraints on the geometry of the geomagnetic field during reversals, *Nature, v. 356*, p. 400–407, 1992.

Vallis, G. K., El Niño: A chaotic dynamical system?, *Science, v. 232*, p. 243–245, 1986.

Valsecchi, G. B., A. Carusi, Z. Knezevic, L. Kresák, and J. G. Williams, Identification of asteroid dynamical families, in R. P. Binzel, T. Gehrels, and M. S. Matthews, eds., *Asteroids II*, p. 368–385, Tucson, AZ, The University of Arizona Press, 1989.

Valtonen, M. J., and K. A. Innanen, The capture of interstellar comets, *Astrophys. Jour., v. 255*, p. 307–315, 1982.

van Dalen, D., *Brouwer's Cambridge Lectures on Intuitionism*, Cambridge, Cambridge University Press, 109 pp., 1981.

van den Bergh, S., The age and size of the universe, *Science, v. 258*, p. 421–423, 1992.

van den Heuvel, E. P. J., and J. van Paradijs, Fate of the companion stars of ultra-rapid pulsars, *Nature, v. 334*, p. 227–228, 1988.

Van Flandern, T. C., Is gravity getting weaker?, *Sci. Am., v. 234*, p. 44–52 (Feb.), 1976.

———, A former asteroidal planet as the origin of comets, *Icarus, v. 36*, p. 51–74, 1978.

———, Do comets have satellites?, *Icarus, v. 47*, p. 480–486, 1981.

Van Flandern, T. C., E. F. Tedesco, and R. P. Binzel, Satellites of asteroids, in T. Gehrels, ed., *Asteroids*, p. 443–465, Tucson, AZ, The University of Arizona Press, 1979.

Veizer, J., Life and the rock cycle (Atmospheric Evolution), *Nature, v. 359*, p. 587–588, 1992.

Verhoogen, J., *Energetics of the Earth*, Washington, DC, National Academy of Sciences, 139 pp., 1980.

References Cited

Vermeij, G. J., Survival during biotic crises: The properties and evolutionary significance of refuges, in D. K. Elliott, ed., *Dynamics of Extinction*, p. 231–246, New York, John Wiley & Sons, Inc., 1986.

Veverka, J., The surfaces of Phobos and Deimos, *Vistas in Astronomy, v. 22*, p. 163–192, 1978.

Veverka, J., and P. Thomas, Phobos and Deimos: A preview of what asteroids are like, in T. Gehrels, ed., *Asteroids*, p. 628–651, Tucson, AZ, The University of Arizona Press, 1979.

Veverka, J., Y. Langevin, R. Farquhar, and M. Fulchignoni, Spacecraft exploration of asteroids: The 1988 perspective, in R. P. Binzel, T. Gehrels, and M. S. Matthews, eds., *Asteroids II*, p. 970–993, Tucson, AZ, The University of Arizona Press, 1989.

Veverka, J., M. Belton, C. Chapman, and the Galileo Imaging Team, 951 Gaspra: First pictures of an asteroid, *Eos, Trans. Am. Geophys. Union, v. 74*, p. 70–71, 1993.

Vickery, A. M., and H. J. Melosh, Melt droplet formation in energetic impacts, *Lunar and Planetary Sci. Conf. XXII, Abstracts of Papers*, p. 1441–1442, Lunar and Planetary Science Institute, Houston, TX, March 18–22, 1991.

Vilas, F., C. R. Chapman, and M. S. Matthews, eds., *Mercury*, Tucson, AZ, The University of Arizona Press, 794 pp., 1988.

Vine, F. J., and D. H. Matthews, Magnetic anomalies over ocean ridges, *Nature, v. 199*, p. 947–949, 1963.

Vogel, A., The irregular shape of the Earth's fluid core: A comparison of early results with modern computer tomography, *Proc. Sixth International Mathematical Geophysics Seminar, Inverse Modeling in Exploration Geophysics, v. 3*, p. 453–463, Braunschweig/Wiesbaden, Friedr. Vieweg & Sohn, 1989.

Vogel, K., Global models and Earth expansion, in S. W. Carey, ed., *The Expanding Earth: A Symposium, Sydney, 1981*, p. 17–27, Hobart, Tasmania, University of Tasmania, 1983.

Vogel, S. N., S. R. Kulkarni, and N. Z. Scoville, Star formation in giant molecular associations synchronized by a spiral density wave, *Nature, v. 334*, p. 402–406, 1988.

Vogt, P. R., Asthenosphere motion recorded by the ocean floor south of Iceland, *Earth Planet. Sci. Lett., v. 13*, p. 153–160, 1971.

———, Evidence for global synchronism in mantle plume convection, and possible significance for geology, *Nature, v. 240*, p. 338–342, 1972.

———, Changes in geomagnetic reversal frequency at times of tectonic change: Evidence for coupling between core and upper mantle processes, *Earth Planet. Sci. Lett., v. 25*, p. 313–321, 1975.

———, Global magmatic episodes: New evidence and implications for the steady-state mid-oceanic ridge, *Geology, v. 7*, p. 93–98, 1979.

———, Volcanogenic upwelling of anoxic, nutrient-rich water: A possible factor in carbonate-bank/reef demise and benthic faunal extinctions?, *Geol. Soc. Am., Bull., v. 101*, p. 1225–1245, 1989.

Volk, T., When climate and life finally devolve, *Nature, v. 360*, p. 707, 1992.

von Baeyer, H. C., *Taming the Atom: The Emergence of the Visible Microworld*, New York, Random House, 223 pp., 1992.

Wagner, C. A., and B. C. Douglas, Perturbations of existing resonant satellites, *Planet. Space Sci., v. 17*, p. 1505–1517, 1969.

Wahr, J. M., The effects of the atmosphere and oceans on the Earth's wobble and on the seasonal variations in the length of day — II. Results, *Geophys. Jour. Roy. Astron. Soc., v. 74*, p. 451–487, 1983.

Walcott, R. I., Late Quaternary vertical movements in eastern North America: Quantitative evidence of glacio-eustatic rebound, *Rev. Geophys. Space Phys., v. 10*, p. 849–884, 1972.

Walker, J. C. G., and K. J. Zahnle, Lunar nodal tide and distance to the Moon during the Precambrian, *Nature, v. 320*, p. 600–602, 1986.

Walker, S. C., Address to the American Association for the Advancement of Science on the subject of a new analogy in the periods of rotation of the primary planets, discovered by Daniel Kirkwood, Esq., of Pottsville, Pennsylvania, *Proc. Amer. Assoc. Advance. Sci. (Held at Cambridge, August, 1849), Second Meeting*, p. 207–220, 1850.

Walzer, U., and R. Maaz, On intermittent lower-mantle convection, in S. W. Carey, ed., *The

Expanding Earth: A Symposium, Sydney, 1981, p. 329–340, Hobart, Tasmania, University of Tasmania, 1983.

Wang, K., Glassy microspherules (microtektites) from an Upper Devonian limestone, *Science, v. 256*, p. 1547–1550, 1992.

Wang, K., and H. H. J. Geldsetzer, A late Devonian impact event and its association with a possible extinction event on eastern Gondwana, *Papers Presented at the International Conference on Large Meteorite Impacts and Planetary Evolution, 31 August–2 September, 1992, Sudbury, Ontario, Canada*, p. 77–78, Lunar and Planetary Institute, Houston, TX (LPI Contribution No. 790), 1992.

Wang, K., C. J. Orth, M. Attrep, Jr., B. D. E. Chatterton, H. Hou, and H. H. J. Geldsetzer, Geochemical evidence for a catastrophic biotic event at the Frasnian/Famennian boundary in South China, *Geology, v. 19*, p. 776–779, 1991.

Ward, P. D., The Cretaceous/Tertiary extinctions in the marine realm: A 1990 perspective, in V. L. Sharpton and P. D. Ward, eds., *Global Catastrophes in Earth History; An Interdisciplinary Conference on Impacts, Volcanism, and Mass Mortality*, p. 425–432, Boulder, CO, The Geological Society of America, Special Paper 247, 1990a.

——, A review of Maastrichtian ammonite ranges, in V. L. Sharpton and P. D. Ward, eds., *Global Catastrophes in Earth History; An Interdisciplinary Conference on Impacts, Volcanism, and Mass Mortality*, p. 519–530, Boulder, CO, The Geological Society of America, Special Paper 247, 1990b.

Ward, W. R., Comments on the long-term stability of the Earth's obliquity, *Icarus, v. 50*, p. 444–448, 1982.

Wasson, J. T., The building stones of the planets, in F. Vilas, C. R. Chapman, and M. S. Matthews, eds., *Mercury*, p. 622–650, Tucson, AZ, The University of Arizona Press, 1988.

——, Climate and the impact formation of silicate melts (Abstract), *Eos, Trans. Am. Geophys. Union, v. 71*, p. 1425, 1990.

——, Planetesimal heating by FU-Orionis-type events (Abstract), *Eos, Trans. Am. Geophys. Union, v. 73* (Supplement, 27 Oct.), p. 336, 1992.

Wasson, J. T., and W. A. Heins, Tektites and climate, *Jour. Geophys. Res., v. 98*, p. 3043–3052, 1993.

Wasson, J. T., and G. W. Wetherill, Dynamical, chemical, and isotopic evidence regarding the formation locations of asteroids and meteorites, in T. Gehrels, ed., *Asteroids*, p. 926–974, Tucson, AZ, The University of Arizona Press, 1979.

Watts, A. W., R. Greeley, and H. J. Melosh, The formation of terrains antipodal to major impacts, *Icarus, v. 93*, p. 159–168, 1991.

Weaver, H. A., and L. Danly, eds., *The Formation and Evolution of Planetary Systems*, New York, Cambridge University Press, 344 pp., 1989.

Webb, S. D., On two kinds of rapid faunal turnover, in W. A. Berggren and J. A. Van Couvering, eds., *Catastrophes and Earth History: The New Uniformitarianism*, p. 417–436, Princeton, NJ, Princeton University Press, 1984.

Weems, R. E., and W. H. Perry, Jr., Strong correlation of major earthquakes with solid-earth tides in part of the eastern United States, *Geology, v. 17*, p. 661–664, 1989.

Weidenschilling, S. J., The distribution of mass in the planetary system and solar nebula, *Astrophysics and Space Sci., v. 51*, p. 153–158, 1977.

——, Progress toward the origin of the Solar System, *Rev. Geophys. Space Phys., v. 21*, p. 206–213, 1983.

——, Formation processes and time scales for meteorite parent bodies, in J. F. Kerridge and M. S. Matthews, eds., *Meteorites and the Early Solar System*, p. 348–371, Tucson, AZ, The University of Arizona Press, 1988.

——, A plurality of worlds, *Nature, v. 352*, p. 190–192, 1991.

Weidenschilling, S. J., P. Paolicchi, and V. Zappalà, Do asteroids have satellites, in R. P. Binzel, T. Gehrels, and M. S. Matthews, eds., *Asteroids II*, p. 643–658, Tucson, AZ, The University of Arizona Press, 1989.

Weinstein, S. A., Catastrophic overturn of the Earth's mantle driven by multiple phase changes and internal heat generation, *Geophys. Res. Lett., v. 20*, p. 101–104, 1993.

References Cited

Weiss, N., Magnetic geometry of sunspots, *Nature, v. 362*, p. 208–209, 1993.

Weissman, P. R., Terrestrial impact rates for long and short-period comets, in L. T. Silver and P. H. Schultz, eds., *Geological Implications of Impacts of Large Asteroids and Comets on the Earth*, p. 15–24, Boulder, CO, The Geological Society of America, Special Paper 190, 1982.

———, The Oort cloud and the galaxy: Dynamical interactions, in R. Smoluchowski, J. N. Bahcall, and M. S. Matthews, eds., *The Galaxy and the Solar System*, p. 204–237, Tucson, AZ, The University of Arizona Press, 1986.

———, The cometary impactor flux at the Earth, in V. L. Sharpton and P. D. Ward, eds., *Global Catastrophes in Earth History; An Interdisciplinary Conference on Impacts, Volcanism, and Mass Mortality*, p. 171–180, Boulder, CO, The Geological Society of America, Special Paper 247, 1990a.

———, The Oort Cloud, *Nature, v. 344*, p. 825–830, 1990b.

———, Why did Halley hiccup?, *Nature, v. 353*, p. 793–794, 1991a.

———, Dynamical history of the Oort Cloud, in R. L. Newburn, Jr., M. Neugebauer, and J. Rahe, eds., *Comets in the Post-Halley Era, Volume 1*, p. 463–486, Boston, MA, Kluwer Academic Publishers, 1991b.

———, The discovery of 1992 QB_1, *Eos, Trans. Am. Geophys. Union, v. 74*, p. 257, and 262–263, 1993.

Weissman, P. R., and J. Wasson, Naked stars and hot meteorites, *Nature, v. 345*, p. 208–209, 1990.

Weissman, P. R., M. F. A'Hearn, L. A. McFadden, and H. Rickman, Evolution of comets into asteroids, in R. P. Binzel, T. Gehrels, and M. S. Matthews, eds., *Asteroids II*, p. 880–920, Tucson, AZ, The University of Arizona Press, 1989.

Wetherill, G. W., Occurrence of giant impacts during the growth of the terrestrial planets, *Science, v. 228*, p. 877–879, 1985.

———, Accumulation of Mercury from planetesimals, in F. Vilas, C. R. Chapman, and M. S. Matthews, eds., *Mercury*, p. 670–691, Tucson, AZ, The University of Arizona Press, 1988.

———, Cratering of the terrestrial planets, *Meteoritics, v. 24*, p. 15–22, 1989a.

———, Origin of the Asteroid Belt, in R. P. Binzel, T. Gehrels, and M. S. Matthews, eds., *Asteroids II*, p. 661–680, Tucson, AZ, The University of Arizona Press, 1989b.

———, The formation of the Solar System: Consensus, alternatives, and missing factors, in H. A. Weaver and L. Danly, eds., *The Formation and Evolution of Planetary Systems*, p. 1–30, New York, Cambridge University Press, 1989c.

———, Formation of the Earth, *Ann. Rev. Earth Planet. Sci., v. 18*, p. 205–256, 1990.

———, End products of cometary evolution: Cometary origin of Earth-crossing bodies of asteroidal appearance, in R. L. Newburn, Jr., M. Neugebauer, and J. Rahe, eds., *Comets in the Post-Halley Era, Volume 1*, p. 537–556, Boston, MA, Kluwer Academic Publishers, 1991a.

———, Occurrence of Earth-like bodies in planetary systems, *Science, v. 253*, p. 535–538, 1991b.

Wetherill, G. W., and C. R. Chapman, Asteroids and meteorites, in J. F. Kerridge and M. S. Matthews, eds., *Meteorites and the Early Solar System*, p. 35–67, Tucson, AZ, The University of Arizona Press, 1988.

Wetherill, G. W., and E. M. Shoemaker, Collision of astronomically observable bodies with the Earth, in L. T. Silver and P. H. Schultz, eds., *Geological Implications of Impacts of Large Asteroids and Comets on the Earth*, p. 1–13, Boulder, CO, The Geological Society of America, Special Paper 190, 1982.

Whalley, P., Ancient residents in resin, *Nature, v. 360*, p. 714, 1992.

Whelan, N. D., D. A. Goodings, and J. K. Cannizzo, Two balls in one dimension with gravity, *Phys. Rev. A, v. 42*, p. 742–754, 1990.

White, O. R., ed., *The Solar Output and Its Variation*, Boulder, CO, Colorado Associated University Press, 526 pp., 1977.

White, R., and D. McKenzie, Magmatism at rift zones: The generation of volcanic continental margins and flood basalts, *Jour. Geophys. Res., v. 94*, p. 7685–7729, 1989.

Whitmire, D. P., and A. A. Jackson, II, Are periodic mass extinctions driven by a distant solar companion?, *Nature, v. 308*, p. 713–715, 1984.

References Cited

Widdel, F., S. Schnell, S. Heising, A. Ehrenreich, B. Assmus, and B. Schink, Ferrous iron oxidation by anoxygenic phototrophic bacteria, *Nature, v. 362*, p. 834–836, 1993.

Widmann, P. J., M. Gorman, and K. A. Robbins, Nonlinear dynamics of a convection loop II. Chaos in laminar and turbulent flows, *Physica D, v. 36*, p. 157–166, 1989.

Wildi, T., *Units and Conversion Factors: A Handbook for Engineers and Scientists*, New York, The Institute of Electrical and Electronics Engineers (IEEE) Press, 80 pp., 1991.

Wilhelms, D. E., *The Geologic History of the Moon*, U.S. Geol. Survey, Prof. Paper 1348, Washington, DC, U.S. Gov't Printing Office, 302 pp., 24 Plates, 1987.

Willeboordse, F. H., Time-delayed map phenomenological equivalency with a coupled map lattice, *Internat. Jour. Bifurcation and Chaos, v. 2*, p. 721–725, 1992.

Williams, G. E., Geological evidence relating to the origin and secular rotation of the Solar System, *Modern Geol., v. 3*, p. 165–181, 1972.

———, ed., *Megacycles: Long-Term Episodicity in Earth and Planetary History*, Woods Hole, MA, Hutchinson Ross Publishing Co., 435 pp., 1981.

———, The Acraman impact structure: Source of ejecta in Late Precambrian shales, South Australia, *Science, v. 233*, p. 200–203, 1986.

———, Late Precambrian tidal rhythmites in South Australia and the history of the Earth's rotation, *Jour. Geol. Soc. London, v. 146*, p. 97–111, 1989a.

———, Precambrian tidal sedimentary cycles and Earth's paleorotation, *Eos, Trans. Am. Geophys. Union, v. 70*, p. 33, and 40–41, 1989b.

Wilson, E. O., The current state of biological diversity, in E. O. Wilson and F. M. Peter, eds., *Biodiversity*, p. 3–18, Washington, DC, National Academy Press, 1988.

Wilson, J. T., Evidence from islands on the spreading of the ocean floor, *Nature, v. 197*, p. 536–538, 1963.

———, Did the Atlantic close and then re-open?, *Nature, v. 211*, p. 676–681, 1966.

Winograd, I. J., T. B. Coplen, J. M. Landwehr, A. C. Riggs, K. R. Ludwig, B. J. Szabo, P. T. Kolesar, and K. M. Revesz, Continuous 500,000-year climate record from vein calcite in Devil's Hole, Nevada, *Science, v. 258*, p. 255–260, 1992.

Winston, D., M. Arora, J. Maselko, V. Gáspár, and K. Showalter, Cross-membrane coupling of chemical spatiotemporal patterns, *Nature, v. 351*, p. 132–135, 1991.

Wisdom, J., The resonance overlap criterion and the onset of stochastic behavior in the restricted three-body problem, *Astron. Jour., v. 85*, p. 1122–1133, 1980.

———, The origin of the Kirkwood gaps: A mapping for asteroidal motion near the 3/1 commensurability, *Astron. Jour., v. 87*, p. 577–593, 1982.

———, Chaotic behavior and the origin of the 3/1 Kirkwood gap, *Icarus, v. 56*, p. 51–74, 1983.

———, Meteorites may follow a chaotic route to Earth, *Nature, v. 315*, p. 731–733, 1985.

———, Chaotic behavior in the Solar System, in M. V. Berry, I. C. Percival, and N. O. Weiss, eds., *Dynamical Chaos*, p. 109–129, Princeton, NJ, Princeton Univ. Press, 1987a.

———, Chaotic behavior in the Solar System, *Proc. Roy. Soc. London, v. A 413*, p. 109–129, 1987b.

———, Urey Prize Lecture: Chaotic dynamics in the Solar System, *Icarus, v. 72*, p. 241–275, 1987c.

———, Is the Solar System stable? *and* Can we use chaos to make measurements?, in D. K. Campbell, ed., *CHAOS/XAOC: Soviet-American Perspectives on Nonlinear Science*, p. 275–303, New York, American Institute of Physics, 1990.

Wolbach, W. S., I. Gilmour, and E. Anders, Major wildfires at the Cretaceous/Tertiary boundary, in V. L. Sharpton and P. D. Ward, eds., *Global Catastrophes in Earth History; An Interdisciplinary Conference on Impacts, Volcanism, and Mass Mortality*, p. 391–400, Boulder, CO, The Geological Society of America, Special Paper 247, 1990.

Wolf, A., Quantifying chaos with Lyapunov exponents, in A. V. Holden, ed., *Chaos*, p. 273–290, Princeton, NJ, Princeton University Press, 1986.

Wolfe, J. A., Paleobotanical evidence for a marked temperature increase following the Cretaceous/Tertiary boundary, *Nature, v. 334*, p. 665–669, 1990.

———, Paleobotanical evidence for a June 'impact winter' at the Cretaceous/Tertiary boundary, *Nature, v. 352*, p. 420–423, 1991.

References Cited

Wolszczan, A., and D. A. Frail, A planetary system around the millisecond pulsar PSR1257 + 12, *Nature, v. 355*, p. 145–147, 1992.

Wones, D. R., and H. R. Shaw, Tidal dissipation: A possible heat source for mare basalt magmas, *Lunar Sci. Conf. VI, Abstracts of Papers*, p. 878–880, Lunar Science Institute, Houston, TX, 1975.

Wood, F. J., *Tidal dynamics: Coastal Flooding, and Cycles of Gravitational Force*, Boston, MA, D. Reidel Publishing Company, 558 pp. 1986.

Wood, J. A., *The Solar System*, Englewood Cliffs, NJ, Prentice-Hall, Inc., 196 pp., 1979.

——, Moon over Mauna Loa: A review of hypotheses of formation of Earth's Moon, in W. K. Hartmann, R. J. Phillips, and G. J. Taylor, eds., *Origin of the Moon*, p. 17–56, Houston, TX, Lunar and Planetary Inst., 1986.

——, Chondritic meteorites and the solar nebula, *Ann. Rev. Earth Planet. Sci., v. 16*, p. 53–72, 1988.

Wood, J. A., and G. E. Morfill, A review of solar nebula models, in J. F. Kerridge and M. S. Matthews, eds., *Meteorites and the Early Solar System*, p. 329–347, Tucson, AZ, The University of Arizona Press, 1988.

Wood, K. D., Sunspots and planets, *Nature, v. 240*, p. 91–93, 1972.

Wood, R. M., *The Dark Side of the Earth*, London, George Allen & Unwin, 246 pp., 1985.

Woodard, M., and H. S. Hudson, Frequencies, amplitudes and linewidths of solar oscillations from total irradiance observations, *Nature, v. 305*, p. 589–593, 1983.

Wright, I. P., Interstellar diamonds: Rich pickings for astronomers, *Nature, v. 360*, p. 20, 1992.

Wright, S., *Evolution: Selected Papers* (edited by W. B. Provine), Chicago, The University of Chicago Press, 649 pp., 1986.

Wright, T. L., D. L. Peck, and H. R. Shaw, Kilauea lava lakes: Natural laboratories for study of cooling, crystallization, and differentiation of basaltic magma, in *The Geophysics of the Pacific Ocean Basin and its Margins*, p. 375–392, Washington, DC, Am. Geophys. Union, Geophys. Monograph 19, 1976.

Wright, T. L., T. J. Takahashi, and J. D. Griggs, *Hawaii Volcano Watch: A Pictorial History, 1779–1991*, Honolulu, HI, University of Hawaii Press, 162 pp. (143 Figs.), 1992.

Wu, X., and Z. A. Schelly, The effects of surface tension and temperature on the nonlinear dynamics of the dripping faucet, *Physica D, v. 40*, p. 433–443, 1989.

Xu, D.-Y., Zhang, Q.-W., and Sun, Y.-Y., Mass extinction — A fundamental indicator for major natural divisions of geological history, *Acta Geologica Sinica, v. 1*, p. 1–12, 1988.

Xu, D.-Y., Zhang, Q.-W., Sun, Y.-Y., Yan, Z., Chai, Z.-F., and He, J.-W., *Astrogeological Events in China*, New York, Van Nostrand Reinhold, 264 pp. 1989.

Yamaji, A., Periodic hot-spot distribution on Io, *Science, v. 254*, p. 89–91, 1991.

——, Periodic hotspot distribution and small-scale convection in the upper mantle, *Earth Planet. Sci. Lett., v. 109*, p. 107–116, 1992.

Yeates, C. M., Initial findings from a telescopic search for small comets near Earth, *Planet. Space Sci., v. 37*, p. 1185–1196, 1989.

Yeomans, D. K., *Comets: A Chronological History of Observation, Science, Myth, and Folklore*, New York, John Wiley & Sons, Inc., 485 pp., 1991a.

——, Killer rocks and the celestial police: The search for near-Earth asteroids, *The Planetary Report, v. 11*, p. 4–7 (Nov./Dec.), 1991b.

Yoder, C. F., J. G. Williams, J. O. Dickey, B. E. Schutz, R. J. Eanes, and B. D. Tapley, Secular variation of Earth's gravitational harmonic J_2 coefficient from Lageos, and nontidal acceleration of Earth's rotation, *Nature, v. 303*, p. 757–762, 1983.

Yoshikawa, M., Motions of asteroids at the Kirkwood Gaps: II. On the 5:2, 7:3, and 2:1 resonances with Jupiter, *Icarus, v. 92*, p. 94–117, 1991.

Young, C. J., and T. Lay, The core-mantle boundary, *Ann. Rev. Earth Planet. Sci., v. 15*, p. 25–46, 1987.

Young, G. M., Late Proterozoic stratigraphy and the Canada-Australia connection, *Geology, v. 20*, p. 215–218, 1992.

Zahnle, K., J. B. Pollack, D. Grinspoon, and L. Dones, Impact-generated atmospheres over Titan, Ganymede, and Callisto, *Icarus, v. 95*, p. 1–23, 1992.

References Cited

Zaslavsky, G. M., The simplest case of a strange attractor, *Phys. Lett, v. 69A*, p. 145–147, 1978.

Zeeman, E. C., General discussion, in M. V. Berry, I. C. Percival, and N. O. Weiss, eds., *Dynamical Chaos*, p. 199, Princeton, NJ, Princeton University Press, 1987.

Zener, C., *Elasticity and Anelasticity of Metals*, Chicago, IL, The University of Chicago Press, 170 pp., 1948.

Zewail, A. H., Small is beautiful, *Nature, v. 361*, p. 215–216, 1993.

Zhang, K., and D. Gubbins, On convection in the Earth's core driven by lateral temperature variations in the lower mantle, *Geophys. Jour. International, v. 108*, p. 247–255, 1992.

Zhou, L., F. T. Kyte, and B. F. Bohor, Cretaceous/Tertiary boundary of DSDP Site 596, South Pacific, *Geology, v. 19*, p. 694–697, 1991.

Ziegler, A. M., C. R. Scotese, and S. F. Barrett, Mesozoic and Cenozoic paleogeographic maps, in P. Brosche and J. Sündermann, eds., *Tidal Friction and the Earth's Rotation II*, p. 240–252, New York, Springer-Verlag, 1982.

Zimmermann, R. E., The anthropic cosmological principle: Philosophical implications of self-reference, in J. L. Casti and A. Karlqvist, eds., *Beyond Belief: Randomness, Prediction and Explanation in Science*, p. 14–54, Boston, MA, CRC Press, 1991.

Zirin, H., *Astrophysics of the Sun*, New York, Cambridge University Press, 433 pp., 1988.

Zolensky, M., and H. Y. McSween, Jr., Aqueous alteration, in J. F. Kerridge and M. S. Matthews, eds., *Meteorites and the Early Solar System*, p. 114–143, Tucson, AZ, The University of Arizona Press, 1988.

Zonenshain, L. P., J. Verhoef, R. Macnab, and H. Meyers, Magnetic imprints of continental accretion in the U.S.S.R., *Eos, Trans. Am. Geophys. Union, v. 72*, p. 305, 310, 1991.

Index

Index

In this index an "f" after a number indicates a separate reference on the next page, and an "ff" indicates separate references on the next two pages. A continuous discussion over two or more pages is indicated by a span of page numbers, e.g., "57–59. " *Passim* is used for a cluster of references in close but not necessarily consecutive sequence.

Abraham, F. D., 134, 408, 464, 546; on psychology and nonlinear dynamics, 234, 412
Abraham, R. H., 29, 183, 234, 236, 392, 408; on manifolds, 80; on mechanics and nonlinear dynamics, 134; monumental textbooks by, 134, 406; on bifurcation, 134; on phase/state space, 157, 422; and "blue sky catastrophes," 226; on phase space portraiture, 420; on "Big Picture," 456; on separatrices, 457, 471; on nested attractors, 473
Accretionary continents, 36, 264, 277, 526
Acraman Crater, South Australia, 62, 119, 128, 271
AES, *see* Artificial-Earth satellites
"Africa Master APWP," 366 (App. 20)
Ager, D. V., xliii, 207, 282, 284ff, 292, 507; on catastrophes, 53, 288–89, 316, 364, 529; and biostratigraphic code/record, 53, 55, 316; on self-similarity in catastrophes, 288, 294, 301; on biostratigraphy, 292, 507
Aggarwal, H. R., 110, 212, 261f, 376, 528
Ahlquist, J. E., 512–13; on DNA-based phylogeny, 295, 299ff, 511
Ahrens, T. J., 198, 201, 205, 212, 252; on discoveries of asteroids, 175, 547; on deflection of impacts, 303; on 1991 BA, 307; on near-Earth objects, 310, 495

Airburst, xliv, 172, 308, 559. *See also* Bolide events; Tunguska event
Aki, K., 239, 393, 399, 535
Aksnes, K., 171, 496
Alfvén, H., 74, 81, 201, 209, 252, 454, 474; on plasma dynamics, 9, 221, 516; on Earth-Moon system resonance, 43, 52, 173, 345, 476f; on meteroid streams/delivery, 50, 429; on resonant-level states, 91, 206, 477; on resonant interactions, 133, 326; on particle aggregation in space, 170; on retrograde orbits, 199; on matter-antimatter, 247, 436; on cosmology, 302, 449, 453, 492, 540
Alfvén-Arrhenius resonance model, 174
Algorithmic complexity, 54, 425, 452, 460; incompressibility rule of, 1; examples of reduction in, 2, 28; definition of, 415
Alien (motion picture), 24
Alignment, 191; and mean-motion resonances, 192–93
Allan, D. W., 218, 232, 341
Allan, R. R., 3, 35, 92, 174, 185, 190, 195f, 232, 310, 327, 428, 436, 467, 495; on spin-orbit resonances, 34, 94 (Fig. 20), 96 (Fig. 21), 149, 166, 168f, 180, 189, 199, 325, 497; theory of, 91, 172–73, 178, 324, 498; on natural satellites, 173; on eccentricity, 496

Index

Allen, T., 13, 559
Allometry, 302
Alt, D., 212f, 240f, 260, 280, 352
Alvarez, L. W., xxxii, xxxiii; on impact-extinction hypothesis, 56, 285, 304, 507
Alvarez, W., xxxiii, 48, 212, 284, 293; on impact-extinction hypothesis, 57, 285, 304, 507; on iridium anomalies, 99, 210, 214, 262, 507
Amber, 539, 547
Ammonites at K/T boundary, 427
Amor asteroid group, 78, 139, 144, 147, 150, 216, 487
Ancestral Pacific Basin/Ocean, 52, 265, 276
Anders, E., 210; on diamonds in meteorites, 210–11, 414; on meteorites, 435, 490, 548
Anderson, A. J., 17, 34, 178
Anderson, D. L., 36, 45, 229, 240, 245, 266, 275, 279, 352, 506, 515, 523, 525
Anderson, O. L., 249, 251, 342, 516; on thermal feedback, 27, 408, 413; on thermal runaway, 401, 491
Andrews, J. A., 37, 40, 44, 47f, 266, 367
Andreyev, Y. V., 15, 134, 234, 388
Antimatter, 247, 436
"Aperiodic crystal," 19, 23
Apparent polar wander, 47, 225
Apparent polar wander paths, 53, 203, 364 (App. 19), 366 (App. 20), 368 (App. 21), 369 (App. 22), 372 (App. 23), 519, 527, 561; definition of, 38; cusps, hairpins in, 39, 42, 56; correlations with Geologic Column, 41f, 56; compared with cratering patterns, 225, 227
Appenzeller, I., 551; on contracting star systems, 433f
Appenzeller, T. on coronal mass ejections, 252, 254, 257
Apollo asteroid group, 78, 82, 139, 144f, 147, 150f, 216, 487
Approach scenarios of impactors, 35f, 91 (Fig. 19), 91–93, 129, 166, 194f, 199, 206, 419
APW, *see* Apparent polar wander
APWP, *see* Apparent polar wander paths
Arnold, V. I., 134–35, 392
Arrhenius, G., 74, 81, 170, 199, 209, 436, 454; on plasma dynamics, 9, 221, 516; on Earth-Moon system resonance, 43, 52, 173, 345, 476f; on resonant-level states, 91, 206, 477; on resonant interactions, 133, 326
Arrhenius, S., 305
Arthur, M. A.: on periodic concept in geology, 256, 508
Artificial-Earth satellites, 17, 34, 132, 166, 169, 172, 175, 195f, 198, 270, 445, 485–86, 495, 498
Asaro, F., xxxiii, 57, 285, 507
Asteroid Belt, 50f, 74, 80 (Fig. 12), 81–83 (Fig. 13), 133, 137, 140, 150–54, 166, 168, 175, 345, 433f, 487, 490; dead/decaying comets in, 129, 494; dynamics of, 132; resonances in, 135, 166, 176, 182, 185, 191, 313, 432; Kirkwood Gaps in, 143ff, 148f, 170, 176, 185, 313, 432, 440; "filling" and "emptying" in, 149, 182, 310; bin model of, 149; and Jupiter, 153, 185, 432; as reservoir, 165, 313; and sine-circle map, 218; Flora region of, 308; commensurabilities in, 440. *See also* Main-Belt Asteroids
Asteroids, 2f, 9, 50, 129, 190, 194, 262, 454, 485; Amor group of, 78, 82, 139, 144, 147, 150, 216, 487; Apollo group of, 78, 139, 144f, 147, 150f, 216, 487; Aten group of, 78, 139, 144, 147, 150, 216, 487; 1991 BA, 172, 307f; 4769 Castalia, 170; Ceres, 9, 74, 78, 147, 150, 216, 260, 440–41, 445, 494; 2060 Chiron, 82, 145; Cybeles group, 149, 487; 951 Gaspra, 308; Hidalgo, 82; Hildas, 150; Hungaria group, 149, 487; 2201 Oljato, 82, 145; Pallas, 74; 5145 Pholus, 82; Thule, 150; Trojan, 150; 4 Vesta, 74, 216; optical searches for, 11; growth rate of discoveries of, 55, 131, 146–47, 167, 247, 311, 453, 548; delivery to vicinity of Earth, 127, 191, 226, 487; as distinct from comets, 129, 140, 494; orbital distributions of, 129; and fast-moving objects, 147, 310; ballistic agglomeration of, 170; discoveries of, 170f, 175, 200, 308; "contact binary," 170; rubble pile, 170, 490f, 494; satellites of, 170, 309; as parent bodies of meteorites, 216, 487, 548; as satellites of planets, 309; formation of, 490
Asteroids II (conference), 148
Asthenophere, 279f, 524
Astrodynamics, 2
Astrogeology, 212
Astronomical date of reference, 31
Astronomical phenomena, 161; "guidance effects" in, 168
Astronomical search strategies, 175–76, 482–85, 494–96
Astronomical units, 132, 470, 477
Astronomy, xxii, xxvii, 4, 247, 311, 398, 433, 441–43, 444, 453, 469–70, 482, 484, 492, 499f, 548–49
Astrophysical processes, 234, 236, 283
Astrophysicists, 168, 174, 247, 447, 465; and capture scenarios, 168

646

Index

Astrophysics, 309, 311, 468f, 492, 549; and dissipation, 275, 436
Aten asteroid group, 78, 139, 144, 147, 150, 216, 487
Atkins, P. W., 381, 398, 412
Atomic clock, 19, 21f, 185, 317, 417, 430, 560
Atomic decay, 20
Attractor basins, *see* Basins of attraction
Attractors and attractor systems, 146, 202, 313, 391, 407, 460–61, 518, 527; as celestial strange attractors, 4; as Solar System and cosmological structures, 4; "universal" properties of, 157; uncountable infinities of, 163; degrees of complexity, 188; memory capacities, 188; in weather-climate system, 201; definition of, 388–89. *See also* Fixed-point attractors; Strange attractors
Atwater, T.: on "hot spots," 48, 240, 352
AU, *see* Astronomical units
Auditory illusions, xxviii
Avak structure, Point Barrow, Alaska, 49, 64, 122f, 261, 328, 334
Averaging (dimensional scaling), 22
Axelrod, R., 209; on evolutionary cooperation, 303, 464, 537
Aztex, Ovenden's "missing planet," 489

Babcock, K. L., 161
Babloyantz, A., 13–14
Background extinction, 203, 285, 295, 316, 544
Bacteriophage, 23–24
Badjukov, D. D., 64, 121
Bagby, J. P., 35, 92, 170, 196; on near-Earth satellites, 270, 309, 486, 495, 497, 548, 559; on near-Earth objects, 496
Bahcall, J. N., 477
Bailey, M. E., 86, 129, 132f, 136, 140, 151, 167f, 213, 226, 435, 454f, 477–79, 493
Bai-Lin, H., *see* Hao, B.-L.
Bak, P., 141, 156, 209, 231, 421, 448; on self-organized criticality, 2, 140, 161f, 225, 302, 397, 503; on Devil's Staircase, 85, 143, 149
Baker, D. N., 462
Baker, G. L., 134, 389, 391; his tutorial on nonlinear dynamics, 29–30; on equation of pendulum, 60 (Table 1), 473; on pendulum and orbital mechanics, 86; on phase space, 396, 422; on chaotic dynamics, 420; on winding numbers, 432; on universality parameters, 436
Baker, R. M. L., Jr., 35, 170, 196, 309, 495, 559
Baksi, A. K., 57; on Deccan Traps, 66; on Siberian Traps, 66, 529

Baldwin, B., 123, 322, 427, 467, 559
Baldwin, R. B., 3, 53, 138, 212, 252, 260, 263, 265, 325, 489, 540
Ballistic agglomeration, 170
Bambach, R. K., 112, 279, 367, 376, 526
Bandermann, L. W., 50, 81, 263, 490, 559
Bargar, K. E., 275, 352, 535
Barnes, J. E., 14, 135, 448, 453
Barucci, M. A., 487
Barycenter, 479
Base of the upper mantle, 278–79
Basins of attraction, 2, 80, 158, 168, 182, 225f, 233, 309, 457f, 471, 473, 475; of cometary objects within Galaxy, 152; definition of, 389
Bate, R. R., 32, 81, 91, 192, 327f, 464ff, 469f, 479; his equation on acceleration and mass distribution, 60 (Table 1); on Earth's gravitational influence, 168
Baum, S. K.: on paleogeographic reconstruction, 376, 527
Baxter, D. C., 9, 50, 81, 170, 201, 429, 474
Beatty, J. K., xliv, 435, 454; his test on Solar System, 213
Beaverhead structure, Montana, 334
Beckmann, P., 397f
Belbruno, E., 132, 144, 169; and "fuzzy route," 2, 192, 195; and transfer orbits, 2, 309, 415, 465–66, 468, 495; on "fuzzy" boundaries, 35, 144, 472–73, 478; and Japanese space program, 465
Bell, J., 441; on pulsars, 441–42
Bell, J. F., 50, 414, 548
Benson, S. V., 108; on route to turbulence, 108, 502, 541; on period-doubling route, 386; on quasiperiodic route, 387
Benton, M. J., 33, 364, 427, 513
Benz, W., 252, 489; on collisional formation of Moon, 53, 75, 92, 263, 265, 325
Bergé, P., 134, 157, 389, 420, 456
Berger, A., 32, 47, 53, 96, 130, 132f, 201, 253, 266, 466, 528, 536
Berggren, W. A., 511
Berry, M. V., xxix, 29, 132, 134, 213, 389; on quantum chaology, 20, 399; on quantum chaos, 22, 182
Besse, J., 32, 45, 47, 506, 562; on TPWPs, 40–41, 341; on hairpins, 42; on geomagnetic studies, 48, 341; on APWPs, 364, 366, 368f, 372
Bice, D. M., 33, 213, 262, 334, 364, 427, 513
Bifurcation, 5, 15, 134, 158, 232, 256, 294, 386, 392, 394, 411, 541; synapses, 388, 452; ring-vortex, 395, 412; bubbles, 452

647

Index

Big Bang theory, 247, 302, 437, 447, 451, 540, 551
"Binary planet," 137, 494
Binary-star-catastrophe model of Solar System, 73–75 (Fig. 9b Inset), 137f, 346–48, 414, 433–36, 548–51
Binary-star convective-advective oscillations, 549
Binary-star systems, 137, 313, 434; and star formation, 74, 548–51
Bin model of Asteroid Belt, 149
Binzel, R. P., 50, 168, 213, 260, 263, 309f, 435, 454, 548; on asteroids, 78, 81f, 144, 146, 171f, 175, 216, 487f, 490f, 494; on Aten, Apollo, Amor groups, 139; on Main-Belt Asteroids, 147, 150, 345; on Asteroid Belt, 148, 434; on 4769 Castalia, 170; on 1991 BA, 307; on near-Earth objects, 474, 495; on rotation data, 500
Bioenergetic feedback, 304
Biogenetic, *see* Genetic
Biological adaptation, 302
Biosphere, 25, 53, 163, 178, 239, 246, 283, 285, 302; threats to, 3; and geosphere, 294, 304
Biostratigraphic code, 53, 56, 59, 203, 287, 312, 530
Biostratigraphic column, 53
Biostratigraphic discontinuities, *see* Mass extinction
Biostratigraphic record, 29, 284, 304
Biostratigraphy, 3, 235, 283, 288, 292, 312
Birds, 316; survival across K/T boundary, 207ff, 300; and refuges, 209; DNA-based phylogeny for, 295, 301, 511; study of Pelecaniformes and Passeriformes, 114–15 (Fig. 36b), 299–300; evolution of, 301, 512–13
Black holes, 23, 419, 449, 451, 491
Blagg, M. A., 444
Blagg-Richardson relationship, 342 (App. 6 Inset B), 344, 444
Blazars, 400–401, 419, 421
"Blazing sky effect," 393, 441, 446–47, 448, 451
Bleil, U., 40, 352, 506
Blekhman, I. I.: on synchronization, 165, 190, 407, 429, 438
Bloxham, J., 502; on flux lobes, 102, 219, 220–22, 226, 271, 499; on geomagnetism, 103, 342; on virtual geomagnetic poles, 218; on perturbation of core, 231
"Blue sky catastrophes," 226, 306
Bock, W. J., 544; on evolution, 295; on paleontology, 542

Bode, J. E., 444
Bode's Law, 444. *See also* Ovenden, M. W.; Titius-Bode relationship
Bogue, S. W., 218, 222, 224, 342
Bohm, D.: on "implicate order," xxix, 20
Bohr, T., 161f, 181, 225, 387, 397, 400, 403, 450, 461f, 497, 503f, 562
Bohor, B. F., 49; on "fireball phase," 205
Bolide events, 177, 205–6, 208, 210, 303, 307, 326, 428. *See also* Tunguska event
Bonnet, C., 442f
Boss, A. P., 433, 490, 551; on collisional origin of Moon, 92, 110, 263, 265, 489; on planet formation, 434
Bostrom, R. C., 40, 477; on Earth's rotation, 273, 501
Bosumtwi Crater, Ghana, 120
Bottleneck effect in evolution, 209, 510
Bottomley, R. J., 118
Bowell, E., 50, 133, 145, 170; on discovery of asteroids, 55, 82, 147f, 171–72, 308, 441; on minor planets, 484
Bowin, C. O., 34, 178, 269f, 272
Boynton, W. V., 64, 213; on Chicxulub Crater, 119, 122, 128; on Caribbean impact site, 212
Brachiopod, *see* Clam and brachiopod genera
Bradley, W. H., 541; on ring-vortex bifurcation, 395, 412
Brahic, A., 2, 9, 169, 173, 263, 309, 428, 436, 454, 475
Brookins, D. G., on past nuclear reactions, 484
Brosche, P., 92; on Earth's tides, 477, 481–82
Brouwer, L. E. J.: on mathematical intuitionism, 411
Brown, J., 451; on superstring theory, 399, 450
Brown, W. K., 50, 81, 263, 559
Brush, S. G., 3, 53, 212
Buland, R., 533f
Bullard, E. C., 341; on disk-dynamo, 218, 232; on Euler pole, 364; on paleogeographic reconstruction, 368
BUM, *see* Base of the upper mantle
Burdon, J. J., 544, 547
Burek, P. J., 342, 506f; on mass impacts, 212, 240
Burgess, A., 301f
Burke, K., 49
Burnet, M., 16; on molecular genetics, 16, 293; on evolution, 543–45
Burrows, A., 491
Bursts, *see* Chaotic bursts
Busse, F. H., 222; on core, 222; on geomagne-

Index

tic field, 341; on magnetohydrodynamics, 390
Butler, R. F., 227, 342, 498; on Earth's magnetic field, 31, 38, 563; geocentric axial dipole (GAD) hypothesis of, 37, 40, 225; on Euler pole, 39, 211; on polar wander paths, 42, 364, 561; on geomagnetic reversals, 218; on virtual geomagnetic poles, 219, 422
Butterfly effect, xi, xxiv, xxxiv, 177, 387f, 544
Butterfly strange attractor, 233; sunspot resemblance to, 517
Byl, J., 74, 140, 143, 180, 344, 455, 489, 494

Cadek, O., 40, 501
Cahalan, R. F., 460, 473; on dripping-faucet system, 21, 141, 413, 462
Cairns, J.: on "directed mutation," 544–45
Cairns-Smith, A. G., 246; on essentials for life, 246; on "mineral origins" of life, 416; on evolution, 542
Calculus, xxxvii, 404; as applied to linear systems, 14
Caldeira, K., 186, 256, 291, 303, 352, 356, 358, 360, 374, 507, 535; on "hot spots," 242, 520
Calloway, C. B., 55, 114, 208, 289, 294f, 537
Camargo, Z. A., 119, 122, 261
Cameron, A. G. W., 252, 489; on collisional formation of Moon, 53, 75, 92, 263, 265, 325; on Mercury, 433–34, 520
Campbell, D. B., 242, 520; on statistics on crater counts, 260
Campbell, I. H., 57, 66, 201, 285, 506, 529; on catastrophic Siberian volcanism, 529
Campo del Cielo craters, Argentina, 210
Cande, S. C., 48; on geomagnetic time scale, 505; on seafloor spreading, 506–7
Candle flame, Faraday's study of, 412–13
Cannizzo, J. K., 434, 551
Cantor, G., xxxvi, 412; Kronecker's treatment of, xxvi, 411; transfinite set theory of, 392, 411
Cape-Fold-Belt-Karroo system, S. Africa, 272
Capra, F., 7; on Tao of Pooh, xxx, 551
Capture of satellitic objects, 177, 199, 465–66; of natural satellites, 2, 123; conditions for, 169, 193, 309; effect of resonance, 192–93, 467
Capture scenarios, 78, 168, 176, 192, 309
Carey, S. W., 454; on expanding Earth theory, 376, 492
Caribbean impact site, 212
Carlisle, D. B., 262, 435; on diamonds at K/T boundary, 49, 210, 414
Carnevale, G. F., 170

Carr, M. H.: on impact cratering, 242, 260
Carroll, R. L.: on radiation of land vertebrates, 427
Carroll, T. L., 88, 106, 150, 158, 183, 223, 226, 233, 254, 459, 518; on spin-wave experiments, 4, 160, 224, 231; on fractal dimension, 90; on chaotic bursts, 138–39, 161, 257; on synchronization, 190, 407, 429–30, 438, 463
Carter, W. E., 17, 482, 562
Carusi, A., 50, 435; on observation of Solar System objects, 201; on comets, 454, 494
Casdagli, M.: on model of time-series data, 460
Casenave, A., 17, 34, 178
Cassen, P., 434
Cassen, P. M., 92, 174f, 490, 517; on Io, xxxiv, xxxv, 325, 374, 401
Cassidy, W., 215, 490
Cassidy, W. A., 35, 210, 309, 336, 428, 467
Castellan, G. W., 13
Casti, J. L., xxvi, 29; on self-referential relativity/context, 411, 417
Catastrophes, xxxviii, 3, 29, 56, 294, 316, 364, 529–30; scaling of, 53, 301; and global biostratigraphic correlations, 284; self-similarity in, 288, 427; effect on evolution, 304; vs uniformitarianism, 543; and CRFH, 559. *See also* "Blue sky catastrophes"
Catastrophic implications of large-body impactors, 3
Cause and effect, 231, 292f, 504, 513f, 543; classical issues of, 57; vs effect and effect, 424–25
Cavalli-Sforza, L. L., 209; on linguistics/genetics, 286
CCD, *see* Charged-coupled device
Celestial attractors, *see* Attractors and attractor systems
Celestial dynamics, xxvii, 11, 18, 143, 188, 275, 389, 468f, 482, 485, 503, 544
Celestial mechanics, vii, 4, 13, 404, 415, 436ff, 471, 479, 482, 527, 538
"Celestial pin ball" (metaphor), 159
Celestial poles, 32
Celestial reference frame, 33, 43, 48, 53, 58, 125, 165, 178, 191, 194, 203, 236, 242, 248, 278, 289, 516, 528; definition of, xxvii, 32; invariance, 34, 37, 214, 221, 314f, 497, 499; reservoir, 50, 52, 201, 210, 238, 264, 270, 311, 314, 497; for geologic processes, 59; trajectories, 198; and satellitic resonances, 259, 313; resonances, 280; as test framework, 281; as nonlinear clock, 316

649

Index

Celestial reference frame hypothesis, 10, 41ff, 47f, 49–50, 131, 191, 198, 235, 262ff, 290, 310, 333, 476, 526, 540; and patterns of mass anomalies, 36; and geocentric axial dipole hypothesis, 42; and GAD and hot spot reference frame models, 46; and latitudes and longitudes, 51; and Geologic Column, 55–56; and near-Earth asteroids, 148; and proximal flight control, 165, 309; predictions of, 170, 311–12, 323; observational problems in, 170; and magmatic-tectonic effects, 202; and reconciliation with boundary events, 207; tests of, 210, 217; and geographic patterns of meteorite falls, 217; and impact-extinction hypothesis, 237, 382, 529; and MGK anomaly, 278f; and paleogeographic models, 281; and coupled processes, 283; issues stimulated by, 312–16; for Earth-Moon system, 319, 322; and pulsars, 419; and path of GFBP, 558

Celestial sphere, xxvii, 51, 171, 175, 497, 499; and projection of Earth's geographic coordinates, 31ff, 194; and approach scenarios, 36, 419; and condition for alignment in, 193

Cell death, 301f

Cellular automata, 156, 463, 503

Center of mass, 31, 479–80; barycenter of system, 479

Center of mass of Earth-Moon system, 179, 480

Center of mass of Solar System, 31, 78, 179, 183f, 432, 469, 480–81

Ceres (asteroid), 9, 74, 78, 133, 147, 150, 216, 260, 494; discovery of, 440–41, 445, 485; Gauss on orbit of, 441

CFD, *see* Critically fast dynamo

CGMN, *see* Critical golden-mean nonlinearity

Chaikin, A., 435, 454; his text on Solar System, 213

Chaitin, G. J., 29, 288, 505; on Gödel, xxv, 26; on algorithmic complexity, 1, 415; on randomness, 18, 545

Chamberlin, T. C. on "greenhouse warming," 540

Champion, D. E., 342; on geomagnetic field, 37, 225, 563; on subchrons, 38, 47; on paleomagnetic reversals, 507

Chandler wobble, 518–19, 562–63

Chandrasekhar, S.: on magneto- and hydrodynamics, 390, 502

Chandrasekhar mass, 491

Chang, L.-M. on ring-vortex bifurcation, 395, 541

Chant, C. A., 35, 91, 309; on Great Fireball Procession, 334, 337–38 (App. 3), 428, 467, 558–59

Chao, B. F., 243, 259, 482; on Chandler wobble, 519, 562

Chao, E. C. T., 212

Chaos, vii, 16, 54, 391, 394, 410, 540f; history of, viii; definition of, ix, 15, 187; "classical," xxix, 20, 399; nonlinear dynamics and science of, 4; in popular media, 12; edge of, 148, 152, 162, 181, 183f, 188, 448, 497, 504; statistical properties of, 188. *See also* Controlled chaos

Chaos Cabal, xxxii, 508

Chaotic, definition of, 391

Chaotic bursts, xxiv, xxv, 138, 145, 160, 163f, 175, 226, 254, 306, 395, 458, 474, 518; in Asteroid Belt, 145, 182, 313; in ferromagnetic resonance, 160; and experimentation, 394

Chaotic crises, xxv, xxx, xxxvii, 47, 139f, 156ff, 159, 163, 177f, 182f, 216, 225f, 254, 283, 294, 302, 316, 395, 421, 472, 503, 526, 528, 533, 540, 549; in systems of diverse type, 21, 161, 185; geophysical examples of, 106–7 (Fig. 31), 108–9 (Fig. 32), 160; and self-organized criticality and CGMN, 161; and delivery of asteroids, 191; relative to orbital frequencies, 197; in geomagnetic phenomena and cratering record, 203; and intermittencies, 217; in geophysical-astrophysical resonances, 244; definition of, 457. *See also* Interior crises

Chaotic intermittency, 124, 138, 156, 158, 170, 183, 248, 282, 309, 406, 412, 457, 510, 522

Chaotic transition, *see* Edge of chaos

Chapman, C. R., 50, 198, 213, 263, 491, 520, 548; on collisional origin of Moon, 53, 265, 325, 489; on asteroids, 74, 81, 129, 170, 262, 309, 414, 434, 454, 490, 494; on Tunguska event, 172, 308, 559; on impacts on Mercury, 350; on near-Earth objects, 495

Characteristic paleomagnetic poles of epoch as distinct from paleomagnetic pole positions (PPPs) and virtual geomagnetic poles (VGPs), 47

Charged-coupled device, 547–48

Charlevoix Crater, Quebec, 66

Charon (satellite of Pluto), 133, 136–37, 145, 184, 483, 494; discovered by J. W. Christy, 136

Chemistry, 381, 398, 406, 410, 423, 525; and cosmic abundances, 414, 435

Chen, C. H.: on ring-vortex bifurcation, 395, 541

Index

Chen, K., 156, 209, 225, 231, 421, 448; on self-organization criticality, 2, 140, 161, 302, 503
Chen, P., 209
Chicxulub Crater, Yucatan, 49, 55, 119, 121f, 128, 206, 257, 260f, 334, 426–27
Children's story, 551–57
Chillingworth, D. R. J., 108; on magnetic field, 232, 252, 341
Chirikov, B. V., 135; on theories of electron beams, 390
Chondrites, *see* Meteorites
Chouet, B., xxii, 2, 74, 76, 135, 140, 162, 184–209 *passim*, 225–45 *passim*, 253–57 *passim*, 275–91 *passim*, 387, 395–407 *passim*, 421–23, 440–49, 457–63 *passim*, 497–508 *passim*, 532–33, 546, 561; on seismic tremor, 10f, 56, 162–63, 393, 399–401, 419–20, 535; on Hawaiian magma transport, 163, 218, 234, 244, 403; on CGMNs, 161, 256, 302, 397, 400, 448, 450, 503; on Hawaiian-Emperor volcanism, 534–35
Christy, J. W.: on discovery of Charon, 136
Church, W.: on past nuclear reactions, 484
Chyba, C. F., 209, 269, 305, 427; on Tunguska event, 99, 308, 322, 333, 338, 428, 558; on heterogeneity of upper mantle, 262; on geochemical history of Earth, 266; on estimate of meteoritic material, 267; on "mineral origins" of life, 416
Circle map, *see* Sine-circle map
Claeys, P., 528; on impact craters, 213, 364; on microtektites, 262, 334
Clam and brachiopod genera, 208f, 294, 301, 316, 537; study based on Gould and Calloway (1980), 114–15 (Fig. 36a), 208, 295–99
Clarke, C.: on star formation, 548–51
Clarke, D., 32, 194; on orbital mechanics, 91, 342, 464f, 470, 479; on precession, 197, 497; on gravitational sphere of influence, 466; on Kepler's third law, 469
Clemens, W. A.: on extinctions, 511
Clement, B. M., 218, 342, 502; on virtual geomagnetic poles, 103, 219, 271, 422; on VGP and core-flux patterns, 218, 220, 226; on helical flows, 223
Clock, 286; as perfect, 19f; atomic, 19, 21f, 185, 317, 417, 430, 560; as nonlinear, 55, 316, 432; and sine-circle map, 304; astronomical, 316. *See also* Mechanical clock
Clockwork, 559–61
Cloos, E., xxxi
Cloos, H., xxxi

Clube, S. V. M., 84, 139, 209, 213, 454f, 488, 507; on giant comets, 9–10, 130, 151, 167–68, 177, 326; on Galactic sources of comets, 78, 129, 136, 477, 493; on Tunguska event, 98, 200, 428, 558; on "comet factories," 152, 478; on cometary aggregates, 170; on catastrophic impacts/bolides, 201, 205, 252, 338; on Gould's Belt, 479
Clube-Napier hypothesis, 130, 477
Clustering, 246, 447ff, 453
CM, *see* Center of mass
CMB, *see* Core-mantle boundary
CMem, *see* Center of mass of Earth-Moon system
CMss, *see* Center of mass of Solar System
Coe, R. S., 38, 47, 342; on geomagnetism, 31, 422, 499; on paleomagnetism, 37, 47, 225, 507, 563
Coffey, S. L., 14, 196, 199, 327, 389, 394, 496; on spin-orbit resonance, 34, 498; on artificial satellites, 92, 132, 486, 495; on resonances, 177, 195, 467, 495, 497; on critical inclination, 196, 199, 327, 496, 498
Coin-flipping machine, *see* Perfect coin-flipping machine
Coin tosses, 10, 187, 404, 417; and concept of "fair," 19ff, 26
Coleridge, S. T., 451–52
Collisional exchanges/processes, 92, 138, 433
Colombia Basin structure, 122
Columbia River basalts, 213; and Hagstrum-Turrin model, 213
Colwell, J. E., 436; on satellites, 2, 263; on rings in Solar System, 9, 309, 475; on Saturn's rings, 173
Comet Halley, 82–83
Comets, 3, 9, 50, 151–53, 170, 199, 262, 485, 489; "giant" size of, 9, 129–30, 151, 167; optical searches for, 11; evolution of, 86–87 (Fig. 15b), 154; as showers, 124, 129, 138, 153f, 164, 190, 289, 550; as contributors to formation of craters, 128; reservoirs of, 129, 132f, 140, 144, 152, 154, 165f, 313, 396, 458, 487; and Oort Cloud, 129, 151, 153, 190, 289; as distinct from asteroids, 129, 140, 494; and storage within Solar System, 129; and Galactic tide, 129; and Hills Cloud, 129, 151, 313; and growth rate of discoveries, 131, 167, 311, 453, 548; short-period type of, 144, 151, 153, 167f, 176, 190, 195; long period type of, 151, 153, 195; and Kuiper Belt, 151, 313; Monte Carlo simulation of, 84, 86–87, 152, 154–55, 167; delivery to inner Solar System, 191, 313; discoveries of, 200; satellites of,

651

Index

309; estimates of number of, 311; carbon-poor class of, 414; references on sources of, 454; hypothetical source-concepts of, 455; trajectories of, 477–78

Commensurabilities, 143f, 180, 189f, 256, 292, 438, 440, 508; of satellite orbits, 173

Complexity, 387, 401; undecidable propositions, 1; fabric of, 11, 285; periodic-aperiodic mixture of states in, 15; and compression, 23, 25, 28; in framework of Gödelian arrays, 29; in Solar System, 134; from synthetic viewpoint, 415. *See also* Algorithmic complexity

Computer experiments, ix; in nonlinear dynamics, 162, 408–9

Computers, 162; as self-referential extensions, 26

Computer software, 29–30

Conjunction as condition of alignment, 191–92; and synodic periods, 192

Conjunctions, 192–93, 195, 432, 498–99

Connection machine (computer metaphor), xxxvi, 14

Considine, D. M., 8, 76, 216, 308, 350, 442f, 446, 454

Constable, C., 218f, 226, 342, 422

Continental drift, xxxi, 35ff, 39, 42, 264, 278, 281f, 314f, 514, 526

Control function, 386f

Controlled chaos, 257, 297, 463, 540, 543, 546; and inverse of, 164

Cook, G. E., 196, 270, 486

Coordinate frames, 40

Copernican Period of Moon, 53, 319

Coppersmith, K. J., 534

Core, 3, 46f, 53, 163, 216, 224, 264, 267, 283, 499, 502; motions in, 41, 44, 112, 191, 221, 239, 248, 500; formation of, 42, 52, 249, 251, 490; flux patterns in, 59, 219, 221, 226; convection in, 218; and virtual geomagnetic pole paths, 226; as heat source, 223, 244, 247–51, 490; and suppression of magnetic reversals, 230–31; and perturbation of core flow, 231; volumes of inner and outer core, 249; spallation in, 240, 515; resonance ratio in, 500

Core attractor, 146, 158ff, 163, 190, 526

Core dynamo, 46f, 139, 228, 230, 238, 243, 252, 423, 492, 514, 519; resonant behavior of, 42, 44; two-disk model of, 108–9, 217–18, 219, 232; convection in, 203; self-organized configurations in, 227; and magma-tectonic oscillator, 229; and other geological phenomena, 506

Core-mantle boundary, xxxvi, 40, 220f, 223, 238f, 243, 248f, 271, 273, 278f, 314, 490, 505, 514; and mantle plumes, 45, 245; and D" layer, 251, 271, 274, 500, 515f

Core-mantle coupling, 46; and global heat flow, 46; and geomagnetic signatures, 244

Cosmic abundances, 414, 435, 477, 525

Cosmic imprint on Earth, vii, 29

Cosmic plasmas, 9, 436

Cosmic world order, *see* Children's story

Cosmological interface, 236–37

Cosmology, 8, 18, 26, 28, 142, 302, 390, 404

Cosmos, xxii, xxvi, xxvii, xxx, xxxvii, xli, 142, 248, 316, 437, 442, 505; interstellar reaches and processes in, 2; attractor systems in, 4; resonances with, 7; and governing principle of evolution, 28, 425; clustering in, 246, 447; chemical heterogeneity of, 414, 435

Cosmosynaptic interface, 452

Counting, 22

Coupled geodynamic and orbital-dynamic resonances, 30

Coupled-oscillator systems, xxx, 36, 182–83, 188, 224, 229, 239, 257, 317, 387, 396, 405, 450, 452, 504, 528; and mechanical clock, 13; and celestial reference frame hypothesis, 43

Coupled phenomena references, 208–9

Coupling between impact, geo-, and tidal dynamics, xxxvi

Courtillot, V., 32, 422, 506f, 562; on geomagnetic field, 37, 48, 225, 342, 515, 563; on paleomagnetic subchrons, 38; on TPWPs, 40f, 45, 47, 341; on hairpins, 42; on APWPs, 364, 366, 368–69, 372

Courtillot, V. E., 509; on volcanic "winters," 57, 201; on volcanic phenomena as cause, 285

Cox, A., 36; on plate tectonics, 36; on Euler poles, 39, 211; on geomagnetic fields, 341, 423; on geomagnetic time scale, 505f

Cox, K. G.: on correlations with paleomagnetic record, 341

CPPEs, *see* Characteristic paleomagnetic poles of epoch

Craig, H., 436; on chemical heteorgeneity, 414, 525

Cratering, *see* Impact cratering; Impact cratering patterns

Cratering nodes, 31, 112, 168, 201, 211, 229, 240, 354 (App. 13); additional, 58, 211, 220; and conditions of alignment, 193. *See also* Phanerozoic cratering nodes

Craters and similar structures, 334–35; Acraman Crater, South Australia, 62, 119, 128,

652

Index

271; Avak structure, Point Barrow, Alaska, 49, 64, 122f, 261, 328, 334; Beaverhead structure, Montana, 334; Bosumtwi Crater, Ghana, 120; Campo del Cielo craters, Argentina, 210; Caribbean impact site, 212; Charlevoir Crater, Quebec, 66; Chicxulub Crater, Yucatan, 49, 55, 119, 121f, 128, 206, 257, 260f, 334, 426–27; Colombia Basin structure, 122; Dellen structure, Sweden, 118; Falkland Plateau structure, 57, 272, 312; Flynn Creek Crater, Tennessee, 66; Gusev structure, Russia, 121–23; Haughton (crater), N.T., Canada, 328; Jänisjärvi Crater, Russia, 118; Kaluga Crater, Russia, 66; Kamensk structure, Russia, 121–23; Kara Crater, Siberia, 64, 121ff; Manicouagan Crater, Quebec, 121f; Manson structure, Iowa, 49, 121ff, 206, 427; Mishina Gora Crater, Russia, 66; Montagnais Crater, North Atlantic, 64, 120, 312; Piccaninny Crater, Western Australia, 66; Popigai Crater, Siberia, 49, 64, 121f, 206, 328, 426–27; Rio Cuarto crater field, Argentina, 210, 334, 340; Sääksjärvi Crater, Finland, 118f; Siljan Crater, Sweden, 66; Taihu Lake structure, China, 128, 334; Ust-Kara Crater, Siberia, 121ff. *See also* Precambrian craters
Cratonic terranes, 30, 36f, 314, 378–79
CRB, *see* Columbia River basalts
Creer, K. M., 342, 506, 520; on geomagnetic dipole axis, 37f; on geomagnetic excursions, 47, 225, 563; on reversal frequencies, 69, 126, 218; on spallation, 240, 515; on impacts and polarity reversals, 562
Cretaceous normal superchron, 228, 232, 242, 244, 253
Cretaceous/Tertiary extinction, *see* K/T boundary
CRF, *see* Celestial reference frame
CRFH, *see* Celestial reference frame hypothesis
CRF/PCN, *see* Phanerozoic cratering nodes–celestial reference frame
Crichton, M., 539; on amber, 547
Crick, F. H. C., 209; on genetic code (DNA), 23, 288, 416, 532, 542; on "panspermia," 305
Crisp, J. A., 352; on magmatic energy, 258
Critical golden-mean nonlinearity, 161, 256, 302, 397, 399–400, 445, 448, 450, 503f
Critically fast dynamo, 225, 231
Critically self-organized regimes, 1; of *cellular automata*, 156
Critical self-organization, 42, 170, 178, 216, 225–27, 231, 234f, 254, 283, 286, 302, 306, 421, 458, 474, 503, 534, 546; in Solar System, 78–79 (Fig. 11). *See also* Self-organization
Croquette, V.: on period-doubling route, 387
Crough, S. T.: on "hot spots," 352
Crowell, J. C.: on paleogeographic reconstruction, 376, 527
Crowley, T. J., 281; on glaciation, 281, 326, 527; on paleogeographic reconstruction, 376, 527
Cruikshank, D. P., 2, 263, 309, 475
Crust, 29, 40, 198, 218, 227, 239, 243, 259, 267f, 275, 279, 520
Crutchfield, J. P., xxvi, 29, 209, 236, 390, 407, 460, 463; on nonlinear time-series analysis, xxxii–xxxiii; on linear dynamics, 134, 540; on bifurcation, 158, 452; on nonlinearly coupled oscillators, 403; on audio-visual feedback, 404; on chaos, 413, 508; on chaotic dynamics, 414; on synchronization, 429, 438; on spatial intermittency, 457–58
CSI, *see* Cosmosynaptic interface
Cvitanovic, P.: papers on nonlinear dynamics, 134, 420
Cybeles group (asteroid), 149, 487
Cybernetic self-tuning, 161, 178, 400
Cyclic-feedback process, 52
Czamanske, G. K., 57, 285, 506; on Siberian volcanism, 66, 529; on volcanic "winters," 201

Dalrymple, G. B., 266, 269, 352; on atomic decay, 20; on age of Moon, 32, 35, 525; on age by means of heat transfer, 251; on age of Earth and Moon, 259–60; on age of meteorites, 267; on research in Hawaii, 275; on age of universe, 453; on geomagnetic time scale, 505
Dalton, J.: on "atomic theory," 381
Daly, R. A., 252; on collisional origin of Moon, 53, 138, 263, 265, 325, 489, 540; on cratering phenomena, 212
Dalziel, I. W. D., 37; on "supercontinents," 37, 376, 526; on paleogeographic reconstructions, 264, 278
Damon, P. E., 256; on orogenic-epeirogenic episodes, 256; on geologic periodicities, 508; on variation in solar activity, 518
Danielsson, L., 50, 429, 474; on asteroid/meteor "streams," 9, 170, 201; on stream-like focusing of asteroids, 81
Dark matter theory, xxiv, 447
Darwin, C., xxv, 285f, 317; and neo-Darwinian models, 16; "10,000 wedges"

653

Index

of, 56, 293f, 304–5, 537–38; challenge to Neo-Darwinism, 544–47
Darwin, G. H., 92, 491; on birth-scar of lunar origin, 265; on rotational stability of Earth, 345
Data sets, 460–64
Dauben, J. W., xxx, 143; on completed infinities, 411; on Georg Cantor, 411; on transfinite numbers, 412; on universe as wheel of fortune, 448
Davies, M. E., 8; on satellites of planets, 8; on Io, 373
Davies, P., 7
Davies, P. C. W., 451; on superstring theory, 399, 450
Davis, D. R., 50, 454; on collisional origin of Moon, 53, 138, 252, 263, 265, 325, 489; on asteroids, 74, 309, 434; on cratering phenomena, 212, 260; on Phobos and Deimos, 311
Davis, J. L.: on artificial satellite geodesy, 17, 178
Davis, M.: on undecidable propositions, 411
Davy, H., 381
Dean, D. R.: on James Hutton, 423, 540
Deccan flood basalt province, India, 66, 121, 213f
Deep-mantle P-wave anomaly, *see under* Mantle
Deep-mantle reference frame, 40f, 43, 45
Deep mass anomalies, 32, 35, 52, 264, 270
Deformation and flow far from equilibrium, 45
Degn, H., 209
Degrees of freedom, 160–61, 281, 421–22, 460, 463–64, 501, 504, 526f
Deimos (satellite of Mars), 309, 311
Deino, A. L.: on age of Popigai Crater, 64, 121, 426–27
Delany, J. M.: on volcanism on Io, 401
Delayed-feedback coupling phenomena, 57, 202, 229, 257, 514
Delayed-feedback systems, 202, 229, 297, 301
Dellen structure, Sweden, 118
DeMets, C.: on mantle reference frame, 48
Dence, M. R., 30, 119, 213
Dermott, S. F., 8
Derry, D. R., 130; on craters, 130, 321, 379; his *World Atlas of Geology and Mineral Deposits*, 379
Des Marais, D. J., 504
De Sousa Vieira, M., 257; on synchronization, 407, 430, 438, 463
Deterministic chaos, xi, xxi, 4, 8, 27f, 43, 164, 286, 391, 396, 399, 401, 420, 482

Deutsch, D., xxviii; on auditory illusions, xxviii; on language, 286
Devaney, R. L., 12; his text on nonlinear dynamics, 12, 134; on phase space map, 386; on period-doubling route, 387; on chaotic bursts, 394f
De Vaucouleurs, G., xxix, 416, 477, 533; on astronomical discoveries, 55, 147; on power-law density distribution, 73; on mass distributions, 74, 135; on age of universe, 246; cosmological model of, 246–47, 492, 540; his paper "The Case for a Hierarchical Cosmology," 401; on "blazing sky" effect, 448; on fractal dimension of galactic processes, 449; on microwave background, 449; on Hubble's law, 451; on "growth rate of astronomical progress," 453
"Devil's Staircase," 82, 143, 148–49, 176, 180, 395; in Asteroid Belt, 13, 85 (Fig. 14), 150, 349
Dewdney, A. K., 134, 209, 234; on frustration, 4, 183, 233, 293, 463
De Wit, M. J., 37, 264, 526
Diamond, J. M., 209; on linguistics, 286
Diamonds, 48; at K/T boundary, 48, 210, 414; in interstellar medium, 210, 414; in meteorites, 414
Dickey, J. O., 481–82; on rotation speed of Earth, 259, 516, 519, 562
Dickman, S. R., 501
Dietz, R. S., 112; on paleogeographic reconstruction, 112, 279, 281, 376; on "astrogeology," 212, 507
Dimensional analysis, 198
Dimensional scaling, 21f, 386, 399, 463
Ding, E. J.: on winding numbers, 181f, 432, 440
Dinosaurs, xxxvii, 316, 329, 427, 513
Dirac, P. A. M., 393; on dimension of spherical surface, 393; on cosmological constants and expansion, 492
Discontinuity (mantle), 264, 272f, 278, 524
Disease concept of phenomena, 25
Dissipative feedback, xxviii, 5, 514; and diffusive transport, 27; and transfer of work, heat and mass, 27
Dissipative systems, 10, 168, 183, 188, 389f, 394, 410, 415, 420, 503, 542
Ditto, W. L., 90, 108, 232, 290, 394, 458, 463; and magnetoelastic experiment, 88–89 (Fig. 17), 106–7, 139, 157–60, 459; on chaotic bursts, 88, 139; on chaotic crises, 140, 421, 533f; on strange nonchaotic attractors, 161, 517; on negative feedback, 224, 226

Index

Dixon, T. H: on artificial satellite geodesy, 17, 178
D″ layer, 251, 271, 274, 500, 515f
DNA, structure of, 23, 163, 416
DNA-hybridization methods, 295, 299, 301f
Dobzhansky, T.: on micro- and macro-evolution, 542
Dodd, R. T., 262, 286, 414, 434f, 490
Doehne, E.: on iridium anomalies, 49, 210
Doell, R. R.: on paleomagnetic reversal scale, 505f
Doherty, P., xxvi, 209, 390, 403, 452, 460, 518; on attractors, 157, 463, 517; on synchronization, 407, 429, 438
Dohnanyi, J. S., 50, 81
Dones, L., 50, 436; on angular accelerations of Earth, 259; on ring arcs in Solar System, 309, 475; on impulses of torque in Solar System, 481–82
Donnison, J. R., 50; on sizes and shapes of asteroids, 263, 490
Dormand, J. R., 454; on Solar System structure, 436
Draconis (star), 32
Drainage-basin (metaphor), 309, 471–72, 476
Drake, C. L., 249, 251, 266f; on endogenous cause of extinctions, 57, 285; on IEH debate, 509
Dripping-faucet system, xxxii, 21, 141, 413, 462
Drummond, J. D., 474, 488; on meteor/meteoroid/asteroid streams, 9, 50, 81, 170, 201, 429; on Apollo group, 144–45
Druyan, A., 137, 433, 455
Duba, A.: on density of core, 249, 251, 515
Duffield, W. A.: on foundering/subduction, 277, 520, 522
Duncan, M., 78, 136, 138, 164, 257, 469, 479, 488, 493f; on comets, 75, 140, 199, 414, 454–55; on Monte Carlo simulation of comets, 84, 86, 152, 437; on galactic perturbations, 129, 154, 177, 245; on cometary reservoirs, 132, 151, 166; on short-period comets, 151, 153, 168, 190; on orbital inclinations, 153, 498
Duncan, R. A., 48, 352; on true polar wander, 40; on "hot spots," 229, 240, 352, 357 (App. 16), 506, 519; on paleogeographic reconstructions, 112, 279, 281
Dunham, D. W., 490, 494
Dunning, G. R., 33; on possible Triassic/Jurassic impact, 33; on age of Manicouagan, 121; on end-Triassic extinctions, 364, 427, 513
Dust bands, 9

Dynamical dimension, 393, 459–60, 463
Dynamical resonance, 429
Dynamical universalities, 24
Dynamics, xxxvii, 12, 48, 191, 396, 405–6, 407f, 410
Dynamo theory, 503
Dziewonski, A. M., 269, 275; on deep-mantle P-wave anomaly, 103, 111f, 220, 271; on core-mantle boundary, 223, 278; on discontinuity, 264; on earthquakes and Chandler wobble, 518–19, 562
Dzurisin, D., 477; on tidal deformation, xxxvi; on scales of volcanotectonic effects, 275; on Hawaiian-Emperor volcanics scaling, 535

Earth, xxvii, 1f, 17, 24, 48, 133, 136f, 150; rotational deceleration in, xxxv; longitudinal mass anomalies, 3, 51, 173, 201, 476; "ratcheting" of surface and interior motions of, 3; early history of and origin of moon, 5, 42, 249, 250–51, 259–60, 282; legend of beginning of, 6–7; weather-climate system of, 25, 163, 179, 253, 517, 527–28; hot spots in, xxxvi, 29, 43–44, 45, 203, 213f, 229, 240, 242, 244, 275, 375; magnetic field of, 31, 37, 160, 227, 250, 252, 500–501, 518; spin-axis, 31ff, 36f, 42, 44ff, 47f, 227, 264, 270, 279f, 498, 519, 527, 561–64; geomagnetic axis, 31; precession of, 31–32, 51, 166, 194, 270, 497f; virtual poles mapped on celestial sphere, 32; deep mass anomalies in, 32, 35, 52, 264, 270f; mass distribution of, 32f, 37–38, 44, 51, 110–11, 273, 314, 498; moments of inertia in, 33, 40, 42, 44, 46, 201, 243, 264, 266, 270, 520; gravitational field of, 34f, 270; departures from symmetry, 34, 51; equatorial bulge of, 34, 46, 264, 270; mass anomalies of, 34f, 52, 103, 215, 221, 480; mass distribution, evolution of, 35, 263, 269; near-surface mass anomalies in, 36, 52, 273; rotational pole of, 37f, 40, 278; coordinated frames and outer shells of, 39–40; mass heterogeneity in, 45; plumes/plume heads in, 45f; shear-wave attenuation mapping of, 45; ancestral Pacific Basin/Ocean of, 52, 110–11, 265, 276; gravitational sphere of influence of, 168, 466; tidal bulge of, 174, 480; sphere of influence, 177, 179, 195, 197, 266, 466, 474; orbital-rotational motion of, 217, 527; thermal evolution of, 251; age of, 259–60; meridional geodetic keel, 271–74, 276–81, 527; Chandler wobble of, 518–19, 562–63

655

Index

Earth-crossing objects, 309, 320, 323, 326; delivery of, 11, 127, 140; discovery of, 11

Earth-crossing orbits, 131, 146, 193, 200; of comets, 129, 226; of asteroids, 129, 144, 191, 226, 487; delivery of objects in, 226

Earth-Moon system, 34f, 125, 171, 201, 274, 307, 319; and tidal phenomena, xxxiv–xxxv, 136, 477, 480; and spin-orbit resonance, 36, 51, 166, 169, 173, 215, 311; capture of satellitic objects, 50, 226, 322; and "geocentric" system, 51; resonance states of, 173ff, 477; loss of Earth's K. E. to, 174; longitudinal mass anomalies in, 176, 178, 184; and center of mass, 176, 179, 184; and satellitic reservoirs, 179; and artificial satellites, 195; gravitational sphere of influence, 195; Lagrangian points of, 474–75

Earthquakes, xxiii, 25, 198, 243, 261, 395, 421, 490f, 515, 519, 533–34, 562; and mass impacts, 244; unpredictability of, 257; scaling of, 258, 289

Earth science, xxiii, 539; and interdisciplinary outlook, xxx, xxxi; and "periodicity controversy," 256

East African rift, 241

Eccentricity, 196f, 199, 215, 263, 467, 471, 496f; of asteroid orbits, 194, 487

Ecliptic plane, 31–32, 190, 195, 199

Ecologists and biological diversity, 303

Eddy, J. A., 253; on Maunder Minimum, 253, 257; on sunspot cycles, 481; on solar activity, 517–18

Edelman, G. M., xxviii, 209, 538, 543; on neuronal group selection, xxvi–xxvii; on neurophysiology, 234; on population dynamics, 302; on brain-mind plexus, 408; on therapeutic application, 464; on Neo-Darwinian molecular genetics, 546

Edge of chaos, 148, 152, 162, 181, 183, 188, 448, 504; and Solar System, 184, 497

Ediacaran fauna, 282

Eigen, M., 29, 237; on essentials for life, 246; on genetic code, 288, 542; on living matter, 545

Eighth Day of Creation (book), 23

Einstein's theory of relativity, 451

Eldredge, N., 209, 511; on punctuated equilibria, 237, 304; on evolution, 285–86, 295, 540–41, 543–44; on paleontology, 542; on fossil record, 547

Elitzur, M., xxiv, 421

Elliott, C. G., 36, 264, 376, 526

Elsner, J. B., 10f, 161, 163, 188, 423, 460, 462, 561

"Empty" and "filled" resonances, 50, 149, 176f, 182

Encke, R.: on quasiperiodic route to chaos, 387

Energy, 252, 306; dissipation of, 26, 250, 432, 474; as potential, 244, 249f, 468, 470f; conservation of, 249, 294; tidal dissipation of, 250, 258; cascade of, 252; of impact cratering, 257–59; and global ecology, 304; in astrophysics, 469; as specific, 470. *See also* Kinetic energies

Engebretson, D. C., 276; "lithospheric graveyards" of, 35, 264, 272, 277f; on subduction, 40; on plate motions, 45, 178

Entropic feedback, 27

Epimenides' Paradox, 18

Equations and nonlinear systems, 14–15, 385–86

Ernst, W. G.: on "accretionary" and "suspect" terranes, 37, 264, 527

Erwin, D. H., 57

Escher, M. C., xxviii

ESOCs, *see* Extra-Solar Oort clouds

Esposito, L. W., 436; on satellites, 2, 263; on rings in Solar System, 9, 309, 475; on Saturn's rings, 173

Essex, C., 163; on attractor states in climate systems, 163, 462; on time series in climate, 460

Eugster, H., x

Euler poles, 39f, 211

Europa (satellite of Jupiter), xxxiv, 314, 494

Everhart, E.: on chaotic orbits in Solar System, 11

Evernden, J. F., 256; on orogenic-epeirogenic episodes, 256, 277; on periodic concept in geology, 508

Evolution, xxv, xxvii, 286, 290, 293f, 316, 504, 508, 538f, 542; immunological, 208; genetic/linguistic, 208; neurophysiological, 208; punctuated equilibria in, 237; Ediacaran fauna, 282; first identifiable organism in, 282; first reproducing organism in, 282, 423; multicellular organism in, 282; and environmental change, 285; of birds and vascular plants, 301, 512; and mass impacts, 303; politico-economic context of, 305; latency in, 509, 513, 530; bottleneck effect in, 510; founder effect in, 510; genetic drift in, 510, 546f; "common ancestor" in, 540–41; gradual vs punctuational, 543; vortex-tree model, 544. *See also* Mutation; Natural selection

Exclusive Economic Zones, 64, 214, 312

Index

"Expanding-Earth theory," 492; and Dirac-Jordan-Carey theory, 393
Experiments and experimentation, xxxii, 394; magnetoelastic ribbon, 88–89 (Fig. 17), 106 (Fig. 31), 139, 157–60; yttrium-iron garnet, 90–91 (Fig. 18), 160, 226; by J. Sommeria on Great Red Spot, 222–23; on chaotic bistable fluid oscillator, 104 (Fig. 29), 224. *See also* Computer experiments
Exponential divergence, 4, 27, 33
Exponential divergence from initial conditions, *see* Sensitive dependence on initial conditions
Extinction patterns, 209; 26 m.y. -period of Raup and Sepkoski, 54, 291; background vs mass, 203, 285, 295. *See also* Mass extinction
Extra-Solar Oort clouds, 478
Ezzell, C., 12, 134, 234, 464

Falkland Plateau structure, 272; proposed crater site on, 57, 312, 334, 528–29
Fanale, F., 169, 428, 436
Faraday, M., 381; on hypotheses, 381–82, 540; experimental-experiential method of, 410–11, 439; on candle flame, 412–13
Farley, K. A., 414; on chemical heterogeneity, 414, 525; on neon in mantle, 436
Farmer, D., xxxii–xxxiii, 460, 508, 540
Farmer, J. D., xxxiii; on predictive methods, 188; on dynamical dimensions, 459–60; on nonlinear periodicity, 507–8; on chaotic-time series, 561
Farrar, E., 57, 66, 529
Fast dynamo, 503
Fast-moving objects, 78, 147, 171, 175, 310, 485, 497
Fastovsky, D. E.: on extinctions, 511
Feder, J.: on fractal geometry, 393
Feedback coupling, 59, 314; between impact dynamics and geodynamics, 42; among impacts, magmatic cycles and geomagnetic reversals, 243
Feedback systems, xxii, xxviii, 178, 459; pattern-generating effects of, 12; computations of, xxiii, 385–87. *See also* Dissipative feedback; Geodynamic feedback; Negative feedback; Nonlinear feedback systems
Feigenbaum, M. J., xxxi, 14, 420, 439, 502; on mathematical universality, xxxi; on logistic map/equations, 187, 446, 463, 539; on tuning, 385; on period-doubling route, 386–87, 413, 541; on Feigenbaum ratios, 396; on metric universality, 396, 461; on bifurcation, 411

Fein, A. P., 161, 254, 403, 445, 504; on sine-circle map, 60, 397; on golden mean as parameter, 162, 316, 399, 444, 458, 497; on power-law distribution of frequencies, 184, 188; on tuning, 385, 458; on coupled-fluid convection studies, 400; on winding numbers, 432, 440; on models of chaotic dynamics, 502; on quasiperiodic route, 503; on event frequencies, 534f
Feit, S. D., 135
Fernández, J. A.: on comets, 488, 494
Ferromagnetic resonance of YIG, 160
Feudel, U.: on fast dynamo, 225, 503
F-F, *see* Frasnian-Famennian boundary/ extinction
Fink, J. H.: on impact cratering, 198
Fink, U., 262; on heterogeneity of impactors, 262, 414, 478, 494; on carbon-poor comets, 414; on comets, 453; on cosmic abundance of elements, 477, 525
Fireball trajectories, 34–35; across Canada and N.Y., 308
Fischer, A. G., 209, 326; on accretionary continents, 264, 526; on periodic concept in geology, 256, 508
Fish, F. F., Jr.: on orbital energies of planets, 76, 344
Fisher, J., 510; on diversity of avian families, 114–15 (Fig. 36), 300; on avian evolution, 295; on avian classification, 299, 511–13
Fisher, Rev. O., 265; on birth-scar of lunar origin, 265; on Earth's moments of inertia, 266; on thermal contraction of Earth, 401; on crustal foundering, 520
Fixed-point attractors, 46, 388, 394f, 409, 413, 464, 541, 543
Flam, F., 14; on system of galaxies, 14; on cosmological/galactic phenomena, 135, 437; on microwave background, 449, 533
Fleming, J. R.: on "greenhouse warming," 540
Fleming, R. F., 207
Flessa, K. W.: on extinction horizon, 511
Flood-basalt provinces, 66–67 (Fig. 4), 99 (Fig. 24), 100 (Fig. 25), 101 (Fig. 26), 124, 258, 272; at British-Arctic-Tertiary Province, 66; at Columbia River, 66, 213; at Deccan volcanics, India, 66, 121, 213f; at Ethiopian Province, 66; at Karroo field, South Africa, 66, 119, 272; at Keweenawan Volcanic Province, U. S. & Canada, 66; at Serra Geral (Pararn) Province, S.A., 66; at Siberian Traps, Russia, 66; at North Atlantic, 121; impact origin of, 213, 244, 258, 260; and impact craters, 119, 210, 229, 240,

657

Index

528; correlations with mass impact and mass extinctions, 211–12; and antipodal position to impact sites, 213, 260; and Hagstrum-Turrin model, 213
Flow, dissipative, 27, 402–3
Flux lobes, *see* Magnetic flux lobes
Flybys, 91–93 (Fig. 19), 168, 193, 199
Flynn Creek Craters, Tennessee, 66
FMOs, *see* Fast-moving objects
Folded feedback, 17, 236, 342, 386, 413, 449f, 503, 522
Foote, M.: on Cambrian (beginning of Phanerozoic) fauna, 28, 282, 537
Force and tensor mathematics, 407
Ford, J., 22, 29; on classical chaos and nonlinear dynamics, xxix, 182, 399; on random chance/coin flipping, 18, 20; on quantum chaos, 20; on stochastic resonance, 437
Forrester, J. W., 209; on "industrial dynamics," xxii; on cybernetic self-tuning, 178; on lack of economic balances, 305; on nonlinear feedback computations, 385
Fossil record, 53, 174, 286, 293, 300f, 303, 457, 508, 539, 543; "rare" events in, 3; of birds, 207; latent-acceleration effect in, 510; Signor-Lipps effect in, 510, 512f; extinction horizons in, 511
Founder effect in evolution, 510
Fractal, definition of, ix
Fractal-attractor basin boundaries, 144, 169, 177, 188, 195, 204, 451, 468, 471–73; and Belbruno's "fuzzy boundaries," 2, 168–69, 466, 472; of Earth-Moon system resonances, 35. *See also* Attractors and attractor systems
Fractal dimension, 73–74, 90, 135, 146f, 347, 393, 395; "Multi" scaling of, 22; in universe, 533
Fractal geometry, xxxvii, 74, 392, 409, 541; statistical techniques in, 11; and Kilauean lava prominences, 393
Fractal potential well, 4. *See also* Gravitational potential well
Fractal singularity analysis, 188. *See also* Multifractal singularity spectrum; Singularity spectrum
Franceschini, V., 385; on nonlinear feedback computations, 385; on period-doubling route, 386; on turbulence, 502
Francheteau, J., 48, 352
Frank, L. A., 170; on search for near-Earth objects, 170, 175, 309, 495–96; his reports of natural-Earth satellites, 548
Frasnian-Famennian boundary/extinction, 33, 56, 58, 124, 212, 255, 288f, 297, 528, 529–30; McLaren Line, 66, 124, 290; multiple events at, 427
Freeman, W. J., xxvii, xxix, xxxvi, 105, 134, 209, 228, 234, 299, 546; on neurophysiology, 15, 229f, 342, 405, 438; on nonlinear synchronization, 43, 407, 430; on delayed-feedback systems, 105, 202, 229, 257, 297, 425, 514; on article "How Brains Make Chaos . . . ," 405, 409, 438; on therapeutic applications, 464; on information processing of brain, 408
French, B. M.: on "astrogeology," 212
French, L. M., 50; on asteroid Chiron, 82, 145; on charge-coupled device camera, 548
Frey, H., 212; on Precambrian cratering events, 212; on early evolution of Earth and Moon, 260
Friedman, H.: on coupling of solar wind and coronal ejections, 252
Friesen, L. J., 495f; on spin-orbit resonance, 34; on geosynchronous distance, 35, 189; on orbital processes/motions, 132, 415; on natural-Earth satellite-resonance families, 195
Friis-Christensen, E., 562; on weather patterns and solar cycles, 253; on sunspot cycles, 481
Froeschlé, C., 494; on Kirkwood Gaps, 81, 440; on mean motion resonances, 165, 190, 192, 195; on comet-asteroid orbital evolution, 166, 488; on commensurabilities, 438; on Asteroid Belt resonances, 191
Frouzakis, C. E.: on resonance conditions, 150
Frustration, 5, 54, 299, 431, 463, 538, 546; and Solar System, 183; and social landscape, 233–34, 293; definition of, 233
Fryxell, B. A., 491
Fujiwara, A., 50, 170, 197, 216, 263, 490
Fukao, Y., 45; on "hot spots," 45, 229, 240; on discontinuity, 220, 264, 278; on thermal structure of upper mantle, 245, 279
Fuller, M., 131; on production of ocean crust, 131; on correlation with paleomagnetic record, 341, 506
Fundamental matter entities, 12–13; splitting and lumping of, 13; structural units of, 27
FU-Orionis type of T-Tauri stars, 138, 313, 548–50
"Fuzzy boundary," 2, 144, 151, 168f, 451, 464–66, 468, 472–74
Fuzzy route, 2, 188, 192, 195, 309, 327, 468, 495; discovery by Belbruno of, 2, 465–66

GAD, *see* Geocentric axial dipole
Gaffey, M. J., 50; on origins of asteroids/

Index

meteorites, 434; on parent bodies of meteorites, 216, 487
Gaffin, S., 223, 252, 287, 326; on geomagnetic reversal record, 218, 460; on scaling dimensions, 462; on sea level and paleomagnetic record, 506; on periodicity and paleomagnetic record, 507
Gaffin, S. R., 287; on correlations of natural phenomena, 287, 506; on glaciation and weather patterns, 326, 460; on scaling dimensions, 462
Gaia paradigm, 246, 504
Galactic disk, 151
Galactic plane, 153; and ecliptic plane of Solar System, 153
Galactic tide, 84, 129, 152–54, 164, 178
Galactic year, 245, 257
Galaxy, xli, 14, 137, 155, 170, 248, 418, 436; Solar System in, 2, 136, 183, 238, 305; and black holes, 23; and outer reservoirs, 139; and Oort Cloud, 152, 184; basins of attractions of comets in, 152; "rotation curve," 179; comet formation in, 477; and "Extra-Solar Oort clouds," 478
Galer, S. J. G.: on magma ocean, 267, 277
Galileo: on falling objects, 5
Gallagher, J. S., III, 209; on galaxies/cosmological phenomena, 14, 135; on star formation and supernova, 449, 549
Gallagher, W. B., 209; on selective extinction, 209; on markers and extinction horizons, 287, 511
Gallant, R., 212; on "astrogeology," 212, 507; on impact phenomena, 213; on comets, 454; on shift of spin axis, 562–63
Galle, J. G., 443
Galloway, D. J., 222; on complex flows, 222–24; on fast dynamo, 225, 503; on geomagnetic field, 341–42
Gamow, G., 5; on pendulum, 344; on law of gravity, 396; on calculus, 404
Ganapathy, R.: on cratering and meteoritic material, 213, 262
G & C, see Gould and Calloway (1980)
Ganymede (satellite of Jupiter), xxxiv, xxxv, 8, 484
Garey, M. R.: on challenge of computability, 411
Garfinkel, A., 161; on applications of nonlinear dynamics, 134, 234, 407, 464; on negative feedback, 226; on synchronization, 429–30, 438, 463
Gartner, A. E.: on spatiotemporal scaling, 533f
Garvin, J. B.: on age of Popigai Crater, 426–27

Garwin, L., 265; on collisional origin of Moon, 265; on early geochemistry of Earth, 266; on magma ocean, 520, 523
Gas, structured, 10, 547
Gatlin, L. L., 288; on linguistic complexity, 288; on genetic information/language, 532, 542
Gault, D. E., 198; on capture mechanisms, 123, 169, 210; on antipodal effects, 213, 237, 240; on impacts and craters, 198, 212, 427f, 515, 520; on grazing impacts, 322, 338, 428, 467
Gauss, K. F., 410f; on Ceres orbits, 441f
Gehrels, T., 50; on Aten, Apollo, Amor groups, 139; on "future" discoveries of asteroids, 483; and "Spacewatch," 548
Geldsetzer, H. H. J., 364, 528
Genetic code, xxi, 2, 19, 288, 416, 532, 542
Genetic drift, 510, 546f
Genetics, 16, 293, 542
Geocentric axial dipole, 38f
Geocentric axial dipole–hot spot reference frame model and CRFH–hot spot reference frame model, 46
Geocentric axial dipole hypothesis, 37, 40–41, 43f, 225; and celestial reference frame hypothesis, 42; and TPWPs, 45
Geocentric/near geocentric orbits, 3, 123, 171, 192f, 200, 215, 497
Geocentric reservoir objects, see Reservoirs
Geodesy, 17, 34, 392
Geodynamic feedback, 35, 178
Geographic coordinates, see Celestial reference frame; Celestial sphere
Geographic invariances, 498
Geoid, see Earth
Geological history, 35, 178, 423–24; and principles of organization, 28
Geological time series, 54, 113 (Fig. 35), 290–91
Geologic Column, 3, 286f, 315–16, 508f, 530; correlations with APWPs-TPWPs, 41–42; and nonlinear dynamics, 54; and CFRH, 55–56
Geologic Time Scale, 41, 54, 287f, 315; as multidimensional calibration chart/code, 3–4
Geology, xxi, xxiv, xxxiv, 24, 26, 248, 278, 387, 390, 406, 423f, 492, 508, 540, 542
Geomagnetic, see also entries beginning with Magnetic
Geomagnetic code, 56
Geomagnetic axial dipole, 40
Geomagnetic field, 38, 225, 250, 252, 341;

659

Index

secular variation of, 38; westward drift of, 223, 499ff
Geomagnetic phenomena, 161, 217; suggested references on, 218
Geomagnetic reversal frequencies, 69 (Fig. 6), 70 (Fig. 7), 71 (Fig. 8), 131, 227–29
Geomagnetic reversals, 29, 126, 160, 191, 253, 314, 423; and impact-cratering, 104–5 (Fig. 30), 106–7 (Fig. 31), 227, 242; 150-m.y. record of, 218; and critically fast dynamic model, 225; and thirty-m.y. cycle, 228, 243, 255; and perturbations at core-mantle boundary, 243. *See also* Paleomagnetic reversals
Geomagnetic time scale, xxi, 2, 532; Cretaceous normal superchron of, 228, 232, 242, 244, 253; and Vine-Matthews-Morley hypothesis, 505; and Jaramillo event, 505; and variability of seafloor spreading rates, 505
Geometric progressions, xxxvii, 54, 187f, 190, 290–91, 395, 412, 444, 533, 541; and Solar System, 185, 446, 488
Geometry-from-time series, 508
Geophysical phenomena, 191
Geophysical processes, 59, 234, 236, 283
Geophysicists, 168, 174, 401, 408
Geosynchronous distance, 35
Geothermal energy, x
Gerola, H., 14, 135, 209, 449, 549
Gerstenkorn, H., 92; on Roche limit, 92; on tidal evolution, 477
Gestalt, 389, 407; and "scientific method," 404
GFBP, *see* Great Fireball Procession
Ghil, M., 130; on glaciation cycles, 130, 266, 326, 536; irradiances and solar cycles, 238, 253
"Ghost Binary," 73 (Fig. 9b Inset), 75, 76–77 (Fig. 10a), 137, 344, 346 (App. 7), 346–48, 549
Ghyka, M.: on golden mean in art and nature, 397–98
Giant comets, hypothesis of, 129–30, 151, 167, 177, 326
"Giant piston core," 214
Gibbons, A., 510; on "microscopic" phenomena, 286
Gierasch, P. J., 222; on Jupiter's atmosphere, 222, 238; on Great Red Spot, 395, 499
Gilbert, A. D., 4, 236; on complex flows, 222–24; on fast dynamo, 225, 503; on geomagnetic field, 341–42
Gilbert, G. K., 212; pioneering studies by, 212; on "random" source of impactors, 322

Gillespie, J. H., 541, 544, 547; on models of molecular genetics, 16, 539, 546; on neutral drift, 293, 538
Gilmour, I., 209, 262; on diamonds at K/T boundary, 48–49, 210, 414; on "wildfires" at K/T boundary, 207–8
Giovanelli, R., 238; on Jupiter's atmosphere, 238; on solar flares, 254, 393; his illustrations of cloud structures, 392
Glaciation, 130, 229, 326, 527, 536
Glacier-ocean-atmosphere-magnetosphere, 53, 163, 238–39, 243, 252, 282; and patterns of weather and climate, 201
Glass, B. P., 326, 507; on tektites, 99, 262; on mass impacts and paleomagnetic reversals, 506
Glass, L., 234, 464
Gleick, J., xxxii, 461; on Butterfly Effect and strange attractor, xxxiv, 233, 407, 517; his historical and popular review, 12, 134, 387, 420; on Great Red Spot, 222
Glen, J., 42; on correlation of APWP cusps and Geologic Column, 42; on VGP patterns, 101 (Fig. 26), 227
Glen, W., xxi, 36, 49, 121, 387, 389f, 404, 407, 526, 532; his influence on H. R. Shaw, xi-xii, xxxiii, 506f; his study of debates on mass extinction, xi, xxiii-xxiv, 285, 507, 509; on plate tectonics, xxiii, 240; on peer-review in science, xxvi, 392, 412, 445; on mass extinction hypothesis, xxxiii; on history of IEH and related viewpoints, xxxiii, 57, 202, 457, 509; on Jaramillo event, 505; on gestalt in evolutionary biology, 537
"Glen Line," 49, 121
Gliese, W.: on nearest star, 477
Global magma-tectonic oscillator, 229
GLORIA (Geologic LOng Range Inclined Asdic) survey, 122, 215; of U.S. Exclusive Economic Zones in Pacific Ocean, 214, 312
GOAM, *see* Glacier-ocean-atmosphere-magnestophere
Gödel, K., xxv; and principle of self-reference, xxv, xxxvi; Incomplete Theorem of, xxv, xxvi, xxviii
Gödelian arrays, 29
Gödelian state vs random state, 26
Goedecke, G. H.: on Eötvös force, 281
Gold, T., 40, 367; on Gold's "beetle," 44; on Gold's conjecture, 45; on Earth's axial moments of inertia, 266
Goldberger, A. L., 29; on applications to neurophysiology, 134, 209, 234; on therapeutic applications, 464
Golden mean ($), 142, 316, 396, 401, 432,

Index

444, 497; and pi (π), 142, 398–99; and inverse of, 161f, 184, 504; and method of continued fractions, 397; definition of, 397. *See also* Critical golden-mean nonlinearity "Golden Spike," 287

Goldreich, P., 477; on polar wander, 37, 40, 367; on Gold's "beetle," 44; on Earth's axial moments of inertia, 266; on mantle reference frame, 276

Gollub, J. P., 134, 387, 389, 391; his tutorial on nonlinear dynamics, 29–30; on equation of pendulum, 60 (Table 1), 86, 473; on "routes to turbulence," 108, 502, 541; on period-doubling route, 386, 541; on phase space, 396, 422; on chaotic dynamics, 420; on winding numbers, 432; on universality parameters, 436

Goodfellow, W. D., 55, 284, 292, 304, 334, 338, 364, 507; on Frasnian-Famennian extinction, 33, 58, 66, 124, 297, 427; on impact-extinction hypothesis, 57, 285

Goodwin, B. C., 302; on macromorphogenetic field, 302, 543; on micro- and macroevolution, 543–44

Gordon, R. G., 37, 211; on TPW, 40–41, 367; on cusps, 42; on geomagnetic studies and pole positions, 48, 278; on plate motions, 273, 281, 501

Gorman, M., 104; on chaotic bursting, 104, 160–61; on reversing-flow systems, 223–24

Gostin, V. A.: on Acraman Crater, 62, 119, 128

Gould, S. J., 7, 29, 55, 209, 305, 404, 550; on Darwin, 16, 294, 538; on "perfect determinism," 18; on early Cambrian fauna, 28, 282, 537; on clam and brachiopod diversity, 114, 208, 296–99, 537; on punctuated equilibria, 237, 304; on evolution, 285–86, 295, 543–44; on Permian-Triassic transition, 289; on biological clocks, 293; on Signor-Lipps effect, 510; on extinction horizon, 511

Gould and Calloway (1980 publication), 55, 208, 296–99, 537

Gould's Belt, 479

GPC-3, *see* "Giant piston core"

Gradie, J. C., 50; on Aten and Amor groups, 144

Graham, J.: on protist-protoctist evolution, 547

Graham, R. L., 29

Grand unified theory, 452

Grant, B. R., 541; on "common ancestor," 540–41; on hybridization, 547

Grant, P. R., 541; on "common ancestor," 540–41; on hybridization, 547

Grantz, A., 428; on Avak structure, 64, 122, 261, 328, 334, 338, 360; on low-incidence angle, 123

Grassberger, P., 163; on attractors in weather, 163; on dynamical dimensions, 459–60, 462

Gravitational lenses, xxiv

Gravitational potential wells, 2, 4, 159, 169, 181, 468, 472, 485

Gravitational sphere of influence, 168, 466

Gravity anomalies, 34, 78, 112

Gray, D. R., 36, 264, 376, 526

Grazing impacts, 3, 169, 308, 322, 338, 467

G-R Catalog, *see* Grieve, R. A. F.

Great Fireball Procession, xliii, 91, 206–7, 333–34, 336–38, 428, 467, 557–59

Great Red Spot, 47, 226–27, 395; stationarity of, 221; vortex structures in, 222; stroboscopic image, 222, 499f

Grebogi, C., 139, 157ff, 160f, 232, 421, 457ff, 474, 503

Greeley, R., xxxiv, 520; on tidal deformation, xxxv–xxxvi, 401; on magmatic consequences of impacts, 241–42; on tidal dissipation and magmatism, 476–77; on magma as irreversible process, 517

Green, D. H., 212, 259f

Green, H. W., II: on deep-focus earthquakes, 515, 525

Greenberg, J. H.: on genetic/linguistic evolution, 209, 286

Greenberg, R., 2, 9, 169, 173, 263, 309, 428, 436, 454, 474f, 548; on delivery of impactors, 11; on Kirkwood Gaps, 81, 440; on Earth-crossing orbits, 82, 226; on mean motion resonances, 165, 190, 192, 195; on Asteroid Belt resonances, 191; on commensurabilities, 438; on orbital delivery of asteroids, 487; on asteroid-comet evolution, 488

Grew, P. C.: on extinction, 303

Grieve, R. A. F., 30, 49, 55, 62, 64, 66, 121, 241, 257, 260; his G-R Catalog, 30, 55, 118f, 122–28 *passim*, 213, 335, 337, 340, 359, 369, 379; on Precambrian impacts, 212; on age of Beaverhead structure, 334; on age of Popigai structure, 426

Grimm, R. E.: on radiogenic heating, 548

Grolier, M. J., 119, 213; on bibliography on impact structures, 119; his compendium on impact sites, 213

Grommé, C. S., 281; on history of Arctic

Index

Ocean, 281; on paleogeographic reconstruction, 376
Gross, R. S., 243, 259, 482; on Chandler wobble, 519, 562
Grossman, J. N.: on origin of chrondridic meteorites, 434, 436
GRS, *see* Great Red Spot
Gruntfest, I. J., 352, 374, 509; on Earth's tides, xi, xxxv–xxxvi, 250, 401, 476–77, 501; on material science/rheology, xi, xxii, xxxi, 92, 387; on exponential divergence, xxiv; on thermomechanical feedback, xxiv, xxxiv, 27, 502; on computer experiments in feedback systems, 385, 408–9; on thermal feedback, 413; on geological recurrence patterns, 508; on magma as irreversible process, 517
Gu, Y., xxvi, 157, 407, 429, 438, 463; on computer experiments, xxvi; on phase space, 157, 429; on synchronization, 407, 438; on logistic equations, 463
Gubbins, D., 223, 249, 502, 506; on geomagnetism, 31, 103, 342, 422, 499; on flux lobes, 102, 219, 220–22, 226, 271; on virtual geomagnetic poles, 218, 422; on structure of core, 222; on perturbation of core, 231; on thermal convection and magnetic field, 250
Guckenheimer, J.: text on linear dynamics, 12, 134
Gudlaugsson, S. T.: on impact sites, 213
Guidance mechanisms, 49, 166, 191, 194
Gusev structure, Russia, 121–23
GUT, *see* Grand unified theory
Gwinn, E. G., 139, 158, 161f, 397; on pendulum, 86, 145, 157, 239, 389, 396, 439, 462, 473; on intermittency, 138, 456f, 510; on attractor-basin boundaries, 144, 474; on singularity spectra, 188, 504; on critical self-organization, 225, 421; on electronic transport, 400, 403; on chaotic/interior crises, 421, 457f, 503; on numerical dynamics, 431

Haase, R.: on Kepler, 142
Hackman, R. J., 3
Hagee, V. L., 506; on core and core dynamo, 222f, 230–31, 243; on thermal plumes, 275; on geomagnetic fields, 342
Hager, B. H.: on mass anomalies/geoid and satellite orbits, 34, 178
Hagstrum, J. T., 39, 42, 212, 520; on flood basalts antipodal to impacts, 213, 237, 260, 358, 361, 514; on impacts and magmatic propagation, 240f; on possible crater sites, 312; on "hot spots," 352, 374; on APWPs, 364, 561; on craters filled with lava, 379
Hagstrum-Turrin model, 213ff, 217
Hairpins, 39, 42, 56
Hajduk, A.: on decaying comets, 494
Haken, H., 12; on slaved systems/mechanisms, 223, 386, 421, 546
Hallam, A., 507; on endogenous cause for extinction, 57, 285; on extinctions across boundaries, 207; on multiple end-K extinctions, 513
Halley's Comet, 82–83
Halliday, D., 405; physics texts of, 405, 470; on dynamics and mechanics, 406; on why sky is blue, 410; on "Maxwell's Equations," 411; on energies of mechanical systems, 471
Halsey, T. C., 161; on sine-circle function/equation/map, 161f, 397, 461; on singularity/analysis spectrum, 184, 188, 287, 504; on chaotic dynamics, 502
Hamilton, W. D., 209; on evolutionary cooperation, 303, 463, 537
Hamiltonian models/systems, 134–35, 275, 390, 394, 406, 420, 484, 503, 538; as conservative gestalt, 389, 404, 527; conservative-energy equations in, 471
Hanneman, D. L.: on Cenozoic stratigraphy, 287, 511
Hao, B.-L., 134; on anthologies on linear dynamics, 134, 420; on phase space, 157
Harbert, W., 281; on Arctic geologic and paleomagnetic data, 281; on paleogeographic reconstruction, 376
Hardy, G. H.: on irrational numbers, 186, 397
Hargittai, I.: on symmetry, 288
Hargraves, R. B., 213; on true polar wander, 40; on paleogeographic reconstruction, 112, 279, 281; on Beaverhead structure, 334
Harland, W. B., 30, 55, 63, 120, 229, 282f, 286, 364, 514, 528; on fossil record, 3; on Geologic Time Scale, 41, 54, 114, 287, 296, 508; on age of Frasnian-Famennian, 56, 58, 66, 288, 529; on polarity chrons, 225, 423; on polarity events, 227–28, 255, 279
Harper, C. L., Jr., 249; on collisional origin of Moon, 249; on core formation, 251; on 4.5 Ga differentiation of mantle, 260; on geochemical history of Earth, 266f
Harris, A. W., 175; on discoveries of asteroids, 175, 547; on deflection of impactors, 303; on 1991 BA, 307; on near-Earth

Index

objects, 310, 495; on asteroid "light curves," 548
Harrison, C. G. A.: on hot-spot reference frame, 506
Harrison, E. R.: on "blazing sky" effect, 447–49
Harry, D. L.: on Serra Geral Province, 66
Hart, R. B., 36; on plate tectonics, 36; on Euler poles, 39, 211
Hart, S. R.: on model of mantle melting, 525
Hartmann, L.: on FU-Orionis type stars, 138, 433, 548
Hartmann, W. K., 53; on collisional origin of Moon, 53, 138, 252, 263, 265, 325, 489; on cratering, 212, 260
Hartnady, C. J. H., 37; on continental drift, 36–37; on cratering, 213; on paleogeographic reconstruction, 264, 278, 376, 526
Harvey, P. H., 538f
Hartung, J. B.: on Manson Crater, 122
Hassell, M. P., 209; on biological population dynamics, 405; on evolution of cooperation, 463
Haucke, H.: on quasiperiodic route, 387
Haughton (crater), N.T., Canada, 328
Hawaii, xxxi; ponded basaltic lavas of, xxxi, 524; and "hot spots," 46; crustal foundering of lava, 520
Hawaiian-Emperor Bend, 122
Hawaiian-Emperor volcanism, 241, 255, 534–37
Hawaiian mantle-melting anomaly, 214
Hawaiian Volcano Observatory, x–xi, 275
Head, J. W.: on impacts/"hot spots" on other planets, 242, 260, 520
Heat flow, 46, 506
Heat transfer, 522, 524
H-E Bend, *see* Hawaiian-Emperor Bend,
HED meteorites, 216
Heezen, B. P., 262, 506f
Heisler, J., 129; on comet flux and comet showers, 87–88 (Fig. 16), 154; on outer cometary reservoirs and perturbations, 129, 136, 139–40, 151–52, 177; on Monte Carlo simulations, 167, 437
Held, G. A., 141; on sandpile model, 141; on self-organized criticality, 161
Held, I. M., 130; on glacial cycles, 130, 266, 536; on glaciation and bolide impact events, 326
Heliker, C.: on Hawaiian volcanoes, 261, 275
Helleman, R. H. G., 157; on dissipative phase space/systems, 157, 389; on period-doubling route, 386–87; on education and nonlinear dynamics, 390, 406; on universality parameters, 396; on nonintegrable forms, 408; on standard map, 562
Helz, R. T.: on lava lakes, 521
Hemmer, P. C.: on winding numbers, 181f, 432, 440
Henbest, N.: on Titius-Bode relation, 133–34
Henderson, M. E., 460; on coupled nonlinear oscillations, 145; on pendulum, 157, 239
Henkel, H.: on cratering in Fennoscandia, 213, 379
Hennig, W.: on dynamical bauplans, 509
Hénon, M., 159; on "celestial pin ball," 159; on chaotic scattering, 474
Herbert, F., 313; on FU-Orionis type stars, 313; on "differentiated" meteors, 548
Herbig, G. H., 433; on planetary system formation, 433; on binaries, 434
Herman, Y., 200; on duration of K/T event, 200, 210; on extinction horizon, 511
Hernquist, L., 14, 135, 448, 453
Herrick, S.: on asteroid orbital evolution, 488
Herschel, W., 442
Herzel, H.: on solar flares and oscillations, 162–63, 460, 462
Hess, H. H.: on "hot-spot tracks," 240
Hess Rise, 64, 122, 214f
Hewins, R. H.: on origin of asteroids, 434, 490
Hewitt, G. M., 209
Hi, J. F.: on Eötvös force, 281
Hide, R., 482; on torque in Solar System, 481–82; on Earth's rotation rate, 519; on ocean-atmosphere torque, 562
Higgins, M. on evidence of cratering, 213
Hilbert, D.: on his twenty-three problems, xxv–xxvi
Hildas asteroids, 150
Hildebrand, A. R., 213; on Chicxulub Crater, 64, 119, 122, 128, 260f, 334; on Caribbean impact site, 212
Hillhouse, J. W., 281, 376; on history of Arctic Ocean, 281; on paleogeographic reconstruction, 376
Hills, J. G., 138–39, 177, 180; on calculations of comets, 88, 154; on "new" comets, 454; on "missing" planet, 489; on comet showers/"Nemesis," 550
Hills Cloud, 129, 151, 167, 182, 190, 313, 418, 454, 479, 493
Hinderer, J., 482, 518, 562
Hiroshima, Japan, nuclear device at, 308
History of science, xxi, xxii, 410; uncertainty-of-precedence principle, xxv, 392, 539; fortuitous discoveries in, 418, 445

663

Index

HMMA, *see* Hawaiian mantle-melting anomaly

Hodge, P. W.: source books on comets/galaxies, 454

Hodges, A., xxvi; on computability of undecidable propositions, 411

Hodych, J. P., 33; on possible Triassic/Jurassic impact, 33; on age of Manicouagan, 121; on end-Triassic extinctions, 364, 427, 513

Hoecksema, J. T., 252; on coronal mass ejections, 252, 257; on magnetospheric substorms, 254

Hoff, B.: on Pooh-Tao, xxx, 551

Hoffman, A.: on cause-effect and IEH, 457

Hoffman, K. A., 218; on virtual geomagnetic pole patterns, 218f, 271; on geomagnetic fields, 341–42; on dynamo families, 422

Hoffman, P. F., 36; on "accretionary continents," 36; on paleogeographic reconstruction, 264, 278, 376; on Pangaea, 526

Hoffmann, M., 50, 396, 488, 493; on asteroids, 82, 154, 170, 487, 494; on origins of asteroids, 434, 454; on particle aggregation in space, 170

Hofstadter, D. R., xxv, xxvi, xxix, xxxii, xxxvi, xxxviii, 14, 29, 385, 409, 420; his *Gödel Escher Bach: an Eternal Golden Braid*, xxviii, 316; on Endlessly Rising Canon, xxxviii, 316; on "Strange Loop," xxviii, xxx, xxxvi, 18, 230, 304, 316, 505, 509, 539; on fundamental and irreducible mechanisms, 12; on period-doubling route, 386–87, 413, 541; on theories of chaos, 391; on bifurcation, 411; on logistic map, 429

Hohmann, W., 465

Holcomb, R. T., 64; on Exclusive Economic Zones, 64, 214, 312; on Hawaiian volcanoes, 261, 275

Holden, A. V., 134; source book on linear dynamics of, 134, 420; on medical application, 234; on dynamical diseases, 464

Holden, J. C.: on paleogeographic reconstruction, 112, 367, 376

Holland, H. D.: source book on comets, 454

Hölldobler, B., 405; on "chain of command," 405; on synergy in evolution, 537; on colonial insect communities, 547

Holle, A.: on chaotic hydrogen atom, 381

Holmes, P.: texts on linear dynamics, 12, 134

Holmes, P. J., 108; on magnetic field, 232, 252, 341

Holser, W. T.: on impact-extinction debates, 285

Honda, S., 275; on models of mantle convection/"avalanches," 275, 422, 525; on phase-change instabilities, 515

Hones, E. W., Jr.: on model of Earth's magnetotail, 462

Hoover, W. G., 10; on "structured gas," 10, 546–47; on kinetic theory of gases, 21; on "celestial pin ball," 159

Hora, H.: on expanding Earth hypothesis, 492

Horn, L. J., 2; on satellites of planets, 2, 263, 475; on satellitic rings, 9, 173

Horton, C. W., Jr., 134, 213, 389f, 461

Hoshi, M., 106, 342; on Rikitake two-disk dynamo, 108 (Fig. 32), 219, 232

Host systems, 24–25; and concept of alien entities, 24; and self-organization, 25

Hot spot reference frame, 45ff, 48, 59, 217, 500f, 506, 520

"Hot spots," 29, 43–44, 45f, 229, 240, 242, 275, 352–56 (App. 12–15), 362, 375 (App. 25); on Io, xxxv, 204, 325; definition of, xxxvi; invariant angular relationship of, 44; antipodal to impact sites, 203, 213, 215, 227, 244, 357–63 (App. 16–18), 514, 519–20; at Reunion, 214; and paleomagnetic reversals, 506

Hourglass, 1, 21, 140–41, 148f, 161; model, 2, 78–79 (Fig. 11), 140, 216, 289, 504; avalanches in, 231

House, M. R., 53, 287

Howard, K. A., 265f

Howell, D. G., 264, 526; on "suspect terranes," 36f, 376, 527

HSRF, *see* Hot spot reference frame

Hsü, K. J., 48; on iridium anomalies, 48; on impact loci, 210

Hubbert, M. K., 198, 502

Hubble, E. P., 451; on cosmological indeterminancy, 451; his *Realm of the Nebulae* (1936), 533

Hubble's law, 451

Huffman, A. R., 49, 58, 99, 262

Hughes, D. W., 17, 392, 445, 486

Hughes, H. G., 213; on antipodal effects of impacts, 213, 237, 240, 520; on influence of impacts on core, 515, 520

Hughes, T. J., 376, 492; on glaciation, 266, 326, 338; on history of Arctic Ocean, 281

Hungaria group (asteroid), 149, 487

Hunt, E. R., 223, 459; on transient bursting phenomena, 88, 161; on synchronization, 190, 463; on negative feedback, 224, 226

Hunter, D. A., 209; on galaxies/cosmological phenomena, 14, 135; on star formation, 449; on supernova, 549

Index

Huntley, H. E.: on golden mean, 398
Huss, G. R.: on meteorites, 490
Hut, P., 169; on K/T multiple impacts, 49, 121, 200, 210; on "comet showers," 124; on K/T multiple extinction events, 207, 427, 513; on Nemesis hypothesis, 138, 550
Hutchinson, D. R., 62; on flood basalt provinces, 62, 66; on correlations between impacts and volcanism, 240; on rifting and impacts, 241
Hutchinson, G. E., 395; on rapid sedimentation rates, 395; on E. Sewell, 412, 433, 531; address by, 433
Hutton, J., 285, 317; on uniformitarianism, 6, 56, 423–25, 540
Hyde, R., 481–82; on rotation speed of Earth, 259, 516, 519, 562

Iapetus (satellite of Saturn), 8
IEH, *see* Impact-extinction hypothesis
Iidaka, T.: on deep-focus earthquakes, 515, 525
Immune system, 24–25
Immunological evolution, 208
Impact cratering, 130, 178, 201, 204–6, 429, 529; large-body and related scars, 3, 128; records of, 10f, 29, 49, 126–29, 258, 271, 325; database of, 30–31; grouping by ages of, 30–31, 33, 63 (Fig. 2), 64–65 (Fig. 3), 120, 312, 336 (App. 2); and discoveries of new craters, 33, 55; cumulative energy of, 35, 46, 252, 257–59; history of, 49; multiple events, intermittency of, 49; from CRF reservoir, 52; and magma production/processes, 52, 239f, 244, 341 (App. 5); and geomagnetic reversals, 104–5 (Fig. 30), 227–29, 341 (App. 5); low to grazing angle of incidence, 123, 206, 428; ratio of potential to documented, 127–28; and nonlinear resonances, 133; dynamics of crater-forming process, 197; and rifting, 211, 241, 280; potential oceanic sites of, 211, 214–15; and magnetic flux lobe patterns, 220–21; thirty-m.y. cycle, 228, 243f, 255; and magmatic-tectonic-geomagnetic interactions, 242; and moments of inertia, 243; and dissipated energies, 244, 514; ratio between Earth and Moon of, 258; comparative histories of Earth and Moon, 319. *See also* Impact cratering patterns; Impact craters *under* Moon; K/T cratering swath
Impact cratering patterns, 16, 32, 36, 48, 57, 59, 100, 112, 118, 194, 198, 200, 217f, 315, 370–72, 427–28; and spatiotemporal records of, 3, 29f, 52; on planets, satellites, and larger asteroids, 8; on Io, 11; on Mercury, 11, 166, 204, 350 (App. 10), 375 (App. 25), 525; on Moon, 11, 29, 49, 116–17 (Fig. 37), 130, 166, 204, 215, 258, 259, 260, 265f, 319–26, 375 (App. 25), 490, 525; sizes of, 49, 197; dominance in Northern Hemisphere of, 58, 66, 123, 206, 215, 340 (App. 4); as evolving, 178; invariances of, 217, 497; symmetry in, 220–21; and patterns of "hot spots," 362, 519–20
Impact dynamics, 47; and geodynamics, coupling between, xxxvi, 42, 237, 314; and tidal dynamics, coupling between, xxxvi, 130; and terrestrial dynamics, 231; and biological dynamics and geodynamics, 284
Impact-extinction hypothesis, 56, 209, 237, 311, 483, 492, 540, 543; debates on, xi, xxiv, xxxiii, 167, 201, 212, 285, 293, 454, 457, 507, 509, 543, 550; and opposing volcanists, xi, 201–2, 285, 529; and CRFH, 382, 529
Impactite-tillite hypothesis, 110, 212, 528
Impact phenomena, xxvii, 490; symposia, texts, etc. on, 212–13; and locations of lakes, 338–39 (App. 3b). *See also* Impact cratering; Impact extinction hypothesis; "Impact winter"; Mass impacts
Impacts, *see* Impact cratering; Impact cratering patterns; Mass impacts
"Impact winter," 201, 205, 252; one or several at K/T boundary, 210
Inclination, 153, 194, 196, 199, 215, 263, 467, 471, 487, 497f
Inclined axial dipole, 39, 47; and nonlinear resonances, 41, 47
Incompressibility, 1
Independent systems, 10
"Industrial dynamics," xxii
Ineffable transfinite, xxx, 557
Ingersoll, A. P., 238; on Great Red Spot, 221f, 395, 499f
Inland Whale (book), 7–8
Innanen, K. A., 245; on Galactic year, 245; on comet orbital evolution, 488
Inside-out, 248, 251, 284
Intelligent creature and scaling, 22
Interior crises, 138, 153, 158f, 161ff, 225, 230ff, 233f, 248, 253f, 305–6, 458, 474, 534. *See also* Chaotic crises
Intermittency, xxxii, 36, 158, 177, 190, 217f, 229–31, 233, 239, 242, 244, 256–57, 283, 289, 313, 453, 540; 100,000-year type of, 130; as bistable, 145; of paleomagnetic-reversal phenomena, 191, 203; as used by author, 456–57; Pomeau-Manneville Types

665

Index

II and III, 474. *See also* Choatic intermittency
Introductory references, 29
I-0, *see* Inside-out
Io (satellite of Jupiter), xxxiv, 11, 314, 325, 494; and system of resonances, xxxiv; synchronous rotation relative to Jupiter, xxxiv; volcanism on, xxxiv, 375 (App. 25), 401, 525; Voyager I observations of, xxxiv; and tidal dissipation, xxxv, 8, 519; and interactions between satellites of Jupiter, xxxv, 325; and hot spots, xxxv–xxxvi, 325, 373–74 (App. 24), 519–20
Iooss, G.: on turbulence, 386f, 502
Ip, W.-H., 344, 520; on comets, 50, 132, 488, 494; on planetesimals, 251–52
Ipatov, S. I., 50; on asteroid-comet orbits, 78, 81, 226, 488; on commensurability ratios, 144
Iridium anomalies, 48, 58, 99 (Fig. 24), 203, 210f, 214, 262, 507
Irving, E., 37; on polar wander, 37; on hairpins, 42; on paleogeographic reconstruction, 264, 376, 526
Irving, G. A., 37; on polar wander, 37; on hairpins, 42; on paleogeographic reconstruction, 376
Irwin, A. J.: on stochastic resonance, 155, 437
Ishi, 7–8
Ito, E.: on deep-focus earthquakes, 515, 525
Ito, K., 106, 341; Rikitake two-disk dynamo of, 219, 232; minimum entropy region of, 108–9 (Fig. 32)
Izett, G. A., 81, 206, 427; on tektites, 48–49, 99, 262; on K/T boundary sites, 205

Jablonski, D., 208; on marine bivalve end-K extinctions, 208–9; on extinction horizon, 511; on background extinctions, 544
Jackson, A., 218f, 231, 342, 422
Jackson, A. A., II, 138, 550
Jackson, E. A., 134, 420
Jackson, E. D., 266, 280, 352, 401, 422, 533; on transport of magma, xxii, xxxvi; on plume models, 45f, 275; on hot spot reference frame, 48, 501; on volcano-tectonic code, 56; on magmatic propagation, 240, 242, 250; on Hawaiian-Emperor magmatism, 241, 535–37; on periodicity, 507–8; on magma-related phase-change instability, 516, 524
Jacob, F.: on viruses, 23
Jacobs, J. A., 223, 341, 492, 506, 562; on geomagnetic reversals, 69, 218; on perturbation of core/core-mantle, 231, 239; on properties of core, 249, 251, 502, 516; on spallation, 515
Jacobsen, S. B., 249; on collisional origin of Moon, 249; on core formation, 251; on 4.5 Ga differentiation, 260; on geochemical history of Earth, 266f
Jahreiss, H.: on nearest star, 477
Jakubick, A. T.: on past nuclear reactions, 484
James, G., 397; on golden mean, 397; mathematical definition (glossary) by, 399; on Fibonacci sequences, 400
James, R. C., 397; on golden mean, 397; mathematical definition (glossary) by, 399; on Fibonacci sequences, 400
Jänisjärvi Crater, Russia, 118
Janle, P.: on impact/"hot spots" on other planets, 242, 520
Jansa, L. F., 64, 120
Jantsch, E.: on "autopoiesis," 161, 397
Japanese space program and Moon, 465
Jaramillo event, 505, 514
Jeanloz, R., 40, 223, 239f, 249, 251, 342, 490, 515f
Jeffreys, H., xxxv, 477
Jensen, M. H., 161, 188, 225, 413, 462, 497, 502; on sine-circle map/equation, 60 (Table 1), 162, 181, 397, 461, 503; on Devil's Staircase, 85, 395; on period doubling, 387; on sine-circle function and CGMN, 400; on electronic forms of transport, 403; on singularity spectrum, 504
Jensen, R. V., 134, 158, 181, 452, 461, 471; on quantum chaos and effects, 20, 22, 182, 288; on kicked rotor, 60 (Table 1), 396, 430, 562; on chaotic hydrogen, 381–82; on phase-space map, 386, 420; on dissipative systems, 389, 394, 421; on quantum dynamics, 390, 399; on boundary crises, 458, 473
Jensen, W. B.: on periodic table of elements, 288, 435
Jessberger, E. K., 33, 121, 364, 427
Jewitt, D. C., 2, 9, 173, 263, 309, 436, 475; and 1992 QB1, 492
Jin, R.-S., 37f, 482, 562
Jin, S., 37f, 482, 562
Johnson, D. S.: on challenges of computability, 411
Joly, J.: on pulse of Earth, 507
Jordan, P.: on expanding Earth hypothesis, 492
Jose, P. D., 31, 76, 96, 168, 179, 201, 247, 253, 347, 480f
Judson, H. F., xxi, 19, 288, 387, 416, 532; his *Eighth Day of Creation*, 23

Index

Jupiter, 36, 136, 145f, 148, 150, 173, 189, 191, 440; and Io, xxxiv–xxxv, 8, 314, 325, 401, 494, 519–20; and Europa, xxxiv, xxxv, 314, 494; and Ganymede, xxxiv, xxxv, 8, 484; and interactions between satellites, xxxv; and Shoemaker-Levy 9 Comet, xliii–xliv; atmosphere of, 10, 222, 238, 392; and Great Red Spot, 47, 221–22, 226–27, 395, 499f; and Asteroid Belt, 144, 153, 185, 432; orbital period of, 144, 184, 248; and Trojan asteroids, 150; celestial volume inside orbit of, 171; as relative to Saturn, 184–85, 486; as relative to Pluto, 185–86; gravitational field of, 486

Jurdy, D. M., 37; on polar wander, 37; on plate motions, 273, 501; on "hot spots," 352

Kadanoff, L. P., 2, 140f, 161f, 225, 397, 421, 503
Kaitchuck, R. H., 434, 551
Kaku, M., xxiv, 399, 450ff
Kaluga Crater, Russia, 66
KAM, *see under* Kolmogoroo, A. N.
Kamensk structure, Russia, 121–23
Kanamori, H., 198, 243, 519, 534, 562; on scaling of great earthquakes, 198, 258
Kaneko, K., xxvi, xxxiii, 56, 209, 390, 407, 458, 460, 463; on nonlinearly coupled oscillators, 403; on synchronization, 429, 438
Kaplan, H., 10, 14, 187, 387, 446
Kara Crater, Siberia, 64, 121ff
Karroo field, South Africa, 66, 119, 272
Kasting, J. F., 303
Kauffman, E. G. on extinctions, 207, 511, 513
Kaufmann, W. J., 72, 76, 110, 213, 398, 401, 421, 454, 477, 493, 500, 520, 548; on quasars, xxiv, 418, 421; on collisional origin of Moon, 53, 263, 265, 325; on binary stars, 137, 455; on "Roche" condition, 345; on blazars, 418; on black holes, 419; on "Big-Bang" and age of universe, 451, 453; standard text by, 470, 549
Kaula, W. M., 34, 134, 170, 325, 444, 454, 488; on tides of Earth and Moon, xxxiv–xxxv; as data source on Solar System, 76, 133, 242, 479, 520; on Apollo group, 82, 144f, 487; on equations of motion, 86, 473; on geosynchronous lunar rotation, 116, 326, 476; on commensurabilities, 143; on "new" and "old" comets, 151, 477f; on orbital-rotational balance, 173, 177; on Earth's rotation, 269–70
K. E., *see* Kinetic energies
Keller, G., 213, 262, 511

Kennett, J. P., 506, 537; on "hot spot" volcanism, 506; on Circumpacific volcanism, 537
Kent, D. V., 48, 262, 513; on geomagnetic time scale, 505; on seafloor spreading, 506–7
Kenyon, S. J.: on FU-Orionis type stars, 138, 433, 548
Kepler, J., 5, 143, 447, 452; and numerical regularities in universe, 142, 442
Kepler's Third Law, 181, 185, 189, 342, 345, 469, 488
Kerguelen Island, south Indian Ocean, 215; predicted impact site, 215
Kerr, R. A., 4, 53, 82, 201, 213, 287, 303–13 *passim*, 329, 427, 429, 443, 474, 497; on critical-state processes in Solar System, 2–3, 420; on collisional origin of Moon, 110, 263, 265, 489, 520; multiple impacts at K/T, 121, 204–5, 210; on Chicxulub Crater, 122, 206, 426; on glaciation cycles, 130, 266, 326; on near-Earth objects (NEOs), 310, 547; on Asteroid Belt, 344, 415; on Planet X, 483, 485; on 1991 QB1, 492–93; on Precambrian/Cambrian events, 528, 537
Kerridge, J. F., 491, 548; on meteorites, 50, 435f, 490; on diamonds in meteorites, 210–11, 414; textbooks by, 212–13; on compositional heterogeneity, 262, 414; on planetary evolution, 433, 435
Kiang, T., 50
Kieffer, S. W., xxxiv; on volcanism on Io, xxxiv, xxxv, 373, 401; on geology of meteor impacts, 130, 198, 307–8; on Meteor Crater, Arizona, 307–8; on geology of cratering, 321, 323, 379
Kilauea volcano, Hawaii, 261, 535–36; and fractal geometry, 393; Mauna Ulu episode, 393, 524
Kinematics, 33, 41, 394, 405f, 410, 516
Kinetic energies, 189, 197, 467–68, 470f, 481, 519; in Solar System, 76–77 (Fig. 10), 136, 177, 344–45, 434; of particles at depth inside Earth, 189; of impacts, 241, 244, 259
Kinetic theory of gases, 21, 416
Kinetics, 406, 410
King-Hele, D., 112, 178; on artificial satellites, 17, 34, 270, 486, 495; his *Tapestry of Orbits*, 392, 445
Kinzel, W., 4, 233, 463
Kippenhahn, R., 435, 440; on meteor showers, 165, 167–68, 477; textbooks on comets, 212–13, 454; on Lagrangian points, 474–75

667

Index

Kirby, S. H., 491; on deep-focus earthquakes, 491, 515; on mantle transitions, 491, 525

Kirkwood, D., 50, 454; on "empty" resonances, 81; on commensurabilities, 143; on Kirkwood ratios, 144; on rotation states in Solar System, 189, 345, 500; on asteroid-comet orbital evolution, 487–88

Kirkwood Gaps, 80 (Fig. 12), 82–83 (Fig. 13a), 143ff, 148, 170, 176–77, 185–86, 216, 313, 349, 432, 440

Kirschner, C. E.: on Avak structure, 64, 122–23, 261, 328

Kistler, R. W., 255; on Sierra Nevada plutonism, 255, 535; on orogenic-epeirogenic episodes, 256, 277; on periodic concept in geology, 507–8

Klacka, J., 488

Klein, F. W., 275; on earthquakes and Earth tides, xxxvi; on Earth-Moon system tides, 477; on Hawaiian-Emperor volcanism, 535

Knight, D. M., 288, 381, 382

Knobloch, E., 385f; on bifurcation bubbles, 452

Knoll, A. H., 28, 256, 282, 301, 512, 537

Koeberl, C., 121

Kolmogorov, A. N., 134–35; KAM theorem, 134, 389f

Kolvoord, R. A., 2, 9, 173, 263, 309, 436, 475

Kono, M., 106, 108, 232, 342; on Rikitake two-disk dynamo, 108–9 (Fig. 32), 219

Kot, M., 209

Koyanagi, R. Y., 239; on seismic tremor, 239, 393, 399–400; on volcanotectonic scales, 275, 535

Kresák, L., 50, 145, 488, 494

Krinov, E. L., 309, 490; on Tunguska event, 98, 338, 428–29, 558

Krinsky, V. I., 14, 29, 390, 505; on dissipative feedback systems, 27, 403–4, 413, 450, 491, 539

Kristian, J., xxiv

Kroeber, A. L., 8

Kroeber, T., 7; her *Inland Whale*, 7; her biography of Ishi, 7–8

K/T boundary, 33, 121, 123, 127, 178, 201, 255, 312, 528; extinctions at, 3, 207ff, 296–97, 427; diamonds at, 48, 210, 414; iridium anomalies, 48, 58, 99 (Fig. 24), 210f, 507; references on, 48–49; multiple impact events at, 124, 200, 206, 301, 427; duration of events at, 200; at sites in western North America, 205; at Teapot Dome, WY, 205, 207f; and fossil record of birds, 207, 300; and wildfires, 207f; selective effects in survival of, 208f; and Deccan flood-basalt province, 214; ammonites at, 427

K/T cratering swath, 57–58, 63, 65 (Fig. 3), 67 (Fig. 4), 68–69 (Fig. 5) 120, 122, 124, 199, 214, 279, 312, 329, 334–38, 426f; and Phanerozoic cratering swath, 48, 529; "cloudburst" of impacts, 48; and magnetic flux lobes, 219

Kuiper, G. P., 151, 435, 455, 493

Kuiper Belt, 51, 83f, 86, 133, 150–54, 165, 190, 454, 479, 493; resonances in, 135, 182; reservoir of comets, 313

Kupers, R., 135, 393, 451, 492, 533

Kurths, J.: on solar flare and oscillations, 162–63, 460, 462

Lada, C. J.: on star formation, 433

Lagrangian points, 176, 328, 474–75, 495

Laj, C., 69, 103, 112, 218f, 226, 253, 271, 276, 342, 422, 507; on VGP paths and seismic P-waves, 220

Lake Acraman impact site, South Australia, 119, 128

Lambeck, K., 92, 249, 259, 477, 482, 501, 515f, 518f, 562

Landscape models, 149–50, 233–34, 293, 305, 457, 471f, 544

Landscheidt, T., 168, 345, 562; on sun and center of mass of Solar System, 76, 78, 96, 179, 247, 347, 476, 480–81; on weather and climate, 201, 326, 517–18; on sunspots, 253, 311, 518, 563

Langford, W. F.: on turbulence, 386–87, 502

Language, 3, 22f, 423, 531–33; of nonlinear dynamics, xxxvii, 5, 54, 56, 286; of dynamical uniformitarianism, 293; numerical structure of, 412; logons and cognons in, 531

LaPaz, L., 35, 91, 309, 334, 467

Laplacian ideal, xxxvi, 404, 484

Larimer, J. W.: on chemical elements, 288, 435f

Larson, R. L., 247, 506f; on magma production, 71, 126, 131, 242; on core/core dynamo, 223, 230–31, 243, 249; on thermal plumes, 245, 275

Lasers, xxvi, 13, 223, 419, 421

Laskar, J., 2, 5, 168, 317, 443, 466, 489; on chaotic motions of planets, 20, 96, 132, 183, 481–82; on Solar System as chaotic, 133, 175, 483, 492, 500, 527, 560; on equations of motions, 390, 420, 528

Lassen, K., 562; on weather patterns and solar cycles, 253; on sunspot cycle, 481

Latency in evolution, 509, 513, 530

Index

Lay, T., 239, 264, 278f, 490, 515
LDEF, *see* Long-Duration Exposure Facility
Leake, M. A., 84, 263, 348, 434, 484, 489f
Legend of the Beginning Written in the Earth, 6–7, 550
Le Mouël, J.-L., 37f, 47, 225, 342, 507, 515, 562f
Length-of-day, 174, 185, 243, 259, 481, 519; and chaotic bursts, 175
Length-of-month, 174, 481; and chaotic bursts, 175
Length-time feedback loop, 17
Lenski, R. E.: on mutation and natural selection, 538, 544f
Lerbekmo, J. F., 48, 207, 210
Lerner, E. J., 302, 436f, 447, 449, 453, 492, 540
Levine, A. J., 23
Levine, D. S., xxviii
Levinton, J. S., 28, 209, 537
Lewin, R.: on genetics and linguistics, 209, 286, 541
Li, G.-X., 108, 169, 182, 394, 472
Li, T.-Y.: on "Period Three Implies Chaos," 187, 445–46
Lianza, R. E., 35, 169, 210, 213, 309, 322, 334, 336, 340, 428, 467
Liapunov, A. M.: on divergence, 135, 187, 420
Liapunov functions/numbers, 187–88, 317, 394, 420
Lighthouse model, *see under* Pulsars
Light year, 477
Lim, T. T.: on ring-vortex bifurcation, 395, 412, 541, 543
Lima-de-Faria, A., 29, 237, 246, 288, 505, 542
Lin, E. C. C., 24
Lindblad, B. A., 429, 474; on meteoroid "streams," 9, 50, 81, 170, 201
Lindgren, W., 424
Lindh, T.: on hot-spot reference frame, 506
Lindley, D., 5, 303, 307, 547
Lindström, M., 213, 379
Linear resonance, 430–32, 439
Linear systems, 12; differences between nonlinear and, 12, 14, 430; lack of synergy in, 12; and closed-system conceptions of nature, 13; as nonexistent idealization, 15
Linear thermodynamics, 13
Linguistics, xxii, xxviii, 531; and nonlinear dynamics, 438, 532
Lipps, J. H., 510
Lipschutz, M. E., 50, 75, 129; on meteorites, 215f, 262, 324, 436, 487, 490

Lissauer, J. J., 454; on cratering of Saturn's satellites, 8
"Lithospheric graveyards," 35, 264, 272, 277f
Lithospheric westward drift, 273
Living systems/states, 229, 246, 290, 302, 305, 416, 425, 451, 530, 532, 539
Lobes, *see* Magnetic flux lobe
Lockner, D. A.: on "earthquake lights," 421
Logistic map/equation, 187, 429, 446, 463
Logons and cognons, 531
Long-Duration Exposure Facility, 486
Longitudinal mass anomalies, 3, 34, 51, 173, 178, 180, 184, 201, 314, 324, 476
Lord Kelvin (William Thomson), 448
Lorenz, E. N., xxiv, 11, 108, 233, 391, 404, 439, 460, 462, 502; and computer experiments, xi, xxv, 26, 385–86, 503, 528; on meteorology, xxx–xxxi, 163, 503; on butterfly effect, xxxiv, 387, 517; on sensitive dependence on initial conditions, 26–27, 387, 408; on degrees of freedom, 161, 422–23
Lorenz attractor, 108, 233, 409
Lorenz effect, xxiv
Lorenz equations, 30
Lovelock, J. E., 209, 246, 303, 504
Lowell, P., 482, 492, 494; established Lowell Observatory, 443
Luedke, R. G., 56, 449, 533, 536
Lunar and Planetary Science Conference, 22nd, 121; and papers on K/T boundary, 121
Luo, X., 449; on fractal scaling in universe, 135, 448, 451, 492, 533
Lupishko, D. F.: on asteroid "light-curves," 548
Lyell, C., 545
Lyrae (Vega) (star), 32

Maasch, K. A., 326, 460, 462; on correlations of natural phenomena, 287, 506
MacArthur, R. H., 209
MacDonald, G. J. F., 92; on tidal friction, xxxv, 177, 249–50, 477, 481–82; on rotation of Earth, 259, 501, 516, 518–19, 561–63
Mackey, M. C., 134; on psychology/physiology, 134, 234; on "dynamical diseases," 208–9, 464
Macropaleontologic code, 53
Magma and molten, definitions of, 516–17
Magma/magmatism, x, xxii, xxxi, 47, 218, 245–46, 248, 249, 261, 507, 509; production induced by mass impacts, 239, 240–41, 514, 520; geomagnetic/impact frequen-

669

Index

cies/production of, 71 (Fig. 8), 242; and thirty-m.y. cycle, 243f, 255; transport and seismicity of, 244, 399–400, 403, 420f; production induced by tidal strain energy feedback, 401

Magma ocean, 266f, 525; crustal foundering of, 276f, 520–24; solidification time of, 523

"Magmasphere," 53, 163, 239, 283

Magmatic-tectonic events, 36; and geomagnetic interactions, 242; and celestial reference frame hypothesis (CRFH), 202

Magnetic, *see also entries beginning with* Geomagnetic

Magnetic dipole axis, 37, 42, 419

Magnetic flux lobes, 53, 219, 244, 278, 314, 505; and cratering patterns, 102 (Fig. 27), 103 (Fig. 28), 218, 219–21; and K/T swath, 219; and VGP patterns, 220–21, 264; and Phanerozoic cratering nodes, 271; stationarity of, 499–500

Magnetic-pole historic record, 38

Magnetoelastic ribbon experiment, 88–89 (Fig. 17), 139, 157–60, 459

Main-Belt Asteroids, 137, 147f, 150, 185, 216, 345, 487; orbital frequencies of, 144

Makaopuhi lava lake, Hawaii, x; foundering process of, x–xi, 521–22

Mallet, R.: on Earth's dissipation of energy, 401–2

Mallove, E. F., 2, 420, 483, 489

Mandelbrot, B. B., xxxvi, 73, 391, 393, 408, 412; on self similarity in Solar System, 9; on fractal geometry, 11, 392, 409; on de Vaucouleurs' observations, 135; on "blazing sky effect," 447f; on turbulence, 502; on power-law (fractal) function, 533

Mangan, M. T.: on lava lakes, 521

Manicouagan Crater, Quebec, 121f

Manifolds, 80, 134, 163, 182, 190, 254, 392, 456, 474

Mankinen, E. A., 37f, 225, 342, 507

Mansinha, L.: on earthquakes and Chandler wobble, 518–19, 562

Manson structure, Iowa, 49, 121ff, 206, 427

Mantle, 3, 29, 42, 262, 266, 283, 523f; convection of, 44, 47, 203, 217f, 244f, 247f, 267, 271, 275, 422; instabilities of, 45; reference frames of, 45ff, 48, 276, 516; shear-wave attenuation mapping of, 45; deep-mantle P-wave anomaly, 103, 112, 220, 271–72, 278, 314; as heat source, 239, 247, 249f; "thermal plumes" and/or "hot spots" in, 245, 275, 422; density anomalies in, 264; chemical heterogeneity of, 525; evolution of protomantle, 525. *See also* Core-mantle boundary; Core-mantle coupling

Maran, S. P., xxiv, 8, 76, 78, 137, 238, 345, 418, 421, 441, 449, 454f, 470, 499f, 516, 548f

Marcus, P., 394; on Great Red Spot, 221–22; on Jupiter's atmosphere, 238

Margolis, S. V., 49, 210

Margulis, L., 209, 246, 544, 547; on Gaian-symbiogenetic concepts, 504; on interaction often enhances, 539; on common ancestor in evolution, 540–41; Mars, 137, 150, 313, 520; and Phobos and Deimos, 309, 311; as source of meteoritic material on Earth, 324

Mars-crossing orbits, 144, 146

Marsden, J. E., 134, 392, 406

Marshall, J. R., 110, 212, 261f, 376, 528

Masers, xxiv, xxvi, 223, 419, 421

Mason, S. L., 423f, 540

Mass anomalies, 34f, 40, 78, 110, 179, 221, 259, 262f, 271, 278, 476; and motions of cratonic terranes, 36; and distribution of satellitic resonances, 215. *See also* Longitudinal mass anomalies; Meridional band; Meridional geodetic keel; *see also under* Earth; Moon

Mass extinction, 292f, 550; and mass impacts, 3, 211–12, 303–5, 316, 457; caused by volcanic eruptions, 201; and background extinction, 203, 285, 295, 316, 544; correlation with flood-basalt volcanism, 211–12; and biological evolution, 294; "cure" for, 303. *See also* Impact-extinction hypothesis

Mass impacts, xxvii, 1, 46, 236, 240, 292, 294, 519, 538; correlation with massive extinctions, 3, 301, 303f; spatiotemporal patterning in, 3; and influence on core, 47, 515; and "impact winter," 201, 252; and geological processes, 211–12, 237; and magmatic propagation, 211, 239f; and geomagnetic reversals, 243, 506; energy dissipation of, 250f, 514, 519; and global signatures, 284; and Darwin's organic evolution, 294. *See also* Grazing impacts

Masters, T. G., 45, 229, 240, 245, 264, 278f

Mateo, M., xxiv, 447

Matese, J. J., 488; on Planet X, 483, 492

Mathematical illusions in Gödel's Theorem, xxviii

Mathematical incompleteness, xxv, xxix, 12, 412, 425

Mathematical set theory and self-reference, xxix

Mathematical universality, xxxi, xxxvii
Mathematics, xxi, xxv, 409ff, 439, 441; as handicap, xxxvii; evolution of, 18; tensor, 407
Mather, K. F., 423f, 540
Matter, *see* Fundamental matter entities
Matthews, D. H., 505
Matthews, M. S., 454, 491, 548; on meteorites, 50, 435f, 490; text books by, 212–13; on compositional heterogeneity, 262, 414; on planetary evolution, 433, 435
Matthews, R., 3, 175, 303, 307, 310, 443, 483, 492, 495, 547
Mauger, R. L., 256, 508
Mauna Loa volcano, Hawaii, 535
Maunder, E. W., 253–54, 517
Maxwell, J. C., 411
May, R. M., xxx, 12, 14, 158, 209, 286, 385, 408, 411, 420, 439, 452f, 461, 539; on biological populations, xxxi, 302, 387; on chaotically self-selected number sequences, 10; on standard statistical test, 10, 39; on logistic map/equations, 187, 429, 446, 463; on Sarkowskii's theorem, 187–88; on evolutionary cooperation, 303, 463; on period-doubling route to chaos, 386, 413, 541; his paper on mathematical models, 387–88; on science education, 390, 405f; on metric universality, 396; on time-series data (natural phenomena), 460, 561; on "spatial game" theories, 537
Mayr, E., 542; on Darwin, xxv, 538–39, 545
Mazaud, A., 69, 253, 341f, 507
Mazurs, E. G.: on chemical elements, 288, 435
MBAs, *see* Main-Belt Asteroids
McDonald, C.: on attractor basin boundaries, 144, 169, 182, 457–58, 473–74
McElhinny, M. W., 218, 228f, 341, 518; on polarity superchrons, 514
McFadden, L. A., 82, 144f, 454, 487f, 494
McFadden, P. L., 422, 507; on geomagnetic reversals, 69, 218, 227–28; on magnetic field, 221, 225, 341–42; on core fluid, 222, 224, 502
McIntosh, B. A., 9, 50, 81, 145, 170, 195, 199ff, 226, 429, 474, 488
McKay, D. S., 50, 490
McKenzie, J. A.: on impact loci, 210
McLaren, D. J., 53, 284, 296, 301, 304, 334, 338, 364, 507; on Frasnian-Famennian extinction, 33, 56, 58, 66, 124, 212, 214, 288f, 297, 427, 530; on biostratigraphic record, 55; on causes of extinction, 56, 294, 530; on impact-extinction hypothesis, 57, 285;
on "sudden" events in evolution, 114; on background extinction, 285; his landmark paper, 292, 509
McLaren Line, 66–67 (Fig. 4), 124, 290, 336–38, 528, 529–30; and Taihu Lake structure, China, 338
McLean, D. M., 57, 201, 207, 285, 293, 507
McSween, H. Y., Jr., 434, 436, 490; on radiogenic heating, 548
Mean-motion resonances, 51, 165f, 190–95 *passim*, 215, 309, 311f, 322, 440, 473; of comets, 169; and conditions of alignment, 191
Mean North Pole of whole Earth rotation, 37f
Mebane, A. D., 35; on fire-ball trajectories, 34–35, 309, 467; on Great Fireball Procession, 91, 334, 428, 558–59
Mechanical clock, 13, 19, 560
Mechanics, xxxvii, 134, 390, 405–10 *passim*
Meiss, J. D., 135, 155, 389f, 394, 407, 421, 429, 437, 473, 495
Melosh, H. J., 50, 309, 454, 477, 520, 523; on Tunguska event, 99, 322, 428–29, 467; on impacts, 123, 197–98, 212–13, 324; on collisional origin of Moon, 138, 263, 265
Meneveau, C., 161, 387, 462, 502
MER, *see* Minimum entropy region
Mercury, 9, 137, 313, 484, 520; impact patterns on, 11, 166, 204, 350 (App. 10), 375 (App. 25); collisional model for evolution of, 75, 138, 434, 489; antipodal tectonic effects on, 213; Mariner 10 flyby of, 350; strange density of, 433
Meridional band, 45, 264, 270–71, 272, 277, 279, 314
Meridional geodetic keel, 112, 271–74, 276–81; and distribution of continents, 272, 526–27; and metaphor of sailing vessel, 273–74
Merrill, R. T., 342, 518; on geomagnetic reversals, 69, 218, 227–28; on magnetic field, 221, 225, 341–42; on core fluid, 222, 224, 502
Metamorphic rocks, 392; petrographic fabrics of, 392; as migmatites, 392
Meteor Crater, Arizona, 307–8, 559
Meteorites, 204, 262, 267; on Antarctica, 215; HED type derived from 4 Vesta, 216; parent bodies of, 216, 487, 490; and transfer (from Moon to Earth), 324; diamonds in, 414; source reservoirs of, 414; carbonaceous chondrite type, 414, 434–35, 491; elements in, 435; chemical reactivity in water of, 436

Index

Meteor/meteroid/asteroid streams, 9, 50, 81, 168, 170, 195, 201, 429
Meteoroid impacts, xxvii, 192, 211, 238, 303, 428. *See also* Mass impacts
Meteoroids, 50, 132, 205; sources of, 50, 216; delivery from MBA to Earth, 216; delivery rates, 429
Meteorology, xxx, xxxi
Meteor radiants, 167f, 170, 201, 311
Meteors, 171; ballistic agglomeration, 170
Meteor showers, 165, 167f, 311
Metric universality, 4, 156, 397, 409, 452, 461; definition of, 396. *See also* Universality parameters
Meyer, C., 256; on episodic mineralization, 256; on ore deposition, 319, 424; on spigot method, 424
MGK, *see* Meridional geodetic keel
Microscopic reversibility, 13
Microwave background, 452, 533
Middleton, G. V., 29f, 134
Milani, A., 2, 390, 420, 443, 466; on stable chaos, 5, 104, 180, 185, 197, 546; on chaotic motions of planets, 20, 132–33, 183, 388, 481; on Solar System as chaotic, 175, 483, 489, 492, 560
Milankovich, M., 238; on glaciation cycles, 130
Milankovich-like oscillations, 130
Miles, J., 86; on pendulum, 86, 145, 239, 439, 473; on spherical pendulum, 153, 157, 164, 187, 194
Milgrom, M., 447; on modification of Newton's Laws, 5, 344
Milky Way Galaxy, xxvii, xli, 25, 51, 130, 418, 437, 453, 549; and intergalactic collision, 290
Millis, R. L., 490, 494
Milton, J. G., 234; on "dynamical diseases," 208–9, 464
Mimas (satellite of Saturn), 8
Mind-body system, 464
Minimum entropy region, 108–9 (Fig. 32); definition of, 232–33
Minor planets, *see* Asteroid Belt
Mirror Theorem, 96, 154, 195, 438
Mishina Gora Crater, Russia, 66
"Missing planet," 442, 444, 488f
Mitrovica, J. X., 266, 326, 338
Mittler, J. E.: on mutation and natural selection, 538, 544f
Mode-locked resonance, *see* Phase-locked resonances
Molecular biology, 23; language and semantics of, 23; and immunological research, 24

Molnar, P., 376, 506; on "hot spots," 48, 240, 352; on hot spot reference frame, 48, 501, 520
Moments of inertia, *see under* Earth
Montagnais Crater, North Atlantic, 64, 120, 312
Montanari, A., 285
Monte Carlo simulations, 437; for comets, 84, 86–87, 152, 154–55, 167; for extinctions, 295
Moon, 1f, 51, 116, 284, 495, 507; tides on, xxxv; low-amplitude seismicity in, xxxv; "fuzzy" route to, 2, 465, 468; formation of, 5f, 32, 35, 42, 52, 122, 259, 263, 282; collisional origin of, 11–12, 53, 75, 92, 138, 249, 251, 258f, 263, 265, 277, 313f, 434, 489f, 525, 540; impact craters on, 11, 29, 49, 130, 166, 204, 215, 258ff, 265f, 319–26, 375 (App. 25), 490, 525; geocentric distance, 35; spin-orbit resonance of, 43, 52, 173f, 324, 327, 476; "Orbit-spin" resonance of, 52; mass anomalies, 52, 215, 476; Procellarum Basin on, 52, 110, 261–62, 265, 278; Copernican Period of, 53, 116–17 (Fig. 37), 319f, 326; Pre-Nectarian Period of, 53, 110, 261; tidal bulge of, 173–74; librations of, 174; rotational synchronous resonance of, 174, 177; synodic periods of, 192; antipodal tectonic effects on, 213; basin formations on, 259, 267; Tycho on, 320; Imbrian Period of, 268, 320; Eratosthenian Period of, 320; ratio of craters on Earth/Moon, 258, 320–21; age of, 260; Apollo 15 KREEP basalts of, 269; Copernicus on, 319f; transfer of meteoritic material to Earth, 324; Hayn and Bel'kovich K craters on, 116, 325; selenocentric reservoir, 326–27
Moon, F. C., 389, 459; on routes to chaos, 108, 387; on text on nonlinear dynamics, 134, 407; on phase space, 157, 429–30; on fractal basin boundaries, 169, 182, 457, 472; on synchronization, 190, 429, 438, 463; on negative feedback, 224, 226; on chaotic dynamics, 394, 420
Moore, D. R., 385; on nonlinear feedback computations, 385–86; on route to turbulence, 386–87; on bifurcation bubbles, 452
Moore, J. G., 163, 214, 499
Moore, P., 11, 134, 175, 184, 196, 441ff, 482f, 485, 492, 495
Moores, E. M., 36; on "accretionary" terranes, 36; on paleogeographic reconstruction, 264, 278, 376

Index

Moran, B., 10; on "structured gas," 10, 547; on kinetic theory of gases, 21; on "celestial pin ball," 159
Morell, V., 539, 547
Morfill, G. E., 490
Morgan, W. J., 40, 47; on "hot spots"/thermal plumes, xxxvi, 275, 352, 366; on "hot-spot tracks," 240
Morris, D. E., 326, 342, 514; on correlation of impacts and geomagnetic reversals, 243, 506, 562; on Muller-Morris model, 252
Morris, S. C., 28, 282, 537
Morrison, D., xxxiv, 11, 53, 172, 198, 213, 263, 265, 308, 325, 373, 435, 454, 483, 495, 520, 559
Morrison, F., 12, 15, 385, 389, 397, 412, 439, 559; on mechanical clock, 13
Mosekilde, E., 4, 233, 305, 463
Moser, J., 135
Mullen, M. W., 123, 428; on Avak structure, 64, 122, 261, 328, 334, 338, 360
Müller, N., 118
Muller, R. A., 138, 252, 326, 342, 454, 514, 550; on correlation of impacts and geomagnetic reversals, 243, 506, 562
Muller and Morris model, 252
Multifractal singularity spectrum, 184, 400, 485, 546. *See also* Fractal singularity analysis; Singularity spectrum
Multiple folded feedback, 342
Munk, W. H., xxxv; on tidal friction, xxxv, 177, 249, 477, 481–82; on Earth's rotation, 259, 501, 516, 518–19, 561–63
Murphy, J. B., 37, 264, 376, 526–27
Murray, C. D., 5, 104, 180, 183, 185, 263, 388, 434, 482, 500, 546
Music, 433
Music of the spheres, 142, 317, 452
Mutation, 530, 544–47
Myers, C. R.: on boundary crises, 458, 473

Nagel, E., 409; on Gödel's Theorem, xxv
Nance, R. D., 37, 256, 264, 376, 526–27
Napier, W. M., 84, 140, 209, 213, 454f, 507; on giant comets, 9–10, 130, 151, 167–68, 177, 326; on Galactic source of comets/objects, 78, 129, 136, 477, 493; on Tunguska event, 98, 200, 428, 558; on catastrophic impacts/bolides, 201, 205, 252, 338; on Gould's Belt, 479
Natural algorithm, 395–96
Natural-Earth satellites, 3, 34, 43, 49, 123, 176, 229, 270, 280, 428, 485, 497f, 548, 558; search for, 92, 175, 196, 495–96; reservoir of, 177, 235, 311, 326; spin-orbit resonances of, 309
Natural hazards and disease concept, 25. *See also* Catastrophes; Earthquakes; Volcanism
Natural properties of numbers, 291
Natural selection, xxv, 285, 303, 530, 544–47; and Darwin's metaphor of wedges, 56, 293f, 537–38, 546; and "uncertainty-of-precedence principle," 539
Natural systems/data, 460–64, 504, 532; time series of, 460
Natural vs unnatural processes, 11
Nazarov, M. A., 64, 121
NDCP, *see* Nonlinear-dynamical comparator program
Near-Earth asteroids, 78, 147, 216; and CRFH, 148; discoveries of, 171–72; search for, 171, 195
Near-Earth objects, 3, 43, 78, 145, 172, 229, 270, 307, 309, 311, 323, 328, 475; approach scenarios of, 91 (Fig. 19); of cometary origin, 168; search for, 171, 195f, 485–86, 495–96; reservoir of, 235; and fast-moving objects, 310; deflection of, 310; new discoveries of, 547–48
NEAs, *see* Near-Earth asteroids
Negative feedback, 164, 224, 226, 257, 303, 386, 403, 413, 429, 449, 491, 540, 543
Negentropy, 27
Nelson, S. A., 27; on thermal feedback, 27, 408, 413; on feedback systems computation, 385; on strain-energy feedback, 401
Nemesis hypothesis, 138, 550
NEOs, *see* Near-Earth objects
Neptune, 75, 136, 443, 445, 482f, 485; ring arcs of, 9, 170, 173, 309, 475; six new satellites, 170, 173, 309
Neptune-crossing distances, 152
NES, *see* Natural-Earth satellites
Nese, J. M., 135, 188, 386, 391, 420
Neukum, G., 242, 260, 350, 434, 520
Neurophysiological model, xxvii, xxviii–xxix, 230; and delayed feedback, 229, 257; and nonlinear dynamics, 234, 405, 438
Neville, K., 397; on mystery of number theory, 397; on "music of the spheres," 452
Newell, A. C., 162
Newell, N. D.: on geologic periodicities, 256, 508
Newman, J. R., xxv, 409
Newsom, H. E., 213, 435; on origin of asteroids, 434, 490
Newton, I.: mathematical methods of, 404
Newtonian mechanics, 5, 450
Newton's Laws of Motion, 5, 136, 410f

673

Index

NGCs, *see* Nodal great circles
Nickels, T. B.: on ring-vortex bifurcation, 395, 412, 541, 543
Nicolis, C., 163; on dimension and attractor states, 460, 462; and climate analysis, 163, 508
Nicolis, G., 1, 27, 29, 401; on self-organization, 13–14; on randomness, 18, 21; on dissipative systems, 389, 391, 450; on dimension and attractor states, 460, 462; and climate analysis, 163, 508
Nichols, D. J., 207
Nicola, N., 301
Niehoff, J., 483
Nieto, M. M., 134, 344, 442ff, 446, 454
Nishenko, S. P., 533f
Nobili, A. M., 2, 50, 81, 226, 349, 390, 420, 443, 466, 487; on stable chaos, 5, 104, 180, 185, 197, 546; on chaotic motions of planets, 20, 132–33, 183, 388, 481; on Solar System as chaotic, 175, 483, 489, 492, 560
Nodal great circles, 98 (Fig. 23a), 116, 210, 213, 333–34, 338–40, 353–57, 558; and Phanerozoic cratering nodes, 125, 198
Nolan, M. C., 81, 474, 488, 548; on delivery of impactors/asteroids, 11, 487; on earth-crossing orbits, 82, 226
Nonintegrability, 5, 143, 390
Nonlinear, definition of, 12
Nonlinear cosmic proportion, 398–99
Nonlinear-dynamical comparator program, 495
Nonlinear-dynamical coupling phenomena, 204
Nonlinear dynamics, xxi, xxx, xxxvii, 6, 41, 57, 132, 381, 405, 408, 436, 482, 538, 540; and chaos science, ix, 4; interdisciplinary principles of, xxiv; universal properties of, xxvii–xxviii, 163; and theories of relativity, 5; definition of, 12, 15, 28, 156, 388, 408–10; mystification of, 12, 409; mathematical rigor in, 12; and search for universality parameters, 15; phenomena as increasingly familiar in, 15; surprise in, 15, 409, 438; and recognizability, 16, 28; recursive patterns in, 21; and Geologic Column, 54; useful texts and symposia on, 134; as applied to hourglass models, 140; uncertainty principle of, 144; and quantum dynamics, 182; as applied to psychology and sociology, 233–34; and Darwin, 294; of organic evolution, 303; historical overview, 387; and superstring theories, 399, 450, 547; references on application to diverse disciplines, 407; as applied to motions of planets, 415; references on terminology of, 420–21; and Hutton's uniformitarianism, 424–25; and "literary" sense of phenomena, 438; therapeutic applications of, 464
Nonlinear feedback systems, xxiii, xxxi, 13, 14–15
Nonlinear laser printer (metaphor), 3, 50
Nonlinear resonances, 46, 160, 432, 438, 499, 501, 514; phenomena of, 11, 25, 43; and impacts on Earth, 133, 165; distinction between linear and, 430–32, 439–40; and transfer orbits, 466
Nonlinear systems, xxvi, xxxii, 14–15, 246; differences between linear and, 12, 14; experiental understanding in, 15. *See also* Nonlinear dynamics
North, G. R.: on glaciation, 326, 527
North American lakes, 338–39 (App. 3b)
North Atlantic flood-basalt province, 121
North Geographic Pole, 31f, 58
Northern Hemisphere, 199; K/T impact localities in, 48, 58, 206; dominance in cratering pattern, 58, 69, 123, 199, 206, 215, 324
Notes, *see* Topic of Chapter Notes, xix–xx
Nowak, M. A., 209, 286; on evolutionary cooperation, 303, 463; on "spatial game" theories, 537
Nuclear reactions, 484
Numerical dynamics, 390, 408, 431, 539
Nutation, 501, 518–19, 561

Oberbeck, V. R., 128, 376; on impactite-tillite hypothesis, 110, 212, 528; on craters filled with magma, 260–61, 379; on impact products, 262, 374; on Falkland Plateau, 272, 334, 529
Oceanus Pacificus, 110–11, 262, 265f, 268f, 274, 276, 277–78, 525
Officer, C. B., 57; on endogenous cause of extinctions, 57, 285; on IEH debates, 509
O-I, *see* Outside-in
Okamura, R., 521
O'Keefe, J. D., 198, 201, 205, 212, 252, 262
Olbers, H. W., 441, 445, 453; on "blazing sky effect," 446–47
Olinger, D. J., 161f, 188–89, 225, 254, 287, 302, 387, 397, 400, 403, 458, 461f, 497, 503f, 534f
Olsen, L. F., 209
Olson, P., 245, 249, 275, 342, 506; on core/core dynamo, 222–23, 230–31, 243
Olsson-Steel, D., 50, 81, 129f, 151, 199f, 226, 429, 474, 488; on meteor and debris

Index

streams, 9, 170, 195, 201; on breakup of "giant comets," 167, 177, 326; on planet-crossing asteroids and comets, 176
Onsager reciprocity relations, 13
Oort, J. H., 488, 494; on Oort Cloud, 139, 151, 414, 454; on orbital energy, 469; on "new" comets, 477
Oort Cloud, 124, 129, 133, 140, 150–54, 257, 289, 310, 418, 455, 493; as reservoir of comet-like objects, 51, 167, 177, 190; Hills Cloud in, 129, 190; perturbation of, 138, 151, 177, 234, 245; Monte Carlo simulations of comet flux in, 87–88 (Fig. 16), 167; cometary aggregates in, 170; resonances in, 135, 182, 313; and Galaxy, 152, 184; and planets, 414; thermal processing of comets, 477
Opdyke, N. D., 505
Open-system nonlinear process, 2
Öpik, E. J., 438, 454
Optical illusions, xxviii
Orbital eccentricity, 173, 196, 467, 528; of asteroids, 145f, 437
Orbital energy, 137, 182, 189, 469, 471
Orbital evolution, 4, 56, 139, 145, 164, 174, 177, 194, 200, 203, 206, 315, 444, 472, 491, 497; of Earth-Moon system, 51; asteroid-comet references on, 488
Orbital instabilities, 133, 138, 164
Orbital resonances, 10, 35, 41, 149–50, 156, 168, 176, 183, 419. *See also* Spin-orbit resonances
Orbital stabilities, 150, 178
Orbital synchronization of meteor showers, 165
Orbital system, 51, 437
Orbital transfer, *see* Transfer orbits
Orbital transfer route, 2, 195; and synodic periods, 192
Orbiting-object reservoirs, *see* Reservoirs
Orbits, 152; circular, 152, 173, 194, 196, 467, 496; geocentric, 171, 176, 192f, 466, 497; heliocentric, 171, 192, 199, 466, 496; of satellitic objects, 173; concentric, 181; independent, 181; chaotic, 183, 195, 494; of planets, 183, 415; "strange," 197; selenocentric, 466
Order, viii, 17–19, 181, 286, 421, 446, 482; as nonlinear, 143
Order of magnitude, 398–99
Ore deposition, 319, 424
Organic forms recognition of, 28, 517–18
Organic molecules, 23. *See also* Viruses
Orgel, L. E., 209; on "pansperamia," 305; on genetic code, 542

Originations (biological), 3, 53, 292, 294ff, 316, 427, 512f
Orion (constellation), 130, 549
Orrery, 19f
Orth, C. J., 49, 210, 262
Ortoleva, P., 209, 392
Oscillations, 29, 43, 53, 157f, 226, 234, 239, 256, 342, 400, 432, 462, 540; between entropy and negentropy, 27; of Earth's magnetic field, 160; of chaotic bistable fluid, 104 (Fig. 29), 224; of cratering and geomagnetic reversal frequencies, 228–29; of seismovolcanic signals, 419–20. *See also* Milankovich-like oscillations
Ostlund, S., 161, 387, 400, 503f
Ott, E., 187f, 224, 226, 236, 394, 413
Ottino, J. M., 236, 390, 394, 414
Ouspensky, P. D., xxx
Outer gaseous planets, 136f, 344, 489, 499f, 549
Outside-in, 244, 248, 251f, 284
Ovenden, H. G., 96, 143, 455, 489; and principle of least interaction action, 180
Ovenden, M. W., 29, 74, 140, 142f, 344, 455, 463, 482, 494, 538; his *Bode's Law – Truth or Consequences?*, xlv, 4; and celestial strange attractors, 4; on Newton's Laws of Motion, 5; and Mirror Theorem, 96, 154, 195; and principle of least interaction action, 96, 180, 438, 475; and "missing planet," 489
Overstreet, H. A., 16
Owen, H. G., 376, 492
Owen, T., 414, 436
Oxburgh, E. R., 402

Pachner, J., 385f, 404
Pacific Basin/Ocean, 262, 268, 314; and "ring of fire" and/or subduction zone, 277, 504; search for impact traces, 312. *See also* Oceanus Pacificus
Packard, N. H., xxxii, xxxiii, 460, 540; on geometry-from-time series idea, 508
Padian, K., 207, 209, 237, 513
Pal, P. C., 342; on reversal frequencies, 69, 126, 218
Paleogeographic models, 112 (Fig. 34), 280–81
Paleogeography, 203, 235, 454
Paleomagnetic pole positions, 37, 41, 46f, 225; paleomagnetic record, 38; distinctions between VGPs, CPPEs, and, 47
Paleomagnetic poles of epoch, *see* Characteristic paleomagnetic poles of epoch
Paleomagnetic reversals, 203, 225; correla-

675

Index

tions with other Earth processes, 505; research papers on correlations, 506–7. *See also* Geomagnetic reversals
Palmer, R. G., 4, 233, 463
Pangaea, 279, 367, 529; collision of Laurasia and Gondwana, 526
Panisset, M., 539, 544, 547
Pannella, G., 53, 92, 345, 477; fossil studies of, 174
"Panthalassa," 277
Papaelias, P. M., 140; on antimatter, 436, 455
Paradox of alternative states, 182
Parent-star Genesis, 6
Parmenter, R. H., 22, 182, 288, 382
Parrish, J. M., 209, 367, 376, 527
Parsec, 477
Particle-wave duality, 5
Particulate/particle rings, 9, 309
Partridge, L., 538f
Pasachoff, J. M., xxiv, 8, 35, 76, 83, 133, 136f, 238, 345, 350, 446, 449, 451, 453ff, 470, 477, 485, 494, 500, 507, 548f; on Pluto, 72, 484; source book on astronomy, 213; on decimal scaling, 398; on quasars, 418–19; on pulsars, 441–42; on "blazing sky effect," 447
Pattern periodicities, 138, 461, 486, 517
PCN, *see* Phanerozoic cratering nodes
PCN-CRF, *see* Phanerozoic cratering nodes–celestial reference frame
PCFM, *see* Perfect coin-flipping machine
Peale, S. J., 92, 263, 265, 325, 345, 350, 374, 475, 477, 489, 517; on volcanism on Io, xxxiv
Peat, F. D., xxix, 20
Peck, D. L., 521, 524
Pecora, L. M., 257; on synchronization, 190, 407, 429–30, 438, 463
Peltier, W. R., 130, 253, 266, 326, 338, 515, 525, 536
Pendulum, ix, 21, 60, 86, 145, 148, 239, 389, 396, 439, 458, 473. *See also* Spherical pendulum
Penfield, G. T., 64; on Chicxulub Crater, 119, 122, 128, 261
Perfect clock, 19f
Perfect coin-flipping machine, 20
"Perfect determinism" vs "perfectly random chance," 18
Perfect orrery, 19f
Period-doubling, 187, 394, 445, 447, 461
Period-doubling bifurcations, 256, 386
Period-doubling cascade, 411–12, 445, 541
Period-doubling route to chaos, 386f, 413, 541. *See also* Quasiperiodic route to chaos

Periodic attractors, 46, 156, 158ff, 183, 226, 236, 395, 457, 458; in Asteroid Belt, 146. *See also* Attractors and attractor systems
Periodicities, xxiii, 38, 53, 130, 138, 223, 256, 396, 471, 504, 506ff; in geological record, 186, 444, 508; definition of, 421; and concept of intermittency, 457. *See also* Pattern periodicities
Periodic Table of the Elements, 287, 435
Periodic windows, 182f, 187, 231f, 305, 412
"Period Three Implies Chaos," 187, 445–46
Perkins, P. C., 27; on thermal feedback, 27, 408, 413; on thermal runaway, 401, 491
Permian/Triassic boundary/extinction, 3, 33, 255, 272, 279, 289, 297, 528–29
Perry, W. H., Jr., xxxvi
Pesonen, L. J., 30, 118, 121, 334, 426; on cratering in Fennoscandia, 213, 379
Peterson, I., 2, 12, 134, 234, 436, 464
Phanerozoic cratering nodes, 33, 36, 41, 44, 120, 203, 213, 236, 270, 323, 498f; invariance of, 31f, 34, 264, 280, 315, 500; and paleomagnetic poles, 37; patterns in Northern Hemisphere, 48, 312, 340 (App. 4); search for "fourth," 57f; as "benchmarks," 58; and nodal great circles, 125, 210, 558; and magnetic flux lobes, 271; and continental masses, 279; and spin axis, 280
Phanerozoic cratering nodes–celestial reference frame, 41, 45f, 52; invariance of, 35, 37, 42, 204, 527; coordinate system of, 41
Phanerozoic cratering swath, 34, 57f, 66–67 (Fig. 4), 68–69 (Fig. 5) 124, 214, 236, 334–36, 338; and K/T cratering swath, 48, 529
Phase-locked resonances, 13, 137, 148, 162, 181, 183, 187, 305, 390, 432, 497
Phase portraits, 54, 146, 204, 396, 420, 460; "stroboscobic" type, 223, 459; definition of, 395; in Hawaiian earthquake data, 395
Phase space, 4, 46, 55, 153, 156, 182, 188, 246, 396, 437; and immune system, 24; volume of, 26; manifolds in, 134, 456; as three-dimensional, 146; as dissipative, 157; a.k.a. state space, 204, 422; regions of, 218; diagram of, 287; nonlinear partitions of, 305; fixed points in, 386; return map of, 395, 422; definition of, 422, 430; and degrees of freedom, 460; blind sample of, 461
Phase space maps, definition of, 386
Phase-space trajectories, *see* Trajectories in phase space
Phillips, R. J., 242, 260, 520, 523
Phobos (satellite of Mars), 309, 311
Phoebe (satellite of Saturn), 309
Physics, xxi, 12, 390f, 396, 405f, 409f, 423,

439, 441, 453; and "allowable" problems, 407–8
Pi (π), 396f, 444; and golden mean ($), 398–99
Piazzi, G., 445; on Ceres, 441
Piccaninny Crater, Western Australia, 66
Pierce, J. R., 54
Pietronero, L., 22, 135, 393, 451, 492, 533
Pike, R. J., 242, 260, 350, 520
Pilcher, F., 145, 148, 216, 485
Piper, J. D. A., 37, 376
Planet formation theories, 434, 489–92
Planet O, 443
Planet X, 190, 443, 482f, 485, 494, 550
Planetary resonances, 488–89
Planetary system, 2, 179, 181, 184, 248, 263, 475f; orbital evolution of, 177, 491; and Galactic system, 191
Planetesimals, 9, 25, 74–75, 132, 138, 252, 313, 415, 433; and 1992 QB1, 494; and "pluton" of Stern, 494
Planets, 288; chaotic motions of, 20, 96, 132, 183, 390; discoveries of, 442–43, 482–85; stabilities of, 489–92. *See also* Terrestrial planets; Titius-Bode relationship
"Plant pathology," 204
Plasma, 75, 224, 516. *See also* Cosmic plasmas
Plate motion, 47, 53, 112, 273, 281–82, 315, 520; speeds of polar wander vis-à-vis speeds of, 45
Plate subduction, *see* Subduction
Plate tectonics, xxi, xxiii, 35, 42ff, 48, 217, 240, 250, 264, 314, 402, 424, 501, 525; episodic character of, 36; cratering node obliterated by, 57; inception of rift systems, 241; and MGK, 276; flat-world effect in, 406; and "hot spot" patterns, 520
Pleistocene glaciation cycles, 130
Plumes, 45f, 245, 274–76
Pluto, 9, 135, 180, 189; and Charon, 8, 75, 133, 136–37, 184, 190, 483, 494; discovery of, 11, 184, 443, 445, 482–83, 484–85, 492, 494; as relative to Jupiter, 185–86
Pluton (geologic context), 136
"Plutons" (small planets), 75, 489; objects predicted by Stern, 136f, 190, 484; term coined by Weidenschilling, 136, 494
PMR, *see* Paleomagnetic reversals
PO, *see* Perfect orrery
Poag, C. W., 64, 213
Poe, E. A., 448
Poetry, 438, 531; order and disorder numerical realms of, 437; and S. T. Coleridge, 451–52; and E. Sewell, 410, 433, 437

Poinar, G. O., Jr., 539, 547
Poincaré, H., xxxvi, 26, 387, 391, 404, 412, 439, 482; on mathematical methods of prediction, xxv
Poincaré sections, 80–81, 88–89, 146, 157, 223, 395, 459
Point Barrow structure, Alaska, 49, 64, 122f, 328
Poirier, J. P., 249, 251, 342, 502
Poitou, C.: on period-doubling route, 387
Polaris (star), 31f, 58
Polar wander (paths), 37, 42–43, 59, 131, 204, 225, 505, 562
Polarity reversals, 108 (Fig. 32d), 160, 232, 543
Pollack, J. B., 169, 428, 436
Pool, R., 12, 134, 234, 464
Pope, K. O., 119, 122
Popigai Crater, Sibera, 49, 64, 121f, 206, 328, 426–27
Porco, C. C., 2, 173, 263, 310; on Neptune's satellites, 9, 170, 309, 475
Poreda, R. J., 414; on neon in mantle, 436
Post, E. L., 1, 411
Potential well, 145, 157–58, 164, 181, 473, 476
Power-law functions, 146f, 159, 164, 533
Power-law scaling (relationships), 161f, 188, 231, 286, 302, 399f, 504
Power-law statistics, xxxvii, 23, 260, 447
PPP, *see* Paleomagnetic pole positions
Precambrian basement rocks, 204, 378–79 (App. 27)
Precambrian craters, 33, 62 (Fig. 1), 120, 212, 528; at Sudbury, Canada, 128; at Vredefort, South Africa, 128; possibly at Bohemian Massif, Czech Republic, 379
Precambrian/early Cambrian biota, 28
Precambrian midcontinent rift systems of NA, 241
Precessing inclined axial dipole, 39
Precession, 31–33, 51, 166, 194, 270, 497f
Pre-Nectarian Period of Moon, 53, 261
Prévot, M., 342; on geomagnetic field, 37; on paleomagnetism, 38, 47, 225, 507, 563
Prigogine, I., 1, 27, 29, 209, 237, 246, 381, 390, 401, 403, 413, 460, 491, 505, 539, 562; on linear thermodynamics, 13; on "Thermodynamics of Evolution," 13–14; on nonlinear systems, 14; on randomness, 18, 21; on dissipative systems, 389, 391, 450
Prigogine-Stengers definition, 13
Principle of incompleteness, 21

677

Index

Principle of least interaction action, 96, 180, 438, 475, 491
Principle of self-reference, xxv, 21
Pringle, M. S., 37
Probability, 2; and distribution for orbital transfers, 2; of chance events, 10; and views of nature, 21
Procaccia, I.: on dimensions, 460, 462
Procellarum Basin on Moon, 52, 261–62, 265, 278
Proctor, M. R. E., 222; on complex flows, 222–24; on fast dynamo, 225, 503; on geomagnetic field, 341–42
Prograde motion/orbits, 31, 199
Project SPACETRAP, xliv, 486, 495
Proto-continental masses, 259, 267f, 276, 525
Proto-Earth, 6, 12, 53, 251, 265, 267f, 325; Stage One, 110–12 (Fig. 33), 265–68; Stage Two, 110–12 (Fig. 33), 268–69; and collisions, 520
Protostar (now called Sun), 6
Protostars and Planets III (conference), 138
Proxima Centauri (star), 477
Pulliam, R. J., 41, 274
Pulsars, xxiv, 47, 226f, 238, 400–401, 417, 495; signal from as measure of time, 20; lighthouse model of, 39, 41, 418; periodic to chaotic resonance effects in, 418; and CRFH, 419; discovery of by Jocelyn Bell, 441–42
Pumped systems far from equilibrium, 13, 223, 246, 306, 386, 476, 542
Punctuational dynamics, 237

Quantum chaology, 20
Quantum chaos, 20, 182, 184, 288, 381
Quantum dynamics, xxix, 5, 20, 390, 399; and nonlinear dynamics, 182
Quasars, xxiv, 418f, 421, 451
Quasiperiodic route to chaos, 387, 413, 503

Radioactive-waste isolation, xxii, 261
Rajlich, P., 213, 379
Rampino, M. R., 57, 62, 66, 110, 128, 205, 213, 253, 256, 260ff, 285, 291, 334, 352, 356, 358, 360, 374, 376, 379, 506f, 514, 535; on flood-basalt provinces, 119; on geologic time series, 186; on volcanic "winters," 201; on correlations between impacts, volcanism, etc., 211–12; on "hot spots," 242, 520; on Falkland Plateau, 272, 312, 528–29
Ramsey, W. H., 240; on phase-change instabilities, 240, 490–91, 492, 515; on implosion mechanism, 491, 515, 525

Random gas, 21
Random model, 1, 27–28; definition of, 1
Randomness, 1–2, 19, 29, 415, 417; and order, viii, 18; not "natural state, 16–17; and models of molecular genetics, 16; and relativity, cyclicity, and compressibility, 25; and Gödelian state, 26; statistical properties of patterns in, 188; in mutations, 544–45
Random null model, 16f, 38
Random-number generator, 10, 19. *See also* Perfect coin-flipping machine
Rapp, P. E., 134, 209, 234, 463f
Raup, D. M., xxv, xxxiii, 25, 29, 53, 55, 138, 207, 237, 253, 256, 303f, 454, 510, 543, 550; on periodicity of extinction, 54, 291, 507f; landmark paper, 56; on marine bivalve end-K extinction, 208–9; on background extinction, 295, 544; on impact extinction, 295; on "killing," 301; on extinction horizon, 511
Raup-Sepkoski analysis, 507
Rawal, J. J., 9, 151, 177, 414, 455, 493
Recognition, 23; of living and fossil forms, 28, 517–18
Reconstruction, paleogeographic, 39, 59, 112 (Fig. 34), 264, 272, 278, 280–81, 314, 368, 376–77 (App. 26), 454, 492, 527, 529
Recursive lifespans, 302
Regenerative feedback systems, 449
Renne, P. R., 66, 506
Reservoirs, 1, 49, 51, 196, 201, 210, 235, 312, 325f, 328; of asteroids, 2, 396; of comets, 2, 132f, 140, 144, 152, 165f, 167, 313, 396, 458, 479, 487; of potential impactors, 3, 49, 78 (Fig. 11), 124, 139, 141, 218, 238–39, 314; and celestial reference frame, 50, 210; "filling" and "emptying" in, 50, 176f; references on objects in, 50; storage of meteroids in, 50; and hourglass model, 78–79; transfer types in, 139; disruption of and multiple impacts, 177, 206; delivery of objects to, 178; searches for, 200
Resnick, R., 411, 471; physics texts by, 405, 470; on dynamics and mechanics, 406; on why sky is blue, 410
Resonance ratios, 181, 185–86, 189, 191, 193, 223, 351, 440, 488, 497, 500
Resonances, 9, 30, 34, 36, 42, 48, 51, 150, 168, 170, 197, 203, 222, 259, 263, 407, 467, 495; of planetary satellites, 8; in Solar System, 95–96 (Fig. 21), 173, 182, 184–85, 238, 313, 488; as synchronous, 173f, 177, 186, 189, 327, 429, 488; exterior in Solar System, 186, 190; interior in Solar System, 186, 190; and secular method of analysis,

Index

190; as sideral, 190, 193; as linear, 430–31; as experiential phenomena, 438–39; of planets, 488–89. *See also* Dynamical resonance; Mean-motion resonances; Nonlinear resonances; Phase-locked resonances; Stochastic resonances
Retrograde motion/orbits, 31, 199
Return map, *see* Phase space
Reunion "hot spot," 214
Rhea (satellite of Saturn), 8
Rheology, xi, xxii, xxiv, 189, 198, 402, 410; of basaltic magma, x, xxxi; definition of, xxxi, 406, 421; of mantle instabilities, 45; on Io and Europa, 314
Ricard, Y., 40, 273, 501
Rice, A., 506, 520; on spallation, 240, 515; on impacts and polarity reversals, 562
Richards, D., 22, 150, 181f, 288, 381, 390
Richards, M. A., 35, 48, 264; on true polar wander, 40; on "hot spots," 229, 240, 352, 506; on plate motions, 178
Richardson, D. E., 444
Richardson, G. P., 12, 178, 209, 385
Rifting and impact cratering, 211, 241, 280, 514
Rikitake, T., 219
Rikitake two-disk dynamo, 106–7 (Fig. 31 Inset), 108–9 (Fig. 32), 219, 232
Ring arcs, 9, 170, 173, 309, 313
Ringwood, A. E., 110, 137, 435, 523; on collisional origin of Moon, 92, 138, 249, 251–52, 263, 265, 489; on geochemistry of proto-Earth, 92, 266–67
Rio Cuarto crater field, Argentina, 210, 334, 340
Rio Grande rift, 241
Robbins, K. A., 104, 108, 223f, 232, 502
Roberts, P. H., 390
Robertson, D. S., 17
Robertson, P. B., 30, 62, 64, 66, 118f, 213, 335, 337, 359, 369
Robin, E., 49
Robutel, P., 443, 482, 500, 527
Rocchia, R., 49, 98f, 200, 309, 558
Roche condition/limit, 75, 92, 345
Roddy, D. J., 49, 212
Rogers, R. R., 33, 256, 364, 427, 513
Rollins, R. W., 223, 459; on transient bursting phenomena, 88, 161
Roosen, R. G., 50, 201
Ross, M. N., 9, 173
Rössler, O. E., 411
Rotation and/or revolution, 179, 193, 273, 500f
Rotman, B., 5, 26, 405, 409

Roux, J. C., 224, 237, 412, 420, 435, 551
Roy, A. E., xii, 32, 192, 389, 428, 436, 444, 454, 558; his *Principle of Planetary Claustrophobia*, 4; on orbital mechanics, 91, 342, 464f, 470, 479; and Mirror Theorem, 96, 154, 195, 438; on commensurabilities, 143; on precession, 194, 497; on gravitational sphere of influence, 466; "n-body problem," 469, 485; on Lagrangian points, 474–75; on mirror configurations, 476; on center of mass, 479–80; his *Orbital Motion* (1988), 482
Rubin, V. C., 14, 130, 135, 179, 245, 345, 549
Ruelle, D., 10, 18, 26, 29, 416f, 460, 508, 541; on algorithmic complexity, 1, 396, 415; on strange attractors, 391–92, 541; on comparative chaotic dynamics, 394; on tuning, 395; on "mathmetical understanding," 410–11; on synthetic viewpoint, 415; on turbulence, 502
Ruelle-Takens on strange attractors, 391f
Ruhlen, M.: on genetic/linguistic evolution, 209, 286
Runcorn, S. K., 40, 252, 267ff; on cratering on Moon, 110, 265f, 284, 320, 476; on "true equatorial wander" of Moon, 525
Russell, B., xxv; his *Principia Mathematica*, xxv
Russell, C. T.: on satellites of planets, 2, 9, 173, 263, 475
Russell, D. A., 57, 329, 454, 510f
Russell, S. S., 210, 262, 414, 434f
Ruthen, R., xxvi, 152, 162, 183, 209, 305, 420, 448
Ruzmaikina, T. V., 50
Ryan, M. P., 275
Ryder, G., 266, 268–69, 525

Sääksjärvi Crater, Finland, 118f
Safronov, V. S., 50, 129, 455, 488, 494
Sagan, C., 137, 209, 269, 305, 433, 455; on geochemical history of Earth, 266; on "mineral origins" of life, 416
Sager, W. W., 37, 40, 352, 506
Sampson, J. R., 23f, 541, 547; on genetic code, 288, 532, 542
Sanders, R. H., 5, 344, 447
Sandpile model, 141, 163, 225, 231, 504
Sarkovskii's theorem, 187f
Satellite orbital stabilities, 34, 178. *See also* Natural-Earth satellites; Near-Earth objects
Satellite resonance code, 53
Satellite rings, 9, 173, 313
Satellites of planets, 2, 8–9, 309; and synchronous resonance, 173, 189

679

Index

Sato, H.: on deep-focus earthquakes, 515, 525
Saturn, 136, 145, 443; and Iapetus, 8; and Mimas, 8; and Rhea, 8; and Titan, 8, 484; rings of, 173; as relative to Jupiter, 184–85, 486; and Phoebe, 309
Savage, H. T., 139, 157, 159, 161, 232, 421, 458f, 533
Sawyer, D. S.: on Serra Geral Province, 66
Scaling, 22, 24, 198, 202, 290f, 398–99, 400–401, 419, 437, 458, 461, 533–36, 543; of catastrophes, 53, 301; as universal, 163; of events in geological records, 282
Schaber, G. G., xxxv, 242, 260, 373, 520, 523
Schaffer, W. M., 209
Schelly, Z. A., 21, 141, 413, 460, 462
Schlegel, R., 29, 182
Schmidt, G., 135, 389ff, 421, 429, 437, 471, 473, 503, 562
Schmidt-Kaler, 130, 179, 477, 513, 549
Schmidt, M., 139, 488, 494
Schmitt, R. A., 57, 212
Schneid, B. D., xxxiv, 520; on tidal deformation, xxxv–xxxvi, 401, 476–77; on magma, 241–42, 476–77, 517
Schneider, N. M., xxxiv, xxxv, 373
Schneider, S. E., 5, 14, 135, 344
Schneider, S. H., 205, 208
Scholl, H., 81, 190f
Schopf, T. J. M., 57, 510f
Schramm, D. N., 449; on fractal scaling in universe, 135, 448, 451, 492, 533
Schrödinger, E., 23, 29, 505; on "aperiodic crystal," 19; his *What is Life?*, 19, 416
Schrödinger's cat, 182, 473
Schubert, G., 9, 173, 267, 277, 520
Schultz, P. H., 35, 40, 260, 266, 268f, 280, 309, 334, 336, 340, 350, 454, 509ff; on Mars, 110, 525; on capture mechanisms, 123, 169, 210; on antipodal effects, 213, 237, 240; on impacts, 198, 241, 427f, 515, 520; on grazing impacts, 322, 338, 428, 467
Schwartz, D. P., 534
Science education, 390, 405
"Scientific method," 16, 404
Scientific research, xxii–xxiii, 404; and principles of self-organized chaos, xxvi
Scintillation in astronomical radio sources, 442
Scott, D. W., 196, 270, 486
Scott, E. R. D., 50, 434, 490
Seafloor spreading, 57, 59, 241, 276–77, 281, 505–6
Sea level, 506, 519
Searls, D. B., 286, 288, 438, 532
Sears, D. W. G., 262, 414, 435, 490

Secular resonances mechanisms, 190–91
Seiden, P. E., 14, 135, 209, 449, 549
Seismic tremor, 239, 401; as magma-transport related, 10, 234; in volcanic systems, 162, 399–400, 535
Self-organization, xxvi, xxx, 9, 14, 16, 25, 36, 160, 190, 204, 254, 283, 303, 313, 416–17, 434, 440, 445, 519, 534, 538; in infinite-dimensional systems, 163
Self-organized criticality, xxx, 21, 161, 225, 275, 289, 306, 397, 449, 503, 534; in "shower activity," 140; as open-system nonlinear process, 2; parallelisms between chaotic crises, CGMN, and, 161; in nature, 162; in Solar System, 78–79 (Fig. 11), 183; in convective motions of Earth's core, 218; in stellar processes, 549
Self-organized nonlinear coupled oscillations, 342
Self-organized systems, xxxvii, 151, 201, 313; of extraterrestrial objects, 50; in nature, 162
Self-referential phenomena/system, xxvii, 425; in framework of universe, xxix–xxx, 17; and concepts of length and time, 17; folded-feedback loop in, 17; and computers, 26; in framework of Gödelian arrays, 28–29; "Strange Loop" in, 316
Self similarity, ix, xxx, xxxvii, 9, 163, 180, 203, 256, 275, 288, 290, 401, 419, 450
Sensitive dependence on initial conditions, xxiv, xxix, xxxiv, xxxv, 5, 26–27, 251, 387, 408, 414, 424, 504, 527, 538, 551
Separatrices, 225, 457, 471f
Sepkoski, J. J., Jr., xxxiii, 33, 237, 256, 304, 457; on periodicity of extinctions, 54, 291, 507f; on landmark paper, 56
Sergin, V. Ya., 130, 253, 266, 326
Sewell, E., 410, 531; on numerical structure of language/poetry, 41, 433; on order and disorder, 437; on linguistics, 438
Sexl, H., 5, 405; on Titius-Bode relation, 133–34; "Jupiter is almost a star!," 433, 484; on white dwarf star, 348, 484
Sexl, R., 5, 405; on Titius-Bode relation, 133–34; "Jupiter is almost a star!," 433, 484; on white dwarf star, 348, 484
Seyfert, C. K., 36, 212f, 217, 240f, 280, 319, 374, 376, 506f, 514; on impact epochs, 211, 256
Shafto, M., xxviii, 234, 411
Shannon, C., 54; and information theory, xxxii, 420
Sharpton, V. L., 49, 119, 122, 213, 257, 260, 334, 454, 509ff, 543

Shaw, C. D., 80, 134, 236, 392, 420, 422, 456f, 471, 473; on bifurcation, 134

Shaw, H. R., xxiii, xxv, xxviii, xxx, xxxvi1, 5–66 *passim*, 74–92 *passim*, 104–226 *passim*, 233–57 *passim*, 266–305 *passim*, 316–52 *passim*, 374–449 *passim*, 460–77 *passim*, 491–518 *passim*, 531–60 *passim*; his "Periodic Structure of Natural Record," vii; vita of, ix–xi; on geothermal energy, x; on studies of magma (in Hawaii), x, xi, xxxi, 46, 163, 218, 234, 243–44, 253, 403, 515, 520–24, 534–35; and rheology, xi, xxii, xxxi, 44, 92, 402; on tidal dissipation, xi, xxxiv–xxxv, 250, 258, 401, 476–77; on radioactive-waste isolation, xxii; on thermomechanical feedback, xxiv, 14, 27, 241, 502–3; computer experiments in nonlinear dynamics, xxvi, 385, 408–9; on continental volcanism, xxxi; on hourglass model, 2, 140–41, 231; on seismic tremors, 10f, 56, 162–63, 393, 399–401, 419–20, 535–37; on thermal plumes, 45f, 275; on geological time series, 113 (Fig. 35), 290–91; on cratering history, 129f; on pattern periodicities, 138, 461, 486; on wheel of fortune, 142, 163, 286; on critical golden-mean nonlinearity, 161, 256, 302, 397, 399f, 448, 450, 503; on attractor states, 169; on survival of birds over K/T boundary, 207–8, 512–13; on submarine volcanism, 214; on intermittency, 242, 457–58; computer experiment and Kirkwood Gaps, 349; on bifurcation synapses, 388, 452; on nonlinear resonance, 432; NDCP proposal of, 495

Shaw, R. S., 15, 21, 29, 187f, 236, 413, 420, 456, 460, 462, 508, 540; on model of dripping faucet, xxxii, 141; his paper on strange attractors, xxxii

Sheaffer, Y., 123, 322, 427, 467, 559

Shearer, P. M., 45, 229, 240, 245, 264, 278f

Shear rate, 403; in deformation of solids and liquids, 402, 502

Shear stress, 403; in deformation of solids and liquids, 402, 502

Shear-wave attenuation mapping of Earth, 45

Sheridan, R. E., 231, 506

Shinbrot, T., 463, 546

Shocked quartz, 48, 58, 99 (Fig. 24), 210f, 214, 262, 334

Shoemaker, E. M., 3, 49f, 81, 84, 126, 130, 139, 166f, 170, 175, 198, 258, 308, 321ff, 427, 455, 488, 490, 507, 563; on astronomical/optical search strategies, 11, 485; on completeness of cratering catalog, 127–29; on impactors, 127–28, 130–31, 206; on cometary sources, 128; on impact phenomena, 128, 212, 260, 262; on Trojan asteroids, 150; on K/T boundary sites, 205; as pioneer in "astrogeology," 212; on energy scaling, 257; on Meteor Crater, Arizona, 307, 559

Shoemaker-Levy 9 Comet, xliii–xliv

"Shower activity" of meteors, 7, 201

Shu, F. H.: on star formation, 433

Shull, J. M., 129, 133, 136, 140, 151, 177, 455, 477f, 488, 550

Sibley, C. G., 512–13; on DNA-based phylogeny, 295, 299ff, 511

Sideral periods, 184, 192, 499f

Sidorowich, J. J., xxxiii, 188, 460, 508, 560

Signor, P. W., III, 510

Signor-Lipps effect, 510, 512f

Sigurdsson, H., 49, 256, 262

Sikhote-Alin trajectories, 340, 360

Siljan Crater, Sweden, 66

Silk, J., 449, 533

Silver, L. T., 213, 241, 280, 454, 509ff

Simberloff, D. S., 209, 511

Simonds, C. H. on geology of impacts, 130, 198, 321, 323, 379

Simpson, G. G., 542

Sine-circle function, 13, 30, 149, 161, 163–64, 180, 183, 218, 400, 432, 504, 512

Sine-circle map, xxx, 60 (Table 1), 143, 146, 148, 153, 159, 187, 245, 255, 390, 396, 432, 437, 446, 452, 461f, 503f; and structure of Asteroid Belt, 218

Singer, J., 104, 161, 164, 224, 226, 463

Singularities, definition of, 237

Singularity spectrum, 76, 161–62, 287, 293, 316, 399, 504. *See also* Fractal singularity analysis; Multifractal singularity spectrum

Sirkin, L. A., 36, 212f, 217, 240f, 280, 319, 374, 376, 506f, 514; on impact epochs, 211, 256

Siva, 7, 550

Skarda, C. A., xxvii, xxix, xxxvi, 15, 134, 209, 234, 299, 408, 464, 546; on neurophysiology, 229f, 342, 405; on article "How Brains Make Chaos . . . ," 405, 409, 438

Slaved systems/mechanisms, 223, 386, 421, 546

Smale, S., 391f

Smit, J., 48, 99, 212, 262, 511

Smith, B. A., 9, 170, 173, 309, 475

Smith, D. E., 17

Smith, D. G., 53, 287f, 291

681

Index

Smith, L. A., 135, 389ff, 394, 407, 421, 429, 437, 456, 471, 473, 503

Smith, R. L., 517, 536; on geothermal energy, x; on volcano-tectonic code, 56; on lineaments in western U.S., 56; on styles of intermittent magmatism, 242, 258, 449, 507f; on rates of magma evolution, 258; on rate of production of silicic magma, 268; on invariant spectrum of volcanic frequencies, 449; on geologic periodicity, 507; on igneous recurrence patterns, 508; on organic complexity of volcanic systems, 517; on logarithmic scaling of magmatic phenomena, 533

Smoking gun vs smoking Gatling gun, 124

Smoluchowski, R., 151, 226, 454, 477, 488; on comet reservoirs, 2, 132, 152, 166, 479; on Kuiper Disk, 86; on orbital instabilities, 138; on "short-period" comets, 493

Smylie, D. E., 238, 249, 251, 342, 515f; on kinematic viscosity of core, 502; on earthquakes and Chandler wobble, 518–19, 562

Snow, J. T., 395, 429

Social phenomena, nonlinear-dynamical studies of, 463

Sokal, R. R., 209, 286

Solar System, xxvii, xli, 1, 25, 36, 47f, 56, 72, 129f, 132, 135, 147, 170f, 201, 248, 257, 456, 542; rotational decelerations in, xxxv; critical-state processes in, 2; showers of asteroids, comets, NEOs in, 3; as attractor system/basin, 4, 458, 476; and early binary-star stage, 6, 137–38, 433, 549–51; cratering records in, 8, 10; and increasing self-organization, 9; as self-similar, 9; chaotic evolution of, 9, 305; spectrum of object classes in, 9, 348 (App. 8); discovery of objects in, 10; structure of, 14, 59, 131, 143, 312, 436–38; center of mass in, 31, 179, 183; kinetic energies in, 76–77 (Fig. 10), 136; resonances in, 95–96 (Fig. 21), 173, 175, 176–77, 178, 181f, 185; as chaotic, 132, 175, 182f, 474, 481, 483, 485, 489, 492, 500, 527, 560; chaotic orbits in, 133; dynamical complexity of, 133–34; objects in outer part of, 133, 150–51; mass distributions in, 136; and "Ghost Binary," 137, 346–48 (App. 7), 549; distal reaches of, 151; ecliptic plane of, 153; as critically self-organized, 163; implications of unstable phase-locked resonances in, 183; as multifractal singularity spectrum, 184; resonant ratios in, 189, 351 (App. 11), 446; symposia, texts, etc., on impact phenomena, 212–13; planet formation theories, 433–34; orbital attractor structures in, 436–38; Titius-Bode relation in, 444–46; n-body interactions in, 486; as clockwork, 559

Solar-terrestrial cycles, 32

Solar wind, xxxii, 9, 217, 252, 519

Solheim, L. P., 515, 525

Solitons (e.g. stadium wave), xxvi

Solomon, S. C., 242, 520, 524

Sommerer, J. C., 139, 157, 159ff, 188, 232, 421, 457ff, 474, 503, 533; on Liapunov numbers, 394

Sommeria, J., 221–23, 224, 395, 429; on Jupiter's Great Red Spot, 222–23

Sonea, S., 539, 544, 547

Sonett, C. P., 198, 238, 393, 434f, 503, 518, 563; on impacts on oceans, 212; on FU-Orionis type star, 313; on "differentiated" meteorites, 548

Sornett, A., 460

Southern Hemisphere, 199; search for "fourth" PCN in, 58; and additional sites, 211, 215

Southworth, R. B., 429, 474; on meteoroid "streams," 9, 50, 81, 170, 201

Space and time, 17–18, 407

Space program (American and Soviet), 2; Voyager I and Io, xxxiv, 373; transfer orbits in, 2, 464–65; spacecraft launch of, 192; spacecraft return of, 192; and monitoring satellite debris, 196, 495; and Sputnik, 196, 507; Apollo 15 KREEP basalts from Moon, 269; Galileo spacecraft, 308; Mariner 10 flyby (of Mercury), 350; and space probes, 415; and Hohmann transfer orbit, 465; and Long-Duration Exposure Facility, 486; and NORAD's SPACETRACK system, 486; and Russian communication satellite, 496. See also Capture of satellitic objects

SPACETRAP, xliv, 486, 495

Spacewatch at University of Arizona, 548

Spada, G., 40, 501

Spallation, 240, 515

Sparrow, C., 233; on Lorenz attractor, 233, 409, 517; on nonlinear feedback computations, 385f

Spencer, J. H., 29

Spencer-Brown, G., 26, 29, 504, 540; on principle of self reference, xxv, xxxvi, 409; on self reference in universe, xxix–xxx; on mathematics as handicap, xxxvii; on "logic," 16, 447; his *Laws of Form* and "truth," 410f

Spera, F. J., 53, 138, 212, 252, 263, 265, 325, 489, 540

Index

Sphere of influence, 158, 177, 179, 195, 197, 266, 388, 458, 464, 471f, 485
Spheres of attraction, 2
Spherical pendulum, 153, 157, 164, 187, 194
Spiegel, E. A., 135, 223f, 389ff, 394, 407, 421, 429, 437, 456, 471, 473, 503
Spin-axis, 31ff, 36, 42, 45, 47f, 227, 264, 279f, 498, 527; toroidal convective resonance with, 37; geologically "instantaneous" changes in, 44; redistribution of mass relative to, 44; and pole normal to plane of equatorial bulge, 46; wobble of, 259, 270; nutations of, 519, 561–63
Spin-glass theory, 4; a theory of complexity, 4; and Solar System, 183; model of electromagnetic phenomena, 223; application to social dynamics, 463
Spin-orbit resonances, 11, 35–36, 49, 59, 94–96, 149, 156, 166, 168, 172–73, 175, 194f, 197, 215, 262, 311, 315, 323, 325; and pattern of impacts, 32; of satellitic objects, 33, 112, 169; and mass anomalies, 34f, 179, 201, 314, 429; and positioning of cratering patterns, 36; of Moon, 43, 52, 324, 327; in vicinity of Earth-Moon, 51, 175; in Solar System, 180; universality in hierarchical system of, 180; as function of distance from Earth, 189; of natural Earth satellites, 309, 497
Spin-wave experiments/system, 224, 231, 233
Spin-wave instabilities, 160, 226
Sreenivasan, R. R., 161f, 188–89, 225, 254, 287, 302, 387, 397, 400, 403, 458, 461f, 497, 502ff, 534f
Sridhar, S., 92, 265, 345, 477, 491
Stable chaos, 4, 180, 185, 197, 387–88, 481, 546
Stacey, F. D., 34, 238, 249, 266, 562
Standard statistical test, 10, 39
Stanley, S. M., xxv, 57, 285, 510f
Star formation, 74, 302, 403, 433, 449, 456, 548–51
Stark, L. E., 53, 138, 212, 252, 263, 265, 325, 489, 540
Stark, P. B., 41, 274
Stationarity, 221, 223, 498–99
Statistics, 39. *See also* Standard statistical test
Stauffer, R. C.: on Darwin, 545
Steel, D., 3, 81f, 129, 168, 171, 175, 307, 310, 429, 474, 547; on NEOs, 172, 495
Stein, D. L., 4, 183, 224, 233, 463
Stellar evolution, 72–75 (Fig. 9), 76–77 (Fig. 10)

Stengers, I., 14, 27, 29, 390, 413, 450, 491, 539; on linear thermodynamics, 13
Stern, S. A., 2, 73, 75, 78, 82, 84, 129, 132, 140, 151, 169, 177, 313, 344, 443, 454f, 483, 485, 488f, 550; on Pluto-Charon system, 8, 137; on planetesimals, 10, 252, 414–15, 484, 493; on objects in outer part of Solar System, 133, 135–36, 154, 170, 184, 484; on "plutons," 137, 190; on extra-Solar comets, 152, 477–78
Stevenson, D. J., 53, 92, 138, 252, 263, 265f, 325, 433, 489; on magma oceans, 523
Stewart, G. R., 110, 138, 263, 434, 489; on Mercury, 520
Stewart, H. B., 134, 157, 226, 387, 389, 420, 429, 459
Stewart, I., 7, 134; on orbital phenomena, 433
Stochastic resonances, 155, 437
Stock, J. M., 48, 240, 352, 501, 506, 520
Stöffler, D., 50, 263, 490
Story of the Beginning, 551–57
Stothers, R. B., 3, 62, 66, 69, 119, 130, 211f, 240, 242, 253, 256, 260, 352, 356, 358, 360, 374, 379, 506f, 514, 535, 563; on cratering on Moon, 319
Strain-energy feedback, xxxv, 401; applied to tidal energy, xxxv
Strange attractors, ix, xxviii, xxx, xxxii, 4f, 10, 25, 43, 46, 236, 305, 395, 409f, 413, 447, 452, 541; in Asteroid Belt, 150; as orbital trajectories in Solar System, 197; definition of, 391; physical examples of, 392–94; on tidal pond, 393–94
Strange Loop, xxviii, xxx, xxxvi, 18, 304, 316, 505, 509, 539
Stratigraphy, 283f, 316
"Stroboscopic portrait," 223
Strom, R. G., xxxiv, xxxv, 242, 260, 263, 350, 373, 434, 489, 520
Suarez, I. M., 4, 134, 183, 209, 233f, 293, 463
Subchrons, 38, 47, 227, 423. *See also* Geomagnetic time scale
Subduction, 35, 40, 57, 264, 272, 276–77, 279, 281, 314
Suetsuga, D.: on deep-focus earthquakes, 515, 525
Sugden, R. A., 50; on sizes and shapes of asteroids, 263, 490
Sugihara, G., 188; on standard statistical test, 10, 39; on time-series data (natural phenomena), 460, 561
Sugihara-May approach, 188
Sun, 9, 17, 36, 50f, 162, 195, 248, 503, 549; and motion about Solar System center of mass, 31, 78, 179f, 182, 184, 189, 247, 253,

683

Index

311, 347, 480–81, 519; and vortical invariances in, 47; and Nemesis hypothesis, 138; as pre-main-sequence T-Tauri star, 138; sunspot activity, 179, 224, 247f, 253–55, 257, 311, 393, 480–81, 500, 517–18, 563; and Jupiter, 184–85; and nonlinear dynamics, 217; and solar-flare activity, 247, 254, 257, 393, 524; and tides raised by planets, 247–48; coronal mass ejections of, 252, 254, 257; and Maunder-Minimum sunspot activity on, 253–54, 257, 517; gravitational potential well of, 485

Sündermann, J., 92; on Earth's tides, 477, 481–82

Supercontinent cycles, 37, 314, 526–27

Supernova, 6, 302, 449, 477, 491, 549

Superstring theory, xxiv, 452; and nonlinear dynamics, 399, 450, 547

Suspect terranes, 36f, 526f

Sussman, G. J., 2, 5, 20, 133, 317, 388, 390, 420, 443, 466, 481, 483; on Solar System as chaotic, 132, 175, 183, 489, 492, 527, 560

Swanson, D. A., 66, 241, 352, 508, 522

Sweet, A. R., 49, 207, 210

Swindle, T. D., 50

Swinney, H. L., 224, 237, 412, 420, 435, 551; on turbulence, 108, 502; on period-doubling route, 386, 541

Swisher, C. C., III, 49, 119, 122, 128, 260, 262, 334, 426–27

Sykes, M. V., 50, 147, 201

Synchronization, 165, 190, 407, 429; suggested references on, 438

Synchronous distance, 189, 327f

Synergy, 12, 303, 317, 537, 539

Synodic periods, 192

Tackley, P. J., 275, 422, 516, 525

Taff, L. G., 32, 91f, 175, 192, 196, 464, 479, 485; on tracking systems, 486, 495

Taihu Lake structure, China, 128, 334

Tait, L.: on evidence of cratering, 213

Takahashi, E.: on deep-focus earthquakes, 515, 525

Takens, F., 460, 502; on strange attractor, 391–92; on geometry-from-timeseries, 508

Tam, W. Y., 237, 412

Tang, C., 2, 140f, 161, 225, 302, 421

Tassoul, J. L., 345, 500

Tassoul, M., 345, 500

Tatum, J. B., 78, 82, 172, 175, 308, 310; on fast-moving objects, 147, 171, 497

Tautologies vs immutable constructs, 5

Taxonomy, 148; and Precambrian/early Cambrian biota, 28; avian, 512–13

Taylor, G. I., xxxv; on tidal friction, xxxv

Taylor, G. J., 50, 81, 548

Taylor, S. R., 260, 266f, 490, 492, 523, 540; on collisional origin of Moon, 249, 251, 265, 489

Teapot Dome, WY, K/T site at, 205, 207f

Tebaldi, C., 385f, 502

Tedesco, E. F., 50, 133, 145, 147f, 172, 216, 260, 441

Tektites, 48, 99, 211, 262, 426

Terranes, *see* Cratonic terranes

Terrestrial and cosmological mechanisms, 28

Terrestrial cratering, *see* Impact cratering

Terrestrial planets, 75, 82, 136f, 186, 248, 313, 344–45, 433, 525, 549; references on evolution of, 489

Terzian, Y., 5, 14, 135, 344

Theories of relativity, 5, 451

Theory of Nothing, 5

Thermal feedback, 27, 402f, 408, 413, 422, 502f; in structural organization (astronomical), 14; in structural organization of Earth, 14

Thermodynamics, 13, 21, 404–10 *passim*, 437, 450; Second Law of, 27, 468, 539

Thermomechanical feedback, xxiv, xxxi, 27, 241, 502

Thermosyphon loop/model, 104 (Fig. 29), 108, 224, 250

Thirty-m.y. cycle, 228, 243, 255ff, 535, 550

Tholen, D. J., 50, 487

Thomas, E., 49

Thomas, L., 230, 298, 539

Thomas, P., 2, 311

Thompson, J. M. T., 134, 387, 389, 420, 429, 459

Thompson, J. N., 544, 547

Thompson, R., 20, 22

Thompson, W. B., 9, 50, 81, 170, 201, 429, 474

Thouveny, N., 37f, 47, 225, 342, 563

Thule asteroids, 150

Tidal dissipation, 136, 249f, 258, 476, 520; model of Shaw and Gruntfest, xi, xxxiv, 250; influence on Earth, xxxiv–xxxv, 174, 480; influence on Io, xxxiv–xxxv, 8, 519

Tilling, R. I., 64, 122, 214, 275, 393

Tillites, 212; possible impact origin of, 212, 528; correlation with flood basalts, 528

Tilton, G. R., 490

Time, 17, 561; concepts of, 17, 20; measure of, 17, 19ff, 185, 430; "telling" of, 20; "shallow" vs "deep," 303

Titan (satellite of Saturn), 8, 484

Titius, J. D., 442f

Titius-Bode relationship, 133, 135, 186, 313, 346, 444–47, 482, 485, 488; alternative expression of, 342–45 (App. 6)
Tittemore, W. C., 132
Tombaugh, C. W., 134, 170, 175, 195, 442, 484f, 495, 559; and discovery of Pluto, 11, 184, 441, 443, 445, 482–83, 494; and trans-Neptunian planet searches, 11, 196, 443, 482, 489, 492; on astronomical searches, 496–97
Tonks, W. B., 523
Toomre, A., 44, 266, 276; on polar wander, 37, 40, 367
Torbett, M. V., 151, 226, 488; on comet reservoirs, 2, 132, 152, 166, 479; on Kuiper Disk, 86; on orbital instabilities, 138; on "short-period" comets, 493
TPW, see True polar wander
TPWPs, see True polar wander paths
Trainer, J., xxiv, 399, 450ff
Trajectories, 68, 182, 200, 204, 232, 308, 336–40 (App. 3–4), 420, 466–67, 472f; of orbital motions, 5, 153, 197, 485–86; of natural-Earth satellites, 49; "random" set of, 166; and celestial reference frame, 497; of Great Fireball Procession, 467, 558
Trajectories in phase space, 134, 156–57, 194, 283, 447, 456, 540; divergence vs convergence of, 187; transient type of, 457
Transfer of mechanical work, heat, and mass, 27
Transfer orbits, 2, 139, 166, 176f, 188, 192, 309, 327, 415, 464–68
Transformation of thought, 18–30; First, 18–26, 208; Second, 26–30
Trans-Neptunian planet, search for, 11, 196, 443, 482, 489, 492
Transport, diffusive and chemical reactions, 27
Trego, K. D., 242, 260, 322
Tremaine, S., 92, 138, 140, 177, 265, 345, 477, 491, 550; on angular accelerations of Earth, 259; on impulses of torque in Solar System, 481–82
Trendall, A. F., 454
Triassic/Jurassic extinction, 33
Trieloff, M., 33, 121, 364, 427
Tritton, D. J., 86, 145, 153, 157, 194, 224, 239, 439, 473
Trojan asteroids, 150
True polar wander, 40, 43, 47, 225, 500, 506; a.k.a. "roll" of the mantle, 40; speeds of, 45, 341 (App. 5); and differences determined from APW and hot spot reference frame (HSRF), 47

True polar wander paths, 40, 203, 278, 366–67; cusps, hairpins in, 41–42, 56; correlations with Geologic Column, 42, 56; and GAD hypothesis, 45
"Truth" and circular reasoning, 25
Tryon, E. P., 492
Tschudy, B. D., 48, 207
Tschudy, R. H., 48, 207
Tsonis, A. A., 10f, 161, 163, 188, 423, 460, 462, 561
T-Tauri stars, 138; as FU-Orionis type, 76, 138, 313, 414, 433, 548
Tufillaro, N. B., 29f, 187, 386, 391, 420
Tunguska event, xliv, 322, 467, 558; trajectory of, 98–99, 200, 337–38; as explosive runaway and blast wave, 307–8, 428–29
Tuning, 274, 385f, 395, 430, 458–59, 463, 518, 541f; in magnetoelastic experiment, 158; and cybernetic self-tuning, 161, 178, 400. See also Control function
Turbulence, 391f, 394, 403, 502, 541
Turcotte, D. L., 50, 263, 348, 402, 490, 523, 559
Turing, A. M., 541
Turrin, B. D., 212, 312; on flood basalt antipodal to impacts, 213, 237, 260, 358, 361, 514; on impacts and magmatic propagations, 240–41, 379; on "hot spots," 352, 374
Tyndall, J., 410

Umbgrove, J. H. F.: on pulse of earth, 507
Uncertainty-of-precedence principle, xxv, 392, 539
Uncertainty principle of nonlinear dynamics, 144
Undecidable propositions, 1, 411, 451
Unidentified near-Earth objects, 172
Unidentified object discovered in 1992, 492–93
Uniformitarianism, 6, 246, 285, 293, 423, 540; vs catastrophism, 543
Universality parameters, xxxvii, 54, 162, 245, 286, 317, 397, 450, 456, 458, 461, 543; search for, 15; of nonlinear dynamics, 134, 316, 436; and scaling, 234, 289; definition of, 396; and seismic tremors in Hawaii, 400. See also Dynamical universalities
Universality principles, xxvii-xxviii, 203, 425; definition of, xxxi-xxxii
Universal order, 142, 396–97
Universe, xxxviii, 34, 142, 492, 505; and superstring theory, xxiv, 399, 450, 452; conception of, xxix, 21; as dissipative system, 10, 168, 183, 397, 415; of experience, 26f;

685

Index

entropy of, 27; numerical regularities in, 142; clustering in, 246, 448, 453; size and/or age of, 246–47, 404, 416, 453; order vs disorder in, 446, 482; stringed-oscillator model of, 452; living systems/states, 530; microwave background of, 533
University of California, Berkeley, ix, xxxii, xxxiii, 8, 424, 506
University of California, Lawrence Berkeley Laboratory, xxxii, xxxiii
University of California, Santa Cruz, xxxii, 227, 508
Uranus, 75, 133, 136, 152, 442f, 482, 483, 485
Urey, H. C., 262, 528
Ust-Kara Crater, Siberia, 121ff

Valet, J.-P., 218f, 226, 342, 422
Vallis, G. K., 163
Valsecchi, G. B., 50, 435; on observation of Solar System objects, 201; on comets, 454, 494
Valtonen, M. J., 488
Van Couvering, J. A., 511
Van den Bergh, S., 416, 449, 451, 453, 533, 540
Van Flandern, T. C., 50, 74, 140, 309, 344, 455, 488–94 *passim*; on asteroids having satellites, 170
Vega (star), 32
Venus, 137, 269, 314, 485, 520, 524; tectonic transition on, 33, 525
Verhoogen, J., 341; on energy and core, 249f
Vermeij, G. J., 209, 511
Verne, Jules, 424, 563–64
Veverka, J., 2, 308, 311, 490, 494
VGPs, *see* Virtual geomagnetic pole
Vickery, A. M., 198
Vilas, F., 11, 435, 454
Vine, F. J., 505
Vine-Matthews-Morley hypothesis, 505
Virtual geomagnetic pole (paths), 29, 43, 47, 53, 59, 271, 278, 315, 492, 505, 519; clustering and VGP paths as sensitive tests, 47; distinctions between PPPs, CPPEs, and VGPs, 47; and deep mantle seismic P-waves, 103, 220, 271, 314; compared with core flux patterns, 218f, 226, 264; references on patterns of, 218; and core dynamo, 227; and cratering, 227, 244; and mantle-convection, 227; definition of, 422; and studies of paleomagnetic reversal, 423
Viruses: and intelligence, 23–24
Viscosity, 273, 402, 501–3; of Earth's outer core, 224, 502; in gaseous systems, 403

Vogel, A., 220, 264, 278, 342
Vogel, T. A., 57
Vogt, P. R., 57, 201, 209, 285, 341, 506f; on core-mantle processes, 223; on plumes, 275; volcanism, 537
Volcanism, xii, xxxi, 25, 29, 191, 243, 399, 529, 534–37; on Io, xxxiv–xxxvi; global fabric of, 42; and induced "winters," 201; induced by mass impacts, 240; correlation with impacts and paleomagnetic reversals, 506. *See also* Flood-basalt provinces; Hot spot reference frame; "Hot spots"; Magma; "Magmasphere"
Volcano-tectonic code, 56
Volk, T., 57, 201, 285, 303, 507
von Baeyer, H. C., 381
von Zach, F. X., *see* Zach, F. X. von
Vortex-tree model, 544
Vulcanoids/"Vulcanoid region," 347–48, 484f

Wahr, J. M., 518, 562
Walker, J. C. G., 92
Walker, S. C., 189, 345, 500
Wallace, A. R., xxv, 545
Walter, M. R., 28, 537
Wang, K., 49, 128, 213, 262, 334, 338, 364, 528
Wänke, H., 212, 240, 342, 506
Ward, P. D., 49, 213, 427, 454, 509ff, 543
Ward, W. R., 256, 482, 527f
Wasserburg, G. J., x
Wasson, J., 76, 138, 313, 433; on FU-Orionis type events, 548
Wasson, J. T., 50, 129, 262, 326, 350, 434, 488
Watson, J.: on genetic code (DNA), 23, 416
Watts, A. W., 213, 237, 240, 515, 520
Weather and climate, 25, 163, 179, 253, 503, 517, 527f; and nonlinear dynamics, 201
Weaver, W., 54
Weeks, R., 131, 341, 506
Weems, R. E, xxxvi
Wegener, A., xxxi
Weidenschilling, S. J., 50, 73, 84, 132, 151, 309, 313, 415, 434, 454, 485, 489f, 493, 548; on plutons, 136, 190, 484; on asteroids having satellites, 170
Weinstein, S. A., 516, 525
Weissman, P. R., 2, 50, 73, 76, 82, 84, 88, 124, 126, 128f, 133, 135f, 140, 145, 151f, 154, 167, 177, 260, 313, 344, 433, 454, 479, 487f, 493f, 548; on cometary sources, 128–29, 477–78; on protostar conference, 138; on comet-asteroid dynamics, 166
Westervelt, R. M., 139, 158, 161f, 397, 431,

458; on pendulum, 86, 145, 157, 239, 389, 396, 439, 462, 473; on intermittency, 138, 456–57, 510; on singularity spectra, 188, 504; on critical self-organization, 225, 421; on attractor-basin boundaries, 144, 474; on electronic transport, 400, 403; on chaotic crises, 421, 457, 503
Wetherill, G. W., 50, 129, 145, 252, 350, 434, 455, 488ff, 520, 548; on collisional origin of Moon, 53, 263, 265, 325; on planetesimal accretion, 74–75; on cometary source of asteroids, 84, 494; on collisional exchange, 92, 138; on nebular accretion, 136
What is Life? (book), 19
Wheel of fortune, 142, 163, 286, 448
Whitehead, A. N., xxv
Whitmire, D. P., 138, 483, 488, 492, 550
Wideman, C. J.: on Cenozoic stratigraphy, 287, 511
Widmann, P. J., 104, 108, 161, 224, 502
Wildi, T., 398, 470
Wilhelms, D. E., xxxv, 3, 138, 212f, 252, 267, 284, 325, 476, 489, 540; on impact craters on Moon, 11, 116, 215, 259f; on collisional origin of Moon, 53, 263; on Procellarum Basin on Moon, 110, 261, 265, 278; his *Geologic History of the Moon* (monograph), 262, 319; on Imbrian Period on Moon, 268; on Copernican Period on Moon, 319–20
Willeboordse, F. H., xxvi, 390, 407, 430, 438, 460, 462f
Williams, G. E., 53, 62, 92, 119, 128, 174, 245, 256, 433, 454, 477, 540; his *Megacycles*, 508; on Earth's spin axis, 527
Wilson, E. O., 303, 511; on refuges, 209; on "chain of command," 405; on synergy in evolution, 537; on colonial insect communities, 547
Wilson, J. T., 240; on "hot-spot tracks," 240; on continental drift, 526
Wilson-Morgan hypothesis, 240
Winding numbers, 55, 85, 154–55, 161, 166, 181f, 187, 222, 228, 245, 253, 255, 432, 518; in chaotic regime, 159; definition of, 400; and rotations, 501; and orbital frequencies, 440
Winograd, I. J., 53, 130, 253, 266, 287f, 326
Wisdom, J., 29, 78, 86, 106, 138, 146, 190, 317, 389, 443, 474, 493, 548; on planetary system, 2, 466; on stable chaos, 5; on delivery of impactors/asteroids/meteoroids, 11, 191, 216, 226; on chaotic motions of planets, 20, 388, 481, 483; on "fast moving objects," 78; on asteroid orbital instabilities, 80–81 (Fig. 12), 97 (Fig. 22), 145, 153, 196, 487, 497; on equations of motion, 86, 390, 473; on Solar System as chaotic, 132–33, 175, 183, 420, 489, 492, 500, 527, 560; on Asteroid Belt, 144, 153, 166, 185, 479
Wolbach, W. S., 49, 207f
Wolf, A., 135, 188, 420
Wolf, A. R., 253
Wolfe, J. A., 49, 201, 204–8, 210; on "impact winter" at K/T site, 205, 329
Wolfe, R. F., 84, 129, 166
Wollman, E. L.: on viruses, 23
Wones, D. R., xxxiv, xxxv, xxxvi, 27, 92, 325, 374, 385, 401, 409, 413, 476f, 517
Wood, F. J., 480, 489
Wood, J. A., 213, 262f, 265, 414, 434, 454, 490
Wood, K. D., 168, 179, 201, 248, 253, 311, 481
Wood, R. M., 265, 520, 526
Woodhouse, J. H., 269, 275; on deep-mantle P-wave anomalies, 103, 111f, 220, 271
Woolfson, M. M., 454; on Solar System structure, 436
Woolum, D. S., 75, 436, 490
Wright, E. M.: on irrational numbers, 186, 397
Wright, I. P., 210, 435
Wright, S., 234, 305, 414
Wright, T. L., 393, 508, 521f, 524; on Hawaiian volcanoes, 261, 275
Wu, X., 21, 141, 413, 460, 462

Xu, D.-Y., 49, 454, 528
Xu, S., 50, 216, 434, 487

Yamaji, A., 242, 325, 352, 356, 358, 360, 373f, 520; on "hot spots" on Io, xxxv–xxxvi
Yeates, C. M., 175, 309, 488, 495, 548
Yeomans, D. K., 82ff, 165, 167f, 171, 175, 213, 307, 310, 435, 440f, 454, 474, 547
YIGs, *see* Yttrium iron garnet
Yorke, J. A.: "Period Three Implies Chaos," 187, 445–46
Yoshikawa, M., 50, 81, 144, 226
Young, C. J., 239, 490, 515
Yttrium iron garnet, 160; ferromagnetic resonance in, 160; spin-wave instabilities in, 90–91 (Fig. 18), 226, 233
Yu, L. Y., 22, 182, 288, 382

Zach, F. X. von, 441, 445
Zahnle, K. J., 92, 169, 436

687

Index

Zaslavsky, G. M., 391
Zen, xxix; and ineffable transfinite, xxx
Zener, C., 403
Zero nonlinear coupling, 180–81. *See also* Tuning
Zewail, A. H., 381
Zhang, K., 222f
Zhou, L., 49, 58, 99
Zimmermann, R. E., 480

Zircon, 260, 267; derived from protocontinents, 267
Zirin, H., 238, 253, 393, 516f, 524
Zodiacal dust, 9, 170, 201
Zolensky, M., 434, 436
Zonenshain, L. P., 264, 281, 367, 376, 379; on "accretionary continents," 36, 526
Zvjagina, E. V., 50
Zwart, P. A., 262, 506

Library of Congress Cataloging-in-Publication Data

Shaw, H. R.
   Craters, cosmos, and chronicles : a new theory of Earth / Herbert R. Shaw.
   p. cm.
Includes bibliographical references and index.
ISBN 0-8047-2131-9 (alk. paper) :
1. Earth. 2. Celestial mechanics. 3. Orbits. 4. Impact.
5. Cratering. 6. Nonlinear theories. I. Title.
QB631.S528 1994
551.7′001′175—dc20
93-31725   CIP

⊚ This book is printed on acid-free paper
It has been typeset in 10/12½ Times Roman by Keystone Typesetting, Inc.